VALVES MANUAL
International

Handbook of Valves and Actuators

Edited by

Brian Nesbitt

Elsevier
Butterworth-Heinemann is an imprint of Elsevier
Linacre House, Jordan Hill, Oxford OX2 8DP, UK
30 Corporate Drive, Suite 400, Burlington, MA 012803, USA

First edition 2007

ISBN: 978-1-85617-494-7

Transferred to digital print 2008

Printed and bound by CPI Group (UK) Ltd, Croydon, CR0 4YY

Published by Elsevier in association with Roles & Associates Ltd

ELSEVIER

Amsterdam Boston Heidelberg London New York Oxford
Paris San Diego San Francisco Singapore Tokyo

ROLES & associates

Foreword

Valves are one of the most fundamental components in any process plant. Engineering design relies on a host of mechanical devices to enable the transportation of product to take place under controlled conditions and valves are at the heart of it.

The global control valve market is seeing unprecedented growth as a result of new greenfield plants in developing countries and current high investment in oil and gas exploration and production. A 250,000 bpd refinery might have as many as 2500 of them installed in its pipes and vessels.

Life cycle cost analysis shows the initial purchase price represents around 20% of a valve's total cradle-to-grave cost. Almost three-quarters of the cost is expended on maintenance, major overhauls and repairs, which offers a huge opportunity for cost reduction.

Process plants have seen a dramatic shift in their control system architecture, with advanced strategies being implemented, but critical process flows are still controlled by valves that differ little in function or appearance from those in plants that operated 50 years ago.

Incorrectly specified or underperforming control valves can cost the process plant operator dearly. And that does not mean simply losses from incorrect maintenance. Factor in poor product quality, operating disruptions, safety compliance and other indirect costs, and the figure can grow considerably.

The most effective way to cut control valve costs is to use preventive and predictive maintenance to reduce the number of valves that require turnaround maintenance. A diagnostic system, capable of evaluating the performance of control valves without taking them off line saves plant downtime and provides a far more certain analysis, because it avoids a possible misdiagnosis that can result from removing the valve from its working environment.

The effective management of control valves leads to cost-reduction. Installing correctly specified valves can improve plant performance and product quality, while extending scheduled maintenance periods. Preventive maintenance measures, such as regular calibration and physical inspection, can identify areas of concern before valves require costly repair and disrupt operations.

Predictive maintenance helps an operator see into the device, pinpoint hidden problems and eliminate the need to isolate good valves. With maintenance consuming the lion's share of a valve's life cycle cost, initial selection and an effectively managed maintenance programme are essential in a plant operator's drive for cost reductions.

Valves Manual International is therefore a welcome aid to engineers and users of all disciplines. It combines, in one convenient volume, chapters on fluid theory, terminology, valve and actuator types and selection, materials, QA, testing, installation and maintenance, together with valve applications and solutions and a Buyers' guide. Users and specifiers will find sensible and practical information to help them make informed decisions which will impact positively on whole-life costs. This is not simply another textbook on valves an actuators.

Ian Leitch

Commercial Director

Energy Industries Council
(London, Houston, Singapore, Rio de Janeiro, Macae, Aberdeen, Dubai and Indonesia)

The editor

Brian Nesbitt is a highly respected consultant specialising in valve and pump technology. He works with valve and pump manufacturers, equipment manufacturers, valve and pump users and others who require specialist assistance. Brian regularly publishes articles, presents papers, conducts seminars and workshops, and undertakes work for industry ranging from theoretical research projects and equipment design studies, pump and valve duty evaluations, to site visits to investigate problem valves, pumps, or problem systems. As a pump and valve designer, with experience from 10 to 2048 barg he is admirably suitable to discuss all aspects of valves and actuators. He is the current chairman of the British Standard MCE/6/6 and European CEN/TC197/SC5 subcommittees which have formulated the draft European Standards for rotary and reciprocating pd pumps and is the UK Principal Technical Expert to ISO/TC 67/SC6 Joint Working Group 2 for "Oil & Gas" reciprocating pumps. Brian was one of the UK delegates on the API task-force looking at modifications for the 3rd edition of API 674 and its conversion to ISO 13710.

Brian Nesbitt's working career began as a technical apprentice in the turbo-generator division of CA Parsons, gaining hands-on experience of building machines up to 500MW. From the shop floor, his promotion led to periods in drawing and design offices working on rotary and reciprocating compressors and gas turbines. Large rotary compressors included gas circulators for nuclear reactors. Many machines included specially designed ancillary systems and complicated lube and seal oil systems. Mechanical seals were designed for specific compressor applications. Machine installation and interconnection was an important feature on some contracts. Process piping was considered during machine design. A brief spell in industrial refrigeration, with reciprocating and screw compressors and considerable system design and site/installation exposure, was followed by an introduction to pumps at Ingersoll-Rand where he worked until 1985. Brian was recruited to provide engineering and application support for reciprocating pumps manufactured within Europe. Conversion from reciprocating compressors to reciprocating pumps was accomplished by an extended visit to the "parent" factory in the USA. During this time, Brian assisted with ongoing design work on current contracts, including critical valve components on a batch of 2.0 MW pumps.

Once settled in the UK, Brian provided guidance to Sales for pump selection and choice of accessories and visited potential customers to discuss applications. Special pump designs were implemented for duties unsuitable for standard pumps. Emphasis was directed towards easier maintenance, low NPIPr, high pressure, high viscosity and solids handling applications. New valve designs were produced for arduous applications. North Sea pump applications and associated quality requirements proved to be a great source of development.

Many pumps required special ancillary systems for the crankcase and the stuffing boxes. On-skid process pipework featured on a considerable number of pumps where multiple pumps were assembled to create a single unit. Close liaison with the test department was essential to ensure test rigs were capable of providing the required accuracy for a wide range of operating conditions. Brian was very fortunate in that the test department had an enviable range of equipment and facilities available, together with very experienced staff. Extensive tests were conducted, in parallel with contractual tests, to explore the capabilities of the pump ranges. Novel rig designs were developed to cope with unusual operating conditions.

Brian also provided support for the after-sales departments with site visits to advise on pump and system operational problems. Exposure to complicated system problems, such as acoustic resonance, provided opportunities to work with consultants developing leading-edge technologies and investigation methodologies.

While at Ingersoll-Rand, he was surrounded by some of the world's most eminent rotodynamic pump specialists and met lots of others who visited the plant regularly. Although not knowing everything about valves and pumps, Brian does know who will know the answer to those insoluble problems!

Using this book

Written specifically as a practical reference book for valve and actuator users, *Valves Manual International* is intended to provide useful information about the outline design, selection and installation of valves and actuators and how these affect performance. *Valves Manual International* is not intended to be just "another" textbook on valves; rather it seeks to address the problems that exist at the interface between valve manufacturers and users. It has been compiled with the help of and benefit from the practical experience of valve users; it is aimed at everyone who has technical problems as well as these wanting to know who supplies what, and from where.

Valves Manual International can be used in a variety of ways depending on the information required. For specific problems it is probably best used as a reference book. The detailed Contents section at the front of the book, combined with the Index at the end, will simplify finding the appropriate topic. The "Useful references" at the end of most Chapters also provide helpful guidance, useful information and suggestions for further reading. As a textbook though, *Valves Manual International* may be read from cover to cover to obtain a comprehensive understanding of the subject. Of course, individual Chapters may be studied separately.

Chapter 1 is an important and probably the most-often referred to Chapter. It is an A-Z of commonly used terms, definitions and abbreviations frequently encountered in the daily use of valves and actuators and their associated systems. Technical terms, unique expressions, concepts, topics, synonyms and acronyms are all defined, along with an explanation of a number of examples of the misuse of valve, actuator and piping terminology. With the valve and actuator market being global and with English being used more and more as a universal engineering language, the need for such clarification of important technical terms and definitions is all the more important. Chapter 1 can be used as a straightforward dictionary or, perhaps more effectively, in conjunction with the reading and study of individual chapters. A number of cross references to the other chapters in the book are also given and the comprehensive Index at the end of the book is will also be of help in locating topics.

The properties of fluids are discussed in Chapter 2. Chapters 3 to 7 are devoted to the main valve types, grouped into Isolating, Non-return, Control and Safety relief. Actuators are covered specifically in Chapter 12.

The book then follows a logical pattern with Chapters devoted to valve and pipe sizing and then piping and connectors through to sealing arrangements. Ancillary products and services are also discussed. Testing and quality assurance is dealt with in Chapter 15. Chapters 16 and 17 are devoted to valve and actuator standards. Chapter 19 is concerned with installation, commissioning and maintenance as well as efficiency, economics and selection.

Chapter 18 provides details of a number of interesting valve and actuator applications and solutions that illustrate some of the problems encountered in the practical use of valves and how these have been solved. This Chapter embraces a number of valve and actuator designs and uses, showing some of the diverse uses for these products outside the more traditional areas. Chapter 20 gives useful guidance and information on many fluid properties and brings together data on liquids, gases as well as important units and conversions used in valve selection and in making preliminary design decisions.

The Buyers' Guide, Chapter 21, summarises the various valve types, divided them into Isolating, Non-return, Control, Safety relief and Actuators, based on the individual Chapters in the book devoted to each type. The Guide has been grouped in this way to simplify representation of the major generic valve groups. This is due to the difficulty in trying to impose tight boundaries on various valve and actuator type descriptions used in various parts of the world.

This is followed by Ancillary products and services. Trade/brand names are comprehensively listed too. It is preceded by the names and addresses and contact details of all companies appearing in the Guide. They are listed alphabetically, by country.

Although each Chapter describes in detail all the valve types that have been placed within that generic group, it is strongly recommended that direct contact with the relevant companies is made - initially perhaps via their websites - to ensure that their product ranges are clearly identified from the broad classification and that these details can be clarified wherever necessary.

Contents

Definitions and abbreviations 1

This Chapter is an A-Z of commonly used terms, definitions and abbreviations frequently encountered in the daily use of valves and actuators and their associated systems.

Technical terms, unique expressions, concepts, topics, synonyms and acronyms are all defined, along with an explanation of a number of examples of the misuse of valve, actuator and piping terminology.

With the valve and actuator market being increasingly global and with English being used more and more as a universal engineering language, the need for such clarification of important technical terms, abbreviations and definitions is all the more important.

The Chapter can be used as a straightforward dictionary or, perhaps more effectively, in conjunction with the reading and study of individual chapters. A number of cross references to other chapters in the book are also given and the comprehensive Index at the end of the book is will also be of help in locating topics.

Definitions and abbreviations

A

3-A

3-A Sanitary Standards, Inc. (of USA). An American company which writes voluntary "accepted practices" to maintain good hygiene within food handling processes.

AA

Aluminium Association (of America)

Absolute pressure

The actual pressure, not a pressure compared to the local atmospheric pressure. Gauge pressure refers the measured pressure to the local atmosphere, nominally 101.325 kPa (absolute).

ABNT

Associação Brasileira de Normas Técnicas, the National Standards Authority of Brazil.

Absolute pressure regulator

A regulator which maintains a sub-atmospheric pressure rather than a positive gauge pressure.

Accuracy

The variation in the controlled variable between the minimum controllable flow and the rated flow. The value is equal to 100% minus the offset, see Figure 1.2.

Accumulation

The pressure increase above the system design pressure or system maximum allowable working pressure (MAWP) during safety valve operation. If the accumulation is limited to 10% and the overpressure is 25% then the set pressure must be 110/125 = 0.88 = 88% of system design pressure, maximum.

ACI

Alloy Casting Institute, a division of the Steel Founders' Society of America, has a series of identification codes for alloy and stainless steels, see Table 1.1.

AISI	ACI	Cr/Ni/Mo
303	CF-16F	18/9/0.6
304	CF-8	19/9
304L	CF-3	19/10
310	CK-20	25/20
316	CF-8M	17/12/2
316L	CF-3M	17/12/2
317	CG-12	19/13/3
347	CF-8C	18/11
410	CA-15 CA-15M CA-6NM	12/1 12/1/1 12/4/1
17-4PH	CB-7Cu	17/4
420	CA-40	12/1

Table 1.1 ACI identification codes

Acme thread

A thread form similar to square threads but the thread flanks form an included angle of 29°. The Acme thread is stronger than a square thread and is easier to machine. Thread is rolled rather than cut to improve strength and surface finish. Used on the stems of some linear valves.

Acoustic resonance

Severe pressure pulsations caused when an exciting frequency coincides with an organ pipe natural frequency of the pipework. Sound waves travel successfully around corners hence pipe bends do not divide piping systems into a collection of straight lengths.

Actuator

The device fitted to a valve for moving the sealing element. The actuator can be a handwheel for manual operation or a diaphragm or piston powered by compressed air or hydraulics.

Actuator stiffness

The force per unit length, N/mm, required to move the actuator stem without the positioner driving the actuator. When a pneumatic diaphragm actuator is set in a specific position the diaphragm casing contains a fixed volume of air. The air volume is elastic because air is compressible. If a force is applied to the actuator stem it will deflect as the air compresses or expands. The stiffness of fluid powered actuators is dependent upon the fluid volume and therefore the stroke position. Electromechanical actuators have fairly constant stiffness. Actuator stiffness for various types is listed in Table 1.2.

Most stiff	Electro-mechanical
	Electro-hydraulic
	Double-acting pneumatic piston
	Single-acting pneumatic piston
Most flexible	Pneumatic diaphragm

Table 1.2 Actuator stiffness

AECMA-STAN see ASD-STAN

AENOR

Asociación Española de Normalización y Certificación, the National Standards Authority of Spain.

AFNOR

Association française de normalisation, the National Standards Authority of France

Air loaded regulator

Another name for an External pressure loaded regulator.

AISI

American Iron and Steel Institute

Al

Aluminium is used as a light, corrosion resistant metal and also as an alloying element in steels. Aluminium is the strongest deoxidiser and also combines with nitrogen, present in air, to reduce strain ageing. It is also useful in inhibiting grain growth and improves scale resistance for hot applications.

Allotropy

The property possessed by certain elements to exist in two or more distinct forms that are chemically identical but have different physical properties. In the case of iron the crystal structure has one form at room temperature and another at high temperature. When heated above 910°C the atomic structure changes from body-centred cubic to face -centred cubic but reverts again when cooled. The allotropy of iron modifies the solubility of carbon, and it is because of this that steel can be hardened.

Altitude valve

An accurate pressure reducing regulator combined with a reverse differential pressure regulator used to maintain the level in a liquid reservoir. When system pressure is high liquid is supplied to the reservoir until the maximum level is reached. If the system pressure falls below the reservoir pressure liquid is released into the system. Used extensively in water distribution systems with local water towers.

Amplifier valve

Another name for the pilot valve in a **Pilot-operated regulator**.

AMS

Aerospace Materials Specifications (from SAE specifications)

Anaerobic

Usually applied to liquid gaskets, thread seals and impregnation compounds to indicate the hardening or thickening process commences after assembly or application when air is excluded.

Analogue signal

A signal which varies in intensity. Typical analogue signals are pneumatic; 0.2 to 1.0 barg; electric; 0 to 10V or 4 to 20 mA.

Angle valve

A valve body style which has the process connections at 90°C.

Annealing

Heating steel to, and holding at a suitable temperature, followed by relatively slow cooling. The purpose of annealing may be to remove stresses, to soften the steel, to improve machinability, to improve cold working properties, to obtain a desired structure. The annealing process usually involves allowing the steel to cool slowly in the furnace.

Anti-static design

A design which provides electrical continuity for the moving valve components and across the valve body to the connected piping.

AOD

Argon-oxygen decarburisation, (a stainless steel refining process)

Approved materials

When valves are to be used with fluids for human consumption, e.g.. water, milk, food products; or valves are to be used in manufacturing processes where contamination would degrade or destroy a critical fluid constituent, only tested and approved materials of construction are allowed. Several bodies have lists of approved materials or will make a judgement on a specific material. Table 1.3 is indicative:

Country	Approvals Body	
Belgium	ASEAU	
Denmark	VA	
France	AFNOR	Association Françiase de Normalisation Beaureau Veritas
Germany	DVGW BI	Deutscher Verein vor Gas-und Wasserfachmannern Bau Institut
Italy	UNI	Ente Nazionale Italiano de Unificazione
The Netherlands	KIWA	Keurings Instituut voor Waterleiding-Artikelen
Norway	GNFS	
Sweden	PV	
Switzerland	SVGW	
UK	DWI WFB WRc	The Drinking Water Inspectorate The Water Research Council
USA	FDA USDA	Food and Drink Administration The US Department of Agriculture

Table 1.3 Learned bodies approving materials of construction

For the specific area of drinking water there is no European Standard. This is because the member countries have adopted different approaches to the problem and rationalisation is not possible without enormous expenditure in some countries. CEN is considering the problem at present. National Standards prevail in this area.

ANSI

The American National Standards Institute, the National Standards Authority of the USA.

API

The American Petroleum Institute

API pipe thread see **NPT**

Arc welding

An electric welding process where the necessary heat is provided by striking an electric arc between two conductors. One of the conductors is the workpiece, the other can be a non-consumable electrode or a consumable weld filler material.

ARSO

African Regional Standards Organisation

ASD-STAN

Aerospace and defence industries association of Europe - Standardisation

Asbestos fibre

A natural mineral fibre which has good mechanical properties and chemical resistance at high temperatures. No longer a popular material due to the possible health risk but can be used for difficult applications as a last resort.

ASME

American Society of Mechanical Engineers

Aspirator

Another name for a **Pitot tube**.

ASRO

Asociatia de Standardizare din Romania, the National Standards Authority of Romania.

ASTM

American Society for Testing and Materials

Atmospheric pressure

Atmospheric pressure is standardised, technically, at 101.325 kPa(absolute), equivalent to 14.696 psi(a). Remember! This is the pressure at sea level. Equipment working at altitude will not experience the same atmospheric pressure. Also the local atmospheric pressure changes with the weather. Local values are likely to vary from 94 to 108 kPa at sea level.

Austempering

Quenching from a temperature above the transformation range to a temperature above the upper limit of martensite formation, and holding at this temperature until the austenite is completely transformed to the desired intermediate structure, for the purpose of conferring certain mechanical properties.

Austenite

A face centred cubic crystal structure found in chrome-nickel stainless steels.

Austenitic stainless steel

Steels containing high percentages of certain alloying elements such as manganese and nickel which are austenitic at room temperature and cannot be hardened by normal heat-treatment but do work harden. They are also non-magnetic in the annealed condition and very corrosion resistant. Chromium content usually falls between 16 and 26% and nickel content less than 35%. Typical examples of austenitic steels include the 18/8 stainless steels and 14% manganese steel. Work hardening or welding may induce some magnetism.

Autoclavable

The ability of a valve to be heated in an oven for sterilisation. The time and temperature being dependent upon the organism to be destroyed. Lower temperatures are usually applied for longer periods. A typical low temperature regime could be 105°C for 60 minutes. A higher temperature regime could be 121°C for 5 minutes.

Automatic high pressure inlet valve

A regulator used in two pressure systems to admit automatically a high pressure source when a set minimum pressure has been reached. It is used in systems where a low pressure high volume supply is used to fill equipment or provide low power actuation against reduced loads. When the equipment is filled or the load increases, the pressure within the low pressure supply rises as the flow reduces. The automatic high pressure inlet valve senses the pressure rise and admits the high pressure supply. The high pressure supply is prevented from pressurising the low pressure supply by a non-return valve.

It is similar in construction to a piston pressure reducing regulator, see Figure 5.2 in Chapter 5, but very rugged. It is a push-down-to-close regulator rather than the push-down-to-open as shown. The high pressure supply would be connected to the left hand connection. Low pressure systems are typically 70 to 140 barg with high pressure systems up to 550 barg.

Auto-tuning

A feature found on electronic PID process controllers, see **Control algorithms**. During the operation of the controller, the controller monitors the response of the system to control changes. The controller uses this information to optimise its control algorithm. Auto-tuning can be single-shot or continuous optimisation.

AWN

Arc welding noise

AWR

Arc welding robot

AWS

American Welding Society

B

Boron

Boron is used in very low concentrations to improve the depth of heat treatment in steels and can increase the strength of stainless steels at the expense of corrosion resistance.

Backseat bush

A seal found on some valves to isolate the packing box from the valve internal pressure. In rising stem valves the stem can be physically prevented from lifting out by fitting a collar which engages in a seating when the valve is wide open. The seating can be a good metal-to-metal contact or can incorporate a resilient seal. Fluid is prevented from escaping up the stem by the surface contact. Valves in which the internal pressure generates an axial thrust in the stem or spindle can have a pressure loaded backseat bush. The use of a backseat bush allows the packing box to be maintained while the valve is pressurised.

Back pressure

The pressure at the safety valve outlet. This can be expressed as a percentage of the fluid inlet pressure or as a specific pressure. Back pressure is very important and has a significant effect on valve performance.

Back pressure regulator

A regulator which maintains a preset pressure upstream of the regulator. The regulator may be direct-acting or pilot operated by the process fluid. Also known as a pressure retaining valve.

Balanced safety valve

A safety valve which is designed to minimise the effects of back pressure on the valve performance. Dead weight and spring-loaded safety valves can be balanced by fitting a bellows or a sliding seal.

Balanced steel

Steels in which the deoxidisation is controlled to produce an intermediate structure between a rimmed and **Killed steel**. Sometimes referred to as semi-killed steels, they possess uniform properties throughout the ingot and amongst their applications are boiler plate and structural sections.

Barbed fitting

A type of coupling used with small hoses. A tail fits in the bore of the hose. The tail is not smooth but covered with angled serrations which grip the hose. A compression clip around the hose may complete the assembly for pressurised applications.

Barrel nipple

Also called a "full coupling", a short piece of material with a female pipe thread in each end used to connect two pieces of pipe.

Barstock valve

A valve with a body machined from standard rolled bright bar. Small isolating valves with screwed connections are made from square section, 1.25", 1.5" or 1.75", or hexagon bar. Combination valves are made from rectangular bar, typically 1.25" by 2.5". Common materials are carbon steel and austenitic stainless steel. The steel quality may be general, to BS 970/ASTM A668, or pressure vessel quality to BS 1502/ASTM A105.

Basic steel

Steel produced in a furnace in which the hearth consists of a basic refractory such as dolomite or magnesite, as opposed to steel melted in a furnace with an acid lining. The basic process permits the removal of sulphur and phosphorous and in this respect is superior. Present day BOS (Basic Oxygen Steel) and electric arc furnaces use basic linings.

BAT

Best Available Technology

BASMP

Institute for Standards, Metrology and Intellectual Property of Bosnia and Herzegovina

BDS

Bulgarian Institute for Standardization, the National Standards Authority.

Be

The rare element beryllium, is alloyed with copper to make a superior spring material.

Belleville washer

A disc spring, effectively a conical washer, which can supply high forces in confined spaces.

Bellows safety valve

A safety valve fitted with a bellows seal to prevent process fluid leaking into the bonnet or spring housing.

Bellows sealed bonnet

A bonnet which uses a bellows to eliminate leakage from the valve stem. The bellows may be sealed with the body by an "O"-ring or gasket or may be welded. The bonnet may incorporate a secondary seal as a back-up in the event of bellows failure.

BELST

Committee for Standardization, Metrology and Certification of Belarus, the National Standards Authority.

Bessemer Process

A method of producing steel, first introduced in the last century, where air is blown under pressure through molten iron to remove the impurities by oxidation. The development of the process has led to the present day Basic Oxygen Steel making plants that account for bulk production of commercial quality steels in the UK.

BHN

Brinell hardness number

Billet

A section of steel used for rolling into bars, rods and sections. It can be a product of the ingot route, or increasingly today produced directly by continuous casting.

Binary number

A number represented by a series of 0s and 1s. Binary numbers are usually described by how many digits or bits are required. Popular sizes used in measurement, communications and computing have 4, 8, 10, 12, 16, 32 and 64 bits. Table 1.4 shows the relationship between decimal numbers and 8 bit binary; the most popular format. The accuracy of an 8 bit system is 1 part in 255, 0.4%. Electronic process sensors often use 10 or 12 bit data communications for increased accuracy.

0	00000000
1	00000001
2	00000010
3	00000011
4	00000100
5	00000101
6	00000110
7	00000111
8	00001000
9	00001001
10	00001010
16	00010000
32	00100000
64	01000000
128	10000000
255	11111111

Table 1.4 Equivalent binary numbers

Blast furnace

A cylindrical, refractory lined furnace for the production of pig iron or hot metal for direct conversion into steel.

Blind

Blank flange

Bloom

A large square section of steel intermediate in the rolling process between an ingot and a billet. Blooms are now also being produced by the continuous casting process eliminating the necessity of first producing an ingot.

Blowdown

The difference between the set pressure and the reseat pressure, usually expressed as a percentage of the set pressure. The values quoted by manufacturers are based on the valve starting from full lift and passing 100% flow. If the valve does not achieve full lift, or is not required to pass 100% flow, the blowdown will be shorter.

Blowdown valve

A valve which permits pressurised fluid to be released to the environment.

Bolted gland

A gland which is attached to the bonnet by two or more bolts or studs. Because of the difficulty of maintaining exact alignment on adjustable versions some form of spherical bearing on the packing follower is recommended.

Bonnet

The component attached to the valve body through which the stem or spindle passes and has locations for mounting the actuator. The bonnet includes the packing box which is equivalent to the stuffing box on a pump or compressor. The bonnet can be threaded, flanged, union, clamped, breechlock or welded to the body. Special valves can have the bonnet retained by shear rings.

Booster

A pneumatic amplifier used to convert a pneumatic controller output signal to a higher pressure. The normal controller maximum output of 1.03 barg can be increased to 2.4 barg for higher pressure diaphragm actuators or 10 barg and higher for piston actuators. An air supply about 0.5 bar higher than the maximum actuator pressure is necessary.

Bottom flange

A flange on the valve body opposite the bonnet. The bottom flange may be the extra connection on a three port valve. The bottom flange may be fitted for access to the valve internals or to house a bottom guide bush.

Bourdon tube

A curved tube which expands to a slightly larger radius when internal pressure is applied. Used to detect fluid pressure in higher pressure mechanical systems, when bellows would be unsuitable.

bps

Bits per second — the speed of electric digital communications in bits per second.

BPS

Bureau of Product Standards, the National Standards Authority of the Philippines.

Braided packing

Strands or filaments, all running in the same longitudinal direction, that are twisted together to form square or rectangular sections. Various types of braiding are possible to produce harder, stiffer or softer, more pliable, packing. The braiding can be applied over a soft or hard central core. Reinforced corners can be incorporated. The strands can be coated with solid or fluid lubricant to extend operating conditions. Metallic strands can be included for reinforcement and to improve heat transfer. Braided packing can be extruded to compress the section and control the finished size. Braided packing is generally supplied as a continuous length, 5 to 20 m, depending upon the cross-section. Sealing rings are produced by cutting the required length from the coil.

Brazing

A method of joining metal parts together by fusing a layer of brass between the adjoining surfaces. A red heat is necessary and a flux is used to protect the metal from oxidation.

Breakaway force

The force required on a linear motion sealing element to open the valve against maximum differential pressure.

Breakaway torque

The torque required on a rotary motion sealing element to open the valve against maximum differential pressure.

Bright annealing

An annealing process that is carried out in a controlled atmosphere furnace or vacuum in order that oxidation is reduced to a minimum and the surface remains relatively bright.

Bright drawing

The process of drawing hot rolled steel through a die to impart close dimensional tolerances, a bright, scale free surface, and improved mechanical properties. The product is termed bright steel.

BSI

British Standards Institution, the National Standards Authority of the United Kingdom.

BSMI

Bureau of Standards, Metrology and Inspection, the National Standards Authority of Taiwan.

BSN

Badan Standardisasi Nasional, the National Standards Authority of Indonesia.

BSP(P)

British Standard Pipe {thread} (Parallel), the Whitworth thread form, 55°, pipe thread used in the UK. Originally defined by BS 2779 for pipes from 1/16" to 6" nb but redefined in metric by ISO/R228. Sealing of the process fluid is not on the threads. Table 1.5 shows the main dimensions. Standard sizes are defined up to 18" nb, but 6" is the practical limit for process applications. Many users will not accept threaded fittings over 3".

Pipe nominal bore in	Approx pipe od in	Diameter at top of thread in	Root diameter of thread in	Number of threads per in	Minimum length of thread in
1/8	0.406	0.383	0.337	28	0.375
1/4	0.531	0.518	0.451	19	0.437
3/8	0.687	0.656	0.589	19	0.500
1/2	0.844	0.825	0.734	14	0.625
3/4	1.062	1.041	0.950	14	0.750
1	1.344	1.309	1.193	11	0.875
1 1/4	1.687	1.650	1.534	11	1.000
1 1/2	1.906	1.882	1.766	11	1.000
2	2.375	2.347	2.231	11	1.125
2 1/2	3.000	2.960	2.844	11	1.250
3	3.500	3.460	3.344	11	1.375
4	4.500	4.450	4.334	11	1.625
5	5.500	5.450	5.334	11	1.750
6	6.500	6.450	6.334	11	2.000

Table 1.5 Main dimensions of BSP pipe threads

BSP(T)

British Standard Pipe {thread} (Taper), the Whitworth form, 55°, pipe thread used in the UK. Originally defined by BS 21 for pipes from 1/16" to 6" nb, but redefined in metric by ISO7/1 and ISO7/2. Sealing of the process fluid is on the threads. The thread taper is 0.75" per foot on diameter. Standard sizes are defined up to 18" nb, but 6" is the practical limit for process applications. Many users will not accept threaded fittings over 3".

BSTI

Bangladesh Standards and Bangladesh Standards and Testing Institution, the National Standards Authority.

Buckling pin safety valve

A safety valve which is held closed by a strut. When the set pressure is exceeded the strut is loaded above its critical load and fails by buckling, allowing the valve to open. The valve does not close when operating conditions return to normal but must be reset, with a new pin, manually.

Built-up back pressure

The pressure experienced at the safety valve outlet when passing 100% flow, due to friction losses in attached pipework and any static back pressure.

Burst pressure

Hoses are designed with a theoretical maximum pressure at which the hose ruptures. Samples are tested to destruction regularly to ensure manufacturing quality. Working pressures are related to the burst pressure by a factor of safety. The minimum factor of safety, for low pressure hoses, is 3. If the hose may be subject to chemical degradation the safety factor should be increased to 4. For hot or high pressure applications, such as steam, the factor of safety can be as high as 10.

Bursting disc

A pressure relief device which utilises a thin membrane to enclose the fluid. The device relieves when the membrane fails. A bursting disc cannot close and seal after operating. Special versions are available for protection against severe shock waves caused by explosions. Bursting discs are also called rupture discs.

Bus

A multi-core cable for digital communications in computers. All the bits for a binary number are transmitted simultaneously. An 8 bit bus will have at least nine cores. Very fast data rates are possible. One multi-core cable can have more than one bus and other digital services. Analogue signals should be routed through different cables. All the services in one multi-core cable should be at similar voltages; a maximum difference of 50V is specified by some users. Most specifications require all cores to be terminated properly in the terminal boxes; and have no coiled ends. Multi-cores and terminal boxes should have at least 10% spare capacity for expansion and re-routing broken cores. See also fieldbus

Butt fusion weld

A thermal welding technique used with some non-metallic piping systems. Materials such as polypropylene can be welded by using electric heaters. Machines can join 1200 mm diameter pipe.

Butt weld joint

A welded joint between two components usually of the same diameter and with identical weld preparations; for example, a connection between two pipes or a pipe and a fitting. Welded pipe can be manufactured with a butt weld joint. A pipe or a fitting can be butt welded to a flat surface without a weld preparation but this form is not acceptable to many purchasers. Butt weld pipe joints can be the highest integrity connection, if proved by suitable **NDE**, or **NDT**, such as 100% radiography.

C

C

The chemical symbol for the element carbon; alloyed with iron to produce cast iron and steel. Carbon increases the strength of iron and steel and can increase the hardness.

Cage

A hollow cylindrical component, part of the trim, which guides and aligns the plug with the seat. The cage may have profiled openings or multiple small holes to create the required characteristic or impart cavitation resistance or for noise attenuation.

Capacity

The flow through a regulator under specified operating conditions with a set regulator opening.

Capillary fittings

A type of soldered connection used with copper pipe. The fitting is made slightly larger than the pipe to provide a socket for location. The bore of the socket is treated with flux and contains a reservoir of solder. After the pipe is fitted the connection is heated and a perfect soldered joint is produced.

Carbon steel

A steel whose properties are determined primarily by the amount of carbon present. Apart from iron and carbon, manganese up to 1.5% may be present as well as residual amounts of alloying elements such as nickel, chromium, molybdenum, etc. It is when one or more alloying elements are added in sufficient amount that it is classed as an alloy steel.

Carbo-nitriding

A case-hardening process in which steel components are heated in an atmosphere containing both carbon and nitrogen.

Carburising

The introduction of carbon into the surface layer of a steel that has a low carbon content. The process is carried out by heating the components in a solid liquid, or gaseous carbon containing medium. The depth of penetration of carbon into the surface is controlled by the time and temperature of the treatment. After carburising it is necessary to harden the components by heating to a suitable temperature and quenching.

Case-hardening

The process of hardening the surface of steel whilst leaving the interior unchanged. Both carbon and alloy steels are suitable for case-hardening providing their carbon content is low, usually up to a maximum of 0.2%. Components subject to this process, particularly in the case of alloy steels, have a hard, wear-resistant surface with a tough core.

Cast iron

A definition can be applied that cast iron is an alloy of iron and carbon in which the carbon is in excess of the amount that can be retained in solid solution in austenite at the eutectic temperature. Carbon is usually present in the range of 1.8% to 4.5%. In addition, silicon, manganese, sulphur and phosphorus are contained in varying amounts. Various types of cast iron are covered by many of the National Standard authorities and include grey, malleable and white irons. Elements such as nickel, chromium, molybdenum, vanadium can be added to produce alloy cast irons.

CDA

Copper Development Association of the United Kingdom

CDS

Cold Drawn Seamless; generally small diameter pipe and tube.

Cellulosics

Natural vegetable fibres used for packing of which cotton, flax, jute and ramie are the most popular. Natural fibres often have poor resistance to acids but can be acceptable for alkalis. Natural fibres can be much cheaper than man-made fibres depending upon the coatings applied and the compounds added to the braid.

Cemented joint

A joint used on non-metallic pipe. Similar in concept to soldered connections but the sealing compound is an adhesive and provides the structural strength. Manufacturers of the pipe and fittings qualify their pressure ratings and chemical compatibility, by stipulating that their solvent or adhesive must be used. It is essential the manufacturers instructions are followed. Some adhesives require a priming coat before the adhesive is applied.

CE

Carbon equivalent, an assessment made of carbon and carbon manganese steels used in a sour environment (H_2S present), according to the percentages of the constituents:

$PRE_N = Cr + 1.5 \times (Mo + W + Nb)$ for stainless steel

$PRE_N = 3.3 \times Mo + 16 \times N_2$ for super austenitic stainless steel

$PRE_N = Cr - 0.8 \times Cu + 1.5 \times (Mo + W)$ for nickel alloy weld wire

$$PRE_N = Cr + 3.3 \times Mo + 30 \times N_2$$

CEN

Comité Européen de Normalisation (the European Standards authority). CEN is the European regional Standards Authority which issues Standards for the European Union. It is made up of the following National members:

Austria, Belgium, Bulgaria, Cyprus, Czech Republic, Denmark, Estonia, Finland, France, Germany, Greece, Hungary, Iceland, Ireland, Italy, Latvia, Lithuania, Luxembourg, Malta, The Netherlands, Norway, Poland, Portugal, Romania, Slovakia, Slovenia, Spain, Sweden, Switzerland, United Kingdom.

CENELEC

European Committee for Electrotechnical Standardisation, (a branch of CEN).

Certified performance

Learned bodies, such as ASME, and classification organisations, such as Lloyd's and Underwriters Laboratories Inc, have regulations for sizing safety valves for applications. If the nozzle area and the mechanical construction of the valve comply with the regulations, the performance of the safety valve can be certified.

Certifying Authority

An organisation, retained by the purchaser, which is entrusted with ensuring that all relevant statutory requirements are fulfilled. This is usually an insurance company but it can be a firm of consultants or an accredited test house.

CFS

Cold Finished Seamless; pipe and tube which is made hot but finished when cold to improve dimensional stability and surface finish.

Change-over valve

A valve with three connections which allows one circuit to be connected to one of a possible two circuits. The circuit not connected is isolated. Valves of this type are frequently built into equipment such as duplex filters. The change-over valve allows one filter element to be used while the other is cleaned or replaced. In the case of duplex filters, two change-over valves are ganged together. An inlet change-over valve is ganged with an outlet change-over valve so that two filters work with one inlet and outlet process connection. Very special change-over valves are available to be used with safety relief valves. One process connection may be connected to one of two safety relief valves (srvs). The valve ensures that at least one srv is in circuit at all times. Both srvs cannot be isolated simultaneously.

Characteristic

The characteristic of a control valve is the relationship between valve movement and flow through the valve at constant differential pressure. Several characteristics have become standard; quick opening (QO), linear (L), modified parabolic (MP)

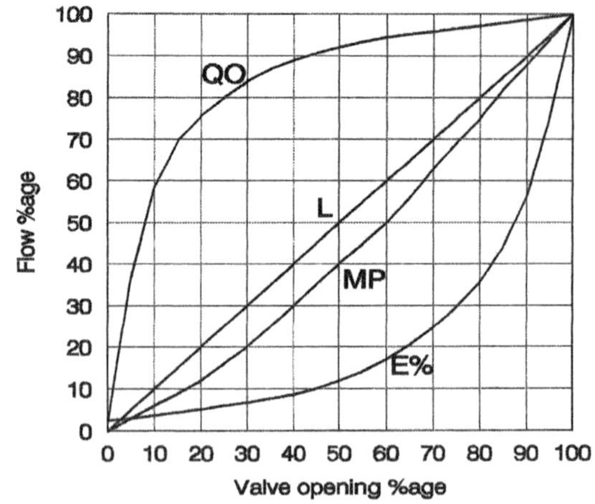

Figure 1.1 Comparison of control valve characteristics

and equal percentage (E%). Modified parabolic is also known as exponential. Figure 1.1 indicates these relationships graphically. Some manufacturers will produce a special characteristic to suit a specific application.

Charpy test

A test to measure the impact properties of steel. A prepared test piece, usually notched, is broken by a swinging pendulum. The energy consumed in breaking the test piece is measured in Joules. The more brittle the steel the lower the impact strength. CEN and ISO provide test methods.

Chatter

The rapid oscillation of the disc against the seat caused by unstable flow conditions and resulting in rapid wear of the nozzle, disc and guides. This condition can occur when the safety valve is only required to pass a small percentage of the rated flow. Depending upon the characteristics of the fluid, chattering can be a problem at flows below 30% of rated.

Check valve

Non-return valve

Chokes

A manual valve used for continuous throttling applications; a heavy duty needle valve. An American term used mainly in connection with crude oil production.

Clamped bonnet

Large valves can have the bonnet attached to the body by a proprietary clamped connection. These connections provide a very rigid joint with high accuracy positioning due to the shaped metal seal ring used.

CI

Cast iron

CIP see **Clean-In-Place**

Clamped connection

A pipe connection which relies on the wedging action of tapers to provide the axial force to close the joint. Gaskets can be metallic or non-metallic. A very rigid, quick, pipe connection.

Classification Societies

Learned bodies, and usually insurance companies, who have rules governing the use of valves and/or associated equipment in certain industries, environments or applications. Typical areas of influence include drinking water, hygienic applications, marine and offshore installations. Table 1.6 indicates some of the most commonly used societies.

Body	Field of influence
American Society of Mechanical Engineers (ASME)	Stressing of pressure vessels, valve sizing (USA)
American Petroleum Institute (API)	Valve installation and sizing (USA)
Associated Offices Technical Committee (AOTC)	Commercial and industrial compressed air and heating systems (UK)
Bau Institut (BI)	Hygienic applications (Germany)
Beaureau Veritas	Marine, offshore, general process (France)
Det Norske Veritas (DNV)	Marine, offshore, general process (Norway)
Factory Mutual Systems (FM)	Commercial and industrial fire protection systems (UK)
Germanischer Lloyd	Marine, offshore, general process (Germany)
Institute of Electrical Engineers (IEE)	Electrical safety and wiring (UK)
Keurings Instituut voor Waterleiding-Artikelen (KIWA)	Drinking water (The Netherlands)
Lloyd's Register	Marine, offshore, general process (UK)
Loss Prevention Council (LPC)	Commercial and industrial fire protection systems (UK)
TUV ad-Merkblatt	Pressure vessels (Germany)
Underwriters Laboratory (UL)	Electrical equipment, fire safety (USA)
Underwriters Laboratory (UL)	Electrical equipment, fire safety, fire protection systems (Canada)
Water Research Council (WRc)	Drinking water (UK)

Table 1.6 Classification societies

Clean-in-Place, CIP

A requirement for some equipment used in hygienic applications. Complete systems are cleaned by washing out, sterilising or steaming without any disassembly. The externals of equipment are hosed down or washed with caustic, detergent, chlorine or alcohol. Equipment design must be optimised to remove dead volumes where product may accumulate and decay; systems must drain completely to allow thorough flushing. See also **Steam-in-place**.

Clearance flow

Any flow less than the minimum dictated by the valve rangeability.

Closed-loop control system

A control system in which the measurement of process variables corrects, if necessary, the position of a control element. Consider a householder sitting in the living-room of a house with gas/water central heating. The living-room is fitted with a room thermostat for temperature control. The householder goes to the kitchen leaving the living-room door open. Cold air from the hall enters the living-room reducing the temperature. The thermostat detects the fall in temperature and sends a signal to the boiler to maintain the water temperature. If the radiator was fitted with a thermostatic control valve, the sensor would detect the fall in temperature. The thermostatic valve would open and allow more hot water to flow through the radiator. The hot water temperature would fall prompting the boiler to turn on or increase the fuel flow.

CMA

Chemical Manufacturers Association (of USA).

Co

Chemical symbol for Cobalt. It may be present in steels as an impurity added with nickel. It is added specifically to tool steels to improve hot strength. Cobalt becomes highly radioactive when irradiated; specifications for nuclear applications restrict the cobalt concentration to very low levels.

Cock

A small manual valve used for low pressure applications. Usually a ¼ turn ball or plug valve. An imprecise term subject to interpretation and should not be used.

Code of Practice

A document, very much like a standard, issued by any of the standards organisations, which explains the best way to do something to achieve satisfactory results. A Code of Practice can only be mandatory if precise instructions are given and there are no alternatives. Phrases using "should" and "may" can only be used for giving guidance and advice.

Cogging

An intermediate rolling process when a hot ingot is reduced to a bloom or slab in a cogging mill.

COGUANOR

Comisian Guatemalteca de Normas, the National Standards Authority of Guatemala.

Coil bound

The condition of helical springs when compressed until all the coils touch. The spring length when coil bound is called the solid length. Some spring designs are not suitable for compression to solid length and permanent plastic deformation occurs rendering the spring useless for its intended purpose. See also **Permanent set**.

Cold differential set pressure

When a spring-loaded valve which will normally operate hot is preset on a cold test rig, the cold set pressure must be increased slightly so the valve will operate at the correct pressure when hot. Typical corrections are shown in Table 1.7.

Operating temperature	% age increase in cold set pressure
-18 to 120°C	0
121 to 315°C	1
316 to 425°C	2
426 to 530°C	3

Table 1.7 Cold differential set pressures

Cold drawing

A technique for improving the surface finish and correcting the sizes of pipes, tubes, bars, shapes or wire. The semi-processed material is drawn through a series of dies which polish the surface and reduce the diameter/shape slightly to the correct size. The material is work-hardened and the enhanced mechanical properties can be utilised provided it is not welded or heated.

Cold pull

Tension applied to a piping system during erection when cold. Thermal expansion, on heating, relieves the initial stresses.

Cold-setting compound

An adhesive, liquid gasket, thread seal or impregnation compound which hardens or thickens at ambient temperature without the application of external heat.

Cold water finish see **Mirror finish**

Cold working

Altering the shape or size of a metal by plastic deformation. Processes include rolling, drawing, pressing, spinning, extruding and heading. It is carried out below the recrystallisation point usually at room temperature. Hardness and tensile strength are increased with the degree of cold work whilst ductility and impact values are lowered. The cold rolling and cold drawing of steel significantly improves surface finish.

Column separation

When a non-return valve closes, the liquid travelling towards the valve may create water hammer. The inertia of the liquid travelling away from the valve may create column separation; that is the liquid may divide and create a vapour pocket. Dissolved gas can also be released in these situations. If, when the inertia energy is dissipated, the liquid column reverses and returns to the non-return valve, a classic cavitation condition is created.

Compliance

Requirements in specifications and standards must be verifiable in order to establish compliance. Any requirement which cannot be verified is worthless. CEN editorial policy dictates that requirements must be verifiable at the manufacturer's works prior to dispatch. A simple example may clarify the situation.

An EN standard may state that cast steel globe valves shall be painted with a total paint thickness of 100 µm. A third-party inspector can inspect a finished valve, measure the paint thickness, and state categorically that the valve does or does not comply with the painting requirements. An EN standard cannot state valve packing shall have a minimum life of 2 years and the leakage rate shall not increase by more than a factor of 2. This requirement is not verifiable in the manufacturer's works, before shipment, and cannot, sensibly, be considered as part of a purchase agreement.

The requirement is also application specific and not suitable for inclusion in a standard. Standards can contain notes which are explanatory or describe policy. A note could be as follows, "Valves complying with standard are intended to be heavy-duty and packing should be suitable for a 2 year working life." This would be a policy statement, not an enforceable, verifiable requirement.

Compressibility factor Z

A factor used in gas applications to convert the ideal gas formula to real gas. The basic gas equation, $pV = mRT$, becomes $pV = ZmRT$. Values for Z are obtained from charts depending upon the operating pressure and temperature.

Compression coupling

A pipe fitting which relies on the compression of a ferrule to seal the pipe and locate the pipe axially. Compression fittings with one or two metal ferrules are limited to 1" or 2" od Imperial tube but are capable of high to very high pressures; fittings for thin wall copper tube are available up to 108 mm. Non-metallic ferrules can be used for lower pressure applications with tubes up to 125 mm od.

CONACYT

Consejo Nacional de Ciencia y Tecnología, the National Standards Authority of El Salvador.

Concentric serrated

A flange finish which uses a similar straight V groove to the spiral serrated finish but consists of a set of concentric grooves, not a continuous spiral groove.

Contact corrosion

When two dissimilar metals are in contact without a protective barrier between them and they are in the presence of liquid or other conducting material, an electrolytic cell is created. The degree of corrosion is dependent on the area in contact and the electro-potential voltage of the metals concerned. The less noble of the metals is liable to be attacked, i.e. zinc will act as a protector of steel in seawater whereas copper or brass will attack the steel in the same environment.

Continuous casting

A method of producing ingot, blooms, billets and slabs in long lengths using water cooled moulds. The castings are continuously withdrawn through the bottom of the caster whilst the teeming of the metal is proceeding. The need for primary and intermediate mills and the storage and use of large numbers of ingot moulds is eliminated. The continuous cast product is generally cleaner than batch cast material and of a more homogenous quality.

Control algorithms

Controllers, especially intelligent electronic process controllers, can decide, on the basis of the process variables measured, the best way to return the process to the desired set point. Process systems can have mechanical and thermal inertia. If a control valve opens to increase the flow rate on a long liquid pipeline, nothing happens immediately. The mass of liquid in the pipeline must be accelerated and this can take some time, often minutes rather than seconds. The valve could be opened until the maximum working pressure was achieved and this would accelerate the liquid as fast as possible. The speed of heating or cooling materials depends upon the mass and the specific heat. The controller can apply some heat and measure the effect over a short period. The controller can then decide whether to apply more or less heat to return to the set point.

Mechanical and pneumatic controllers can be purely proportional controllers i.e. the control signal issued is in proportion to the deviation from the set point. Maximum restoring effort can only be applied once the maximum set deviation has been reached. As the variable returns to approach the set point, restoring effort is reduced to avoid over-shoot. Pneumatic controllers can have an integration function added to complement the proportional mode, PI. The integral function is the integral of the deviation from the set point. This means the controller output changes as long as the deviation exists. Very small deviations can produce full controller output if they persist long enough.

Pneumatic controllers with proportional control can have a derivative function added, PD. The derivative effect is to anticipate the system response to the control signal. Pneumatic controllers can incorporate a derivative function to complement the proportional and integral functions. The derivative function measures the rate of change of the deviation and anticipates how the process is going to react to the controller output. The three term controller, PID, is very useful for processes with large inertias. Electronic controllers are usually of the PID type.

Controlled atmosphere

A gas or mixture of gases in which steel is heated to produce or maintain a specific surface condition. Controlled atmosphere furnaces are widely used in the heat treatment of steel as scaling and decarburisation of components is minimised by this process. In the case of steel this refers to a component that has been case-hardened where the centre is softer than the hard surface layer or case. It can also be applied to the central part of a rolled rimming steel.

Control valve

A valve which opens and closes in response to an external signal and is powered by an external source. Similar to a regulator but requires an external supply of electricity, compressed air or pressurised liquid. The control signal may be electric analogue, electric digital or fluid analogue.

Controlled bore pipe

Pipe which is specially manufactured to close tolerances. Ordinary pipes may have a 12.5% tolerance on the wall thickness which leads to considerable variation in the bore size. Controlled bore pipes use strict manufacturing controls to ensure the bore size varies from 1.5% or 3.2 mm in the larger sizes.

Controller

A device which measures a process variable against a set point and issues an adjusting signal to correct any deviation. It may be mechanical, pneumatic or electric/electronic. The smallest controllers measure one process variable. Large controllers can supervise many process variables.

Conventional safety relief valve

A spring-loaded valve with the bonnet or spring case vented to the outlet side of the valve.

COPANIT

Comisión Panameña de Normas Industriales y Técnicas, the National Standards Authority of Panama.

Cover

Another name for the bonnet.

CPRU

Construction Planning and Research Unit, Ministry of Development, the National Standards Authority of Brunei Darussalam.

Cr

The chemical symbol for the element chromium, sometimes abbreviated to "chrome". Chrome is added in small quantities to steel, less than 3%, to improve the effectiveness of heat treatment and hardenability and to increase the strength. When the content is increased to between 10 and 30%, the steel acquires stain resistance or becomes stainless, i.e. less likely to oxidise.

Creep

The gradual permanent change in dimensions of solid materials when exposed to long term tensile stress. Non-metallic materials tend to creep much faster than metallic materials. High temperature is not necessary but it will accelerate the effect.

Crossbar

The plate attached to the outer ends of the pillars, on linear valves, which supports the handwheel.

Crow-foot guided

A method of guiding valve plugs from the seat bore. The plug is manufactured with three, four or five slender extensions which slide in the seat bore and centralise the plug. The plug may be characterised by shaping the ports between the extensions.

CSNI

Czech Standards Institute, the National Standards Authority of the Czech Republic.

CSSN

China Standards Information Center, the National Standards Authority of China.

Cu

The chemical symbol for the element copper, which is usually present in steels as a trace impurity. It is added to some chrome-nickel alloys to impart precipitation hardening properties. It is added to some stainless steels to improve the strength and corrosion resistance at the expense of elasticity, and possibly castability.

C_v see **Flow coefficient**

Cyanide hardening

A process of introducing carbon and nitrogen into the surface of steel by heating it to a suitable temperature in a molten bath of sodium cyanide, or a mixture of sodium and potassium cyanide, diluted with sodium carbonate and quenching in oil or water. This process is used where a thin case and high hardness are required.

D

Damper

A hydraulic device which can be fitted to power-actuated valves to eliminate spindle/stem vibration and/or control actuation speed.

Dashpot

A device which relies on fluid friction to produce a force which opposes the motion. This can be a piston in a cylinder or a disc in a close-fitting housing. The opposing force increases as velocity increases.

Data-logging

The recording of process variables for subsequent review and analysis.

Data sheet

A formatted sheet or sheets which fully describes the operating conditions the valve will experience. Operating conditions are not limited to the internal process fluid. External environmental factors can play a decisive part in the valve selection and the design of auxiliary equipment. See Chapter 19 for a full list of essential information.

dB

deciBel; a tenth of a Bel, is the ratio between two sound power or pressure levels.

dB(A)

A noise scale which is weighted to approximate the characteristics of the human ear. The weighting adjusts the sampling time and the amplitude of the octave bands. Table 1.8 shows some relative sound pressure levels.

Noise source	Sound pressure level dB(A)
Ordinary conversation	60
Heavy traffic	80
A party	90
An underground train	100
A loud pop group concert	100 to 130
A jet plane at close range	150

Table 1.8 Typical sound pressure levels

DBB

Double block and bleed; a single valve or two valves which seal process fluid from two directions. The space between the seals is vented to indicate leakage.

DCS

Distributed Control System; a control system which uses more than one computer, PLC or controller. The logic for control decisions is distributed over the control system.

Dead band

The range through which the controlled variable can reverse and change in magnitude without initiating a regulator response.

Decarburisation

The loss of carbon from the surface of steel as a result of heating in a carbon weak atmosphere. During the rolling of steel hot surfaces are exposed to the decarburising effects of oxygen in the atmosphere and as a result the surface is depleted of carbon. In steels where the components are to be subsequently heat treated it is necessary to remove the decarburised surface by machining.

Degassing

The release of trapped gases from metallic and non-metallic materials when exposed to much lower pressures than previously experienced.

Delamination

The separation of a surface layer due to over-stressing the bond or failure of the adhesive.

Deoxidation

The addition of silicon and/or aluminium to molten steel to combine with free oxygen and form stable solid oxides. The solubility of oxygen in steel is reduced as temperature is lowered during solidification and the excess oxygen combines to form carbon monoxide. If the molten metal is not deoxidised the effervescence produced by the evolution of carbon monoxide during solidification would result in blow holes and porosity. Steel treated in this way is termed, "**Killed Steel**".

DFI

Design, fabrication and installation

DGN

Dirección General de Normas, the National Standards Authority of Mexico.

DGSM

Directorate General for Specifications and Measurements, the National Standards Authority of Oman.

Diameter Nominal (DN)

A European metric size which approximates the valve or pipe size. Similar to nominal bore, see **nb**.

Diamond Pyramid Hardness Test see **Vickers Hardness**

Diaphragm bonnet

A style of bonnet which utilises a flexible diaphragm, metallic or non-metallic, to isolate completely the operating mechanism from the components in contact with the process fluid. The diaphragm does not influence the flow of process fluid as in the case of weir type and full-bore diaphragm valves. The diaphragm may be sealed with the body by an "O"-ring or gasket or may be welded. The diaphragm bonnet may incorporate secondary seals as a back-up in the event of diaphragm failure.

Diaphragm spring

A disc spring with multiple radial slots which converts the solid single disc spring into multiple finger springs. Usually used individually; not built up into stacks.

Die-formed rings

Most stems/spindles are sealed by soft packing — rectangular or square rings braided from strands or filaments. Rings are cut from a straight length and fitted individually into the packing box. The rings can shrink 35 to 50% in the packing box when subjected to high compressive loads. The amount of shrinkage can be reduced to 10 to 15% by pre-compressing the rings before fitting. Pre-cut rings are placed in dies which are subjected to loads which exceed the normal loads experienced in the packing box. Once die-formed, the rings maintain their volume without the rapid initial reduction usually found during bedding in. Long term volume reduction is greatly reduced.

Differential pressure reducing regulator

A differential pressure regulator which maintains a constant pressure difference between the regulator outlet connection and an external pressure signal.

Differential pressure regulator

A regulator which maintains a constant differential pressure, i.e. maintains a constant difference between upstream and downstream pressures.

Differential pressure relief regulator

A differential pressure regulator which maintains a constant pressure difference between the regulator inlet connection and an external pressure signal. An extension of the back pressure regulator principle.

Digital signal

A signal, usually electrical but not a prerequisite, which carries digital information in a series of pulses. A sequence of on-off pulses transmits a binary number.

DIN

Deutsches Institut für Normung, the German National Standards Authority.

Direct-acting actuator

Some actuators can be fitted to valves in more than one way. This allows the actuator action to be changed for a constant control signal. Control valves can be made to open or close in response to the same signal. Direct-acting usually refers to the effect of negative feedback; on increasing control signal the valve closes. See also **Reverse-acting**.

Direct-acting regulator

A regulator which utilises the process fluid directly without amplification. The size of these valves is limited because the mass and friction of the control element becomes large in comparison to the fluid pressure forces available for actuation. Increasing inertia leads to reduced accuracy. The process sensing connection is usually internal and external pipework is unnecessary.

Direct-operated

An American term used to describe the action of a thermostatic system on a control valve.

Disc

The moving element in a butterfly valve. The disc can be plain; a parallel plate; or specially profiled to reduce operating torque and increase rangeability.

Disc spring

A linear spring produced by deforming a disc to form part of a cone. The spring force is generated when attempts are made to flatten the disc. The disc spring has very high spring rates and can withstand very high loads. Disc springs can be stacked together to increase or decrease the overall spring rate. When stacked to increase the spring rate, the influence of friction between the discs provides inherent damping.

Diverting valve

A valve with at least three connections which can divert flow from one circuit to another.

DOM

Drawn Over Mandrel. Welded pipe and tube can be straightened and the bore adjusted to close tolerances by subsequent cold drawing over a mandrel.

Dome

The pressurised chamber above the piston in a process pilot-operated safety relief valve.

Dome-loaded regulator

Another name for a **Pressure-loaded regulator**.

Double-beat construction

A type of globe valve which has two seats. The fluid flow divides and flows through the seats in opposite directions balancing the dynamic forces on the plug.

Double flange body

The normal valve body construction; two standard flanges which require bolts/nuts or short stud-bolts/nuts for installation; which is expected when no other description is specified. The description may be applied to valve types which are normally associated with flangeless or wafer bodies.

Double-port construction

The American description for a double-beat valve. Double-port is confusing because in this context port is used to refer to seats. In "three-port", port refers to process connections. It is more correct to use double-beat to avoid confusion.

Drift

The slow change in set point over an extended period of time.

Droop

Another name for offset.

DS

Dansk Standard, the National Standards Authority of Denmark.

DSSU

State Committee on Technical Regulation and Consumer Policy of Ukraine, the National Standards Authority of Ukraine.

DTNM

Dirección de Tecnología, Normalización y Metrología, the National Standards Authority of Nicaragua.

Duplex stainless steels

A high strength, corrosion resistant group of stainless steels, iron-chrome-nickel alloys, where the crystal structure is a mixture of body-centred cubic ferrite and face-centred cubic austenite. The balance between the two crystal groups is usually about 50-50 but heat treatment can change the proportions. The ferrite content is usually controlled to avoid cracking or lack of toughness or susceptibility to stress corrosion cracking but a minimum percentage is required to ensure the strength.

Ductility

The property of materials which permits it to be reduced in cross-sectional area, while experiencing tensile stress, without fracture. In a tensile test, ductile metals show considerable elongation eventually failing by necking, with consequent rapid increase in local stresses.

Ducting

Square or rectangular pipes constructed from thin gauge sheets or low pressure hose usually designed for handling air. One or more metal spiral wires are incorporated in the hose to maintain the shape. The wires can also act as conductors to allow anti-static applications.

Dust ignition-proof

A style of enclosure design that excludes dust in sufficient quantities which might affect performance with regard to internal arcs, conduction, heating or ignition, and to restrict the enclosure surface temperature to prevent the ignition of external dust accumulation or clouds.

Duty rating

For valves whose components react to changes in the process system, it is very important to assess the system variations, the impact on the valve service life and the maintenance interval. Valves with poor bearing designs will wear quickly which may

Duty description	Running hours/day
Intermittent	0 to 3
Light	3 to 8
Continuous	8 to 24
Cyclic	User to quantify
Irregular	User to quantify

Table 1.9 Duty ratings

affect valve sealing dramatically when experiencing unforeseen cycles. Duty ratings are indicated in Table 1.9.

Non-return valves (nrvs) are frequently used in conjunction with compressors and pumps. When used with a machine rated as intermittent, the valve would only be expected to open/close once per day, and perhaps not every day. The same would apply to light duty, but with everyday operation expected. With a continuous operation machine the valve may open and not close for weeks or months. Conversely, the standby machine and valve may lie idle for weeks or months, dry or with stagnant liquid, waiting for something to happen. Cyclic operation can be the most arduous of conditions. Machines can operate in response to pressure or level signals. A typical operating regime could be:

- operating conditions 10 barg @ 80°C
- operating cycle for 48 weeks
- 7 days per week
- pump start at 07.30, run for 1 hour
- 09.00 to 20.00, pump run for 5 min, stop for 15 min
- 20.15 pump start for flush out, run for 10 min

Reciprocating pumps and dosing pumps can be used to inject liquid into pressurised systems. A non-return valve can be fitted to eliminate any backflow. A typical operating cycle could be:

- operating conditions 100 barg @ 20°C
- operating cycle for 16 weeks
- 7 days per week
- 24 hours per day
- valve opens and closes 100/min

Dye penetrant inspection

Colloquially known as "dye-pen", a method for detecting surface porosity or cracks in metal. The part to be inspected is thoroughly cleaned and coated with a dye which penetrates any surface flaws that may be present. The surface is wiped clean and coated with a white powder. The powder absorbs the dye held in the defects indicating their location.

Dynamic back pressure see **Built-up back pressure**

Dynamic unbalance

The force on a plug, or torque on a disc, at any open operating condition with specific liquid properties and pressures.

DZNM

State Office for Standardization and Metrology, the National Standards Authority of Croatia.

E

Eccentric disc

A modified type of disc used in control butterfly valves. The disc is mounted so that it moves axially away from its seat as it opens. See Figure 3.38 in Chapter 3, for examples of disc mounting and design.

ECISS

The European Committee for iron and steel standardisation, an advisory committee to CEN.

EFW

Electric Fusion Welded; pipe and tube which is rolled from plate and has a welded seam with filler metal.

Electrical equipment classification

There are various methods of manufacture for safe electrical equipment. Not all types are acceptable in all zones. Table 1.10 indicates where equipment may be used.

Zone	Equipment type	Full description
Zone 0	Ex ia	Intrinsically safe with 2 faults
Zone 1	Ex ia	Intrinsically safe with 2 faults
Zone 1	Ex ib	Intrinsically safe with 1 fault
Zone 1	Ex d	Flameproof
Zone 1	Ex p	Pressurised
Zone 1	Ex e	Increased safety
Zone 1	Ex q	Powder filled
Zone 1	Ex m	Encapsulated
Zone 2	Ex N	(constructed to avoid sparks and hot spots)
Zone 2	Ex o	Oil immersed + all equipment suitable for Zone 1

Table 1.10 Electrical equipment for use in hazardous zones

EHEDG

The European Hygienic Equipment Design Group, makes recommendations on the design of hygienic equipment. Both EHEDG and the American 3-A Sanitary standards state a surface finish of 0.8 μm or better is suitable for food production. Rougher surfaces maybe acceptable if cleanability can be proved. Materials must be wear resistant and corrosion resistant to the product(s) and cleaning solutions.

Elastic limit

The maximum stress, tensile or compressive, that can be applied to a metal without producing permanent deformation. When external forces act upon a material they produce internal stresses within it which cause deformation. If the stresses are not too great the material will return to its original shape and dimension when the external force is removed.

Elasticity

The property which enables a material to return to its original shape and dimensions after stressing by external forces.

Electrical zones

Electrical safety from explosions is primarily defined by a zone type which describes the probability of a flammable gas or dust hazard, see Table 1.11.

European designaton	Definition	USA designation
Zone 0	An area in which an explosive gas-air mixture is continuously present or present for long continuous periods. > 1 000 hours/annum	Class 1 Division 1
Zone 1	An area in which an explosive gas-air mixture is likely to be present for only a short period during normal operating conditions. > 10 hours/annum	Class 1 Divison 1
Zone 2	An area in which an explosive gas-air mixture is not likely to be present during normal operating conditions, but if it was present it would only be for a short time. > 0.1 hour/annum	Class 1 Divison 2
Zone Z	An area in which a combustible dust-air mixture may be present during normal operating conditions	
Zone Y	An area in which a combustible dust-air mixture may be present during abnormal operating conditions	
Safe area	An area in which an explosive gas-air or dust-air mixture will not be present	

Table 1.11 Electrical area classifications

Electronic process controller

A controller which evaluates one or more process variables and issues corrective signals to restore the process to a predetermine state. Controllers can have pre-programmed control algorithms, such as **PID**, to return the process variable to the set point as quickly and as smoothly as possibly. Electronic controllers can have more than one algorithm and select the best logic to manage the situation. Controllers can auto-tune themselves to the process conditions; that is, the controller remembers what is necessary to produce the desired results. Auto-tune can be single-shot or continuous optimisation. Controllers using many process variables and continuous optimisation are sometimes called multi-variable predictive controllers, **MVC**.

Controllers can have independent alarm settings; these can be based on absolute values or deviation from a set point. Some electronic process controllers have data-logging memory fitted and recent process history can be recalled for review. Electronic process controllers can be used to operate control valves; the process variables being wired directly to the controller. The process input data may be interrogated remotely and the controller settings adjusted remotely via a serial or parallel communications network, thus forming a distributed control system, **DCS**. Electronic process controllers can have enclosures to IP65 and better; the controller can be mounted in the field, close to the relevant equipment.

Some controller manufacturers have developed particular models or designed special control algorithms to suit specific common process control situations. Manufacturers should be consulted before embarking on costly in-house solutions for standard processes.

Electro-pneumatic positioner

A pneumatic valve positioner which responds to an electrical control signal, typically 4-20 mA, rather than a 0.2 to 1.03 barg pneumatic signal. An air supply is necessary to actuate the diaphragm or piston. See **Positioner**.

Electro-pneumatic transducer

This is a misnomer used by some manufacturers, see **I/P converter**.

Electro-polished

A good surface finish produced by electrochemical machining. Used to enhance the surface texture of components which have been mechanically polished by reducing the variations in peaks and troughs but also polishing at a macro level. Electro-polishing is effectively, the reverse of electroplating and produces a very uniform, smooth surface finish with the added benefit, when used on chrome alloys, of improving the integrity and strength of the oxide layer essential for corrosion resistance.

Electroslag refining

A specialised steel making process in which a rolled or cast ingot in the form of an electrode is remelted in a water cooled copper mould. The melting is activated by resistive heat generated in a conductive slag. The resulting product has a similar basic chemical composition to the original ingot, but is characterised by high purity and low inclusion content.

Elongation

A measure of the ductility of a material; how much a material can stretch before it fractures.

ELOT

Hellenic Organization for Standardization, the National Standards Authority of Greece.

End quench test see **Jominy Test**

Endurance limit

The maximum stress which can be applied, repeatedly, to a material without suffering fatigue failure.

EO

Egyptian Organization for Standardization and Quality Control, the National Standards Authority of Egypt.

Equal percentage

A flow characteristic at constant differential pressure which results in equivalent increases in flow for increases in valve travel. A change of 10% travel produces a change of 10% in the flow.

ESA

European Sealing Association

ERW

Electric Resistance Welded. Pipe and tube which is rolled from plate and has a welded seam without filler metal.

ESD (1)

Electrostatic discharge; the discharge of static electricity. Operators can carry a 10 kV potential. ESD can destroy integrated circuits, such as computer chips, and erase memory chips. Electronic equipment must be protected from static electricity.

ESD (2)

Emergency shutdown system; an independent control system within a process plant dedicated to stopping production safely. The ESD uses independent process measurements to detect faults which would cause serious problems. The ESD can also be initiated manually by strategically located pushbuttons or key-switches. The ESD system can use control and isolating valves fitted for normal operation. ESD commands override all normal process control instructions. The ESD system is usually individually piped or hardwired to all its actuators. Electric circuits may be double wired via different routes for higher reliability. Isolating and control valves with pneumatic actuators can have the ESD pneumatic signal injected via a solenoid valve which would normally be closed.

ET

Eddy-current testing (**NDE**)

EVS

Eesti Standardikeskus, the National Standards Authority of Estonia.

Explosion relief valve

These valves are designed to react very quickly and relieve large quantities of low pressure gas or vapour. The name is a misnomer, the valves should be called safety valves because the action is pop. On detecting a very small pressure rise the valve opens wide to pass excess gas or vapour. The valve recloses when operating conditions have returned to normal.

Explosive decompression

A problem which can occur with liquefied gases which destroys gaskets and seals. Many liquids are handled which would normally be a gas at atmospheric pressure. During normal operation the liquid can be absorbed by the gasket/seal material. If a sudden loss of pressure occurs, a broken pipe connection for example, the liquid flashes to vapour as it escapes. Liquid absorbed in gaskets/seals expands rapidly and also changes to vapour. The rapid expansion and sudden increase in volume can destroy the gasket/seal. An associated problem is the sudden drop in temperature as the liquid cools to the boiling temperature at atmospheric pressure. If explosive decompression is a risk, then special attention must be given to gasket materials and elastomers and the low temperature ductility of metallic parts.

Explosion-proof

A type of enclosure which is designed to withstand the explosion of an internal flammable gas/vapour mixture without igniting and flammable gas/vapour mixture which might surround the enclosure.

Exposure limits

Fluids and dusts which are hazardous to personnel are regulated by controlling the amount of fluid or dust in the local air. Allowable concentrations are based on long term exposure over an 8 hour day and short term exposure, isolated incidents, over 10 minutes. The values shown in Table 1.12 are taken from the UK's Health and Safety Executive document EH40 and show the variation in allowable exposure for extreme examples. It is the users responsibility to monitor and control the exposure levels experienced by personnel. Routine sampling must be performed, and results recorded, to ensure compliance with local regulations.

Substance	Formula	Long term exposure (8 hours)		Short term exposure (10 minutes)	
		ppm	mg/m^3	ppm	mg/m^3
Carbon dioxide	CO_2	5 000	9 000	15 000	27 000
Osmium tetraoxide	OsO_4	0.0002	0.002	0.0006	0.006

Table 1.12 Typical personnel exposure limits

Extended bonnet

A bonnet which is lengthened to separate the packing box from the bonnet bottom for hot or cold applications.

External pressure loaded regulator

A regulator which uses an external pressurised fluid supply as a reference signal. The external fluid can be applied to a diaphragm or piston. Changing the pressure of the external fluid supply changes the regulator set point.

F

Fatigue

The effect on materials of repeated cycles of stress. The dangerous feature of fatigue failure is that there is not necessarily an obvious warning, a crack forms without appreciable deformation of structure making it difficult to detect the presence of growing cracks. Fractures usually start from small surface nicks or scratches or abrupt changes in shape which cause localised stress concentrations. Failure can be influenced by a number of factors including size, shape and design of the component, condition of the surface or the overall operating environment.

Fault tolerance

Control systems need to be reliable. Overall reliability can be improved by installing hot standby redundant equipment. Electronic equipment can be designed so that the standby equipment assumes immediate control if a fault is detected in the primary device. Stored information is held in at least two devices.

FCAW

Flux Cored Arc Welding. An electric arc welding method where the arc is struck with the consumable which has a flux core. The flux provides the shielding but additional shielding, by a gas blanket, may be provided in some cases.

FDA

The Food and Drug Administration (of America); the organisation which sets the regulations for hygienic applications.

Feedback

A control signal which originates after the control element. A room thermostat for a central heating system sends feedback signals to the control system. Feedback can be negative or positive. Negative feedback reduces the input as the output is detected to be increasing. Positive feedback increases the input as the output increases.

Feedforward

A control signal which originates before the control element and allows pre-emptive action before the critical set-point has changed. Consider a house with central heating and an external thermostat. When the outside air temperature falls to 2°C the external thermostat can turn on the central heating to prevent the water pipes from freezing. The heating would start before the internal temperature had fallen to an unsafe level.

Ferrite

A body-centred cubic crystal structure found in chromium alloys.

Ferritic stainless steel

A term usually applied to a group of stain resistant steels with a chromium content in the range of 12 to 30% and whose structure consists largely of ferrite. Alloys with chromium content at the bottom of the range will not be "stainless" in the generally accepted sense of the term. Such steels possess good ductility and are easily worked but do not respond to any hardening or tempering processes and have poor high temperature properties. Types of applications include automotive trim and architectural cladding.

Ferrule

A small cylindrical component, used in a compression fitting, which becomes attached to the pipe od after fitting. Compression fittings can have one or two ferrules depending upon the design.

Fieldbus

A serial communications network with two or four cores. Some fieldbus specifications use an extended version of the RS485 multi-drop system. Many proprietary fieldbus specifications currently exist. Profibus, P-net and World-FIP are European Standards in EN 50170. An international Standard, IEC 1158, has combined what is claimed to be the best features of Profibus and World-FIP. It is hoped that IEC 61158 will become the world Standard in the future.

Fire-safe

A valve which is suitable for installations where the risk of being engulfed by fire cannot be ignored. Valves are tested to a specification and qualified, see Chapter 15. Most valves which qualify are ball and butterfly. Linear valves are often said to suffer from stem thermal problems but some wedge gate valves have qualified. Most wafer and lug valves are generally not suitable.

Flame hardening

A surface hardening process in which heat is applied by a high temperature flame followed by quenching jets of water. It is usually applied to medium to large size components such as large gears, sprockets, slide ways of machine tools, bearing surfaces of shafts and axles, etc. Steels most suited have a carbon content within the range 0.40 to 0.55%.

Flameproof

An enclosure design which does not allow sparks, arcs, hot gas or hot components to ignite any surrounding flammable gas mixtures.

Flanged bonnet

A bonnet which is attached to the valve body by a flange and bolts or studs and nuts. The flange need not necessarily be integral with the bonnet. High pressure or large diameter valves may use a loose flange, of a different material, to hold the bonnet in position.

Flanged connection

A pipe joint which uses two discs held together by bolts. Many flange designs are available for high and low pressure. Flanged connections are usually the preferred method of joining pipes for 3" nb, DN 80 and larger.

Flare coupling

A coupling which forms a seal with the pipe wall. The pipe end is shaped to match a tapered adapter and then clamped to the adapter.

Flash welding

A type of electric resistance welding.

Flow coefficient C_v, K_v

A factor, determined experimentally, which indicates the flow capacity of incompressible fluid during non-choked, non-flashing flow of a valve with a unit differential pressure. Two flow coefficients are in common use; the American C_v and the metric K_v. Equations 1.1 and 1.2 relate to liquid flow. Liquid flow coefficients are based on turbulent flow conditions for a water-like liquid.

$$C_v = Q\sqrt{\frac{sg}{\Delta p}} \qquad \text{Equ 1.1}$$

where:

C_v = US flow coefficient (USgpm/psi)

Q = liquid flow (USgpm)

sg = liquid specific gravity (non-dimensional)

Δp = differential pressure (lb/in^2)

$$K_v = Q\sqrt{\frac{\rho}{\Delta p}} \qquad \text{Equ 1.2}$$

where:

K_v = metric flow coefficient ($m^3/h/bar$)

Q = liquid flow (m^3/h)

ρ = liquid density (kg/dm^3)

Δp = differential pressure (bar)

Other definitions of liquid flow coefficient exist and are in use. If a flow coefficient value seems dubious check with the manufacturer.

Equations 1.3 and 1.4 show the approximate flow coefficients for valves handling gases. Equations 1.5 and 1.6 show the modified versions for low pressure air at 20°C when the density is approximately 1.29 kg/m^3. Equations 1.7 and 1.8 show the approximate flow coefficients for steam valves.

NOTE: See Section 8.4.7, in Chapter 8, for a detailed explanation of compressible flow.

Valid

when $p_2 > \frac{p_1}{2}$

and $\Delta p < \frac{p_1}{2}$

$$K_v = \frac{Q_n}{514}\sqrt{\frac{\rho_n \cdot T_1}{\Delta p \cdot p_2}} \qquad \text{Equ 1.3}$$

Valid when $p_2 < \frac{p_1}{2}$

and $\Delta p > \frac{p_1}{2}$

$$K_v = \frac{Q_n}{257 p_1}\sqrt{\rho_n \cdot T_1} \qquad \text{Equ 1.4}$$

Valid when $p_2 > \frac{p_1}{2}$

and $\Delta p < \frac{p_1}{2}$

$$K_v = \frac{Q_n}{\sqrt{26.4\,\Delta p \cdot p_2}} \qquad \text{Equ 1.5}$$

Valid when $p_2 < \frac{p_1}{2}$

and $\Delta p > \frac{p_1}{2}$

$$K_v = \frac{Q_n}{13.2\, p_1} \qquad \text{Equ 1.6}$$

Valid when $p_2 > \frac{p_1}{2}$

and $\Delta p < \frac{p_1}{2}$

$$K_v = \frac{G}{31.6}\sqrt{\frac{v_2}{\Delta p}} \qquad \text{Equ 1.7}$$

Valid when $p_2 < \frac{p_1}{2}$

and $\Delta p > \frac{p_1}{2}$

$$K_v = \frac{G}{31.6}\sqrt{\frac{2v_x}{p_1}} \qquad \text{Equ 1.8}$$

where:

K_v = gas flow coefficient ($m^3/h/bar$)

p_1 = inlet pressure (bara)

p_2 = outlet pressure (bara)

Δp = differential pressure (bar)

Q_n = normalised flow (1) (m^3/h)

ρ_n = gas density (1) (kg/m^3)

T_1 = absolute inlet temperature (K)

G = mass flow (kg/h)

v_2 = specific volume (2) (m^3/kg)

v_x = specific volume (3) (m^3/kg)

(1) at 0°C and 760 mm Hg

(2) from steam tables at p_2 and T_1

(3) from steam tables at $p_1/2$ and T_1

The flow coefficient is measured in a test rig using inlet and outlet piping which is the same size as the valve connections. Using pipework which is larger or smaller, with reducers for diameter transition, will modify the effective valve maximum C_v/ K_v. The installed valve flow coefficient may vary from 30% to 160% depending upon the valve design and the pipework configuration.

NOTE: See Section 8.4.5 in Chapter 8 for effects on the flow coefficient caused by choked flow.

Flow control valve, FCV

An automatic regulating valve which is actuated by a control signal derived from a flow sensing element.

Flow regulator

An application of a constant differential pressure regulator. The regulator maintains a constant differential pressure across an orifice which ensures a constant flow.

Flow ring

A ring which can be fitted to some ball valves, in place of the seats, which is more resistant to erosion damage but does not seal as well.

Flow-to-close, FTC

The direction of flow which tends to close the valve. In single-seat valves, it is the direction in which fluid flows over the sealing element and then out through the seat.

Flow-to-open, FTO

A linear motion valve fitted in a system so that the fluid flows through the seat and passed the plug to the outlet. The hydraulic process forces tend to open the valve.

Flowing pilot

A pilot valve used with a pilot-operated safety valve which relies on a flow of process fluid, while the safety valve is relieving, for proper functioning.

Fluid

The term fluid is a generalisation. It describes a group of "states" — liquids, vapours, gases and plasmas. The term could be used to describe mixtures of these states and also mixtures with solids. Fluid is a vague term and should be used with great caution. It is not, as many people seem to think, synonymous with liquid.

Flutter

The rapid oscillation of the disc resulting in significant changes of lift. The disc does not touch the seat or the lift stop. Significant wear of the valve guides may occur. Flutter may be caused by high fluid friction losses in the inlet or outlet pipework creating unstable flow conditions with consequent pressure pulsations in the system.

FONDONORMA

Fondo para la Normalizaci6n y Certificaci6n de la Calidad, the National Standards Authority of Venezuela.

Foot valve

A non-return valve installed at the bottom of a pump suction pipe to prevent the suction pipe and pump draining when the pump is stopped. Fitting a foot valve removes the requirement to vent and prime the pump before every start.

Forging

A process of working metal to a finished shape by hammering or pressing and is primarily a "hot" operation. It is applied to the production of shapes either impossible or too costly to make by other methods or needing properties not obtainable by casting, such as inherent integrity. Categories of forgings include hammer, press, drop or stamping.

FSA

Fluid Sealing Association (International, based in USA)

FTC see **Flow-to-close**

FTO see **Flow-to-open**

Fugitive emissions

These are leaks of hazardous fluids from pressurised systems. Leaks can be from gland seals, pipe connections or through porous metal components.

Full bore

A valve which has the same flow area as the attached pipework. This concept is not as simple as it seems. Pipework is produced in discrete sizes and a valve will be matched to a particular wall thickness with a similar pressure/temperature rating. Always check actual sizes. Refer to **Reduced-port** for an indication of problems.

Fusible plug safety device

A device, usually a screwed plug, of suitable low temperature melting point material which seals the fluid. On rising temperature, the fusible material weakens and is forced from the tapped hole. The plug cannot close after the fault condition has been rectified and must be replaced with a new plug.

G

Gas

A state of matter, achieved by heating beyond the superheated vapour phase to a temperature higher than the critical temperature, in which the molecules move randomly to fill the volume available. A gas cannot be converted to liquid by simply increasing the pressure. The volume of a gas changes easily to accommodate changes in the confining volume. The pressure and temperature of an ideal or perfect gas can be calculated by using the characteristic equation and Boyle's and Charles' Laws. Ideal gas properties can be converted to real gas properties by including compressibility factors in the characteristic equation. Gases can exist in mixtures, air, and can be dissolved in liquids, like fresh water or seawater. Compressors are designed to handle gases. Special compressors, like refrigeration compressors, are designed to handle gases and vapours, sometimes with trace quantities of liquid.

Gas carburising

A heat treatment method used in the case-hardening of steel. Carbon is absorbed into the outer layers of the components by heating in a gas flow, rich in carbon compounds. The process is more versatile than some other methods as the depth of the case and the limiting carbon content of the case can be controlled by the composition of the atmosphere, the dew point and the temperature.

Gas groups

Flammable gases are divided into groups depending upon their ignition temperature. See BS EN 60079-14 or IEC 79-4.

Gateway

A protocol converter for bus communication systems. A gateway allows information to be transmitted to a another system which uses a different protocol. Multiple gateways can be connected to a primary link. For example, gateways to several different serial fieldbus networks can be connected to a single computer via one RS232 or RS485 serial link.

Gauge plate

A low alloy tool steel supplied in flat and square section with the surfaces ground to close limits. It is also known as Ground Flat Stock and is used for manufacturing accurate brackets and mounting assemblies.

Gauge pressure

A pressure measurement relative to the local atmospheric pressure. Normal pressure gauges indicate this relative pressure. Units are usually suffixed with "g", but this common practice is frowned upon by the Standards Authorities.

GDBS

Grenada Bureau of Standards, the National Standards Authority of Grenada.

Geared handwheel

Both linear and rotary motion valves can be operated directly by handwheels or levers attached to the stem or spindle. As valve sizes and operating pressures increase the torque required to operate them also increases. At some point it is necessary to fit torque multiplication so that personnel can operate a valve relatively easily. A bevel gearbox or worm drive may be fitted. A geared handwheel may be fitted as an option on small valves to reduce water hammer effects.

Nominal Value R_a	micrometre	50	25	12.5	6.3	3.2	1.6	0.89	0.8	0.64	0.51	0.4	0.28	0.2	0.1	0.05	0.025
	microinch	2000	1000	500	250	125	63	35	32	25	20	16	11	8	4	2	1
Roughness number		N12	N11	N10	N9	N8	N7		N6			N5		N4	N3	N2	N1
Polish number								#4		#5	#6		#7				
Grit number								150		180	240		320				

Table 1.13 Different surface finish designations

Gland

The flange or special nut which retains and/or compresses the packing in the packing box.

GMAW

Gas Metal-Arc Welding, another name for **MIG welding**, Metal Inert-Gas welding.

GNBS

Guyana National Bureau of Standards, the National Standards Authority of Guyana.

GOST-R

Federal Agency on Technical Regulating and Metrology, the National Standards Authority of Russia.

Grey iron

Grey cast iron, also known as "flake iron", is the normal cast iron used for many applications and is available in many grades (strengths).

Grit number

A number which quantifies the number of silicon carbide grains per square inch of polishing tape or disc. As the number of grains per square inch increases the grain size must reduce and the maximum quality of the surface finish increases. Grit numbers have been superseded by actual surface finish designations. See Table 1.13.

Grooved coupling

A type of clamped pipe coupling which relies on a groove cut in each pipe end for axial restraint. An elastomer across the pipe ends provides the fluid seal.

GRP

Glass-fibre reinforced plastic

GTAW

Gas Tungsten-Arc Welding, another name for **TIG welding**, Tungsten Inert-Gas welding.

Guide bushing

A replaceable bushing, considered as a stationary trim component, within a valve to align the moving element with the seat.

Guidelines

Guidelines are advice and must be written in an appropriate form, using "should" and "may". Guidelines cannot be enforced or verified as with Standards and Regulations.

H

Hammer lug coupling

A heavy duty threaded coupling used for oilfield applications. Pressure ratings start at 345 barg and increase to 1380 barg.

Hard facing

A method of increasing the wear resistance of a metal by the deposition of a hard protective coating. Alloys such as Stellite or Colmonoy, or a metallic carbide or a ceramic oxide are most often used for the coating. Increased corrosion resistance is an additional benefit.

HAZ

Heat affected zone; the parent material immediately next to a weld; the material which could have been "heat-treated" by the welding process. Welding can change the local hardness and/or modify the grain structure of the parent material.

Hazardous fluids

These are fluids, which by nature of their chemical properties, create a potential for human injury, damage to property, to the environment or a combination of these. There is no European or International agreement on which fluids should be classed as hazardous. The UK Health and Safety Executive has a document, EH40, regarding the safety of personnel. In America there is a useful document produced by the National Fire Protection Agency, NFPA 325M. Another good source of information is the requirements for the transportation by road of dangerous goods, commonly known as ADR. A similar agreement applies to rail transportation, RID. These documents are published in the local language, not just English, French and German, and cover over 20 countries. For flammable fluids useful data can be found in IEC 79-12 which identifies gases and vapours for potentially explosive atmospheres for electrical equipment. Dust clouds can create a potential hazard and these should be checked rather than assuming dust to be safe.

HAZOP

Hazard and Operability Study

Heat-curing compound

An adhesive, liquid gasket, thread seal or impregnation compound which must be heated to above the ambient temperature to start curing.

Helical spring

The type of compression coil spring used in packing boxes and actuators. A linear motion spring coiled from round wire or square section strip. Very accurate springs can be machined from the solid.

Hex nipple

A piece of hexagon bar with both ends turned and threaded with a male pipe thread. Used to connect two fittings.

HFS

Hot Finished Seamless; pipes of all sizes, but usually having thicker walls.

High pressure change-over valve

A type of regulator which automatically selects between two feed systems depending upon the downstream pressure. For rapidly filling vessels a low pressure high flow supply, from a centrifugal pump, can be used. As the vessel pressure approaches the pump closed valve head, the change-over valve changes supplies to a low flow high pressure source, possibly from a positive displacement pump. See Figure 5.2, Chapter 5.

High recovery valve

A valve design which only dissipates a small amount of energy and has a low pressure drop.

High Speed Steel (HSS)

The term "high speed steel" was derived from the fact that it is capable of cutting metal at a much higher rate than carbon tool steel and continues to cut and retain its hardness even when the point of the tool is seen to glow red. Tungsten is the major alloying element but it is also combined with molybdenum, vanadium and cobalt in varying amounts. Sometimes used for valve seats and other small components prone to wear.

HIP

Hot Isostatic Pressing; a technique using high pressure and high temperature to convert casings to forgings.

HIPPS

High Integrity Pressure Protection System

HMI

Human, machine interface; another name for **MMI**.

Hot finished

A type of pipe which is made when the metal is hot and does not have any manufacturing operations after the metal has cooled. Usually covered by a thin scale and the surface finish may be a little uneven.

Hot-melt compound

An adhesive, liquid gasket, thread seal or impregnation compound which is heated to a molten phase before application.

HRc

Rockwell hardness on Scale "C"

H_2S trim

A valve with all contact materials in compliance with NACE MR-01-75, latest edition.

HT

Hardness testing (**NDE**)

HV

Vickers Hardness

Hydraulic lock

A situation usually encountered in gas systems when liquid has become trapped in a confined volume which should only contain gas, and prevents the movement of components.

Hygienic

A general term used to designate equipment suitable for food production; animal and human food. CEN uses the term "agrifoodstuffs" as an all encompassing description for any product consumed orally. Hygienic does not imply equipment is suitable for Biotechnology.

Hygienic quality

Some valves can be manufactured for processes which require cleanliness to be maintained at the highest levels. Initial cleanliness is ensured by mechanical design, material choice and polished surface finishes. Mechanical design must remove crevices where product can be trapped and decay. Also the valve must not contain dead volume, that is, volume which is not constantly exposed to moving product. Special connectors are required. Only approved materials may be used. All surfaces are polished to reduce the likelihood of product sticking.

Hysteresis

The greatest difference in the controlled variable in relation to a single set of operating conditions after increasing, then decreasing, the control signal. The effect of the dead band is subtracted.

I

IANOR

Institut algarien de normalisation, the National Standards Authority of Algeria.

IBN

The Belgian Institution for Standardization, the National Standards Authority of Belgium.

IBNORCA

Instituto Boliviano de Normalizaci6n y Calidad, the National Standards Authority of Bolivia.

ICONTEC

Instítuto Colombiano de Normas T6cnicas y Certificaci6n, the National Standards Authority of Columbia.

ID

Internal diameter; the bore of a pipe or tube.

IDF

The International Dairy Federation; a trade association which has standardised hygienic pipe union fittings.

IEC

The International Electrotechnical Commission, the electrical branch of ISO.

IEEE

Institute of Electrical and Electronic Engineers (of America)

IIW

International Institute of Welding

ILC

A proprietary hygienic fitting based on the best features of IDF and RJT fittings; capable of much higher pressures.

Impregnation

The sealing of a collection of very small pores using a chemical compound or the binding of electrical windings with an insulating compound.

Impulse line

A small bore pipe which carries a pressure signal from a process pipe to an instrument. The detailed design of the impulse line depends upon the process fluid, the size and the operating requirements. If the process fluid is hazardous the design specification may exclude the use of screwed connections. In this context compression fittings are not screwed as the union nut thread is not in contact with the fluid. If the distance between the process pipe and the instrument is great, a primary isolation valve, at the junction with the process pipe, may be specified. The primary isolation valve may be specified as flanged. At the instrument end, facilities for block and bleed, double block and bleed, equalise and/or calibration may be required. Bleeding can be accomplished by a plug or a valve. Hazardous fluids must be piped into a closed drains system, so plugs would be unacceptable. Because of the number of facility variations and permutations it is essential for the designer or safety officer to define the detailed requirements precisely.

Inclusions

Usually non-metallic particles contained in metal. In steel they may consist of simple or complex oxides, sulphides, silicates and sometimes nitrides of iron, manganese, silicon, aluminium and other elements. In general they are detrimental to the mechanical properties but much depends on the number, their size, shape and distribution.

Increased safety

A design approach which reduces the probability that unusual heating or a dangerous arc can be produced which may ignite a flammable gas/vapour mixture which may be present during unusual operating conditions.

INDECOPI

Instituto Nacional de Defensa de la Competencia y de la Protección de la Propiedad Intelectual, the National Standards Authority of Peru.

IndiaBIS

Bureau of Indian Standards, the National Standards Authority of India.

Induction hardening

A widely used process for the surface hardening of steel. The components are heated by means of an alternating magnetic field to a temperature within or above the transformation range followed by immediate quenching. The core of the component remains unaffected by the treatment and its physical properties are those of the bar from which it was machined, whilst the hardness of the case can be within the range of 37 to 58 HRc. Carbon and alloy steels with a carbon content in the range 0.40/0.45% are most suitable for this process.

INEN

Instituto Ecuatoriano de Normalización, the National Standards Authority of Ecuador.

Ingot

The mass of metal that results from casting molten steel into a mould. An ingot is usually rectangular in shape and is subsequently rolled into blooms and billets for rods, bars and sections and slabs for plates, sheet and strip. With the increasing use of the continuous casting process the ingot route is less used as the molten steel is now directly cast into a bloom or billet.

INN

Instituto Nacional de Normalización, the National Standards Authority of Chile.

INTECO

Instituto de Normas Técnicas de Costa Rica, the National Standards Authority of Costa Rica.

Intercrystalline corrosion

Chromium-nickel austenitic stainless steels are prone to this form of corrosion when they are welded and subsequently in contact with certain types of corrosive media. When the metal is heated to within a temperature range of 450 to 800°C precipitation of the chromium carbides takes place at the grain boundaries in the area of the weld and these areas no longer have the protection of the chromium on the peripheries of the grains. This type of corrosion is also known as Weld Decay and Intergranular Corrosion. The most common way to avoid the problem is to select a grade of steel that is very low in carbon i.e. 0.03% or less, or one that is stabilised with niobium or titanium.

Intrinsically safe (IS)

A type of electric circuit which limits the amount of electrical energy available for heating or the production of arcs so that surrounding flammable gas/vapour mixtures cannot be ignited.

I/O

Input/Output; the abbreviation used to describe the connection facilities of PLCs. 16 I/O would indicate a total of 16 connections for inputs and outputs. Large PLCs have thousands of I/O.

Inherent flow characteristic

The same as **Characteristic** when the differential pressure is maintained constant.

Inherent rangeability

The ratio of maximum to minimum flow within a specified tolerance band of the specified rangeability.

Inside screw

Used on linear motion valves to move the sealing element. The screwed part of the stem is below the packing and exposed to the process fluid. The nut is built into the lower part of the bonnet or the sealing element.

Installed flow characteristic

The control valve characteristic is defined at constant differential pressure. When the valve is fitted in a system the differential pressure is unlikely to remain constant over the control flow range. Consider a control valve with inlet and outlet pipework between a constant pressure supply and a constant pressure demand; the valve is partially open and a stable flow regime exists. There is a frictional pressure drop in the inlet and outlet pipework. The control valve opens to increase the flow rate. The frictional pressure drop in both inlet and outlet pipes increases. The total differential pressure of the system is constant so the pressure drop across the control valve is reduced. The increase in flow is not as great as expected because the control valve differential pressure is reduced; the valve must open wider. The frictional components of practical piping systems together with flow-pressure relationships of the supply and demand modify the valve characteristic.

Interlocking

The function of restricting valve operation depending upon the state of another valve or piece of equipment. Control systems can be interlocked electrically by requiring switches mounted on several pieces of equipment to be in certain positions before a circuit can be made. Hand-operated valves can be interlocked mechanically by using special keys. Mechanical and electrical equipment can be interlocked this way. See Chapter 14.

Internal safety valve

A spring-loaded relief valve which is designed for mounting inside pressure vessels. The complete valve assembly is mounted within the vessel and all components are exposed to the fluid.

I/P converter

A device which converts an analogue electric signal into an analogue pneumatic signal. Typical electrical input of 4-20 mA dc would produce an output of 0.2 to 1.03 barg (3 to 15 psig). The converter must have an air supply; usually about 0.4 bar above the maximum output pressure. Some I/P converters include a voltage circuit so that 1 to 10 V dc can also be converted to pneumatic.

IP

Institute of Petroleum

IPPC

Integrated Pollution Prevention Control

IPQ

Instituto Português da Qualidade, the National Standards Authority of Portugal.

IRAM

Instituto Argentino de Normalización y Certificatión, the National Standards Authority of Argentina.

ISIRI

Institute of Standards and Industrial Research of Iran, the National Standards Authority of Iran.

ISO pipe threads

Metric versions of BSP.

ISO

International Standards Organisation

Isolating valve

A valve which is intended to be wide open or closed, with no flow regulation functions.

ISS

International Sanitary Standard; a largely obsolete standard for hygienic pipe unions. For most practical purposes ISS has been replaced by, and is interchangeable with, IDF unions. 4" unions are not interchangeable because the threads are different.

ISSM

Institution for Standardization of Serbia and Montenegro, the National Standards Authority of Serbia and Montenegro.

IST

Icelandic Council for Standardization, the National Standards Authority of Iceland.

ITCHKSAR

Innovation and Technology Commission, the National Standards Authority of Hong Kong.

Izod impact test

A test specimen, usually of square cross-section is notched and held between a pair of jaws, to be broken by a swinging or falling weight. When the pendulum of the Izod testing machine is released it swings with a downward movement and when it reaches the vertical the hammer makes contact with the specimen which is broken by the force of the blow. The hammer continues its upward motion but the energy absorbed in breaking the test piece reduces its momentum. A graduated scale enables a reading to be taken of the energy used to fracture the test piece.

To obtain a representative result the average of three tests is used and to ensure that the results conform to those of the steel specification the test specimens should be machined and polished accurately. Charpy testing is now more popular.

J

JBS

Bureau of Standards, Jamaica, the National Standards Authority of Jamaica.

JISC

Japan Industrial Standards Committee, the National Standards Authority of Japan.

JIS

Jordan Institution for Standards and Metrology, the National Standards Authority of Jordan.

Jominy Test

Sometimes known as an **End quench test**, it is used to assess the hardening ability of steels.

K

KATS

Korean Agency for Technology and Standards, the National Standards Authority of the Republic of Korea.

KAZMEMST

Committee for Standardization, Metrology and Certification, the National Standards Authority of Kazakstan.

KEBS

Kenya Bureau of Standards, the National Standards Authority of Kenya.

Killed steel

The term indicates that the steel has been completely deoxidised by the addition of an agent such as silicon or aluminium, before casting, so that there is practically no evolution of gas during solidification. Killed steels are characterised by a high degree of chemical homogeneity and freedom from porosity.

Knoop hardness test

A micro hardness test in which a pyramid shaped diamond is pressed into the surface.

KOWSMD

Public Authority for Industry, Standards and Industrial Services Affairs, the National Standards Authority of Kuwait.

K_v see **Flow coefficient.**

KYRGYZST

State Inspection for Standardization and Metrology, the National Standards Authority of Kyrgyzstan.

L

Lamination

A problem which can occur with rolled raw materials, like steel plate. The bulk material is not homogeneous and can separate into layers.

Laminated packing

Packing can be formed by laminating fabric layers and curing with suitable bonding agents such as rubber. Many fabric and bonding agent combinations are possible to suit operating conditions.

Lantern ring

A spacer/bush used in packing boxes to allow the injection of lubricant, the collection of leakage and the application of a water seal, to the spindle/stem. Lantern rings are usually metallic but plastic or ceramic can be used. A short spacer with material removed from the central portion to provide storage or collection volumes in the bore and around the od. Internal and external volumes are connected by a series of holes or slots.

LCO_2

Liquid carbon dioxide

Leaded steels

When added to steel, lead does not go into solution but exists in a very finely divided state along the grain boundaries. It greatly assists machinability as it acts as a lubricant between the steel and the tool face. Lead is normally added in amounts between 0.15 to 0.35% and when combined with similar amounts of sulphur, optimum machinability is attained. The suffix Pb may be added to the steel designation.

Leakage rates

For substances hazardous to the environment the allowable leakage rates are controlled by local, national or country grouping legislation. There is no complete list of hazardous substances but a list entitled Red List Chemicals or NFPA 325, is available for guidance. Local inspectors evaluate the site and issue an operating licence which specifies maximum emissions. More than one licence may be necessary for a site.

For substances hazardous to personnel, but not to the environment, the allowable leakage rate is dependent upon local conditions; it is the concentration in the air which is important. Totally enclosed processes, with little or no air movement, will achieve high concentrations with low seal leakage. Open processes, with very large air volumes and continuous air movement, will be able to tolerate larger leakage rates without creating dangerous concentrations. Leakage will generally be dispersed downwind.

There is a problem in defining environmental hazards which are not direct personnel hazards. No complete European list is available and the American NFPA 325 standard includes a reactivity hazard rating based on a scale of 0 (zero) to 4. Substances with a rating of 1 or higher would be considered unacceptable for unmonitored emissions. Discussions with local pollution inspectors are necessary to clarify the containment requirements.

Level regulator

An ambiguous term which should not be used. In some applications level regulator is used to describe a float valve. It can also be used to describe an altitude valve.

LIBNOR

Lebanese Standards Institution, the National Standards Authority of Lebanon.

Limiting ruling section

The maximum diameter of cross-section of a bar or component in which certain specified mechanical properties can be achieved after heat treatment. Figures are often quoted in standards for alloy steels, fastener material standards for example. Stainless steel standards and duplex steels, frequently omit this data and this can create manufacturing and operational problems for larger components. Corrosion problems can be experienced when solution annealing, for maximum corrosion resistance, or tempering, for strength, cannot be performed correctly.

Linear

A flow characteristic which yields flow increases proportional to the valve opening, at constant differential pressure. If the valve opens to 50%, the flow will be 50%.

Linear valve

A valve style in which the moving element travels in a straight line when opening and closing.

Liquid

A liquid is a phase of matter which can be considered, for many practical applications, as a fixed volume with a variable shape. When subjected to significant pressure or temperature changes, liquids do change volume slightly. Liquids of low viscosity, like water, change shape rapidly. Liquids of significant viscosity, like syrup or molasses, change shape much more slowly. Liquids, like water, are Newtonian. Viscosity is independent of the rate of shear, and decreases with increasing temperature and is slightly influenced by pressure. Some liquids are "shear-sensitive" or "pseudo-plastic" and do not flow easily until a critical shear stress is applied, and then the properties of the liquid may change with time, like custard. Liquid is a very important phase of matter; pumps are designed specifically to handle liquid, not fluid. Many valves are designed specifically for liquids.

Liquid trim

Some manufacturers have design variations specifically for liquids.

Live-loaded gland

A gland which uses springs to maintain a load on the gland when the packing compresses. See Figure 11.12, Chapter 11.

Live-loaded packing

Packing which uses a spring to apply the load when fitted in a non-adjustable packing box. See Figure 11.11, Chapter 11.

LN_2

Liquid nitrogen

Loader

An American name for a **Pressure reducing regulator**.

Lockable valve

A manual valve which is fitted with additional lugs on the yoke or gland to allow a padlock to be fitted to lock the valve open or closed.

Locked-in

The condition experienced by pipes and equipment between two closed isolating valves. If the ambient temperature rises significantly, due to a fire for example, the internal pressure may rise to an unacceptable level. Thermal relief valves are fitted to remove hazards. If the ambient temperature fell significantly, a partial vacuum maybe created internally. If a vacuum was undesirable, a vacuum breaker could be fitted. If it was not permissible to introduce air, then a suitable fluid would have to be supplied.

Lockshield valve

A manual valve which is operated with a removable handle, sometimes called a key. The top of the gland is extended to surround the stem completely and thus prevent unauthorised adjustment with a spanner. The top of the stem is sometimes machined to a triangle or special rounded shape to prevent a spanner from gripping.

Lockup

The deviation in the controlled variable when the flow is reduced from the minimum controllable flow to zero flow, see Figure 1.2.

LST

Lithuanian Standards Board, the National Standards Authority of Lithuania.

Lubricator

A connection on the packing box to facilitate packing lubrication.

Lug body

A valve body which is designed to be clamped between two flanges with long stud-bolts. The valve body has lugs through which the stud-bolts pass. Some lug body designs do not allow the use of stud-bolts and the two lug faces are tapped and have studs fitted. Valves with studded connections require much more space for fitting and removal. Drawings must be checked for actual design. Tap bolts could be used in place of studs, but these will wear the holes in the valve if regular dismantling is planned.

LVS

Latvian Standard, the National Standards Authority of Latvia.

M

Mach Number M

The relationship of the fluid velocity to the velocity of sound in the fluid. Most liquid applications operate at low velocity and Mach Number is irrelevant. However, pressure waves generated by water hammer and surge travel through the liquid at the acoustic velocity. As a rough guide the local velocity of sound in liquids can be taken as 1500 m/s. Some gas and vapour applications operate at high velocity and the local velocity of sound can be important for step changes in characteristics. The local velocity of sound in gases is given by Equation 1.9. The Mach Number is a simple ratio as shown in Equation 1.10.

$$a = 91\sqrt{\frac{\gamma \cdot T}{m}} \qquad \text{Equ 1.9}$$

where:

a = velocity of sound (m/s)

γ = ratio of specific heats (non-dimensional)

T = absolute temperature (K)

m = molecular weight (kg/kmol)

The velocity of sound in low pressure air can be approximated from $a = 20.02\sqrt{T}$. For approximations using dry steam use γ = 1.135, for superheated steam use 1.3.

$$M = \frac{c}{a} \quad \text{Equ 1.10}$$

where:

M = Mach Number (non-dimensional)

c = fluid velocity (m/s)

a = velocity of sound (m/s)

NOTE: The velocity of sound in gases and liquids varies with both temperature and pressure. The presence of solids or gas/vapour bubbles reduces the acoustic velocity very significantly.

Machined seals

Some materials which make excellent seals are very difficult to mould into suitable cross-sections. Most of these materials can be extruded or cast into tube or hollow bar. Seal profiles can be machined from the raw material which is a useful option for the production of seals to special dimensions. Both one-piece and split rings can be produced.

Machined springs

Coil springs, or helical springs are usually manufactured by winding round, or square, wire around a mandrel. Coil springs can be machined from solid bar or a tube to produce a more accurate spring, without the constraint of standard wire gauges or wire materials.

MACT

Maximum Achievable Control Technology

Magnetic crack detection

The material to be tested is magnetised by passing a heavy current through it or by making it the core of a coil through which a heavy current is passed. Cracks or inclusions cause the magnetic flux to break the surface forming free magnetic poles. When the component is sprayed with a suspension of finely divided magnetic particles they collect at the free poles to visibly show the presence of surface defects.

Major repair

During the production of special valves the purchaser and the manufacturer agree definitions of major and minor repairs which are permitted. A major repair is classed as a hold point on the quality plan and reclamation cannot proceed without the purchasers agreement. The purchaser must agree firstly that the component may be repaired. Secondly the purchaser and the manufacturer agree the repair method. The distinction between major and minor is usually defined by:

- location —

  - is the flaw in an accessible location for pre- and post inspection as well as repair?

  - is the flaw in a functionally critical location?

  - is the flaw in a highly stressed region?

- size —

  - the depth of the flaw as a percentage of the pressure retaining wall thickness

  - the length of the flaw, as a percentage of a descriptive dimension of the component

  - the area of the flaw, as a percentage of the surface area of a region of the component

  - the weight of repair material required, as a percentage of the weight of the component

- material —

  - some materials are very easy to repair, by welding for example, and others are very difficult to repair

- component —

  - repairs to certain components may be classed as major irrespective of the type of flaw

Malleable iron

Grade of cast iron where the carbon content is largely converted to graphite to provide better ductility.

MAM

A metric version of **AMS**.

Mandatory

National government legislation is mandatory; it must be observed and obeyed. Certain sections of standards are mandatory for compliance.

Man-made fibres

Braided packing can be manufactured from man-made fibres or in combination with natural fibres. Popular man-made fibres include: Aramid carbon (carbonaceous), glass, graphite, metal and PTFE.

Martensitic stainless steel

An alloy of iron, carbon and chromium which can be hardened by heat treatment, are magnetic but are not as corrosion resistant as other grades of stainless steel. The chrome content is usually between 10 and 18% with a carbon content over 1%. Carbon allows the material to be very hard and provides a degree of wear resistance.

Matter

What everything is made of. Matter can exist in different phases, depending upon the temperature and pressure: solid, liquid, vapour, gas or plasma. Note especially, that fluid is not a state of matter. Fluid covers the collection — liquid, vapour, gas and plasma. Fluid is vague and should be used carefully and only when appropriate. Fluid is not interchangeable with liquid!

Mechanical resonance

Severe vibration caused by the coincidence of an exciting frequency with a pipework natural frequency. Unacceptable levels of vibration may be transmitted to building structures or through foundations. Pipework failure due to fatigue may result if the resonance persists for an extended period of time.

Meter valve

A type of needle valve. Most needle valves are small and the needle has a sharp point. Meter valves are larger, of heavier construction and the needle has a rounded end to reduce wear problems. Meter valves are suitable for continuous throttling and should be capable of isolation.

Metric threads

A metric thread to ISO R/262 and French Standard NFE 03013 is popular for connecting pipe fittings to equipment. This is a parallel thread and uses a flat gasket for sealing. Table 1.14 shows the important dimensions.

Thread designation	Minimum facing mm	Diameter at top of thread mm	Root diameter of thread mm	Thread pitch	Minimum length of thread mm
M10 x 1	18	9.97	8.92	1.0	8.5
M12 x 1.5	20	11.97	10.34	1.5	9.0
M14 x 1.5	22	13.97	12.34	1.5	10.0
M16 x 1.5	24	15.97	14.34	1.5	11.0

Thread designation	Minimum facing mm	Diameter at top of thread mm	Root diameter of thread mm	Thread pitch	Minimum length of thread mm
M18 x 1.5	28	17.97	16.34	1.5	11.0
M20 x 1.5	30	19.97	18.34	1.5	12.0
M22 x 1.5	32	21.97	20.34	1.5	12.0
M24 x 1.5	34	23.97	22.34	1.5	13.0
M27 x 2	37	26.96	24.8	2.0	14.0
M33 x 2	42	32.96	30.8	2.0	16.0
M36 x 2	49	35.96	33.8	2.0	16.0
M45 x 2	58	44.96	42.8	2.0	18.0
M48 x 2	58	27.96	45.8	2.0	20.0

Table 1.14 Main dimensions of metric threads for pipe fittings

MF

Mechanical flowsheet, now generally called **P&ID**

MIG welding

Metal arc Inert-Gas welding is an electric arc welding process where the arc is struck with the filler metal. The weld site is protected from atmospheric contamination by an inert gas shield, argon, helium or carbon dioxide being the most popular.

MIL

American Military specifications

Minimum controllable flow

The lowest flow at which the control variable can be maintained at a constant value, see Figure 1.2.

Minor repair

Read **Major repair** first. If a repair is not prohibited or major then it is minor. The manufacturer inspects and repairs the component but informs the purchaser of his actions. All documentation regarding the inspection and repair must be available for review by the purchaser.

Mirror finish

A very smooth surface finish applied to flange facings. Usually obtained directly by turning with a large radius tool at high speed. The actual surface finish should be between 0.8 and 1.8 µm. Also called **Cold water finish** because it is popular with water companies.

MMA

Manual Metal Arc welding. The arc is struck between the workpiece and the consumable which is covered (shielded) by a flux coating. The consumable and the flux melt in the weld pool. The melted flux covers the weld to prevent contamination. This is the most common type of manual arc welding.

MMI

Man, machine interface. A keypad or keyboard, with or without text/graphical display, which allows operators to communicate with the control system.

Mn

The chemical symbol for manganese, a very important element for steel alloys. Manganese is present in the lowest alloys of steel and is added to improve hot working (during steel manufacture), and to increase strength, toughness and hardness. Manganese refines the steel grain structure by forming austenite. It improves wear resistance and facilitates welding.

Mo

Molybdenum is alloyed with high strength steel to improve the high temperature physical properties. It is added to stainless steels to increase resistance to pitting corrosion. "Moly" has a favourable influence on the welding qualities.

Modified parabolic

A flow characteristic which is very close to linear. 50% opening produces slightly less than 50% flow at constant differential pressure. Also known as an exponential characteristic.

Modulating

To adjust the valve position to any point between closed and wide open in response to a control signal. The alternative to modulating is on-off or step control, see the Special note on Digital valves in Chapter 6.

Modulating-action

Once a safety valve has started to open because the set pressure has been reached, the valve opens sufficiently to pass the required flow. As the valve flow changes the lift adjusts accordingly. This type of valve action produces small pressure waves when the valve opens and closes but not usually sufficient to cause problems.

Moulded seals

Finished one-piece seal rings can be moulded from most elastomers, plastic or leather. Dies are required for every size, incurring tooling costs and an initial investment which increases production costs. They are an ideal production method for large quantities of seals. Seals can include fabric reinforcement to strengthen the seal and increase rigidity. Seals can be moulded to shapes which allow the rings to be self-energised. The tolerances of finished seals can be a problem and must be strictly controlled. "O"- rings, square rings and lobed rings, also called X rings or QUAD© rings, are moulded and can be used as seals.

MOLDST

Department of Standardization and Metrology, the National Standards Authority of Moldova.

MSA

Malta Standards Authority, the National Standards Authority of Malta.

MSB

Mauritius Standards Bureau, the National Standards Authority of Mauritius.

MSS

The Manufacturers Standardization Society of the Valve and Fittings Industry, Incorporated of America. MSS produces specifications and guidelines for valves, fittings and piping installations.

MSZT

Magyar Szabványügyi Testület, the National Standards Authority of Hungary.

MT

Magnetic particle testing (**NDE**)

MTBR

Mean Time Between REpairs

N

N

Nitrogen, the chemical symbol is more usually written N_2. Nitrogen increases the austenitic stability of chrome-nickel stainless steels and greatly increases the strength, hardness and improves pitting corrosion resistance. Via a nitriding process, nitrogen can produce very hard surfaces while maintaining a tough core.

NACE

National Association of Corrosion Engineers (of America). Although the NACE organisation provides guidance on all types of corrosion its name has become synonymous with sour service applications where hydrogen sulphide may be a problem.

NAMUR

Dimensions for interfacing valves, actuators and auxiliaries. Became a German Standard as VDE/VDI 3845 and now an International Standard as ISO5211.

Natural frequency of a spring

The frequency at which the spring resonates. A spring design should be selected so that the natural frequency is not a harmonic of any significant frequency present in the fluid system. The spring must not be considered in isolation. The spring will always have a moving mass associated with it. The natural frequency of the spring-mass system must be calculated. Include of the spring mass for accurate calculations.

Nb

The element niobium (used to be known as Columbium) is added to some steels to stabilise carbon, in a similar way to titanium, but has the added benefit of improving elevated temperature strength.

nb

Nominal bore, the nominal size of a valve or pipe in inches. Nominal bore pipework has a fixed outside diameter and variable wall thickness to accommodate different pressure ratings, up to 12". Pipe sizes over 12" are designated by the outside diameter. In low pressure pipe the bore size may be larger than the nominal bore. In high pressure pipe the bore size can be considerably smaller than the nominal bore and the actual diameter must be checked when calculating flow velocities. The wall thickness of pipe is subject to manufacturing tolerances and ±12.5% is fairly typical. The pipe bore is therefore variable to some extent. The pipe wall thickness available for stressing and corrosion is also variable. **DN** is the metric equivalent nb.

NC

Oficina Nacional de Normalizaci6n, the National Standards Authority of Cuba.

NDE

Non-destructive examination

NDT

Non-destructive testing

NEN

Nederlandse Normalisateltsituut, the National Standards Authority of The Netherlands.

NFPA (1)

National Fire Protection Association (of America).

NFPA (2)

National Fluid Power Association Inc (of America).

Ni

The chemical symbol for nickel. An important element in alloying steel. In small quantities, nickel improves toughness. In quantities of 8% and greater, it works with chromium to produce the austenitic stainless steels typified by AISI 304 and AISI 316, probably the most popular stainless steels.

Nipple

A short length of pipe, usually of small diameter, 2" nb, DN50, or smaller.

NISIT

National Institute of Standards and Industrial Technology, the National Standards Authority of Papua New Guinea.

NIST

National Institute for Standards and Technology (of America).

Nitriding

A case hardening process that depends on the absorption of nitrogen into the steel. All machining, stress relieving, as well as hardening and tempering are normally carried out before nitriding. The parts are heated in a special container through which ammonia gas is allowed to pass. The ammonia disassociates into hydrogen and nitrogen and the nitrogen reacts with the steel penetrating the surface to form nitrides. Nitriding steels offer many advantages: a much higher surface hardness is obtainable when compared with case-hardening steels; they are extremely resistant to abrasion and have a high fatigue strength. Stainless steels, such as 17-4PH, can be nitrided to provide extreme wear resistance with only a slight loss of corrosion resistance.

Non-destructive testing

All those forms of testing that do not result in permanent damage or deformation to the part being tested. Typical examples include magnetic crack detection, ultrasonic inspection, X-ray inspection and gamma radiography. Some types of hardness testing do not leave any damaging impressions.

Non-flowing pilot

A pilot valve used with a pilot-operated safety valve which only passes sufficient process fluid to relieve the holding pressure above the disc.

Non-rising stem

Used on linear motion valves with inside screws. The nut is built into the sealing element which moves due to rotation only of the stem.

Non-rotating tip

The disc, plug or needle of a linear valve is attached to the stem but does not necessarily rotate. It is a good way to reduce wear when tightening the tip against the seat.

Non-shock

Usually applied to pressure ratings but applies equally to temperature. Equipment manufacturers design for maximum and minimum applied pressure and temperature. The designer expects the equipment to operate at steady-state conditions with only slow changes in pressure or temperature. In this context the term slow cannot be quantified exactly as it is dependent upon the detail design and the size of the equipment. Rapid pressure changes may be caused by surge, water hammer or cyclic flow variations from compressors or pumps. Mechanical shock may have the effect of doubling the material stresses.

Rapid temperature changes can be caused by using cold standby equipment without preheating or warming up. Thermal shock can create such high uneven stresses that the material cracks. A temperature change of 4°C/minute may be considered acceptable for most equipment. Whenever operating conditions are liable to change quickly the equipment ratings should be discussed with the manufacturer.

Normalising

A heat treatment process that has the object of relieving internal stresses, refining the grain size and improving the mechanical properties. The steel is heated to 800 to 900°C according to analysis, held at temperature to allow a full soak and cooled in still air.

Normally closed

An isolating or control valve which requires control pressures above atmospheric to open. If the actuating system fails and the control pressure is reduced to atmospheric, the valve will close. A control valve can also be called direct-acting.

Normally open

An isolating or control valve which requires control pressures above atmospheric to close. If the actuating system fails and the control pressure is reduced to atmospheric, the valve will open wide. A control valve may also be called reverse-acting.

Notched bar test

A test to determine the resistance of a material to a suddenly applied stress, i.e. shock. A notched test piece is employed in an Izod or Charpy machine impact test machine.

Notified Body

An organisation retained by the manufacturer to confirm that requirements such as the European PED or other regulations have been fulfilled.

Nozzle

The inlet portion of a safety relief valve which is the only pressurised component when the valve is closed. The nozzle has the valve seat which is sealed by the disc when closed. See also semi-nozzle.

Nozzle area

Also called orifice area. Safety valves are manufactured in standard sizes with specific nozzle areas listed by API. Special nozzle sizes can be supplied to match specific applications. Table 1.15 indicates the areas available from various manufacturers.

API Letter designation	Area in^2	Area cm^2
	0.080	0.516
	0.084	0.542
	0.096	0.619
	0.107	0.609
D	0.110	0.710
	0.131	0.845
	0.135	0.871
	0.176	1.13
	0.191	1.23
E	0.196	1.26
	0.210	1.35
	0.212	1.37
	0.231	1.49
	0.237	1.53
	0.300	1.93
F	0.307	1.98
	0.329	2.12
	0.368	2.37
	0.349	2.54
	0.490	3.16
G	0.503	3.24
	0.517	3.33
	0.766	4.94
H	0.785	5.06
	1.257	8.11
J	1.287	8.30
K	1.838	11.9
L	2.853	18.4
M	3.600	23.2
N	4.340	28.0
P	6.379	41.2
Q	11.5	71.2
R	16.00	103
T	26.00	168
V	42.19	272
W	60.75	387
Y	82.68	533
Z	90.95	587
Z_2	108.86	702
AA	136.69	882
BB	168.74	1089
BB_2	185	1193

Table 1.15 Safety relief valve nozzle areas

NPS

Nominal pipe size (see **nb**)

NPT

National Pipe Thread (of America), is a tapered pipe thread which seals the fluid. Thread tape or sealant is applied to the thread during assembly. The basic 60° thread form is cut on a taper of 0.75" per foot on diameter. The standard provides sizes up to 24" nb, but in practice 6" is about the largest used in process applications, see Table 1.16. Many users will not accept threaded fittings over 3".

Pipe nominal bore in	Approx pipe od in	Pitch diameter at beginning of thread in	Pitch diameter of perfect thread in	Number of threads per in	Minimum length of thread in
1/8	0.405	0.363	0.380	27	0.392
1/4	0.540	0.477	0.502	18	0.595
3/8	0.675	0.612	0.637	18	0.600
1/2	0.840	0.758	0.792	14	0.781
3/4	1.050	0.968	1.002	14	0.793
1	1.315	1.214	1.256	11.5	0.984
1 1/4	1.660	1.557	1.601	11.5	1.008
1 1/2	1.900	1.796	1.841	11.5	1.025
2	2.375	2.269	2.316	11.5	1.058
2 1/2	2.875	2.719	2.791	8	1.571
3	3.5	3.341	3.416	8	1.634
4	4.5	4.334	4.416	8	1.734
5	5.563	5.391	5.479	8	1.840
6	6.625	6.446	6.541	8	1.946

Table 1.16 Main dimensions of NPT threads

NSAI

National Standards Authority of Ireland, The National Standards Authority of the Republic of Ireland.

O

Oblique valve

A valve body style for linear valves which has in-line process connections and the valve actuating axis not at 90°.

Observed activity

During the production of special valves the purchaser or his representative or the Certifying Authority may express a desire to observe certain activities being performed. These activities may include valve assembly or routine production testing. The observer is informed of the timing of the activity and the activity is carried out whether the observer is present or not. Observed activities are not hold points on the quality plan.

Obturator

The moving sealing element in a valve.

Octave bands

Complex noise signals can be analysed and separated into discrete frequency signals or harmonics; each frequency would have a specific amplitude. Rotating and reciprocating machinery have the signal analysed without grouping discrete frequencies into sets. The raw harmonics can indicate specific problems with the machine. Noise sources which should not contain important harmonics have the discrete frequencies grouped into standardised octave bands. Each octave band would have an amplitude indicating whether the noise source was low frequency, high frequency or random. The standardised octave bands are designated by their centre frequencies.

Table 1.17 shows the octave bands generally used with their top and bottom cut-off frequencies.

31.5 Hz	63 Hz	125 Hz	250 Hz	500 Hz	1 kHz	**2 kHz**	4 kHz	8 kHz	16 kHz	
22.2 Hz	44.5 Hz	89 Hz	177 Hz	354 Hz	707 Hz	1.414 kHz	2.828 kHz	5.657 kHz	11.31 kHz	22.63 kHz

Table 1.17 Standard octave band frequencies

od

outside diameter; the outside diameter of pipe or tube.

Olive

Another name for a **Ferrule**

ON

Austrian Standards Institute, the National Standards Authority of Austria.

Open-loop control system

A control system in which the measurement of process variables does not influence directly the adjustment of the control system. Consider a gas fire in the living-room of a home. The householder sitting in the living-room feels cold and turns the gas fire up. Later, the householder feels hot and turns the gas fire down. There is no physical connection between the instrumentation and the operating mechanism.

Outside screw

Used on linear motion valves to move the sealing element. The screwed stem is exposed to the atmosphere above the gland. The nut is built into the handwheel or crossbar.

Overpressure

The maximum pressure attained when the safety valve is passing 100% flow. When the system pressure exceeds the set pressure the safety valve will open. The system pressure may increase or decrease depending upon the volume flowing through the valve and the valve action. The valve nozzle area is a function of the allowable overpressure. Higher overpressures mean smaller valves.

BS 5500 requires overpressures up to 110% of set pressure for gases and vapours and up to 125% for liquids. ASME, ANSI and API safety valves can be oversized to reduce the overpressure. Compressors and pumps with safety relief valves fitted may have the set pressure and overpressure selected to reduce driver power. Because safety relief valves are generally built in discrete sizes the actual overpressure for 100% flow should be confirmed in the valve quotation.

Over-torqueing

The over-tightening of fasteners. Over-torqueing is most common when trying to prevent leaks from flanged connection. Leaks are usually due to incorrect gasket materials, the use of old gaskets or warped flange faces.

Oxidation

The most common form of chemical reaction which is the combining of oxygen with various elements and compounds. The corrosion of metals is a form of oxidation, rust on iron and steel for example is iron oxide. Refined metals can be considered as unstable compounds; the natural stable state is their original mineral ores.

P

P

The element phosphorous, usually considered as a impurity in steels and refined to very low concentrations, 0.05% and lower.

P&ID

Process and Instrumentation Diagram, a schematic arrangement of a process showing all the equipment and all the associated instruments.

P & S

Phosphorous and sulphur are often quoted together as impurities in steel. Electric arc furnaces can reduce the concentration to less than 0.025%. Vacuum remelted products can be expensive to manufacture but have higher than normal cleanliness and integrity.

Packing box

The part of the bonnet which contains the seal or packing to reduce leakage along the valve stem.

Packing flange see **Gland**

Packing follower

A plain, close tolerance metal ring fitted on the atmospheric side of the packing and used to apply an even gland load.

Packing lubricator

An optional fitting on the bonnet to allow the packing to be lubricated while working.

Packing tape

Packing boxes can be sealed using thin tape which is rolled to fill the box. Usually only PTFE and graphite is available in this form.

Pallet valve

Another name for a dead-weight safety valve.

Panel mounting

A facility supplied on smaller valves which enables the valve to be mounted with the connections behind the panel. Usually applied to valves with screwed glands, extra nuts are provided to clamp the panel.

Parkerising

A chemical treatment applied to ferrous metals to improve their corrosion resistance. The process is based on a manganese phosphate solution which produces a fairly thick coating. This can subsequently be painted or impregnated with oil.

Patenting

A heat treatment process often applied to high carbon spring wire. The steel is heated to a suitable temperature well above the transformation range, followed by cooling in air or a bath of molten lead or salt. A structure is produced suitable for subsequent cold drawing and which will give the desired mechanical properties in the finished state.

Pb

The chemical symbol for lead. Lead is added to steel to improve machining qualities, sometimes in combination with sulphur, to make grades called "free-machining". The combination of lead and sulphur can increase the susceptibility to corrosion.

PC

Personal Computer, a desktop or laptop computer.

PD

Proportional and Derivative; see **Control algorithms**.

Permanent set

The difference in length of a spring after deflection to a specified length and being released.

Permeability

The ability of gases, vapours and liquids to pass through solid materials. Many materials have pores through which molecules can pass. Permeability can be a very serious problem with very hazardous fluids and elastomer seals.

Physical protection

The physical protection of an electrical enclosure is described by an IP rating. The rating consists of two numbers which define the protection against solids and liquids, see Table 1.18.

First number Protection against solid objects	Second number Protection against liquids
0 No protection	0 No protection
1 Protection against solids ≥ 50 mm	1 Protection against vertically falling drops of liquid
2 Protection against solids ≥ 12 mm	2 Protection against liquid falling at up to 15° from the vertical
3 Protection against solids ≥ 2.5 mm	3 Protection against liquid spray at up to 60° from the vertical
4 Protection against solids ≥ 1.0 mm	4 Protection against splashing liquid
5 Protection against large dust	5 Protection against water jets
6 Protection against dust	6 Protection against heavy seas
	7 Protection against intermittent immersion
	8 Protection against continuous immersion

Table 1.18 Electrical enclosure physical protection

IP ratings comply with IEC 529 for static equipment and IEC 34-5 for rotating equipment.

PI (1)

Pressure indicator

PI (2)

Proportional and Integral; see **Control algorithms**.

PID

Proportional, Integral, Derivative; see **Control algorithms**.

Pillars

Usually two round bars attached to the valve cover or bonnet which support the yoke, handwheel or actuator and guide the outer end of the stem.

Pilot-operated regulator

A regulator which uses a small independent valve to sense and regulate the flow of control process fluid into the main valve. By using a small pilot valve the main valve can be large without loosing control precision. The process sensing connection may be internal or external to a remote point. Multiple pilots can be used to sense different control parameters or a single parameter with different set points.

Pilot-operated safety relief valve

A safety relief valve which uses the process fluid pressure to hold the valve closed rather than a spring. Pilot-operated valves can be pop or modulating action. A pilot-operated valve may be set at a lower pressure, closer to normal operating pressure, and possibly reduce the system design pressure.

Also, the overpressure requirements may be reduced. The use of pilot-operated valves, in conjunction with positive displacement compressors and pumps, may result in smaller driver requirements.

Pilot regulator

A small regulator which senses the controlling pressure and supplies process fluid to actuate the main regulator.

Pipe

A conduit for carrying fluid which is designated by its nominal bore, nb, or its nominal diameter, DN. Pipes below 14" nb have unusual outside diameters i.e. 6.625", 10.75", etc. Pipes 14" and above use the od as the size. Pipe with an appropriate wall thickness can be threaded with standard pipe threads.

Pipe anchor

A pipe support which prevents pipe movement and rotation in all three planes.

Pipe flexibility

An assessment of the forces and moments applied by the pipe on its supports and end connections. Loads imposed by the pipework can vary considerably as temperature and/or pressure change. A very rigid pipe system can impose tremendous loads on valve connections. Some equipment, such as compressors, pumps and vessels, must be protected from high pipe loads and special pipe anchors must be incorporated in the pipe design to divert high loads directly into the support structures. Dynamic forces, from the movement of vibrating equipment, pulsating internal pressures or flow surges, can create unacceptable levels of pipework vibration if the flexibility is too high. The addition of pipe bends in the layout, to increase flexibility, can create unacceptable fluid flow patterns. The prime consideration of the piping design — to carry the process fluid safety and appropriately, can be compromised, unwisely, by poor details.

Pipe pup

The short lengths of pipe, welded to a valve body prior to assembly, used as a heat sink for field welds.

Pipe supports

A device for holding the pipe in one or two planes while allowing movement in the other plane(s). Most supports hold the pipe at a specific elevation but allow movement laterally and longitudinally. Pipe supports can be spring loaded to allow movement as pipe loads change. Non-metallic pipes are much more flexible than metal pipes and require more supports. As operating temperatures increase, non-metallic piping requires even more supports as the material becomes more elastic.

Piping Geometry Factor FP

A factor which adjusts the valve flow coefficient to accommodate process piping which is larger or smaller than the valve. Depending upon the installed pipework the valve can appear to have a flow coefficient between 30% and 160% of the test value.

Piping kick

The sudden movement of piping, quite often at a bend, in response to a pressure wave or a flow surge.

Piston

A special moving element in a linear motion valve which creates the variable area geometry in conjunction with the guiding cage. A piston seals on its circumference against the cage.

Pitot tube

A tube for measuring the static and dynamic pressure of a fluid. The tube end faces upstream or downstream in the fluid flow, not perpendicular to the flow, as a standard pressure tapping.

PKN

Polish Committee for Standardization, the National Standards Authority of Poland.

Plasma welding

Similar to **TIG welding**, where a tungsten electrode is used to create the electric arc which initiates the process. The parent metal is heated by a stream of argon plasma at about 20000°C and flows freely to form the joint. No filler material is used. The weld pool and the hot gas plasma stream are protected from atmospheric contamination by an argon gas shield. Very narrow welds are possible with small heat affected zones, HAZ.

Plastic flow

The gradual permanent change in dimensions of solid materials when exposed to long term compressive stress. Non-metallic materials tend to plastic flow much faster than metallic materials. High temperature is not necessary but it will accelerate the effect.

PLC

Programmable Logic Controller; a computer specifically designed for process control applications. A computer with input and output, I/O, connections for analogue and digital electrical signals and communication ports for other PLCs and computers. Most PLCs have an expansion port to allow more memory and I/O to be added. Special expansion modules, sometimes including another processor, can be added for communications with proprietary fieldbuses. The program is usually stored in a memory chip rather than disc as with commercial computers. Usually all the process logic, the program, must be supplied by the user. Programming can be by means of a handheld console or via a PC running special software. Some PLCs include self-diagnostic routines to validate the integrity of the hardware and software; this facility can be invaluable when problems occur.

Many manufacturers of PLCs still use their own programming language; programs are not generally transferable. Some large multinational corporations, who purchase numerous PLCs, have co-operated and insisted on standardisation to allow much easier cross platform migration. IEC 61131 defines five standard languages for PLCs. PLCs are generally programmed by flowchart, ladder logic or statement list. There are over 60 PLC manufacturers covering a very broad spectrum, Table 1.19 indicates some important variables.

Variable	Range
Memory kb	0.7 to 8000
I/O capacity	10 to 18 000
Speed of reading programme ms/1000 instructions	0.1 to 100

Table 1.19 Programmable logic controller limitations

PLCs are available in PC card format. Standard PCs have ISA expansion slots to allow additional hardware to be fitted. PLCs on a card can be fitted inside a normal PC. The PC becomes an MMI. The PLC can be powered from the ISA slot (when the PC is switched off the PLC is switched off) or from an external 24 V dc supply. The PLC programme is stored in FLASH RAM and is not volatile. The PLC can be programmed across the ISA slot from the host or externally via an RS 232 port on the card. The PLC can utilise standard external expansion cards via an interface.

Plug

A special moving element in a linear motion valve which creates the variable area geometry in conjunction with the seat or seat and cage. A plug seals on its end face or an angled portion on the end.

PN

Pressure nominal, a European designation, in barg, of the maximum working pressure at ambient temperature. Pressure-temperature rating tables or graphs must be reviewed for the actual material at specific operating temperatures. Metallic materials are usually suitable up to 65°C without derating. Some non-metallic materials may need derating over 20°C.

Polish numbers

A number which describes the surface finish of a component, #7 being better than #4. A fairly loose specification which has been replaced by **Grit numbers**.

Pop-action

Once a safety valve has started to open because the set pressure has been reached, the valve opens fully very quickly due the combined effect of pressures and fluid forces. Also known as snap action, this type of action can create pressure waves within the system leading to piping kick and pump/compressor surge.

Power-actuated safety valve

A safety valve which is actuated by an external power source. Pneumatic and hydraulic power are the most usual, electric is possible in some cases. Care must be taken to ensure fail-safe operation or back-up equipment must be installed.

Positioner

A control valve actuator used for accurate control applications. The positioner supplies air to the diaphragm or piston actuator to adjust the valve in response to control signals. The positioner is coupled to the valve/actuator stem and monitors the valve position to ensure the valve does adjust in response to the control requirement. The feedback of valve stem position allows very precise valve control.

Positive Material Identification, PMI

Certain critical components may have the material of construction identified beyond any doubt. A piece of the component can be removed and given to a witnessing inspector who carries the sample to a laboratory and watches the analysis. Alternatively, the component is analysed by a portable instrument which confirms the chemical composition.

Powder metallurgy

A technique using high pressure and high temperature to convert metal granules into solid, high integrity components. Very useful for mass producing flanges and "funny" shapes.

PRE

Pitting resistance equivalent, calculated from a formula, depending on the material considered:

$PRE_N = Cr + 1.5 \times (Mo + W + Nb)$ for stainless steel

$PRE_N = 3.3 \times Mo + 16 \times N_2$ for super austenitic stainless steel

$PRE_N = Cr - 0.8 \times Cu + 1.5 \times (Mo + W)$ for nickel alloy weld wire

$PRE_N = Cr + 3.3 \times Mo + 30 \times N_2$

Precipitation-hardening steels

There are relatively few precipitation-hardening steels compared to other alloy categories. A group of iron-chrome-nickel alloys with relatively small nickel proportions. In the annealed condition, alloys can be austenitic or martensitic. Heat treatment hardens the martensitic crystal structure.

Preferred mounting orientation

Some valve designs work better or last longer when installed in a specific orientation. Most linear valves work best when the stem is vertical with the packing box on the top. Rotary valves can usually have the spindle horizontal or vertical. Check the manufacturers instructions before proceeding with detailed piping designs. Ensure sufficient space and access is available around actuators.

Pressure balanced regulator

A regulator which compares two pressure signals and tries to equalise the pressures. A special type of Differential pressure regulator where the differential is zero.

Pressure containing part

A valve part which is in physical contact with the process fluid or any actuating fluid. These parts may be subject to corrosion or erosion by the fluid. Parts include: valve body, bonnet, stem and spindle.

Pressure control valve, PCV

A valve which is actuated by a control signal derived from a pressure sensor. The controlling pressure may be upstream or downstream of the control valve.

Pressure-loaded regulator

A regulator which uses a fixed mass of compressible fluid to replace the spring. The fluid charge may be contained behind the diaphragm or in an accumulator or in a sealed bellows capsule. Also called a **Dome-loaded regulator**.

Pressure reducing regulator

A regulator which maintains a preset pressure downstream of the regulator. The regulator may be direct-acting or pilot operated by the process fluid.

Pressure registration

A term used to describe the control pressure sensing. Pressure registration is either internal or external. Internal registration indicates that the control pressure is sensed directly from the regulator valve body; no control pipework is required. External registration indicates that the regulator is fitted with a port which must be connected to a point in the system where the control pressure can be sensed. Pilot-operated regulators with external pilots can be considered as external registration even though the sensing point is piped from the main valve body.

Pressure relief valve

A valve designed to open at a specific pressure and prevent dangerous excess pressures occurring. The valve recloses when normal operating conditions have been restored.

Pressure retaining part

A valve part which is stressed by internal fluid pressure due to its function of holding one or more pressure containing parts in position. Pressure retaining parts are not in direct physical contact with the process or actuating fluids and are not subject to corrosion or erosion by them. Pressure retaining parts may be subject to corrosion or erosion by atmospheric pollution. Parts include: bonnet fasteners, glands, gland fasteners, stems, spindles and stem nuts.

Pressure retaining valve

An alternative name for a **Back pressure regulator**.

Pressure-temperature ratings

Metallic and non-metallic materials used for valve manufacture usually weaken and become more elastic as temperature increases. Some materials become brittle as temperature decreases. These factors are well known and have been studied extensively by classification societies and standards authorities. Allowable stress values have been correlated with working temperature by pressure vessel standards and codes. The most popular Standards are EN 13445, ASME VIII and ANSI B16.34. BS 5500 has been replaced by PD 5500 and is popular outside Europe. These Standards use different formulae to calculate the stresses arising from fluid pressure; the allowable stress values are not interchangeable. Pressure-temperature ratings are usually qualified as non-shock, may be time limited or have specified full-cycle pressure limitations.

Pressure-temperature regulator

A regulator which uses a pressure and a temperature sensor for control. A pilot-operated regulator with two independent pilots.

Pressure shut-off regulator

A regulator which senses pressure and operates an isolating valve. The regulator may operate with falling or rising pressure only or both.

Pressure sustaining valve

Another name for a **Back pressure regulator**.

Pressure switching regulator

A regulator which acts as an automatic diverting valve in response to a pressure signal. Flow from an inlet connection is directed to one of two outlet connections depending upon the pressure signal.

Process controller

This can be a mechanical, pneumatic or electric/electronic device. Mechanical and pneumatic controllers would normally measure one or two variables and adjust a single output. Electric/electronic controllers can measure many variables and select the correct output logic to manage the situation. Controllers can have pre-programmed control algorithms, such as PID, to return the process variable to the set point as quickly and as smoothly as possibly.

Profiler

Another name for an electronic process controller. Profilers are usually used for processes whose set point changes with time. A sequence of events is initiated and the profiler ensures the process variable follows the correct time/value curve. Multiple curve profiles can be pre-programmed and memorised by the profiler, each curve containing a number of time/value settings.

PSI

Palestine Standards Institution, the National Standards Authority of Palestine.

PSQCA

Pakistan Standards and Quality Control Authority, the National Standards Authority of Pakistan.

PT

Penetrant testing (dye penetrant inspection) (**NDE**)

Pump governor

Another name for a **Pump pressure regulator**.

Pump pressure regulator

This term is misleading. Pumps handle liquids. This regulator is a steam regulator controlling the steam flow to a turbine driving a pump. The pump pressure can also be controlled using a **Back pressure regulator** to dump excess pump flow to the suction source. The pump pressure regulator is also called a **Pump governor** in some publications.

Push-down-to-close

A style of globe valve where the plug is pushed into the body to close the flow port(s). Sometimes called direct-acting.

Push-down-to-open

A style of globe valve where the plug is pushed into the body and opens the flow port(s).

Q

QQ

The prefix for American Federal Specifications

Quick-opening

A valve characteristic which allows the passage of large flows with relatively small openings from closed.

QSAE

Quality and Standards Authority of Ethiopia, the National Standards Authority of Ethiopia.

R

Ra

The symbol used to denote surface finish or surface roughness in accordance with BS 1134, ISO R/468, ISO 4287/1, DIN 4768/1 and DIN 4762/1. The standards are not identical but are

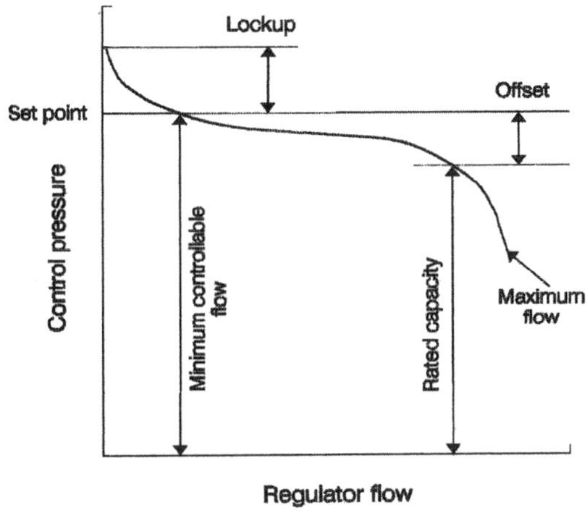

Figure 1.2 Typical regulator performance characteristics

based on the same principle, the Centre Line Average, (CLA) roughness. The method finds the arithmetic average of the deviation in peaks and troughs from a centreline over a measured length. The following formula is used.

$$R_a = \frac{1}{/m} \int_{x=0}^{x=/m} |y| dx \qquad \text{Equ 1.11}$$

In order to reduce confusion in global manufacturing companies, the concept of a roughness number was introduced for surfaces. The relationship between inch and metric values is shown in Table 1.13 together with Polish and Grit numbers.

For comparison purposes the surface finish of a good investment casting should be about 4.5 µm. A satin polished finish would be about 2 µm.

RAL

The German organisation, Reichsausschuss für Lieferbedingungen, was instigated to standardise precise technical terms to allow rationalisation of purchasing (long before the EU Single Market Project). Now the RAL prefix is used for colour shades in accordance with international standards such as BS 381C.

Rangeability

The effective flow control range of a control valve. The result obtained when the maximum flow is divided by the minimum flow. Usually expressed as a ratio; 50:1.

Rated capacity

The flow through a regulator when operating under specified conditions and deviating from the set point pressure by the limiting offset, see Figure 1.2.

Reduced-opening valve see **Reduced-port**

Reduced-port

Always described as reduced-bore, a valve in which the seat area or the ports leading to the seat are smaller than the nominal pipe area in which it is fitted. In globe valves the ports may be smaller and the seat may be larger than the nominal pipe area. Gate valves may have 50% flow area, globe valves even less. The increased velocity in the port or seat can result in increased erosion, corrosion and noise. Accelerated material loss is not limited to solids handling. Increased liquid velocities can damage the protective oxide layer resulting in unexpected corrosion. Some valves do not have circular ports. Sudden changes in area can create violent turbulence leading to greatly increased erosion and corrosion.

Reducer

A piping fitting which allows pipes of different sizes to be connected together. Concentric and eccentric styles are available. Sometimes eccentric are preferred to prevent the formation of gas/vapour pockets or sludge traps.

Reflux

The name given to reverse flow fluid passing through a non-return valve before the valve closes.

Regulation

The ability of a valve to operate at intermediate positions between wide open and closed and to effectively adjust the flow rate. Most valve types can regulate but not always effectively. When regulation is achieved by creating very high velocities, wear, noise and cavitation can be expected.

Regulations

Legally enforceable requirements imposed by national governments. Regulations are enacted by law in the country of origin.

Regulator performance

The performance of a regulator is best shown on a flow/pressure graph which can indicate the important reference points, see Figure 1.2.

Relaxation

The reduction in induced stress due to a fixed deformation when the material begins to creep or exhibit creep or plastic flow. Helical compression springs can suffer a reduction in the force applied after prolonged compression.

Relay valve

Another name for the pilot valve in a **Pilot-operated regulator**.

Relief valve

A valve which opens at the set pressure and modulates the lift in response to the pressure and the volume of fluid flowing. This type of valve is used mainly for liquids rather than gases or vapours.

Repairs

During the normal manufacturing process, errors occur. The errors can be small, that is easily rectified without impairing the function or performance of the component, or major, requiring substantial rework, or critical and unrecoverable. During the production of standard valves, the valve manufacturer imposes his standard repair requirements on components. When valves are produced specially to order, the purchaser has the opportunity to decide which flaws are repairable and which are not. The manufacturer and the purchaser must agree on a policy before the contract starts. See **Major** and **Minor repair**.

In certain instances some types of component have restrictions placed on the nature of repairs allowed. For example, it may be advantageous not to permit weld repairs on forged valve bodies. Forged valve bodies are used because a casting is not considered to have sufficient inherent integrity. If a flaw is found in the bore, the most highly stressed region, it may be easy to excavate and weld repair. The reclaimed forging, however, does not have the same integrity as a good forging. If a flaw is discovered on the outside of the valve body, the lowest stressed region, other repair techniques may be appropriate. Proprietary specifications for pressure containing parts frequently prohibit repairs by peening, plugging, burning-in or impregnating. Limitations on repairs must be agreed before manufacturing commences.

Repeatability

The tolerance on the controlled variable value obtained when the operating conditions are returned to a specific condition after operating at other conditions.

Requirements

Standards and regulations should contain requirements. That is to say statements specifying how objectives must be achieved. Requirements must be verifiable or there is no way of establishing compliance.

Reseat pressure

The pressure at which the safety valve will close after opening and allowing fluid to pass. The reseat pressure is usually between the system normal operating pressure and the safety valve set pressure but can be below the normal operating pressure.

Resolution sensitivity

The smallest change in the controlled variable which will initiate a regulator response.

Reverse-acting

Some actuators can be fitted to valves in more than one way. This allows the actuator action to be changed for a constant control signal. Control valves can be made to open or close in response to the same signal. Reverse-acting usually refers to the effect of positive feedback i.e. on increasing the control signal the valve opens. See also **Direct-acting**.

Reynolds Number, Re

In the same way that Reynolds Number can be calculated for pipework, it can be calculated for valves to determine whether operation is lamina or turbulent. Lamina operation can occur when viscosity is high or the differential pressure is low or with small valves. If operating conditions vary over a wide range it is advisable to check Re at the extremes. The valve capacity may be drastically reduced in the lamina region. The following formula has been used by one manufacturer to calculate the valve Re.

$$Re_v = \frac{76\,000 F_d \cdot Q}{m\sqrt{F_p \cdot F_L \cdot C_v}} \times \left\{ \frac{(F_p \cdot F_L \cdot C_v)^2}{0.002\,14\, d^4} + 1 \right\}^{0.25} \quad \text{Equ 1.12}$$

where:

Re_v = valve Reynolds Number (non-dimensional)

F_d = valve design characteristic (non-dimensional)

Q = flow (m^3/h)

μ = fluid viscosity (centiStoke)

F_p = pipe geometry factor (non-dimensional)

F_L = pressure recovery factor (non-dimensional)

C_v = valve flow coefficient ($m^3/h/barg$)

d = valve inlet diameter (mm)

F_d = 1 for ball valve, modified ball valve, plug valve, eccentric plug valve, single seat globe valve

= 0.71 for butterfly valve, double-beat globe valve

= 0.5 for special low noise and anti-cavitation trims

F_P = 1 when inlet and outlet piping are the same size as the valve, (Fp can vary from 0.6 to 3 depending upon the valve design and the piping sizes)

F_L = 1 to 0.9 for anti-cavitation trims

= 0.8 to 0.9 for globe valves

= 0.5 to 0.7 for special rotary valves

Rimmed steel

A steel which is not fully deoxidised before pouring; gas blowholes can be a problem.

Rising handwheel

On linear motion valves when the handwheel is directly attached to the stem and the stem rises when the valve is opened.

Rising stem

On linear motion valves the stem rises through the yoke when the valve is opened.

RJT

Ring Joint Type; a hygienic union pipe connector. Not to be confused with RTJ.

Rockwell hardness testing

A method for testing the hardness of metals by determining the depth of penetration of a steel ball or a diamond sphero-conical indentor. The value is read from a dial and is an arbitrary number related to the depth of penetration. For testing hard steels, a sphero-conical diamond is used with a 150 kg load, the result is read from the black scale on the dial and is prefixed with the letter C. A hardened tool steel would typically give a reading of 62 HRc. For softer metals Scale B is used with a 1/16" diameter steel ball and a standard load of 100 kg.

Rotary valve

A valve style in which the moving element rotates about an axis when opening and closing.

Rotating tip

The disc, plug or needle of a linear valve is part of the stem and rotates when the valve is adjusted. Rubbing rotation can cause wear when tightening the tip against the seat.

Rough finish

An intermediate flange finish between smooth and the serrated finishes. The surface roughness should be between 3.2 and 6.3 μm. Sometimes known by the inch designation 250 AARH.

RTU

Remote Terminal Unit; an operator interface to a control system. Another name for an **MMI**.

Rupture disc see **Bursting disc**

S

S

Sulphur is usually an impurity in steels, creating brittleness. But it may be added to improve machinability although it can create problems during hot working.

SA

Standards Australia, the National Standards Authority of Australia.

SABS

South African Bureau of Standards, the National Standards Authority of South Africa.

SAC

Standardization Administration of China, the National Standards Authority of China.

SAE

Used to be The Society of Automobile Engineers (of America) but is now called The Society of Automotive Engineers, advancing mobility in Land, Sea Air and Space.

SAE pipe thread

SAE J475 and ISO 725 is an American, 60°, parallel pipe thread used extensively in automotive applications. Sealing is accomplished by an "O"-ring or a flat washer. Table 1.20 indicates the important dimensions.

Thread designation	Minimum boss od in	Diameter at top of thread in	Root diameter of thread in	Number of threads per in	Minimum length of thread in
7/16-20	0.830	0.432	0.375	20	0.450
1/2-20	0.910	0.494	0.437	20	0.450
9/16-180	0.970	0.557	0.493	18	0.500
3/4-16	1.187	0.743	0.672	16	0.560
7/8-14	1.375	0.868	0.785	14	0.660
11/16-12	1.625	1.054	0.959	12	0.750
15/16-12	1.912	1.304	1.209	12	0.750
15/8-12	2.270	1.617	1.521	12	0.75
17/8-12	2.562	1.867	1.771	12	0.75

Table 1.20 Main dimensions of SAE pipe threads

Safety valve

A valve which opens fully once the set pressure has been exceeded irrespective of the volume of fluid flowing. These valves usually have a pop-action and are used mainly on gases and vapours which expand considerably as the pressure drops through the valve.

Safety relief valve

A valve which can have either a pop or modulating action.

Sanitary

American terminology for **Hygienic**

SARM

National Institute of Standards and Quality, the National Standards Authority of Armenia.

SASO

Saudi Arabian Standards Organization, the National Standards Authority of Saudi Arabia.

SASMO

Syrian Arab Organization for Standardization and Metrology, the National Standards Authority of the Syrian Arab Republic.

SAW

Submerged Arc Welded, a type of electric welding process used on thick wall pipe and tube. The submerged arc process produces a very high quality, clean weld. Powdered flux is poured into a dam which is clamped either side of the seam. The end of the welding electrode is covered by the flux and air is effectively excluded. The powdered flux melts to protect the molten filler metal. Both flux and filler metal solidify on cooling. Unused powdered flux is collected for re-use. When cold, the flux forms a solid slag which peels of the weld metal. Typical pipe sizes would be in the range of 400 mm diameter and larger with wall thickness of 6 to 32 mm. For corrosive applications the filler metal should be the same chemical composition as the parent metal or slightly more noble. For critical applications clarification will be worthwhile.

SBS

Seychelles Bureau of Standards, the National Standards Authority of the Seychelles.

SCADA

Supervisory, Control And Data Acquisition; an intelligent control system for large scale processes. Usually implemented by employing DCS, Distributed Control System, with extensive process data measurement. Small systems may only have 40 data points; large systems may have 2000 and over.

Scanner

An electronic device which has multiple measuring points. Each point is measured consecutively at a preset rate. Scanning rate can be as low as 4/s. Each measurement can have at least two associated alarm or trip settings. If a measurement is outside the alarm settings a local or remote annunciator can be initiated or a signal can be transmitted over a network, RS 232 or RS 485. A scanner can act as a data-logger by printing measured values at preset time intervals and when an alarm has been initiated. Sophisticated scanners can calculate mean values and remember maximums and minimums. Small scanners have six inputs, larger scanners can have over 700.

SCC

Standards Council of Canada, the National Standards Authority of Canada.

Screened cable

A cable with one or more cores where the cores are completely surrounded by a conducting screen which is usually earthed. The earthed screen prevents interference from stray magnetic fields close to the cable.

Screwed bonnet

A bonnet which is threaded and screwed directly into the valve body.

Screwed gland

A gland which screws into a female thread in the bonnet. It can be either adjustable or non-adjustable.

Se

The element Selenium, added to some steels to improve machinability.

Seal weld

A threaded pipe connection, usually a taper thread but could be applied to a parallel thread, which is made dry without sealants. A fillet weld is added at the external joint face to prevent any leakage. This type of construction is only acceptable when the process fluid is not corrosive. The pipe threads, which provide the strength for the joint, are not protected and are susceptible to crevice corrosion or galvanic corrosion if the metals are significantly different.

Seamless

Pipe or tube which does not have a welded seam. Some rolled plate and strip products have no seam because the edges are joined by pressure welding when the metal is hot.

Seat

The pressure-retaining contact seal between the stationary and moving parts of the valve. A valve may have more than one seat.

Seat leakage

The leakage through the seat when the valve is closed. Leakage is standardised in many cases, see Chapter 16.

SEE

Service de l'Energie de l'Etat, Organisme Luxembourgeois de Normalisation, the National Standards Authority of Luxembourg.

Self-energised packing

Packing which uses the fluid pressure to create the sealing forces to prevent leakage. Higher pressures result in higher sealing forces. See Figure 11.9 in Chapter 11.

Semi-lug wafer body

A valve body designed to clamp between two flanges with several lugs with clearance holes. The lugs allow the valve body to be held in place against one flange using extra nuts on the long stud-bolts.

Semi-nozzle construction

A safety relief valve style where the nozzle does not form all the valve inlet. The valve body forms part of the inlet and is exposed to the process fluid. See Figure 7.7 in Chapter 7.

Serial transmission

Sending digital information bit by bit. An 8 bit binary number would be sent as eight discrete pulses one after another. The type of digital data transmission over two-core cable and twisted pairs.

Servo-loaded relief valve

A conventional spring-loaded relief valve which is normally held closed by fluid pressure supplied by a control valve. When the set pressure is reached, the control valve unloads the servo so the relief valve can open and reseat as normal. At a predetermined low pressure, the control valve replaces the fluid pressure on the servo to hold the valve closed again. Although more complicated, this approach allows successful operation very close to the set pressure and prevents prolonged simmering which can wiredraw seats. The valve is still fail-safe; the servo pressure is insufficient to hold the valve closed without the spring force.

Servo-operated

Applied to solenoid valves to indicate the main valve is fluid-powered and actuated by a small valve in a servo or pilot circuit. Identical in principle to pilot-operated.

Servo-regulator

An alternative name for a **Pilot regulator**.

Set point

The value of the controlled variable at the minimum controllable flow, see Figure 1.2.

SFS

Finnish Standards Association, the National Standards Authority of Finland.

SG iron

An abbreviation for Spheroidal Graphite Cast Iron. As the name implies, graphite is present in spheroidal form instead of flakes and compared with Grey Cast Iron it has higher mechanical strength, better ductility and increased impact resistance.

Shear ring

As working pressures increase it becomes more difficult to design a satisfactory bolted flanged connection. The pressure and gasket loads must be supported by the studs. Stud stresses can become irrelevant, as flange thicknesses increase, and the elasticity of the stud becomes important. Increasing the stud diameter has a finite limit. Space must be allowed for spanner clearance. Cap nuts can be used but again a limit is reached. A rectangular metal ring, a shear ring, set in a groove in the body can provide much more load carrying area exactly where it is needed. Bolted retainers are used to secure the shear ring until pressure loads are applied. Bonnets of this style should have radial seals against the bore, not face seals.

Shore scleroscope

An instrument that measures the hardness of a sample in arbitrary terms of elasticity. A diamond tipped hammer is allowed to fall freely down a graduated glass tube on to the sample under test. The hardness is measured by the height of the rebound. In another form the rebounding hammer actuates the pointer of a scale so that the height of the rebound is recorded. The Shore test can only be used on "large" components.

Shutdown valve, SDV

An isolating valve, part of the ESD system, which closes to isolate equipment or a process during the shutdown sequence.

Si

Silicon, is added to steels, although usually present anyway from the iron ore, as a deoxidiser to improve quality and reduce porosity. It acts in a similar manner to aluminium but produces a harder steel.

SII

Standards Institution of Israel, the National Standards Authority of Israel.

SIL

Safety Integrity Level

Simmer

A condition which occurs with spring-loaded valves when the system operating pressure is very close to the set pressure. The closing force on the disc is much reduced allowing the safety valve to weep. No measurable volume of fluid escapes but it can be visible and audible. Some manufacturers state this condition only occurs with gases and vapours; however it is also applicable to high pressure liquid valves. The valve may make a noise very similar to a steam whistle.

Single flange wafer body

A short wafer body which incorporates a single, central flange. Valve installation requires long stud-bolts with an extra nut to clamp the valve body to one flange. Some designs have tapped holes at the spindle centreline. Drawings should be checked for exact construction.

Single-seat globe valve

A standard globe valve with one seat.

SIP

Steam-in-place

SIRIM

Malaysian Standards Institution, the National Standards Authority of Malaysia.

SIS

Swedish Standards Institute, the National Standards Authority of Sweden.

SIS

Safety Instrument System

SIST

Slovenian Institute for Standardization, the National Standards Authority of Slovenia.

Skirt-guided

A method of guiding the valve plug from the seat bore. The plug bottom is fitted with a tube which extends into the seat bore. Flow ports through the skirt are positioned immediately beneath the plug lower surface.

SLBS

Saint Lucia Bureau of Standards, the National Standards Authority of Saint Lucia.

SLSI

Sri Lanka Standards Institution, the National Standards Authority of Sri Lanka.

Slug (1)

A method of fitting soft packing in the packing box. The packing is not cut into individual rings which are fitted separately, but is fitted as a single continuous length. The ends of the packing are cut at an angle so they can blend in and form a square face.

Slug (2)

A change in consistency of a fluid flow stream; a slug of liquid in a gas/vapour stream or a slug of gas/vapour in a liquid stream.

SMAW

Stick Metal Arc Welding or Shielded Metal Arc Welding, the simplest and most common type of manual arc welding, also known as **Manual Metal Arc welding, MMA**. The arc is struck between the workpiece and the consumable which is covered (shielded) by a flux coating. The consumable and the flux melt in the weld pool. The melted flux covers the weld to prevent contamination.

SMS

Swedish Manufacturing Standard; a Swedish standard for hygienic pipe unions.

Smooth finish

A flange facing finish which is smooth but not polished. The finish would normally lie between 1.8 and 3.2 µm. No visible tool marks should be discernible.

SMYS

Specified minimum yield strength

SN

Standards Norge, the National Standards Authority of Norway.

SNIMA

Service de normalisation industrielle marocaine, the National Standards Authority of Morocco.

Snubber

A non-adjustable high fluid friction resistance to attenuate pressure pulsations.

SNV

Swiss Association for Standardization, the National Standards Authority of Switzerland.

SNZ

Standards New Zealand, the National Standards Authority of New Zealand.

Socket fusion weld

A thermal welding technique used with some non-metallic piping systems. Materials such as polypropylene can be welded by using electric heaters.

Socket weld fittings

Fittings for connecting pipes, which use a single fillet weld around the od of the pipe/tube for mechanical attachment and sealing. The pipe/tube is inserted into the fitting, then welded. It should only be used with safe fluids. The weld cannot be X-rayed to verify integrity.

Soft PLC

Software which provides a standard PC with the programming facilities normally found on an industrial PLC.

Soft-seat

A soft insert of plastic, elastomer or other readily deformable material, in the valve body or the moving element which deforms on contact with the other sealing surface to provide low or zero seat leakage.

Soldered connection

A type of connection used mainly with copper tube. The tube is inserted into a socket in the fitting and the clearance around the tube is filled with solder. The solder can be supplied externally during heating or can be part of the fitting, see **Capillary fittings**.

Solution annealing

A heat treatment to allow precipitated constituents to return in to the solution. Quenching "freezes" all the constituents in the required state.

SON

Standards Organisation of Nigeria, the National Standards Authority of Nigeria.

Sound power

The rate at which sound energy is radiated by a source; Watts per unit time.

Sound Power Level, SPL

Measured in dB, ten times the logarithm to base 10 of the ratio of sound power radiated by the operating source to the reference power level.

Reference level = 1pW º 10-12W.

Sound Power Level = SPL =

$$= 10\log_{10}\left\{\frac{\text{Power Level W}}{10^{-12}}\right\} \qquad \text{Equ 1.13}$$

An increase of 3 dB is double the sound power level; an increase of 10 dB is ten times the sound power level.

Sound pressure

The sound pressure, at a specified position near the source, when the source is operating under specific conditions, mounting and reflecting planes, in the absence of background noise and reflections from surfaces other than the plane below the equipment.

Sound Pressure Level, SPL

Measured in dB, ten times the logarithm to base 10 of the square of the emission sound pressure to the square of the reference sound pressure, measured with a particular time weighting and a particular frequency weighting. The reference sound pressure level is 20 mPa. Suitable weightings are listed in IEC 651; dB(A) is the most used. Sound pressure levels are usually specified at 1m from the source and 1m above ground level or at the height of the centreline.

Sound Pressure Level = SPL =

$$= 10\log_{10}\left\{\frac{\text{Pressure Level Pa}}{20^{-6}}\right\}^2 \qquad \text{Equ 1.14}$$

$$= 20\log_{10}\left\{\frac{\text{Pressure Level Pa}}{20^{-6}}\right\} \qquad \text{Equ 1.15}$$

An increase of 6 dB is double the pressure level; an increase of 20 dB is ten times the pressure level.

Sour

A description of the environment or the process fluid; hydrogen sulphide, H_2S, is present.

Specification

A document issued as part of a contractual agreement for equipment. Most specifications refer to standards and amend or extend requirements to suit the particular application. Some specifications are "vague" in the sense that requirements are aspirations and cannot be verified. Before writing a specification, authors could be usefully guided by consulting Standard writing rules, like BS 0 (Standards, specifications for structure, drafting and presentation.). A specification which is vague or ambiguous will be interpreted to the suppliers" advantage. A specification which is capable of interpretation, is of little value to the purchaser.

Spill area

This is the area formed by the circumference of the nozzle bore and the disc lift. Some standards refer to this as the curtain area. The spill area need not be equal to the nozzle area.

Spindle

The shaft which rotates in rotary valves and actuators.

Spiral serrated

A very similar finish to "stock finish" but a straight V groove is used rather than a radiused one. The finished grove is 1/64" deep and the feed rate is 1/32" for all flange sizes.

Spiral-welded

A type of pipe which is made by rolling a narrow strip around a mandrel and welding the edges together. It is an effective way of making large diameter pipe.

Spring

When used without qualification, spring is construed to mean a helical coil spring wound from round wire. There are other spring designs and other manufacturing methods. Springs should be described to avoid confusion.

Spring range

The change in force of a diaphragm return spring when the applied air pressure varies between 0.21 and 1.03 barg.

Spring rate

The force per unit length deflection applied by the spring. The spring rate may be a linear function of deflection or may increase as deflection increases. Variable spring rates can be helpful in eliminating spring resonance.

Spring resonance

This occurs when a spring is excited at its natural frequency. Resonance usually results in uncontrolled movement of individual coils causing material damage and early failure.

SPRING SG

Standards, Productivity and Innovation Board, the National Standards Authority of Singapore.

Squish

A pseudo technical term used in fluid dynamics to denote a special type of compression process. Fluid is compressed by a closing tapered enclosure and forced to move in a specific direction. This type of compression can be used to slow the impact of discs on seats, a type of fluid damping. The fluid must possess a minimum viscosity for the technique to be viable. End-of-travel cushioning can be accomplished in the same way.

SSC

Sulphide stress cracking; a type of corrosion mechanism which attacks components made of high strength materials which derive their strength from heat treatment to relatively high hardness levels. Sulphide stress cracking is usually associated with hydrogen sulphide in connection with crude oil and natural gas production.

SSPC

Society for Protective Coatings (of America).

Stability

The capability to maintain a constant controlled variable value within the limits of the accuracy of regulation.

St

Steel

Stagger-set

Applicable to multiple safety valve installations when the set pressures are all slightly different.

Stainless steel

An alloy of iron and chromium which is usually, at least, 50% iron and up to 30% chromium. Carbon is usually present from very low concentrations, 0.03% to 1.0% or more. Stainless steel can be divided into five groups: martensitic, ferritic, austenitic, duplex and precipitation-hardening.

Standard

A specification, issued by a National Standards Authority, a government sponsored organisation, which is characterised by strict editorial control and verifiable requirements. Standards should consist of clauses which mainly use "shall" to indicate a "requirement" which must be met. Compliance with standards must be verifiable, otherwise the effort is wasted. "Should" is used to describe recommendations; "may" indicates a course of action which is acceptable. Clauses which require an agreement should be kept to an absolute minimum.

Standard flow

Some linear and rotary valves have a preferred, or mandatory, direction of flow due to the valve construction.

Static back pressure

The pressure at the safety relief valve outlet connection when the valve is closed and not operating. Also called **Superimposed back pressure**.

Static unbalance

The force on a plug, or torque on a disc, when closed with specified operating conditions.

Steam-in-place, SIP

A method of cleaning and sterilising hygienic equipment after production. The equipment is not disassembled but cleaned as a system. Low pressure saturated steam, 1 to 2.4 barg, 120 to 137°C, is circulated for 45 minutes or up to 1 hour.

Steel

An alloy of iron and carbon with small quantities of silicon, manganese, moly or nickel. Phosphorous and sulphur are usually present, in quite small quantities, less than 0.05%, as impurities which impair the physical and chemical properties. If the carbon content is between 0.4 and 1.2% approximately the steel can be hardened by heat treatment although steels with lower carbon contents can be heat treated, inadvertently, during welding.

Stem

The shaft which reciprocates in linear valves and actuators.

Stem connector

A clamp, usually in two halves, to connect the valve stem to the actuator stem.

Stem leakage

Leakage through the stem (or spindle) seal when the valve is open or closed. Some valve designs seal the packing box when open; others when closed. Rotary valves generally are unable to seal the packing box.

Sterilise-In-Place, SIP

The procedure of sterilising complete hygienic systems after production without disassembling any equipment. Sterilisation can be accomplished with low pressure saturated steam, 1 to 2.4 barg, 120 to 137°C, or proprietary liquids. Low pressure steam has the added benefit of being able to loosen and soften product which may have become stuck.

Sticktion

American spelling of **Stiction**.

Stiction

Short for static friction; the physical phenomenon which creates the very high force necessary to start something moving; friction to prevent movement rather than resist movement. It can be considered like a "tack weld'; something which holds in place but is quite easily broken. It can happen in valves, handling viscous material, when the plug becomes stuck to the seat.

Stick welding

Short for Stick Metal Arc Welding or Shielded Metal Arc Welding, the simplest and most common type of manual arc welding, also known as **Manual Metal Arc welding**, MMA. The arc is struck between the workpiece and the consumable which is covered (shielded) by a flux coating. The consumable and the flux melt in the weld pool. The melted flux covers the weld to prevent contamination.

STLE

Society of Tribologists and Lubrication Engineers (of USA)

Stock finish

The standard surface finish applied to flange facings by the flange manufacturer. Stock finish is a shallow spiral groove machined as one continuous cut across the face. The angled crests created apply a strong grip to the gasket to resist blow-outs. Flanges up to 12" have a 1/16" radius groove cut with a feed rate of 1/32" per revolution. Larger flanges use a 1/8" radius tool and 3/64" per revolution.

Stress relieving

A heat treatment including heating and soaking at a suitable temperature e.g. 600 to 650°C, followed by cooling at an appropriate rate in order to reduce internal stresses without substantially modifying the steel's structure. This treatment may be used to relieve stresses induced by machining, quenching, welding or cold working.

STRI

Icelandic Council for Standardisation, the National Standards Authority of Iceland.

St St

Stainless steel

Stud-bolts

Continuously threaded studs supplied with two nuts used mainly for pipe flanges.

Surface finish

Raw materials are machined to size and with a specified degree of surface roughness or surface finish, to accommodate the function and adjacent parts. In some cases it is necessary for fairly rough surfaces to be polished to a bright finish so that visual inspection can easily detect contamination or surface flaws. Other surfaces may be polished to a mirror finish. Three grades of polished finish are standard for hygienic applications; Ra 6.3, Ra 1.0 and Ra 0.1; finishes quoted in micrometres, µm, a millionth of a metre. Better finishes are sometimes available. Mechanical polishing can achieve finishes to about 0.28 µm.

Surge

An operational mode of some rotodynamic machines when installed in systems with flat HQ characteristics, when the head and flow oscillate between two sets of values and create strong, possibly damaging, vibration.

Superimposed back pressure see **Static back pressure**

SUTN

Slovak Standards Institute, the National Standards Authority of Slovakia.

Sweet

A description of the environment or the process fluid; hydrogen sulphide, H_2S, is not present.

Sympathetic ratio

The rate of change of outlet pressure for a given change in inlet pressure at a specific flow.

T

Ta

Tantalum is a rare metal and is used in some applications for its special corrosion resistance. It is used as an alloying element in steels and may be used instead of niobium.

TA-Luft

Technische Anleitung zur Reinhaltung der Luft, the German Regulatory Body for air pollution and control regulation.

TBS

Tanzania Bureau of Standards, the National Standards Authority of Tanzania.

TCVN

Directorate for Standards and Quality, the National Standards Authority of Vietnam.

Tempering

A heat treatment applied to ferrous products after hardening. It consists of heating the steel to some temperature below the transformation range and holding for a suitable time at the temperature, followed by cooling at a suitable rate. The object of tempering is to decrease hardness and increase toughness to produce the desired combination of mechanical properties.

Test gag

A bolt with a locknut screwed into the cap which can be used to lock the valve spindle down and prevent opening. This optional feature is useful when hydro-testing equipment and piping after installation. Proper safety procedures must be implemented, to ensure gags are released, before this feature can be used prudently.

Temperature regulator

An alternative name for a **Thermostatic control valve**.

Thermal relief safety valve

A small spring-loaded valve fitted in a pipe run between isolating valves to ensure that excess pressure is not created by the expansion of fluid due to expected temperature rises.

Thermal safety valve

A safety valve which is held closed by relatively low melting temperature material. When the temperature of the valve exceeds the set temperature the material is sufficiently weakened to allow the valve to open. The valve does not close when operating conditions return to normal but must be reset, with a new material insert, manually.

Thermostatic control valve

A temperature sensitive valve which utilises the expansion properties of a heated fluid as the motive power for actuation.

Thermowell

A small leak-proof container used to house thermometers or temperature sensors. Thermowells can be screwed or flanged. Flanged thermowells are preferred for hazardous fluids. The highest quality thermowells are machined from solid bar; usually stainless steel.

Third-party inspection

An independent inspection agency employed to ensure compliance with all relevant standards, specifications and regulations.

Threaded fittings

A way of connecting pipes by screwing each pipe into a fitting with at least two female threads. Threads can be parallel or tapered. Parallel threads need additional components for sealing. Tapered threads can seal on the threads by using PTFE tape or compounds.

Threaded unions

A pipe coupling, using threaded union nuts to hold the components together, into which the pipes are threaded.

Ti

Titanium, a metal element, used in its own right because of its strength, lightness and corrosion resistance, and also as an alloying element in steels. Titanium is used to stabilise carbides in stainless steels to prevent intergranular corrosion.

TIG welding

Tungsten arc Inert-Gas welding is an electric arc welding process, where the arc is struck with a dedicated tungsten electrode across to the workpiece. A separate filler rod or wire is fed into the weld pool which is protected by an inert gas shield.

TISI

Thai Industrial Standards Institute, the National Standards Authority of Thailand.

Tool steel

A generic term applied to a wide range of steels, both plain carbon and alloys with tungsten and vanadium.

Torsion spring

A spring which produces torque rather than linear movement. The spring is wound-up with rotary movement. The type of spring used in mouse traps and twin disc non-return valves.

Toughness

The ability of a metal to rapidly distribute within itself both the stress and strain caused by a suddenly applied load, or more simply expressed, the ability of a material to withstand shock loading. It is the opposite of "brittleness" which carries the implication of sudden failure. A brittle material has little resistance to failure once the elastic limit has been reached.

Tramp elements

An American expression describing undesirable inclusions found in slabs, billets and blooms during subsequent rolling operations.

Transducer

A measuring instrument which produces an electrical output in proportion to the measured variable. The electrical output may be specific to the manufacturer rather than an industrial standard. Additional equipment may be required to condition the signal for further use. A transducer may include a display which indicates the current value of the measured variable. This is a useful facility which eliminates the need for a separate instrument with additional process connections. Transducers are high speed measuring instruments which can capture transients. Transducers are usually used for test purposes.

Transmitter

A measuring instrument which produces a standard electrical output in proportion to the measured variable. The electrical output may be analogue, 4-20 mA for example, or digital. A transmitter may include a display which indicates the current value of the measured variable. This is a useful facility which eliminates the need for a separate instrument with additional process connections. Transmitters are usually damped or include averaging algorithms. Transmitters are unsuitable for capturing brief transients. Transmitters are used with process controllers for system automation.

Travel

The linear or angular movement of the actuator expressed in millimetres, degrees or as a percentage.

Travel indicator

A pointer, attached to the valve or actuator stem, which shows how much the valve is open or closed.

Trim

The internal components of a valve in contact with the process fluid.

Trip amplifier

An electric device which monitors, and sometimes conditions electric analogue control signals. The trip amplifier can be adjusted to provide multiple high/low alarm or trip signals. If the analogue signal produced by a transducer or transmitter is outside the preset limits the trip amplifier completes another circuit and initiates an alarm or trip sequence. Most trip amplifiers have two low and two high set points allowing alarm before trip operation.

TTBS

Trinidad and Tobago Bureau of Standards, the National Standards Authority of Trinidad and Tobago.

TSE

Türk Standardlari Enstütü, the National Standards Authority of Turkey.

Tube

A conduit for carrying fluid which is designated by its outside diameter. The bore size is determined by the wall thickness. Both inch and mm tube are readily available. Tube cannot be threaded with standard pipe threads. If necessary, tube with an appropriate wall thickness can be threaded with standard bolting parallel threads.

Twin-seat valve

A valve which can seal upstream and downstream simultaneously.

Twisted pair

A two core cable used for analogue signals and digital communications when the data bits are transmitted one by one, as in Serial transmission. Not as fast as a parallel data bus but can be much cheaper. Screened or plain twisted pair, analogue and digital signal cables may be routed together. Power cables should be at a safe minimum distance to avoid interference.

Two-stage regulator

Two direct-acting pressure reducing regulators connected in series in one body. Used for applications when the outlet pressure is very much smaller than the inlet pressure; typically gas regulators used with bottled gas. The inlet pressure may be 200 barg and the outlet pressure 2 barg.

U

UL

Underwriters Laboratory Inc. (of USA and Canada), an accreditation society.

Ultrasonic

Sound above the range of human hearing; typically frequencies greater than 20 kHz. These high frequency sounds can be detected by piezo-electric crystals.

UNBS

Uganda National Bureau of Standards, the National Standards Authority of Uganda.

UNI

Ente Nazionale Italiano di Unificazione, the National Standards Authority of Italy.

Union bonnet

A style of bonnet which is attached to the valve body by a single nut.

Union gland

A gland which screws onto a male thread on the bonnet; it can be either adjustable or non-adjustable.

UNIT

Instituto Uruguayo de Normas Técnicas, the National Standards Authority of Uruguay.

UNS

Unified Numbering System for Metals and Alloys; a joint effort between ASTM and SAE to produce unique identifiers for materials based on the chemical composition. Metal groups are prefixed by a letter denoting the general category of material:

- A = aluminium and alloys
- C = copper and alloys
- D = steel with specified mechanical properties
- F = iron alloys
- G = AISI & SAE carbon steels
- J = cast alloys
- K = miscellaneous steels and alloys
- N = nickel and nickel alloys
- S = heat and corrosion resistant stainless steels
- T = tool steels
- W = weld filler metals
- Z = zinc and zinc alloys

U-section wafer body

A short valve body with two flanges. The flanges are so close together there is insufficient space for two nuts. The valve must be installed using extra long stud-bolts. Some designs have two or four tapped holes in each flange at the spindle centreline. Studded flanges require extra space for installations; tapbolts wear the threads in the casting.

UT

Ultrasonic testing (**NDE**)

V

V

The chemical symbol for vanadium, added in small quantities to tools steels and high speed steel to improve hot hardness and restrict grain growth.

V-notch ball

A ball, for a control valve, which has a specially profiled V passage through it rather than the normal circular hole. When correctly sized gives a linear valve characteristic.

Vacuum break valve

A pressure relief valve which is designed to prevent a vacuum forming in a process system. The valve opens when the set vacuum is exceeded and admits fluid, usually air, into the system to increase the pressure. The valve closes when the vacuum is reduced.

Vacuum regulator

A regulator which maintains a sub-atmospheric pressure rather than a positive gauge pressure.

Vacuum remelt

A process used for producing advanced steels to the most demanding and critical specifications, particularly in such areas as aerospace applications or high temperature/pressure. The steel is first produced to a very close analysis and the resulting ingot is slowly remelted in a vacuum arc remelting furnace for up to 14 hours. Such steels are expensive to manufacture but have higher than normal cleanliness and integrity.

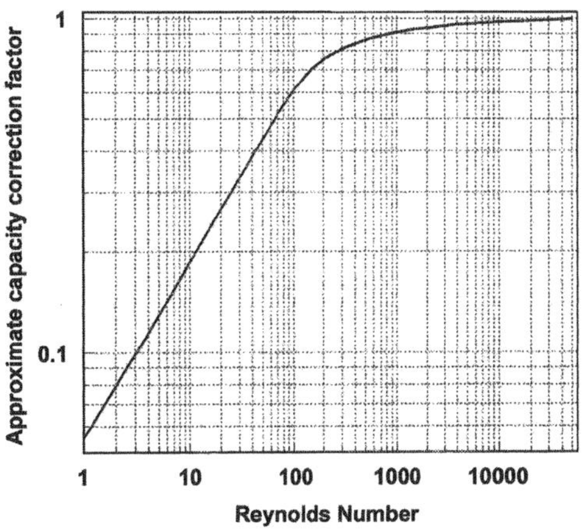

Figure 1.3 Valve viscous correction factor

Valve block

Another name for a valve manifold; an assembly with two or more valves.

Valve body

The major pressure-containing part of a valve which has the process connections.

Valve schedule

A tabulation of similar valves used to identify common construction features, materials and sizes. See Tables 8.1 and 8.2 in Chapter 8.

Valve twist

A special form of braided packing which is rope-like in construction. Strands can be removed to reduce the cross-sectional area for fitting in smaller packing boxes.

Vapour

If a liquid is heated, eventually it will boil to produce vapour. Vapour is a gas-like state in that the volume and shape change easily. A vapour can expand to fill a container, but if the container is too big, a proportion of the vapour will condense back into liquid. Saturated vapour is a mixture of vapour and liquid droplets and exists when vapour coexists above boiling liquid. Vapour is a "complicated" gas and may not obey the perfect gas laws. Vapour is important when considering cavitation.

Vickers hardness test

This test, more commonly known as the Vickers test, finds greater use in the laboratory than the workshop. It employs a pyramid shaped diamond with an included angle of 136° which is impressed into the specimen using loads of 5 to 120 kg making a small square impression. This test is used for finished or polished components because the impression can be very small. The diamond pyramid hardness number is obtained from a calculation based on measuring the diagonals of the impressions in the material. Test results are usually quoted as VPN, Vickers Pyramid Number, or just HV.

Viscous capacity correction, FR

The flow coefficients stated for valves are based on water-like liquids operating with fully developed turbulent flow. For operation in the lamina range the valve capacity can be reduced by using the graph in Figure 1.3.

Viscous correction

Basic safety valve sizing for liquids is conducted for water-like liquids. If the liquid is viscous, a correction factor, based on the application Reynolds Number, is included in the sizing equation. Increased viscosity usually requires larger valves.

VOC

Volatile organic compound; compounds which are normally handled as liquids but which vaporise quickly at atmospheric pressure and temperature. Valves handling VOCs do not show external visible signs of leakage as the liquid flashes to vapour when passing through the seal and escapes in the vapour or gas phase.

VOD

Vacuum-oxygen decarburisation (a stainless steel refining process).

VPN

Vickers Pyramid Hardness

VT

Visual inspection (**NDE**)

W

W

The chemical symbol for tungsten, which used to be called Wolfram. Tungsten can greatly improve the hot properties of tool steels. It is added to some chrome-nickel alloys to aid carbide formation.

Wafer body

A valve body which is intended to be clamped between two flanges with long stud-bolts. The valve body diameter is small enough to fit inside the bolt circle. Large diameter wafer valves may have two or four lugs, with clearance holes, to assist alignment during installation. Some designs have two or four tapped holes on the spindle centreline. Some very large valves are described as wafer when the valve body has integral flanges with blind tapped holes and the body is slimmer than normal. Looking at a drawing is the best way to be certain of the exact construction.

Water hammer

This is the high pressure generated in liquid systems when a large mass of liquid or high velocity liquid has its velocity reduced suddenly. The mobile energy of the liquid is converted into pressure energy. Water hammer is not a kinetic energy problem, it is an acoustic problem. The wave speed will be a lot faster than the liquid velocity. The magnitude of the pressure increase is a function of the liquid properties, especially density and acoustic velocity, liquid velocity and the pipe elastic properties. The acoustic velocity of a fluid is a function of the bulk modulus and the density.

$$c = \sqrt{\frac{K}{\rho}} \qquad \text{Equ 1.16}$$

where:

c = acoustic velocity (m/s)

K = liquid bulk modulus (Pa)

ρ = liquid density (kg/m^3)

The acoustic velocity defined by Equation 1.16 applies to fluids which are unconfined. Placing a fluid in a pipe modifies the acoustic characteristics and a wave speed is used instead of acoustic velocity, see Equation 1.17.

$$a = \sqrt{\frac{g/w}{\frac{1}{k} + \frac{1}{8E}(d_o^4 - d_i^4)}} \qquad \text{Equ 1.17}$$

where:

a = wave speed (m/s)

g = acceleration due to gravity (m/s^2)

w = specific weight (N/m^3)

E = Young's Modulus of Elasticity of the pipe (N/m^2)

d_o = pipe outside diameter (m)

d_i = pipe inside diameter (m)

The elasticity of the pipe wall has a significant effect on the wave speed. Air or gas bubbles have a much more severe effect on the wave speed and can reduce it to 100 m/s. The pressure wave is reflected at the end of the pipe as a negative pressure wave, returns to the valve, is reflected again and so on. In this way a process of pressure oscillation is set up in the pipeline and continues, even after the valve is completely closed, until the oscillations are damped out by friction. The hydraulic gradient lines H_{tmax} and H_{tmin} connect the highest and lowest pressures which occur at different points along the pipeline during the oscillation process. The greatest pressure variations are to be found in the valve and occur when the valve is closed rapidly. The greatest amplitude of pressure surge is:

$$\Delta p_{max} = \rho \cdot a \cdot v \qquad \text{Equ 1.18}$$

where:

Δp_{max} = max. pressure variation (Pa)

ρ = density of the liquid (kg/m^3)

a = speed of wave propagation (m/s)

v = velocity of flow before valve begins to close (m/s)

For water in steel pipes $\rho \cong 1000$ kg/m^3, $a \cong 1100$ m/s, and a flow velocity of 1 m/s creates, for rapid valve closing, a water hammer pressure of $1.1 \cdot 10^6$ Pa = 11 bar $\cong$ 110 m. The pressure process at the valve cannot be affected before the pressure wave has been reflected from the end of the pipe and returned to the valve. This time is referred to as the pipeline reflection time and is designated $t_{(s)}$.

$$t_r = 2 \cdot \frac{L}{a} \qquad \text{Equ 1.19}$$

where:

L = the pipe length (m)

a = wave speed (m/s)

If for example the length of pipe is 1100 m, then $t_r \cong 2$s. All valve closing times ≤ 2s thus give maximum water hammer pressure.

For a plastic pipe $\Delta p_{max} = 3 \cdot 10^5$ Pa $\cong$ 30 m is obtained for valve closing times ≤ 7.36s. Increasing the closing time reduces the magnitude of water hammer and very slow closing valves result in no line shock at all.

Some manufacturers control valve selection programmes include a water hammer evaluation procedure. Valve closure times can be adjusted and the effect on peak pressure reviewed. Check for this useful facility before purchasing software.

Water hammer does not just affect the pressure rating of the piping system. If water hammer occurs regularly the piping system may have to be designed to resist fatigue failure rather than to standard code stress levels. Pressure vessel codes are based on steady state operating conditions and excursions which may cause fatigue are limited, in some codes, to 7000 cycles. The pressure pulse can create axial forces in the system where the pipework changes direction, resulting in a "kick". The axial forces can excite pipe vibration or damage or fracture pipe supports and anchors resulting in unexpected loads being transferred to other locations. If vessel or machine nozzles are subjected to increased loading the useful life or operational times between overhauls can be severely reduced. Pumps and compressors can be made to seize; mechanical seal rotating parts can wear rapidly; pressure vessels can be permanently deformed. Water hammer is a complex phenomenon and should be investigated fully in systems involving large fluid masses, high velocity fluids or quick-acting valves. The consequences of water hammer can be catastrophic.

Water hammer suppression

The effects of water hammer can be alleviated at source by valve design. Ensuring the valve closes slowly, rather than quickly, can dramatically reduce the pressure rise. In the specific case of non-return valves the valve should close when the fluid velocity is zero. If the valve closes when reverse flow has commenced there is a chance of creating water hammer. A non-return valve should modulate with the flow and close slowly. If water hammer generation cannot be avoided then suppression devices should be used to protect system integrity. Surge columns, surge tanks, air tanks and accumulators can be fitted to reduce the magnitude of the pressure pulse.

Wave spring

A linear spring made from a ring of material which is deformed to produce a series of peaks and troughs. The spring force is generated by trying to crush the spring and level out the peaks and troughs. Useful for loading packing when space is restricted.

Welded pipe

Pipe can be manufactured by rolling strip or plate and then welding the seam. Nominal diameters between 20 and 1000 mm, ¾" to 40" nb, with wall thicknesses up to 55 mm, 2.165", are regularly produced. Manufacturing can be accomplished in two ways so either a single longitudinal straight weld is formed or a spiral weld is produced. Both methods can produce a high integrity product depending on the welding method and the quality controls implemented. Welding can be: manual arc, submerged arc, MAG, SMAW with coated electrodes, tungsten inert gas (TIG), plasma, LASER or electron beam. Weld beads can be dressed. After welding, the pipe lengths can be heat treated to remove welding residual stresses, adjust the molecular structure in the heat-affected-zone (HAZ) or modify the physical properties of all the material. Typical heat treatments include annealing, solution annealing and tempering. Pipes can be finished by descaling or pickling. The integrity of the weld can be checked by ultrasonic examination or real-time radiography. Final inspection checks include air leak testing under water and hydrostatic testing. Hydrostatic testing can be combined with cold-forming to improve circularity.

Weld preparation

Usually abbreviated to weld prep, this is the machining performed on a component or pipe end, to allow a specific type of weld to be used.

Welded bonnet

For applications requiring very high integrity the bonnet can be welded to the body. Care must be taken to ensure the weld can be inspected adequately to verify the soundness.

Weldolet

A specially-designed, mass-produced, forged fitting for making tees in pipe runs. It produces a better result, from stressing and fluid flow criteria, than butting a branch directly to the run pipe. The trade name has passed into common usage to represent fittings of this style produced by other manufacturers. Versions are available for connections on bends, branches at an angle and adding screwed connections.

Wing-guided

A method of guiding valves from the seat bore. The valve plug is manufactured with three, four or five very large flutes, which extend down into the valve seat, so that only a narrow land is remaining at the periphery. The lands centralise the plug within the seat and maintain alignment.

Wiredrawing

The erosion effect of high velocity fluid jets across metal surfaces. This can be experienced on valve discs, plugs and seats and in packing boxes and the spindle/stem. It can look like a saw cut across the surface.

Witnessed activity

A production activity which is witnessed by the purchaser or his representative or the Certifying Authority. The activity may not be performed if the agreed witnessing parties are not present. A witnessed activity is a hold point on the quality plan.

WPS

Welding procedure specification

WQC

Welder qualification certificate

Y

Yoke

The structure which attaches some actuators to the bonnet. The yoke is usually open with two columns; totally enclosing circular yokes are possible.

Z

Zr

Zirconium, a rare element which is used in very small quantities in some steels as a carbide former, a deoxidiser, a sulphur remover and a nitrogen remover.

Properties of fluids

2

2.1 Explanation of terms

2.1.1 Introduction

Liquids, together with solids and gases, are the forms in which substances occur in nature: the solid state, the liquid state and the gaseous state. The three physical states are sometimes called phases. Liquids and gases can be combined in the general group called fluids. Fluids differ from solids in that they will readily change shape to suit the container.

A solid body subjected to a small shear force undergoes a small elastic deformation and returns to its original shape when the force is removed. When subjected to larger shear force the shape may be permanently changed due to plastic deformation.

A fluid when subjected to an arbitrarily small shear force undergoes a continuous deformation. This happens regardless of the inertia of the fluid. For a fluid the magnitude of the shear force and the speed of deformation are directly related. In a solid body it is the deformation itself which is related to the shear force.

A fluid may be either a liquid or a gas. A gas differs from a liquid in that it will expand to completely fill the container. A gas at conditions very close to boiling point or in contact with the liquid state is usually called a vapour. Fluids are compressible; gases being much more compressible than liquids.

A substance can exist in all three states. A typical example of this is ice, water and steam. When ice is heated at constant pressure, the ice converts to water at the melting point and to steam at the boiling point. If the steam pressure is increased at constant temperature, the steam converts to water at the saturation (vapour) pressure.

Solid particles can be suspended or mixed in a liquid. Such a combination, liquid plus particles, is called a suspension. When the particles distribute themselves evenly through the liquid, we speak of a homogeneous mixture. When concentration gradients occur, we speak of a heterogeneous mixture.

The word "solution" refers to an otherwise pure liquid in which a solid, another liquid or gas has been dissolved.

Two liquids which are not soluble in each other, can be mixed by mechanical action. Such a mixture is called an emulsion. Emulsions can be very difficult to separate.

The properties of liquids vary greatly. For the purposes of pumping the following characteristics of liquids should generally be known:-

- changes of state
- viscosity
- density
- compressibility
- pH value
- hazards

2.1.2 Changes of state

2.1.2.1 Melting point

SI unit °C

The melting point is the temperature at which a substance changes from a solid to a liquid state and also solidifies from a liquid to a solid. The melting point in most substances is pressure dependent only to a very limited degree. In those cases where the pressure dependence has to be taken into account, the boundary between solid and liquid state is shown by the melting curve in a pressure-temperature diagram.

2.1.2.2 Boiling point

SI unit °C

The boiling point is the temperature at which a liquid converts to vapour or gas at a particular local pressure. The boiling point is usually stated at a standardised atmospheric pressure, 101.325 kPa (760 mm Hg). The boiling point of water at this pressure is 100°C. The boiling point of all liquids is heavily dependent upon pressure.

2.1.2.3 Vapour pressure

SI unit Pa, kPa, MPa

Preferred unit bar

All liquids have a tendency to evaporate. Vapour or gas accumulates above the free surface of a liquid by reason of interchange of molecules. The partial pressure of the vapour rises to a point at which as many molecules are being returned to the liquid as are leaving it. At this equilibrium state the partial pressure of the liquid is called the vapour or saturation pressure.

Vapour pressure depends on temperature alone and increases with the temperature of the liquid. At a particular temperature, the equilibrium pressure above the surface of a liquid can never be less than the vapour pressure. Any attempt to lower the pressure of the vapour (by means of a vacuum pump for example) immediately results in increased evaporation, i.e. the liquid boils. If the pressure in a liquid decreases locally to the vapour pressure at the actual temperature, vapour bubbles are generated in the liquid. In a pump installation, the formation of vapour bubbles (cavitation) can cause serious mechanical damage and can seriously diminish the performance of the installation.

Various liquids display widely differing values of vapour pressure as a function of temperature. Some substances in a solid state can also change directly to the gaseous state without passing through the liquid state. As an example, a complete pressure-temperature diagram, showing the phases for water is shown in Figure 2.1. At the triple point, all three states may exist simultaneously. In practice the only substance which exhibits the solid-gas transformation is carbon dioxide.

NOTE: The SI unit of pressure is very small. In most practical applications it may be necessary to use Pa or kPa for suction pressures and MPa for discharge pressures. The pump industry is concerned that the change in prefix could confuse operators and lead to potentially dangerous situations. The MPa is a very large unit; small numbers representing large pressures. The bar is a much more suitable unit and does not require prefixes.

Whenever a pressure value is quoted it must be qualified as absolute or gauge.

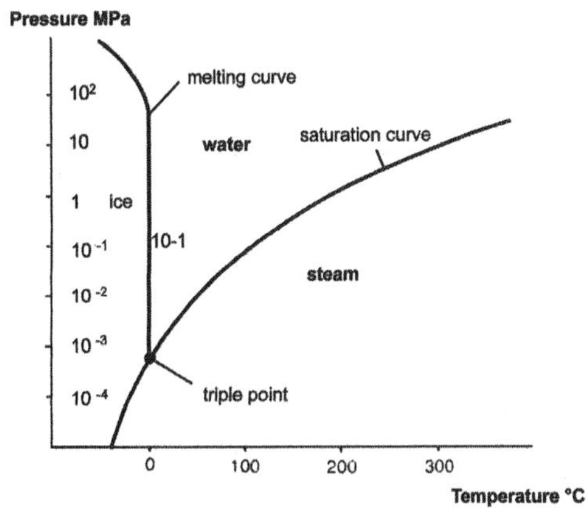

Figure 2.1 Phase diagram for water

2.1.2.4 Comments

Melting point (freezing or solidifying point) must be known when assessing the pumpability of a liquid and the risk of blockage, freezing or solidifying.

Information regarding the vapour pressure is required for calculating the permitted suction lift and $NPSH_a/NPIP_a$. Vapour pressure is a critical factor in for example the choice of pump type, speed and shaft sealing requirements.

2.1.3 Viscosity

Viscosity (the ability to flow) is a property of liquids treated under the heading of rheology. The word rheology derives from the Greek "rheos" meaning flow.

Between two layers of liquid flowing at different speeds, a tangential resistance, a shear stress, is developed because of molecular effects. We say that the shear stress is caused by the internal friction of the liquid, or conversely that the liquid transmits shear forces by reason of its internal friction.

A liquid in motion is continuously deformed by the effects of the shear forces. The magnitude of the stress depends on the rate of shear deformation and the sluggishness of the liquid, i.e. the viscosity.

2.1.3.1 Newton's Law of Viscosity

Viscosity is defined for flow in layers, laminar flow, by Newton's Law of Viscosity, see Figure 2.2.

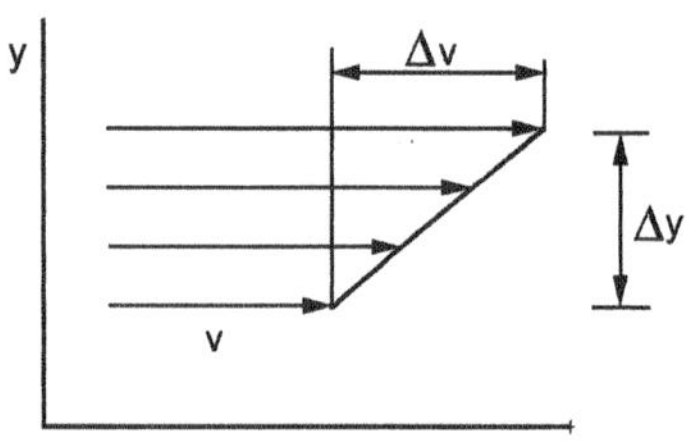

Figure 2.2 Definition of viscosity

$$\tau = \mu \frac{\Delta v}{\Delta y} \qquad \text{Equ 2.1}$$

where:

τ = shear stress (N/m^2)

μ = dynamic viscosity (kg/ms)

Δv = change in viscosity (m/s)

Δy = distance between layers (m)

2.1.3.2 Absolute viscosity

In the SI system the unit is:

1 kg/ms = 1 Ns/m^2

Other units are:

1 Poise = 1P = 0.1 kg/ms

or

1 centipoise = 1 cP = 0.01 P = 0.001 kg/ms

2.1.3.3 Kinematic viscosity

In viscous flow equations the dynamic viscosity divided by the density of the liquid is given the symbol ν. This parameter is called kinematic viscosity.

$$\nu = \frac{\mu}{\rho} \qquad \text{Equ 2.2}$$

where:

ν = kinematic viscosity (m^2/s)

μ = dynamic viscosity (kg/ms)

ρ = density (kg/m^3)

The SI unit for kinematic viscosity is 1 m^2/s.

Sometimes, for convenience, the following units are used:

Stoke = 1 St = 0.0001 m^2/s

or more usually

1 cSt = 0.01 St = 0.000001 m^2/s = 1 mm^2/s

Water at 20°C and 0.1 MPa has a kinematic viscosity of 1 cSt.

2.1.3.4 Newtonian liquids

A liquid which follows Newton's Law of Viscosity in laminar flow and has constant viscosity, regardless of shear rate and time, is known as a Newtonian liquid, see Figure 2.3.

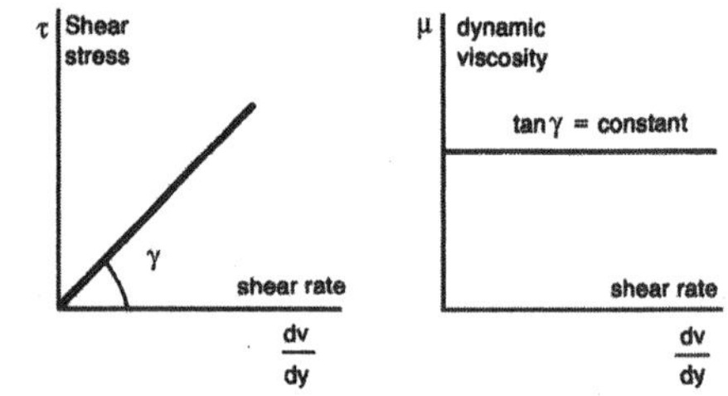

Figure 2.3 Newtonian liquid

Examples: water, aqueous solutions, low molecular liquids, oils and oil distillates. Black liquor, liquid resin and resinous sebacic acid also behave like Newtonian liquids.

2.1.3.5 Non-Newtonian liquids

Liquids which do not fulfil the requirements for Newtonian liquids are called non-Newtonian liquids. Most high molecular liquids, suspensions and emulsions display non-Newtonian properties. Non-Newtonian liquids usually fall into three main groups:

- **Non time-dependent**
  pseudo-plastic
  dilatant
  plastic
- **Time-dependent**
  thixotropic
  rheopectic
  irreversible
- **Visco-elastic**

Non time-dependent liquids are not affected by the length of time of the flow process, see Figure 2.4. The shear stress in the case of laminar flow at a given temperature is determined entirely by the shear rate. In analogy with Newtonian liquids, it may be stated:

$$\tau = \mu_1 \frac{dv}{dy} \qquad \text{Equ 2.3}$$

where:

μ_1 = apparent dynamic viscosity

In the case of pseudo-plastic liquids, the apparent viscosity decreases with increasing shear rate.

Examples: high molecular solutions, rubber, latex, certain molten materials, mayonnaise.

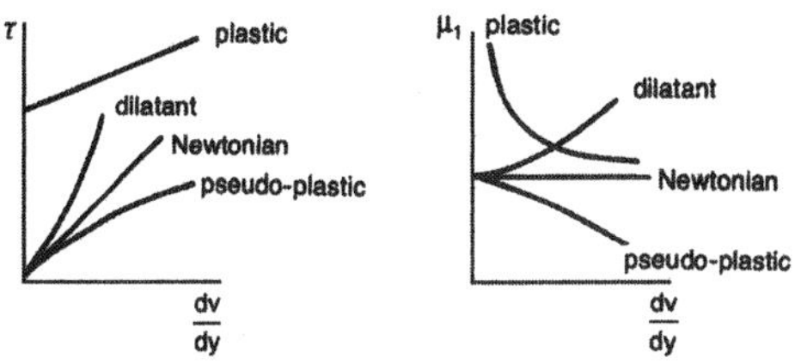

Figure 2.4 Non time-dependent non-Newtonian liquids

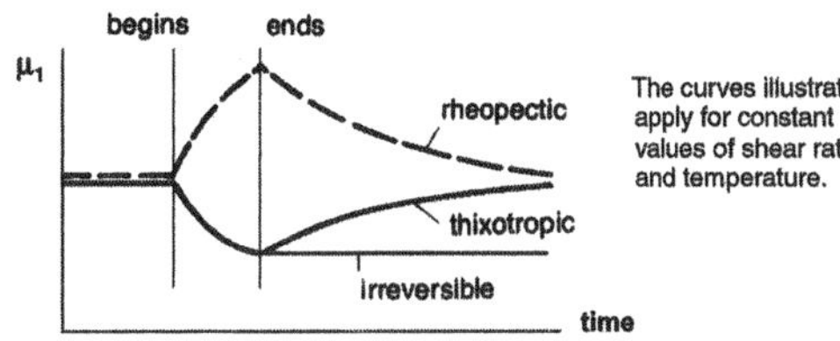

Figure 2.5 Illustrations of time-dependent non-Newtonian liquids

In the case of dilatant liquids the apparent viscosity increases with increasing shear rate.

Examples: base in oil paints, suspensions with high concentrations of small particles: cement, lime, sand, starch.

Plastic liquids require a certain minimum shear stress - yield stress - in order to initiate flow. The apparent viscosity decreases from an infinitely high value as the shear rate increases.

Examples: toothpaste, ointments, grease, margarine, paper pulp, printers ink, emulsions.

With time-dependent liquids the apparent viscosity is affected not only by the shear rate but also by the length of time during which the flow continues, see Figure 2.5.

In the case of thixotropic liquids, μ_1 diminishes when flow commences. When the flow ceases, the liquid returns to its original viscosity after a certain period of time.

Examples: paint, gelatinous foodstuffs.

Rheopectic liquids display increasing viscosity under mechanical influence and resume their original viscosity when the influence ceases.

Examples: certain suspensions of gypsum.

Irreversible liquids do not resume their original viscosity at all or perhaps only after a very long time after the removal of the influence. These liquids must be pumped carefully.

Examples: cheese coagulates, yoghurt, marmalade.

Visco-elastic liquids, contain liquids exhibiting both elastic and viscous properties. Visco-elastic liquids undergo both elastic and viscous deformations. When the flow ceases there is a certain reversal of elastic deformation.

Examples: asphalt, liquid nylon, rubber, polymer solutions.

2.1.3.6 Comments

- Dynamic viscosity is temperature dependent. Increasing temperature results in diminishing viscosity. Certain liquids can be pumped easier when heated.
- Viscosity is also an indication of the lubricating qualities of the liquid. At low values, less than 1 cSt, additional lubrication may be necessary.
- Temperature dependence on density has also to be considered in kinematic viscosity. At pressures of 10 MPa a certain pressure dependence can be observed in the viscosity.
- Loss coefficients in pipe flows are dependent upon Reynold's Number which is in turn dependent upon dynamic viscosity.
- The performance of rotodynamic pumps depends upon Reynold's Number. Standard pump data is always for water and this has to be corrected when pumping other liquids. Rotary positive displacement pumps are better for high viscosities but there is no specific value to aid selection. The viscosity effects are dependent upon the pump size. For rotodynamic pumps, increased viscosity reduces efficiency and increases NPSHr. The effects of viscosity on positive displacement pumps are dependent upon the pump type as well as size.
- Viscosity is defined as the ratio of shear stress and shear rate in laminar flow. In the case of turbulent flow, this ratio is affected by exchange of momentum between the layers caused by the random motion of the liquid particles.
- In the past viscosity has been specified by the results of proprietary test methods such as Engler, Saybolt and Redwood. These tests measured the time taken for an oil sample to flow through an orifice. In Chapter 20, Fluid properties & conversion, the necessary basis is given for conversion to SI units.

2.1.3.7 Penetration

For some non-Newtonian liquids, for example grease and asphalt, determination of viscosity does not provide sufficient information about flow properties and supplementary information regarding consistency is required. This information is obtained from determination of penetration.

For greases, penetration is a measure of consistency. This is the depth, in tenths of a millimetre, to which a cone forces its way down into a test receptacle containing a grease sample heated to 25°C. The penetration depends on whether the consistency is changed by stirring, shaking and so on. Hard greases have low penetration figures whilst soft greases have high penetration. Manufacturers of greases state the penetration figure for each quality, e.g. 240 - 325. Two lubricant greases, furthermore, may have the same penetration figure but may still have different flow capabilities depending upon the viscosity of their respective component oils. The oil grades and thickeners should also be specified.

Determination of penetration for asphalt is carried out in principle at 25°C in the same way, as with lubricant grease, but using a loaded pin instead of a cone. The values measured are used to classify asphalt, typical penetration figures being 10 - 50. See BS 2000-50 - which is equivalent to ISO 2137.

2.1.4 Density and relative density

2.1.4.1 Density

Unit, kg/m^3, symbol ρ

The density of a liquid is the relationship between the mass of the liquid and its volume. The density of liquids changes slightly with temperature and very slightly with pressure unless the liquid is very compressible. The density of a liquid is to a very slight extent affected by the quantity of dissolved gases. Since the ability to dissolve gases is temperature and pressure dependent, there is an indirect dependence on these parameters. The affect of dissolved gases on density can generally be neglected however.

2.1.4.2 Relative density

Unit, non-dimensional, symbol d

Relative density, formerly called specific gravity, is the ratio of a liquid's density to water at standard conditions, atmospheric pressure, 101.325 kPa and 4°C.

NOTE: Information about the density of a liquid, or its relative density, is required when converting pressure to head. Rotodynamic pumps for example, calculate power using differential head and specific gravity. Positive displacement pumps calculate power using differential pressure therefore density or specific gravity has no influence.

2.1.5 Compressibility

All liquids are compressible to some extent. It may be noted however that water is about 100 times more elastic than steel and about 0.012 times as elastic as air. Compressibility is very temperature dependent and slightly pressure dependent. Any values used must relate to the operating conditions. Classically, compressibility is expressed in terms of the bulk modulus defined by the relationship:

$$\text{Compressibility} = \frac{1}{K} \qquad \text{Equ 2.4}$$

$$K = \rho \cdot \frac{\Delta p}{\Delta \rho} = -V\frac{\Delta p}{\Delta V} \qquad \text{Equ 2.5}$$

where:

K = bulk modulus of the liquid (N/m^2)

p = liquid pressure (Pa)

ρ = density of the liquid (kg/m^3)

V = volume of liquid (m^3)

Δ = change of magnitude

The change in volume due to a change in pressure can be calculated directly from the definition

$$\Delta V = -\frac{V\,\Delta p}{K} \qquad \text{Equ 2.6}$$

where the minus sign indicates that the volume decreases with increasing pressure.

Densities and specific volumes of water under a wide range of pressures and temperatures are readily available in good steam tables. However compressibility data for other liquids can be difficult to obtain. Some data is available in the form of "contraction per unit volume per unit pressure increase" which is designated by χ. In the cases where definitive data is not available the compressibility can be approximated as a multiple of water compressibility.

2.1.5.1 Acoustic velocity

In the case of acoustics however, compressibility is of critical importance. The acoustic velocity, or wave speed, is directly related to the bulk modulus and compressibility. If acoustic resonance occurs in the pipework the acoustic velocity must be known to affect a successful cure. Acoustic resonance can be a very serious problem creating destructive piping vibrations and large pressure pulsations. In suction pipework the resonance can create cavitation conditions leading to valve and pump damage and reduced performance.

The acoustic velocity calculated from the bulk modulus applies to pure clean liquid. If the liquid contains gas bubbles or solid particles the acoustic velocity will be greatly reduced from the theoretical value. Testing may be the only approach to find the true value. See Table 2.1 for indicative values of bulk modulus and acoustic velocity.

2.1.6 pH value

The concentration of hydrogen ions in an aqueous solution is a measure of the acidity of the solution and is expressed as the pH value.

The definition of the pH value is

$$pH = \log_{10}\frac{1}{H^+} \qquad \text{Equ 2.7}$$

where:

H^+ = concentration of hydrogen ions (mol/l).

If, for example, $H^+ = 10^{-4}$ mol/l, then the pH value = 4.

The hydrogen exponent is another name for the pH value.

The numerical value of pH in Figure 2.6 varies between the limits 0 and 14. Acid solutions have pH values between 0 and 6.5, neutral solutions pH = 6.5 to 7.5 and the alkaline or base solutions pH = 7.5 to 14. Acid solutions turn blue litmus red and alkaline solutions turn red litmus blue.

All aqueous solutions reacting as acids contain a surfeit of hydrogen ions H^+. All aqueous solutions reacting as alkalis have a surfeit of hydroxide ions OH^-. The product of the hydrogen ion concentration and the hydroxide ion concentration is always a constant at any fixed temperature.

At 22°C for example the ion product is:

$$H^+ \times OH^- = 10^{-14} \text{ mol/l}$$

When the hydrogen ion concentration and the hydroxide ion concentration are equal, i.e.

$$H^+ = OH^- = 10^{-7} \text{ then}$$

pH = 7 and the solution reacts neutrally.

Pure water is a very poor electrical conductor. Water only acquires good conductivity in solutions of so-called electrolytes such as salts, acids or bases. When dissolved in water, these substances split into two components with opposing electrical charges, by means of which an electric current may be carried. This act of splitting is called electrolytic dissociation. The

Liquid	Bulk modulus N/m^2	Wave propagation speed m/s	Estimated wave propagation speed in pipe	
			Steel	PVC
Acetone	$0.8 \cdot 10^9$	1000	850	280
Petrol	$1.0 \cdot 10^9$	1170	1000	320
Ethylene glycol	$2.7 \cdot 10^9$	1560	1150	300
Glycerol	$4.5 \cdot 10^9$	1890	1400	380
Chloroform	$1.0 \cdot 10^9$	820	700	230
Mercury	$24 \cdot 10^9$	1330	430	-
Methanol	$0.8 \cdot 10^9$	1000	850	280
Oil	$1.5 \cdot 10^9$	1300	1000	300
Turpentine	$1.4 \cdot 10^9$	1270	1000	300
Water, distilled	$2.2 \cdot 10^9$	1480	1100	300
Water, sea	$2.4 \cdot 10^9$	1530	1100	300

Table 2.1 Elastic properties of some liquids at 25°C and 0.1 MPa (1bar)

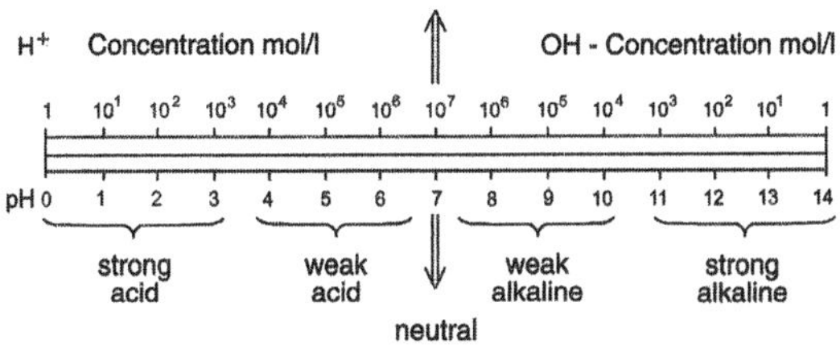

Figure 2.6 Ion concentration, pH value, degree of acidity and alkalinity

charged particles so formed are called ions. Distinction is made between the positively charged cation particles and the negatively charged anion particles. The positively charged particles are symbolised by $^+$ and the negatively charged particles by $^-$. Metals and hydrogen in acids produce cations by electrolytic dissociation. The acid remnants and the alkaline hydroxyl groups on the other hand generate negatively- charged anions.

For liquids with pH values of less than 4, the hydrogen ion concentration is a powerful factor affecting the speed of the dissolving process for most metals. The positively-charged hydrogen ions relinquish their charge at the surface of the metal to the metal atoms and break down their solid connections, whilst they themselves, robbed of their ion construction, return to the atomic state and unite into molecules leaving the metal surface in the form of gas bubbles. The dissolving process is faster, therefore, when the concentration of hydrogen ions is greater, i.e. the lower the pH value of the liquid.

Within the pH range of about 4 to 9, the oxygen present in the liquid, from air for example, has an influence on the corrosion process, say, for iron. The discharged hydrogen does not leave in the form of gas bubbles, as is the case with pH values in the region of 0 to 4, but combines immediately with the oxygen to produce water. At the same time the iron is oxidised to rust. The affect of pH on non-metallic materials must be considered on a case by case basis, see Figure 2.7.

Although pH is an important factor in the attack of metals, this is by no means the only factor which needs to be considered. The concentration (as distinct from pH) of the chemical, particularly of acids and alkalis, has an important bearing on the degree of corrosion or attack. Dissolved oxygen plays an important part in corrosion of metal components. There is now a whole range of organic chemical compounds and biological compounds which will give rise to chemical attack, and many of these are completely unrelated to pH. Even very small concentrations of some modern chemicals can create serious hazards both from the point of view of chemical attack, leakage and toxic hazards.

2.1.7 Hazards

The word "hazard" is in common use in the English language. Its use in *Valves Manual International* is defined as follows:

Hazard: — a physical situation with a potential for human injury, damage to property, damage to the environment or some combination of these.

It can be seen from the definition that three distinct types of affect are considered but in some cases one hazard may lead to others. Fire, for example, can be a serious health hazard.

A hazardous substance: — a substance which, by virtue of its chemical properties, constitutes a hazard.

2.1.7.1 Health hazards

The effects of the liquid and its vapour on the health of the operators and employees must be considered. Most countries have legislation limiting the exposure of employees to substances judged to be hazardous. If the liquid to be handled is listed in local regulations the type of health hazard must be specified.

Another health hazard, sometimes not recognised as such, is noise. This is discussed in detail in Chapter 10. Some countries have regulations stipulating the acceptable noise levels and exposure times. Some equipment is inherently noisy and large equipment, in general, is noisy. Noise levels can be attenuated by fitting acoustic enclosures however these tend to drastically diminish the maintainability of the equipment by hindering access. In some instances, costly acoustic enclosures have been removed at site and scrapped in order to achieve acceptable access. One easy solution to this hazard is to declare certain areas as "Ear Protection Zones".

2.1.7.2 Physical hazards

Physical hazards include fire and explosions as well as corrosion and temperature. The degree of risk attached to the hazard is dependent upon the properties of the liquid and the vapour; if the liquid evaporates quickly at ambient temperature and whether the vapour is lighter or heavier than air. A light vapour can disperse easily in an open installation whereas a heavy vapour can become trapped in sumps, depressions and pipe trenches. The upper and lower limits of flammability are important when considering allowable concentrations.

Within the EU, the Machinery Directive, 89/392/EEC, amended 98/37/EEC, places the responsibility for safety on the machine designer. The machine must be designed to be safe in all aspects; installation, commissioning, operation and maintenance. If the designer is unable to devise a completely safe machine the areas of concern must be documented and recommended precautions communicated to the user. Because this is a legal requirement in all EU countries the machine designer may not be relieved of the obligations by a third party.

Looking to the future, the new Machinery Directive 2006/42/EC has now been published in the Official Journal of the European Union and will be effective from 29th December 2009.

2.1.7.3 Environmental hazards

The Earth's resources and waste-disposal capabilities are finite. Stricter limitations are being imposed gradually on the amount of pollutant which can be released while the list of pollutants is becoming longer. The valve user must be aware of the full consequences of leakage of liquid/vapour from the valve and installation. The environment can be considered in two separate identities:

- local
- global

If the site is surrounded or close to a town what risk is likely to the population, structures or habitat in the event of a failure? In the global sense, what are the likely cumulative affects of product leakage?

2.1.7.4 Installation hazard assessment

The user and system designer are in full possession of all the relevant available facts regarding the liquid and the installation. Any assumptions made should be passed to the valve manufacturer and identified as such.

Auto ignition point	The temperature above which a substance will start to burn without an ignition source being necessary
Flash point	The lowest temperature at which a liquid gives off sufficient vapour to burn if an ignition source is present
Atmospheric boiling point	The temperature at which the liquid boils at atmospheric pressure, 101.325 kPa
Vapour specific gravity	Specific gravity is the ratio of a vapour's density to air at standard conditions, atmospheric pressure, 101.325 kPa and 4°C
Surface tension	The surface tension at process temperature will indicate how difficult the liquid is to seal

Table 2.2 Liquid properties reviewed during hazard assessment

The user must assess the risks attached to all the possible hazards and decide what, if any, leakage is acceptable. Liquid

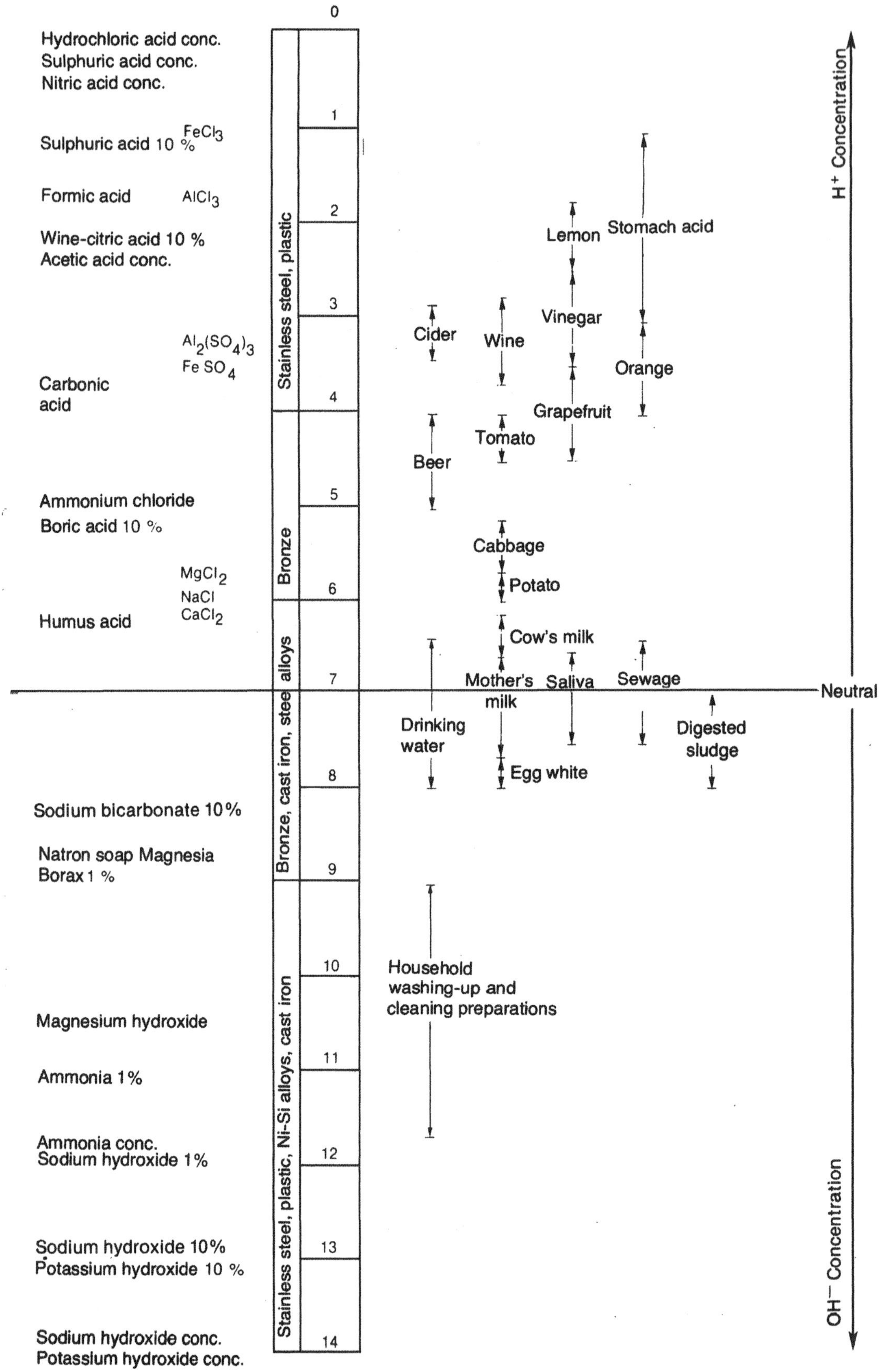

Figure 2.7 Effect of pH values on non-metallic materials

properties reviewed during the assessment, should include those set out in Table 2.2.

The nature of the hazards will also dictate the type of pipe connections to be used; screwed, flat-face flanged, raised-face flanged, ring-type joints. Process upset conditions must be considered as part of the assessment. Upset conditions which last for more than one or two hours may have a significant impact on pump and ancillary equipment selection.

The physical location of the valve, indoor or outdoor, will decide the behaviour of the leakage once outside the valve. Will any vapour cloud quickly disperse on a breeze which always blows over the un-manned site or will a manned enclosure gradually build up a dangerous concentration of vapour? Only the user can assess these questions and specify the necessary precautions.

It is the responsibility of the user to define exactly what the valve is intended to do. It is the responsibility of the valve manufacturer to supply equipment to meet the required performance.

2.1.7.5 Useful sources of information

Basic information on plant and system design, together with risk assessment, can be obtained from various sources. See Section 2.7 for more complete information. For example the British Institution of Chemical Engineers publishes several useful texts as well as the American Institute of Chemical Engineers.

Information regarding trade names and manufacturers of dangerous chemicals is available from the UK Chemical Industries Association.

BS 5908 summarises the statutory requirements, storage and movement of materials, process plant piping, ventilation, fire protection etc. This publication also lists a wide range of reference literature including: UK Health Protection Agency, American Petroleum Institute (API), American Society of Mechanical Engineers, (ASME), the now defunct US Bureau of Mines, and publications of the Oil Companies' European Organisation for environment, health & safety in refining and distribution (CONCAWE).

BS 5908 classifies substances according to the operating temperature and flash point temperature, see Table 2.3, agreeing with the UK Energy Institute.

	Flash point °C	Operating temperature °C
Class I	< 21	
Class II(1)	21 to 55	below flash point
Class II(2)	21 to 55	at or above flash point
Class III(1)	55 to 100	below flash point
Class III(2)	55 to 100	at or above flash point

Table 2.3 Classification of substances

American practice is outlined in ANSI B31.3. The UK Health & Safety Executive also produces a wide range of literature on all aspects of safety. Work with chemical substances that are classed as hazardous to health is covered in the UK by The Control of Substances Hazardous to Health (Amendment) Regulations 2003. General advice on control measures can be found in the COSHH Approved Code of Practice. (Health & Safety Executive (HSE) Books Reference number: L5). COSHH covers the following significant areas:

Regulation 6 assessment of exposure,

Regulation 7 control of exposure,

Regulation 8 & 9 use and maintenance of control measures,

Regulation 10 monitoring exposure.

Separate regulations cover the hazards of carcinogenic substances, lead and asbestos. COSHH does not apply to underground mining installations or to the hazards posed by micro-organisms.

Another useful source of information is The European Agreement concerning the International Carriage of Dangerous Goods by Road (ADR) on 30 September 1957, under the auspices of the United Nations Economic Commission for Europe, and it entered into force on 29 January 1968.

A set of new amendments came into force on 1 January 2005, and consequently, a third consolidated "restructured" version was published as document ECE/TRANS/175, Vol.I and II (ADR 2005).

The new structure is consistent with that of the United Nations Recommendations on the Transport of Dangerous Goods, Model Regulations, the International Maritime Dangerous Goods Code (of the International Maritime Organization), the Technical Instructions for the Safe Transport of Dangerous Goods by Air (of the International Civil Aviation Organization) and the Regulations concerning the International Carriage of Dangerous Goods by Rail (of the Intergovernmental Organisation for International Carriage by Rail).

Conditions are laid down in the UK Petroleum (Consolidation) Act 1928 and in particular The Highly Flammable Liquid and Liquefied Gases Regulations SI (Statutory Instrument) 1972 No. 917, governing the storage, handling and conveyance of gases and liquids constituting a fire hazard.

According to these regulations:

- **Highly flammable liquid** means any liquid, liquid solution, emulsion or suspension which, when tested in the manner specified in Schedule 1 to the Regulations, gives off a flammable vapour at a temperature of less than 32°C and when tested in the manner specified in Schedule 2 to the Regulations, supports combustion;
- **Dangerous concentration of vapours** means a concentration greater than the lower flammable limit of the vapours.

Fire hazardous liquids are listed in the Petroleum (inflammable Liquids) Order SI 1968 No 570 under Part 1 of the Schedule and any solution or mixture containing any of those substances specified in Part 1 of the Schedule, if it gives off an inflammable vapour at a temperature below 22.8°C.

NOTE: The Fire Hazard categories specified in the Liquid Table in Section 2.6.2 are classified according to the following:

CLASS 1 Liquids having a flash point below +21°C.

CLASS 2a Liquids with a flash point above +21°C but not exceeding +30°C.

CLASS 2b Liquids with a flash point above +30°C but not exceeding +60°C.

CLASS 3 Motor fuel and heating oil with a flash point in excess of +60°C.

Below 32°C (73°F):

Abel Apparatus: Statutory Method - BS 2000: Part 33:1982 (IP 33/59(78))

-18°C to 71.5°C (0°F to 160°F):

Abel Apparatus: Non-Statutory Method - BS 2000: Part 170:1982 (= IP 170/75(81))

-7°C to 371°C:

Pensky-Martens Apparatus - BS 6664: Part 5:1990 (= IP 34/67 and = ISO 2719)

American literature can be checked for additional information. NFPA (Fire) HAZ-01 published by the US National Fire protection Association maybe useful.

ACGHI® documentation of threshold limit values for substances in workroom air is available from American Conference of Government Industrial Hygienists.

Gas constituting a fire hazard is not classified according to a flash point, the criteria being the way in which the gas can be ignited and the flame propagated. An explosive atmosphere occurs if the percentage concentration by volume of an explosive gas or vapour in air is such that the air will burn when ignited and flame propagation occurs. The upper and lower explosive limits are referred to as the explosive range or flammability limits.

Regulations governing such installations so as to reduce the risk of explosion caused by electrical equipment are laid down and certified by National bodies, European and International agreements. This aspect of hazards and safety is fully covered in Chapter 10 but some basic information is given here.

BS 5345 sets out a code of practice for selection, installation, and maintenance of electrical apparatus for use in potentially explosive atmospheres (other than mining applications or explosive processing and manufacture), gives guidance and references for all aspects of electrical installations.

European - EN 50 014 to EN 50 020, plus EN 50 028 and EN 50 039 are the relevant European standards issued by the European Committee for Electrotechnical Standardization (CENELEC) and adopted by all EU member countries. The International Electrotechnical Committee (IEC) has issued Standard IEC 79, in 18 sections for world use.

Accredited test houses check for standard compliance and issue certificates for suitable equipment. The UK, Germany and USA have the most popular test houses, see Chapter 15.

2.2 Water

2.2.1 Demineralized water

Demineralized water is chemically pure water and can be produced by various methods :

- Multiple distillation
- Multi-bed ion exchange
- Reverse osmosis
- Electro-dialysis

Chemical impurities and dissolved gases are removed which cause the water to become chemically active. At reasonable temperatures, glass and austenitic stainless steels are completely corrosion resistant. Cast iron and bronze are both attacked resulting in contamination of the water. At higher temperatures austenitic stainless steels suffer from intergranular corrosion.

The hydraulic properties of demineralized water can be determined with very great accuracy. The viscosity and vapour (saturation) pressures given in Tables 2.4 to 2.7, conform to the international standard values accepted at the ICPS-8 (Eighth International Conference on Properties of Steam, 1975). The values stated for density agree with ICPS-6 (1964).

Temperature°C	Dynamic viscosity μ kg/m.s			
	0.1 MPa	1 MPa	10 MPa	100 MPa
0	$1.793 \cdot 10^{-3}$	$1.789 \cdot 10^{-3}$	$1.768 \cdot 10^{-3}$	$1.651 \cdot 10^{-3}$
25	$0.891 \cdot 10^{-3}$	$0.891 \cdot 10^{-3}$	$0.889 \cdot 10^{-3}$	$0.891 \cdot 10^{-3}$
50	$0.547 \cdot 10^{-3}$	$0.547 \cdot 10^{-3}$	$0.549 \cdot 10^{-3}$	$0.568 \cdot 10^{-3}$
75	$0.377 \cdot 10^{-3}$	$0.378 \cdot 10^{-3}$	$0.380 \cdot 10^{-3}$	$0.404 \cdot 10^{-3}$
100	-	$0.282 \cdot 10^{-3}$	$0.285 \cdot 10^{-3}$	$0.300 \cdot 10^{-3}$
150	-	$0.182 \cdot 10^{-3}$	$0.185 \cdot 10^{-3}$	$0.207 \cdot 10^{-3}$
200	-	-	$0.136 \cdot 10^{-3}$	$0.156 \cdot 10^{-3}$
250	-	-	$0.108 \cdot 10^{-3}$	$0.128 \cdot 10^{-3}$
300	-	-	$0.087 \cdot 10^{-3}$	$0.110 \cdot 10^{-3}$
350	-	-	-	$0.096 \cdot 10^{-3}$

Table 2.4 Dynamic viscosity of chemically pure water at various temperatures and pressures

Temperature °C	Density ρ kg/m³			
	0.1 MPa	1 MPa	10 MPa	100 MPa
0	999.8	1000.2	1004.7	1045.5
50	988.0	988.5	992.4	1027.4
100	-	958.6	962.8	1000.0
150	-	917.1	922.2	965.3
200	-	-	871.1	924.2
250	-	-	806.0	876.7
300	-	-	715.4	823.2
350	-	-	-	762.5

Table 2.5 Density of chemically pure water at various temperatures and pressures

Temperature °C	Dynamic Viscosity μ kg/m.s	Kinematic Viscosity ν m²/s	Density kg/m³	Remarks
-25	$5.842 \cdot 10^{-3}$	$5.889 \cdot 10^{-6}$	992.1	
-20	$4.342 \cdot 10^{-3}$	$4.365 \cdot 10^{-6}$	994.7	Super-
-15	$3.342 \cdot 10^{-3}$	$3.353 \cdot 10^{-6}$	996.7	cooled
-10	$2.650 \cdot 10^{-3}$	$2.655 \cdot 10^{-6}$	998.3	water
-5	$2.156 \cdot 10^{-3}$	$2.158 \cdot 10^{-6}$	999.3	
0	$1.793 \cdot 10^{-3}$	$1.793 \cdot 10^{-6}$	999.8	
5	$1.518 \cdot 10^{-3}$	$1.518 \cdot 10^{-6}$	1000.0	
10	$1.306 \cdot 10^{-3}$	$1.306 \cdot 10^{-6}$	999.8	
15	$1.137 \cdot 10^{-3}$	$1.138 \cdot 10^{-6}$	999.2	
20	$1.003 \cdot 10^{-3}$	$1.005 \cdot 10^{-6}$	998.3	NB: All
25	$0.891 \cdot 10^{-3}$	$0.895 \cdot 10^{-6}$	997.2	values are
30	$0.798 \cdot 10^{-3}$	$0.801 \cdot 10^{-6}$	995.7	for a
35	$0.720 \cdot 10^{-3}$	$0.724 \cdot 10^{-6}$	994.1	pressure of
40	$0.654 \cdot 10^{-3}$	$0.659 \cdot 10^{-6}$	992.3	0.1 MPa
50	$0.547 \cdot 10^{-3}$	$0.554 \cdot 10^{-6}$	988.0	(1bar).
60	$0.466 \cdot 10^{-3}$	$0.474 \cdot 10^{-6}$	983.2	
70	$0.403 \cdot 10^{-3}$	$0.412 \cdot 10^{-6}$	997.7	
80	$0.354 \cdot 10^{-3}$	$0.364 \cdot 10^{-6}$	971.6	
90	$0.315 \cdot 10^{-3}$	$0.326 \cdot 10^{-6}$	965.2	
95	$0.298 \cdot 10^{-3}$	$0.310 \cdot 10^{-6}$	961.7	
99	$0.285 \cdot 10^{-3}$	$0.297 \cdot 10^{-6}$	958.9	

Table 2.6 Dynamic viscosity, kinematic viscosity and density for chemically pure water in the temperature range -25°C to 99°C and at a pressure of 0.1 Mpa (1 bar)

Temperature°C	Vapour pressure MPa	Vapour pressure, metres, column of water
0	0.00061	0.06
10	0.00127	0.13
20	0.00234	0.24
30	0.00424	0.43
40	0.00738	0.76
50	0.01234	1.27
60	0.01992	2.07
70	0.03116	3.27

Temperature°C	Vapour pressure MPa	Vapour pressure, metres, column of water
80	0.04736	4.97
90	0.07011	7.41
100	0.10133	10.8
110	0.14327	15.4
120	0.19854	21.5
150	0.47600	52.9
200	1.5549	183
250	3.9776	508
300	8.5927	1229
350	16.535	2810

Table 2.7 Vapour (saturation) pressure for chemically pure water

Various characteristics of chemically pure water are shown in Table 2.8.

Molecular weight	18.015
Melting point, ice to water	0°C at 101 kPa
Latent heat of fusion, ice to water	335 kJ/kg
Specific heat	4.182 kJ/kg K at 101 kPa and 20°C
Boiling point	100°C at 101 kPa
Latent heat of steam	2260 kJ/kg at 100°C
Coefficient of expansion g =	0.000207 per deg°C at 101 kPa and 20°C
Bulk modulus K =	2.2×10^{-9} N/m^2 at 101 kPa and 20°C

Table 2.8 Chemically pure water characteristics

2.2.2 Fresh water

2.2.2.1 General

Fresh water is the water derived from rivers, streams and wells. Generally containing less than 1% salt, sodium chloride, it can be "hard" or "soft". Hard water is high in calcium and/or magnesium salts and results in poor lather when using soap. Dissolved gases are always present, usually oxygen and carbon dioxide. The oxygen content should be about 5 mg/l.

During treatment for drinking water chlorides are added. Unless stated otherwise, fresh water is considered to be clean; this is taken as meaning no measurable concentration of solids over 10 microns. Untreated water should have micro-organisms 0.5 to 1.0 microns.

Fresh water has many uses, among other things as raw water for drinking purposes, as cooling and process water within various industries and for irrigation.

For the purposes of pipework installations, the most interesting aspect is the corrosive action of water on the commonly used construction materials:

- steel
- grey cast iron
- bronze

The characteristics of water which contribute to corrosion are:

- pH value
- hardness and carbonic acid content
- content of various chemicals, primarily salts
- acidity

Apart from these, the temperature of the water and the following factors will influence installation corrosion and wear:

- Velocity of flow; normally several m/s in pipes and 10-40 m/s in a centrifugal pump
- NPSH available and cavitation
- Content of solid bodies, e.g. sand and sludge from various sources

Not all materials are suitable for higher velocity service. Not all materials have equal cavitation resistance. If the absolute pressure of fresh water is reduced to approximately 0.4 bara the dissolved gases will start to evolve. If this occurs in pipework, suction cavitation symptoms and failures may occur. Small quantities of solids can cause high wear rates and accelerated corrosion in valves designed and selected for clean water.

Consideration should be given to the rate of corrosion. Popular valve types can be supplied with an internal corrosion allowance providing an estimated component life. It is very difficult to make general rules when even an insignificant quantity of salt may magnify an attack of corrosion. Water containing chlorides can be especially troublesome. The corrosive action on steel, for example, is increased by a factor of 8 at 50 mg/litre of Cl_2, which means that steel is not practicable for this purpose.

2.2.2.2 pH value and choice of material

Natural waters usually have pH values of between 4 and 8. They are divided into two main groups according to their acid content.

- **Ground water** from deep sources:— this contains very little acid and it is thus the hydrogen ion concentration which is the decisive factor in the aggressiveness of the water. It should be pointed out that iron is attacked noticeably at pH values of 6 to 7 in this low acid water.
- **Surface water**, which is acidic:— here the pH value is no absolute measure of aggressiveness although it is important to know what it is.
- **Drinking water**: — this may be low acid ground water or high acid surface water which has, furthermore, been treated chemically and filtered. Waterworks can supply the necessary information in this respect. Special conditions apply for lime-containing water, which is dealt with in the following Section.

Since valves and pumps used for handling fresh water are for the most part constructed of cast-iron, some general aspects are given below about the use of this material with particular reference to the pH value of the water.

- Grey cast iron may be used without any real problems within the pH range 7 to 10. If chlorides are present, it may be that cast iron is not adequate.
- Grey cast iron can often be used within the pH range 5 to 7, but the affects of those factors arising from the lower pH values can be great. Within this pH range, cast iron is superior to steel as regards resistance to corrosion. The high carbon content of grey cast-iron (3 to 4 %) means that at moderate speeds of flow the graphite, together with corrosion products, can build up an anti-corrosive film, so-called graphitization.

For pH values at which cast-iron is not resistive, bronze, steel or stainless steel has to be used. Aluminium bronze is particularly suitable for fresh water, although nickel aluminium bronze may be preferred for longer life. (See also Chapter 13, Materials).

Steel can be used successfully under the right conditions. High velocity areas can be coated. 11-13Cr steel is popular for boiler feed applications although stainless steel is sometimes used to protect against poor water quality control. It should be emphasised that the pH value of water can always be adjusted by suitable chemical treatment before pumping. Care must be taken to ensure any chemical treatment is performed correctly and that additional corrosion problems are not introduced.

When handling corrosive liquids, particularly liquids supporting electrolytic cells, the correct choice of material combinations for valves is critical. Materials for shafts, sleeves, bushes and wear rings must also be considered. Material properties at clearances can be varied by coating or plating one or both components.

2.2.2.3 Hardness of water

The degree of hardness of water depends upon the presence of impurities, mainly calcium (Ca) and magnesium (Mg) in the form of carbonates, although non-carbonates, for example sulphates, nitrates and chlorides, also have an effect. The quantities in which these impurities occur are a measure of the hardness of the water. Hardness can be expressed in terms of the specific substance such as calcium hardness (Ca-H), magnesium hardness (Mg-H) and so on. The total hardness comprises the sum of the individual hardnesses.

Distinction can also be made between permanent and temporary hardness. Temporary hardness consists of alkali ions bound to carbonates, and permanent hardness of alkali ions bound to non-carbonates. Temporary hardness is so called because it disappears with heating. Distribution under the various headings is according to the chemically equivalent content of alkali ions. Since the various alkaline metals have different atomic weights, the unit for hardness - 1 milli-equivalent per litre (meq/l) is defined as:

$$1\ \text{meq}/\text{l} = \frac{1\text{m mol}/\text{l}}{\text{chemical valency}} \qquad \text{Equ 2.8}$$

The hardness unit 1 meq/l corresponds to the following ion contents in mg/l:

1 meq calcium hardness = 20.04 mg/l Ca++

1 meq magnesium hardness = 12.16 mg/l Mg++

1 meq strontium hardness = 43.82 mg/l Sr++

1 meq barium hardness = 68.68 mg/l Ba++

The so-called English degree of hardness, Eng°, is expressed as the equivalent content of calcium carbonate ($CaCO_3$) per Imperial gallon of water, thus:

$$1\text{Eng}° = 14.2\text{mg}\,CaCo_3\,/\text{l} \qquad \text{Equ 2.9}$$

For the various oxides of mineral alkaline metals:

$$1\text{Eng}° = 8.0\text{mg}\,CaO/\text{l}$$

$$1\text{Eng}° = 5.7\text{mg}\,MgO/\text{l}$$

$$1\text{Eng}° = 14.78\text{mg}\,SrO/\text{l}$$

$$1\text{Eng}° = 21.88\text{mg}\,BaO/\text{l}$$

and the relation in terms of meq/l is:

$$1\text{Eng}° = 0.285(\text{meq/l})$$

$$1\text{meq}/\text{l} = 3.5(\text{Eng}°)$$

It is usual in the UK to state the hardness of water in English hardness degrees. See Table 2.9.

English Hardness degrees	Classification
3	very soft
3 - 6	soft
6 - 10	medium hard
10 - 15	rather hard
15 - 24	hard
- 24	very hard

Table 2.9 English hardness degrees

Other units of concentration are used in other countries; for conversion of various hardness units, see Table 2.10.

Units	Alkali ions meq/l	German hardness degrees °dH	English hardness degrees	French hardness degrees	ppm $CACO_3$ (USA)
1 meq/l alkali ions	1.00	2.8	3.5	5.0	50.5
1 German hardness degree °dH	0.356	1.0	1.25	1.78	17.8
1 English hardness degree	0.285	0.80	1.0	1.43	14.3
1 French hardness degree	0.20	0.56	0.70	1.00	10.0
1 ppm $CACO_3$ (USA)	0.02	0.056	0.07	0.10	1.0

Table 2.10 Conversion factors for various degrees of hardness

Example:

1°dH = 1.78 French hardness degrees

1 Eng° = 0.80 German hardness degrees °dH

Soft water in general is more suitable for the majority of household tasks, than hard water. When washing, a high bicarbonate content in the water is damaging because soaps consisting of a mixture of sodium stearate and palmitate generate insoluble calcium salts of organic acids with calcium bicarbonate. Hard water always contains calcium carbonates whose solubility diminishes with increase of temperature.

That is why calcium carbonate is deposited as fur and scale in boilers, heat exchangers and other heating vessels. In order to combat the deposition of scale, which can lead to local overheating in steam boilers, soft water has to be used as a source of supply. Water intended for such applications has to be softened, or dehardened.

2.2.2.4 Carbonic acid and carbonate equilibrium

Precipitation seeping down through the ground to become ground water, absorbs carbon dioxide (CO_2) from the air in the soil, generated there by the oxidisation of organic material or by the action of various acids on limestone. Carbon dioxide and water from carbonic acid H_2CO_3 converts the carbonates $CaCO_3$ (limestone) and $MgCO_3$, both difficult to dissolve in water, into soluble bicarbonates $Ca(HCO_3)_2$ and $Mg(HCO_3)_2$. The latter contains some CO_2 from the original carbonate (bonded carbonic acid), and some CO_2 from the carbonic acid which converted the carbonate into bicarbonate (semi-bonded carbonic acid) (see Figure 2.8). In order to keep the bicarbonate dissolved, a certain extra amount of CO_2 is required (free attached carbonic acid).

If there is enough carbonate in the ground, and if all the CO_2 is used up in the conversion of this to bicarbonate and in keeping the bicarbonate in solution, then the water is in a state of equilibrium as regards carbonate-carbonic acid. Thus a special condition of equilibrium arises between lime and the attached free carbonic acid. If the free carbonic acid content is less than that required for equilibrium, lime is separated. If the carbonic acid content increases the lime is re-dissolved. If, on the other hand, there is a surplus of CO_2, this is called "free surplus carbonic acid" or "aggressive carbonic acid". It is this part of the carbonic acid content which usually causes corrosion.

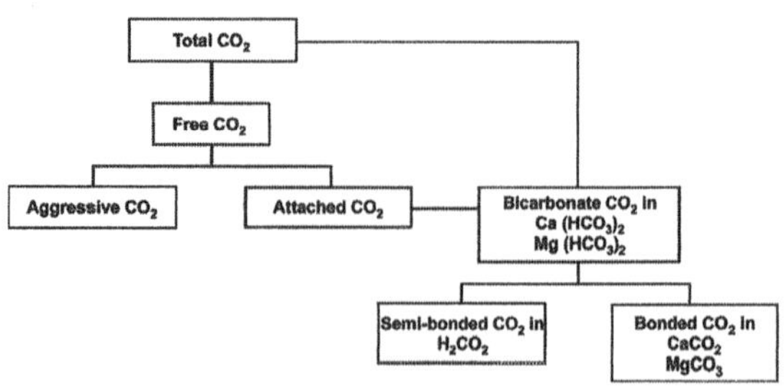

Figure 2.8 Various states for carbon dioxide (CO_2) in water

Water without free carbonic acid, if oxygen is present at the same time, will easily generate carbon-containing protective films which prevent corrosion on exposed valve/pump surfaces, vessels and pipes. If there is free "aggressive carbonic acid" in the water, the build-up of the protective layer is hampered, particularly if there is a lack of oxygen. If this kind of water comes into contact with lime, concrete, treated lime or soda, the lime is dissolved until its chemical equilibrium is restored. Free carbonic acid is thus aggressive to lime.

Iron, steel and light metals are attacked by all free content of carbonic acid, so that the lower the value of pH, the greater is the speed of corrosive breakdown. Ground water often contains bicarbonates with "attached carbonic acid" and perhaps also "aggressive carbonic acid". Water containing aggressive carbonic acid can be aerated and/or dosed with a suitable alkali (lime, soda, sodium hydrate) to bond the carbonic acid and to increase the pH value. At the same time, however, it must be observed that the so-called lime-saturated pH value should not be overstepped. If that happened, troublesome precipitations might result.

For practical purposes, the pH for natural carbonated water depends almost entirely on the relationship between the "bonded" and the "free carbonic acid" and, according to Klut, is determined by the relationship:

$$pH = 6.82 + \log\left(\frac{\text{bonded } CO_2}{\text{free } CO_2}\right) \qquad \text{Equ 2.10}$$

If the equilibrium value pH is set in relation to the carbonate hardness, then relationships are obtained as shown in Figures 2.9 and 2.10. Using Figure 2.10 distinction can also be made between "attached" and "aggressive" carbonic acid.

Carbonate precipitation can cause trouble in certain assemblies in pumps, for example shaft seals and water-lubricated plain bearings.

Figures 2.9 and 2.10 can be used to measure the aggressivity of natural water when the pH value and the carbonate hardness are known. Oxygen-rich carbonated water is aggressive only when the pH value is less than the values along the equilibrium curve. In other cases, a lime protective film is built up. If the pH value is considerably below the values on the equilibrium curve, the affect of other pH reducing substances may be suspected. Corrosion is then caused by these.

In the case of waters having low oxygen content there will be no build-up of protective film. All free content of carbonic acid adds to the aggressivity of the water. The corrosion rate increases with decreasing pH value.

The oxygen content of raw water, which is so important, can vary greatly. In both spring water and ground water emanating from upper strata, the oxygen content is almost always sufficient to build up the natural lime rust-protective layer for carbonate hardnesses of 6°dH. Soft surface water, because of lack of lime, cannot build up lime rust-protective layers and is, therefore, always more or less aggressive.

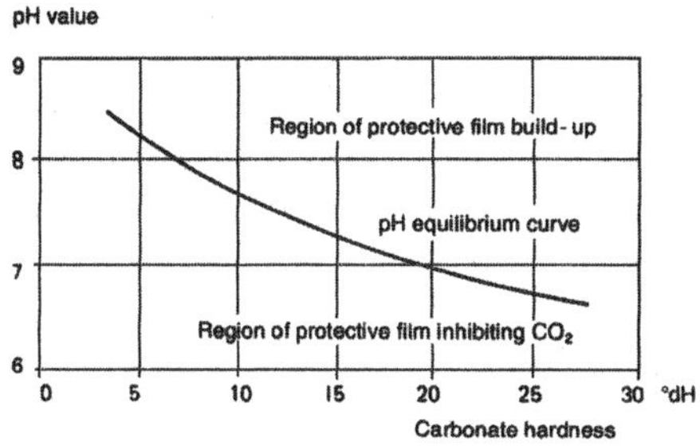

Figure 2.9 Condition of equilibrium for the creation of carbon content protective films in acidic waters

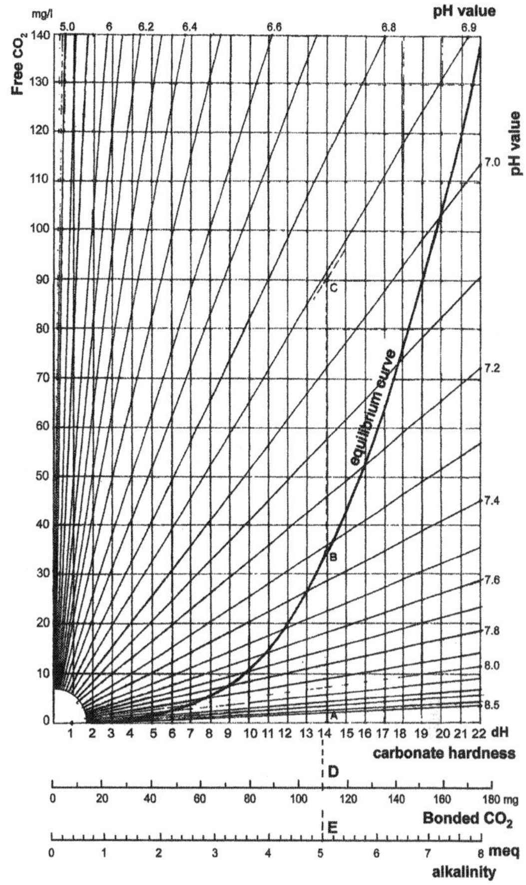

Example:

The following values are obtained after chemical analysis of raw water:

Carbonate hardness	14 °dH	(A)
Free carbonic acid	90 mg/l	(AC)
According to the nomogram:		
Free attached carbonic acid	35 mg/l	(AB)
Free aggressive carbonic acid	55 mg/l	(BC)
Bonded carbonic acid	110 mg/l	(D)
pH value	6.91	(C)
Alkalinity	5.0 meq	(E)

Figure 2.10 Carbonate hardness relative to carbonic acid content

Fresh water is frequently converted to drinking water, called technically, potable water.

Potable water is strictly controlled in developed countries which have built the infrastructure to supply it. Potable water should be clear, bright and have no discernable taste or odour. The actual chemical analysis, at the point of use, may vary slightly. Table 2.11 shows a typical recommendation:

Conductivity	50 to 100 S/cm
pH	6.5 to 8.0
Anionic	detergents 0.2 mg/litre
Coliforms	zero in 95% of samples
Pesticides	0.5 mg/litre
PAH	0.2 g/litre
Phenols	0.5 g/litre
Aluminium	0.2 mg/litre
Ammonia	0.5 mg/litre
Arsenic	0.05 mg/litre
Barium	0.1 mg/litre
Calcium	100 mg/litre

Cadmium	5 g/litre
Chloride	200 mg/litre
Chromium	0.05 mg/litre
Copper	0.1 mg/litre
Cyanide	0.05 mg/litre
Fluoride	1.5 mg/litre
H_2S	0
Iron	0.2 mg/litre
Lead	0.05 mg/litre
Magnesium	50 mg/litre
Manganese	0.05 mg/litre (0.1 mg/litre affects taste)
mercury	1 g/litre
Nitrate	50 mg/litre
Oil	25 ppm maximum
Selenium	0.01 mg/litre
Sodium	150 mg/litre
Soluble lubricants	100 mg/litre
Suphate	250 mg/litre
Zinc	5 mg/litre

Table 2.11 Chemical analysis of potable water

2.2.3 Brackish water

Brackish water occurs at the mouths of rivers where fresh water and seawater mix. The salt and chlorides content are diluted to approximately 1 to 2.5% and 4000 ppm respectively, giving a pH range of 6 to 9. Because of the turbulent flow regime brackish water contains suspended solids, typically silt and sand. Particle sizes would range from 1 to 200 microns. The percentage of solids, and the size distribution, could only be quantified by sampling.

River water may be contaminated with industrial waste such as nitrogen compounds or caustic. Ammonia, sodium carbonate and sodium hydroxide are some of the most popular wastes. If nitrogen compounds are present, the oxygen content may be reduced due to the activity of bacteria and fungi. Small micro-organisms, as in fresh water, may be present.

Cast iron, steel and 11-13Cr steel will generally not be suitable. The presence of ammonia or caustic will cause stress corrosion cracking problems with brass. The addition of 0.25% to 4% tin to copper alloys improves corrosion resistance and resistance to stress corrosion cracking. Aluminium bronzes, with 10% aluminium, and nickel aluminium bronzes, with 4% nickel, are generally corrosion resistant. Nickel aluminium bronze has proved superior where mud or silt deposits accumulate. Austenitic stainless steels, such as AISI 316, have given adequate service on brackish water with 0.2% ammonia compounds up to 35°C. The solids content must be quantified to allow proper valve selection.

2.2.4 Seawater

Seawater is used continuously in two important processes:

- crude oil secondary recovery
- desalination

Secondary recovery may require pressures up to 350 barg. Desalination, by reverse osmosis, requires only 70 barg. These two processes account for the recent developments in valve materials.

2.2.4.1 General

Seawater contains a mixture of inorganic salts. Cooking salt, sodium chloride NaCl, forms about 75% of the salt content. The salt content at great depths varies insignificantly between the oceans of the world. At a depth of 500 m the salt content is about 3.5%. At the surface, however, the salinity is affected considerably by the climatic and other factors. In the Northern Baltic, the salinity is almost 0%, and in the Red Sea it is about 4%. Evaporation, rain, polar ice, rivers and industrial/chemical effluents are all factors which affect the properties of seawater.

The temperature at the surface of the sea varies from about -2°C in the Arctic Ocean and all around Antarctica to about +37°C in the Persian Gulf. The sea bed temperature is generally between + 2°C to + 5°C. At temperatures between 10 and 15°C the oxygen content varies between 8 and 10 mg/l. Chlorides are present up to 25000 ppm. Unless polluted there is no hydrogen sulphide. Seawater properties vary throughout the year and care must be taken to establish the full range. At one location temperatures ranged from 3.3 to 27°C, pH 7.9 to 8.4, whilst oxygen varied from 72 to 100% saturation. The physical properties of seawater vary little from chemically pure water.

2.2.4.2 The qualities of seawater and corrosion

The quality of seawater is of great significance in resistance to corrosion. There are examples of identical ships where the one vessel has had serious problems with its seawater pump systems those in the sister ship have remained intact. This phenomenon has usually been explained by the fact that the ships have been plying in different types of seawater. By and large, those ships engaged in coastal and river traffic are attacked more by corrosion than are those which sail the open seas.

As regards coast-based industries and power stations, the quality of the water may vary greatly, depending upon where the installation is located. In such cases however, the quality of water to be dealt with is usually already known at the design stage, enabling the choice of materials to be made accordingly. This is not usually possible in the case of marine pump installations. Coastal installations and others subjected to saline atmospheres require special attention to external construction, particularly where dissimilar metals are used, to reduce the affects of galvanic corrosion. The variations in the properties of seawater and the affects these variations have on corrosion rates can be summarised as follows:

- The temperature of seawater is of relatively great importance with regard to corrosion rates. In general, higher temperatures increase the corrosion risks
- Solid contaminants damage the protective-film on the material, thus giving corrosion a chance to attack. The rate of mechanical wear can of course, also be increased
- Chemical contaminants of various types can also lead to increased rate of corrosion
- Tests have shown that oxygen contents over 1 mg/l will promote stress corrosion cracking. If pre-treatment is considered, oxygen reduction is of prime importance. The pH value of seawater is normally about 8
- The dosing of additives may perhaps be regarded as a deliberate chemical contamination of seawaters. Properly conducted, such dosing should not bring any problems from the corrosion point of view. Too strong a dose of sodium hypochlorite, sometimes used to inhibit algae, can on the other hand cause serious pitting corrosion

Summarising, it can be confirmed that clean and cold seawater with normal oxygen content gives the least corrosion problems.

2.2.4.3 Choice of materials

Care must be taken with material combinations involving seawater. An electrolytic cell is set up between dissimilar metals due to the high conductivity of the seawater. The less noble material should have a much larger surface area than the noble material. The problem is exacerbated by changes in the electrolytic potential caused by high flow velocities.

Steel is not very good in moving seawater; corrosion of steel increases by a factor of 4 from static to 4.5 m/s. When suitable, cast iron castings with stainless steel components last longer overall than nickel aluminium bronze components. Ni-resist and ductile ni-resist have been used for some applications; ductile ni-resist is difficult to cast and both are difficult to weld repair.

Copper alloys give good service lives in seawater and are generally easy to cast and machine. Nickel aluminium bronze can be difficult for some machining operations, like tapping. Special tools may be required. Nickel aluminium bronze and magnesium aluminium bronze are better than aluminium bronze and gunmetal. Aluminium bronze is resistant to stress corrosion cracking provided metallurgy and heat treatment are correct.

Nickel aluminium bronze has exceptional resistance to cavitation damage compared to some stainless steels; BS1400 AB2 = 0.025mm; stainless steel AISI 321 = 0.305 mm in 3% NaCl. Welding of nickel aluminium bronze is possible, but heat treatment after is necessary to reduce corrosion possibilities. Aluminium and nickel aluminium bronze are limited to 18 to 25 m/s to eliminate erosion problems. Aluminium bronze can be used in combination with stainless steel and titanium without problems. Seawater piping 90/10 Cu Ni has adequate service life.

Stainless steels are the most popular materials for pumping seawater. Care must be taken when reviewing test results and when specifying materials. The chemical composition, and therefore corrosion resistance, of austenitic stainless steels varies with material production; cast AISI 316 is different to wrought AISI 316.

Ferritic steels suffer from local corrosion and pitting in seawater. Stainless steels without molybdenum are generally poor. AISI 304 is generally regarded as the lowest quality austenitic stainless steel. AISI 304 is not used for seawater. AISI 316 has proved successful in tests up to 90°C on synthetic seawater although when tested at 20°C with bubbled chlorine gas pitting occurred. The general consensus is that cast 316L casings with 316 shafts is the minimum working combination. Cast 316L has 10 to 15% ferrite and a higher chromium content than wrought 316L which has no ferrite structure.

316L castings are as good as Alloy 20 for pitting and crevice corrosion resistance, have good stress corrosion cracking resistance and are easier to cast and weld repair. The ferrite content improves corrosion resistance and aids weldability. Pitting can be a problem with mechanical seals.

Research over a number of years tended to show that molybdenum played an important part in the corrosion resistance. An early theory suggested the Cr + Mo percentage should be over 30%.

This was followed by an Uddeholm equation for Pitting Resistance Equivalent (PRE).

$$PRE = (Cr\ weight\%) + (3.3 \times Mo\ weight\%) \qquad \text{Equ 2.11}$$

Uddeholm thought that PRE should equal 28 minimum. More research and testing. Later, nitrogen was found to have an important role in the corrosion resistance. A new equation was proposed.

$$PRE = (Cr\ weight\%) + (3.3 \times Mo\ weight\%) + (16 \times N_2\ weight\%) \qquad \text{Equ 2.12}$$

At one time it was suggested that the PRE_N value be set at 35. Some users have requested PRE_N values of over 40. Nitrogen strengthened steels had been available since the early 70s. Now more exotic varieties are compounded. Another benefit provided by nitrogen hardened stainless steels was improved cavitation damage resistance. Table 2.12 shows the range of chromium and nickel contents of current popular materials.

Designation	Composition %	
	Cr	Ni
AISI 316	16-18	10-14
ACI CF-8M	18-21	9-12
AISI 316L	16-18	10-14
ACI CF-3M	17-21	9-13
Avesta 254SMO	20	18
Avesta 254SLX	20	25
ACI CN-7M	19-22	27.5-30.5
Rex 734	21.6	9
Armco Nitronic 50	20.5-23.5	11.5-13.5
Zeron 25	24-26	5.5-7.5
Zeron 100	24-26	6-8
Ferralium 255	24-27	4.5-6.5
Ferralium 288	26-29	6-9

Table 2.12 Popular materials for use in seawater

Some exotic stainless steels have a low tensile strength combined with high ductility. More costly, stronger materials may be better resulting from thinner casings and smaller corrosion allowances. Materials, such as titanium, may cost less than proprietary stainless steels.

2.2.5 Produced water

Produced water is underground water which comes to the surface as a result of mining or drilling operations. Produced water may be brought to the surface with solid mined products or flow with natural gas and crude oil from a "well" — do not confuse this with "well water". Produced water is a natural product but usually brought to the surface as a result of 'man's' efforts. As a natural product it is very variable but usually needs to be considered as a pollutant and an environmental hazard. When thinking about valves, pumps and piping, the following should be considered:

- temperature, pH, total alkalinity, calcium hardness, magnesium hardness, sodium, potassium, sulphates, chlorides, chlorine, bicarbonate, suspended solids, sulphides, sulphate reducing bacteria and other micro-organisms, dissolved gases, entrained gases.

Table 2.13 gives a very generalised composition:

pH	3.5 to 7.7
Chlorides	12 000 to 190 000 ppm
Dissolved hydrocarbons	Carboxylic acid, acetic acid, propionic acid, BTEX (= benzene, toluene, ethyl benzene, xylene), PAH (poly aromatic hydrocarbons)
Dissolved solids	Sodium, calcium, magnesium, barium, strontium, ferrous, ferric, aluminium; carbonates, bicarbonates, sulphates, sulphides, bromides, iodides, silicates
Dissolved gases	O_2, CO_2, H_2S
Partially soluble organics	Aliphatic & aromatic hydrocarbons, phenols
Suspended solids, micro-organisms	Algae, bacteria, fungi; typically 0.2 to 10 micron
Suspended solids, inorganic	Clay, sand, calcium carbonate, iron sulphides
Dispersed oil	Typically 4 to 6 micron droplets, total oil up to 40%
Radioactive	Radium 226, radium 228

Table 2.13 Generalised composition of produced water

The very high chloride content makes the brine very corrosive; exotic metallic materials should be considered and non-metallic materials. Glass-fibre lined carbon steel pipes have been used successfully when the sand content is low. Some bacteria can feed on the dispersed oil and produce H_2S. The dispersed oil increases the viscosity and the surface tension.

Produced water can be very difficult to dispose of if not reinjected in a well. Even reinjection water requires treatment to protect the well, **and** the equipment:

- remove all solids larger than 3.5 micron
- remove 98% of solids larger than 2.0 micron
- oil, 5 ppm maximum
- oxygen, 50 ppb maximum

Produced water is usually considered pollution and spills must be reported to the relevant local authority. A continuously flowing water course, is the best option, with sufficient flow to provide a 10:1 dilution. Even so, elaborate pre-treatment may be necessary before the local authority will allow limited disposal. But, the chemicals used to treat the water are hazardous and subject to restrictions. Disposal flow must be measured accurately and recorded. Typical treatments and location restrictions are shown in Table 2.14:

Removal of iron and manganese precipitates	to avoid discolouration
Dissolved solids concentration	less than 5 g/litre
Total suspended solids	less than 25 mg/litre
pH	between 6.5 and 9.0
Chlorides	less than 1500 mg/litre
Boron	less than 1.0 mg/litre
Oil total	maximum, 5 ppm
Trace elements	less than 10 times the water course concentration
PAH	less than 10 times the water course concentration
BTEX	less than the local drinking water regulation limits
Dissolved oxygen	greater than 3.0 mg/litre
Temperature	not higher than 5°C above water course
Test toxicity	to local marine life
Not closer than 5 km	to a drinking water extraction point
Not closer than 2 km	to an irrigation water extraction point

Table 2.14 Typical treatments and location restrictions

Produced water can be allowed to soak-away, providing:

- dissolved solids are compatible with the dissolved solids in the existing ground water
- dissolved solids concentration is not greater than double the existing concentration
- dissolved solids concentration is less than 5 g/litre
- suspended solids is less than 25 mg/litre
- sodium content is reduced to an acceptable level (local soil conditions)
- pH is between 6.5 and 9.0
- volatile hydrocarbons is less than 15.0 mg/litre
- extractable hydrocarbons is less than 5.0 mg/litre
- BTEX is less than the local drinking water regulation limits
- not closer than 2 km to a drinking water well

2.3 Oils

2.3.1 General

Oils are classified according to their origins:

- Mineral
- Animal
- Vegetable

In the context of valves they can all be treated the same. Along with mineral oils, we can include such petroleum products as solvents, petrol, kerosene and the like, which should be considered when pumping oil stocks. When handling oil, the upper and lower operating temperature limits, viscosity, cloud point, lowest flow (pour point), solidifying temperature and the vapour pressure should all be established. The flow capability of oils follows Newton's Law. In common with water, they have constant viscosity independent of shear rate and time. The viscosity is temperature dependent, oil flowing more easily when heated. The viscosity falls as the temperature rises. In order to assess the needs of a pump installation, the viscosity-temperature relationship must be known and the way the oil behaves with variations in operational temperatures must also be clarified.

2.3.1.1 Cloud point, pour point and solidifying point

Mineral oils transform gradually from the liquid to the solid state, as opposed to other liquids (water, for example, which has an exact freezing point). When the oil is chilled, it goes cloudy at a certain temperature because of precipitation of paraffin crystals, i.e. generation of wax. This temperature is called the **cloud point** (cold filter plugging point).

The **pour point** is reached if the temperature is further reduced. A few degrees below this temperature, the oil changes to a completely solid form, the **solidifying point**.

Because of waxing, it is considered that mineral oils can only be handled at a temperature of at least 10°C above the pour point. When considering high viscosity oils with solidifying temperatures near to or above the ambient temperature (for outdoor installations for example), notice must be taken of the pour point and the installation must be designed so that the pipes and pumps or valves can be heated. Pipes which are not heated must be able to be emptied in order to prevent stoppages building up if flow is interrupted in ambient temperatures below the lowest flow temperature. Low sulphurous heating oils have higher pour point temperatures than high sulphurous oils.

Complex petroleum products, petrol for example, have a vapour pressure range which is dependent upon the most easily flowing components. This property affects the calculation of the $NPSH_a/NPIP_a$ for pumps. Mineral oils are classed as dangerous liquids in fire hazard Class 3. Light distillates, such as petrol and photogene are in fire hazard Class 1, see the Liquid Table in Chapter 20, Section 20.2.2.

2.3.2 Viscosity

2.3.2.1 Burner oils

Low sulphur oils (max 0.8% by mass - EEC Directive 25/716/EEL) are covered by classes C, and C2 have viscosities of between 1 - 2 cSt at 37°C and together with Class D oils can be stored and handled at ambient temperatures likely to be encountered in the UK. Class E - H are residual blended oils for atomising burners and normally require preheating before atomisation.

Class of fuel	Min. temperature for storage	Min. temperature for outflow from storage and handling
E	7°C	7°C
F	20°C	27°C
G	32°C	38°C
H		(special purpose fuels)

Table 2.15 Fuel classes

The four lines on the chart in Figure 2.11 show average viscosity/temperature relationships for fuel of Classes E to H at the maximum viscosity allowed by the specifications. Table 2.12 gives details of the class. The approximate viscosity/temperature relationship for any petroleum fuel within these classes, for which viscosity at one temperature is known, can be determined by drawing through the known viscosity/temperature in-

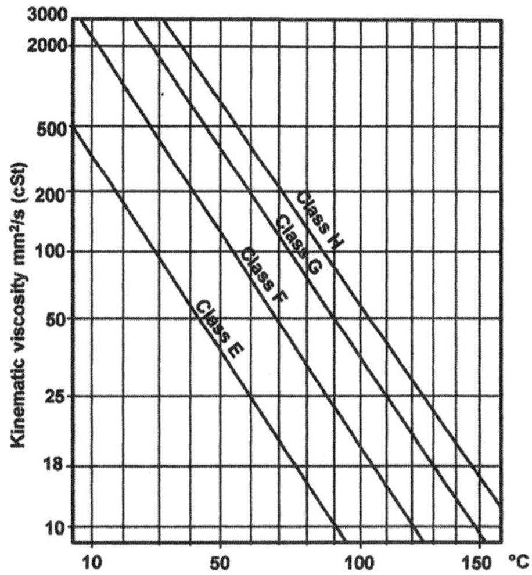

Figure 2.11 Kinematic viscosity/temperature chart

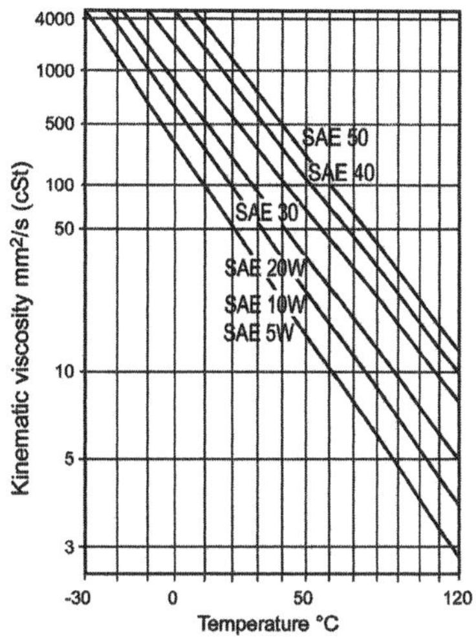

Figure 2.12 Kinematic viscosity for engine oils, SAE 5W to SAE 50

tersection a line parallel to those shown. From this line can be read the temperature required for any desired viscosity e.g. that specified by burner manufacturers for proper atomisation. As the temperature is lowered towards the pour point (lowest flow temperature) there is an increasing upward deviation from the viscosity indicated on the chart. This deviation is of such a magnitude that in no case shall the chart be used within 15°C of the pour point, information on which can be obtained from the suppliers. BS 2869 may be useful in this respect.

2.3.2.2 Engine oils

The SAE system was devised in America and laid down by the Society of Automotive Engineers (SAE) in 1926. The Society's aims are "to develop technical information on all forms of self-propelled vehicles including automobiles, trucks and buses, off-highway equipment, aircraft, aerospace vehicles, marine, rail, and transit systems" and is usually known by the abbreviation SAE.

Lower numbers in the SAE system mean thinner oils and the letter W after the number indicates that the oil is suitable for use in winter. The SAE system is accepted and used internationally.

Below the cloud point temperature, there is no rectilinear relationship because of waxing.

Pour point: -20°C to -30°C (lower for certain special qualities).

Lowest recommended handling temperature: 10°C to 15°C above pour point, see Figure 2.12.

2.3.2.3 Gearbox oils

Below the cloud point temperature, there is no rectilinear relationship, because of waxing.

Pour point: -20°C to -30°C (lower for certain special qualities).

Lowest recommended handling temperature: 10°C to 15°C above pour point. See Figure 2.13.

2.3.2.4 Industrial oils

The International Organisation for Standardisation, (ISO), has developed a system of viscosity classification for lubricants for industrial use which came into use effectively from 1977. The system consists of 18 viscosity categories stated in centi-Stokes at 40°C. Each class of viscosity is identified by an ISO VG (viscosity grade) number which in general coincides with the mean value in accordance with Figure 2.14 from BS 4231:1992, ISO 3448:1992.

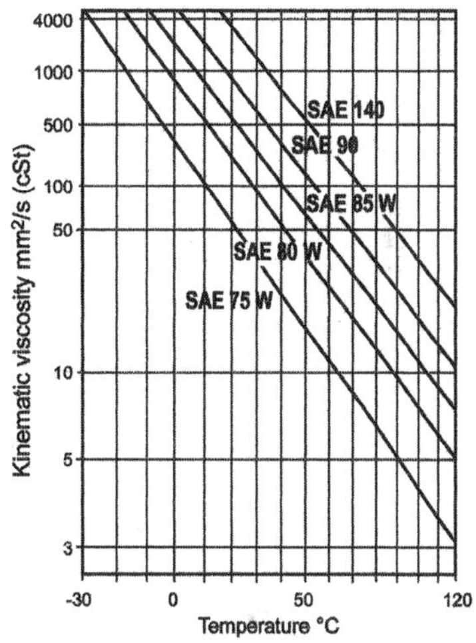

Figure 2.13 Kinematic viscosity for gearbox oils, SAE 75W to SAE 140

ISO Viscosity Class	Kinematic viscosity at 40°C Mean value	Minimum	cSt (mm2/s) Maximum
ISO VG 2	2.2	1.98	2.42
ISO VG 3	3.2	2.88	3.52
ISO VG 5	4.6	4.14	5.06
ISO VG 7	6.8	6.12	7.48
ISO VG 10	10	9.00	11.0
ISO VG 15	15	13.5	16.5
ISO VG 22	22	19.8	24.2
ISO VG 32	32	28.8	35.2
ISO VG 46	46	41.4	50.6
ISO VG 68	68	61.2	74.8
ISO VG 100	100	90.0	110.0
ISO VG 150	150	135	165
ISO VG 220	220	198	242
ISO VG 320	320	288	352
ISO VG 460	460	414	506
ISO VG 680	680	612	748
ISO VG 1000	1000	900	1100
ISO VG 1500	1500	1350	1650

Table 2.16 Viscosity classes in accordance with ISO 3448:1992, BS 4231:1992

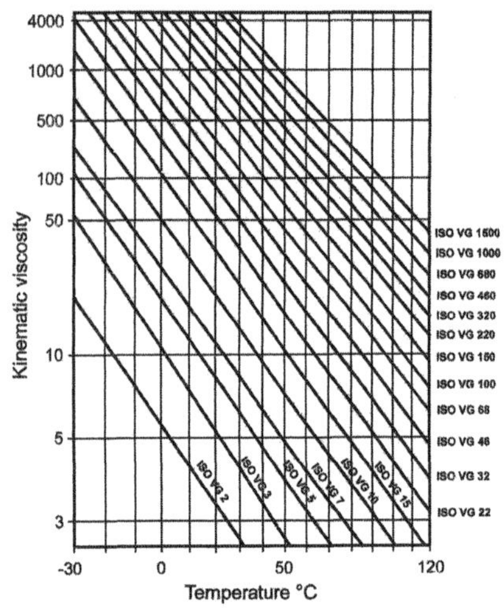

Figure 2.14 Kinematic viscosity for industrial oils in accordance with ISO 3448:1992, BS 4231:1992

The classification system has special advantages:

- The ISO VG number gives information on the viscosity of the oil
- ISO 3448 is fully supported by the leading national standardisation organisations such as ASTM, DIN, BSI, JIS making it easier to compare the viscosity of the oils with that specified by the machine manufacturers
- ISO 3448 is directly comparable with BS 4231:1982 for classifications at 40°C

Pour point: 20°C to 50°C (lower values for certain special qualities).

Lowest recommended handling temperature: 10°C to 15°C above pour point.

The scales in the viscosity diagrams are themselves not linear but are adjusted so that the viscosity relationships are linear. The diagram may be used to construct the viscosity relationship for other oils if the associated viscosity temperature values are known for two points. These relationships will also be linear. The Kinematic viscosity for various oils is given in Table 2.17.

Liquid	Kinematic viscosity ν mm²/s (cSt) at a given temperature	
Aviation fuel	0.7/0°C	0.6/38°C
Castor oil	300/38°C	42/54°C
Corn oil	29/38°C	9/100°C
Groundnut oil	42/38°C	23/54°C
Kerosene	2.0/20°C	1.6/38°C
Linseed oil	31/38°C	19/54°C
Olive oil	43/38°C	24/54°C
Petrol	0.7/20°C	0.6/38°C
Soya oil	35/38°C	19/54°C
Turpentine	2.1/20°C	2.0/38°C
Valtran	37/38°C	22/54°C

Table 2.17 Kinematic viscosity for various oils

2.3.2.5 Other liquid data

Liquid	Density kg/m³ at 20°C	Region of boiling°C at 0.1 MPa	Vapour pressure kPa at 20°C	Fire hazard class*
Aviation fuel	720	40-150	25-35	1
Burner oils —				
BS 2869 Class D	840			3
Class E	940			3
Class F	950			3
Castor oil	960		(0)	
Corn oil	920		(0)	
Oil SAE 5W-50	900		(0)	3
Oil SAE 75W-140	910		(0)	3
Groundnut oil	910		(0)	
Oil ISO VG2-1500	880-905		(0)	3
Kerosene	780	170-250	0.1	2b
Linseed oil	930		(0)	
Liquid resin	970		(0)	
Olive oil	910		(0)	
Petrol	730	40-180	25-70	1
Soya oil	940		(0)	
Turpentine	860		(0)	2b
Valtran	920		(0)	

Table 2.18 Some liquid properties

*Refers to classification in the Liquid Table, Chapter 20, Section 20.2.2

2.4 Liquid-solid mixtures

2.4.1 General

Particles of inorganic and organic solid matter are to be found, more or less finely distributed, suspended in liquids either as contaminants or for the purpose of transportation. Liquid-solid mixtures belong to the non-Newtonian liquids. The characteristics of the mixture depend upon:

- Pure liquid properties
- Size of the solid particles
- Density of the solid particles
- Shape, hardness and abrasiveness of the solid particles
- Deformability of the solids
- Friability of the solids
- Concentration of particles in the liquid

These parameters are of decisive importance in the choice of valve. It is usual in valve catalogues to state the suitability of valves for various kinds of liquids, e.g. clean, no solid contamination, slightly contaminated, contaminated, slurry, mud, transport of pulps, solids, etc. without closer definition. However, all applications for liquid-solid mixtures should be discussed in detail with manufacturers.

Small quantities of solids can have a serious affect on wear. Vetter, Thiel and Störk presented test results for valve wear with low concentrations of 15 μm quartz with a Miller Number of 135, see Figure 2.15. These test results can be taken as representative of fine sand. A concentration by weight of only 0.25% doubled the wear rate. A concentration of 1% solids with a Miller Number of 50 in the same valves would increase valve wear by 80%.

It is helpful in equipment selection to use the following parameters for the particle content of the liquid:

- Sizes <0.1, 0.1 to 1.0, 1.0 to 10 and 10 to 100 mm
- Deformable
- Abrasive
- Concentration approx. <1%, >1%

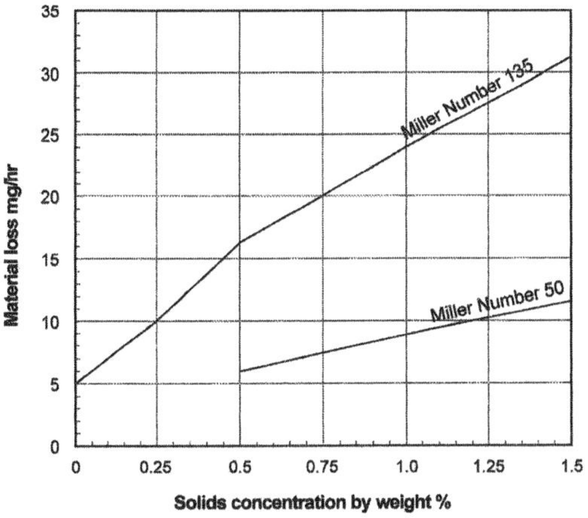

Figure 2.15 Valve material loss

2.4.1.1 Liquid properties

The physical properties of the carrier liquid influence the properties of the liquid-solid mixture. Solids in the liquid tend to increase the viscosity. The clean liquid viscosity is necessary to evaluate the mixture viscosity. Also the abrasiveness of a mixture is affected by the liquid viscosity. Sand in oil is much less abrasive than sand in water. Solids in liquids tend to increase the relative density. Liquid relative density is required to calculate the mixture relative density.

2.4.1.2 Size of particle

Particles less than 1 μm occur in clean fresh water at very low concentrations. Solids in chemical process applications are usually small, up to 1 mm. Solids transported in liquids can vary over a wide range, dependent upon the working pressure and the choice of pump, up to 25 mm is not uncommon. Larger hard solids are encountered with dredging, quarrying and building site applications.

A standard requirement is to pass a 100 mm sphere. Large soft solids, such as fruit, fish and sewage waste up to 140 mm are pumped on a regular basis. Examples of particle sizes are shown in Figure 2.16.

2.4.1.3 Particle density

The particle density is important for power consumption. Energy is expended moving the particles and the liquid. The power required is a function of the mixture density which is dependent upon the solids density.

2.4.1.4 Shape, hardness and abrasiveness

The shape of manufactured or processed solids can be defined. Naturally occurring solids may be difficult to describe. The hardness of solids can be measured and should be specified.

Because of the difficulties caused by shape, shape and hardness can be combined to give abrasiveness. Abrasiveness is quantified by testing representative samples of the mixture. The term "representative" is critically important. Extrapolating data from one mixture to another is unreliable. A popular test method is the Miller Number; originally intended for reciprocating piston pumps and now standardised as ASTM G75. The test measures the weight loss of a reciprocating metal block, 28% chrome iron, due to the affect of a 50% concentration by weight in distilled water.

The Miller Number is calculated from the test results to produce a linear abrasive scale; a Miller Number of 200 produces twice as much wear as a Miller Number of 100 all other conditions being equal. Miller Numbers less than 50 are considered non abrasive; over 50 consideration should be given to wear problems. Table 2.19 indicates typical values of Miller Number.

Solid	Minimum values	Maximum values
Bauxite	9	134
Coal	6	57
Copper ore	20	135
Iron ore	28	157
Limestone	22	46
Magnetite	64	134
Mud, drilling	10	10
Sand	51	246
Sewage (digested)	15	15
Sewage (raw)	25	25
Shale	53	59
Tailings	24	644

Table 2.19 Typical Miller Number Values

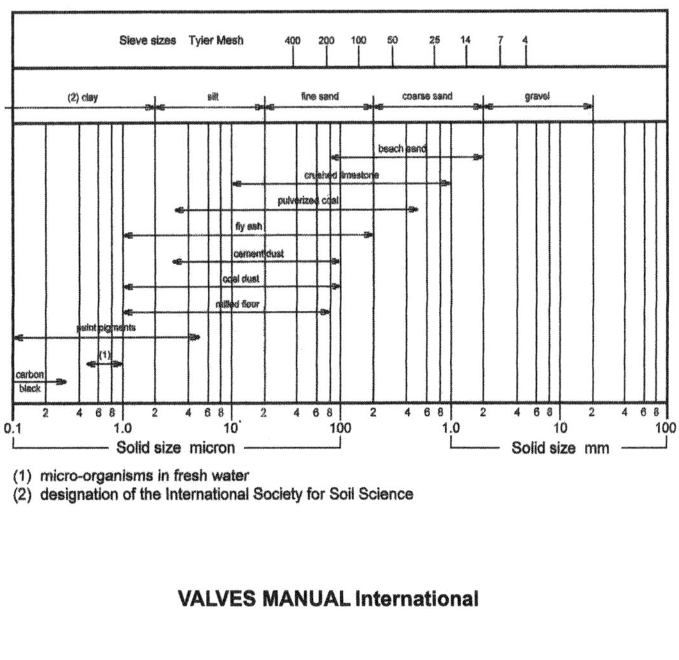

Figure 2.16 Particle sizes

2.4.1.5 Deformable particles

Particles of organic matter, soft, fibrous particles which can be compressed, deformed or may be subject to damage. For example, dispersions of plastic or natural rubber in water - latex. Latex can coagulate under pressure and therefore requires careful handling as do other soft solids — fruit, vegetables and fish.

2.4.1.6 Friable particles

Friable means easily crushed or easily broken. Hard solids, like some coals, can be crushed or broken when trapped by certain system components. Friable solids can be handled by gear pumps, some screw pumps and reciprocating pumps with valves.

2.4.1.7 Solids concentration

Concentration refers to the quantity of particles in suspension (dry substance = DS content) and is expressed as a percentage by weight or volume. Equations 2.13 to 2.16 show the volumetric relationships between liquid, particles and mixture. Equations 2.17 to 2.20 show the mass relationships.

$$\rho_M = \rho_L + \frac{C_V\,(\rho_s - \rho_L)}{100} \qquad \text{Equ 2.13}$$

$$C_V = \frac{100\,(\rho_M - \rho_L)}{(\rho_S - \rho_L)} \qquad \text{Equ 2.14}$$

$$\rho_L = \frac{\left(\rho_M - \dfrac{C_V\,\rho_S}{100}\right)}{\left(1 - \dfrac{C_V}{100}\right)} \qquad \text{Equ 2.15}$$

$$\rho_S = \frac{100}{C_V}\left(\rho_M - \rho_L + \frac{C_V\rho_L}{100}\right) \qquad \text{Equ 2.16}$$

$$\rho_M = \frac{1}{\left(\dfrac{C_M}{100\rho_S} + \dfrac{1}{\rho_L} - \dfrac{C_M}{100\rho_L}\right)} \qquad \text{Equ 2.17}$$

$$C_M = \frac{\left(1 - \dfrac{\rho_M}{r_L}\right)}{\left(\dfrac{\rho_M}{100\rho_S} - \dfrac{\rho_M}{100\rho_L}\right)} \qquad \text{Equ 2.18}$$

$$\rho_L = \frac{\left(1 - \dfrac{C_M}{100}\right)}{\left(\dfrac{1}{\rho_M} - \dfrac{C_M}{100\rho_S}\right)} \qquad \text{Equ 2.19}$$

$$\frac{1}{\rho_S} = \frac{100}{C_M}\left(\frac{1}{\rho_M} - \frac{1}{\rho_L} + \frac{C_M}{100\rho_L}\right) \qquad \text{Equ 2.20}$$

where:

C_M = percentage of solids in mixture by weight

C_V = percentage of solids in mixture by volume

ρ_L = density of liquid kg/m^3

ρ_M = density of mixture kg/m^3

ρ_S = density of solids kg/m^3

The relationship between C_M and C_V is given by :

$$\frac{\rho_L C_M}{\rho_S} + \frac{C_M C_V}{50} + \frac{C_V \rho_S}{\rho_L} - \frac{C_M C_V \rho_S}{100\rho_L} - \frac{C_M C_V r_L}{100\rho_S} - C_M - C_V = 0 \qquad \text{Equ 2.21}$$

Table 2.20 indicates typical values of solids concentration found in some mixtures that are handled.

Mixture	Solids concentration by weight %
Polluted municipal sewage water	0.1
Paper pulp	0.5 to 6
Dense pulp	6.0 to 20
Centrifuged sewage	up to 35
Copper concentrate pipeline	up to 45
Coal pipeline traditional	13 to 50
Activated sewage sludge	20 to 50
Carbonation slurry	50 to 60
Limestone slurry	50 to 60
Iron ore slurry	20 to 65
Coal water fuel	up to 75
Sand	20 to 75
Clay, raw cement sludge	50 to 75
Mine backfill slime	up to 78

Table 2.20 Solids content in mixtures

It is not possible to generalise the effect of solids concentration on valve selection, performance or component life.

Figure 2.17 shows a nomogram for estimating the mixture density when the carrier liquid is water.

The example shows that a suspension having a particle content of 40% by weight and a particle density of 2700 has a total density of about 1340.

2.4.1.8 Velocity

One important aspect of solids handling has not been mentioned and that is velocity. Abrasive solids, and large solids, are handled at much lower velocities than clean liquids.

Experience has shown that abrasive wear in valves and pumps is related to flow velocity raised to a power between 2.5 and 5 depending upon the component. If the velocity is increased by 10% the wear rate will increase by a minimum of 27% up to 61%. The velocity-wear relationship can be used with the Miller Number to estimate the effects of duty changes.

2.4.2 Sewage

Sewage is a generic term for:

- **Soil sewage**: discharge from water-closets, urinals, slop-sinks including waste water from domestic baths etc. and even industrial effluent.
- **Surface water**: rainfall and storm water.
- **Drain water**: drainage from building sites, fields, leaks in broken pipes etc.

Waste water from households, businesses, hotels, offices and restaurants serves as a carrier for the contaminants from water closets and other sources of waste in the sewage system. The size of the waste thereafter is limited in principle to the area of the intake connected to the pipeline. It is not easy to define the limits of length of a soft and flexible objects. These objects, such as sheets of plastic, sanitary towels, etc., pass unchecked down the water closet and into the public sewer.

It is forbidden to discharge into any open drain or sewer, material which is likely to cause damage, create a hazard or which could have a prejudicial effect upon the treatment of their contents (i.e. waste matter which should be dealt with by commercial waste disposal companies). Such waste could be dirty engine oil, volatile liquids producing flammable or toxic vapour or biological waste. The legal regulations applying to local sew-

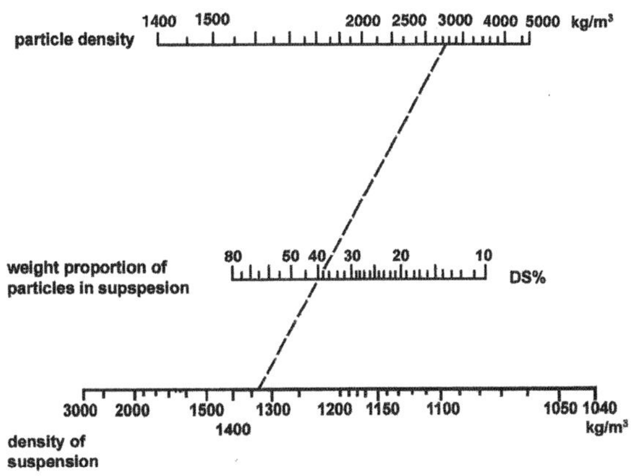

Figure 2.17 Mixture density nomogram for water

age works are enforced by the local water authorities supervised by government bodies such as river authorities and environmental protection agencies.

The quantity of unwanted or prohibited objects collected at rain water and sewage grills continues to increase. Especially difficult, and containing a high content of textile matter amongst others, is the sewage from densely populated city centres. The heart of the city with its restaurants, offices and department stores presents a far greater sewage problem than do the surrounding dormitory suburbs.

The demand placed on transporting daily sewage should be assessed according to the contaminants (permitted or not) which are contained in the sewage. Different types of industry produce varying types of effluent which often contain substances which can cause damage to the sewage treatment works and the recipient water installation.

In the UK the law regarding the discharge of industrial waste into sewers is restricted by the Public Health (Drainage of Premises) Act 1937, amended in 1961. The responsibility for administering and enforcing this legislation is in the hands of the water authorities. Provisions for the restriction of pollution of tidal waters and estuaries appear in sections 2-5 of the Rivers (Prevention of Pollution) Act 1951 and amended by the 1961 Act. The Clean Rivers (Estuaries and Tidal Waters) Act 1960 also applies with limitations specified in the Schedule to the Act 1960.

The UK Control of Pollution Act 1974 sections 43 and 44 extends legislation with respect to specific pollutants. Limiting values for the discharge of substances which are toxic, corrosive, explosive and/or damaging to systems and system materials etc. are the subject of consent which must be obtained from the local water authority. (Many of these liquids appear in the Liquid Table in Chapter 20, Section 20.2.2). Cooling water, surface water and drain water display properties as in Section 2.2. Solid precipitations from the circulating water can accumulate in cooling tower installations in the form of lime furring, sludge and concentrations of minerals giving rise to corrosion, hot spots and blockage in pumps and piping.

2.4.3 Sludge

Sludge is the name given to the residue which forms when inorganic and organic particles are separated when a liquid is cleaned. The following mechanical and chemical cleaning methods are used:

- **Sedimentation-settling:** particles fall to the bottom in a separator under their own weight
- **Flotation:** particles are made to float by the injection of small air bubbles
- **Centrifuging:** particles are separated by centrifugal force
- **Filtering:** liquid is passed through a filter which allows the liquid to pass through whilst trapping the particles, the filter may consist of a grill, porous material or one or more beds of filtering media
- **Precipitation-flocculation:** by the addition of chemical reagents, the particles form insoluble combinations -flocs- which can be separated by sedimentation or flotation

The dry solids or substance (DS) content of the sludge is of interest as is also the size and hardness of the particles. Since sludge always occurs in the treatment of water and in industrial processes, there can be no general definition. The values obtained in the treatment of sludge in a typical municipal sewage works can be used however as a guide. See Table 2.21.

Material	DS-content%
Before thickening	
Chemical sludge (scum) after precipitation or flotation	0.5 to 1
Mechanical sludge, sedimentation or flotation	2 to 3
After thickening	
Mechanically agitated sludge	6 to 10
Activated sludge	2 to 3
Bio-filtration	4 to 8
Mechanical and activated sludge	5 to 8
After straining	
In the centrifuge, filter screen press, vacuum filtering etc.	> 30

Table 2.21 DS-content after various treatment stages

In the bottom, or compression zone, of sedimentation basins, sand traps, oil tanks and other containers of stationary liquids, the bottom layer, during sedimentation is subjected to mechanical pressure due to the weight of sludge lying above it. This causes liquid to be compressed out of the bottom layer thus increasing its density. Other names given to sludge are sediment and slurry.

2.4.4 Pulps

2.4.4.1 General

Pulps are used in the cellulose and paper industries. It is a name for fibre suspensions in water. Pulps belong to the non-Newtonian liquids; group 1, the time-independent, plastic liquids.

The density of a pulp can be calculated from Equations 2.13 and 2.17 with reservation for possible air content. Often the effects of the fibres and air cancel each other out. The density then coincides with that of the water.

Pulps exhibit widely varying pH values, depending on the production method and the bleaching process. Valves made of stainless steel are usually needed.

2.4.4.2 Pulp quality

The characteristics of pulp depends, in the first instance, upon the raw material used, the additives and the method of production. Examples of raw materials used are coniferous (soft) wood (fibre length 3 to 4 mm), deciduous (hard) wood (fibre length 1 to 1.5 mm) and rag (fibre length 25 to 30 mm). Basically the methods of production can be divided into chemical and mechanical methods. The chemical method is most common. Sulphate and sulphite pulps are produced both bleached and unbleached. The most important pulp to be produced mechanically is groundwood pulp.

Aside from the concentration - the dry solids substance (DS) content - the flow characteristics of the pulp are also affected by the fibre ratio - length/diameter and the degree of pulverisation.

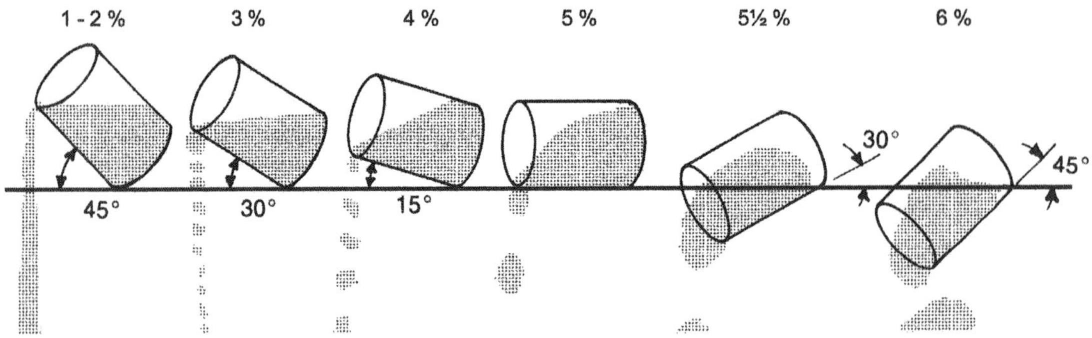

Figure 2.18 Illustration of fluidity of paper pulp

An illustration of the fluidity of paper pulp is shown in Figure 2.18. The percentage figures given indicate the DS-content.

2.4.4.3 Air content in pulp

Pulp differs from many other suspensions in that it consists of three phases; water, solid-fibre and air.

Air in the pulp occurs either in the form of bubbles or in a combination. Air occurs as bubbles either in a free form or attached to the fibre. Air in the combined form occurs in solution in the water or absorbed in the fibre. The content of air in the pulp depends upon; quality, concentration, additives, pulverisation, temperature and time; and also the pulp handling process. Groundwood pulp contains more air than sulphate or sulphite pulp. Better sizing increases the air content considerably as does pulverising.

An increase in temperature increases the content of free air whilst the ability of air to dissolve in water decreases. Generally speaking, the air content of pulp reduces in storage. The air content of pulp increases quickly to a certain level specific to each type of pulp when mixed with air. This happens typically when the pulp is allowed to fall freely into a vessel or cistern.

2.4.4.4 Pulp properties which create complications

- **Content of solid particles**: — The pulp fibres build up a network, the density of which increases with increased concentration. It is relatively difficult to get a high concentration pulp moving, since when high energy pulses are transmitted locally in the network, it is probable that the pulp will break up and movement will take place only locally.
- **Air content**: — Air in the form of bubbles in the pulp is very inconvenient from the point of view of pumping. An air content of only 1% to 2% is enough to change the characteristics.
- **Pressure drop**: — For flow at low speeds the pressure drop for pulp is much higher than for water. In general, pulp which is hot or which contains additives is easier to handle than pure, cold pulp. It can be somewhat of a problem to get pulp moving again in a pipe system after a shut-down, especially in the case of high-concentration pulp. This is because the starting resistance, "stiction", is greater than the pressure drop at low flow speeds. Dilution with water can assist re-starting.
- **Blockage characteristics**: — Flocculation of the pulp, sticks, twigs and synthetic fibre, can cause blockage in the pump. A total blockage of both valve, pump and pipe is not an unknown occurrence.
- **The tendency to thicken in reducing sections**:— Rapidly reducing conical sections on the suction side of the pump, e.g. reduction in the suction pipe from 400 mm to 150 mm, can cause flocculation at concentrations of only 3 to 4%.

2.5 Liquid-gas mixtures

2.5.1 General

Liquid-gas mixtures cover areas such as:

- Small oil and gas wells
- Boiler feed pumps
- Aircraft fuel systems
- Chemical, pharmaceutical and food processing

Small oil and gas wells: Especially offshore, where it may not be economic to lay two pipelines — one for crude oil and one for natural gas — or fit components with associated drivers and controls.

Boiler feed pumps: Generally high speed multi-stage centrifugal pumps operating with small NPSH margins. Process upsets initiating cavitation, can cause severe costly damage, resulting in unscheduled shut-downs.

Aircraft fuel systems: Prone to air entrainment due to manoeuvring. Loss of fuel pressure and reduced flow rates can be disastrous.

Chemical, pharmaceutical and food processing operations: The handling of gas and liquids together are seen as a process simplification resulting in cost savings.

2.6 Gas laws and gas properties

2.6.1 Thermodynamic laws and basic gas laws

Thermodynamics is the study of heat, work, the properties of matter and the properties of systems. Matter may be solid, liquid, vapour or gas or a mixture. The First Law of Thermodynamics is based upon the principle of the Conservation of Energy — energy cannot be created or destroyed. The principle will be valid providing all of the systems studied exclude nuclear reactions.

The First Law of Thermodynamics states:

When a system undergoes a thermodynamic cycle the net heat supplied is equal to the net work done.

The First Law is based on the work of Joule who found by experiment that work and heat are mutually convertible. The Second Law of Thermodynamics recognises that any real system cannot be 100% efficient and that losses are inevitable.

The Second Law of Thermodynamics states:

It is impossible for a system to produce net work in a thermodynamic cycle if it only exchanges heat with sources/sinks at a single fixed temperature.

The Second Law is based on a principle proposed by Clausius that heat flows unaided from hot to cold but cannot flow, unassisted, from cold to hot. Lord Kelvin used the proposal to show that work may be completely transformed into heat but only a proportion of heat could be transformed into work. If a gas is heated at constant volume there will be no work done but the energy level of the gas will be increased, see Equation 2.22.

$$Q = mc_v(T_2 - T_1) \qquad \text{Equ 2.22}$$

$$= m(u_2 - u_1)$$

where:

Q	=	heat transferred (kJ)
m	=	mass of gas (kg)
c_v	=	specific heat capacity at constant volume kJ/(kg K)
T_2	=	final absolute temperature (K)
T_1	=	initial absolute temperature (K)
u_2	=	final specific internal energy (kJ/kg)
u_1	=	initial specific internal energy (kJ/kg)

NOTE: There is no degree symbol associated with the absolute temperature. Absolute temperatures in Kelvin can be converted to degrees Celsius by subtracting 273.15. Specific heat capacity is normally abbreviated to specific heat. It is easy to see that specific internal energy, u, is equal to the product of c_v and the absolute temperature. Internal energy is an intrinsic property of a gas and is dependent upon the temperature and pressure. In this case it would have been possible to use degrees Celsius to obtain the same result. However it is worthwhile working in absolute temperatures consistently to avoid problems with ratios. If a gas is restrained and heated at constant pressure there will be work done, Equation 2.23.

$$Q = mc_v(T_2 - T_1) + W \qquad \text{Equ 2.23}$$

$$= mc_p(T_2 - T_1)$$

$$= m(h_2 - h_1)$$

so that:

$$W = m\left((h_2 - u_2) - (h_1 - u_1)\right)$$

also:

$$W = p(V_2 - V_1)$$

and:

$$h = u + pv$$

where:

W	=	work done (kJ)
c_p	=	specific heat capacity at constant pressure kJ/(kg K)
h_2	=	pecific enthalpy (kJ/kg)
h_1	=	specific enthalpy (kJ/kg)
p	=	absolute gas pressure (kPa)
V_2	=	final gas volume (m^3)
V_1	=	initial gas volume (m^3)
v	=	gas specific volume (m^3/kg)

Absolute pressures are gauge pressures plus 101.325 kPa. The International Standard Atmosphere, at sea level, is 101.325 kPa. The actual local sea level atmospheric pressure is not constant and will vary with the weather by +/- 4%. Some locations which experience severe weather conditions may experience larger variations. The atmospheric pressure will reduce at altitudes above sea level; there is no international agreement regarding pressure reductions for altitude.

2.6.1.1 Enthalpy

Enthalpy is an intrinsic property of a gas and is dependent upon the temperature, pressure and volume. The total enthalpy in a system, H, is the product of gas mass, m, and the specific enthalpy, h. Equation 2.23 can be rewritten as shown in equation 2.24 when it is known as the Non-flow energy equation. U is the product of m and u.

NOTE: The specific heat capacities, c_v and c_p, are variables not constants. The values for dry air, not real air, at atmospheric pressure and 275 K are 0.7167 and 1.0038; at 1000 K the values increase to 0.854 and 1.411.

$$Q = (U_2 - U_1) + W \qquad \text{Equ 2.24}$$

For heat to be transferred into or out of a system a temperature differential must exist. The general equation for heat transfer by conduction is given by equation 2.25.

$$q = \frac{ka\,(T_h + T_c)}{l} \qquad \text{Equ 2.25}$$

where:

q	=	energy transfer (kW)
k	=	thermal conductivity kWm/(m^2K)
a	=	area (m^2)
T_h	=	hot absolute temperature (K)
T_c	=	cold absolute temperature (K)
l	=	length of conductive path (m)

The thermal conductivity, k, will not be a simple value based on the boundary material. The conductivity value used must take account of the inside and outside boundary layer films and, if necessary, an allowance for the reduction in conductivity due to surfaces being coated with deposits or modified by corrosion.

It will be appreciated that the rate of heat transfer due to conduction is proportional to the temperature differential. If the heat source cools as transfer proceeds it will take an infinite length of time to transfer all the heat available providing there are no losses. Energy losses usually occur via convection and radiation and by heating the system as well as the gas. Perfect systems are massless; only the mass of the working fluid is considered.

2.6.1.2 Entropy

Entropy is another intrinsic property of gases. Entropy is very unusual when compared to other gas properties; entropy only changes when heat transfer occurs. Entropy is not dependent upon temperature, pressure or volume. A change in entropy is defined as:

$$ds = \frac{dQ}{T} \qquad \text{Equ 2.26}$$

where:

ds	=	change in entropy (kJ)
dQ	=	heat transfer (kJ)
T	=	absolute temperature (K)

The units for specific entropy, s, are kJ/(kg K). Values of intrinsic properties; u, h, s; are quoted in gas tables and appear on the axes of gas charts. It is very important to verify the base temperature of printed data before starting calculations. Some gases use 0 °C and some, like refrigerants, use -40° C.

2.6.2 The Gas Laws

The state of perfect gases can be predicted by several laws generally grouped together as the Gas Laws.

Boyle's Law states:

The volume of a mass of gas at constant temperature is inversely proportional to the absolute pressure.

Boyle's Law provides the following relationship.

$$pV = \text{constant} \qquad \text{Equ 2.27}$$

$$p_1V_1 = p_2V_2$$

p can be in Pa or kPa but both must be the same.

Charles' Law states:

The volume of a mass of gas at constant pressure is proportional to the absolute temperature.

Charles' Law provides the following relationship:

$$\frac{V}{T} = \text{constant} \qquad \text{Equ 2.28}$$

$$\frac{V_1}{T_1} = \frac{V_2}{T_2}$$

The Pressure Law states:

The pressure of a mass of gas at constant volume is proportional to the absolute temperature.

The Pressure Law provides the following relationship:

$$\frac{p}{T} = \text{constant} \qquad \text{Equ 2.29}$$

$$\text{or } \frac{p_1}{T_1} = \frac{p_2}{T_2}$$

The three preceding laws may be combined to provide one of the most useful perfect gas relationships commonly known as the Combined Gas Law.

$$\frac{p_1v_1}{T_1} = \frac{p_2v_2}{T_2}$$

$$\text{or } \frac{p_1V_1}{T_1} = \frac{p_2V_2}{T_2} \qquad \text{Equ 2.30}$$

The Combined Gas Law can be used with the specific volume of the gas, v, when considering the effects on a single kg of gas or with V the volume of a known mass of gas.

2.6.3 Fundamental equations

The processes of flow occurring in nature are usually very complex and difficult to deal with. In many technical applications, however, perfectly acceptable results can be obtained from calculations based upon simplified considerations. Some of the concepts and conditions germane to this concept are discussed in this chapter.

A flow process is steady if all flow parameters - pressure, velocity, etc. - at any specific point in the flow field are independent of time. According to this definition practically all flow processes are unsteady. In practice, many can, with sufficient accuracy, be treated as steady using suitable mean time values. The fundamental equations presented in later sections are valid for steady flow.

Extra care should be taken with certain installations where steady state conditions may not exist. Low head, high flow centrifugal pumps with wide impeller tip widths produce velocity variations across the flow stream. Some positive displacement pumps, with a small number of pumping elements, produce cyclic flow and pressure variations. The amplitude of these variations is dependant upon the piping system as well as the pump construction.

Flow is generally three-dimensional, i.e. the flow parameters vary with all three co-ordinates defining some point in space. In many practical cases of flow the number of dimensions studied may be reduced whilst still maintaining accuracy. A common example is pipe flow, where the parameters are assumed to vary in only one direction, i.e. the longitudinal dimension of the pipe.

One-dimensional flow assumes that flow parameters can be described by mean values across the flow through-section, see Figure 3.1. In principle, different mean values should be applied when studying continuity, momentum and energy. In the case of pipe flow, the mean velocity in the pipe is defined as the volume flow divided by the cross-sectional area of the pipe (v = Q/A). This mean velocity can as a rule be used with adequate accuracy for most purposes.

An important property of a flowing medium is its density and the changes in density which occur during flow. A gas is compressed, i.e. the density increases, when the pressure somewhere in the flow increases. This type of flow is called compressible flow. In a liquid system density variations are very small under great changes of pressure. Liquid flow can thus, with sufficient accuracy, be treated as incompressible. This is also the case with gases at low flow velocities when the pressure changes are small. Liquid compressibility may be important for valve selection. Hot liquids and liquefied gases may require special consideration depending upon the differential pressure. Power is required to compress liquids. Ignoring liquid compressibility can result in low flow and/or undersized pump drivers.

A streamline is a curve to which the velocity vector is tangential at any point. In the case of steady flow, the streamlines remain unchanged with time and describe the path of a liquid particle passing through the flow field. The streamlines through all points on a closed curve in the flow field constitute a flow tube. See Figure 2.19. No mass will pass through the circumferential boundary surface of a flow tube. The flow tube is thus reminiscent of an ordinary pipe. In the case of a pipe, however, there are strong frictional effects associated with the wall of the pipe which do not necessarily occur in a flow tube.

2.6.3.1 The continuity equation

The continuity equation is a statement for the condition that mass is not created or destroyed during a flow process.

Assuming that the flow is steady, the mass flow $\dot{m}$ must be of equal magnitude everywhere along the pipe or the flow tube. In the case of one-dimensional flow in Figure 2.20,

$$\dot{m} = r_1 \cdot v_1 \cdot A_1 = r_2 \cdot v_2 \cdot A_2 \qquad \text{Equ 2.31}$$

or for an incompressible liquid flow,

$$Q = v_1 \cdot A_1 = v_2 \cdot A_2 \qquad \text{Equ 2.32}$$

where:

Q	=	volume flow (m^3/s)
v	=	Q/A = flow velocity (m/s)
A	=	cross-sectional area (m^2)
$\dot{m}$	=	mass flow (kg/s)

When the cross-sectional area in a pipe reduces, then, according to the continuity equation, the flow velocity increases. If the area is halved, the velocity doubles and so on.

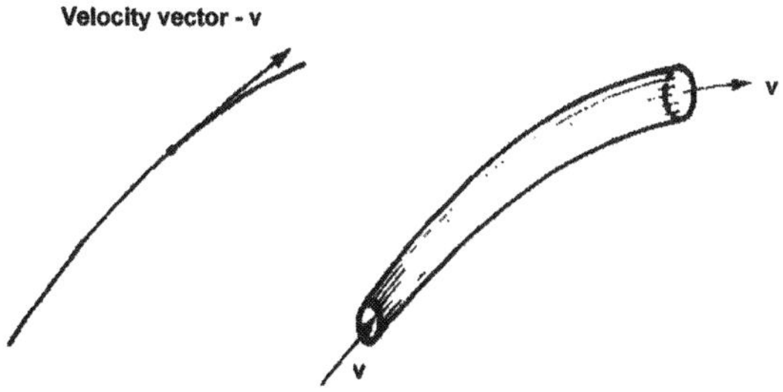

Figure 2.19 Streamline and flow tube

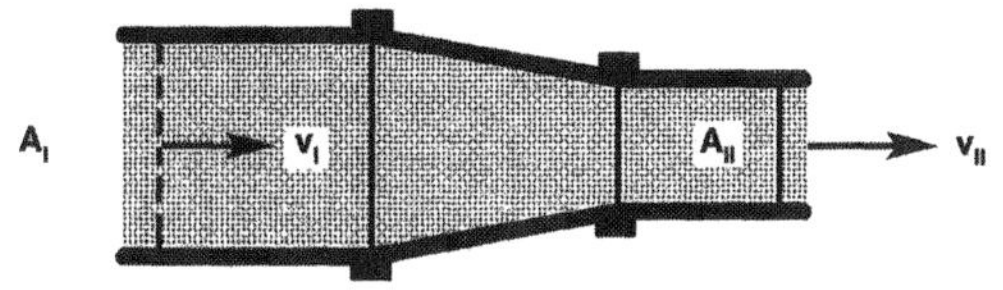

Figure 2.20 Example of one-dimension flow

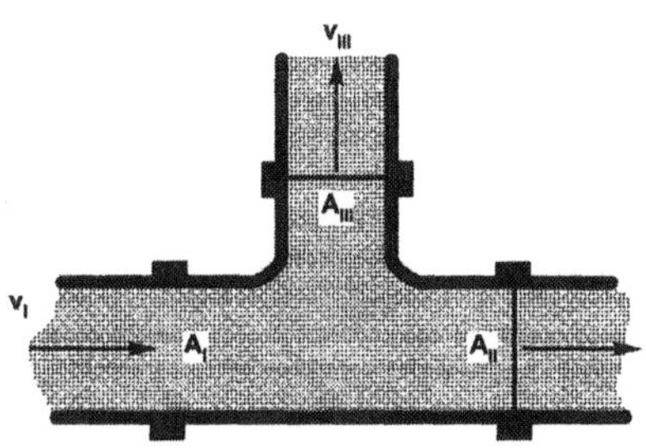

Figure 2.21 Branching

Since there is no increase in mass at the point of branching, the mass flow entering will equal the total mass discharging per unit time. Using the symbols of Figure 2.21:

$$Q_1 = Q_2 + Q_3$$

or

$$v_1 \cdot A_1 = v_2 \cdot A_2 + v_3 \cdot A_3 \qquad \text{Equ 2.33}$$

2.6.3.2 Bernoulli's equation

Bernoulli's equation is an equation of motion for incompressible flow based on Newton's Second Law of Motion adapted for fluids and Euler's Equation of Motion. The equation expresses the relationship between:

- static pressure
- potential energy
- kinetic energy
- friction losses

Using Figure 2.22, Bernoulli's equation, for steady, one-dimensional, constant temperature and incompressible flow between stations I and II, becomes:

$$\frac{p_I}{r} + \frac{v_I^2}{2} + gz_I = \frac{p_{II}}{r} + \frac{v_{II}^2}{2} + gz_{II} + \frac{\Delta p_f}{r} \qquad \text{Equ 2.34}$$

where:

p	=	static pressure (gauge)(Pa)
ρ	=	density (kg/m^3)
v	=	velocity (m/s)
g	=	gravitational acceleration (m/s^2)

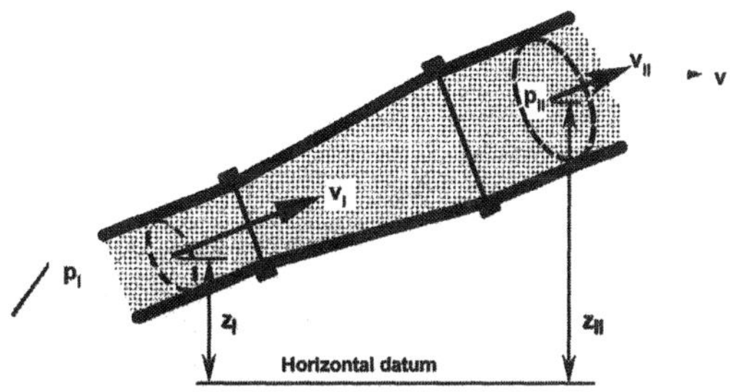

Figure 2.22 Symbols used in Bernoulli's equation

z	=	height above datum (m)
Δp_f	=	frictional pressure loss (Pa)

where:

p	=	static pressure (gauge) (Pa)
ρ	=	density (kg/m^3)
v	=	velocity (m/s)
g	=	gravitational acceleration (m/s^2)
z	=	height above datum (m)
Δp_f	=	frictional pressure loss (Pa)

Equation 2.34 expresses the various energy terms per unit mass of fluid, J/kg. The gravitational acceleration, g, is not a constant but a variable. For most applications the general value of 9.80665 m/s^2 will be acceptable. However, for applications at altitude or where high accuracy is required the local value of g can be calculated from:

$$g = 9.7803\left(1 + 0.0053\text{SIN}^2\varphi\right) - 3.1\text{x}10^{-6}z \qquad \text{Equ 2.35}$$

where:

φ	=	latitude (degrees)
z	=	altitude (m)

In rotodynamic machine technology it is practical to express the Bernoulli equation in terms of "head" i.e. metres of gas column. If all terms in Bernoulli's equation are divided through by g, we get:

$$\frac{p_I}{w} + \frac{v_I^2}{2g} + z_I = \frac{p_{II}}{w} + \frac{v_{II}^2}{2g} + z_{II} + \frac{\Delta p_f}{w} \qquad \text{Equ 2.36}$$

where:

w	=	specific weight (N/m^3)
w	=	ρg

The various terms are then called:

$\frac{p}{\rho g}$ or $\frac{p}{w}$	=	static head (m)
$\frac{v^2}{2g}$	=	velocity head (m)
z	=	potential head (m)
$\frac{\Delta p_f}{w}$	=	h_f = head loss (m)

Since all the terms in Equation 2.35 refer to height they are easy to illustrate graphically. See Figure 2.23.

Total head is sometimes designated by H. The relationship between pressure and head is p/w = h.

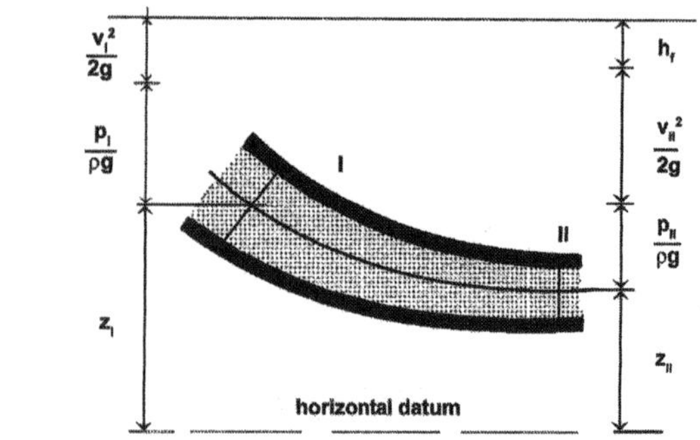

Figure 2.23 Graphic illustration of Bernoulli's equation

The three head terms can be converted to equivalent pressure terms. Static pressure plus velocity, or rotodynamic pressure is called total pressure or stagnation pressure.

$$\text{total pressure} = p + \frac{\rho v^2}{2} = wh + \frac{wv^2}{2} \qquad \text{Equ 2.37}$$

Total and static pressure are measured in different ways and illustrated in Figure 2.24. A pressure tap at right angles to the direction of flow will sense the static pressure. Ahead of a Pitot tube the velocity is reduced to zero and the static pressure rises to stagnation pressure.

As a rule, it is easy to measure the static pressure, whereas the total pressure can easily be affected by measurement errors.

A Prandtl tube or Pitot-static tube measures both total and static pressure. The differential pressure, the dynamic pressure, $p_{dyn} = wv^2/2g$ is obtained directly.

2.6.3.3 The momentum equation

The product of mass times flow velocity for a liquid particle is called its momentum. The momentum equation for steady flow reads:

$$F = \frac{d}{dt} \cdot \dot{m}$$

The resultant F of all external forces on a control volume is equal to the changes in momentum for the outflowing and inflowing mass per unit time, or,

$$F = \dot{m}\left(\bar{v}_2 - \bar{v}_1\right) \qquad \text{Equ 2.38}$$

where:

$\dot{m}$ = mass flow dm/dt $\left(\frac{kg/s}{(m/s)}\right)$

$\bar{v}$ = velocity vector

The momentum equation is illustrated in the following example. The problem is to determine the force required to hold a pipe bend in position, as illustrated in Figure 2.25.

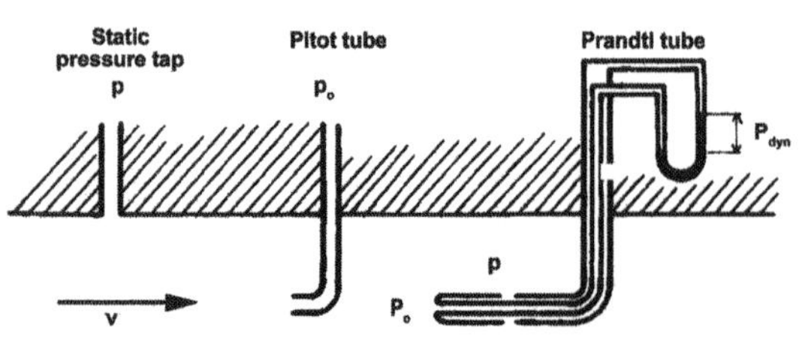

Figure 2.24 Measuring of static, total and dynamic pressure

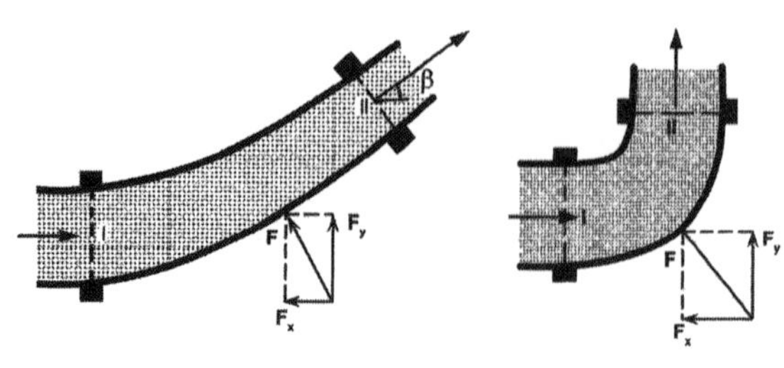

Figure 2.25 Pipe bends, arbitrary (left) and 90° bend

The control volume is defined by the interior of the pipe bend. Then, for the arbitrary bend in the x axis:

$$-F_x + p_I \cdot A_I - p_{amb} \cdot A_I - p_{II} \cdot A_{II} \cdot \cos\beta + p_{amb} \cdot A_{II} \cdot \cos\beta = \dot{m}\left(v_{II} \cdot \cos\beta - v_I\right)$$

or:

$$F_x = \dot{m} \cdot v_I + \left(p_I - p_{amb}\right) \cdot A_I - \left[\dot{m} \cdot v_{II} + \left(p_{II} - p_{amb}\right) \cdot A_{II}\right] \cdot \cos\beta$$

Similarly, in the y axis:

$$F_y = \left[\dot{m} \cdot v_{II} + \left(p_{II} - p_{amb}\right) \cdot A_{II}\right] \cdot \sin\beta$$

For a 90° bend β = 90°, therefore:

$$F_x = \dot{m} \cdot v_I + \left(p_I - p_{amb}\right) \cdot A_I$$

$$F_y = \dot{m} \cdot v_{II} + \left(p_{II} - p_{amb}\right) \cdot A_{II}$$

$$F = \sqrt{F_x\,2 + F_y^{\,2}} \qquad \text{Equ 2.39}$$

NOTE: It is a gauge pressure and not the absolute pressure in the pipe which determines the magnitude of the force. Note also that the forces can be determined without detailed knowledge of the internal flow process through the bend. The momentum equation applies regardless of whether the process has losses or not.

It is often the case that the terms m and v are small compared with the others. For a 90° degree bend with A_I being equal to A_{II}.

$$F = \sqrt{2} \cdot \left(p - p_{amb}\right) \cdot A \qquad \text{Equ 2.40}$$

where:

p = pressure in pipe (absolute) (Pa)

p_{amb} = ambient pressure (absolute) (Pa)

A = pipe cross-sectional area (m^2)

F = fixing force (N)

The fixing force has an angle of 135 degrees to the incoming direction of flow.

2.6.3.4 The energy equation

The energy equation is an extension of the energy principle which states that energy cannot be created or destroyed but can only be converted to other forms. The energy equation for steady state one-dimensional machine flow is:

$$W_{in} = \left(h_{II} - h_I\right) + \frac{\left(c_{II}^2 - c_I^2\right)}{2} + g\left(z_{II} - z_I\right) + Q_{out} \qquad \text{Equ 2.41}$$

The term h, specific enthalpy, is made up from two components; u, (u = Tc_p) internal energy and pv, which represents the mechanical energy due to pressure, often called flow work. The energy equation can be rewritten:

$$W_{in} = \left(T_{II}\, c_{v2} - T_I\, c_{v1}\right) + \left(p_{II}\, v_{II} - p_I v_I\right) +$$

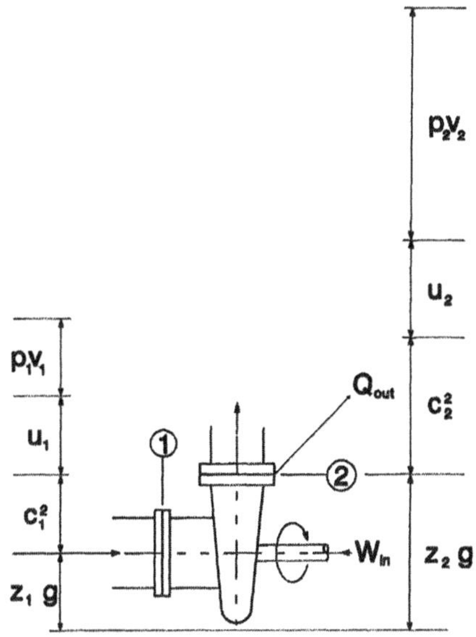

Figure 2.26 Illustration of the energy equation

$$\frac{\left(c_{II}^{2} - c_{I}^{2}\right)}{2} + g(z_{II} - z_{I}) + Q_{out} \qquad \text{Equ 2.42}$$

where:

T	=	temperature (°C)
c_v	=	specific heat (J/kg/°C)
u	=	Tc_v
p	=	pressure (Pa)
v	=	specific volume (m^3/kg)
c	=	velocity (m/s)
z	=	height above datum (m)
g	=	gravitational acceleration (m/s^2)
W_{in}	=	work input to the shaft (J/kg)
Q_{out}	=	heat rejected (J/kg)

NOTE: In these equations the velocity is represented by c not v. Engineers frequently use the same symbol for different quantities. The context is important and should be explanatory. All the terms in the energy equation are calculated per kg of medium flowing. The terms $c^2/2$ and gz respectively represent the kinetic and potential energy of the medium per kg. The effect of a machine on the energy equation is shown diagrammatically in Figure 2.26.

Mechanical work is input through the rotating shaft, all the variables change and some heat is rejected from the machine casing. Notice particularly that a temperature rise has occurred, u_2 is greater than u_1. The change from suction conditions to discharge conditions usually takes place very quickly within the machine so the ideal process is isentropic, constant entropy, not constant temperature. The ideal process is never realised in practice due to friction, heat conduction along or through the machine walls and heat losses or gains due to convection and radiation. Machine inefficiencies will generally appear as increased temperature rise and heat rejection. Noise and vibration will take their energy as portions of heat rejection.

2.6.4 Perfect and real gases

Perfect gases are characterised by the following relationship:

$$pv = RT \qquad \text{Equ 2.43}$$

remembering that V = mv

$$\text{so that } pV = mRT \qquad \text{Equ 2.44}$$

and as

$$\rho = \frac{1}{v}$$

then

$$p = \rho RT$$

where:

R	=	Characteristic Gas Constant kJ/(kg)
V	=	volume (m^3)
ρ	=	gas density (kg/m^3)

Another very important relationship is given by:

$$pv_0 = MRT \qquad \text{Equ 2.45}$$

where:

v_0	=	volume of 1 Mol of gas (m^3)
M	=	mass of 1 Mol of gas (kg)

A Mol of gas is defined as the product of the gas molecular weight and 1 kg. For example; the molecular weight of oxygen, O_2, is 2 x 15.9994 which is 32 for most practical purposes. A Mol of oxygen is 32 kg.

Avogadro's Hypothesis states that equal volumes of gases at the same conditions will contain the same number of molecules. Therefore:

$$\frac{pv_0}{T} = MR = \text{a constant for all perfect gases}$$

$$= \tilde{R}$$

$$= 8.31451\ kJ/(kmol\ K) \qquad \text{Equ 2.46}$$

remembering that 1J = 1N x 1m and 1pa = $1N/1m^2$

$$\tilde{R} = 8.31451\ kPa\ m^3/(kmol\ K)$$

$$\text{or} = 8314.51\ Pa\ m^3/(kmol\ K)$$

$$\text{or} = 0.0831451\ bar(a)\ m^3/(kmol\ K)$$

where:

v_0	=	volume of 1 Mol of gas (m^3)
R_0	=	Molar Gas Constant

From Equation 2.46 this extremely important relationship can be derived:

$$R = \frac{\tilde{R}}{M}\ J/(kgK) \qquad \text{Equ 2.47}$$

Providing the molecular weight of a gas, or mixture, is known the Characteristic Gas Constant, R, can be calculated. Using R in Equation 2.44 provides a theoretical starting point for all compression calculations. Other useful perfect gas relationships include:

$$\gamma = \frac{c_p}{c_v} \qquad \text{Equ 2.48}$$

$$R = c_p - c_v$$

$$a = \sqrt{\gamma RT}$$

so that

$$Ma = \frac{c}{a}$$

where:

γ = isentropic exponent

a = sonic velocity (m/s)

Ma = Mach Number (dimensionless)

The sonic velocity, the velocity of sound waves, which are of course pressure waves, is useful when dealing with high speed rotodynamic machines and when considering the properties of piping.

The perfect gas characteristic equation as shown as Equation 2.44, can be modified to cope with real gases by the addition of a "compressibility factor", Z, as shown in Equation 2.49. The compressibility factor modifies the perfect gas volume to a real gas volume.

$$pV = ZmRT \qquad \text{Equ 2.49}$$

For preliminary calculations Z can be assumed to be unity. Values of Z are usually obtained from charts of "Generalised Compressibility Factors", see Figure 2.27, but charts for specific gas groups, such as natural gas and mixtures, see Figure 2.28 are available. The compressibility factor can vary from 0.3 to about 4 and must be included if accurate results are required. The compressibility factor is obtained from charts after the reduced pressure and the reduced temperature have been calculated, see Equation 2.50.

$$p_r = \frac{p}{p_c}$$

$$T_r = \frac{T}{T_c} \qquad \text{Equ 2.50}$$

where:

p_r = reduced pressure

p_c = critical pressure

T_r = reduced temperature

T_c = critical temperature

The working pressure, p, and the critical pressure, p_c, must be in the same units and the value of R must be compatible.

2.6.4.1 Examples of using generalised compressibility factor

NOTE: Compressibility factors are used to convert ideal gas properties to real gas properties. If thermodynamic tables or a gas chart are available there is no requirement to use compressibility factors as the data provided is "real". Gas data available from computer programs which use equations of state such as van der Waals, Redlich-Kwong, Soave Redlich-Kwong (SRK), Benedict-Webb-Rubin-Starling (BWRS), and Peng Robinson, is also "real".

Example 1:

10 kg of propane at 120 °C and 2000 kPa(g);
what volume does the gas occupy?

$$T = 120 + 273.15 = 393.15\ \text{K}$$

$$p = 2000 + 101.325 = 2101.325\ \text{kPa(a)}$$

Data for propane is taken from Chapter 20, Section 20.4.6

M = 44.097 kg,

p_c = 4240 kPa (a)

T_c = 369.77 K

$$R = \frac{8.31451}{44.097} = 0.18854\ \text{kJ/(kg K)}$$

$$p_r = \frac{2101.325}{4240} = 0.4956$$

$$T_r = \frac{393.15}{369.77} = 1.0634$$

Z = 0.845 from Figure 2.27

$$V = \frac{mZRT}{p} = \frac{10 \times 0.845 \times 0.18854 \times 393.15}{2101.325}$$

$$= 0.298\ \text{m}^3$$

Example 2:

100 kg of methane at 40 °C and 1000 kPa (g);
what volume does the gas occupy?

$$T = 40 + 273.15 = 313.15\ \text{K}$$

$$p = 1000 + 101.325 = 1101.325\ \text{kPa(a)}$$

Data for methane is taken from Chapter 20, Section 20.4.3

M = 16.043 kg,

p_c = 4599 kPa(a)

T_c = 190.56 K

$$R = \frac{8.31451}{16.043} = 0.51826\ \text{kJ/(kg K)}$$

$$p_r = \frac{1101.325}{4599} = 0.23947$$

$$T_r = \frac{313.15}{190.56} = 1.64331$$

Z = 0.975 from Figure 2.27

$$V = \frac{mZRT}{p} = \frac{100 \times 0.975 \times 0.51826 \times 313.15}{1101325}$$

$$= 14.4\ \text{m}^3$$

2.6.5 Diagrams of gas properties

The properties of gases can be represented diagrammatically. Most diagrams do not just show the gas state but also show the liquid, wet vapour, superheated vapour and the gas state. Diagrams are based on "real" data and consequently the complications of compressibility factors, **Z**, are avoided. It is essential to know where the diagram thermodynamic datum is; where enthalpy and entropy are zero. The datum is essential if any "off-diagram" property calculations are to be performed.

The most notable of the property diagrams is the Mollier diagram. Mollier was the first to realise how useful a diagram could be to visualise the thermodynamic processes. A Mollier diagram is a plot of enthalpy against entropy. Figure 2.29 shows the general style of a diagram.

The Saturation line separates the wet vapour state from both the pressurised liquid region and the superheated vapour region. The constant vapour lines, which indicate partial evaporation of the liquid, are also called constant dryness or quality. 90% vapour, 90% dryness or 90% quality indicates that 90% of the latent heat of evaporation has been supplied. Considering entropy as a gas property at a particular condition, is extremely

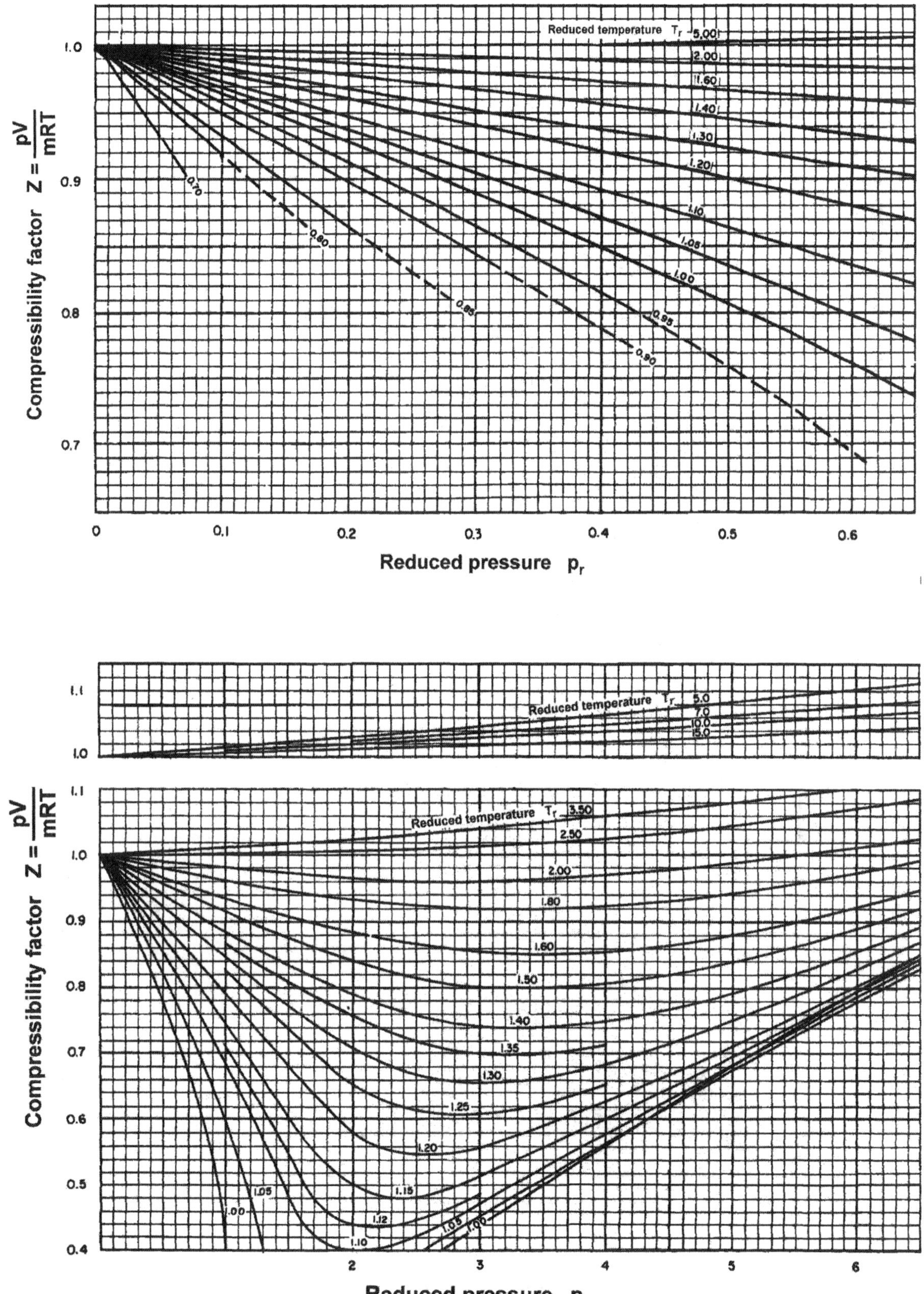

Figure 2.27 Generalised compressibility factors

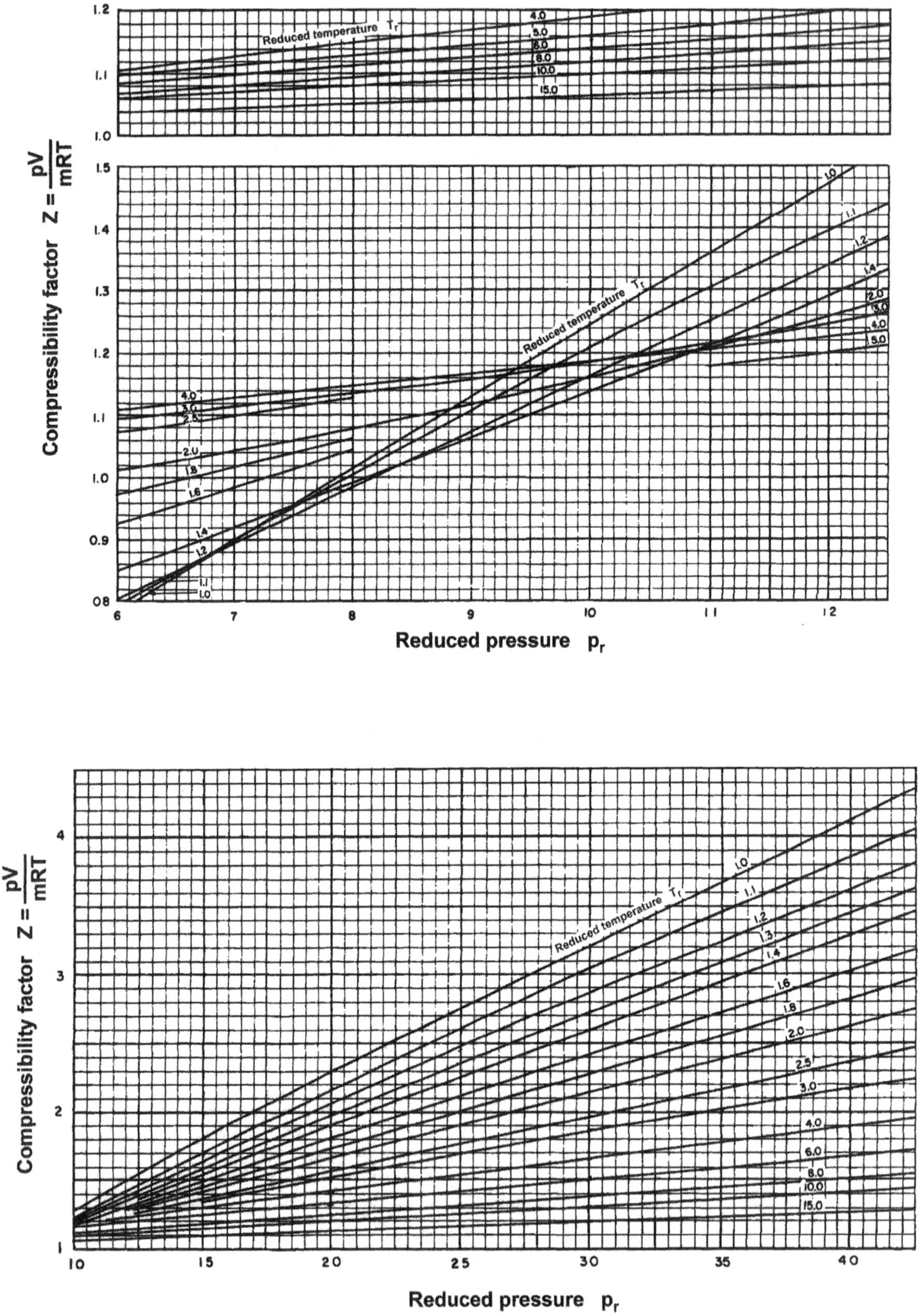

Figure 2.27 Generalised compressibility factors *Continued*

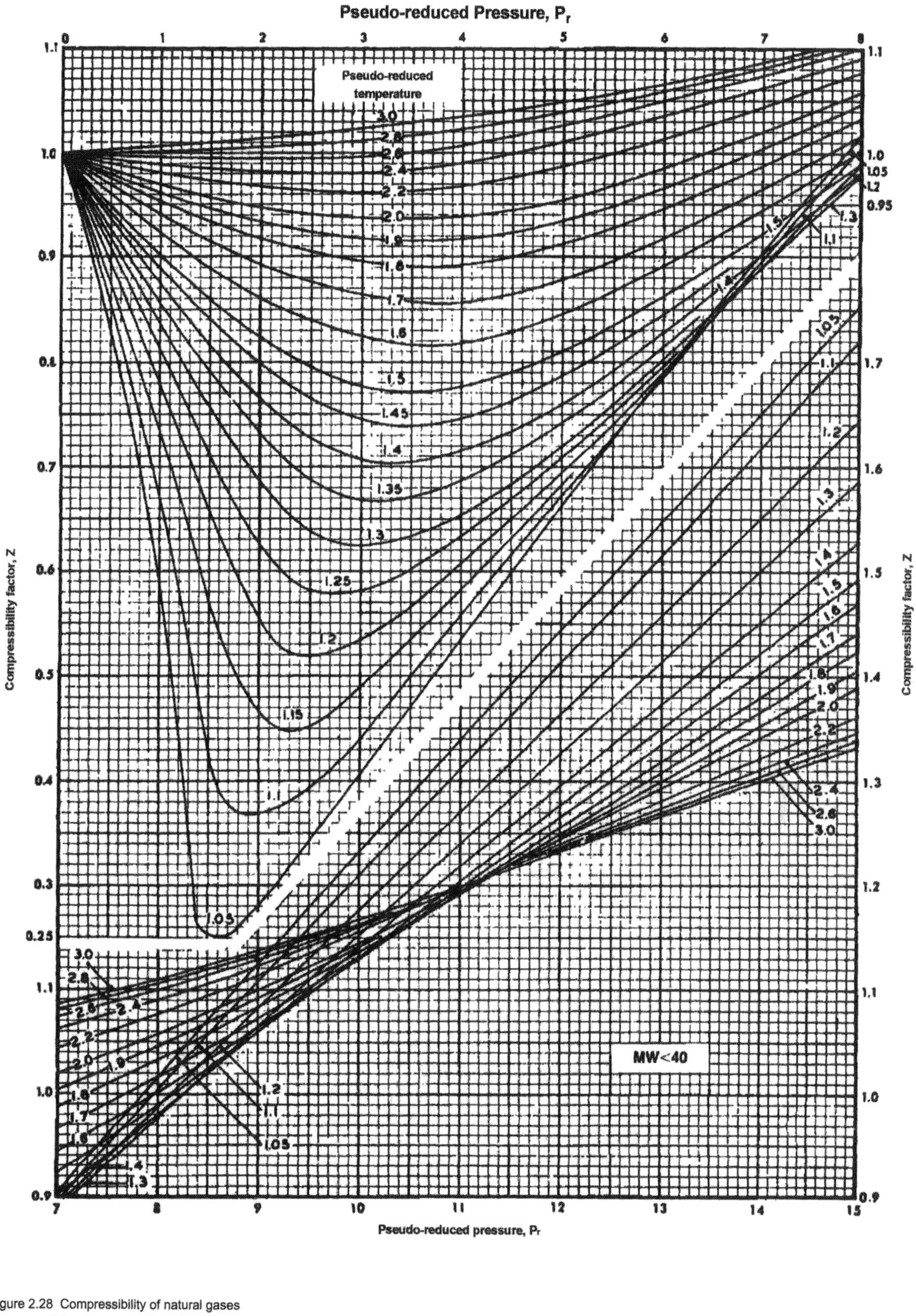

Figure 2.28 Compressibility of natural gases
Courtesy of Gas Processor Suppliers Association

useful for ideal processes which do not involve heat transfer. If heat transfer is zero the change in entropy is also zero; therefore a process line must be a vertical straight line. Inefficiencies, which transform an ideal process into a practical process, normally include heat transfer and/or friction which results in a change in entropy. The difference in enthalpy, for the two process lines, can be used to calculate an isentropic efficiency.

It is possible to draw other property diagrams — temperature-entropy diagrams are very common. Figure 2.30 shows the important features. These diagrams are sometimes, mistakenly, called Mollier diagrams.

Temperature-entropy diagrams are very useful for processes which involve evaporation and/or condensation. Evaporation and condensation take place at constant temperature and constant pressure; both processes follow a horizontal straight line between the saturated liquid line and the saturated vapour line. The latent heat is easily determined.

Pressure-enthalpy diagrams, see Figure 2.31, are used extensively in refrigeration and air-conditioning applications. When the suction and discharge pressures and temperatures are known the enthalpy change is easily found. The effects of changes in operating conditions can be quickly converted to changes in ideal power requirements; the product of mass flow and enthalpy increase will provide theoretical power in kJ/s. If the type of compression process is unknown the discharge temperature must be predicted by selecting an appropriate value for the polytropic exponent, **n**.

Figure 2.32 shows a psychrometric chart for air at normal atmospheric pressure. Air is a mixture of gases and usually contains water vapour. The psychrometric chart cannot be used for visualisation of compression processes because the chart is constructed for a single, fixed pressure, 101.325 kPa(a). However, the chart is very useful for potential users of air compressors.

Firstly, it is very easy to assess the change in specific volume with variations in humidity. Secondly, the amount of moisture in the air, kg/kg dry air, is easily read from the chart; this allows the water condensation problem in the discharge to be evaluated. Provisions can be considered to accommodate the condensate or to remove it. Thirdly, the additional compression power over the theoretical air power can be approximated. The difference in dry air enthalpy and humid air enthalpy can be read from the chart. The theoretical dry air compression power can be increased by the same proportion to give an approximation of compression power for "real" air.

Table 2.22 shows useful data relating to normal atmospheric pressure.

Relative humidity at 20°C	ppm by volume water vapour
100	23733
53	12319
26	6097
11	2583
4.4	1018

Table 2.22 Water content by volume in humid air

The use of thermodynamic property diagrams is often superseded by the calculation of individual properties using computer software but it is helpful to see the theory behind the programs!

NOTE: Some software produces "ideal" gas properties rather than "real" gas properties so **Z** must still be evaluated. If individual properties are calculated it is worthwhile sketching a Mollier or T-s or p-H diagram to visualise the process. The diagram is especially useful if process conditions are close to a saturation condition and a change of phase may occur due to small variations in the process.

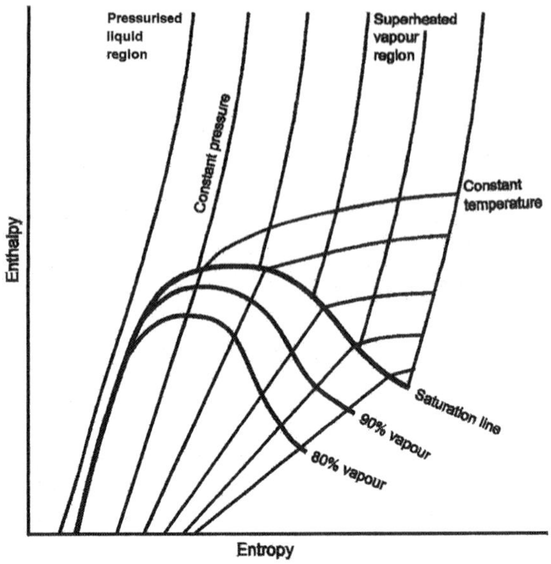

Figure 2.29 Characteristic thermal properties shown on a Mollier diagram

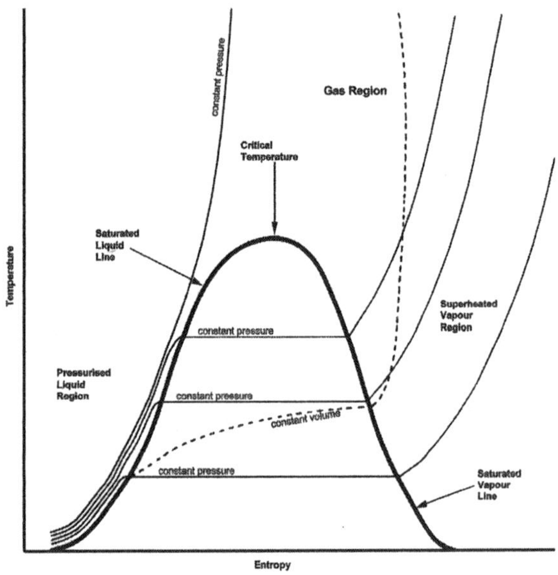

Figure 2.30 Characteristic thermal properties shown on a temperature-entropy diagram

Diagrams of gas properties are available from a number of sources.

2.6.6 Properties of mixtures

The methods given in the following paragraphs can be used to determine those properties of a gas mixture required for compressor calculations, with sufficient accuracy for most purposes associated with the selection and outline design of a compressor.

2.6.6.1 Mean molar mass

The mean (effective) molar mass of a mixture of gases can be calculated from the following formula:

$$\bar{M} = \sum M_x m_x \qquad \text{Equ 2.51}$$

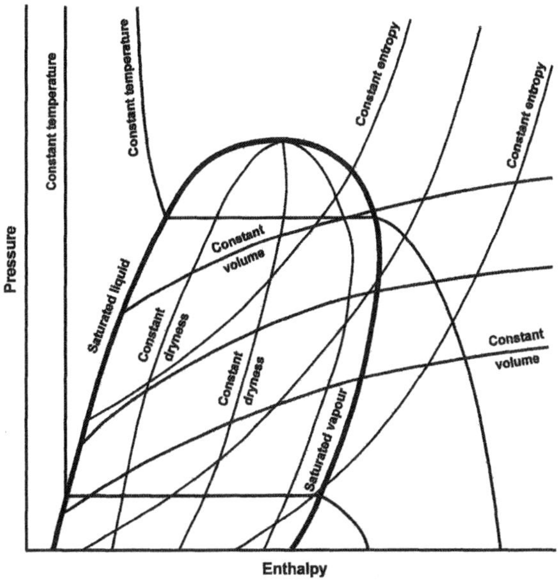

Figure 2.31 Characteristic thermal properties shown on a pressure-enthalpy diagram

where:

$\overline{M}$ = effective mean molar mass of the mixture

M_x = molar mass of constituent x

m_x = molar fraction of constituent x

2.6.6.2 Effective value of the gas constant

The gas constant for the mixture is calculated from the following formula:

$$\overline{R} = \frac{\overline{\overline{R}}}{\overline{\overline{M}}} \qquad \text{Equ 2.52}$$

where:

$\overline{R}$ = effective gas constant kJ/(kgK)

2.6.6.3 Mean molar specific heat

The mean molar specific heats can be calculated from:

$$\overline{c}_p = \sum m_x (M.c_p)_x \qquad \text{Equ 2.53}$$

$$\overline{c}_v = \sum m_x (M.c_v)_x \qquad \text{Equ 2.54}$$

where:

$(M.c_p)_x$ = molar specific heats of constituent x

$(M.c_v)_x$ = molar specific heats of constituent x

2.6.6.4 Isentropic exponent of the mixture

The isentropic exponent of the mixture is calculated from:

$$\overline{\gamma} = \frac{\overline{c}_p}{\overline{c}_v} = \frac{\overline{c}_p}{\overline{c}_p - \overline{R}} \qquad \text{Equ 2.55}$$

2.6.6.5 Compressibility of the gas mixture

The compressibility factor (Z) for the mixture, at absolute temperature T and absolute pressure P, can be obtained as follows:

Firstly the "pseudo-critical pressure" and the "pseudo-critical temperature" are calculated from the following formulae:

$$\overline{p_c} = \sum p_{cx} m_x \qquad \text{Equ 2.56}$$

$$\overline{T_c} = \sum T_{cx} m_x \qquad \text{Equ 2.57}$$

where:

$\overline{p_c}$ = pseudo-critical pressure of the mixture

p_{cx} = critical pressure of component x

$\overline{T_c}$ = pseudo-critical temperature of the mixture

T_{cx}= critical temperature of component x

Secondly the pseudo-reduced pressure and temperature for the gas mixture are calculated from:

$$\overline{p_r} = \frac{p}{\overline{p_c}} \qquad \text{Equ 2.58}$$

$$\overline{T_c} = \frac{T}{\overline{T_c}} \qquad \text{Equ 2.59}$$

Finally the value of the compressibility factor $\overline{Z}$ may be obtained from Figure 8.23, Chapter 8, using the values of $\overline{P_r}$ and $\overline{T_r}$, determined as above.

2.6.6.6 Mixture specific volume

The specific volume of gas mixture, $\overline{V}$, can be calculated from the following formula:

$$\overline{v} = \frac{\overline{Z}\overline{R}T}{p} \qquad \text{Equ 2.60}$$

Example:

Assuming a gas mixture of 60% methane, 20% ethane, 15% propane and 5% iso-butane by volume (mol percentages). See Table 2.23.

Hence:

- the mean molar mass of the gas mixture is 25.161 kg/kmol
- the gas constant $\overline{R}$ for the gas mixture is

  $$\frac{8314 \times 10^{-5}}{25.161} = 330.4 x 10^{-5} \frac{bara.m^3}{kg.K}$$

- the mean specific heat (c_p) at 15°C, 1.013 bara is $\frac{46.89}{25.161} = 1.864$
- the effective isentropic exponent for the gas mixture at 15°C, 1.013 bara is

  $$\frac{1.864}{(1.864 - 0.3304)} = 1.215$$

Component	Mole fraction m	Molar mass M kg/kmole	m.M	Spec heat at 15°1.013 bar(a) Cp kJ/kg	m.M.Cp	Critical pressure Pc (bar(a))	m.Pc	Critical temperature Tc (K)	m.Tc
Methane	0.6	16.043	9.626	2.204	21.21	45.99	27.59	190.56	114.34
Ethane	0.2	30.070	6.014	1.705	10.25	48.80	9.76	305.41	61.08
Propane	0.15	44.097	6.615	1.624	10.74	42.40	6.36	369.77	59.52
Iso-butane	0.05	58.123	2.906	1.616	4.69	36.40	1.82	407.82	20.39
TOTALS	1.00		25.161		46.89		45.53		255.33

Table 2.23 Gas mixtures

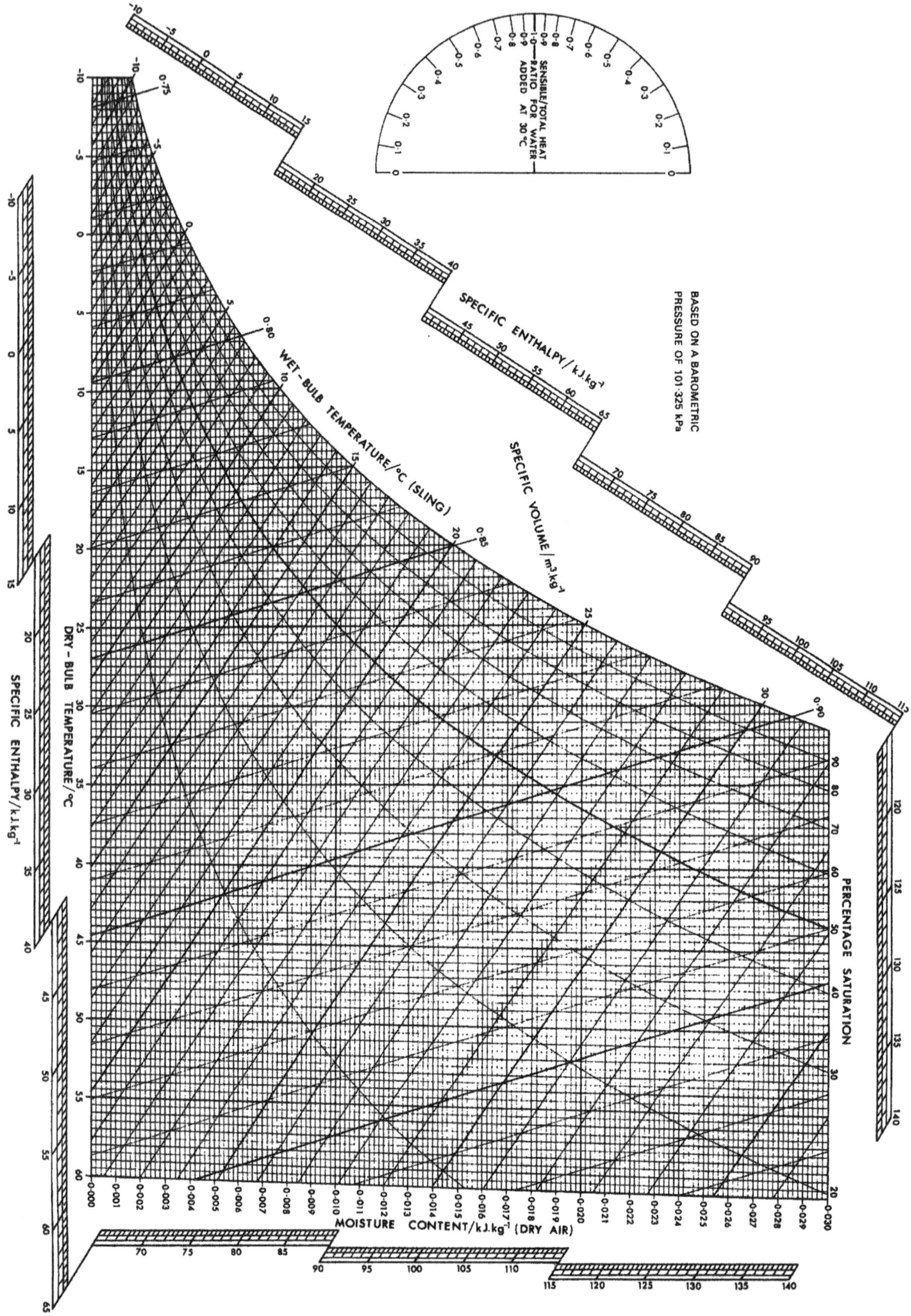

Figure 2.32 A psychrometric chart for air

Courtesy of CIBSE

Guidance on use of Figure 2.32

An ordinate is drawn vertically from the dry bulb temperature to the point of intersection with the relevant diagonal wet bulb temperature line. The "real" air properties; specific volume, specific enthalpy, percentage saturation and moisture content; can be read from the chart. Percentage saturation is also known, incorrectly. as relative humidity. The two parameters can be considered as interchangeable for industrial purposes. Most users will probably know the dry bulb temperature and the relative humidity. The 'state point' can be found at the intersection of the vertical dry bulb temperature line and the percentage saturation curve.
Notice there are two different sets of diagonal lines; one for wet bulb temperature and one for specific volume.

NOTE: The value of the isentropic exponent will vary a little with temperature and pressure. However for most initial calculations this value may be used.

- the pseudo mean critical pressure for the gas mixture is 45.53 bara
- the pseudo mean critical temperature for the gas mixture is 255.33 K
- assuming gas conditions of 10 bar(a) and 300 K, the reduced pressure and temperature for the mixture are 0.22 and 1.17 respectively
- from the psychometric chart (Figure 2.32) the compressibility factor for the gas mixture is approximately 0.95
- if gas conditions of 100 bar(a) and 400 K are assumed, the reduced pressure and temperature are 2.2 and 1.56 respectively

hence from Figure 2.28 the compressibility factor is about 0.835.

2.6.7 Mass/volume relationship

If the required duty is specified in terms of mass flow, then the actual volume flow at the inlet can be determined from the following formula:

$$V_1 = W \frac{Z\tilde{R}T_1}{Mp_1} \frac{3600}{10^2} \qquad \text{Equ 2.61}$$

where:

V_1 = inlet volume flow (m^3/h)

W = mass flow (kg/s)

Z = compressibility factor (for this purpose may generally be assume to be unity)

$\tilde{R}$ = universal gas constant

T_1 = absolute inlet temperature (K)

M = molar mass (g/mol)

p_1 = absolute inlet pressure (barg)

Alternatively, for pure gases, the gas density can be determined from a pressure-enthalpy chart, if available, (commonly but strictly incorrectly, referred to as a Mollier chart) and hence the volume flow calculated.

2.6.8 Principles of expansion

When a gas is contained within a pressurised system, it will try to return to the lowest possible energy state; this is usually to the conditions of the surrounding atmosphere. Leaks generally tend to be small, unless a major fault has developed, so a continuous process occurs. During process control, it may be necessary to reduce the pressure of a gas stream. A control valve can vary the rate of throttling to adjust the expansion ratio. An orifice can be used to provide a fixed degree of throttling. Figures 2.33, 2.34 and 2.35, show the expansion processes plotted on diagrams.

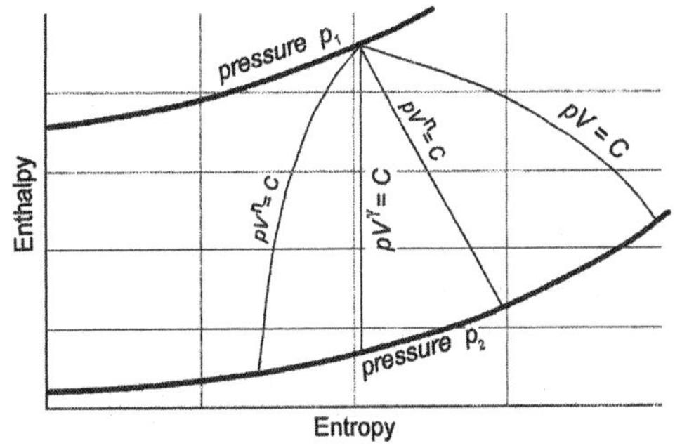

Figure 2.33 Expansion processes on a Mollier diagram

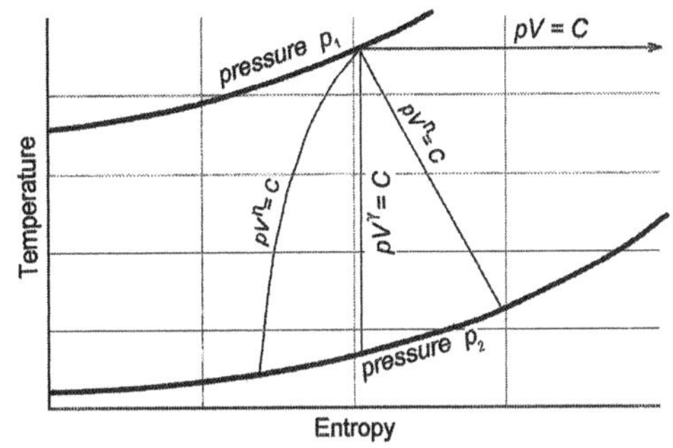

Figure 2.34 Expansion processes on a temperature/enthalpy diagram

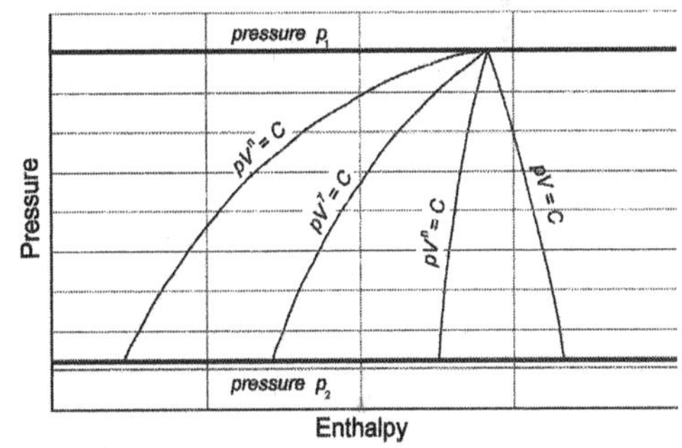

Figure 2.35 Expansion processes on a pressure/enthalpy diagram

Throttling is, theoretically, a process in which there is no heat transfer and so is adiabatic. Due to the high velocities involved, frictional heating occurs leading to an increase in entropy. Throttling is not an isentropic process but, in some cases, may be approximated as such, without inducing major errors. If testing is possible, with accurate temperature measurement, an isentropic efficiency can be evaluated leading to accurate future predictions;

$$\eta_{IS} = \frac{T_1 - T_2}{T_1 - T_{IS}} \qquad \text{Equ 2.62}$$

where T_{IS} is the temperature found from

$$T_{IS} = T_1 \left(\frac{p_2}{p_1}\right)^{\frac{\gamma-1}{\gamma}} \qquad \text{Equ 2.63}$$

Throttling is generally an inefficient process, in as much as the frictional heat generated is not usually recovered usefully. If large flows and/or large differential pressures are involved, then a more efficient expansion method should be evaluated. Power recovery turbines are becoming increasingly popular and the payback period is continuously shortening as energy costs increase.

2.6.9 Polytropic head

Remember that in a compression process, n can be less than or greater than γ. If the gas compression process is assumed to be a polytropic process, that is a process which follows the law:

$$pv^n = \text{constant} \qquad \text{Equ 2.64}$$

Then the specific compression work is given by the following equation:

$$H_p = \frac{Z_{av} R T_1}{m}\left(r^m - 1\right) \qquad \text{Equ 2.65}$$

where:

- H_p = specific compression work (kJ/kg)
- Z_{av} = average compressibility factor
- R = the gas constant $(kJ.kg^{-1}.K^{-1}) = \tilde{R}/M$ (M= molar mass (kg/kmol))
- T_1 = inlet temperature (absolute) (K)
- r = pressure ratio = P_2/P_1 (P_1 = inlet pressure (bara)) (P_2 = discharge pressure (bara))
- m = (n-1)/n (n = polytropic exponent)

The "specific compression work" is commonly referred as the "head" or, in the case of polytropic compression, the "polytropic head". The term "head" is analogous to the concept of head in pump technology and also comes from the units in the Imperial system, which are: "foot-pounds per pound", often shortened to "feet".

Strictly speaking, simplifying the units in this way is incorrect since the units are: "foot-pounds (force) per pound (mass)". However the term "head" is in common use.

2.6.10 Polytropic efficiency

In an ideal rotodynamic compressor the compression process would be isentropic, which means that it would obey the law:

$$pv^\gamma = \text{constant} \qquad \text{Equ 2.66}$$

where:

- γ = the isentropic exponent of the gas

The polytropic exponent is related to the isentropic exponent by the expression:

$$\frac{n-1}{n} = \frac{\gamma - 1}{\gamma \eta_p} \qquad \text{Equ 2.67}$$

where:

- η_p = polytropic efficiency

It is normally assumed that for a given compressor operating at a given speed, the polytropic efficiency is a function of the volume flow only.

This assumption is reasonably correct provided that the gas velocities in the compressor are significantly less than the speed of sound in the gas.

2.6.11 Power consumption

Compressor power consumption is given by the following equation:

$$G = w\frac{H_p}{\eta_p} + F \qquad \text{Equ 2.68}$$

where:

- G = compressor power consumption (kW)
- w = mass flow rate (kg/s)
- F = mechanical power losses (kW) including frictional losses etc.
- H_p = polytropic specific compression work (kJ/kg)
- η_p = polytropic efficiency

The mechanical losses F comprise all losses, the heat from friction which is not carried away by the compressed gas. Such losses include bearing and seal frictional losses and any losses in the mechanical power transmission. The heat from these losses will typically be carried away by the lubricating oil or by heat transfer to the surroundings. Internal losses where the consequent heat is carried away by the gas being compressed, will result in an increase in gas discharge temperature and will be allowed for in the compressor efficiency.

2.7 Useful references

Institution of Chemical Engineers (IChemE), Davis Building, 165-189 Railway Terrace, Rugby CV21 3HQ UK, Tel: 01788 578214, Fax: 01788 560833, www.icheme.org.

American Institute of Chemical Engineers (AIChE), 3 Park Avenue, New York, NY 10016-5991 USA, Tel: 212 591 8100, Fax: 212 591 8888, www.aiche.org.

Chemical Industries Association (CIA), Kings Buildings, Smith Square, London SW1P 3JJ UK, Tel: 020 7834 3399, Fax: 020 7834 4469, www.cia.org.uk.

UK Fire Protection Association, London Road, Moreton-in-Marsh, Gloucestershire GL56 0RH UK, Tel: 01608 812500, Fax: 01608 812501, www.nfpa.org.

National Fire Protection Agency (NFPA), 1 Batterymarch Park, Quincy, Massachusetts 02169-7471 USA, Tel: 617 770 3000, www.nfpa.org.

Health Protection Agency, 7th Floor, Holborn Gate, 330 High Holborn,London WC1V 7PP UK, Tel: 020 7759 2700, Fax: 020 7759 2733, www.hpa.org.uk

American Petroleum Institute (API), 1220 L Street, NW, Washington, DC 20005-4070 USA Tel: 202 682 8000, www.api.org.

American Society of Mechanical Engineers (ASME), Three Park Avenue, New York, NY 10016-5990 USA, Tel: 973 882 1167, www.asme.org.

CONCAWE, Boulevard du Souverain 165, B-1160 Brussels Belgium, Tel: 02 566 9160, Fax: 0 2 566 9181, www.concawe.be.

Energy Institute, 61 New Cavendish Street, London, W1G 7AR, UK,Tel: 020 7467 7100, Fax: 020 7255 1472, www.energyinst.org.uk.

American National Standards Institute (ANSI), 25 West 43rd Street, (between 5th and 6th Avenues), 4th floor, New York, NY 10036 USA, Tel: 212 642 4900, www.ansi.org.

Health and Safety Executive (HSE), Rose Court, 2 Southwark Bridge, London SE1 9HS UK, Tel: 0845 345 0055, www.hse.gov.uk.

United Nations Economic Commission for Europe (UNECE), Palais des Nations, CH - 1211 Geneva 10, Switzerland, Tel: 022 917 1234, Fax: 0 22 917 0505, www.unece.org.

ACGIH®, 1330 Kemper Meadow Drive Cincinnati, Ohio 45240, USA, Tel: 513 742 6163, Fax: 513 742 3355, www.acgih.org.

BSI British Standards, 389 Chiswick High Road, London W4 4AL UK, Tel: 020 8996 9000, Fax: 020 8996 7001, www.bsi-global.com.

CENELEC, 35, rue de Stassartstraat, B-1050 Brussels, Tel: 02 519 6871, Fax: 02 519 6919, www.cenelec.org

IEC, 3 rue de Varembé, PO Box 131, CH-1211 Geneva 20, Switzerland, Tel: 022 919 0211, Fax: 022 919 0300, www.iec.ch.

SAE 400 Commonwealth Drive, Warrendale, PA 15096-0001 USA, Tel: 724 776 4841, www.sae.org.

Gas Processors Suppliers Association, 6526 E. 60th Street, Tulsa, OK 74145 USA, Tel: 918 493 3872, www.gpsa.gasprocessors.com.

EN 50 014 Electrical apparatus for potentially explosive atmospheres- General requirements.

BS 5345 :1980 Code of Practice for Selection, Installation. & Maintenance of Electrical. Apparatus for Use in Potentially Explosive Atmospheres (save mining applications or explosive processing & manufacture).

BS 2869:2006 Fuel oils for agricultural, domestic and industrial engines and boilers. Specification.

ISO 3448:1992 Industrial liquid lubricants - ISO viscosity classification.

BS 4231:1992 Classification for viscosity grades of industrial liquid lubricants

BS 5908:1990 Code of practice for fire precautions in the chemical and allied industries.

ANSI B31.3 (Chemical Plant and Petroleum Refinery Code)

Process Plant Design and Operation, D Scott, F Crawley, Institution of Chemical Engineers, 1992, ISBN 0 85295 278 3.

Nomenclature for Hazard and Risk Assessment in the Process Industries, David Jones Institution of Chemical Engineers, ISBN: 085295297X.

Reciprocating Pump Valve Design, G Vetter, E Thiel, U Störk, TAMU Pump Show Conference paper, page 39, 1989.

Internal Flow Systems, DS Miller, Crane Company, USA , Technical Papers Nos. 409 and 410, 1988.

The Two K Method Predicts, William B Hooper, Chemical Engineering, August 24, 1981, pp 96-99.

Calculate Head Loss Caused by Change in Pipe Size, William B Hooper, Chemical Engineering, November 7, 1988, pp 89-92.

BS 2000-50:1993 Methods of test for petroleum and its products. Determination of cone penetration of lubricating grease.

ISO 2137:1985 Petroleum products - Lubricating grease and petrolatum - Determination of cone penetration.

ASTM G75-01 Standard Test Method for Determination of Slurry Abrasivity (Miller Number) and Slurry Abrasion Response of Materials (SAR Number).

Isolating valves 3

3.1 Isolating valves and the system

Isolating valves are fitted to systems for a variety of reasons. Most are probably fitted for maintenance or safety purposes. Equipment which is supported by installed standbys will have isolating valves to permit maintenance and inspection without shutting down the process and loss of production. Long instrument impulse lines may be fitted with primary isolation valves so that in the unlikely event of pipe failure the fluid source can be sealed to prevent further loss of product. Isolating valves are also often fitted at the blank end of lines or manifolds to allow future connection of additional equipment without shutting down operating processes. Figure 3.1 shows a typical rotodynamic pump installation with multiple pumps. Each pump has isolating valves. Additionally each pump outlet is fitted with a non-return valve to prevent reverse flow through stationary pumps with consequential recirculation back to the suction system.

Many small valves are installed as drain and vent valves. When equipment or pipelines are isolated there is frequently a need to drain the contents before any work is commenced. Draining can simply be the putting of dirty lube oil into a bucket or water on to open ground. Hazardous liquids must be drained into a proper enclosed drainage and treatment system or a closed container. In large piping systems drain valves are essential at all low points. When systems are prepared for production it is often necessary to remove any internal air.

Vent valves are provided at strategic locations. Venting and priming pumps are a typical example. Many pumps must be full of liquid prior to start up. Some pump designs are not self-venting and vent valves must be fitted to the pump casing. The vent valves are opened and the incoming liquid forces the air out. At the first sign of liquid at the valve it is closed. Automatic vent valves can be fitted in those situations when gas or vapour may accumulate regularly. Vent valves may be necessary at high points in pipework when gas/vapour cannot be carried away by the liquid flow. Some liquids contain dissolved or entrained gases which may require venting a regular intervals.

When choosing a valve for a particular purpose the valve type should be selected on the basis of operating conditions. Operating conditions are more than merely the normal operating pressure and temperature.

The following points should also be considered:

- How often will the valve be cycled from open to closed?
- How quickly will process conditions change?
- Will the valve temperature change significantly when wide open or closed tight?
- Will inspection of the valve be possible when operating?
- Will the seals need to be changed on-line?

The location of manual valves is very important. Good access is obviously important but sometimes overlooked. An isolating valve fitted for safety purposes is not very effective if scaffolding must be erected for access. Further points to consider:

- Valves can be very heavy. Those fitted in the centre of a span create pipe deflection and impose additional stresses on the pipework. Valves should be supported or be very close to a pipe support. If a good location is not possible then a review of the optional extras which can be fitted to alleviate the situation, should be made.
- Valves operate more effectively with uniform flow patterns. Placing a valve immediately after a short radius elbow is not a good idea. High wear rates may shorten the useful life of seats and seals and impose unacceptable maintenance costs.

3.2 Isolating valve design

Most valves are designed for pressure applications; that is to say, the internal pressure is higher, even if only slightly, than atmospheric pressure. Valves are designed to prevent the contents leaking out. Some are designed specifically for vacuum applications when precautions must be taken against the atmosphere leaking in. Not all packing arrangements are suitable for both conditions and specifications must be checked.

Valves are designed in a similar way to pressure vessels. The internal or external pressure produces a stress in the material. The designer decides on a suitable safe stress and adjusts the wall thickness to suit. Manufacturing techniques, such as casting, may mean the wall thickness must be thicker than necessary. The spare material can be used as a corrosion allowance. When a pressure vessel code is applied to the design the allowable working pressure may increase or decrease to accommodate the code allowable stress. When a valve is fitted with flanges the body should be designed to match the flange ratings. Pressure vessel codes and flange specifications are very useful because the pressure-temperature relationships are tabulated for many materials. When a valve has non-metallic components, such as seats and seals, these must also be considered when deciding extreme operating conditions.

Valves are designed for steady-state conditions. Temperature and pressure change slowly. If a valve is to be subject to rapid temperature changes thermal shock may occur causing cracking. If pressure is going to change rapidly the shock load may break components or spring joints. If the pressure is going to pulsate cyclically then fatigue may result. Operation under non steady-state conditions should be discussed with the designer.

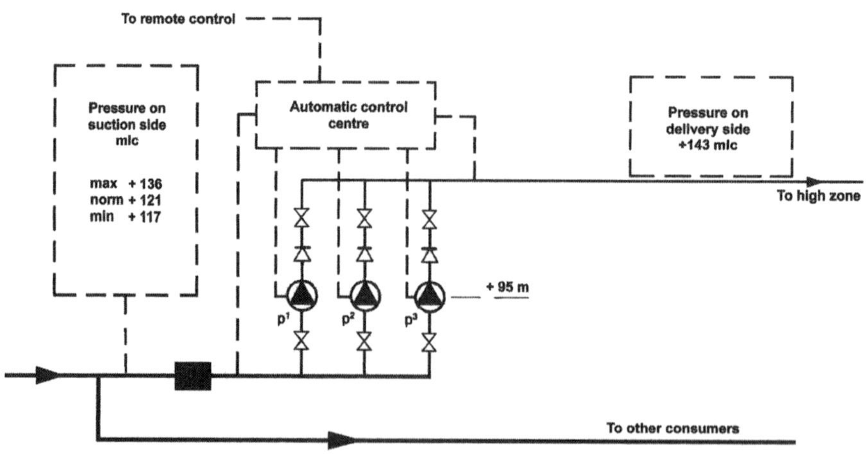

Figure 3.1 A typical rotodynamic pump installation

Valves can be designed to seal in both flow directions or only in one. Those only suitable for one direction will normally have an arrow on the body indicating the direction of flow. When sealing in both directions is essential, consult the specification. Some valves achieve sealing by some form of wedging action. When open, one or more components may be "loose" and easily rattled by vibration. This can be a serious problem with pulsating fluid flow from compressors and pumps. Not only positive displacement and reciprocating machines produce these effects. Some centrifugal pumps have flow variations which produce orifice meter differential pressure variations of greater than ±6%. Obviously this aspect of operating conditions must be considered when selecting a valve type.

Non-metallic materials are very useful for low pressure corrosive applications. However, great care must be taken if junctions between metallic and non-metallic systems are inevitable. Non-metallic components and cemented junctions can be very intolerant of pulsating flow described previously. Electrical continuity must be ensured. Anti-static precautions may be necessary.

Some valves are designed for specific applications. Some can be modified to possess specific qualities, such as hygienic cleanliness or NACE material compatibility for sour service. The user must understand the full application environment — design, installation, use and maintenance — if the extra qualities are to be realised in practice.

3.3 Isolating valve types

3.3.1 Linear motion valves

Most linear motion valves can have rising or non-rising stems. When a rising stem is fitted, a portion of the stem which is inside the valve or outside in the local environment, is moved to the packing area each time the valve is cycled. The fluid in the valve may not corrode the stem inside, without the presence of air, but may corrode in the packing when air is available. Conversely the stem may be attacked by the environment and not the product. Both situations lead to packing leakage problems when the stem is at extremes of its travel. A good stem material, or corrosion/wear resistant coating, may prove very worthwhile in the long-term. Screw threads can be cut or rolled. Rolled threads are stronger and have a better surface finish than cut threads. Rolled threads are less prone to shearing and require less friction torque. Specifying rolled threads can also be a long-term benefit.

Parallel gate valve

The parallel gate valve is manufactured in three gate designs which have different characteristics. The gate is usually made in two halves which are spring-loaded against the seats. One design relies on the initial spring load plus the line pressure to create the seal. This type of valve can only seal on one seat. The other valve style includes a wedge between the spring-loaded halves, see Figure 3.2. The initial seat force is provided by the springs. As the stem is screwed in further a wedge forces the halves apart and increases the seat force on both seats. This type of valve seals both seats simultaneously. The gate can be made solid, see the later description of compact parallel gate valves.

Figure 3.2 shows a parallel gate valve with wedged gate discs. The valve has a bolted bonnet and bolted gland. It can be difficult to maintain even-loading and alignment on two bolt glands and good designs have a gland with a self-aligning follower. Phosphor bronze gland nuts for hot applications are much better than steel ones. The valve shown has a separate yoke. The yoke is clamped to the bonnet. Separate yokes make it much easier to repack the box. Valves for outdoor applications should also have lubrication facilities for the handwheel nut.

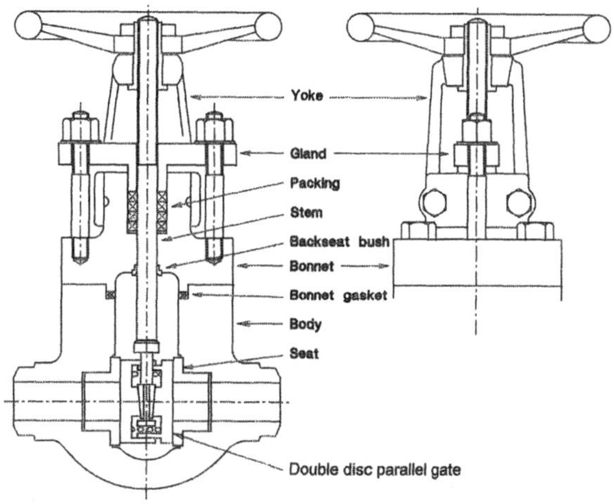

Figure 3.2 A high pressure parallel gate valve

Parallel gate valves are suitable for clean fluids. Entrained solids increase wear and leakage results. The spring-loaded version is suitable for applications where significant temperature changes occur and where valve body distortion may be a problem. As only one seat is loaded when the valve is closed the gate cannot be jammed. The wedged version can be jammed closed by temperature changes. Because both seats are sealed, significant temperature increases may create high pressures in the valve cavity. Manufacturers should be consulted about cavity relief.

Parallel gate valves have an unobstructed bore when wide open. Low friction losses and high C_v values result. Some valves may have reduced ports which will make "pigging" impractical, so ensure sizes are checked before purchasing. Flow regulation is only possible by inducing very high local velocities which can result in rapid wear and erosion. These valves should only be used for isolation. Gate valves are slow to operate manually, resulting in no water hammer problems.

Figure 3.3 shows a small parallel gate valve with bolted bonnet and bolted gland.

Parallel gate valves are very popular for power generation applications. Cast and forged steel and alloy steel valves are available in many sizes, see Table 3.1. Valves larger than DN40 for pressures over ANSI Class 900lb may have bonnets secured by shear rings.

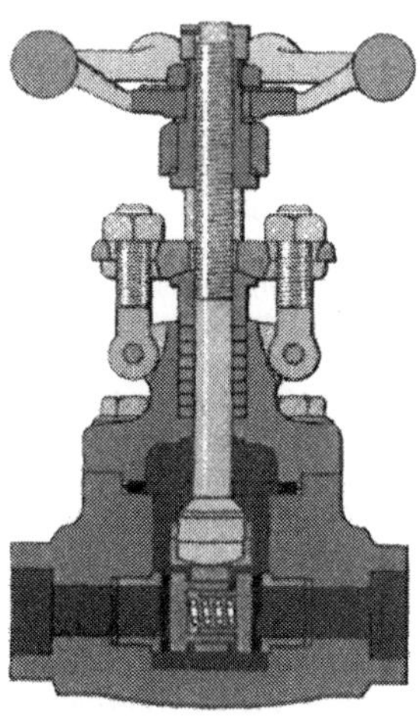

Figure 3.3 A typical small parallel gate valve
Courtesy of Dresser-Masoneilan

Body and bonnet material	Pressure rating	Size	Connections
Cast carbon steel	ANSI 150lb	DN50 to DN500	Flanged Butt weld
	ANSI 300lb	DN50 to DN500	
	ANSI 600lb	DN50 to DN500	
Cast carbon steel	ANSI 600lb	DN15 to DN50	Flanged Butt weld Spocket weld
Cast carbon steel Cast alloy steel	ANSI 900lb	DN125 to DN600	Butt weld
	ANSI 1 500lb	DN125 to DN600	
	ANSI 2 500lb	DN125 to DN600	
	500 barg	DN125 to DN600	
Forged carbon steel Forged alloy steel	ANSI 900lb	DN50 to 100	Butt weld
	ANSI 1 500lb	DN50 to 100	
	ANSI 2 500lb	DN50 to 100	
	500 barg	DN50 to 100	
Forged carbon steel Forged alloy steel	ANSI 2 500lb 500 barg	DN15 to DN40 DN15 to DN40	Flanged Butt weld

Table 3.1 Popular parallel gate valves

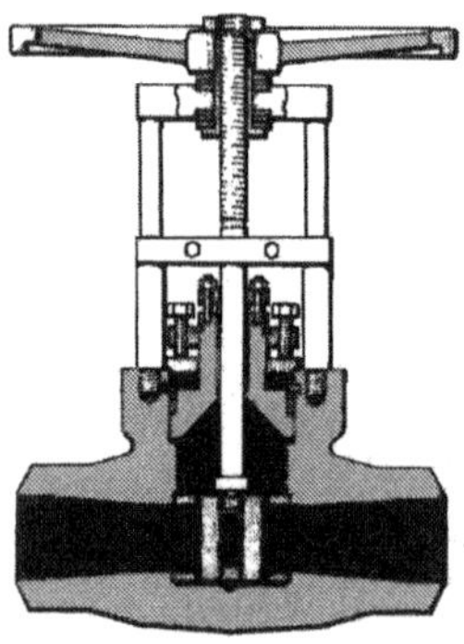

Figure 3.4 A high pressure high temperature parallel gate valve
Courtesy of Dresser-Masoneilan

Figure 3.4 shows a valve design for up to DN100 with a bonnet secured by shear rings.

Notice that the yoke columns are fabricated from turned bar, rather than cast and become guide pillars, used to guide the stem and prevent rotation.

Parallel gate valves can be built as compact shorter versions to save space. These valves are sometimes described as "wafer gate valves". In this context "wafer" is used to mean "thin". These valves are not clamped between two pipe flanges as the name would normally imply. Valve length is reduced by incorporating the flange facings directly into the body, see Figure 3.5.

The flanges are facings only, studbolts with two nuts or bolts and nuts cannot be used. The valve facing can be fitted with studs in those installations where the valve body is unlikely to be removed. Alternatively bolts must be used. When bolting is used, for valves which may be removed routinely, thread inserts in the valve body are a sound investment. The body can have a bottom clean out cover. The gate is solid. Sealing is effected by spring-loaded seats. The seats may have elastomer inserts and "O" ring seals. Cast carbon steel body and bonnets are standard for low pressure ANSI 150lb valves with 11/13 Cr trim. Gates are usually plated carbon steel. Sizes range from DN100 to DN1 200, and larger valves can can also be built.

Wedge gate valve

The wedge gate valve is one of the first choices of isolating valve, with good sealing capabilities and low fluid friction losses. Leakage rates should be of the order of 10 ml/hr/25 mm seat bore for metal-to-metal seals. The wedge is effectively clamped between two full-bore seats and seals on both sides.

Figure 3.5 A very large wafer gate valve

The taper of the wedge can compensate for wear on both the gate and the seats. Low pressure valves up to DN350 will be full-bore but some larger sizes may have a slight bore reduction. Check the design detail carefully if "pigging" is required. The wedge gate must be lifted clear of the seat bores to open the valve fully. With manual valves, this can take some time. Water hammer is not usually a problem. Wedge gate valves can be suitable for liquid-solid mixtures. Problems can be experienced when closing the valve if sediment accumulates in the bottom between the seats.

The wedge gate is available in three different styles each with slightly different characteristics. The styles are:

- solid wedge
- flexible wedge
- double disc wedge

The solid wedge is popular and is easy to manufacture in small sizes. Valves from ¼ ", in brass and bronze, are used in many

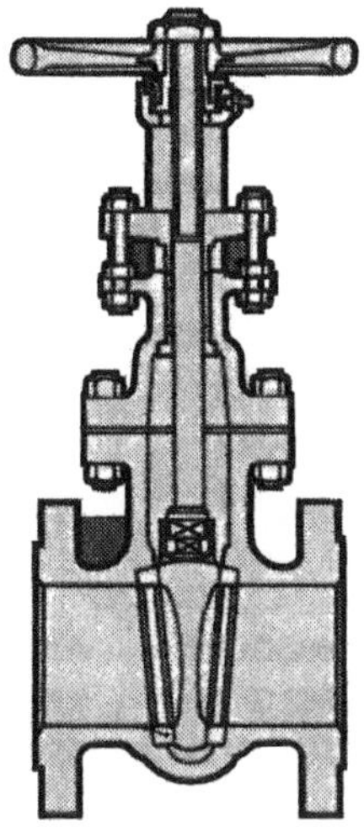

Figure 3.6 Section through a high pressure solid wedge gate valve

water applications. Cast steel valves are available in sizes from DN50. Forged steel valves can be as small as ¼". When closed the solid wedge can become jammed between the seats if there is a significant drop in temperature or the valve body is distorted by pipe forces and moments. Figure 3.6 shows a high pressure solid wedge valve with handwheel.

The flexible wedge is designed to overcome these problems. The wedge is made in one piece but the sealing faces are sprung on cantilevers from the centre. When the valve is closed it is the spring load which is applied to the seat faces, not the wedge force directly. Thermal contraction or body distortion increases the spring load without jamming the gate. Some flexible wedge gates can be fitted with PTFE inserts to enhance sealing but operating temperatures are limited.

Valves with solid wedges may have flexible wedge gates as an option. The double-disc design or expanding wedge gate valve has a gate made up of two separate, but connected discs. The discs are aligned but allowed some angular freedom. When the discs are in position over the seats, an inter-spaced central spherically-mounted wedge attached to the stem forces the discs apart on to the seats. The spherical mounting allows the discs to comply with the seat faces and clearances assure balanced seating forces. This valve design seals on both the upstream and downstream seats, see Figure 3.7, which shows the internal construction and method of operation. The central valve body cavity can be fitted with a drain to verify sealing integrity. For some applications the discs can be coated with an elastomer to improve sealing and reduce corrosion. Seat faces can be hard faced to reduce wear, corrosion, galling and friction. For applications with significant entrained solids, purge connections can be fitted to the seats for cleaning prior to opening or closing.

Figure 3.7 Construction and operation of expanding wedge gate valves
Courtesy of YPS Valves Ltd

The double-disc design cures two other problems associated with solid wedge designs. If unexpected forces and moments are applied to the valve body, attempting distortion, the gate can still be freed because the seat sealing loads are removed before gate withdrawal is attempted. The independent discs can follow their respective seats if the valve body distorts slightly. Temperature changes, while the valve is closed, do not cause the gate to jam solid. This valve design is available in sizes from DN50 to DN600, 2" to 24" nb, with pressure ratings up to 100 barg on 600 lb flanges. Body materials include steel, stainless steel, duplex stainless steel and nickel alloys to cope with a temperature range of -200 to 816°C. The bonnet can be extended to permit near ambient operating conditions for the stem seal and actuator, which can be powered.

All styles of wedge gate valves are prone to "chattering" when used for flow regulating or throttling. The wedge is loosely guided by cast passages in the body when not in contact with the seats. Both the wedge and seats can wear or be damaged. The diameter and unsupported length of the stem, plus the design of the wedge attachment to the stem, determine the valve's susceptibility to vibration. Wedge gate valves are recommended for isolation only.

The wedge gate lifts out of the seat bores to open the valve. The upper portion of the valve body must be wide enough to accommodate the gate. This means the bonnet attachment can be almost elliptical. Some designs use circular bonnets which makes the valve body longer than strictly necessary but greatly improves the access to the internals. Bolted bonnets are standard on steel valves rated for 16 barg and over. Small brass and bronze valves, for 10 barg applications, will have screwed bonnets and integral seats. Carbon steel valves generally have screwed 11/13 Cr seats and some designs include "O" ring seals. Integral seats may be optional for valves in applications where crevice corrosion can cause problems. Suppliers should be consulted about facilities to maintain the seat faces.

Seat designs can be modified for fire-safe applications. Valve bodies can be partially lined by fitting long seat inserts through the connections. The location of the bonnet on the body is critical for the gate to be centrally placed between the seats. If the gate is offset axially, the stem will be bent when the valve is closed, possibly leading to fatigue. A good, positive bonnet location is essential for long trouble-free operation; a spigot is probably best for low pressure valves, a ring-type joint performing a double function on high pressure valves. A spiral-wound gasket under the spigot is a good method of sealing low pressure bonnets. Large valves may have a clamped bonnet and so alignment is assured. The bonnet can be extended for cold and hot applications.

The bottom of the packing box should be fitted with a backseat bush of reasonable length. This provides stem guidance and alignment. If it is too short, rapid wear can occur if the valve is operated often. The bottom of the stem should form a good seal with the backseat bush to allow repacking with the valve on-line. Both components may be described as "stainless steel" when they are made of 11/13 Cr, AISI 410. Ensure that corrosion will not be a problem or else the back seal will be ineffective when required. Small brass and bronze valves will have screwed glands. Steel valves are generally bolted. A form of graphite packing is the popular standard. Standard valves can have the packing box extended for vacuum applications. Most valves are of the outside screw, rising stem type. Depending upon the pressure rating, valves over DN200 may have geared handwheels; valves over DN350 will always have geared handwheels.

Brass/bronze is popular for small low pressure water applications. Cast and forged carbon steel are popular for larger installations. Valves with closed die-forged bodies may have reduced ports. Some forged body valves have flanges attached by butt welding or friction welding; check the quality assurance on the joint. Forged body valves can have welded bonnets for increased integrity. The weld can be difficult to inspect so integrity may not be verified.

A few manufacturers make cast iron valves for PN16 ratings. These valves feature an inside screw/non-rising stem with gunmetal seats and gunmetal inserts in the cast iron wedge. Aimed at waterworks applications, the valves can operate between -10 to 220°C. Table 3.2 indicates the sizes and ratings of some popular wedge gate valves.

Body and bonnet material	Pressure rating	Size	Connections
Cast carbon steel	ANSI 150lb	DN 50 to DN900	Flanged Butt weld
Cast low temp steel	ANSI 300lb	DN 50 to DN750	
Cast 1.25% Cr	ANSI 600lb	DN 50 to DN600	
Cast 5% Cr	ANSI 900lb	DN 50 to DN300	
Cast 18% Cr 10% Ni	ANSI 1 500lb	DN 50 to DN300	
Forged carbon steel	ANSI 150lb	DN10 to DN40	Flanged
	ANSI 300lb	DN10 to DN40	
	ANSI 600lb	DN10 to DN40	
	ANSI 1 500lb	DN10 to DN25	
Forged carbon steel	ANSI 800lb	DN10 to DN 50(1)	Screwed
		DN10 to DN50	Socket weld
	ANSI 1 500lb	DN10 to DN50	
	ANSI 2 500lb	DN10 to DN25	

(1) Bellows sealed packing box with butt weld connections option

Table 3.2 Popular wedge gate valves

Because the wedge gate valve seals effectively on both seats it can be used for process "double block and bleed" applications; the soft-seated versions proving very efficient. The body cavity can be bled at the bottom, like a drain, or through the bonnet. The bleeding can be automatic, triggered by the valve closure, or manual. Standard valves, from ANSI 150lb to 1 500lb, can be adapted.

A special version of the wedge gate valve is available in small sizes for use in instrument impulse lines and similar applications. Sometimes called a plug valve it is better described as a circular wedge valve. It is a linear valve and looks similar to a needle valve but the ports are in-line allowing the valve to be cleaned with rods if necessary. The circular wedge is attached to the stem and does not rotate as it is pressed into the seat. The rising stem is sealed by a screwed gland with PTFE packing. The seat may be metallic or non-metallic and replaceable. Barstock valves, with reduced ports, in stainless steel with Delrin™ seats for pressures up to 414 barg at temperatures between -29 to 121°C are made in screwed pipe sizes from ¼" to ½" female and ¼" to ¾" male. Socket and butt weld ends are optional.

Knife gate valve

Knife gate valves are a derivative of the parallel gate valve. The thin gate is completely smooth and positively guided and sealed by elastomers or flexible metal seals mounted in the body. The guides and seals are circular to match the bore of the valve. The bottom of the gate seals along the edge on an elastomer or sprung insert. The bottom seal is flush with the valve body to eliminate pockets where solids could collect. The seals are spaced axially to provide positive interference, with the gate providing a bubble tight seal under all operating conditions. Sealing in both directions is standard. Deflection cones can be fitted to some designs to direct solids away from the seals.

The knife gate valve is different from other gate valves in that the gate is not totally enclosed within the body, see Figure 3.8. The gate lifts out of the body as the valve opens. The packing box is not circular but rectangular. The packing seals the gate itself against the body. Glands cannot be of screwed or union types and are always bolted. Complex packing arrangements are difficult, if not impossible, to fit and these valves are only suitable for products which are not hazardous.

Figure 3.8 Typical knife gate valve

The gate is usually thin which allows the valve body to be slim. The cast body is similar in style to a lugged body. Bolting cannot pass through all the lugs because the gate rises through the top of the body. In order to make all the fasteners identical the lugs are tapped. The body is so slim that there is no space between the lugs for nuts. The body slimness increases the ruggedness.

Because the bore of the valve is smooth and full diameter, the fluid friction losses incurred are low resulting in high C_v values. The valves are ideal for very viscous materials and fluids carrying solids. The gate is designed to cut through pastes, wax, fibres and pulps and to seal effectively. The gate is completely surrounded by seals and does not touch the valve body. Distortion of the body, caused by extreme forces and moments, may tighten the grip of the seals but will not cause the gate to seize.

The gate valve is not intended for flow regulation. Pressure losses across the valve can only be increased by creating high velocities. Some gate valves can "chatter" when opening/closing due to fluid forces moving the gate. The knife gate valve is effectively damped by the seals. Valves are of the rising screw design, generally with outside screw. Valve opening and closing is slow, unless actuated, and water hammer effects should not be a problem.

Knife gate valves are mass-produced in sizes from 2" to 24" although valves up to 72" are not uncommon. Popular body materials include cast iron, cast steel and stainless steel. Cast iron lined with austenitic stainless steel is possible for sizes between 2" and 24". Valves with fabricated steel bodies are made in sizes from 2" to 42". Exotic alloys, such as Alloy 20, Hastelloy B™ and Hastelloy C™, are also available. Valves can be fully lined with an elastomer to reduce corrosion and erosion. These valves may have a shaped knife edge and a preferred flow direction. The gate is generally of stainless steel. The yoke is generally cast or fabricated in steel although stainless steel may be an option. A bronze bearing may be fitted.

Seat seals include: 316, AISI 317, 17-4PH, Neoprene™, ABS, Hypalon™, Viton™, EPDM.

Glands can be of ductile iron, carbon steel or stainless steel. Ductile iron and carbon steel can be plastic coated. Valves can be fitted with standard flange bolt patterns but pressure limits may be much less than the nominal flange rating. Valves with ANSI 150lb flange facings can be limited to 3.5 to 10 barg. Differential pressures for solids handling, slurries and pulp stock as well as dry solids, may be restricted to 50% of the body pressure rating.

Open frame versions of knife gate valves are manufactured in large sizes to be incorporated into concrete tanks and pits. The pressure ratings are usually only slightly above atmospheric pressure. This type of valve is also called a "penstock valve".

Conduit gate valve

Conduit gate valves are characterised by the smooth regular bore presented to the fluid when the valve is wide open and an absence of pockets or crevices in which solids or sediment can accumulate. They are a derivative of the parallel gate valve with extra features for more arduous applications. Because the flow passage is full-bore and smooth, the fluid friction losses are very low with consequently high C_v values. The smooth full-bore ports of these valves do not lend themselves to flow regulating applications without creating very high velocities. Conduit gate valves are not recommended for flow regulation or throttling, but are suitable for solids handling applications. These valves can be cleaned mechanically and "pigged". Conduit gate valves are also called "ported gate valves" and "conduit pipeline valves".

Figure 3.9 Typical conduit gate valve
Courtesy of Tyco Flow Control

The conduit gate valve is unlike other valves in that it does not simply uncover a seat and allow fluid to flow. It is however, similar in principle to the line blind valve described in Section 3.3.5. The solid portion of gate sealing the seats is replaced by a portion of gate with a hole the same size as the seat bore. The seats are always in contact with the gate. To accomplish constant seat contact, the gate must be much longer than normal and this can be appreciated by the size and shape of the body, see Figure 3.9.

The gate is about twice as long as normal. When the valve is closed the portion of the gate containing the flow port is lowered below the pipe centre-line. An extension to the body is fitted to cater for this. In large valves the body may be cast in three pieces — a central section and an upper and lower section bolted on. Valves generally have rising stems. Manual operation is slow so water hammer effects should not be noticeable.

The gate is plain and smooth and is guided solely by the seats. Replaceable seats are standard and can be metallic or non-metallic. Metal seats can have elastomer inserts, such as PTFE, to aid sealing and reduce friction. Metal seats can be loaded against the gate by coil springs, disc springs or the "O" rings used to prevent leakage around the seat. Gate "chatter" should not be apparent when the spring loading is correct. As the seats are spring-loaded against the gate there is no tendency for the gate to bind due to temperature changes. On opening, the plain portion of the gate is lifted into the upper part of the valve body allowing the bore in the gate to line up with the seats. Bodies are normally cast and have flanged or butt weld connections. A drainage connection is usually fitted in the bottom section.

Special designs are made with the gate built in three pieces. Figure 3.10 shows such a valve for high temperature, high pressure power generation applications.

Standard materials are cast steel with 11/13 Cr trim. Seats are 11/13 Cr with PTFE inserts. The gate may be chrome plated. Valves larger than DN250 may be fitted with geared handwheels depending upon the pressure rating. Valves to ANSI pressure ratings 150lb, 300lb and 600lb are available in sizes from DN80 up to DN900 as standard. Larger sizes can be supplied to order. Valves from DN125 to DN600 can be suitable for pressures up to 500 barg.

Low pressure versions are available where the gate withdraws through the packing box in a similar manner to knife gate valves

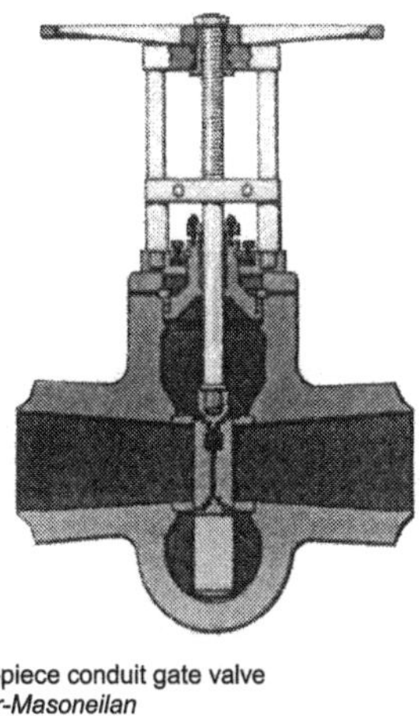

Figure 3.10 A three-piece conduit gate valve
Courtesy of Dresser-Masoneilan

Figure 3.11 A low pressure conduit gate valve

but the bottom of the gate is not semicircular to seal on the seat, see Figure 3.11. The valves are fitted with elastomer linings which also act as the seat and flange gasket. These valves, rated for 10 barg, are available from DN50 to DN900. Standard body materials include ductile iron and AISI 316 austenitic stainless steel.

Lining elastomers include: natural rubber, Buna N™, EPDM, Neoprene™ and chloro-butyl rubber.

The flange facings are machined directly on the ends of the body. Through-studbolts are possible in some holes, others are tapped for bolts or studs.

Small flanged valves, DN15 to DN50, are available with ANSI pressure ratings of 150lb up 2 500lb. High pressure versions are available for isolation in connection with oil exploration and production, see Figure 3.12. Raw crude oil may contain corrosive ingredients, such as CO_2, H_2S and HCl, together with erosive agents like sand. A smooth flow passage eliminates corners and edges where erosion can commence. The gate is totally enclosed within the valve body. Valve bodies and bonnets are closed die forgings in high tensile steel and full stainless steel versions are also available. The popular stainless steels, AISI 304 or 316, are not strong enough to replace high tensile carbon steel. A close inspection of specifications shows the replacement material to be 11/13 Cr. This material is not stainless but stain-resistant or rust-resistant. The versions with 4% nickel have better corrosion resistance but not equivalent to the austenitic stainless steels.

Bolted bonnets are standard but clamped versions may be optional. The bonnet flange may be sealed with a trapped spiral wound gasket or a metal seal ring similar to those used with clamped connections. The replaceable seats may be of surface

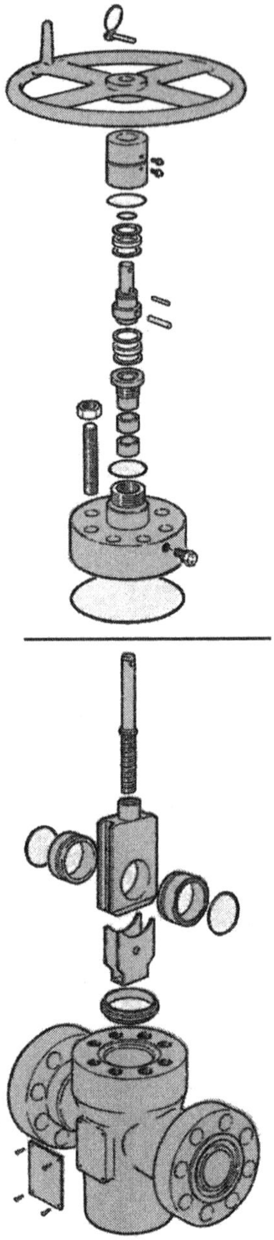

Figure 3.12 High pressure forged conduit gate valve
Courtesy of Anson Ltd

hardened carbon steel or treated hardened stainless steel. Hard facing could be optional and tungsten carbide coatings have proved successful as well. The gate can be made of the same materials as the seat.

Galling is generally not a problem at the low speeds encountered. Valves will generally be of the inside screw type with a non-rising stem. The screw should be totally enclosed in the gate in a pocket which can be used as a lubricant reservoir. The nut in the gate should be renewable.

Because of the high axial loads on the stem, due to the high operating pressures, ball or roller bearings may be used as thrust bearings. The high loads encountered require the use of a relatively fine pitch screw; between 12 to 60 turns may be required from "open" to "closed". The gland may be screwed or bolted with back-up "O" ring seals. The packing box should be deep

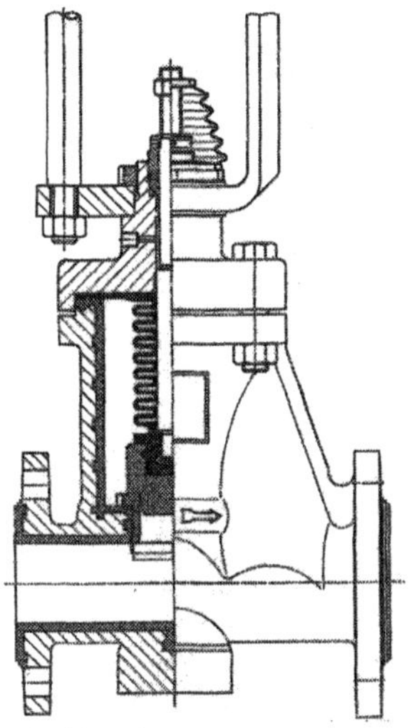

Figure 3.13 A lined-body globe valve
Courtesy of Richter Chemie-Technik GmbH

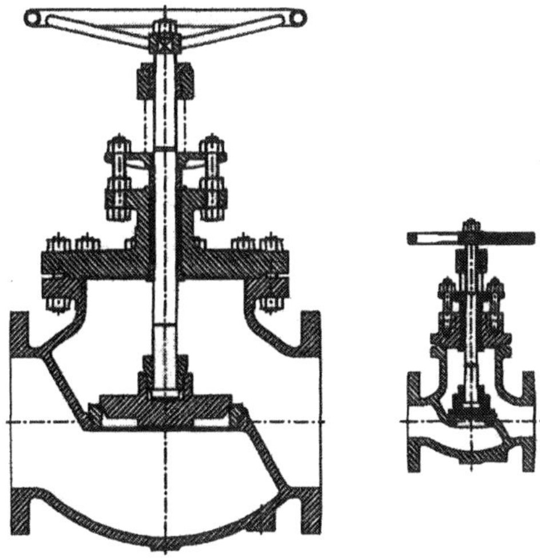

Figure 3.14 A typical globe valve construction
Courtesy of H H Valves Ltd

enough for a lantern ring with a connection for lubrication or venting.

Valves are available with various connections including API flanges, proprietary screwed and clamped fittings, as well as butt welding. Standard sizes of 1", 2", 2½", 3" and 4" are made for pressures up to 689 barg.

Straight metal globe valves can be lined with corrosion resistant material to permit a wider range of fluid applications. The metal body and bonnet provides the strength and integrity to contain the pressurised fluid. Body materials include ductile cast iron and steel and these can be coated with PFA (per fluoro alkoxy), FEP (fluorinated ethylene polypropylene) or PVDF (poly vinyl idene fluoride). The internal coating is usually 2 to 6 mm thick and moulded to the body profile. The valve body can be coated externally with protective epoxy, or, in extreme cases, the valve body and bonnet can be made of stainless steel. For applications with hazardous fluids, these valves can have a PTFE bellows sealed bonnet to eliminate fugitive emissions. Valve sizes between DN25 to DN100, with pressure ratings up to 16 bar(g) can cope with operating temperatures between zero and 180°C. Figure 3.13 shows the general construction with a bellows sealed bonnet.

Straight globe valve

These globe valves, with flat discs and seats, are sometimes called "screw down stop valves". Straight globe valves are called "straight" because the connections are in line. However the flow path is far from straight and is fairly tortuous. The flow must change direction through 90° in order to pass through the seat and then another 90° to return to the original direction. The shape and area of the passages in cast valves varies considerably with valve size and pressure rating. A typical construction is shown in Figure 3.14.

Valves with closed die-forged bodies or machined from barstock can have severe limitations placed on the port sizes. Port machining is only possible through the connections and the bonnet opening. Inlet and outlet passages must be drilled at an angle through the connection openings. The small diameters possible increase the fluid flow losses. Additionally, the sharp corners formed by manufacture may induce cavitation in liquid applications.

Some valves are described as "full-bore" and "reduced port" but this is irrelevant. Check fluid losses on a case by case basis. Globe valves can be used for flow regulation as well as isolation but this extra facility is at the cost of increased pressure drop. Flow losses are higher than gate valves but sealing can be better. Because of the internal passage construction, globe valves cannot be cleaned with rods. They are also not recommended for solids handling applications or dirty fluids. Globe valves should only be operated with flow passing up through the seat. Flow in the other direction will create much higher flow losses.

Straight globe valves can have cast or closed die-forged bodies or be machined from barstock. Some forged valves have flanges attached by friction or butt welding, so check weld quality assurance. The seat is at right angles to the centre-line of the connections so that all assembly can be performed through the bonnet opening. Seats are usually screwed in. A special key is required to engage with flats in the seat bore.

The disc can be one of two types — flat or crow-foot guided with shaped ports. The flat disc type has lower pressure losses but is not as good for flow regulation. The flat disc creates high velocities across the full seat periphery when throttling. The crow-foot guided disc divides the flow stream into multiple small streams with consequently higher losses at lower velocities. The crow-foot guided disc is bottom guided, as well as top guided from the stem, and is a more rigid assembly. The flat disc tends to deflect the stem away from the seat centre due to fluid circulating from the inlet side to the outlet. This circulation is worst when the valve is throttling, creating high velocities.

Guided discs can be fitted with an isolation seating which is separate from the throttling portion. The disc should be attached to the stem with a self-aligning connection so that closing forces are applied to the seat evenly. The disc does not rotate against the seat and there are no galling tendencies between similar materials. Small brass and bronze globe valves may have an integral hemispherical tip on the stem and no disc. Some steel valves may have an option for a needle-shaped disc for better regulation characteristics.

Steel valves will generally have a bolted bonnet with low pressure bonnets located by a spigot and high pressure ones by a ring-type joint. Special valves may have a clamped connection. A substantial backseat bush is necessary to provide good guidance for the stem. Some valves can achieve a good seal between the backseat bush and the stem when wide open to allow repacking the box on-line. Bolted glands are common with reinforced graphite packing. Live-loaded glands, using disc spring stacks, may be optional. Outside screw types with rising stems are standard. Geared handwheels may be standard over DN100 depending upon the pressure rating.

Brass and bronze valves, with screwed bonnets/glands and connections, suitable for 10 barg are used extensively for clean water applications. Popular cast and forged steel straight globe valve ranges are listed in Table 3.3.

Body and bonnet material	Pressure rating	Size	Connections
Cast carbon steel	ANSI 150lb	DN50 to DN250	Flanged Butt weld
	ANSI 300lb	DN10 to DN200	
	ANSI 600lb	DN10 to DN150	
	ANSI 900lb	DN10 to DN150	
	ANSI 1 500lb	DN10 to DN150	
Forged carbon steel	ANSI 150lb	DN10 to DN40	Flanged Butt weld
	ANSI 300lb	DN10 to DN40	
	ANSI 600lb	DN10 to DN50	
	ANSI 900lb	DN10 to DN50(1)	
	ANSI 1 500lb	DN10 to DN25	
Forged carbon steel	ANSI 800lb	DN10 to DN50(1) DN10 to DN50	Screwed Socket weld
	ANSI 900lb (454 °C max)	DN10 to DN50	
	ANSI 1 500lb	DN10 to DN50	
	ANSI 2 500lb	DN10 to DN25	
Forged stainless steel	ANSI 2 500lb	DN10 to DN40(2)	Butt weld

(1) Welded bonnet (2) Extended cooled bonnet for 800°C operation

Table 3.3 Popular cast and forged steel straight globe valve sizes and ratings

Larger valves are available, and alternative materials can be used. The standard material trim is 11/13 Cr but some valves specifically intended for continuous regulation, such as those for balancing circuits, can have 17 Cr precipitation-hardened seats and disc. The best versions of these valves will have skirt-guided or crow-foot guided discs.

Small barstock globe valves can be over-centre toggle operated, see Figure 3.15. The stem is spring-loaded from the bonnet to close the valve. Unidirectional operation is specified with the incoming fluid entering under the flat plug which is fitted with a trapped PTFE soft seal. Shut-off capabilities are limited by the spring strength. Maximum pressure is only 20 barg. There is no packing fitted. The plug is sealed in the burnished bonnet by a Viton™ "O" ring. Toggle operated valves can be bellows sealed.

Slightly larger globe valves with hardened stainless steel ball tips for pressures up to 414 barg are made in sizes from $\frac{1}{8}$" to $\frac{1}{2}$" NPT, $\frac{1}{4}$" to $\frac{3}{4}$" od compression fitting $\frac{1}{4}$" to $\frac{1}{2}$" nb socket weld. Small valves can utilise metal diaphragms to seal the bonnet. Figure 3.16 shows a straight globe valve with bellows sealed bonnet.

Figure 3.15 Barstock globe valve with compression fittings
Courtesy of Swagelok Company

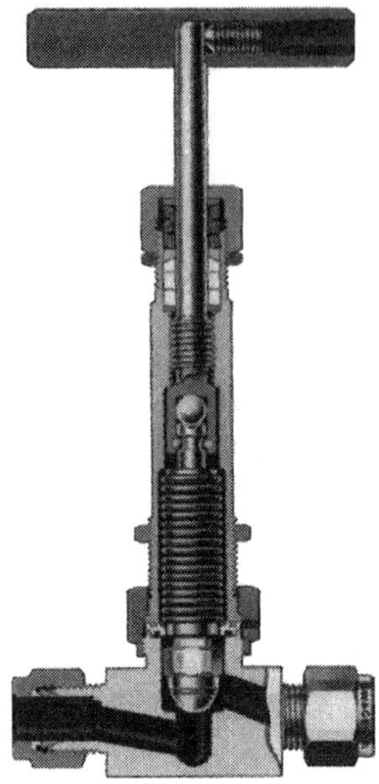

Figure 3.16 Straight globe valve with bellows sealed bonnet
Courtesy of Swagelok Company

High pressure globe valves, compatible with $\frac{1}{4}$", $\frac{3}{8}$" and $\frac{9}{16}$" od heavy wall tube, are available for small scale extreme pressure applications to 3100 barg. The valve body is machined from a solid block of wrought stainless steel. The angled seat is machined directly in the body. The inlet and outlet screwed connections are in-line but the internal reduced bore ports are drilled at an angle. The non-rotating stem is in hardened stainless steel and may have a Stellite™ tip. The PTFE or Grafoil™ packing is compressed by the phosphor bronze screwed bonnet via a shaped follower. The handwheel stem outside screw is located in the bonnet bore. Operation at up to 454°C is possible but at reduced pressure. Various connectors can be fitted to suit the application.

Figure 3.17 shows the valve assembly, using clamped connections, designed specifically to allow in-process cleaning, without disassembly, and to minimise spoil accumulation. To facilitate the long stroke necessary, these valves are actuated by pneumatic cylinders, piston actuators.

Angle globe valve

The angle globe valve, also called a "crown valve", is a very close relative of the straight globe valve. The angle valve was probably developed first and the straight version introduced later to include all the best features of an in-line valve. The straight version is more popular but the angled valve has some distinct advantages. Because of the close similarities, the preceding Section on the straight globe valve should be read first.

The angle globe valve allows the flow to change direction by 90°, see Figure 3.18. Flow always enters under the seat. The flow path is much more open and less tortuous than the straight version with consequently less pressure drop. The valve is slightly less prone to body erosion by solids but clean fluids are preferred. Disc options include crow-foot and skirt-guided for better regulation characteristics. The valve body will be subjected to fluid reaction forces due to the change in flow direction. These forces will normally be insignificant but as valve size and fluid density increase so will the magnitude of the forces.

Small screwed valves in brass and bronze are used extensively in clean water applications. Most valves for industrial use are made in cast steel with bolted bonnets. Bronze, stainless steel and duplex materials are readily available. Trim in 11/13 Cr is standard but hard facing is regularly applied to combat erosion.

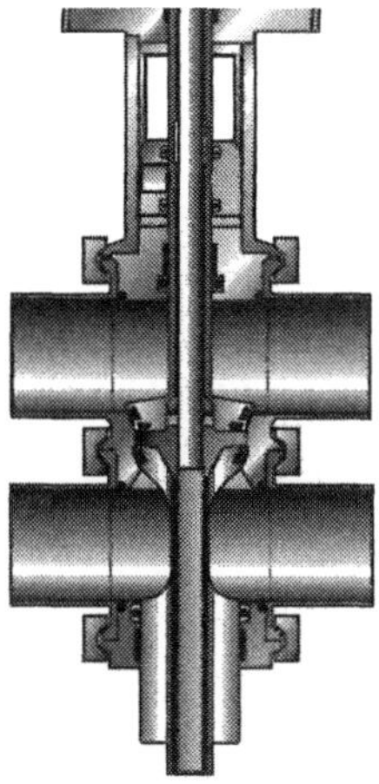

Figure 3.17 Mix-proof straight globe valves with CIP facilities
Courtesy of Waukesha Cherry-Burrell

Popular sizes are from DN50 to DN250 with ratings for BS 10 Table E and F, ANSI 150lb, 300lb and 800lb. Temperatures up to 425°C are common. Valves for higher pressures can be fitted with a balance piston to reduce the axial thrust on the stem.

Small barstock globe valves can be over-centre toggle operated. The stem is spring-loaded from the bonnet to close the valve. Unidirectional operation is specified with the incoming fluid entering under the flat plug which is fitted with a trapped PTFE soft seal. Shut-off capabilities are limited by the spring strength. Maximum pressure is only 20 barg and there is no packing fitted. The plug is sealed in the burnished bonnet by a Viton® "O" ring.

Slightly larger globe valves with hardened stainless steel ball tips for pressures up to 414 barg are made in sizes from ⅛" to ½" NPT, ¼" to ¾" od compression fittings to ¼" to ½" nb socket weld.

Small bellows sealed valves are manufactured for difficult conditions. They are only available in stainless steel with socket or butt weld connections, with the bellows clamped to the body by the bonnet and sealed with a silver plated metal "O" ring. A hemispherical tip seals and regulates using the integral seat. Stem rotary movement is isolated from the non-rotating bellows assembly by a ball bearing. With outside screw and screwed bonnet the valve is compact even though it is full bore for ¾" and 1" od tube and ¾" nb pipe. Operating pressures are up to 172 barg with temperatures to 650°C with Grafoil® secondary packing.

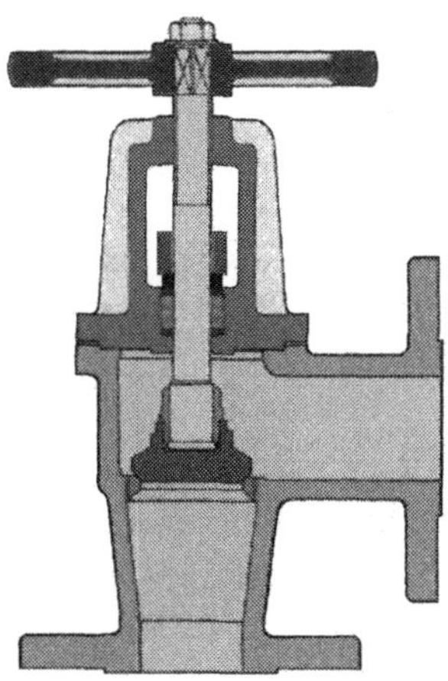

Figure 3.18 A typical cast angle globe valve

Small diaphragm-sealed valves are available for less arduous applications. A stainless steel diaphragm isolates the process fluid from the actuating mechanism which may be quarter turn or outside rising screw. A return spring under the diaphragm ensures the tip lifts when the valve is opened.

High pressure angle globe valves, compatible with ¼", ⅜" and $^{9}/_{16}$" od heavy wall tube, are available for small scale extreme pressure applications to 3 100 barg, see Figure 3.19. The valve body is machined from a solid block of wrought stainless steel. The angled seat is machined on the end of the inlet connection adaptor which is screwed into the bottom of the body. The non-rotating stem is in hardened stainless steel and may have a Stellite™ tip. The PTFE or Grafoil® packing is compressed by the phosphor bronze screwed bonnet via a shaped follower. The handwheel stem outside screw is located in the bonnet bore. Various connectors can be fitted to suit the application. Maximum temperature is 454°C at reduced pressure.

Larger high pressure angle globe valves, or choke valves, are used for continuous throttling in crude oil and natural gas production. Heavy duty needle valves were adequate for this application but operating pressures continue to increase and operating conditions deteriorate due to sand, H_2S and CO_2 content. Solids erosion combined with cavitation erosion produced rapid material loss requiring frequent valve adjustment. A control valve style of globe valve has proved to be more successful for the latest operating environments, and is shown in Figures 6.14 and 6.17 in Chapter 6. An angle globe valve with a piston sliding inside a perforated cage is used. Unlike other angle valves the flow enters the valve through the branch and not under the seat.

Figure 3.20 shows a typical angle choke valve for oil field applications. The incoming flow is directed around the periphery of the cage. Flow passes through the cage wall via multiple holes or slots. The flow area available through the cage is controlled by the piston position. The high velocity jets self-destruct in the bore of the cage, away from the cage wall, and do not create any cavitation-induced metal erosion. The fluid travels down the cage, through a lined portion of body, and exits the valve. A

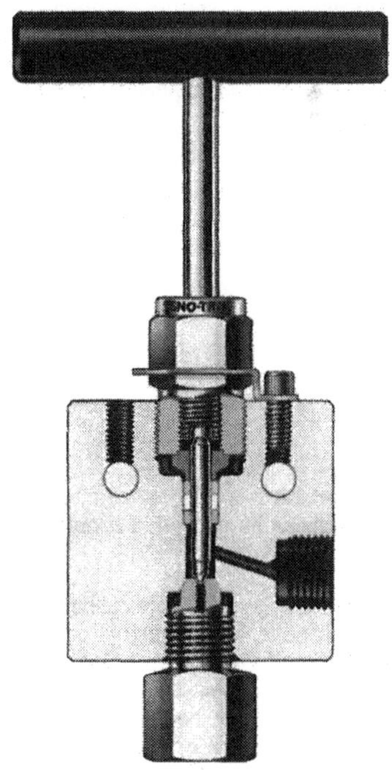

Figure 3.19 High pressure angle globe valve
Courtesy of Swagelok Company

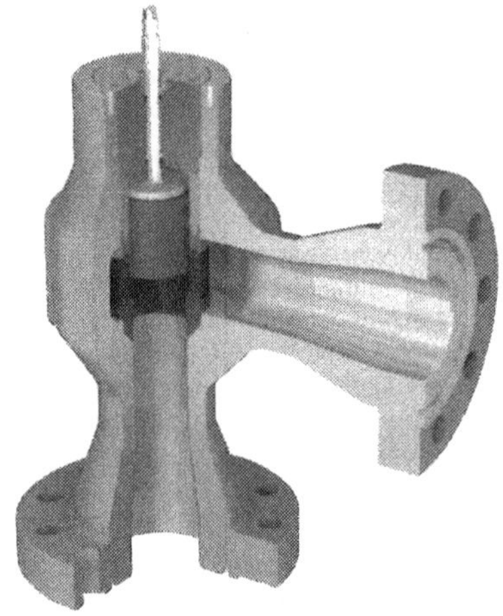

Figure 3.20 High pressure angle choke valve
Courtesy of Mokveld Valves BV

considerable amount of pressure energy can be continuously destroyed, so pressure recovery is very low. To reduce operating forces the piston is pressure-balanced using the downstream process pressure. Noise can be a problem. Special attention is given to noise reduction techniques for valves used on high differential pressures.

Different styles of cage trim are available to suit specific applications, such as:

- crude oil pressure reduction
- natural gas re-injection
- brine injection
- steam injection
- water injection

Valve bodies are cast or forged in one piece in a wide range of materials including:

- carbon steel, high tensile steel, 11/13 Cr
- austenitic stainless steel, duplex, high nickel alloys

The valve trim is selected specifically to suit the operating conditions; tungsten carbide coatings and inserts are available. Typical valve sizes range from DN25 to DN200 with pressure ratings covering ANSI 900lb through to 4500 lb. API pressure ratings include 3000 through to 15 000.

Process connections can be:

- ANSI flange
- API flange
- proprietary clamp
- butt weld

Operating temperatures would normally be -50° to 205°C. Valves up to DN400 can be supplied to order.

Oblique globe valve

The oblique globe valve is sometimes called a "Y-globe valve". It is a compromise between the straight and angled versions. The normal straight valves have a tortuous flow path. By setting the stem and seat at an angle to the body the flow path is straightened somewhat and the flow losses reduced. The oblique style is more popular for general applications than the angled, which is largely used for steam. Solids handling capabilities are improved but should be tested with caution. Flow is still only in one direction. Full-bore and reduced port versions

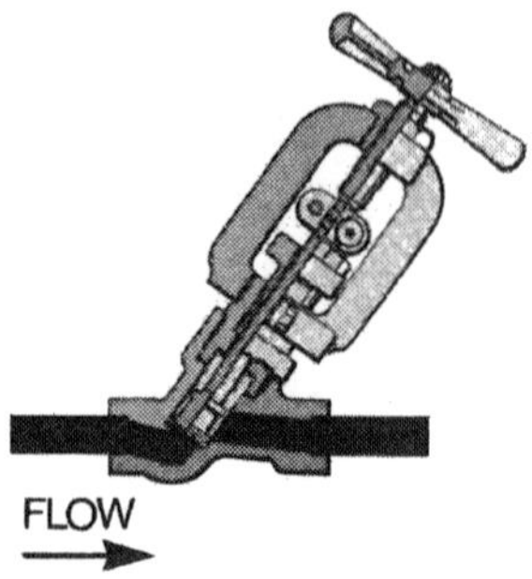

Figure 3.21 High pressure oblique globe valve
Courtesy of Dresser-Masonelian

are available. Cleaning with rods is not possible unless the bonnet is removed.

Bodies are normally cast or die-forged in one piece. Special designs are available with two piece bodies. A double-sided replaceable seat is clamped between the body flanges. This feature gives excellent maintenance access but also provides an extra leak path for the fluid. The benefits must be weighed against the potential problems. One-piece body valves may have integral seats. Valves for high pressure throttling may have the seating surface hard-faced to extend their useful life. Seats can be machined, lapped and polished in situ and integral seats can be useful in applications where crevice corrosion is a problem. Clamped seats can be acceptable for crevice corrosion applications depending upon the materials, the seals and the detailed design.

Discs should be plain, crow-foot guided or needle contoured to suit the application and differential pressure. Needle profiles can be multiple tapers to produce primary and secondary throttling. To combat erosion both seat and disc can be hard-faced. Valves with plain and crow-foot discs can be fitted with wipers to clean the seat before face-to-face sealing contact is made; or seats can be fitted with an elastomer seal.

The most popular material is cast carbon steel, with forged carbon steel for high pressure valves. Duplex stainless steel valves are available for special applications. Normal ranges include valves up to DN250 for ANSI 150lb, up to DN150 for ANSI 300lb. Valves to DIN 3790 cater for 26 barg up to DN32 and 10 barg up to DN200. High pressure valves, with integral bonnet in closed die-forged high tensile steel, and bolted gland, can operate at high temperatures. Figure 3.21 shows a typical high pressure oblique globe valve with screwed bonnet and separate yoke. Table 3.4 indicates typical performance of the pressure ratings available.

Body and bonnet material	Pressure rating	Size	Connections
Forged carbon steel	248 barg at 38 °C 103 barg at 565 °C	DN10 to DN65	Socket weld
	414 barg at 38 °C 172 barg at 565 °C	DN10 to DN65	
	745 barg at 38 °C 310 barg at 565 °C	DN10 to DN50	
Forged carbon steel	ANSI 1500lb to ANSI 4500lb	DN20 to DN100	Socket weld Butt weld
Fabricated: stainless steel duplex nickel alloys titanium	ANSI 150lb to ANSI 2500lb	up to DN300	Flanged Butt weld

Table 3.4 High pressure/temperature oblique globe valve performance

Double-beat globe valve

The double-beat globe valve is an old design which is very useful for special applications. Figure 3.22 shows the basic construction.

Globe valves are very good for low leakage shut-off and regulation. As the valve size and the operating pressure increase, the hydraulic load on the stem becomes greater making the valve difficult to operate manually. This problem was first experienced with steam engines as sizes and steam pressures increased to produce more power. The double-beat globe valve reduces the hydraulic load by balancing the forces on the discs.

The upper disc has inlet pressure on the top; the lower disc has inlet pressure on the bottom. The areas are not quite equal, because of the difference between the inner and outer seat diameters. This difference is very small and produces only a small thrust for large high pressure valves. Operation by ungeared handwheel with rising stem valves can be considered for applications impossible with other valves. They are also very good for use on sites without power or for emergency back-up systems.

As with other globe valves there is some flow resistance, but this may not be as high because two seats are used. Double-beat globe valves cannot be cleaned with rods or "pigged". Operation with clean fluids is preferred to reduce the possibility of differential seat/disc wear.

Essentially the valve has two pressure casings. The outer casing being at inlet pressure and the inner at outlet pressure. The incoming flow circulates around the casing and flows through both seats into the inner casing. Flow in either direction is possible; sealing in both directions is equal. The discs require a degree of flexibility to ensure both seat properly. The body is relatively complicated and casting is the preferred method of construction. However fabrication and forging is possible if necessary. Connections do not necessarily have to be in-line. They can be at 90° without significantly affecting the flow losses.

Double-beat globe valves are reserved for special applications and probably made to order. Typical applications include main isolating valves for steam and water turbines.

Three-port globe valve

The three-port globe valve is usually used as a diverting valve in high pressure applications. Figure 3.23 shows a typical valve for hot, high pressure water used for boiler feed water. Flow is generally confined between two ports for the majority of operation; change-over is normally confined to start-up, shut-down or fault conditions.

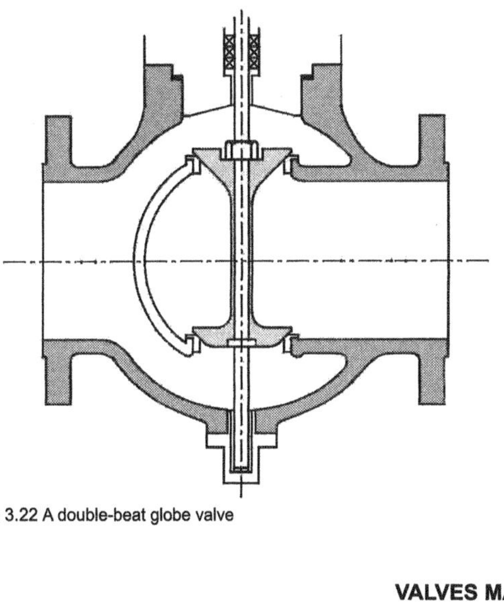

Figure 3.22 A double-beat globe valve

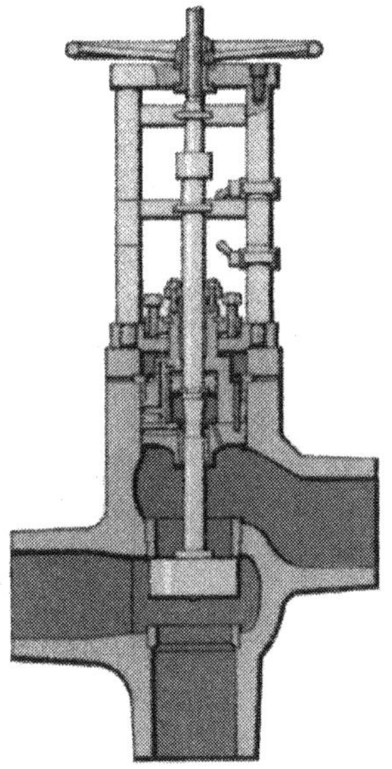

Figure 3.23 A typical three-port globe valve
Courtesy of Dresser-Masoneilan

Another popular application is change-over valves for pressure relief systems. Two relief valves can be mounted on one process connection. One valve is operational while the other is isolated and maintained. Two change-over valves, on inlet and outlet, can be ganged together to act as one. Some flow resistance is apparent due to the connection configuration. The changes in direction of the fluid may create considerable reaction forces on large liquid valves.

Valve bodies are generally cast in steel or alloy steel. Valves for power generation would normally have butt weld connections to overcome flange problems. Trim in 11/13 Cr would be standard. Hard-facing on seats, disc, backseat bush and the stem local to the bush, is a well-used option. Operating pressures up to 200 barg are common with temperatures at 350°C.

Axial piston valve

Figure 3.24 shows a novel concept in isolating valve design. This valve design embodies the best qualities of the globe valve but smooths out the tortuous flow path very considerably. The fluid flow does have to change direction radially, twice, but it is accomplished quite gently. This may look like a plug valve, but sealing is in the bore of the seat, see Figure 3.25. When wide open, the soft replaceable seal is withdrawn inside the body casting. The piston is moved via a novel twin rack mechanism which translates the actuator linear motion into a similar motion in a perpendicular direction. It is capable of fast operation, for emergency shut-down applications and Safety Instrumented Systems (SIS) applications. The piston is pressure balanced and this allows relatively small actuators to be used. Considerable weight savings can be achieved by using this design approach. Valves of this style are available in sizes from DN50 to DN1 200 with pressure ratings up to ANSI 2500 and API 10 000. Valve bodies can be machined for taper clamp connections as an alternative to flanges.

Weir type diaphragm valve

The diaphragm valve is different from all the valves previously described because the bonnet and the stem are always completely isolated from the product. A diaphragm valve is almost equivalent to a bellows sealed linear valve. The diaphragm fulfils two functions. Firstly it separates the inlet from the outlet when the valve is closed and secondly it separates the operating mechanism from the product. The diaphragm forms a part of the pressure containment.

Figure 3.24 A high pressure axial piston valve
Courtesy of Mokveld Valves BV

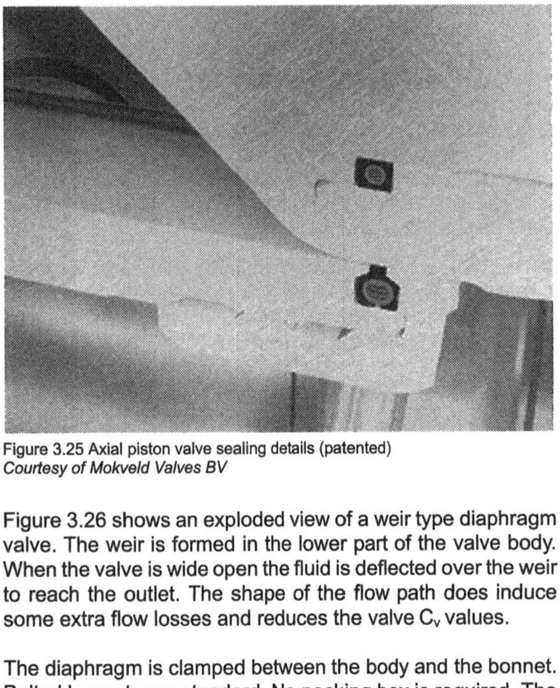

Figure 3.25 Axial piston valve sealing details (patented)
Courtesy of Mokveld Valves BV

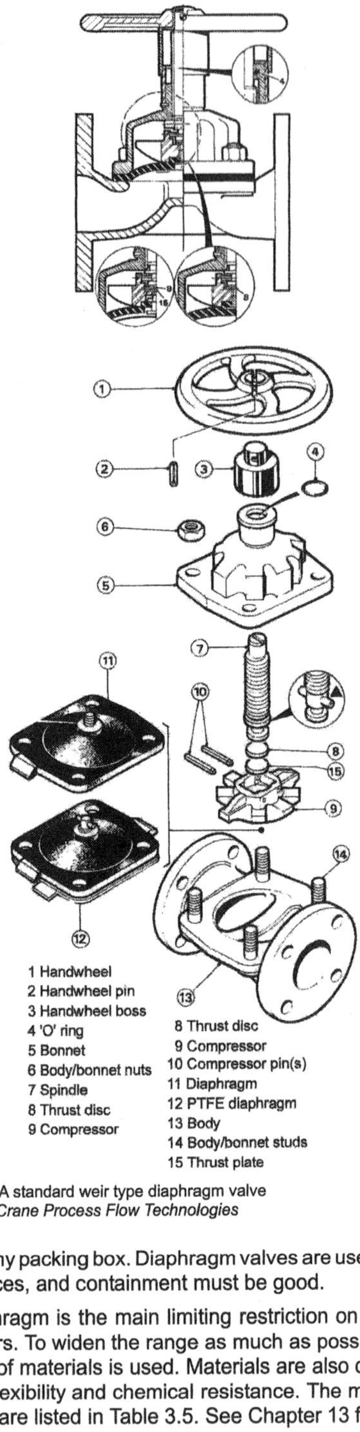

Figure 3.26 A standard weir type diaphragm valve
Courtesy of Crane Process Flow Technologies

Figure 3.26 shows an exploded view of a weir type diaphragm valve. The weir is formed in the lower part of the valve body. When the valve is wide open the fluid is deflected over the weir to reach the outlet. The shape of the flow path does induce some extra flow losses and reduces the valve C_v values.

The diaphragm is clamped between the body and the bonnet. Bolted bonnets are standard. No packing box is required. The bonnet has an inside rising screw, which can have a sealed lubrication reservoir. A shield on the handwheel is sealed by a lip seal to prevent ingress of moisture or dust into the thread area. The operating mechanism is sealed both ways for long life. Valve maintenance is easy once the line is depressurised and all parts can be removed from under the bonnet.

Valve shut-off capabilities are excellent because of the elastomer diaphragm. Solids handling is quite good, with 15% maximum recommended concentration. Diaphragm valves are a sensible choice for hazardous fluids because there is no packing box to worry about. Routine diaphragm maintenance will ensure almost total containment and diaphragm life can be optimised by the fitting of limit stops to the actuator. Over-tightening is eliminated thus reducing stresses in the elastomer. The precise level of containment is dependent upon the fluid. Elastomers are permeable; although permeation through a diaphragm should be considerably less than the leakage through any packing box. Diaphragm valves are used on hydrogen services, and containment must be good.

The diaphragm is the main limiting restriction on operational parameters. To widen the range as much as possible a broad selection of materials is used. Materials are also combined to improve flexibility and chemical resistance. The most popular materials are listed in Table 3.5. See Chapter 13 for details of material and product compatibility.

Diaphragm materials	Min temperature °C	Max temperature °C
Gum rubber	-50 to -29	70 to 100
EPDM	-34	149
PTFE (1)	-34	176
Butyl rubber	-29 to -18	107 to 120
Neoprene™	-20	90
Nitrile rubber	-12	82
Hypalon™	-10	100
Viton™	-5	150
PTFE + EP rubber	-30	150

Diaphragm materials	Min temperature °C	Max temperature °C
PTFE + Butyl rubber	-20	150
PTFE + Hypalon™	-10	120
PTFE + Viton™	-5	175
PTFE + EPDM	-34 to -20	160 to 176

(1) Specially processed PTFE to reduce entrained air and permeability and increase dimensional stability

Table 3.5 Diaphragm materials for weir type valves

Valve size	Max pressure barg	Lower limit temp °C	Upper limit temp °C	Max temp °C	Pressure barg @ max temp
DN10 to DN50	16	-50 to 5	55	175	5
DN65 to DN150	10	-50 to 5	55	175	4
DN200 to DN350	6	-50 to 5	55	175	2.5

Table 3.6 Weir type diaphragm valve pressure/temperature limits

The size of the valve influences the stresses imposed on the diaphragm. Table 3.6 indicates typical pressure/temperature limits.

Weir type diaphragm valves are also suitable for vacuum services. The maximum vacuum depression in practice depends upon the valve size, the temperature and whether the bonnet is evacuated. Vacuums down to 762 Torr are possible. The pressure and temperature data shown indicate the extremes for the materials. If the process fluid reacts with the diaphragm the operating limits may be considerably reduced. When operating at extremes the valve manufacturer should be consulted, giving the full operating conditions to allow material selections to be confirmed and the useful life predicted.

The most popular industrial valves are made with cast iron or SG iron bodies. The body can be coated or lined with various compounds to improve corrosion resistance.

Coatings and linings include:

- hard rubber
- butyl rubber
- borosilicate vitreous enamel
- ethylene chlorotrifluorothylene (ECTFE)
- polypropylene (PP)
- polyvinylidene fluoride (PVDF)
- ethylene tetrafluoroethylene (ETFE)
- perfluoroalkoxy (PFA)
- polytetrafluoroethylene (PTFE)

Figure 3.27 shows a typical valve with a corrosion resistant thermoplastic lining.

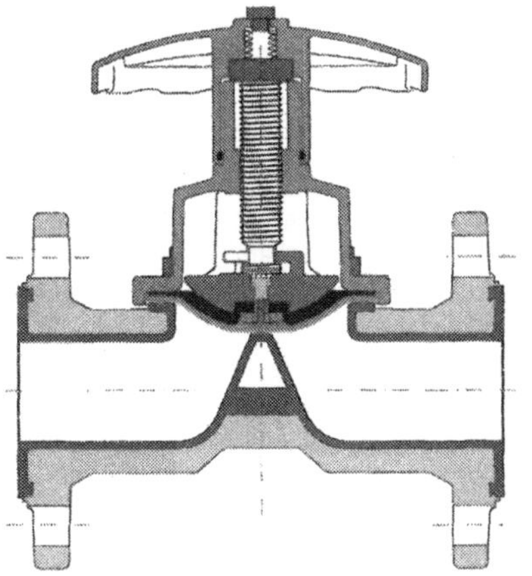

Figure 3.27 A typical lined weir type diaphragm valve
Courtesy of Richter Chemie-Technik GmbH

The PTFE lining is attached by isostatically pressing the material to the body. The body material and lining have an impact on operating temperatures. SG and cast iron bodies would not normally be used below -10°C. Stainless steel and copper alloys can be used down to -30°C. PTFE lined valves with PTFE diaphragms can operate between -40 and 176°C. Valves up to DN80 can have threaded connections. All sizes can be flanged. Brass valves for 10 barg, 90°C are mass-produced in DN10 to DN15 sizes. Stainless steel valves up to DN300 can operate at 10 barg at temperatures up to 150°C.

Full-bore diaphragm valve

The full-bore diaphragm valve is a logical extension to the weir type diaphragm valve. The flow path is straight and the fluid

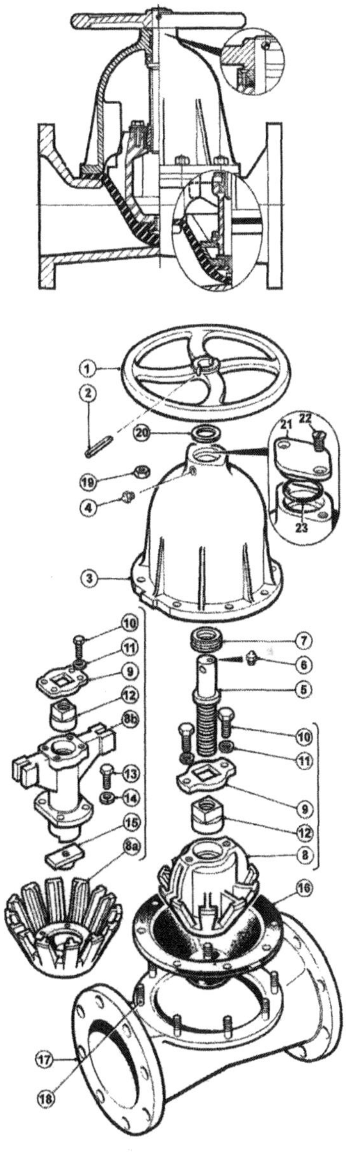

1. Handwheel
2. Handwheel pin
3. Bonnet
4. Bonnet grease nipple
5. Spindle
6. Spindle grease nipple
7. Thrust race
8. Compressor
8a. Compressor base
8b. Compressor guide
9. Compressor plate
10. Compressor plate screws
11. Spring washers
12. Spindle nut
13. Compressor base screws
14. Compressor base washers
15. Diaphragm stud nut
16. Diaphragm
17. Body
18. Body/Bonnet stud
19. Body/Bonnet nut
20. Felt washer
21. Sealing plate
22. Sealing plate screws
23. 'O' Ring

Figure 3.28 Typical full-bore diaphragm valve
Courtesy of Crane Process Flow Technologies

losses consequently smaller. The full bore valve has most of the advantages of the weir type for fluid handling and sealing, plus better solids handling. An exploded view is shown in Figure 3.28. Abrasion is less of a problem because of the straighter flow pattern. The diaphragm rolls up into the bonnet out of the flow stream. It is therefore recommended that information on the weir type diaphragm valve is read first. The extra benefits are paid for by increased diaphragm deflection and stressing resulting in lower pressure ratings. Size for size, full-bore valves are capable of about 60% of the working pressure of weir type valves.

Construction, body materials and diaphragm options are similar to weir type valves, see Table 3.5. Pressure ratings are shown in Table 3.7. Full-bore diaphragm valves can be used for dry solids.

Valve size	Max pressure barg	Lower limit temp °C	Upper limit temp °C	Max temp °C	Pressure barg @ max temp
DN15 to DN 100	10	-40 to 5	50	140	4.5
DN125 to DN150	6	-40 to 5	50	140	1
DN200 to DN300	3.5	-40 to 5	50	140	0.5
DN350	1.75	-40 to 5	50	140	0

Table 3.7 Full-bore diaphragm valve pressure/temperature limits

Standard valves are mass-produced in cast iron, SG iron, brass and stainless steel. Valves up to DN80 can have screwed connections, all sizes may be flanged.

Pinch valve

The pinch valve can be considered as a diaphragm valve taken to the extreme. The flat diaphragm is replaced by an elastomer tube and the bore of the tube is the only component in contact with the product. Exotic metal alloys or sophisticated surface coatings are unnecessary. Figure 3.29 shows the enclosed type of pinch valve. When wide open the bore of the tube is completely clear and presents a 100% area circular passage. Friction losses are very low and C_v values very high. The construction of the valve is extremely simple and therefore reliable. There is no packing to adjust or replace. Sealing is equally effective in both directions. The correct tube material selection should also allow effective throttling.

Isolation is achieved by compressing the elastomer tube between two rollers. The rollers are actuated by a screwed stem. The action of the thread is applied to one roller whilst the other is loaded by the nut reaction. The open frame design has aluminium structural parts to reduce weight. The enclosed design uses cast iron for the body. The stem is of the inside screw, non-rising type, so that all important components are protected.

The tube is available in several compounds for broad chemical compatibility and abrasion resistance:

- gum rubber, Neoprene™, Hypalon™, polyurethane
- chloro-butyl rubber, Buna N™, Viton™, EPDM

Figure 3.29 Enclosed pinch valve

Elastomers can be permeable to gases. The extent of product loss should be evaluated during the valve selection process. The thermoplastic PTFE can be used as a tube when other materials are suspect. PTFE does not achieve the same level of sealing as the elastomers. The tube is unsupported in the region of the rollers and must be allowed to flex, pressure ratings being limited by the strength of the elastomer.

Table 3.8 indicates the pressure/size relationships for popular pinch valves.

Size	Max pressure barg
DN15 to DN20	12
DN25 to DN150	10
DN200	8.5
DN250 to 300	6.9
DN350	5.1
DN400 to DN600	3.4

Table 3.8 Size and pressure ratings of pinch valves

Pinch valves are suitable for a wide range of liquid and liquid-solid mixtures. The acceptable size of solids being dependent upon the valve size and the precise design and construction of the tube. Pneumatic and hydraulic powered versions are suitable for pressures up to 49 barg, see Section 3.3.7.

Straight needle valve

Straight needle valves are manual valves intended for continuous throttling to achieve flow regulation. Valves are fitted in circuits to allow the flow to be adjusted to suit variable operating conditions. Straight needle valves are also used in instrument impulse lines to attenuate pressure pulsations to limit gauge reading fluctuations. These "meter" valves are of a more robust construction. Some valves have accurate calibration, micrometer style scales, to permit repeatability. Some needle valves are not recommended for isolation purposes, so product specifications must be checked carefully. Like globe valves there is a preferred direction of flow; fluid enters the bottom of the seat, travels up the taper and exits from the top.

The stem of a needle valve is machined to an accurate taper and small valves can have a very sharp point on the end. The seat, a much longer seat than in other valves, is machined to the same taper.

On opening, the valve creates a long narrow parallel flow passage. As the stem is withdrawn the flow passage enlarges and reduces the flow losses. Straight needle valves impose high losses in a circuit and have low C_v values compared to other valve types. The nature of the flow passage excludes solids handling, and clean fluids are essential for acceptable component lives.

In general, straight needle valves are small, up to DN25 is usual. Valve bodies can be forged with integral bonnets to remove a potential leak path and solve the connection problem. Most valves have rising stems. The needle portion may or may not rotate depending upon the detailed design. Some valves do not have a backstop which allows the stem to come out of the bonnet. This is not a good idea with hazardous fluids. Stem sealing can be by adjustable packing or by "O" ring. The stem thread may be under the packing or above it. Typical specifications are for valves up to DN15 to be suitable for 689 barg, up to DN25 for 414 barg.

Valves with closed die-forged bodies or machined from barstock can have severe limitations placed on the port sizes. Port machining is only possible through the connections and the bonnet opening. Inlet and outlet passages must be drilled at an angle through the connection openings, see Figure 3.30.

The small diameters may increase the fluid flow losses. Additionally, the sharp corners formed during manufacture may induce cavitation in liquid applications.

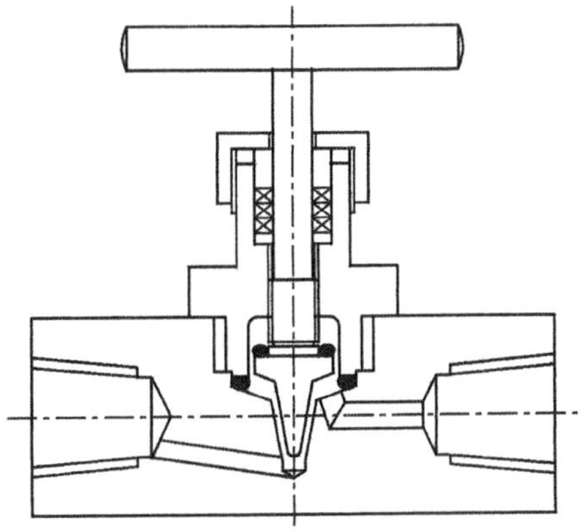

Figure 3.30 Typical straight needle valve construction

Valves with stainless steel bodies machined from barstock use a tapered needle with an angled collar for shut-off. These valves have a stem in hardened stainless and an inside screw protected from the elements. The union bonnet has a metal-to-metal seating to maintain alignment.

Deep PTFE packing, high in the bonnet, gives a temperature range of -29 to 232°C and pressures of up to 345 barg. Connections can be compression fittings for ¼" od tube and ⅛" or ¼" NPT. Small barstock valves with integral bonnets and union glands are suitable for 414 barg at temperatures between -29 and 232°C. The inside screw is protected by the PTFE packing. Available in ¼" or ½" NPT. Some valves are available with packing "live-loaded" by disc springs inside the packing box. Austenitic stainless steel barstock valves for steam applications up to 205°C in sizes from DN10 to DN25 can be used on cold applications up to 689 barg.

Small valves with precision cast bodies in brass, Monel™ and stainless are fitted with stainless steel bellows for difficult fluids. The bellows and bonnet can be sealed to the body with a metal "O" ring or brazed/welded. Maximum pressure for cold fluids is 70 barg reducing to 14 barg at 482°C. A normal handwheel can be replaced by calibrated micrometer adjustment.

Connections can be ¼" socket weld or compression fittings. Similar designs of packed valves are available with panel mounting options. However the panel is clamped to the screwed gland. Before the gland can be adjusted the valve must be loosened at the front of the panel before the gland can be adjusted behind. Smaller cast brass and 316 stainless valves, ⅛" NPT, are suitable for 35 barg at temperatures up to 82°C or 121°C depending upon the packing.

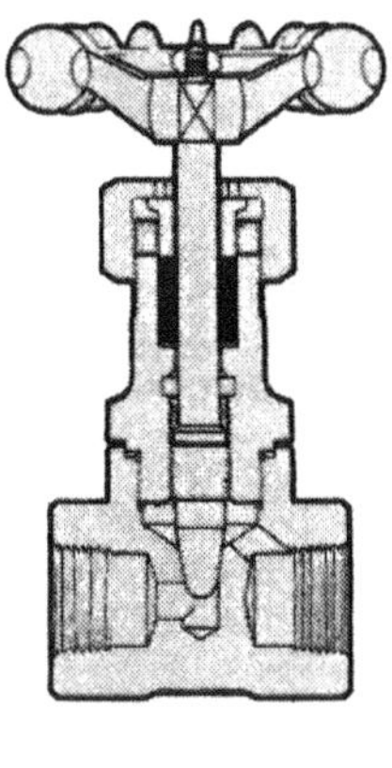

Figure 3.31 Forged needle valve for meter applications
Courtesy of H H Valves Ltd

A range of straight needle valves for meter applications is available in forged carbon steel for applications up to 345 barg, see Figure 3.31. The seat is integral with the body but lapping and polishing is easily accomplished to repair slight damage. Clean fluid applications are preferred. The stem, in 11/13 Cr, has an inside screw with fine threads for good adjustment. The screwed bonnet is machined from the solid and has a long thread for good attachment. A backseat is machined directly in the bonnet rather than fitting a bush. The packing is adjusted by a screwed gland. Sizes range from DN10 to DN25 with female screwed connections. The valve is not specified to a code pressure/temperature rating. At 232°C the working pressure reduces to 276 barg.

Small carbon steel flanged valves are manufactured by friction welding the flanges to a closed die-forged body. The valve uses a bolted bonnet and gland to achieve an ANSI 800lb rating. The bonnet is sealed by a spiral wound gasket. The 11/13 Cr rising stem is outside screw and matches an 11/13 Cr seat. Sizes range from ¼" to 2" with options for a welded bonnet, screwed or socket weld connections. Valves with screwed bonnets, ¼" to 1", can operated up to 345 barg at ambient temperatures.

Angle needle valve

Angle needle valves are very similar in internal construction to straight needle valves, so that Section should be read first. The angled construction removes the tortuous flow path and eases considerably the manufacturing difficulties of the inlet and outlet passages, see Figure 3.32. Passages in closed die-forged and barstock bodies can be drilled directly in line with the connections. Small valves are easily manufactured from square, rectangular or hexagonal barstock or simple solid forgings.

Screwed connections are straightforward to produce, but compression fittings are slightly more complicated. Closed die forgings are not as popular in angle needle valves as they are in straight versions, but integral bonnet versions are produced for pressures up to 689 barg. The bonnet assemblies can usually be fitted to both straight and angled designs. Angled valves can be fitted with bleed plugs to perform block and bleed/vent duties. Valves up to DN15 should be capable of 689 barg and up to DN25 of 414 barg. The smallest popular valves are ⅛" NPT.

A reaction force is generated by the change in fluid direction. The force should be small and insignificant due to the low flow rates normally encountered. Small valves without stem packing are often used as bleed or vent valves. The needle taper may have an included angle of 60° or 90° and be lapped into the seat to provide a good seal. This style of valve, and some more standard types, do not have backstops to prevent the stem from being removed.

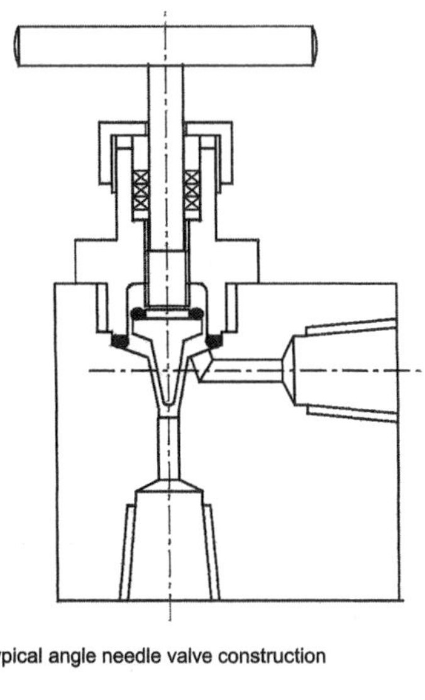

Figure 3.32 Typical angle needle valve construction

Valves with stainless steel bodies machined from barstock use a tapered needle with an angled collar for shut-off. These valves have a stem in hardened stainless and an inside screw protected from the elements. The union bonnet has a metal-to-metal seating to maintain alignment. Deep PTFE packing, high in the bonnet, gives a temperature range of -29° to 232°C and pressures up to 345 barg. Connections can be compression fittings for ¼" od tube and ⅛" or ¾" NPT. Precision cast valve bodies in brass and 316 stainless steel are suitable for pressures up to 35 barg at temperatures up to 82° or 121°C.

Special angle needle valves can easily be constructed for unusual applications. High pressure needle valves are regularly built into blank flanges to provide vent and bleed facilities. The flange forms the valve body and seat with the bonnet assembly attached to the outer face. The outlet port is drilled radially to the rim of the flange. ½" valves, rated up to 690 barg, built into API flanges are popular for oil & gas applications.

Solenoid valve

Solenoid valves are usually small magnetically operated isolating valves. Some special versions can be modulated, see the **NOTE: Solenoid valves** in Chapter 6. Solenoid valves can be considered as a sealed globe valve with electric actuation.

Figure 3.33 shows a typical direct-acting valve. The metal body, shown with threaded female connections, is cast in one piece. The body has a flange facing on the top where the bonnet would normally be found. A light stainless steel, non-magnetic, pressing is attached which forms a pressure boundary. Inside, the solenoid armature slides to open and close the valve; the bottom face sealing on the body seat.

Valves which are normally-open, (NO), have a spring under the armature to hold the valve open. In normally-closed, (NC), valves the spring is on top. The electric coil, which creates the operating magnetic field, is located around the stainless steel enclosure. Coils are normally energised for the whole period of operation against the spring. Small valves may be bidirectional or unidirectional, so the manufacturers' instructions should be consulted. Some valves require a minimum inlet pressure to work; so specifications should be checked carefully.

The effective size of solenoid valves can be increased by servo or pilot operation. The solenoid valve operates in a small bore fluid circuit which actuates the main valve, see Figure 3.34. These valves will be unidirectional as the inlet and outlet pressures are essential for the correct function.

The main valve, a crow-foot plug, is isolated by a diaphragm which also acts to amplify the servo forces. The servo feed is taken from the valve inlet and directed above the diaphragm. The servo feed is filtered by a mesh gauze in the bore of the main valve body. The solenoid valve acts in the servo outlet path to the main valve outlet. When closed, the valve inlet pressure acts over the diaphragm holding the main valve closed. When the solenoid opens the servo fluid is allowed to escape faster than the inlet can maintain; servo pressure is therefore reduced and the main valve opens. Some valves can be water hammer damped. Water hammer is the damaging effect created by changing the velocity of liquid too quickly. This can be caused by opening or closing valves too quickly. Small solenoid valves can open and close in 10 ms. Larger servo-operated valves can be slowed down to 5 or 10 s. In Figure 3.34 the orifice in the servo inlet can be clearly seen.

Larger pilot-operated valves are integral-piston operated globe valves. The piston has the globe valve's disc facing on its bottom surface and the upper portion of the piston is subjected to pilot pressure. Elastomer seals can be fitted for both piston and seat leakage control. Large pistons can be bottom-guided to assist alignment. Main valve damping can take two forms:

- an orifice in the pilot control supply
- viscous damping incorporated in the bottom guide

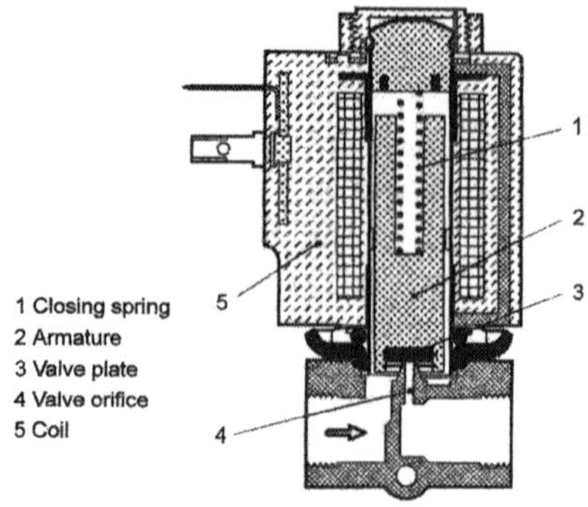

Figure 3.33 Direct-acting solenoid valve
Courtesy of Danfoss A/S

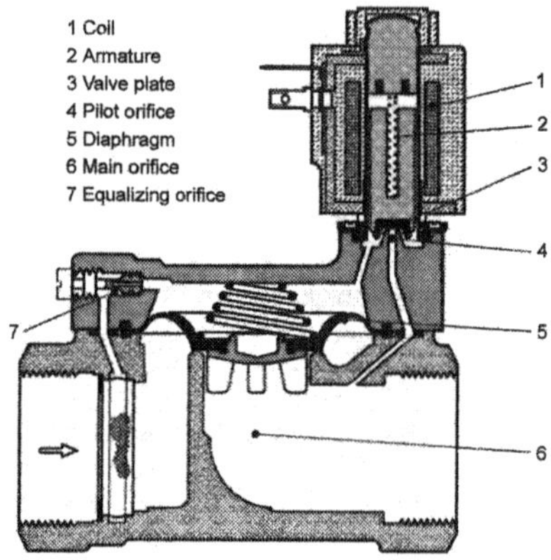

Figure 3.34 Servo-operated solenoid valve
Courtesy of Danfoss A/S

Solenoid valves are normally isolating valves with two ports, an inlet port and an outlet port. These valves may be called a "two-way" solenoid valve and designated "2/2" followed by NC or NO. Solenoid valves can be designed with three ports for diverting applications, "three-way" valves, designated "3/2" followed by NC or NO.

They can also be constructed as three-port diverting valves. Energising the coil swaps outlet ports. Multi-valve manifolds are produced with up to six valves which allow up to six inlets to discharge into one outlet.

The coils for solenoid valves are very important. Because coils are electrical the usual electrical parameters require consideration:

- physical protection
- insulation
- duty rating
- power supply

Physical protection ranges from IP00, no protection, to IP67, fully protected against dust and intermittent immersion in water. The quality of the coil insulation determines how hot the coil can operate without damage. Class "F" insulation: 40°C ambient air, 100°C temperature rise, is fairly standard. Class "H" insulation: 40°C ambient air, 125°C temperature rise, is available on some heavy duty industrial valves. The duty rating is crucial for successful long term operation. If the coil is to be energised for long periods then a continuous or 100% rating is essential. Some coils are only rated for 30%, 18 minutes per hour. It is possible to build voltage reduction into the solenoid which reduces power consumption, and heating after the power has been applied for a specific time.

The power supply for the coil can vary greatly. The following are very common; 12 Vdc, 24 Vdc, 110 Vdc, 220 Vdc, 24 Vac, 36 Vac, 48 Vac, 110 Vac, 220 Vac, 240 Vac, 380 Vac. Frequency of alternating current supplies is important. Coils for dc can have a rectifier built-in to use on an ac supply. Coils do not consume much power, typically between 10 and 20 W. Coils can be intrinsically safe. In some circumstances solenoid valves require venting. This requirement must be discussed before ordering.

A variation of the normal solenoid design is the zero-power pulsed valve. The valve is "toggled" or "latched" by applying a pulse of power. These valves can save energy by not requiring the coil to be energised for long periods. Some solenoid valves can be fitted with a local manual override.

Solenoid valves are available in many body materials and sizes. Table 3.9 lists typical sizes and ratings.

Body material	Connection sizes	Pressure rating barg
Brass	1/8" to 3/8" BSP	30
Brass	1/2" NPT	50
Brass	3/8" to 2" BSP	10
Brass	1/4" to 2"	21
Brass	2 1/2" to 4" flange	10
Zinc free bronze	1/8" to 3/8" BSP	30
Zinc free bronze	3/8" to 2" BSP	10
Stainless steel	1/8" to 1/2"	24
Stainless steel	1/8" to 3/8" BSP	30
Stainless steel	3/8" to 2" BSP	10
Stainless steel	1/4" NPT	100
Stainless steel	3/8" NPT	15

Table 3.9 Typical solenoid valve sizes and pressure ratings

Valves with brass bodies can be supplied in alloys to resist de-zincification by liquids such as demineralised water.

3.3.2 Common optional extras

As can be seen from Section 3.3.1, linear valves are very diverse to cater for wide ranging operating conditions. However, most valves can be adapted with small modifications or additions to improve the performance and/or facilities offered. Numerous options are available, a number of which are discussed further.

Some simple additions to valves are easy to supply. Position indicators can be fitted. Special nameplates, with process and identification information, can be fitted after the valve is finished.

Probably the most requested options relate to connections. Screwed valves will normally have BSP taper or NPT threads. Changing from one to the other is usually not a problem. Some applications may require parallel threads, BSP or SAE, or even a bolt thread. These threads may be more difficult to supply depending upon where the seal is to made. The end facings of valves with female screwed connections may require special machining to produce a smooth square surface.

There are many flange specifications in current use. Changing the drilling pattern is not normally a problem. Some users may require the raised face on steel flanges to be removed to allow the connection of flat cast iron flanges. Cast iron can be used with raised face flanges with care and skill. Removing the raised face removes the doubt. Special flange machinings can often be supplied if there is material on the casting or forging. Flange facings can be machined to various surface finishes; most can be applied to finished valves if essential. See Chapter 9, Section 9.7, regarding flange details.

Material upgrades are very popular requirements. Compressed fibre gaskets can be replaced by metal/fibre spiral wound gaskets or proprietary metal gaskets such as Armco iron. Steel valves can be modified to remove any copper alloy components. Specific areas of valves can be hard-faced to improve erosion/corrosion resistance. Seats, discs and portions of the stem are normally involved. Changing the packing material, which does not involve any packing box modifications, can be easily accomplished.

Valves for sour service should comply with NACE requirements. The material changes required can usually be easily accommodated. However the design of the valve should be scrutinised carefully to ensure simple material changes will have the desired effect. The lower hardness values imposed by NACE mean some materials cannot develop their usual strength. This is most notable with fasteners. Detailed reviews of designs are necessary to ensure bolted connections are not compromised. The NACE requirements embody a spirit as well as specifics. Screwed threads should not be exposed to sour fluids. Select the correct valve design first before applying NACE.

Phosphor bronze gland nuts have proved to be much better than steel nuts for hot applications. Manufacturers do not offer this as a "catalogue option". Slightly longer gland studs may be required. Inquiries on a case-by-case basis should be considered.

The operating mechanism can be modified to suit site conditions and safety requirements. Grease nipples can easily be fitted for outside screws. The stem can be fitted with limit switches to provide remote indication or electrical interlocking. The facility to fit a padlock, to lock a valve open or closed, can usually be provided.

Proprietary interlocking systems, see Chapter 14, can usually be fitted to ensure valves can only be operated under controlled circumstances. Normal handwheels can be replaced by geared handwheels. Specific design requirements, bevel or spur gears, may not be available. Some handwheels can be replaced by a lever. Differential pressure capabilities may be reduced. Care should be exercised with valves which can be operated quickly and water hammer implications investigated.

For valves to be situated in elevated positions above working floor levels, chainwheels and chains may be available. Some valves however, must be positioned below working floor levels and for this extended spindles with universal joints can be fitted. For valves that need to be buried in earthworks, totally enclosed yokes with extended spindles can be supplied to accommodate most situations.

Valve bodies can have extra small connections for specific duties. Standard cast bodies can have bosses strategically placed to allow extra connections to be used without requiring pattern modifications.

The most popular addition is a drain. Depending upon the valve body design two drains may be necessary. Large valves can be fitted with a by-pass. By-passes can be used for preheating equipment during start-up or to maintain hot standby equipment at a temperature close to operating. A by-pass can also be used to balance a valve by relieving differential pressure loads on guides and seats. By-pass valves can be manual or powered to allow remote operation. Some valves can be fitted with a body cavity relief connection. The simplest auxiliary connections are screwed.

Large valves may have an option for flanged connections. Bosses suitable for screwed connections should be able to accept socket welded connections. The highest integrity connection would be butt welded. Auxiliary connections should have the same level of sealing integrity as the main process connections. If a valve is fitted with RTJ flanges or clamped proprietary connections to eliminate hazardous leaks it does not make sense to fit a screwed drain plug or screwed auxiliary piping. Bodies can be fitted with heating jackets to use steam or heat transfer fluids. This requirement must be established at the start of the inquiry stage.

Valve bodies and internals can have the surface finish improved to comply with hygienic requirements. The external surfaces can be similarly treated. As always these can be costly and time consuming processes, so investigate the impact before committing a purchaser order. Bonnet internals for some valve types can be improved to permit hygienic applications.

Bonnet and packing box modifications can be considerable. Bonnets can be extended to cope with low temperature, high temperature or the thickness of insulation applied to the pipework. Extended bonnets may be used to house bellows seals which eliminate the packing box. The packing box may be modified to accommodate vacuum service. Adding a water seal may be possible. Extra packing sets with a lantern ring for leak-off may be required. Packing which is live-loaded, to reduce routine maintenance, may be a worthwhile extra for use on sites where operator skill levels are low. See Chapter 11 for a complete view of packing box design.

Optional extras must be considered as part of the initial valve selection and costing. Some options cannot be fitted just before the valve is painted and valuable time and money may be wasted.

3.3.3 Rotary motion valves

Rotary motion valves operate on different principles to linear motion valves and can offer some advantages. With some designs of linear valves, with outside nut and rising stem, the stem withdraws through the packing box to open the valve. If the valve remains open for extended periods the external portion of the stem can be subject to atmospheric corrosion while the inner portion is protected. When the valve is subsequently closed the packing will not seal on the corroded portion.

The converse can apply satisfactorily when internal corrosion is more severe than external. When the valve is closed, the packing seals on to an uncorroded stem. The spindles on rotary valves do not usually move axially so the seals always work on the same portion of spindle. External corrosion agents do not affect the packing sealing function unless good access to the top of the packing is provided.

Internal corrosion can always attack the same portion of the spindle and careful material selection may be required for some applications. When the spindle and active member are separate components, corrosion and wear can increase backlash to an extent where the valve does not open or close completely even though the handle is moved to its stops. Under these conditions rapid corrosion and wear of the active member is possible due to high velocity fluid and turbulence. Because the same portions of spindle are always inside the body, erosion due to solids can also be a problem. Butterfly valves, and some types of plug valves, can fail due to spindle material loss. Valves with soft seats and non-metallic seals can insulate all the moving components from the body.

Static electricity generation and retention of charge can become a problem too. An operator can transfer a potential of 10 kV. Electrical continuity must be maintained to avoid arcing problems. Rotary motion valves are included in the group of valvs known as "quarter-turn valves". The ball/plug/butterfly valve only needs to rotate 90° to change from wide open to closed. Many of these valves are lever operated which allows an operator to quickly open or close a valve. In some applications rapid valve closure can cause water hammer, see Chapter 4, Section 4.1. If water hammer is likely to be a problem then operating levers should be removed and replaced by geared handwheels.

Rotary motion valves have fairly simple bodies when considering their function and, if one ignores material and production costs, makes them ideal for one-off specials. Valve bodies can easily be fabricated from thick wall tube or cast cored bar. High pressure valves can be machined from solid forgings. High tensile steel; BS 970 708M40, BS 1503 271-560 and AISI 4140; is an ideal material for high strength combined with good machining properties. It does have a drawback, inasmuch as it is tedious to weld. Pre-heat and post weld heat treatment are necessary. Final heat treatment for hardening and tempering to achieve the desired strength may be required.

Welding problems can be circumvented by design and selection of connectors. Flange facings can be machined directly on the body and studded. If studding is precluded because of access problems then bolts can be used in tapped holes fitted with thread inserts. Profiles for clamp connectors are easily machined from the solid. This option can provide the highest integrity pressure containment and remove any doubts regarding weld strength, fatigue and corrosion.

Ball valves in general

The ball valve can have very low fluid resistance and consequently very high C_v values. These properties depend upon the size of the port through the ball. Most ball valve designs can be both full-bore and reduced bore and specifications must be examined in detail to check the ball bore. Even with reduced bores, valves can be cleaned mechanically. If, in the larger sizes, "pigging" is envisaged then care must be taken to ensure full-bore valves are purchased. Ball valves should be used as isolating valves and not for throttling. The ball and seat materials must be compatible for rubbing contact under pressure. The rubbing velocity is low but the interface pressure can be very high. Self-lubricating materials, such as PTFE, are popular for seats for a wide range of applications. Designs with metal balls and metal seats can be used for solids handling applications. Valves with metal seats are more costly than soft seat valves but they are expected to last at least twice as long.

The ball itself can be made in many materials and can also be coated to improve wear and corrosion resistance and impart lubricating properties. Balls are regularly coated with:

- FEP
- PFA
- PTFE
- Rislan™
- hard facing such as Stellite™
- tungsten carbide
- ceramic, usually aluminium oxide

Stainless steel balls can be surface treated to increase hardness and eliminate galling tendencies. Coated balls should not be used when erosion is liable to occur. Erosion damage starts at corners and edges and progresses upstream as well as downstream. Erosion damage will commence where the bore emerges through the ball surface. The erosion will remove the coating from the corner and expose the substrate to accelerated corrosion/erosion damage. In these situations a hardened ball has proved superior to a coated ball. 17-4PH and AISI 440C can be used when corrosion resistance equivalent to AISI 304 is acceptable. A solid aluminium oxide ceramic, Al_2O_3, is available from a few manufacturers.

Several different seat design principles are used to achieve good sealing and low operating torque requirements. The ball usually has a very good surface finish to reduce friction forces and wear. Soft seats conform to the ball contours to establish a seal. Hard and metallic seats can be flat or curved. Flat seats tangential to the ball surface have a theoretical line contact but very quickly bed-in to create a narrow land. Some manufacturers call the initial transition "burnishing". Friction is very low and consequently operating torque is low. Hard and metallic seats with wide contact surfaces must be lapped-in to provide a good seal. Friction forces are higher requiring higher operating torque. Fire-safe valves may lose soft seats when the valve is hot. Back-up seals, of graphite or machined on the metal carriers, form temporary seals until the valve can be maintained. Valves with metal seats and graphite primary seals can survive a fire without damage.

Ideally the ball and spindle should be in one piece. This is a very costly design alternative and only used in high performance, expensive valves. In most ball valves the spindle is a separate component which engages with the ball to drive. To avoid backlash, the method of connecting the spindle to the ball must be positive so that the assembly functions as a single unit. Free play between the ball and the spindle will cause wear and perhaps accelerate corrosion. The most favoured alternative solution is a splined joint between spindle and ball. Some manufacturers use multiple keys to transmit the drive which are almost as good as splines.

The most popular method used is a tongue engaging in a slot in the ball. This design is useful in that it allows the ball to be fitted in the body axially rather than from the top or the bottom. The actual method chosen depends upon the body design philosophy adopted and the operating conditions. If a valve is to be opened and closed frequently a good connection is essential. Also consider the number of opening/closing cycles between possible maintenance access. The ability to repack the gland without removing the valve from the line or depressurising may be essential. When judging the integrity of a valve design the number of potential leak paths to atmosphere should be considered.

Some ball valves only have one seat and will only seal in one direction. These valves will have an arrow indicating the direction of flow. If sealing in both directions is essential this must be specified as a requirement to avoid costly surprises later.

One-piece body ball valve

The one-piece body ball valve is an inherently high integrity valve. The number of leak paths to atmosphere is reduced to a minimum. Standard valves have a single body casting which can include the bonnet. Small valves with screwed connections can be machined from barstock. Special purpose high integrity valves can be manufactured by machining from a solid forging, internal access being gained through the bonnet and bottom flange when fitted.

Standard valves are assembled through one of the end connections, as shown in Figure 3.35. The body casting is not symmetrical. Although the flange connections are identical in size the body passage from one flange is larger. A seat is fitted into its location first, then the spindle is fitted through the bore. Some designs have pressure energised packing and no gland. The packing assembly is fitted to the spindle beforehand. The ball is actuated by a tongue on the stem. The stem must be rotated to the correct position to allow the ball to slide in. The second seat is fitted to an insert which is assembled in the valve bore. The closing insert may be screwed into the body or attached by small screws. The insert is sealed by a joint but also forms all or part of the flange facing of the connection. The ball is located by the two seats and floats to equalise the seating pressure.

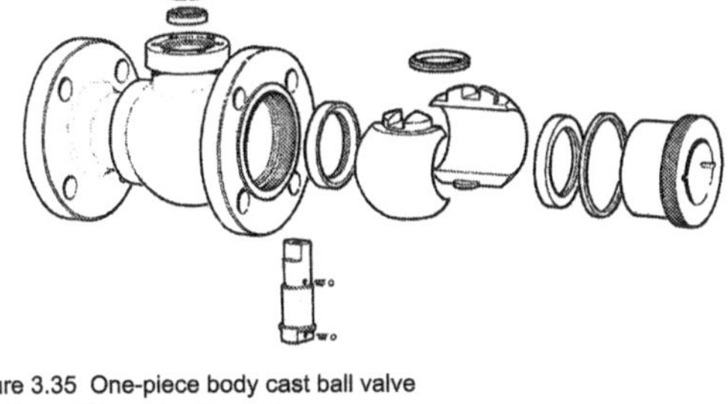

Figure 3.35 One-piece body cast ball valve
Courtesy of Tyco Flow Control

These valves will work in either direction but a good case can be made for using them in one particular way. If the seat which is fitted directly to the body is placed downstream, all seat loads when closed are carried by the casting. However the high pressure fluid is constantly exposed to the insert joint. If the valve is used in the other direction the closed seat load is carried by the insert.

Small barstock valves are assembled by inserting all the internals through the bonnet fixing. Screwed bonnets and glands are standard. The small ports involved result in low stresses. Valves made in austenitic stainless steel can be capable of 690 barg.

Special valves are machined from solid forgings. The bore for the bonnet must be large enough to pass the ball. Connections can be machined directly on the end of the forging or added by welding. Clamped connectors for all openings, including bonnet and bottom flange, ensure a very high integrity assembly.

A technique had been developed to make the ball in two halves. The bottom of the ball is located in position by a shaped lug. The top of the ball is located and held together by a ring around the driving slot.

One-piece body valves are available in a wide range of sizes and materials. Popular reduced bore valves, suitable for pressures of 20 barg or 50 barg, are available from ½". These can be cast in ductile iron, steel, stainless steel, Monel™, Alloy 20 and aluminium bronze.

Temperature ranges for soft seats are -50° to 230°C. For valves over 6" to about 16" the standard materials are steel and stainless steel. Popular full bore valves are available in sizes up to 14".

Small stainless steel valves, ¼" to 2", with PTFE seats can operate at pressures up to 100 barg with a temperature range from -40° to 230°C.

Nickel plated brass valves, ¼" to 4", with reduced bores are suitable for low pressure applications up to 20 barg for temperatures between -30° and 180°C.

A range of barstock valves in brass or AISI 316 stainless steel, stainless steel ball and PTFE coated metal seats, is produced with screwed and compression fitting connections. Full and reduced bore patterns are assembled by an internally screwed combined bonnet/gland. Female connections, ⅛" to ½" NPT, compression fittings ⅛" to ¾" od tube, are suitable for pressures up to 207 barg for temperatures between 10° and 65°C.

Special one-piece ball valves, with integral metal seats, are suitable for cryogenic applications and hot services to more than 1100°C with operating pressures from vacuum to over 3000 barg. The metal seats make these valves suitable for solids handling applications.

A wafer style ball valve made in titanium is also available. Sizes range from DN15 to DN200 with ANSI pressure ratings of 150lb and 300lb.

A special type of ball valve, the rising stem ball valve, is available from a few manufacturers and is claimed to offer superior sealing and better packing box facilities. The quarter turn motion of the valve ball is produced by a rising stem which has a profiled slot engaged with a fixed pin in the bonnet. The quarter

Figure 3.36 Tilt and turn ball valve
Courtesy of Cameron's Valve & Measurement Group

turn rotary motion is only commenced after the ball has been tilted to unload the metal seats. The rising stem is actuated by a screw in a similar manner to a gate valve. Figure 3.36 shows a cut-away view of a typical tilt and turn valve.

Two-piece body ball valve

Small barstock valves for instrument and allied applications can be made using the two-piece body construction. One seat is inserted down the bore of the body. Then the spindle is fitted up through the body into the integral bonnet. The ball must slide into the body and engage the tongue on the spindle. The second seat is fitted. The end of the body is closed by a screwed union which forms the other process connection. These valves, in sizes from 1/8" to 1/2" NPT, can be capable of 414 or 690 barg.

In small cast valves the two-piece body design enables the bonnet to be made integral with the body by replacing the bonnet junction by a body junction. The ball and seat assemblies are located and compressed by a bolted end piece, sometimes called an adaptor. In larger valves the end opening is the only access to fit the ball. A seat assembly is fitted into the body first. The ball is turned by a separate spindle which has a tongue to engage with a groove in the top of the ball. The spindle, with its thrust or sealing washer, is fitted next followed by the ball. Enough space is left for a seat to fit into the body next to the ball. The loose end piece can now be bolted up to hold all the internals in place. The bonnet can be fitted to larger valves. The packing and gland can then be assembled, followed by the handle.

A variation on the two-piece design uses two identical body halves. The body is split down the spindle centre-line. Bolting the halves together locates the ball and the seats. A separate bolted bonnet is used to retain the spindle. Some valves can have spring-loaded metal seats fitted; these would be suitable for solids handling and high temperature applications.

Small valves in ductile iron or brass, 1/4" to 2", with PTFE seats and seals can operate between -20° and 160°C with pressures up to 64 barg. Similar valves in cast steel are suitable for -40° to 230°C at pressures up to 100 barg.

A restricted size range, 6" to 10", is offered in a wide range of body and seat materials. Bodies are cast in:

- ductile iron, carbon steel, aluminium, bronze

Figure 3.37 A lined two-piece body full-bore ball valve
Courtesy of H H Valves Ltd

- aluminium bronze, Monel™, stainless steel, Alloy 20

The reduced-bore valves with integral bonnet and gland are suitable for ANSI 150lb and 300lb pressure ratings and operating temperatures can span -200° to 370°C with metallic and non-metallic seats. All valves except aluminium bronze can be fire-safe.

Stainless steel full-bore valves are mass-produced in sizes from DN25 to DN600 in pressure ratings of PN10, 16, 25 and 40. Bolted bonnet and gland are standard with PTFE packing. Temperature range is -50° to 230°C. Metal seats are an option. Carbon steel versions up to DN300 can have a temperature range of -200° to 450°C.

Valves split around the spindle centre-line are produced in steel and stainless steel from 2" to 36". Pressure ratings cover ANSI 150lb, 300lb and 600lb. Various seat and seal options can be fitted, including metallic and graphite, which cover operating temperatures from -200° to 600°C.

Two-piece body valves are available in both full-bore and reduced-bore styles. These can be fully lined with an elastomer for corrosion protection. FEP, PFA and UHMWPE are popular compounds. Lined valves are mass-produced in sizes from DN25 to DN150 for cold working pressures up to 19 barg. Figure 3.37 shows a full-bore ball valve with integral bonnet and spring-loaded chevron packing.

Three-piece body ball valve

Three-piece body ball valves can be both full-bore and reduced-bore and specifications must be examined in detail to check the ball bore. The three-piece body construction is very versatile and allows many variations in detail design. An ideal method for the manufacture of large valves (over DN1 000 is claimed by some companies), but this method can also be suitable for small valves. The basic design concept is a ball mounted in a central portion, with two identical end connectors clamping the seats in position. These valves are suitable for liquids and gases.

Small metallic valves, up to 3" or DN80, follow a common pattern. The central portion is designed as a wafer or semi-lug body and includes an integral bonnet and gland. The spindle is assembled through the main bore and passed up through the bonnet. The spindle may be sealed by "O" ring, pressure energised chevron packing or disc spring-loaded plain packing. The spindle is retained in the body by a nut and possibly loaded by a wave spring. The spindle is made "blow-out proof" by including a collar which cannot pass through the packing box. The spindle motion is transmitted to the ball by a tongue engaging in a slot. The spindle must be positioned so that the ball can slide into the body and along the tongue.

A seat is placed in each end of the body and the end pieces bolted up. The seats locate the ball in the body. Sealing is accomplished by fluid forces moving the ball slightly towards the downstream seat. Pipe connections can be: female screwed, compression fittings, socket and butt weld and flanged. Handles are usually made of flat bar with plastic grips. Round and shaped handles are available. Bolted gland versions with pressure energised packing are also available. Some manufacturers offer inserts to completely fill the body cavity. As the valve size increases variations in packing box and seat design develop.

Larger valves have the end sections attached independently to the centre section by studs and nuts. Round flanges can be used as there is no requirement to extend beyond the centre section diameter. These studs are mostly protected and are not prone to the severe heating problems in the fire-safe versions, an example of which is shown in Figure 3.38.

Metal valves can be lined with non-metallic materials for corrosion resistance. FEP, PFA and UHMWPE are popular choices. These materials exhibit the chemical resistance of PTFE but possess superior moulding qualities. The lining can be separate and replaceable or actually moulded to the valve body. With the latter the lining is attached by plain or dovetail grooves. In some designs the lining extends through the packing box. The generation of static electricity can be a problem and a complete electric circuit must be ensured to avoid arcing.

Figure 3.39 shows the construction of a three-piece body wafer ball valve with a thermoplastic lining. The wire for electrical continuity is clearly visible. Be vigilant when selecting valve manufacturers! Small valves of this style are full-bore but larger sizes might be reduced-port. Site staff might get a nasty shock during commissioning!

Non-metallic valves are built in a similar manner. Valves up to 2" may have union ends rather than four bolt flanges or sockets for cement connections. Small brass valves, for domestic and commercial applications, are available in union construction. These valves, with screwdriver slots instead of handles, are used as maintenance isolation valves in low pressure systems.

The basic design can have weaknesses. The valve body has two joint faces which must be sealed. The compression of the faces is supplied by the bolts holding the valve body together. The same bolts apply the compression to the seats.

The design of the flanges and the sizing of the bolts are crucial. The bonnet and gland can be made very short to save material in the body and make the valve low profile. Insufficient bearing area for the spindle can result in rapid wear of the spindle, body, or both and lead to short seal/packing life. If the valve is not operated regularly then this will not be a problem. If the slot in the

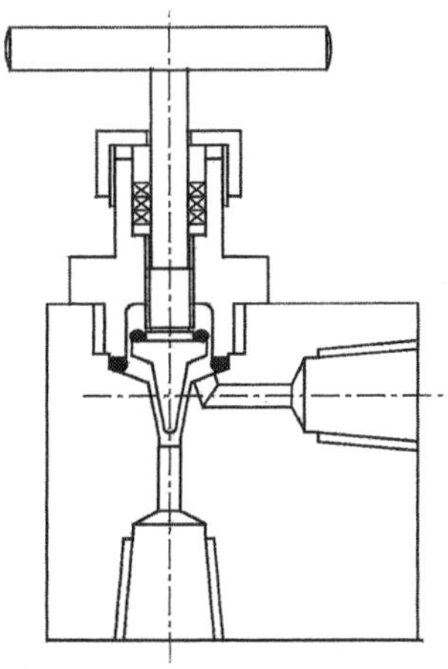

Figure 3.38 A fire-safe three-piece body ball valve
Courtesy of Swagelok Cpmpany

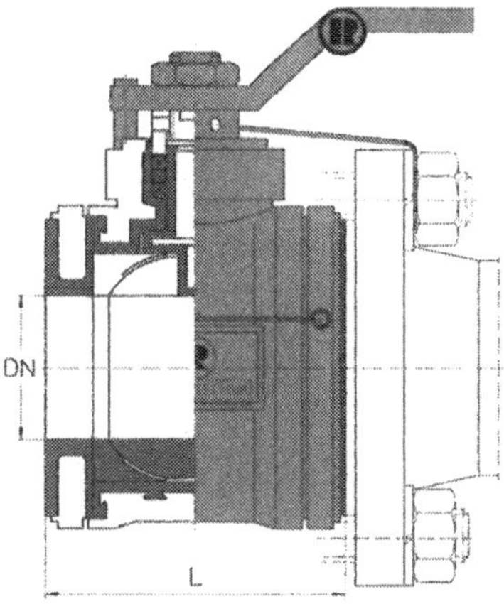

Figure 3.39 A lined three-piece body valve
Courtesy of Richter Chemie-Technik GmbH

ball is considerably wider than the tongue thickness, excessive backlash will occur. Erosion of the ball bore can result with possible damage to the seat especially in reduced bore valves. Seats which have too little bearing area on the ball can wear rapidly with use.

Depending upon the design of the seats, the valve body may always be pressurised; a vent hole may be drilled in the ball for equalisation. To depressurise the body the valve can only seal in one direction. When closed the vent hole must point downstream. The tie bolts which clamp the ends together are critical for the successful operation of the valve. Valves may be fitted with carbon steel or austenitic stainless steel bolts. In critical applications, and especially with fire-safe versions, it may be advisable to upgrade to alloy steel, high tensile steel, or a stronger grade of stainless steel. Because the bolts are almost completely unprotected they can become very hot in an external fire.

Some valves only have one set of ends which are machined to provide screwed, socket or butt weld connections. Flanged connections are provided by screwing in or welding on separate flanges. The type of construction should be evaluated when use with corrosive or toxic fluids is considered. In metallic valves the ball is generally made of austenitic stainless steel irrespective of the body material. Localised galvanic corrosion may be a problem with some material combinations.

Seat designs vary. Plain elastomer seats can be self-loaded due to compression by the bolted ends. Elastomer inserts in metal carriers are used on some higher pressure valves. Metallic seats and some elastomer seats are loaded by springs. Fire-safe valves, in sizes from 2" to 42" have spring-loaded seats. The rigidity and temperature limitations of elastomer seats has a marked effect on valve capabilities. Table 3.10 indicates the variation on a stainless steel valve.

Seat material	Low temperature	High temperature
PEEK	172 barg @ 175°C	138 barg @ 260°C
UHMWPE	207 barg @ 38°C	17 barg @121°C
filled PTFE	152 barg @ 38°C	55 barg @ 150°C
virgin PTFE	100 barg @ 93°C	55 barg @ 150 °C

Table 3.10 Pressure and temperature variations of elastomer seats

Valves are produced in a wide range of materials and pressure ranges as indicated by Tables 3.11 and 3.12

Body	Ball	Seat	Seal
SG iron	Stainless steel	Virgin PTFE	Grafoil™
Carbon steel	Ceramic	Filled PTFE	Graphite
Stainless steel	Tungsten carbide coated	Acetal	Vespel

Body	Ball	Seat	Seal
Duplex stainless steel	Ceramic coated	UHMWPE	Viton™
Brass	Carbon steel	PEEK	EPDM
Aluminium bronze	PFA coated	EPDM	Tungsten carbide coated
Nickel aluminium bronze		Viton™	
PVC		Grafoil	
PP		Tungsten carbide coated	
PVDF			

Table 3.11 Three-piece body ball valve materials

Sizes	Ball bore size	Rated presssure barg at 38°C
DN10 to DN50	full	138
DN10 to DN150	full	345
DN10 to DN150	reduced	207
¼" to 2½"	full	70
¼" to 2½"	reduced	100
¼" to 4"	full	100
DN50 to DN1 000	full	20 to 414
DN80 to DN200	reduced	50
⅜" to 2" (1)	reduced	8.5

(1) Union construction with PVC, PP or PVDF bodies and balls

Table 3.12 Sizes and pressures of popular ball valves

Trunnion-mounted ball valve

The ball valve designs discussed previously have floating balls, the ball being located by the seats. The trunnion-mounted ball valve has two journal bearings to locate the ball within the body. Trunnion-mounted valves are more costly than other designs because of the extra components and machining involved. Small low pressure valves, do not need the extra support of the trunnion bearings. The loads applied to the ball can be adequately supported by the seats. Trunnion-mounted ball valves are intended to seal equally in both directions. The body may be fitted with a bleed port to monitor seat leakage; allowing double block and bleed operation. Some manufacturers fit safety plugs to prevent the body being pressurised above a certain pressure. If the body cavity is required to sustain full line pressure then this should be specified.

Trunnion-mounted ball valves can be built with one, two or three-piece bodies depending upon the pressure and temperature requirements and the severity of the application. The valves have spring-loaded or elastically-mounted seats to ensure good sealing contact. Because the seat loading is largely independent of the body the ball cannot become jammed due to body distortion caused by excessive pipe forces and moments. This type of valve offers the highest mechanical reliability when designed with dependable closures and joints. With suitable material selections the fire-safe versions can exceed minimum standard requirements.

The highest quality valves will have the trunnions integral with the ball. This ensures true alignment under all operating conditions. The actual bearing surface of the trunnion may be a replaceable sleeve. The bearings will be replaceable inserts in the bonnet and bottom flange. Popular bearing materials are austenitic stainless steel which has been surface treated to eliminate its galling tendencies and PTFE. Various options have been introduced to eliminate the high cost of integral trunnions. Short stub shafts have been keyed and/or shrunk into blind holes in the ball. These can cause problems due to differential expansion if thermal transients occur. Dissimilar materials can magnify the problem. The bottom bearing can be reversed so that the bearing is in a blind hole in the ball and the trunnion is a stub shaft attached to the bottom flange. This is a design device and reduces the mounting problems by 50%.

Figure 3.40 One-piece forged body valve
Courtesy of H H Valves Ltd

One-piece body valves are sometimes called "top entry valves". All major components are fitted through the top of the body before the bonnet is attached. These valves usually have a smaller bottom flange to house the bottom bearing. This style of construction is suitable for valves of the highest integrity. Valve bodies can be machined from solid forgings, plain rectangular or shaped, rather than castings, see Figure 3.40. Notice, the valve has ring-type joint flanges, the highest integrity flange connection commonly used in the process industries.

The highest pressure ratings required can be accommodated using this construction. Valves up to about DN150 or DN200 are available for API 10 000 pressure rating. Process connections can be flanged directly to studded machined facings, or standard flanges can be attached by butt or socket welding. Socket welded fittings are difficult to inspect for weld integrity. Radiographs are often difficult to interpret and should not be used for the highest integrity requirements. Profiles for clamped connections are easily machined on forgings, also for bonnet and bottom flange, and can offer extremely high integrity joints.

For high integrity pipeline duties, where the valve is permanently installed, the process connections can be butt weld, see Figure 3.41. The butt weld stubs are machined directly on to the forging. Heating, during welding, can create distortion problems for precision components. To avoid this problem, the valve manufacturer welds a short section of pipe to the valve body **prior** to final machining. At this stage, the fabrication can be heat-treated if necessary. The site welds are now remote from the valve body, the previously fabricated pipe length provides a heat sink for site welds. Valves of this style, which are welded in, can be maintained, in some cases, by pipe freezing rather than

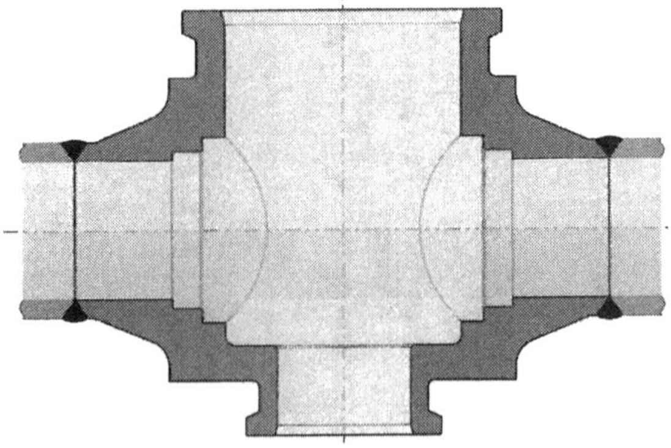

Figure 3.41 A forged valve body with butt weld process connections

draining a section of the pipeline. Figure 3.41 shows the top and bottom entries machined for a clamped attachment

Seats for these valves may be required to have more axial movement than normally expected and will probably have a spring-loaded metal carrier. The springs may be multiple helical coil springs in pockets or stacked disc springs. The carrier must be sealed in its counter-bore to prevent leakage. Elastomers are suitable for temperatures up to 230 - 260°C. For higher temperature applications graphite is used. Graphite seals and packing rings, combined with metal seats allow these valves to operate at temperatures over 850°C. Valves of this type can be used for dirty fluid or solid-fluid mixtures. Mildly abrasive mixtures are accommodated by fitting shields over the carrier springs and using metal seats. For very abrasive or erosive applications the seat is replaced by hard facing on the carrier. Valves with forged bodies will have to be made to order. Whilst this makes them more costly and longer to deliver they can obviously be customised to suit individual project specifications.

Severe-duty trunnion-mounted ball valves can be built in a similar manner to two-piece body ball valves, Figure 3.42. One spring-loaded seat is passed into the main body and located in its counter-bore. The ball is then fitted from the top followed by the bearings which centralise the ball. The bottom flange should be fitted next to retain the bottom bearing. The second seat is fitted to the other body half which can then be bolted up. The bonnet and gland assemblies complete the valve.

When fitted with spring-loaded packing the gland assembly may be eliminated; the fixed length packing space being incorporated in the bolted bonnet. This style of construction is ideally suited to valves with metal seats and metallic seals. Valves of this type are produced routinely in cast and forged carbon steel and austenitic stainless steel as indicated in Table 3.13. Operating temperatures from -200 to 1000°C can be accommodated.

Pressure rating ANSI	Nominal bore in
150lb / 300 lb	1/2 to 42
600lb / 900lb / 1 500lb	2 to 20
2 500lb / 4 500lb	2 to 12

Table 3.13 Sizes and ratings of two-piece body trunnion-mounted ball valves

Three-piece body valves provide the capability of producing the largest valves economically. Valves up to DN1000 are produced to standard designs e.g. two identical cast or forged end sections bolted to a central core with provision for the bearing mountings and the bonnet. The design is also suitable for

Figure 3.42 A trunnion-mounted ball valve with two-piece body
Courtesy of H H Valves Ltd

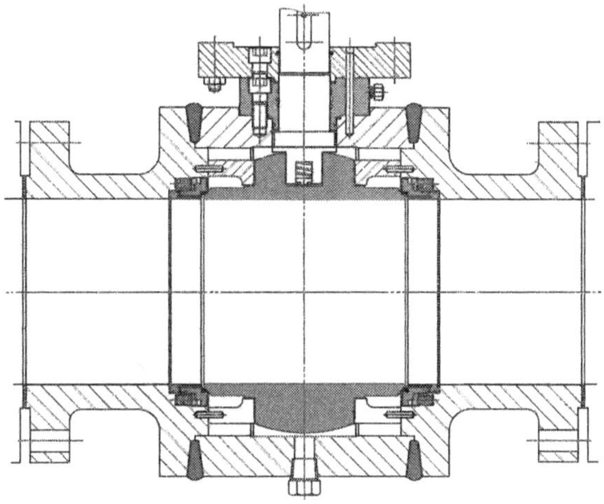

Figure 3.43 A valve design to avoid!

smaller valves and is available in sizes down to DN50. The larger sizes, even at low pressure, produce such high axial forces on the ball that trunnion bearings are necessary to reduce seat loading pressures to acceptable levels. Valves are usually flanged to allow easy removal from the pipework. Some manufacturers allow the upstream end of the valve to be butt welded directly to the pipework for increased integrity. Valves are available in all popular materials for pressure ratings from ANSI 150 lb to 2500 lb.

Some users have experienced problems with sediment accumulating behind the seat carriers and preventing the seats from moving correctly. If the fluid is clean there can be no solids that will drop out and jam the seats. If the fluid is specified as "dirty", the valve manufacturer must take precautions against this contingency.

The valve design shown in Figure 3.43 was offered recently for a high integrity application. The design assumes its "high-integrity" qualifications comes from the fact the body is welded and has no seals to leak; notice the valve cannot be dismantled for maintenance! However, in this case, the design is inconsistent and is not high integrity. A high integrity process would not use flat-face flanges and simple joints. Nor would a threaded taper plug be acceptable. The designers assumed that removing the two major joints would be sufficient. They were wrong!

High integrity processes require all potential leak paths to be eliminated, or at the very least, to use the highest quality seals. The process connections could easily be converted to ring-type joints. The drain connection could be converted to a parallel thread with a metal seal protecting the thread or be replaced by a flanged connection. Oversights like this should alarm potential purchasers. What else have they forgotten?

Small valves with a 4.8 mm port through the ball are available in stainless steel. The trunnion bearings are larger than the ball to allow assembly through the bottom bore, see Figure 3.44. The bearings include a central "O" ring to seal the body. The two seats are located in metal carriers which are spring-loaded by disc springs. The seat carriers are fitted through the inlet and outlet ports and retained by screwed adaptors which provide the system connections; 1/8" of 1/4" NPT female, 1/4", 3/8" or 1/4" compression fittings. The blow-out proof stem is restrained by a step in the integral bonnet and fitted with a non-metallic bearing and "O" ring seal. Operating pressures up to 414 barg for temperatures between -29° to 121°C are possible when Kel-F seats are fitted. PTFE seats increase the maximum temperature to 232°C but the pressure is reduced to 100 barg maximum. All the threads in these valves are protected by seals. The compression fitting versions will be suitable for NACE without modifications.

Figure 3.44 Small trunnion-mounted ball valve
Courtesy of Swagelok Company

A special type of trunnion design has very limited availability. The ball is formed by a central section with the trunnions and two almost hemispheres attached by a snap ring. Top entry bodies are possible allowing high pressure capabilities. Sizes from DN50 to DN300 can be supplied for ANSI pressure ratings from 150 lb to 2500 lb and API ratings up to 15 000 psi.

Multi-port ball valve

Ball valves can be used for mixing and diverting applications. In this context the valves described are not intended for proportioning; a control valve is required. A multi-port ball valve can divert the full flow from the inlet to one of two or more outlet connections. Conversely a valve can accept flow from two or more inlets and direct them to a single outlet.

Trunnion-mounted designs with independently sprung seats are preferred for these applications. Seat wear due to ball movement should be evenly distributed on all the seats. However the chemical action of the process fluid and the possibility of drying-out of some seats can lead to significant differences in seat condition and volume. Body designs vary considerably and should be judged on each application. Cast bodies using standard components are economical for production-run batches. Special valves can be fabricated or machined from the solid depending upon the material and the pressure rating.

Small three port valves, DN10 to DN50, are suitable for 69 barg at temperatures up to 120°C. These valves can be made in steel, stainless steel and brass with virgin PTFE, filled PTFE or UHMWPE seats. Connections can be screwed, socket weld or butt weld.

Valves with three, four or five ports suitable for 50 barg are made in various materials:

- carbon steel, stainless steel and nickel aluminium bronze

Seats can be of virgin PTFE or glass filled PTFE. Sizes from DN10 to DN65 can be screwed, socket weld or butt weld. Sizes from DN25 to DN150 are flanged.

Two-piece body valves can be adapted to three port valves. Available in various materials in sizes from DN15 to DN200, these valves are rated for PN16 and ANSI 150lb.

Special versions of multi-port ball valves are available for water supply applications where a supplementary supply is possible when the pressurised mains supply fails. A three port mixing valve is fitted with a fourth atmospheric port which allows any trapped supplementary water to be drained prior to reconnecting the mains supply. The mains supply port is fitted with a non-return valve to prevent reverse flow. These facilities prevent contamination of the mains supply. Valves with screwed unions for ½" and ¾" are made in bronze with PTFE seats. Maximum water pressure is 10 barg at up to 65°C.

Butterfly valve

One of the earliest uses for butterfly valves was the throttle on petrol engines. A simple valve, with low friction torque and relatively low fluid losses was required. A brass/bronze disc was used inside an aluminium or zinc alloy casting. High performance engines however, which require the lowest possible inlet losses, use a style of conduit gate valve.

The butterfly valve has become very popular as a process valve because it is smaller and much lighter than gate or globe valves. Wafer and lug designs are the favourite body styles for valves between DN50 and DN300. Single flange and double flange body styles are also available. Simple construction allows large valves to be easily manufactured, typically between 6 and 10 m diameter. The basic idea is very simple but design variations can be introduced to improve performance. Figure 3.45 indicates the fundamental differences in the basic design approaches.

3.45(a) shows the basic simple design concept; a symmetrical disc is mounted centrally in a symmetrical plain bore. This design is know as "concentric" or "centric". Sealing in both direc-

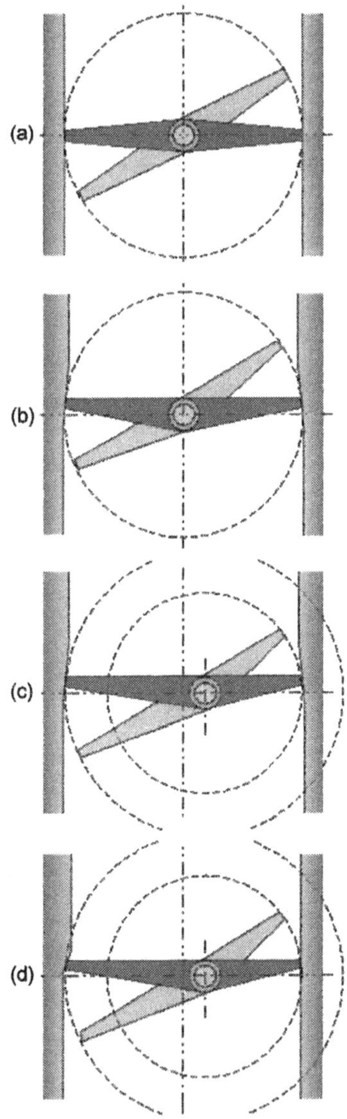

Figure 3.45 Butterfly valve — various operating principles

tions is identical; flow in both directions is identical. Valve rotation can be in either direction. This style of design is useful with soft seats in the valve body. The disc profile can be modified to a "fish-tail" to improve regulation but this makes the valve uni-directional. This style has a significant drawback when used with soft body seats, or lined bodies, to provide a tight seal. The tips of the disc are dragged across the seal surface during opening and closing. Rubbing wear can be a problem when valves are opened/closed frequently. The design modifications which can be implemented try to reduce or eliminate the rubbing wear.

3.45(b) shows a butterfly valve with the disc mounted eccentrically on the central spindle. This design style is known as "single eccentric". The valve body incorporates a symmetrical taper for the seating surface. The disc tips tend to lift off the body seat as the disc rotates. The disc usually incorporates a wedge profile to improve sealing; disc rotation is uni-directional. The valve does not have identical flow patterns. The flow direction upwards, in the figure, will produce less turbulence and losses than downwards.

3.45(c) shows a further variation on the eccentric approach; a "double eccentric" valve. The spindle is moved from the centreline of the valve body. The valve body taper seat is symmetrical. The offset spindle exaggerates the tip lift off and reduces the rubbing wear even further. This style is promoted for applications which require frequent opening/closing. Rotation is uni-directional because of the disc wedge profile. The flow is also better in one direction compared to the other.

3.45(d) shows the design known as, the "triple eccentric". The body taper seat is not symmetrical. Using this design approach it is possible to eliminate all disc/seat rubbing. The disc touches the seat as it seals. This design allows the construction of "all metal" valves which do not require soft seats. The body shape produces much higher losses than the other designs. Concentric valve designs may have C_v values which are 50% to 100% larger than the "eccentric" valves.

Depending upon the seat style, butterfly valves can be used for isolation and regulation of clean fluids. Body linings and disc coatings can be applied to resist corrosion and erosion. Valves can be fitted with replaceable elastomer seats or flexible metal seats. Flexible metal seats are more costly and more difficult to set up to achieve a good seal but they do last longer. Metal seats can be extended to eliminate the short, high velocity fluid paths created just before closing. Metal seats for liquid-solid mixtures can be shielded from the main flow path to extend useful life. Flexible seals, elastomers and metallic, can be built into the disc edge to allow the valve bore to be perfectly smooth. The butterfly valve with resilient seals is free from jamming problems caused by thermal or physical distortion. Fire-safe valves can have elastomer and metal back-up seats.

A common method of reducing leakage is to line the valve body with rubber or other soft elastomer, as shown in Figure 3.46, against which the valve disc presses when closed. The shaft can be located eccentrically or obliquely through the disc in such a way as to obtain a continuous periphery which seals against the valve body.

Alternatively, the disc can be fitted with a moulded ring which seals against the body, as seen in Figure 3.47. A butterfly valve, for example, when fitted with a soft-sealing ring passed a leakage of less than 10^{-7} cm^3 of helium per second when tested at full differential pressure, this corresponds to about 30 bubbles per year. A standard test requirement is for no gas leakage to be visible during a 15 second or 1 minute test. High integrity valves would have to be tested for much longer to assess the seat leakage accurately.

The method of connecting the shaft to the disc must be positive without backlash so that the assembly functions as a single unit. Free play between the disc and the shaft may encourage

Figure 3.46 Lined butterfly valve
Courtesy of Tyco Flow Control

corrosion. Ideally the shaft and disc should be manufactured as a single component, however this tends to be costly. The most favoured alternative solution uses splines, keys or taper pins. Splines are preferable for hygienic applications. Care must be taken to ensure corrosion does not weaken the connection by selective attack. The actual method chosen often depends upon the operating conditions. The spindle does not have to go right through the disc. It can be in two pieces and fit in blind

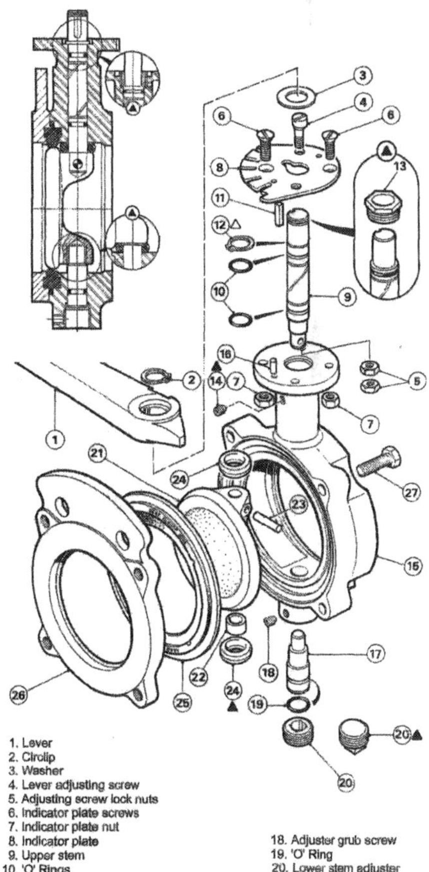

1. Lever
2. Circlip
3. Washer
4. Lever adjusting screw
5. Adjusting screw lock nuts
6. Indicator plate screws
7. Indicator plate nut
8. Indicator plate
9. Upper stem
10. 'O' Rings
11. Stem key
12. Stem clip
13. Stem adjuster
14. Adjuster grub screw
15. Body
16. Indicator plate locking dowel (where fitted)
17. Lower stem
18. Adjuster grub screw
19. 'O' Ring
20. Lower stem adjuster
21. Disc
22. Lower stem bearing
23. Disc pin
24. Sealing ring(s)
25. Seat
26. Seat clamp
27. Seat clamp screw

Figure 3.47 Butterfly valve with external bolted retaining flange

holes. The blind holes can be square to provide very good torque transmission. The bottom spindle-half can be locked in the body and a bearing fitted in the disc, as can be seen in Figure 3.41. Most valves are operated by a spring-loaded lever which can be locked in intermediate positions for regulation.

The valve can be assembled in a number of ways depending upon the detailed design. Valves with solid one-piece bodies must be assembled through the packing box and a bottom flange if the spindle is split. Pins may be required to locate the disc. Complete bore inserts can be fitted in the body prior to the disc and spindle, see Figure 3.40. The body may be split about the horizontal centre-line to produce top and bottom halves. The spindle can be integral with the disc and no bottom flange is necessary. The body must be long enough to provide space for external clamping bolts. These valves can have deep spherical seatings machined in the body. Valves can be assembled using an external or internal screwed or bolted ring to hold the seat in place, see Figures 3.47 and 3.48.

Spindles can be one or two piece. Pins are used to locate the disc. Valve bodies can also be built in two halves around the vertical spindle centre-line. Wafer style valves can be made very compact. The problem of spindle erosion is overcome in some designs by using "O" rings. Usually the spindle bearings are of the plain sliding type, but large valves may use roller or needle roller bearings. The typical bearing materials include bronze, nylon, graphite and PTFE composites. The manufacturer may advise installing the valve with the spindle horizontal to ensure the weight of the disc/spindle is supported by the bearings rather than the valve body or the body seat.

The packing box can be extremely simple — a single "O" ring, or more complex. The primary seal may be an "O" ring or a PTFE labyrinth seal followed by spring-loaded packing. Provisions can be made for lubrication, venting or purging. Extended packing boxes are available for hot and cold applications.

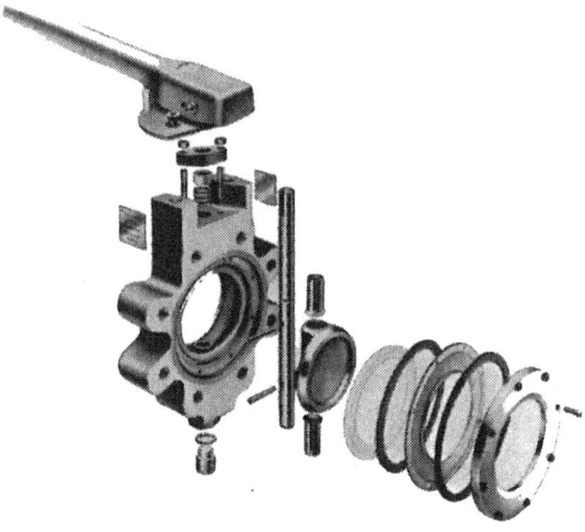

Figure 3.48 Butterfly valve with internal bolted retaining flange
Courtesy of Tyco Flow Control

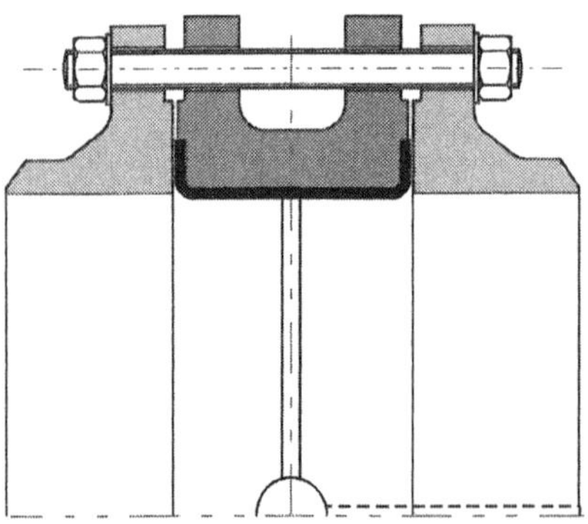

Figure 3.49 A lined double flange butterfly valve

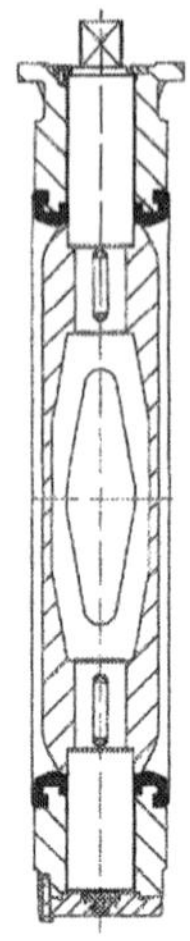

Figure 3.50 A lined wafer-style butterfly valve
Courtesy of AMRI Inc

Figure 3.49 shows the construction and installation of a lined double flange valve. Note that the disc still extends into the pipework even with a wider body. Figure 3.50 shows a cross-section through the other plane; the construction where the spindles pass through the lining is obvious. Figure 3.51 shows a typical example with a geared handwheel.

Figure 3.52 shows a large lined double-flange butterfly valve. The potential problems of the disc interfering with adjacent fittings/equipment are obvious.

Butterfly valves are manufactured in a wide range of materials, Table 3.14 indicates some popular options.

The body and the disc can be coated with various surface treatments to enhance properties:

- epoxy resin, Halar™, Kynar™, Nylon 11™ and Rislan™

Figure 3.51 Typical double flange butterfly valve with geared hand wheel
Courtesy of Tyco Flow Control

Figure 3.52 A large lined double-flange butterfly valve
Courtesy of AMRI Inc

Body material	Disc material	Elastomer seat	Metal seat
PVC	Cast iron	Nitrile rubber	Inconel™
Cast Iron	Ductile iron	Viton™	
Ductile iron	Aluminium	EP rubber	
Carbon steel	Aluminium bronze	EPDM	
Aluminium bronze	Steel	PTFE	
Stainless steel	Stainless steel	UHMWPE	
Duplex		PEEK	
Titanium			
Monel™			

Table 3.14 Butterfly valve materials of construction

Large valves may have a two-pack epoxy coating as the standard finish. Wafer and lug body valves can have liners which extend around the end facings to provide a gasket. Popular lining materials include:

- Buna N™, EPDM, UHMWPE, Hypalon™, Neoprene™, PTFE and Viton™.

PTFE linings can be backed by an elastomer to provide some "spring". Lined hygienic valves, up to DN80, can be used up to 10 barg between -40° and 120°C. Standard butterfly valves are mass-produced in a very wide range of sizes and pressure ratings. Table 3.15 indicates the more popular choices.

Size	Pressure rating
DN25 to DN100	ANSI 600lb, 900lb
DN50 to DN900 lug or wafer	ANSI 150lb
DN80 to DN1200	ANSI 150lb, 300lb, 600lb
DN50 to DN2000	PN10
DN1600 to DN3700	PN6
DN50 to DN600	ANSI 300lb or PN25 or PN40
DN80 to DN800 (-200 to 260°C)	PN10, PN16, PN25, PN40
DN80 to DN1500 (-200 to 700°C)(1)	PN10, PN16, PN25, PN50

(1) Fire-safe

Table 3.15 Butterfly valves sizes and pressure ratings

Cryogenic applications have become larger in scale over the recent past. LNG, liquefied natural gas, is carried by ships all over the globe to countries experiencing shortages of natural resources. Liquefaction, storage and transportation are now conducted on a very large scale. Figure 3.53 shows a butterfly valve specifically designed for butt-welding into an LNG piping system. It can be seen that the design includes access for internal maintenance in-situ. These valves have replaceable metal seals clamped to seat. Figure 3.54 shows a cryogenic gate valve being tested in the factory before dispatch.

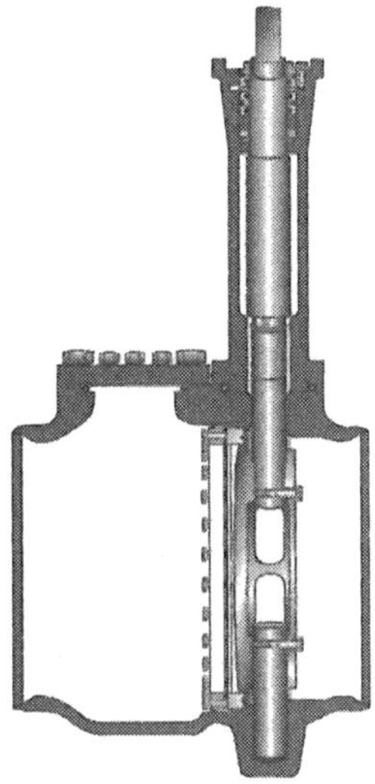

Figure 3.53 A cryogenic fire-safe butterfly valve
Courtesy of AMRI Inc

Some manufacturers produce butterfly valves with inflatable linings. In the open position the lining is clear of the disc tip and avoids rubbing wear. In the closed position the lining is extended to touch the disc tip and form a good seal. The act of inflation dislodges particles which may have become attached to the seat area. Also the moving lining compensates for wear. Air

Figure 3.54 Cryogenic valve testing
Courtesy of YPS Valves Ltd

pressure, up to 11 barg, is required to inflate the seat when closed to ensure a good seal. A differential pressure of 10 bar can be maintained across the closed disc. Butterfly valves of this style are available from DN50 to DN750.

Two or three butterfly valves can be linked to a common actuator to create "change-over" or "diverting" valves. Fabricated pipework is necessary to connect the valve bodies. Complete packages are available from a few manufacturers. Butterfly valves with lined bodies can be built as double block and bleed valves. These are described in Section 3.3.5.

Parallel plug valve

The parallel plug valve is a quarter-turn valve manufactured in several styles. This valve has the capability of having very low fluid resistance and high C_v values. This is dependent upon the port size and profile through the plug. Small valves can have full-bore circular ports which are of low loss and allow the valve to be cleaned mechanically without disassembly. Larger valves have rectangular ports with reduced area. The transition from full-size circular pipe bore to plug rectangular port is important for fluid losses. Sharp corners and abrupt changes in section and area can create turbulence, and cavitation in liquid applications, which will result in high fluid losses and possible metal erosion. Valves used with solid-containing fluids will have the sharp corners removed and abrupt changes blended. The area of reduced ports must be known so that fluid velocity can be evaluated.

The basic valve has a solid plug rotating in a close fitting cylindrical body. When direct metal-to-metal contact is used for sealing, considerable torque may be required to operate the valve. Coatings may be applied to the body and/or the plug to reduce friction and improve sealing. Typical coatings include PTFE, Rislan™ and ceramics. Plugs may be hard faced to reduce wear and galling tendencies. A complete PTFE sleeve can be fitted between the plug and the body to seal and lubricate. The sleeve can easily be replaced when worn. Valves without sleeves should have much harder bodies than plugs; the plug can be replaced when worn. No adjustment for wear is possible, and parts replacement is inevitable. The plug can be fitted with a top "O" ring to eliminate the need for a packing box.

Small full-bore valves are mass-produced in brass and austenitic stainless steel. A one-piece barstock body is used for sizes from $\frac{1}{8}$" to $\frac{1}{2}$" NPT and compression fittings from $\frac{1}{8}$" to $\frac{1}{2}$" od tube. The bore for the plug is machined straight through the body, see Figure 3.55. The plug is made in the same material as the body and is coated with PTFE and is fitted with "O" rings around the port to assist sealing and top and bottom around the plug. No bonnet is fitted. The standard "O" rings are PTFE coated Viton™. The plug is located by a circlip on the bottom and the operating lever on the top. Operating pressure is up to 207 barg at temperatures between -23° and 204°C. A silicone-based grease lubricant, supplied by the manufacturer, is recommended for good service life. Larger flanged valves in cast steel, with a PTFE sleeve, are suitable for 50 barg at temperatures up to 200°C.

A modification to the solid, parallel plug design results in an adjustable version with very good maintainable sealing qualities. The solid plug is replaced by two partial shells which have a tapered slot in the bore. A wedge placed in the slots is adjusted to maintain the interface pressure with the valve body. The wedge is formed on the bottom of the spindle and drives the plug shells. Each partial shell has a PTFE insert seal. A bolted bonnet with PTFE gasket carries the non-adjustable PTFE chevron stem packing. An adjusting plate, which looks like a two-bolt gland, is used to compensate for wear without the valve having to be dismantled. This type of valve is only available in duplex stainless steels and exotic alloys. Sizes from DN15 to DN100 are rated at ANSI 150 lb.

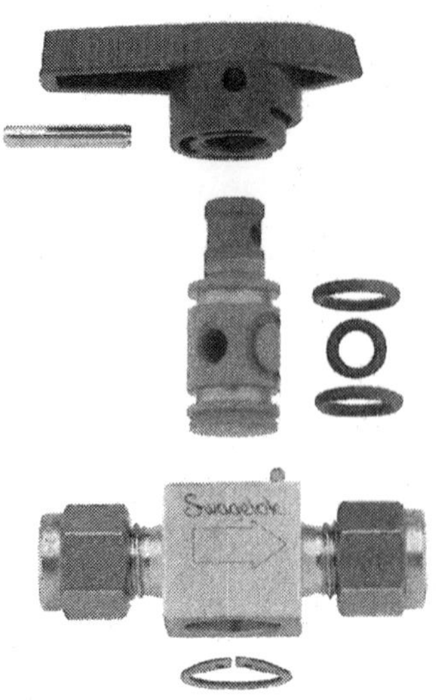

Figure 3.55 Small parallel plug valve with "O" ring seals
Courtesy of Swagelok Company

Another design variation is one in which the plug is manufactured in three pieces, see Figure 3.56. The central portion, attached to the spindle, has a shaped port. Full bore and reduced bore versions are also available. The two faces at right angles to the port are tapered and carry separate shoes which together form a parallel sealing shell. When open, the spindle is raised, reducing the outside diameter of the loose shoes, allowing the spindle to rotate without the shoe peripheries contacting the valve body. The top and bottom edges of the shoes are constrained by the bonnet and bottom flange. To close, the plug first rotates 90° and presents the two loose shoes to the process connections. The spindle and central portion then descend and expand the outside diameter of the sealing shoes. The sealing shoes are wedged against the process ports. The ports are wedged closed against each other to form a very rigid mechanical system.

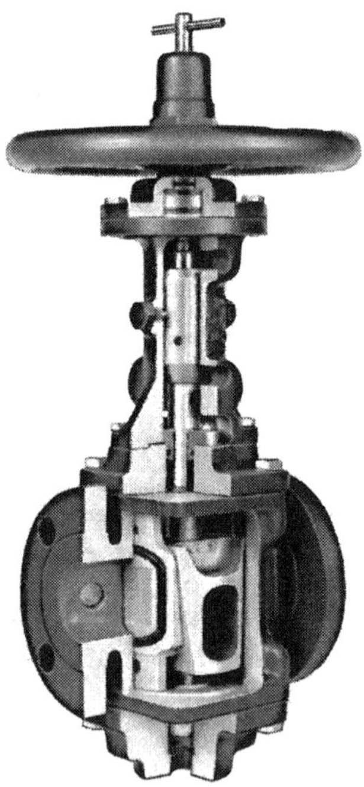

Figure 3.56 Plug valve with three-piece wedge plug

The loose shoes may be fitted with elastomer or graphite seals to reduce leakage rates. The spindle is guided in bearings in both the bonnet and bottom flange to ensure the loose shoes are always clear of the body ports when rotation occurs. The shoes can be removed through the bottom flange, for inspection or replacement, without disturbing the rest of the valve. As both process ports are sealed independently these valves can be used for double block and bleed applications. Generally made in carbon steel or low temperature carbon steel they are made in a broad range of sizes and ratings, Table 3.16.

Pressure rating ANSI	Nominal bore in
150lb	2 to 42
300lb	2 to 20
600lb	2 to 16
900lb	2 to 10
1 500lb	2 and 3

Table 3.16 Sizes and ratings of three-piece wedge plug valves

A special version is also available which enables the shoes to be changed without draining the pipeline. These valves can be certified as fire-safe. All sizes can be fitted with geared handwheels to eliminate water hammer problems.

Tapered plug valve

The tapered plug valve is a derivative of the parallel plug valve. Slight wear of the surface of the plug can be accommodated by lowering the plug in the body. The position of the plug is usually adjusted automatically by spring-loading. Tapered plug valves can be full-bore or reduced bore. The port area of reduced port valves may only be 50% of the nominal pipe size. Also the port may not be circular. Tapered plug valves have the potential to be low loss isolating valves but port areas must be checked. Most valves have two independent seats so that bi-directional sealing is good. These valves are intended for clean fluids including low pressure steam, although special adaptations can be used for solids handling. Tapered plug valves can be fire-safe. The plug valve is a quarter turn valve. Valves up to DN100 will probably be fitted with a lever. Rapid opening and closing can create water hammer in liquid systems and should be taken into consideration.

Generally, these valves are constructed with a one-piece cast body and all assembly is through the bonnet. Replaceable seats are easily fitted through the bonnet aperture. Seats can be fairly loose inside the body; once the plug is fitted the seats are positively located. The seats may be fitted with seals. The plug may not touch the body casting internally and may be completely supported by the seats. The wiping action of the plug against the seats may help to keep the seats clean. Any solids trapped on the seat are just as likely to fall into the body cavity as well as fall into the ports and be flushed away. A disc spring stack is a convenient way to load the plug, and the spring reaction need not be wasted. If the bonnet and gland are integrated and spring-loaded packing is used, one spring stack can perform both functions, see Figure 3.57. The internal spring stack, assisting the pressure loading, is preferred to external spring configurations. Because the internal spring is exposed to the process, fluid corrosion must be considered. Disc springs can be provided in various materials to ensure compatibility such as: AISI 316L, Alloy 20, Hastelloy C™ and Inconel X-750™.

Figure 3.57 Tapered plug valve with integral gland
Courtesy of SMG Valves

The plug surface can be coated to reduce wear and tungsten carbide coatings have been used successfully. Bolted bonnets are common, sealed with a spiral wound or compressed fibre gasket. End connections can be threaded, flanged, socket weld or butt weld. Because these valves seal independently on each seat they are suitable for double block and bleed applications. A bleed valve can be fitted to the body to check seat leakage.

An alternative approach is to machine a complete matching taper in the valve body and line it with a lubricating/sealing material. The plug is spring-loaded into the liner which forms a very effective seal. In these designs a packing box is almost unnecessary, a lip seal can be used as a secondary seal. The gland can be loaded by external springs and used to ensure good plug contact with the liner. PTFE and PFA are popular liner materials. It is also possible to coat the plug with a suitable material rather than use a separate liner. PFA on a ductile iron plug is available. Valves are made in various materials to suit a wide range of applications, see Table 3.17.

Body	Plug	Seat	Packing
Ductile iron	Ductile iron	Virgin PTFE	PTFE
Carbon steel	Stainless steel	Filled PTFE	Graphoil™
Stainless steel	Alloy 20	UHMWPE	
Alloy 20	Monel™	Graphite	
Monel™	Hastelloy™	PFA liner	
Hastelloy™			

Table 3.17 Materials for tapered plug valves

Graphite seats do not seal as well as the other options; seat leakage is reduced from ANSI/FCI 70-2 Class VI to Class IV.

Popular valve sizes are from DN15 to DN200 with pressure ratings from PN10 to PN40. Valves up to 6" can be ANSI 300lb. Valves up to 18" nb are produced regularly with pressure ratings up to ANSI 300lb. Valves over DN100 will have a geared handwheel. The temperature range is largely controlled by the seat material, see Table 3.18.

Seat material	Operating temperature range °C
UHMWPE	-45 to 80
PTFE	-45 to 232
Reinforced PTFE	-45 to 260
Graphite	-45 to 540

Table 3.18 Taper plug valve seat material temperature range

Lubricated plug valve

Lubricated plug valves can be parallel and tapered plug, and the preceding Sections should be read first. The bore of the body and the plug surface are furnished with shallow grooves to distribute the lubricant. This is intended to cover the whole interface surface to reduce friction and wear and assist in sealing. The lubricant selected must have a high viscosity at the operating temperatures and be insoluble in the process fluid. Grease is the most usual. Petroleum based greases are limited to temperatures from -30° to 90° or 120°C. Synthetic based greases can be used down to -73°C and up to 232°C. High temperature synthetic greases can be very costly, up to 30 times the cost of average petroleum based products.

Grease quality should not be compromised if the valve is required to last and operate well. The grease should have corrosion and oxidation inhibitors as well as extreme pressure additives. The grease must be mechanically stable and not

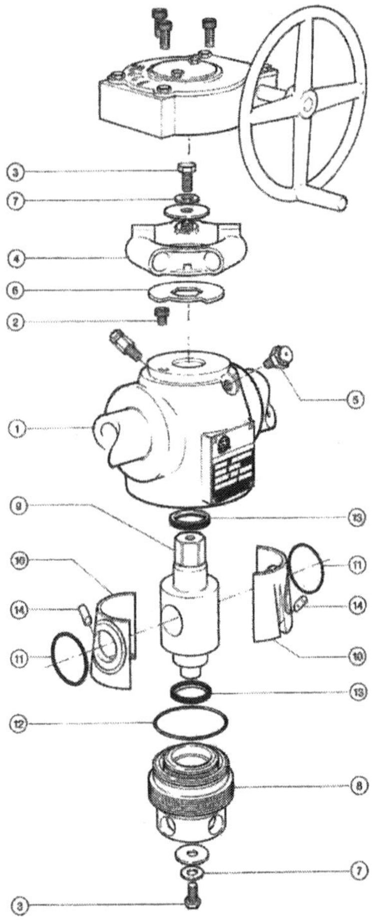

Figure 3.58 Typical oil-field high pressure lubricated plug valve
Courtesy of Anson Ltd

tend to separate into its constituents. Hexagon head slide-on grease nipples are preferred for injection. A pressure relief valve should be fitted to indicate when sufficient grease has be injected. The plug may be coated with PTFE or similar products to reduce friction and aid sealing and a complete elastomer sleeve can still be used. Plug valves in cast steel and cast low temperature steel are produced in sizes from DN15 to DN600 in all ANSI pressure ratings from 150lb to 2500lb. Both standard and pressure balanced designs can be supplied.

High pressure plug valves are used extensively in oil field production and exploration. These valves can have pressure ratings of 414, 689, 1034 and 1379 barg. Figure 3.58 shows a typical example. The valve bodies are machined closed die forgings or solid forged blocks of high tensile steel. The transverse bore of the valve is machined tapered to accept a replaceable seat insert. Various materials are used; aluminium bronze and ductile iron are typical. The outside of the insert may have "O" rings to seal around the flow ports in the body. The insert is adjustable in the body by a screwed bottom cover which also locates the parallel plug relative to the insert. The plug can be made of case hardened carbon steel or coated stainless steel. The top and bottom stems need to be sealed, but leakage is not a problem when the plug and seat are in good condition.

Grease is introduced into the clearance space at the end of the plug. Hexagon head slide-on grease nipples are preferred to "lemon" heads because of the better location and grip provided. Some stem seals offered as standard may seem inadequate for H_2S contaminated applications. Valves may be fitted with a tee-bar, a lugged wheel to accept a tommy-bar or a geared handwheel.

Connections can be provided to suit the installation standard. The range would normally be:

- hammer lug union, API flange
- hubs for clamped connections, Metlock™ connections,
- butt weld ends

These valves are manufactured in standard sizes according to the high pressure pipework used; 1", 2", 2½", 3" and 4".

Eccentric plug valve

The eccentric plug valve is a type of parallel plug valve with important modifications. The plug is only in contact with the body or seat when closed, and the surface of the plug is not used as a support or guide bearing.

Consider a parallel plug with its port, then remove half the plug so that the port is half in the plug and half in the body, this is an eccentric plug valve, see Figure 3.59. When the valve is wide open the half plug port effectively becomes part of the valve body. The flow path through the valve is very smooth. Because the plug has only one side it can only seal on one seat. The preferred direction of flow presses the plug against the seat. For liquid applications, sealing in both directions is quoted as "drip-tight".

The eccentric plug valve has the potential to have extremely low losses and high C_v values. Most valves have reduced port areas but losses are still low. The ports may not be circular; this factor must be checked if "pigging" is envisaged. Most of the eccentric plug valves produced by various manufacturers have the standard product with reduced port area. A manufacturer now claims a range of 3" - 36" valves that have 100% port area. These are easily pigged due to the enlarged port area. Eccentric plug valves are available for regulation and solids handling applications. The partial plug can wipe solids from the seat during closing. Valves below DN150 can have lever actuators, to reduce cost, but maximum pressure differentials may be reduced. The handwheel is generally geared to allow intermediate positioning; water hammer should not be a problem.

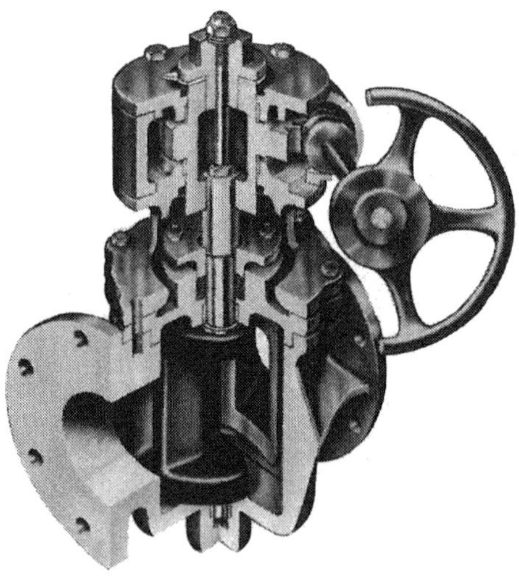

Figure 3.59 Typical eccentric plug valve
Courtesy of DeZurik Water Controls

The valve body is usually a single casting very similar to a parallel plug valve body. Bolted glands are standard so that all assembly can take place through the bonnet aperture. The body seating can be hard faced. The plug has top and bottom spindle extensions to allow the use of dedicated bearings to ensure accurate plug locations. Although the body seating can be hard faced, many are of high purity nickel, and the plugs are lined with a variety of elastomers to suit various applications, and to provide bubble tight shut off to full rated pressure. At least two manufacturers claim to have seating that have sealed bearings; hence the bearings are not exposed to the media. The bearing material is exposed to the process fluid and must change to prevent excessive corrosion. Typical bearing materials include bronze, stainless steel, nickel, Alloy 20, Monel™ and Hastelloy™. The plug can be fitted with various soft inserts to enhance sealing qualities. All-metal plugs are recommended for throttling applications. A separate bolted gland holds and adjusts the spindle packing.

Eccentric plug valves are available in many sizes from DN 15 to DN 1800. Popular materials include cast iron, Ni-Resist, bronze (less common now), aluminium, carbon steel and stainless steel. Exotic materials are also available. Coated plugs are possible. Valves fully lined with hard or soft rubber can also be supplied. Spindle packing can be a single "U" ring for low pressure applications, soft square rings or self energised chevron packing. Materials range from Buna "N" to Grafoil™. One manufacturer offers packing adjustable with chevron packing and packing glands.

Eccentric plug valves are essentially a low pressure type of valve although some small valves can be capable of 30 barg. Valves up to DN900 may be rated for PN10, PN16, ANSI 125 lb or ANSI 150 lb. Larger valves will be PN10. Connections can be screwed, flanged or grooved.

Multi-port plug valve

The tapered plug valve with separate seats, described earlier, is available as a three-way mixing or diverting valve, and is a typical example of a multi-port plug valve. An extra port is added to the bottom of the body. Various plug configurations are possible which allow all ports to be isolated in turn, any port to be connected to any other port and all ports to be connected simultaneously. Sizes are DN15 to DN150 in pressure ratings PN10, PN16, PN25 and PN40. Body materials include cast steel, austenitic stainless steel, Alloy 20, Hastelloy C™ and Monel™.

The three-piece wedge plug valve as shown in Figure 3.56 is available as a four-port diverter valve.

3.3.4 Common optional extras

As with linear valves, rotary valves can be customised by fitting options to extend facilities. Connections and flange variations are similar to linear valves, see Section 3.3.2. Packing box options are legion, see Chapter 11. Valves can have a polished surface finish or approved elastomer lining to assist in hygienic applications, see **Special Note** in Section 3.3.6.

Actuators can be converted from ¼-turn lever to geared handwheel. Spindles can be extended for remote operation and to cater for high/low temperature. Rotary valves have become popular for sub-sea oil and gas production applications, with special actuator options available. Material combinations can be adjusted and modified by using hard facings and coatings. Materials to comply with NACE are a popular option for oil field sour service. Bodies can be fitted with full or partial heating/cooling jackets.

Some valves can be fitted with more obvious position indicators to allow confirmation from a distance; these can be fairly flimsy and easily damaged. Also, some ball and plug valves can be fitted with facilities to allow the injection of sealing compounds, typically PTFE paste, to seal damaged seats, but this is not a common option.

3.3.5 Special purpose valves

Line blind

Positive isolation of pipework or vessels has long been undertaken manually by introducing a spade (blank-plate) between two flanges or rotating a spectacle-plate, and rarely considered an aspect of process or maintenance where safety or productivity improvements can be achieved. A spectacle-plate consists of two inserts: one, a full-bore orifice to allow unrestricted fluid flow, and the other, a solid blank, to prevent all flow. The two inserts are constructed like a "lugged valve" to be clamped between two standard pipe flanges. This traditional practice of spading lines is labour intensive and often requires lifting or jacking equipment to "spread" and lift pipework. This type of activity is no longer possible on many process systems due to the better understanding of risk and pollution and the regulation of site activities by Health and Safety and Environmental Protection agencies. Hazardous substance piping is isolated by appropriate valves, sometimes multiple valves.

Some valve designs do not offer "positive" isolation, since these types of valves have an acceptable leak-rate given the internal seal design. Yet they are still considered in large pipework where isolation is frequently required. The larger the pipe, the more difficult isolation becomes, as the mechanical issues increase with the number of nuts and bolts to loosen and tighten (or replace each changeover), weight and lifting requirements, and the considerable effort required to spread heavy pipes apart to replace flange face gaskets and introduce a spade.

Line blind technology is strictly speaking not a valve, as there is no possible leak-path downstream and pipes must be depressurised before "blinding", yet process engineers place the line blind in the valve category. Line blind technology has developed considerably over the last thirty years as its acceptance by the major oil companies world-wide has grown with the realisation of its fast payback, as plant owners and operators look beyond the initial investment in equipment which offer safety and productivity benefits in-service. A line blind investment provides a healthy return since this otherwise labour intensive operation of line-isolation is reduced to a single-operator role taking less than 5 minutes without need for any tools.

Whilst positive isolation of small bore process pipework is easily achieved with double-block and bleed, the greatest line blind benefits are seen in isolating DN400, 16" nb piping and larger, effortlessly in minutes without the usual sight of scaffolding towers or maintenance crews huddled around pipes. The line blind mechanism enables the operator to spread heavy pipework quickly and effortlessly by utilising the principles of leverage in translating the modest force of a single person into a force suffi-

Figure 3.60 Typical line blind
Courtesy of MV Fluids Handling

Figure 3.61 Combi-system line blind
Courtesy of MV Fluids Handling

cient to spread ANSI 2500 rated pipe or large diameter pipe up to 48" in a controlled and parallel manner. When a line blind spreads the flanges, axial movement is only 5 mm total which is sufficient to release the blind plate from its clamped state and enable change-over.

A large blind plate can weigh up to 750 kg, hence the line blind mechanism incorporates a supporting guide-roller system which automatically centralises the plate evenly between the flange faces as the line blind spreads open. By centralising the plate at all times it prevents the risk of "O"-ring damage during change-over. As the "O"-ring or gaskets are located on the plate this facilities ease of regular quality inspection and rapid maintenance. Figure 3.60 shows a typical smaller line blind.

The line blind is designed to open pipework to atmosphere to allow a blind plate to be put into position, replacing the orifice plate, which means the process has to be depressurised and drained/vented before positive isolation takes place. However, "live" lines can also be positively isolated by including ball-valves directly into the line blind body to provide a single or double-block and bleed possibility before the pipes are opened to atmosphere to facilitate the isolation. When ball valves are integrated into the line blind it is significantly safer, as a mechanical interlock can be incorporated between the ball valve and blind ensuring the ball valves are 100% closed-off before an operator can spread the pipes and risk a release to atmosphere. Figure 3.61 shows a combined blind and double ball valve system. Figure 3.62 shows a line blind assembly with limit switches and hydraulic jacking.

Application possibilities include pipework up to DN1 900, 76" nb, pressure rating from ANSI 150 to ANSI 2500, and temperatures ranging from -104° up to 760°C. Plus the option for full automatic operation, and ganged configuration in parallel process lines. Line blinds can be locked and fitted with limit switches for control room indication of blind position.

Fire hydrant valve

The fire hydrant valve is the standard water system outlet valve, as specified in BS750, which is set below cast iron covers in roads and pavements. The valve body is made of cast iron and has a bottom flanged inlet for the supply, PN16 to BS EN 4504. The valve is a flat disc globe valve with a nitrile rubber seal. The stem terminates in a square drive to suit a standard tee key. The body has a bolted bonnet which includes a bolted stem guide and replaceable stem nut. The outlet connection is a male taper screwed spigot in gunmetal. The valve body is hydro-tested to 24 barg and the seat to 16 barg. Standard finish is bitumen black paint. Hydrants with wedge gate valves are also defined in the Standard.

Sample valve

Sample valves are used to remove small quantities of product from pressurised systems. The main feature is the absence of any dead volume in the valve where product could accumulate and deteriorate or be contaminated. These valves are used in hygienic applications or biotechnology. Most sample valves rely on a plunger or piston sliding in a close fitting bore. The use of a piston or plunger is why these valves are sometimes called "ram valves". Strictly speaking, these are piston valves.

A primary seal is placed close to the product end of the piston to prevent any product entering the valve body, see Figure 3.63. The seal is mounted on the piston and wipes the bore of the valve. Pistons have moving seals. Plungers only have stationary seals. When closed the end of the piston/plunger is flush with the pipe or vessel bore. A secondary seal is usually fitted to prevent product escaping to the environment; many combinations are possible, see Chapter 11. Special versions are available for high viscosity liquids and dry solids. Valves can be designed to dispense a measured sample rather than random amounts controlled by the operator.

Most valves are lever-operated and fit directly to tank bottoms or flush with the bottom surface of horizontal pipe runs. Double ball valves have been used with some success for sample applications. Poppet valves have been used when the poppet can open into the vessel.

Piston valves with hygienic grade PTFE seals are made in many metals and suitable for operating at ANSI pressure ratings from 150 lb to 2500 lb. Metal seals can be provided to allow a temperature range from -212° to 648°C.

A portable sample valve is available for fitting to liquid containers to permit the safe transfer of the contents to a smaller vessel. The valve is similar in principle to the quick release couplings for hoses. Half the valve is fitted to both vessels.

Figure 3.62 A line blind assembly with limit switches and hydraulic jacking
Courtesy of MV Fluids Handling

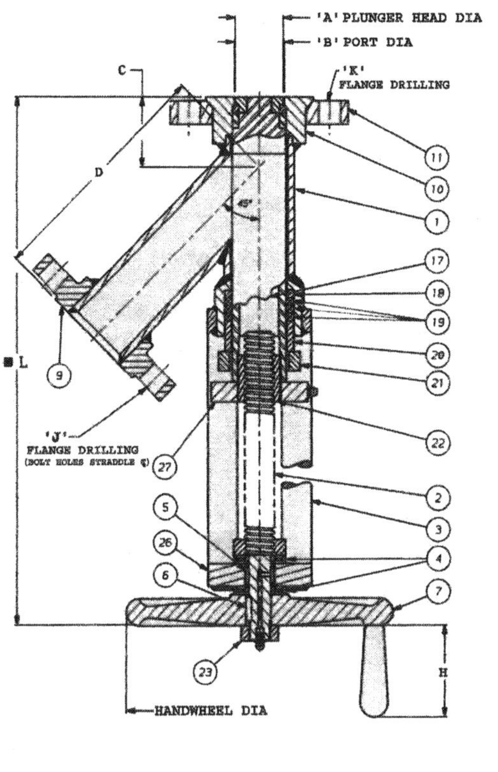

Figure 3.63 Typical piston sample valve
Courtesy of Strahmann Valves France

When both halves are connected together the valve on the larger vessel can be opened. The halves can only be separated when the valve is closed. When separated, both halves are sealed.

Flush tank valve

A flush tank valve is used to drain tanks completely in those situations where it is not feasible to shape the tank bottom to include a gravity drain point. Most tanks can have a sloped bottom so that all sludge/sediment and product can be drained. Tanks of this type have dead volumes which cannot be stirred and product cannot be circulated. These operating restrictions make this style of tank construction unsuitable for hygienic and biotechnology applications. Equipment must be designed to allow all product to circulate freely and pockets, where the consistency may vary from the bulk, must be eliminated. Flush tank valves are designed to be fabricated into torispherical dished ends or circular shells. The complete contents of the tank can be drained without inhibiting the internal flow patterns.

The valve body can be flanged to bolt to a facing fabricated into the tank. Alternatively, the valve body can be shaped and weld prepared, or "weld prepped", for butt welding into the shell. The most popular valve types are piston type sample valves. Care must be taken because of the very close clearances between the piston and the body. The valve has an extremely poor tolerance to pipe strain and imposed forces and moments and so the design of the outlet pipe is critical.

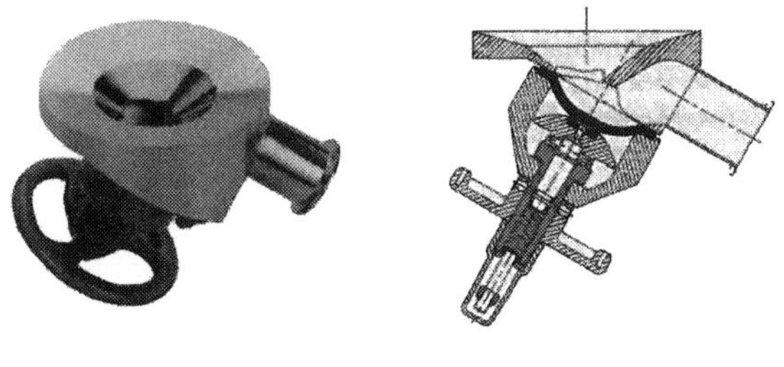

Figure 3.64 Weir type diaphragm valve for hygienic tanks
Courtesy of ITT Corporation

Three-piece body ball valves have been adapted to tank bottom applications. One end piece is fabricated into the tank and the rest of the valve attached by long studs. See Section 3.3.3 for full details of their construction and material options.

The weir type, hygienic diaphragm valve, described in Section 3.3.1, can be adapted to flush tank bottom applications. The standard valve is manufactured in AISI 316L to ASME VIII Division 1 requirements. Full material traceability is maintained. The valve body is designed for direct fabrication into the tank bottom as shown in Figure 3.64. Sizes cover DN15 to DN100 outlets.

Float valve

The float valve is a simple mechanical liquid level controller. A float detects changes of elevation in the liquid free surface and opens or closes a valve accordingly. Large numbers of small float valves are used in domestic water systems. A float valve maintains the level in the cold water header tank which regulates the pressure of the hot water system. A smaller float valve is fitted to toilet cisterns to measure the flush water. Float valves for domestic cold water applications are made to suit 15 and 22 mm copper pipe. These valves are designated "low pressure" and "high pressure" to cover varying local conditions. The high pressure float valve has a smaller seat area and is capable of working up to 16 barg. In some areas the water supply pressure can be as low as 2 barg. Domestic valves are cast in brass or bronze or injection moulded in plastic.

Most valves close on a rising liquid level, but this can be reversed on industrial valves. The valve is mounted on the tank side above the liquid level to allow maintenance without draining the tank. The liquid level on low cost valves is adjusted by bending the ball arm. Superior valves have a bolted connection in the arm to facilitate adjustment. If the valve outlet is not piped below the liquid level the incoming liquid can be aerated to a great extent. The aeration can create serious problems with pumps operating very close to the NPIP or NPSH limit. The effective NPIPa or NPSHa is reduced by the air coming out of solution. At best, pump flow is reduced; at worst, cavitation damage can drastically reduce operating life.

Float valves can create waves on the liquid surface which in turn can cause surge problems. Waves cause the float valve to modulate and vary the flow rate of the liquid. The flow modulation can make the problem worse. If modulation occurs close to valve closure then the disc and the seat can be damaged and water hammer symptoms induced. The inlet process system can suffer significant disturbance problems due to surge. Pressure gauges vibrate and the needles fall off. Pressure transducers produce erratic readings and flowmeters give incorrect readings. The liquid surface must be stabilised to eradicate the problem. Piping the outlet under the surface and diffusing the flow can work. Baffles across the tank or stilling tubes, to prevent the surface movement, always work.

The mounting of float valves on tank walls can create problems. The ball arm can produce considerable torque to close the valve. The torque must be resisted by the tank wall. Thin wall tanks and glass fibre tanks may need reinforcing to cope with the loads. The problem increases as both valve size and inlet pressure increase. The torque can be approximated using the seat area and the inlet pressure. Add 20% to this value because the actual torque is greater than the approximation. Do not skimp on the reinforcing. Spread the load as far as possible. If the valve causes the tank wall to crack, there will be serious consequences.

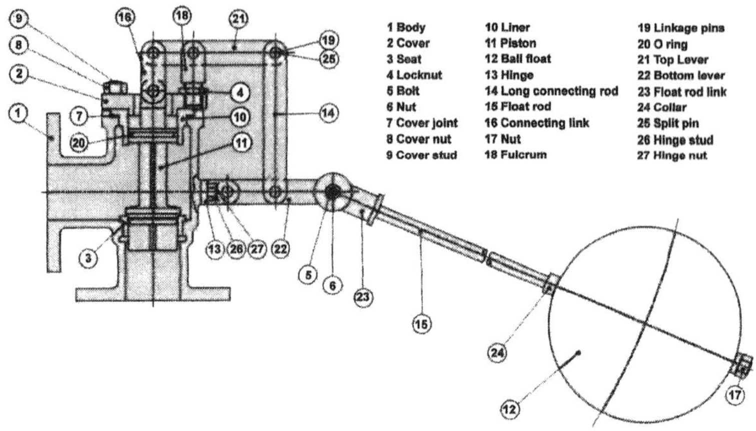

Figure 3.65 A balanced float valve
Courtesy of Taylor Shaw

Float valves are generally available in sizes from DN15 to DN300. Standard materials include cast iron, gunmetal and stainless steel. More exotic materials can be used but the problem of float manufacture must be considered. Working pressures up to PN20 are normal.

The problem of high operating torque can be solved by using a balanced valve. Some manufacturers call balanced valves "equilibrium float valves". Balanced valves work in a similar manner to double-beat globe valves, see Figure 3.22. The balanced float valve does not have two valves, but one valve and a balance piston. The balance piston has the same area as the valve but is loaded in the opposite direction by the same inlet pressure. The operating torque is reduced significantly although a little friction is introduced by the balance piston seal. The friction can be very beneficial as damping to reduce modulation induced by waves. Figure 3.65 shows a balanced float valve with an "O" ring seal on the balance piston. Note the replaceable sleeve for the balance piston, the replaceable seat and the bolt adjustment for ball position. The wing-guided plug can be fitted with a soft seal for cold applications.

Float valves can be used to prevent overfilling or overflow of tanks when the liquid needs to be diverted rather than dumped in a drain. Suitable valves can be mounted on the base of the tank with the outlet piped through the tank wall. The valve communicates with the float arrangement by suitable links. Large float valves, for filling or overflow control, can be constructed in a similar manner to the poppet style controlled closure non-return valve shown in Chapter 4, Figure 4.13. Valves of this style, up to 12m diameter for reservoir applications, can be mechanically operated or hydraulically actuated via a small pilot float valve.

Figure 3.66 shows a combination float valve used for water level control in steam condensers. The valve caters for "supply" and "dump" depending upon the water level. The valve body bolts to the condenser wall and all parts are subject to condenser vacuum. The combination consists of two double-beat piston valves on different systems. The top right hand connection is an overflow. If everything goes wrong it is important that the condenser does not flood. In this context "flood" means covering tubes intended to condense steam with condensate already formed. If the condenser loses capacity due to flooded tubes the consequences can be disastrous and very costly.

The valve is shown in the "water level high" position. The bottom valve is open allowing condensate to drain into the feed tank or hot well. As the condensate level stabilises both valves will close. This condition means that condensate extracted by the extraction pump is the same as steam condensed. If the extraction rate is too high the condensate level will fall, opening

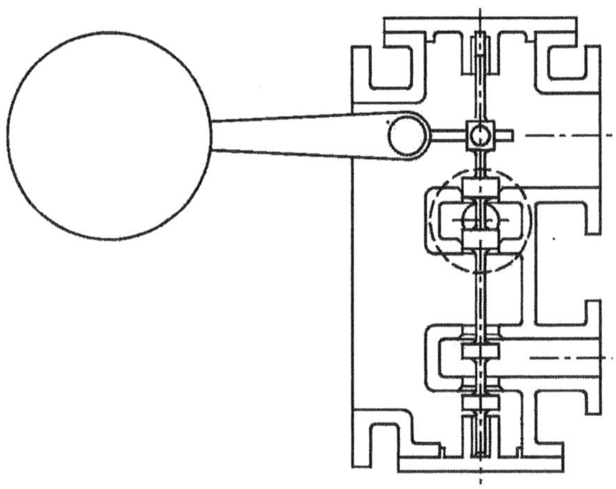

Figure 3.66 Combination float valve

the top port on the back of the valve. The top port is a recirculation connection from the extraction pump discharge.

Excess pump flow is returned to the condenser to maintain the level. A minimum condensate level is very important to provide sufficient NPSHa to the extraction pump. In an efficient system the condensate is very close to its boiling temperature. The extraction pump NPSHa is created solely by static head above the pump impeller. If flows stabilise, the level will rise causing both valves to close. When a fixed speed extraction pump is used the capacity can only match one set of condenser operating conditions. The combination float valve allows operation above and below rated capacity.

Automatic air release valve

The automatic air release valve is used in those situations where air is liable to accumulate and create an "air lock". If a pocket of air becomes trapped in a liquid pipe system the air can create an orifice effect and severely restrict the flow capacity. The problem is not necessarily caused by air drawn into the system, or by remnants of air left behind after venting. Many liquids carry air in solution. If the liquid pressure is reduced, for whatever reason, the air can be released and exist as a confined volume.

The most likely sites for air accumulation are high points in systems which do not "fall" consistently. Pipework in industrial and refinery installations rises or falls in one direction so that any gas can escape. In pipework installations which cannot be so well ordered, such as domestic installations or large scale dis-

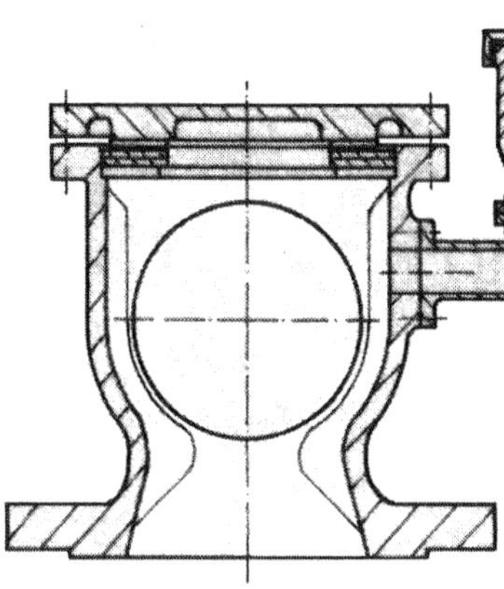

Figure 3.67 Combined automatic air valves

tributions systems which follow ground contours, pipework must rise and fall to suit local conditions. Air pockets are inevitable. In domestic installations the householder does not want to vent water systems regularly. In large scale distribution systems it is not feasible to manually vent pipelines regularly. The automatic air release valve can fulfil this function when necessary and not based on arbitrary timescales.

The automatic air release valve relies on a float to hold the valve closed. The float is mounted below a simple valve. The automatic valve unit is mounted on top of the pipeline at a high point. Initially when the pipeline is filled the float has no liquid to support it so it must rest on the bottom stop and allow air in the pipe to escape. When the pipe fills with liquid the valve body will also fill up and lift the float until it closes the air valve. The pipeline is now sealed. As long as only liquid exists in the pipeline the release valve will remain closed. If there is a sudden demand, or if a pump fails, the pressure of the liquid may fall to a point where air comes out of solution. Because the release valve is at the highest point the air will accumulate in the valve body. When sufficient air has accumulated to displace the liquid the float will drop allowing the valve to open and release the air. As air escapes the liquid level rises again and the valve closes. The automatic air release valve operates whenever necessary.

Most valves are designed for water systems and made in cast iron, ductile iron, brass or bronze. Most popular sizes are small, DN15 to DN50 although valves over DN150 are listed. Cold working pressures are normally PN16, PN20 or PN25. Non-metallic valves, up to DN50, can work at pressures up to PN16.

Large low pressure pipework distribution systems can take a very long time to fill and displace very large volumes of air. Larger air release valves are manufactured to cater for large systems. Figure 3.67 shows a large valve combined with a smaller high pressure valve. The larger ball valve on the left allows large volumes of low pressure air to escape from the pipework during filling. When the ball floats on the liquid surface the valve closes. This valve also allows air into the pipework to prevent vacuum formation. The small high pressure ball valve, mounted on the right-hand side, releases small volumes of air which may accumulate during normal operation. The high pressure valve releases air while the pipework is pressurised.

The system designer or pipework designer must consider all the operating modes, including filling and emptying, and decide on the type of protection required. It may not be desirable to allow air into the pipework. Rapid closure of a large release valve may cause surge problems. Many pipework systems contain hazardous gases rather than air. These must be piped to an appropriate system for recovery or safe disposal.

Combined block and bleed valve

The combined block and bleed valve is the simplest of the valve combinations in common use. One valve is used to isolate a circuit and another is used to bleed fluid slowly from the circuit to allow inspection, calibration or maintenance. The isolating valve may be used for throttling to damp pressure pulsations. In some situations the bleed valve can be replaced by a bleed plug. Bleed plugs cannot be used with hazardous fluids when the escaping fluid must be totally contained and directed to a closed drains system. For very hazardous fluids the block and bleed valve may be inadequate and facilities may be required to

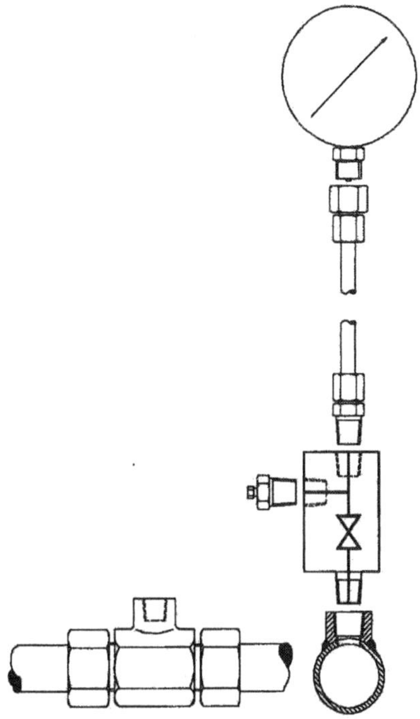

Figure 3.68 Pressure gauge installation for safe fluids

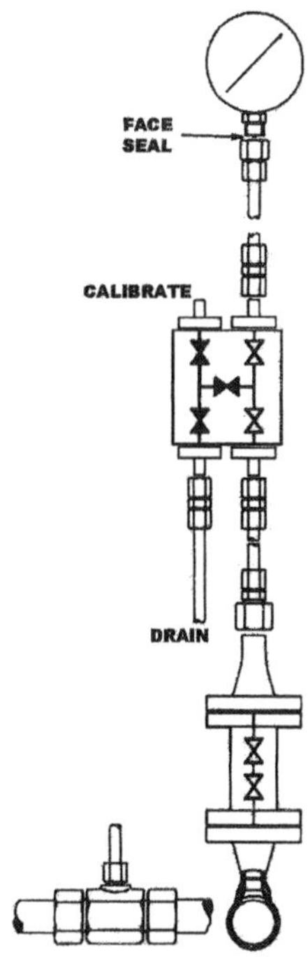

Figure 3.69 Pressure gauge installation for hazardous fluids

evacuate the circuit completely prior to splitting any connections. Most block and bleed valves are small and are used in instrumentation pipework. Figure 3.68 shows a pressure gauge impulse line for safe fluids such as air, nitrogen, water or lube oil, whilst Figure 3.69 shows an impulse line for hazardous fluids, such as sour crude oil or sour natural gas.

Notice the style of installation changes dramatically for the hazardous fluid. In general, screwed connections should not be used for hazardous fluids. If process pipes use ring-type joint flanges or clamped connections to reduce the risk of leakage, it does not make sense to use screwed connections on instruments. For sour services, where H_2S is present, screwed connections and screwed bonnets should not be used unless the thread is completely isolated from the process fluid.

The combined block and bleed valve is designed to save space, pipe fittings and reduce the risk of leaks from connections. The functions can be easily duplicated using separate valves and plenty of pipe fittings. A slightly longer valve body can readily accommodate two valve bonnet assemblies. Flow through the valve is not anticipated under normal operating conditions, the valve is designed to allow the transmission of pressure waves. The port sizes are generally not important and C_v values are not quoted. Valves are selected to match pipe or fitting sizes.

Figure 3.70 shows the diagrammatic arrangements of combined valves as used in P & IDs (Piping/Process & Instrument Diagrams). The left-hand illustration shows a valve with a bleed plug. The right-hand illustration shows a valve with a closed bleed valve. The dotted line indicates an optional plugged connection for calibration. Figure 3.71 shows a barstock block valve with a bleed plug and an extra plugged connection which can be used for calibration. The bleed plug is fitted with a 90° seat screwed release bolt.

Figure 3.72 shows a barstock block valve with a bleed valve. In both cases all process connections are screwed. Both valves use the same bonnet assemblies. Needle valve or plug valve assemblies can be used. The bonnet assembly is screwed into the body and may be locked with a spiral pin. The valves are of the outside screw with rising stem type. The packing is adjusted by a screwed gland. The valves shown have a plastic dust cap

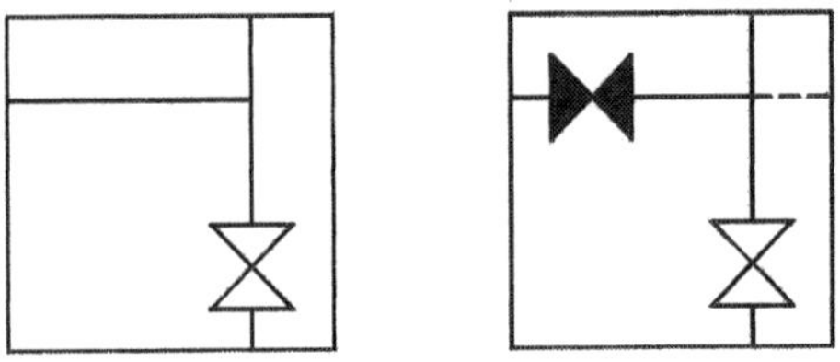

Figure 3.70 Diagrammatic representation of block and bleed valve

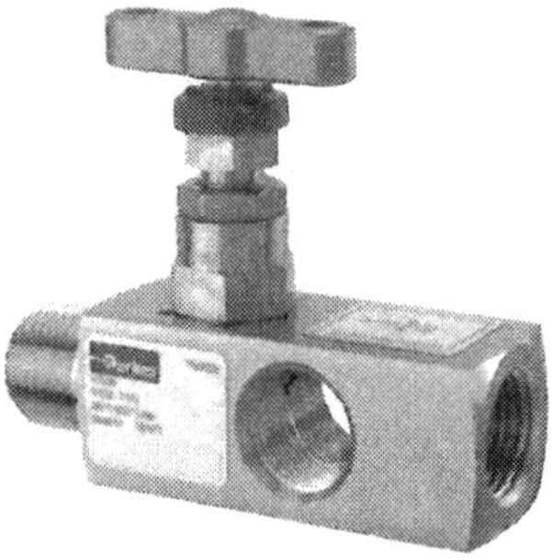

Figure 3.71 A barstock block valve with bleed plug
Courtesy of Instrumentation Products Division, Parker Hannifin Ltd

Figure 3.72 Barstock block and bleed valves
Courtesy of Instrumentation Products Division, Parker Hannifin Ltd

fitted to the gland. The dust cap can be colour coded to identify valve functions. Various types of handle can be fitted to suit the installation. Valve bodies can be square, rectangular or hexagonal and forged bodies are available. Connections can be male or female pipe threads. Some versions can have socket weld connections.

A few manufacturers can supply valves with special two bolt flanges or "kidney flanges" which can be socket or butt welded to the pipework or used with compression fittings. Other styles of valve can have bolted glands and flanged connections. Compression fittings are also available. Valve bodies can also be shaped like a normal blank flange to allow direct connection to flanged process pipework. The inlet is a flanged connection, the outlet can be flanged, screwed or socket weld.

The flanged in/out version is mounted like a "lug" pattern valve between two process flanges. The valve bonnets and auxiliary connections are positioned around the periphery of the flange. This style of construction is sometimes called a "monoflange" valve. Most valves can be used with tubing from $\frac{1}{4}$" od to $\frac{3}{4}$" od although larger valves up to $1\frac{1}{2}$" NPT can be found. Barstock valves are generally suitable for 414 barg, forged versions for 689 barg.

For secure installations valves can be fitted with padlock facilities. Proprietary interlocks can be fitted to some styles. The valves can sometimes be panel mounted using the threads on the bonnet for the gland. Other valves can have tapped or clearance holes for bolts.

Linear valves can have rotating or non-rotating tips. Spherical tips can be optional with some body designs. Seats can be integral with the body or replaceable in which case they can be non-metallic. Some stems are not fitted with backstops to prevent accidental removal. Some of those with backseats can isolate the packing box while on-line. Bellows sealed bonnets are available in some sizes.

There are ball valve versions of combined block and bleed valves. These valves are assembled through the end of the body and the ball and seats are retained by an insert. The spindle may not have packing but one or two seal rings.

Valves are generally available in carbon steel or stainless steel. A replacement for the 90° seat bleed bolt uses a spring-loaded ball in an angled seat. Materials include steel, brass, stainless steel, Monel™, aluminium and PTFE. Metal ball versions can operate up to 275 barg, PTFE ball versions to 14 barg.

Straight barstock needle valves, described in Section 3.3.1, can be fitted with bleed or vent plugs and used for block and bleed applications. The valve must be capable of tight shut-off and may include a dedicated seating or have an "O" ring in the seat. Angle needle valves, also described in this Section, can be used as well.

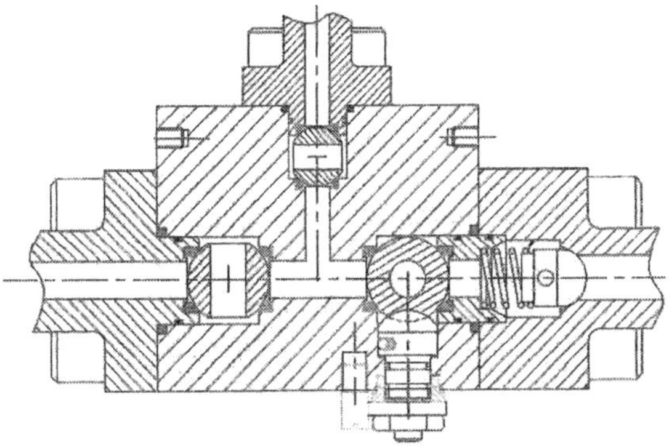

Figure 3.73 Cross-section through a typical process block and bleed valve
Courtesy of Sabre Instrument Valves Ltd

A special version of the wedge gate valve is better described as a "circular wedge valve". It is a linear valve and looks similar to a needle valve but the ports are in-line allowing the valve to be cleaned with rods if necessary. The wedge is attached to the stem and does not rotate as it is pressed into the seat. The seat may be metallic or non-metallic and replaceable. Barstock valves in stainless steel with Delrin™ seats for pressures up to 414 barg at temperatures between -29° to 121°C are made in screwed pipe sizes from ¼" to ½" female and ¼" to ¾" male. Socket and butt weld ends are optional and bleed plugs are fitted. An optional screwed connection is possible for calibration.

The valves described so far are small and intended for instrumentation applications. Block and bleed valves for process applications are available in sizes up to DN150 or 6" nb. Both valve functions can be accommodated by ball, globe or gate valves. Needle valves can be fitted for bleed functions. Non-return valves can be incorporated if required. Figure 3.73 shows a cross-section through a double block and bleed valve with primary isolation, secondary isolation and bleed ball valves and an in-line check valve. Process valves are available for ANSI 2500lb, 4500lb and API 5000, 10 000 and 15 000 psi ratings. Valves can be manufactured in a wide range of materials to suit the application with appropriate customer specified connections.

Combined double block and bleed valve

"Double block and bleed" can mean different things to different people. The effect of scale is very important. Small valves provide certain facilities; larger valves **may** provide different facilities. This Section must be read completely to ensure a full understanding of the different applications.

Small combined double block and bleed valves are used in instrumentation circuits to allow either portion of the circuit to be bled or to protect the open instrument portion with two isolating valves. Figure 3.74 shows diagrammatically the arrangement of the valve for normal operation; two blocks open, bleed closed. A single valve body houses two independent isolating valves and either a bleed plug or valve. The individual valves can be needle, plug or ball. Figure 3.75 shows a current version of the combined double block and bleed valve.

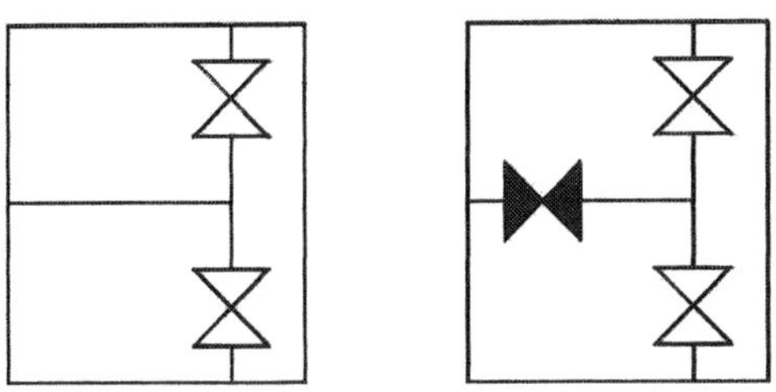

Figure 3.74 Diagrammatic representation of double block and bleed valve

Some instrumentation specifications require the primary isolation of instrument impulse lines to be flanged. The valve illustrated in Figure 3.75 fulfils this requirement and can, when necessary, have a ring-type joint flange. This valve is manufactured in carbon steel and stainless steel with other materials to special order. The flanged inlet can be from ½" to 2" with ANSI ratings from 150 lb to 2500 lb. The screwed outlet connections can be parallel or taper in sizes ¼", ⅜" and ½". Some versions are forged in one piece others are fabricated. If compliance with NACE is specified then socket welding should not be used. The valve shown is also fire-safe. The instrument double block and bleed valves follow similar types of construction as the block and bleed valves described earlier in this Section and should be reviewed.

Process versions of double block and bleed valves are manufactured in sizes up to DN150 and 6" nb. These valves can be rated up to ANSI 4500 and API 15 000psi. Figure 3.76 shows their construction using a double block and bleed valve with primary and secondary isolation gate valves and a needle bleed valve. The needle valve shown has a threaded bonnet construction; this style would not be acceptable for NACE applica-

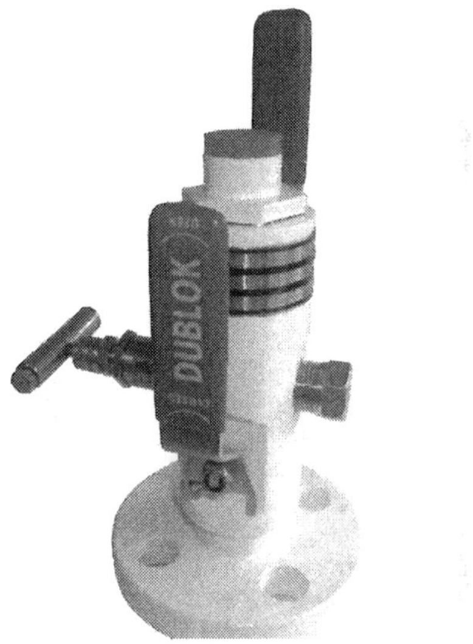

Figure 3.75 Special combined double block and bleed valve
Courtesy of Sabre Instrument Valves Ltd

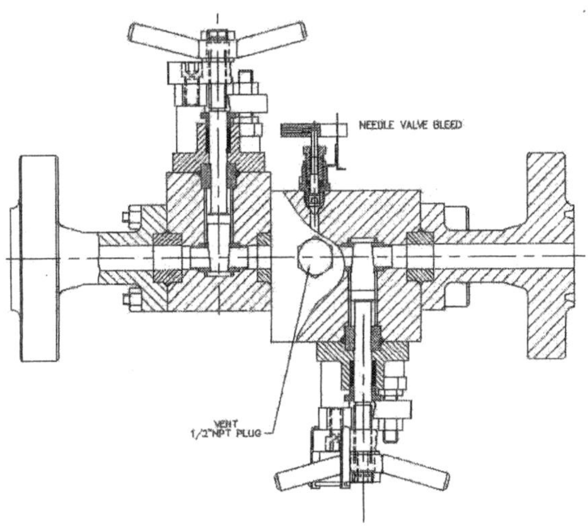

Figure 3.76 A larger double and bleed valve
Courtesy of Sabre Instrument Valves Ltd

tions unless the thread was isolated from the process medium. Various valve styles — ball, gate, globe and needle can be combined to maximum the assembly effectiveness for the individual application.

Valves larger than DN50, 2" nb, are advertised by manufacturers as "double block and bleed" for a completely different application. Large pipework installations have isolating valves placed at strategic locations to allow pipe sections to be closed down. Product may be re-routed or the pipe could be isolated to allow hydro testing. Valves which offer very good seat-sealing qualities are used to ensure pressure containment. The following valve types are often used:

- wedge gate valve
- weir type diaphragm valve (hygienic applications)
- trunnion-mounted ball valve
- butterfly valve
- parallel plug valve
- tapered plug valve
- line blind

To indicate that the valve is sealing correctly the valve body is fitted with a drain connection, called the "bleed indicator". If either of the valve seats leaks the product will be seen to bleed from the valve body. Both pipes attached to the valve are sealed. The bleed valve cannot be used to drain or vent one section of pipe. The application is different but the terminology is the same. The type of valve required can only be determined by the context of the requirement. Typical sizes and ratings of valves used are shown in Table 3.19.

Valve type	Sizes	Pressure rating
Tapered plug valve	½" to 6"	ANSI 150lb ANSI 300lb
Flexible wedge gate	DN50 to DN600 DN50 to DN450 DN50 to DN300	ANSI 150lb ANSI 300lb ANSI 600lb
Elastomer lined butterfly	DN50 to DN300	PN10

Table 3.19 Typical double block and bleed process valve sizes and ratings

Combined stop and non-return valve

The combined stop and non-return valve is a special version of the globe valve. Both straight and angle patterns are made so these types described in Section 3.3.1 should be reviewed first. Flat discs are used which close on flat seats. The disc is not attached to the stem and is fitted with top and bottom guide stems. The top stem slides in a bore in the screw-down stem while the bottom stem is guided by a bush in the valve body. The non-return function is accomplished by the disc acting as a piston non-return valve, see Chapter 4, Section 4.3.

The disc is normally not spring-loaded, using gravity effects only, but there is no theoretical reason why a spring cannot be used to enhance the non-return valve operation. A spring can be designed to prevent the valve slamming wide open if an increase in flow occurs. A spring can also help the valve to modulate with changing flows. It can increase the pressure drop across the valve if the initial compression and spring rate is too high. As with all spring-loaded mechanisms the detailed spring design is crucial for successful operation. Valves must generally be mounted so the disc operates on the vertical axis. Operation in any other position should be confirmed by the manufacturer. The screw-down stem can be used to limit the valve opening as well as preventing it from opening altogether. Globe valves can be used for flow regulation depending upon the nature of the fluid and the materials used for the disc and seat.

The combined stop and non-return valve performs two functions in one valve body. The concept allows a reduction in the use of pipe fittings and saves space. The benefits do not seem to appeal to many users; very few valves of this type are used in modern installations, except steam. Alternatively, it is possible that most valve users are unaware of the existence of this valve type and therefore specify two valves where one could be used.

Valves are currently produced in cast iron, bronze or gunmetal and carbon steel in sizes from DN15 to DN300. Larger sizes can be made to order. Pressure ratings up to ANSI 300lb or PN50 are normal for valves up to DN200. Valves have outside screw rising stems. Packing boxes can be modified to handle most sealing requirements.

Valve manifolds

Valve manifolds are used for special instrument interfaces or when multiple small connections need to be made to one pipe. The valve manifold is an extension of the combined block and bleed valve concept. The manifold is an extended valve body which allows many bonnet assemblies to be fitted to control individual circuits. Mass-produced manifolds cater for two to 12 connections. The valve manifold can save considerable space and reduces the number of pipe fittings and potential leak paths.

A two valve manifold is used for some differential pressure instruments. The low and high pressure signals travel through the valve manifold via two completely independent ports. Each port is fitted with its own isolating valve.

The next size has three valves. The valve is similar to the two valve manifold but the third valve is connected between the two ports, shown diagrammatically in Figure 3.77. The interconnecting valve is shown in the normal operating position, closed.

Figure 3.78 shows the construction of a typical barstock valve manifold. The valve bonnets can be oriented in several ways to suit instrument and panel arrangements. The third valve is used as an equalising valve. Equalising is a frequent requirement on some differential pressure instruments to allow calibration. If the valve is accidentally left open, or the seat is damaged/worn and does not seal effectively, the instrument will

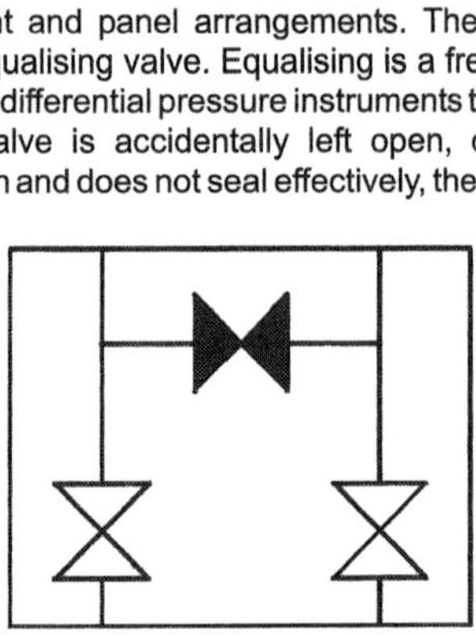

Figure 3.77 Diagrammatic representation of three valve manifold

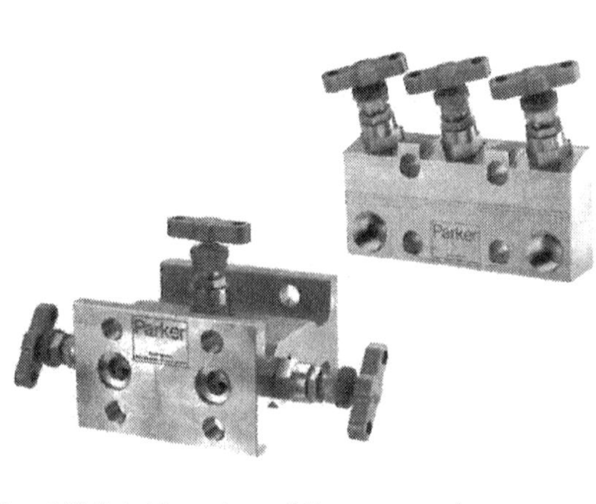

Figure 3.78 Typical three valve manifold
Courtesy of Instrumentation Products Division, Parker Hannifin Ltd

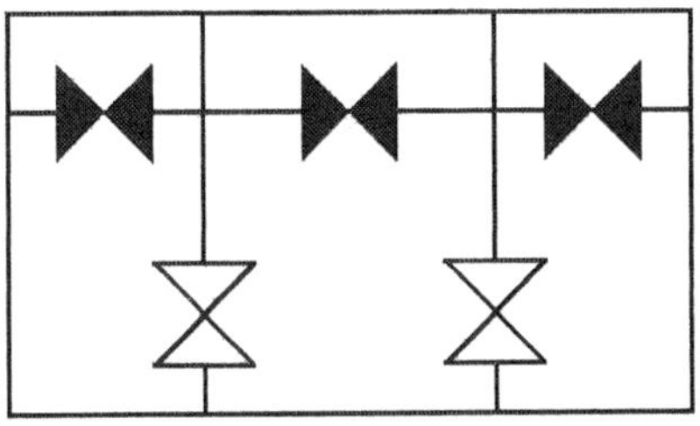

Figure 3.79 Diagrammatic representation of five valve manifold

Figure 3.80 Typical five valve manifold
Courtesy of Instrumentation Products Division, Parker Hannifin Ltd

show incorrect readings. Some users prefer the two valve manifold described earlier to eliminate possible equalising valve problems. Some body styles allow the equalising valve to be replaced by a solenoid valve; this facility permits remote equalisation and calibration. Two and three valve manifolds can be fitted with drain/bleed plugs.

Five valve manifolds are very similar to three valve versions. Two extra valves are added for drain/bleed functions. Figure 3.79 gives a diagrammatic representation and Figure 3.80 shows an actual manifold.

In Figure 3.79 the equalising and bleed valves are shown closed. The five valve manifold is used with differential pressure instruments when the process fluid must be contained and the bled fluid is piped away. The five valve manifold with "kidney" flange connections can be used in high integrity pressure instrument applications, as shown in Figure 3.69. It may be possible to replace all or some of the valve bonnets with solenoid valves to permit remote operation. Figure 3.81 shows a typical installation of a five valve manifold and a differential pressure transmitter used with an orifice plate for flow measurement.

Two, three and five valve instrument manifolds usually have female screwed connections or are flanged to suit instrument facings. Figure 3.81 shows the differential pressure transmitter bolted directly to the valve manifold. Some valves can have socket weld connections. The valves can be needle, plug or ball. The bleed valves can be lockshield bonnets to prevent unauthorised adjustment. Alternatively, all valves can be interlocked to ensure operation in the correct sequence. Some valve stems do not have backstops to prevent accidental removal. Some have sealing backstops to allow the packing box to be re-packed on-line. Some are sealed by an "O" ring. Barstock valves are generally suitable up to 414 barg. Valves with forged bodies can operate up to 689 barg.

Distribution manifolds are made with two to 12 connections. The valve body can be barstock or forged. Working pressure is generally up to 414 barg. The inlet connections can be screwed or socket weld. All outlet connections are screwed in a choice of sizes. The largest inlet connection is 1". The manifold has inlet

Figure 3.81 Typical five valve manifold installation
Courtesy of Emerson Process Management

connections on both ends and can be piped into the middle of a run. All distribution manifolds have holes for mounting. Figure 3.82 shows the diagrammatic arrangement. Figure 3.83 shows a typical manifold.

Valve manifolds can be used for multiple solenoid valve installations. The manifold incorporates the piping connections and

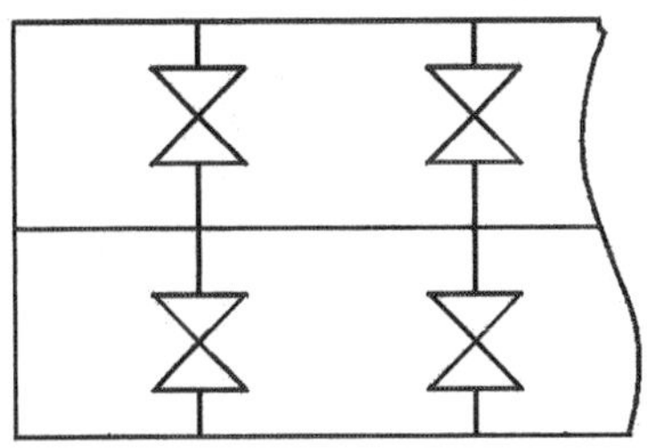

Figure 3.82 Diagrammatic representation of distribution valve manifold

Figure 3.83 A typical distribution valve manifold
Courtesy of Instrumentation Products Division, Parker Hannifin Ltd

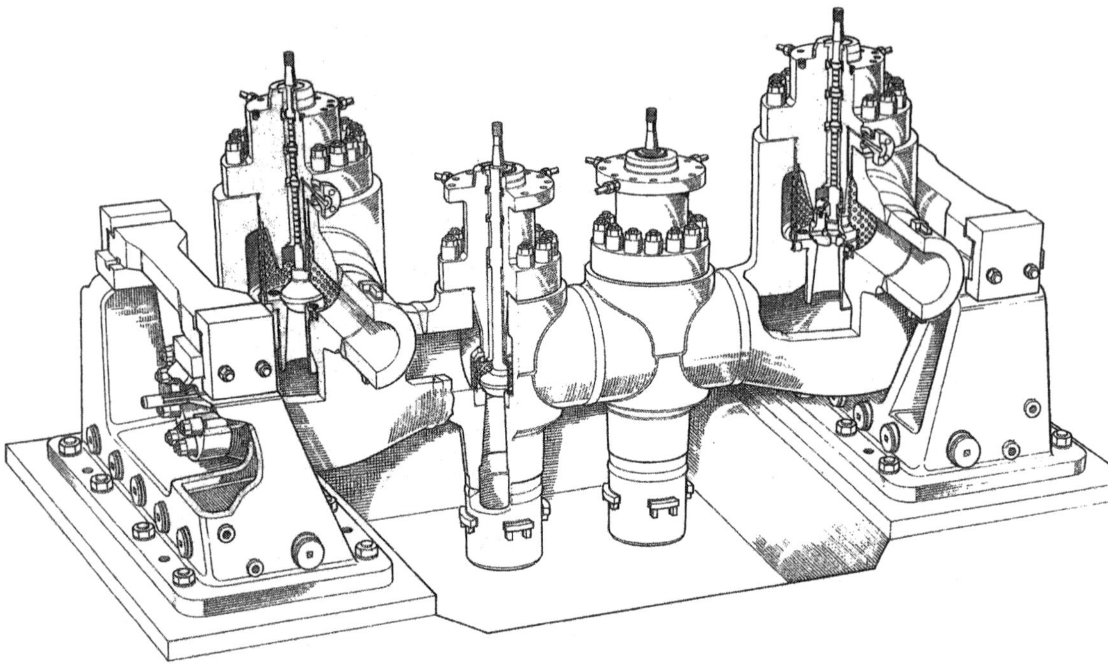

Figure 3.84 Turbine steam chest
Courtesy of Siemens AG, Power Generation

locations for the coil assemblies. Coils can be replaced without disturbing pipework. Non-metallic manifolds are available for special applications. PTFE manifolds, to accept two to six solenoid valves, can operate from 0.13 bara to 2 barg at 70°C.

Figure 3.84 shows what is possibly the ultimate in valve manifolds, a turbine steam chest. The steam chest houses the stop valves and control valves which allow steam from the boiler into the turbine. Fairly typical inlet steam conditions would be 158 barg at 560°C. Turbines used for electricity generation, 500 MW or 660 MW, would have four steam chests. More recent larger turbines may have up to eight separate steam valves rather than a combined chest as shown. Only the steam valves are shown. The hydraulic actuators, which would be mounted alongside, behind the chest, are not shown. The steam chest is quite large, up to 4 m long. It is mounted directly on the foundation block on its own sole plates. Because of the large temperature difference between cold and running conditions the steam chest is "keyed" to its pedestals and retained by clamps which allow the chest to expand and float to balance steam pipe expansions. The two outer valves are pilot-balanced stop valves. The inner valves are speed or load control valves; generally called "governor valves" because they respond to speed changes detected by the governor.

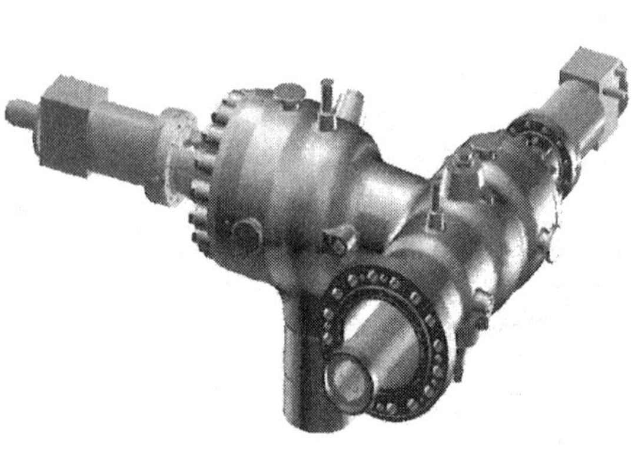

Figure 3.85 A modern large steam valve
Courtesy of Siemens AG, Power Generation

The basic valve construction is interesting because of the differences compared to standard mass-produced valves. All the internals are replaceable; having to remove a steam chest would be a disaster. Notice how deep the bonnet is and the lack of a packing box. Leakage is controlled by very tight clearances around the stem and a long labyrinth created by grooves machined in the stem. Sealing steam is injected through the small connections in the bonnet top flange. When the stop valves are wide open the backseat bush seals the stem.

Pilot-balancing serves two different functions. When the stop valve is cracked open only the pilot flows; this allows very small quantities of steam to be introduced for pre-heating and low speed off-load running. Large steam turbines take a considerable amount of time to warm-up properly and the pilot may be used for a week during barring. Barring large rotating equipment involves driving the rotor at very low speeds, typically 10 to 20 rpm, while warming-up or cooling down. Pilot-balancing used this way removes the need for an external by-pass valve. Once the pilot is flowing, the differential pressure loads on the main valve are considerably reduced allowing the use of a smaller actuator. This style of pilot-balancing removes the need to use larger and more complex valves such as double-beat globe valves. Flanges and connections are kept to an absolute minimum to avoid leaks.

The chest is built up from four valve bodies which are butt welded together. The main steam pipes are butt welded to the chest. All welds are 100% radiographed to ensure integrity. The blank flanges on the end of the chest are used for inspection and maintenance purposes. All the flanged connections use capnuts indicating the high loads to be restrained. Clamped connections or shear rings should be considered as alternatives for these types of applications. Each valve is protected by a steam strainer. In large, hot installations there is always a chance of debris being carried with the fluid.

Valve manifolds like steam chests are not mass-produced to standard designs. Each manifold is specifically engineered to match the application. Very few companies have the expertise to design and manufacture equipment on this scale.

Figure 3.85 shows a modern single valve arrangement. There is only one major valve body component and hence no requirement to weld large, heavy components together. At site, the steam pipes are butt-welded to the valve so that non-destructive examinations can prove the integrity of the welding.

Special Note: Pig launchers

A very special type of rotary valve is used to launch and receive "pigs". These are multi-port valves to enable the pig to be fitted into or retrieved from a branch to the main line. The branch would be fitted with vent and drain valves and a quick release blank cover. Once inserted into the valve, rotation through 90° allows the process fluid to drive the pig through the system. Various pressure ratings are available in sizes from DN200 to DN1000.

3.3.6 Valves for dry solids

Solids can be transported by various means. Over short distances light solids can be carried by air or gas streams. A wide range of dry solids are handled routinely, see Table 3.20.

Animal feed	Flour	Phosphate rock
Bauxite	Fly ash	Plastic granules
Bentontite(1)	Frozen food	Precipator dust
Bone meal	Granulated sugar	Sand
Brewers grain	Grass meal	Sawdust
Cement	Gypsum powder	Snuff
Ceramic powder	Ice	Soap noodles
Cereals	Icing sugar	Soda ash
Coal dust	Iron ore	Soya mela
Coal singles	Kaolin (2)	Starch
Cocoa beans	Lime	Tea
Cocoa powder	Limestone powder	Thermoplastic powder
Detergent powder	Malt	Titanium dioxide
Dried vegetables	Malt culm	Tobacco
Fertilisers	Meat meal	Whiting powder
Fish meal	Milk powder	Wood chips
(1) Mineral clay powder		
(2) China clay powder		

Table 3.20 Dry solids using isolating valves

After storage in silos, compacted solids need to drained out. Valves are required for on/off and flow control. Both linear and rotary valve styles can be used. Most systems considered flow under gravity, but valves can be suitable in preventing reverse gas flow under adverse differential pressure conditions.

Dry solids isolating valve

Some parallel gate valves, described in Section 3.3.1, can be suitable for dry solids provided the particle size is relatively large compared to the seat diameter. Valve orientation is important. If the valve is mounted normally, stem vertically upwards, small particles can accumulate in the bottom of the body. Mounting with the stem horizontally will work more reliably. Very small particles may leak past the gate and fill the bonnet. An adaptation of the parallel gate valve, for small dry solids handling in gas or air may be called a "slide valve" or a "parallel slide valve". This is very similar in most respects to a gate valve but great attention is paid to ensuring the valve bore is smooth and without pockets, for solids to accumulate. Intermittent guides may be positioned on the side walls but the bottom of the bore is smooth. When the valve is open the slide is removed completely from the flow path. Seals and wipers on both sides prevent solids from migrating into the body.

Some styles are open, i.e. the slide is withdrawn through the packing box and out of the valve, as shown in Figure 3.7. To assist sealing when handling small particles, the slide edge may be fitted with a replaceable elastomer insert. This is available in aluminium, carbon steel and stainless steel in round and square versions from 100 mm bore to 420 mm square. Larger sizes can be operated by rack and pinion.

A variation on the parallel slide valve eliminates the side seals and wipers by enclosing the slide in a rolling membrane. When wide open the membrane rolls into the body leaving an unobstructed flow path. As the valve closes the membrane unrolls until it completely seals the flow path. Leakage of fine powder into the valve mechanism or out of the valve is eliminated. The use of a membrane to isolate the working parts of the valve from the product removes all the problems inherent with tight seals and dry operation. The forces required to move the slide are reduced dramatically. However, elastomers can fail due to excessive bending. Inquiries should request data regarding routine replacement of membranes and predicted operational life. This style of valve can be made for square and rectangular ducts as well as pipes. Standard materials include AISI 304 and 316. Standard valve sizes are from DN50 to DN450. The valves can be fitted with a quick release mechanism to allow removal from the system for cleaning.

Standard knife gate valves, described in Section 3.3.1, may be suitable for dry solids; the manufacturers' guidance should be sought when the solids properties are known. Special versions of knife gate valves are manufactured to operate with pulverised coal, sometimes called "pulverised fuel", on coal-fired boilers. The pulverised coal is blown into the boiler by a pressurised air stream. Knife gate valves are used to isolate fuel lines during maintenance. Boilers can occasionally become unstable and blow-backs down the fuel lines are not unknown. Isolating valves must be capable of withstanding a specified blowback pressure, normally 3.4 barg. Valve bodies can be cast or fabricated in carbon steel or stainless steel. Purge connections are standard to allow steam to be admitted. The replaceable seat can be carbon steel or AISI 304 stainless steel. Hard facing can be applied to extend the seat life by limiting abrasion damage. Flange facings can be ANSI 125 lb or 150 lb, NFPA or manufacturers proprietary dimensions. Valves are normally outside screw rising stem. Some sizes can be modified to be non-rising stem where space is restricted. Standard valves start at DN150 and range up to DN600. Knife gate valves can be capable of differential pressures up to 5.1 bar for safe dry solids.

The conduit gate valve, See Section 3.3.1, also known as a "conduit pipeline valve" and "ported gate valve" is eminently suitable for solids handling applications. The absence of pockets and crevices removes the obvious features which result in blockage and jamming. Spring-loaded metal seats with elastomer inserts or spring-loaded "plastic" seats should provide the highest reliability. Hardened 11/13Cr gates or thick chrome plate, 1 mm, should prove to be wear resistant against abrasive solids. Valve sizes range from DN80 to DN1200 with pressure ratings up to 10 barg.

A variation of the parallel slide valve could be considered as a rotary valve. The slide rotates about a hinge pin and slides radially out of the pipe bore, Figure 3.86. The bore is completely clear when open. A stainless steel slide is totally contained within a cast aluminium body and supported by nylon inserts. Designed primarily for hygienic applications these valves are for 150 mm to 300 mm bore systems.

The full-bore diaphragm valve, in Section 3.3.1, has been used successfully on aggregate and powders. Standard valves are used with diaphragm compounds formulated to resist abrasion. Soft rubber, butyl rubber and Neoprene™ can be used. Cast

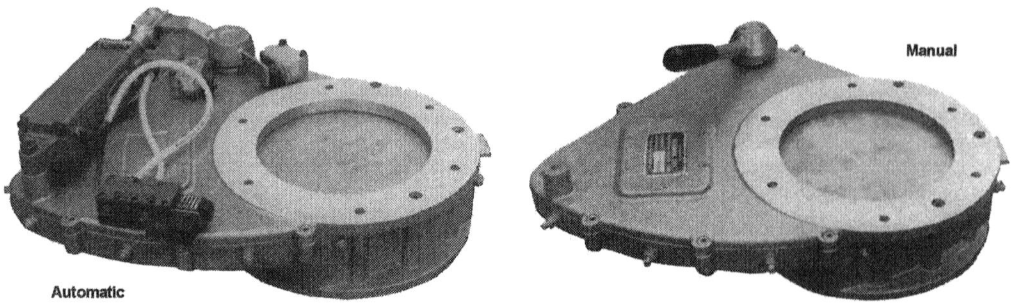

Figure 3.86 A radial parallel slide valve
Courtesy of Rotolok Ltd UK

iron valve bodies can be lined with glass, as well as rubbers, to increase valve life. Table 3.21 indicates the size, pressure and temperature limits of standard valves.

Valve size	Max pressure barg	Lower limit temp °C	Upper limit temp °C	Max temp °C	Pressure barg @ max temp
DN15 to DN100	10	-40 to 5	50	140	4.5
DN125 to DN150	6	-40 to 5	50	140	1
DN200 to DN300	3.5	-40 to 5	50	140	0.5
DN350	1.75	-40 to 5	50	140	0

Table 3.21 Full-bore diaphragm valves for solids

Other full-bore diaphragm valves which may be suitable for solids are available in non-metallic materials such as — PVC, PVDF, PP, PVDF, ABS and CPVC

Sizes range from DN15 to DN100 with pressure ratings of 6 and 10 barg. Some valves are only available as pneumatic operation.

The pinch valves, described earlier, are also suitable for dry solids up to pressures of 12 barg. Powders can be handled without problems, but handling larger solids depends upon the valve size and the tube material/construction. High pressure powered versions up to 49 barg are available. Another version of the pinch valve is even better. Pneumatic powered versions of the pinch valve are available in sizes from DN15 to DN2 100 for pressures between 10 barg and 1.7 barg. The pneumatic pinch valve applies a uniform contraction pressure over the majority of the valve length and can produce a long sealed section. Solids can be easily contained without the fear of high localised forces creating high stresses. Pneumatic pressure is normally 1.7 bar greater than the product pressure. See Section 3.3.7 for more details.

The simplest form of rotary valve is very similar in concept to a swing disc non-return valve with external loading, see Chapter 4, Section 4.3. These valves are for the vertical flow of solids. A flap is held closed by a dead weight or a spring load. The flap is opened by rotating the spindle. A single flap valve works well for gravity flow but does not provide a gas seal. This can be maintained by using two flap valves. Motorised double flap valves are sometimes called "double dumps". The flaps are opened alternately by a rotating cam. For closing, the cam is stepped to provide a rapid movement creating vibration which assists in the solids movement and clears the seats. Standard valves are of carbon or stainless steel in sizes from 150 mm to 300 mm round or square, with larger sizes to order. Double flap valves can sustain differential pressures up to almost 0.7 bar. Double flap valves can also be operated pneumatically and controlled by fluid logic. Valves can be cast or fabricated depending upon the size and material, see Figure 3.87.

Special one-piece body ball valves, see Section 3.3.5, with integral metal seats, are suitable for solids handling applications. A variation on the two-piece cast body design ball valve, this cvalve uses two identical body halves. The body is split down

Figure 3.87 A typical double dump flap valve
Courtesy of Rotolok Ltd UK

the spindle centre-line. Bolting the halves together locates the ball and the seats. A separate bolted bonnet is used to retain the spindle. Valves with spring-loaded metal seats fitted would be suitable for solids handling applications.

Some butterfly valves can be applied to dry solids applications. Wafer and lug designs with coated discs and lined bodies have been used successfully. Butterfly valves with inflatable seats are very useful for solids handling. The ability to withdraw the seat from the disc edge removes the problems associated with solids trapped on the seat. The movement of the inflatable seat would tend to dislodge large particles and be able to seal around small particles. Valves of this style are manufactured in sizes from DN50 to DN750.

A style of butterfly valve sometimes called an "eccentric disc butterfly valve" or a "ball segment valve", is a hybrid of the eccentric plug valve, Section 3.3.3, and the eccentric rotating plug control valve, described in Chapter 6, Section 6.3.2, see also Figure 3.59 and Figure 6.52 in Chapter 6. Instead of the sealing element being circular in one plane and rectangular in the other, it is spherical. It is similar in concept to the characterised ball valve shown in Figure 6.42, Chapter 6, but the disc, a partial sphere, is mounted on longer extensions from the spindle. When the valve is open the partial sphere swings out of the flow path to reveal a full-bore seat. The construction is similar to a ball valve but only enough of the ball is used to cover the seat when sealing. Only one seat is fitted. Flow in one direction is preferred.

These valves are quite special and only built by a few manufacturers. They are useful for isolation and regulation. The valves are available in sizes from DN25 to DN100 with pressure ratings to ANSI 600lb. Some sizes may be suitable for 900lb. Standard material is AISI 316 austenitic stainless steel and surface texture can be made good enough for hygienic applications.

A very special version of the valve is built for abrasive solids handling applications. The seat includes an inflatable seal. During normal valve open operation the seal is flush with the metal seating surface. When the valve closes the seat is inflated and can seal around trapped particles. This style of valve is made in sizes from DN50 to DN600 and is suitable for working pressures up to 43 barg. Maximum temperature is limited to 480°C. Tapered plug valves, Section 3.3.3, have been used with some success on dry solids. Sample valves, described in Section 3.3.5, can be manufactured specifically for dry solids applications.

A valve unique to dry solids handling is the iris valve or iris diaphragm valve. A fabric iris can be adjusted from closed, to any intermediate bore size, to wide open. When wide open the valve bore is formed by a fabric tube, see Figure 3.88. A valve operation similar in principle to the pinch valve. These valves can be used for regulation as well as isolation. The valve is constructed as a slim flanged package for easy mounting in pipework. Normal operation is by lockable lever which has 180° travel around the periphery of the valve body. An all-enclosing circular handwheel can be fitted optionally to allow operation from all sides. The valve action is described as "gentle but sure".

The diaphragm can be made in various materials to suit the application; elastomers, fabrics and elastomer coated fabrics. Elastomers and fabrics provide a wide range of chemical resistance, abrasion resistance, strength and the ability to seal the smallest particle sizes. For extremely abrasive products the diaphragm can be fitted with a replaceable liner. To reduce the effects of flying dust and loss of product the diaphragm can be extended through the valve body to allow attachment to containers when used for filling applications. Metal components can be aluminium, bronze, steel and stainless steel. PTFE coated metal parts can be supplied. All materials of construction can be suitable for hygienic applications. Sizes start at 50 mm bore and extend to 450 mm. Valves can be fitted with a special quick-release mechanism to allow the valve to be removed quickly. These valves can also be used for special liquid-solid mixtures, such as live fish.

Another valve unique to solids handling applications is sometimes called a "rotary airlock". This is a rotary valve for isolation and flow control which looks more like a vane compressor, see Figure 3.89. A rotor with straight vanes is encased in a tight fitting body. Product falls into the rotor through the top connection. The rotor rotates continuously and transfers the product from the top of the body to the bottom where the product can fall out through the bottom connection. The rotor vanes maintain multiple tight clearances with the valve body to prevent loss of higher pressure air or gas. These valves can support a differential pressure of 1.3 barg. Flow is isolated when the rotor stops. Flow regulation is achieved by varying the speed of the rotor. Valves typically run at speeds up to 30 rpm.

The rotor is supported by ball bearings. Internal "sealed-for-life" bearings can be used with some products and removes the need for stuffing boxes. External bearings can be fitted when necessary. Soft packed stuffing boxes are standard. Shaft sleeves under the packing should be fitted to prevent shaft wear. Some designs allow a clean gas flush in front of the packing to extend packing life. Gas is required at 0.07 to 0.14 bar above the line pressure. Valves for both circular and rectangular duct systems are available. Standard body and end cover materials include: cast iron, aluminium, carbon steel and stainless steel.

Rotors can be fabricated from carbon steel or stainless steel. Vanes can be replaceable and be non-metallic, PVC or PTFE, as well as metallic. Metallic vanes can be tipped with hard fac-

Figure 3.88 Operation of the iris diaphragm valve
Courtesy of Kemutec Powder Technologies Ltd

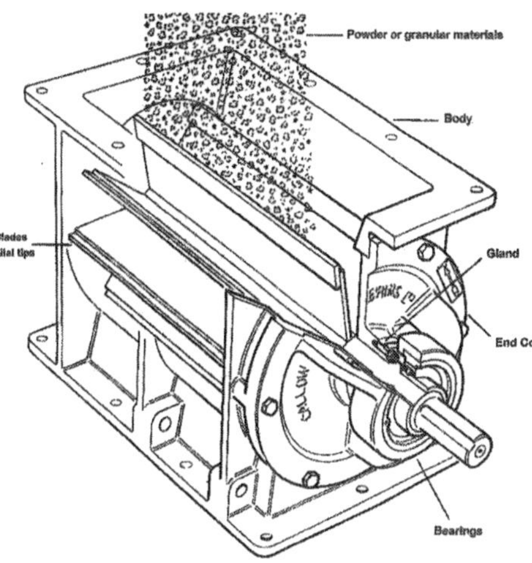

Figure 3.89 Rotary valve for dry solids
Courtesy of Callow Engineering Ltd (A Division of J Fletcher Engineers Ltd)

ing to reduce wear or with an elastomer to improve sealing. The detail design of the rotor can be optimised to suit the precise properties of the bulk solids. Body bores and end cover internal faces can be plated or surface treated. Construction and materials can be adapted to suit any hygienic requirements including fulling machined external surfaces and quick release covers. Materials and construction can be adapted to cope with operating temperatures up to 400°C. The rotor can be driven via a chain or toothed belt from a standard geared motor.

Valves are built with rotor diameters from 50 mm to 750 mm which can pass up to 300 to 442 m^3/h depending upon the flowing nature of the product. Standard valves can accept square ducts up to 750 by 750 mm and circular ducts up to 600 mm diameter. Connections can be adapted to accommodate rectangular ducts. Large valves with 610 mm by 1130 mm inlet ports have been designed to cope with 13 tonnes per hour of wood strands. The valve itself weighed 3 tonnes and used a 15 kW geared motor.

As well as the isolating valves described above, mixing and diverting valves are manufactured in various styles to suit piping and ductwork systems.

Special Note: Non-metallic valves

Non-metallic parallel gate valves, Section 3.3.1, are available in uPVC in sizes from DN15 to DN100. These valves can have female threaded BSP or plain socket connections and can be used up to 10 barg between 5° and 50°C. Parallel gate valves, similar in style to the open construction knife gate and conduit gate valves, can be used, from DN40 to DN300 for low pressure applications up to 0.7 barg with clean fluids. The stainless steel gate is withdrawn through a seal out of the body. The body can be ABS or PVC with Buna N seals. Connections are cement sockets.

Straight globe valves, Section 3.3.1, in PP are available in ¾" and 1" BSP. These valves are suitable for 10 barg up to 80°C. Non-metallic full-bore diaphragm valves are available in PVC and PVDF for 6 barg in sizes from DN15 to DN50. Pneumatically operated on/off valves are produced in a wider range, see Section 3.3.7. Non-metallic straight needle valves are produced in PTFE. The stem, made of Kel-F, has an integral needle. Available only with screwed connections in sizes from ⅛" to ⅜" the valves are limited to 5 barg working pressure. Solenoid valves, can have non-metallic bodies. Typical materials include; Delrin™, nylon, PVC, cPVC, PEEK, PTFE, PVDF and polypropylene. Valve sizes from ⅛" to 3" are popular. Valves are specifically produced for high vacuum applications. Table 3.22 indicates the limitations of popular solenoid valves.

Body material	Connection sizes	Pressure rating barg
PVC	⅛" to ½" UNF	8.2
PVDF	2 to 6 mm	1.5 to 6
PVC	10 to 50 mm spigot	6
Nylon 66™	¾" to 3" BSP	9
Delrin™	¼" to 1½"	16

Table 3.22 Non-metallic solenoid valves limitations

Solenoid valves can also use the pinch principle. Valves can be normally-open or normally-closed. A single solenoid can operate on one, two, four or eight tubes. One solenoid can be fitted with a rocking mechanism which allows some tubes to be open while others are closed. Soft flexible tubes are recommended to reduce the solenoid force required. Silicon tube is popular allowing operating temperatures between 0° and 90°C.

Non-metallic one-piece body ball valves, Section 3.3.3, in PVC or ABS are manufactured in sizes from DN15 to DN150. Up to DN65 the valves are rated at 10 barg, larger valves are only 6 barg. One-piece body valves are also available in moulded from vinyl ester resin which is reinforced with glass fibre or carbon fibre. The ball and spindle are made of the same material. The valve internals are assembled from one end and retained by a threaded insert in a similar manner to that shown in Figure 3.35. Seats can be virgin PTFE or glass reinforced. All sizes have reduced-ports. Sizes from DN25 to DN150 are suitable for 19 barg, DN200 for 11 barg, at temperatures up to 120°C.

For abrasive applications the valve body can be fitted with two ceramic bore inserts and a ceramic ball is fitted. Soft seats are not fitted; the ceramic ball seals against the ceramic inserts which are diamond lapped to provide an accurate profile. A stronger spindle is required and this is accomplished by coating a Hastelloy™ C insert with vinyl ester resin. Non-metallic three-piece body ball valves are built in a similar manner. Valves up to 2" may have union ends rather than four bolt flanges or sockets for cement connections.

Butterfly valves, from DN15 to DN300, are available with a PVC body and a PVC or PP disc mounted on a stainless steel spindle. Small valves can cope with 10 barg but the larger sizes are limited to 4 barg.

Special Note: Hygienic valves

A type of knife gate valve, Section 3.3.1, which is a variation on the parallel slide valve, eliminates the side seals and wipers by enclosing the slide in a rolling membrane. When wide open the membrane rolls into the body leaving an unobstructed flow path. As the valve closes the membrane unrolls until it completely seals the flow path. Leakage into the valve mechanism, which could spoil, or out of the valve is eliminated. The use of a membrane to isolate the working parts of the valve from the product removes all the problems inherent with tight seals and forces required to move the slide are reduced dramatically. However, elastomers can fail due to excessive bending. Inquiries should request data regarding routine replacement of membranes and predicted operational life. This style of valve can be made for square and rectangular ducts as well as pipes. Valves are suitable for regulation. Standard materials include AISI 304 and 316 which are electro-polished. Standard valve sizes are from DN50 to DN450.

For hygienic applications, two straight globe valves can be built into a single unit, one above the other. The two seats are arranged back-to-back with a small flush connection to the interspace, see Figure 3.17. Both circuits can operate independently when the valves are closed. Either circuit can be flushed by opening a valve and injecting a solution through the seat interspace; a power-operated flush isolating valve is controlled by the main valve actuator. Diverting can be achieved by opening both globe valves simultaneously. Angle globe valves, are easily adapted for hygienic applications.

Figure 3.90 shows a valve fabricated from standard stainless tube and machined rings. Approved pipe clamps are used for the bonnet connection as well as the pipework. Valves of this style are available in sizes from DN40 to DN100 for pressures up to 12.5 barg. Flush tank valves can be fabricated on long radius bends. Standard valve bodies can be stacked, as shown in Figure 3.91 to produce complicated diverting configurations.

Versions of the weir type diaphragm valve, described in Section 3.3.1, are specifically manufactured for hygienic, pharmaceutical and bioprocessing applications. Standard body material is AISI 316L. The surface finish of the valve can be specified down to 0.28mm and electro-polishing is an option. These valves can be lined with glass, PVDF or polypropylene approved by the American FDA. All diaphragm options — rubber, nitrile rubber, butyl rubber, EPDM and PTFE, comply with FDA requirements. Temperature limits shown in Table 3.5 apply.

Bonnets in various materials, metallic and non-metallic, can be fitted to suit specific operating environments. PAS, polyarylsulphone, bonnets can operate continuously up to 149°C. Cleaning-in-place and steaming-in-place can be performed without valve disassembly. These valves are manufactured

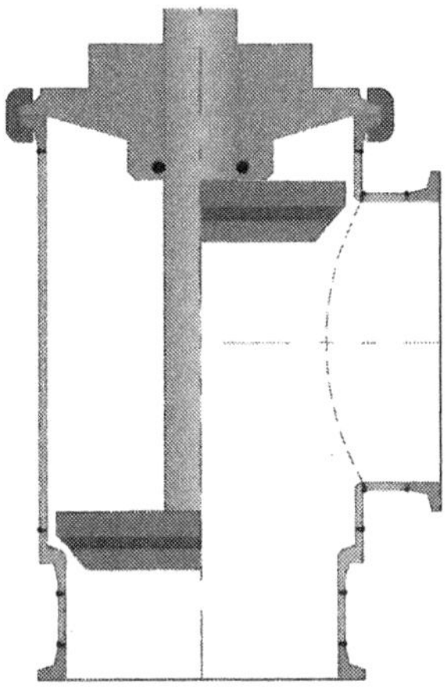

Figure 3.90 A fabricated hygienic angle globe valve

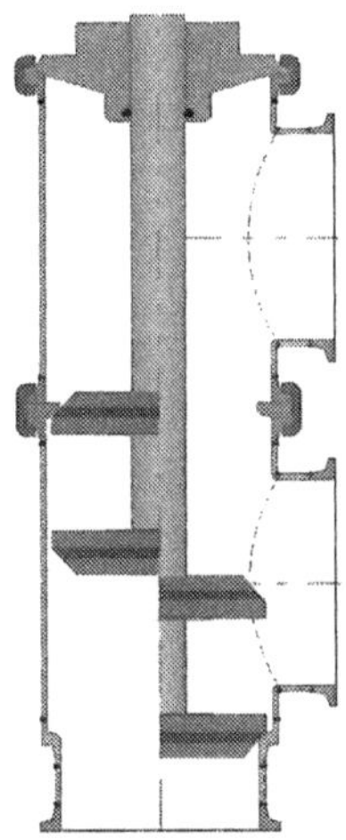

Figure 3.91 A fabricated hygienic diverting valve

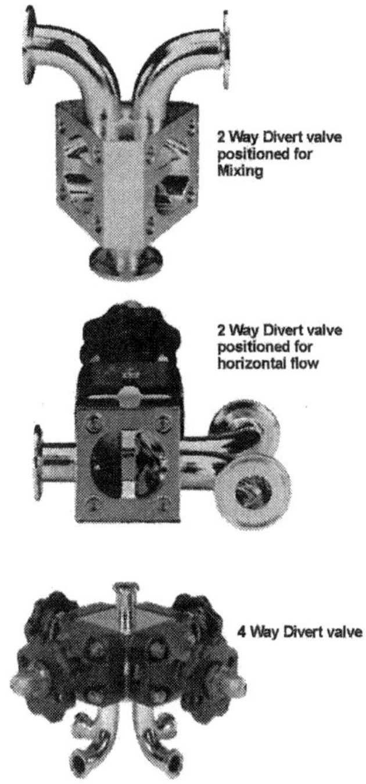

Figure 3.92 Weir diaphragm valves for diverting applications
Courtesy of ITT Corporation

with a very wide range of end connections to suit every eventuality:

- butt weld (to suit many pipe thicknesses)
- butt weld ends to ISO for automatic welding
- proprietary clamp fittings
- factory fitted tube extensions

Weir diaphragm valves are available in sizes from DN10 to DN100 for pressures up to 13.7 barg cold and 6 barg at 176°C.

Some manufacturers will fabricate standard valves into sub-assemblies. A typical example would be four valves connected to form a double block and sterilise/bleed assembly. Standard valves can be modified to include a continuous by-pass or circulation connection. This connection is used to prevent product stagnating or cooling down while the valve is closed. Some valves up to DN80 can be fitted with an integral small bore by-pass. Special versions of hygienic weir type valves can be built for diverting applications. Standard configurations include two, three, four and five way in sizes from DN10 to DN80. Figure 3.92 indicates some possibilities.

The full-bore diaphragm valve, see Section 3.3.1, is also available as a hygienic valve. Valves from DN6 to DN80 can be fitted with approved clamp, screw or butt weld ends. Bonnets can be stainless steel or plastic. As an option to normal screw actuators some valves can be fitted with quarter turn levers. Full-bore diaphragm valves can be configured for "non-dead leg" operation, with tangential branch valves, sample points and flush tank bottom installation.

Pinch valves can be constructed for hygienic applications. As the elastomer sleeve is the only component in contact with the product a sleeve manufactured from an FDA approved material is all that is required. Pneumatically actuated valves are suitable for pressures over 30 barg. Low pressure systems utilising small bore plastic tubes can be constructed without valves. Clamps can be applied anywhere on the tubes to provide isolation.

Solenoid valves can also use the pinch principle. Valves can be normally-open or normally-closed. A single solenoid can operate on one, two, four or eight tubes. One solenoid can be fitted with a rocking mechanism which allows some tubes to be open while others are closed. Soft flexible tubes are recommended to reduce the solenoid force required. FDA approved tubes, in many sizes, are readily available. Peristaltic pump tubes may be suitable.

One-piece, two-piece and three-piece body ball valves, are available in both full-bore and reduced bore styles. These valves can be fully lined with an approved elastomer or polished. Valves are mass-produced in sizes from DN6 to DN100 for cold working pressures up to 19 barg.

Butterfly valves, Section 3.3.2, from DN25 to DN150, for up to 16 barg working pressure, can be upgraded to hygienic quality by use of approved coatings and/or polished stainless steel/duplex stainless steel discs. Standard metallic bodies, with approved connectors, can be coated with FDA approved EPDM. Note that there are many EPDM "alloys" available and the operating temperature restrictions vary widely. Pairs of butterfly valves can be linked together to provide interlocked isolation or flow. This arrangement can be particularly useful for introducing sterilising or CIP flows into pipework when tanks/vessels are isolated.

Plug valves can be constructed in stainless steel for low pressure applications such as dairy products. Three-port diverting valves are also available in sizes from DN25 to DN80.

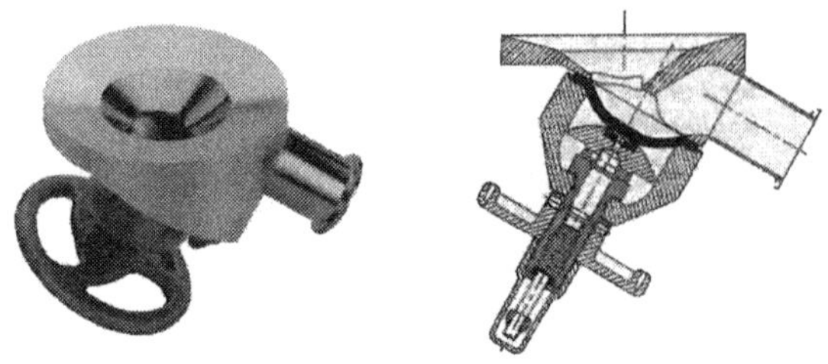

Figure 3.93 Weir type diaphragm valve for hygienic tanks
Courtesy of ITT Corporation

Sample valves, Section 3.3.5, can be made suitable for hygienic and biotechnology applications. The basic design lends itself to these applications because there are no dead volumes. Valves are upgraded by using stainless steel throughout, polished surfaces inside and out, and by fitting approved elastomers for seals.

Flush tank valves of the weir diaphragm or full-bore diaphragm type or three-piece body ball valves, are suitable for all types of hygienic applications. Figure 3.93 shows a weir diaphragm valve suitable for flush butt welding into a tank bottom.

3.3.7 Power-actuated valves

Valves for isolating functions can usually be operated by power. This feature is useful for complete automatic control of processes and also for those situations where physical access to a valve is difficult or impossible. Actuated valves have the advantage of being very easy to interlock to ensure the correct sequence of operation.

The steam chest described earlier and shown in Figure 3.84 has four actuated valves, the valves being too big for effective manual operation. Smaller steam chests, for turbines up to 60 MW, would normally have, what appears to be, a manual main stop valve. The stop valve was actually hydraulically actuated by lube oil pressure. The manual handwheel allowed the valve to open but lube oil pressure opened the steam valve against a retreating stem.

This type of actuation reduces the manual effort required and interlocks the steam valve with the lube oil system. If the lube oil system is not running the stop valve will not open, even though the handwheel can be screwed right out, thus preventing costly bearing damage. If the lube oil system fails during normal running the stop valve closes to limit bearing damage. The handwheel incorporates a reset mechanism. Once the valve has tripped the handwheel must be screwed right down to reset the hydraulic power circuit.

Although most valves can be power actuated some small valves may prove difficult. Mounting points for the actuator are necessary and some valve body designs cannot accommodate extra drilled holes. Mounting on the bonnet, using the gland thread, may be possible but can be unreliable if positive locking against rotation is impractical. Solenoid valves are the obvious choice for small on/off applications, see Section 3.3.1.

When purchasing an actuated isolating valve, buy both assemblies from the same supplier. Do not buy the two separately and assemble the unit. If the package does not work correctly, the valve manufacturer will say it is an actuator problem and vice versa. If you buy both from one source, then the supplier has the responsibility to ensure the package works correctly.

Sizing the actuator is critical. If the actuator is too small the valve will never operate correctly. If the actuator is too big, depending upon the type, the valve may operate very quickly causing water hammer problems. Pneumatic and hydraulic actuators can be speed controlled using orifices or needle valves in the control lines. Electric actuators may be more difficult, since a new gearbox may be necessary. The stem force and spindle torque must be predicted with some accuracy. Special packing boxes, extra seal rings or extra sets of packing, will increase the force/torque required. Manually adjusted packing can be over-tightened by unskilled staff resulting in poor actuator performance. If the skill of site personnel skill is in doubt, opt for spring-loaded or non-adjustable packing, see Chapter 11.

Some actuated isolating valves are intended to work fast. Emergency shut-down (ESD) valves, are fitted to isolate process circuits in the event of impending disaster. These valves are usually held open against a spring and close on power failure, a normally closed valve. Pneumatic and hydraulic actuation is also possible using small local high pressure reservoirs. Because these valves are intended to operate quickly, the effects of water hammer and hydraulic shock cannot be ignored. There is little point in closing a valve to isolate a system if the pipe ruptures due to water hammer. All the possible operating cases, under which the valve may close, must be examined with regard to pipe integrity. Applied axial forces may fracture pipe restraints but these can be repaired later. If the excessive pipe movement applies unbearable forces and moments to fixed connections, which subsequently fail, the impending disaster is made worse rather than limited or confined.

Linear motion valves, Section 3.3.1, can be fitted with various types of actuator, the popular choices are:

- electric
- pneumatic
- hydraulic

If the stem screw mechanism is retained, an electric motor can operate the valve through a gearbox. If the stem screw mechanism is removed, the valve can be operated directly by a pneumatic/hydraulic piston or diaphragm depending upon the stroke required. Globe and diaphragm valves have fairly short strokes and can use a diaphragm actuator. Electric actuators for wedge gate valves, some globe valves and needle valves should have a "hammer-blow" effect built in to ensure the valve disengages from the seat immediately.

Small parallel gate valves can be used for remote vent and drain applications with pressures up to 500 barg. Actuation is via a pneumatic cylinder, see Figure 3.87.

Knife gate valves can be power-operated by removing the screwed stem and fitting a pneumatic or hydraulic cylinder. Double-acting cylinders, for air-to-open/air-to-close operation, are normally fitted to isolating valves. Working air pressure can be 3.4 or 5.5 barg to suit the supply. If necessary, a totally enclosed yoke can be supplied to protect the stem. Knife gate valves can also be motorised using various manufacturers systems.

Small straight globe valves, ¼" nb, in cast AISI 316L have a bellows sealed bonnet to allow operation up to 240 barg at temperatures between -40° and 65° or 204°C depending upon whether Kel-F™ or Vespel™ seats are used. Leakage rate is quoted as 4E-9 std ml/s using a non-rotating ball tip. The valve has a union bonnet and can have screwed, compression or butt weld connections. Actuation is via a pneumatic piston.

Some angle globe valves are specifically designed for pneumatic operation. Valves in either bronze or stainless steel, from DN15 to DN150, can operate between 3 and 35 barg at temperatures up to 130°C. Seats have PTFE inserts and seals are EPDM as standard. The valves are normally closed, NC, flanged and have screwed bonnets.

The high pressure choke valves or angle globe valves described in Section 3.3.1, can be power-actuated to allow remote adjustment. In addition to the actuators listed above, electro-pneumatic and electro-hydraulic actuators are commonly employed.

Oblique globe valves, in sizes from DN15 to DN80, are similar to the angle globe valves for 3 to 35 barg described above. Oblique valves in polypropylene, DN15 to DN80, can be used up to 9 barg between 0° and 50°C. Valve bodies are suitable for cement socket ends. Actuation is by pneumatic diaphragm.

Air-operated weir diaphragm valves, in PVC or PVDF, are available for plastic piping systems. The valves are rated at 6 or 10 barg in sizes from DN15 to DN50. Other valves have bodies in polypropylene and PTFE for 7 barg working pressure.

Weir diaphragm valves have been highly successful as control valves when fitted with actuators and positions. Both plain and lined valves have been used for applications involving abrasion and corrosion. The weir style valve has approximately double the flow capacity of an equivalent double-beat globe valve without the possible problems of the packing box. Valves in cast iron and SG iron can be lined with glass or fluorocarbon polymers. Valve sizes from DN15 to DN250 are suitable for 16 barg.

Hygienic quality weir diaphragm valves can be fitted with actuators for remote operation. Pneumatic diaphragm actuators, with various diaphragms including PTFE, can have PAS, polyarylsulfone thermoplastic, casings to allow cleaning with caustic, chlorine or alcohol. The casing openings are sealed with Viton™ "O" rings. A stop to limit opening can be fitted.

This actuator is suitable for valves from DN10 to DN50 and is "autoclavable". Larger actuators with vinyl ester thermoplastic casings are suitable for DN65 to DN100 valves. Operating air pressure is up to 6.2 barg at 149°C. The actuators can be selected from:

- air-to-open/air-to-close
- air-to-open/spring-return
- air-to-close/spring-to-open

Actuators can be fitted with proximity detectors or limit switches for remote position indication.

Weir type diaphragm valves can be motorised. Small valves, DN15 to DN50, can be fitted with electric actuators, for various ac voltages, and usually in a choice of speeds. Fast operation can be 15 to 20 seconds and slow, up to 50 seconds. Positional feedback may be optional to allow control functions. Tank bottom versions can be pneumatically actuated.

Full-bore diaphragm valves can be fitted with air actuation for automatic control. Actuators can be fail-safe open or closed, or double-acting. Non-metallic versions for on/off isolation are available in — PVC, PP, PVDF, ABS and CPVC.

Valves from DN15 to DN100 can have diaphragms in EP rubber, Viton™ and PTFE, and are suitable for working pressures up to 10 barg. Pneumatic air can be up to 6 barg depending upon the process pressure.

The pinch valves described in Section 3.3.1 can be power operated. The operating screw is replaced by an air or hydraulic cylinder. The cylinder casing is attached to one roller and the piston to the other. Valves from DN25 to DN900 are catalogue items. Valves over DN600 are only available as "open frame" style. A high pressure version with nylon, polyester or Kevlar™ fabric reinforcement is available in sizes from DN25 to DN300. Maximum working pressure is 49 barg at temperatures up to 38 °C.

The powered pinch valve is also manufactured in another style which cannot be manually operated. The elastomer tube is mounted within a pressure vessel, of cast iron or aluminium, and compressed by air pressure directly, see Figure 6.19, in Chapter 6. Operating air pressure should be 1.7 barg greater than the process pressure.

These valves are available in a wide range of sizes and can be fitted with various designs of tube for different applications. Special "double wall" tubes for abrasive applications are claimed to outlast Stellited metal components used for slurries. Sizes and pressure ranges are listed in Table 3.23.

Size	Max pressure barg
1/8" to 3/8"	5.1
DN15 to DN25	5.1 to 10
DN32 to DN50	3.4 to 10
DN65 to DN150	10
DN200	8.6
DN250 to DN300	6.9
DN350 to DN400	5.1
DN450 to DN600	3.4
DN750/900/1 200/1 500	3.4
DN1 800/2 100	1.7

Table 3.23 Size and pressure ratings of direct-acting pneumatic pinch valves

Multi-turn electric actuators have special ac squirrel cage electric motors with low inertia, producing high torque. Limit switches can be fitted so that the valve only opens to a set point. Torque sensors can be fitted to allow the valve to be closed to a prescribed torque value. A handwheel, for manual operation, is a worthwhile investment for emergency operation if power supply or control system is lost. Open and close local push buttons are usually optional to allow local powered operation. Position indicators must be fitted to the actuator if the valve mechanism is not visible. Limit switches, when fitted, can also provide remote indication of valve status.

Wedge gate valves, or valves which only operate infrequently, can benefit from a "hammer-blow" effect to ensure valves operate correctly. The effect permits the use of smaller electric motors with consequential savings on starters and cabling. The speed of operation can be selected from a wide range to suit all types of installation.

Pneumatic and hydraulic piston actuators can be used for all valves. Pneumatic and hydraulic diaphragm actuators are limited to short stroke valves. Both types can include speed control and end of stroke cushioning. Auto-close by spring return, on power failure, may be optional or the actuator can be "stay-put". Electro-hydraulic linear actuators can provide thrusts of 220 kN to open or close gate valves up to DN900. Water hammer prevention controls can be included in the mechanism.

Rotary motion valves can also be power actuated. The possible modes of operation are similar to those listed for linear valves but the detailed design is different in some cases to include motion translation from linear to rotary.

Ball and butterfly valves are often power actuated to facilitate remote operation. Most of the valves described in Section 3.3.3 can have actuators fitted. Some small butterfly valves are specifically designed for powered operation. Pneumatically actuated PVC valves, DN15 to DN50, can operate up to 6 barg at temperatures up to 50°C. Seals can be EPDM or Viton™ and the valve bodies have sockets for cement connections. Maximum air pressure is 6 barg. The same valves can be electrically actuated; 24 V, 100 V or 240 V ac. The actuator is physically protected to IP65.

Further information on all actuator types can be found in Chapter 12.

3.4 Useful references

NACE (originally known as "The National Association of Corrosion Engineers") 1440 South Creek Drive, Houston, Texas 77084-4906 USA Tel: 281 228 6200, www.nace.org.

DIN EN 1503-1 Valves Materials for bodies, bonnets and covers - Part 1: Steels specified in European standards.

BS 750 : 2006 Specification for underground fire hydrants and surface box frames and covers.

BS EN 4509 : 2006 Screws, 100 degrees countersunk normal head, offset cruciform recess, threaded to head, in titanium alloy, anodized, with aluminium pigmented coating, metric series. Classification: 1100 MPa (at ambient temperature)/315 C°.

ANSI (American National Standards Institute), 1819 L Street, NW(between 18th and 19th Streets), 6th floor, Washington, DC, 20036 USA, Tel: 202 293 8020, Fax: 202 293 9287, www.ansi.org.

AISI (American Iron and Steel Institute), 1140 Connecticut Avenue, NW Suite 705, Washington, DC, 20036 USA, Tel: 202 452 7100, www.steel.org.

NFPA (National Fire Protection Association), 1 Batterymarch Park, Quincy, Massachusetts, 02169-7471 USA, Tel: 617 770 3000, www.nfpa.org.

FDA (US Food and Drug Administration), 5600 Fishers Lane, Rockville, MD 20857-0001 USA, Tel: 888 463 6332, www.fda.gov.

Non-return valves

4

4.1 Non-return valves and the system

Non-return valves, or nrvs as they are known, prevent fluid from travelling the wrong way in systems. A popular application is to prevent circulation in rotodynamic compressors and pumps, described in Chapter 3, Section 3.1.

P&IDs (Process/Piping &Instrumentation Diagrams), show the engineer how process equipment and instrumentation are connected. Different types of compressors and pumps are indicated by different symbols allowing system designers to decide what piping components are required. P&IDs do not show the orientation of pipework. Many items of equipment can have connections in a variety of orientations to suit the local installation. The physical orientation of the pipe run must be known before the final decision can be made regarding nrv style.

Rotodynamic machines are not always fitted with nrvs. A nrv is not required in a system with a single rotodynamic machine when all the pressure generated by the machine is dissipated as pipe friction losses. There is no opportunity for reverse flow. This situation rarely occurs in practice, because most pipe systems include changes in pipe elevation, which require the rotodynamic machine to impart some static head to the fluid which pressurises the system. When the machine stops the static head decays as a result of the fluid returning to the lowest points in the system. A non-return valve can prevent the fluid from returning to the suction source.

In some installations the fluid returning to the suction source does not cause a system problem. However the reverse flow can cause serious machine problems. Many compression machine types can run very successfully as an expansion machine, i.e. a turbine. Uncontrolled reverse flow can cause a compression machine to overspeed and damage itself. Some machines will not run successfully in reverse at all. Lube oil pumps driven from the machine spindle may not supply any lube oil when running backwards. Running at any speed in reverse can cause costly damage to bearings and gearing.

In complex systems, where good reliability and continuity of production must be assured, multiple rotodynamic machines are installed. Some machines will run continuously while others will be on stand-by in case a fault develops. In some processes the stand-by machines must take over almost instantaneously. There is no time for an operator to manually open isolating valves or even wait for power-actuated valves to open. The stand-by machine must be ready to run as soon as an operator presses the start button or the process control system initiates the start sequence. A non-return valve provides a good solution.

The running machines discharge into the system. The higher pressure fluid is prevented from returning to the suction source through stationary machines by nrvs. When a stand-by machine starts it must initially generate a slightly higher pressure than normal to open the nrv; but then the machine follows its HQ characteristic to find the point of equilibrium with the system curve. The failed running machine can then be isolated and inspected or repaired while production continues.

Some positive displacement machines need to be fitted with nrvs in a similar way to rotodynamic machines. The following machine types should be fitted with non-return valves when used in a stand-by situation:

- gear
- lobe
- screw
- rotary piston
- progressing cavity
- vane

Peristaltic machine installations should be reviewed with the manufacturer. Positive displacement machines which use internal valves do not require a nrv for simple stand-by functions.

Positive displacement machines with internal valves may be fitted with a non-return valve to prevent reverse flow in a particular situation. These positive displacement machines often require a substantial torque for starting on-line; that is starting at full discharge pressure. The starting torque can be as high as 150% of the normal running torque. This high value is only required for a very short time, normally less than a second, but it must be applied to start the machine. Only direct-on-line DOL starting, of ac squirrel cage fixed speed motors can supply this level of torque. Some motor designs do not develop a starting torque of this magnitude, even with DOL. Be warned!

With larger motor sizes, star-delta starting, S-D, is preferred to reduce the voltage drop in the electrical system. Star-delta starting produces very little starting torque, so a method of reducing the driven machine torque requirements is necessary. If started with a discharge pressure very close to suction pressure the starting torque can be reduced to 10 or 20% of normal running torque. A by-pass, returning discharge fluid to the suction system, can accomplish the required pressure reduction. If the by-pass is opened on a running system the high pressure fluid will be short-circuited to suction. A nrv is fitted to prevent this happening.

Mounting the nrv remote from the compression machine can create additional problems. Seawater pumps for offshore platforms are submerged, but the control equipment and non-return valves are on the platform. For some reason, foot valves are not used in the suction pipework. A discharge nrv is fitted at platform level. The discharge pipe from the pump to the nrv is empty at start-up. Beyond the nrv the discharge pipework is full of seawater. When the new seawater hits the nrv, tremendous pressure spikes are created. A back pressure regulator can be used to automatically bleed air out of the system and control the seawater pressure. The regulator sizing is critical.

Offshore operating staff have experienced many problems including over pressurised riser tubes, stretched flange bolts, damaged pump bearings, corrosion, erosion, broken bursting discs in downstream systems and other damage generally attributed to pressure and flow transients. According to D Fitzgerald, simulation of many of these problems is possible, using powerful computers.

The manner in which the fluid starts and stops to flow, under normal operating conditions, can be important for the long term durability of the nrv. Small rotodynamic machines, started DOL, will "run up" very quickly causing a rapid fluid velocity increase. The same machines will stop quickly when the power is switched off. These operating conditions can lead to the moving valve components bouncing off stops and seats. Larger rotodynamic machines will be started with closed discharge valves so the increase in fluid velocity can be controlled by the valve speed. Small positive displacement machines will react in a similar manner to small rotodynamic machines. Large positive displacement machines will probably be started on a by-pass, so the increase in fluid velocity can be controlled by the by-pass valve operation. The non-return valve required for a specific installation may require resilient "bump" stops and a resilient seat insert, or damping, or end-of-travel cushioning, to prevent shock loading.

Special versions of double non-return valves can be fitted in domestic drinking water systems to prevent contamination of the system. These valves are often called "backflow preventers". When reverse flow tries to enter the system the downstream nrv opens a branch port and allows the suspect liquid to be dumped.

Non-return valves impose a friction loss on the system. A straight swing disc nrv, wide open, is generally assumed to be similar to a square elbow in the pipe. If the nrv is not wide open the losses will be higher. Swing disc nrvs are manufactured in an oblique style, see Section 3.3.1 in Chapter 3.

These valves will have slightly lower friction losses, everything else being equal. Piston nrvs generally have higher losses. Special streamlined versions of straight piston valves are manufactured to reduce losses to a minimum. Friction losses by piping and pipe fittings are a constant drain on energy. Whether or not overall efficiency is important from a corporate viewpoint, it is always important from a global perspective to reduce energy consumption, depending upon the source and cost of the energy. Compression machines frequently have small discharge connections. The first item a piping designer fits in the pipework is a reducer*, to increase the pipe bore and reduce the flow velocity to an acceptable level. The velocity through the nrv should also be considered to ensure proper valve selection.

NOTE: A fitting to change the pipe size is always called a "reducer". Even if it is fitted backwards to increase the pipe size it is still a "reducer".

The type and size of the non-return valve which must be fitted in a system can be greatly influenced by the necessity of cleaning or inspecting the pipework with a "pig". If "pigging" is a prerequisite then the non-return valve must be able to present a completely clear bore of approximately the same diameter as the pipe. A slightly larger bore would be acceptable, but not a smaller one.

Finally, it may be necessary to fit protection for the non-return valve or instrumentation to indicate correct function or valve status. Clean fluid systems are only clean after flushing and some time of normal operation. Systems which operate with a degree of solids content frequently have unexpected solids in the worst possible locations. Crude oil, specified to contain sand, sometimes contains pebbles. Equipment selected to pass sand up to 2mm may be damaged or disabled by 10mm pebbles.

Nrvs fitted to stand-by machines can be a considerably drain on energy if the valve is permanently "cracked-open" by sand on the seat. Permanent instrumentation may not be necessary if maintenance staff perform routine inspections with modern equipment. The high frequency acoustic signal created by a high velocity fluid passing through a small aperture should be easily detected.

4.2 Non-return valve design

The non-return valve is an automatic, self-powered valve which tries to ensure fluid only moves in one direction. The moving element normally rests on the seat to form a seal. A small pressure must be applied to the moving element to open the valve initially. Once open, fluid-dynamic forces are generated which hold the valve open or increase the opening. Fluid flow must usually stop before the valve will close. The fluid-dynamic forces created by any fluid flowing across the seat will generally prevent all valves closing. Springs may, or may not, be used to control opening and assist closing. Some valves rely solely on gravity to provide the closing forces. Valves which rely on gravity must be installed according to the manufacturers' instructions. If level pipe is specified then the local pipe fall must be modified for a short distance.

Swing disc valves are gravity-powered. As the valve opens, the force required to hold the valve open increases. If the balance between disc mass and fluid-dynamic forces is incorrect the valve will not open completely. Increased fluid velocities may result in unexpected corrosion or erosion damage. Gravity-powered valves must be matched to the fluid and operating conditions.

When valves open fully, the disc or piston travel must be limited by a stop. Valves which open fully, but are not stop limited, are liable to "flutter". Flutter can cause rapid wear of hinge pins or piston guides. Valves which use springs can suffer from early spring failure due to fatigue. Flutter can be caused by the shedding of eddies, or turbulence. Damping can be used to restrict the flutter movement. Fluid damping, using squish, can be effective when the fluid has some viscosity. Valves using springs can have variable rate springs fitted. If the full travel stop incorporates squish, to prevent rebound after rapid opening, this can be an effective flutter damper.

Squish can be incorporated into the seat and disc/piston design to prevent the valve slamming shut. Extra material is added around the seat contact area to create two squish zones. Trying to squeeze the fluid out of these zones during rapid closure slows the valve down. But there is a penalty to be paid. The increased area of limited clearance is an ideal site for trapping small solids. Squish protection for damped closure can lead to more problems caused by trapped solids, unless there is adequate clearance for the squish action to eject the solids. Valves with narrow seats can crush friable solids, such as coal. The squish zones tend to broaden the effective seat width and reduce the valve's capability to crush solids. This effect must be considered, taking into account the nature of any pertinent solids. Ball valves usually have very narrow seats and can clear most solids enabling the seating to be effective. (See Chapter 3, Section 3.3.3.)

The problem of flutter may be restricted to small valves. As valves become larger the inertia of the moving parts becomes much greater. The increased inertia can effectively dampen flutter and lead to delayed closure, after reverse flow has commenced. Damping at the seat therefore becomes very important.

As with all valves, the flow areas must be checked and velocities calculated for design operating conditions. Areas around discs and pistons are as important as the main port areas. Flow areas which are smaller than others will be the zones where erosive, and possibly cavitation wear, will occur.

Non-return valve bodies can incorporate extra connections for special functions such as venting and draining. Valves for hot applications can sometimes be fitted with an external by-pass to allow system preheating at low flows.

4.3 Non-return valve types

Selecting the correct type of non-return valve is essential for process reliability. Frequent unscheduled interruptions can be very costly and also result in product delivery delays. The following descriptions should assist in informed valve selection.

Swing disc non-return valve

The swing disc non-return valve, also commonly known as the "swing check", is the standard basic nrv usually supplied by a manufacturer unless a specific type is requested. Figure 4.1 shows a typical arrangement of a valve with a one-piece cast body.

All access to the internals is through the top cover. The valve has a replaceable seat. The hinge pin is located so that gravity

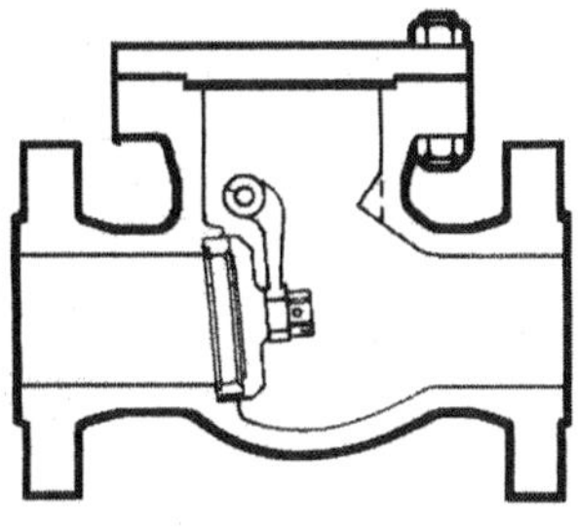

Figure 4.1 A typical on-piece body cast swing disc non-return valve

holds the disc against the seat. Some of these valves are suitable for vertical operation. The hinge pin is fitted through a boss in the body wall and retained by a screwed plug. This hole and the top cover are the only leak paths for process fluid. For hazardous applications the nature of the plug seal must be reviewed. A seal which isolates the plug threads is preferable.

Hinge pins are generally of 11/13Cr. The end-of-travel stop is cast into the body allowing plenty of scope for squish design. Cast valves of this style can usually be "pigged" without problems. Valves of this type are widely available, as shown by Table 4.1.

Body material	Seat material	Disc material	Pressure rating	Sizes
Cast carbon steel	Hard faced carbon steel	11/13 Cr	ANSI 150lb	2" to 4"
		11/13 Cr on carbon steel		6" to 12"
	11/13 Cr			14" to 24
	Hard faced carbon steel	11/13 Cr	ANSI 300lb	2" to 4"
		11/13 Cr on carbon steel		6" to 12"
	11/13 Cr on carbon steel			14" to 20"
	Hard faced carbon steel	11/13 Cr	ANSI 600lb	2" to 4"
		11/13 Cr on carbon steel		6"
	11/13 Cr on carbon steel			8" to 20"
	11/13 Cr	11/13 Cr	ANSI 900lb	3" & 4"
		11/13 Cr on carbon steel		6"
	11/13 Cr on carbon steel			8" to 20"
	11/13 Cr	11/13 Cr	ANSI 1 500lb	2" & 4"
		11/13 Cr on carbon steel		6"
	11/13 Cr on carbon steel			8" to 12"

Table 4.1 One-piece body, swing disc non-return valve materials and ratings

Butt weld connections are available as an option instead of flanges. Valves for pressure ratings over 300lb may have corrugated soft metal gaskets or ring-type joints for the body cover. Elastomer inserts in either seat or disc is possible. Valves of this type are available for pressures up to ANSI 2500 lb in sizes up to 48". Alternative body materials include:

- cast iron, (for 16 barg)
- low temperature carbon steel
- $1\frac{1}{4}$% Cr $\frac{1}{2}$% Mo steel
- 5% Cr $\frac{1}{2}$% Mo steel
- 18% Cr 10% Ni 2% Mo stainless steel

The design can be modified slightly by increasing the angle of the seat; this is claimed to reduce water hammer effects. Cast valves can be of the oblique style. Both modifications may prevent "pigging".

The seat is normally screwed in to the body; therefore check how it is locked. The hinge pin should be locked in the body and the disc rotates about the pin. This construction ensures the body is not a wearing part. When the seat and hinge pin are of different materials to the body, selective corrosion or galvanic corrosion may be a problem; it is advisable to seek the manufacturers' advice. For austenitic stainless steel valves, a hardenable stainless steel should be used for the hinge pin. 17-4PH could be a good choice.

Closed die forged bodies are produced as standard for valves of ANSI 900lb and 1 500lb. Sizes include DN6 to DN50. Connection options include screwed, socket weld, butt weld and flanged. The flanged versions may have the flanges butt-welded to a forged body and so it is important to check the

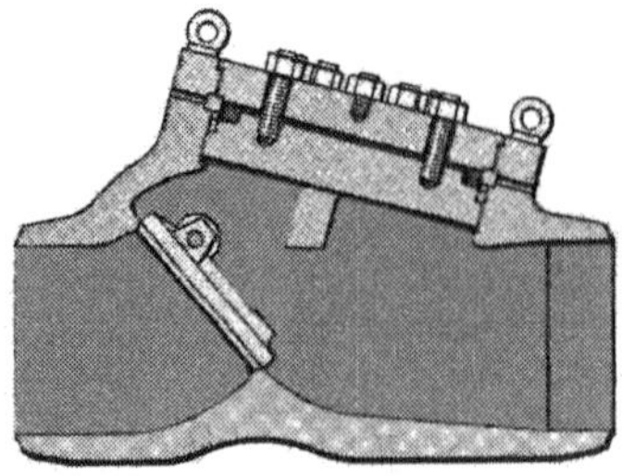

Figure 4.2 An oblique style swing disc non-return valve
Courtesy of Dresser-Masoneilan

weld quality. Some small high pressure valves may not be full bore; and so check the seat size if this is important. Forged body valves with hammer lug connections are specifically manufactured for oilfield applications.

The style of construction shown in Figure 4.1 can be used for special materials if the complete body is machined from the solid. The hinge pin can be fitted from the top, through the cover, and clamped in position. The facing for the cover seating is made by machining a flat on the circular body. The integrity of the valve is improved by removing the hinge pin retaining plug. Valves are available up to DN200 in the following materials:

- aluminium, Alloy 20, bronze, duplex stainless steel
- Hastelloy™, Inconel™,Monel™, nickel, Nitronic™
- tantalum, titanium, zirconium

The design of the swing disc valve can be modified slightly by moving the position of the hinge relative to the disc centre-line. The disc can be pivoted from a point within the disc diameter, called a "tilting disc valve". Moving the centre of rotation closer to the disc reduces the mass and the inertia, although it makes the seat and seal design more complicated.

Figure 4.2 shows an oblique style valve for steam. Valves of this style are available up to DN1000 for pressures up to 60 barg. Valves for liquid can be up to DN350 with pressure capabilities to 500 barg. Oblique style valves can be fitted with pneumatic actuation to override the fluid dynamic forces. These valves are not full bore.

The swing disc can be made in two other styles:

- two-piece body
- wafer

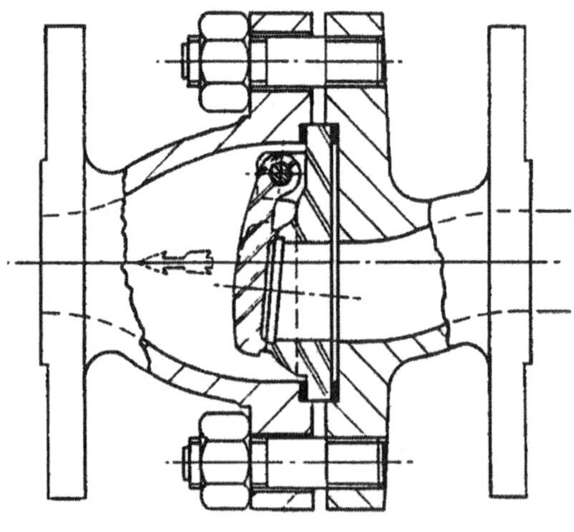

Figure 4.3 A two-piece body swing disc non-return valve
Courtesy of YPS Valves Ltd

The two-piece body version is similar to that shown in Figure 4.1. The body is split vertically next to the seat location. The body flange is used to hold the seat and the hinge pin in position, see Figure 4.3. One possible leak path is removed, increasing the valve integrity. Valves of this style are cast in exotic alloys, as listed above, for sizes from DN15 to DN150 in pressure ratings up to PN50.

The wafer version is useful for locations where space is restricted. Wafer valves are built in sizes from DN50 to DN900 in pressure ratings up to ANSI 2 500lb. Carbon steel is standard with a stainless steel option. Some wafer valves have an "O" ring seat insert as standard.

Full bore wafer non-return valves, from DN15 to DN100, are suitable for 100 barg at temperatures up to 200°C. Reduced-bore stainless steel or gunmetal valves, for 40 barg maximum, with Viton™ seat seals are available from DN50 to DN400. The bore of the DN50 valve is 31 mm; DN400 is 305 mm. Maximum operating temperature is 105°C.

A derivative of the simple flat swing non-return valve is the curved plate non-return valve. The flat disc is replaced by a curved plate which seals the bore. The valve body has a bore diameter slightly larger than the pipe bore and is fitted with a replaceable insert profiled to a shaped metal seating. The curved plate is designed to lift into the horizontal position and rest in the top of the valve bore. The flow path is completely unhindered and results in very low pressure drops and allows the valve/pipeline to be cleaned or inspected by a "pig".

Flat plates are inefficient for supporting evenly distributed loads and need to be disproportionately thick. End caps for pipework are therefore elliptical. The curved plate is much stronger than a flat disc and can be made much thinner and consequently lighter. The reduced weight of the curved plate allows the valve to open fully at lower fluid velocities and accept lower density fluids. This valve style is ideal for gases. In fact, the design was originally developed for use in the pulverised coal-feed lines on large boilers. The valves were fitted to eliminate the danger of explosions in the feed lines due to blow-backs from the boiler. As the curved plate is virtually invisible to the flow when the valve is wide open, these valves are very suitable for solids handling and abrasive applications.

Swing disc non-return valve with external loading

The swing disc non-return valve, with external loading is a logical development from the previous valve design. The disc is attached to the hinge pin which now rotates in bearings in the valve body. A shaft seal is required where the hinge pin is extended through the body wall. Externally the hinge pin is fitted with a radial arm and adjustable weight. The external weight can be used to increase the effective weight of the valve disc and reduce opening, or reduce the effective weight and increase opening. In both cases it must be remembered that the inertia of the valve moving parts has been increased and the valve response to similar forces will be slower. The hinge pin seal can provide a degree of damping.

When a standard swing disc nrv is mounted horizontally, the force required to hold the valve open increases as the valve opens. If it is mounted vertically the force reduces. An external weight on a radial arm can be used to counterbalance or partially counterbalance the disc weight effect. Alternatively, the arm can be oriented so that the external weight complements the disc weight. The arm can be set at any angle and the weight adjusted until the desired valve performance is achieved.

An external dashpot can be fitted to limit the disc angular velocity and angular acceleration. Also a quadrant can be attached to the valve body to allow the valve to be locked open or closed or to limit the opening. The external lever can be held in the open position permanently by a wire and fusible link. This arrangement enables the valve to prevent reverse flow in the event of an external fire.

Because the external loading is a fairly simple modification of this valve type, the availability is similar to those shown in Table 4.1. The external lever can be replaced by a pulley or sprocket which allows a constant external torque to be applied to the hinge pin.

Twin disc non-return valve

The problem of valve inertia has just been mentioned above. The twin disc non-return valve, significantly reduces the problem. The single disc described previously is divided in two and supported across the centre-line of the valve bore. The twin disc nrv has two nominally semi-circular discs. The inertia of each disc is less than half that of the normal swing disc because the axis of rotation has been moved and the "half" disc can be thinner. The linear movement required to achieve full opening is much shorter. The two discs are pivoted from a single central hinge pin or a central column.

Because this valve has an obstruction in the centre it cannot be cleaned or inspected by a "pig". This shortcoming must be considered but the basic design concept is eminently suitable for large valves. A well-designed twin disc non-return valve will have a flow coefficient as high as, or higher than, a conventional swing disc design. Some manufacturers do not offer valves for applications with pulsating flow. Installations with reciprocating machines, and some other positive displacement machines, should therefore be reviewed with the valve manufacturer. Some rotodynamic machines can be prone to surge when operated with certain system characteristics. The possibility of surge should be evaluated in these circumstances.

The valves are installed with the hinge pin vertical. On opening, the discs rotate towards each other and move to a position pointing downstream. Physical stops should be provided to prevent over travel.

These non-return valves are most popular as a wafer pattern, see Figure 4.4, in all pressure ratings. The wafer construction permits considerable space saving and is much lighter than that of flanged valves. Higher pressure versions can be similar to the lug style, with the body od the same as the flange od. Clearance holes allow the use of one set of studbolts. With this design the bolting is around the outside of the valve, still however maintaining the "wafer" face-to-face dimension (API 594 standard). The lug (also known as a wafer lug sometimes) and double flanged types are equally as popular, if not more so, in the

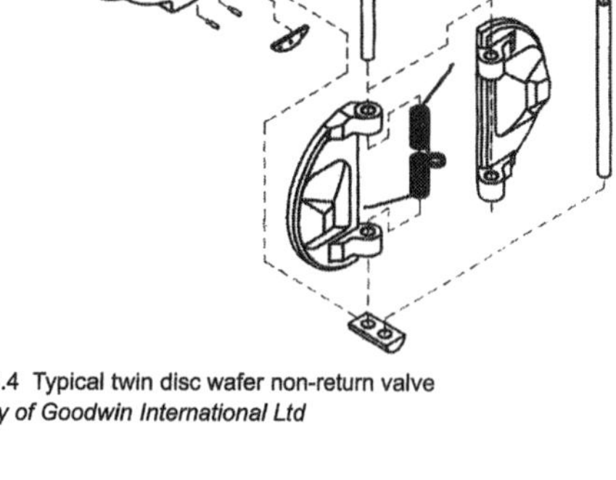

Figure 4.4 Typical twin disc wafer non-return valve
Courtesy of Goodwin International Ltd

hydrocarbons industry. This type of valve is also sometimes referred to as a dual plate check valve (API 594 standard).

Twin disc nrvs are available in three designs:

- conventional
- retainerless
- cartridge

The conventional approach is to fit the hinge pin through the valve body and provide the pressure containment by a screwed plug or flange. The retainerless design fits all the internals axially from one end of the body. Internal stops secure the hinge pin without piercing the body pressure containment. The cartridge valves have all the internal components, including the seats, mounted in a sleeve which is locked axially in the valve body. Again, the valve body is not pierced with potential leak paths. Spare cartridges can be stocked to allow rapid repair in emergency situations. The cartridge can be serviced and inspected on a bench rather than on site. All designs should include a fixed spring stop to allow each disc to be independently sprung and balanced for synchronised operation.

Popular valves have bodies and cartridges of cast steel with 11/13 Cr discs. Seats can be integral and hard-faced or have resilient inserts. Valves from DN50 to DN600 can have pressure ratings from ANSI 150lb to 1500lb; valves up to DN300 have ratings of 2500lb. Valves up to DN2100 and valves with pressure ratings up to ANSI 4500lb are also available. Some metal wafer non-return valves have a lining. Stainless steel valves with PFA lining are mass-produced in sizes from DN15 to DN100. Maximum pressure is 100barg at temperatures up to 200°C. Cast iron wafer valves, with aluminium bronze discs, are available from DN50 to DN600 for 16 barg. Bodies can be painted with epoxy for corrosion protection.

Because these valves have two sealing elements, the elements can respond individually to variations in flow velocity. If the valve is subject to uneven flow distribution across the bore the discs will open to different positions. Wear may be localised in one half. If the discs are sprung against each other, rather than a fixed stop, travel stop damage and early spring failure may result. The detailed design of the valve should be fully investigated if uneven flow distribution, possibly caused by close proximity of bends to the inlet, is likely.

Piston non-return valve

The piston non-return valve, belongs to a general group of nrvs called "lift nrvs" or "lift checks". These are linear motion valves similar in concept to globe valves. The ball non-return valve, described in the next Section, belongs to the same group but has different properties. The piston non-return valve can be used for the most arduous clean applications, such as dosing and metering. Spring-loaded valves can be cycled up to 1000

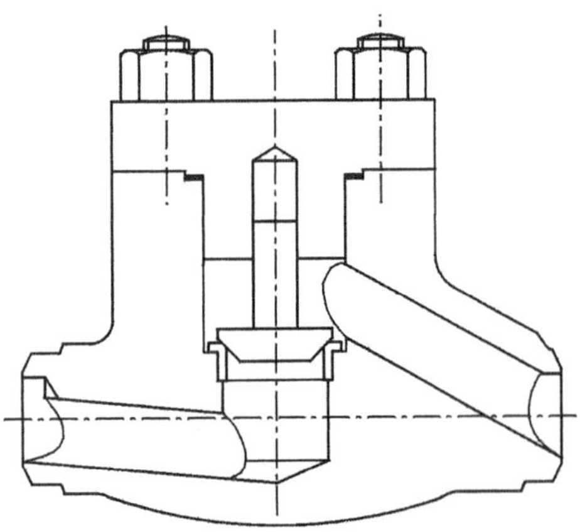

Figure 4.5 A typical closed die-forged piston non-return valve

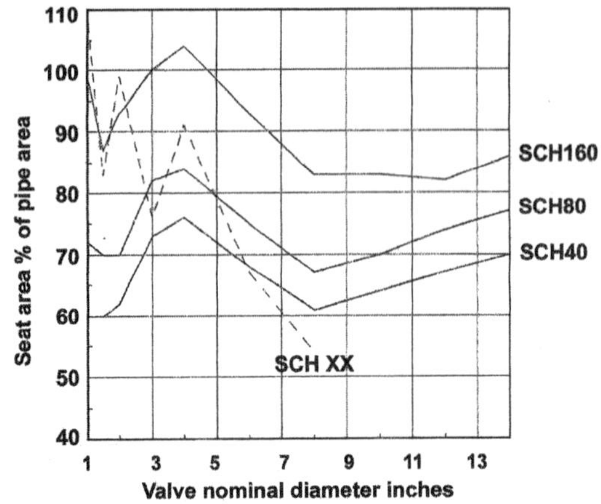

Figure 4.6 Seat areas of piston non-return valves

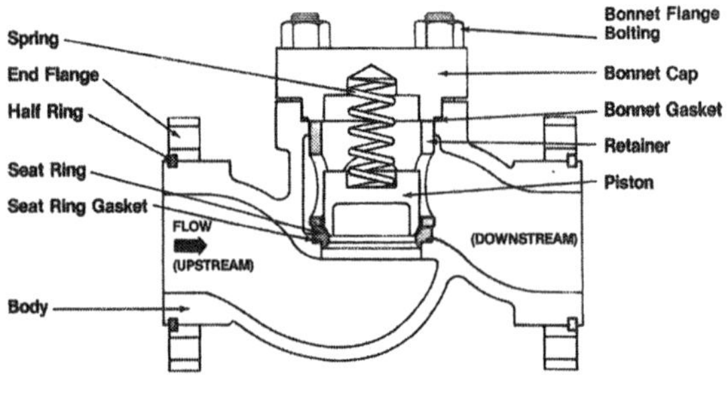

Figure 4.7 A typical cast cage-trim piston non-return valve

times per minute. Figure 4.5 shows the construction of a typical high pressure non-return valve for sizes up to DN80.

The valve shown has butt weld connections and a bolted, spigotted cover. Small valves of this type may have a screwed cover. The valve has a replaceable seat and a top-guided piston. Guiding can be bottom, crow-foot or wing-guided, or top, or both. The valve relies on gravity alone to close. Spring-loading is also popular. When springs are fitted it is important that the springs do not become coil bound. For valves which cycle regularly, the springs must be stressed to avoid fatigue. The valve performance can be optimised by adjusting the spring design, spring rate and initial compression.

The flow passages in small valves can be tortuous when machined from the solid. The velocity in the passages should be checked as well as through the seat. The tortuous flow path makes these valves unsuitable for solids handling applications. Forged body valves are mass-produced in carbon steel and austenitic stainless steel. Typical pressure ratings include PN20, PN50, PN100, PN150 and PN250. Specialist manufacturers produce valves for ANSI2500lb and 4500lb. Some flanged valves are made by welding on flanges, so the weld quality should always be checked. Some small valves have an angle pattern but are not produced in such large quantities.

Flow passage problems are eliminated in larger valves by casting the whole body. All valves are not full bore, so check seat areas when evaluating the pressure drop. Figure 4.6 shows the typical seat area as a function of pipe area for common pipe schedules.

Some designs are very similar to cage-trim globe control valves, see Chapter 6, Section 6.3.1. Figure 4.7 shows a cage-trim piston non-return valve. Notice the flanges are not integral but mounted and retained by shear rings. The valve seat is retained by the cage. All components are assembled through the top cover. The valve can be maintained in situ without special tools. Squish can be built into the valve design to soften end-of-travel and closing. End-of-travel squish can be accomplished by the machining of the back of the piston and the stop. Closing squish is adjusted by the positioning of the ports through the cage relative to the full diameter of the piston .

Valves of this style and angle pattern, with various end connections, are produced in a range of sizes and pressure ratings, see Table 4.2.

Body material	Pressure rating	End connections	Sizes
Cast carbon steel Cast low temperature steel Cast stainless steel	ANSI 150lb ANSI 300lb ANSI 600lb ANSI 1500lb ANSI 2500lb	Screwed Flanged Socket weld Butt weld	up to DN50
Cast carbon steel Cast low temperature steel Cast stainless steel	ANSI 150lb ANSI 300lb ANSI 600lb ANSI 1500lb ANSI 2500lb	Flanged Butt weld	DN80 to DN600

Table 4.2 Sizes and ratings of cage-trim non-return valves

Special high pressure versions, to ANSI 4500lb, machined from solid forgings are available in suitable materials. Pistons and seats would normally be made from austenitic stainless steel. The piston cage can be carbon steel or stainless steel, both with the bore surface treated to eliminate friction and galling. The valve body can be cast in more exotic materials, such as those used in swing disc non-return valves.

The tortuous flow passage problems can be largely eliminated by constructing an in-line valve, see Figure 4.12. General purpose cast iron valves, rated at PN10, are produced in sizes from DN50 to DN400 and incorporate a flat Buna "N" seal ring on the piston. These valves are very popular as foot valves for pumps; matching strainers are available.

Wafer pattern valves, in carbon steel or stainless steel, suitable for 48 barg and 150°C, are available in sizes from DN15 to DN150. Soft seats, of neoprene or Viton™, are also an option.

Small non-return valves can be fitted with compression fittings or screwed ends. Because of the small seat sizes involved, the fluid pressure forces are small, and the piston or disc can be made quite thin and therefore light. Valve discs can have bonded soft seals to act against integral metal seats. Valves up to ½" or for 12mm od tube may be assembled by internal screw threads, see Figure 4.8. These valves are capable of operating up to 414barg at temperatures up to 204°C. Slightly larger valves, for ¾" and 1" tube and pipe, may only be rated for 139barg. Any internal threads, not protected by seals, would render the valve unsuitable for NACE applications.

Valves for cold water system applications with brass bodies and screwed or single ferrule compression fittings will be suitable for 10barg and temperatures up to 95° or 105°C. Sizes will generally be suitable for use with copper pipe of 8mm to 54mm od.

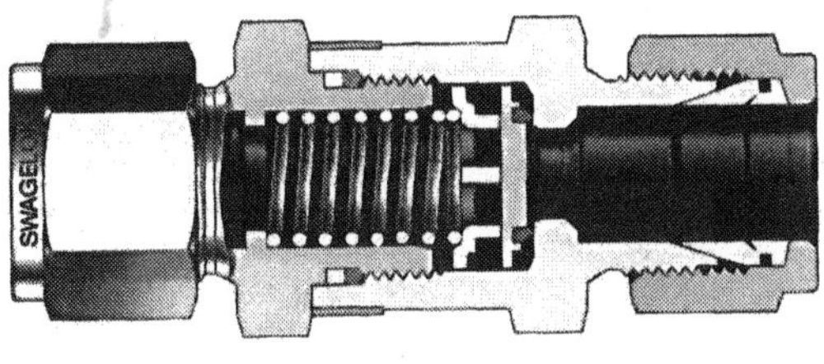

Figure 4.8 Small piston non-return valve with compression fittings
Courtesy of Swagelok Company

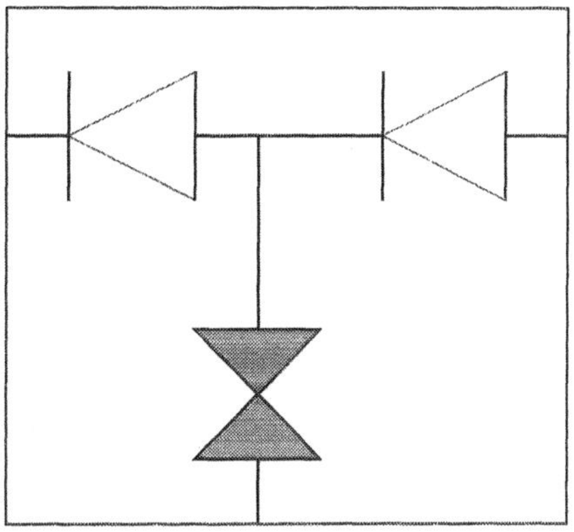

Figure 4.9 Diagrammatic arrangement of a backflow preventer for drinking water system

The backflow preventers used in drinking water systems follow a similar type of construction as the valve shown in Figure 4.8 but incorporate three separate valves, two nrvs and an automatic dump valve. Figure 4.9 shows a diagrammatic arrangement. Under normal operating conditions, both nrvs are open and the dump valve is closed, and flow travels from the supply to the consumer. If the pressure differential reverses and water attempts to flow into the supply, both nrvs close. The downstream non-return valve has a moveable seat which slides back and opens the dump valve. Water, which could contaminate the supply system, is diverted to the dump connection which can be piped to a drain. When the normal differential pressure conditions are restored, the dump valve is closed and both nrvs open.

Backflow preventers of this type should be fitted in the consumer's local pipework and not in the distribution system. Valves are small, DN15 and DN20, suitable for pressures up to 10barg and temperatures up to 65°C. Bodies in bronze can have screwed connections or be fitted with single ferrule compression fittings for standard copper pipe. The sliding seat which actuates the dump valve may be fitted with a rubber bellows or membrane. Valves of this type must be approved by the appropriate authority before fitting to the system.

A special derivative of the piston non-return valve, a type of plate valve, is used exclusively on gas applications. Special plate valves have been developed for reciprocating gas compressors. A non-return valve is required at the inlet and outlet of each cylinder to ensure the gas only travels downstream. These valves generally have the plate and spring combined in a single component. The plate flexes to allow the valve to open. Figure 4.10 shows an exploded view of a typical valve. The bottom component is the valve seat. The od of the seat is stepped to allow clamping in the compressor cylinder.

Many valves are retained by a cage in a similar manner to cage-trim control valves and the piston nrv shown in Figure 4.7. The flow passages are shaped curved slots. The valve plate consists of concentric rings connected by slender spokes to a central boss. The flow passages are generally narrow to allow the plate to be very thin. A thin, light plate has very low inertia and can respond quickly to changes in gas velocity. Compressors frequently operate faster than 1000rpm and valve inertia would create serious problems. The complete valve assembly is held together by a central stud. Additional small helical compression springs may be fitted near the periphery of the valve plate to supply additional spring load. Simple versions of this style of valves, with a single port, are sometimes called "reed valves" and are used in small air and refrigeration compressors.

The compressor valves described can be used for process applications with gas. The valve can be mounted in a body or as-

Figure 4.10 A typical compressor plate valve
Courtesy of Hoerbiger Holding UK Ltd

sembled with spacers to form a wafer unit. Unlike most others these valves can be mounted in any position. Pulsating gas flow is not a problem. Valves for process applications are available for working pressures from 10 barg to 320 barg. Sizes range from DN25 to DN500. Standard valve materials are 11/13Cr but stainless steel and non-metallic materials, such as PEEK, are also available.

Ball non-return valve

The ball non-return valve, is suitable for the most arduous, dirty or clean applications. The nature of the ball shape produces smooth fluid streamlines which reduce the tendency for erosive wear considerably. Great attention must be paid to the detail design if the ball nrv is to achieve its real potential. Failure to deal with these detail design problems will result in high fluid losses and rapid wear.

The main problem to be understood and resolved is the weight and inertia of the ball. The volume of the ball is much greater than the equivalent piston and when made of the same material is much heavier. The increased inertia means the ball is much slower to respond to fluid and spring forces. The fluid forces generated are smaller than those of a piston due to the balls better drag coefficient. Selecting the right material can assist with the ball weight. Alumina ceramic is approximately half the weight of steel, and such balls are routinely manufactured. Ceramics are very hard, abrasion resistant and have extremely broad chemical compatibility. If the ball fails to lift successfully, high velocity jets will travel across the seat and emerge to erode the valve body.

For the longest possible life the ball must be balanced. If the ball does not rotate equally about all three axes, localised wear will occur which severely reduces useful life. A balanced ball will wear evenly and gradually reduce in diameter. Considerable material removal is possible however before the ball requires replacement. Seats can have elastomer inserts which improve the sealing for clean fluids and fluids with very small solids.

Squish can be designed into the seat and ball lift stop to cushion the ball. Because of the increased weight, squish can be very important in reducing impact damage. If good lift is not achieved damping is unlikely to be a problem.

Ball non-return valves are generally small and are similar in style to piston nrvs, see Figures 4.5 and 4.8. The ball may, or may not, be spring-loaded. The spring must not rest on the ball directly as this will inevitably prevent the ball from rotating. A light follower must be used. The most popular commercial ball is a standard ball bearing. These are manufactured in SAE52100 carbon steel. Balls are also routinely made in AISI440C, tool steel, tungsten carbide and bronze.

Metal ball non-return valves are generally restricted to sizes between DN8 and DN50 with pressure ratings from ANSI150lb to 2500lb.

A special version, the double ball non-return valve, is produced in the smaller sizes. Two nrvs are built into one body. These valves follow the in-line, straight construction shown in Figure 4.8. Valves can have screwed or compression fittings. The valves are used for high pressure lubrication in conjunction with single-shot lubricators. A very small quantity of oil is injected at regular intervals into machine elements such as:

- compressor cylinders
- engine cylinders
- compressor stuffing boxes
- pump stuffing boxes

The double ball is a safety feature to ensure combustion products or process product does not escape in the event of lube oil line fracture. These valves are used routinely to 414barg; higher pressures are possible.

Diaphragm non-return valve

Diaphragm non-return valves, are made in two distinctly different designs:

- flat disc
- "duck-billed"

The flat disc diaphragm valve uses the flexibility of an elastomer disc against a multi-port plate to produce the valve action. Flow in the correct direction bends the elastomer disc away from the ported plate to allow the fluid to flow. If the fluid tries to change direction the elastomer disc straightens out and seals off the ports. This style of valve is recommended for clean fluids. Valve bodies are of cast iron or cast iron coated with PTFE. The steel ported plate can be Rislan™ coated or PTFE coated. Various elastomers can be used to suit the fluid. Sizes range from DN10 to DN200 for PN16, valves up to DN150 can be PN25. Maximum operating temperature is 100°C.

The "duck-billed" nrv relies on a moulded diaphragm to seal against itself. A tube is moulded and one end is formed and flattened to seal the bore. The slightest increase in internal pressure opens the flattened portion and allows flow. If the internal pressure falls below the surrounding pressure the flattened portion closes and seals. The length of the flattened portion is significant and the valve can seal around solids. The tube seal is very effective and the valve can be used for critical applications. The flow path is unobstructed when open and large solids can be passed. The basic tube is available in a range of materials:

- natural rubber, chlorobutyl rubber, Neoprene™, Buna N,
- polyurethane, Hypalon™, Viton™, EPDM.

The valve can be used by attaching the moulded tube to the end of a pipe. Tubes for this purpose are made in sizes to suit pipes from ½" to 120"nb. Alternatively the moulded tube can be built into a cast iron or fabricated steel body to form a flanged valve assembly. This style of valve is made in sizes from 3" to 48" nb. Standard small valves can withstand back pressures of 10 barg and large standard valves can withstand 2 barg. Special high pressure tubes can be moulded to special order.

Non-return foot valve

Foot valves are a special type of non-return valve, intended for the suction pipes of pumps. Most pumps require priming before starting. Priming involves filling the pump casing with liquid and expelling any trapped air. Some pumps are self-priming and these pumps will be specifically described as such. Even these pumps need some liquid in the pump casing to be able to work properly.

The foot valve is designed to maintain the suction pipework and the pump casing full of liquid. Foot valves are used on pump installations where the pump lifts liquid from a source below the pump.

The pump casing and the suction pipe can be filled manually by pouring liquid into the pump. Alternatively the liquid can be induced into the pipework and casing by applying a vacuum to the casing. In both cases the foot valve prevents the liquid running straight out and emptying the casing. It must be remembered that a foot valve is a resistance in the suction system and results in a loss of suction pressure and NPSHa/NPIPa for the pump. The resistance must be considered when designing the pipework and selecting a valve, see Chapter 8, Section 8.2.

There are many types of non-return foot valve to suit many applications:

- plate
- poppet
- ball
- flat diaphragm
- tube diaphragm

All valve types may be fitted with a suction strainer; the design of the strainer being dependent upon the pump application. Contractors' pumps, i.e. those pumps used on building sites to drain ditches and pits, may be suitable for handling solids up to 100mm. The strainer must limit the solid intake to the pump's capacity. A heavy duty wire cage is necessary. Process pumps generally operate with clean liquids. A perforated metal tube or woven wire mesh, of adequate flow area, is adequate.

The plate versions are very similar in style to the small valve shown in Figure 4.8 with different body constructions. Small valves can have three-piece bodies similar to the ball valves in Section 3.3.3, Chapter 3. Small and large valves can have one-piece bodies with all the internals fitted through the downstream end of the body. The spring retainer holds all parts in position. Plate versions are available in cast iron, brass and bronze. Sizes range from DN15 to DN200 with pressure ratings from PN6 to PN40. Valves with elastomer sealing inserts may be limited to 110°C but all metal valves can be suitable for 350°C. Plate foot valves are recommended for clean liquids. A derivative of the plate version uses a crow-foot guided valve. This design is available from DN175 to DN 800 for pressures up to 10barg.

The poppet style valves are very similar to the controlled closure nrv shown in Figure 4.13. The body is a one-piece casting and all components are fitted through the downstream end. The spring retainer holds all parts in place. These valves are made in cast iron, bronze and austenitic stainless steel. Sizes range from DN50 to DN400 with pressure ratings from PN10 to PN40. Elastomer seals are used to prevent leakage and temperature limits will be 80°/110°/150°C depending upon the compound used. These foot valves are not recommended for liquids carrying solids.

The ball style foot valve is not like the ball non-return valve described earlier. The non-return foot valve version is designed for vertical or horizontal mounting and the ball is not spring-loaded. The ball is lifted by hydrodynamic forces and is pushed into a pocket alongside the liquid flow path. When the valve is wide open most of the ball is retracted, almost providing a full-bore passage. The ball version of the foot valve is intended for liquids carrying solids. These valves are made with cast iron bodies and balls can be of cast iron or non-metallic. Nylon and polyacetal balls have proved useful. Valve sizes range from DN50 to DN350 for pressures up to PN10. Working temperatures cover -10° to 80°C.

The flat diaphragm valve uses the flexibility of an elastomer disc against a multi-port plate to produce the valve action. When the pump creates a suction depression the elastomer disc is bent away from the ported plate to allow the liquid to flow. This style of valve is recommended for clean liquids. Valve bodies are cast iron or cast iron coated with PTFE. The steel ported plate can be Rislan™ coated or PTFE coated. Sizes range from DN10 to DN200 for PN16, valves up to DN150 can be PN25. Maximum operating temperature is 100°C.

The tube diaphragm foot valve is a logical development of the flat diaphragm version. The tubular diaphragm lies inside the perforated plate strainer. When the pump creates a suction depression, the tube collapses away from the strainer and opens the flow path. If the depression is removed the tube expands to regain its normal shape and form a seal against the strainer. This style of valve is only effective with clean liquids. Tube diaphragm foot valves are available in sizes from DN40 to DN300 with a maximum operating pressure of 6barg. Standard body material is cast iron with a natural rubber diaphragm.

Controlled closure non-return valve

The problem of water hammer generation due to rapid valve closure has already been emphasised. Peak pressures experienced through water hammer can cause considerable damage with consequential loss of production. Some non-return valve designs can be modified to regulate the velocity of the moving element(s). The modifications are an improvement on squish which can only function close to the end-of-travel. A form of damping is included to limit valve element velocity. These valve types are also called "non-slam valves".

The swing disc non-return valve can be modified to include external loading, and was described earlier. Damping, in the form of a dashpot, can be attached to the external arm. Depending upon the forces to be restrained, damping can be achieved using atmospheric air or oil. Atmospheric air is best when possible since there are no consumables, no shortage of supply, and no maintenance refills. Oil is used when air becomes impractical. Needle valves should be fitted to allow adjustment at site to suit actual operating conditions. If orifices are used, a good selection of taper pin reamers will be required for final adjustment. Dashpots with fixed orifices may require seasonal oil changes to cope with viscosity variations.

The cage-trim version of the piston nrvs described earlier can be modified to include damping, see Figure 4.11. The two valves are basically the same. The piston in the controlled closure version is modified to include viscous damping by the product. The volume of fluid above the piston is trapped and only allowed to enter or exit via controlled ports. To open, the fluid above the piston must flow through the spring-loaded ball

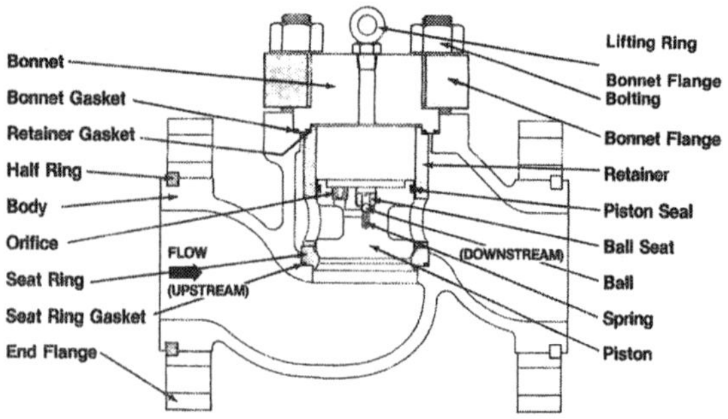

Figure 4.11 A controlled closure piston non-return valve
Courtesy of Flowserve Flow Control

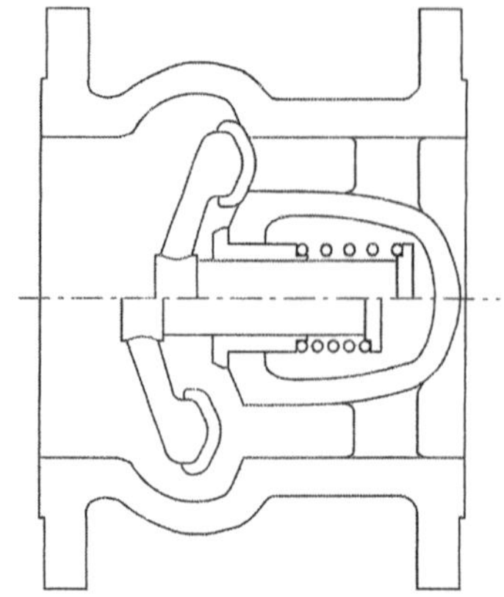

Figure 4.12 Annular plate controlled closure non-return valve

check valve. To close, fluid must flow from below the piston to above, via the removable orifice. Both flow paths are adjustable to optimise performance. Squish is still used to cushion the extremes of travel. These valves are available in a range of sizes and ratings and angle pattern, see Table 4.3. Special high pressure versions, to ANSI4500lb, machined from solid forgings are available in suitable materials. The comments regarding seat areas and Figure 4.7 apply to these valves also.

The valve shown in Figure 4.11 is constructed with a standard globe valve style body. Other designs straighten the flow path to avoid corners. Figures 4.12 and 4.13 show two different styles:

- annular plate
- poppet valve

The annular plate style is used extensively in high pressure hydraulic systems where the liquid is very clean. The poppet valve style is very popular for oilfield applications where the fluid may contain solids such as sand. Figure 4.14 shows a detail of the poppet/seat arrangement. Care is taken with both designs to maintain smooth streamlines. The shape of the moving element, plus squish built into the seat design, can control the valve closure. Damping can be incorporated when necessary.

Both designs are capable of having full-bore seats, but in practice this is rarely achieved. Full-bore seats require plenty of space which makes the valve body large and long.To comply with standardised valve lengths the seat area is compromised. The seat flow area, and the areas around the moving element, should be checked and the velocities evaluated. Erosion corrosion can occur with clean fluids when the protective oxide layer is removed. More exotic materials of construction or surface treatment may be required to obtain satisfactory service life. Alternatively, the pipe size can be increased locally to permit a larger valve to be installed.

Body material	Pressure rating	End connections	Sizes
Cast carbon steel Cast low temperature steel Cast stainless steel	ANSI 150lb ANSI 300lb ANSI 600lb ANSI 1500lb ANSI 2500lb	Screwed Flanged Socket weld Butt weld Proprietary clamp	up to DN50
	ANSI 150lb ANSI 300lb ANSI 600lb ANSI 1500lb ANSI 2500lb	Flanged Butt weld Proprietary clamp	DN80 to DN300
	ANSI 600lb	Flanged Butt weld Proprietary clamp	DN350 to DN900

Table 4.3 Sizes and ratings of controlled closure piston non-return valves

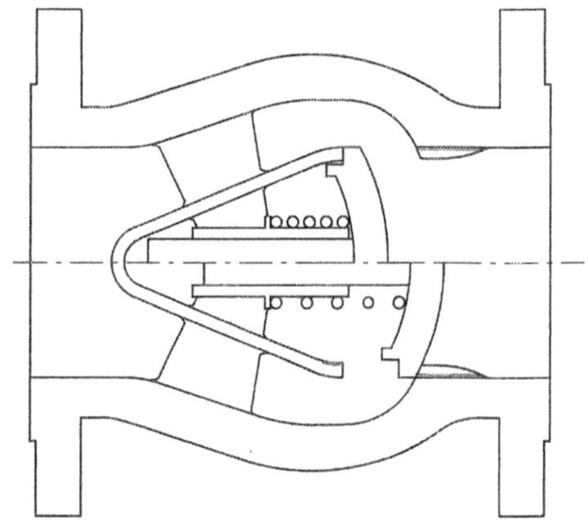

Figure 4.13 Poppet valve controlled closure non-return valve

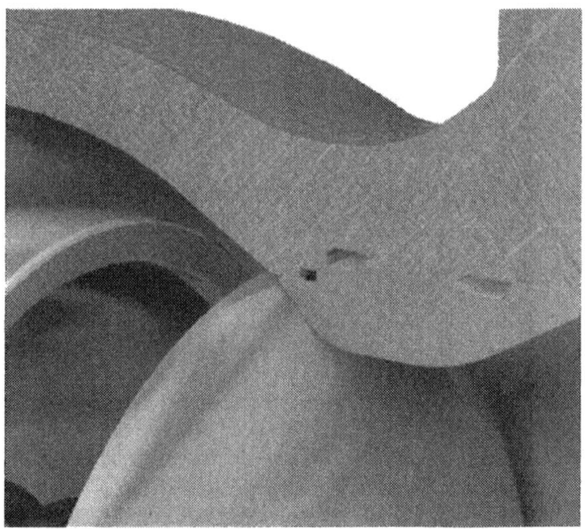

Figure 4.14 A high pressure poppet non-return valve seat detail
Courtesy of Mokveld Valves BV

Valves of these styles are manufactured in large sizes. Up to DN1800 are already installed and DN2500 are advertised. Valves smaller than DN150 are rare. Pressure ratings up to ANSI2500lb are standard. The poppet style valve is also manufactured to API pressure ratings: 2000, 5000 and 10 000.

The gas compressor plate valve shown in Figure 4.10 can be considered as a controlled closure valve. The very light plate design and very low inertia, enables the valve to respond rapidly to changes in velocity. The background history of the use of these compressor valves means that small valves must be able to operate at frequencies in excess of 17Hz. Pulsating flow and repeated changes in gas direction are not problems.

Special Note: Pump valves

Reciprocating pumps, like reciprocating compressors, require inlet and outlet valves to ensure the liquid always travels from inlet to outlet. The valves are non-return valves, for a very dynamic application. Small pumps for clean liquids can operate at speeds up to 1500rpm which require valves to operate successfully at frequencies up to 25Hz. Reciprocating pumps are used for a wide variety of applications not just high pressure. A diversity of valve types is available to cope with extreme requirements. If a mass-produced nrv is not suitable for a particular application, a pump valve, used in a similar manner to the compressor plate valve shown in Figure 4.10, may be a viable solution. Figure 4.15 shows some pump valves in common use.

Special Note: Valves for dry solids

Non-return valves are generally unsuitable for dry solids applications. The ball valve without spring for vertical mounting or the duck-billed nrv, could be suitable depending upon the specific operating conditions. System design should remove the necessity for non-return valves. The use of power actuated isolating valves and an appropriate control system would provide the most reliable installation.

Special Note: Non-metallic valves

Reduced bore wafer swing-disc nrvs, in PVC or polypropylene are available in sizes from DN50 to DN350 for pressures up to 8 barg. Viton™ seat seals are standard. Operating temperature is limited to 90°C. Bore size may be considerably smaller than pipe bore; DN50 valve has 26mm bore; DN350 valve has 240mm bore. Full-port valves are available from DN15 to DN200. PVC valves, with cement socket bodies, in sizes from DN15 to DN100 are suitable for 5 barg. The hinge and disc seal are Buna "N".

Piston non-return valves can be non-metallic. Valves from DN15 to DN100, in PVC or polypropylene are suitable for 10 barg up to 50°C. Non-metallic valves in polyacetal, limited to DN10 to DN25, can be suitable for 95°C. The piston can be fitted with EP or Viton™ seals. Valves are fitted with cement socket connections. Some polypropylene valves are available with female BSP connections. Small valves, DN6 to DN30, are manufactured in POM and polypropylene for 8 or 5 barg. Wafer pattern bodies, for valves in polypropylene and uPVC, are suitable for 10 barg for sizes from DN15 to DN300. A Viton™ soft-seat limits maximum temperature to 90°C. Similar designs are also available in PVDF.

Non-metallic ball nrvs, in PVC or ABS with union connectors, ⅜" to 4c, have EPDM seals and are suitable for a maximum working pressure of 10barg.

Special Note: Hygienic valves

It is possible that the wafer versions of all-metal swing-disc non-return valves, may be suitable for hygienic installations. The crevice created by the hinge pin could be removed by the fitting of a suitable plastic bearing. Manufacturers should be contacted regarding third-party testing and certification.

In-line piston non-return valves, in stainless steel are popular for hygienic applications. The two-part valve body is clamped by an external nut. Valves from 1" to 3" are supplied with butt weld tails for direct fabrication in the pipework.

4.4 Useful references

Start-up dynamics of offshore water winning pumps, D Fitzgerald, World Pumps no. 331, pp. 31-37, 1994 Elsevier Ltd.

Water Hammer Control in Centrifugal Pump Systems, H Kinno, Journal of the Hydraulics Division, May 1968, pp 619-639.

Water hammer analysis, J Parmakian, Dover Publications 1963, New York.

Water Hammer Charts for Centrifugal Pump Systems, Kinno & Kennedy, ASCE Journal of the Hydraulics Division, 1965

Fluid transients in systems, E B Wylie, V L Streeter,1993, Prentice Hall, New York.

Hydraulic analysis of unsteady flow in pipe networks, J A Fox, 1987: MacMillan Press Ltd, London.

ANSI/API 594 Check Valves: Wafer, Wafer-Lug, and Double Flanged Type.

Regulators 5

5.1 Regulators and the system

Regulators can be used in place of control valves when set point deviation is not critical. Regulators are a self-powered control valve but usually on a small scale. Regulators can be used where control valves would be used but external power is not available, such as in portable applications. Most control valves are pneumatic but increasing numbers of electric powered valves are being installed. A minority of control valves are hydraulically powered.

All control valves require external power:

- compressed air
- electricity
- hydraulic fluid

In some locations external power is not available and the cost of supply power is prohibitive. In some hazardous locations it may not be desirable to introduce electric supplies. A regulator can overcome all of these difficulties providing the process fluid is suitable.

One of the most popular regulator applications is the reduction in pressure of bottled gas. Bottled gas is widely used for cooking, heating, industrial and medical purposes. Gases include:

- acetylene
- butane
- carbon dioxide
- LNG
- LPG
- nitrogen
- nitrous oxide
- oxygen
- propane

Gas bottles usually contain pressurised liquid gas. The liquid evaporates when a depression occurs. The liquid pressure in the bottle is generally too high for the user and a regulator is fitted to allow a suitable pressure to be selected.

Regulators are used extensively by the water and gas supply utilities. Fluid is transported over long distances in pipelines and the demand, ie. flow rate, can vary significantly. The variation in frictional losses would result in large pressure changes at the consumer outlets. Regulators are placed at strategic locations in local supply branches to minimise pressure excursions.

Back pressure regulators, described in Section 5.4, are also used as low pressure safety valves in some applications. Vessels which store flammable liquids, and also compressor, engine and reciprocating pump crankcases, are often pressurised or purged with a safe gas such as nitrogen. The nitrogen, at a pressure slightly above atmospheric, displaces the air and forms a non-combustible safety blanket. The gas blanket also prevents the ingress of dirt and moisture which could promote corrosion. In storage vessels the gas pressure can be used to increase pump suction pressure and, theoretically, increase NPSHa/NPIPa. The predicted increase in NPSHa/NPIPa may not be realised due to dissolved gas evolving from solution. The vapour pressure of the liquid is effectively raised to the pressure at which dissolved gas begins to evolve. High pressure nitrogen is reduced in pressure locally by a suitable regulator.

If the liquid level is lowered due to liquid use, the regulator supplies more gas to maintain the pressure. If liquid is admitted to the vessel the gas blanket is compressed and the pressure increases. To prevent over pressurisation, a back pressure regulator is used as a relief valve. As pressure increases the regulator opens and discharges gas to a vent system. Back pressure regulators can be used in a similar manner on vessels with volatile liquids. If the vapour pressure increases, due to liquid input or temperature rise, the back pressure regulator can release vapour to a flare, vent or recovery system. This type of application is sometimes called "vapour recovery".

Regulators are widely used in refrigeration applications. One compressor can be connected to several evaporation systems at different temperatures, and therefore different pressures. Back pressure regulators can maintain the different evaporation pressures irrespective of compressor suction conditions. Large complex industrial refrigeration systems can use separate compressors to perform multi-stage compression. Back pressure regulators and regulator relief valves can be used to control inter-stage pressures.

The differential pressure reducing regulator may be unfamiliar under this name but is the basis of all constant flow regulators. Frequently used in heating and ventilating systems, the constant flow regulator can balance individual circuits while coping with varying system conditions. The differential pressure reducing regulator is a popular accessory used with centrifugal pumps. Centrifugal pumps can have a very flat H-Q characteristic which results in significant changes of flow for relatively small changes in differential head. In most centrifugal pumps the absorbed power increases as the flow increases. If the pump flow may increase above the rated flow, because of transient operating condition variations, a larger driver must be installed to provide the extra power. This driver "oversize", results in a reduction of pump unit efficiency at the rated conditions and increased power costs.

The differential pressure reducing regulator can maintain a fixed differential between suction and discharge heads and thus restrict the pump flow range. The driver for rated conditions can be fitted, with consequential reduction in operating costs. The differential pressure reducing regulator is used mainly in water distribution systems and district heating schemes.

5.2 Regulator design

Regulators can be divided into two basic groups:

- diaphragm
- piston or spool valves

Both groups can be direct-acting and pilot-operated to cope with size, pressure and accuracy.

The easiest to understand is the direct-acting diaphragm regulator, see Figure 5.1 The regulator is shown partially open and fluid enters from the left. The downstream port of the regulator communicates with the underside of the diaphragm via a small port. The spring case above the diaphragm is vented to atmosphere to ensure a constant pressure, theoretically. The regulator set point is adjusted by modifying the spring force using the bolt shown. Increasing the spring force increases the set point and vice-versa. If the downstream pressure is low the spring force overcomes the diaphragm force and opens the regulator. The lower the pressure the wider the regulator opens. As the downstream pressure approaches the set point the diaphragm

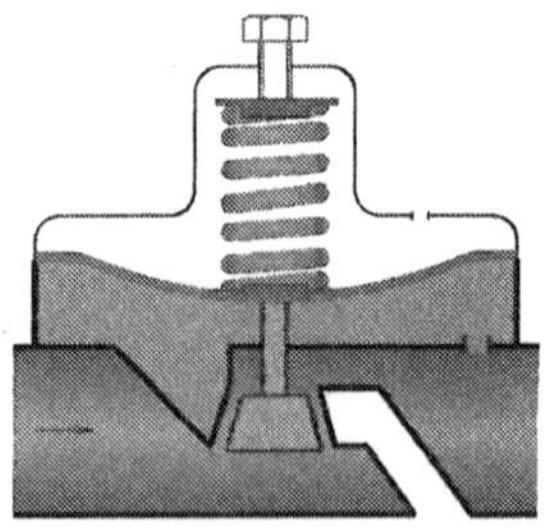

Figure 5.1 Direct-acting diaphragm pressure reducing regulator

force increases and gradually compresses the spring. If the set point is reached the regulator closes and prevents further flow.

The diaphragm regulator can be modified to use an external sensing connection rather than an internal port. This can be implemented by reversing the valve operation; changing push-down-to-open to push-down-to-close. The spring and adjustment is fitted under the diaphragm and the sensing connection piped to the top. External sensing can be very useful for controlling the pressure in vessels.

The diaphragm regulator is a very simple device. The diaphragm is clamped between the "valve" body and the spring case. Diaphragms can be metal as well as elastomer. The valve spindle can be guided by a bush in the body or the plug can be guided by the seat. The communication port can be incorporated in the guide bush. Friction of the guidance and stiffness of the diaphragm tend to reduce accuracy but impart damping to reduce "hunting". The diaphragm area can be increased to compensate for friction and stiffness; but this introduces increased inertia which creates opening "over-shoot" and valve closing impacts, unless porting and spring modifications are incorporated. Double-beat globe valve designs can be used to reduce the spring and diaphragm forces required to operate the regulator, see Table 3.4 in Chapter 3, and Figure 6.7 in Chapter 6.

The set point range can be extended by fitting a choice of springs. If the set point ranges overlap, the lighter spring should be used to obtain the best accuracy. The accuracy of direct-acting regulators should not be expected to be better than ± 5% of set point. The regulator characteristic can be modified in various ways. The valve plug can be profiled as in a control valve. Standard helical coil springs have a constant spring rate. An increasing spring rate can be produced by winding the end coils at reduced pitch. This would reduce regulator opening with pressure variation. The communication port can be adapted to suit varying operating conditions. Throttling, fixed or adjustable, can be added. Multiple ports can be used, some with non-return valves; this allows faster opening or closing.

The diaphragm can be replaced by a metal bellows. Both elastomer/metal diaphragms and metal bellows can suffer from fatigue problems. The fluid and operating conditions must be fully evaluated when selecting materials and the type of construction. These styles of regulators can be used for inlet pressures up to 415barg in the smaller sizes. Sizes extend to DN50.

For clean fluids, a piston or spool pressure reducing regulator may be used. Depending upon the nature of the fluid it may be possible to eliminate all non-metallic materials to allow high temperature operation. Figure 5.2 shows a typical arrangement. The piston regulator is very similar to the diaphragm version.

When piston seals are used they increase the friction forces, tending to impair movement and also increase damping. Metal pistons have more inertia than diaphragms and this effect can exacerbate any "over-shoot" or hard closing problems. Modern plastics can alleviate this problem. Some designs utilise a smaller piston diameter on top. This piston is exposed to up-

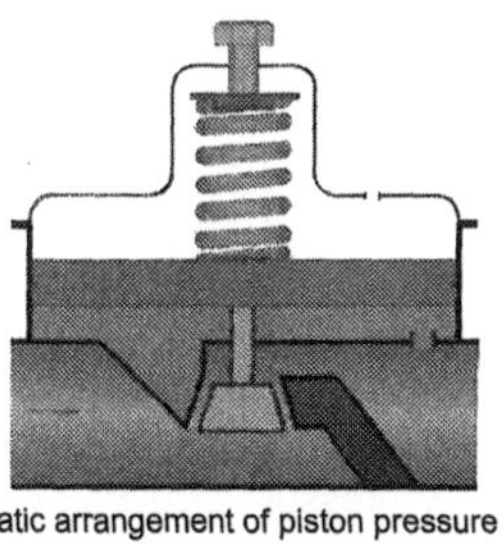

Figure 5.2 Diagrammatic arrangement of piston pressure reducing regulator

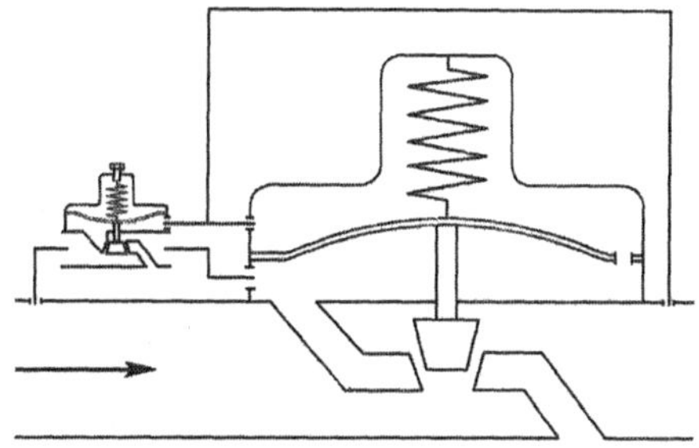

Figure 5.3 A pilot-operated diaphragm pressure reducing regulator

stream pressure in the "reducing" mode. The upstream pressure force complements the spring force and allows the use of smaller springs. Varying the relationship between the two piston diameters can increase the overall pressure range. Piston regulators can be used for higher pressures and higher differentials between upstream and downstream. Piston regulators can be modified in a similar manner to diaphragm versions to allow external pressure tappings.

As direct-acting regulators become larger the problems of inertia and diaphragm stiffness increase and accuracy is sacrificed. The problems can be overcome by using a pilot-operated regulator. A small regulator, of the type shown in Figure 5.1, is used to sense the control signal. The flow from the regulator is directed to the underside of the much larger diaphragm in the main valve, see Figure 5.3.

The small control regulator is sometimes called a "relay" or "amplifier". The pilot-operated regulator is push-down-to-open while the main regulator is push-down-to-close. The downstream pressure is applied under the pilot and over the main diaphragm. If the downstream pressure falls the pilot valve opens allowing upstream process fluid to flow under the main diaphragm. This unbalances the main diaphragm and opens the main valve. Process fluid flows downstream to attempt to maintain the set pressure. If the downstream pressure recovers the pilot valve closes. The pressure under the main diaphragm decays by leaking through a port in the diaphragm and flowing downstream through the sensing line.

Pilot valves can be fitted in a similar manner to piston regulators. Pilot-operated regulators should be used when an accuracy of better than ± 5% of set point is required or the valve size is over DN50. DN150 is the largest popular pilot-operated regulator size for general process applications. Cast iron regulators, specifically for water applications, are readily available up to DN300.

The pilot can be incorporated in the main valve so that it is almost impossible to distinguish between direct-acting and pilot-operated versions. Alternatively the pilot valve can be mounted externally with all connections between the two being piped. Both styles have advantages and disadvantages. Whilst the external pilot pipework can be prone to leakage the pipework allows the fitting of optional filters and coolers to extend the operational range. Also multiple external pilots can be fitted to allow complex control functions to be implemented by one valve. Pressure and temperature control is possible. Multiple pressure settings can be implemented by individual pilots connected via remote controlled solenoid valves.

Regulators can be sensitive to the mounting position. A few can be mounted in any position without affecting operation. Low pressure regulators may require fitting at a specific attitude. The manufacturer's specification should be checked before finalising the pipework design. The regulator will generally be smaller than the nominal pipe size used, so allow space for reducers.

Figure 5.4 Sliding gate valve

The set point of a regulator is adjusted by varying the spring compression. This is normally accomplished locally using the bolt and locknut fitted as standard. Some regulators can be fitted with a motor drive to allow remote set point adjustment.

The regulator valve body is designed for pressure containment. The top of the body is extended to form the diaphragm mounting or the piston cylinder. Direct pressure communication, from either valve port, to the underside of the diaphragm/piston is simple. When the spring case is vented to atmosphere there is no pressure containment requirement. The spring case can be a light steel pressing or die cast aluminium or zinc alloy. The connecting flange which clamps the diaphragm must have sufficient rigidity to make a good seal. When the spring case is pressurised, in differential pressure regulators for example, a heavier style of construction may be required depending upon the pressure involved.

The regulators discussed so far have used a type of globe valve to control the fluid; this is the most popular arrangement. An alternative style of valve, the sliding gate valve, is also available. The sliding gate valve relies on a ported seat and a matching ported plate to regulate and isolate the flow. The plate slides over the seat to uncover the ports to increase the flow area and is illustrated in Figure 5.4.

The sliding action can clear the seat of solids and the body does not have pockets where solids can accumulate. The plate and seat are lapped to optical flatness and ensure shut-off leakage to better than 0.01% of rated capacity. The upstream pressure and retainer guides hold the plate against the seat. The plate is undercut at strategic locations to reduce the fluid pressure loading. Fluid pressure does not exert variable unbalanced forces on the actuator which improves accuracy.

The standard plate and seat material is stainless steel. The plate is chrome plated to improve wear resistance and remove galling tendencies. Alternative materials and coatings can be supplied for specific applications. Various port areas can be fitted to each valve body size to vary the C_V. Some manufacturers offer customised port sizing to suit user applications. The sliding gate construction requires a very short stroke from closed to wide open; this promotes a rapid response to changing control signals.

Regulators modulate in response to changing system pressures. Some regulators can be modified to perform as pressure sensitive isolating valves. The set point defines a "switch setting" or "toggle setting". Reaching the "switch setting" causes the regulator to open or close. The regulator does not have any intermediate positions. This style of regulator can be useful in certain conditions.

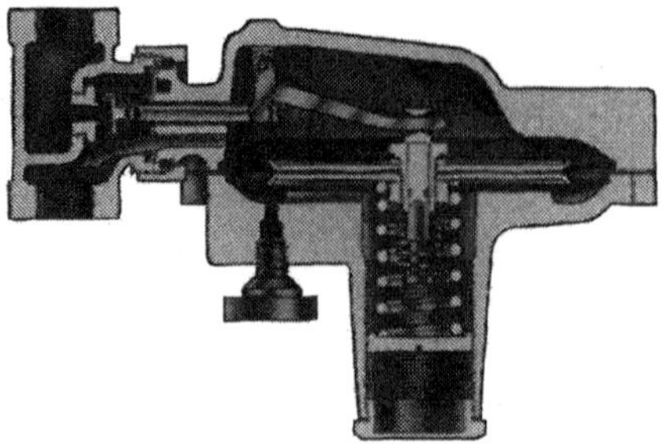

Figure 5.5 Low pressure reducing regulator
Courtesy of Emerson Process Management

5.3 Regulator types

Constant pressure regulator

The most popular of all constant pressure regulators is the pressure reducing version, as discussed in Section 5.2 and illustrated in Figures 5.1, 5.2 and 5.3. Fluid is supplied from a higher pressure source. A regulator is fitted to reduce the pressure locally to the desired value. Figure 5.5 shows a low pressure reducing gas regulator used for nitrogen blanket and purge applications.

The regulator shown is intended for mounting in vertical pipes with the bottom connection piped to the downstream low pressure. The operation of low pressure regulators of this type can be influenced by the weight of internal components and the manufacturer's installation instructions must be consulted regarding acceptable attitudes. Regulators of similar design are mounted with the spring on top of the diaphragm. The diaphragm is large compared to the valve body. The downstream pressure is low resulting in small diaphragm forces. The diaphragm area is increased to produce as high a force as is practical. Even so, the force is magnified by using a modified bell crank. The diaphragm deflection is quite large but the valve movement is only about 30% of this, due to the force magnification. The lower diaphragm surface is vented to atmosphere and a filtered vent is fitted. The regulator downstream pressure is sensed via a Pitot tube in the valve body and transmitted to the diaphragm upper surface.

Regulators such as these are suitable for gas inlet pressures of 10barg. The outlet pressure can be controlled over a range from 12mm water gauge to 0.7 barg. Sizes range from DN20 to DN50 although direct-acting regulators up to DN100 can be made for low pressure applications.

Pilot-operated regulators are produced for similar low pressure applications. Minimum set pressure can be as low as 6mm water gauge for sizes up to DN150. The inlet pressure capabilities are increased to 13.8 barg.

These gas regulators usually have cast iron, steel or stainless steel bodies to cope with all gas and environmental conditions. Some models may also be available in aluminium and zinc.

The low pressure regulators described above are used for specific gas applications. As working pressures increase the range of general purpose pressure reduction regulators becomes very broad, as shown in Table 5.1.

In some designs of regulator the maximum differential pressure across the regulator is limited. This restriction must be considered at the system design stage when selecting suitable equipment.

Diaphragms are available in nitrile rubber, EPDM, Neoprene™ and Viton™. High pressure regulators may use a stainless steel diaphragm. Small low pressure regulators may have nitrile rubber or nylon valve discs.

Set pressure range barg	Maximum inlet pressure barg	Size range	Operation	Fluids (1)	Body materials (2)
0.02 to 2.0	8	DN8	Direct	A	B, BRZ
0.14 to 2.0	21	DN8 to DN25	Direct	A, G, O, S, W	CI, CS, SS
0.7 to 4.1	13.7	DN50	Direct	A, G, L, S (3)	SS
0.3 to 5.1	10	DN8 to DN15	Direct	L	PVC, PTFE, polypropylene
0.14 to 5.5	25	DN15 to DN300	Pilot (4)	W	CI
0.2 to 6.9	17.2	DN8	Direct	A, G	A, BRZ, SS
0.7 to 6.9	13.7	DN40	Direct	A, G, L, S (3)	SS
0.015 to 6.9	69	DN50	Pilot	A, G	CI, CS
3.5 to 6.9	207	DN8	Two stage	A, G	A, SS
0.015 to 8.6	17.2	DN8	Direct	A, W	Glass filled nylon
0.14 to 9.3	24	DN20 to DN25	Direct	A, G, L, S (3)	SS
0.14 to 10	21	DN25 to DN100	Pilot	S, W	CI, CS
0.3 to 10	21	DN15 to DN25	Pilot	A, G, S	CI, CS
0.3 to 10	21	DN8 to DN50	Direct	A, G, S, W	CI, CS, SS
1.0 to 10	21	DN8 to DN50	Direct	A, G, O, S, W	CI, CS, SS
0.2 to 13.7	17.2	DN15	Direct	A, G	A
0.005 to 20.5	27.5	DN25 to DN150	Pilot	A, G	CI, CS
0.14 to 25	25	DN15 to DN300	Pilot	W	CI
0.2 to 35	100	DN25 to DN50	Direct	A, G	CI, CS
0.35 to 35	138	DN20 to DN50	Direct	A, G	DI, CS
0.7 to 35	414	DN8	Direct	A, G, O, W	B, BRZ, SS
0.012 to 41	72	DN25 to DN150	Pilot	A, G	CI, CS
0.1 to 41	138	DN15	Direct	G, L	A
0.7 to 46.5	48	DN15 to DN50	Pilot	A, G, L, O, S	DI

(1) A = air, G = gas, L = liquids, O = oil, S = steam, W = water
(2) A = aluminium, B = brass, BRZ = bronze, CI = cast iron, CS = carbon steel, DI = ductile iron, SS = stainless steel
(3) Hygienic applications with CIP or SIP
(4) Altitude valve

Table 5.1 Typical pressure reduction regulators

All the pressure regulators described previously use the movement of a diaphragm or piston to sense the change in control pressure and physically modulate the regulator — a purely mechanical system. Another type of air regulator utilises the accuracy and speed of electronics to sense and control the regulator — in effect a pneumatic control valve. The regulator is an air operated diaphragm valve. The downstream pressure is sensed by a transducer and compared electronically with the set point. Two small, high speed solenoid valves control the air flow in to and out of the diaphragm case.

If the system pressure is low the inlet solenoid valve opens more often to allow the pilot air pressure to increase, and so the diaphragm opens the regulator wider. If the system pressure is high the inlet solenoid valve remains closed and the outlet solenoid valve opens to increase the vent rate. At steady-state conditions, when both solenoid valves are closed, control air, filtered to 5 μm, is required to preserve a minimum pressure in the control circuits. The benefits include higher control accuracy with smaller lockup, offset and hysteresis.

Pressure transducers can measure to within ± 0.1% of full scale deflection so a fairly constant regulator performance can be expected. Direct control of the set point, electronically, allows automatic adjustment and integration into PLC and computer control systems.

Set pressure range barg	Size range	Operation	Fluids (1)	Body materials (2)
0.002 to 0.5	DN25 to DN150	Pilot	G	CS, CI, SS
0.005 to 0.5	DN20 to DN50	Direct	G	CI, DI, SS
0.27 to 2.0	DN60 to DN100	Direct	G, L, S, W	BRZ, CS, DI, SS
0.7 to 4.1	DN50	Direct	G, L, S, (3)	SS
1.7 to 5.5	DN60 to DN100	Direct	G, L, S, W	BRZ, CS, DI, SS
0.14 to 5.8	DN15 to DN40	Direct	G, L, S, (3)	SS
0.14 to 10.3	DN25 to DN50	Direct	G, L, S, W	BRZ, CS, DI, SS
0.14 to 12.7	DN15 to DN20	Direct	G, L, S, W	BRZ, CS, DI, SS
0.14 to 12.7	DN25 to DN 150	Pilot	A, G, L, S, W	CS, DI, SS
0.14 to 13.8	DN6 to DN10	Direct	G, L, S, W	BRZ, CS, DI, SS
0.14 to 25	DN15 to DN300	Pilot	W	CI
3.4 to 31	DN15 to DN50	Direct	G, L, S, W	BRZ, CS, DI, SS

(1) A = air, G = gas, L = liquids, O = oil, S = steam, W = water
(2) A = aluminium, B = brass, BRZ = bronze, CI = cast iron, CS = carbon steel, DI = ductile iron, SS = stainless steel
(3) Hygienic applications with CIP or SIP

Table 5.2 Typical back pressure regulators

NOTE: In the context of using solenoid valves as regulators or control valves read **Special Note: Digital valves** in Section 6.3.1, Chapter 6. Although this type of regulator is advertised as a regulator it is in fact a control valve; external power, electricity, is required.

Back pressure regulators maintain the upstream pressure at a constant value. The regulator acts as a dump valve when the pressure is too high, and fluid is allowed to flow to a lower pressure system. If the pressure is low the regulator closes to try to maintain the pressure. Standard back pressure regulators work on systems where the regulator is always required to pass some flow; the regulator does not shut-off and the lockup effects are not considered. The regulator is selected so that the "minimum controllable flow" is smaller than the minimum dump flow. Back pressure regulators are not as popular as pressure reducing regulators so the available range is smaller, as shown by Table 5.2.

Pilot-operated back pressure regulators can be fitted with an additional pilot and function as a pressure reducing regulator simultaneously.

Back pressure regulators can be used as relief valves. Figure 7.1 in Chapter 7 shows the normal relationships between the various operating pressures for different types of relief valve. Some of the operating conditions deviate from the normal operating pressure by a considerable margin. A back pressure regulator may permit a much tighter operating band than a standard spring-loaded relief valve. The characteristics of the back pres-

Set pressure range barg	Size range	Operation	Fluids (1)	Body materials (2)
0.14 to 2.6	DN6 to DN25	Direct	A, G, L, S, W	CI, CS, SS
0.7 to 6.9	DN25	Pilot	A, G	A
0.004 to 9.0	DN25 to DN150	Direct	A, G, L, W	CI, CS, SS
0.05 to 10	DN25 to 100	Direct	A, G, L, S, W	CI, CS, SS
0.2 to 20.5	DN50 to DN150	Pilot	L, S, W	CI CS
0.14 to 25	DN15 to DN300	Pilot	W	CI
1.0 to 26	DN6 to DN50	Direct	A, G, L, S, W	CI, CS, SS
7 to 83	DN25 to DN150	Pilot	A, G, L, W	CS

(1) A = air, G = gas, L = liquids, O = oil, S = steam, W = water
(2) A = aluminium, B = brass, BRZ = bronze, CI = cast iron, CS = carbon steel, DI = ductile iron, SS = stainless steel

Table 5.3 Typical regulators used as relief valves

sure regulator are optimised to reduce lockup effects when the regulator is closed. The scope of regulators used as relief valves does not extend to very high pressures but can be effective up to about 80barg, see Table 5.3. Low pressure performance is much better than conventional spring-loaded valves.

Depending upon the exact design of the valve body these regulators can have varying degrees of seat leakage when closed. The worst should be ANSIB16.104 Class IV and best can be Class VI.

Differential pressure regulator

The differential pressure regulator tries to maintain a constant differential pressure between the regulator inlet and outlet pressures. This type of regulator is used extensively in the water distribution industry. Cast iron globe valve bodies are extended to include the lower diaphragm casing. The valve disc is also extended to provide support for, and a connection to, the diaphragm. An external differential pilot controls the water flow above the diaphragm. Water industry regulators use copper pipework for the pilot and bronze or brass fittings. These regulators are available in sizes up to DN300 for pressure ratings up to PN25. The differential pressure can be adjusted between 0.14 barg and 17.24 barg. Figure 5.6 shows a schematic arrangement of a typical valve.

The three pilot circuit connections are fitted with isolating valves to allow pilot maintenance without disturbing the main valve. The pilot circuit is protected by a fine filter. This is the water industry standard. It should be noticed that no differential

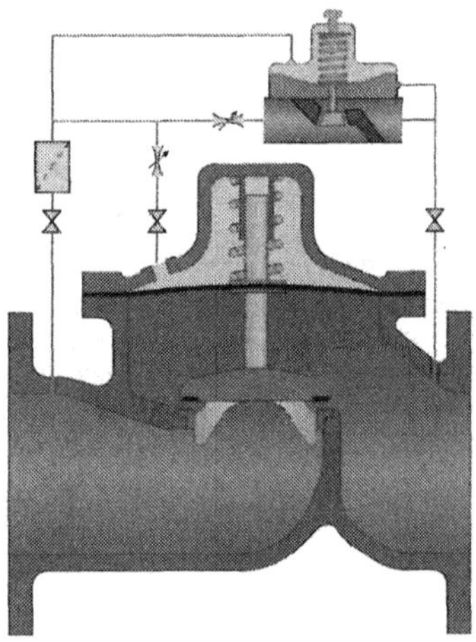

Figure 5.6 Typical differential pressure regulator for water

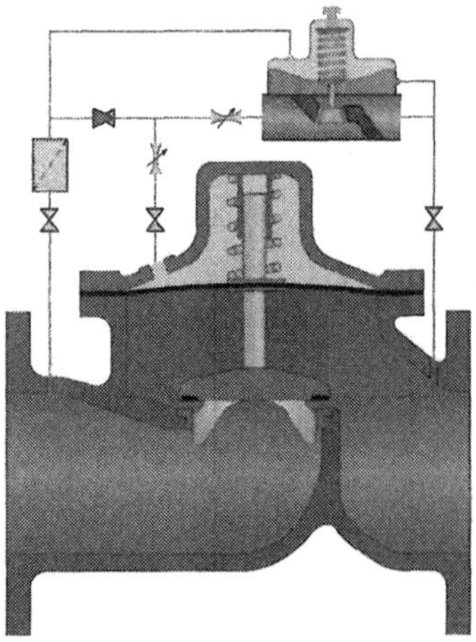

Figure 5.7 Differential pressure regulator with non-return valve in pilot circuit

pressure indication for the filter is fitted; how does one know when the filter is dirty? Whenever a filter is fitted, differential pressure measurement must be included. The two variable orifices shown, usually needle valves, allow speed adjustment of valve operation. The differential pressure regulator shown in Figure 5.6 does not close if the downstream pressure increases and the flow reverses. A non-return valve function can be added to the regulator by including a small non-return valve in the pilot circuit, see Figure 5.7.

The differential pressure reducing regulator tries to maintain a constant differential pressure between the regulator outlet pressure and an external pressure source. The regulator outlet pressure is sensed internally and the pressure signal applied to the underside of the diaphragm or piston. The external pressure signal and the adjustable spring load is applied to the upper surface of the diaphragm or piston. The set differential pressure is increased by increasing the spring compression and hence the spring load. The use of heavier springs, for higher differentials, can be avoided by employing a piston with two diameters. Diaphragm regulators can be fitted with two independent diaphragms to create a fixed pressure differential.

Table 5.4 shows some of the more popular pressure reducing regulators available. Pilot-operated regulators, similar to that shown in Figure 5.7, but with remote upstream pressure registration, are also used.

Set differential pressure range barg	Maximum pressure barg	Size range	Operation	Fluids (1)	Body materials (2)
0.004 to 10	13.5	DN25 to DN150	Direct	G, L	CI, CS, SS
0.14 to 10	17.2	DN3 to DN50	Direct	G, L, S	CI, CS, SS
0.14 to 13.7	41.3	DN5 to DN20	Direct	A, G, L, S, W	BRZ, CS, DI, SS
0.14 to 17.24	25	DN25 to DN300	Pilot	W	CI
5.2 to 23.4	41.3	.DN15 to DN50	Direct	A, G, L, S, W	BRZ, CS, DI, SS

(1) A = air, G = gas, L = liquids, O = oil, S = steam, W = water
(2) A = aluminium, B = brass, BRZ = bronze, CI = cast iron, CS = carbon steel, DI = ductile iron, SS = stainless steel

Table 5.4 Typical differential pressure reducing regulators

The differential pressure **relief** regulator is similar in concept to the differential pressure **reducing** regulator but the internal pressure registration is from the valve inlet rather than the outlet. Table 5.5 shows some of the sizes available.

Some designs of differential pressure relief regulators may not close if the outlet pressure exceeds the inlet pressure. This can be a problem in some systems. When necessary the problem can be cured by fitting small non-return valves in the pilot circuit.

Set differential pressure range barg	Maximum pressure barg	Size range	Operation	Fluids (1)	Body materials (2)
0.004 to 9.0	17.2	DN25 to DN150	Direct	A, G, L	CI, CS, SS
0.14 to 13.7	34.5	DN15 to DN150	Pilot	A, G, L, S, W	CI, CS, DI, SS
0.14 to 17.2	25	DN15 to DN300	Pilot	W	CI
5.2 to 23.4	41.3	DN15 to DN50	Direct	A, G, L, S, W	BRZ, CS, DI, SS
0.14 to 25.0	25	DN15 to DN300	Pilot	W	CI
0.14 to 25.8	27.6	DN6 to DN25	Direct	A, G, L, S	CI, CS, SS

(1) A = air, G = gas, L = liquids, O = oil, S = steam, W = water
(2) A = aluminium, B = brass, BRZ = bronze, CI = cast iron, CS = carbon steel, DI = ductile iron, SS = stainless steel

Table 5.5 Typical differential pressure relief regulators

A pressure balanced regulator is a special version of the differential pressure regulator. "Pressure balanced" means equalising the pressure at the two sensing points. The diaphragm actuator has no spring. The upstream and downstream pressures are sensed externally and piped to the diaphragm casing. One pressure connection is directed to either side of the diaphragm. The orientation of the connections is dependent upon whether the regulator is push-down-to-close or push-down-to-open. The system pressure, which may be higher, must load the diaphragm in the opening direction. The pressure balanced regulator only balances in one direction. If the system pressure expected to be low is unexpectedly high, the regulator will close. Two regulators can be piped across the systems with the diaphragm connected in reverse on one; this arrangement will balance in both directions.

The pressure balanced regulator is not a common device and choice is limited. Maximum differential pressure is 12bar for sizes from DN25 to DN150. Valve bodies can be cast iron, carbon or stainless steel. Seat leakage at shut-off can be poor, ANSI Class II, depending upon the valve detail design.

Constant flow regulator

As mentioned earlier the constant flow regulator is a differential pressure regulator. Flow sensing is added as the controlling differential pressure. Usually an orifice plate is fitted to the valve inlet. Pressure taps before and after provide the differential signal for control. The orifice size is selected to provide a reasonable signal at the required flow. Both direct and pilot types are available. Pilot constant flow regulators are very similar to that shown in Figure 5.7. The differential pressure signal from the orifice plate is applied directly across the pilot diaphragm.

These regulators are most common in water systems but chemical processing can be accommodated by the use of suitable materials. Viscosity correction may be necessary on smaller regulators when the operating viscosity range is wide. For water applications the regulator body will be cast iron for pressure ratings up to PN25. Fabric reinforced nitrile rubber is standard for diaphragms. The valve plug generally has a rubber insert to improve sealing qualities. Sizes range from DN15 to DN300. Pilot operated regulators can cope with maximum flow rates between 19.5 to 1180m^3/h.

Small direct-acting constant flow regulators are produced specifically for high pressure water applications. Sizes are usually up to DN50 for maximum pressures of 345 barg. The differential pressure across the regulator is limited to between 1.7 and 3.4 bar. The standard body material is carbon steel.

Excess flow shut-off valve

An excess flow shut-off valve is an isolating valve which is flow sensitive. Under normal operating conditions the valve is open and fluid flows with a slight pressure drop. In the event of process control failure or pipe/vessel fracture the normal fluid flow rate can increase dramatically and very serious consequential damage can ensue. Incorrect proportions of chemicals being mixed can lead to devastating results. Fluids flowing from open pipes and vessels can create dire personnel hazards or permanent environmental damage. Costly products can be lost forever. The excess flow shut-off valve senses fluid velocity higher than required and isolates the system. The higher fluid velocities create increased fluid drag within the valve and produce sufficient force to close the valve. Figure 5.8 shows a typical liquid excess flow design.

Valves up to DN600 have been produced. Pressure ratings are not a problem as the valve body has no openings around the periphery. Valves can be manufactured in most materials to suit the process. Once closed, the excess flow shut-off valve must be reset before it will open. To reset the valve, the pressure on both sides must be equalised; this can easily be achieved by a small by-pass.

Low pressure shut-off valve

The low pressure shut-off valve is a pressure sensitive isolating valve. When the control pressure falls below a preset value the valve closes. When the control pressure increases to a value above the set pressure the valve re-opens. These valves can have internal or external pressure registration and the sensing point can be either on the inlet or outlet. The low pressure shut-off valve is very versatile and can be used in a similar manner to the excess flow show-off valve. It can also be applied as a type of non-return valve. Table 5.6 indicates the range of popular valves.

Some low pressure shut-off valve designs use two valve bodies actuated by a single diaphragm. These valves have been popular in gas distribution systems, working with larger pilot operated valves, as safety isolating valves

Switching valve

A switching valve is similar in concept to the low pressure shut-off valve but the valve can be configured to open or close on rising pressure. The globe valve body can be built as push-down-to-close or push-down-to-open. The external pres-

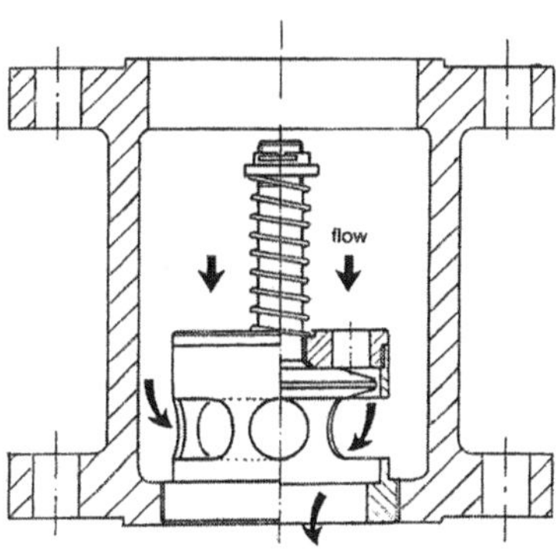

Figure 5.8 A typical liquid excess flow shut-off valve

Set pressure range barg	Maximum inlet pressure barg	Size range	Operation	Fluids (1)	Body materials (2)
0.007 to 0.012	0.07	DN20 to DN35	Direct	G	CI
0.007 to 0.34	5.2	DN20	Direct	F	CI
0.03 to 5.7	17.2	DN15 to DN50	Direct	L, G	CI, CS, SS
0.03 to 5.7	41	DN25 to DN100	Pilot	L, G	CI, CS, SS
0.3 to 6.9	8.6	DN20	Direct	G	CI

(1) A = air, G = gas, L = liquids, O = oil, S = steam, W = water

(2) A = aluminium, B = brass, BRZ = bronze, CI = cast iron, CS = carbon steel, DI = ductile iron, SS = stainless steel

Table 5.6 Popular low pressure shut-off valves

Set pressure range barg	Maximum inlet pressure barg	Size range	Operation	Fluids (1)	Body materials
0.2 to 10	10.3	DN20 to DN30	Direct	A, G, L	Cast iron
0.2 to 13.8	17.2	DN5 to DN15	Direct	A, G	Aluminium

(1) A = air, G = gas, L = liquids, O = oil, S = steam, W = water

Table 5.7 Typical switching valves

Set pressure range barg	Maximum inlet pressure bage	Size range	Operation	Fluids (1)	Body materials
0.2 to 6.9	17.2	DN5 to DN15	Direct	A, G	Aluminium
0.2 to 10.3	10.3	DN20 to DN25	Direct	A, G, L	Cast iron

(1) A = air, G = gas, L = liquids, O = oil, S = steam, W = water

Table 5.8 Typical 3-way switching valves

Figure 5.9 A non-metallic pressure regulator

sure signal is applied to the underside of the diaphragm. Set point adjustment is via a compression spring acting on the upper surface of the diaphragm. These valves can be used on liquids and gases. The external pressure signal must be compatible with the diaphragm; neoprene and nitrile rubber are most common. Table 5.7 shows typical valve parameters.

3-way switching valve

The 3-way switching valve is the mixing/diverting version of the switching valve. Fluid flow is controlled by an external pressure signal. Some valves are specifically designed for pneumatic control and are really a power operated 3-way valve. The basic characteristics for 3-way switching valves are listed in Table 5.8.

Special Note: Valves for dry solids

Because regulators use the process fluid as the driving medium regulators are unsuitable for dry solids applications.

Special Note: Non-metallic valves

Constant pressure regulators, are available in uPVC, PP and PVDF with EPDM or Viton™ diaphragms. The Viton™ diaphragm may be faced with PTFE for improved chemical compatibility. These valves are similar in construction to the safety relief valves shown in Chapter 3, Figure 3.26. Valves are suitable for inlet pressures up to 10 barg. Connections via solvent cement socket or fusion spigot are suitable for pipes from 16 to 110mm od.

Non-metallic back pressure regulators are constructed in a similar way to the regulator shown in Figure 5.9. Small regulators, for pipe of 16 to 63 mm od are suitable for pressures up to 10 barg. Larger valves, for 110 mm od pipe are suitable for 4 barg. Pilot-operated non-metallic back pressure regulators are also available.

Special Note: Hygienic valves

Constant pressure regulators of hygienic quality have the internal surfaces polished and only approved materials are used. Optionally, the external surfaces can be polished to allow effective washdown procedures. Table 5.1 should be reviewed for typical regulator capablities.

5.4 Useful references

Compressed Gas Association (CGA), 4221 Walney Road, 5th Floor, Chantilly, VA 20151 USA, Tel: 703 788 2700, Fax: 703 961-1831, www.cganet.com.

The LP Gas Association, (LPGA) Pavilion 16, Headlands Business Park, Salisbury Road, Ringwood, Hampshire BH24 3PB, UK, www.lpga.co.uk.

National Propane Gas Association (NPGA), 1150 17th Street, NW, Suite 310 Washington DC 20036-4623,USA, Tel: 202 466 7200, Fax: 202 466 7205, www.npga.org.

Control valves

6

6.1 Process control valves and the system

An automatic control system consists of:

- the fluid to be controlled
- a sensor for the process variable
- a controller which affects the actuator
- an actuator which modulates the valve
- a control valve to control the flow

The interaction of these elements provide the basis of any control system, which can, in principle, be used to control any fluid which flows in pipes. The system can adopt many different forms, depending upon the particular process requirements.

The system shown in Figure 6.1 shows a coolant storage tank in which a constant coolant level (1) should be maintained, irrespective of the volume of liquid entering (2). The transmitter (3) measures the level in the tank, converts the change in level into a pneumatic signal, with a standard range of 0.2 to 1.0 barg (or 3 to 15 psi), and transmits this signal to the controller (4). This in turn transmits a control signal with the same standard range. The controller is reverse acting, as the input from (3) rises its output to the control valve (5) falls causing it to open when the level rises and to close if the level tends to drop.

The control valve regulates the quantity flowing out (6) so that the level (1) is maintained at the desired value, usually called the "set-point" of the controller.

6.2 Control valve design

A control valve regulates the rate of flow. To do this, it should be so constructed as to be unaffected by the working conditions - no matter how difficult or unique they may be. This does not imply that all valves are suitable for all fluid conditions. Valves are supplied to handle with the operating conditions specified.

Many types of control valve have been produced over the years. Some are used in great numbers covering a wide range of processes, whilst others have been developed to fulfil unique and/or special requirements.

The design and operational features of control valves may be summarised as follows:

- operating function
  - control (modulating)
  - on/off (in some instances)
- fluid to be controlled (liquid, gas, steam, liquid/gas mixture)
  - density
  - temperature
  - chemical composition, if unusual
  - physical properties (clean/contaminated, viscosity, abrasive etc.)
  - corrosion of materials
  - personnel hazards
  - environmental hazards
- inlet pressure of the valve for various rates of flow
- pressure drop
  - pressure drop in operation
  - pressure to be sealed when closed
- flow
  - minimum flow to be controlled
  - normal flow through valve
  - maximum flow
- maximum permitted noise level

- sound power level or
  - sound pressure level
- leakage
  - seat tightness, leakage class
  - stem/spindle leakage
- valve size
  - nominal pipe size
  - pressure class
  - connection type

6.3 Process control valve types

The design of a particular control valve depends upon the specific process requirements and the need to provide a suitable economic solution. Control valves can be grouped into two main categories:

- Valves which control the flowing volume by the reciprocating linear motion of a valve plug attached to a stem -
  - globe valves
  - cage trim valves
- Valves which control the flowing volume by means of rotating a disc or sphere -
  - butterfly
  - ball valves

Similarly, actuators can be divided into two groups:

- linear
- rotary

Figure 6.2 shows the relationship of both major valve and actuator types. The classification shown applies to control valves with external actuators. Any particular control valve could be fitted with a choice of actuators to match a specific application.

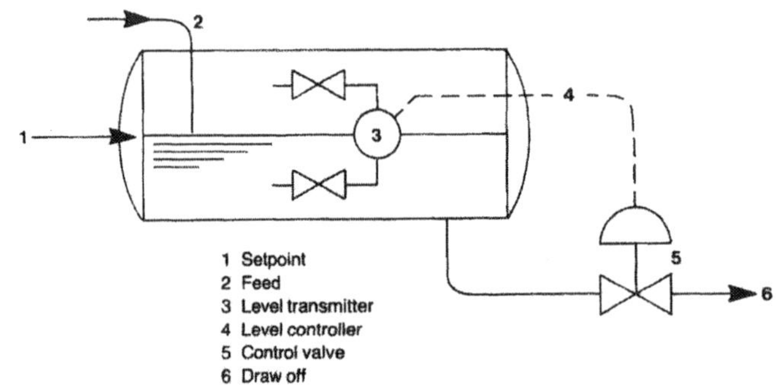

Figure 6.1 Schematic arrangement of level control by modulating outflow

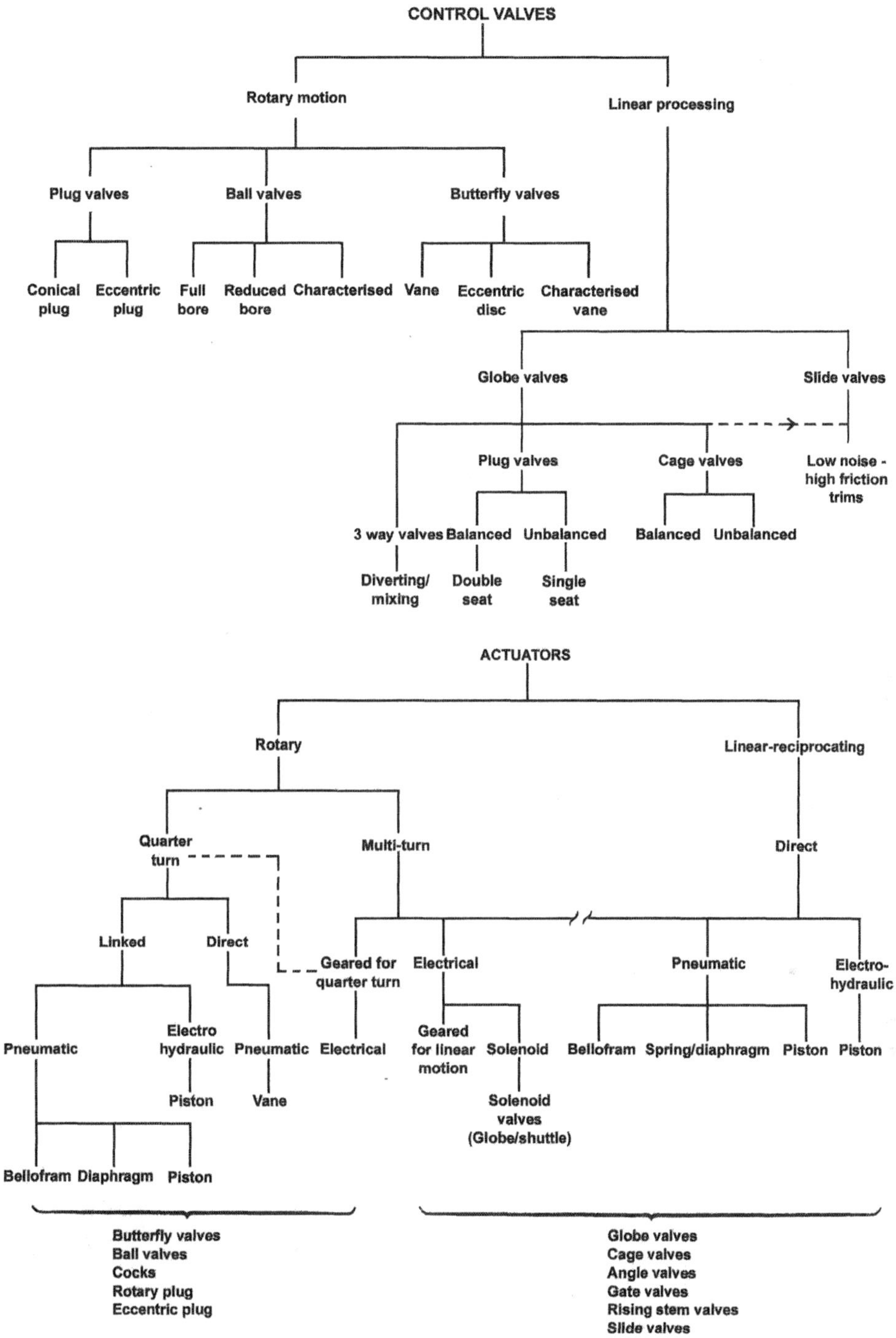

Figure 6.2 Classification of control valves and actuators

There is a special control valve which has an internal actuator and is always powered by the process liquid — the automatic pump recirculation valve, which is described in Section 6.3.3. Another special control valve type, which can be considered as self-powered, but not powered by the process fluid, is the thermostatic valve, which is also in Section 6.3.3.

Linear and rotary valve types are manufactured in various sizes and ratings. The table in Table 6.1 indicates the range available and the suitability for various operating conditions.

6.3.1 Linear motion valves

Figure 6.3 shows the standard arrangement of a linear motion globe valve with pneumatic diaphragm actuator. The valve positioner is mounted on the yoke.

Single seat globe valve

Linear motion valves for general service are operated by a pneumatic diaphragm. Figure 6.4 shows a typical straight cast body globe control valve Air pressure opens the valve and the spring, clearly shown, closes the valve when air pressure is reduced. This valve would "fail closed". The plug is top guided by

	Linear Reciprocating	Rotary Ball Valve	Rotary Butterfly Valve
Size			
nominal bore	6 to 1 800	12 to 1 200	50 to 3 600
flow coefficient for 100mm	C_v = 240	C_v = 530	C_v = 400
max C_v for valve type	60 000	14 000	400 000
Pressure rating			
ISO 7005	PN 420	PN 110	PN 420
ANSI	6 600 LB	600 LB	2 500 LB
API	20 000 PSI		
Special connections	3 447 barg		
Max differential pressure, barg (1)	690, 3 447 **(2)**	**102**	**145**
Operating temperatures			
maximum °C	38 to 815	79 to 427	149 to 815
minimum (normal trim) °C	17 to -73	-30 to -46	-7 to -80
minimum (cryogenic trim) °C	-198 to -269		-198 to -254
Materials			
iron, steel, stainless steel	✓	✓	✓
Duplex stainless steel	✓	✓	✓
non-metallic	✓	✓	
Hastelloy, Carpenter and titanium	✓	✓	✓
NACE MR-01-75 materials	✓	✓	✓
Connections			
screwed	✓		
compression fittings	✓		
flanged	✓	✓	✓
proprietary clamped fittings	✓		
socket weld	✓		✓
butt weld	✓	✓	✓
wafer	✓	✓	✓
lug		✓	✓
Valve characteristic flow in relation to valve opening	E%, L, MP, QO	E%, MP	E%, MP QO
Flow range (Rangeability) max flow/min flow	**50:1 to 200:1**	**150:1 to 300:1**	**100:1**
Seat leakage (N)ormal, (R)educed, (Z)ero	**N, R, Z**	**N, R, Z**	**N, R, Z**
Actuator			
pneumatic diaphragm	✓	✓	✓
pneumatic piston	✓	✓	✓
electro-hydraulic	✓	✓	✓
electro-mechanical	✓	✓	✓
Action on control failure			
return to wide open	✓	✓	✓
return to closed	✓	✓	✓
stay-put	✓	✓	✓
Installation options			
space required	large	smaller	smallest
trim servicing	in-situ	**(3)**	after removal
actuator mounting	fixed	variable	variable
Environment			
atmospheric corrosion	**(4)**	**(4)**	**(4)**
noise attenuation	in valve & pipe	in valve & pipe	in pipe
Operating conditions			
corrosive	✓	✓	✓
abrasive	limited	✗	✗
flashing	✓	✓	✓
cavitating	✓	✗	✗
hazardous emissions	bellows seal	double packing	double packing
pulp	✗	✓	✓
hygienic (CIP)	✓	✗	✓

(1) Carbon steel; ASTM A105, ASTM A216 Gr.WCB, BS 1503-161 Gr. 32, BS 1504-161 Gr. 480; at temperatures beween -29 and +38 °C. See Chapter 9 for more information on pressure-temperature ratings.
(2) Alloy steel
(3) Top entry ball valves can be serviced in-situ, others after removal.
(4) Correct selection of materials or surface treatment.

Table 6.1 Process control valve availability and capabilities

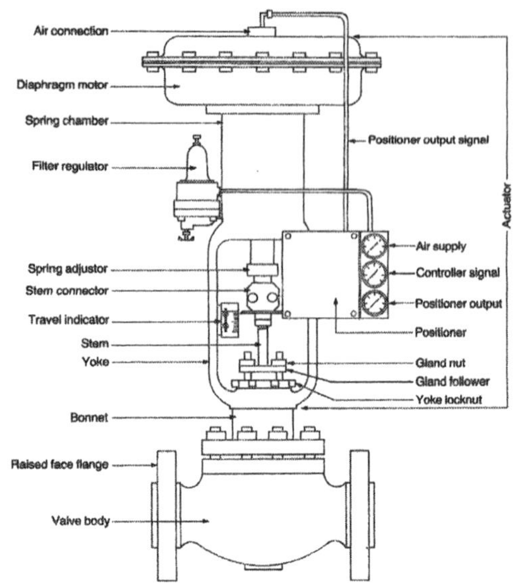

Figure 6.3 A typical linear control valve assembly

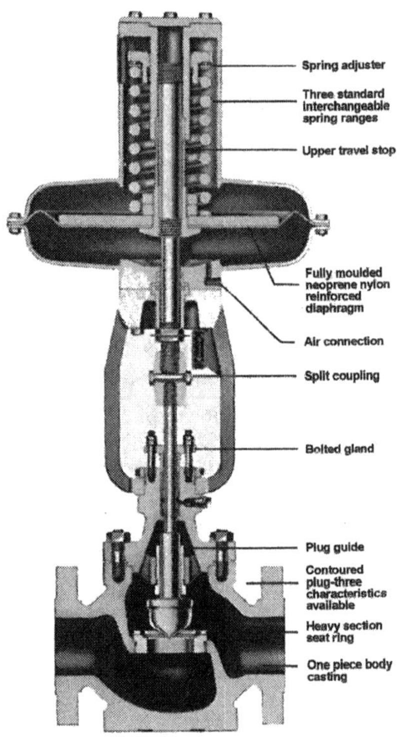

Figure 6.4 Single seat globe valve with pneumatic diaphragm

a replaceable bush. The replaceable seat is screwed into the cast body. The bonnet is bolted to the body. The gland is bolted and a lantern ring is provided in the packing. The valve is designed for relatively low pressures. Note the valve body wall thickness, the number of bolts and the valve trim proportions.

As a result of the pressure difference acting upon the unbalanced valve plug area, the stem area is not pressurised by the process fluid, and so there is always an upward force. The direction of flow is normally from beneath the seat, or called "flow to open". A more powerful actuator is therefore required when the upward force is large due to higher pressures. Figure 6.5 shows another cast valve but for higher pressure.

Figure 6.6 shows a variation on the unbalanced plug design which incorporates additional bottom guidance to reduce stem vibration. This type of construction is used for higher pressure valves when the pressure drop across the plug and seat is liable to create significant turbulence. The bolted gland, with external Belleville washers, is clearly shown. Review Chapter 11 for more sealing options.

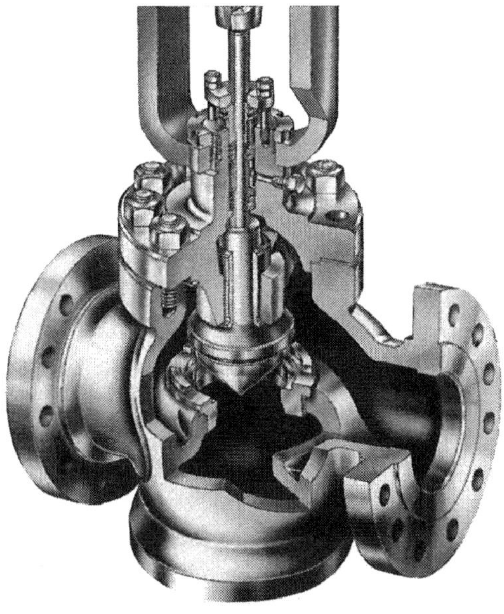

Figure 6.5 Single seat globe valve for higher pressures

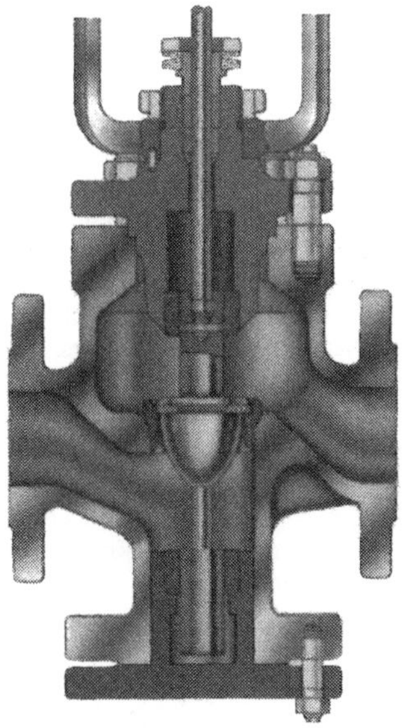

Figure 6.6 A top and bottom guided plug valve

As an alternative to screwed replaceable seats it is possible to have press-fit seats or to clamp the seat in position. A style of cage is fitted which sits on top of, or underneath, the seat and is held in position by the bonnet or a bottom flange. These cages do not guide the plug, as in cage-trim globe valves, described later, but merely clamp the seat in position. Clamped seats are preferable to screwed seats for some corrosive applications.

The valves shown are designed to be maintained in-situ. All wearing parts can be inspected/replaced when the bonnet, and the bottom flange, are removed. For some applications with hazardous fluids it may be desirable to fit a drain valve to the bottom of the control valve. The drain valve would allow fluid trapped in the control valve to be removed safely before the control valve is dismantled for service.

The valve characteristic is a consequence of the plug profile. Standard characteristics are stocked as standard parts, see definition of Characteristic in Chapter 1. Special characteristics, to suit the installed system, can be provided for a small increase in price and an extension in delivery time.

Metal globe valve bodies can be lined with a thermoplastic, PFA (per fluoro alkoxy) or FEP (fluorinated ethylene polypropylene), to improve corrosion resistance. Popular valve sizes, DN15 to DN100, are produced in ductile cast iron or steel then lined. These valves can be rated for 16 barg at temperatures between -60° and 180°C; extremes of both not necessarily simultaneously. Figure 6.7 shows a lined valve with a PTFE bellows sealed bonnet to eliminate fugitive emissions. The seat and plug are available in several combinations including reinforced PTFE, Hastelloy© or tantalum.

The valves shown are classed as "full-bore", the seat area is comparable to the process pipe area. It is not unusual to see installations where the control valve is a much smaller nominal size than the pipework; a 100 mm valve in a 150 mm pipeline.

To avoid problems with reducing fittings in the pipework, extra radiography of welds or extra screw threads to seal, a valve can be fitted with a reduced port. Figure 6.8 shows the typical arrangement of a relatively small port in a cast body. The valve characteristic is created by the profile of the groove machined

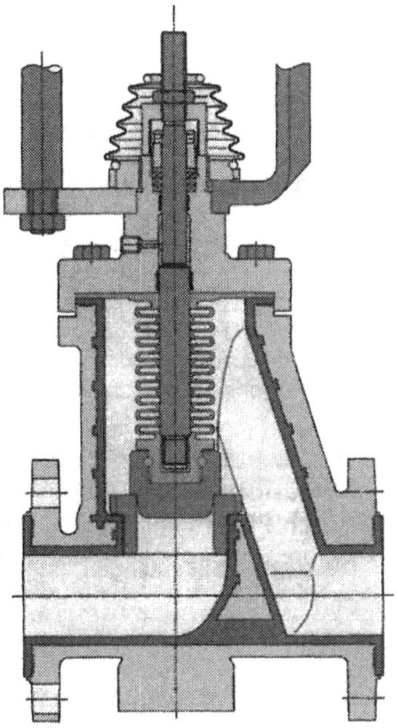

Figure 6.7 A lined single-seat globe valve
Courtesy of Richter Chemie-Technik GmbH

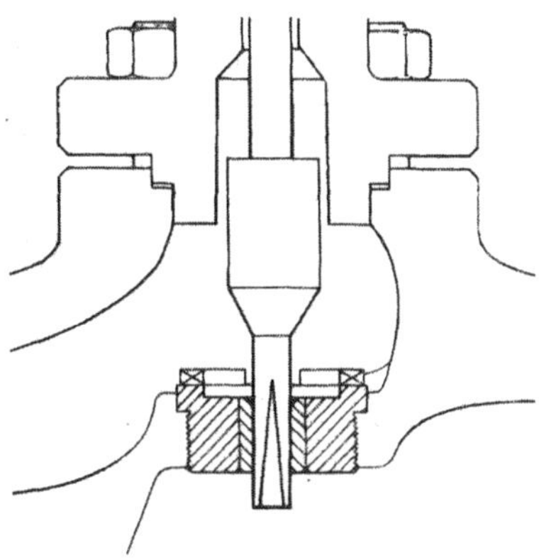

Figure 6.8 Globe control valve with reduced port

in the parallel plug. In effect, a needle valve is created with a parallel needle and a parallel seat.

Profiled plugs, as shown in Figures 6.4 and 6.5, are usually only top guided but some designs are also bottom guided. The guidance provides plug/seat alignment and lateral stiffness to resist vibration. A top guided valve stem is at its most flexible, when considered in bending, just as the plug touches the seat. The overhang from the top guide bush is at its maximum. This means the force required to deflect the stem radially is at its lowest making the plug vulnerable to vibration. Unbalanced radial flow forces are caused by uneven flow distribution around the plug and vortices created by throttling and cavitation when operating at high differential pressures. The reduced port construction shown has an added advantage in that the parallel plug is bottom guided by the seat.

There is a style of single seat plug valve which incorporates a fullbore seat and seat guidance to complement the top guide. A plug similar to that shown in Figure 6.13 is designed to slide through the seat, no cage is necessary. Profiled slots are machined through the wall of the plug and the shape of these determines the valve characteristic. This style of construction is limited to valves up to 50 mm and pressures of 50 barg.

Single seat valves are also produced in oblique and angle designs. Angle valves in some designs have larger outlet connections. Various packing box options can be fitted to suit emission requirements. Small valves can be machined from solid bar. Higher pressure valves can be manufactured by machining from solid forgings. Some valve bodies are a fabricated assembly. Where welds are used the quality assurance should be available for scrutiny. Some valve bodies are only cast as screwed versions and flanges are provided by socket or butt welding. Screwed process connections and socket weld assemblies are not suitable for sour services when NACE (see Chapter 13) requirements are imposed. Additionally, socket weld assemblies are difficult to radiograph and are therefore not suitable for the highest integrity requirement applications.

When necessary, some valve designs can be fitted with steam jackets to prevent heat loss and assist warm-up during hot process start-up. If the fluid is liable to form crystals or deposit residue on cold surfaces the manufacturer must be informed.

Single seat valves are used for applications demanding low shut-off leakage. Soft seats are available on some designs for very low leakage rates but these seats tend to reduce temperature capabilities. On all metal valves, leakage can be reduced by lapping the plug to the seat. Typical shut-off leakage is at worst 0.01% of rated flow, ANSI B16.104 Class IV, when tested on air or water. ANSI B16.104 Class VI is possible on some sizes and designs. See Chapter 16 for details of standard leakage definitions.

Single seat valves are mass-produced in sizes from 6 mm to 200 mm. Larger sizes, up to 1200 mm, are available to special order. Some designs restrict the operating pressure to 10 barg, even in the smallest sizes. Generally valves can be purchased to operate at ISO 7005 Part 1 PN 100 rating; 100 barg at ambient temperatures. High pressure valves are generally rated for PN 420 but valves for 760 barg are available. 25 mm and 50 mm valves have been developed for special applications; these sizes are available in API 5000, 10 000 and 15 000 psi ratings and for 3447 barg.

Throttling through a seat creates high fluid velocities which can produce erosion (wire-drawing) and noise. The velocity can be reduced by arranging multiple seats in series. A tortuous flow path is made and similar pressure drops can be produced with lower velocities. Figure 6.9 show the arrangement of a high pressure valve with four "seats" in series; only the bottom seat seals on closure. High differential pressures are possible but the plug is not balanced; large actuator forces are required to locate the plug axially.

Figure 6.9 A high pressure angle valve for large differential pressures
Courtesy of Dresser-Masoneilan

Double-beat globe valve

Double-beat globe valves are balanced, i.e. the pressure differential acts both upwards and downwards on the respective plug areas so that hydrostatic and hydrodynamic forces tend to cancel each other. The intermediate stem between the two plugs is subject to tensile forces. The hydrostatic forces resulting from the valve stem passing through the packing box to the atmosphere remain unbalanced. The net resulting upward force, which is to be overcome by the actuator thrust, is however considerably less than for single seat valves. The valve shown in Figure 6.10 is designed for high pressure duty. Note the valve body wall thickness, the number of bolts and the diameter of the top and bottom plug guides.

Figure 6.11 shows a cross-sectional arrangement of a double-beat globe valve. The screwed seats and the top and bottom plug guides can clearly be seen. The drawing shows a bolted bonnet and gland.

Figure 6.10 Double-beat globe control valve

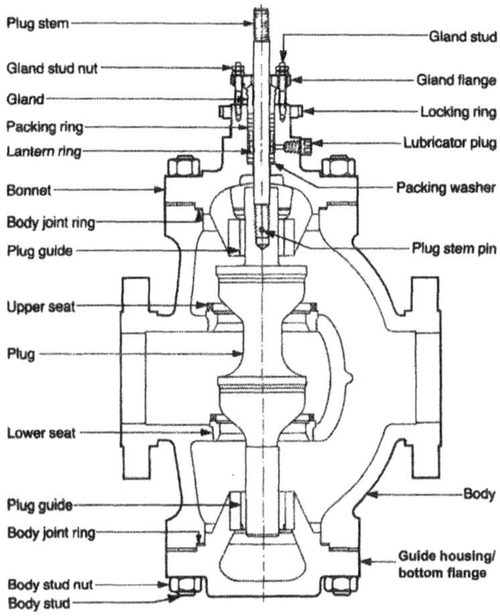

Figure 6.11 Cross-section through cast double-beat globe control valve

The screwed seats are identical in the bore but have different diameter threads. The bottom seat must usually pass through the top seat location in the body. If only the bottom seat needs replacing the top seat must still be removed. Most valves are designed for top access; some though, are designed for access through the bottom flange. Inspection and replacement of seats can be performed with the valve mounted in the pipework.

The valve characteristic can be adjusted by modifying the profile of the plugs. An alternative type of guidance is available on a few designs. The top and bottom guides are replaced by seat guidance, see Figure 6.20. The shaped plugs, shown in the illustrations, are replaced by wing-guided poppets connected by the stem. Seat-guided valves can be built with a bottom flange to allow the stem and the seats to be withdrawn from the bottom without disturbing the actuator.

Double-beat globe valves are only available in the straight pattern as shown. Valve bodies are normally cast. Various packing box options can be fitted and steam jackets are possible on some models. These globe valves can be used for fluids containing small quantities of solids or fluids which will leave a deposit. The valves are manufactured from 35 mm to 300 mm. Three different pressure ratings are popular, 50 barg, 103 barg and 259 barg. Contrary to popular belief, double-beat valves can seal very well. Seat leakage varies from ANSI B16.104 Class II to Class VI depending upon the size and the pressure rating.

Cage-trim globe valve

The natural development of control valves continues towards increased capacity for a given body size, a limited number of standard body types for normal applications and valve designs that are less prone to vibrate. Valve design is also aimed at reducing cavitation, cavitation damage and limiting noise emission, whilst increasing serviceability.

One design that meets most of these design objectives is the cage-guided valve shown in Figure 6.12. This valve type has a greater capacity for a given body size. Close examination of the components shown indicates that the valve plug is always guided by the cage as it travels up and down, thus giving maximum lateral stability with a greatly reduced tendency to vibrate. The plug is sometimes balanced by drilling holes to allow process pressure to act on both sides of the plug. A seal must then be fitted to the plug to provide proper shut-off in the closed position. Consequently it is possible to use them as an alternative to double-beat valves. Because this type of valve is balanced it is also possible to use smaller, less expensive actuators with the added benefit of more rapid response. See Figure 6.13.

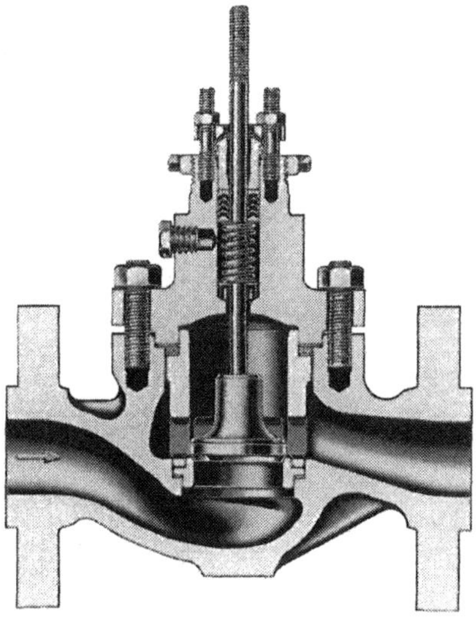

Figure 6.12 Cage-trim plug valve with solid plug
Courtesy of Emerson Process Management

In both valve types illustrated, the spring-loaded chevron stem packing is important. The packing is loaded by an internal spring which assists the fluid pressure unlike external springs which must overcome fluid pressure. This style of packing is excellent for sealing and high pressure applications. See Chapter 11 for more detailed information.

The desired valve characteristic is achieved by varying the shape of the cage ports. It is also possible to design special cages to reduce noise and cavitation, see Chapter 10. Valve capacity and characteristics can easily be changed on site by replacing a number of standard components. Capacity is reduced by cutting smaller ports in the cage. The shape of the ports modifies the characteristic. Figure 6.14 shows the essential components for replacement.

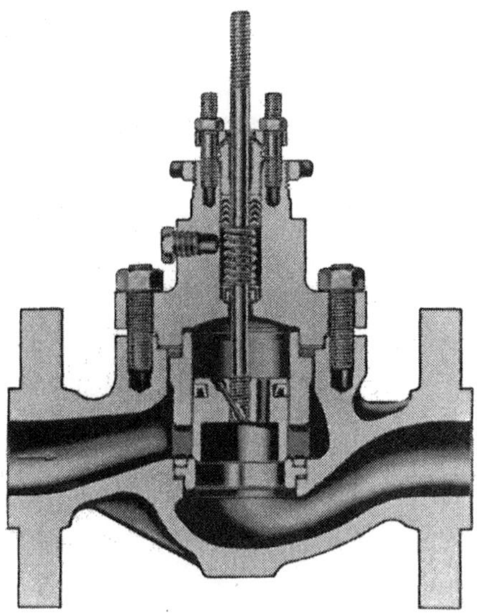

Figure 6.13 Cage-trim globe valve with balanced cylindrical plug
Courtesy of Emerson Process Management

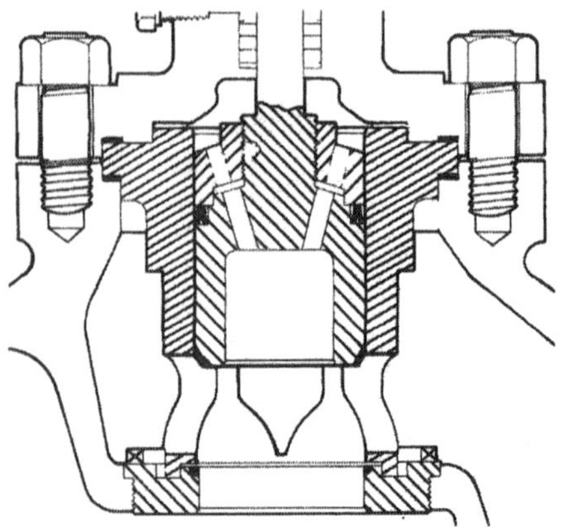

Figure 6.14 Essential components for capacity/characterisation modifications

This type of valve is easily serviced as there are no screwed components within; the parts simply lift out. The seat is clamped in position by the cage which is loaded by the bonnet. Screwed seats are an option with some designs. Cage-guided valves have acquired a good reputation in service. For certain applications such as intermittently condensing steam, the gasketing has to be designed and fitted with great care if an effective seal is to be maintained.

The direction of fluid for cage-guided valves is normally from below the plug and out through the cage ports.

Two exceptions to this rule are:

- balanced valves
- some valves with cavitation-resistant trim

Part of the cavitation resistance is achieved by arranging jets of liquid to hit each other, as shown in Figure 6.15, and so destroy the energy in the liquid/vapour mixture. The cage is perforated with small holes which guide the liquid flow so that it does not impinge upon the inner valve components; thus avoiding valve wear.

NOTE: Cavitation is only a problem in some liquid valves. See Chapter 8 for more information on cavitation and cavitation damage. Cage-guided valves can also be designed specifically for noise reduction. The fluid is separated into a number of fine jets, as shown in Figure 6.16, which effectively reduce noise in the valve.

Special cages can be constructed by stacking suitably machined or cast rings. Many small, distinctly shaped passages can be formed between adjacent rings; the DRAG™ valve and Tiger-Tooth™ cages being typical examples. Another technique is to build a cage from concentric cylinders, each cylinder having precisely positioned and sized drilled holes. The HUSH Trim™ is typical. Noise control and cavitation resistance can be combined in one valve.

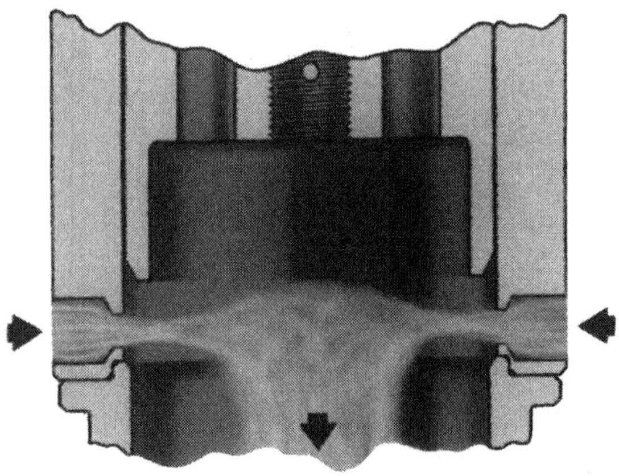

Figure 6.15 Cavitation-resistant cage design

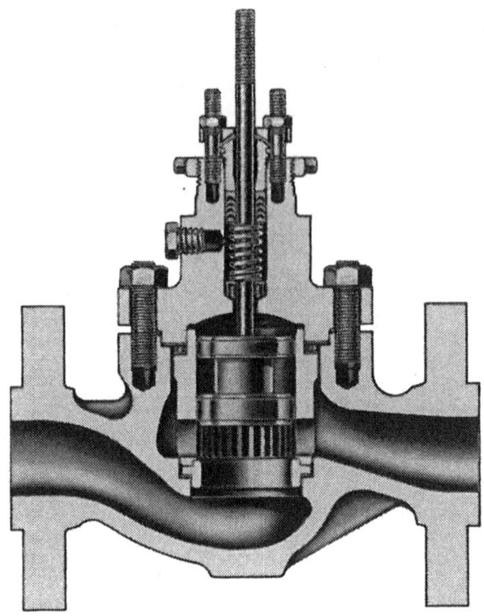

Figure 6.16 Low noise cage design

Reverse-acting, that is push-down-to-open, cage-guided valves are particularly suitable for applications requiring frequent inspection/service of the valve trim. Their advantage is in the plug being able to be removed without dismantling the actuator or other parts of the control system, see Figure 6.17.

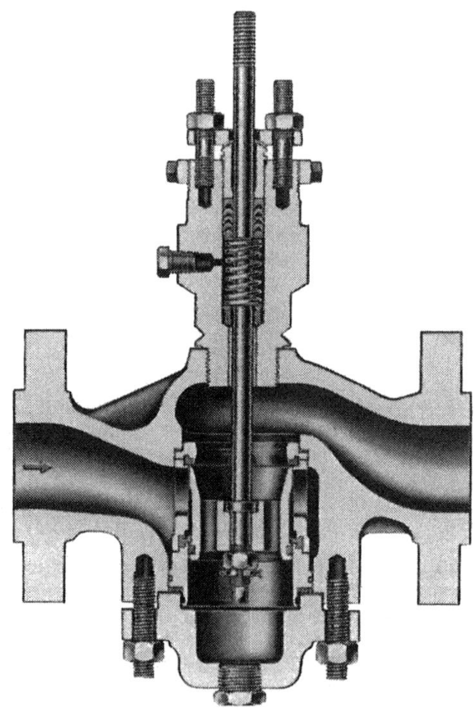

Figure 6.17 Reverse-acting cage-trim globe valve
Courtesy of Emerson Process Management

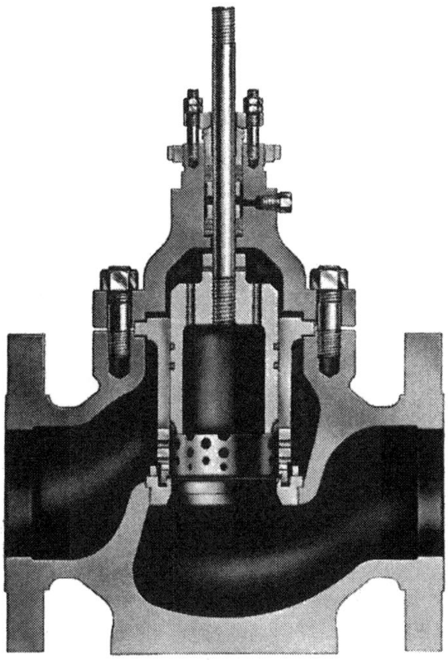

Figure 6.18 Zero-leakage soft seat with low-noise cage

Zero-leakage cage-trim valves can be constructed by fitting a soft elastomer seat. This type of seal is, however, only suitable for temperatures up to 200°C and pressure drops of up to 100 barg. Figure 6.18 shows such a valve with a "drilled" low-noise cage.

Some cage-trim globe valves are manufactured as oblique and angle patterns. Applications for all clean fluids are standard, but high viscosity can be a problem with some designs. Valves from 40 mm to 300 mm are mass-produced; and others up to 1000 mm are made to order. Pressure ratings up to ISO 7005 Part 1 PN 420 at 900 mm are possible. Valves up to 200 mm can be rated for ANSI 4500 LB, 765 barg at ambient temperatures. Seat tightness varies from ANSI B16.104 Class II to Class VI.

Multi-port globe valve

Multi-port globe valves are an adaptation of the single seat globe valve for flow diverting or mixing applications. Valves always have two seats and may have a double-ended single plug or two separate plugs. Cage-guided plug designs are also available. Cage-guided plug valves have clamped seats using two cages. Stem-guided valves normally have screwed seats. Press fit seats may be an option when thermal expansion is not a problem. Cage-guided plug and wing-guided plug valves will usually have higher capacities than profiled plug valves. Double-ended profiled plugs may require bottom guidance to prevent stem vibration and maintain valve accuracy.

An example of this valve type is the three-port globe valve, described in Chapter 3, Section 3.3.1. The valve body has three process connections. For diverting duties there is one inlet and two outlets. For mixing duties, two inlets and one outlet. Valve bodies with a bottom flange are easily converted; the bottom flange being replaced by a process connection. Two connections will normally be in-line, the third on the valve bottom, see Figure 6.19. Valves with a double-ended plug can have the third connection on the side of the body, at right angles to the plane of the other two connections. Some designs do not have in-line connections. The process connections on opposite sides of the body are at different elevations.

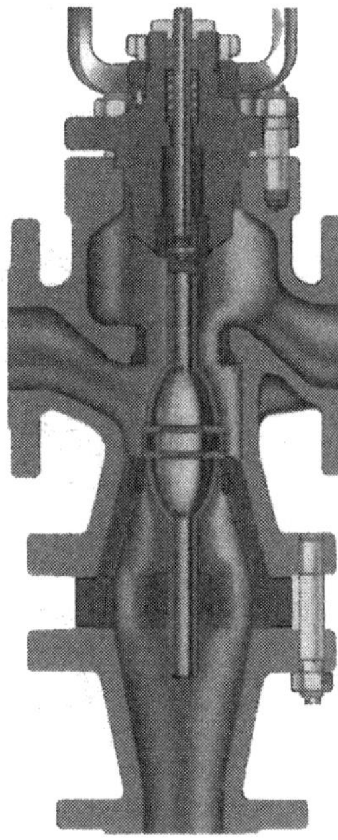

Figure 6.19 Three-port globe valve with double-ended profiled plug

The valve body in Figure 6.19 is very deep to accommodate the bottom stem guide which is clamped between the valve body and the bottom connection extension. A three-port globe valve can be made considerably smaller by the use of wing-guided plugs which are seat-guided, see Figure 6.20.

Valve bodies can be cast, machined from bar or forgings. Small sizes have screwed connections. Flanges may be added by butt welding. Valves machined from the solid can have flange facings and studs incorporated and those for hydraulic applications may use SAE flanges. SAE flanges are useful because of the small space requirements at relatively high pressures.

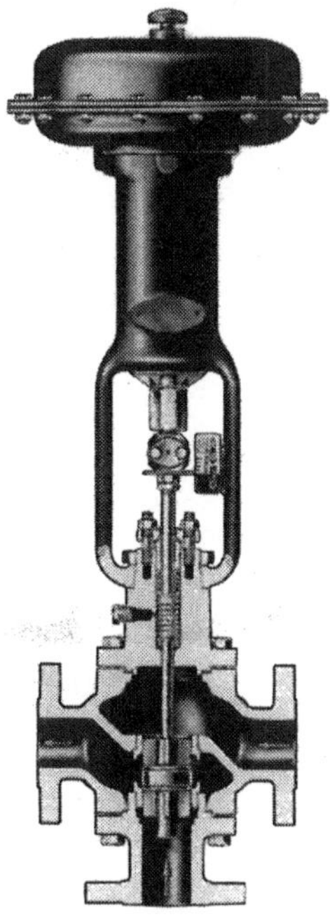

Figure 6.20 Three-port globe valve with double-ended wing-guided plugs
Courtesy of Emerson Process Management

Cage-guided valves can have balanced plugs to reduce the actuator force required. Stem-guided valves are unbalanced. The actuator force required being dependent upon the relationship of the three pressures involved. Because the flow is controlled by two independent devices the characteristic of each does not have to be identical.

Valves designed for liquids and gases are usually restricted to clean fluids. General purpose valves are available from 12 mm to 300 mm with pressure ratings up to ISO 7005 Part 1 PN 110, 102 barg. Some high pressure valves have very limited throttling capabilities, only 7 barg in some cases. Single plug style valves, designed specifically for oil and water-oil emulsions in 20 mm to 100 mm sizes, are capable of 240 barg as standard; with special versions to 690 barg. Seat leakage varies widely and is very dependent upon the style of construction, ANSI B16.104 Class II to Class VI.

Axial-cage control valve

This interesting design approach is more easily appreciated by looking at a picture, see Figures 6.21 and 6.22. This is an outside-guided or skirt-guided plug valve; sealing for closure is affected by the replaceable soft ring on the end of the plug. The term "axial" is used here because the cage is parallel to, and concentric with, the process piping. This style of construction is not a "needle" valve as described in some manufacturers' literature. It can be seen from Figure 6.21 that the plug motion is generated by a crank on the spindle and transmitted via a connecting rod. Pressure across the piston is balanced. The fluid flow path for this design style looks a lot smoother, and simulation proves the flow to be smoother, than the conventional globe-body control valve, but, the fluid still has to negotiate two radial changes in direction, albeit gently. Performance is the key comparator; flow coefficients can be significantly larger than globe valve designs .... but not always! There are different trim styles in both designs and detailed specifications must be examined closely. The axial-cage style of valve is very likely to be lighter than the conventional globe valve equivalent. The radial inward flow, in front of the plug, is a good way to avoid cavitation damage in liquid applications; the vapour bubbles implode in a body of liquid and not next to a solid surface, see Figure 6.15. Valves of this style are produced from DN100 to DN1800 with pressure ratings from PN10 to PN100. Standard materials of construction include cast iron, ductile cast iron, steel and alloy steel all with stainless steel internals. Figure 6.23 shows a typical valve with a two-piece body.

Figures 6.24 and 6.25 show a slightly different approach to the axial-cage design; this is a caged piston valve. This design is different to the earlier style because sealing is around the periphery of the piston, not on the end. This style of construction could be considered better for high pressure applications because of the greater area afforded to sealing. The piston style can include soft replaceable seat seals, as shown in Figure 6.25. A seal located in the cage, in this manner, would not be exposed to high velocity fluid. Various styles of cage trim can be fitted to a single body design including multi-stage cascade trims, to provide a wide range of application suitability. The par-

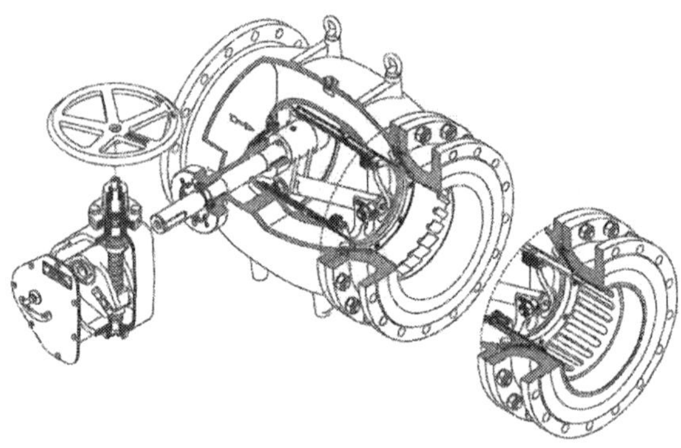

Figure 6.21 Schematic of a medium pressure axial-cage control valve
Courtesy of Erhard Armaturen GmbH & Co KG

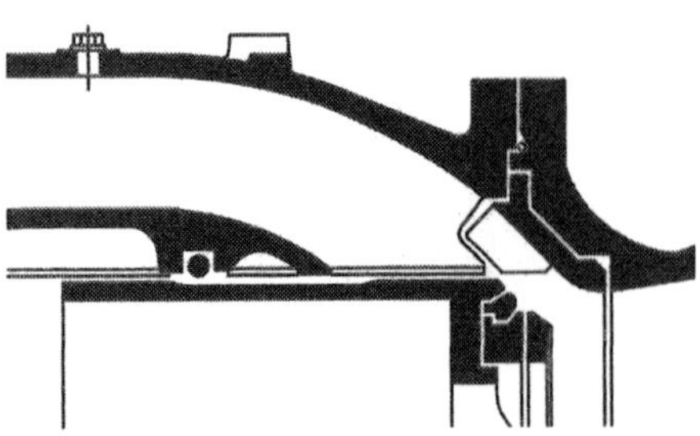

Figure 6.22 Plug guidance and sealing
Courtesy of Erhard Armaturen GmbH & Co KG

Figure 6.23 A medium pressure axial-cage control valve
Courtesy of Erhard Armaturen GmbH & Co KG

Figure 6.24 A high pressure axial-cage piston valve
Courtesy of Mokveld Valves BV

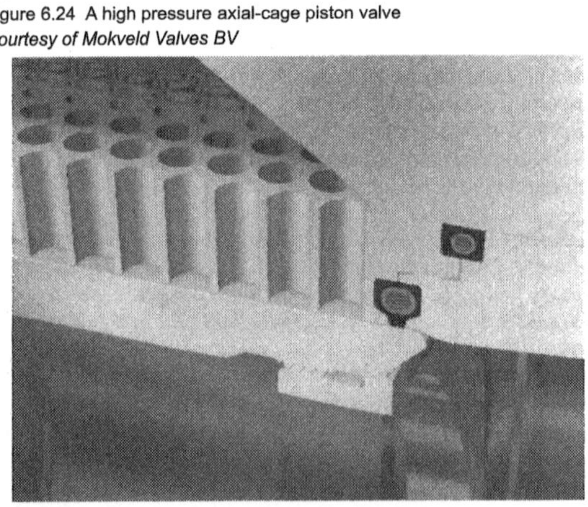

Figure 6.25 A high pressure axial-cage piston seal (patented)
Courtesy of Mokveld Valves BV

ticular valve design shown uses a novel twin rack mechanism to to produce the piston movement. It translates the actuator linear motion into a similar motion in a perpendicular direction. Fast motion is possible but water hammer and surge effects should be considered carefully. Valves are available in a wide range of sizes, DN50 to DN1800, with pressure ratings up to ANSI 2500 or API 10 000. Flanges can be replaced by clamped connectors. These valves are available in a wide range of steels, stainless steels and exotic metals.

Valves of a similar style are manufactured in very large sizes specifically for water applications in transportation, reservoir management and hydroelectricity, see Figure 6.26. The larger valves are suitable for throttling and shut-off and rely on creating a pressure drop across the tapered plug section without the use of a "perforated cage" although a characterised cage can be fitted for specific applications. The turbulence and flow disturbance, associated with control cages, are avoided but are

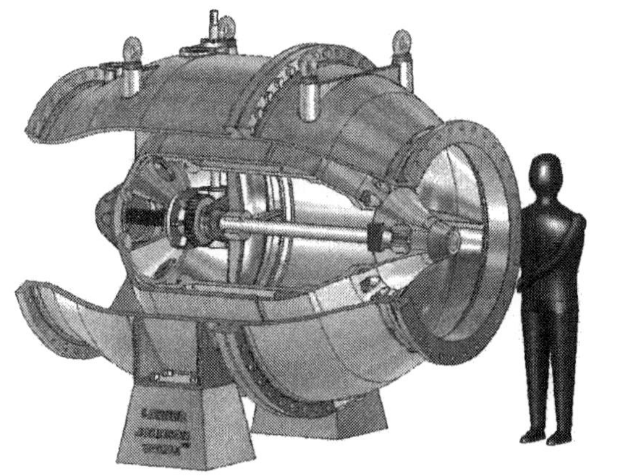

Figure 6.26 A Larner-Johnson® control valve for water
Courtesy of Blackhall Engineering Ltd

replaced by high velocities as the liquid accelerates along the taper of the body and plug. Material erosion, caused by long exposure, can be reduced by material selection. Reducing the exposure time obviously reduces erosion problems. The Larner-Johnson® valve is unusual in that it is pilot-operated; initial opening/closing movement opens/closes a pilot port in the main valve. All valves are fitted with scour ports to facilitate solids/debris removal prior to plug movement. Plug movement can be effected by different mechanisms to permit the use of linear or rotary actuators.

General construction is cast iron, cast steel or aluminium bronze to avoid corrosion problems. The cage may be fitted with a cast iron "bearing" liner, which provides a very good, hard wearing bearing surface, but smaller ones have an option of a bronze liner. Sealing is effected on the end of the plug. Both the plug and the body have replaceable seals. Valves up to DN4000 have been manufactured for pressures of 30 barg; smaller diameter valves are operating up to 90 barg. Figure 6.26 shows the general construction of a large valve.

Pinch valve

Pinch valves for control purposes are available with two types of operating mechanism:

- fluid
- mechanical

Also the basic valve construction can be one of two types:

- supported sleeve
- unsupported sleeve

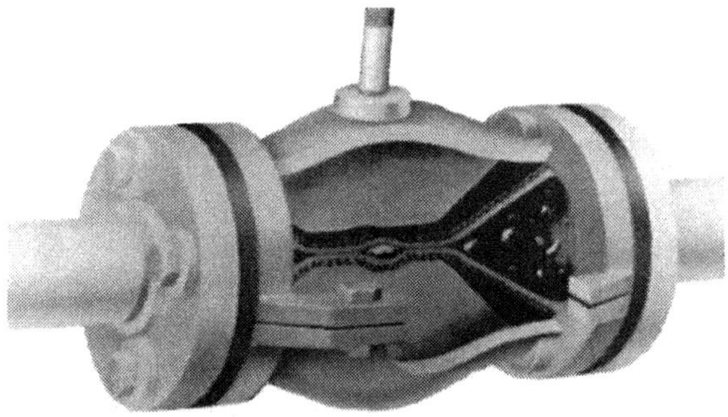

Figure 6.27 Unsupported sleeve pinch control valve

The unsupported sleeve pinch valve, is the simplest valve construction, see Figure 6.27. An elastomer sleeve is housed inside a split metal body and the only component exposed to the fluid is the sleeve. The sleeve is designed with the same bore as the nominal bore of the pipework. When fully open there is effectively no pressure drop. An annular reservoir around the sleeve is pressurised with the operating medium — compressed air, hydraulic liquid or process fluid.

In some applications the pinch valve can be self-powered and in effect becomes a regulator. The operating medium compresses the tube and creates a flow restriction, introducing a pressure drop. When the valve closes the sleeve bore touches for approximately 60% of the valve length. This type of pinch valve is extremely effective for fluids carrying solids. The sleeve, when compressed by a fluid, can close around solids and still make a good seal. The elastomers available have a wide range of chemical compatibility and good abrasion resistance for solids:

- Buna-N™, Chlorobutyl, EPDM, Hypalon™
- Neoprene™, Polyurethane, pure gum rubber, Viton™

The flow control range is not as wide as other valve types; rangeability will be from 8:1 to 26:1 depending upon the design

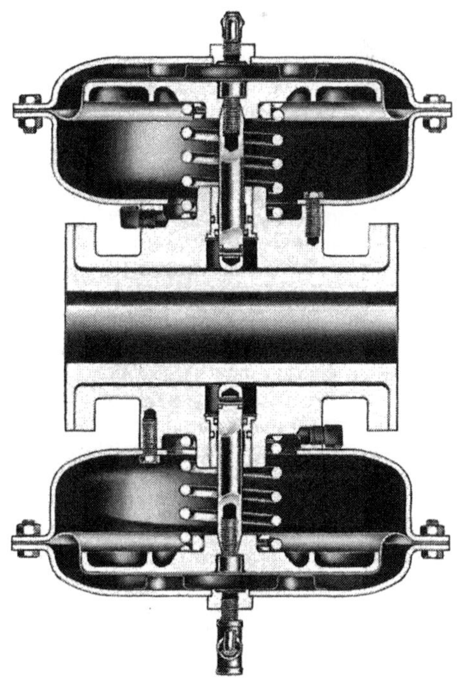

Figure 6.28 Mechanically operated unsupported sleeve pinch control valve
Courtesy of Emerson Process Management

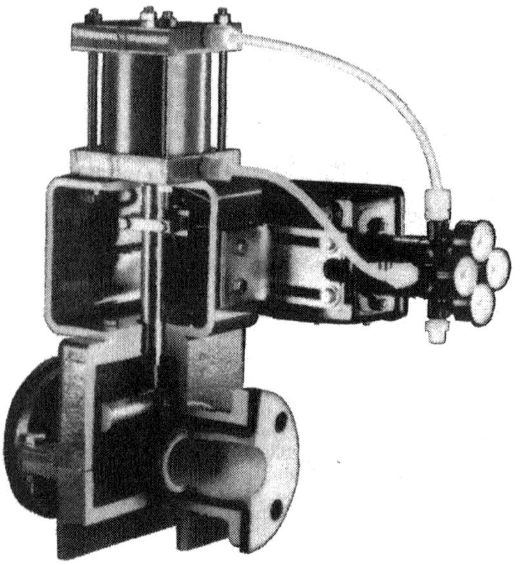

Figure 6.29 Mechanically operated unsupported sleeve pinch control valve — open version

of the sleeve. Differential pressure is restricted to 30% of the valve rating. The differential pressure restriction can be circumvented by connecting valves in series. The sleeve can be closed to bubble-tight requirements, even when handling solids. The fluid-operated, unsupported sleeve pinch valve is available in sizes from 25 mm to 1800 mm. Working pressure on the smaller sizes is limited to 10 barg reducing to 2 barg for the largest sizes. The pressure required to operate the valve is low, just over 4 barg. Valves can be operated from normal compressed air systems.

The unsupported sleeve valve can also be operated mechanically, the sleeve being crimped between two rollers or one roller and the valve body, as shown in Figure 6.28. The crimping action of the roller(s) varies the cross-sectional area creating a pressure drop. The pinching roller can be actuated by pneumatics or electrically. These valves are available in sizes from 25 mm to 1200 mm. Valves up to 300 mm can be suitable for pressures of 50 barg.

The mechanically operated unsupported sleeve valve is available as an "open" version; the sleeve is not housed in a metal casing, see Figure 6.29. The sleeve is fitted with metal flanges at the ends for connections, the flanges being held at the correct length by two spacers. The sleeve is crimped between two balanced rollers which are actuated pneumatically or electrically. Valves up to 100 mm can be rated for 10 barg; 600 mm are rated for 3.5 barg.

Unsupported sleeve pinch valves are very good for fluid streams handling solids. Also the valve design is inherently leak-free. There are no shaft or spindle seals as leak paths, or maintenance worries. The enclosed version of the pinch valve can contain leakage in the event of sleeve failure. Obviously the open pattern valves can fail with serious consequences. The type of valve used must be based on the possible hazards and likely consequences.

Supported sleeve pinch valves are a completely different variety. The elastomeric sleeve is supported by a contoured and slotted tube and enclosed in a pressure casing. The support tube is divided into inlet and outlet sections by a shaped blank which prevents any fluid flowing straight through the tube. Fluid entering the valve is forced to flow radially out of the tube through slots and in so doing lifts the elastomeric sleeve off its circumferential seat. The fluid must then flow radially inwards,

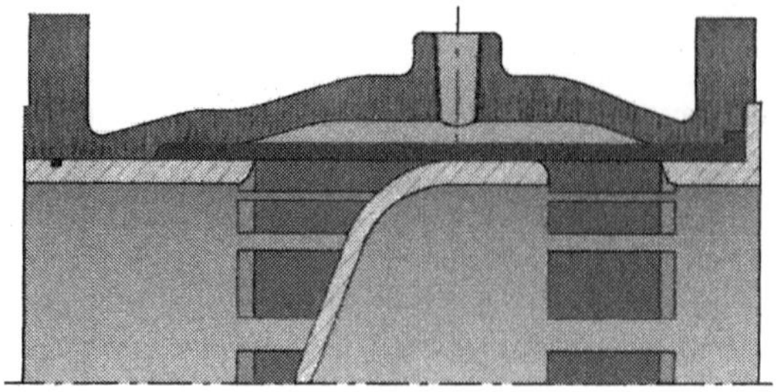

Figure 6.30 Supported sleeve pinch control valve

back inside the tube, and out of the valve, see Figure 6.30. Fluid forces lift the sleeve off the support tube seat. The fluid forces can be counterbalanced by control fluid pressures as well as the elastic properties of the elastomer. Fluid around the outside the sleeve provides an external control influence.

The control fluid can be any compatible fluid including the process fluid. A small pilot valve responds to the process changes and adjusts the control fluid accordingly. Increasing the control pressure tends to reseat the sleeve increasing the pressure drop. Supported sleeve valves are best for clean fluids. Valves of this style are available up to 300 mm for pressures up to 100 barg.

All pinch valves rely on an elastomeric element as the control element. The temperature and pressure capabilities are restricted by the elastomer. Continuous operating temperatures over 200°C are possible.

Special Note: Sliding gate valves

A special version of gate valve is made for control purposes. Unlike gate valves for isolation, described in Chapter 3, Section 3.3.1, the control version does not have a virtually unobstructed flow path which is the size of the pipe connection. The sliding gate control valve uses matched ports in two plates to vary the flow area. One plate is fixed in the valve bore. The other plate slides, to cover or uncover the ports, see Figure 5.4 in Chapter 5. The ports are arranged as horizontal slots in the plates. When wide open, the ports are revealed fully to the fluid flow. During closing, the ports are gradually covered until metal-to-metal sealing around each port is achieved, resulting in very low shut-off leakage. Leakage rates of 0.0001 % Kv and ANSI Class IV are possible.

The port slots are generally narrow in comparison with the pipe bore. The moving plate travel from "open" to "closed" is very short, resulting in very fast response times. A very short stroke permits the use of metal diaphragms in actuators. However the narrowness of the slots results in drastic capacity restrictions, as viscosity increases. The slot profiles can be adjusted to create various characteristics. Rangeability is generally 40:1 to 100:1.

The force required to move the plate is low in comparison with plug valves; plate movement is not resisted by hydrodynamic forces. The sliding action of the plates tends to clean and "wear-in" rather than "wear-out" the plates. Sliding gate valves are suitable for high differential pressures when cavitation may occur. The collapse of the cavitation bubbles occurs downstream, outside the valve, and does not affect control or sealing.

Sliding gate valves are available with various packing box options to suit most applications. Materials include cast iron, ductile iron, bronze, steel and stainless steel. For some applications the fixed plate can be PTFE coated. Sizes range from DN6 to DN 200. Valves up to DN50 can be suitable for PN100, larger valves PN40 or PN16. Special versions, with extended bonnets, can cope with temperatures down to -200°C.

Special Note: Solenoid valves

Solenoid valves are usually considered for on-off control. Small solenoid valves in stainless steel and Viton™ are available for modulating control applications. Valves respond linearly to dc voltages between 0 and 24. Interfaces are available to convert 4-20 mA and 0-5 V. Port sizes are 0.25" NPT female and operating pressures are up to 34.5 barg. Valves rated at 10 barg, in brass with a PTFE diaphragm, are available from 0.25" BSP to 0.75". Valves respond to either 4 to 20 mA or 0 to 10 V dc. The valves require a power supply of 21 to 30 V full wave rectified unsmoothed ac.

Special Note: Digital valves

At first glance this term seems incorrect; the actuator or positioner may be digital but the valve is still a valve. However a digital valve can be constructed from standard components. Multiple solenoid valves can be connected between one supply and one demand. The smallest programmable logic controller, PLC, will have at least eight digital outputs. That is eight switches which can be programmed to be on or off. If eight identical solenoid valves were used, and the differential pressure remained constant, a linear control valve would result with flow increments of 12.5%. In practice, the overall characteristic will not be linear due to the influence of the system characteristic. But a very simple and effective control valve has been produced.

With isolating valves on each solenoid valve, the control valve can be maintained without being 100% off-line. The solenoid valves do not have to be identical; the control valve could be characterised to match specific system requirements. With five valves sized for; 5%, 10%, 20%, 40% and 80%; it is possible to control between 0% and 100% in 5% increments. By adding a sixth valve at 2.5% the increment is reduced to 2.5%.

If eight digital outputs are available, very fine adjustment is possible, or spare valves could be fitted for the solenoids which would cycle the most. If the solenoid valves were fitted in the pilot circuit of larger valves, then very large flows could be controlled this way.

6.3.2 Rotary motion valves

The general style and construction of popular rotary valves is shown by Figure 6.31 The upper illustration shows a wafer style, profiled port ball valve actuated by a pneumatic diaphragm with spring return. Access to the valve internals is only possible after removal from the pipework. The lower illustration shows a butterfly valve actuated by a pneumatic cylinder. It can be seen that much less space is required for the butterfly valve.

The advantages of rotary valve designs which include butterfly valves, ball valves and also rotary plug valves of various types arise partly from hydraulic balance which minimizes fluid forces on the trim and partly from the economy of constructional methods that can be employed.

Balanced fluid forces impose small loads on actuators. The valve bodies themselves can be machined from barg, cast or fabricated from rolled plate. Flangeless designs, wafer and lug, are quite safe in many applications and save weight and cost, especially for the larger sizes.

Because of the unobstructed flow passage presented by a fully open full-bore ball valve these have a large capacity-to-size ratio, closely followed by butterfly and eccentric plug types. If they are fitted with leak free seats they exhibit rangeabilities in excess of 100:1, although the extent to which such rangeabilities can be realized in practice will depend on the quality of the actuator and the process hydraulics of the particular application.

The unobstructed design copes well with slurries and solids in suspension; special rotary designs are available for fibrous materials such as paper stock.

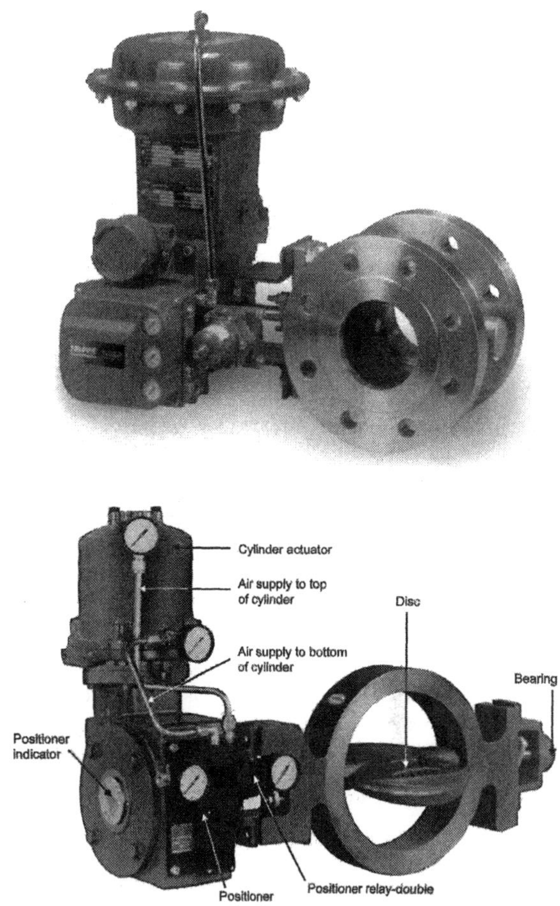

Figure 6.31 General arrangement of rotary valves with pneumatic actuators
Courtesy of Emerson Process Management

A disadvantage of most rotary designs, that may preclude their use in high pressure drop applications or when flashing or cavitation is likely to occur, is the "high recovery" shown by all of these valves. The term is explained later in Chapter 8, Section 8.4.2, but briefly it means that the dynamic pressure drop in the valve can be between 3 and 8 times as great as the design differential pressure with correspondingly high velocities in the valve. These high velocities can cause environmental noise, and accelerated wear of the valve.

Ball valve

The ball valve body design is conventional and flow is controlled by the position of a hole through the ball. This type of valve is used to control flow, pressure control in gas distribution systems and pressure reduction in connection with gas storage. These valves are also often used as isolating valves since they give rise to very little pressure drop in the fully open position. Ball valves can be:

- full-bore
- reduced-port

Full-bore valves are useful in large scale piping systems, oil and gas distribution for example, because the valve will pass a "pig". A pig can merely be a special plug pushed through the pipe system by the fluid to clean the pipe or the pig can be an intelligent capsule for conducting internal examinations and non-destructive testing. The advantage of cleaning and inspection, without disassembly and without total loss of production, can not be underestimated in large systems.

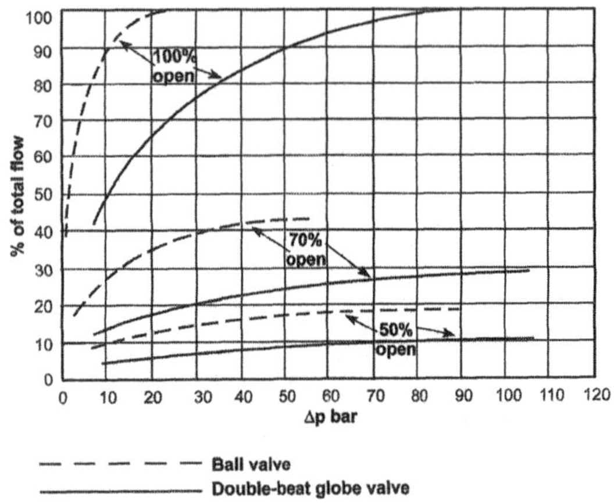

Figure 6.32 Flow v pressure drop for ball and double-beat globe valves

The disadvantage of the full-bore ball valve is its poor control characteristics. Figure 6.32 compares the inherent flow characteristic of a ball valve with a traditional double-beat globe valve of the same size. For a given pressure drop the capacity of the ball valve is clearly greater. The diagram also shows the earlier onset of "choked" flow in the ball valve. It can be seen how the ball valve passes considerably more flow at a specific opening with the same pressure drop. This means it does most of its controlling when nearly closed. This creates high velocities, subsequent wear and noise. Accurate positioning is necessary for good control. This is why characterised ball valves (described next) are better for control applications.

Ball valves are manufactured in all straight variations; threaded, flanged, wafer, lug and welding. The body can be manufactured in many styles to suit various applications; machined from solid forgings is an option for high pressure critical valves. The ball is replaceable. Generally chromium plated stainless steel or tungsten carbide coating is available, and will seal against pressures of just over 100 barg. Hardened stainless steel is preferred for the ball in applications where erosion is likely to be a problem. The ball can be guided by the seats or supported in top and bottom bearings. Seats can be non-metallic and metallic. The packing box is usually short for simple seals. Some designs include facilities for longer boxes to house dual packing sets.

Some valves have the differential throttling pressure restricted to 70 barg. Seat leakage can be very low; down to 0.0001 % of rated flow. Valves between 25 mm and 400 mm are used routinely for control purposes. Figure 6.33 illustrates the compact construction and the ball supported in its PTFE seat rings. Ball valves are not generally quoted for temperatures above 450°C because of the proximity of the shaft seal to the main flow, but they can be fire-safe.

Characterised ball valve

Ball valves are characterised in two ways:

- specially profiling the hole through the ball
- adding additional elements to a round bore valve

Characterised ball valves, of which the Vee-ball™ valve in Figure 6.34 is an example, are especially suited for many applications, not only combining the properties of both the linear and rotary valve, but having their own unique qualities. This type of valve has a large control range and is very resistant to clogging, making it specially suitable for paper stock, fibrous material and liquid-solid mixtures. For applications where erosion may be a problem, and some leakage can be tolerated, the ball seats can be replaced by flow rings.

Figure 6.33 Flanged ball valve with actuator

Figure 6.34 Vee-ball™ characterised ball valve
Courtesy of Emerson Process Management

Figure 6.35 shows the relationship between valve opening and flow. Note that the curves representing water (incompressible) and air (compressible) diverge. This illustrates why valves for different fluids are sized according to different rules. These valves may be used in any process requiring a large control range with low pressure drop. They can be used for control and on-off applications and can be equipped for manual and/or automatic operation.

Round port ball valves can have perforated plates fitted in the ball bore or the single bore can be replaced by multiple small bores. Both these approaches have two advantages:

- increased pressure drop, therefore increased rangeability
- reduced noise levels

The ball can be made of alloy steel, stainless steel or titanium, with or without chromium plate or cobalt alloy coating, to make the surface extra resistant to corrosion and erosion. Some designs allow the ball to have a replaceable spherical sector in the sealing zone. Seats can be stainless steel, cobalt alloy, titanium or PTFE. A flat ring can be used in ball valves to seal against the

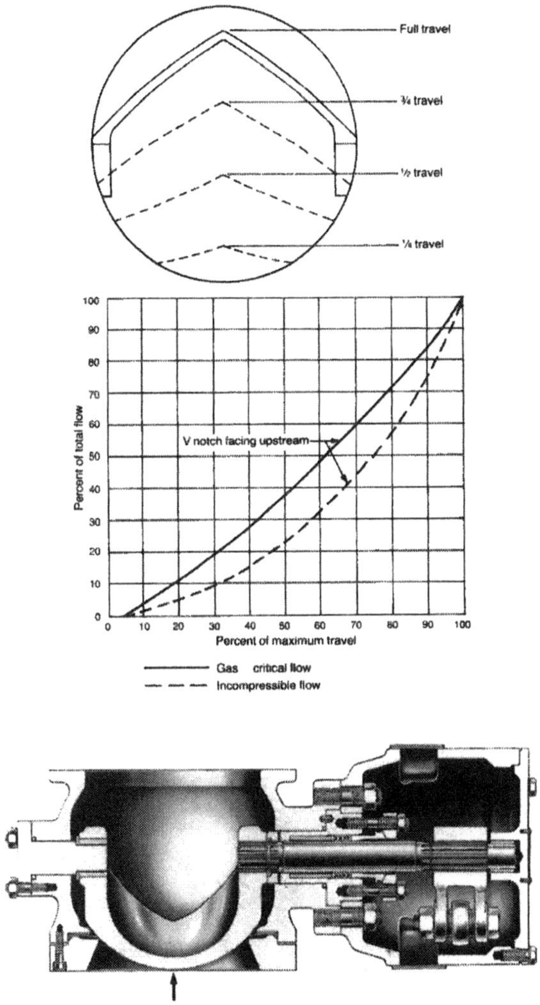

Figure 6.35 Vee-ball™ characteristic
Courtesy of Emerson Process Management

ball periphery. This seal ring is made of stainless steel or some other "soft" metal, depending upon the particular process requirements. A special inlet nozzle is often fitted to protect the seal ring from the fluid flow, Figure 6.36.

As with ball valves for isolation, the ball can be supported in two ways:

- "floating" between the seats in the valve body
- carried on two trunnions with plain or roller bearings

"Floating" is the cheapest but suffers from the disadvantages of requiring very high actuator torque if the pressure drop is of any magnitude and being prone to wear and loss of sealing efficiency. Separate bearings are preferable in high pressure

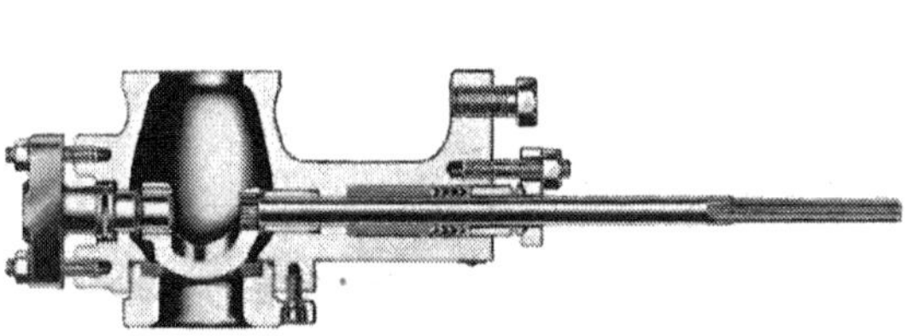

Figure 6.36 Profiled ball valve with protecting inlet nozzle
Courtesy of Emerson Process Management

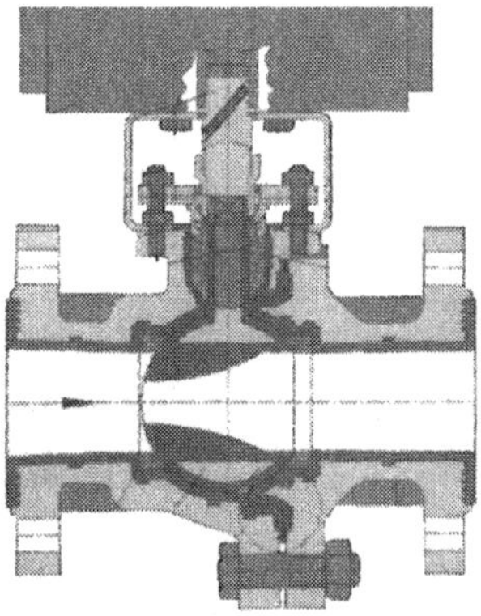

Figure 6.37 A lined body ball control valve
Courtesy of Richter Chemie-Technik GmbH

duties. Large shaft diameters are required to reduce bearing pressures.

To avoid backlash, the method of connecting the shaft to the ball must be positive so that the assembly functions as a single unit. Free play between the ball and the shaft will cause poor response to small actuator movements and may increase corrosion. Ideally the shaft and ball should be manufactured as a single component. This makes the ball costly and is only used on high pressure valves or valves in exotic materials. Trunnion mounted balls should be made this way. The most favoured alternative solution for separate shafts is a splined joint. A flanged shaft with driving pins is a popular construction, as is a slot in the top of the ball to accept a tongue on the shaft. The actual method chosen often depends upon the operating conditions and the body construction.

Characterised ball valves are popular in sizes from DN25 to DN400 for pressure ratings up to PN100. Carbon or stainless steel valve bodies are options. Hardened stainless steel balls, with chrome plate or Stellite™ inlays, can be fitted. Shut-off sealing can be as good as ANSI Class VI with soft seats.

Metal two-piece body ball valves, with characterised ports, can be lined with a thermoplastic to improve corrosion resistance. The metal body provides the structural integrity and is separated from the fluid by a bonded lining. Ductile cast iron is protected by PFA or PVDF with a minimum thickness of 3 mm. Valves of this style are manufactured from DN15 to DN200 for pressures up to 16 barg while coping with -60° to 200°C. The valve shown in Figure 6.37 has a single packed box but double packing is available as an optional extra.

Special Note: Multi-way valves

As with linear valves, ball valves can be adapted for mixing and diverting. Three-way valves are available which have connections at 90° or 120°. Specialist manufacturers produce four and five-way valves.

Butterfly valve

Traditional butterfly valves now work at high pressure drops across the disc which can be both metallic and "soft". Upper and lower temperature limits are the same, by and large, as those for globe valves, depending on duty and material of construction. The butterfly construction is especially suitable for high temperatures. Bodies can be fabricated from bar and plate and the seals can be mounted on cooling extensions away from the main flow. The upper temperature limit can be extended by using a refractory lining. A standard butterfly valve is shown in Figure 6.38.

Butterfly valves can be used as a control valve and also as a shut-off valve, as discussed in Chapter 3, Section 3.3.3, against high pressure drops of regularly up to 415 barg. Depending upon the materials of construction and the seat design a butter-

Figure 6.38 Standard wafer butterfly valve with actuator

fly control valve may have very limited shut-off pressure drops. Some 100 barg valves are only rated for 4 barg shut-off differential.

A butterfly valve should have a range of possible shaft diameters for each nominal valve size in order to handle the variation in torque due to various operating pressure conditions and packing box friction. Shafts should not be made of material prone to creep, such as some austenitic stainless steels. In these situations a precipitation hardening stainless steel such as 17-4PH is preferred. The corrosion resistance of such materials, equivalent to AISI 304, must be borne in mind. The disc must withstand high differential pressures. Some valves do have restrictions on the maximum throttling differential pressure, 35% of pressure rating in some cases.

Figure 6.39 shows the pressure distribution caused by the fluid flowing through a standard butterfly valve. The disc can be considered as an aerofoil, where greater forces are applied on the upper side than on the lower. The pressure is therefore relatively low where the velocity is high and relatively high where the velocity is low. These dynamic pressures generate an unbalanced torque which tends to close the valve. This torque varies from zero when the valve is closed, to a maximum at about 80° open, returning to zero again when the valve is fully open. It is this torque which imposes the pressure drop limitations which can be tolerated by the valve. It also determines the required actuator thrust. Furthermore, unbalanced torque reduction in these valves increases their range of applications.

Special Note: Fishtail™ valves

Detailed studies have been carried out to determine the effects of different disc profiles on the fluid flowing through valves. This has led to the development of hydrodynamically balanced disc profiles, an example of which is given in Figure 6.40. The valve can operate at all angles between 0 and 90°, the control range being greatly increased because of the considerable torque reduction. The hydrodynamically balanced characteristic is achieved entirely by virtue of the disc profile. Figure 6.39 shows how the fluid flow is divided equally above and below the disc. The flow velocities and dynamic pressures above and below the disc are thus also equal. The valve is therefore always balanced and only low operating torques are required. The use of this type of disc has several advantages:

- reduced pressure drop means lower pumping costs
- a large control range, or rangeability can now be achieved

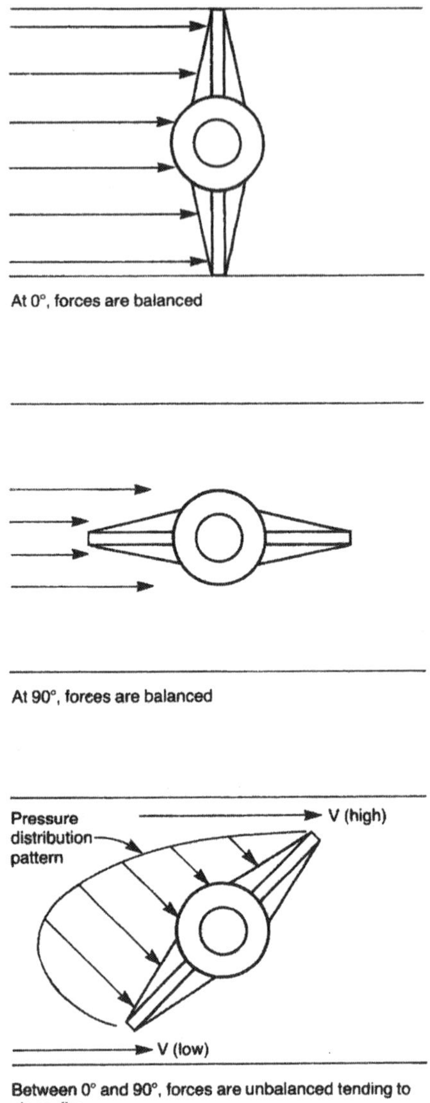

Figure 6.39 Pressure distribution on a standard butterfly valve disc

Figure 6.40 Fishtail™ butterfly valve
Courtesy of Emerson Process Management

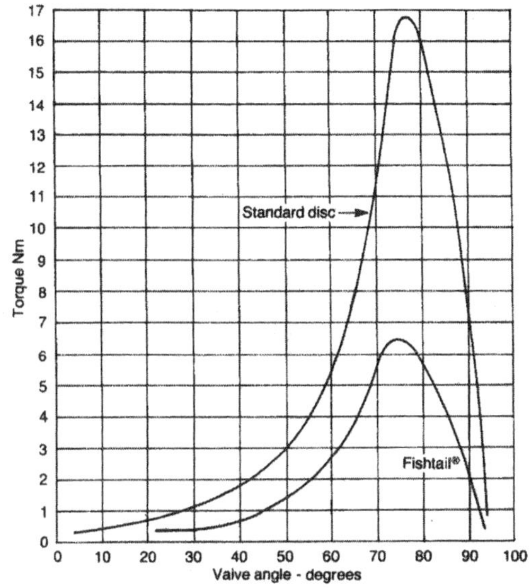

Figure 6.41 Shaft torque for standard and Fishtail™ butterfly valves

- a conventional valve with a maximum travel of 60° can be replaced by a smaller Fishtail™ disc valve, possibly saving installation costs
- the operating torque can be reduced by about 50%, depending upon the valve size, with consequential reductions in actuator force enabling cost savings

Figure 6.41 shows the torque requirement for conventional disc valves compared to a corresponding Fishtail™ disc valve at various opening angles. Figure 6.42 indicates the characteristics of butterfly valves. Special note should be made of the fact that the control range of a Fishtail™ disc valve is much greater than that of a conventional butterfly valve.

Another method of modifying the characteristic of butterfly valves is to add orifices to the edge of the disc. A short sector of a thin perforated sphere can be attached to the disc edge, see Figure 6.43, to modify the valve characteristic. The valve characteristic can be modified individually by adjusting the orifice configuration. Figure 6.44 shows the standard Rotrol™ characteristic which is very similar to equal percentage. This modification to the basic butterfly valve design does not increase the valve length; these valves are available as standard wafer pattern.

A variation on the Rotrol™ principle is possible by adding a perforated plate to the valve body bore. The perforated plate is place in the high velocity fluid jet as the disc closes. This type of construction is said to assist in torque reduction and provide noise attenuation up to 8 dB(A).

Other manufacturers have extended the Rotrol™ principle to partial hemispheres of perforations attached to the disc. Perforations can be uniform or graduated. Valves of this style are longer than normal and not usually offered in wafer patterns. These additions to the disc do increase rangeability and can reduce operating torque. However, one of the main advantages of butterfly valves is severely weakened — clear flow passages.

All fluid systems suffer occasional transients and process upsets which can result in solids carryover. Standard butterfly valves present a clean profile to the fluid and offer few opportunities for solids to collect. Perforated plates can be clogged, solids may accumulate in the corners where the perforated plate is attached to the disc and the valve characteristic will be modified. The full range of potential operating conditions must be considered when specifying valve types.

The method of connecting the spindle to the disc must be positive without backlash so that the assembly functions as a single unit. Free play between the disc and the spindle will cause poor response to small actuator movements and may lead to increased corrosion. Ideally the spindle and disc should be manufactured as a single component, however this tends to be costly. The most favoured alternative solution is splines or keys; taper pins can be effective but replacing either disc or

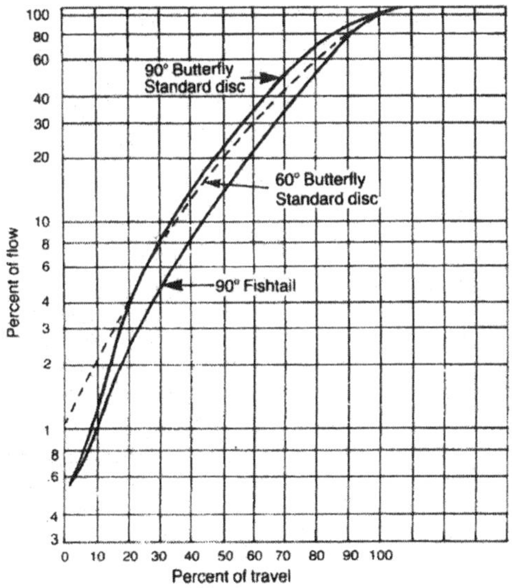

Figure 6.42 Characteristic of standard and Fishtail™ butterfly valves

Figure 6.43 Rotrol™ wafer butterfly valve
Courtesy of Koso Kent Introl Ltd

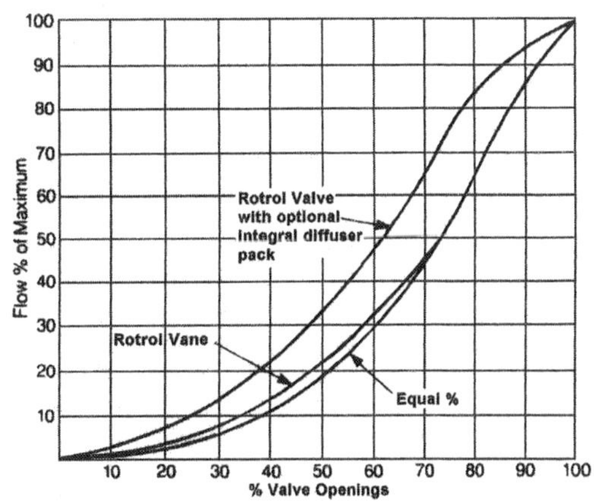

Figure 6.44 Characteristic of Rotrol™ butterfly valve

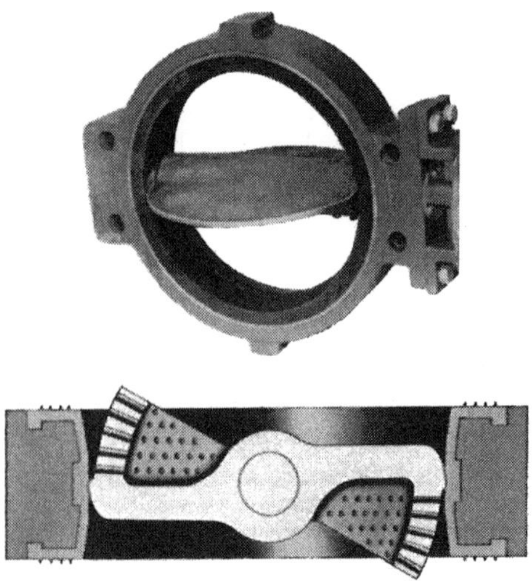

Figure 6.45 Lined butterfly valves for reduced leakage

spindle can be a problem. Care must be taken to ensure corrosion does not weaken the connection by selective attack. The actual method chosen often depends upon the operating conditions.

All metal butterfly valves with solid seats tend to leak a little. Leakage rates will vary from 0.06% to 4.6% of rated flow. Specially shaped floating metal seats, like NELDISC™ can reduce leakage to ANSI 816.104 Class V, better than 0.01 %. Valves with soft seats can be gas tight, on air, to 1 bubble per minute. Lined valves, as shown in Figure 6.45, have claimed leakage rates of 30 bubbles per year when tested on helium. Valves up to ANSI 2500 LB rating, 420 barg, are readily available. Most popular valves are in the 20 to 50 barg range. Butterfly valves are produced in sizes from DN50 to DN900 in a very wide range of materials including nickel and titanium.

Special Note: Three-way valves

Some manufacturers produce special tees and mounting kits with connecting linkage to allow two butterfly valves to be mounted to form a three-way valve. Both valves are operated by one actuator and act as a single unit.

Eccentric rotating plug valve

This valve design combines the advantages of a full-bore ball valve, of an unobstructed passage and therefore a high capacity when fully open, with a metal-to-metal seating or soft seating to provide bubble tight shut-off of the single seated globe valve, see Figure 6.46.

Reservations as to safety in a fire have been expressed for wafer versions. This type of installation can be made adequately fireproof by suitable lagging. Flanged versions of these valves are available which are certified fire-safe.

A modified version of this style of plug valve includes perforated plates, similar to those described in connection with characterised ball valves. The effect of the perforated plates is to modify the characteristic, from modified parabolic to almost linear, and to reduce the noise generated. Noise attenuation is claimed up to 18 dB(A).

The valves are suitable for flows in both directions. Flow past the plug and out through the seat, FTC, is recommended for erosive and flashing applications. FTC generates slightly more noise than FTO, about 2 to 4 dB(A). The valves, specifically designed for control applications, have a rangeability of about 100:1.

Standard eccentric rotating plug valves are used within the range of -80°C to +600°C; soft seat valves are limited to 260°C.

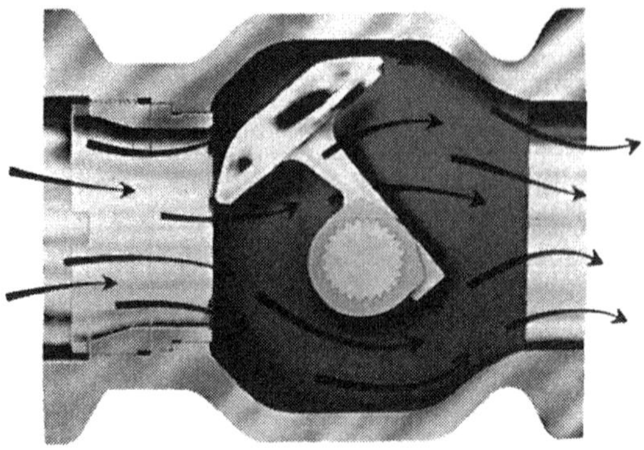

Figure 6.46 Eccentric rotating plug valve
Courtesy of Tyco Flow Control

Valves are available up to ANSI Class 600 body rating, from 25 mm to 300 mm and with flange ratings for Classes 150, 300 and 600 LB. Wafer design is an option above 80 mm. Combined with a good actuator, this type of valve will allow most of the claimed rangeability of 100:1 to be realised in practice.

Unlike ball and butterfly valves that have the same shut-off ratings regardless of body size, because of their pressure balanced construction, the eccentric rotating plug valve is not fully balanced as the name suggests; and the shut-off rating decreases as bodies get larger. Seat tightness is better than ANSI Class IV.

6.3.3 Special purpose control valves

Automatic pump recirculation valves could be theoretically classed as regulators and should be in Chapter 5. Regulators, however, tend to be small. The valves in this category can be some of the largest valves built and are very special.

Automatic pump recirculation valve

Centrifugal pumps are the most popular type of pump for industry and process applications. One good reason why these pumps are so popular is the ease of flow regulation. A valve placed in the pump outlet pipework can throttle the flow, i.e. increase the operating pressure of the pump, so that the pump will operate further back on its curve, see Figure 6.47.

The upper graph shows a system with pure friction losses, the lower graph shows a system with static head and friction losses. It can be clearly seen that the pump operating point is moved closer to the zero flow condition. Many pumps do not suffer at low flow conditions. High power pumps, and pumps operating with low density, low specific heat liquids, can however have serious problems. As the pump flow is reduced:

- pump efficiency reduces passing more heat to the liquid
- vibration levels increase
- NPSH margin may be eroded
- the pump can suffer low flow instability

The conditions listed are capable of seriously damaging the pump during extended running. It is therefore essential to protect the pump by preventing it from operating below a set flow limit even though the system requirements are less. This is easily accomplished, using control valve types already described, as shown in Figure 6.48.

The process demand is controlled by FCV1. As process demand is reduced, FCV1 closes in response to a control signal and the pump flow is reduced. The pump flow is monitored, via an orifice plate in the inlet line, by the DPC.

When the flow is reduced below a preset limit the DPC opens the bypass valve, FCV2, and allows some flow to return to the suction source. If the process demand is reduced to zero the to-

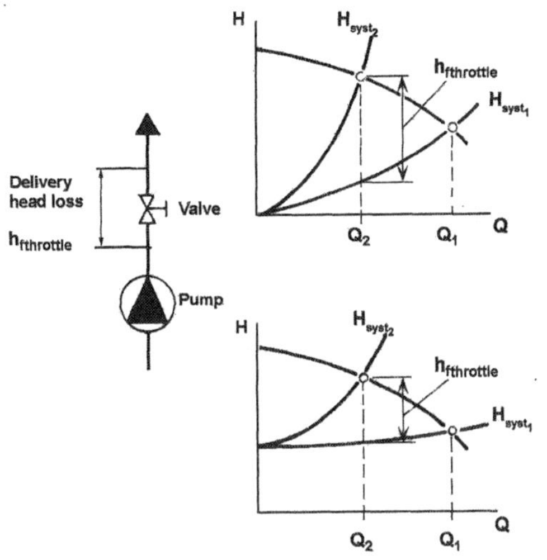

Figure 6.47 Effect of throttling on a centrifugal pump

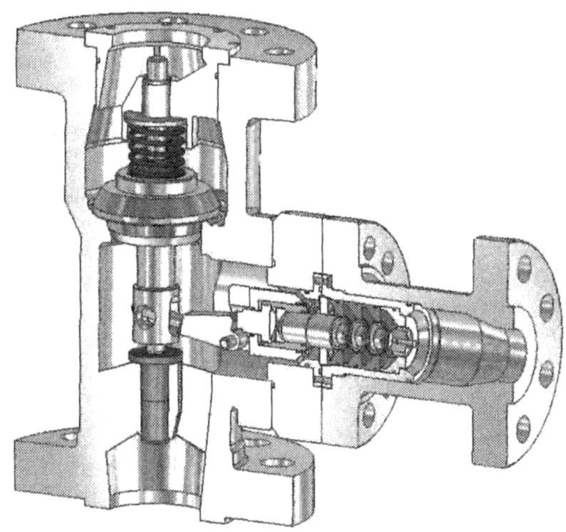

Figure 6.50 Automatic pump recirculation valve, straight-through version
Courtesy of Blackhall Engineering Ltd

tal minimum flow will be passed through the by-pass. The pump is fitted with isolating valves and a non-return valve to prevent "turbining" in the event of power failure. The control valves each have three valve bypasses for maintenance and manual control.

Figure 6.49 shows an equivalent system with an automatic recirculation valve. The recirculation valve is shown as a strange combination of two non-return valves joined together. The system demand is still controlled by FCV1. The recirculation valve senses the pump flow passing through it so external flow measuring devices are unnecessary. If the flow is reduced to a preset level the by-pass port is opened allowing flow to return to the suction source. The recirculation valve is self-powered. Very large valves can be pilot-operated, but external power is not required. The valve incorporates a non-return valve in the main process feed, as shown in Figure 6.50, and so does two jobs. A non-return valve can be incorporated in the by-pass outlet. The bypass outlet from the valve is specially designed to throttle the pump discharge pressure to much lower pressures without cavitation damage.

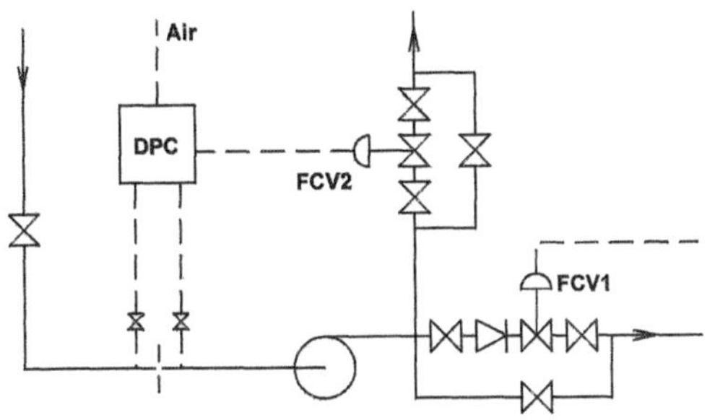

Figure 6.48 Centrifugal pump with flow control and minimum flow protection

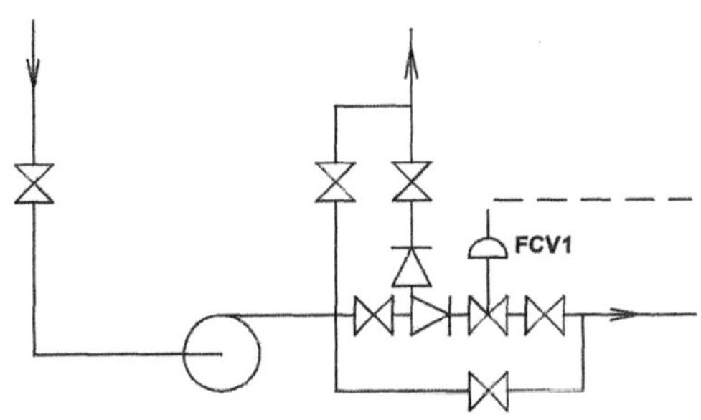

Figure 6.49 Centrifugal pump with automatic recirculation valve

Figure 6.51 shows a different style of automatic recirculation valve; this style of construction could be considered as "Oblique" rather than "straight-through". The bypass connection is on the back of the valve body. Figure 6.51 shows the valve in the full by-pass condition with the main process flow non-return valve closed. The multi-stage throttling of the by-pass flow is clearly visible. Some automatic recirculation valves can be configured as "on-off" bypass rather than a modulating bypass. The surge implications of "on-off" control must be carefully considered **before** implementation.

The automatic recirculation valve can be custom-built to modify the pump H-Q characteristic. Unstable pump curves can be converted to stable. Extra connections can be fitted for start-up conditions and for flushing. These valves are used on many clean liquids in various processes. Automatic recirculation valves can be used in conjunction with load-unload control systems for applications such as descaling in steelworks.

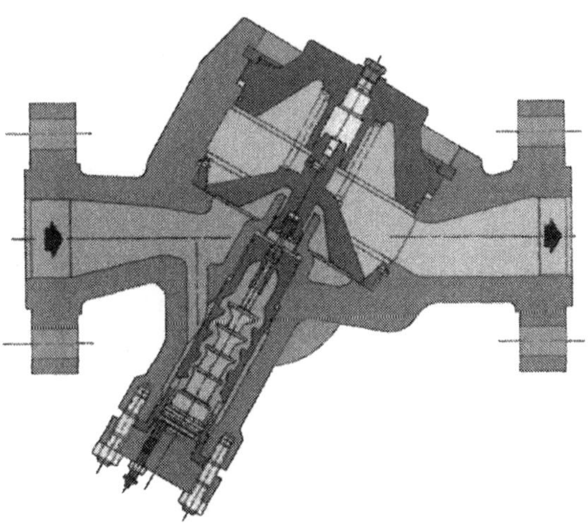

Figure 6.51 Automatic pump recirculation valve, oblique style version
Courtesy of Tyco Flow Control

As with all control valves, it may be prudent to ensure the valve moves periodically to prevent binding. The manufacturer's operating instructions should be consulted for specific information regarding particular installations. Valves are made in sizes from 32 mm to 500 mm, pressure ratings up to ISO 7005 PN420 and ANSI B16.5 Class 2500LB.

NOTE: There is only one major concern regarding the use of automatic recirculation valves — that is **EFFICIENCY !**

If the low flow condition that only occurs for a very short time the increased operating costs may not be significant. If the low flow condition, or variable flow conditions, occur regularly then the loss of operating efficiency may produce significantly increased energy costs. For regular low flow operation a variable speed pump should be considered.

Thermostatic control valve

The thermostatic control valve is a control valve which is self-contained and designed specifically to respond to temperature changes. It is self-powered but the process fluid is not the motive fluid. A sealed system encloses a liquid or vapour or an absorption charge. The valve modulates to find the point of equilibrium.

Thermostatic control valves consist of two major assemblies:

- the thermostatic control system
- the valve body - Figure 6.52

The thermostatic control system is sealed to ensure the integrity of the charge. A temperature sensor, usually cylindrical, is connected by a capillary tube to the actuator, usually a bellows or diaphragm assembly. When an absorption charge is used, the crystals are trapped in the temperature sensor. On rising temperature the pressure within the thermostatic control system increases. In liquid and vapour systems the fluid expands in response to the temperature rise. In an absorption system gas is released from the crystals creating an increase in pressure. The internal pressure on the actuating bellows causes the bellows to distend to provide the control movement. Manual adjustment is usually provided for the set-point.

Thermostatic actuators can be complex and multi-functional. One manual adjustment can be organised to control the effect of two sensors on two different valves. For example, heating and cooling systems can use a single heat exchanger to maintain a process temperature. For some applications the temperature sensor can be built into the actuator. Air temperature sensors for radiator control valves and process liquid temperature sensors for return temperature limiters are typical examples. Remote temperature sensors are normally quite compact, 10 mm diameter by 100 mm long. However on the larger industrial units the sensor increases in size, 22 mm diameter by 250 mm long being common.

Good heat transfer between the process fluid and the sensor is essential for accurate control. If sensors are fitted in thermowells, the well must be filled with a suitable liquid, such as light oil. The valve actuator and the set-point adjustment can be two separate units. This refinement allows the valve body and sensor to be fitted within equipment housings but allows the adjustment to be conveniently panel mounted. The capillary connecting the sensor to the actuator can be up to 2 m long. With separate control and actuator, 8m capillary lengths are possible. Installation instructions must be observed fully and some designs restrict the position and orientation of the sensor. Larger valves can be pilot-operated. The thermostatic actuator controls a process fluid pilot signal which operates the main valve.

The valve body can be in many configurations to suit diverse applications. The most common styles are straight and angled pattern globe valves with a plain or contoured plug. Plugs can incorporate soft seals to enhance shut-off sealing. The valves are opened by an internal spring and closed by the expansion of the control medium. Thermostatic valves are generally small, screwed valves from 10 mm to 25 mm and are rated for working pressures of 10 barg. Some designs limit the maximum differential pressure to 0.2 barg.

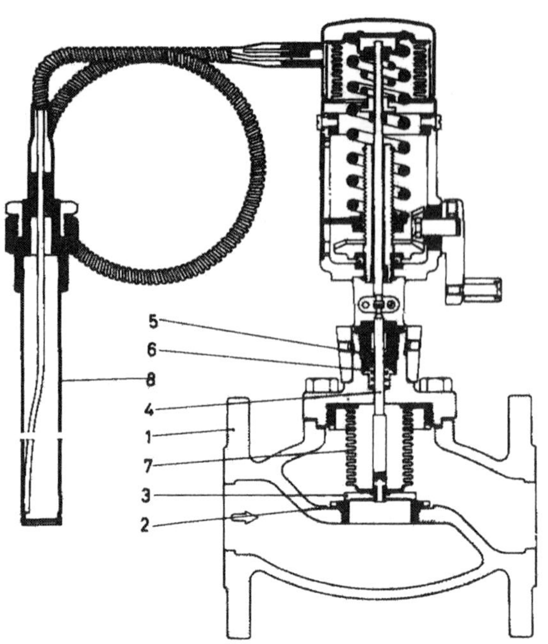

Figure 6.52 Flanged thermostatic control valve
Courtesy of Danfoss A/S

Small valves are unbalanced and are typically used in domestic and commercial central heating systems. Valve bodies are of brass or nickel plated brass and use diaphragms and/or "O" rings for the stem seal. Heavier flanged valves in cast iron, from 15 mm to 50 mm, used for commercial and industrial systems, including steam, have a balancing bellows and are rated for 16 barg with differentials up to 8 barg. Secondary stem sealing is achieved by spring-loaded Teflon™ seal rings.

Three-port globe valve bodies are available for mixing, diverting and bypass applications. Contoured and plain double plug styles are standard. Screwed valves of this type, from 10 mm to 25 mm, are suitable for 10 barg.

The majority of thermostatic valves are used in domestic and commercial heating, ventilating and refrigeration systems. Valves are used for process control, especially useful in electrical hazardous areas or where instrument air is unavailable or unreliable. Valves can also be used as backpressure regulators for constant temperature control in refrigeration systems.

Special Note: Pressure/temperature valves

Thermostatic temperature control valves can be combined with pressure control regulators. The control mechanism of the pressure regulator is adjusted by the movement resulting from temperature changes of the "charge". Both, or either of the control impulses can be pilot amplified. This type of valve is very special but can be extremely useful in situations when external power is not available for a control valve.

Special Note: Bimetallic actuation

A modified type of thermostatic control valve can be produced by making use of the concept of differential expansion of dissimilar metals. A linear motion is produced by connecting an outer bulb, exposed to the temperature varying fluid, to an inner rod. By correct selection of the two metals the rod can produce positive or negative motion on rising temperature. This style of valve is only available in the compact, unit construction form;

sensor, actuator, valve body and set-point adjustment being a single assembly. The fluid being sensed does not necessarily have to be the fluid being controlled.

Special Note: Control valves for dry solids

Whether flow control can be successfully implemented is dependent upon the nature of the dry solids. Flow control of graded solids, solids within tight size limits, can be very accurate when the solids are fairly regularly shaped. When the size of solids varies over a wide range flow control is difficult and accuracy is sacrificed. Even with graded, regularly shaped solids external vibration may be required to ensure a continuous flow. Many standard control valve styles are unsuitable for dry solids. Best results are achieved with valves adapted for dry solids applications, see Chapter 3, Section 3.3.6.

Linear motion control valves i.e. globe, double-beat globe, cage-trim and multi-port sliding gate, are not used for dry solids applications. Simple valve designs have proved very effective. Horizontally mounted parallel slide valves, similar in construction to knife gate valves, Chapter 3, Section 3.3.1, can be fitted with piston actuators. Piston actuators are necessary for the long travel. The control characteristic can be modified by replacing the normal circular port with a vee-notch. The valve C_v value is dramatically reduced by using the vee-notch seat and this must be considered when sizing ducts. Figure 6.53 shows a vee-notch seat valve with a piston actuator; the pneumatic controller is not shown.

Figure 6.53 A knife gate valve with vee-notch seat

Valves of this type are available from DN50 to DN900; valves with circular seats are available up to DN1800. Valve bodies can be of cast iron, carbon steel or austenitic stainless steel. Gates are usually of austenitic stainless steel. Knife gate valves with polyurethane bodies are available from DN100 to DN600. These valves, with standard threaded stems, can be fitted with rotary electric actuators. Parallel gate slide valves are manufactured for square and rectangular ducts up to 420 mm square.

Full-bore diaphragm valves can be power actuated, pneumatic or electric, to provide control functions. Table 3.7 in Chapter 3, lists the operating limitations for the full range of valves from DN15 to DN350. Pneumatically or hydraulically operated pinch valves are possibly the best design of control valve because the portion of the valve which creates the throttle is long and not effectively a knife edge orifice. A long orifice can be of bigger diameter than a knife edge orifice with consequential lower velocity. Lower velocity can be equated to lower wear and longer parts life and hence lower operating cost.

The wearing part of a pinch valve is an elastomer tube which should not be costly compared to metal components which are surface coated with exotic materials. Pinch valves also provide a leak tight seal when closed. Table 3.23 in Chapter 3, lists the available sizes and pressure capabilities for direct-acting pinch valves. The iris diaphragm valve, described in Section 3.3.6 of Chapter 3, is also popular for controlling dry solids. Sizes range from DN50 to DN450.

The rotary slide valve, see Table 3.21 in Chapter 3, could be fitted with controlled actuation to perform as a control valve. These valves are very similar to parallel gate valves and knife gate valves but the motion is radial.

Rotary motion control valves such as butterfly, vee-port ball and eccentric rotating plug valves are used successfully with dry solids. Butterfly valves are manufactured in sizes up to DN2 000. Valves with elastomer linings will provide good shutoff sealing. The disc and the bore of the body can be coated with an elastomer or thermoplastic, such as PTFE, to provide a non-stick surface.

A few butterfly valves are available with inflatable linings. In the open position the lining is clear of the disc tip and avoids rubbing wear. In the closed position the lining is extended to touch the disc tip and form a good seal. The act of inflation dislodges particles which may have become attached to the seat area. Also the moving lining compensates for wear. A differential pressure of 10 barg can be maintained across the closed disc. Butterfly valves of this style are available from DN50 to DN750.

Vee-port ball valves are popular in sizes from DN25 to DN400 for pressure ratings up to PN100. Carbon or stainless steel valve bodies are options. Hardened stainless steel balls, with chrome plate or Stellite™ inlays, can be fitted. For good shut-off, sealing soft seats must be used which may have a short life with abrasive solids.

Rotary eccentric plug valves with soft seats are limited to 260°C. Valves are available from 25 mm to 300 mm and with flange ratings for ANSI Classes 150, 300 and 600 LB. Wafer design is an option above 80 mm. A special version of these valves can be fitted with an inflatable seat. During normal control functions the seat is flush with the valve body and largely protected. For isolation the seat is inflated and seals against the spherical plug face. Rotary eccentric plug valves with inflatable seats are available from DN50 to DN600. Operating temperatures up to 350°C are possible with a water cooled seal and a water cooled plug.

Motorised valves, specifically designed for dry solids applications, see Chapter 3, Section 3.3.6, may not require external vibration assistance. Double dump valves and rotary airlocks can perform flow control functions by varying the speed of the drive motor. Variable frequency inverters are quite inexpensive for small motors and very effective. However, to avoid incompatibility problems the inverter should be purchased from the motor manufacturer or the motor supplier.

Special Note: Non-metallic control valves

Parallel gate valves with uPVC bodies are manufactured in standard sizes from DN15 to DN100 and are suitable for 10 barg between 5° and 50°C. Valve bodies can have female BSP or plain socket connections. Parallel gate valves with ABS or PVC bodies, suitable for 0.7 barg, are available in sizes from DN40 to DN300. These valves have stainless steel gates.

It may be possible to coat the gate with ceramic or a thermoplastic such as FEP, PFA, PTFE or Rislan™. A ceramic coating would probably provide the longest service life as well as the broadest corrosion resistance. Knife gate valves with polyurethane bodies are available from DN100 to DN600. A one-piece moulded polyurethane seat also forms a complete liner for the wafer body. Standard gate materials are AISI 304 or 316 stainless steel. See the comment above regarding coatings.

Valves up to DN150 are rated for 6.9 barg; the largest valves are rated at 3.5 barg. The operating temperature range is limited to -40° to 55°C. Pneumatic piston actuators or electric rotary actuators may be fitted. Remember that gate valves have poor control characteristics and most throttling effect is created when the valve is nearly closed which results in high fluid velocities. Fluids contaminated with solid particles are liable to cause rapid wear.

Full-bore diaphragm valves, with bodies of PVC or PVDF, are available in sizes from DN15 to DN50. Valves are fitted with pneumatic diaphragm actuation and are suitable for pressures up to 6 barg.

Pinch valves should be considered as a non-metallic valve because the only component in contact with the process fluid is an elastomer tube, see Figure 6.54. Pinch valves operated by metal rollers are suitable for system pressures up to 49 barg; pneumatic and hydraulic operated valves can operate up to 10 barg. The range of tube materials available is quite broad to provide wide chemical compatibility:

- Buna-N™
- ChlorobutylEP
- DM, Hypalon™
- Neoprene™
- Polyurethane
- pure gum rubber
- Viton™

The rangeability of pinch valves can be up to 26:1 but the differential pressure is usually restricted to 30% of the pressure rating. The control range can be extended by connecting valves in series. Pinch valves are available in sizes from 25 mm to 1 800 mm.

Solenoid valves are available in a range of non-metallic materials as shown in Table 6.2. Valves with PEEK and polypropylene bodies are also produced.

Body material	Connection sizes	Pressure rating barg
PVC	$\frac{1}{8}$" to $\frac{1}{2}$" UNF	8.2
PVDF	2 to 6 mm	1.5 to 6
PVC	10 to 50 mm spigot	6
Nylon 66™	$\frac{3}{4}$" to 3" BSP	9
Delrin™	$\frac{1}{4}$" to 1$\frac{1}{2}$"	16

Table 6.2 Typical non-metallic solenoid valves

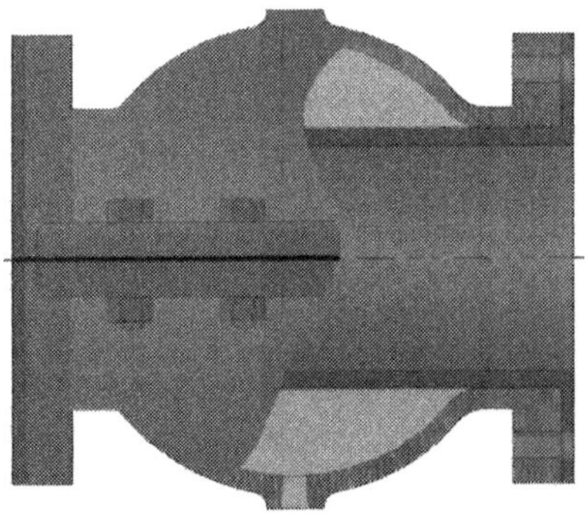

Figure 6.54 Typical pneumatic/hydraulic pinch valve construction

Solenoid valves can be used to construct a digital control valve. At first glance this concept seems incorrect; the actuator or positioner may be digital but the valve is still a valve. However a digital valve can be constructed from standard components. Multiple solenoid valves can be connected between one supply and one demand. The smallest programmable logic controller, PLC, will have at least eight digital outputs. That is eight switches which can be programmed to be on or off. If eight identical solenoid valves were used, and the differential pressure remained constant, a linear control valve would result with flow increments of 12.5%. In practice, the overall characteristic will not be linear due to the influence of the system characteristic. But a very simple and effective control valve has been produced.

With isolating valves on each solenoid valve, the control valve can be maintained without being 100% off-line. The solenoid valves do not have to be identical; the control valve could be characterised to match specific system requirements. With five valves sized for; 5%, 10%, 20%, 40% and 80%; it is possible to control between 0% and 100% in 5% increments. By adding a sixth valve at 2.5% the increment is reduced to 2.5%. If eight digital outputs are available, very fine adjustment is possible, or spare valves could be fitted for the solenoids which would cycle the most. Also remember that solenoid valves can operate on the mechanical pinch valve principle.

Ball valves can be fitted with ceramic balls which are profiled with slots, parallel and variable depth or teardrop, to provide control characterisation. One piece bodies are moulded from vinyl ester resin which is reinforced with glass fibre or carbon fibre. Seats can be virgin PTFE or glass reinforced. Sizes from DN25 to DN150 have reduced-port bores and are suitable for 19 barg, DN200 for 11 barg, at temperatures up to 120°C.

For abrasive applications the valve body can be fitted with ceramic inserts, see Figure 6.55. Valve bodies are predrilled to accept actuator mounting brackets so any rotary actuator may be used. These valves have very small C_V values compared to isolating ball valves. The pressure drop through the valve should be evaluated carefully when designing the system.

PVC wafer butterfly valves, from DN50 to DN400 can be fitted with electric actuators. Valves up to DN80 are rated at 10 barg; DN100 to DN250 at 6 barg and DN300 to DN400 at 3 barg. The discs are of the plain symmetrical type so that low flows are achieved by creating high velocities. Lined body butterfly valves can be considered as non-metallic valves if the disc is non-metallic or coated. Standard linings include:

- Buna N™
- EPDM
- UHMWPE
- Hypalon™
- Neoprene™
- PTFE
- Viton™

Standard options for disc coatings include:

- two pack epoxy resin
- Halar™
- Kynar™
- Nylon 11™
- Rislan™

The metal spindle would have to be adequately isolated from the fluid. Lined and coated butterfly valves are available in a wide range of sizes. Valves up to DN1600 can be rated for 16 barg; larger standard valves are rated for 10 barg. Lined valves with inflatable seats are rated at 10 barg. These valves can

Figure 6.55 Ceramic lined non-metallic ball valve

have nylon discs or cast iron discs coated with nylon or PTFE. Sizes range from DN50 to DN750. Air pressure, up to 11 barg, is required to inflate the seat when closed to ensure a good seal.

Special Note: Hygienic control valves

The cleanliness requirements for hygienic applications places severe limitations on control valve construction. The necessity to remove "dead" volumes and crevices where process material could accumulate restricts the style of valve assemblies which can comply with legal regulations. Experience has shown that some valve designs are eminently suitable. A mixture of Imperial and metric tube specifications are used for hygienic applications; specifying the correct sizes is very important.

Parallel slide valves, sealed by a rolling membrane rather than packing or bellows, can be fitted with electric or pneumatic actuation and suitable controllers. Valves are manufactured in approved materials in sizes from DN50 to DN450. Single seat globe valve bodies can be fabricated, cast or machined from the solid, see Figure 3.88 in Chapter 3. Valve actuation can be electric, pneumatic diaphragm or pneumatic piston for long stroke valves. Actuators with sealed, polished stainless steel housings can be provided to allow hose-down cleaning. Valve plugs can be contoured to provide equal percentage or linear characteristics. Valve bodies, with pressure ratings up to 16 barg at 160°C can be angle or oblique in sizes from DN15 to DN150. A wide range of approved connectors can be supplied; butt weld, Acme threads, clamp, DIN, flange, IDF, RJT or SMS.

Diaphragm valves, both full-bore and weir, can be fitted with pneumatic actuation for control applications. Some valves can have electric actuation. Enclosures can be plastic or stainless steel. Sizes range from DN5 to DN80 and all valves can operate at 10 barg. Some sizes can be uprated to 16 barg depending upon the specific operating conditions. Diaphragm control valves can operate up to 180°C depending upon the choice of diaphragm elastomer.

Pinch valves can be constructed to comply with the strictest hygienic regulations. The normal flat-faced flange connections, shown in Figure 6.54, can be replaced by approved hygienic clamp connectors. Even though the product does not contact the interior of the valve metal body the inner surfaces are polished to a hygienic finish. Valve bodies can be cast iron, aluminium or stainless steel. Non stainless steel valves can be treated with an approved polyester coating. The "hygienic" treatment is normally given to valves between 15 mm and 250 mm bore. Working pressures up to 12 barg are possible.

Solenoid valves can be adapted to the pinch principle. Many approved tube grades are available; peristaltic pump tubes may be suitable. Multiple solenoid valves can be connected in parallel to create a digital control valve. A small PLC could be programmed to control up to eight valves. Incremental control steps, smaller than 2.5% are possible.

NOTE: Diaphragm and pinch valves are especially useful for processes with solids. The valves are able to seal around solids possibly without damaging the solids.

Butterfly valves can be fitted with electric or pneumatic actuators. AISI 316L polished valves can have approved rubber, nitrile, EPDM, Neoprene™, silicone, Viton™ or PTFE seats. The flow control characteristic of plain butterfly discs is poor and moderate throttling is achieved by creating high velocities. Popular valve sizes range from DN25 to DN100.

6.4 Useful references

Society of Automotive Engineers (SAE), World Headquarters, 400 Commonwealth Drive, Warrendale, PA 15096-0001 USA, Tel: 248 273 2494, www.sae.org.

International Organization for Standardization (ISO), 1 rue de Varembé, Case postale 56, CH-1211 Geneva 20, Switzerland, Tel: 22 749 01 11, Fax: 22 733 34 30, www.iso.org.

American National Standards Institute (ANSI), 1819 L Street, NW (between 18th and 19th Streets), 6th floor, Washington, DC 20036 USA, Tel: 202 293 8020, Fax: 202 293 9287, www.ansi.org.

ISO 7005 Part 1 : Steel Flanges; Part 2 : Cast Iron Flanges; Part 3 : Copper alloy and composite.

Safety relief valves

7

7.1 Safety relief valves and the system

In general, pressurised systems are designed to a code. The code specifies allowable material stress levels to provide an acceptable safety margin for the material. The safety margin considers process variations which are likely to occur and the effect on the material stress. Most codes are based on steady-state operating conditions, the effects of fatigue due to continuous pressure oscillations not being considered. When fatigue failure is likely special precautions must be taken.

Low pressure systems — systems with design pressures up to 1barg — can usually be designed to accommodate excess pressure without any material over-stressing. In these systems excess pressure protection may not be necessary. Low pressure relief valves are manufactured for installations which cannot be designed to be inherently safe.

Low pressure systems can be subject to another mode of failure which is peculiar to these systems, namely: elastic instability. If the low pressure is reduced until a partial vacuum is created, some components can deform or buckle. This can be a serious problem for pressure vessels and tanks. Additional protection may be required, by using a vacuum break valve.

High pressure systems may be subjected to excess pressures, that is pressures which exceed the design pressure causing material over-stressing. The nature of the system and its environment must be analysed to evaluate the size of the problem. Rotodynamic pumps and compressors generate a theoretical maximum pressure which cannot be exceeded. Selecting a design pressure slightly above this value removes all problems which could be machinery related.

Problems due to temperature and chemical reactions must be considered independently. Temperature rises can be due to other systems, such as a fire in adjacent equipment. Dangerous chemical reactions can be created by control errors resulting in incorrect proportions of constituents or errors in basic constituents.

Positive displacement pumps and compressors are capable of generating very high pressures. Systems are usually protected against excess pressure. Safety valves, or bursting discs, can be fitted to prevent pressures above a predetermined value being applied.

Figure 7.1 indicates the relationship between the various pressures for a group of safety relief valves:

- conventional (1)
- pilot-assisted (2)
- supplementary loaded (3)
- pilot-operated (4)

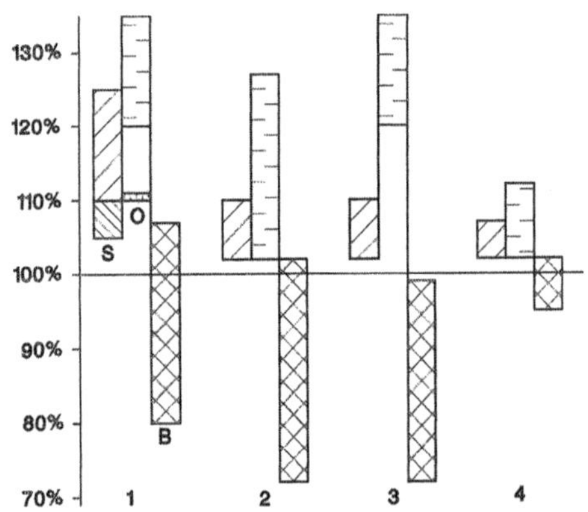

Figure 7.1 Operating pressures for safety relief valves

Normal operating pressure is indicated as 100%. The typical ranges of set pressure (S), overpressure (O) and blowdown (B) can be easily appreciated. Conventional spring-loaded valves are greatly influenced by the fluid properties, and to some extent by the inlet and outlet pipework. More complex, and costly, valve designs allow the minimum set pressure to be moved closer to the normal operating pressure, although a special conventional valve, with a resilient seat, does allow low set and overpressure operation. Some designs allow the overpressure to be drastically reduced. The figure also shows that the blowdown can be adjusted over a very wide range and this point must be carefully considered.

Two aspects of valve operation must be considered in the context of valve cycling:

- Does the valve close at a pressure between the normal and set?
- Does the system pressure have to reduce below the normal pressure for the valve to close?

If the valve closes too soon the fault condition may not have been rectified and the valve will re-open immediately. Valve cycling causes wear and if very rapid pressure changes are involved can damage the nozzle and disc. Valve cycling is dependent upon the process system characteristics and must be evaluated on a case-by-case basis. Pilot-assisted and pilot-operated valves can help to alleviate this problem by holding the valve open for a specific time every time the valve opens. As with controlled closure, the wear and damage problems can be greatly reduced. However, the valve life is extended at the cost of wasted product.

Positive displacement compressors and pumps, without instructions to the contrary, operate intermittently at any pressure above the set pressure. This time period cannot be quantified as it is dependent upon:

- the running conditions immediately prior to fault
- the running time prior to the fault
- the size and type of the driver

Operation above the set pressure is considered as an emergency condition which should be rectified very quickly. If process operation or control dictates extended running over the set pressure, the time period must be evaluated and communicated to the compressor/pump manufacturer and the valve manufacturer.

Positive displacement machines can create pressure pulsations in the process system. The peak-to-peak value of the pulsations is dependent upon the system design and the fluid as well as the machine design and speed. When selecting the relief valve set pressure, the magnitude of the positive peak must be considered and an allowance made to avoid spurious valve operation. Pulsation dampers may be fitted to the machine to reduce peak-to-peak values.

Some rotodynamic machines produce undesirable pressure pulsations at certain operating conditions. This factor must be considered if relief valves are fitted for system protection.

When considering the use of a safety valve, the energy of the system and the rate at which energy is added to the system plays an important part. Some systems can produce very high pressures at low energy levels. The pressures are high enough to fracture pipework and fittings but the energy levels are so low that personnel are not put at risk. Such systems include single-shot lubrication systems and diesel injection systems on small engines.

When two processes react physically but are not normally in contact, for example two fluids in a heat exchanger, the other fluid may influence safety requirements. Some chemical processes place special demands on safety valve sizing. Vessels in which reactions, separation or evaporation take place require detailed study. Precautions must be taken if the fluid may change phase while passing through the valve or if the fluid could be multi-phase when entering.

Dynamic simulation has proved useful and economic when considering the relief and flare systems for fractionating columns. The benefits of comparing accurate system simulations against conservative manual methods are worth noting.

Excess pressure protection can sometimes be unnecessary if the source of pressure can be controlled directly. This type of safety control can be useful in chemical processing and refining when heating a vessel by steam or hot liquid. Additional controls in the heating circuit can act as excess pressure protection.

Any pressure relief device which is fitted must be reliable. Reliability is usually ensured by fitting dedicated safety devices; that is safety devices which do not take part in normal control sequences. A dedicated safety device does nothing for the majority of its working life. But when required to operate it is in good condition, not worn out by control actions, and fulfils its prime objective. Relief valves tend to fail safe. Most safety relief valves are spring-loaded and the spring may relax or, very infrequently, break. Under these conditions the valves release process pressure and must be maintained.

Within chemical processing and refining, dedicated relief devices are the norm. Hydraulic power applications sometimes share the excess pressure protection with control functions by utilising a valve incorporated in a multi-valve assembly. Process pilot-operated unloading or by-pass valves frequently double as safety valves. Some hydraulic power manufacturers shun the use of spring-loaded safety valves on the grounds of size. The user should be aware of the physical protection fitted and the degree to which safety is shared with or compromised by control.

Reliability can be increased by the use of simple equipment. Spring-loaded and dead-weight safety valves are among the simplest machine elements in current use. The pilot-operated safety valve is slightly more complex but relies on the action of a spring-loaded spool valve to initiate the relief cycle. Because many pilot-operated safety valves come from nuclear applications the reliability can be assumed to be extremely high.

Standard spring-loaded relief and safety valves are not suitable for shock wave protection. Explosion relief valves are specifically designed for protection against shock waves but are usually only suitable for pressures close to atmospheric. Shock waves travel through the fluid at acoustic velocity. When passing the valve, the local pressure causes the valve to open. If the shock wave is severe the valve may slam wide open. The pressure is only applied momentarily as the wave passes. There is no fluid flow associated with the shock wave so there are no hydrodynamic forces developed to keep the valve open. The valve closes as violently as it opened. Violent closing can damage the disc and the nozzle leading to leakage of product.

Pilot-assisted valves, or pilot-operated valves, may be able to withstand the worst effects of violent closure by holding the valve open for a fixed period followed by controlled closure. Shear pin, buckling pin safety valves, and bursting discs may open allowing product to escape. All operating conditions, including process upsets and possible shock wave generators, must be properly evaluated when selecting valves. The use of variable speed drivers for compressors and pumps has resulted in increased demands for the more complex relief valves.

The safety relief valve is one component in the safety system. The safety system may be extremely small, for example steam exhausting to atmosphere. At the other extreme, the safety system for hazardous or costly fluids may be complex and large involving closed drains or product recovery. The safety system cannot be designed as a separate entity. Decisions regarding the safety relief valve modes of operation and pressures will have a direct impact on the primary systems. The safety system must be designed concurrently with the primary systems if safety and plant reliability are to be achieved.

The European Commission's Pressure Equipment Directive (97/23/EC), PED, was adopted in 1997 and entered into force on 30 May 2002. The PED is a harmonised directive that replaced legislation based on individual country pressure equipment construction codes by using harmonised European standards. The PED is applicable, with a few exceptions, to systems having a working pressure of 0.5 barg or above. The consequence of the PED is that it is illegal to sell or purchase safety valves for use within the European Union unless they are PED compliant. This applies to safety valves sold individually or as part of an equipment assembly.

The PED has had a major impact on the technical requirements for system design, the choice of materials and the cost of purchasing equipment. Valve users should be fully aware of the safety implications when designing new systems. The national valve manufacturers' associations can provide invaluable assistance. Criticisms of the PED are that it is poorly written and is vague in many areas. The full implications of the PED are still being experienced and may not become apparent until requirements have been tested in Law and precedents have been set.

7.2 Safety relief valve design

Safety relief valves (srvs) only became necessary at the beginning of the Industrial Revolution. Mine water pumping spawned very early mechanisation by utilising steam power. Early engines operated at pressures only slightly above atmospheric. In 1769 the average boiler pressure was less than 0.5 barg. By 1800, boilers operating at 3.4 barg were available. Safety relief valves became essential to protect equipment and personnel from higher operating pressures.

The first relief valves were dead-weight, that is to say — a mass resting on the valve disc held the valve closed, see Figure 7.2.

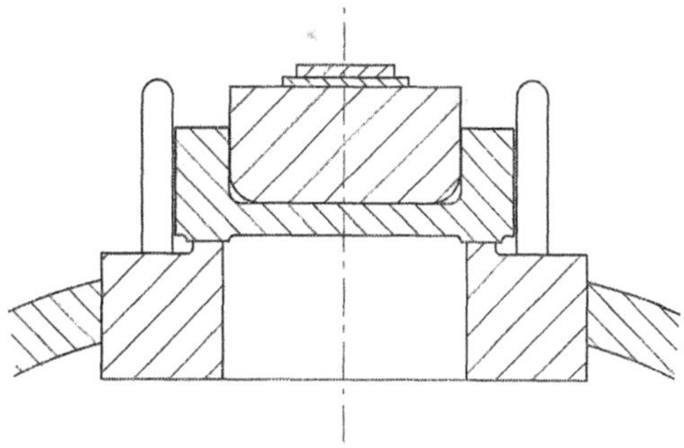

Figure 7.2 Simple dead-weight relief valve

The valve could only be mounted vertically. Steam pressure acting over the nozzle area created a force trying to lift the disc. The weight of the disc, plus additional weights, resisted the steam force. The valve could be easily adjusted to very accurate set points while operating. Reducing the set pressure could be exciting as the steam was not piped away but released to the surroundings!

This type of valve was very reliable and effective. Because the valve only had three components the number of failure modes were very limited. The valve could be adjusted very easily by adding or removing weight. During commissioning and early trials the valve could be set "light" as an extra precaution. As commissioning proceeded and staff became more confident the working pressure could be increased in increments. The valve would modulate and only lift sufficiently to relieve excess fluid.

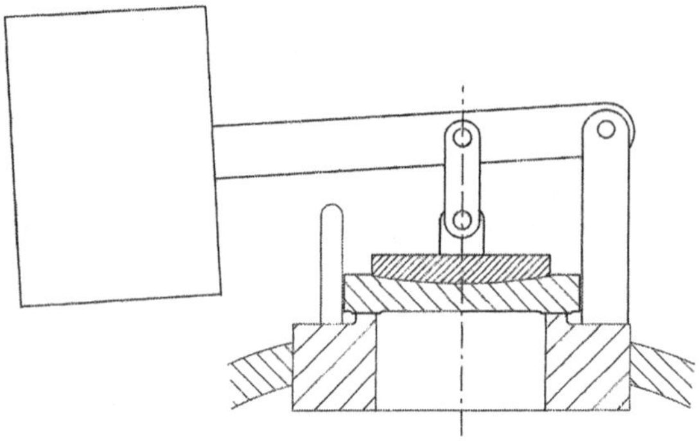

Figure 7.3 Dead-weight lever relief valve

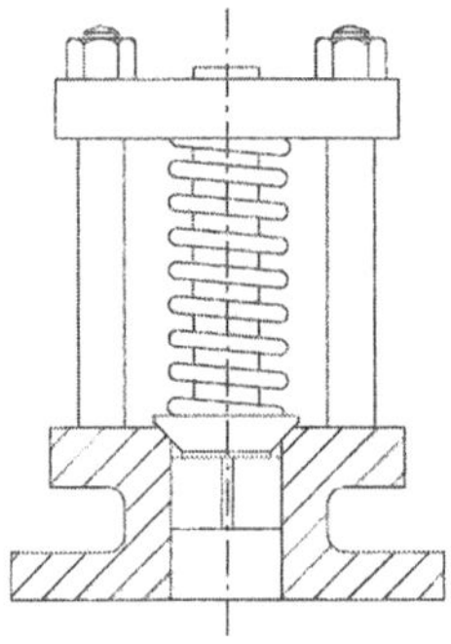

Figure 7.4 Open spring-relief valve

Inspection and maintenance could be readily accomplished when the boiler was shut down. The seating faces could be "blued" and "matched" without difficulty.

The earliest safety relief valves were designed for low pressure steam. As steam pressures increased improvements were necessary and the dead-weight lever valve was developed, see Figure 7.3.

The addition of leverage to the relief valve design allowed higher pressures to be contained while maintaining the nozzle area constant. The valve set pressure could be adjusted by the addition of small weights or by moving the large weight on the lever. The lever could be made any length so that one size of valve could be made suitable for many operating conditions. The basic design was still simple and very reliable. As pressures and valve sizes increased further, the space required for the relief valve also increased. A new style of valve was therefore needed.

Advances in material manufacture allowed springs to be produced. The dead-weight lever relief valve was replaced by an open spring relief valve, Figure 7.4.

The valve disc became an extended poppet which was top and bottom guided. The seating could be angled, as shown, to aid centralisation, or flat. The spring load was adjusted by varying the pre-compression. Shims could be added under the yoke to increase the set pressure. Small increases in set pressure could be accomplished by adding weight on top of the spindle.

The spring-loaded relief valve was very compact and a fixed nozzle area could be used for a wide range of pressures by varying the spring design. Most modern spring-loaded and thermal relief valves, described in the next Section, are derivatives of this simple concept.

7.3 Safety relief valve types

Spring-loaded valves for gas, vapour and steam

The modern spring-loaded valve is similar to that shown in Figure 3.6 with the following exceptions:

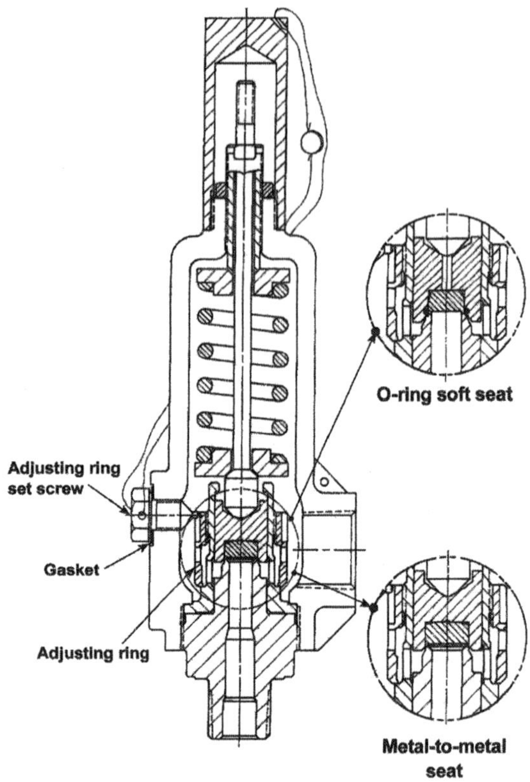

Figure 7.5 Typical small screwed safety relief valve
Courtesy of Tyco Flow Control

- the spring is enclosed
- the valve setting is adjusted by screw
- the disc/nozzle seating is nearly always flat
- the valve outlet is piped away

Valves for steam may still have open springs depending upon the environment. Some manufacturers advocate the use of angled seats for valves used in conjunction with reciprocating compressors when pulsation dampers are not used to attenuate pressure pulsations. Figures 7.5 and 7.6 show examples of a small valve with screwed connections and a larger flanged valve.

The valves are similar in many respects. Notice the nozzles are screwed into the body and the similar construction used to retain and adjust the spring. Large high pressure valves may be fitted with a special thrust bearing at the top of the spring to reduce friction torque on the adjusting screw when setting. Valves may have a choice of springs to increase the range of pressure settings. In both designs care is exercised to ensure the disc seats correctly on the nozzle. Both surfaces are lapped optically flat to provide a good sealing area. Ball bearings or partial spherical surfaces are used to prevent the spindle affecting the angle of contact with the nozzle. The small one has a screwed bonnet, the large has a bolted bonnet.

Both valve types have screwed caps. Large valves may have a bolted cap. Test plugs/gags screwed through the cap are optional on both. In both designs the bonnet vent to the outlet portion of the body can be clearly seen. The small valve has no blowdown adjustment but is standard on the flanged valve. The different approaches to a secondary orifice or huddling chamber is obvious. The small screwed valve can be converted to flanged construction by adding screwed flanges or replacing the nozzle completely with an integral nozzle/flange for higher integrity. The small valve shows the screwed nozzle locked by

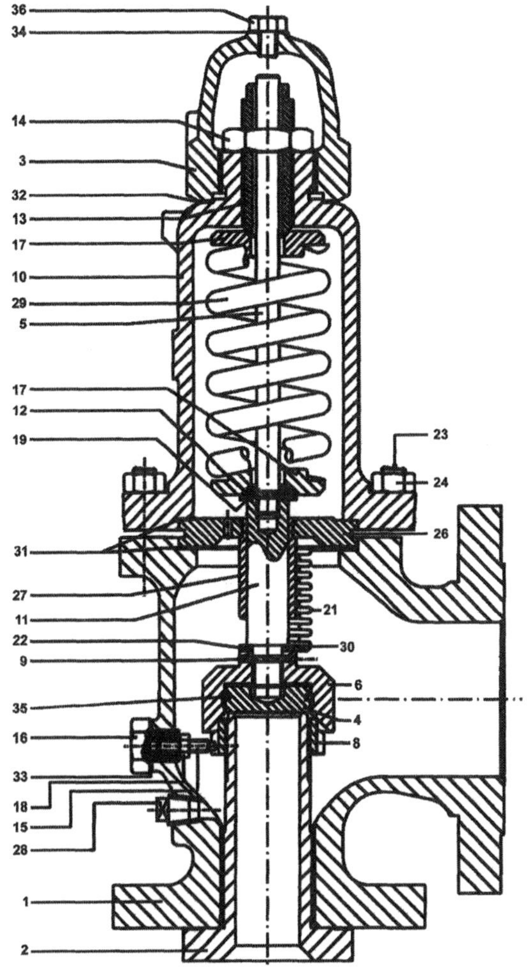

1 Body
2 Nozzle
3 Cap
4 Disc
5 Spindle
6 Disc holder
8 Blowdown ring
9 Pin
10 Bonnet
11 Stem
12 Collar
13 Adjusting screw
14 Locknut
15 Screwed pin
16 Clampscreen
17 Spring carrier
18 Locknut
19 Pin
20 Pin
21 Bellows
22 Washer
23 Stud
24 Stud
26 Guide flange
27 Guide tube
28 Drain plug
29 Spring
30 Joint
31 Joint
32 Joint
33 Joint
34 Joint
35 Circlip
36 Plug/gag

Figure 7.6 Typical flanged safety relief valve
Courtesy of Broady Flow Control Ltd

seal welding. Seal welding of screwed components is not universally acceptable and design changes may be necessary. Most safety relief valves have one outlet slightly larger than the inlet. Bodies with two outlets may be necessary to cope with some very compressible fluids or large differential pressures.

The larger flanged valve has a screwed blowdown ring on the outside of the nozzle. This ring is used to adjust the overpressure and the blowdown pressure. At one extreme the valve can attain high lift with low overpressure but the blowdown is large requiring the system pressure to be considerably reduced before the valve will reseat. At the other extreme, the valve only achieves full lift with maximum overpressure while the blowdown is short with the reseat pressure close to the set pressure.

Short blowdowns can sometimes result in the valve slamming shut. Final adjustment at site can usually achieve a happy medium. Some valve designs have two adjusting rings; one on the nozzle and one on the disc holder, see Figure 7.7. In these cases the adjustment on the disc holder is the primary adjustment. Adjustment of the nozzle ring should only be necessary after major overhauls when the nozzle has been remachined.

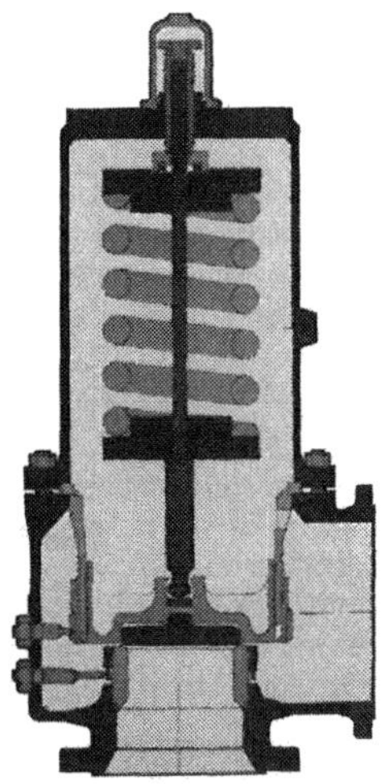

Figure 7.7 Semi-nozzle safety relief valve
Courtesy of Tyco Flow Control

Both designs could suffer from drainage problems. The large valve is not self-draining and a drain plug is provided. A piped drain may be appropriate at site. The small valve may be self draining depending upon the exact details of the body. Process condensation after blow-off could tend to accumulate in the body. The evaporation of the liquid can result in the build up of sediment on top of the nozzle threads. The evaporation process can lead to the concentration of corrosive elements with consequential local corrosion. If used for heavy gases, an accumulation could build up in the bottom of the body; this may be a problem with hazardous and flammable gases.

The type of construction shown does not comply with the spirit of the NACE standard MR-01-75 for sour service. Metallic materials in contact with hydrogen sulphide, H_2S, are liable to corrode and suffer sulphide stress cracking. Hardened materials, like the nozzle and disc, are more prone than soft materials. The spirit of NACE is to avoid hardened materials and crevices where sediment can accumulate. A screwed thread is effectively a long crevice. The problem of the crevice can be solved by fitting a seal to protect the thread. Socket weld construction would not be suitable for NACE applications. NACE gives guidance on suitable hardened materials and heat treatments.

On flanged valves the nozzle may be a clearance fit in the body rather than screwed. This is particularly convenient in situations when the nozzle may have to be replaced regularly due to corrosion or erosion damage.

Both valves shown have full nozzles. Semi-nozzle construction, see Figure 7.7, is used on valves when corrosion is considered unlikely. The seat is screwed into the body and is replaceable. These valves, usually of cast iron or bronze with cast iron/bronze trim, are used for low pressure steam, compressed air and clean gases. Liquid versions with soft seats are available. The semi-nozzle is usually held in place by tack welding or seal welding. Obviously changing nozzles in these valves is not quite as simple or fast. Some valve body designs include a drain hole, at a low point in the outlet connection, to prevent moisture and condensation accumulating inside the valve.

The standard valves shown are suitable for fluids which are not hazardous and which will not corrode the spring assembly or bonnet. The back pressure must be fairly constant because this acts on the unbalanced area of the disc and affects the set pressure. A fixed back pressure is compensated by adjusting the set pressure without any back pressure.

These problems are overcome by fitting a bellows which seals the bonnet from the process fluid. The diameter of the bellows at the disc/discholder is selected to balance back pressure effects. Valves with bellows are often called "Balanced bellows safety relief valves". The bellows seals the process fluid in the valve body allowing standard springs and bonnets to be used. Figure 7.6 shows the bellows construction on the right-hand side of the discholder. Bellows should be fitted when the back pressure is variable or over 10% of the set pressure. The bonnet can be sealed or vented to atmosphere. Bellows can be fitted to valves, with sealed bonnets, discharging to atmosphere for enhanced corrosion protection.

A balancing piston as well as a bellows are fitted by some manufacturers. A bellows can be highly stressed and subject to fatigue. If the bellows failed, the valve could malfunction depending upon the behaviour of the back pressure. By fitting a balancing piston the correct operation of the valve is ensured; possible leakage of product is the only problem.

Some standard designs have the bonnet vented to atmosphere. It is possible to balance the valve by adjusting the spindle diameter relative to the nozzle diameter. This alternative is simpler than bellows construction but, of course, does not have all the advantages and is unsuitable for hazardous fluids.

When the bonnet is vented to atmosphere, care must be taken to avoid the ingress of dust and moisture. A type of filtered trap should be fitted and maintained regularly. Valves for non-corrosive services frequently have carbon steel springs. These springs must be protected from the worst effects of the weather if a long reliable life is to be achieved.

A more recent valve design utilises a sealed bonnet which is intentionally partially pressurised when the valve opens. The bonnet pressure is used to assist in valve closure and is claimed to aid tight shut-off.

An alternative approach to bellows sealing is diaphragm sealing. Figure 7.8 shows a low pressure, up to 4 barg, diaphragm sealed spring-loaded valve suitable for air, gas and powdered solids conveyed by air/gas. A, is the diaphragm. Notice the valve construction is completely different to the other valves considered so far. The spring is totally enclosed and sealed. The bonnet is attached by a circlip, B. The valve outlet is not piped but relieves to atmosphere. It is obviously only suitable for safe gases and solids.

Most spring-loaded valves are fitted with set pressure adjustment. In general terms the valve can be adjusted to ±10% of the set pressure. At the highest pressure the spring is compressed the most and is subjected to the highest stresses when closed. The spring must have sufficient clearance to allow the valve to open fully without becoming coil bound. The static stresses in the spring must be below the creep range to reduce the possibility of relaxation.

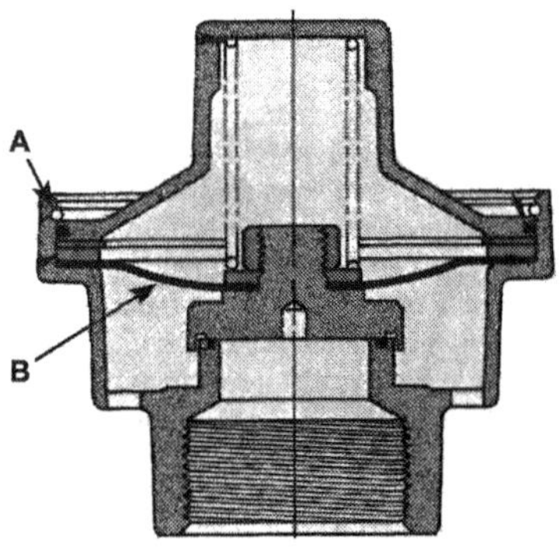

Figure 7.8 Low pressure diaphragm sealed safety relief valve

Valves are tested and set at the factory. The adjusting screw will be locked and wired and may have a lead seal. When overpressure and blowdown adjusting rings are incorporated these too may be wired. If the manufacturer has flow rig capabilities the valve setting can be very accurate. If the valve is set by "popping" without flow it may need adjustment at site when finally installed. Any adjustment at site must be carefully recorded.

Srvs are designed to have stable, repeatable characteristics. This does not mean that valves are capable of handling all flow rates from 1% to 100% of the rated capacity. Spring-loaded valves can suffer instability at low flows. Instability, caused by uneven or erratic flow patterns around the disc, can cause chatter or flutter resulting in, at best, rapid wear or worse damage to the disc and nozzle. If a valve is routinely required to protect against flows below 30% of the rated capacity then damping is essential by pneumatic lift cylinder, and a pilot-assisted or pilot-operated valve should be considered.

Instability can be caused by fluid friction losses in the inlet and outlet pipework. The outlet pipe effects can be alleviated by valve balancing, bellows or spindle. If the inlet pipe pressure drop is too high the valve will cycle causing either flutter or chatter. Short pipework on both inlet and outlet is preferred.

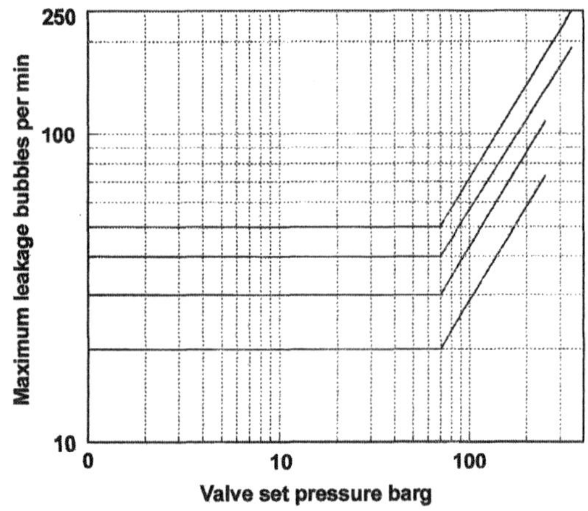

Figure 7.9 Safety relief valve leakage rates

Normal metal-to-metal seats will leak. API 527 has specified maximum leakage rates depending upon the nozzle size and the set pressure for commercial quality valves. Figure 7.9 shows the allowable leakage when tested with air at 90% of the set pressure. This test is conducted after the valve has been "popped" and the leakage is to atmospheric pressure.

The leakage rates, starting with the most demanding, are categorised as follows:

- conventional valves, G orifice and larger
- bellows valves, G orifice and larger
- conventional valves, F orifice and smaller
- bellows valves, F orifice and smaller

A bubble is defined as $0.018in^3$ or 0.295ml. Valves with resilient or soft seats are often quoted as bubble tight at 95% of set pressure.

Safety relief valves are primarily designed around standard nozzle sizes. This standardisation is carried further by fitting the nozzle into a standard body, that is, having fixed inlet and outlet connections for each orifice size. Table 7.1 indicates the connection sizes which can be expected for popular orifice sizes.

Orifice letter	Inlet size DN	Inlet rating barg	Outlet size DN	Outlet rating barg
D	25	20 to 100	50	20

Orifice letter	Inlet size DN	Inlet rating barg	Outlet size DN	Outlet rating barg
D	37	150 to 250	50	20 to 50
D	37	425	63	20 to 50
E	25	20 to 100	50	20
E	37	150 to 250	50	20 to 50
E	37	425	63	20 to 50
F	37	20 to 100	50	20
F	37	150 to 425	63	20 to 50
G	37	20 to 150	63	20 to 50
G	50	250 to 425	76	20 to 50
H	37	20	76	20
H	50	50 to 250	76	20 to 50
J	50	20	76	20
J	53	50 to 150	102	20 to 50
J	76	150 to 250	102	20 to 50
K	76	20 to 150	102	20
K	76	150 to 250	152	20 to 50
L	76	20 to 50	102	20
L	102	50 to 250	152	20 to 50
M	102	20 to 50	152	20
N	102	20 to 150	152	20
P	102	20 to 150	152	20
Q	152	20 to 100	203	20
R	152	20	203	20
R	152	50 to 100	254	20
T	203	20 to 50	254	20
V	254	20 to 50	355	20
W	305	20 to 50	2x 305	20

Table 7.1 Connection sizes for standard orifices

Notice the last valve listed has two outlet pipes. The numbers listed indicate a trend. The inlet pipe areas are larger than the orifice areas and the outlet pipe areas are much larger again. On large orifice valves the area ratios are not so impressive, see Table 7.2.

Orifice	Area cm^2	Inlet pipe	Pipe area cm^2	Inlet area ratio	Outlet pipe	Pipe area cm^2	Outlet area ratio
D	0.71	25	4.61	6.5	50	19	26.8
D	0.71	37	6.13	8.6	63	27.3	38.5
J	8.3	50	21.6	2.6	76	47.6	5.7
J	8.3	63	27.3	3.3	102	74.2	8.9
J	8.3	76	34.9	4.2	102	74.2	8.9

Table 7.2 Relief and safety valve inlet and outlet areas

The orifice areas selected were based on similar inlet and outlet flange ratings indicating similar expansion ratios for similar fluids. The Table shows that relatively smaller pipes are used for larger valves. The outlet pipes show a very marked reduction in area. This may indicate the need for larger outlet pipes than the valve outlet flange. At the very least, more care should be taken in designing inlet and outlet systems for larger valves. Back pressure and built-up back pressure may be more of a problem on larger valves, especially when discharging into closed systems with headers rather than to atmosphere. The outlet area problem can be alleviated by using a valve with two outlets. This course of action has an additional benefit; the axial thrust component of the outlet pipe reaction is balanced if mirror image pipework is used.

Larger valves have much larger moving components than small valves, such as the disc, spindle, spring and spring plate. The mass of these components may be very significant compared to the hydrodynamic forces generated by the fluid. When the moving mass becomes significant it is possible for inertia to result in overshoot. Without sufficient damping overshoot could result in flutter or chatter. The mass effects should be evaluated when the process fluid is "light", e.g. hydrogen, helium, methane, ammonia, steam and air. Perhaps with some fluids it is better to have two 50% valves rather than one 100% valve.

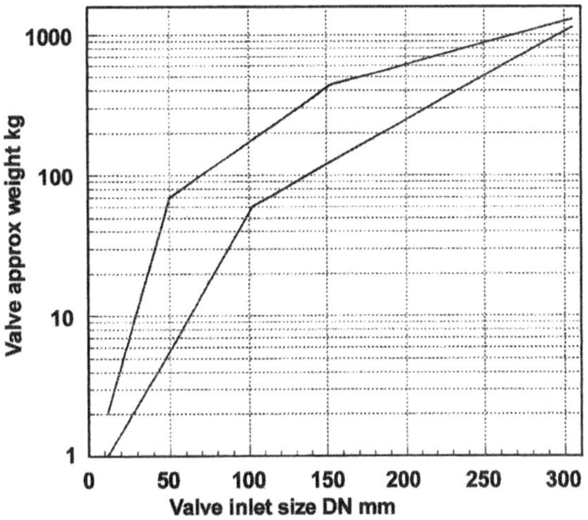

Figure 7.10 Safety relief valve approximate weights

Safety relief valves can be very heavy. Figure 7.10 indicates approximate weights based on the valve inlet connection size. The two lines show the difference between low pressure and high pressure valves. Optional extras, such as pneumatic cylinders, will increase the valve weight significantly. Lifting facilities for maintenance will be required for larger valves.

Safety relief valves are generally constructed with cast bodies. Machined from solid bar or from forgings is possible depending upon the size and the pressure ratings. Most standard designs anticipate the outlet pressure to be low, below 20barg or 50barg, irrespective of the inlet pressure. Flanged valves for high back pressures tend to be special. Some manufacturers use castings which produce flanges which are thicker than the specification requirements. This is good from the point of view of valve body strength and rigidity and being able to withstand angular pipe flange misalignment. However it does mean that special, extra long bolts or studbolts are essential.

Valves for arduous applications, high temperature superheated steam for example, can be supplied with butt weld pipe connections. Valves for hygienic applications can be supplied with various lug union connections. Most valve styles are manufactured in standard material combinations approximating to:

- cast iron with steel/bronze trim
- bronze with bronze/stainless steel trim
- steel with stainless steel trim
- stainless steel with stainless steel trim

NOTE: Some popular materials for trim belong to the 11-13% chrome steels group. These steels are not truly stainless but rust or corrosion resistant.

The bonnet and cap will generally be cast iron, bronze or steel unless operating conditions, corrosion or temperature, make them unsuitable.

Some manufacturers have special material combinations for specific applications such as steam. When NACE requirements are imposed, material changes may be necessary and hard facing may be applied to compensate for softer materials.

Ductile cast iron and steel bodied valves can be lined with corrosion resistant thermoplastic, PFA or PVDF, to broaden the range of applications. The nozzle and disc can be carbon-fibre reinforced PTFE, virgin PTFE or exotic metal alloys. To prevent corrosive fluid entering the spring case, a PTFE bellows is fitted. As well as providing an effective emission seal, this addition imparts the usually pressure balancing functions to render

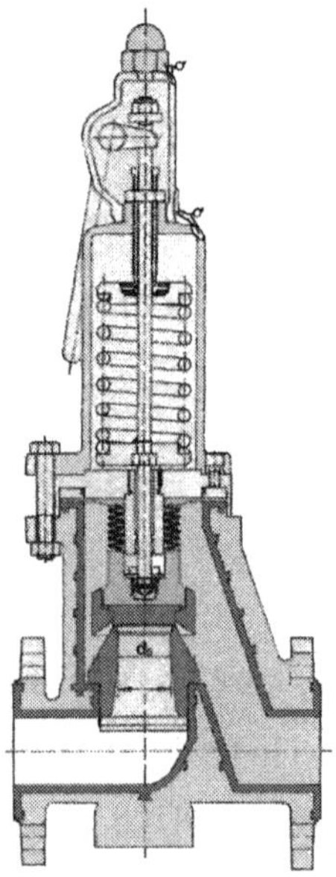

Figure 7.11 A lined metal-body spring-loaded relief valve
Courtesy of Richter Chemie-Technik GmbH

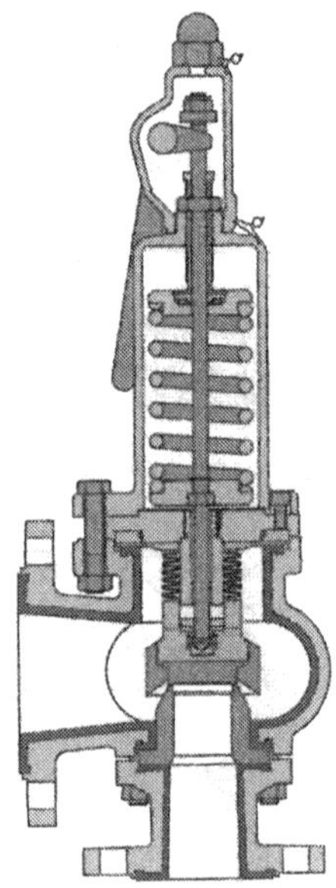

Figure 7.12 A lined metal-body spring-loaded relief valve
Courtesy of Richter Chemie-Technik GmbH

variable back pressures harmless. Valves of this style are manufactured in sizes between DN25 and DN100 and can be rated up to 13 barg for temperatures between -60° and 180°C. Figure 7.11 shows an unusual style for a pressure relief valve; in-line connections.

Notice particularly that there might be limited scope for gas/vapour expansion; the inlet and outlet connections are the same size. Figure 7.12 shows a more conventional valve design which is lined for improved corrosion protection; the scope for fluid expansion is obvious; DN100 inlets have DN150 outlets.

7.3.1 Spring-loaded valve optional extras

Standard spring-loaded valves may be fitted with various optional extras.

- **Lifting lever**

  Most valves can have a lifting lever fitted to the cap. This allows the valve to be opened manually for checking or to depressurise the system. Lifting levers can be open or enclosed with a packing box if the bonnet is pressurised with hazardous fluid. Levers can be adapted to accept chains or linkage to allow remote manual operation. Small mass-produced valves have open levers as standard.

- **Test gag**

  A screwed test gag can usually be fitted to the cap to allow the valve to be locked closed to allow system pressure testing.

- **Resilient seat or soft seat**

  Some valves can be fitted with a modified disc which incorporates an "O" ring or similar seal to reduce leakage when the valve is closed. These can also be useful for sealing against trace solids and coping with modest corrosion of the nozzle seating. Soft seats for flashing hot water applications are sometimes recommended.

- **Cooling spacer**

  For hot applications a cooling spacer can be fitted between the valve body and the bonnet. This feature allows the spring to operate at lower temperatures and removes the necessity for exotic spring materials and possible problems associated with creep and relaxation.

- **Limit switch**

  The cap can be modified to allow a limit switch to be fitted which detects the spindle movement. This enables remote electrical indication that the valve has opened.

- **Pressure switch**

  Valves with bellows and closed bonnets can have a pressure switch coupled to the bonnet. In the event of a bellows failure the pressure switch will register a rise in pressure. For valves handling safe fluids in continuously manned locations, a whistle can be fitted as an alternative.

- **Nozzle steam jacket**

  Some valve designs have protruding nozzles. In these cases a steam jacket can be fitted to the nozzle only, rather than to the complete valve body. This is essential when the fluid properties change dramatically on cooling or if condensation will occur.

- **Steam jacket**

  A steam jacket fitted around the complete valve body.

- **Pneumatic lift cylinder**

  The cap and spindle can be modified to allow a pneumatic cylinder to be fitted on top of the valve. The piston is directly connected to the valve spindle and allows the valve to be opened, irrespective of the process pressure, for testing or depressurisation. The pneumatic lift cylinder can be useful for preventing chatter and flutter under low flow conditions. The speed of valve opening and closing may be controlled by the port sizes or orifices in the pneumatic circuit. This option should not be confused with pilot assistance or supplementary loading.

For low pressure applications, standard spring-loaded valves can be set at pressures between 0.25 to 0.9 barg depending upon the nozzle area. Some low pressure valves can incorporate a vacuum break facility, see Figure 7.19. Valves can be fitted upside down, with the manufacturer's approval, for even lower settings. In some circumstances, the spring may be removed and replaced by a dead-weight. Dead-weight valves may have different operating characteristics to their spring-loaded counterparts; and data sheets should be read carefully.

Balanced bellows safety valves have higher minimum set pressures because the bellows increases the effective spring stiffness. Set pressures are in the range of 0.9 to 1.5 bar gauge.

Valves are available for the highest set pressures. Valves for over 425 barg and suitable for any fluid tend to be special unless very small. Small valves for safe fluids are available for over 2000 barg.

Low pressure protection for large vessels can be incorporated into personnel access hatches. Hatch covers can be spring-loaded and set for pressures up to 0.03 barg. Standard hatches, 508 mm and 610 mm, can be fitted with special sprung hinges and elastomer gaskets for sealing.

Spring-loaded safety relief valve for liquids

Valves for liquids are very similar to valves for gases and vapours and the previous valve type should be reviewed first. Valves for liquids are so similar that some manufacturers use the same valve for both applications. Valves for liquid applications being derated to 60 or 70% of the gas capacity. The aspect of significant difference in dedicated liquid valves is the shape of the nozzle and disc.

Figure 7.13 shows a standard gas/vapour design next to a liquid design. The standard gas/vapour design is flat. The liquid design has a raised ring added to increase the liquid momentum reaction and avoid the necessity of derating and increasing the overpressure to 25%. Some manufacturers use a shaped disc to replace the flat gas disc. Liquid valves use the same standard orifice areas as gas/vapour valves. Resilient or soft seats are available for liquid valves. Some designs use an "O" ring attached to the disc.

The bonnet arrangement is slightly different for liquids, it is always sealed from the liquid and usually vented to atmosphere. A sliding seal is fitted on the spindle where it passes through the guide. The seal can act as a friction damper to reduce the effects of chatter and flutter. The friction of the seal can be modified by making it sensitive to the pressure under the disc.

Figures 7.5 and 7.6 could both be adapted for liquid applications. Both designs could suffer from drainage problems. The large valve is not self draining and a drain plug is provided. A piped drain may be appropriate at site. The small valve may be self draining depending upon the exact details of the body. Liquid can tend to accumulate in the body if the valve leaks slightly. The evaporation of the liquid can result in the build up of sediment on top of the nozzle threads. The evaporation process can lead to the concentration of corrosive elements with consequential local corrosion. Provision is necessary for the removal of hazardous liquids from the valve body prior to inspection or maintenance.

For spring-loaded valves for gas, vapour and steam, the problem of fluid expansion was discussed in relation to the outlet pipe area; this problem is not automatically eliminated when considering liquids. **Liquids are compressible.** Liquids are not nearly as compressible as gases and vapours but some liquids are very compressible when compared to water. The following liquids should not be assumed incompressible unless the differential pressure across the valve is extremely small:

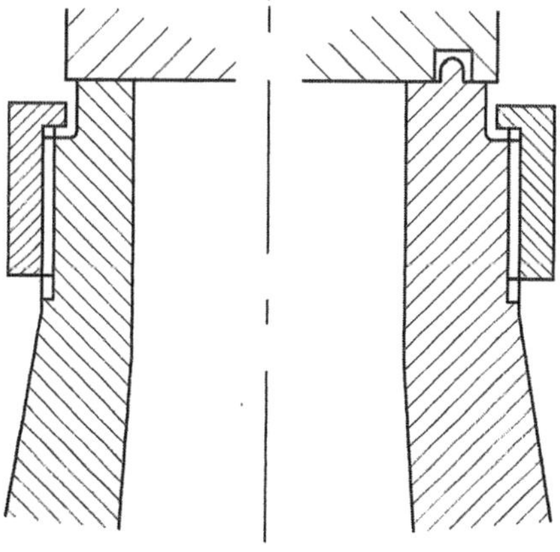

Figure 7.13 Nozzle and disc designs — Left: gas/vapour, right: liquid

- butane, carbon dioxide, cyclopropane, ethane, ethylene
- liquified hydrocarbon gases, LNG (liquified natural gas)
- LPG (liquified petroleum gas), methane, methyl amines
- ammonia, olefins

Liquids can present problems which are not immediately obvious, e.g. flashing to vapour and the release of dissolved/entrained gases. Flashing to vapour can be a problem with hot water and in boiler plants. The pressure in the outlet system is below the saturation pressure and the water is converted to steam. Liquified gases can present similar problems. Liquids which contain dissolved and/or entrained gases can have the outlet pipe choked because of the relatively greater expansion of the gas components. This is a prime example of knowing the exact composition of the process liquid/mixture when specifying equipment.

Another problem that can be associated with liquified gases and ammonia is rapid cooling and freezing. As the liquid changes phase to vapour, the latent heat of vaporisation must be supplied. The liquid temperature reduces locally so that heat transfer from the surroundings can supply the required heat. Cooling of the pipework and the valve itself can cause the moisture in the atmosphere to freeze and build up an ice deposit. As the cooling spacer, as mentioned in Section 7.3.1, can be fitted as a thermal barrier to protect the spring and bonnet. Some valves have no bonnet and external springs to stop the springs freezing up.

The problem of possible low temperatures due to rapid decompression of the system must be considered when evaluating the lower design temperature. Brittle fracture of metallic materials and hardening/crystallisation of elastomers may result in consequential damage unconnected with the primary fault.

Seat leakage is standardised by API 527 as before and the test medium is still air. Some manufacturers advocate the use of angled seats for valves used in conjunction with reciprocating pumps when pulsation dampers are not used to attenuate pressure pulsations.

Standard spring-loaded valves can be set at pressures between 0.25 to 0.9 barg depending upon the nozzle area. In some circumstances, the spring may be removed and replaced by a dead-weight. Balanced bellows safety valves have higher minimum set pressures because the bellows increases the effective spring stiffness. Set pressures are in the range of 0.9 to 1.5 barg. High pressure settings can be over 2000 barg for small valves reducing to 50 barg for larger flanged valves.

Spring-loaded safety relief valve for viscous liquids

Viscous liquids can pose operational problems for safety relief valves. After "popping" the liquid remaining on the seat forms a very good seal when the disc reseats. The sealing effect enhances the "wringing" tendencies of the flat surfaces which can make it very difficult to separate the two parts. This makes the valve opening at set pressure unreliable and leads to higher operating pressures plus surge and pressure waves. These problems are eliminated by creating two or more very small grooves across the nozzle face. The grooves allow process pressure to act over a larger area and increase the force available for the initial opening. The reliability obtained is not without cost; the valve leaks continuously when closed. The leakage is very small and does not usually require consideration.

Safety relief valves of this type are widely used in pressurised lube and seal oil systems. The valve outlet is piped back to the reservoir and a very small quantity recirculates continuously. These valves are only available up to DN50 and for pressures of 20barg. For process applications with viscous liquids use pilot-assistance or pilot-operated valves.

Spring-loaded safety relief valve with pilot assistance

Pilot assistance is almost like an optional extra because it is fitted to standard spring-loaded valves. A pneumatic diaphragm or piston is attached to the top of the valve spindle in a similar manner to a pneumatic lift cylinder, see Figure 7.14. The diaphragm or piston is double-acting because it applies force in both directions. When the valve is closed air is applied on top of the diaphragm/piston which complements the spring force holding the disc down. The process pressure is detected at two locations; under the valve disc as normal and at a point in the process pipework. The pressure in the process pipework is relayed to a sensor which controls the pneumatic power. With a conventional spring-loaded valve the sealing pressure between the disc and the nozzle reduces gradually as the process pressure increases until set pressure is reached when the process fluid pressure equalises the spring force. However, with pilot assistance the pneumatic load is available to supply additional sealing pressure. The pilot-assisted valve can therefore operate closer to the set pressure and still maintain good seat leakage.

When the pressure sensor detects the set pressure in the process pipework the air pressure is removed from the top of the diaphragm/piston and the safety relief valve is allowed to open. After a predetermined period the air is applied to the bottom of the diaphragm/piston and the valve is opened wide. Because the valve is not relying on hydrodynamic forces for lift, the valve can be oversized to achieve very low overpressures. The pneumatic control system can be adjusted to hold the safety relief valve open for a preset time after the process pressure has decayed to the required blowdown. Because the pneumatic diaphragm/piston is double acting the blowdown can be shortened to very small values by closing the valve pneumatically.

Pilot-assisted valves can work successfully with set pressures at +2%, 1 or 2% overpressure and 0% blowdown, see Figure 7.1. Instability, chatter or flutter is not a problem as the pneumatics operate between closed and 100% lift. The speed of opening and closing can be controlled by the port sizes in the pneumatic system. Safety relief valve problems created by the inlet piping can be solved completely by sensing the process pressure in the system rather than in the inlet pipework. The pressure sensing for the pneumatic controller can be placed anywhere.

Pilot-assisted valves can be used in applications when the process fluid properties vary over a wide range and low flow may be possible. The pneumatics rely only on the process pressure. The pilot system can also be triggered remotely irrespective of the process pressure to allow system depressurisation.

The great advantage of pilot-assisted valves is reliability. Even if the pneumatics fail the valve will operate as a conventional spring valve; albeit with wider operating pressures. The reliability of the pneumatic system can be increased by installing a back-up pneumatic controller. The pneumatic system does not have to be air powered; nitrogen and similar clean gases can be used providing a supply at about 3 barg is available. It may be possible to utilise natural gas in those remote locations with few facilities. Very little gas is used.

Because pilot-assistance is effectively a "bolt-on-option" the safety relief valve is available in all the construction and material variations of a conventional valve.

Spring-loaded safety relief valve with supplementary loading

Supplementary loading, also called servo loading, is an optional extra which is fitted to conventional spring-loaded valves, see Figure 7.15. Supplementary loading is identical, in principle, to a pneumatic lift cylinder except it works automatically in the opposite direction. The single-acting diaphragm/piston is pressurised on top and holds the valve closed providing additional sealing pressure. As with pilot-assisted valves, the supplementary loading allows the valve to work well with the set pressure close to the normal operating pressure. When the pneumatic controller senses the process pressure has reached the set pressure, the air is released from above the diaphragm/piston and the valve operates as a conventional spring-loaded valve. The valve continues to operate "conventionally" until it reseats after blowdown. When the pneumatic

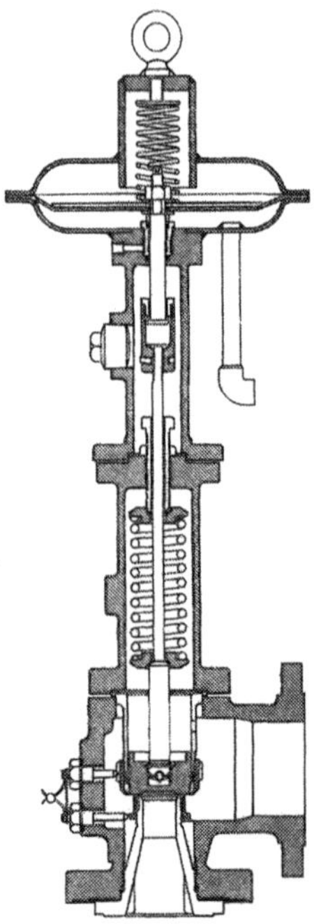

Figure 7.14 Spring-loaded safety relief valve with pilot assistance
Courtesy of Tyco Flow Control

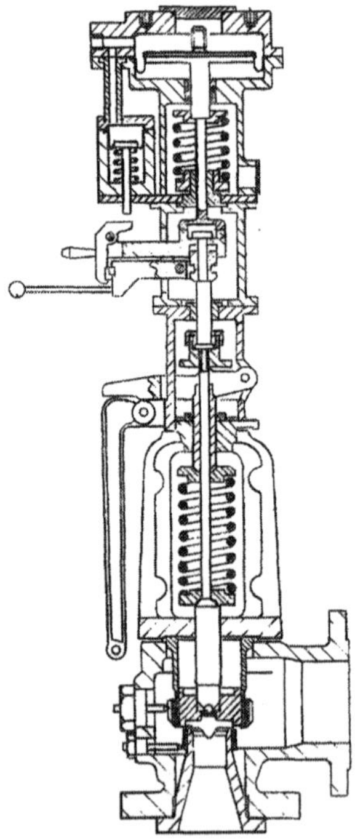

Figure 7.15 A spring-loaded valve with supplementary loading
Courtesy of Tyco Flow Control

controller senses the process pressure has reduced to the blowdown pressure the air is re-applied to the top of the diaphragm/piston.

The supplementary loading spindle can be coupled to the valve spindle via a quick-release fail-safe mechanism. In the event of pneumatic failure the supplementary loading device is automatically uncoupled from the valve and allows the valve to operate "conventionally" unhindered.

Supplementary loading can be used in those applications when operation close to set pressure is essential but normal overpressure and blowdown pressures are acceptable. Supplementary loading is not as complex as pilot assistance but it is not as useful.

Process pilot-operated safety relief valve

Process pilot-operated safety relief valves were first developed for nuclear applications; this means the basic reliability of the valves is assured. Valves for gases and vapours were developed first, the liquid valves followed later. pilot-operated safety valves have more in common with cage-trim control valves, see Chapter 6, Section 6.3.1, than with conventional spring operated relief valves.

The pilot-operated valve consists of two assemblies: the main valve and the pilot valve, see Figure 7.16. The main valve is the safety valve, which is piped into the process pipework, and the pilot valve is effectively an actuator. Pilot-operated valves are suitable for liquids and gases/vapours; variations of pilot valve design are fitted to suit fluid and operating conditions.

Pilot-operated valves function differently to spring-loaded valves. The nozzle is sealed by a piston which is loaded by the process pressure. The differential area between the nozzle bore and the back of the piston creates an unbalanced force which holds the piston down on the nozzle. As the process pressure increases the unbalanced force increases which increases the sealing pressure on the nozzle face. This is exactly the opposite condition to spring-loaded valves which lose sealing pressure as the set pressure is approached.

Pilot-operated valves can operate extremely close to the set pressure, similar to pilot-assisted and supplementary loaded valves, without tending to leak. Most pilot-operated valves have resilient seats to further reduce seat leakage. When the pilot valve senses the set pressure, normally in the nozzle, the pilot vents the process fluid above the piston and the fluid forces from the nozzle bore open the valve.

These valves can be either pop-action or modulating so the main valve either opens fully or partially to match the flow requirements. Pop-action may be preferred for pure safety reasons, modulating causes less problems with pressure waves. Modulating can also waste less product. When the pilot detects the system pressure has decayed to the blowdown value process fluid is re-applied to the top of piston and the valve closes.

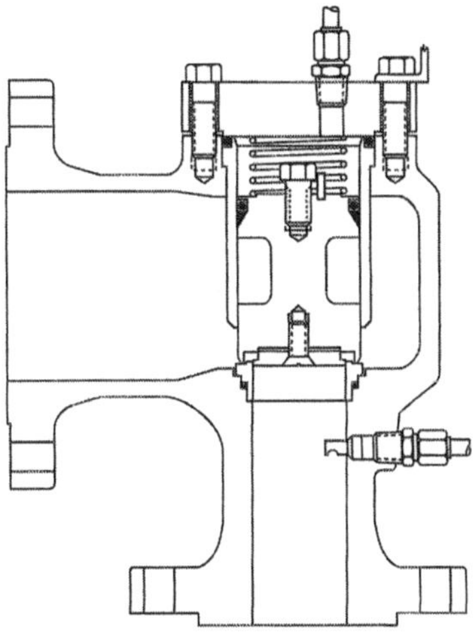

Figure 7.16 Process pilot-operated safety relief valve
Courtesy of Tyco Flow Control

Pilot-operated valves can operate successfully with close set pressures and very small overpressure and blowdown values. They are insensitive to low flow instability; some sizes have friction damping on the piston. The friction damping does mean that valves require a minimum differential pressure to operate correctly, of the order of 1 bar to 40 bar depending upon the size. These valves are good for applications when the fluid properties vary over a wide range and low flow may be possible. The problems associated with inlet pipes can be solved by remote pressure sensing in the process pipework. Variable back pressure and superimposed back pressure are not a problem until greater than the inlet pressure. Pilot-operated valves can be much smaller than conventional valves and space saving can be important with packaged equipment.

All the main valve components are assembled through a top blank flange. The disc and spindle are replaced by a piston, in some cases with a replaceable sealing face, which slides in a cage. The cage style of construction ensures very good alignment of the piston with the nozzle which promotes low seat leakage. The cage is sometimes called a "liner", even though it is not a liner in the usual sense of a sleeve being supported by a heavier cylinder. The cage may be full depth to the top of the nozzle or it may be shortened to provide maximum flow area. When full depth cages are used the nozzle is clamped in the body by the top flange. An "O" ring may be fitted to prevent leakage. This is usually a face seal which allows metal "O" rings to be used for wider temperature and pressure capabilities. When the cage is shortened the nozzle must be attached to the body independently.

Some manufacturers use press fits, and at least one expands the nozzle into the body. A special tool may be required to remove press fit nozzles. Nozzles which are expanded into the body must be returned to the factory for replacement. The nozzle should only require replacement after long service or damage which cannot be "lapped" or machined out in situ. The body is a fairly simple component and usually cast but can be machined from a solid forging for special materials or highest integrity.

Pilot valves are built in several versions to suit different applications. Two distinct design approaches are used:

- flowing - a very small quantity of process fluid passes through continuously and must be exhausted to atmosphere or a low pressure system
- non flowing - a non flowing pilot is a static system, only the fluid pressurising the piston is lost when the main valve opens

The pilot valve is a spring-loaded spool valve which includes the adjustment for set pressure and blowdown. Modulating pilot valves have an inherent overpressure of about 5%. Some include a diaphragm to protect the spring from the process fluid.

Manufacturers have selected two different approaches for varying the size of the main valve. When a full length cage is used and the nozzle is clamped in place a range of nozzle areas can be fitted in a body. In some instances, four nozzle sizes can be fitted to a single body. Special nozzles can easily be made for a particular application. When the nozzle is expanded into the body, changing nozzle areas is impractical. The effective nozzle area is controlled by restricting the lift of the piston. Normally the controlling orifice is the bore of the nozzle. By restricting the piston lift the controlling orifice becomes the spill area formed by the nozzle bore circumference and the lift. The piston lift can be restricted by fitting an adjustable stop or by machining

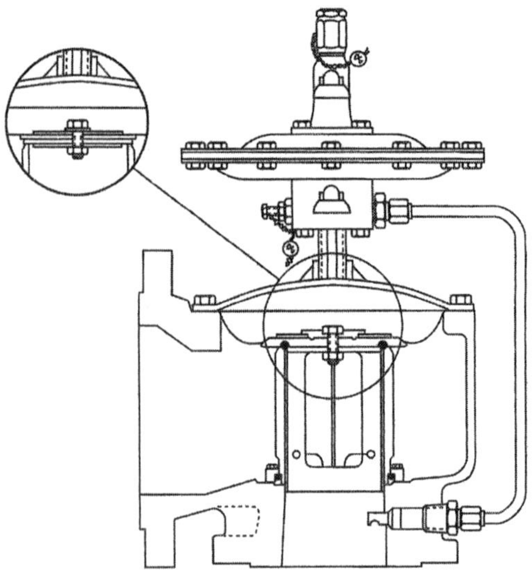

Figure 7.17 Typical low pressure process pilot-operated safety relief valve
Courtesy of Tyco Flow Control

pistons to different lengths. Remachining components at site is not always a good idea or physically possible. Stop adjustments are very easy and quick but must be made by experienced personnel and fully recorded.

A special type of pilot-operated valve is available for low pressure applications, see Figure 7.17. The piston in the main valve is replaced by a disc with a diaphragm. A diaphragm is used to create a larger area differential ratio between the nozzle area and the pilot pressure area holding the valve closed. Diaphragms are used in the pilot, also to magnify the pressure effects. These valves are available for set pressure between 0.07 to 10 barg.

The bellows main valve is a variation on the diaphragm main valve. A stainless steel bellows replaces the diaphragm for sealing the pilot circuit. The stainless steel bellows may have a wider temperature range than the diaphragm. Lower set pressure can be down to 0.035 barg.

Some pilot-operated valves can have back pressure problems. When the back pressure is less than the valve inlet pressure there are no problems, valve lift is unaffected. If the back pressure increases to values greater than the inlet pressure there is a possibility the main valve will open. If the process pressures can change to produce these conditions a back flow preventer can be fitted, sensing the back pressure, and preventing the valve from opening.

As the system pressure is sensed via external pipework the sensing point does not necessarily have to be directly under the main valve body. When very short blowdowns are necessary remote sensing can improve repeatability.

Process pilot-operated safety relief valves are capable of very high flow rates. Some manufacturers claim capacities double the ASME formula rating. The high flow rates can produce very large reaction forces and these must be evaluated to ensure the valve is correctly supported. The high flow rates can also result in fluid expansion problems as explained in the spring-loaded valve sections. Both problems can be solved by using valves with twin outlet.

As most main valves have resilient seats and piston seals, and the pilot valve has resilient seals, there are some temperature limitations not experienced by all-metal conventional valves. Even the best elastomers are limited to about 300°C. The nozzle/piston seal could be simplified to metal-to-metal but this would reduce the benefits of the superior sealing pressure. Resilient seats using "O" rings could possibly be replaced by metal "O" rings. Piston seals in graphite compounds would increase the temperature limit to that of the body material.

A non-flowing pilot valve would not be subjected to the process temperature and a cooling coil in the sensing pipe would give additional protection. The pilot valve may be susceptible to problems when the fluid changes character or deposits solids or wax. The pilot valve must be evaluated, as well as the main valve, for all possible fluid conditions.

Valves are available in sizes from "D" orifice with DN25 inlets up to DN200 inlet, full bore, with two DN250 outlets. Standard materials are carbon steel and stainless steel although main valves can be produced in many materials at extra cost and extended delivery. The piston and cage may be plated to reduce galling. Low pressure valves are available with orifice areas up to 580cm^2.

7.3.2 Pilot valve optional extras

The pilot system can be fitted with various options which can be combined to provide all the functions required.

- **Pilot filter**

  The pilot system can be protected by a filter in the sensing pipe. Not only can solids be removed, but in gas systems liquid droplets can be coalesced and removed.

- **Remote actuator**

  The pilot system can be fitted with a valve or valves which trigger the pilot actuation independent of the process pressure. Any standard isolating valve, compatible with the fluid and operating conditions, can be used and initiated by the most convenient means. This facility can be useful in pre-empting emergencies or depressurising the system.

- **Local test button**

  The pilot valve can be fitted with a push button which allows local triggering of the pilot system irrespective of the process pressure. This is useful for checking the operation of the main valve.

- **Pressure setting test connection**

  A connection can be provided into the pilot system so that an external pressure source can be applied to the pilot. When the external pressure equals the set pressure the main valve will open irrespective of the process pressure. The set pressure can be confirmed on-line.

- **Pilot signal pulsation damper**

  For gas applications only, a pulsation damper can be fitted in the pilot pressure sensing pipe to reduce the chances of spurious valve trips due to pressure waves.

- **Remote indicator**

  A differential pressure switch can be connected across the main valve inlet and actuating pressure above the piston. When the main valve opens a large differential is created and the switch can activate a remote indicator.

- **Multiple settings**

  Some pilots can be fitted with auxiliary spools and diaphragms preset for different set pressures. The main valve setting is adjusted by transferring the pressure sensing pipe to a different pilot.

Shear pin safety valve

All the valves described previously open and close automatically. When the process pressure decays to a predetermined value the valve closes. The shear pin safety valve is different in that it does not close after opening. Valve construction is similar to conventional spring-loaded valves but the spindle does not have provision for a spring. The spindle slides in a sleeve, both are cross drilled to accept a shear pin. The material strength and cross-sectional area of the shear pin dictate the set pressure. When the system pressure reaches the set pressure the fluid force exerted on the disc equals the shear force for failure of the shear pin. The pin fails and the valve opens wide. The valve remains open until the system pressure is reduced to atmospheric.

Because the basic principle of operation is very similar to conventional spring-loaded safety relief valves the shear pin version is subject to set pressure variations due to changing values of back pressure and superimposed back pressure. The valve can be balanced by adjusting the spindle diameter.

Some shear pin valves are designed to use nails as the shear pin. Obviously the reliability of the set point is control by the quality of the nail. Nails are not usually sold with guaranteed mechanical properties so their use may be limited. Shear pins specifically manufactured for a valve are preferred. Some valve designs allow the use of more than one shear pin simultaneously. This can lead to some confusion and this particular type should only be used with extreme caution.

Shear pin valves are available in all common material combinations and can easily be made in special materials.

Buckling pin safety valve

The buckling pin safety valve is similar to the shear pin valve in that it opens but does not close automatically. The valve is held closed by a strut which is loaded in compression by the fluid pressure. The valve opens when the compressive load in the strut reaches a critical value. The critical value is dependent upon:

- the slenderness ratio
- the mechanical properties of the strut
- how the load is applied at the ends

Euler proposed the following formula to calculate the critical load for struts:

$$P_E = \frac{C\pi^2 EI}{l^2} \qquad \text{Equ 7.1}$$

where:

C = constant non-dimensional

P_E = Euler critical load (N)

E = Young's Modulus of Elasticity (MPa)

I = minimum second moment of area (mm^4)

l = length (mm)

The constant takes various values between 0.25 and 4.0 depending upon the exact nature of the end fixings. Safety valves use clamped ends for the strut, a constant value of about 4 would be appropriate. The heat treatment, length and diameter of the strut being varied to achieve a specific set pressure. Factory settings have a tolerance between ± 3% and ± 9%. A typical buckling pin safety valve is shown in Figure 7.18.

The buckling pin valve is slightly different to all the other safety valves. Sealing is not accomplished at a pair of flat faces, or faces at 45°. A seal is made in the bore of the inlet pipe by an "O" ring in a piston. When the buckling pin fails the piston is pushed out of the inlet pipe and opens the relief path. The pin is designed to bend on failure and hold the valve open. Replacement pins are stored in the studs which are used to attach the top flange. Because the seal is made in the bore of the inlet pipe with a resilient seal the buckling pin safety valve is bubble-tight up to set pressure.

As with conventional valves, variable back pressure affects the set pressure. Balanced piston designs are available to eliminate back pressure problems.

The buckling pin concept has been applied to many valve styles including:

Figure 7.18 A buckling pin safety valve
Courtesy of BS & B Safety Systems (UK) Ltd

- inline open outlet
- inline closed outlet
- double-beat
- angle
- oblique

The concept of buckling pin technology has been accept by the ASME VIII pressure vessel code for primary pressure protection. When the valve opens it does so very quickly. The effect of resulting pressure waves should be investigated for repercussions. Valves are available with threaded connections and flanged from DN3 to over DN1 200 with set pressures from 0.006 to over 3000 barg.

Thermal relief valve

Fluid systems, which normally operate at pressures below the design pressure and have no mechanical means to increase the pressure to unacceptable values, may have problems in isolated pipe runs due to external heat sources. Isolating valves may be fitted at both ends of a pipe run for safety and/or maintenance purposes. Fluid trapped in the pipe between the valves may need to expand due to relatively small temperature rises caused by changes in ambient temperature or direct sunlight. Fluid is released at the design temperature to avoid code violations.

Thermal relief valves are small versions of spring-loaded srvs. Due to their size these valves nearly always are of screwed construction and have screwed connections. Special bolted construction valves with flanged or clamped connections should be used for hazardous fluids. Valves for thermal relief are manufactured in carbon steel and stainless steel as standard. Other materials can be supplied to order.

Vacuum break valve

Vacuum break valves are used in low pressure systems which may possibly be subject to partial vacuums. Low pressure systems can be fabricated from thin material. Under normal working conditions the material is subject to low tensile stresses which help to maintain the shape of circular vessels and ducts. Vessels and ducts which consist of flat panels may require strengthening by using ties to opposite panels. In general, ma-

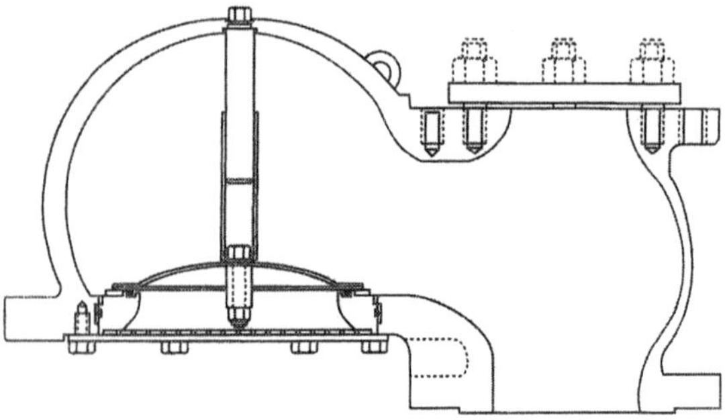

Figure 7.19 A typical vacuum break valve
Courtesy of Tyco Flow Control

terial stresses are low because deflection and deformation are the prime criteria. If any operating conditions exist which allow the low positive internal pressure to decay to sub-atmospheric values the structural integrity of the pressure containment can be seriously compromised.

Thin circular structures are very efficient at containing low internal pressures. The internal pressure tends to stretch the material evenly and maintains the shape. When an external pressure is applied the structure can become unstable. The external pressure tries to crush the vessel and may cause local buckling. The vessel fails due to elastic instability. Failure may be due to the radial pressure on the shell or axial load on the ends.

Flat panels can be braced by ties. When a partial vacuum occurs the tie becomes a strut in compression and can buckle at very low loads. A tie can support very high tensile loads without structural problems. The compressive capacity of a strut is very dependent upon the end conditions and whether the strut is straight or slightly bent. Any horizontal tie will become a bent strut when the loading is reversed. Bent struts are very poor in compression.

The vacuum break valve is a pressure relief valve which contains the atmosphere; the process fluid creates a back pressure on the outlet of the valve. When the back pressure reduces to a preset partial vacuum the valve opens and allows atmospheric pressure to enter and increase the internal pressure. Figure 7.19 shows the arrangement of a typical vacuum break valve.

Because of the low pressures involved vacuum break valves can be of the dead-weight type, the weight of the disc controlling the set pressure. The valve must be mounted as shown to work correctly. Spring-loaded valves are also used. The valve shown has two seat seals, an elastomer "O" ring and a flat sponge rubber gasket. Because atmospheric air is drawn into the process system the valve is fitted with a wire mesh screen to prevent the ingress of solid contaminants. The valve has an extra connection, with a blank flange, which can be used to mount the pressure relief valve when fitted. One connection to the process can be used for two functions. The valve body can be modified to include both valves.

Under normal operating conditions the valve will be subject to a positive back pressure. The back pressure will be the design pressure from a pressure vessel point of view. Small valves, DN50, may be suitable for pressure up to 3.5barg, large valves, DN600, only 1.5barg. Standard materials include aluminium, carbon steel, stainless steel and glass reinforced plastic, GRP. Aluminium may not be suitable for some outdoor environments. Special painting techniques are required for aluminium due to the tough oxide layer. These techniques may not be suitable for the valve internal passages and a material upgrade may be the better option. GRP is resistant to a wide range of chemicals and is a good material selection for all pressure ranges.

Explosion relief valve

An explosion is a sudden violent increase in volume generally caused by the release of a large quantity of heat over a small volume. The violence of the volume increase creates shock waves which usually cause the damage. In engineering fluid systems it is possible to create a violent expansion of gas or vapour during fault conditions. Typical examples are pressurised air ducts for boilers, crank cases of large engines and tanks of oil-cooled transformers.

Boilers work more efficiently when the air/fuel ratio is accurately controlled. Air can be controlled accurately by using a blower to force the air into the boiler. To do this the boiler air pressure must be increased above atmospheric pressure. The pressure increase may be very small, millimetres of water, but an enclosed pressurised system is created. A combustion fault within the boiler can cause a "blowback". A large quantity of heat is injected into the inlet air ductwork causing a sudden pressure increase. The low pressure ductwork must be protected to avoid damage and possible injury to personnel.

Engine crankcases contain the crankshaft and bearings which convert the reciprocating piston motion into rotary motion. The crankcase may be completely isolated from the cylinders by oil scrapers, which are a type of stuffing box, or the cylinder bottoms may enter the top of the crankcase. The crankcase may also function as the oil reservoir. If a separate reservoir is used the bottom of the crankcase will still be covered by oil. When the engine is hot the crankcase will be full of mist consisting of oil droplets. Some fuel droplets or vapour may also be present.

For combustion to start a hot-spot must be present. If the hot-spot warms up gradually the chemistry of the crankcase contents locally may change to produce lower temperature combustion material. The hot-spot may be a tight bearing or a rub of materials not intended to be in contact. Poor piston sealing can lead to hot combustion gases being injected into the crankcase. When the crankcase contents start to burn there is not necessarily an explosion. If the mixture is weak, a slow burning fire will result. If the mixture is optimum, a rapid burning fire will produce a violent increase in pressure.

Oil-cooled transformers can be subject to rapid temperature rises during extreme fault conditions such as short circuit or lighting strikes. Arcing within the coils can cause the oil to vaporise and create a sudden increase in pressure.

If protection is to be afforded against shock waves a valve must react very quickly. Explosion relief valves are constructed with very light discs to reduce inertia to an absolute minimum. Figure 7.20 shows a standard valve assembly for engine crankcases. Sealing when closed is assured by an "O" ring. Explosion relief valves are designed with large flow passages to allow the free passage of gas and vapour. The relief valves are usually mounted on an access hatch or door. This allows easy access for inspection or maintenance without dismantling major assemblies.

A problem associated with explosion relief is the containment of flames. The relief valve shown is fitted with a flame trap which prevents any flames from escaping to the surroundings. Valves can be fitted with local visual indication to show the valve has operated. Also, micro-switches can be fitted to allow a remote signal to indicate valve operation.

Explosion relief valves for crankcases are manufactured in standard sizes from 71cm^2 to 706cm^2. Multiple valves being fitted to achieve the required relief area and to reduce the maximum flame path length. Valves for transformers are available with 137cm^2 nozzles.

Explosion relief protection can be provided by bursting discs. In systems with very large volumes protection can be provided by using spring-loaded hatches. Inspection or access hatches are fitted with spring-loaded clamps rather than screws or nuts and

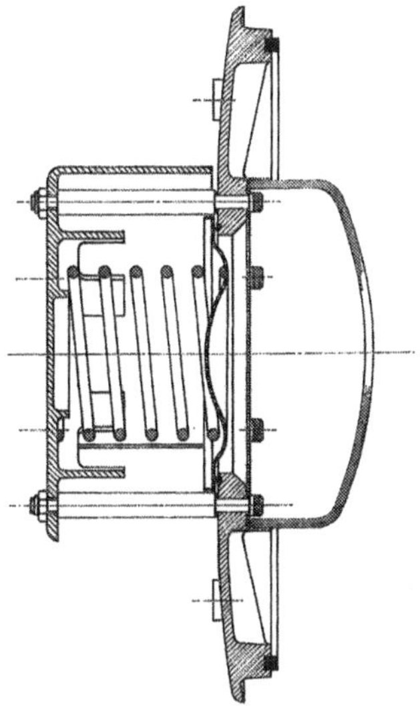

Figure 7.20 Typical engine crankcase explosion relief valve
Courtesy of Pyropress Engineering Co Ltd

bolts. When the set pressure is exceeded the hatch is blown clear of the mounting. Bursting discs and spring-loaded hatches do not reset after operation. Operator intervention is required before normal operation can resume.

Special Note: Bursting discs

Bursting discs, or rupture discs, are thin membranes which act like a blank flange. The membrane is very thin in comparison with the other pressure containing parts. It may be relieved locally to produce stress raisers which dictate the failure mode. On increasing pressure, the membrane fails and allows fluid to escape. Unlike most safety valves it cannot close when the excess pressure condition has past. Before failure bursting discs are completely leak free.

When handling the most hazardous fluids, bursting discs can be used in conjunction with safety valves to provide protection. The bursting disc can be fitted before or after the safety valve. Before the safety valve is best; normally the valve is not pressurised or subject to corrosion. During normal working the bursting disc provides a perfect seal. With excess pressure the disc fails and the safety valve relieves. On returning to normal conditions the safety valve reseats and seals. The bursting disc can be replaced at the earliest convenient opportunity.

Pressure instrumentation can be fitted to the pipework between the bursting disc and the safety valve. Valves normally leak slightly so the pressure under the safety valve will be equal to the safety valve back pressure. When the bursting disc fails the pressure will rise allowing a remote indication or alarm to be triggered.

For safety reasons bursting discs must not break up on failure. The disc must fail and establish a specific flow path for the fluid. All portions of the disc must remain attached to the mounting. When used in conjunction with a safety valve loose fragments must not be available to impede valve closure.

Bursting discs are available in a wide range of materials suitable for low and high pressures. The bursting disc, being considerably cheaper than a safety valve, is ideal for temporary installations such as test rigs.

Special Note: Dry solids applications

Very small solids at low concentrations can be sealed by resilient or soft seats. When the solid size is not small the resilient seat is unable to make a seal around individual particles. When the solids concentration is low the occasional particle travelling through the safety relief valve can not do much damage. When the solids concentration is significant many particles will pass across the nozzle/disc face, at high velocity, with the associated risk of erosive wear. After passing solids most valves will not close successfully. Solids are invariably trapped between the nozzle and the disc and wiredrawing ensues. Nozzle and disc life is short; the valve only seals until the first overpressure incident.

Low pressure powder applications, up to 4 barg, have been accommodated successfully by diaphragm sealed spring-loaded valves, see Figure 7.8. These valves are not adjustable and are preset at the factory. The system designer must make provision to collect the valve output.

Shear pin safety valves are very good for solids handling applications when the solids concentration and size are significant. Serious damage can occur when a spring-loaded valve tries to close with solids moving between the two sealing surfaces. The shear pin valve reduces the "velocity effect" damage by opening wide and remaining open. The nozzle and disc can be manufactured from precipitation hardening stainless steel or be coated with Stellite™ to minimise erosion damage.

The buckling pin safety valve is also very good for solids handling applications because it possesses almost identical qualities to the shear pin valve.

A special version of the full-bore diaphragm valve, described in Chapter 3, Section 3.3.1, can be used for safety relief. The diaphragm is not actuated by a normal threaded stem but is spring-loaded. The spring characteristics can be adjusted to provide the desired valve performance. Valves are available up ANSI 150lb rating which permits maximum pressures up to about 21.6 barg depending upon the temperature. These valves are available in the usual valve materials with a wide range of diaphragm elastomers. Natural rubber and Neoprene probably being the most abrasion resistant options. Valve bodies can be lined; glass may provide very good abrasion resistance.

The pinch valve, see Chapter 3, Section 3.3.1, can be arranged to provide safety relief features. If a pneumatically powered valve is fitted with a constant pressure air supply the valve will open when the solids pressure exceeds a preset value. The valve will close automatically when the solids pressure is reduced. Full-bore valves can handle large particles. The pinch valve could be an ideal processing solution.

Bursting discs can be used with dry solids but a provision for resetting must be considered.

Special Note: Non-metallic safety relief valves

Safety relief valves in uPVC, PP and PVDF are available in sizes from DN15 to DN100, see Figure 7.21. These valves are diaphragm sealed style for pressures up to 10 barg. The PVC valves can operate between 0°C and 50°C, the PP valves -10°C to 90°C and PVDF valves 0°C to 100°C. All valves can have screwed union connectors. UPVC valves can have solvent cement sockets and PP/PVDF can have fusion spigots. The valve design is slightly unusual in that the process flow is transported through the valve; the valve is not fitted to a branch which would normally be static. This style of construction removes the possibility of the most common causes of safety relief valve operational problems; inlet branch too small or too long.

Full-bore diaphragm valves, (Chapter 3, Section 3.3.1), are available in non-metallic materials:

- PVC, PVDF, PP, PVDF, ABS, CPVC.

Sizes range from DN15 to DN100 with pressure ratings of 6 and 10 barg. These valves could be fitted with spring-loaded stems or constant pressure pneumatic/hydraulic loading to perform safety relief valve functions.

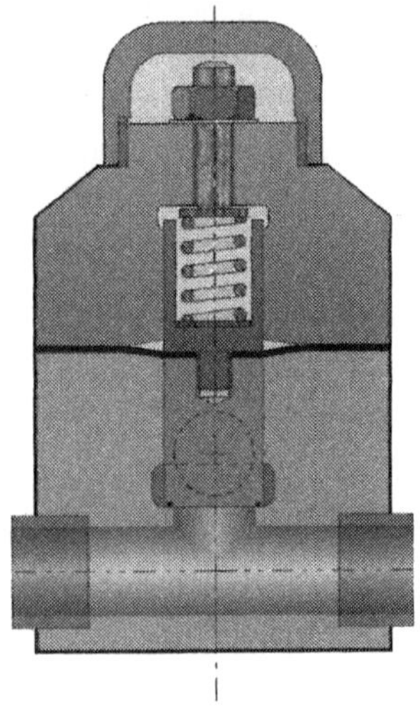

Figure 7.21 Typical non-metallic safety relief valve

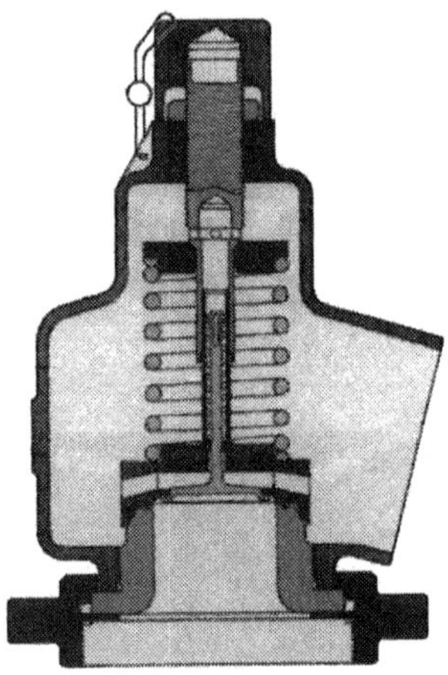

Figure 7.22 Dual function safety relief valve

The pinch valve can be considered as a non-metallic valve as all the process fluid contacting parts are elastomers. Pneumatically/hydraulically actuated valves can also function as safety relief valves.

Special Note: Hygienic applications

Spring-loaded pressure relief valves, similar in construction the the angled globe valve shown in Chapter 3, Figure 3.75, are manufactured in stainless steel. Valves are readily available for butt-welding to od pipework ¾" to 4". Working pressures range from 7.5 to 20 barg.

Figure 7.22 shows the construction of a low pressure valve which performs two important functions. The valve, up to DN100, can protect against overpressures to 8.5 barg. The inner inverted poppet valve is independently sprung and can protect against internal vacuums. The valve shown has a threaded lug nut for attachment to an approved connector.

Weir type and full-bore diaphragm valves, can be fitted with spring-loaded stems or constant pressure pneumatic/hydraulic loading to perform safety relief valve functions.

Pinch valves fitted with sleeves of approved materials and pneumatically/hydraulically actuated can function as safety relief valves.

7.4 Piping reactions

Emphasis has been placed in earlier Sections on the correct mounting of safety relief valves and the operational effects of relieving high flows. The following formulae can be used to evaluate the dynamic effects of flowing fluids.

A force will be created whenever the flowing fluid changes direction. Safety relief valves and their associated pipework operate at high velocities. Consideration should be given to using pipe sizes larger than indicated by the valve connections.

For a normal angle pattern safety relief valve, two basic forces will be applied to the valve when open:

- A vertical force, caused by the reaction of the disc, will tend to lift the valve
- A horizontal force, applied at the centre-line of the outlet connection, will try to push the valve away from the outlet pipe

The horizontal force will generate a moment at the inlet connection which must be restrained.

7.4.1 Gases and vapours

Physical data on gases and vapours is given in Chapter 8, Section 8.5.7.

To calculate the mass flow, the characteristic equation, given in Equation 7.2, can be used:

$$pV = q_m RT \qquad \text{Equ 7.2}$$

$$R = \frac{R_0}{M} \qquad \text{Equ 7.3}$$

where:

p	=	absolute pressure (Pa)
V	=	volume flowing (m^3/s)
q_m	=	mass flow (kg/s)
R	=	specific gas constant (kJ/kg K)
T	=	absolute temperature (K)
R_0	=	universal gas constant (kJ/kmol K)
M	=	molar mass (kg/kmol)

To calculate the velocity through the nozzle based on the gas/vapour volume at overpressure use:

$$v = \frac{10000V}{A} \qquad \text{Equ 7.4}$$

where:

v	=	velocity (m/s)
A	=	nozzle area (cm^2)

The vertical force, F_v in Newtons, can be calculated from:

$$F_v = 2\,v\,q_m \qquad \text{Equ 7.5}$$

For the horizontal force the volume flowing at back pressure conditions is required. Because the gas/vapour has expanded the temperature will be lower. The expansion takes place very quickly and adiabatic conditions can be assumed to apply.

The outlet temperature can be calculated from:

$$T_2 = T_1\left(\frac{p_2}{p_1}\right)^{\left(\frac{\gamma-1}{\gamma}\right)} \qquad \text{Equ 7.6}$$

where:

T_1	=	inlet absolute temperature (K)
T_2	=	outlet absolute temperature (K)

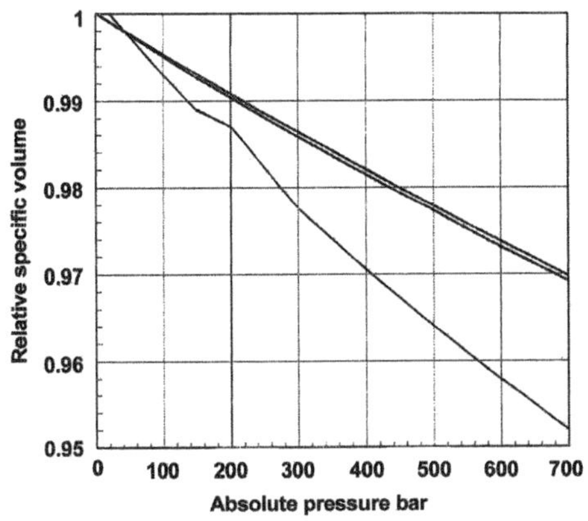

Figure 7.23 Relative specific volume for water at 10, 100 and 200°C

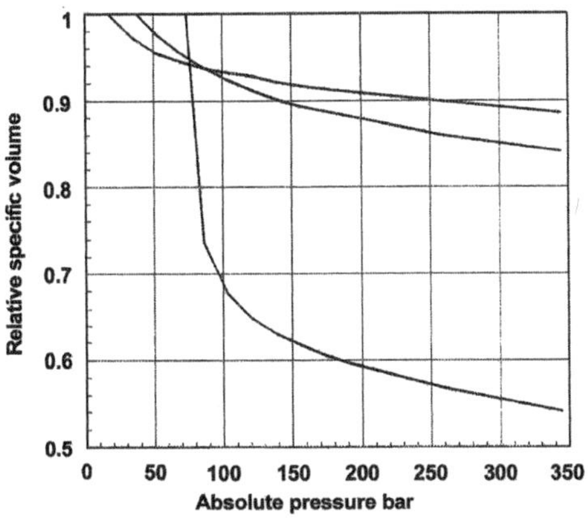

Figure 7.24 Relative specific volume for CO_2 at -29, 1.5 and 31°C

p_1 = inlet absulute pressure (Pa)

p_2 = outlet absolute pressure (Pa)

γ = ratio of specific heats (non-dimensional)

The volume of gas/vapour flowing at p_2/T_2 must be calculated. The velocity through the valve outlet connection can then be calculated.

The horizontal force, F_h in Newtons, can be calculated from:

$$F_h = v\, q_m \qquad \text{Equ 7.7}$$

7.4.2 Liquids

Physical data on liquids is given in Chapter 8, Section 8.5.8.

The approach with liquids is slightly different because the specific volume is not so easily correlated with pressure and temperature. Figure 7.23 shows the variation in specific volume with pressure for water.

Even at 200°C and a 700 bar differential pressure the volume changes by less than 5%. For liquids like water the compressibility effects can be disregarded when evaluating safety relief valves. The flowing volume and hence the velocities can easily be calculated and the forces from Equations 7.5 and 7.7.

At the other extreme the liquified gases are very compressible. Carbon dioxide is one of the most compressible liquids, see Figure 7.24.

A 5% change in volume is easily achieved. As the temperature approaches the critical temperature the liquid becomes even more compressible as can be seen by the curve for 31°C.

For the following liquids the change in specific volume with pressure cannot be ignored:

- butane, carbon dioxide, cyclopropane, ethane, ethylene
- liquified hydrocarbon gases, LNG (liquified natural gas)
- LPG (liquified petroleum gas), methane, methyl amines
- ammonia, olefins

Accurate information for the pressurised liquid phase must be available. The specific volumes at overpressure and back pressure must be evaluated so that accurate velocities can be calculated. The two forces can be calculated with Equations 7.5 and 7.7.

7.5 Useful references

High pressure trip systems for vessel protection, HG Lawley and TA Kletz, Chemical Engineering, May 1975, pp 81-88.

Control the ΔP on the PRV inlet line, Wing Y Wong, Chemical Engineering 2001, Vol. 108, no10, pp 129-132 (3 ref.) McGraw-Hill, New York, NY, USA.

Modeling of tower relief dynamics II, JR Cassata, S Dasgupta, SL Gandhi, Hydrocarbon Processing, 1993, vol. 72, no 11, pp 69-74, Gulf Publishing, USA.

Dynamic simulation improves column relief-load estimates, Hydrocarbon Processing, 1999, vol. 78, no 12, pp 81-86, Gulf Publishing, USA.

ANSI/API 527 (R2002),1991 Seat Tightness of Pressure Relief Valves.

American Society of Mechanical Engineers (ASME), Three Park Avenue, New York, NY, USA, Tel: 973 882 1167, www.asme.org.

European Commission, Pressure Equipment Directive (97/23/EC), (PED) dedicated website in English: http://ec.europa.eu/enterprise/pressure_equipment/ped/index_en.html

Valve and piping sizing 8

8.1 Sizing isolating valves

Isolating valves are usually the same size as the pipe; that is to say a DN 50 valve is fitted in a DN 50 pipeline. Valves have a seat area comparable to the pipe bore area, but globe, ball and plug valves are an exception. Globe valves can have particularly tortuous flow paths through the valve, inducing a significant pressure loss.

These valves can be fitted with reduced ports. This can be useful in particular to reduce costs in situations when the pressure drop is not important, for example in:

- vent and drain lines
- non-flowing instrument lines
- lines oversized for structural reasons

Some pipes require a clear unobstructed full bore, for example for cleaning or inspection by pig. The selection of valve type is then restricted and the minimum port size is defined.

In theory the pipe size can be selected by several different methods. In practice these methods typically require some modifications to cope with practical situations. One popular starting point for pipe sizing is to use velocities which have been tried and tested in particular industries or situations. Tables 8.15 and 8.16 list values for liquids, gases and vapours in some common applications.

Once a pipe size has been selected, the friction losses can be calculated, see Section 8.5.2, and reviewed for acceptability. In the case of reciprocating compressors and pumps, the pipe size may require additional analysis to evaluate the effect of flow variations and the consequential pressure pulsations. Pulsation dampers may be required to attenuate pulsations, or the pipe size may be increased to remove the need for dampers. Pulsation dampers do not remove all the pressure pulsations but only attenuate them to acceptable levels. In some situations the damper cannot attenuate sufficiently and other hardware changes are also necessary. Suction pipework for all pumps must be evaluated for the NPSHa or NPIPa; if insufficient is available for the selected pump, several options are available for modification such as pipe size/length, pump elevation, pulsation damper and pump speed.

When the friction losses are evaluated it may be found that the individual losses due to isolating valves form a large proportion of the overall loss. The isolating valve losses may be reduced by changing the valve type or manufacturer rather than increasing the pipe size. Figures 8.1, 8.2 and 8.3 show the range of available flow coefficient K_v, values for isolating valves. The variation in internal design from different manufacturers generally results in a broad range of available K_v values. If a decision has been made on a valve type which results in unacceptable losses the decision must be questioned.

The speed of valve operation may be an important factor for process control. This aspect must also be considered, including surge/water hammer implications, together with the valve size, when comparing valve and actuator types.

In this book, the float valve is considered as a special purpose valve within the isolating valve categories, see Chapter 3, Section 3.3.5. Float valves for clean liquids should be sized so that the maximum liquid velocity through the seat does not exceed 3m/s. If entrained solids are present then lower velocities should be considered and the trim materials improved.

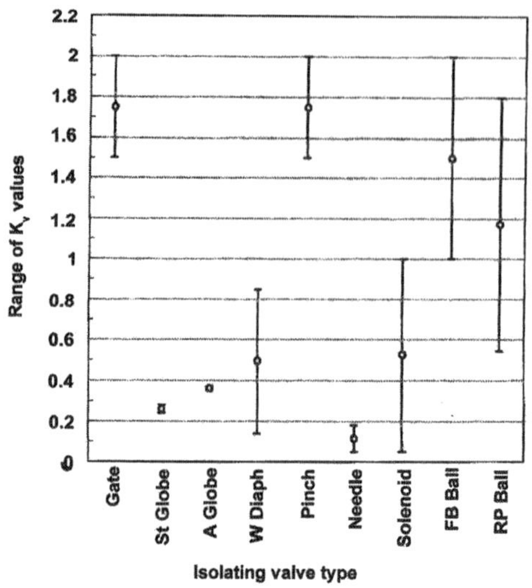

Figure 8.1 K_v values for DN8 isolating valves

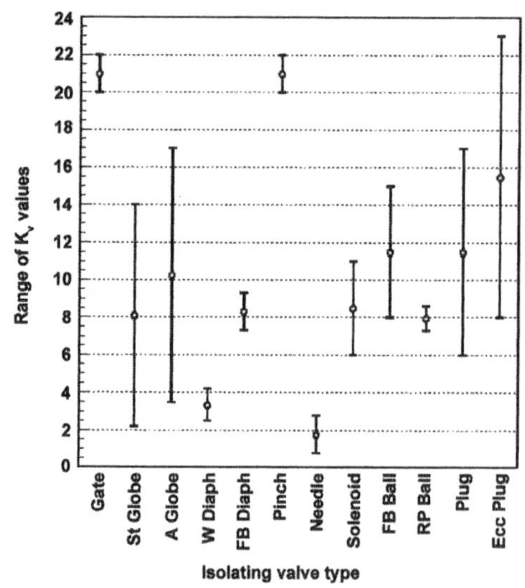

Figure 8.2 K_v values for DN25 isolating valves

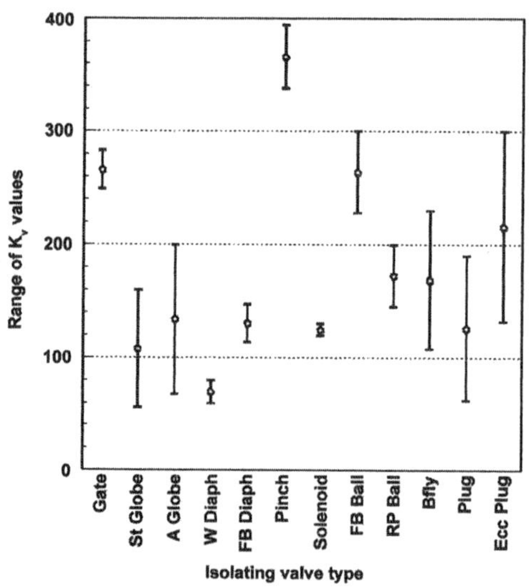

Figure 8.3 K_v values for DN100 isolating valves

8.2 Sizing non-return valves

Non-return valves, or nrvs, are very similar to isolating valves when considering sizing. A non-return valve of the same size as the pipe is usually fitted. If the pipe is to be cleaned or inspected by a pig the choice of non-return valve is restricted to the swing disc style of valve. Some swing disc valves have facilities to lock the valve wide open to allow unhindered passage of the pig.

There are considerable differences in the pressure losses created by the various styles of non-return valve. If the overall pressure loss within a system is too high, and the non-return valve is a significant contributor, then the manufacturer's actual data should be consulted rather than typical approximations. A change of non-return valve design may reduce losses considerably.

Non-return valves with springs can be adjusted to some extent. If the required flow is smaller than the maximum flow recommended by the manufacturer a slightly weaker spring will allow the valve to open wider at the required flow.

Figures 8.4, 8.5 and 8.6 show the K_v values for three popular sizes of non-return valve for different styles.

Notice how the relationship between K_v values vary as the size varies. It is not possible to make a policy decision to select one style to achieve the lowest friction losses for all sizes. The viscous correction based on Reynolds Number can be applied to non-return valves.

During the non-return valve sizing and type selection procedure the dynamic response of the valve **must** be considered. To select the valve correctly the true nature of the process system and the operating regime must be known. If the valve is to operate infrequently, for example with a pump which runs continuously except for a six month inspection, there should be little wear on the seating surfaces. Conversely, a valve fitted to a dosing pump injection system may open and close 100 times per minute. The rate of change of the flow rate is very important. Sudden changes in flow can cause the non-return valve to slam wide open or slam shut. Sudden changes in flow can be created in various ways:

- **compressor or pump starting** — run-up times are very short even on star-delta starting
- **compressor or pump power failure** — modern machines have comparatively little inertia and therefore stop very quickly when the power is removed
- **changing diverting or mixing valve position** — this can create a sudden increase and decrease simultaneously in two systems
- **pipe fracture** — the loss of process flow will be very abrupt

Many more fluid machines are being driven by variable frequency/speed ac motors. Normal starting and stopping can be controlled by changing the speed gradually. The problem of power failure, however, is not overcome.

If the fluid flow rate is going to change suddenly, or the risk of a sudden change is high, the non-return valve and system dynamics should be considered carefully. Specialist consultants can use analogue or digital techniques to analyse the system dynamic responses.

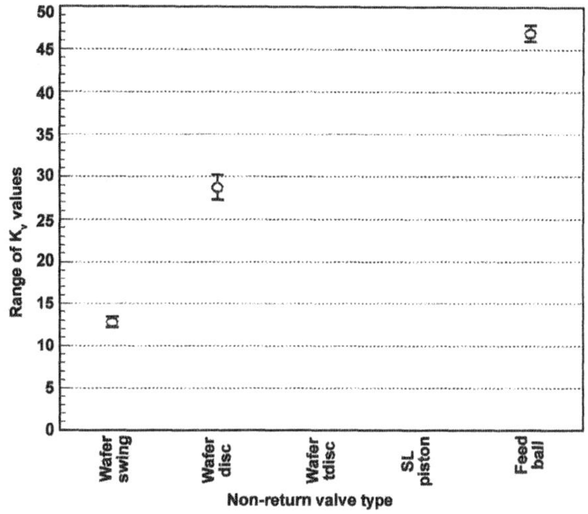

Figure 8.4 Comparative K_v values for DN40 non-return valves

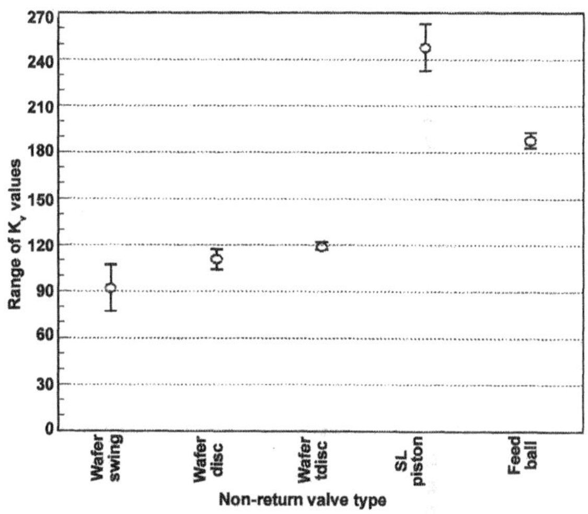

Figure 8.5 Comparative K_v values for DN80 non-return valves

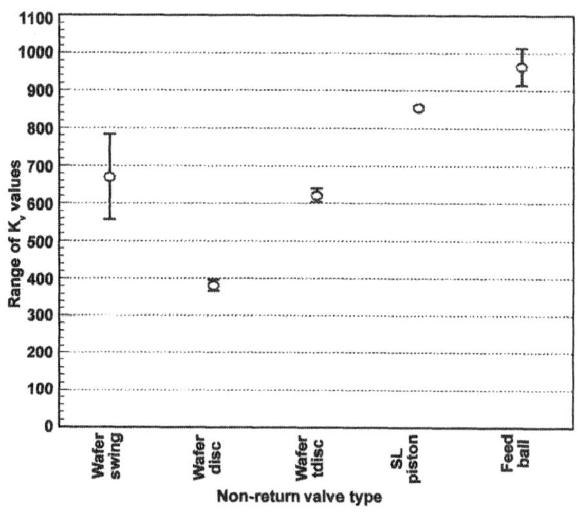

Figure 8.6 Comparative K_v values for DN150 non-return valves

8.3 Sizing safety relief valves

8.3.1 General

For oil and gas related processes API RP520 gives various formulae for calculating the required nozzle area.

The prospective purchaser is well advised to place the onus for correct sizing of the valve on the manufacturer by fully disclosing the physical conditions under which the valve is to operate. Dedicated slide rules and proprietary computer software are available for sizing valves. International organisations, such as API and ASME, and classification societies, such as Lloyd's Register, publish formulae and rules for sizing valves. In practice, to avoid guarantee problems, most safety valves are sized by the manufacturer and its agents.

The flow range over which a safety valve is expected to work is very important. This factor applies particularly to variable speed positive displacement pumps. When the pump flow varies by more than a factor of three, manufacturers should be consulted regarding the suitability of standard spring-loaded valves and whether a pilot-operated or air-assisted valve should be fitted to prevent low flow chatter and instability.

Safety valves are usually fitted on a tee in a process line. The pressure drop in the branch feeding the safety valve is very important. When closed, the safety valve experiences the static pressure in the process line less any static elevation differences. When the valve is open, friction losses and entry losses in the tee become important. The operating losses in the pipe leading to the safety valve should be small in comparison to the set pressure.

The flow to be relieved can be easy or difficult to define depending upon the process requirement. Fixed speed positive displacement compressors and pumps are the easiest. The flow rate varies only slightly over a wide pressure range. The required relieving capacity can be accurately calculated. Protection for other process applications can be difficult to quantify. Some typical, awkward operational conditions include:

- **control system malfunction** — predicting the results of all possible control system combinations can be an extremely long and complicated process
- **undesirable chemical reactions** — producing gas or vapour
- **fire hazards** — systems operating at much higher temperatures than normal

Valves for thermal expansion protection can be sized from the volume contained and the predicted rate of temperature rise.

The first step to be taken in sizing safety relief valves is to decide the set pressure and its relationship to the normal working pressure. The set pressure is generally 10% above the normal working pressure. If the normal working pressure is much lower than the pipework design pressure, the set pressure can be increased. The installed driver power of compressors and pumps should be checked.

The second step relates to overpressure. Liquid, gas and steam valves are generally set at 10% although the standard liquid equations calculate the size for a 25% over pressure. Pilot-operated and pneumatic-assisted valves can allow set pressure and overpressure to be very close to the normal operating pressure. There is much scope for adjustment depending upon the process requirements, pipework design pressure and the type of valve selected, see Figure 7.1 in Chapter 7. The absorbed power of compressors and pumps should be checked again once an overpressure is decided.

8.3.2 Sizing equations

Various sizing equations are available for gases and vapours, steam and liquids.

Equations for gases and vapour for 10% overpressure:

$$A = \frac{W}{P_1\ (C \times K) K_b \sqrt{\frac{M}{TZ}}} \quad \text{Equ 8.1}$$

$$A = \frac{W\sqrt{TZ}}{P_1\ (C \times K) K_b \sqrt{M}} \quad \text{Equ 8.2}$$

$$W = AP_1\ (C \times K) K_b \sqrt{\frac{M}{TZ}} \quad \text{Equ 8.3}$$

$$A = \frac{SCFM\sqrt{TGZ}}{1.175\ (C \times K) P_1 K_b} \quad \text{Equ 8.4}$$

$$SCFM = \frac{1.175\ (C \times K) P_1 K_b}{\sqrt{TGZ}} \quad \text{Equ 8.5}$$

where:

A	=	minimum nozzle area (in^2)
W	=	capacity (lb/h)
P_1	=	set press + over press + atm (lb/in^2 absolute)
(C ×K)	=	factor (non-dimensional)
K_b	=	back pressure correction (non-dimensional)
M	=	molecular weight (non-dimensional)
T	=	absolute temp (deg Rankine)
SCFM	=	capacity (standard ft^3/min)
G	=	relative density (non-dimensional)
Z	=	compressibility factor(non-dimensional)

Equations for steam:

$$A = \frac{W}{50 P_1 K_{sh} K_N H_b} \quad \text{Equ 8.6}$$

where:

A	=	minimum nozzle area (in^2)
W	=	capacity (lb/h)
P_1	=	set press + overpressure +atm (lb/in^2 absolute)
K_{sh}	=	superheat capacity correction (non-dimensional)
K_N	=	capacity correction (non-dimensional)
H_b	=	back pressure correction (non-dimensional)

(K_N is required by ASME for P_1 values above 1500 psia.)

Equations for liquid with standard trim:

$$A = \frac{Q\sqrt{G}}{24.32 K_p K_w K_v \sqrt{1.25p - p_b}} \quad \text{Equ 8.7}$$

where:

A	=	minimum nozzle area (in^2)
Q	=	capacity (USgpm)
G	=	relative density (non-dimensional)
K_p	=	overpressure correction (non-dimensional)
K_w	=	back pressure correction (non-dimensional)
K_v	=	viscous correction (non-dimensional)
p	=	set pressure (lb/in^2 gauge)
p_b	=	back pressure (lb/in^2 gauge)

Equations for liquid with liquid trim:

$$A = \frac{Q\sqrt{G}}{28.14 K_w K_v \sqrt{\Delta P}} \quad \text{Equ 8.8}$$

where:

A	=	minimum nozzle area (in^2)
Q	=	capacity (USgpm)
G	=	relative density (non-dimensional)
K_w	=	back pressure correction (non-dimensional)
K_v	=	viscous correction (non-dimensional)
Δp	=	differential pressure (lb/in^2)

The viscous correction factor can be applied to safety relief valves for approximate sizing.

The API RP520 equation for liquid valves is:

$$A = \frac{Q_L}{36.4\,K_p K_g K_v K_1 \sqrt{\Delta p}} \qquad \text{Equ 8.9}$$

where:

A = minimum nozzle area (in^2)

Q_L = capacity (USgpm)

Δp = differential pressure (lb/in^2)

K_p = overpressure correction (non-dimensional)

K_g = relative density correction (non-dimensional)

K_v = viscous correction (non-dimensional)

K_1 = back pressure correction (non-dimensional)

API RP520 contains procedures for the sizing of relieving devices for insulated vessels which may be subject to external heating. The values given for the relevant factors are not universally accepted. Advice should be sort when sizing valves or bursting discs for this application.

The ASME VIII Design Standard gives the following equation for gases and vapour, including a 0.9 safety factor:

$$A = \frac{Q}{0.852 C P K_2 K_3}\sqrt{\frac{TZ}{M}} \qquad \text{Equ 8.10}$$

where:

A = minimum nozzle area (in^2)

Q = capacity (lb/h)

C = gas constant (non-dimensional)

P = set press + overpress + atm (lb/in^2 absolute)

K_2 = back pressure correction (non-dimensional)

K_3 = back pressure correction (non-dimensional)

T = absolute temperature (R)

Z = compressibility factor (non-dimensional)

M = molecular weight (non-dimensional)

The ASME VIII equation for steam, including 0.9 safety factor:

$$A = \frac{Q_v}{43.89 P K_S K_2} \qquad \text{Equ 8.11}$$

where:

A = minimum nozzle area (in^2)

Q_v = capacity (lb/h)

P = set press + over press + atm (lb/in^2 absolute)

K_S = super heat correction (non-dimensional)

K_2 = back pressure correction (non-dimensional)

8.3.3 Explosion relief valves

Explosion relief valves that are used for the protection of engine, compressor and pump crankcases can be sized according to simple rules based on the crankcase internal volume. Three rules in common use provide the seat area directly:

BICERI	700 cm^2/m^3 internal volume
Lloyd's Register	120 cm^2/m^3 internal volume
Pyropress	350 cm^2/m^3 internal volume

Explosion relief valves are sometimes fitted to oil cooled transformers. Pyropress uses the simple relationship:

137 cm^2 for small transformers

274 cm^2 for large transformers

Some control valve software packages include a module to size pressure relief valves.

8.4 Sizing regulators and control valves

8.4.1 General

The initial selection of control valves poses problems unique to control valves — where to put the valve:

- Should the valve be positioned at the beginning of the pipework, exposed to the highest pressure, and rely on the outlet pipe pressure drop to prevent cavitation and flashing?
- Should the valve be positioned at the end of the pipework, at lowest pressure, and take precautions against cavitation and flashing?
- Should it be positioned at some optimum point within the length of the pipe which provides a low operating pressure combined with adequate downstream pressure drop to prevent cavitation and flashing? This may be the optimum hydraulic and superficial "low cost financial position", but are there any services available such as electrics, instrument air, hydraulics, to operate the valve?

These decisions must be made before valve sizing can be undertaken.

The prospective purchaser is again well advised to place the onus for correct sizing and functioning of the valve on the manufacturer by fully disclosing the physical conditions under which the valve is to operate. In practice, to avoid guarantee problems, most control valves are sized by the manufacturer and its agents. Computer software is available for regulator and control valve sizing. Independent software houses have collated data from many manufacturers and the selection program can evaluate different valve types. For a complete understanding of the complexity of regulator/control valve sizing the European and International Standards IEC/EN 60534-2-1 and IEC/EN 60534-2-2 governing sizing should be reviewed:

Regulator Tag No.	Line number	Line Size	Fluid	Max Flow	Min Flow	Pressure Rating	Max Diff Pressure	Min/Max Temp	Material	Connections

Table 8.1 Typical layout for a regulator schedule

Control Valve Tag	Line No.	Line Size	Fluid	Max Flow	Min Flow	Pressure Rating	Max Diff Pressure	Min/Max Temp	Material	Connections	Actuator Type	Control Signal

Table 8.2 Typical layout for a control valve schedule

Data sheets are the best way to communicate operational data for regulators and control valves. Manufacturers, and trade associations such as the Instrument Society of America, ISA, have produced standardised formatted data sheets. (See Chapter 16 for information on standardised data sheets.) This type of form is very useful especially when large numbers of valves have to be ordered for a project, as it will highlight any missing data. Regulators and control valves, with their important parameters, should be listed in a schedule. This allows comparison of duties and sizes and can reduce the overall number of different sizes/materials required with resulting reductions in spares inventory. Typical layouts for schedules are shown in Tables 8.1 and 8.2.

Manufacturers have a great deal of experience in the art of sizing regulators and control valves, acquired partly by flow testing representative valves in development laboratories. Even more importantly they will keep in touch with their clients, and learn how their valves perform in day-to-day industrial processes and utilities.

Major manufacturers publish results of research and development work in the form of sizing tables for the valves they manufacture. These tables cover the standard ranges of globe valves from DN25 to DN600 and for the various types of valve body available. Manufacturers also list the capacities of valves fitted with reduced trims, i.e., where the valve seat is reduced in size relative to full bore. Sometimes the capacity of a control valve is reduced by its installation in the process piping, this too can be considered when sizing.

The choice of a particular size of valve to perform a given duty is almost always a compromise. Valves are built with standard body sizes, and a standard body with a full size or reduced trim must be chosen with a capacity slightly more than the capacity calculated from design data. If the valve is too large, and this can happen if design margins and trim sizes are chosen independently and applied cumulatively, it may be impossible to provide accurate control at lower rates of flow. If on the other hand the valve capacity is very close to maximum, there is nothing in hand to cover deterioration of pump performance or the fouling of pipes or heat exchangers. The control valve that is fully open under those conditions will have ceased to control altogether.

The purpose of this Section is not to provide a complete guide to do the sizing, but to explain how the manufacturer would size the valve on the basis of operating data supplied by the purchaser. However, the most important job of an design engineer is to ensure that the valves tendered by a supplier in response to an inquiry are adequately and economically sized. Each and every valve must be reviewed in detail, and its capacity calculated. It is too late to discover that a valve is too small, or too large when the control loop is being commissioned.

Whilst a smaller trim can often be fitted to an oversized globe or cage design, an undersized valve needs to be replaced completely and usually also needs piping changes. Undersize trims are not available for butterfly valves. Reduced ports may be possible on ball and eccentric rotary plug valves. In any event, commissioning would have to be interrupted for a time to permit major changes to be made.

It is not quite as difficult to review the choice of a suitable actuator, but unless the operating pneumatic signal range is specified, it is sometimes found that the manufacturers fit the smallest possible actuator in order to remain competitive and then to rely on excessive "bench setting" in order to make the assembly perform "in theory". In extreme cases this may mean that the effective control signal of the controller is cut to less than one third of its output range. This is totally unacceptable and must be avoided. Quoted operational data must include details of instrument air supplies. When hydraulically-powered valves are required, utilising site power pack outputs, the flow and pressure limitations must be specified.

Normally the actuator should be large enough to overcome the hydraulic and frictional forces acting on the valve, without the need to bias the actuator by more than 10% by bench setting. An undersized actuator will give rise to uncontrolled movements of the valve stem as the balance of hydraulic forces on the trim changes. If a valve is found to be difficult to commission because of an undersized actuator, a valve positioner could be fitted to remedy this without having to shut down the process.

Very often control valves are fitted with positioners for reasons other than actuator force e.g., to obtain a certain characteristic or to increase the operating speed. The connections of actuators to valve bonnets are standardized, it is therefore relatively easy to replace a deficient actuator with the next larger standard size. Actuator sizing should be reviewed when valve sizes are checked.

8.4.2 Throttling and pressure recovery

In practically all processes the flow of fluids in pipes is fully turbulent. Under these conditions the rate of flow through a throttling device such as a control valve is proportional to the square root of the differential pressure across it. This is sometimes represented by the Bernoulli equation and defines the "normal" condition to which the sizing equations apply.

Bernoulli's equation, for steady, one-dimensional, constant temperature and incompressible flow between stations I and II is given by:

$$\frac{p_I}{\rho} + \frac{v_I^2}{2} + gz_I = \frac{p_{II}}{\rho} + \frac{v_{II}^2}{2} + gz_{II} + \frac{\Delta p_f}{r} \qquad \text{Equ 8.12}$$

where:

p	=	static pressure (gauge) (Pa)
ρ	=	density (kg/m^3)
v	=	velocity (m/s)
g	=	gravitational acceleration (m/s^2)
z	=	height above datum (m)
Δp_f	=	frictional pressure loss (Pa)

When the flow velocity is low, or the viscosity is high, or the valve is small, the flow regime may be laminar. It is always worthwhile checking the Reynolds Number, Re, to confirm the flow regime before proceeding. Manufacturers have correction factors to cope with laminar flow conditions, see Chapter 1.

The positions at which the differential pressure is thought to act are defined in EN 60534-2-3/IEC 534-2-3. The upstream tapping is in a straight run of pipe, at least 20 pipe diameters long at a distance of 2 pipe diameters from the valve inlet. The downstream tapping is 6 pipe diameters downstream of the valve, at a point where the pipe flow has largely resumed its usual velocity profile and is no longer disturbed by the presence of the valve. The differential pressure between these two points or between two equivalent points in each and every control valve installation is the one included in the sizing calculation.

Bernoulli's equation, which relates rate of flow and pressure drop is an idealised one which describes a reversible change. A reversible change is one in which energy is not lost but merely converted between the potential energy of the fluid represented by its pressure and its kinetic energy associated with its flow velocity.

If the inlet and outlet piping were of the same diameter and if there were no change in density across the valve or of head, then the inlet pressure and the outlet pressure measured in the positions indicated would be identical; and there would be no pressure drop at all. But in the actual throttling that takes place in a control valve, some of the kinetic energy of the fluid at the most constricted part of the flow passage, the vena contracta, is converted into heat and is lost, not to be recovered as pressure. Exactly what proportion of the kinetic energy is lost and what is recovered as pressure depends largely on the construction of the valve restriction and seat.

Gases and vapours exhibit large changes in density with pressure. Liquids are usually assumed to be of constant density irre-

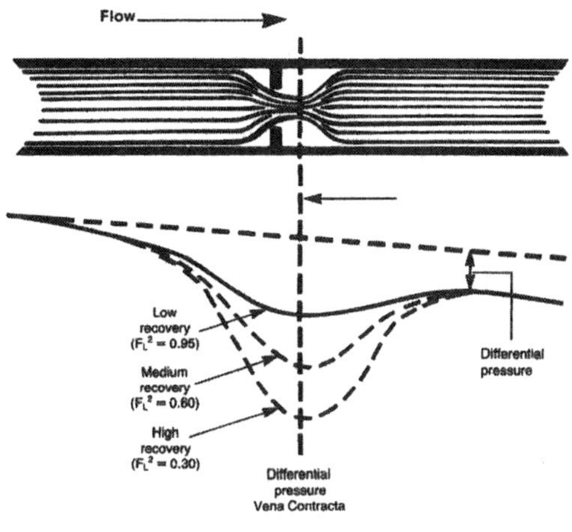

Figure 8.7 Pressure recovery after throttling

spective of pressure. This assumption can be valid for most liquids provided the pressure change is small. All liquids are compressible. If the liquid is water and the pressure differential is small, say 10barg, the density change is usually insignificant. Modern processes run routinely at pressures over 350 barg and valves are required to impose large pressure drops, so density changes of liquids can not be ignored. Liquified gases such as LPG, LNG and CO_2, are much more compressible than water.

In general, if a large jet of fast moving fluid impinges on a comparable mass of fluid moving more slowly, or indeed if the large jet is decelerated in a conical diffuser e.g., the venturi outlet of an angle valve, then much of its kinetic energy will be recovered in the form of pressure, providing the profile is carefully designed. Butterfly valves, ball valves, eccentric rotating plug valves and some angle valves have one or two comparatively large openings, which discharge a large jet or jets downstream. These types of valve are called high recovery valves.

Figure 8.7 indicates the pressure along the fluid path in three recovery rate valves. In the upstream pipe, fluid pressure drops slowly under the influence of pipe friction. Between the inlet of the valve and inlet of the actual trim the fluid accelerates and pressure, as a result, drops more sharply. The pressure then drops rapidly to the point just beyond the exit of the trim, this point is the vena contracta, the smallest area of the fluid stream. As the fluid stream expands downstream of this point, pressure will rise until it resumes the normal downstream value and then continues to decay by pipe friction.

The liquid pressure recovery factor of a valve F_L, is defined as the ratio of the actual choked flow rate to a theoretical flow rate calculated on the assumption that the differential pressure is the difference between the inlet pressure and the super-cooled vapour pressure at the vena contracta.

F_L is determined experimentally by testing with water at 40°C. F_L is sometimes called C_f in American publications. F_L varies with the degree of opening of a control valve. Depending upon the valve detail design F_L can increase or decrease as the valve closes. This can have important consequences for the noise generated. For comparison, the same figure includes a similar curve for a medium pressure recovery valve.

The differential pressure is the same for both valves. Indeed it must be, because this is one of the design conditions. It is not immediately obvious that the pressure at the vena contracta has not dropped nearly as much. This leads to the conclusion that the velocity at the vena contracta, in a high recovery valve, is much greater than that of a comparable medium recovery valve; one reason why high recovery valves are noisier. Typical medium recovery valves of this type are single seat and double-beat globe valves with pressure recovery factors F_L=0.8 to 0.9.

Finally, there are valves in which the frictional element is deliberately designed to destroy almost all of the kinetic energy of the vena contracta, exhibiting pressure recovery factors F_L=0.95 or more. Valves of this type are employed when conventional valves would produce excessive flow noise and for the control of flashing and cavitating media. Typical applications include pump and compressor testing, to create and adjust the total differential pressure, and pressure let-down valves for processes with subsequent sections at much lower pressures.

In pump and compressor testing applications the let-down valve is sometimes called a DRAG™ valve. Energy conservation demands have considerably reduced the use of let-down valves in continuous processes; power recovery turbines are a better and cost-effective solution.

8.4.3 Sizing equations for incompressible fluids

It is essential to verify whether the assumption of incompressible flow is valid. For liquids similar to water, standard steam tables are a good source of data. Only steam tables which have data on pressurised water are useful. Data at the saturation line is only useful for checking flashing. Assume a constant enthalpy throttling process and evaluate the outlet liquid properties. For very compressible liquids such as LPG, LNG, CO_2, the volume change could be up to six times that of water. Check the saturation line data to make sure the outlet liquid is still liquid.

Much of the early empirical analysis of experimental results of hydraulic tests on control valves was carried out in the United States and formed the basis early standards of the Fluid Controls Institute. They have been withdrawn and are replaced by ANSI/ISA755.19.01.

The capacity of a regulator or control valve is expressed in terms of its flow coefficient. In the original definition which employs US units, this flow coefficient is the rate of flow as measured in US gallons per minute of water divided by the square root of the differential pressure, 2D upstream and 6D downstream of the valve in full size pipework, expressed in pounds per square inch.

A useful rule of thumb, that gives results within 10% for most single seat and double-beat globe valves, is that this flow coefficient, C_v, is 12 times the square of the body size expressed in inches or:

$$C = 12\,d^2 \qquad \text{Equ 8.13}$$

where:

C_v = approx flow coefficient (US gpm/psi)

d = nominal valve diameter (in)

Values of the flow coefficient C_v are listed by all valve manufacturers using the inch system. Some European manufacturers and their customers prefer use of a flow coefficient based entirely on metric units.

This flow coefficient K_v, is defined as the rate of flow measured in cubic metres per hour of water divided by the square root of the differential pressure measured in bar. In this context it should be remembered that K_v was originally defined in terms of kgf/cm^2 and only more recently in terms of bar. This is relevant to some of the older Standards such as BS 4740: Part 1: 1971, which was withdrawn.

Example:

Figure 8.8 shows part of a process schematic representing a centrifugal pump discharging into a heat exchanger and then into a pressure vessel which is maintained at a constant pressure of 4barg. The flow rate through the valve required to maintain a constant level in the vessel, is between a minimum of 5 and a maximum of 50m³/h. The total pipe resistance at full flow amounts to 2bar, at the minimum flow it is 1/100th of the full value and can be ignored. The pressure vessel is elevated at 10m creating a static pressure of 1barg. The liquid has a density of 1000kg/m³ or 1kg/dm³. Figure 8.9 shows the pump system characteristics together with the various pressures in the control system, at design flow and at full turndown.

Using Equation 6.2 and the conversion factor at maximum flow conditions:

$$K_v = Q\sqrt{\frac{\rho}{\Delta p}} = 50\sqrt{\frac{1}{2}} = 35$$

$$= 50\sqrt{\frac{1}{2}}$$

$$= 35$$

Therefore:

$$C_v = 35 \times 1.156 = 41$$

Applying the rule of thumb to the C_v value, a body size of 2" or 50mm is required. This is confirmed by reference to the capacity table in Table 8.3, which shows this capacity to correspond to a valve travel of 83%.

Valve Opening - percent of total travel	Coefficients - C_v (Liquid) Nominal body size mm 25	40	50	65	80	100	150	200
10	0.783	1.54	1.74	4.05	4.05	6.56	13.20	18.80
20	1.29	2.52	3.15	7.19	6.84	11.40	24.60	33.60
30	1.86	3.57	4.72	10.60	10.00	17.30	41.10	53.60
40	2.71	4.94	5.91	14.50	15.00	27.00	62.50	79.80
50	4.18	7.41	10.60	21.20	23.80	42.20	97.10	114.00
60	6.44	11.60	16.30	31.60	37.80	58.40	155.00	168.00
70	9.54	17.20	25.00	45.50	59.00	103.00	223.00	242.00
80	13.10	23.50	36.70	64.20	87.10	148.00	286.00	345.00
90	15.70	28.70	47.80	77.70	110.00	184.00	326.00	467.00
100	17.40	33.40	56.20	82.70	121.00	203.00	357.00	570.00

Table 8.3 Valve flow coefficients for equal percentage characteristics

When choosing a body size, it is important to ensure that the valve is not fully stroked under design conditions, otherwise there would be nothing in reserve should a sudden increase in demand require to be corrected. To cater for this, a valve is usually arranged to control at an opening of 70% to 80% under maximum design conditions. The most convenient way of achieving this, without lengthy reference to sizing tables, is to multiply the calculated C_v by a factor of 1.2 to 1.4 for linear trims and by a factor between 1.4 and 1.6 tor equal percentage trims.

The corrected C_v is then used in choosing the valve body. In this case this corrected maximum flow coefficient is between:

$$C_{v\ max} = 41 \times 1.4 = 57$$

$$C_{v\ max} = 41 \times 1.6 = 66$$

If the 50 mm valve body were installed in a 50 mm pipe it would be reasonable to choose to apply a 50 mm valve. If, on the other hand, the pipe upstream of the valve were 80 mm diameter, necessitating the use of reducers to accommodate the valve, BS 5793, now replaced by BSEN 60534-4:2006, recommended a further increase in the flow coefficient by a factor of 1.06 to allow for the additional pressure drop.

The maximum calculated C_vs now lie between:

$$C_{v\ max} = 57 \times 1.06 = 60$$

$$C_{v\ max} = 66 \times 1.06 = 70$$

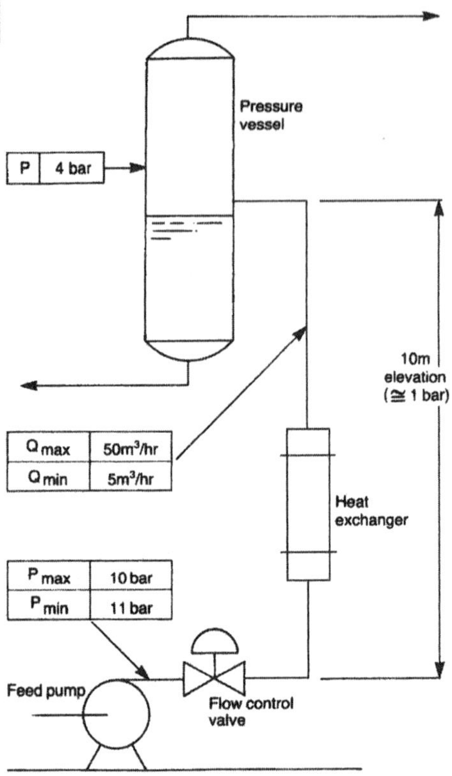

Figure 8.8 Simple process schematic

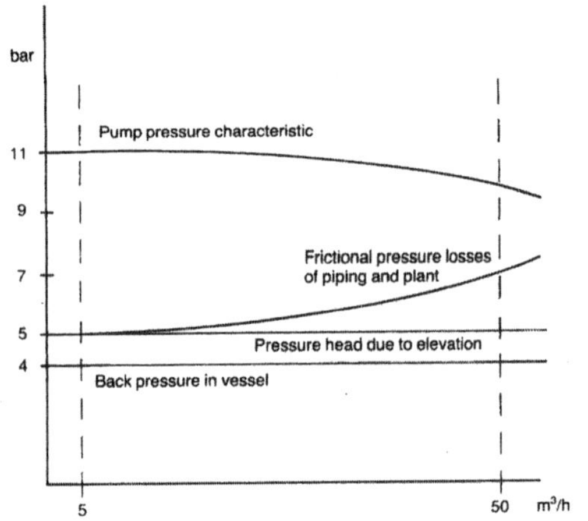

Figure 8.9 Pump and system characteristics

Reviewing the data in Table 8.3 the higher of these C_vs corresponds to a 65mm, 2½" valve at 85% of full travel. If the calculated valve size is less than the pipe size such an allowance should always be made. Most manufacturers include tables of correction factors for installation conditions, sometimes called Piping Geometric Factors, in their sizing data. The valve would then actually control at a travel of 69% at which it has a flow coefficient of 44.

The flow coefficient of a fully open 65mm valve with equal percentage trim listed in the table is 82.7. There is plenty of capacity to allow for deterioration of the pump, fouling of the heat exchanger and other influences. It is worthwhile checking that the valve will still adequately control at the minimum rate of flow.

Referring to Figure 8.9, at the minimum flow rate of $5m^3/h$, the available pressure drop is 6 barg.

$$K_v = Q\sqrt{\frac{\rho}{\Delta p}} = 5\sqrt{\frac{1}{6}} = 2.04$$

$$C_v = 2.04 \times 1.156 = 2.36$$

If the 50 mm valve were chosen, this flow coefficient is achieved at a travel of 14%, and the 65mm valve would travel approximately 6% to provide this capacity. Clearly if satisfactory control at the minimum rate of flow is essential, better results with the smaller valve body would be expected.

However, if the maximum flow of the specification is more important one should use a 65mm body with some deterioration in control at lower rates of flow. The example is typical in so far as the final choice is frequently a compromise between conflicting design criteria.

8.4.4 Inlet velocity

There are two practical limits to the liquid capacity of a control valve, even if sufficient differential pressure were to be available to drive the liquid through the valve. One limit is set by the tendency of liquids to flash into vapour in the vena contracta, and this is discussed in the next Section. The other is set by the mechanical robustness of the valve itself and is expressed as an upper limit to the inlet velocity permissible.

Manufacturers specify limiting inlet velocities for their valves and these different values are given in Tables 8.4, 8.5 and 8.6. One globe valve manufacturer states a general maximum of 15m/s outlet velocity for liquids.

Valve style and rating	Trim	Size	Carbon steel body m/s	Alloy steel body m/s
All body styles Ratings up to ANSI 600 lb ISO PN100	Contoured plug	DN15, 20, 25	12.5	14.0
		DN40	11.9	12.8
		DN50	11.3	11.6
		DN80	10.4	10.4
		DN100	9.75	9.75
		DN150	9.15	9.15
		DN200	8.85	8.85
		DN250	8.25	8.25
		DN300, 350	7.30	7.30
		DN400,450	6.70	6.70
		DN500	5.50	5.50
		DN600	3.70	3.70
All body styles Ratings of ANSI 900lb ISO PN160 and over	Contoured plug	DN15, 20, 25	12.5	14.0
		DN40	12.5	14.0
		DN50	12.5	14.0
		DN80	12.5	12.8
		DN100	11.9	11.9
		DN150	11.3	11.3
		DN200	11.0	11.0
		DN250	10.1	10.1
		DN300, 350	9.45	9.45
		DN400,450	8.25	8.25
		DN500	6.7	6.70
		DN600	4.3	4.30
Angle bodies All pressure ratings		All sizes	14.6	17.3

Table 8.4 Control valve inlet liquid flow velocity limits

Body material	Service conditions	Max velocity m/s
Cast iron Ductile iron Carbon steel	Continuous modulation Δp > 35 bar	6.0
	Intermittent modulation or Δp < 35 bar	9.0
	2% duty cycle	12.0
Low alloy steel Stainless steel	Continuous modulation	13.7
	Intermittent modulation	18.3
	2% duty cycle	27.5

Table 8.5 Globe valve liquid velocity limits

Valve type	Continuous	Intermittent
Ball, plug, eccentric plug, anti-cavitation trim	10 m/s	12 m/s
Butterfly	7 m/s	8.5 m/s

Table 8.6 Rotary valve liquid velocity limits

The outlet velocity, based on the downstream density, should also be checked. In general, a maximum velocity of 16m/s should not be exceeded. If flashing occurs, the maximum velocity may be increased to 150m/s or Mach Number=0.2 for valves in continuous operation. Valves which only suffer flashing intermittently can operate with mixture velocities up to 450m/s or Mach Number=0.3 providing materials are selected accordingly.

8.4.5 Choked flow

Flow is "critical" or "choked" when the rate of flow through a control valve is no longer observed to follow the square root relationship with the differential pressure. Ultimately then, under constant upstream conditions, a further decrease in downstream pressure no longer produces any increase in the flow. The choking effect is caused by bubbles of vapour in the Vena Contracta.

It now depends on the subsequent pressure recovery as to whether the vapour bubbles formed by flashing in the Vena Contracta persist downstream of the valve or if they collapse altogether as pressure rises towards the outlet of the valve. If the vapour persists downstream of the valve, flashing flow is said to occur.

The capacity of the valve can be calculated either from the formula in BS EN 60534-2-1 or by calculating separately, the flow coefficients required to pass the vapour produced by the flashing process, and the liquid remaining. Then adding these coefficients, as in a procedure for sizing valves for mixed phase fluids. Neither of these methods is suggested with much confidence when there is a high percentage of flashing. The choked flow formula gives results that are slightly too small for liquids with little flashing.

$$K_v = \frac{Q}{F_L}\sqrt{\frac{r}{p_1 - F_F \cdot p_v}} \qquad \text{Equ 8.14}$$

where:

F_F = liquid critical pressure ratio factor (non-dimensional)

p_v = liquid inlet vapour pressure (bar)

F_F can be approximated by the following relationship, plotted in Figure 8.11 where p_c is the liquid critical pressure in bara:

$$F_F = 0.96 - 0.28\sqrt{\frac{p_v}{p_c}} \qquad \text{Equ 8.15}$$

See Table 8.17 in Section 8.7.1 for critical pressure information.

Example:

Steam condensate, SG 0.92, leaves a flash drum at a pressure of 6 barg and a temperature of 145°C. The design flow rate is 80 m^3/h and the pressure differential is a constant 3 bar irrespective of the flow. At this stage it is appropriate to remember that pressures, vapour pressures and similar used in thermodynamic calculations are measured on an absolute scale.

The water temperature of 145°C creates a vapour pressure of 3 barg equal to approximately 4 bara. The standard atmosphere is 14.696 psia which is equal to 1.01325 bara. If flashing is likely to occur, as it is in this application, one would select a medium recovery valve with a pressure recovery factor of F_L = 0.87.

The maximum allowable differential pressure for sizing purposes is:

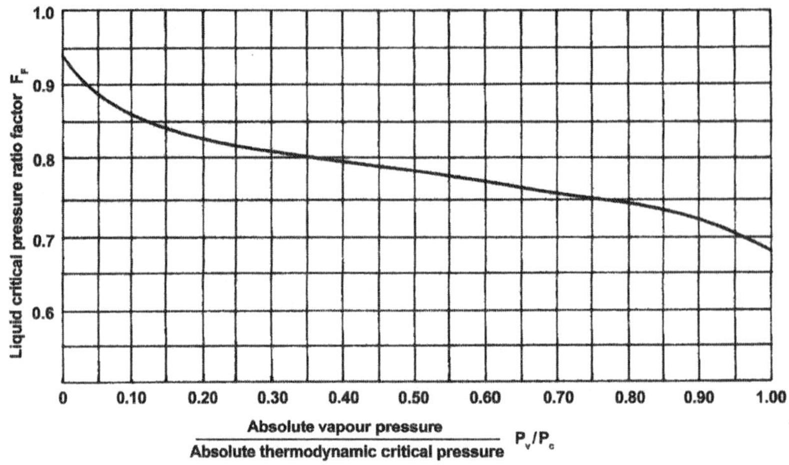

Figure 8.10 Liquid critical pressure ratio factor, F

$$\Delta p = F_L^2 \, (p_1 - F_F \cdot p_v) \qquad \text{Equ 8.16}$$

$$p_1 = 6+1 = 7 \text{ bara}$$

$$p_v = 3+1 = 4 \text{ bara}$$

0

$$F_L = 0.87$$

$$F_L^2 = 0.76$$

$$\Delta p = 0.76\,(7 - 0.93 \times 4) = 2.48 \text{ bar}$$

$$\rho = 0.92 \times 1 = 0.92 \text{ kg} / \text{dm}^3$$

Using the standard equation for K_v:

$$K_v = Q\sqrt{\frac{\rho}{\Delta p}} = 80\sqrt{\frac{0.92}{2.48}} = 48.7 \qquad \text{Equ 8.17}$$

$$C_v = 48.7 \times 1.156 = 56$$

Because upstream and downstream pressures in this example are independent of the flow rate, a linear valve trim is chosen to obtain the desired linear relationship between the control signal to the valve and the flow of liquid. Correction factors for this type of trim are 1.2 to 1.4, and this adjusts the flow coefficient by:

$$C_v = 1.2 \text{ x } 56 = 68$$

$$C_v = 1.4 \text{ x } 56 = 79$$

A 65mm single seated globe valve with a maximum flow coefficient of 85 is selected from the manufacturer's standard sizes. The corresponding 80mm valve has a C_v of 110.

In this particular example the pressure drop was not particularly great and would not impose unusually large forces on the trim. If it had been greater, a trim specially designed for flashing and cavitating conditions would have been chosen, with $F_L = 0.99$. The fitting of such a valve allows the use of a slightly larger pressure drop in calculating the valve size:

$$\Delta p = 0.98\,(7 - 0.93 \times 4)$$

$$\Delta p = 3.2 \text{ bar}$$

$$C_v = 51$$

$$C_{v\ corr} = 60 \text{ to } 70$$

An 80 mm single seated globe valve with a flash flow trim has a flow coefficient =76. When flashing occurs in a valve a certain fraction of vapour will persist downstream of the valve. This

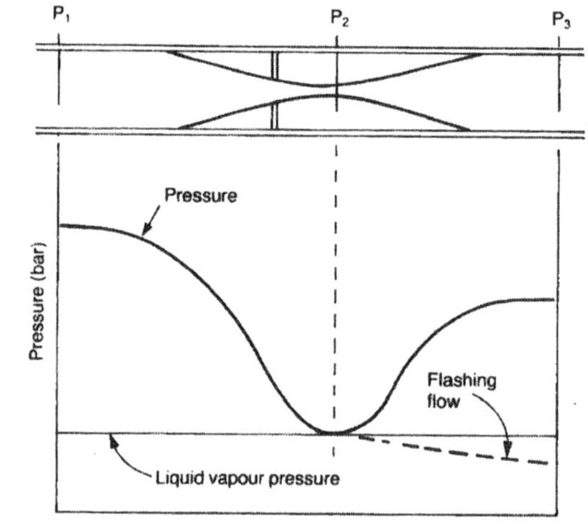

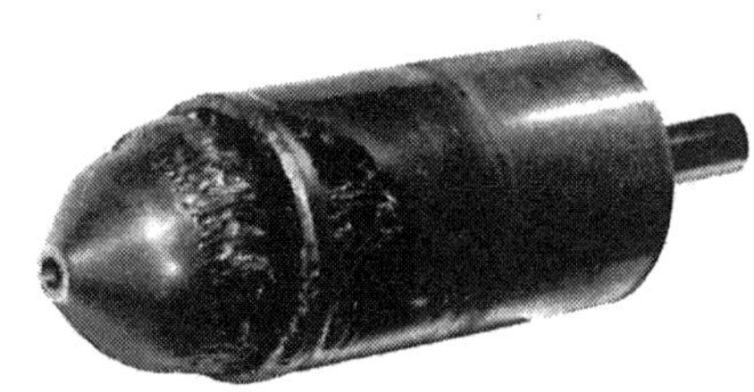

Figure 8.11 A valve plug showing typical flashing damage

fraction of vapour can be calculated from temperatures and pressure upstream and downstream of the valve. This is usually done on the assumption that the throttling action of the valve is performed at constant enthalpy, $h_1 = h_2$. Flashing can lead to high local velocities in the valve which may lead to rapid erosion of standard trim materials unless these are protected by hard coating, see Figure 8.11.

8.4.6 Cavitation

Cavitation can only occur in liquid valves because cavitation requires changes of phase. When cavitation occurs in a valve, there are two complete changes in phase, first from liquid to vapour near the Vena Contracta, and the second change from vapour back to liquid somewhere in the outlet of the valve.

The approach taken to sizing is the same as that for flashing with the proviso that cavitation and its effects, see Figure 8.12, are avoided as much as possible:

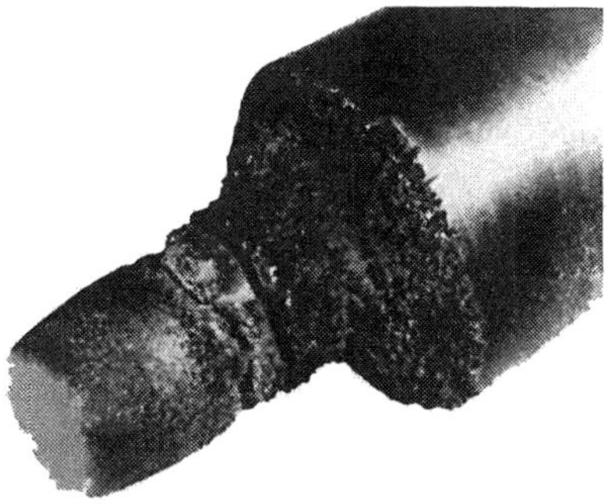

Figure 8.12 A valve plug with typical cavitation damage

- By choosing special cavitation control trims so arranged that the energy of the fluid is made to dissipate itself in the body of fluid downstream of the valve, see Figure 8.9
- By limiting the pressure drop occurring across the valve, or by providing several stages of pressure drop within one valve body, see Figure 8.13
- By surface treatment, hard coating or similar design measures

Valve size depends upon the liquid flow being laminar or turbulent.

8.4.7 Sizing equations for gas/vapour

In Section 8.4.5 the gradual limitation of "choking" flow was treated as part of the hydraulics of incompressible fluids, although choking is caused by the appearance of a vapour phase which itself is compressible. This was done because choking is an important limiting factor that is sometimes, but not always, associated with liquid flow. Only when the differential pressure is a small fraction, say 1/50th, of the inlet absolute pressure, is it sufficiently accurate to ignore compressibility effects and to use the sizing formulae for incompressible flow.

If the fluid is a compressible gas or vapour, then there will always be a choked flow condition which is the maximum rate of flow which the fluid can attain when passing through a valve. Under these conditions and with fixed upstream pressure, a further decrease in downstream pressure, increasing the differential pressure, will not produce an increase in the rate of flow.

The critical differential pressure ratio, X_T, which results in choked flow, is the maximum value of the ratio X, of differential pressure to upstream absolute pressure, that is effective in driving the fluid through the valve.

In the past, based on limited experimental evidence on the performance of steam nozzles, it was thought that the critical differential pressure ratio had a fixed value of 0.5. It is now known that the critical flow occurs in very low recovery, high friction low noise or cavitation control trim, valves at quite a high critical differential pressure ratio, $X_T = 0.8$.

Critical flow in high recovery, ball or butterfly, valves can occur at differential pressure ratios as low as $X_T = 0.2$.

The gas/vapour sizing equations are now listed:

$$C_v = \frac{Q}{417\, p_1 \cdot Y \sqrt{\dfrac{X}{G_g \cdot T_1 \cdot Z}}} \qquad \text{Equ 8.18}$$

$$C_v = \frac{Q}{2250\, p_1 \cdot Y \sqrt{\dfrac{X}{M \cdot T_1 \cdot Z}}} \qquad \text{Equ 8.19}$$

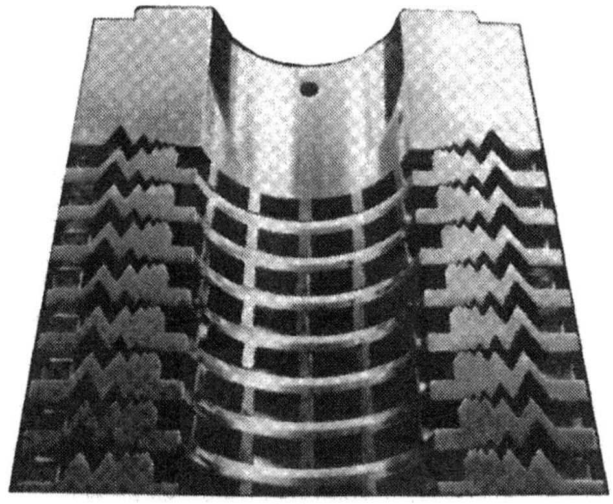

Figure 8.13 Multiple pressure drop stages. Tiger-Tooth™ valve trim
Courtesy of Flowserve Flow Control

$$C_v = \frac{w}{27.3\, Y \sqrt{X \cdot p_1 \cdot r_1}} \qquad \text{Equ 8.20}$$

$$C_v = \frac{w}{94.8\, p_1 \cdot Y \sqrt{\dfrac{X \cdot M}{T_1 \cdot Z}}} \qquad \text{Equ 8.21}$$

$$X = \frac{p_1 - p_2}{p_1} \qquad \text{Equ 8.22}$$

$$F_k F_k = \frac{\gamma}{1.40} \qquad \text{Equ 8.23}$$

$$Y = 1 - \left\{\frac{X}{3\, F_k \cdot X_T}\right\} \qquad \text{Equ 8.24}$$

where:

Q	=	flow (m^3/h)
		(Flow normalised to 1.01325 bara at 15°C)
w	=	mass flow (kg/h)
p_1	=	inlet pressure (bara)
Y	=	expansion factor (non-dimensional)
X	=	differential pressure ratio (non-dimensional)
X_T	=	critical differential pressure ratio (non-dimensional)
G_g	=	gas relative density (non-dimensional)
		(Relative to air at 1.01325 bara at 15°C)
T_1	=	inlet temperature (K)
Z	=	compressibility factor (non-dimensional)
ρ_1	=	inlet density (kg/m^3)
F_k	=	specific heat ratio factor (non-dimensional)
γ	=	ratio of specific heats (non-dimensional)
M	=	gas molecular weight (non-dimensional)

$$C_v = \frac{Q}{1360\, p_1 \cdot Y \sqrt{\dfrac{X}{G_g \cdot T_1 \cdot Z}}} \qquad \text{Equ 8.25}$$

$$C_v = \frac{Q}{7320\, p_1 \cdot Y \sqrt{\dfrac{X}{M \cdot T_1 \cdot Z}}} \qquad \text{Equ 8.26}$$

$$C_v = \frac{w}{63.3\, Y \sqrt{X \cdot p_1 \cdot r_1}} \quad \text{Equ 8.27}$$

$$C_v = \frac{w}{19.3\, p_1 \cdot Y \sqrt{\frac{X \cdot M}{T_1 \cdot Z}}} \quad \text{Equ 8.28}$$

where:

Q = flow (ft^3/h)

(Flow normalised to 14.73 psia at 60°F)

w = mass flow (lb/h)

p_1 = inlet pressure (psia)

G_g = gas relative density (non-dimensional)

(Relative to air at 14.73 psia at 60°F)

T_1 = inlet temperature (R)

ρ_1 = inlet density (lb/ft^3)

Values of X_T are characteristic of the valve design and must be obtained from the valve manufacturer. Piping geometry may affect the X_T value.

Typical values of X_T are listed in Table 8.7.

Valve type	X_T
Globe valve, contoured plug, flow-to-open	0.7 to 0.72
Globe valve, contoured plug, flow-to-close	0.5 to 0.75
Globe valve, plug/cage trim, large ports	0.81
Globe valve, plug/cage trim, multiple small ports	0.68
Globe valve, needle valve trim	0.72
Globe valve, quiet and anti-cavitation trims	0.84 to 1.0
Butterfly valve	≈ 0.25
Eccentric plug valve	≈ 0.3

Table 8.7 Typical values of critical differential pressure ratio

The expansion factor Y expresses the change in density of the compressible fluid on its passage through the valve. Although the value of Y is determined by experiment, it is still useful to review the contributing mechanical influences. When valve body capacity is evaluated experimentally in accordance with IEC 60534-2-3, tests with compressed air are conducted to examine the influence of the ratio of the port area to the body inlet area, of the path taken by the fluid, and of the ratio of pressure differential to the inlet absolute pressure.

These influences are expressed by the limited differential pressure ratio factor. If the inlet pressure p_1 is held constant and the outlet pressure p_2 is progressively lowered until a maximum flow limit, namely choked flow, is reached, at this limit the differential pressure ratio is:

$$X = F_k \cdot X_T \quad \text{Equ 8.29}$$

Most manufacturers quote values of X_T measured with air. For fluids other than air a correction factor, F_k, accommodates the influence of specific heat ratio. Theoretically the expansion factor Y is also affected by Reynolds Number but this is of no consequence in practice because, with the possible exception of hydrogen, which has a high kinematic viscosity and therefore lower Reynolds Number, most gas and vapour flow is in the fully turbulent region.

In the equations, X is not allowed to exceed the value X_T even if the apparent differential pressure ratio is greater. When X is greater than $F_K \times X_T$ then Y=0.667 and X in all the equations should be replaced by $F_K \times X_T$.

There is also an empirical relationship between the liquid pressure recovery factor, F_L, and X_T, which was established by experiment:

$$F_L = \sqrt{\frac{X_T}{0.84}} \quad \text{Equ 8.30}$$

This appears to be correct for most traditional styles of globe body but should be used with caution when applied to cage trim and angle valves.

Example:

A valve is required to pass 50 000kg/h of superheated steam at a line pressure of 80bara and an inlet temperature of 300°C. The valve discharges into a header at 35 bara.

Use the sizing equation to estimate an initial valve, assuming a worst case value of Y = 0.66 for the expansion factor and limiting the maximum differential pressure to 0.6 times inlet pressure. The steam specific volume is approximately $0.02352 m^3/kg$. Take g for steam as 1.3.

The differential pressure ratio can be checked using Equation 8.22:

$$X = \frac{80 - 35}{80} = 0.5625$$

This is smaller than the imposed limit, so does not present a problem. An approximate C_v can be calculated to allow an initial valve selection, using equation 8.20.

$$C_v = \frac{50\,000}{27.3 \times 0.66 \sqrt{0.5625 \times 80 \times \left\{\frac{1}{0.02352}\right\}}} = 63.4$$

A double-beat valve with reduced trim, which has an X_T value of 0.53, is selected from a manufacturer's data. The flow is critical, so the expansion factor should be calculated.

$$Y = 1 - \frac{0.5625}{3 \times \left\{\frac{1.3}{1.41}\right\} \times 0.53} = 0.6163$$

The C_v value can now be calculated more accurately:

$$C_v = \frac{50\,000}{27.3 \times 0.616 \sqrt{0.5625 \times 80 \left\{\frac{1}{0.02352}\right\}}} = 67.9$$

Because upstream and downstream pressures are constant, the valve is fitted with a linear trim. The corrected flow coefficients are in the range:

$C_v = 67.9 \times 1.2 = 81$ to $C_v = 67.9 \times 1.4 = 95$

An 80 mm control valve, with a medium reduced trim, 1 step down, has a C_v of 100 and is chosen for this duty.

After the valve size has been selected the valve body velocities should be checked. Table 8.8 indicates acceptable values.

Figure 8.15 shows the input and output data for computerised control valve selection for a particular gas application.

8.4.8 Multi-phase flow and rheology

Pipework does not necessarily handle only a liquid or a gas/vapour and mixtures are commonplace. The paper making industry handles plenty of water/pulp mixtures. Economic pressures have forced crude oil producers to move crude oil and natural gas in one pipeline. Solids transportation, with a liquid carrier, can be commercially attractive in some situations; coal with methanol has been considered, coal with water is already used on a large scale. Regulators and control valves may be required to handle a diverse range of mixtures.

The most difficult mixture, from a valve sizing consideration, is a liquid with its own vapour. This combination is very complicated to size and is best left entirely to the valve manufacturer.

A liquid with a non-condensing gas component is dealt with by calculating an equivalent specific volume for the mixture and assuming both components travel at the same speed through

Valve style and rating	Trim	Size	Valve inlet m/s	Valve outlet m/s
All body styles Ratings up to ANSI 600LB ISO PN100	Contoured plug	DN15,20,25	104	Globe valve bodies 250 m/s or 0.5/0.65 Mach No.
		DN40	99	
		DN50	99	
		DN80	90	
		DN100	90	
		DN150	81	
		DN200	81	
		DN250	67	
		DN300,350	67	
		DN400,450	58	
		DN500	46	
		DN600	35	
All body styles Ratings of ANSI 900LB and over ISO PN160 and over	Contoured plug	DN15,20,25	130	Intermittent 0.7 Mach No.
		DN40	123	
		DN50	123	
		DN80	112	Angle bodies 1.0 Mach No.
		DN100	112	
		DN150	100	
		DN200	100	
		DN250	84	
		DN300,350	84	
		DN400,450	72	
		DN500	58	
		DN600	46	
Angle bodies All pressure ratings		All sizes	144	

Table 8.8 Maximum valve body velocities for gas flow

the regulator/valve. Both components must be checked individually to see if either is "choked".

Liquids, mixtures and suspensions which are shear sensitive should have their viscous characteristics evaluated during passage through the valve. The viscosity may change significantly depending upon the change in velocity experienced and the valve opening. Products which thicken may suffer considerably higher pressure losses than expected.

8.4.9 Erosion

Erosion can be a costly problem in valves which were defined as "clean fluid" but in actual operation included very small quantities of abrasive solids. Solids in fluids can never be treated as "trace quantities" and as a chemical group. Solids can destroy the valve characteristic and prevent sealing on closure in less than two weeks. Short production runs interspersed with expensive parts replacement can be the unexpected legacy of overlooking entrained solids.

Tests with reciprocating pump valves; non-return valves which open and close continuously; conducted by Vetter and Störk showed a doubling of the valve material loss when exposed to 0.25% concentration of fine sand. These valves would operate with a nominal liquid velocity of about 2m/s. The effect would be much greater at higher velocities. Other research work suggests the erosion rate to be proportional to the velocity raised to the power of 2.5 minimum. In some tests the power was found to be 5. The component shape is a very important controlling factor in erosion.

Material	Relative abrasion resistance
Hard chrome plate	2.06/2.28
Tungsten carbide	1.39/4.12
Stellite 6	0.83/3.31
Ni Hard	0.89/1.11
27 Cr cast iron	1
13 Cr steel 441 BHN	0.32
316L stainless steel	0.26
Carbon steel 195 BHN	0.22
Aluminium bronze	0.13
Nickel aluminium bronze	0.12
Cast iron	0.09

Table 8.9 Relative abrasion resistance to sand in water

All metals are not equally resistant to erosion. Table 8.9 shows the relative performance of various popular materials. The original data has been conditioned to show 27Cr cast iron as unity; this is a standard material for solids handling centrifugal pumps and piston mud pumps.

Solid	Miller Number range
Coal	6/57
Drilling mud	10
Sewage (digested)	15
Limestone	22/46
Sewage (raw)	25
Shale	53/59
Bauxite	9/134
Copper ore	20/135
Iron ore	2 8/157
Sand	51/246
Magnetite	64/134
Tailings	24/644

Table 8.10 Abrasiveness of solids by Miller Number

All solids are not equally abrasive. There are no universally adopted tests for abrasion testing of flowing solids. The Miller Number test is used for reciprocating pumps and appears consistent. Table 8.10 shows various solids in water ranked according to Miller, ASTMG75. Miller Numbers less than 50 are considered to be not abrasive and do not require special material considerations. However, the Miller test evaluates the wear caused to a 27 Cr cast iron test block. Lesser materials, see Table 8.9, may suffer unacceptable material loss with Miller Numbers less than 50. The wear rate is approximately proportional to velocity and Miller Number.

Test facilities specifically for testing valves with solids are available. Up to 5% sand, 0.5mm maximum size, can be circulated with water through valves to evaluate the wear characteristics. Valve opening can be adjusted automatically to simulate site conditions. Continuous tests are conducted to check valve/material suitability.

Solids in fluids, just like trace chemical elements, can have catastrophic effects on valve performance. It is the responsibility of the purchaser to declare the relevant information to allow the correct valve selection. Because velocity is an important factor the actual operating velocities must be checked.

8.4.10 Valve characteristics

The flow characteristic of a control valve is the relationship between the flowing quantity of fluid through the valve and the valve stem travel from closed (0%) to the fully open (100%) position. Normally the relationship is represented as in Figure 8.14, where the valve lift and capacity are stated as percentages of maximum travel and the flow coefficient of the fully open valve.

Many types of inherent flow characteristics have been developed to satisfy the needs of various process requirements. In order to select the optimum requirements of a particular process, knowledge is required of the function of the process with regard to the pressure conditions, flow, control range and precision demands, etc. Of special interest is the installed characteristic, which is a combination of the process and valve characteristics. A basic requirement is to maintain sufficient pressure drop across the valve so that it can control the flow within the full specified control range.

In practice it can be both difficult and/or impractical to obtain such in-depth process information and, for this reason, guide-

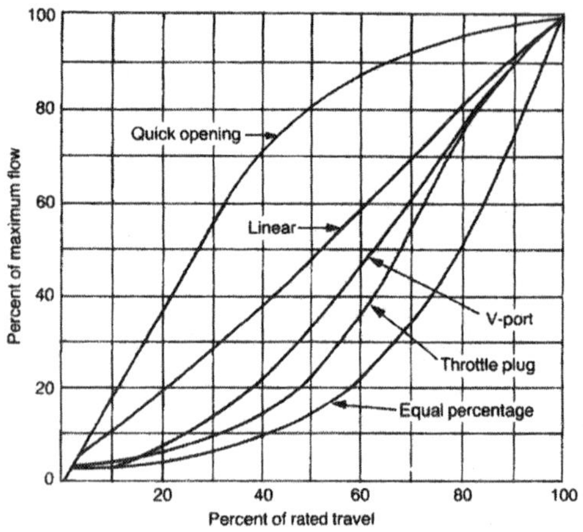

Figure 8.14 Typical control valve characteristics

lines have been developed to assist in the choice of inherent valve characteristics for the various control applications.

Unlike conventional globe valves, most rotary motion valves do not have a mathematically defined flow characteristic, which can be described in graphical or tabular form by the makers. ISA Standard S75.11defines inherent rangeability as the ratio of the largest flow coefficient to the smallest flow coefficient within which the deviation from the specified inherent flow characteristic does not exceed the limits stated. See Section 8.4.12.

The rangeability which is implied by the data in Figure 8.14 is never attained in practice for a variety of reasons:

- In many process applications the available pressure drop increases as the flow decreases, which can reduce the actual rangeability to less than 25% of that quoted in catalogues, values which are determined under constant pressure conditions,
- It is inaccurate to speak of rangeabilities greater than 20:1 if the positioning accuracy of a spring and diaphragm actuator is ± 5% at best, because there can be no effective control at less than 10% of the full flow coefficient,
- Some globe valves with wall thicknesses appropriate to low pressure operation, say Class 150lb, can be distorted by piping strain and show very significant leakage if they are not installed correctly.

Plug style	Port	Body size	Basis for min coefficient	Min coefficient % of Max coefficient
Equal percentage	Single	25mm & above	DC	1.5 to 2.5
Linear and quick opening	Single	All	LF	0.0 to 0.1
	Double	All	LF	0.0 to 0.5
V-port	Single	Below 65 mm	CF	1.0 to 3.0
	Single	65mm & above	CF	0.5 to 1.5
	Double	Below 25 mm	CF	2.0 to 7.0
	Double	25 mm & above	CF	1.0 to 3.0
Throttle	Single	65 mm & below	CF	1.5 to 3.5
	Single	Above 65 mm	CF	1.0 to 2.0
	Double	Below 25 mm	CF	4.5 to 6.0
	Double	25 mm to 65 mm	CF	1.5 to 4.0
	Double	Above 65 mm	CF	1.0 to 3.0
V-ported ball - with seal	Single	All	DC	0.0 to 0.1
- without seal	Single	All	LF	0.5 to 1.5
Butterfly - rotation 0° to 60°	Single	All	LF	0.5 to 3.0
- rotation 0° to 90°	Single	All	LF	0.5 to 1.5

Table 8.11 Minimum sizing coefficients for control valves

The control range is specified preferably in terms of percentages of the maximum flow through the valve. Table 8.11 shows that the valve control range, and by implication the rangeability, varies not only with design but also with the pressure drop which determines the minimum flow.

If one considers a valve in the tightly closed position, it should, theoretically, have an infinitely large control range if it were not for the fact that the minimum valve coefficient is determined by:

- DC — Design Characteristics, the capacity of the trim to control the minimum required flow
- LF — Leakage Flow, leakage in the closed position
- CF — Clearance Flow

The minimum coefficient is the value at which the valve's change in flow, divided by the change in opening, is greater than that determined by the valve characteristic.

In practice there are three types of situation under which the valve will operate:

- Where the pressure drop across the control valve is constant, in this situation the installed characteristic will be the same as the valve's inherent characteristic, see Figure 8.14
- Where the pressure drop is small for large flows and high for small flows, see the valve sizing example in Section 8.4.3
- Where the pressure drop is small for small flows and greatest for large flows, see Figures 8.15 and 8.16 for more information

8.4.11 Installed characteristic

Generalizing, the flow through a valve is a function of the pressure drop and the valve travel. When a valve is tested on a test rig, the pressure drop is held constant so that the flow is only a function of the valve opening. This relationship is called the "inherent flow characteristic", see Figure 8.14. In reality, the valve operates under conditions where the pressure drop also varies and the relationship between the valve opening and fluid is called the "installed characteristic". The installed characteristic is, in other words, dependent upon both the valve characteristic and the system characteristic.

The pressure drop across the valve also governs how much the flow rate through the valve changes as a result of the valve opening. A large pressure drop produces a larger change in flow than a small pressure drop. If the pressure drop in the system changes so that it is greatest at small flows and small for large flows, and this is often found in systems which incorporate centrifugal pumps, the relationship shown in Figure 8.17 for a linear valve, applies.

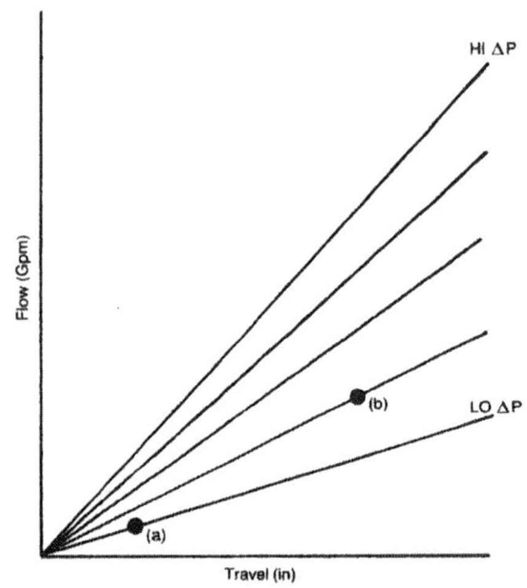

Figure 8.15 Control valve pressure drop with flow

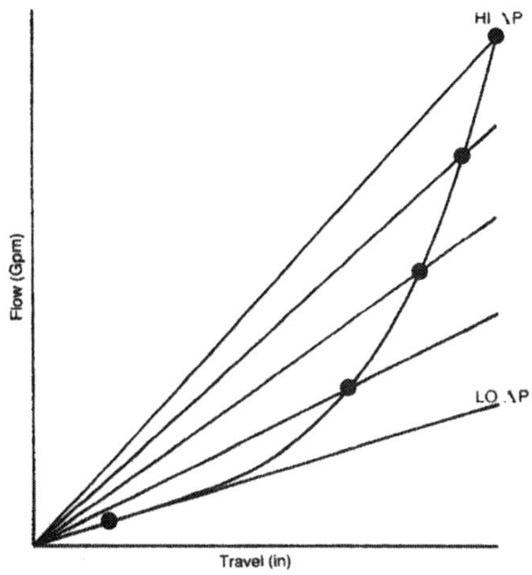

Figure 8.16 Control valve pressure drop with flow

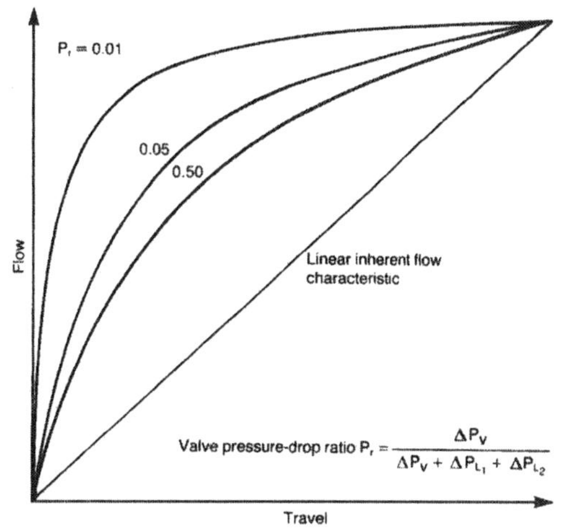

Figure 8.18 Installed flow characteristic for a linear valve

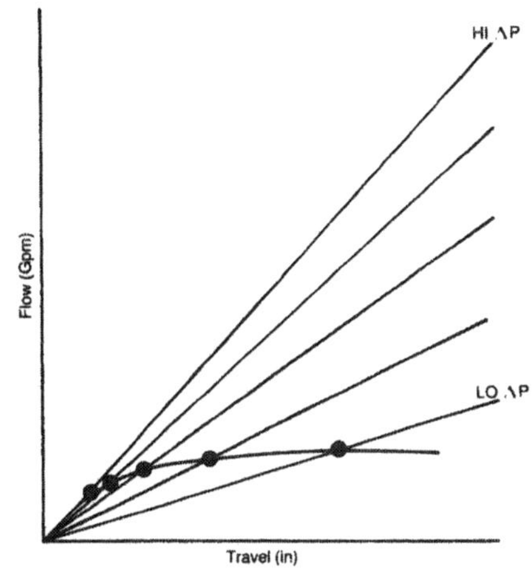

Figure 8.17 A linear valve in a centrifugal pump system

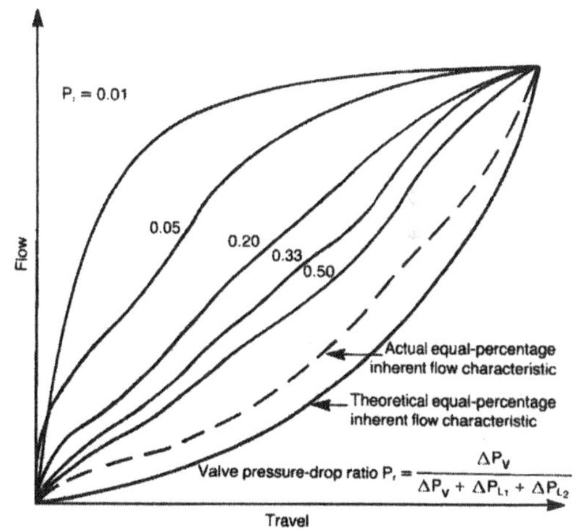

Figure 8.19 Installed flow characteristic for an equal percentage valve

The examples in Figures 8.15, 8.16 and 8.17 demonstrate an important point. There are many different installed characteristics which vary with the system characteristic, depending on the proportion of the pressure drop across the control valve in relation to the pressure available as a result of pump head and pipe friction at that rate of flow.

One of the most important questions to be answered is how large the pressure drop across the control valve should be. Figures 8.18, 8.19 and 8.20 illustrate the relationship between the valve opening and the flow, showing the inherent valve characteristic and the installed characteristic, when the ratios of pressure drop due to the valve, P_r, to the total dynamic pressure drop for the system are: 0.01, 0.05, 0.5 etc.

If, for example, we accept that an installed linear inherent characteristic is desirable, then a closer study of Figures 8.18 and 8.20 shows that inherent linear and inherent quick-opening are not suitable.

The equal-percentage inherent characteristic shown in Figure 8.19 on the other hand, is acceptable, if the valve pressure-drop ratio, P_r, the ratio of valve pressure drop to system pressure drop, equals 0.5. The disadvantage of applying the above reasoning is that it can lead to the selection of a pump having too high a delivery pressure for the actual flow, resulting in increased process running costs. An economical compromise may be necessary when making final assessments.

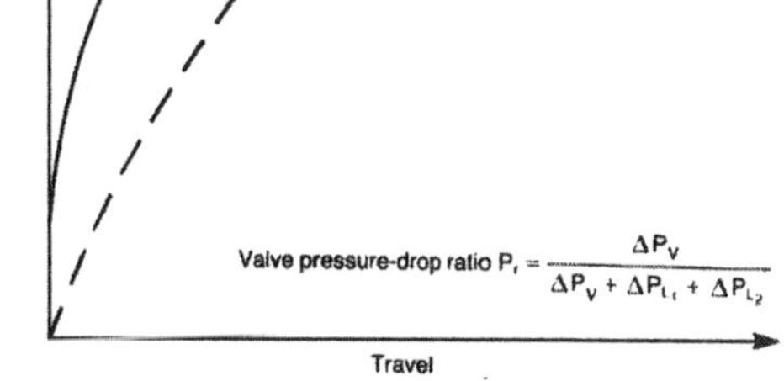

Figure 8.20 Installed flow characteristic for a quick-opening valve

In the case of Figure 8.19 we can see that $P_r = 0.5$ may be too expensive and that $P_r = 0.05$ is perhaps the most non-linear characteristic which can be accepted. By pure estimation, a P_r value between 0.1 and 0.33 may be chosen. Little however, is gained by going as high as 0.33 and so a P_r of 0.1 should be selected for the following reasons:

- **the quality of the data on which the calculations are based** — if the data is accurate then we can afford to approach $P_r = 0.1$
- **pump size** — if the pump is large, then $P_r = 0.1$ should be approached for economical reasons
- **control range requirement** — the greater the control range, the more reason to aim for $P_r = 0.33$

These rules are very generalised and should not be used to replace conclusions based on detailed, dynamic studies of the actual process.

8.4.12 Selecting a characteristic

An insight into the design and construction of the various types of plug valve and their characteristics should be taken into consideration when selecting a characteristic.

Quick-opening

The valve is often used for on/off applications because the full C_v is achieved with a short stroke.

Linear

The relationship between opening and flow is almost a straight line. A given change in opening results in an equally large change in the flow.

Equal percentage

Equal variations in valve travel result in equal changes in the percentage flow rate. When the valve plug is near the seat and the flow is small, then the relative changes in flow will also be small. When the valve plug is near the fully open position and the flow is large, then the relative changes in flow will also be large. This type of characteristic is often used in processes where a large portion of the pressure drop is taken up by the system itself and only a small proportion occurs across the valve, as for example, in a pressure control system. In the case of larger, varying pressure conditions in the system, then this type of valve is often preferred.

In Tables 8.12, 8.13 and 8.14 the guidelines are based on experience, the various characteristics having proved to be advantageous for different conditions.

8.5 Sizing piping

Pipe sizing is accomplished initially by using guidelines for acceptable fluid velocities. If the approximations indicate pipe sizes which fit in with major process equipment, then detailed pressure loss calculations are performed to allow the process design to be finalised. If pressure losses are too high then the pipe size must be increased. If pressure losses are very low then it may be possible to use smaller pipe sizes if the increased fluid velocity does not create other problems.

Erosion by entrained solids will increase proportionally with velocity$^{2.5}$ as a minimum; increasing the flow velocity by 32% will at least double the erosion rate. High velocity clean liquids can effectively erode materials without entrained solids. Water at 40m/s can remove the protective oxide layer from austenitic stainless steel resulting in rapid material loss.

Higher fluid velocities demand better pipework designs and higher quality manufacture and assembly. Poorly fitted slip-on flanges, badly aligned, will cause extra turbulence and pressure losses. Bent pipe is preferable to fabricated bends. In most cases, large radius should be used in place of short. The minimum number of connections, well-aligned, such as weld neck flanges with ring-type joints or clamped connections fittings, will preserve good flow patterns.

8.5.1 Flow velocities

Tables 8.15 and 8.16 indicate the range of velocities normally associated with specific liquid and gas/vapour applications.

Process type	Preferred valve characteristic
Liquid processes, in general as in pumped circulation	Equal percentage
Gas and steam flows at constant upstream pressure, large volume. (Total flow volume in process corresponds with nominal pipe diameter). Reducing pressure drop with increasing demand. Pressure drop at maximum demand greater than 20% of the pressure drop at minimum demand. Liquid flow under gravity, constant head.	Linear
Gas processes, large volume. Reducing pressure drop with increasing demand. Pressure drop at maximum demand less than 20% of pressure drop at minimum demand.	Equal percentage

Table 8.12 Pressure control

Process type		Preferred valve characteristic	
Signal output from transmitter	Location of control valve downstream of the measuring element	Large flow variations	Small flow variations, but large pressure drop changes across the control valve with increased demand
Proportional to the rate of flow (Rotameter, Magnetic Flowmeter)	In series	Linear	Equal percentage
	In bypass	Linear	Equal percentage
Proportional to square of the rate of flow (Orifice, Venturi)	In series	Linear	Equal percentage
	In bypass	Equal percentage	Equal percentage

Table 8.13 Flow control

Process type	Preferred valve characteristic
Constant pressure drop	Linear
Reducing pressure drop with increased demand. Pressure drop at maximum demand greater than 20% of pressure drop at minimum demand.	Linear
Reducing pressure drop with increased demand. Pressure drop at maximum less than 20% of pressure drop at minimum demand.	Equal percentage
Increasing pressure drop with increased demand. Pressure drop at maximum less than 200% of pressure drop at minimum demand.	Linear
Increasing pressure drop with increased demand. Pressure drop at maximum demand greater than 200% of pressure drop at minimum demand.	Quick opening

Table 8.14 Level control

The actual pressure losses must be evaluated to check suitability.

Liquid application	m/s
Dirty water	
Mine water lines	1.75 to 2.5
Hot water	
Condensate	1.0 to 1.5
District heating distribution	1.0 to 2.5
District heating service main	0.75 to 2.0
Domestic heating circulation	0.3 to 0.6
Feed water lines	1.2 to 4.3
Crude oil	
Oil pipelines	0.25 to 1.8
Lube oil	
Drain line (1)	0.15 to 0.75
Feed lines	1.0 to 2.3
Pump discharge line	0.6 to 1.8
Pump suction line (2)	0.45 to 1.5
Pump suction line (3)	0.3 to 1.0
Petrol	

Liquid application	m/s
Pump discharge	1.0 to 1.4
Pump suction	0.45 to 0.75
Refrigerant	
Condenser to receiver	≤ 1.0
Receiver to system (4)	≤ 1.5
Sewage	
Pump discharge, centrif	≥ 2.4
Pump suction, centrif	≥ 1.5
Warm water	
Economiser tubes	0.75 to 1.5
Water	
Pump discharge, centrifugal	8 to 15
Pump discharge, reciprocating	3.0 to 4.5
Concrete suction bays	≤ 0.3
Pump suction headers	≤ 1.0
Pump suction, centrifugal	0.45 to 1.5
Pump suction, reciprocating	≤ 0.6
Consumer supply	1.0 to 1.4
Service mains	0.6 to 1.5
Trunk main	1.0 to 3.0

(1) 1:50 fall minimum, half full, (2) flooded suction, (3) suction lift, (4) 0.115 bar / 25m maximum

Table 8.15 Guideline flow velocities for liquid applications

Gas vapour application	m/s
Compressed air	
Distribution main	5 to 10
Shop ring main	10 to 80
Natural gas	
Distribution mains	0.4 to 0.75
Domestic/commercial systems(1)	0.00125 bar max pressure drop
Reciprocating process compressor	
Suction	12 to 15
Discharge	20 to 25
Refrigerants	
Compressor discharge	0.04 to 011 bar/10m
Compressor suction	
-51 to -7°C	0.005 to 0.05 bar/10m
> -7°C	0.02 to 0.07 bar/10m
NH3 -51 to -7°C	0.005 to 0.025 bar/10m
Steam	
Bleed lines	20 to 30
Boiler outlets	30 to 60
Branch lines	30 to 75
Industrial distribution line	
< 1 barg	20 to 30
1 to 10 barg	30 to 50
> 10 barg	50 to 100
Commercial heating supply line	
< 0 barg	0.003 to 0.006 bar/10m
0 to 1 barg	0.006 to 0.022 bar/10m
2 to 10 barg	0.045 to 0.22 bar/10m
Headers	30 to 40
Superheater tubes	10 to 25
Turbine exhausts	30 to 75
Turbine feeds	30 to 60
Vacuum lines	100 to 200

(1) 0.0344 barg regulated systems

Table 8.16 Guideline flow velocities for gas and vapour applications

8.5.2 Pipe flow losses

8.5.2.1 Reynolds Number

Flow losses occur because of the effect of internal friction. Shear stresses arise as soon as velocity gradients exist.

$$\tau = \mu \cdot \frac{dv}{dy}\left(N/m^2\right) \qquad \text{Equ 8.31}$$

where:

τ = shear stress (N/m^2)

μ = dynamic viscosity (Ns/m^2)

v = flow velocity (m/s)

y = thickness of flow stream (m)

The work done by the shear force generates heat and adds to the internal energy of the liquid. The increase in internal energy means that the static pressure becomes a little less hence the term: pressure loss or pressure drop. The level to which the liquid temperature can rise due to friction heating is dependent upon the heat transfer mechanism to the surroundings.

The flow losses are great wherever the shear stress is great, i.e. where the velocity gradients are great. High velocity gradients occur in the boundary layer, at flow around sharp corners, in strong vortex flows and so on. Viscous flow in pipes is characterised by Reynolds Number Re.

$$Re = \frac{v \cdot d}{\nu} \qquad \text{Equ 8.32}$$

where:

v = flow velocity (m/s)

d = pipe diameter (m)

n = kinematic viscosity (m^2/s)

The following relationship exists between kinematic or flow viscosity and dynamic viscosity:

$$\nu = \frac{\mu}{\rho} \qquad \text{Equ 8.33}$$

where:

ν = flow velocity (m/s)

μ = dynamic velocity (Ns/m^2)

ρ = liquid density (kg/m^3)

In the case of low Reynolds Numbers, Re <2300, the flow is laminar. The liquid flows in layers with different velocities which do not mix with each other. At Re > 2300 the flow is turbulent, that is to say the liquid particles do not flow in a straight line in the direction of motion of the main flow. See Figure 8.21.

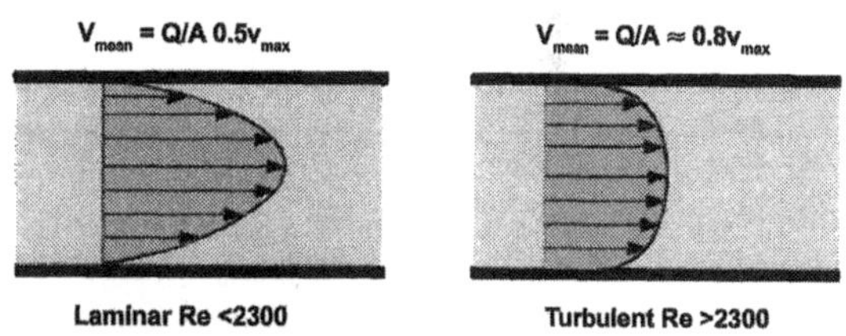

Figure 8.21 Laminar and turbulent flow

Since the velocity gradients differ, the shear stress and flow losses will also differ between the two cases. The critical value of Re Number has been stated as $Re_{cr} \cong 2300$. It should be noted however, that the flow may continue to be laminar at considerably higher values of Re if the flow is very well protected from disturbances and entrance conditions to the section are the optimum. Low values of Re Number occur at low flow velocities, in small pipe diameters or at high viscosity.

8.5.2.2 Head losses in straight pipes

Table 8.17 gives the transition to turbulent flow for liquids, for a range of pipe diameters. It can be seen from the examples that laminar flow is less usual in practice for liquids with viscosity similar to that of water, but is quite common for oils.

The following formula applies to head losses, h_f, for flow in a straight pipe:

Liquid	Kinematic Viscosity n m^2/s	Transition to turbulent flow (for the pipe diameters given) V_{cr} m/s			
		d = 0.001m	d = 0.01 m	d = 0.1 m	d = 1 m
Water, 20 °C	1×10^{-6}	2.3	0.23	0.023	0.0023
SAE 5W 60 °C	1×10^{-5}	23	2.3	0.23	0.023
Burner oil Class E (BS 2869: 1970) 30 °C	1×10^{-4}	230	23	2.3	0.23
SAE 50 18 °C	1×10^{-3}	2300	230	23	2.3

Table 8.17 Transition to turbulent flow in pipes

$$h_f = \lambda \cdot \frac{l}{d} \cdot \frac{v^2}{2g} \quad (m)$$ Equ 8.34

where:

λ = loss coefficient for straight pipes (dimensionless)

l = length of pipe (m)

d = pipe diameter (m)

v = Q/A = flow velocity (m/s)

The loss coefficient is dependent upon Reynolds Number and upon the internal roughness of the pipe. This is shown in Figure 8.22.

The roughness of the pipe can be assessed from the values in Table 8.18.

Pipe and pipe material	Condition of pipe	Factor k Calculation values m
Steel pipe	Seamless, new	0.00003 - 0.00005
	Welded, new	0.00003 - 0.0001
	in use, rusted	0.00015 - 0.0005
	Seamless, galv.	0.0001 - 0.00016
	Welded, galv.	0.0001 - 0.0002
	Heavy asphalt coating	0.0005
	Mature water mains	0.0012
	Stainless, acid resist.	0.000045
Cast iron pipe	Bitumen coated	0.0001-0.0003
	Main conduit	0.0001
	Other conduits	0.0002-0.0003
	in use, rusted	0.0004 - 0.002
Copper pipe	Diameter <200 mm	0.00001
Brass, glass		
Al	Technically smooth	0.000002
Concrete pipe	Water main	0.0001-0.0002
	Waste, drains, new	0.0002-0.0005
	Waste, drains, old	0.001
Asbestos cement pipe	Newly laid	0.00005
Earthenware pipe	Newly laid drains	0.0002
	Older drains	0.001
	Mature foul sewer	0.003
PE, PVC	Dia. <200> mm	0.00001
	Dia. 200 mm	0.00005
GRP	All dimensions	0.0001
Wood	New	0.0002-0.001

Table 8.18 Examples of approximate roughness in pipes

The loss coefficient can also be calculated from the following formulae:

Laminar flow Re < 2300

$$\lambda = \frac{64}{Re}$$ Equ 8.35

Turbulent flow Re > 4000

$$\frac{1}{\sqrt{\lambda}} = -2 \cdot \log_{10}\left(\frac{2.51}{Re\sqrt{\lambda}} + \frac{k/d}{3.71}\right)$$ Equ 8.36

Equation 8.36 is sometimes called the "Colebrook-White" equation; "Colebrook" in France and "Prandtl-Colebrook" in Germany.

In the region Re = 2300 to Re = 4000 laminar and turbulent flows may alternate in various sections of the pipe. The resultant loss coefficient will then assume values between those shown in Equations 8.35 and 8.36.

The expression given in Equation 8.36 is somewhat inconvenient for manual calculation. The following formula, proposed by Miller, can be used for turbulent flow:

$$\lambda = \frac{0.25}{\left[\log_{10}\left(\frac{k}{3.7d} + \frac{5.74}{Re^{0.9}}\right)\right]^2}$$ Equ 8.37

Other formulae which often occur in literature for the calculation of straight pipe losses in the turbulent region are the "Hazen-Williams and Manning" formulae. These, however, do not provide any additional information to the equations presented above and have, therefore, been omitted.

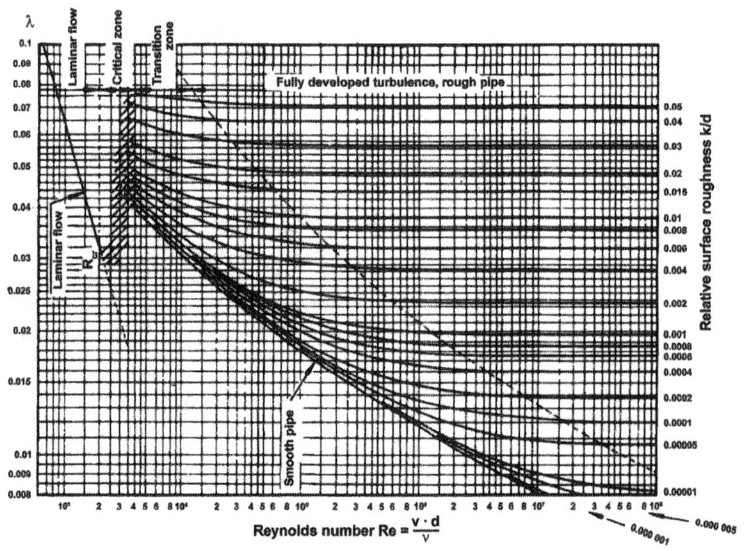

Figure 8.22 Loss coefficient for straight pipes

8.5.2.3 Head losses in fittings

Head losses in bends, valves, etc. can be calculated with the help of this formula:

$$h_f = z \cdot \frac{v^2}{2\,g} \quad \text{(m)} \qquad \text{Equ 8.38}$$

where:

z = loss coefficient of fitting (dimensionless)

v = flow velocity (m/s)

The magnitude of the head losses are in principle influenced, as in the case of straight pipe flow, by surface roughness and by Reynolds Number. Examples of approximate values of loss coefficients for losses in fittings are illustrated in Figure 8.23 All values are those for conditions of normal roughness and at high Reynolds Number, i.e. at fully developed turbulent flow. In cases where the velocity changes it must always be the highest velocity which is used for calculating h_f in accordance with Equation 8.38. In the case of varying diameters, the velocity to be applied is the one resulting from the smallest diameter, and in the case of T-pieces, the velocity related to the total volume flow is applicable.

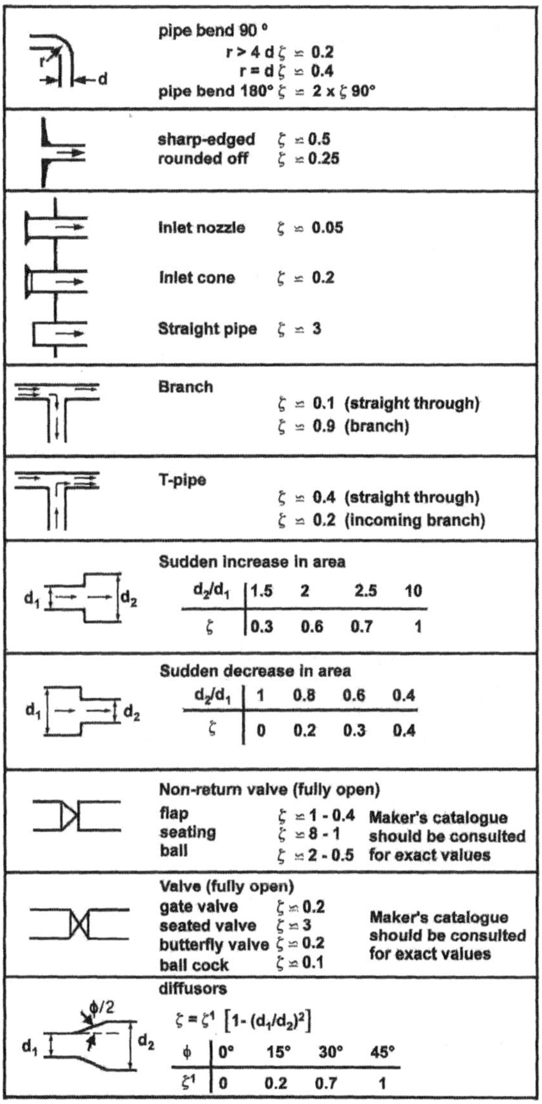

Figure 8.23 Approximate values for loss coefficients for head losses in fittings

An alternative method of expressing the magnitude of head losses in fittings is the use of the concept of equivalent pipe length, l_{eq}, i.e. the equivalent length, (l_{eq}) of straight pipe giving rise to the same losses in pressure as the fitting, at the same flow velocity.

By comparing Equations 8.34 and 8.38:

$$l_{eq} = \frac{\zeta}{\lambda} \cdot d \qquad \text{Equ 8.39}$$

where:

ζ = loss coefficient of fitting (dimensionless)

λ = loss coefficient for straight pipes (dimensionless)

It can be seen from Equation 8.39 that the equivalent pipe length is a function of l, i.e. the surface roughness of a straight pipe and Reynolds Number. It is therefore difficult to quantify l_{eq}, as a specific property for, say, a pipe bend and is thus of an approximate character.

For $\lambda = 0.02$, the equivalent pipe length of a short radius 90° bend (r = d) is $l_{eq} = 20d$.

For additional information and data on losses see Crane Company (USA) Technical Paper 410.

8.5.2.4 Hydraulic diameter

For flow in a partially filled pipe, in non-cylindrical ducts or in open channels the head losses can be calculated in principle in the same way as described in earlier Sections. Instead, however, of using the diameter of the pipe, the hydraulic diameter, d_h, must be used in these cases.

$$d_h = \frac{4 \cdot A}{P} \qquad \text{Equ 8.40}$$

where:

d_h = hydraulic diameter (m)

A = cross-sectional area flowing (m^2)

P = wetted perimeter (m)

For a completely full cylindrical pipe

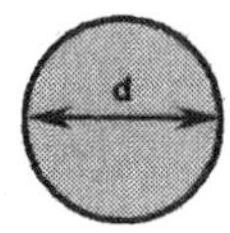

$$d_h = \frac{4 \dfrac{\pi \cdot d^2}{4}}{\pi \cdot d} = d$$

i.e. in this case the hydraulic and geometric diameters coincide.

For a half-filled cylindrical pipe,

$$d_h = \frac{4 \cdot \dfrac{\pi \cdot d^2}{2 \cdot 4}}{\dfrac{\pi \cdot d}{2}} = d$$

For a half-filled rectangular section,

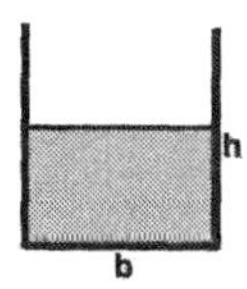

$$d_h = \frac{4 \cdot b \cdot \frac{h}{2}}{b + 2 \cdot \frac{h}{2}} = \frac{2 \cdot b \cdot h}{b + h}$$

With Reynolds Number

$$Re = \frac{v \cdot d_h}{n} \qquad \text{Equ 8.41}$$

and relative roughness k/d_h, Figure 8.22 and the formulae in Equations 8.34 and 8.36 apply with unchanged numerical values.

The value of Re_{cr} will change as the section proportions change and will have to be evaluated by experiment or computer simulation. Pipes flowing entirely due to gravitational head can oper-

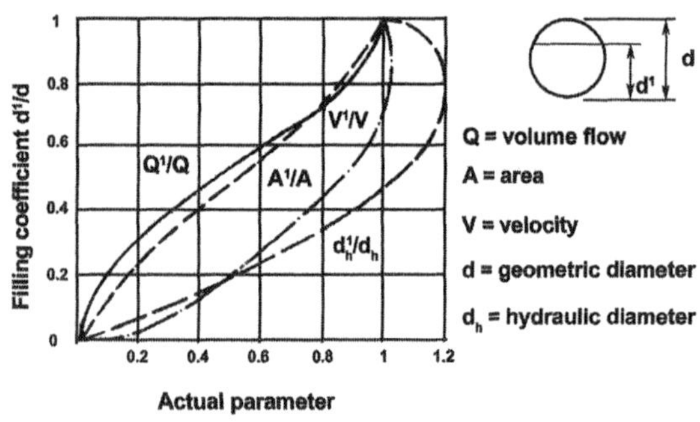

Figure 8.24 Flow in a partially-filled pipe

ate completely filled. The degree of fullness depends, among other things, on the slope of the pipe and the operating conditions.

Pipe fullness is defined by the ratio d^1/d where d^1 is the depth of liquid in the pipe and d is the diameter of the pipe. Figure 8.24 illustrates how various flow parameters change with the degree of fullness of the pipe. Both flow velocity and hydraulic diameter have their greatest value at a filling coefficient of just below 1.

For gravity flow pipes carrying solids and liquids care must be taken with regard to sedimentation; the avoidance of which can place restrictions on the viable configuration and operation of the pipe.

8.5.2.5 Total head losses in the pipe system

The system total losses consist of the sum of the straight pipe head losses, the head losses in fittings and valves etc. and static losses due to changes in elevation.

$$h_f = h_f\,(\text{straight pipe}) + h_f\,(\text{fittings and valves}) + \Delta z \qquad \text{Equ. 8.42}$$

Using the usual symbols, head loss:

$$h_f = \lambda \cdot \frac{1}{d} \cdot \frac{v^2}{2g} + \Sigma\zeta \cdot \frac{v^2}{2g} + \Delta z$$

$$h_f = \left(\lambda \cdot \frac{1}{d} + \Sigma\zeta\right) \cdot \frac{v^2}{2g} + \Delta z$$

$$h_f = \left(\lambda \cdot \frac{1}{d} + \Sigma\zeta\right) \cdot \frac{Q^2}{\left(\frac{\pi \cdot d^2}{4}\right)^2} \cdot \frac{1}{2g} + \Delta z \qquad \text{Equ 8.43}$$

where:

h_f	=	head loss (m)
λ	=	loss coefficient for straight pipe
ζ	=	loss coefficient for fittings, valves etc.
$\Sigma\zeta$	=	sum of all loss coefficients
l	=	length of pipe (m)
d	=	diameter of pipe (m)
Q	=	volume flow (m^3/s)
v	=	flow velocity (m/s)
Δz	=	$(z_{II} - z_I)$ (m)

Another useful method for calculating head losses in fittings is the Two-K method, proposed by William B. Hooper.

Pipe flow losses can also be evaluated by converting fittings and bends into equivalent straight lengths of pipe. This type of data is available in many handbooks produced by valve, compressor and pump manufacturers.

8.5.3 Economics

Another method of selecting the pipe size is to consider the financial implications of the investment involved. Any given pipe size will incur installation and operational costs. The installation costs must include the costs of the pipe and the pipe and connectors, valves and fittings, supports and foundations and labour.

Once the pipework is installed, operational costs will be incurred due to energy consumption, and perhaps physical loss due to corrosion or erosion. A study of pumps used in paper making showed the energy cost during the useful life amounted to seven times the cost of the pump. The same study also indicated the cost of the pipework to be five times the cost of the pump. The cost of pipework and energy should not be treated lightly. The payback period of the investment will be fixed by corporate policy. Pipework may be categorised for "write-off" differently to equipment such as pumps and compressors and may be as short as five years.

The evaluation process commences by selecting a pipe which is too small, that is a pipe size which produces flow velocities which the "Engineering" department would not accept. The installation and operational costs are calculated. The evaluation is repeated for incremental increases in pipe size. The cost of construction and the energy cost can be plotted against the pipe size; this will indicate the optimum values and the rate of change. The payback period can also be plotted against the pipe size. This graph should not be a smooth curve because of all the different parameters which come into effect. A sound financial judgement can be made based on the two graphs.

The financial approach is not quite as simple as the foregoing description might suggest. Small pipework, running fast, will require a different philosophy to larger, slower pipework. Eddies and turbulence created by poorly aligned connections will cause operational problems with compressors, pumps and control valves. Better quality connectors are necessary. Small pipework creating higher pressure drops may need to have thicker walls to withstand the higher pressures. When considering the suction pipework for pumps, the Net Positive Suction Head available (NPSHa) or Net Positive Inlet Pressure available (NPIPa) will be reduced with smaller pipes. This will modify the pump selection as well as increasing the differential pressure.

Very low NPSHr/NPIPr requirements usually mean large pumps running slow; with increased initial cost plus larger space requirements and bigger foundations. Pump efficiency is greatly affected by the pump size which will influence the cost of energy consumption by the pipework. Suction and discharge pipework for compressors and pumps are usually of different sizes, which is a further complication.

Pipework for compressors and pumps must be evaluated as a system; suction pipework plus machine plus discharge pipework. The companion volume to this book, *Pumping Manual International*, may be helpful for pipe and financial interaction.

8.6 Useful references

API RP 520 Sizing, Selection and Installation of Pressure-Relieving Devices in Refineries, 7th American Petroleum Institute 2000.

ISA-75.11.01-1985 (R2002) Inherent Flow Characteristics and Rangeability of Control Valves.

BICERI testing progress, B Wilson, Industrial Lubrication and Tribology, Volume 48, Number 6, 1996, pp. 6-9(4) Emerald Group Publishing Ltd.

BS EN 60534-2-1 Industrial-process control valves - Part 2-1: Flow Capacity - Sizing Equations for Fluid Flow under Installed Conditions.

BS EN 60534-2-5:2003 Industrial-process control valves - Flow capacity - Sizing equations for fluid flow through multistage control valves with interstage recovery.

DIN EN 60534-2-3 Industrial-process control valves - Part 2-3: Flow capacity; test procedures (IEC 60534-2-3:1997).

ANSI/ISA 75.19.01-2001 Hydrostatic Testing of Control Valves 2001.

BS EN 60534-4:2006 Industrial-process control valves. Inspection and routine testing.

IEC 60534-2-3 Industrial-process control valves - Part 2-3: Flow capacity - Test procedures.

Verschleiß selbsttätiger Pumpenventile durch abrasive Suspensionen / Wear of automatic pump valves by abrasive suspensions, G Vetter, U Störk, Chemie Ingenieur Technik, Volume 59, Issue 5, pp 383-392, 2004.

Explicit Design of Commercial Pipes with No Secondary Losses, D H Yoo and V P Singh, Journal of Irrigation and Drainage Engineering, Volume 130, Issue 5, pp 437-440, 2004.

Flow of Fluids through Valves, Fittings, and Pipe, Crane Technical Paper No. 410 (TP-410), Flow of Fluids, 4529 Intelco Loop SE, Suite A, Lacey, WA 98503 USA, Tel: 360 292 4070, www.tp410.com.

Internal Flow Systems, 2nd edition D S Miller, 1990, ISBN 0-87201-020-1.

Handbook of Hydraulic Resistance Idelchik, 3rd edition, Begell House Publishers; 2001, ISBN-13: 978-1-56700-0740, ISBN-10: 1-56700-07-46,

The Two-K Method Predicts Head Losses in Pipe Fittings, W B Hooper, Chemical Engineering, p. 97-100, August 24, 1981.

Correlate Pressure Drops through Fittings, R Darby, Chemical Engineering, pp 101-104, July, 1999.

Guidelines for Pressure Relief and Effluent Handling systems, pp 265-268, AIChE Center for Chemical Process Safety, New York, 1998.

Pumping Manual International, Edited by B Nesbitt, Elsevier, 2006, ISBN-13:978-1-85617-476-3. ISBN-10:1-85617-476-X.

API (American Petroleum Institute),1220 L Street, NW, Washington, DC 20005-4070, USA, Tel: 202-682-8000, www.api.org

ASME (American Society of Mechanical Engineers), Three Park Avenue, New York, NY 10016-5990, USA, Tel: 973 882 1167, www.asme.org.

Lloyd's Register EMEA, 71 Fenchurch Street, London EC3M 4BS, UK, Tel: 020 7709 9166, Fax: 020 7488 4796, www.lr.org.

Piping and connectors 9

9.1 Pipework design principles

The prime function of a piping system is to convey fluid and/or solids. Depending upon the particular nature of the process the piping system may also be required to protect the fluid or protect the environment/personnel from the fluid. Protection of the fluid may:

- simply require the correct selection of materials to avoid corrosion products contaminating the fluid
- involve the elimination of fluid contact with a wide range of chemical and biological agents normally present in the environment.

The second type of protection is usually associated with hygienic or biochemical processes but is also becoming more common in certain chemical processing. Chemical cleaning or sterilising of piping systems can be more arduous than the primary fluid functions. Secondary process requirements, such as cleaning, must not be overlooked when defining operating limitations. Sterilising or steam cleaning may take place at much higher temperatures than primary fluid processing. Protecting the "world" from the fluid usually means providing adequate pressure containment and eliminating potential leak paths.

Pipework systems are designed to "codes" or regulations; a set of rules which define materials, allowable stress values and the type of connections which are permitted. A large proportion of piping design is performed almost automatically by computer software because the "code" decisions necessary are very logical. Dedicated pipework design programs and sub-routines within CAD packages perform the labourious stress calculations and evaluate the pipe flexibility. Pipe wall thickness or material can be changed and the effects on stressing and flexibility can be evaluated very quickly. The pipework designer must be aware of any constraints imposed by individual components and check these areas for over-stressing or distortion.

Sophisticated computer software will evaluate a piping design for mechanical vibration. A pipe length suspended between two supports will have a natural frequency depending upon the pipe, the fluid and the type of supports. It is usually a manual task to check if any of the pipe natural frequencies coincide with a forcing frequency normally present. Compressors and pumps can inject various pulsations into the fluid which may excite one of the pipe natural frequencies. Once excited, a section of pipe may resonate, leading to early fatigue failure. A full understanding of the equipment used is essential in predicting the important exciting frequencies. Heavy weights, such as valves or flanges, at the centre of a pipe span can dramatically reduce the mechanical natural frequency. Adequate supports should be provided to maintain natural frequencies high enough to be unaffected by system pulsations.

Machinery is not the only source of pressure pulsations. The pipework itself can generate specific frequencies due solely to the flowing fluid. Branches into main lines can create eddies and vortices. If the vortices are "shed" in a regular pattern then an exciting frequency is produced. Branches of specific lengths can produce an acoustic tone which may excite other forms of vibration.

Most computer software does not recognize the possibility of acoustic resonance. The pressure pulsations mentioned earlier can excite a pipe system in a similar manner to that of an organ pipe. The pipe length and fluid properties combine to produce a set of acoustic harmonics. When resonating, the acoustic waves can create pipework vibration and cause machinery failure due to the increased levels of pressure pulsa-

Standard	Grade	Major constituents	Min tensile strength MPa	Max temp °C	Remarks
BS 1387		P & S 0.06 max	320/460	260	@ 21 barg
BS 3601 ISO 2604 Pt2 ISO 2604 Pt3	320 TS1 &TS2 TW1 & TW2	0.5Mn 0.16C	320/440	≈400	Welded and seamless
BS 3601 ISO 2604 Pt2 ISO 2604 Pt3	360 TS4 &TS5 &TS6 TW4 & TW5 & TW6	0.6Mn 0.17C	360/480	≈450	Welded and seamless 27J @ -40°C 27J @ -40°C
BS 3602 DIN 17175	HFS 410 ST45.8	0.8Mn 0.21C	410/550 410/530	450	
BS 3603	410LT50	0.9Mn 0.2C	410/530	≈50	27J @ -50°C
BS 3602 DIN 17175	HFS 460 17 Mn 4	1Mn 0.22C	400/600	450	
BS 3602 ISO 2604 Pt2	490 Nb TS18	1.05Mn 0.23C	490/630 490/610	450	
BS 6323	Pt3 HFS5	1.5Mn 0.23C	490		
BS 6323	Pt4 CHS3	0.9Mn 0.2C	450		
DIN 17175	ST35.8	0.6Mn 0.17C	360/480		
DIN 17121	ST52.3	1.6Mn 0.2C	490/630		27J @ -20°C
DIN 1629	ST35 ST35.4	0.18C 0.4Mn 0.17C	343/441		
DIN 1629	ST52	1.5Mn 0.2C	520/620		
API 5L	A B	0.9Mn 0.22C 1.15Mn 0.26C	331 413	205	Seamless and welded
API 5L	X52	1.35Mn 0.31C	455	205	Seamless
ASTM A53	A B	0.95Mn 0.25C 1.2Mn 0.3C	331 413	590	Welded and seamless
ASTM A106	A B C	0.6Mn 0.25C 0.65Mn 0.3C 0.65Mn 0.35C	331 414 485	425	Seamless -29°C min
ASTM A135	A B	0.95Mn 0.25C 1.2Mn 0.3C	331 414	590	ERW pipe
ASTM A155	C45 C50 C53	0.9Mn 0.17C 0.9Mn 0.22C 0.9Mn 0.28C	310/380 350/410 380/450		Large diameter EFW pipe
ASTM A333	1 6	0.7Mn 0.3C 0.65Mn 0.3C	379 414	590	-45°C
ASTMA A334	1 6	0.7Mn 0.3C 0.65Mn 0.3C	379 414		
ASTM A587		0.45Mn 0.15C	331	340	Welded pipe

Table 9.1 Carbon steels for pipes and tubes

Standard	Grade	Major constituents	Min tensile strength MPa	Max temp °C	Remarks
ASTM A335	P1	0.55 Mo 0.55Mn 0.15C	379	≈400	Seamless pipe
BS 3604	660	0.6Mo 0.55Mn 0.45Cr 0.15C	460/610	580	Heat treated alloy steel
DIN 17175	14Mo V 63				Seamless
ISO 2604 Pt2	TS33				
ASTM A335	P2		379		
BS 3604	621	1.25Cr 0.55Mn 0.45Mn 0.12C	420/570	≈570	
ASTM A335	P11		414		Seamless
ASTM A155	1.25CR		410/590		EFW large diameter
BS 3604	622	2.25Cr 1.15Mo 0.55Mn 0.15C	490/640	≈600	Heat treated alloy steel
DIN 17175	10Cr Mo 910				Seamless
ASTM A335	P22		414		
ISO 2604 Pt2	TS34		410/560		
ASTM A155	2.25CR		410/590		EFW large diameter
BS 3604	625	5Cr 0.5Mo	450/600	≈650	
ASTM A335	P5		414		Seamless
ISO 2604 Pt2	TS37		410/560		
ASTM A155	5CR		410/590		EFW large diameter
BS 3603	503LT100	3.5Ni 0.55Mn 0.15C	440/590	≈50	27J @ -100°C
ASTM A333	3				
ISO 2604 Pt2	TS43				
BS 3604	509LT196	9Ni 0.55Mn 0.12C	690/840	≈50	39J @ -196°C
ASTM A335	8		689		
ISO 2604 Pt2	TS45		690/840		
BS 3604	629	9Cr 1Mo 0.45Mn 0.15C	470/620	≈600	
ASTM A335	P9		414		Seamless
ISO 2604 Pt2	TS38		410/560		
BS 3604	762	11.25Cr 1Mo 1Mn 0.55Ni 0.2C	720/870	≈600	
DIN 17175	X20CrMo V 12 1		690/840		
ISO 2604 Pt2	TS40		690/840		
ISO 2605 Pt2	TS40	11.5Cr 1Mo 1Mn 0.55Ni 0.2C	690/840	≈600	Seamless
3CR12		11.5Cr 8Ti 1.5Ni 1.7Mn 0.03C + N_2	460	550	Welded

Table 9.2 Alloy steels for pipes and tubes

tions. Acoustic resonance can remove NPSH/NPIP margins

Standard	Grade	Major constituents	Min tensile strength MPa	Max temp °C	Remarks
BS3605	304S25	18Cr 9.5Ni 2Mn 0.06C	490/690	450	Welded
ISO 2605 Pt2	TS47	18Cr 10Ni 2Mn 0.07C	490/690		Seamless 71J @-196°C
BS 3605	304S18	18Cr 10.5Ni 2Mn 0.06C	490/690	450	Seamless -225°C
ISO 2605 Pt2	TS46	18Cr 11Ni 2Mn 0.03C	460/490		Seamless 71J @-196°C
ISO 2605 Pt2	TS56	18Cr 12.5Ni 2Mn 0.07C	510/710		Seamless
ASTM A269	TP304	19Cr 9.5Ni 2Mn 0.08C	520	≈815	Seamless and welded
ASTM A312	TP304				Seamless and welded
ASTM A430	FP304				Forged seamless -250°C
BS 3605	316S18	17.25Cr 12.5Ni 2.5Mo 2Mn 0.07C	510/710	450	Seamless -225°C
BS 3605	316S26	17.25Cr 11.5Ni 2.5Mo 2Mn 0.07C	510/710	450	Welded
ASTM A269	TP316	17Cr 12.5Ni 2.5Mo 2Mn 0.08C	520	≈815	Seamless and welded
ASTM A312	TP316				Seamless and welded
ASTM A430	FP316				Forged seamless -198°C min
ISO 2605 Pt2	TS63	17Cr 13Ni 2.5Mo 1.5Mn 0.06C	510/710		Seamless
ISO 2605 Pt2	TS60	17.25Cr 12.5Ni 2.25Mo 2Mn 0.07C	510/710		Seamless 63J @ -196°C
ISO 2605 Pt2	TS57	17.25Cr 12.5Ni 2.25Mo 2Mn 0.03C	490/690		Seamless 63J @ -196°C
BS 3605	316S14	17.25Cr 13.5Ni 2.5Mo 2Mn 0.03C	490/690	450	Seamless
BS 3605	316S22	17.25Cr 12.5Ni 2.5Mo 2Mn 0.03C	490/690	450	Welded
ASTM A269	TP316L	17Cr 12.5Ni 2.5Mo 2Mn 0.035C	480	≈815	Seamless and welded
ASTM A312	TP316L				Seamless and welded -198°C
ASTM A269	TP317	19Cr 12.5Ni 3.5Mo 2Mn 0.08C	520	≈815	Seamless and welded
ASTM A312	TP317				Seamless and welded -198°C
ASTM A269	TP317L	19Cr 13.5Ni 3.5Mo 2Mn 0.035C			Seamless and welded
ASTM A312	TP317L				Seamless and welded
ASTM A409	304, 316, 316L 317				EFW 14" to 30" nb sch 5S and 10S
ASTM A269 (21-6-9)	6Mo	19.5Cr 9Mo 6Ni		≈480	Welded and drawn, high pressure tube for hydraulics
254SMO	S31254	20Cr 18Ni 6Mo 0.01C	650/900	≈400	Seamless
ASTM A269	S31254				
ASTM A316	S31254				
ASTM A789	S31803	22Cr 5.5Ni 3Mo 2Mn 0.03C + N_2	680/880	≈300	Seamless
ASTM A790	S32304	23Cr 4.25Ni 2.5Mn 0.3Mo 0.03C + N_2	600	≈300	Welded and seamless
ASTM A790	S32750	25Cr 7Ni 4Mo 1.2Mn 0.03C + N_2	800	≈300	Welded and seamless
ASTM A790	S31200	25Cr 6Ni 1.5Mo 2Mn 0.03C + N_2	690	≈300	Welded and seamless
SAF 2507	S32570	25Cr 7Ni 4Mo 1.2Mn 0.03C + N_2	800/1000	≈315	Seamless and welded
ASTM A790	S32900	25.5Cr 3.75Ni 1.5Mo 1Mn 0.08c	620	≈300	Seamless and welded
ASTM A668	N08028	31Ni 27.5Cr 3.5Mo 2Mn 1.4Cu 0.02C	500/750	≈400	Seamless

Table 9.3 Stainless steels for pipes and tubes

Standard	Grade	Major constituents	Min tensile strength MPa	Max temp °C	Remarks
BS 3074	NA18	63Ni 30Cu 2.7Al 2Fe 1.5Mn	900	≈400	Seamless tube
Inconel 600		74Ni 16Cr 8Fe 1Mn	517/827	≈1150	Seamless pipe and tube
Inconel 625		62Ni 22Cr 9Mo 5Fe 1Co	689/965	≈800	Seamless pipe and tube '-200°C
Inconel X750		70Ni 15Cr 7Fe 2.5Ti 1Mn 1Co	1065	704	Seamless tube
Alloy 200 BS 3074 ASTM B161	 NA 11 N02200	99.2Ni	370	315/600	Seamless -198°C min
Alloy 400 BS 3074 ASTM B165	 NA 13 N04400	63Ni 31Cu 1.75Fe 1.25Mn	450	≈425	Seamless -198°C min
Incoloy 800 ASTM B163 ASTM B407	 N08800 N08800	32Ni 21Cr 1.5Mn	520/700	≈815	Seamless -198°C min
Incoloy 825		42Ni 21.5Cr 3Mo 2Cu	580/720	≈450	
Hastelloy B		65Ni 28Mo 5Fe 1.2Co 0.5Cr	≈950	≈760	
Hastelloy C		56Ni 16Mo 15.5Cr 5.5Fe 4.3W	≈690	≈980	
Titanium ASTM B337	Gr3 R50550	Ti99 0.3Fe	460	≈300	Seamless and welded
Ti3Al 2.5V		94.5Ti 3Al 2.5V	620		CFS
Alloy 20		34Ni 20Cr 3.5Cu 2.5Mo 2Mn 0.06C	1000	≈870	

Table 9.4 Exotic materials for pipes and tubes

from pump suction systems leading to cavitation damage and failure.

In general, computer software does not design the pipework hydraulically, only mechanically. The fluid friction losses are probably evaluated separately and basic size information, 4" sch 80, is used as the initial computer mechanical design. The computer software does not embody "tried-and-tested" rules of thumb such as not siting valves immediately after bends where the distorted flow patterns can create serious problems.

Computers are a great design aid but they have not yet supplanted the experienced, knowledgeable piping designer. After the stress engineer has found a viable "code" design, an experienced engineer, who understands the fluid hydraulic implications, must review the design for overall acceptability.

When designing pipework for liquified gases the possibility of rapid depressurisation, explosive decompression, must be evaluated and a decision made regarding the low temperature design value. All materials must be suitable for the intended operating conditions.

9.2 Rigid piping systems

Sizes and ratings for various types of pipe and tube are listed in the following Sections. The sizes listed are not simply a reiteration of Standards. Where possible, manufacturers' and stockholders' catalogues have been reviewed and the pipes/tubes listed are available.

NOTE: The outside diameter of English and American nominal bore pipes is slightly different. Check sizes carefully when using compression fittings.

9.2.1 Metallic materials

Pipe and tube are mass-produced in many materials to cope with the wide range of fluids handled in modern systems. Cast iron is used for pipes but these are generally specially made for the application. Foundries should be contacted directly to check for diameters, thickness and lengths possible. The availability of fittings must be verified at the same time. Malleable or ductile iron pipe is available in sizes from DN80 to DN1 350 for working pressures up to 24 barg.

Some of these pipes are designed specifically for burying in trenches and may rely on the backfill for support. Ni-Resist is also used for pipes.

See Tables 13.3, 13.4, 13.5 and 13.6 in Chapter 13 for details of standard cast irons. Silicon cast iron, with 14.5% Si and 0.85% C is available in special pipes from DN25 to DN375 for working pressures up to 3.5 barg. Silicon bronze, with 3% Si and 1% Fe is also available for special applications.

Most metallic pipes are in carbon steel, alloy steel or stainless steel, see Tables 9.1, 9.2 and 9.4. Pipes and tubes are available in exotic materials, see Table 9.3.

Tubes are available in high tensile steel. The following materials are available in pipe and tube but are probably not stocked on the shelf:

- 15-5PH
- 17-4PH
- 17-7PH
- 19-9PH

Both welded and seamless in metric and Imperial sizes are produced.

9.2.2 Metallic pipe

The tables which appear in this Section and Section 9.2.3 show the dimensions of pipes and tubes which are generally available. Flow capacity, at a velocity of 1 m/s, and maximum stress determined by Lamé's thick cylinder formula for an internal pressure of 10 barg, are shown for each size.

Tables 9.5, 9.6, 9.7 and 9.8, show the range of sizes available in nominal bore pipes to British and American standards. Other than BS 1387 (replaced by BS EN10255 : 2004) these pipes are available in a wide range of materials. Some Standards confuse pipe and tube and specify them together, see Table 9.9.

nb	WT	Od in	Od mm	WALL in	WALL mm	Id in	Id mm	1m/s m^3/h	10barg N/mm^2
1/8	MED	0.400	10.16	0.080	2.03	0.240	6.10	0.105	2.1
1/8	HVY	0.400	0.161	0.104	2.64	0.192	4.88	0.067	1.6
1/4	MED	0.534	13.56	0.092	2.34	0.350	8.89	0.223	2.5
1/4	HVY	0.534	13.56	0.116	2.95	0.302	7.67	0.166	1.9
3/8	MED	0.672	17.07	0.092	2.34	0.488	12.40	0.434	3.2
3/8	HVY	0.672	17.07	0.116	2.95	0.440	11.18	0.353	2.5
1/2	MED	0.843	21.41	0.104	2.64	0.635	16.13	0.736	3.6
1/2	HVY	0.843	21.41	0.128	3.25	0.587	14.91	0.629	2.9
3/4	MED	1.059	26.90	0.104	2.64	0.851	21.62	1.321	4.6
3/4	HVY	1.059	26.90	0.128	3.25	0.803	20.40	1.176	3.7
1	MED	1.331	33.81	0.128	3.25	1.075	27.31	2.108	4.8
1	HVY	1.331	33.81	0.128	4.06	1.011	25.68	1.864	3.7
1 1/4	MED	1.672	42.47	0.128	3.25	1.416	35.97	3.658	6.1
1 1/4	HVY	1.672	42.47	0.160	4.06	1.352	34.34	3.334	4.8
1 1/2	MED	1.904	48.36	0.128	3.25	1.648	41.86	4.954	7.4
1 1/2	HVY	1.904	48.36	0.160	4.06	1.584	40.23	4.577	5.5
2	MED	2.374	60.30	0.144	3.66	2.086	52.98	7.938	8.2
2	HVY	2.374	60.30	0.176	4.47	2.022	51.36	7.458	6.7
2 1/2	MED	2.991	75.97	0.144	3.66	2.703	68.66	13.33	10.4
2 1/2	HVY	2.991	75.97	0.176	4.47	2.639	67.03	12.70	8.5

nb	WT	Od in	Od mm	WALL in	WALL mm	Id in	Id mm	1m/s m^3/h	10barg N/mm^2
3	MED	3.496	88.80	0.160	4.06	3.176	80.67	18.40	10.9
3	HVY	3.496	88.80	0.192	4.88	3.112	79.04	17.67	9.1
4	MED	4.491	114.1	0.176	4.47	4.139	105.1	31.25	12.8
4	HVY	4.491	114.1	0.212	5.38	4.067	103.3	30.17	10.6
5	MED	5.496	139.6	0.192	4.88	5.112	129.8	47.67	14.3
5	HVY	5.496	139.6	0.212	5.38	5.072	128.8	46.93	13.0
6	MED	6.499	165.1	0.192	4.88	6.115	155.3	68.21	16.9
6	HVY	6.499	165.1	0.212	5.38	6.075	154.31	67.32	15.3

Table 9.5 Seamless and welded steel pipe to BS 1387/BS EN 10255
Note: The od shown is mid tolerance

Table 9.6 Seamless and welded pipe to BS 3601/3602/3603/3604/3605 (Below and continued)
Note: The British Standards are **similar** to ISO 2604/2605

nb	WT	Od in	Od mm	WALL in	WALL mm	Id in	Id mm	1m/s m^3/h	10barg N/mm^2
¼	11G	0.531	13.49	0.116	2.95	0.299	7.59	0.163	1.9
⅜	10G	0.677	17.20	0.128	3.25	0.421	10.69	0.323	2.3
½	12G	0.840	21.34	0.104	2.64	0.632	16.05	0.729	3.6
½	10G	0.840	21.34	0.128	3.25	0.584	14.83	0.622	2.9
½	8G	0.840	21.34	0.160	4.06	0.520	13.21	0.493	2.2
¾	11G	1.059	26.90	0.116	2.95	0.827	21.01	1.248	4.1
¾	9G	1.059	26.90	0.144	3.66	0.771	19.58	1.084	3.3
¾	7G	1.059	26.90	0.176	4.47	0.707	17.96	0.912	2.6
1	10G	1.327	33.71	0.128	3.25	1.071	27.20	2.092	4.7
1	8G	1.327	33.71	0.160	4.06	1.007	25.58	1.850	3.7
1	6G	1.327	33.71	0.192	4.88	0.943	23.95	1.622	3.0
	8G	1.500	38.10	0.160	4.06	1.180	29.97	2.540	4.2
	¼"	1.500	38.10	0.250	6.35	1.000	25.40	1.824	2.6
1¼	8G	1.669	42.39	0.160	4.06	1.349	34.26	3.320	4.8
1¼	5G	1.669	42.39	0.212	5.38	1.245	31.62	2.827	3.5
	¼"	1.750	44.45	0.250	6.35	1.250	31.75	2.850	3.1
1½	8G	1.900	48.26	0.160	4.06	1.580	40.13	4.554	5.5
1½	6G	1.900	48.26	0.192	4.88	1.516	38.51	4.192	4.5
1½	4G	1.900	48.26	0.232	5.89	1.436	36.47	3.762	3.7
	8G	2.000	50.80	0.160	4.06	1.680	42.67	5.148	5.8
	¼"	2.000	50.80	0.250	6.35	1.500	38.10	4.104	3.6
	5/16"	2.000	50.80	0.312	7.92	1.376	34.95	3.454	2.8
	⅜"	2.000	50.80	0.375	9.53	1.250	31.75	2.850	2.3
	¼"	2.250	57.15	0.250	6.35	1.750	44.45	5.586	4.1
	⅜"	2.250	57.15	0.375	9.53	1.500	38.10	4.104	2.6
	½"	2.250	57.15	0.500	12.70	1.250	31.75	2.850	1.9
2	9G	2.375	60.33	0.144	3.66	2.087	53.01	7.945	8.2
2	7G	2.375	60.33	0.176	4.47	2.023	51.38	7.465	6.7
2	6G	2.375	60.33	0.192	4.88	1.991	50.57	7.231	5.7
2	¼"	2.375	60.33	0.250	6.35	1.875	47.63	6.413	4.3
2	5/16"	2.375	60.33	0.312	7.92	1.751	44.48	5.593	3.4
2	⅜"	2.375	60.33	0.375	9.53	1.625	41.28	4.817	2.8
2	½"	2.375	60.33	0.500	12.70	1.375	34.93	3.449	2.0
	¼"	2.500	63.50	0.250	6.35	2.000	50.80	7.297	4.6
	5/16"	2.500	63.50	0.312	7.92	1.876	47.65	6.420	3.6
	⅜"	2.500	63.50	0.375	9.53	1.750	44.45	5.586	2.9
	½"	2.500	63.50	0.500	12.70	1.500	38.10	4.104	2.1
	⅝"	2.500	63.50	0.625	15.88	1.250	31.75	2.850	1.7
	¼"	2.750	69.85	0.250	6.35	2.250	57.15	9.235	5.1
	⅜"	2.750	69.85	0.375	9.53	2.000	50.80	7.297	3.2
	½"	2.750	69.85	0.500	12.70	1.750	44.45	5.586	2.4
	⅝"	2.750	69.85	0.625	15.88	1.500	38.10	4.104	1.8
	7G	3.000	76.20	0.176	4.47	2.648	67.26	12.791	8.5
	5G	3.000	76.20	0.212	5.38	2.576	65.43	12.105	7.1
	¼"	3.000	76.20	0.250	6.35	2.500	63.50	11.401	5.5
	5/16"	3.000	76.20	0.312	7.92	2.376	60.35	10.298	4.4
	⅜"	3.000	76.20	0.375	9.53	2.250	57.15	9.235	3.6
	½"	3.000	76.20	0.500	12.70	2.000	50.80	7.297	2.6
	⅝"	3.000	76.20	0.625	15.88	1.750	44.45	5.586	2.0
	¾"	3.000	76.20	0.750	19.05	1.500	38.10	4.104	1.7
	¼"	3.250	82.55	0.250	6.35	2.750	69.85	13.795	6.0
	⅜"	3.250	82.55	0.375	9.53	2.500	63.50	11.401	3.9
	½"	3.250	82.55	0.500	12.70	2.250	57.15	9.235	2.8
	⅝"	3.250	82.55	0.625	15.88	2.000	50.80	7.297	2.2
	¾"	3.250	82.55	0.750	19.05	1.750	44.45	5.586	1.8
3	8G	3.500	88.90	0.160	4.06	3.180	80.77	18.447	10.9
3	6G	3.500	88.90	0.192	4.88	3.116	79.15	17.711	9.1
3	5G	3.500	88.90	0.212	5.38	3.076	78.13	17.260	8.3
3	¼"	3.500	88.90	0.250	6.35	3.000	76.20	16.417	7.0
3	5/16"	3.500	88.90	0.312	7.92	2.876	73.05	15.088	5.2
3	⅜"	3.500	88.90	0.375	9.53	2.750	69.85	13.795	4.2
3	½"	3.500	88.90	0.500	12.70	2.500	63.50	11.401	3.1
3	⅝"	3.500	88.90	0.625	15.88	2.250	57.15	9.235	2.4
3	¾"	3.500	88.90	0.750	19.05	2.000	50.80	7.297	2.0
3	¼"	3.750	95.25	0.250	6.35	3.250	82.55	19.268	7.5
	⅜"	3.750	95.25	0.375	9.53	3.000	76.20	16.417	4.6
	½"	3.750	95.25	0.500	12.70	2.750	69.85	13.795	3.3
	⅝"	3.750	95.25	0.625	15.88	2.500	63.50	11.401	2.6
	¾"	3.750	95.25	0.750	19.05	2.250	57.15	9.235	2.1
3½	8G	4.000	101.6	0.160	4.06	3.680	93.47	24.703	12.5
3½	6G	4.000	101.6	0.192	4.88	3.616	91.85	23.852	10.4
3½	¼"	4.000	101.6	0.250	6.35	3.500	88.90	22.346	8.0
3½	5/16"	4.000	101.6	0.312	7.92	3.376	85.75	20.790	6.0
3½	⅜"	4.000	101.6	0.375	9.53	3.250	82.55	19.268	4.9
3½	½"	4.000	101.6	0.500	12.70	3.000	76.20	16.417	3.6
3½	⅝"	4.000	101.6	0.625	15.88	2.750	69.85	13.795	2.8
3½	¾"	4.000	101.6	0.750	19.05	2.500	63.50	11.401	2.3
3½	1"	4.000	101.6	1.000	25.40	2.000	50.80	7.297	1.7
	¼"	4.250	108.0	0.250	6.35	3.750	95.25	25.652	8.5
	5/16"	4.250	108.0	0.312	7.92	3.626	92.10	23.984	6.8
	⅜"	4.250	108.0	0.375	9.53	3.500	88.90	22.346	5.2
	½"	4.250	108.0	0.500	12.70	3.250	82.55	19.268	3.8
	⅝"	4.250	108.0	0.625	15.88	3.000	76.20	16.417	3.0
	¾"	4.250	108.0	0.750	19.05	2.750	69.85	13.795	2.4
	1"	4.250	108.0	1.000	25.40	2.250	57.15	9.235	1.8
4	7G	4.500	114.3	0.176	4.47	4.148	105.36	31.39	12.8
4	5G	4.500	114.3	0.212	5.38	4.076	103.53	30.31	10.6
4	¼"	4.500	114.3	0.250	6.35	4.000	101.60	29.19	9.0
4	5/16"	4.500	114.3	0.312	7.92	3.876	98.45	27.40	7.2
4	⅜"	4.500	114.3	0.375	9.53	3.750	95.25	25.65	5.5
4	½"	4.500	114.3	0.500	12.70	3.500	88.90	22.35	4.1
4	⅝"	4.500	114.3	0.625	15.88	3.250	82.55	19.27	3.2
4	¾"	4.500	114.3	0.750	19.05	3.000	76.20	16.42	2.6
	5/16"	4.750	120.7	0.312	7.92	4.126	104.80	31.05	7.6
	⅜"	4.750	120.7	0.375	9.53	4.000	101.60	29.19	5.9
	½"	4.750	120.7	0.500	12.70	3.750	95.25	25.65	4.3
	¾"	4.750	120.7	0.750	19.05	3.250	82.55	19.27	2.8
	6G	5.000	127.0	0.192	4.88	4.616	117.25	38.87	13.0
	¼"	5.000	127.0	0.250	6.35	4.500	114.30	36.94	10.0
	⅜"	5.000	127.0	0.375	9.53	4.250	107.95	32.95	6.2
	½"	5.000	127.0	0.500	12.70	4.000	101.60	29.19	4.6
	⅝"	5.000	127.0	0.625	15.88	3.750	95.25	25.65	3.6
	1"	5.000	127.0	1.000	25.40	3.000	76.20	16.42	2.1
	¼"	5.250	133.4	0.250	6.35	4.750	120.65	41.16	10.5
	⅜"	5.250	133.4	0.375	9.53	4.500	114.30	36.94	7.0
	½"	5.250	133.4	0.500	12.70	4.250	107.95	32.95	4.8
	¾"	5.250	133.4	0.750	19.05	3.750	95.25	25.65	3.1
	1"	5.250	133.4	1.000	25.40	3.250	82.55	19.27	2.2
5	7G	5.500	139.7	0.176	4.47	5.148	130.76	48.34	15.6
5	5G	5.500.	139.7	0.212	5.38	5.076	128.93	47.00	13.0
5	¼"	5.500	139.7	0.250	6.35	5.000	127.00	45.60	11.0
5	5/16"	5.500	139.7	0.312	7.92	4.876	123.85	43.37	8.8
5	⅜"	5.500	139.7	0.375	9.53	4.750	120.65	41.16	7.3
5	½"	5.500	139.7	0.500	12.70	4.500	114.30	36.94	5.1
5	¾"	5.500	139.7	0.750	19.05	4.000	101.60	29.19	3.2
5	1"	5.500	139.7	1.000	25.40	3.500	88.90	22.35	2.4
		5.750	146.1	0.375	9.53	5.000	127.00	45.60	7.7
		5.750	146.1	0.500	12.70	4.750	120.65	41.16	5.3
	¼"	6.000	152.4	0.250	6.35	5.500	139.70	55.18	12.0
	5/16"	6.000	152.4	0.312	7.92	5.376	136.55	52.72	9.6

nb	WT	Od in	Od mm	WALL in	WALL mm	Id in	Id mm	1m/s m³/h	10barg N/mm²
	3/8"	6.000	152.4	0.375	9.53	5.250	133.35	50.28	8.0
	1/2"	6.000	152.4	0.500	12.70	5.000	127.00	45.60	5.5
	5/8"	6.000	152.4	0.625	15.88	4.750	120.65	41.16	4.4
	3/4"	6.000	152.4	0.750	19.05	4.500	114.30	36.94	3.6
	1"	6.000	152.4	1.000	25.40	4.000	101.60	29.19	2.6
	1/2"	6.250	158.8	0.500	12.70	5.250	133.35	50.28	5.8
	3/4"	6.250	158.8	0.750	19.05	4.750	120.65	41.16	3.7
	1/4"	6.500	165.1	0.250	6.35	6.000	152.40	65.67	13.0
	6G	6.625	168.3	0.192	4.88	6.241	158.52	71.05	17.3
	5G	6.625	168.3	0.212	5.38	6.201	157.51	70.14	15.6
6	1/4"	6.625	168.3	0.250	6.35	6.125	155.58	68.43	13.3
6	5/16"	6.625	168.3	0.312	7.92	6.001	152.43	65.69	10.6
6	3/8"	6.625	168.3	0.375	9.53	5.875	149.23	62.96	8.8
6	1/2"	6.625	168.3	0.500	12.70	5.625	142.88	57.72	6.2
6	5/8"	6.625	168.3	0.625	15.88	5.375	136.53	52.70	4.9
6	3/4"	6.625	168.3	0.750	19.05	5.125	130.18	47.91	4.0
6	1"	6.625	168.3	1.000	25.40	4.625	117.48	39.02	2.9
6	1/4"	7.000	177.8	0.250	6.35	6.500	165.10	77.07	14.0
6	3/8"	7.000	177.8	0.375	9.53	6.250	158.75	71.26	9.3
	1/2"	7.000	177.8	0.500	12.70	6.000	152.40	65.67	7.0
	3/4"	7.000	177.8	0.750	19.05	5.500	139.70	55.18	4.2
	1"	7.000	177.8	1.000	25.40	5.000	127.00	45.60	3.1
		7.000	177.8	1.378	35.00	4.244	107.80	32.86	2.2
	6G	7.625	193.7	0.192	4.88	7.241	183.92	95.64	19.9
	5G	7.625	193.7	0.212	5.38	7.201	182.91	94.59	18.0
	1/4"	7.625	193.7	0.250	6.35	7.125	180.98	92.60	15.3
7	5/16"	7.625	193.7	0.312	7.92	7.001	177.83	89.41	12.2
7	3/8"	7.625	193.7	0.375	9.53	6.875	174.63	86.22	10.2
7	1/2"	7.625	193.7	0.500	12.70	6.625	168.28	80.06	7.6
7	3/4"	7.625	193.7	0.750	19.05	6.125	155.58	68.43	4.6
7	1"	7.625	193.7	1.000	25.40	5.625	142.88	57.72	3.4
7	1/4"	8.000	203.2	0.250	6.35	7.500	190.50	102.61	16.0
7	3/8"	8.000	203.2	0.375	9.53	7.250	184.15	95.88	10.7
7	3/4"	8.000	203.2	0.750	19.05	6.500	165.10	77.07	4.9
	1"	8.000	203.2	1.000	25.40	6.000	152.40	65.67	3.6
	6G	8.625	219.1	0.192	4.88	8.241	209.32	123.89	22.5
	5G	8.625	219.1	0.212	5.38	8.201	208.31	122.69	20.3
	3/8"	8.625	219.1	0.375	9.53	7.875	200.03	113.13	11.5
8	3/4"	8.625	219.1	0.750	19.05	7.125	180.98	92.60	5.3
8	1"	8.625	219.1	1.000	25.40	6.625	168.28	80.06	3.9
8	1 3/8"	8.625	219.1	1.375	34.93	5.875	149.23	62.96	2.7
8	1/4"	9.000	228.6	0.250	6.35	8.500	215.90	131.79	18.0
8	3/8"	9.000	228.6	0.375	9.53	8.250	209.55	124.16	12.0
8	1/2"	9.000	228.6	0.500	12.70	8.000	203.20	116.75	9.0
	1"	9.000	228.6	1.000	25.40	7.000	177.80	89.38	4.1
	1 3/8"	9.000	228.6	1.375	34.93	6.250	158.75	71.26	2.9
	4G	9.625	244.5	0.232	5.89	9.161	232.69	153.09	20.7
	9/32"	9.625	244.5	0.281	7.14	9.063	230.20	149.83	17.1
	3/8"	9.625	244.5	0.375	9.53	8.875	225.43	143.68	12.8
9	1/2"	9.625	244.5	0.500	12.70	8.625	219.08	135.70	9.6
9	3/4"	9.625	244.5	0.750	19.05	8.125	206.38	120.42	6.0
9	1"	9.625	244.5	1.000	25.40	7.625	193.68	106.06	4.4
9	1 1/8"	9.625	244.5	1.125	28.58	7.375	187.33	99.22	3.8
9	1 5/16"	9.625	244.5	1.312	33.32	7.001	177.83	89.41	3.2
9	1 1/2"	9.625	244.5	1.500	38.10	6.625	168.28	80.06	2.8
9	3/8"	10.000	254.0	0.375	9.53	9.250	234.95	156.08	13.3
9	1/2"	10.000	254.0	0.500	12.70	9.000	228.60	147.76	10.0
9	1"	10.000	254.0	1.000	25.40	8.000	203.20	116.75	4.6
	1 1/2"	10.000	254.0	1.500	38.10	7.000	177.80	89.38	2.9
	9/32"	10.750	273.1	0.281	7.14	10.188	258.78	189.34	19.1
	5/8"	10.750	273.1	0.625	15.88	9.500	241.30	164.63	8.6
	3/4"	10.750	273.1	0.750	19.05	9.250	234.95	156.08	7.2
10	1"	10.750	273.1	1.000	25.40	8.750	222.25	139.66	4.9
10	1 3/8"	10.750	273.1	1.375	34.93	8.000	203.20	116.75	3.5
10	3/4"	11.250	285.8	0.750	19.05	9.750	247.65	173.41	7.5
10	1 1/2"	11.250	285.8	1.500	38.10	8.250	209.55	124.16	3.3
10	2"	11.250	285.8	2.000	50.80	7.250	184.15	95.88	2.4
	1/4"	12.000	304.8	0.250	6.35	11.500	292.10	241.24	24.0
	3/8"	12.000	304.8	0.375	9.53	11.250	285.75	230.87	16.0
	1/2"	12.000	304.8	0.500	12.70	11.000	279.40	220.72	12.0
	1"	12.000	304.8	1.000	25.40	10.000	254.00	182.41	5.5
	1 1/2"	12.000	304.8	1.500	38.10	9.000	228.60	147.76	3.6
	2"	12.000	304.8	2.000	50.80	8.000	203.20	116.75	2.6
	1/2"	12.250	311.2	0.500	12.70	11.250	285.75	230.87	12.3
	1"	12.250	311.2	1.000	25.40	10.250	260.35	191.65	5.7
	1 1/2"	12.250	311.2	1.500	38.10	9.250	234.95	156.08	3.7
		12.250	311.2	2.250	57.15	7.750	196.85	109.56	2.3
12	9/32"	12.750	323.9	0.281	7.14	12.188	309.58	270.97	22.7
12	1/2"	12.750	323.9	0.500	12.70	11.750	298.45	251.85	12.8
12	1"	12.750	323.9	1.000	25.40	10.750	273.05	210.80	5.9
12	1 7/8"	12.750	323.9	1.875	47.63	9.000	228.60	147.76	3.0
14	1/2"	14.000	355.6	0.500	12.70	13.000	330.20	308.28	14.0
14	1"	14.000	355.6	1.000	25.40	12.000	304.80	262.68	7.0
	1/2"	15.000	381.0	0.500	12.70	14.000	355.60	357.53	15.0
	1"	15.000	381.0	1.000	25.40	13.000	330.20	308.28	7.5
	1 1/2"	15.000	381.0	1.500	38.10	12.000	304.80	262.68	4.6
	2"	15.000	381.0	2.000	50.80	11.000	279.40	220.72	3.3
	2 1/2"	15.000	381.0	2.500	63.50	10.000	254.00	182.41	2.6
16	1"	16.000	406.4	1.000	25.40	14.000	355.60	357.53	8.0
16	2"	16.000	406.4	2.000	50.80	12.000	304.80	262.68	3.6
	1/2"	17.000	431.8	0.500	12.70	16.000	406.40	466.98	17.0
	1"	17.000	431.8	1.000	25.40	15.000	381.00	410.43	8.5
	1 1/2"	17.000	431.8	1.500	38.10	14.000	355.60	357.53	5.2
	2"	17.000	431.8	2.000	50.80	13.000	330.20	308.28	3.8
	2 1/2"	17.000	431.8	2.500	63.50	12.000	304.80	262.68	3.0
18	3/8"	18.000	457.2	0.375	9.53	17.250	438.15	542.80	24.0
18	1"	18.000	457.2	1.000	25.40	16.000	406.40	466.98	9.0

Table 9.6 Seamless and welded pipe to BS 3601/3602/3603/3604/3605 (above) Note: The British Standards are **similar** to ISO 2604/2605

Table 9.7 Seamless and welded ANSI steel pipe (Below and continued)

nb	SCH	od in	od mm	WALL in	WALL mm	id in	id mm	1m/s m³/h	10barg N/mm²
1/8	40	0.405	10.29	0.068	1.73	0.269	6.83	0.132	2.6
1/8	80	0.405	10.29	0.095	2.41	0.215	5.46	0.084	1.8
1/4	40	0.540	13.72	0.088	2.24	0.364	9.25	0.242	2.7
1/4	80	0.540	13.72	0.119	3.02	0.302	7.67	0.166	1.9
3/8	40	0.675	17.15	0.091	2.31	0.493	12.52	0.443	3.3
3/8	80	0.675	17.15	0.126	3.20	0.423	10.74	0.326	2.3
1/2	40	0.840	21.34	0.109	2.77	0.622	15.80	0.706	3.4
1/2	80	0.840	21.34	0.147	3.73	0.546	13.87	0.544	2.5
1/2	160	0.840	21.34	0.188	4.78	0.464	11.79	0.393	1.9
1/2	XXS	0.840	21.34	0.294	7.47	0.252	6.40	0.116	1.2
3/4	40	1.050	26.67	0.133	3.38	0.784	19.91	1.121	3.5
3/4	80	1.050	26.67	0.154	3.91	0.742	18.85	1.004	3.0
3/4	160	1.050	26.67	0.218	5.54	0.614	15.60	0.688	2.0
3/4	XXS	1.050	26.67	0.308	7.82	0.434	11.02	0.344	1.4
1	40	1.315	33.40	0.133	3.38	1.049	26.64	2.007	4.5
1	80	1.315	33.40	0.179	4.55	0.957	24.31	1.671	3.3
1	160	1.315	33.40	0.250	6.35	0.815	20.70	1.212	2.2
1	XXS	1.315	33.40	0.358	9.09	0.599	15.21	0.655	1.5
1 1/4	40	1.315	42.16	0.140	3.56	1.380	35.05	3.474	5.5
1 1/4	80	1.660	42.16	0.191	4.85	1.278	32.46	2.979	3.9
1 1/2	40	1.900	48.26	0.145	3.68	1.610	40.89	4.728	6.1
1 1/2	80	1.900	48.26	0.200	5.08	1.500	38.10	4.104	4.3
1 1/2	160	1.900	48.26	0.281	7.14	1.338	33.99	3.266	3.0
1 1/2	XXS	1.900	48.26	0.400	10.16	1.100	27.94	2.207	2.0
2	40	2.375	60.33	0.154	3.91	2.067	52.50	7.794	7.7
2	80	2.375	60.33	0.218	5.54	1.939	49.25	6.858	5.0
2	160	2.375	60.33	0.344	8.74	1.687	42.85	5.191	3.0
2	XXS	2.375	60.33	0.436	11.07	1.503	38.18	4.121	3.0
2 1/2	40	2.875	73.03	0.203	5.16	2.469	62.71	11.120	7.1
2 1/2	80	2.875	73.03	0.276	7.01	2.323	62.71	9.844	4.8
3	40	3.500	88.90	0.216	5.49	2.323	77.93	17.170	8.1
3	80	3.500	88.90	0.300	7.62	2.900	73.66	15.341	5.4
3	160	3.500	88.90	0.438	11.13	2.624	66.65	12.560	3.6

nb	SCH	od in	od mm	WALL in	WALL mm	id in	id mm	1m/s m³/h	10barg N/mm²
3	XXS	3.500	88.90	0.600	15.24	2.300	58.42	9.650	2.5
4	40	4.500	114.3	0.237	6.02	4.026	102.3	29.57	9.5
4	80	4.500	114.3	0.337	8.56	3.826	97.18	26.70	6.7
4	160	4.500	114.3	0.531	13.49	3.438	87.33	21.56	3.8
4	XXS	4.500	114.3	0.674	17.12	3.152	80.06	18.12	2.9
6	40	6.625	168.3	0.280	7.11	6.065	154.1	67.10	11.8
6	80	6.625	168.3	0.432	10.97	5.761	146.3	60.54	7.7
6	160	6.625	168.3	0.719	18.26	5.187	131.7	49.08	4.2
6	XXS	6.625	168.3	0.864	21.95	4.897	124.4	43.74	3.4
8	20	8.625	219.1	0.250	6.35	8.125	206.4	120.4	17.3
8	30	8.625	219.1	0.277	7.04	8.071	205.0	118.8	15.6
8	40	8.625	219.1	0.322	8.18	7.981	202.7	116.2	13.4
8	80	8.625	219.1	0.500	12.70	7.625	193.7	106.1	8.6
8	XXS	8.625	219.1	0.875	22.23	6.875	174.6	86.2	4.5
8	160	8.625	219.1	0.906	23.01	6.813	173.1	84.7	4.3
10	20	10.75	273.1	0.250	6.35	10.25	260.4	191.6	21.5
10	30	10.75	273.1	0.307	7.80	10.14	257.5	187.4	17.5
10	40	10.75	273.1	0.365	9.27	10.02	254.5	183.1	14.7
10	XS	10.75	273.1	0.500	12.70	9.750	247.7	173.4	10.8
10	80	10.75	273.1	0.594	15.09	9.562	242.9	166.8	9.0
10	XXS	10.75	273.1	1.000	25.40	8.750	222.3	139.7	4.9
10	160	10.75	273.1	1.125	28.58	8.500	215.9	131.8	4.3
12	20	12.75	323.9	0.250	6.35	12.25	311.2	273.7	25.5
12	30	12.75	323.9	0.330	8.38	12.09	307.1	266.6	19.3
12	40	12.75	323.9	0.406	10.31	11.94	303.2	260.0	15.7
12	XS	12.75	323.9	0.500	12.70	11.75	298.5	251.8	12.8
12	80	12.75	323.9	0.688	17.48	11.37	288.9	236.0	9.3
12	XXS	12.75	323.9	1.000	25.40	10.75	273.1	210.8	5.9
12	160	12.75	323.9	1.312	33.32	10.13	257.2	187.0	4.4
14	S	14.00	355.6	0.375	9.53	13.25	336.6	320.3	18.7
14	40	14.00	355.6	0.438	11.13	13.12	333.3	314.2	16.0
14	XS	14.00	355.6	0.500	12.70	13.00	330.2	308.3	14.0
16	S	16.00	406.4	0.375	9.53	15.25	387.4	424.2	21.3
16	XS	16.00	406.4	0.500	12.70	15.00	381.0	410.4	16.0
18	S	18.00	457.2	0.375	9.53	17.25	438.2	542.8	24.0
18	XS	18.00	457.2	0.500	12.70	17.00	431.8	527.2	18.0
20	10	20.00	508.0	0.250	6.35	19.50	495.3	693.6	40.0
20	S	20.00	508.0	0.375	9.53	19.25	489.0	676.0	26.7
20	160	20.00	508.0	1.969	50.01	16.06	408.0	470.6	4.6
24	10	24.00	609.6	0.250	6.35	23.50	596.9	1007.4	48.0
24	S	24.00	609.6	0.375	9.53	23.25	590.6	986.1	32.0
30	10	30.00	762.0	0.312	7.92	29.38	746.2	1574.1	48.1
30	S	30.00	762.0	0.375	9.53	29.25	743.0	1560.7	40.0
36	S	30.00	914.4	0.375	9.53	35.25	895.4	2266.6	48.0

Table 9.7 Seamless and welded ANSI steel pipe (Above)

Table 9.8 Seamless and welded ANSI stainless steel pipe (Below)

nb	SCH	od in	od mm	WALL in	WALL mm	id in	id mm	1m/s m³/h	10 barg N/mm²
⅛	10	0.405	10.29	0.049	1.24	0.307	7.80	0.172	3.7
⅛	40	0.405	10.29	0.068	1.73	0.269	6.83	0.132	2.6
⅛	80	0.405	10.29	0.095	2.41	0.215	5.46	0.084	1.8
¼	10	0.540	13.72	0.065	1.65	0.410	10.4	0.307	3.7
¼	40	0.540	13.72	0.088	2.24	0.364	9.25	0.242	2.7
¼	80	0.540	13.72	0.119	3.02	0.302	7.67	0.166	1.9
⅜	10	0.675	17.15	0.065	1.65	0.545	13.8	0.542	4.7
⅜	40	0.675	17.15	0.091	2.31	0.493	12.5	0.443	3.3
⅜	80	0.675	17.15	0.126	3.20	0.423	10.7	0.326	2.3
½	5	0.840	21.34	0.065	1.65	0.710	18.0	0.920	6.0
½	10	0.840	21.34	0.083	2.11	0.674	17.1	0.829	4.6
½	40	0.840	21.34	0.109	2.77	0.622	15.8	0.706	3.4
½	80	0.840	21.34	0.147	3.73	0.546	13.9	0.544	2.5
½	160	0.840	21.34	0.187	4.75	0.466	11.8	0.396	1.9
½	XXS	0.840	21.34	0.252	6.40	0.336	8.53	0.206	1.4
¾	5	1.050	26.67	0.065	1.65	0.920	23.4	1.544	8.1
¾	10	1.050	26.67	0.083	2.11	0.884	22.5	1.425	5.9

nb	SCH	od in	od mm	WALL in	WALL mm	id in	id mm	1m/s m³/h	10 barg N/mm²
¾	40	1.050	26.67	0.113	2.87	0.824	20.9	1.239	4.2
¾	80	1.050	26.67	0.154	3.91	0.742	18.8	1.004	3.0
¾	160	1.050	26.67	0.218	5.54	0.614	15.6	0.688	2.0
¾	XXS	1.050	26.67	0.308	7.82	0.434	11.0	0.344	1.4
1	5	1.315	33.40	0.065	1.65	1.185	30.1	2.562	10.1
1	10	1.315	33.40	0.109	2.77	1.097	27.9	2.195	5.6
1	40	1.315	33.40	0.133	3.38	1.049	26.6	2.007	4.5
1	80	1.315	33.40	0.179	4.55	0.957	24.3	1.671	3.3
1	160	1.315	33.40	0.250	6.35	0.815	20.7	1.212	2.2
1	XXS	1.315	33.40	0.358	9.09	0.599	15.2	0.655	1.5
1¼	5	1.660	42.16	0.065	1.65	1.530	38.9	4.270	12.8
1¼	10	1.660	42.16	0.109	2.77	1.442	36.6	3.793	7.6
1¼	40	1.660	42.16	0.140	3.56	1.380	35.1	3.474	5.5
1¼	80	1.660	42.16	0.191	4.85	1.278	32.5	2.979	3.9
1¼	160	1.660	42.16	0.250	6.35	1.160	29.5	2.455	2.9
1¼	XXS	1.660	42.16	0.382	9.70	0.896	22.8	1.464	1.8
1½	5	1.900	48.26	0.065	1.65	1.770	45.0	5.715	14.6
1½	10	1.900	48.26	0.109	2.77	1.682	42.7	5.161	8.7
1½	40	1.900	48.26	0.145	3.68	1.610	40.9	4.728	6.1
1½	80	1.900	48.26	0.200	5.08	1.500	38.1	4.104	4.3
1½	160	1.900	48.26	0.281	7.14	1.338	34.0	3.266	3.0
1½	XXS	1.900	48.26	0.400	10.16	1.100	27.9	2.207	2.0
2	5	2.375	60.33	0.065	1.65	2.245	57.0	9.194	18.3
2	10	2.375	60.33	0.109	2.77	2.157	54.8	8.487	10.9
2	40	2.375	60.33	0.154	3.91	2.067	52.5	7.794	7.7
2	80	2.375	60.33	0.218	5.54	1.939	49.3	6.858	5.0
2	160	2.375	60.33	0.343	8.71	1.689	42.9	5.204	3.0
2	XXS	2.375	60.33	0.436	11.07	1.503	38.2	4.121	2.3
2½	5	2.875	73.03	0.083	2.11	2.709	68.8	13.39	17.3
2½	10	2.875	73.03	0.120	3.05	2.635	66.9	12.67	12.0
2½	40	2.875	73.03	0.203	5.16	2.469	62.7	11.12	7.1
2½	80	2.875	73.03	0.276	7.01	2.323	59.0	9.844	4.8
2½	160	2.875	73.03	0.375	9.53	2.125	54.0	8.237	3.4
2½	XXS	2.875	73.03	0.552	14.02	1.771	45.0	5.721	2.2
3	5	3.500	88.90	0.083	2.11	3.334	84.7	20.28	21.1
3	10	3.500	88.90	0.120	3.05	3.260	82.8	19.39	14.6
3	40	3.500	88.90	0.216	5.49	3.068	77.9	17.17	8.1
3	80	3.500	88.90	0.300	7.62	2.900	73.7	15.34	5.4
3	160	3.500	88.90	0.438	11.13	2.624	66.6	12.56	3.6
3	XXS	3.500	88.90	0.600	15.24	2.300	58.4	9.650	2.5
3½	80	4.000	101.6	0.318	8.08	3.364	85.4	20.64	5.8
3½	XXS	4.000	101.6	0.636	16.15	2.728	69.3	13.58	2.7
4	5	4.000	101.6	0.083	2.11	4.334	110.1	34.26	5.4
4	10	4.500	114.3	0.120	3.05	4.260	108.2	33.10	18.8
4	40	4.500	114.3	0.237	6.02	4.026	102.3	29.57	9.5
4	80	4.500	114.3	0.337	8.56	3.826	97.2	26.70	6.7
4	120	4.500	114.3	0.438	11.13	3.624	92.0	23.96	4.7
4	160	4.500	114.3	0.531	13.49	3.438	87.3	21.56	3.8
4	XXS	4.500	114.3	0.674	17.12	3.152	80.1	18.12	2.9
5	5	5.563	141.3	0.109	2.77	5.345	135.8	52.11	25.5
5	10	5.563	141.3	0.134	3.40	5.295	134.5	51.14	20.8
5	40	5.563	141.3	0.258	6.55	5.047	128.2	46.47	10.8
5	80	5.563	141.3	0.375	9.53	4.813	122.3	42.26	7.4
5	160	5.563	141.3	0.625	15.88	4.313	109.6	33.93	4.0
5	XXS	5.563	141.3	0.750	19.05	4.063	103.2	30.11	3.3
6	5	6.625	168.3	0.109	2.77	6.407	162.7	74.88	30.4
6	10	6.625	168.3	0.134	3.40	6.357	161.5	73.72	24.7
6	40	6.625	168.3	0.280	7.11	6.065	154.1	67.10	11.8
6	80	6.625	168.3	0.432	10.97	5.761	146.3	60.54	7.7
6	120	6.625	168.3	0.562	14.27	5.501	139.7	55.20	5.4
6	160	6.625	168.3	0.718	18.24	5.189	131.8	49.12	4.2
6	XXS	6.625	168.3	0.864	21.95	4.897	124.4	43.74	3.4
6	5	6.625	219.1	0.109	2.77	8.407	213.5	128.93	39.6
6	10	6.625	219.1	0.148	3.76	8.329	211.6	126.55	29.1
6	20	6.625	219.1	0.250	6.35	8.125	206.4	120.42	17.3
6	40	6.625	219.1	0.322	8.18	7.981	202.7	116.19	13.4
6	80	6.625	219.1	0.500	12.70	7.625	193.7	106.06	8.6
10	5	10.75	273.1	0.134	3.40	10.48	266.2	200.42	40.1

nb	SCH	od in	od mm	WALL in	WALL mm	id in	id mm	1m/s m^3/h	10 barg N/mm^2
10	10	10.75	273.1	0.165	4.19	10.42	264.7	198.06	32.6
10	20	10.75	273.1	0.250	6.35	10.25	260.4	191.65	21.5
10	40	10.75	273.1	0.365	9.27	10.02	254.5	183.15	14.7
10	80S	10.75	273.1	0.500	12.70	9.750	247.7	173.41	10.8
10	80	10.75	273.1	0.593	15.06	9.564	242.9	166.85	9.1
12	5S	12.75	323.9	0.156	3.96	12.44	315.9	282.20	40.9
12	5	12.75	323.9	0.165	4.19	12.42	315.5	281.39	38.6
12	10	12.75	323.9	0.180	4.57	12.39	314.7	280.03	35.4
12	20	12.75	323.9	0.250	6.35	12.25	311.2	273.74	25.5
12	40S	12.75	323.9	0.375	9.53	12.00	304.8	262.68	17.0
12	40	12.75	323.9	0.406	10.31	11.94	303.2	259.97	15.7
12	80S	12.75	323.9	0.500	12.70	11.75	298.5	251.85	12.8
12	80	12.75	323.9	0.688	17.48	11.37	288.9	235.99	9.3
14	10S	14.00	355.6	0.188	4.78	13.62	346.0	338.59	37.2
14	10	14.00	355.6	0.250	6.35	13.50	342.9	332.45	28.0
14	30	14.00	355.6	0.375	9.53	13.25	336.6	320.25	18.7
14	XS	14.00	355.6	0.500	12.70	13.00	330.2	308.28	14.0
16	10S	16.00	406.4	0.188	4.78	15.62	396.8	445.29	42.6
16	10	16.00	406.4	0.250	6.35	15.50	393.7	438.25	32.0
16	20	16.00	406.4	0.312	7.92	15.38	390.6	431.27	21.3
16	30	16.00	406.4	0.375	9.53	15.25	387.4	424.23	25.6
16	40	16.00	406.4	0.500	12.70	15.00	381.0	410.43	16.0
16	10S	16.00	457.2	0.188	4.78	17.62	447.6	566.59	47.9
18	10	18.00	457.2	0.250	6.35	17.50	444.5	558.64	36.0
18	40S	18.00	457.2	0.375	9.53	17.25	438.2	542.80	24.0
18	XS	18.00	457.2	0.500	12.70	17.00	431.8	527.18	18.0
20	5	20.00	508.0	0.188	4.78	19.62	498.4	702.48	53.2
20	10	20.00	508.0	0.250	6.35	19.50	495.3	693.63	40.0
20	20	20.00	508.0	0.375	9.53	19.25	489.0	675.96	26.7
20	XS	20.00	508.0	0.500	12.70	19.00	482.6	658.52	20.0
24	10	24.00	609.6	0.250	6.35	23.50	596.9	1007.4	48.0
24	20	24.00	609.6	0.375	9.53	23.25	590.6	986.07	32.0
24	XS	24.00	609.6	0.500	12.70	23.00	584.2	964.97	24.0
30	10	30.00	762.0	0.312	7.92	29.38	746.2	1574.1	48.1
30	STD	30.00	762.0	0.375	9.53	29.25	743.0	1560.7	40.0
30	STD	30.00	914.4	0.375	9.53	35.25	895.4	2266.6	48.0

Table 9.8 Seamless and welded ANSI stainless steel pipe (Above)

Table 9.9 Seamless steel pipe/tube to DIN 1629 ST52 (Below)

od mm	od in	WALL mm	WALL in	id mm	id in	1m/s m^3/h	10barg N/mm^2
38.1	1.500	5.0	0.20	28.1	1.11	2.233	3.4
42.4	1.669	5.0	0.20	32.4	1.28	2.968	3.8
42.4	1.669	5.6	0.22	31.2	1.23	2.752	3.4
42.4	1.669	6.3	0.25	29.8	1.17	2.511	3.0
44.5	1.752	5.6	0.22	33.3	1.31	3.135	3.5
44.5	1.752	7.1	0.28	30.3	1.19	2.596	2.7
48.3	1.900	5.0	0.20	38.3	1.51	4.141	4.4
48.3	1.900	5.6	0.22	37.1	1.46	3.885	3.9
48.3	1.900	6.3	0.25	35.7	1.40	3.597	3.4
48.3	1.900	8.0	0.31	32.3	1.27	2.944	2.6
50.8	2.000	5.0	0.20	40.8	1.61	4.707	4.6
50.8	2.000	5.6	0.22	39.6	1.56	4.434	4.1
50.8	2.000	7.1	0.28	36.6	1.44	3.788	3.2
50.8	2.000	8.0	0.31	34.8	1.37	3.424	2.8
50.8	2.000	8.8	0.35	33.2	1.31	3.117	2.5
50.8	2.000	16.0	0.63	18.8	0.74	0.999	1.3
57.0	2.244	4.0	0.16	49.0	1.93	6.789	7.1
57.0	2.244	5.0	0.20	47.0	1.85	6.246	5.2
57.0	2.244	5.6	0.22	45.8	1.80	5.931	4.6
57.0	2.244	6.3	0.25	44.4	1.75	5.574	4.1
57.0	2.244	7.1	0.28	42.8	1.69	5.179	3.6
57.0	2.244	8.1	0.32	40.8	1.61	4.707	3.1
57.0	2.244	8.8	0.35	39.4	1.55	4.389	2.8
60.3	2.375	5.6	0.22	49.1	1.93	6.822	4.9
60.3	2.375	7.1	0.28	46.1	1.82	6.014	3.8
60.3	2.375	8.8	0.35	42.7	1.68	5.160	3.0
60.3	2.375	10.0	0.39	40.3	1.59	4.597	2.6
60.3	2.375	17.5	0.69	25.3	1.00	1.813	1.4
63.5	2.500	5.0	0.20	53.5	2.11	8.093	5.9
63.5	2.500	5.6	0.22	52.3	2.06	7.734	5.2
63.5	2.500	7.1	0.28	49.3	1.94	6.872	4.0
63.5	2.500	8.8	0.35	45.9	1.81	5.957	3.2
63.5	2.500	10.0	0.39	43.5	1.71	5.350	2.8
69.9	2.750	5.6	0.22	58.7	2.31	9.726	5.8
69.9	2.750	7.1	0.28	55.7	2.19	8.756	4.5
69.9	2.750	8.0	0.31	53.9	2.12	8.199	3.9
69.9	2.750	8.8	0.35	52.3	2.06	7.719	3.5
69.9	2.750	10.0	0.39	49.9	1.96	7.026	3.1
69.9	2.750	14.2	0.56	41.5	1.63	4.858	2.1
76.2	3.000	5.6	0.22	65.0	2.56	11.946	6.8
76.2	3.000	7.1	0.28	62.0	2.44	10.869	4.9
76.2	3.000	8.8	0.35	58.6	2.31	9.709	3.9
76.2	3.000	14.2	0.56	47.8	1.88	6.460	2.3
76.2	3.000	17.5	0.69	41.2	1.62	4.799	1.8
76.2	3.000	20.0	0.79	36.2	1.43	3.705	1.6
82.5	3.248	7.1	0.28	68.3	2.69	13.190	5.4
82.5	3.248	8.0	0.31	66.5	2.62	12.504	4.7
82.5	3.248	8.8	0.35	64.9	2.56	11.909	4.2
82.5	3.248	14.2	0.56	54.1	2.13	8.275	2.5
82.5	3.248	17.5	0.69	47.5	1.87	6.379	2.0
82.5	3.248	20.0	0.79	42.5	1.67	5.107	1.7
88.9	3.500	5.6	0.22	77.7	3.06	17.070	7.9
88.9	3.500	7.1	0.28	74.7	2.94	15.777	5.8
88.9	3.500	8.8	0.35	71.3	2.81	14.374	4.6
88.9	3.500	14.2	0.56	60.5	2.38	10.349	2.7
88.9	3.500	17.5	0.69	53.9	2.12	8.214	2.2
88.9	3.500	22.2	0.87	44.5	1.75	5.599	1.7
88.9	3.500	25.0	0.98	38.9	1.53	4.279	1.5
101.6	4.000	17.5	0.69	66.6	2.62	12.541	2.5
101.6	4.000	28.0	1.10	45.6	1.80	5.879	1.5
108.0	4.250	7.1	0.28	93.8	3.69	24.850	7.6
108.0	4.250	8.0	0.31	92.0	3.62	23.905	6.7
108.0	4.250	8.8	0.35	90.4	3.56	23.081	5.7
108.0	4.250	30.0	1.18	48.0	1.89	6.501	1.5
114.3	4.500	7.1	0.28	100.1	3.94	28.331	8.0
114.3	4.500	8.8	0.35	96.7	3.81	26.439	6.0
114.3	4.500	14.2	0.56	85.9	3.38	20.863	3.6
114.3	4.500	17.5	0.69	79.3	3.12	17.780	2.9
114.3	4.500	28.0	1.10	58.3	2.30	9.610	1.7
114.3	4.500	30.0	1.18	54.3	2.14	8.337	1.6
127.0	5.000	7.1	0.28	112.8	4.44	35.98	8.9
127.0	5.000	8.0	0.31	111.0	4.37	34.84	7.9
127.0	5.000	8.8	0.35	109.4	4.31	33.84	7.2
127.0	5.000	14.2	0.56	98.6	3.88	27.49	4.0
127.0	5.000	17.5	0.69	92.0	3.62	23.93	3.2
127.0	5.000	28.0	1.10	71.0	2.80	14.25	1.9
127.0	5.000	30.0	1.18	67.0	2.64	12.69	1.8
127.0	5.000	32.0	1.26	63.0	2.48	11.22	1.7
127.0	5.000	10.1	0.40	106.8	4.20	32.25	5.8
133.4	5.250	5.0	0.20	123.4	4.86	43.02	13.3
133.4	5.250	7.1	0.28	119.2	4.69	40.14	9.4
133.4	5.250	8.0	0.31	117.4	4.62	38.94	8.3
133.4	5.250	8.8	0.35	115.8	4.56	37.88	7.6
133.4	5.250	14.2	0.56	105.0	4.13	31.14	4.3
133.4	5.250	16.0	0.63	101.4	3.99	29.04	3.7
133.4	5.250	17.5	0.69	98.4	3.87	27.35	3.4
133.4	5.250	20.0	0.79	93.4	3.68	24.64	2.9
133.4	5.250	22.2	0.87	89.0	3.50	22.37	2.6
133.4	5.250	30.0	1.18	73.4	2.89	15.21	1.9
133.4	5.250	32.0	1.26	69.4	2.73	13.60	1.7
133.4	5.250	40.0	1.57	53.4	2.10	8.05	1.4
139.7	5.500	7.1	0.28	125.5	4.94	44.53	9.8
139.7	5.500	8.8	0.35	122.1	4.81	42.15	7.9
139.7	5.500	14.2	0.56	111.3	4.38	35.03	4.5
139.7	5.500	16.0	0.63	107.7	4.24	32.80	3.9
139.7	5.500	17.5	0.69	104.7	4.12	30.99	3.6
139.7	5.500	30.0	1.18	79.7	3.14	17.96	2.0
139.7	5.500	35.0	1.38	69.7	2.74	13.74	1.7
139.7	5.500	45.0	1.77	49.7	1.96	6.98	1.3
152.4	6.000	7.1	0.28	138.2	5.44	54.00	10.7
152.4	6.000	8.0	0.31	136.4	5.37	52.60	9.5
152.4	6.000	8.8	0.35	134.8	5.31	51.38	8.7
152.4	6.000	14.2	0.56	124.0	4.88	43.47	4.9
152.4	6.000	17.5	0.69	117.4	4.62	38.97	3.9
152.4	6.000	22.2	0.87	108.0	4.25	32.98	3.0
152.4	6.000	30.0	1.18	92.4	3.64	24.14	2.2

od mm	od in	WALL mm	WALL in	id mm	id in	1m/s m^3/h	10barg N/mm^2
152.4	6.000	35.0	1.38	82.4	3.24	19.20	1.8
152.4	6.000	36.5	1.44	79.4	3.13	17.83	1.7
152.4	6.000	38.0	1.50	76.4	3.01	16.50	1.7
152.4	6.000	45.0	1.77	62.4	2.46	11.01	1.4
158.8	6.250	6.3	0.25	146.2	5.75	60.39	12.6
158.8	6.250	8.8	0.35	141.2	5.56	56.33	9.0
158.8	6.250	10.0	0.39	138.8	5.46	54.43	7.9
158.8	6.250	14.2	0.56	130.4	5.13	48.04	5.1
158.8	6.250	16.0	0.63	126.8	4.99	45.42	4.5
158.8	6.250	17.5	0.69	123.8	4.87	43.30	4.1
158.8	6.250	25.0	0.98	108.8	4.28	33.44	2.8
158.8	6.250	30.0	1.18	98.8	3.89	27.57	2.3
158.8	6.250	32.0	1.26	94.8	3.73	25.38	2.1
158.8	6.250	35.0	1.38	88.8	3.49	22.27	1.9
158.8	6.250	40.0	1.57	78.8	3.10	17.53	1.7
158.8	6.250	50.0	1.97	58.8	2.31	9.76	1.3
168.3	6.625	8.8	0.35	150.7	5.93	64.20	9.6
168.3	6.625	10.0	0.39	148.3	5.84	62.17	8.4
168.3	6.625	14.2	0.56	139.9	5.51	55.32	5.5
168.3	6.625	17.5	0.69	133.3	5.25	50.23	4.4
168.3	6.625	22.2	0.87	123.9	4.88	43.39	3.4
168.3	6.625	30.0	1.18	108.3	4.26	33.15	2.4
168.3	6.625	35.0	1.38	98.3	3.87	27.31	2.0
168.3	6.625	40.0	1.57	88.3	3.48	22.04	1.8
168.3	6.625	50.0	1.97	68.3	2.69	13.18	1.4
177.8	7.000	8.0	0.31	161.8	6.37	74.02	11.1
177.8	7.000	8.8	0.35	160.2	6.31	72.56	10.1
177.8	7.000	14.2	0.56	149.4	5.88	63.11	5.8
177.8	7.000	16.0	0.63	145.8	5.74	60.10	5.1
177.8	7.000	17.5	0.69	142.8	5.62	57.66	4.6
177.8	7.000	28.0	1.10	121.8	4.80	41.95	2.8
177.8	7.000	30.0	1.18	117.8	4.64	39.24	2.6
177.8	7.000	35.0	1.38	107.8	4.24	32.86	2.2
177.8	7.000	40.0	1.57	97.8	3.85	27.04	1.9
177.8	7.000	55.0	2.17	67.8	2.67	13.00	1.3
193.7	7.625	8.8	0.35	176.1	6.93	87.66	11.0
193.7	7.625	12.5	0.49	168.7	6.64	80.45	7.7
193.7	7.625	14.2	0.56	165.3	6.51	77.24	6.8
193.7	7.625	16.0	0.63	161.7	6.37	73.91	5.6
193.7	7.625	17.5	0.69	158.7	6.25	71.19	5.1
193.7	7.625	22.2	0.87	149.3	5.88	63.01	3.9
193.7	7.625	30.0	1.18	133.7	5.26	50.53	2.8
193.7	7.625	35.0	1.38	123.7	4.87	43.25	2.4
193.7	7.625	40.0	1.57	113.7	4.48	36.54	2.1
193.7	7.625	45.0	1.77	103.7	4.08	30.39	1.8
193.7	7.625	50.0	1.97	93.7	3.69	24.81	1.6
203.2	8.000	8.0	0.31	187.2	7.37	99.08	12.7
203.2	8.000	8.8	0.35	185.6	7.31	97.40	11.5
203.2	8.000	10.0	0.39	183.2	7.21	94.89	10.2
203.2	8.000	12.5	0.49	178.2	7.02	89.79	8.1
203.2	8.000	14.2	0.56	174.8	6.88	86.39	7.2
203.2	8.000	16.0	0.63	171.2	6.74	82.87	5.9
203.2	8.000	22.2	0.87	158.8	6.25	71.30	4.1
203.2	8.000	30.0	1.18	143.2	5.64	57.98	3.0
203.2	8.000	35.0	1.38	133.2	5.24	50.17	2.5
203.2	8.000	40.0	1.57	123.2	4.85	42.92	2.2
203.2	8.000	50.0	1.97	103.2	4.06	30.11	1.7
203.2	8.000	55.0	2.17	93.2	3.67	24.56	1.5
216.0	8.504	22.2	0.87	171.6	6.76	83.26	4.4
216.0	8.504	30.0	1.18	156.0	6.14	68.81	3.2
216.0	8.504	45.5	1.79	125.0	4.92	44.18	2.0
219.1	8.625	8.8	0.35	201.5	7.93	114.78	12.4
219.1	8.625	12.5	0.49	194.1	7.64	106.50	8.8
219.1	8.625	14.2	0.56	190.7	7.51	102.80	7.7
219.1	8.625	16.0	0.63	187.1	7.37	98.96	6.8
219.1	8.625	17.5	0.69	184.1	7.25	95.81	5.8
219.1	8.625	30.0	1.18	159.1	6.26	71.55	3.2
219.1	8.625	40.0	1.57	139.1	5.48	54.69	2.4
219.1	8.625	45.0	1.77	129.1	5.08	47.11	2.1
219.1	8.625	50.0	1.97	119.1	4.69	40.09	1.8
219.1	8.625	60.0	2.36	99.1	3.90	27.76	1.5
228.6	9.000	17.5	0.69	193.6	7.62	105.97	6.1
228.6	9.000	40.0	1.57	148.6	5.85	62.44	2.5
228.6	9.000	48.5	1.91	131.6	5.18	48.97	2.0
244.5	9.625	8.8	0.35	226.9	8.93	145.54	13.9
244.5	9.625	14.2	0.56	216.1	8.51	132.01	8.6

od mm	od in	WALL mm	WALL in	id mm	id in	1m/s m^3/h	10barg N/mm^2
244.5	9.625	16.0	0.63	212.5	8.37	127.65	7.6
244.5	9.625	17.2	0.68	210.2	8.27	124.90	7.1
244.5	9.625	30.0	1.18	184.5	7.26	96.23	3.6
244.5	9.625	35.0	1.38	174.5	6.87	86.08	3.1
244.5	9.625	40.0	1.57	164.5	6.48	76.49	2.7
244.5	9.625	50.0	1.97	144.5	5.69	59.02	2.1
244.5	9.625	75.0	2.95	94.5	3.72	25.24	1.4
244.5	10.00	17.5	0.69	219.0	8.62	135.61	7.3
254.0	10.00	28.0	1.10	198.0	7.80	110.85	4.1
254.0	10.00	35.0	1.38	184.0	7.24	95.73	3.2
254.0	10.00	40.0	1.57	174.0	6.85	85.60	2.8
254.0	10.00	55.0	2.17	144.0	5.67	58.63	1.9

Table 9.9 Seamless steel pipe/tube to DIN 1629 ST52 (Above)

9.2.3 Metallic tube

Table 9.10, Table 9.11, Table 9.12 and Table 9.13, indicate the range of popular tube sizes available.

od in	od mm	WALL in	WALL mm	id in	id mm	1m/s m^3/h	10barg N/mm^2
0.250	6.35	0.016	0.41	0.218	5.54	0.087	7.8
0.250	6.35	0.020	0.51	0.210	5.33	0.080	5.8
0.250	6.35	0.035	0.89	0.180	4.57	0.059	3.2
0.312	7.92	0.016	0.41	0.280	7.11	0.143	9.8
0.312	7.92	0.020	0.51	0.272	6.91	0.135	7.8
0.312	9.53	0.016	0.41	0.343	8.71	0.215	11.7
0.375	9.53	0.020	0.51	0.335	8.51	0.205	9.4
0.375	9.53	0.028	0.71	0.319	8.10	0.186	6.7
0.500	12.7	0.020	0.51	0.460	11.7	0.386	12.5
0.500	12.7	0.026	0.66	0.448	11.4	0.366	9.6
0.625	15.9	0.016	0.41	0.593	15.1	0.641	19.5
0.625	15.9	0.020	0.51	0.585	14.9	0.624	15.6
0.625	15.9	0.026	0.66	0.573	14.6	0.599	12.0
0.625	15.9	0.028	0.71	0.569	14.5	0.591	11.2
0.625	15.9	0.032	0.81	0.561	14.2	0.574	9.8
0.625	15.9	0.033	0.84	0.559	14.2	0.570	9.5
0.750	19.1	0.020	0.51	0.710	18.0	0.920	18.8
0.750	19.1	0.028	0.71	0.694	17.6	0.879	13.4
0.750	19.1	0.035	0.89	0.680	17.3	0.843	10.7
0.750	19.1	0.039	0.99	0.672	17.1	0.824	9.6
0.750	19.1	0.052	1.32	0.646	16.4	0.761	7.2
1.000	25.4	0.020	0.51	0.960	24.4	1.681	25.0
1.000	25.4	0.026	0.66	0.948	24.1	1.639	19.2
1.000	25.4	0.035	0.89	0.930	23.6	1.578	14.3
1.000	25.4	0.052	1.32	0.896	22.8	1.464	9.6
1.250	31.8	0.020	0.51	1.210	30.7	2.671	31.3
1.250	31.8	0.032	0.81	1.186	30.1	2.566	19.5
1.500	38.1	0.026	0.66	1.448	36.8	3.825	28.8

Table 9.10 Seamless stainless steel tube to ASTM A269

Figure 9.11 Seamless and welded stainless steel inch tube (Below)

od in	od mm	WALL in	WALL mm	id in	id mm	1m/s m^3/h	10barg N/mm^2
0.0625	1.59	0.010	0.25	0.043	1.08	0.003	2.7
0.0625	1.59	0.012	0.30	0.039	0.98	0.003	2.2
0.0625	1.59	0.014	0.36	0.035	0.88	0.002	1.9
0.0625	1.59	0.016	0.41	0.031	0.77	0.002	1.6
0.0625	1.59	0.020	0.51	0.023	0.57	0.001	1.3
0.125	3.18	0.012	0.30	0.101	2.57	0.019	4.8
0.125	3.18	0.016	0.41	0.093	2.36	0.016	3.5
0.125	3.18	0.020	0.51	0.085	2.16	0.013	2.7
0.125	3.18	0.028	0.71	0.069	1.75	0.009	1.9
0.125	3.18	0.032	0.81	0.061	1.55	0.007	1.6
0.125	3.18	0.035	0.89	0.055	1.40	0.006	1.5
0.125	3.18	0.042	1.07	0.041	1.04	0.003	1.2
0.125	3.18	0.049	1.24	0.027	0.69	0.001	1.1
0.156	3.97	0.016	0.41	0.124	3.16	0.028	4.4
0.156	3.97	0.020	0.51	0.116	2.95	0.025	3.5
0.156	3.97	0.028	0.71	0.100	2.55	0.018	2.4
0.156	3.97	0.035	0.89	0.086	2.19	0.014	1.9
0.188	4.76	0.016	0.41	0.156	3.95	0.044	5.4
0.188	4.76	0.020	0.51	0.148	3.75	0.040	4.2

od in	od mm	WALL in	WALL mm	id in	id mm	1m/s m^3/h	10barg N/mm^2
0.188	4.76	0.022	0.56	0.144	3.64	0.038	3.8
0.188	4.76	0.028	0.71	0.132	3.34	0.032	2.9
0.188	4.76	0.035	0.89	0.118	2.98	0.025	2.3
0.188	4.76	0.049	1.24	0.090	2.27	0.015	1.6
0.188	4.76	0.065	1.65	0.058	1.46	0.006	1.2
0.219	5.56	0.016	0.41	0.187	4.74	0.064	6.8
0.219	5.56	0.047	1.19	0.125	3.17	0.028	2.0
0.250	6.35	0.012	0.30	0.226	5.74	0.093	10.4
0.250	6.35	0.016	0.41	0.218	5.54	0.087	7.8
0.250	6.35	0.020	0.51	0.210	5.33	0.080	5.8
0.250	6.35	0.022	0.56	0.206	5.23	0.077	5.2
0.250	6.35	0.028	0.71	0.194	4.93	0.069	4.0
0.250	6.35	0.032	0.81	0.186	4.72	0.063	3.5
0.250	6.35	0.035	0.89	0.180	4.57	0.059	3.2
0.250	6.35	0.042	1.07	0.166	4.22	0.050	2.6
0.250	6.35	0.049	1.24	0.152	3.86	0.042	2.2
0.250	6.35	0.058	1.47	0.134	3.40	0.033	1.8
0.250	6.35	0.065	1.65	0.120	3.05	0.026	1.6
0.250	6.35	0.095	2.41	0.060	1.52	0.007	1.1
0.281	7.14	0.035	0.89	0.211	5.37	0.081	3.6
0.313	7.94	0.016	0.41	0.281	7.12	0.144	9.8
0.313	7.94	0.020	0.51	0.273	6.92	0.135	7.8
0.313	7.94	0.028	0.71	0.257	6.52	0.120	5.1
0.313	7.94	0.035	0.89	0.243	6.16	0.107	4.0
0.313	7.94	0.049	1.24	0.215	5.45	0.084	2.8
0.313	7.94	0.058	1.47	0.197	4.99	0.070	2.3
0.313	7.94	0.065	1.65	0.183	4.64	0.061	2.0
0.313	7.94	0.083	2.11	0.147	3.72	0.039	1.6
0.313	7.94	0.095	2.41	0.123	3.11	0.027	1.4
0.344	8.73	0.015	0.38	0.314	7.97	0.180	11.5
0.375	9.53	0.016	0.41	0.343	8.71	0.215	11.7
0.375	9.53	0.020	0.51	0.335	8.51	0.205	9.4
0.375	9.53	0.022	0.56	0.331	8.41	0.200	8.5
0.375	9.53	0.028	0.71	0.319	8.10	0.186	6.7
0.375	9.53	0.035	0.89	0.305	7.75	0.170	4.9
0.375	9.53	0.042	1.07	0.291	7.39	0.154	4.0
0.375	9.53	0.049	1.24	0.277	7.04	0.140	3.4
0.375	9.53	0.058	1.47	0.259	6.58	0.122	2.8
0.375	9.53	0.065	1.65	0.245	6.22	0.109	2.5
0.375	9.53	0.083	2.11	0.209	5.31	0.080	1.9
0.375	9.53	0.095	2.41	0.185	4.70	0.062	1.6
0.375	9.53	0.120	3.05	0.135	3.43	0.033	1.3
0.438	11.11	0.028	0.71	0.382	9.69	0.265	7.8
0.438	11.11	0.035	0.89	0.368	9.33	0.246	5.8
0.438	11.11	0.049	1.24	0.340	8.62	0.210	4.0
0.438	11.11	0.058	1.47	0.322	8.17	0.189	3.3
0.438	11.11	0.065	1.65	0.308	7.81	0.172	3.0
0.500	12.70	0.016	0.41	0.468	11.9	0.400	15.6
0.500	12.70	0.020	0.51	0.460	11.7	0.386	12.5
0.500	12.70	0.025	0.64	0.450	11.4	0.369	10.0
0.500	12.70	0.028	0.71	0.444	11.3	0.360	8.9
0.500	12.70	0.035	0.89	0.430	10.9	0.337	7.1
0.500	12.70	0.042	1.07	0.416	10.6	0.316	5.5
0.500	12.70	0.049	1.24	0.402	10.2	0.295	4.7
0.500	12.70	0.058	1.47	0.384	9.75	0.269	3.9
0.500	12.70	0.065	1.65	0.370	9.40	0.250	3.4
0.500	12.70	0.083	2.11	0.334	8.48	0.203	2.6
0.500	12.70	0.095	2.41	0.310	7.87	0.175	2.2
0.500	12.70	0.120	3.05	0.260	6.60	0.123	1.7
0.500	12.70	0.134	3.40	0.232	5.89	0.098	1.5
0.563	14.29	0.016	0.41	0.531	13.5	0.513	17.6
0.563	14.29	0.028	0.71	0.507	12.9	0.468	10.0
0.563	14.29	0.035	0.89	0.493	12.5	0.442	8.0
0.563	14.29	0.049	1.24	0.465	11.8	0.394	5.3
0.563	14.29	0.065	1.65	0.433	11.0	0.341	3.9
0.563	14.29	0.083	2.11	0.397	10.1	0.287	3.0
0.563	14.29	0.120	3.05	0.323	8.19	0.190	2.0
0.563	14.29	0.188	4.78	0.187	4.74	0.063	1.2
0.625	15.88	0.016	0.41	0.593	15.1	0.641	19.5
0.625	15.88	0.020	0.51	0.585	14.9	0.624	15.6
0.625	15.88	0.028	0.71	0.569	14.5	0.591	11.2
0.625	15.88	0.035	0.89	0.555	14.1	0.562	8.9
0.625	15.88	0.042	1.07	0.541	13.7	0.534	7.4
0.625	15.88	0.049	1.24	0.527	13.4	0.507	5.9
0.625	15.88	0.065	1.65	0.495	12.6	0.447	4.4
0.625	15.88	0.083	2.11	0.459	11.7	0.384	3.3

od in	od mm	WALL in	WALL mm	id in	id mm	1m/s m^3/h	10barg N/mm^2
0.625	15.88	0.095	2.41	0.435	11.0	0.345	2.9
0.625	15.88	0.120	3.05	0.385	9.78	0.270	2.2
0.625	15.88	0.156	3.96	0.313	7.95	0.179	1.7
0.625	15.88	0.188	4.78	0.249	6.32	0.113	1.4
0.688	17.46	0.020	0.51	0.648	16.4	0.765	17.2
0.688	17.46	0.035	0.89	0.618	15.7	0.696	9.8
0.688	17.46	0.049	1.24	0.590	15.0	0.634	7.0
0.688	17.46	0.065	1.65	0.558	14.2	0.567	4.8
0.750	19.05	0.016	0.41	0.718	18.2	0.940	23.4
0.750	19.05	0.020	0.51	0.710	18.0	0.920	18.8
0.750	19.05	0.022	0.56	0.706	17.9	0.909	17.0
0.750	19.05	0.025	0.64	0.700	17.8	0.894	15.0
0.750	19.05	0.028	0.71	0.694	17.6	0.879	13.4
0.750	19.05	0.031	0.79	0.688	17.5	0.863	12.1
0.750	19.05	0.035	0.89	0.680	17.3	0.843	10.7
0.750	19.05	0.042	1.07	0.666	16.9	0.809	8.9
0.750	19.05	0.049	1.24	0.652	16.6	0.775	7.7
0.750	19.05	0.058	1.47	0.634	16.1	0.733	6.0
0.750	19.05	0.065	1.65	0.620	15.7	0.701	5.3
0.750	19.05	0.083	2.11	0.584	14.8	0.622	4.1
0.750	19.05	0.095	2.41	0.560	14.2	0.572	3.5
0.750	19.05	0.120	3.05	0.510	13.0	0.474	2.7
0.750	19.05	0.156	3.96	0.438	11.1	0.350	2.0
0.750	19.05	0.188	4.78	0.374	9.50	0.255	1.7
0.813	20.64	0.048	1.22	0.717	18.2	0.936	8.5
0.813	20.64	0.064	1.63	0.685	17.4	0.855	5.9
0.875	22.23	0.028	0.71	0.819	20.8	1.224	15.6
0.875	22.23	0.035	0.89	0.805	20.4	1.182	12.5
0.875	22.23	0.049	1.24	0.777	19.7	1.101	8.9
0.875	22.23	0.065	1.65	0.745	18.9	1.012	6.7
0.875	22.23	0.083	2.11	0.709	18.0	0.917	4.8
0.875	22.23	0.095	2.41	0.685	17.4	0.856	4.2
0.875	22.23	0.120	3.05	0.635	16.1	0.736	3.2
0.875	22.23	0.188	4.78	0.499	12.7	0.454	2.0
0.875	22.23	0.250	6.35	0.375	9.53	0.257	1.5
0.938	23.81	0.035	0.89	0.868	22.0	1.373	13.4
1.000	25.40	0.016	0.41	0.968	24.6	1.709	31.3
1.000	25.40	0.020	0.51	0.960	24.4	1.681	25.0
1.000	25.40	0.028	0.71	0.944	24.0	1.626	17.9
1.000	25.40	0.035	0.89	0.930	23.6	1.578	14.3
1.000	25.40	0.049	1.24	0.902	22.9	1.484	10.2
1.000	25.40	0.058	1.47	0.884	22.5	1.425	8.6
1.000	25.40	0.065	1.65	0.870	22.1	1.381	7.7
1.000	25.40	0.083	2.11	0.834	21.2	1.269	5.6
1.000	25.40	0.095	2.41	0.810	20.6	1.197	4.8
1.000	25.40	0.109	2.77	0.782	19.9	1.116	4.1
1.000	25.40	0.120	3.05	0.760	19.3	1.054	3.7
1.000	25.40	0.125	3.18	0.750	19.1	1.026	3.6
1.000	25.40	0.188	4.78	0.624	15.8	0.710	2.3
1.000	25.40	0.250	6.35	0.500	12.7	0.456	1.7
1.062	26.97	0.156	3.96	0.750	19.1	1.026	3.0
1.125	28.58	0.020	0.51	1.085	27.6	2.147	28.1
1.125	28.58	0.028	0.71	1.069	27.2	2.085	20.1
1.125	28.58	0.035	0.89	1.055	26.8	2.030	16.1
1.125	28.58	0.049	1.24	1.027	26.1	1.924	11.5
1.125	28.58	0.058	1.47	1.009	25.6	1.857	9.7
1.125	28.58	0.063	1.60	0.999	25.4	1.821	8.9
1.125	28.58	0.065	1.65	0.995	25.3	1.806	8.7
1.125	28.58	0.095	2.41	0.935	23.7	1.595	5.5
1.125	28.58	0.120	3.05	0.885	22.5	1.429	4.2
1.125	28.58	0.188	4.78	0.749	19.0	1.023	2.6
1.125	28.58	0.250	6.35	0.625	15.9	0.713	1.9
1.187	30.15	0.028	0.71	1.131	28.7	2.333	21.2
1.187	30.15	0.065	1.65	1.057	26.8	2.038	9.1
1.187	30.15	0.092	2.34	1.003	25.5	1.835	6.0
1.250	31.75	0.020	0.51	1.210	30.7	2.671	31.3
1.250	31.75	0.028	0.71	1.194	30.3	2.601	22.3
1.250	31.75	0.035	0.89	1.180	30.0	2.540	17.9
1.250	31.75	0.049	1.24	1.152	29.3	2.421	12.8
1.250	31.75	0.065	1.65	1.120	28.4	2.288	9.6
1.250	31.75	0.083	2.11	1.084	27.5	2.143	7.5
1.250	31.75	0.095	2.41	1.060	26.9	2.050	6.1
1.250	31.75	0.109	2.77	1.032	26.2	1.943	5.3
1.250	31.75	0.120	3.05	1.010	25.7	1.861	4.8
1.250	31.75	0.123	3.12	1.004	25.5	1.839	4.6
1.250	31.75	0.125	3.18	1.000	25.4	1.824	4.6

od in	od mm	WALL in	WALL mm	id in	id mm	1m/s m^3/h	10barg N/mm^2
1.250	31.75	0.148	3.76	0.954	24.2	1.660	3.8
1.250	31.75	0.188	4.78	0.874	22.2	1.393	2.9
1.250	31.75	0.250	6.35	0.750	19.1	1.026	2.1
1.250	31.75	0.375	9.53	0.500	12.7	0.456	1.4
1.312	33.32	0.028	0.71	1.256	31.9	2.878	23.4
1.312	33.32	0.036	0.91	1.240	31.5	2.805	18.2
1.312	33.32	0.049	1.24	1.214	30.8	2.688	13.4
1.375	34.93	0.016	0.41	1.343	34.1	3.290	43.0
1.375	34.93	0.028	0.71	1.319	33.5	3.174	24.6
1.375	34.93	0.035	0.89	1.305	33.1	3.107	19.6
1.375	34.93	0.049	1.24	1.277	32.4	2.975	14.0
1.375	34.93	0.061	1.55	1.253	31.8	2.864	11.3
1.375	34.93	0.065	1.65	1.245	31.6	2.827	10.6
1.375	34.93	0.094	2.39	1.187	30.1	2.570	7.3
1.375	34.93	0.120	3.05	1.135	28.8	2.350	5.3
1.375	34.93	0.188	4.78	0.999	25.4	1.821	3.2
1.375	34.93	0.250	6.35	0.875	22.2	1.397	2.4
1.437	36.50	0.094	2.39	1.249	31.7	2.846	7.6
1.437	36.50	0.123	3.12	1.191	30.3	2.588	5.4
1.437	36.50	0.281	7.14	0.875	22.2	1.397	2.2
1.500	38.10	0.016	0.41	1.468	37.3	3.931	46.9
1.500	38.10	0.020	0.51	1.460	37.1	3.888	37.5
1.500	38.10	0.022	0.56	1.456	37.0	3.867	34.1
1.500	38.10	0.025	0.64	1.450	36.8	3.835	30.0
1.500	38.10	0.028	0.71	1.444	36.7	3.804	26.8
1.500	38.10	0.035	0.89	1.430	36.3	3.730	21.4
1.500	38.10	0.049	1.24	1.402	35.6	3.586	15.3
1.500	38.10	0.065	1.65	1.370	34.8	3.424	11.5
1.500	38.10	0.083	2.11	1.334	33.9	3.246	9.0
1.500	38.10	0.095	2.41	1.310	33.3	3.130	7.9
1.500	38.10	0.100	2.54	1.300	33.0	3.083	7.5
1.500	38.10	0.109	2.77	1.282	32.6	2.998	6.9
1.500	38.10	0.120	3.05	1.260	32.0	2.896	5.8
1.500	38.10	0.123	3.12	1.254	31.9	2.869	5.6
1.500	38.10	0.125	3.18	1.250	31.8	2.850	5.5
1.500	38.10	0.134	3.40	1.232	31.3	2.769	5.1
1.500	38.10	0.188	4.78	1.124	28.5	2.305	3.6
1.500	38.10	0.250	6.35	1.000	25.4	1.824	2.6
1.500	38.10	0.313	7.95	0.874	22.2	1.393	2.0
1.500	38.10	0.375	9.53	0.750	19.1	1.026	1.7
1.562	39.67	0.028	0.71	1.506	38.3	4.137	27.9
1.562	39.67	0.156	3.96	1.250	31.8	2.850	4.6
1.562	39.67	0.281	7.14	1.000	25.4	1.824	2.4
1.562	39.67	0.344	8.74	0.874	22.2	1.393	1.9
1.562	39.67	0.375	9.53	0.812	20.6	1.203	1.7
1.625	41.28	0.035	0.89	1.555	39.5	4.411	23.2
1.625	41.28	0.049	1.24	1.527	38.8	4.253	16.6
1.625	41.28	0.065	1.65	1.495	38.0	4.077	12.5
1.625	41.28	0.120	3.05	1.385	35.2	3.499	6.8
1.625	41.28	0.188	4.78	1.249	31.7	2.846	3.9
1.625	41.28	0.250	6.35	1.125	28.6	2.309	2.8
1.625	41.28	0.375	9.53	0.875	22.2	1.397	1.8
1.625	41.28	0.437	11.10	0.751	19.1	1.029	1.5
1.688	42.88	0.092	2.34	1.504	38.2	4.126	9.2
1.688	42.88	0.124	3.15	1.440	36.6	3.783	6.8
1.688	42.88	0.281	7.14	1.126	28.6	2.313	2.6
1.688	42.88	0.344	8.74	1.000	25.4	1.824	2.1
1.688	42.88	0.406	10.31	0.876	22.3	1.400	1.7
1.750	44.45	0.020	0.51	1.710	43.4	5.334	43.8
1.750	44.45	0.028	0.71	1.694	43.0	5.235	31.3
1.750	44.45	0.035	0.89	1.680	42.7	5.148	25.0
1.750	44.45	0.042	1.07	1.666	42.3	5.063	20.8
1.750	44.45	0.049	1.24	1.652	42.0	4.978	17.9
1.750	44.45	0.065	1.65	1.620	41.1	4.787	13.5
1.750	44.45	0.095	2.41	1.560	39.6	4.439	9.2
1.750	44.45	0.120	3.05	1.510	38.4	4.160	7.3
1.750	44.45	0.188	4.78	1.374	34.9	3.444	4.2
1.750	44.45	0.250	6.35	1.250	31.8	2.850	3.1
1.750	44.45	0.312	7.92	1.126	28.6	2.313	2.4
1.750	44.45	0.375	9.53	1.000	25.4	1.824	2.0
1.812	46.02	0.281	7.14	1.250	31.8	2.850	2.8
1.812	46.02	0.344	8.74	1.124	28.5	2.305	2.3
1.812	46.02	0.406	10.31	1.000	25.4	1.824	1.9
1.875	47.63	0.020	0.51	1.835	46.6	6.142	46.9
1.875	47.63	0.035	0.89	1.805	45.8	5.943	26.8
1.875	47.63	0.049	1.24	1.777	45.1	5.760	19.1

od in	od mm	WALL in	WALL mm	id in	id mm	1m/s m^3/h	10barg N/mm^2
1.875	47.63	0.065	1.65	1.745	44.3	5.555	14.4
1.875	47.63	0.083	2.11	1.709	43.4	5.328	11.3
1.875	47.63	0.188	4.78	1.499	38.1	4.099	4.5
1.875	47.63	0.225	5.72	1.425	36.2	3.704	3.7
1.875	47.63	0.250	6.35	1.375	34.9	3.449	3.3
1.875	47.63	0.312	7.92	1.251	31.8	2.855	2.6
1.875	47.63	0.313	7.95	1.249	31.7	2.846	2.6
1.875	47.63	0.375	9.53	1.125	28.6	2.309	2.1
1.875	47.63	0.438	11.13	0.999	25.4	1.821	1.8
1.937	49.20	0.123	3.12	1.691	43.0	5.216	7.9
1.937	49.20	0.281	7.14	1.375	34.9	3.449	3.0
1.937	49.20	0.344	8.74	1.249	31.7	2.846	2.4
1.937	49.20	0.375	9.53	1.187	30.1	2.570	2.2
1.937	49.20	0.437	11.10	1.063	27.0	2.061	1.9
1.937	49.20	0.500	12.70	0.937	23.8	1.602	1.6
2.000	50.80	0.016	0.41	1.968	50.0	7.065	62.5
2.000	50.80	0.020	0.51	1.960	49.8	7.008	50.0
2.000	50.80	0.022	0.56	1.956	49.7	6.979	45.5
2.000	50.80	0.025	0.64	1.950	49.5	6.936	40.0
2.000	50.80	0.028	0.71	1.944	49.4	6.894	35.7
2.000	50.80	0.035	0.89	1.930	49.0	6.795	28.6
2.000	50.80	0.049	1.24	1.902	48.3	6.599	20.4
2.000	50.80	0.065	1.65	1.870	47.5	6.379	15.4
2.000	50.80	0.083	2.11	1.834	46.6	6.136	12.0
2.000	50.80	0.095	2.41	1.810	46.0	5.976	10.5
2.000	50.80	0.109	2.77	1.782	45.3	5.793	9.2
2.000	50.80	0.120	3.05	1.760	44.7	5.650	8.3
2.000	50.80	0.125	3.18	1.750	44.5	5.586	8.0
2.000	50.80	0.133	3.38	1.734	44.0	5.485	7.5
2.000	50.80	0.165	4.19	1.670	42.4	5.087	5.6
2.000	50.80	0.188	4.78	1.624	41.2	4.811	4.9
2.000	50.80	0.250	6.35	1.500	38.1	4.104	3.6
2.000	50.80	0.312	7.92	1.376	35.0	3.454	2.8
2.000	50.80	0.313	7.95	1.374	34.9	3.444	2.8
2.000	50.80	0.375	9.53	1.250	31.8	2.850	2.3
2.000	50.80	0.438	11.13	1.124	28.5	2.305	1.9
2.000	50.80	0.469	11.91	1.062	27.0	2.057	1.8
2.000	50.80	0.500	12.70	1.000	25.4	1.824	1.7
2.125	53.98	0.035	0.89	2.055	52.2	7.703	30.4
2.125	53.98	0.049	1.24	2.027	51.5	7.495	21.7
2.125	53.98	0.065	1.65	1.995	50.7	7.260	16.3
2.125	53.98	0.092	2.34	1.941	49.3	6.872	11.5
2.125	53.98	0.188	4.78	1.749	44.4	5.580	5.2
2.125	53.98	0.250	6.35	1.625	41.3	4.817	3.8
2.125	53.98	0.313	7.95	1.499	38.1	4.099	3.0
2.125	53.98	0.375	9.53	1.375	34.9	3.449	2.4
2.125	53.98	0.438	11.13	1.249	31.7	2.846	2.1
2.125	53.98	0.500	12.70	1.125	28.6	2.309	1.8
2.125	53.98	0.625	15.88	0.875	22.2	1.397	1.4
2.250	57.15	0.016	0.41	2.218	56.3	8.974	70.3
2.250	57.15	0.020	0.51	2.210	56.1	8.909	56.3
2.250	57.15	0.035	0.89	2.180	55.4	8.669	32.1
2.250	57.15	0.049	1.24	2.152	54.7	8.448	23.0
2.250	57.15	0.058	1.47	2.134	54.2	8.307	19.4
2.250	57.15	0.065	1.65	2.120	53.8	8.198	17.3
2.250	57.15	0.083	2.11	2.084	52.9	7.922	13.6
2.250	57.15	0.120	3.05	2.010	51.1	7.370	9.4
2.250	57.15	0.125	3.18	2.000	50.8	7.297	9.0
2.250	57.15	0.188	4.78	1.874	47.6	6.406	5.5
2.250	57.15	0.250	6.35	1.750	44.5	5.586	4.1
2.250	57.15	0.312	7.92	1.626	41.3	4.823	3.2
2.250	57.15	0.313	7.95	1.624	41.2	4.811	3.2
2.250	57.15	0.375	9.53	1.500	38.1	4.104	2.6
2.250	57.15	0.406	10.31	1.438	36.5	3.772	2.4
2.250	57.15	0.438	11.13	1.374	34.9	3.444	2.2
2.250	57.15	0.500	12.70	1.250	31.8	2.850	1.9
2.250	57.15	0.625	15.88	1.000	25.4	1.824	1.5
2.250	57.15	0.500	12.70	1.250	31.8	2.850	1.9
2.500	63.50	0.025	0.64	2.450	62.2	10.95	50.0
2.500	63.50	0.028	0.71	2.444	62.1	10.90	44.6
2.500	63.50	0.035	0.89	2.430	61.7	10.77	35.7
2.500	63.50	0.049	1.24	2.402	61.0	10.52	25.5
2.500	63.50	0.065	1.65	2.370	60.2	10.25	19.2
2.500	63.50	0.083	2.11	2.334	59.3	9.937	15.1
2.500	63.50	0.095	2.41	2.310	58.7	9.734	13.2
2.500	63.50	0.120	3.05	2.260	57.4	9.317	10.4

od in	od mm	WALL in	WALL mm	id in	id mm	1m/s m^3/h	10barg N/mm^2
2.500	63.50	0.188	4.78	2.124	53.9	8.229	6.2
2.500	63.50	0.250	6.35	2.000	50.8	7.297	4.6
2.500	63.50	0.313	7.95	1.874	47.6	6.406	3.6
2.500	63.50	0.375	9.53	1.750	44.5	5.586	2.9
2.500	63.50	0.437	11.10	1.626	41.3	4.823	2.5
2.500	63.50	0.500	12.70	1.500	38.1	4.104	2.1
2.500	63.50	0.562	14.27	1.376	35.0	3.454	1.9
2.500	63.50	0.625	15.88	1.250	31.8	2.850	1.7
2.750	69.85	0.035	0.89	2.680	68.1	13.10	39.3
2.750	69.85	0.049	1.24	2.652	67.4	12.83	28.1
2.750	69.85	0.065	1.65	2.620	66.5	12.52	21.2
2.750	69.85	0.105	2.67	2.540	64.5	11.77	13.1
2.750	69.85	0.109	2.77	2.532	64.3	11.69	12.6
2.750	69.85	0.120	3.05	2.510	63.8	11.49	11.5
2.750	69.85	0.188	4.78	2.374	60.3	10.28	7.3
2.750	69.85	0.250	6.35	2.250	57.2	9.235	5.1
2.750	69.85	0.375	9.53	2.000	50.8	7.297	3.2
2.750	69.85	0.438	11.13	1.874	47.6	6.406	2.7
2.750	69.85	0.469	11.91	1.812	46.0	5.989	2.5
2.750	69.85	0.500	12.70	1.750	44.5	5.586	2.4
2.750	69.85	0.531	13.49	1.688	42.9	5.198	2.2
2.750	69.85	0.594	15.09	1.562	39.7	4.451	2.0
2.750	69.85	0.750	19.05	1.250	31.8	2.850	1.5
3.000	76.20	0.022	0.56	2.956	75.1	15.94	68.2
3.000	76.20	0.025	0.64	2.950	74.9	15.87	60.0
3.000	76.20	0.028	0.71	2.944	74.8	15.81	53.6
3.000	76.20	0.035	0.89	2.930	74.4	15.66	42.9
3.000	76.20	0.040	1.02	2.920	74.2	15.55	37.5
3.000	76.20	0.049	1.24	2.902	73.7	15.36	30.6
3.000	76.20	0.065	1.65	2.870	72.9	15.03	23.1
3.000	76.20	0.083	2.11	2.834	72.0	14.65	18.1
3.000	76.20	0.120	3.05	2.760	70.1	13.90	12.5
3.000	76.20	0.188	4.78	2.624	66.6	12.56	8.0
3.000	76.20	0.250	6.35	2.500	63.5	11.40	5.5
3.000	76.20	0.281	7.14	2.438	61.9	10.84	4.9
3.000	76.20	0.313	7.95	2.374	60.3	10.28	4.4
3.000	76.20	0.375	9.53	2.250	57.2	9.235	3.6
3.000	76.20	0.438	11.13	2.124	53.9	8.229	3.0
3.000	76.20	0.500	12.70	2.000	50.8	7.297	2.6
3.000	76.20	0.562	14.27	1.876	47.7	6.420	2.3
3.000	76.20	0.625	15.88	1.750	44.5	5.586	2.0
3.000	76.20	0.750	19.05	1.500	38.1	4.104	1.7
3.500	88.90	0.025	0.64	3.450	87.6	21.71	70.0
3.500	88.90	0.028	0.71	3.444	87.5	21.64	62.5
3.500	88.90	0.035	0.89	3.430	87.1	21.46	50.0
3.500	88.90	0.040	1.02	3.420	86.9	21.34	43.8
3.500	88.90	0.042	1.07	3.416	86.8	21.29	41.7
3.500	88.90	0.049	1.24	3.402	86.4	21.11	35.7
3.500	88.90	0.065	1.65	3.370	85.6	20.72	26.9
3.500	88.90	0.188	4.78	3.124	79.3	17.80	9.3
3.500	88.90	0.250	6.35	3.000	76.2	16.42	7.0
3.500	88.90	0.312	7.92	2.876	73.1	15.09	5.2
3.500	88.90	0.375	9.53	2.750	69.9	13.80	4.2
3.500	88.90	0.438	11.13	2.624	66.6	12.56	3.6
3.500	88.90	0.469	11.91	2.562	65.1	11.97	3.3
3.500	88.90	0.500	12.70	2.500	63.5	11.40	3.1
3.500	88.90	0.531	13.49	2.438	61.9	10.84	2.9
3.500	88.90	0.563	14.30	2.374	60.3	10.28	2.7
3.500	88.90	0.600	15.24	2.300	58.4	9.650	2.5
3.500	88.90	0.625	15.88	2.250	57.2	9.235	2.4
3.500	88.90	0.688	17.48	2.124	53.9	8.229	2.2
3.500	88.90	0.750	19.05	2.000	50.8	7.297	2.0
3.750	95.25	0.028	0.71	3.694	93.8	24.89	67.0
3.750	95.25	0.065	1.65	3.620	91.9	23.90	28.8
3.750	95.25	0.120	3.05	3.510	89.2	22.47	15.6
3.750	95.25	0.154	3.91	3.442	87.4	21.61	12.2
3.750	95.25	0.188	4.78	3.374	85.7	20.77	10.0
3.750	95.25	0.226	5.74	3.298	83.8	19.84	8.3
3.750	95.25	0.250	6.35	3.250	82.6	19.27	7.5
3.750	95.25	0.375	9.53	3.000	76.2	16.42	4.6
3.750	95.25	0.438	11.13	2.874	73.0	15.07	3.8
3.750	95.25	0.500	12.70	2.750	69.9	13.80	3.3
3.750	95.25	0.563	14.30	2.624	66.6	12.56	2.9
3.750	95.25	0.594	15.09	2.562	65.1	11.97	2.8
3.750	95.25	0.625	15.88	2.500	63.5	11.40	2.6
3.750	95.25	0.688	17.48	2.374	60.3	10.28	2.3

od in	od mm	WALL in	WALL mm	id in	id mm	1m/s m^3/h	10barg N/mm^2
3.750	95.25	0.750	19.05	2.250	57.2	9.235	2.1
4.000	101.6	0.035	0.89	3.930	99.8	28.17	57.1
4.000	101.6	0.042	1.07	3.916	99.5	27.97	47.6
4.000	101.6	0.049	1.24	3.902	99.1	27.77	40.8
4.000	101.6	0.065	1.65	3.870	98.3	27.32	30.8
4.000	101.6	0.083	2.11	3.834	97.4	26.81	24.1
4.000	101.6	0.120	3.05	3.760	95.5	25.79	16.7
4.000	101.6	0.188	4.78	3.624	92.0	23.96	10.6
4.000	101.6	0.250	6.35	3.500	88.9	22.35	8.0
4.000	101.6	0.375	9.53	3.250	82.6	19.27	4.9
4.000	101.6	0.438	11.13	3.124	79.3	17.80	4.1
4.000	101.6	0.469	11.91	3.062	77.8	17.10	3.8
4.000	101.6	0.500	12.70	3.000	76.2	16.42	3.6
4.000	101.6	0.562	14.27	2.876	73.1	15.09	3.1
4.000	101.6	0.594	15.09	2.812	71.4	14.42	3.0
4.000	101.6	0.625	15.88	2.750	69.9	13.80	2.8
4.000	101.6	0.688	17.48	2.624	66.6	12.56	2.5
4.000	101.6	0.750	19.05	2.500	63.5	11.40	2.3
4.000	101.6	0.875	22.23	2.250	57.2	9.235	1.9
4.000	101.6	1.000	25.40	2.000	50.8	7.297	1.7
4.500	114.3	0.065	1.65	4.370	111.0	34.84	34.6
4.500	114.3	0.188	4.78	4.124	104.7	31.02	12.0
4.500	114.3	0.250	6.35	4.000	101.6	29.19	9.0
4.500	114.3	0.344	8.74	3.812	96.8	26.51	6.1
4.500	114.3	0.375	9.53	3.750	95.3	25.65	5.5
4.500	114.3	0.438	11.13	3.624	92.0	23.96	4.7
4.500	114.3	0.500	12.70	3.500	88.9	22.35	4.1
4.500	114.3	0.531	13.49	3.438	87.3	21.56	3.8
4.500	114.3	0.563	14.30	3.374	85.7	20.77	3.6
4.500	114.3	0.594	15.09	3.312	84.1	20.01	3.4
4.500	114.3	0.625	15.88	3.250	82.6	19.27	3.2
4.500	114.3	0.688	17.48	3.124	79.3	17.80	2.9
4.500	114.3	0.750	19.05	3.000	76.2	16.42	2.6
4.500	114.3	0.813	20.65	2.874	73.0	15.07	2.4
5.000	127.0	0.065	1.65	4.870	123.7	43.26	38.5
5.000	127.0	0.083	2.11	4.834	122.8	42.63	30.1
5.000	127.0	0.120	3.05	4.760	120.9	41.33	20.8
5.000	127.0	0.188	4.78	4.624	117.4	39.00	13.3
5.000	127.0	0.250	6.35	4.500	114.3	36.94	10.0
5.000	127.0	0.375	9.53	4.250	108.0	32.95	6.2
5.000	127.0	0.500	12.70	4.000	101.6	29.19	4.6
5.000	127.0	0.594	15.09	3.812	96.8	26.51	3.8
5.000	127.0	0.625	15.88	3.750	95.3	25.65	3.6
5.000	127.0	0.750	19.05	3.500	88.9	22.35	2.9
5.000	127.0	0.813	20.65	3.374	85.7	20.77	2.7
5.000	127.0	1.000	25.40	3.000	76.2	16.42	2.1
5.500	139.7	0.250	6.35	5.000	127.0	45.60	11.0
5.500	139.7	0.375	9.53	4.750	120.7	41.16	7.3
5.500	139.7	0.500	12.70	4.500	114.3	36.94	5.1
5.500	139.7	0.750	19.05	4.000	101.6	29.19	3.2
5.500	139.7	0.875	22.23	3.750	95.3	25.65	2.7
5.500	139.7	1.000	25.40	3.500	88.9	22.35	2.4
6.000	152.4	0.065	1.65	5.870	149.1	62.85	46.2
6.000	152.4	0.083	2.11	5.834	148.2	62.09	36.1
6.000	152.4	0.109	2.77	5.782	146.9	60.98	27.5
6.000	152.4	0.120	3.05	5.760	146.3	60.52	25.0
6.000	152.4	0.250	6.35	5.500	139.7	55.18	12.0
6.000	152.4	0.375	9.53	5.250	133.4	50.28	8.0
6.000	152.4	0.500	12.70	5.000	127.0	45.60	5.5
6.000	152.4	0.625	15.88	4.750	120.7	41.16	4.4
6.000	152.4	0.750	19.05	4.500	114.3	36.94	3.6
6.000	152.4	0.875	22.23	4.250	108.0	32.95	3.0
6.000	152.4	1.000	25.40	4.000	101.6	29.19	2.6
6.500	165.1	0.250	6.35	6.000	152.4	65.67	13.0
6.500	165.1	0.500	12.70	5.500	139.7	55.18	6.0
6.500	165.1	0.750	19.05	5.000	127.0	45.60	3.9
7.000	177.8	0.032	0.81	6.936	176.2	87.76	109.4
7.000	177.8	0.500	12.70	6.000	152.4	65.67	7.0
7.500	190.5	0.250	6.35	7.000	177.8	89.38	15.0
7.500	190.5	0.375	9.53	6.750	171.5	83.11	10.0
7.500	190.5	0.500	12.70	6.500	165.1	77.07	7.5
7.500	190.5	0.750	19.05	6.000	152.4	65.67	4.6
8.000	203.2	0.083	2.11	7.834	199.0	112.0	48.2
8.000	203.2	0.109	2.77	7.782	197.7	110.5	36.7
8.000	203.2	0.120	3.05	7.760	197.1	109.8	33.3
8.000	203.2	0.250	6.35	7.500	190.5	102.6	16.0

od in	od mm	WALL in	WALL mm	id in	id mm	1m/s m³/h	10barg N/mm²
8.000	203.2	0.500	12.70	7.000	177.8	89.38	8.0
8.000	203.2	0.625	15.88	6.750	171.5	83.11	5.9
8.000	203.2	0.750	19.05	6.500	165.1	77.07	4.9
8.000	203.2	1.000	25.40	6.000	152.4	65.67	3.6
8.000	203.2	1.500	38.10	5.000	127.0	45.60	2.3
8.000	203.2	2.000	50.80	4.000	101.6	29.19	1.7
9.000	228.6	0.500	12.70	8.000	203.2	116.7	9.0
9.000	228.6	0.625	15.88	7.750	196.9	109.6	7.2
9.000	228.6	0.750	19.05	7.500	190.5	102.6	5.5
9.000	228.6	1.000	25.40	7.000	177.8	89.38	4.1
9.000	228.6	1.500	38.10	6.000	152.4	65.67	2.6
9.000	228.6	2.000	50.80	5.000	127.0	45.60	1.9
10.00	254.0	0.083	2.11	9.834	249.8	176.4	60.2
10.00	254.0	0.120	3.05	9.760	247.9	173.8	41.7
10.00	254.0	0.250	6.35	9.500	241.3	164.6	20.0
10.00	254.0	0.500	12.70	9.000	228.6	147.8	10.0
11.00	279.4	2.000	50.80	7.000	177.8	89.38	2.4
12.00	304.8	0.083	2.11	11.83	300.6	255.5	72.3
12.00	304.8	0.250	6.35	11.50	292.1	241.2	24.0
12.00	304.8	0.500	12.70	11.00	279.4	220.7	12.0

Table 9.11 Seamless and welded stainless steel inch tube (Above)

od mm	od in	WALL mm	WALL in	id mm	id in	1m/s m³/h	10barg N/mm²
6.0	0.236	1.0	0.039	4.0	0.157	0.045	2.6
8.0	0.315	1.0	0.039	6.0	0.236	0.102	3.6
10.0	0.394	1.0	0.039	8.0	0.315	0.181	4.6
12.0	0.472	1.0	0.039	10.0	0.394	0.283	5.5
12.0	0.472	1.5	0.059	9.0	0.354	0.229	3.6
14.0	0.551	1.0	0.039	12.0	0.472	0.407	7.0
14.0	0.551	1.5	0.059	11.0	0.433	0.342	4.2
14.0	0.551	2.0	0.079	10.0	0.394	0.283	3.1
16.0	0.630	1.0	0.039	14.0	0.551	0.554	8.0
16.0	0.630	1.5	0.059	13.0	0.512	0.478	4.9
16.0	0.630	2.0	0.079	12.0	0.472	0.407	3.6
20.0	0.787	1.0	0.039	18.0	0.709	0.916	10.0
20.0	0.787	1.5	0.059	17.0	0.669	0.817	6.2
20.0	0.787	2.0	0.079	16.0	0.630	0.724	4.6
25.0	0.984	1.0	0.039	23.0	0.906	1.496	12.5
25.0	0.984	1.5	0.059	22.0	0.866	1.368	8.3
25.0	0.984	2.0	0.079	21.0	0.827	1.247	5.8
28.0	1.102	1.0	0.039	26.0	1.024	1.911	14.0
28.0	1.102	1.5	0.059	25.0	0.984	1.767	9.3
28.0	1.102	2.0	0.079	24.0	0.945	1.629	7.0
38.0	1.496	1.0	0.039	36.0	1.417	3.664	19.0
38.0	1.496	1.5	0.059	35.0	1.378	3.464	12.7
38.0	1.496	2.0	0.079	34.0	1.339	3.269	9.5
44.5	1.752	1.5	0.059	41.5	1.634	4.870	14.8
44.5	1.752	2.0	0.079	40.5	1.594	4.638	11.1
44.5	1.752	2.6	0.102	39.3	1.547	4.367	8.6
57.0	2.244	1.5	0.059	54.0	2.126	8.245	19.0
57.0	2.244	2.0	0.079	53.0	2.087	7.942	14.3
57.0	2.244	2.6	0.102	51.8	2.039	7.587	11.0
76.1	2.996	1.6	0.063	72.9	2.870	15.026	23.8
76.1	2.996	2.3	0.091	71.5	2.815	14.455	16.5
76.1	2.996	2.9	0.114	70.3	2.768	13.973	13.1

Table 9.12 Seamless and welded ISO tube (Above)

Table 9.13 Metric stainless steel tube (Below)

od mm	od in	WALL mm	WALL in	id mm	id in	1m/s m³/h	10barg N/mm²
3.0	0.118	0.3	0.012	2.4	0.094	0.016	4.6
4.5	0.177	0.5	0.020	3.5	0.138	0.035	4.1
5.0	0.197	0.5	0.020	4.0	0.157	0.045	4.6
5.0	0.197	0.8	0.030	3.5	0.138	0.035	2.9
5.0	0.197	1.0	0.039	3.0	0.118	0.025	2.1
5.0	0.197	1.3	0.049	2.5	0.098	0.018	1.7
5.0	0.197	1.5	0.059	2.0	0.079	0.011	1.4
6.0	0.236	0.5	0.020	5.0	0.197	0.071	5.5
6.0	0.236	1.0	0.039	4.0	0.157	0.045	2.6
6.0	0.236	1.5	0.059	3.0	0.118	0.025	1.7
8.0	0.315	1.0	0.039	6.0	0.236	0.102	3.6
8.0	0.315	1.5	0.059	5.0	0.197	0.071	2.3

od mm	od in	WALL mm	WALL in	id mm	id in	1m/s m³/h	10barg N/mm²
8.0	0.315	2.0	0.079	4.0	0.157	0.045	1.7
10.0	0.394	1.0	0.039	8.0	0.315	0.181	4.6
10.0	0.394	1.5	0.059	7.0	0.276	0.139	2.9
10.0	0.394	2.0	0.079	6.0	0.236	0.102	2.1
12.0	0.472	1.0	0.039	10.0	0.394	0.283	5.5
12.0	0.472	1.5	0.059	9.0	0.354	0.229	3.6
12.0	0.472	2.0	0.079	8.0	0.315	0.181	2.6
14.0	0.551	1.0	0.039	12.0	0.472	0.407	7.0
14.0	0.551	1.5	0.059	11.0	0.433	0.342	4.2
14.0	0.551	2.0	0.079	10.0	0.394	0.283	3.1
14.0	0.551	2.5	0.098	9.0	0.354	0.229	2.4
15.0	0.591	1.0	0.039	13.0	0.512	0.478	7.5
15.0	0.591	1.5	0.059	12.0	0.472	0.407	4.6
15.0	0.591	2.0	0.079	11.0	0.433	0.342	3.3
16.0	0.630	1.0	0.039	14.0	0.551	0.554	8.0
16.0	0.630	1.5	0.059	13.0	0.512	0.478	4.9
16.0	0.630	2.0	0.079	12.0	0.472	0.407	3.6
16.0	0.630	2.5	0.098	11.0	0.433	0.342	2.8
16.0	0.630	3.0	0.118	10.0	0.394	0.283	2.3
17.0	0.669	1.0	0.039	15.0	0.591	0.636	8.5
17.2	0.677	1.6	0.063	14.0	0.551	0.554	4.9
18.0	0.709	1.0	0.039	16.0	0.630	0.724	9.0
18.0	0.709	1.5	0.059	15.0	0.591	0.636	5.5
18.0	0.709	1.6	0.063	14.8	0.583	0.619	5.2
18.0	0.709	2.0	0.079	14.0	0.551	0.554	4.1
18.0	0.709	2.5	0.098	13.0	0.512	0.478	3.2
18.0	0.709	3.0	0.118	12.0	0.472	0.407	2.6
20.0	0.787	1.5	0.059	17.0	0.669	0.817	6.2
20.0	0.787	1.6	0.063	16.8	0.661	0.798	5.8
20.0	0.787	2.0	0.079	16.0	0.630	0.724	4.6
20.0	0.787	2.5	0.098	15.0	0.591	0.636	3.6
20.0	0.787	3.0	0.118	14.0	0.551	0.554	2.9
20.0	0.787	4.0	0.157	12.0	0.472	0.407	2.1
20.0	0.787	5.0	0.197	10.0	0.394	0.283	1.7
21.3	0.839	1.6	0.063	18.1	0.713	0.926	6.2
21.3	0.839	2.0	0.079	17.3	0.681	0.846	3.6
22.0	0.866	1.0	0.039	20.0	0.787	1.13	11.0
22.0	0.866	1.5	0.059	19.0	0.748	1.02	7.3
22.0	0.866	2.0	0.079	18.0	0.709	0.92	5.1
23.0	0.906	1.5	0.059	20.0	0.787	1.13	7.7
25.0	0.984	1.2	0.047	22.6	0.890	1.44	10.4
25.0	0.984	1.5	0.059	22.0	0.866	1.37	8.3
25.0	0.984	2.0	0.079	21.0	0.827	1.25	5.8
25.0	0.984	2.5	0.098	20.0	0.787	1.13	4.6
25.0	0.984	3.0	0.118	19.0	0.748	1.02	3.7
25.0	0.984	4.0	0.157	17.0	0.669	0.82	2.7
25.0	0.984	5.0	0.197	15.0	0.591	0.64	2.1
26.9	1.059	1.6	0.063	23.7	0.933	1.59	8.4
26.9	1.059	2.0	0.079	22.9	0.902	1.48	6.7
26.9	1.059	2.3	0.091	22.3	0.878	1.41	5.4
28.0	1.102	1.0	0.039	26.0	1.024	1.91	14.0
28.0	1.102	1.5	0.059	25.0	0.984	1.77	9.3
28.0	1.102	2.0	0.079	24.0	0.945	1.63	7.0
28.0	1.102	2.5	0.098	23.0	0.906	1.50	5.1
30.0	1.181	1.5	0.059	27.0	1.063	2.06	10.0
30.0	1.181	2.0	0.079	26.0	1.024	1.91	7.5
30.0	1.181	2.6	0.100	24.9	0.980	1.75	5.4
30.0	1.181	3.0	0.118	24.0	0.945	1.63	4.6
30.0	1.181	4.0	0.157	22.0	0.866	1.37	3.3
30.0	1.181	5.0	0.197	20.0	0.787	1.13	2.6
32.0	1.260	1.2	0.047	29.6	1.165	2.48	13.3
32.0	1.260	1.5	0.059	29.0	1.142	2.38	10.7
33.0	1.299	1.5	0.059	30.0	1.181	2.54	11.0
33.7	1.327	1.6	0.063	30.5	1.201	2.63	10.5
33.7	1.327	2.0	0.079	29.7	1.169	2.49	8.4
33.7	1.327	2.6	0.102	28.5	1.122	2.30	6.0
34.0	1.339	1.0	0.039	32.0	1.260	2.90	17.0
34.0	1.339	1.5	0.059	31.0	1.220	2.72	11.3
35.0	1.378	1.5	0.059	32.0	1.260	2.90	11.7
35.0	1.378	2.0	0.079	31.0	1.220	2.72	8.8
35.0	1.378	2.5	0.098	30.0	1.181	2.54	7.0
35.0	1.378	3.0	0.118	29.0	1.142	2.38	5.4
38.0	1.496	1.2	0.047	35.6	1.402	3.58	15.8
38.0	1.496	1.5	0.059	35.0	1.378	3.46	12.7
38.0	1.496	2.0	0.079	34.0	1.339	3.27	9.5
38.0	1.496	3.0	0.118	32.0	1.260	2.90	5.9

od mm	od in	WALL mm	WALL in	id mm	id in	1m/s m³/h	10barg N/mm²
38.0	1.496	3.6	0.142	30.8	1.213	2.68	4.8
38.0	1.496	4.0	0.157	30.0	1.181	2.54	4.3
38.0	1.496	5.0	0.197	28.0	1.102	2.22	3.4
38.0	1.496	6.0	0.236	26.0	1.024	1.91	2.8
40.0	1.575	1.0	0.039	38.0	1.496	4.08	20.0
40.0	1.575	1.5	0.059	37.0	1.457	3.87	13.3
40.0	1.575	2.0	0.079	36.0	1.417	3.66	10.0
40.0	1.575	5.0	0.197	30.0	1.181	2.54	3.6
42.0	1.654	3.0	0.118	36.0	1.417	3.66	7.0
42.4	1.669	1.6	0.063	39.2	1.543	4.34	13.3
42.4	1.669	2.0	0.079	38.4	1.512	4.17	10.6
42.4	1.669	2.6	0.102	37.2	1.465	3.91	8.2
43.0	1.693	1.5	0.059	40.0	1.575	4.52	14.3
44.5	1.752	2.0	0.079	40.5	1.594	4.64	11.1
44.5	1.752	2.6	0.102	39.3	1.547	4.37	8.6
44.5	1.752	2.9	0.114	38.7	1.524	4.23	7.7
48.3	1.902	1.6	0.063	45.1	1.776	5.75	15.1
48.3	1.902	2.0	0.079	44.3	1.744	5.55	12.1
48.3	1.902	2.6	0.102	43.1	1.697	5.25	9.3
50.0	1.969	4.5	0.177	41.0	1.614	4.75	5.1
50.0	1.969	5.0	0.197	40.0	1.575	4.52	4.6
51.0	2.008	1.2	0.047	48.6	1.913	6.68	21.3
51.0	2.008	1.5	0.059	48.0	1.890	6.51	17.0
51.0	2.008	2.0	0.079	47.0	1.850	6.25	12.8
52.0	2.047	1.0	0.039	50.0	1.969	7.07	26.0
52.0	2.047	1.5	0.059	49.0	1.929	6.79	17.3
53.0	2.087	1.5	0.059	50.0	1.969	7.07	17.7
54.0	2.126	2.0	0.079	50.0	1.969	7.07	13.5
57.0	2.244	2.0	0.079	53.0	2.087	7.94	14.3
57.0	2.244	2.9	0.114	51.2	2.016	7.41	9.8
60.0	2.362	5.0	0.197	50.0	1.969	7.07	5.5
60.0	2.362	5.5	0.217	49.0	1.929	6.79	5.0
60.3	2.374	1.6	0.063	57.1	2.248	9.22	18.8
60.3	2.374	2.0	0.079	56.3	2.217	8.96	15.1
60.3	2.374	2.9	0.114	54.5	2.146	8.40	10.4
63.5	2.500	2.0	0.079	59.5	2.343	10.01	15.9
65.0	2.559	5.5	0.217	54.0	2.126	8.24	5.5
69.0	2.717	2.0	0.079	65.0	2.559	11.95	17.3
70.0	2.756	2.0	0.079	66.0	2.598	12.32	17.5
70.0	2.756	5.0	0.197	60.0	2.362	10.18	7.0
76.0	2.992	2.0	0.079	72.0	2.835	14.66	19.0
76.1	2.996	1.6	0.063	72.9	2.870	15.03	23.8
76.1	2.996	2.0	0.079	72.1	2.839	14.70	19.0
76.1	2.996	2.9	0.114	70.3	2.768	13.97	13.1
83.0	3.268	1.5	0.059	80.0	3.150	18.10	27.7
84.0	3.307	2.0	0.079	80.0	3.150	18.10	21.0
85.0	3.346	2.0	0.079	81.0	3.189	18.55	21.3
88.9	3.500	1.6	0.063	85.7	3.374	20.77	27.8
88.9	3.500	2.0	0.079	84.9	3.343	20.38	22.2
88.9	3.500	3.6	0.142	81.7	3.217	18.87	12.3
90.0	3.543	5.0	0.197	80.0	3.150	18.10	9.0
103.0	4.055	1.5	0.059	100.0	3.937	28.27	34.3
104.0	4.094	2.0	0.079	100.0	3.937	28.27	26.0
105.0	4.134	2.5	0.098	100.0	3.937	28.27	21.0
108.0	4.252	4.0	0.157	100.0	3.937	28.27	13.5
114.3	4.500	1.6	0.063	111.1	4.374	34.90	35.7
114.3	4.500	2.0	0.079	110.3	4.343	34.40	28.6
114.3	4.500	2.6	0.102	109.1	4.295	33.65	22.0
114.3	4.500	3.6	0.142	107.1	4.217	32.43	15.9
128.0	5.039	1.5	0.059	125.0	4.921	44.18	42.7
129.0	5.079	2.0	0.079	125.0	4.921	44.18	32.3
131.0	5.157	3.0	0.118	125.0	4.921	44.18	21.8
133.0	5.236	4.0	0.157	125.0	4.921	44.18	16.6
139.7	5.500	2.0	0.079	135.7	5.343	52.07	34.9
139.7	5.500	2.6	0.102	134.5	5.295	51.15	26.9
139.7	5.500	3.6	0.142	132.5	5.217	49.64	19.4
153.0	6.024	1.5	0.059	150.0	5.906	63.62	51.0
154.0	6.063	2.0	0.079	150.0	5.906	63.62	38.5
156.0	6.142	3.0	0.118	150.0	5.906	63.62	26.0
159.0	6.260	4.5	0.177	150.0	5.906	63.62	17.7
168.3	6.626	2.0	0.079	164.3	6.469	76.33	42.1
168.3	6.626	2.6	0.102	163.1	6.421	75.21	32.4
168.3	6.626	4.0	0.157	160.3	6.311	72.65	21.0
203.0	7.992	1.5	0.059	200.0	7.874	113.1	67.7
204.0	8.031	2.0	0.079	200.0	7.874	113.1	51.0
205.0	8.071	2.5	0.098	200.0	7.874	113.1	41.0
206.0	8.110	3.0	0.118	200.0	7.874	113.1	34.3
219.1	8.626	2.0	0.079	215.1	8.469	130.8	54.8
254.0	10.00	2.0	0.079	250.0	9.843	176.7	63.5
255.0	10.04	2.5	0.098	250.0	9.843	176.7	51.0
256.0	10.08	3.0	0.118	250.0	9.843	176.7	42.7
273.0	10.75	2.0	0.079	269.0	10.59	204.6	68.3
273.0	10.75	2.6	0.102	267.8	10.54	202.8	52.5
304.0	11.97	2.0	0.079	300.0	11.81	254.5	76.0
305.0	12.01	2.5	0.098	300.0	11.81	254.5	61.0
306.0	12.05	3.0	0.118	300.0	11.81	254.5	51.0
323.9	12.75	2.6	0.102	318.7	12.55	287.2	62.3
355.0	13.98	2.5	0.098	350.0	13.78	346.4	71.0
356.0	14.02	3.0	0.118	350.0	13.78	346.4	59.3
406.0	15.98	3.0	0.118	400.0	15.75	452.4	67.7
456.0	17.95	3.0	0.118	450.0	17.72	572.6	76.0
506.0	19.92	3.0	0.118	500.0	19.69	706.9	84.3
508.0	20.00	4.0	0.157	500.0	19.69	706.9	63.5
606.0	23.86	3.0	0.118	600.0	23.62	1017.9	101.0
608.0	23.94	4.0	0.157	600.0	23.62	1017.9	76.0

Table 9.13 Metric stainless steel tube (Above)

Table 9.14 lists metric copper tube sizes to BS 12449. These are the pipes used for domestic and commercial hot and cold water systems. They are also used for hot water central heating systems.

Copper pipes are being replaced in some installations by non-metallic pipes. Light gauge stainless steel tube, to BS 4127 (replaced by BS EN 10312), is manufactured in 316L in the same sizes between 6 mm and 42 mm od. This tube is also designed for domestic/commercial mechanical services applications.

Table 9.14 Metric copper tube (Below)

od mm	OD in	WALL mm	WALL in	id mm	id in	1m/s m³/h	10 barg N/mm²
6.0	0.236	0.50	0.020	5.00	0.197	0.071	5.5
6.0	0.236	0.60	0.024	4.80	0.189	0.065	4.6
6.0	0.236	0.80	0.031	4.40	0.173	0.055	3.3
8.0	0.315	0.50	0.020	7.00	0.276	0.139	8.0
8.0	0.315	0.60	0.024	6.80	0.268	0.131	6.2
8.0	0.315	0.80	0.031	6.40	0.252	0.116	4.6
10.0	0.394	0.50	0.020	9.00	0.354	0.229	10.0
10.0	0.394	0.60	0.024	8.80	0.346	0.219	8.3
10.0	0.394	0.80	0.031	8.40	0.331	0.200	5.8
12.0	0.472	0.50	0.020	11.00	0.433	0.342	12.0
12.0	0.472	0.60	0.024	10.80	0.425	0.330	10.0
12.0	0.472	0.80	0.031	10.40	0.409	0.306	7.5
15.0	0.591	0.50	0.020	14.00	0.551	0.554	15.0
15.0	0.591	0.70	0.028	13.60	0.535	0.523	10.7
15.0	0.591	1.00	0.039	13.00	0.512	0.478	7.5
22.0	0.866	0.60	0.024	20.80	0.819	1.223	18.3
22.0	0.866	0.90	0.035	20.20	0.795	1.154	12.2
22.0	0.866	1.20	0.047	19.60	0.772	1.086	9.2
28.0	1.102	0.60	0.024	26.80	1.055	2.031	23.3
28.0	1.102	0.90	0.035	26.20	1.031	1.941	15.6
28.0	1.102	1.20	0.047	25.60	1.008	1.853	11.7
35.0	1.378	0.70	0.028	33.60	1.323	3.192	25.0
35.0	1.378	1.20	0.047	32.60	1.283	3.005	14.6
35.0	1.378	1.50	0.059	32.00	1.260	2.895	11.7
42.0	1.654	0.80	0.031	40.40	1.591	4.615	26.3
42.0	1.654	1.20	0.047	39.60	1.559	4.434	17.5
42.0	1.654	1.50	0.059	39.00	1.535	4.301	14.0
54.0	2.126	0.90	0.035	52.20	2.055	7.704	30.0
54.0	2.126	1.20	0.047	51.60	2.031	7.528	22.5
54.0	2.126	2.00	0.079	50.00	1.969	7.069	13.5
67.0	2.638	1.00	0.039	65.00	2.559	11.95	33.5
67.0	2.638	1.20	0.047	64.60	2.543	11.80	27.9
67.0	2.638	2.00	0.079	63.00	2.480	11.22	16.8
76.1	2.996	1.20	0.047	73.70	2.902	15.36	31.7
76.1	2.996	1.50	0.059	73.10	2.878	15.11	25.4
76.1	2.996	2.00	0.079	72.10	2.839	14.70	19.0
108.0	4.252	1.20	0.047	105.60	4.157	31.53	45.0
108.0	4.252	1.50	0.059	105.00	4.134	31.17	36.0
108.0	4.252	2.50	0.098	103.00	4.055	30.00	21.6

od mm	OD in	WALL mm	WALL in	id mm	id in	1m/s m^3/h	10 barg N/mm^2
133.0	5.236	1.50	0.059	130.00	5.118	47.78	44.3
133.0	5.236	2.00	0.079	129.00	5.079	47.05	33.3
159.0	6.260	1.50	0.059	156.00	6.142	68.81	53.0
159.0	6.260	2.00	0.079	155.00	6.102	67.93	39.8

Table 9.14 Metric copper tube (Above)

9.2.4 Lined metallic pipe and tube

Pipe and tube can be lined to increase corrosion resistance. A thin exotic layer can protect a substantial, structural, low grade material. Both metallic and non-metallic linings are used.

One of the simplest lining techniques is electroplating. Carbon, alloy and stainless steel pipes can be chrome or nickel plated for protection. Great care must be exercised during the preparation of the bore prior to plating. A practical inspection process must be available to evaluate the final result.

Thicker coatings of Stellite™ or Colmonoy™ can be applied. Proprietary techniques can be used to apply coatings to straight and bent pipe.

A metallic pipe and tube can be lined with a separate, thin non-metallic tube. Thin non-metallic tubes are specifically manufactured for this purpose. This approach can be very cost-effective; the lining can be easily replaced if damaged or when attacked.

9.2.5 Non-metallic materials

Non-metallic materials are very useful for low pressure corrosive applications. However great care must be exercised if junctions between metallic and non-metallic systems are inevitable. Non-metallic components and cemented joints can be very intolerant of pulsating flow and vibration. Electrical continuity must be ensured. Anti-static precautions may be necessary. Table 9.15 shows the range of non-metallic materials available for pipes and fittings.

Material	Temperature range °C	Comments
ABS	-40 to +80	Very good for abrasive solids handling, water and hygienic applications
CPVC	+5 to +100	Good for paraffins, mineral acids, alcohols
HDPE	-40 to +60	Very good for solvents, acids, caustic solutions, water and gas
Poly-propylene[(1)]	0 to +90	Good chemical resistance to weak acids and alkalis and many organic solvents. Suitable for hygienic applications
Poly-ethylene[(1)]	-40 to +60	Good impact resistance and exceptional abrasion resistance
PVC	-18 to +60	General purpose material
PVDF[(1)]	+40 to +140	Good impact and abrasion resistance properties. Very broad chemical compatibility including hydro-carbons, organic and inorganic acids and alcohols. Dedicated compounds for hygienic applications
PVDF-HP (zero conta-mination)	-40 to +140	For high purity process system, demineralised water, etc.
uPVC	0 to +60	Good chemical compatibility, suitable for potable water
uPVC (zero contamination)	0 to +50	For high purity process system, demineralised water, etc.
Centrifugally cast glass reinforced plastic	max temp from 80 to 260 depending upon compound	DN300 to DN2 500 Gravity flow to PN24

[(1)] Fusion welded, butt or socket

Table 9.15 Non-metallic materials for piping systems

As with metallic materials, non-metallic materials are subject to temperature-pressure ratings.

Figures 9.1 to 9.7 show the pipework derating factors. Dotted lines in the curves indicate intermittent operation.

Fittings are readily available, for all the pipe materials shown, to allow complex systems to be constructed.

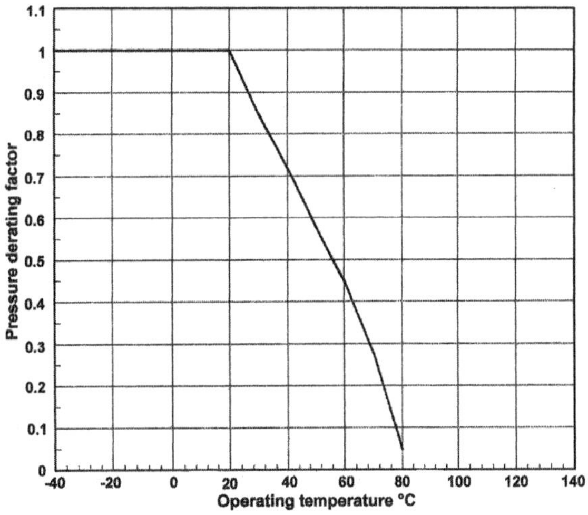

Figure 9.1 Temperature derating for ABS pipes and fittings

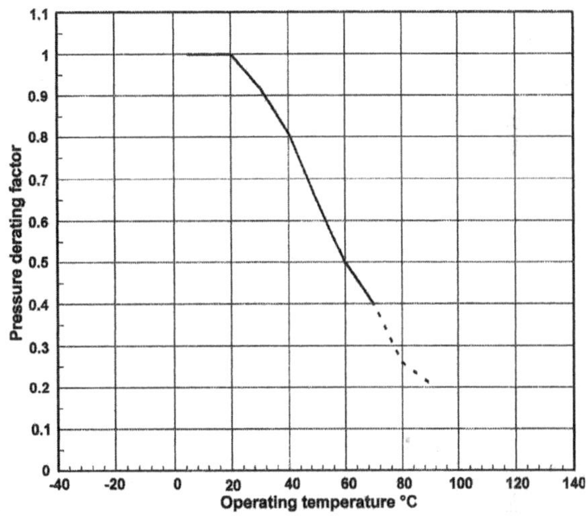

Figure 9.2 Temperature derating for CPVC pipes and fittings

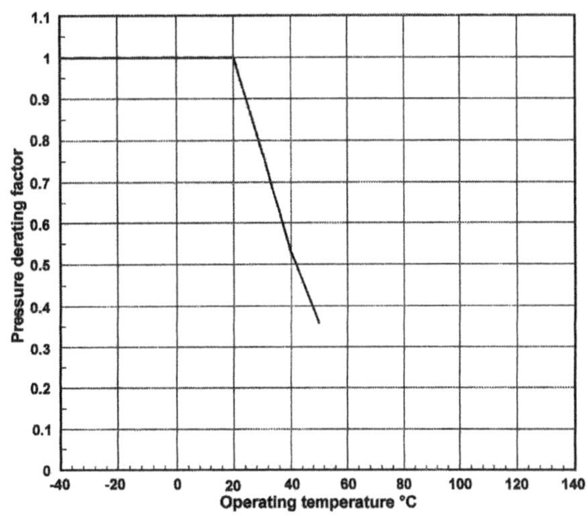

Figure 9.3 Temperature derating for HDPE pipes and fittings

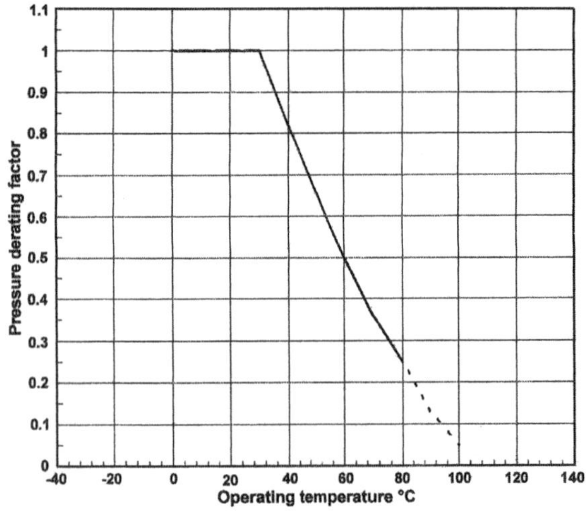

Figure 9.4 Temperature derating for polypropylene pipes and fittings

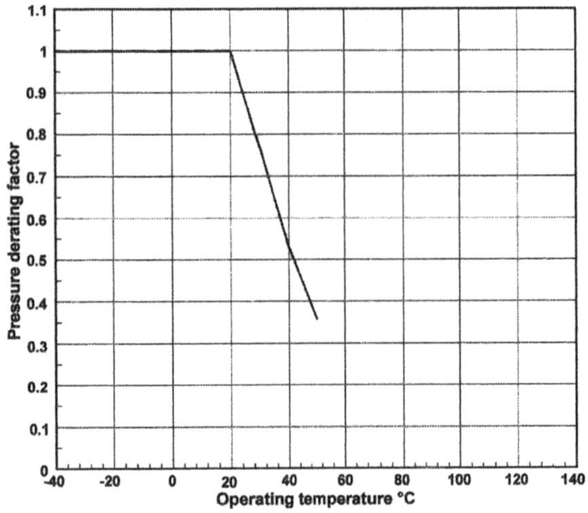

Figure 9.5 Temperature derating for polyethylene pipes and fittings

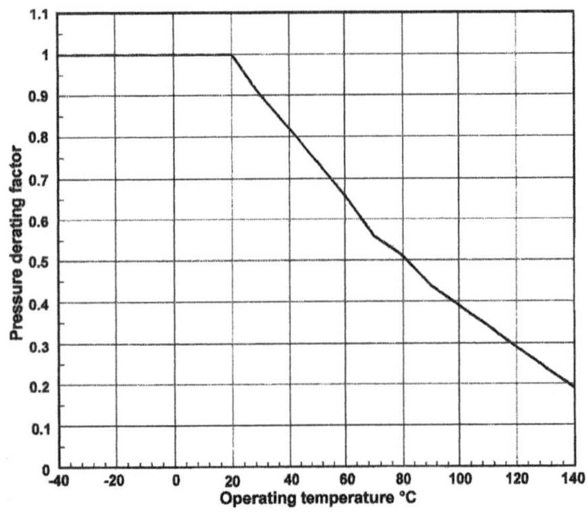

Figure 9.6 Temperature derating for PVDF pipes and fittings

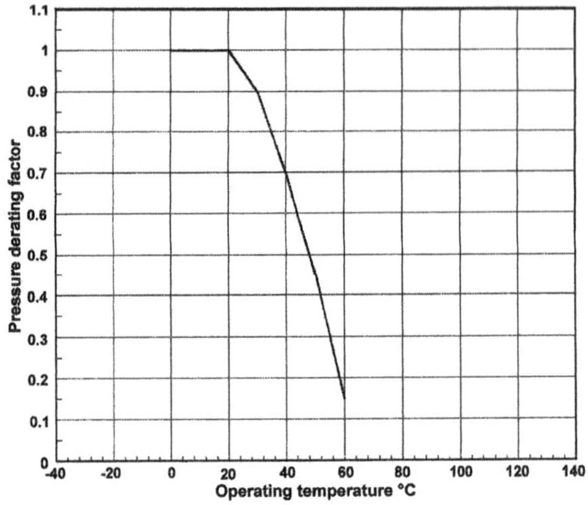

Figure 9.7 Temperature derating for UPVC pipes and fittings

9.2.6 Non-metallic pipe

Tables 9.16 to 9.20 indicate the availability of rigid pipes and tubes. These are normally available in 3, 5 or 6 metre lengths. Some polypropylene pipes are only intended for gravity feed or gravity drain applications. These pipes are shown with a working pressure of 0 barg.

nb	CLASS	od in	od mm	WALL in	WALL mm	id in	id mm	1m/s m³/h	WP barg
⅜	E	0.675	17.15	0.067	1.70	0.541	13.75	0.534	15.0
½	E	0.840	21.34	0.079	2.00	0.683	17.34	0.850	15.0
½	T	0.840	21.34	0.142	3.60	0.557	14.14	0.565	12.0
¾	E	1.050	26.67	0.098	2.50	0.853	21.67	1.328	15.0
¾	T	1.050	26.67	0.142	3.60	0.767	19.47	1.072	12.0
1	C	1.315	33.40	0.079	2.00	1.158	29.40	2.444	9.0
1	E	1.315	33.40	0.122	3.10	1.071	27.20	2.092	15.0
1	T	1.315	33.40	0.169	4.30	0.976	24.80	1.739	12.0
1¼	C	1.660	42.16	0.098	2.50	1.463	37.16	3.905	9.0
1¼	E	1.660	42.16	0.154	3.90	1.353	34.36	3.339	15.0
1¼	T	1.660	42.16	0.209	5.30	1.243	31.56	2.817	12.0
1½	C	1.900	48.26	0.110	2.80	1.680	42.66	5.146	9.0
1½	E	1.900	48.26	0.177	4.50	1.546	39.26	4.358	15.0
1½	T	1.900	48.26	0.236	6.00	1.428	36.26	3.717	12.0
2	C	2.375	60.33	0.142	3.60	2.092	53.13	7.980	9.0
2	E	2.375	60.33	0.220	5.60	1.934	49.13	6.823	15.0
2	T	2.375	60.33	0.283	7.20	1.808	45.93	5.963	12.0
2½	C	2.875	73.03	0.197	5.00	2.481	63.03	11.23	9.0
3	C	3.500	88.90	0.205	5.20	3.091	78.50	17.42	9.0
3	E	3.500	88.90	0.327	8.30	2.846	72.30	14.78	15.0
4	C	4.500	114.3	0.264	6.70	3.972	100.9	28.79	9.0
4	E	4.500	114.3	0.417	10.60	3.665	93.10	24.51	15.0
6	C	6.625	168.3	0.390	9.90	5.845	148.5	62.33	9.0
6	D	6.625	168.3	0.504	12.8	5.617	142.7	57.56	12.0
8	C	8.625	219.1	0.500	12.7	7.625	193.7	106.1	9.0

Table 9.16 Inch ABS pipe to BS 5391

nb	SCH	od in	od mm	WALL in	WALL mm	id in	id mm	1m/s m³/h	WP barg
¼	80	0.540	13.72	0.118	3.00	0.304	7.72	0.168	78.0
⅜	80	0.675	17.15	0.126	3.20	0.423	10.75	0.326	63.0
½	40	0.840	21.34	0.110	2.80	0.620	15.74	0.700	41.0
½	80	0.840	21.34	0.146	3.70	0.549	13.94	0.549	59.0
¾	40	1.050	26.67	0.114	2.90	0.822	20.87	1.232	33.0
¾	80	1.050	26.67	0.154	3.90	0.743	18.87	1.007	48.0
	40	1.315	33.40	0.134	3.40	0.047	26.60	2.001	31.0
1	80	1.315	33.40	0.181	4.60	0.953	24.20	1.656	43.0
1¼	40	1.660	42.16	0.142	3.60	1.377	34.96	3.456	25.0
1¼	80	1.660	42.16	0.193	4.90	1.274	32.36	2.962	36.0
1½	40	1.900	48.26	0.146	3.70	1.609	40.86	4.721	23.0

nb	SCH	od in	od mm	WALL in	WALL mm	id in	id mm	1m/s m³/h	WP barg
1½	80	1.900	48.26	0.201	5.10	1.498	38.06	4.096	32.0
2	40	2.375	60.33	0.154	3.90	2.068	52.53	7.801	20.0
2	80	2.375	60.33	0.217	5.50	1.942	49.33	6.879	28.0
2½	40	2.875	73.03	0.205	5.20	2.466	62.63	11.09	21.0
2½	80	2.875	73.03	0.276	7.00	2.324	59.03	9.851	29.0
3	40	3.500	88.90	0.217	5.50	3.067	77.90	17.16	18.0
3	80	3.500	88.90	0.299	7.60	2.902	73.70	15.36	26.0
4	40	4.500	114.3	0.236	6.00	4.028	102.3	29.59	15.0
4	80	4.500	114.3	0.339	8.60	3.823	97.10	26.66	22.0
6	40	6.625	168.3	0.280	7.10	6.066	154.1	67.12	12.0
6	80	6.625	168.3	0.433	11.00	5.759	146.3	60.50	19.0
8	40	6.625	219.1	0.323	8.20	7.979	202.7	116.1	11.0
8	80	6.625	219.1	0.500	12.70	7.625	193.7	106.1	17.0
10	40	10.75	273.1	0.366	9.30	10.02	254.5	183.1	10.0
10	80	10.75	273.1	0.594	15.10	9.561	242.9	166.8	16.0
12	40	12.75	323.9	0.406	10.30	11.94	303.3	260.0	9.0
12	80	12.75	323.9	0.689	17.50	11.37	288.9	235.9	16.0

Table 9.17 Inch CPVC pipe to ASTM F441

nb	od mm	od in	WALL mm	WALL in	id mm	id in	1m/s m³/h	WP barg
¼	13.50	0.531	2.3	0.091	8.90	0.350	0.224	15.0
⅜	17.00	0.669	2.3	0.091	12.40	0.488	0.435	15.0
½	21.20	0.835	2.9	0.114	15.40	0.606	0.671	15.0
¾	26.60	1.047	3.6	0.142	19.40	0.764	1.064	15.0
1	33.40	1.315	3.7	0.146	26.00	1.024	1.911	12.0
1¼	42.10	1.657	4.6	0.181	32.90	1.295	3.060	12.0
1½	48.10	1.894	5.3	0.209	37.50	1.476	3.976	12.0
2	60.10	2.366	3.5	0.138	53.10	2.091	7.972	6.0
2	60.10	2.366	6.6	0.260	46.90	1.846	6.219	12.0
3	88.60	3.488	5.2	0.205	78.20	3.079	17.29	6.0
3	88.60	3.488	7.5	0.295	73.60	2.898	15.32	9.0
4	113.9	4.484	6.6	0.260	100.7	3.965	28.67	6.0
4	113.9	4.484	9.6	0.378	94.70	3.728	25.36	9.0
6	167.8	6.606	5.0	0.197	157.8	6.213	70.41	3.0
6	167.8	6.606	9.8	0.386	148.2	5.835	62.10	6.0
8	218.3	8.594	6.4	0.252	205.5	8.091	119.4	0
8	218.3	8.594	12.7	0.500	192.9	7.594	105.2	6.0
9	243.4	9.583	6.4	0.252	230.6	9.079	150.4	0
10	271.1	10.67	6.4	0.252	258.3	10.17	188.6	0
10	271.1	10.67	15.8	0.622	239.5	9.429	162.2	6.0
12	322.3	12.69	6.4	0.252	309.5	12.19	270.8	0
12	322.3	12.69	18.8	0.740	284.7	11.21	229.2	6.0

Table 9.18 Inch polypropylene pipe

nb	SCH	od in	od mm	WALL in	WALL mm	id in	id mm	1m/s m³/h	WP barg
½	80	0.840	21.34	0.147	3.73	0.546	13.88	0.544	40.0
¾	80	1.050	26.67	0.154	3.91	0.742	18.85	1.005	32.0
1	80	1.315	33.40	0.179	4.55	0.957	24.30	1.670	29.0
1½	80	1.900	48.26	0.199	5.05	1.502	38.16	4.117	22.0
2	80	2.375	60.33	0.218	5.54	1.939	49.25	6.857	18.0
3	80	3.500	88.90	0.300	7.62	2.900	73.66	15.34	17.0
4	80	4.500	114.3	0.337	8.56	3.826	97.18	26.70	15.0
6	80	6.625	168.3	0.432	11.0	5.761	146.3	60.55	13.0

Table 9.19 Inch PVDF pipe to ASTM D1785 Note: WP for fusion welded joints; screwed joints approx 50% WP

Table 9.20 Inch UPVC pipe to ASTM D2241 (Below)

nb	STD	od in	od mm	WALL in	WALL mm	id in	id mm	1m/s m³/h	WP barg
⅛	D1785/40	0.405	10.29	0.067	1.70	0.271	6.89	0.134	55.5
⅛	D1785/80	0.405	10.29	0.094	2.40	0.216	5.49	0.085	84.5
¼	D1785/40	0.540	13.72	0.087	2.20	0.367	9.32	0.245	53.8
¼	D1785/80	0.540	13.72	0.118	3.00	0.304	7.72	0.168	77.9
⅜	3505/6E	0.675	17.15	0.075	1.90	0.525	13.3	0.504	15.0
⅜	D1785/40	0.675	17.15	0.091	2.30	0.494	12.5	0.445	42.8
⅜	D1785/80	0.675	17.15	0.126	3.20	0.423	10.7	0.326	63.5
½	D2241/13.5	0.840	21.34	0.063	1.60	0.714	18.1	0.930	21.7
½	3505/6E	0.840	21.34	0.083	2.10	0.675	17.1	0.830	15.0
½	D1785/40	0.840	21.34	0.110	2.80	0.620	15.7	0.700	41.4
½	D1785/80	0.840	21.34	0.146	3.70	0.549	13.9	0.549	58.6
½	3050/6T	0.840	21.34	0.157	4.00	0.525	13.3	0.503	12.0
½	D1785/120	0.840	21.34	0.169	4.30	0.501	12.7	0.459	70.0
¾	D2241/21	1.050	26.67	0.059	1.50	0.932	23.7	1.584	13.8
¾	3505/6E	1.050	26.67	0.098	2.50	0.853	21.7	1.328	15.0
¾	D1785/40	1.050	26.67	0.114	2.90	0.822	20.9	1.232	33.1
¾	D1785/80	1.050	26.67	0.154	3.90	0.743	18.9	1.007	47.6
¾	3506/6T	1.050	26.67	0.165	4.20	0.719	18.3	0.944	12.0
¾	D1785/120	1.050	26.67	0.169	4.30	0.711	18.1	0.923	53.3
1	D2241/26	1.315	33.40	0.059	1.50	1.197	30.4	2.613	11.0
1	D2241/21	1.315	33.40	0.063	1.60	1.189	30.2	2.579	13.8
1	3505/6E	1.315	33.40	0.106	2.70	1.102	28.0	2.217	15.0
1	D1785/40	1.315	33.40	0.134	3.40	1.047	26.6	2.001	31.0
1	D1785/80	1.315	33.40	0.177	4.50	0.961	24.4	1.683	43.5
1	3505/6T	1.315	33.40	0.189	4.80	0.937	23.8	1.602	12.0
1	D1785/120	1.315	33.40	0.201	5.10	0.913	23.2	1.522	49.5
1¼	D2241/26	1.660	42.16	0.063	1.60	1.534	39.0	4.293	11.0
1¼	D2241/21	1.660	42.16	0.079	2.00	1.503	38.2	4.118	13.8
1¼	3505/6D	1.660	42.16	0.106	2.70	1.447	36.8	3.822	12.0
1¼	3505/6E	1.660	42.16	0.126	3.20	1.408	35.8	3.616	15.0
1¼	D1785/40	1.660	42.16	0.142	3.60	1.377	35.0	3.456	25.5
1¼	D1785/80	1.660	42.16	0.193	4.90	1.274	32.4	2.962	35.9
1¼	3505/6T	1.660	42.16	0.205	5.20	1.251	31.8	2.853	12.0
1¼	D1785/120	1.660	42.16	0.217	5.50	1.227	31.2	2.746	41.0
1½	D2241/26	1.900	48.26	0.075	1.90	1.750	44.5	5.589	11.0
1½	D2241/21	1.900	48.26	0.091	2.30	1.719	43.7	5.390	13.8
1½	3505/6D	1.900	48.26	0.118	3.00	1.664	42.3	5.050	12.0
1½	3505/6E	1.900	48.26	0.146	3.70	1.609	40.9	4.721	15.0
1½	D1785/40	1.900	48.26	0.146	3.70	1.609	40.9	4.721	22.8
1½	D1785/80	1.900	48.26	0.201	5.10	1.498	38.1	4.096	32.4
1½	3505/6T	1.900	48.26	0.217	5.50	1.467	37.3	3.925	12.0
1½	D1785/120	1.900	48.26	0.224	5.70	1.451	36.9	3.842	37.0
2	D2241/26	2.375	60.33	0.091	2.30	2.194	55.7	8.780	11.0
2	D2241/21	2.375	60.33	0.114	2.90	2.147	54.5	8.406	13.8
2	3505/6C	2.375	60.33	0.118	3.00	2.139	54.3	8.344	9.0
2	3505/6D	2.375	60.33	0.146	3.70	2.084	52.9	7.920	12.0
2	D1785/40	2.375	60.33	0.154	3.90	2.068	52.5	7.801	19.3
2	3505/6E	2.375	60.33	0.177	4.50	2.021	51.3	7.448	15.0
2	D1785/80	2.375	60.33	0.217	5.50	1.942	49.3	6.879	27.6
2	3505/6T	2.375	60.33	0.236	6.00	1.903	48.3	6.603	12.0
2	D1785/120	2.375	60.33	0.252	6.40	1.871	47.5	6.386	32.5
2½	D2241/26	2.875	73.03	0.110	2.80	2.655	67.4	12.85	11.0
2½	D2241/21	2.875	73.03	0.138	3.50	2.599	66.0	12.33	13.8
2½	3505/6C	2.875	73.03	0.142	3.60	2.592	65.8	12.25	10.0
2½	D1785/40	2.875	73.03	0.205	5.20	2.466	62.6	11.09	20.7
2½	3505/6E	2.875	73.03	0.220	5.60	2.434	61.8	10.81	16.0
2½	D1785/80	2.875	73.03	0.276	7.00	2.324	59.0	9.851	29.0
2½	D1785/120	2.875	73.03	0.299	7.60	2.277	57.8	9.454	32.1
3	3505/6B	3.500	88.90	0.122	3.10	3.256	82.7	19.34	6.0
3	D2241/26	3.500	88.90	0.134	3.40	3.232	82.1	19.06	11.0
3	3505/6C	3.500	88.90	0.161	4.10	3.177	80.7	18.41	9.0
3	D2241/21	3.500	88.90	0.165	4.20	3.169	80.5	18.32	13.8
3	3505/6T	3.500	88.90	0.165	4.20	3.169	80.5	18.32	12.0
3	3505/6D	3.500	88.90	0.209	5.30	3.083	78.3	17.33	12.0
3	D1785/40	3.500	88.90	0.217	5.50	3.067	77.9	17.16	17.9
3	3505/6E	3.500	88.90	0.256	6.50	2.988	75.9	16.29	15.0
3	D1785/80	3.500	88.90	0.299	7.60	2.902	73.7	15.36	25.5
3	D1785/120	3.500	88.90	0.350	8.90	2.799	71.1	14.29	30.6
3½	D2241/26	4.000	101.6	0.154	3.90	3.693	93.8	24.88	11.0
3½	D2241/21	4.000	101.6	0.189	4.80	3.622	92.0	23.93	13.8
3½	D1785/40	4.000	101.6	0.224	5.70	3.551	90.2	23.00	16.6
3½	D1785/80	4.000	101.6	0.319	8.10	3.362	85.4	20.62	24.1
4	3505/6B	4.500	114.3	0.142	3.60	4.217	107	32.43	6.0
4	D2241/26	4.500	114.3	0.173	4.40	4.154	106	31.47	11.0
4	3505/6C	4.500	114.3	0.205	5.20	4.091	104	30.52	9.0
4	2241/21	4.500	114.3	0.213	5.40	4.075	104	30.29	13.8
4	D1785/40	4.500	114.3	0.236	6.00	4.028	102	29.59	15.2
4	3505/6D	4.500	114.3	0.268	6.80	3.965	101	28.67	12.0
4	3505/6E	4.500	114.3	0.327	8.30	3.846	97.7	26.99	15.0
4	D1785/80	4.500	114.3	0.339	8.60	3.823	97.1	26.99	22.1
4	D1785/120	4.500	114.3	0.437	11.1	3.626	92.1	23.98	29.7
5	3505/6B	5.563	141.3	0.157	4.00	5.248	133	50.24	6.0
5	D2241/26	5.563	141.3	0.213	5.40	5.138	131	48.15	11.0
5	3505/6C	5.563	141.3	0.248	6.30	5.067	129	46.83	9.0
5	D1785/40	5.563	141.3	0.260	6.60	5.043	128	46.40	13.1
5	D2241/21	5.563	141.3	0.264	6.70	5.035	128	46.25	13.8
5	3505/6D	5.563	141.3	0.327	8.30	4.909	125	43.97	12.0
5	D1785/80	5.563	141.3	0.374	9.50	4.815	122	42.29	20.0
5	3505/6E	5.563	141.3	0.398	10.1	4.768	121	41.47	15.0

nb	STD	od in	od mm	WALL in	WALL mm	id in	id mm	1m/s m³/h	WP barg
6	3505/6B	6.625	168.3	0.189	4.80	6.247	159	71.19	6.0
6	D2241/26	6.625	168.3	0.256	6.50	6.113	155	68.17	11.0
6	D1785/40	6.625	168.3	0.280	7.10	6.066	154	67.12	12.4
6	3505/6C	6.625	168.3	0.295	7.50	6.034	153	66.43	9.0
6	D2241/21	6.625	168.3	0.315	8.00	5.995	152	65.56	13.8
6	3505/6D	6.625	168.3	0.390	9.90	5.845	148	62.33	12.0
6	D1785/80	6.625	168.3	0.433	11.0	5.759	146	60.50	19.3
6	3505/6E	6.625	168.3	0.480	12.2	5.664	144	58.53	15.0
6	D1785/120	6.625	168.3	0.563	14.3	5.499	140	55.16	25.6
8	3505/6B	8.625	219.1	0.220	5.60	8.184	208	122.2	6.0
8	D1785/40	8.625	219.1	0.323	8.20	7.979	203	116.1	11.0
8	D2241/26	8.625	219.1	0.331	8.40	7.964	202	115.7	11.0
8	3505/6C	8.625	219.1	0.346	8.80	7.932	201	114.8	9.0
8	D2241/21	8.625	219.1	0.409	10.4	7.806	198	111.2	13.8
8	3505/6D	8.625	219.1	0.457	11.6	7.712	196	108.5	12.0
8	D1785/80	8.625	219.1	0.500	12.7	7.625	194	106.1	16.6
8	3505/6E	8.625	219.1	0.555	14.1	7.515	191	103.0	15.0
10	3505/6B	10.75	273.1	0.276	7.00	10.20	259	189.7	6.0
10	D1785/40	10.75	273.1	0.366	9.30	10.02	254	183.1	9.7
10	D2241/26	10.75	273.1	0.413	10.5	9.923	252	179.6	11.0
10	3505/6C	10.75	273.1	0.429	10.9	9.892	251	178.5	9.0
10	D2241/21	10.75	273.1	0.512	13.0	9.726	247	172.6	13.8
10	3505/6D	10.75	273.1	0.563	14.3	9.624	244	169.0	12.0
10	D1785/80	10.75	273.1	0.594	15.1	9.561	243	166.8	15.9
10	3505/6E	10.75	273.1	0.689	17.5	9.372	238	160.2	15.0
12	3505/6B	12.75	323.9	0.323	8.20	12.10	307	267.3	6.0
12	D1785/40	12.75	323.9	0.406	10.3	11.94	303	260.0	9.0
12	D2241/26	12.75	323.9	0.492	12.5	11.77	299	252.5	11.0
12	3505/6C	12.75	323.9	0.508	12.9	11.73	298	251.2	9.0
12	3505/6D	12.75	323.9	0.669	17.0	11.41	290	237.5	12.0
12	D1785/80	12.75	323.9	0.689	17.5	11.37	289	235.9	15.9
12	3505/6E	12.75	323.9	0.819	20.8	11.11	282	225.2	15.0
14	3505/6B	14.00	355.6	0.354	9.00	13.29	338	322.3	6.0
14	D1785/40	14.00	355.6	0.437	11.1	13.13	333	314.3	9.0
14	D2241/26	14.00	355.6	0.539	13.7	12.92	328	304.6	11.0
14	3505/6C	14.00	355.6	0.555	14.1	12.89	327	303.1	9.0
14	3505/6D	14.00	355.6	0.732	18.6	12.54	318	286.6	12.0
14	D1785/80	14.00	355.6	0.752	19.1	12.50	317	284.8	15.2
14	3505/6E	14.00	355.6	0.898	22.8	12.50	310	271.7	15.0
16	3505/6B	16.00	406.4	0.402	10.2	15.20	386	421.3	6.0
16	1785/40	16.00	406.4	0.500	12.7	15.00	381	410.4	9.0
16	D2241/26	16.00	406.4	0.614	15.6	14.77	375	398.0	11.0
16	3505/6C	16.00	406.4	0.638	16.2	14.72	374	395.5	9.0
16	3505/6D	16.00	406.4	0.831	21.1	14.34	364	375.0	12.0
16	D1785/80	16.00	406.4	0.843	21.4	14.31	364	373.8	15.2
16	3505/6E	16.00	406.4	1.024	26.0	13.95	354	355.1	15.0
18	D2241/41	18.00	457.2	0.441	11.2	17.12	435	534.5	6.9
18	3505/6B	18.00	457.2	0.469	11.9	17.06	433	531.1	6.0
18	2241/26	18.00	457.2	0.693	17.6	16.61	422	503.5	11.0
18	3505/6C	18.00	457.2	0.717	18.2	16.57	421	500.7	9.0
18	3505/6D	18.00	457.2	0.937	23.8	16.13	410	474.4	12.0
18	D1785/80	18.00	457.2	0.937	23.8	16.13	410	474.4	15.2
20	D2241/41	20.00	508.0	0.488	12.4	19.02	483	660.2	6.9
20	3505/6B	20.00	508.0	0.520	13.2	18.96	482	655.8	6.0
20	D2241/26	20.00	508.0	0.768	19.5	18.46	469	621.9	11.0
20	3505/6C	20.00	508.0	0.795	20.2	18.41	468	618.2	9.0
24	D2241/41	24.00	609.6	0.587	14.9	22.83	580	950.5	6.9
24	3505/6B	24.00	609.6	0.618	15.7	22.76	578	945.3	6.0
24	D2241/26	24.00	609.6	0.921	23.4	22.16	563	895.6	11.0
24	3505/6C	24.00	609.6	0.949	24.1	22.10	561	891.1	9.0

Table 9.20 Inch UPVC pipe to ASTM D2241 (Above)

9.2.7 Non-metallic tube

Non-metallic tube is manufactured in a wide range of sizes, see Tables 9.21 to 9.23. PVDF tube is available for high purity, non-contamination systems.

A special grade is produced specifically for these applications, PVDF-HP, see Table 9.24. UPVC tube is manufactured specifically for potable water applications, see Table 9.25.

od mm	od in	WALL mm	WALL in	d mm	id in	1m/s m³/h	WP barg
16.0	0.63	1.50	0.059	13.0	0.512	0.478	10.0
20.0	0.79	1.65	0.065	16.7	0.657	0.789	10.0

od mm	od in	WALL mm	WALL in	d mm	id in	1m/s m³/h	WP barg
25.0	0.98	1.95	0.077	21.1	0.831	1.259	10.0
32.0	1.26	2.15	0.085	27.7	1.091	2.169	10.0
40.0	1.57	2.70	0.106	34.6	1.362	3.385	10.0
50.0	1.97	3.30	0.130	43.4	1.709	5.326	10.0
63.0	2.48	4.25	0.167	54.5	2.146	8.398	10.0
75.0	2.95	5.00	0.197	65.0	2.559	11.946	10.0
90.0	3.54	6.00	0.236	78.0	3.071	17.202	10.0
110.0	4.33	7.20	0.283	95.6	3.764	25.841	10.0

Table 9.21 Metric ABS tube to DIN 8062 and ISO 161/1 (Above)

Table 9.22 Metric polypropylene tube to BS 5556 and DIN 8077/8078 (Below)

od mm	od in	WALL mm	WALL in	Id mm	id in	1m/s m³/h	WP barg
16.0	0.630	1.80	0.071	12.40	0.488	0.435	10.0
20.0	0.787	1.80	0.071	16.40	0.646	0.760	6.0
20.0	0.787	1.90	0.075	16.20	0.638	0.742	10.0
25.0	0.984	1.80	0.071	21.40	0.843	1.295	6.0
25.0	0.984	2.30	0.091	20.40	0.803	1.177	10.0
32.0	1.260	1.90	0.075	28.20	1.110	2.248	6.0
32.0	1.260	3.00	0.118	26.00	1.024	1.911	10.0
40.0	1.575	1.80	0.071	36.40	1.433	3.746	4.0
40.0	1.575	2.30	0.091	35.40	1.394	3.543	6.0
40.0	1.575	3.70	0.146	32.60	1.283	3.005	10.0
50.0	1.969	1.80	0.071	46.40	1.827	6.087	2.5
50.0	1.969	2.00	0.079	46.00	1.811	5.983	4.0
50.0	1.969	2.90	0.114	44.20	1.740	5.524	6.0
50.0	1.969	4.60	0.181	40.80	1.606	4.707	10.0
63.0	2.480	1.80	0.071	59.40	2.339	9.976	2.5
63.0	2.480	2.50	0.098	58.00	2.283	9.511	4.0
63.0	2.480	3.60	0.142	55.80	2.197	8.804	6.0
63.0	2.480	5.80	0.228	51.40	2.024	7.470	10.0
75.0	2.953	1.90	0.075	71.20	2.803	14.33	2.5
75.0	2.953	2.90	0.114	69.20	2.724	13.54	4.0
75.0	2.953	4.30	0.169	66.40	2.614	12.47	6.0
75.0	2.953	6.90	0.272	61.20	2.409	10.59	10.0
90.0	3.543	2.20	0.087	85.60	3.370	20.72	2.5
90.0	3.543	3.50	0.138	83.00	3.268	19.48	4.0
90.0	3.543	5.10	0.201	79.80	3.142	18.01	6.0
90.0	3.543	8.20	0.323	73.60	2.898	15.32	10.0
110.0	4.331	2.70	0.106	104.6	4.118	30.94	2.5
110.0	4.331	4.30	0.169	101.4	3.992	29.07	4.0
110.0	4.331	6.30	0.248	97.40	3.835	26.82	6.0
110.0	4.331	10.0	0.394	90.00	3.543	22.90	10.0
125.0	4.921	3.10	0.122	118.8	4.677	39.90	2.5
125.0	4.921	4.90	0.193	115.2	4.535	37.52	4.0
125.0	4.921	7.10	0.280	110.8	4.362	34.71	6.0
125.0	4.921	11.4	0.449	102.2	4.024	29.53	10.0
140.0	5.512	3.50	0.138	133.0	5.236	50.01	2.5
140.0	5.512	5.40	0.213	129.2	5.087	47.20	4.0
140.0	5.512	8.00	0.315	124.0	4.882	43.47	6.0
140.0	5.512	12.8	0.504	114.4	4.504	37.00	10.0
160.0	6.299	3.90	0.154	152.2	5.992	65.50	2.5
160.0	6.299	6.20	0.244	147.6	5.811	61.60	4.0
160.0	6.299	9.10	0.358	141.8	5.583	56.85	6.0
160.0	6.299	14.6	0.575	130.8	5.150	48.37	10.0
180.0	7.087	4.40	0.173	171.2	6.740	82.87	2.5
180.0	7.087	7.00	0.276	166.0	6.535	77.91	4.0
180.0	7.087	10.2	0.402	159.6	6.283	72.02	6.0
180.0	7.087	16.4	0.646	147.2	5.795	61.26	10.0
200.0	7.874	4.90	0.193	190.2	7.488	102.3	2.5
200.0	7.874	7.70	0.303	184.6	7.268	96.35	4.0
200.0	7.874	11.4	0.449	177.2	6.976	88.78	6.0
200.0	7.874	18.2	0.717	163.6	6.441	75.68	10.0
225.0	8.858	5.50	0.217	214.0	8.425	129.5	2.5
225.0	8.858	8.70	0.343	207.6	8.173	121.9	4.0
225.0	8.858	12.8	0.504	199.4	7.850	112.4	6.0
225.0	8.858	20.5	0.807	184.0	7.244	95.73	10.0
250.0	9.843	6.10	0.240	237.8	9.362	159.9	2.5
250.0	9.843	9.70	0.382	230.6	9.079	150.4	4.0
250.0	9.843	14.2	0.559	221.6	8.724	138.8	6.0
250.0	9.843	22.8	0.898	204.4	8.047	118.1	10.0
280.0	11.02	6.90	0.272	266.2	10.48	200.4	2.5
280.0	11.02	10.8	0.425	258.4	10.17	188.8	4.0
280.0	11.02	15.9	0.626	248.2	9.772	174.2	6.0
280.0	11.02	25.5	1.004	229.0	9.016	148.3	10.0

od mm	od in	WALL mm	WALL in	id mm	id in	1m/s m³/h	WP barg
315.0	12.40	7.70	0.303	299.6	11.80	253.8	2.5
315.0	12.40	12.2	0.480	290.6	11.44	238.8	4.0
315.0	12.40	17.9	0.705	279.2	10.99	220.4	6.0
315.0	12.40	28.7	1.130	257.6	10.14	187.6	10.0
355.0	13.98	8.70	0.343	337.6	13.29	322.3	2.5
355.0	13.98	13.7	0.539	327.6	12.90	303.4	4.0
355.0	13.98	20.1	0.791	314.8	12.39	280.2	6.0
355.0	13.98	32.3	1.272	290.4	11.43	238.4	10.0
400.0	15.75	9.80	0.386	380.4	14.98	409.1	2.5
400.0	15.75	15.4	0.606	369.2	14.54	385.4	4.0
400.0	15.75	22.7	0.894	354.6	13.96	355.5	6.0
400.0	15.75	36.4	1.433	327.2	12.88	302.7	10.0
450.0	17.72	11.0	0.433	428.0	16.85	517.9	2.5
450.0	17.72	17.4	0.685	415.2	16.35	487.4	4.0
450.0	17.72	25.5	1.004	399.0	15.71	450.1	6.0
450.0	17.72	41.0	1.614	368.0	14.49	382.9	10.0
500.0	19.69	12.2	0.480	475.6	18.72	639.6	2.5
500.0	19.69	19.3	0.760	461.4	18.17	601.9	4.0
500.0	19.69	28.3	1.114	443.4	17.46	555.9	6.0
500.0	19.69	45.5	1.791	409.0	16.10	473.0	10.0
560.0	22.05	13.7	0.539	532.6	20.97	802.0	2.5
560.0	22.05	21.6	0.850	516.8	20.35	755.2	4.0
560.0	22.05	31.7	1.248	496.6	19.55	697.3	6.0
630.0	24.80	15.4	0.606	599.2	23.59	1015	2.5
630.0	24.80	24.3	0.957	581.4	22.89	955.7	4.0
630.0	24.80	35.7	1.406	558.6	21.99	882.3	6.0
710.0	27.95	17.4	0.685	675.2	26.58	1289	2.5
710.0	27.95	27.4	1.079	655.2	25.80	1214	4.0
710.0	27.95	40.2	1.583	629.6	24.79	1121	6.0
800.0	31.50	19.6	0.772	760.8	29.95	1637	2.5
800.0	31.50	30.8	1.213	738.4	29.07	1542	4.0
800.0	31.50	45.3	1.783	709.4	27.93	1423	6.0
900.0	35.43	34.7	1.366	830.6	32.70	1951	4.0
1000.0	39.37	38.5	1.516	923.0	36.34	2409	4.0

Table 9.22 Metric polypropylene tube to BS 5556 and DIN 8077/8078 (Above)

Table 9.23 Metric polyethylene tube (Below)

od mm	od in	WALL mm	WALL in	id mm	id in	1m/s m³/h	WP barg
20.0	0.787	1.90	0.075	16.20	0.638	0.742	10.0
20.0	0.787	2.30	0.091	15.40	0.606	0.671	12.5
20.0	0.787	2.80	0.110	14.40	0.567	0.586	16.0
25.0	0.984	1.80	0.071	21.40	0.843	1.295	6.0
25.0	0.984	2.30	0.091	20.40	0.803	1.177	10.0
25.0	0.984	2.80	0.110	19.40	0.764	1.064	12.5
25.0	0.984	3.50	0.138	18.00	0.709	0.916	16.0
32.0	1.260	1.90	0.075	28.20	1.110	2.248	6.0
32.0	1.260	3.00	0.118	26.00	1.024	1.911	10.0
32.0	1.260	3.60	0.142	24.80	0.976	1.739	12.5
32.0	1.260	4.50	0.177	23.00	0.906	1.496	16.0
40.0	1.575	2.00	0.079	36.00	1.417	3.664	4.0
40.0	1.575	2.30	0.091	35.40	1.394	3.543	6.0
40.0	1.575	3.70	0.146	32.60	1.283	3.005	10.0
40.0	1.575	4.50	0.177	31.00	1.220	2.717	12.5
40.0	1.575	5.60	0.220	28.80	1.134	2.345	16.0
50.0	1.969	2.00	0.079	46.00	1.811	5.983	4.0
50.0	1.969	2.90	0.114	44.20	1.740	5.524	6.0
50.0	1.969	4.60	0.181	40.80	1.606	4.707	10.0
50.0	1.969	5.60	0.220	38.80	1.528	4.257	12.5
50.0	1.969	6.90	0.272	36.20	1.425	3.705	16.0
63.0	2.480	2.00	0.079	59.00	2.323	9.842	3.2
63.0	2.480	2.50	0.098	58.00	2.283	9.511	4.0
63.0	2.480	3.60	0.142	55.80	2.197	8.804	6.0
63.0	2.480	5.80	0.228	51.40	2.024	7.470	10.0
63.0	2.480	7.00	0.276	49.00	1.929	6.789	12.5
63.0	2.480	8.70	0.343	45.60	1.795	5.879	16.0
75.0	2.953	2.00	0.079	71.00	2.795	14.25	2.5
75.0	2.953	2.40	0.094	70.20	2.764	13.93	3.2
75.0	2.953	2.90	0.114	69.20	2.724	13.54	4.0
75.0	2.953	4.30	0.169	66.40	2.614	12.47	6.0
75.0	2.953	6.90	0.272	61.20	2.409	10.59	10.0
75.0	2.953	8.40	0.331	58.20	2.291	9.577	12.5
75.0	2.953	10.4	0.409	54.20	2.134	8.306	16.0
90.0	3.543	2.20	0.087	85.60	3.370	20.72	2.5
90.0	3.543	2.80	0.110	84.40	3.323	20.14	3.2
90.0	3.543	3.50	0.138	83.00	3.268	19.48	4.0
90.0	3.543	5.10	0.201	79.80	3.142	18.01	6.0
90.0	3.543	8.20	0.323	73.60	2.898	15.32	10.0
90.0	3.543	10.0	0.394	70.00	2.756	13.85	12.5
90.0	3.543	12.5	0.492	65.00	2.559	11.95	16.0
110.0	4.331	2.70	0.106	104.6	4.118	30.94	2.5
110.0	4.331	3.50	0.138	103.0	4.055	30.00	3.2
110.0	4.331	4.30	0.169	101.4	3.992	29.07	4.0
110.0	4.331	6.30	0.248	97.40	3.835	26.82	6.0
110.0	4.331	10.0	0.394	90.00	3.543	22.90	10.0
110.0	4.331	12.3	0.484	85.40	3.362	20.62	12.5
110.0	4.331	15.2	0.598	79.60	3.134	17.92	16.0
125.0	4.921	3.10	0.122	118.8	4.677	39.90	2.5
125.0	4.921	3.90	0.154	117.2	4.614	38.84	3.2
125.0	4.921	4.90	0.193	115.2	4.535	37.52	4.0
125.0	4.921	7.10	0.280	110.8	4.362	34.71	6.0
125.0	4.921	11.4	0.449	102.2	4.024	29.53	10.0
125.0	4.921	13.9	0.547	97.20	3.827	26.71	12.5
125.0	4.921	17.3	0.681	90.40	3.559	23.11	16.0
140.0	5.512	3.50	0.138	133.0	5.236	50.01	2.5
140.0	5.512	4.40	0.173	131.2	5.165	48.67	3.2
140.0	5.512	5.40	0.213	129.2	5.087	47.20	4.0
140.0	5.512	8.00	0.315	124.0	4.882	43.47	6.0
140.0	5.512	12.8	0.504	114.4	4.504	37.00	10.0
140.0	5.512	15.6	0.614	108.8	4.283	33.47	12.5
140.0	5.512	19.4	0.764	101.2	3.984	28.96	16.0
160.0	6.299	3.90	0.154	152.2	5.992	65.50	2.5
160.0	6.299	5.00	0.197	150.0	5.906	63.62	3.2
160.0	6.299	6.20	0.244	147.6	5.811	61.60	4.0
160.0	6.299	9.10	0.358	141.8	5.583	56.85	6.0
160.0	6.299	14.6	0.575	130.8	5.150	48.37	10.0
160.0	6.299	17.8	0.701	124.4	4.898	43.76	12.5
160.0	6.299	22.1	0.870	115.8	4.559	37.91	16.0
180.0	7.087	4.40	0.173	171.2	6.740	82.87	2.5
180.0	7.087	5.60	0.220	168.8	6.646	80.56	3.2
180.0	7.087	7.00	0.276	166.0	6.535	77.91	4.0
180.0	7.087	10.2	0.402	159.6	6.283	72.02	6.0
180.0	7.087	16.4	0.646	147.2	5.795	61.26	10.0
180.0	7.087	20.0	0.787	140.0	5.512	55.42	12.5
180.0	7.087	24.9	0.980	130.2	5.126	47.93	16.0
200.0	7.874	4.90	0.193	190.2	7.488	102.3	2.5
200.0	7.874	6.20	0.244	187.6	7.386	99.51	3.2
200.0	7.874	7.70	0.303	184.6	7.268	96.35	4.0
200.0	7.874	11.4	0.449	177.2	6.976	88.78	6.0
200.0	7.874	18.2	0.717	163.6	6.441	75.68	10.0
200.0	7.874	22.3	0.878	155.4	6.118	68.28	12.5
200.0	7.874	27.6	1.087	144.8	5.701	59.28	16.0
225.0	8.858	5.50	0.217	214.0	8.425	129.5	2.5
225.0	8.858	7.00	0.276	211.0	8.307	125.9	3.2
225.0	8.858	8.70	0.343	207.6	8.173	121.9	4.0
225.0	8.858	12.8	0.504	199.4	7.850	112.4	6.0
225.0	8.858	20.5	0.807	184.0	7.244	95.73	10.0
225.0	8.858	25.0	0.984	175.0	6.890	86.59	12.5
225.0	8.858	31.1	1.224	162.8	6.409	74.94	16.0
250.0	9.843	6.10	0.240	237.8	9.362	159.9	2.5
250.0	9.843	7.80	0.307	234.4	9.228	155.3	3.2
250.0	9.843	9.70	0.382	230.6	9.079	150.4	4.0
250.0	9.843	14.2	0.559	221.6	8.724	138.8	6.0
250.0	9.843	22.8	0.898	204.4	8.047	118.1	10.0
250.0	9.843	27.8	1.094	194.4	7.654	106.9	12.5
250.0	9.843	34.5	1.358	181.0	7.126	92.63	16.0
280.0	11.02	6.90	0.272	266.2	10.48	200.4	2.5
280.0	11.02	8.70	0.343	262.6	10.34	195.0	3.2
280.0	11.02	10.8	0.425	258.4	10.17	188.8	4.0
280.0	11.02	15.9	0.626	248.2	9.772	174.2	6.0
280.0	11.02	25.5	1.004	229.0	9.016	148.3	10.0
280.0	11.02	31.2	1.228	217.6	8.567	133.9	12.5
280.0	11.02	38.7	1.524	202.6	7.976	116.1	16.0
315.0	12.40	7.70	0.303	299.6	11.80	253.8	2.5
315.0	12.40	9.80	0.386	295.4	11.63	246.7	3.2
315.0	12.40	12.2	0.480	290.6	11.44	238.8	4.0
315.0	12.40	17.9	0.705	279.2	10.99	220.4	6.0
315.0	12.40	28.7	1.130	257.6	10.14	187.6	10.0
315.0	12.40	35.0	1.378	245.0	9.646	169.7	12.5

od mm	od in	WALL mm	WALL in	id mm	id in	1m/s m³/h	WP barg
315.0	12.40	43.5	1.713	228.0	8.976	147.0	16.0
355.0	13.98	8.70	0.343	337.6	13.29	322.3	2.5
355.0	13.98	11.1	0.437	332.8	13.10	313.2	3.2
355.0	13.98	13.7	0.539	327.6	12.90	303.4	4.0
355.0	13.98	20.1	0.791	314.8	12.39	280.2	6.0
355.0	13.98	32.3	1.272	290.4	11.43	238.4	10.0
355.0	13.98	39.5	1.555	276.0	10.87	215.4	12.5
355.0	13.98	49.0	1.929	257.0	10.12	186.7	16.0
400.0	15.75	9.80	0.386	380.4	14.98	409.1	2.5
400.0	15.75	12.4	0.488	375.2	14.77	398.0	3.2
400.0	15.75	15.4	0.606	369.2	14.54	385.4	4.0
400.0	15.75	22.7	0.894	354.6	13.96	355.5	6.0
400.0	15.75	36.4	1.433	327.2	12.88	302.7	10.0
400.0	15.75	44.5	1.752	311.0	12.24	273.5	12.5
400.0	15.75	55.2	2.173	289.6	11.40	237.1	16.0
450.0	17.72	11.0	0.433	428.0	16.85	517.9	2.5
450.0	17.72	14.0	0.551	422.0	16.61	503.5	3.2
450.0	17.72	17.4	0.685	415.2	16.35	487.4	4.0
450.0	17.72	25.5	1.004	399.0	15.71	450.1	6.0
450.0	17.72	41.0	1.614	368.0	14.49	382.9	10.0
450.0	17.72	50.0	1.969	350.0	13.78	346.4	12.5
450.0	17.72	62.1	2.445	325.8	12.83	300.1	16.0
500.0	19.69	12.2	0.480	475.6	18.72	639.6	2.5
500.0	19.69	15.5	0.610	469.0	18.46	621.9	3.2
500.0	19.69	19.3	0.760	461.4	18.17	601.9	4.0
500.0	19.69	28.3	1.114	443.4	17.46	555.9	6.0
500.0	19.69	45.5	1.791	409.0	16.10	473.0	10.0
560.0	22.05	13.7	0.539	532.6	20.97	802.0	2.5
560.0	22.05	17.4	0.685	525.2	20.68	779.9	3.2
560.0	22.05	21.6	0.850	516.8	20.35	755.2	4.0
560.0	22.05	31.7	1.248	496.6	19.55	697.3	6.0
560.0	22.05	51.0	2.008	458.0	18.03	593.1	10.0
630.0	24.80	15.4	0.606	599.2	23.59	1015	2.5
630.0	24.80	19.6	0.772	590.8	23.26	986.9	3.2
630.0	24.80	24.3	0.957	581.4	22.89	955.7	4.0
630.0	24.80	35.7	1.406	558.6	21.99	882.3	6.0
630.0	24.80	57.3	2.256	515.4	20.29	751.1	10.0
710.0	27.95	17.4	0.685	675.2	26.58	1289	2.5
710.0	27.95	22.1	0.870	665.8	26.21	1253	3.2
710.0	27.95	27.4	1.079	655.2	25.80	1214	4.0
710.0	27.95	40.2	1.583	629.6	24.79	1121	6.0
800.0	31.50	19.6	0.772	760.8	29.95	1637	2.5
800.0	31.50	24.9	0.980	750.2	29.54	1591	3.2
800.0	31.50	30.8	1.213	738.4	29.07	1542	4.0
800.0	31.50	45.3	1.783	709.4	27.93	1423	6.0
900.0	35.43	22.0	0.866	856.0	33.70	2072	2.5
900.0	35.43	28.0	1.102	844.0	33.23	2014	3.2
900.0	35.43	34.7	1.366	830.6	32.70	1951	4.0
900.0	35.43	51.0	2.008	798.0	31.42	1801	6.0
1000.0	39.37	24.4	0.961	951.2	37.45	2558	2.5
1000.0	39.37	31.1	1.224	937.8	36.92	2487	3.2
1000.0	39.37	38.5	1.516	923.0	36.34	2409	4.0
1000.0	39.37	56.6	2.228	886.8	34.91	2224	6.0
1200.0	47.24	29.3	1.154	1141.4	44.94	3684	2.5
1200.0	47.24	37.3	1.469	1125.4	44.31	3581	3.2
1200.0	47.24	46.2	1.819	1107.6	43.61	3469	4.0

Table 9.23 Metric polyethylene tube (Above)

Table 9.24 Metric PVDF and PVDF-HP tube (Below)

od mm	od in	WALL mm	WALL in	id mm	id in	1m/s m³/h	WP barg
16.00	0.630	1.50	0.059	13.00	0.512	0.478	16.0
20.00	0.787	1.90	0.075	16.20	0.638	0.742	16.0
25.00	0.984	1.90	0.075	21.20	0.835	1.271	16.0
32.00	1.260	2.40	0.094	27.20	1.071	2.092	16.0
40.00	1.575	2.40	0.094	35.20	1.386	3.503	16.0
50.00	1.969	3.00	0.118	44.00	1.732	5.474	16.0
63.00	2.480	3.00	0.118	57.00	2.244	9.186	16.0
75.00	2.953	3.00	0.118	69.00	2.717	13.46	10.0
75.00	2.953	3.60	0.142	67.80	2.669	13.00	16.0
90.00	3.543	3.00	0.118	84.00	3.307	19.95	10.0
90.00	3.543	4.30	0.169	81.40	3.205	18.73	16.0
110.0	4.331	3.40	0.134	103.2	4.063	30.11	10.0
110.0	4.331	5.30	0.209	99.40	3.913	27.94	16.0
125.0	4.921	3.90	0.154	117.2	4.614	38.84	10.0
140.0	5.512	4.40	0.173	131.2	5.165	48.67	10.0

od mm	od in	WALL mm	WALL in	id mm	id in	1m/s m³/h	WP barg
160.0	6.299	4.90	0.193	150.2	5.913	63.79	10.0
200.0	7.874	6.20	0.244	187.6	7.386	99.51	10.0
225.0	8.858	7.00	0.276	211.0	8.307	125.9	10.0

Table 9.24 Metric PVDF and PVDF-HP tube (Above)

Table 9.25 Metric UPVC tube to DIN 8061/2 and KIWA 49 (Below)

od mm	STD	od in	WALL mm	WALL in	od mm	id in	1m/s m³/h	WP barg
12.0	8061/2	0.472	1.00	0.039	10.0	0.394	0.283	16.0
12.0	49	0.472	1.10	0.043	9.80	0.386	0.272	12.5
12.0	49	0.472	1.20	0.047	9.60	0.378	0.261	16.0
12.0	8061/2	0.472	1.40	0.055	9.20	0.362	0.239	20.0
16.0	49	0.630	1.10	0.043	13.8	0.543	0.538	12.5
16.0	8061/2	0.630	1.20	0.047	13.6	0.535	0.523	16.0
16.0	49	0.630	1.30	0.051	13.4	0.528	0.508	16.0
16.0	8061/2	0.630	1.80	0.071	12.4	0.488	0.435	20.0
20.0	49	0.787	1.10	0.043	17.8	0.701	0.896	10.0
20.0	49	0.787	1.30	0.051	17.4	0.685	0.856	12.5
20.0	49	0.787	1.50	0.059	17.0	0.669	0.817	16.0
20.0	8061/2	0.787	1.50	0.059	17.0	0.669	0.817	16.0
20.0	8061/2	0.787	2.30	0.091	15.4	0.606	0.671	20.0
25.0	49	0.984	1.30	0.051	22.4	0.882	1.419	10.0
25.0	8061/2	0.984	1.50	0.059	22.0	0.866	1.368	10.0
25.0	49	0.984	1.60	0.063	21.8	0.858	1.344	12.5
25.0	8061/2	0.984	1.90	0.075	21.2	0.835	1.271	16.0
25.0	49	0.984	2.00	0.079	21.0	0.827	1.247	16.0
25.0	8061/2	0.984	2.80	0.110	19.4	0.764	1.064	20.0
32.0	49	1.260	1.70	0.067	28.6	1.126	2.313	10.0
32.0	8061/2	1.260	1.80	0.071	28.4	1.118	2.280	10.0
32.0	49	1.260	2.00	0.079	28.0	1.102	2.217	12.5
32.0	8061/2	1.260	2.40	0.094	27.2	1.071	2.092	16.0
32.0	49	1.260	2.50	0.098	27.0	1.063	2.061	16.0
32.0	8061/2	1.260	3.60	0.142	24.8	0.976	1.739	20.0
40.0	8061/2	1.575	1.80	0.071	36.4	1.433	3.746	6.0
40.0	8061/2	1.575	1.90	0.075	36.2	1.425	3.705	10.0
40.0	49	1.575	2.00	0.079	36.0	1.417	3.664	10.0
40.0	49	1.575	2.50	0.098	35.0	1.378	3.464	12.5
40.0	8061/2	1.575	3.00	0.118	34.0	1.339	3.269	16.0
40.0	49	1.575	3.20	0.126	33.6	1.323	3.192	16.0
40.0	8061/2	1.575	4.50	0.177	31.0	1.220	2.717	20.0
50.0	8061/2	1.969	1.80	0.071	46.4	1.827	6.087	6.0
50.0	49	1.969	1.90	0.075	46.2	1.819	6.035	5.0
50.0	49	1.969	2.10	0.083	45.8	1.803	5.931	7.5
50.0	8061/2	1.969	2.40	0.094	45.2	1.780	5.777	10.0
50.0	49	1.969	2.60	0.102	44.8	1.764	5.675	10.0
50.0	49	1.969	3.20	0.126	43.6	1.717	5.375	12.5
50.0	8061/2	1.969	3.70	0.146	42.6	1.677	5.131	16.0
50.0	49	1.969	3.90	0.154	42.2	1.661	5.035	16.0
50.0	8061/2	1.969	5.60	0.220	38.8	1.528	4.257	20.0
63.0	49	2.480	1.90	0.075	59.2	2.331	9.909	5.0
63.0	8061/2	2.480	1.90	0.075	59.2	2.331	9.909	6.0
63.0	49	2.480	2.10	0.083	58.8	2.315	9.776	7.5
63.0	49	2.480	2.60	0.102	57.8	2.276	9.446	10.0
63.0	8061/2	2.480	3.00	0.118	57.0	2.244	9.186	10.0
63.0	49	2.480	3.20	0.126	56.6	2.228	9.058	12.5
63.0	49	2.480	4.00	0.157	55.0	2.165	8.553	16.0
63.0	8061/2	2.480	4.70	0.185	53.6	2.110	8.123	16.0
63.0	8061/2	2.480	7.00	0.276	49.0	1.929	6.789	20.0
75.0	8061/2	2.953	1.80	0.071	71.4	2.811	14.41	4.0
75.0	49	2.953	1.90	0.075	71.2	2.803	14.33	5.0
75.0	49	2.953	2.10	0.083	70.8	2.787	14.17	6.3
75.0	8061/2	2.953	2.20	0.087	70.6	2.780	14.09	6.0
75.0	49	2.953	2.30	0.091	70.4	2.772	14.01	7.5
75.0	49	2.953	3.10	0.122	68.8	2.709	13.38	10.0
75.0	8061/2	2.953	3.60	0.142	67.8	2.669	13.00	10.0
75.0	49	2.953	3.80	0.150	67.4	2.654	12.84	12.5
75.0	49	2.953	4.70	0.185	65.6	2.583	12.17	16.0
75.0	8061/2	2.953	5.60	0.220	63.8	2.512	11.51	16.0
75.0	8061/2	2.953	8.40	0.331	58.2	2.291	9.577	20.0
90.0	8061/2	3.543	1.80	0.071	86.4	3.402	21.11	4.0
90.0	49	3.543	1.90	0.075	86.2	3.394	21.01	5.0
90.0	49	3.543	2.30	0.091	85.4	3.362	20.62	6.3
90.0	8061/2	3.543	2.70	0.106	84.6	3.331	20.24	6.0
90.0	49	3.543	2.90	0.114	84.2	3.315	20.05	7.5

od mm	STD	od in	WALL mm	WALL in	od mm	id in	1m/s m³/h	WP barg
90.0	49	3.543	3.70	0.146	82.6	3.252	19.29	10.0
90.0	8061/2	3.543	4.30	0.169	81.4	3.205	18.73	10.0
90.0	49	3.543	4.50	0.177	81.0	3.189	18.55	12.5
90.0	49	3.543	5.60	0.220	78.8	3.102	17.56	16.0
90.0	8061/2	3.543	6.70	0.264	76.6	3.016	16.59	16.0
90.0	8061/2	3.543	10.0	0.394	70.0	2.756	13.85	20.0
110.0	8061/2	4.331	2.20	0.087	105.6	4.157	31.53	4.0
110.0	49	4.331	2.40	0.094	105.2	4.142	31.29	5.0
110.0	49	4.331	2.90	0.114	104.2	4.102	30.70	6.3
110.0	8061/2	4.331	3.20	0.126	103.6	4.079	30.35	6.0
110.0	49	4.331	3.50	0.138	103.0	4.055	30.00	7.5
110.0	49	4.331	4.40	0.173	101.2	3.984	28.96	10.0
110.0	8061/2	4.331	5.30	0.209	99.4	3.913	27.94	10.0
110.0	49	4.331	5.50	0.217	99.0	3.898	27.71	12.5
110.0	49	4.331	6.80	0.268	96.4	3.795	26.28	16.0
110.0	8061/2	4.331	8.20	0.323	93.6	3.685	24.77	16.0
110.0	8061/2	4.331	12.3	0.484	85.4	3.362	20.62	20.0
125.0	8061/2	4.921	2.50	0.098	120.0	4.724	40.72	4.0
125.0	49	4.921	2.70	0.106	119.6	4.709	40.44	5.0
125.0	49	4.921	3.30	0.130	118.4	4.661	39.64	6.3
125.0	8061/2	4.921	3.70	0.146	117.6	4.630	39.10	6.0
125.0	49	4.921	4.00	0.157	117.0	4.606	38.70	7.5
125.0	49	4.921	5.10	0.201	114.8	4.520	37.26	10.0
125.0	8061/2	4.921	6.00	0.236	113.0	4.449	36.10	10.0
125.0	49	4.921	6.30	0.248	112.4	4.425	35.72	12.5
125.0	49	4.921	7.40	0.291	110.2	4.339	34.34	16.0
140.0	8061/2	5.512	2.80	0.110	134.4	5.291	51.07	4.0
140.0	8061/2	5.512	4.10	0.161	131.8	5.189	49.12	6.0
140.0	8061/2	5.512	6.70	0.264	126.6	4.984	45.32	10.0
140.0	8061/2	5.512	10.4	0.409	119.2	4.693	40.17	16.0
160.0	8061/2	6.299	3.20	0.126	153.6	6.047	66.71	4.0
160.0	49	6.299	3.40	0.134	153.2	6.031	66.36	5.0
160.0	49	6.299	4.20	0.165	151.6	5.969	64.98	6.3
160.0	8061/2	6.299	4.70	0.185	150.6	5.929	64.13	6.0
160.0	49	6.299	4.90	0.193	150.2	5.913	63.79	7.5
160.0	49	6.299	6.50	0.256	147.0	5.787	61.10	10.0
160.0	8061/2	6.299	7.70	0.303	144.6	5.693	59.12	10.0
160.0	49	6.299	8.00	0.315	144.0	5.669	58.63	12.5
160.0	49	6.299	10.1	0.398	139.8	5.504	55.26	16.0
160.0	8061/2	6.299	11.9	0.469	136.2	5.362	52.45	16.0
180.0	8061/2	7.087	3.60	0.142	172.8	6.803	84.43	4.0
180.0	8061/2	7.087	5.30	0.209	169.4	6.669	81.14	6.0
180.0	8061/2	7.087	8.60	0.339	162.8	6.409	74.94	10.0
180.0	8061/2	7.087	13.4	0.528	153.2	6.031	66.36	16.0
200.0	8061/2	7.874	4.00	0.157	192.0	7.559	104.2	4.0
200.0	49	7.874	4.20	0.165	191.6	7.543	103.8	5.0
200.0	49	7.874	5.20	0.205	189.6	7.465	101.6	6.3
200.0	8061/2	7.874	5.90	0.232	188.2	7.409	100.1	6.0
200.0	49	7.874	6.20	0.244	187.6	7.386	99.51	7.5
200.0	49	7.874	8.00	0.315	184.0	7.244	95.73	10.0
200.0	8061/2	7.874	9.60	0.378	180.8	7.118	92.42	10.0
200.0	49	7.874	12.3	0.484	175.4	6.906	86.99	16.0
200.0	8061/2	7.874	14.9	0.587	170.2	6.701	81.91	16.0
225.0	8061/2	8.858	4.50	0.177	216.0	8.504	131.9	4.0
225.0	8061/2	8.858	6.60	0.260	211.8	8.339	126.8	6.0
225.0	8061/2	8.858	10.8	0.425	203.4	8.008	117.0	10.0
225.0	8061/2	8.858	16.7	0.657	191.6	7.543	103.8	16.0
250.0	8061/2	9.843	4.90	0.193	240.2	9.457	163.1	4.0
250.0	49	9.843	5.20	0.205	239.6	9.433	162.3	5.0
250.0	49	9.843	6.40	0.252	237.2	9.339	159.1	6.3
250.0	8061/2	9.843	7.30	0.287	235.4	9.268	156.7	6.0
250.0	49	9.843	7.60	0.299	234.8	9.244	155.9	7.5
250.0	49	9.843	10.0	0.394	230.0	9.055	149.6	10.0
250.0	8061/2	9.843	11.9	0.469	226.2	8.906	144.7	10.0
250.0	49	9.843	12.3	0.484	225.4	8.874	143.6	12.5
250.0	49	9.843	15.2	0.598	219.6	8.646	136.4	16.0
250.0	8061/2	9.843	18.0	0.709	214.0	8.425	129.5	16.0
280.0	8061/2	11.02	8.20	0.323	263.6	10.38	196.5	6.0
280.0	8061/2	11.02	13.4	0.528	253.2	9.969	181.3	10.0
280.0	8061/2	11.02	20.8	0.819	238.4	9.386	160.7	16.0
315.0	8061/2	12.40	6.20	0.244	302.6	11.91	258.9	4.0
315.0	49	12.40	6.30	0.248	302.4	11.91	258.6	5.0
315.0	49	12.40	8.10	0.319	298.8	11.76	252.4	6.3
315.0	8061/2	12.40	9.20	0.362	296.6	11.68	248.7	6.0
315.0	49	12.40	9.60	0.378	295.8	11.65	247.4	7.5
315.0	49	12.40	12.5	0.492	290.0	11.42	237.8	10.0

od mm	STD	od in	WALL mm	WALL in	od mm	id in	1m/s m³/h	WP barg
315.0	8061/2	12.40	15.0	0.591	285.0	11.22	229.7	10.0
315.0	49	12.40	15.5	0.610	284.0	11.18	228.0	12.5
315.0	49	12.40	19.3	0.760	276.4	10.88	216.0	16.0
315.0	8061/2	12.40	23.4	0.921	268.2	10.56	203.4	16.0
400.0	49	15.75	8.20	0.323	383.6	15.10	416.1	5.0
400.0	49	15.75	10.2	0.402	379.6	14.94	407.4	6.3
400.0	8061/2	15.75	11.7	0.461	376.6	14.83	401.0	6.0
400.0	49	15.75	12.1	0.476	375.8	14.80	399.3	7.5
400.0	49	15.75	15.8	0.622	368.4	14.50	383.7	10.0
400.0	8061/2	15.75	19.1	0.752	361.8	14.24	370.1	10.0
400.0	49	15.75	20.3	0.799	359.4	14.15	365.2	12.5
400.0	49	15.75	24.6	0.969	350.8	13.81	347.9	16.0
400.0	8061/2	15.75	29.7	1.169	340.6	13.41	328.0	16.0
500.0	49	19.69	10.3	0.406	479.4	18.87	649.8	5.0
500.0	49	19.69	12.8	0.504	474.4	18.68	636.3	6.3
500.0	8061/2	19.69	14.6	0.575	470.8	18.54	626.7	6.0
500.0	49	19.69	15.2	0.598	469.6	18.49	623.5	7.5
500.0	49	19.69	20.1	0.791	459.8	18.10	597.8	10.0
500.0	8061/2	19.69	23.9	0.941	452.2	17.80	578.2	10.0
500.0	49	19.69	25.2	0.992	449.6	17.70	571.5	12.5
630.0	49	24.80	16.3	0.642	597.4	23.52	1009.1	6.3
630.0	8061/2	24.80	18.4	0.724	593.2	23.35	994.9	6.0
630.0	49	24.80	20.0	0.787	590.0	23.23	984.2	7.5
630.0	49	24.80	25.4	1.000	579.2	22.80	948.5	10.0

Table 9.25 Metric UPVC tube to DIN 8061/2 and KIWA 49 (Above)

9.2.8 Non-metallic double wall systems

Double wall piping systems are available for those corrosive and hazardous applications where fluid leakage of any magnitude is unacceptable. The primary containment pipe is completely surrounded by a secondary containment which can direct leakage to a suitable collection point. Straight lengths and fittings can be fusion welded. Breakable connections are via special flanges which allow the primary and secondary to flow. This style of piping has very limited use and consequently the manufacturing range is greatly restricted. Material choice is limited to polypropylene or PVDF. Mass-produced sizes include DN25, DN50 and DN100. The pressure rating of all sizes is about 5 barg.

9.3 Flexible piping systems

Flexible piping systems are manufactured in many materials including metal, to provide a wide range of pressure ratings, operating temperatures and chemical compatibility. Flexible pipes can be used in conjunction with rigid pipes and tubes to fulfil particular functions or used separately for maximum manoeuvrability. Flexible pipes can be used to connect machinery or vessels to rigid piping. The use of a flexible pipe removes the possibility of the rigid piping imposing forces and moments due to thermal expansion. Flexible piping can also act as a vibration isolator.

Flexible piping does have bending limitations. Hose specifications should include a minimum bend radius. If hose is bent to smaller radii then flow restrictions should be expected as the hose flattens. Some hoses are designed specifically to resist collapse due to internal vacuum. Minimum pressure ratings should be checked if sub-atmospheric operation is expected. Multi-core hoses are produced to a limited extent, with 6 to 10 mm bore.

Some flexible pipes can be equipped with electric trace heating. This facility is useful if liquid products may tend to solidify, separate or deposit wax on contact with cooler surfaces. In some installations it is not necessary for the pipe itself to be flexible. The required degree of flexibility can be achieved by using rigid pipe with spherically mounted couplings. This approach can be useful for long range outdoor applications such as in quarrying or mining. High pressures can be combined with reasonable bore sizes to allow high flow rates. Piping of this type can be used for large scale hydraulic power or high pressure water descaling ring mains. Care must be exercised with the location and installation of this style of piping as it does try to "straighten out" when pressurised.

9.3.1 Plain hoses

Hoses can be manufactured from a combination of compounds. The lining in contact with the fluid can be selected for good chemical compatibility. The lining can be surrounded by a strong compound to provide the tensile strength to allow higher internal pressures. The hose can be encased in a cover of another compound to provide good protection from ambient conditions or rough handling. Plain hoses are made in many compounds and thicknesses, see Table 9.26. Hoses are available to international Standards, regarding direct contact with foodstuffs, see Table 9.27.

Bore mm	Material	Temperature range °C	Working pressure barg	Approx weight kg/m
1.47 to 20.65	Nylon	-40 to +80	50 to 11	0.01 to 0.19
3 to 50	Clear PVC	-10 to +50	2	Light duty
13 to 25	PVC	-10 to +30	3.5	
38 to 102 (1)	PVC	-10 to +30	atmospheric	0.2 to 0.6
6, 12, 25	PTFE (2)	-53 to 232	207, 138, 69	

(1) Lay flat for storage, (2) With stainless steel braid cover

Table 9.26 Typical plain hoses

Bore mm	Material	Temperature range °C	Working pressure barg	Approx weight kg/m
3 to 19	PVC	-20 to +65	5(+10° to +50°C)	0.03 to 0.14
5 to 44	PVC	-20 to +65	5(+10° to +50°C)	0.11 to 0.9
32 and 38	PVC	-20 to +65	5(+10° to +50°C)	0.78 and 0.9
50	PVC	-20 to +65	5(+10° to +50°C)	1.01
2.5 to 7.5	Silicone rubber	-100 to +260	1 mm wall	
3.0 to 10	Silicone rubber	-100 to +260	2 mm wall	
3.2 to 25.4	Silicone rubber	-100 to +260	3.2 mm wall	
3 to 50	PVC	+10 to +50	5 to 2	.027 to 1.04

Table 9.27 Plain hoses for hygienic applications

Special perforated hose is available for horticultural and agricultural applications. When buried in the ground the hose may irrigate over an area between 15 mm and 450 mm either side. The low pressure rubber hose, 0.5 barg maximum, has a 12 mm bore and is available in standard lengths of 7.5, 15, 30, 50 and 100 m. Other applications include golf courses and ornamental flower beds and public parks.

Hoses can be preformed into spring coils to allow greater axial movement. Nylon has proved a good material for air hoses. Bore sizes of 9 and 12 mm with 2 mm wall thickness are popular for truck applications. Table 9.28 shows the range for general purpose applications.

Bore mm	Material	Temperature range °C	Working pressure barg	Approx weight kg/m
4.3 to 12.5	Nylon	-40 to +85	17	3.8, 7.6, 15.2
5 to 11	Nylon	-40 to +85	10	2, 4, 6, 8

Table 9.28 Typical plain spring coil type hoses

Plain hoses can be covered by a protective braiding. Non-metallic coverings, such as polyester, are available as well as stainless steel. Plain PTFE hose overbraided with 304 stainless steel with bores from 3.5 to 50.8 mm can operate at pressures from 290 barg to 40 barg. Corrugated PTFE hose, with external steel reinforcing and stainless steel overbraid, is produced in a range of bore sizes, 15 to 150 mm, covering pressure ratings from 5 to 41 barg.

These hoses can be fitted with many types of couplings, including proprietary hygienic fittings, to cope with applications from -70° to +260°C. Hygienic hoses can have the overbraid itself coated, with EPDM for example, for easier external cleaning.

9.3.2 Reinforced hoses

Hoses, as described in Section 9.3.1, can be strengthened by embedding fibres or metal strands within the wall during manufacture. The reinforcing can be applied between the various cores and bonded during curing. Reinforced hose can be general purpose or constructed for specific applications such as oxygen and acetylene hoses for cutting/welding torches. Tables 9.29 to 9.37 list the variety of sizes and materials of mass-produced hose.

The linings for air hoses are oil-mist resistant to allow proper lubrication of air tools. Depending upon the bore size, maximum standard lengths vary from 20 m to 100 m.

Bore mm	Lining	Cover	Reinforcing	Working pressure barg	Approx weight kg/m
5 to 25 (1)	SBR	SBR	Synthetic yarn	12	0.14 to 0.5
13 to 152 (2)(3)	NBR	CR	Synthetic yarn	14 to 5	0.61 to 6.38
13 to 76	SBR	SBR	Textile yarn	14	0.4 to 3.0
6 to 10	SBR	NR	Synthetic yarn	16	0.38 to 0.54
6 to 10	SBR	SBR	Synthetic yarn	20	0.4 to 0.57
13 to 63 (4)	SBR	SBR	Synthetic yarn	20	0.46 to 2.55
6.3 to 50 (5)	PVC	PVC	Synthetic yarn	20	0.08 to 0.5
13 to 102 (6)	SBR	EPDM	Textile yarn	28 to 10	0.62 to 4.1
13 to 38	SBR	EPDM	High tensile wire	40	0.63 to 2.1
51 to 127	NBR	EPDM	High tensile wire	40 to 20	2.35 to 7.83
10 and 13 (7)	EPDM	EPDM	Polyester yarn	15	0.3 and 4.15
102 and 127 (8)	NR	CR	Canvas + galvanised wire	2	3.33 and 4.15
50 to 152	Silicone rubber	Silicone rubber	Nomex™ or glass fibre fabric	10 to 2	-50° to +220°/250°C
19 to 380 (9)	Rubber	Rubber	Cotton fabric and galvanised steel wire	0.3	0.45 to 12.9

1) To BS 5118 Type 1 (2) Anti-static (3) Fire retardant (4) To BS 5118 Type 2 (5) -15° to +60°C (6) To BS 5118 Type 3 (7) To BS AU110/5
(8) Air temperature -20° to 100°C continuous
(9) Air with solids or dust, -20° to +100°C

Table 9.29 Reinforced hoses for air at ambient temperature

Bore mm	Lining	Cover	Reinforcing	Working pressure barg	Approx weight kg/m
13 to 25	SBR	NR	Cotton yarn	7	0.3 to 0.67
13 to 25	SBR	SBR	Polyester yarn	10	0.25 to 0.82
13 to 135	SBR	SBR	Textile yarn	10	0.39 to 5.83
25 to 152	SBR	SBR	Textile yarn	5.5 to 3.5	0.36 to 2.45
25 to 152 (1)	PVC	PVC	Synthetic yarn	10 to 4	0.19 to 1.6
13 to 152	SBR	SBR	Textile yarn	17 to 7	0.37 to 6.27
38 to 152 (2)	PVC	PVC	Polyester yarn	16 to 7	0.45 to 2.5
25 to 305 (3)	SBR	SBR	Textile yarn and high tnesile wire	9 to 4	1.07 to 25.4
19 to 152 (4)	SBR/NBR	CR	Polyester yarn and high tensile steel	5	0.61 to 9.46
13 to 19	EPDM	EPDM	Synthetic yarn	6 up to 125°C	0.23 to 0.34
6.5 to 102	Silicone rubber	Silicone rubber	Polyester yarn	8.5 to 2	-50° to 250°C

(1) Temperature range -5° to +60°C, (2) Temperature range -15° to +45°C (3) Non-collapsible for suction applications, (4) Lloyds and DNV approved marine hose

Table 9.30 Reinforced hose for water

Bore mm	Lining	Cover	Reinforcing	Working pressure barg	Approx weight kg/m
19 and 25 (1)	SBR	SBR	Textile yarn	15	0.48 and 0.7
19 and 25 (2)	SBR/NR blend	SBR	Textile yarn	40	0.62 and 0.75
38 to 89 (3)	PVC/NBR blend	PVC/NBR blend	Textile yarn	24.1	0.28 to 0.83
76 to 140 (4)	SBR	SBR	Textile yarn and high tensile steel wire	7.6	3.28 to 7.67

(1) Complies with BS 3169 Class B Type 1 (2) Complies with BS 3169 Class A Type 2 (3) Complies with BS 3169 Type 3

Table 9.31 Reinforced hose for fire hose applications

Bore mm	Lining	Cover	Reinforcing	Working pressure barg	Approx weight kg/m
6 to 15	Polyester	PVC	Poyester yarn	20 (1)	0.14 to 0.38
13 to 25	PVC	PVC	Synthetic yarn	20 (2)	0.18 to 0.45
6.3 to 25	PVC	PVC	Polyester yarn	40 (2)	0.1 to 0.58
8 to 19	PVC	PVC	Polyester yarn	80 (2)	0.18 to 0.5
13 to 50	NBR	NBR	Polyamide	18 up to +95°C 6 up to +164°C	0.35 to 2.35
13 to 76	EPDM	EPDM	Textile yarn	7 up to 170°C	0.4 to 2.76
10 to 19	Polyester	Poly-urethane	Textile yarn	250 to 95 (3)	0.22 to 0.37
13 to 50	SBR	Glass fibre fabric (4)	Synthetic yarn	12 to 8	0.5 to 2.2
25 to 254	NBR/NR blend	SBR	Textile yarn with copper conductor	17 to 3	0.45 to 4.6
51 to 305	SBR/NR blend	EPDM	Textile yarn with high tensile steel wire	10	2.07 to 27.4
25 to 152	SBR	NR	Textile yarn	40	0.9 to 10.4
32 to 140	SBR/NR blend	NR	Brass coated high tensile steel	83	2.2 to 11.8

(1) Temperature range -10° to +80°C, anti-static
(2) Temperature range -15° to +60°C
(3) Temperature range -40° to +80°C
(4) Cover can withstand external temperatures of 300°C
Table 9.32 General purpose reinforced hose

Bore mm	Lining	Cover	Reinforcing	Working pressure barg	Approx weight kg/m
10 to 152 (1)	PVC	PVC	Carbon steel	12 to 3	0.18 to 6.75
32 to 203 (2)	PVC	PVC	Copper	9 to 3	0.7 to 10.3
38 to 152 (3)	Poly-urethane	PVC	PVC	9 to 3.3	0.8 to 6.3

(1) Temperature range -15° to +65°C, hygienic, anti-static
(2) Temperature range -15° to +65°C, anti-static
(3) Temperature range -25° to +60°C, hygienic approved
Table 9.33 Steel helix non-collapsing hose

Bore mm	Lining	Cover	Reinforcing	Working pressure barg	Approx weight kg/m
10 to 25	SBR	SBR	Textile yarn	200	0.45 to 0.95
10	NBR	SBR	Wire braid	400 up to 150°C	0.61
12	SBR	SBR	High tensile steel braid x 6	1 000 -40° to +80°C	1.2
6 to 19	Poly-amide	Poly-ester	Brass plated high tensile steel	310 to 135 -40°C to +120°C	0.16 to 0.43
13 to 51 (2) (3)	EPDM	EPDM	Nylon yarn	7 up to 170°C	0.63 to 1.76
13 to 102	EPDM	EPDM	Brass coated steel wire	17 up to 235°C	0.62 to 7.21

(1) Temperature range -35° to +70°C
(2) Available with a white cover for hygienic installations
(3) Available in approved EPDM compounds for hygienic use
Table 9.34 High pressure water/steam reinforced hose

Table 9.35 Reinforced hose for fuel and oil applications (Below)

Bore mm	Lining	Cover	Reinforcing	Working pressure barg	Approx weight kg/m
32 to 102 (1)	NBR	CR	Tyre yarn and high tensile steel wire	6	1.37 to 5.2
19 to 152	NBR	CR	Tyre yarn and high tensile steel wire	17 to 4	0.62 to 9.0
19 to 152	NBR	CR	Tyre yarn and high tensile steel wire	10	0.64 to 8.0

Bore mm	Lining	Cover	Reinforcing	Working pressure barg	Approx weight kg/m
25 to 100 (2)	Polymer yarn coated with polymer film	PVC	Inner and outer galvanised steel wire	10.3	0.9 to 5.1
75 to 200 (2) 200 (2)	Polymer yarn coated with polymer film	PVC	Inner and outer galvanised steel wire	13.8 10.3	3.75 to 15 21
16 to 25 (3)	NBR	PVC	Textile yarn + 2 electric conductors	10	0.54 to 0.87

(1) Anti-static
(2) Complies with BS 5842, -25° to +82°C
(3) Petrol pump hose to BS 3395 Type 1
Table 9.35 Reinforced hose for fuel and oil applications

Bore mm	Material	Temperature range °C	Reinforcing	Working pressure barg	Approx weight kg/m
3 to 50	Clear PVC	-20 to +65	Polyester yarn	32 to 2.9	0.06 to 1.61
6.3 to 50	Clear PVC	-15 to +60	Polyester yarn	0.15 to 6.0	0.09 to 1.48
16 to 25.4	Translucent silicone rubber	-100 to +175	Polyester yarn	7.2 to 3.8	
2.0 to 12.5	Translucent silicone rubber	-100 to +175	Polyester yarn	12 to 8.6	
13 to 102	Silicone rubber	-60 to +180	Polyester yarn and high tensile steel wire	10 to 3	0.3 to 2.75
13 to 152 (1)	SNR/NR blend lining EPDM cover		Textile yarn	20 to 7	0.57 to 5.84
13 to 100	NR lining SBR cover		Textile yarn	16 to 10	0.3 to 7.3
51 and 63	EPDM lining EPDM cover		Textile yarn, nylon and electric conductor	17	1.04 and 1.26

(1) Carbon free lining
Table 9.36 Reinforced hygienic quality hoses

Hoses are specifically manufactured for hydraulic applications. The elastomer materials are specially compounded to resist the effects of modern hydraulic oils over a temperature range of -40° to +125°C. End connections are usually swaged on and must be fitted using the correct equipment. The local supplier will normally have facilities to make up finished hoses to a particular length. Table 9.37 indicates typical hydraulic hose characteristics.

Bore mm	Lining	Cover	Reinforcing	Working pressure barg	Approx weight kg/m
4.8 to 50.8 (1)	Synthetic rubber	Synthetic rubber	High tensile steel wire	207 to 26	0.18 to 1.83
4.8 to 50.8 (2)	Synthetic rubber	Synthetic rubber	High tensile steel wire	345 to 78	0.35 to 2.75
6.4 to 25.4 (3)	Synthetic rubber	Synthetic rubber	High tensile steel wire	450 to 290	0.63 to 2.1
2 to 6	Nylon	Nylon	ARAMID fibre	500	0.02 to 0.065

(1) Complies with SAE 100R1AT
(2) Complies with SAE 100R2AT
(3) Complies with DIN 20023-4SP/EN 856
Table 9.37 Reinforced hose for hydraulic applications

In the same way that hydraulic hose is manufactured specifically for vehicle and mobile power applications, refrigeration quality hose follows the same pattern. Hoses with synthetic rubber linings, Rayon™ reinforcing and weatherproof synthetic rubber covers are mass-produced for fluorinated hydrocarbon; R12, R22, R502, etc.; in sizes from 6 mm to 16 mm. Refrigerant quality hoses are suitable for a maximum working pressure of 25 barg at temperatures between -30° and +125°C.

9.3.3 Ducting

Hoses of large diameter can be manufactured and the wall thickness can be made relatively thin. This type of hose is suitable for low pressure applications. Table 9.38 indicates the range of sizes and materials available.

Bore mm	Wall material	Reinforcing	Working pressure barg	Approx weight kg/m
102 and 127 (1)	CR coated canvas	Galvanised wire	2	3.33 and 4.15
19 to 380 (2)	Rubber coated cotton fabric	Galvanised steel wire	0.3	0.45 to 12.9
51 to 610 (3) (4)	Vinyl coated glass fibre	Metal wire	0.5 to 0.03	0.23 to 1.55
77 to 610 (4) (5)	Aluminised glass fibre	Metal wire	0.43 to 0.02	0.29 to 1.56
12 to 315 (6)	Neoprene coated glass fibre	Glavanised spring steel wire	1.5 to 0.2	0.06 to 2.5
12 to 203 (7) (8)	Silicone coated glass fibre	Spring steel wire	1.5 to 0.5	0.08 to 1.75
25 to 508 (9)	Polyurethane	Spring steel wire	0.55 to 0.06	0.13 to 2.56
125 to 760 (10)	Neoprene coated glass fibre	Glavanised spring steel wire	0.55 to 0.04	0.58 to 4.8

1) Air temperature -20° to 100°C continuous
(2) Air with solids or dust, -20° to +100°C
(3) Air temperature -5° to +90°C
(4) Complies with BS 476 Parts 6, 7 & 20
(5) Air temperature -40° to +200°C
(6) Air temperature -50° to +150°C
(7) Air temperature -60° to +300°C
(8) CAA approved
(9) Hygienic quality -40° to +100°C
(10) Air temperature -40° to +150°C
Table 9.38 Typical ducting sizes and materials

Ducting can be supplied covered with insulation. This is useful in heating and cooling applications to prevent heat transfer to the surroundings, reducing the process efficiency

9.3.4 Metallic hose

Metal hose can be manufactured in a similar manner to bellows or to corrugated PTFE hose. Metal corrugated hose can be supplied bare or with a cover or metal overbraid. Hoses in two grades of stainless steel are mass-produced:

- 18Cr 10.5Ni Ti stabilised (321),
- 17.5Cr 12.5Ni 2.5Mo 0.03C (316L).

A small range of hoses is made in AISI 321 stainless steel. Popular sizes range from 6 mm bore to 300 mm. Small bore hoses can operate at pressures over 400 barg; the largest sizes are restricted to 0.7 barg. All metal hose assemblies can operate at temperatures from -200° to +600°C. Standard lengths range from 3 m to 6 m. Corrugated metal hose can be supplied with a protective stainless steel over braid; this feature tends to reduce flexibility.

Bore mm	Tube material	Gasket material	Temperature range °C	Working pressure barg	Approx weight kg/m
20 to 400	Galvanised steel	Asbestos	-20 to +230	28 to 4	0.86 to 60
20 to 400	304 SS	Asbestos	-20 to +230	42 to 7	0.75 to 60
16 to 300	304 SS	Soft copper	Suitable for engine exhausts	Atmospheric	
100	304 SS	Rubber	Ambient	Atmospheric	Used for granulated sugar

Table 9.39 Interlocking folded metal hose

Metal hose can also be formed by helically winding interlocking metal strips. The metal strips can seal solely by the action of the folded profile, or a gasket material can be incorporated during the forming process. This type of construction precludes the manufacture of very small hose, 16 mm bore being the approximate lower limit. Hose length can be up to 6 m. Table 9.39 indicates the standard variants available. Hoses in bronze can be made to special order. Flexible metal hose is manufactured specifically for oil and gas production applications. This hose, like most other equipment concerned with oil production, is heavy duty, suitable for high pressures, over 400 barg, and expensive. Materials can be suitable for the fluids generally encountered in oilfield services:

- sweet crude oil
- sour crude oil
- liquified natural gas
- carbon dioxide
- water and brine
- cement-acid
- mud
- well-kill liquids

This type of hose is manufactured specifically for applications and not mass-produced. The hose manufacturer should be consulted on a case-by-case basis with definitive operating requirements.

9.3.5 End connections

Hose lengths can be cut from a drum or coil and fitted over the ends of rigid pipework. In these situations the hose can be clamped in placed by a threaded band clamp; the most notable of these being the Jubilee Clip™ whose name is usually used as the product description. This style of assembly is acceptable for permanent installation. If a hose is to be connected and disconnected routinely then other types of connector may be more suitable.

Hoses can be supplied with a wide range of end connectors which can be fitted by the hose supplier. The connectors may be permanently attached to the hose, by crimping or adhesive, or secured by a threaded band clamp.

Hoses for higher pressures, reinforced hose for hydraulic applications for example, will use end fittings which have a tube insert for the hose bore and an external sleeve slightly larger in diameter than the hose od. Once in place, the connector sleeve is crimped to clamp the hose between the bore tube and the sleeve. Crimping is performed hydraulically with appropriate dies to ensure the hose is clamped adequately but not crushed.

Lower operating pressure hose may be supplied with clamped connectors secured by a threaded band clamp. These connectors are fitted with barbed inserts for the hose bore rather than smooth tube. A sealant or lubricant may be used to assist assembly and improve final integrity.

Non-metallic connectors are becoming more popular as major equipment and rigid piping systems in non-metallic materials increase in use. Favourite materials include:

- Acetal™ (*)
- CTFE
- Delrin™ (*)
- ETFE
- PEEK
- PFA
- polypropylene (*)
- PTFE
- Tefzel™

(*) used for quick-disconnect connectors, without automatic valves.

Many styles of connectors are available to suit all process and assembly requirements as indicated below:

- solid male threaded ends — simple, reliable, but only suitable for short hoses as the hose must rotate during assembly/disassembly
- solid female threaded ends — as above
- female union nut ends — simple, reliable, hose not required to rotate
- ¼ turn flat face unions — simple, sealing faces must be protected when not in use
- over-centre toggle unions — simple, only suitable for low pressures
- spring-loaded detent couplings — simple, limited sizes, 2½" instantaneous coupling
- flanged ends — any standard flange specification
- clamped connectors with screwed or toggled clamps
- compression fittings
- hammer-lug fittings
- hygienic couplings
- quick release couplings — large range of various designs, need to verify number and rating of isolating valves

9.4 Expansion joints

The majority of pipework systems in use are rigid. Solid metal or non-metallic pipes/tubes connect various pieces of equipment, including valves, to form a complete system. Few fluid systems operate at a single, constant temperature and, depending upon the pipe lengths involved, thermal expansion can create problems. Unless the pipe/tube is in one straight length and rigidly anchored at each end, the problems are not experienced by the pipe. When the pipework expands or contracts, forces and moments and deflections are generated at the anchors or terminations. Well-designed pipework will have anchors to absorb excessive forces/moments/deflections and equipment will only experience modest effects.

Pipework design philosophy often appears to have a prime objective of lowest cost. This requires elimination of costly anchors and results in compressors, pumps, valves and vessels, etc. having to sustain very high loads. In the case of compressors, pumps and vessels the high loads can, in some cases, be transmitted directly through the feet to the foundations. The equipment becomes an anchor. Most valves however do not have feet and any imposed loads must be absorbed by the body without undue distortion and stressing. Failure to assess imposed loads can result in early failures or poor valve performance.

Expansion joints can be used to accommodate the pipework expansion/contraction while allowing the loads on equipment connections to remain low. The expansion joint is a very flexible section of pipe which can absorb axial/angular deflection without creating high loads. Expansion joints can also be used to attenuate the transmission of vibration between equipment and pipework or adjacent pipework sections.

An expansion joint can be manufactured in many materials to suit the operating conditions, both metallic and non-metallic. The od of the joint is normally exposed to ambient conditions and some heat transfer through the joint is probable. Any heat transfer which takes place will complicate the assessment of the operating temperature. With hot fluids, the joint may be appreciably cooler than the fluid; with cold fluids, it may be considerably warmer. The actual operating temperature is very important for correct material selection. Tables 13.21 and 13.22 in Chapter 13 list metallic and non-metallic materials which may be suitable.

A special type of expansion joint can be used to seal bulkheads with pipework passing through. One end of the joint is attached to the od of the pipework, the other end to the bulkhead wall. The joint forms a gas-tight seal. In this type of installation a single bellows can accommodate linear, lateral and angular displacement.

A thorough knowledge of the predicted pipe movement is required in order to specify an expansion joint correctly. In general, expansion joints are not capable of accommodating rotation about their longitudinal axis. The bellows will be the weakest link in the piping system. When handling hazardous fluids a bellows failure can result in extremely high costs. The following Sections describe and outline the capabilities of various bellows types.

9.4.1 Axial bellows

The axial bellows is the simplest type of expansion joint. A bellows is fitted between two guided pipe sections. The bellows allows each pipe section to move axially relative to the other. The allowable deflection is quoted as $\pm$ x about the free length. When the bellows is to accommodate both expansion and contraction the free length must be set correctly for installation. An initial deflection in either direction could significantly reduce the operational life. If the bellows is only required to accommodate movement in one direction, then provision must be made to set it at the other extreme. The success of this type of bellows is dependent upon the alignment and guidance of the two pipe sections.

9.4.2 Double axial bellows

This type comprises two bellows units connected by a short straight pipe. It is very similar in most respects to the axial bellows described in Section 9.4.1 but the double version is capable of accommodating some lateral misalignment. The two guided pipe sections should be parallel but need not be inline. Again, initial setting is very important.

9.4.3 Self-guided axial bellows

This is an axial bellows unit which incorporates an internal sleeve or an external sleeve or both to maintain the alignment. The two adjoining pipe sections do not require guidance, only support. The direction of fluid flow may be important and if so should be indicated by a permanently attached arrow.

9.4.4 Tied bellows

Single and double axial bellows can suffer from problems of over-extension and over-compression. Due to the natural expansion of a bellows when internal pressure is applied, the axial bellows may over-extend. If excessive axial loads are applied by the pipework, the axial bellows is liable to compress beyond the design limit and create extreme stresses. Both of these damaging operating conditions can be avoided by "tieing" the bellows. Two tie rods, with spherical bearings, can be fitted with limit stops to prevent both over-travels. Using two tie rods can allow lateral movement; using more tie rods can eliminate lateral movement.

9.4.5 Single hinged bellows

A single hinged bellows unit is designed to allow relative angular movement in one plane without any variation in axial position about the hinge centre-line. A single bellows unit is centred about the hinge pin and provides the angular flexibility. Movement in all other axes are prevented.

9.4.6 Double hinged bellows

This is a version of the double axial bellows unit in Section 9.4.2, fitted with two hinges to allow angular and lateral movement in one plane. The hinge pins are located at the centre of each bellows unit. The hinges are connected by a rigid link. Movement is totally restricted to the plane perpendicular to the hinge pins. The axial length of the bellows unit will shorten if lateral movement takes place.

9.4.7 Gimbal bellows

The gimbal bellows unit combines a single bellows and two hinges. The hinge pins are attached to a central "gimbal" ring and set at 90° to each other. The hinge assembly prevents angular rotation about the longitudinal axis and any lateral parallel deflection. Setting the two hinges at 90° does permit angular deflection in any direction. The axial length of the bellows unit is maintained by the hinge assembly.

9.4.8 Double gimbal bellows

Two gimbal bellows units are connected by a short straight pipe length. The double gimbal bellows unit permits lateral deflection in any plane and angular deflection in all planes but not rotation about the longitudinal axis. The axial length of the unit is fixed and additional facilities must be adopted to accommodate this movement.

9.5 Threaded connections

Threaded connections are very popular for small diameter pipework. In discussions involving pipework the term "small" is relative and varies from industry to industry. It is generally agreed that threaded connections up to 1½" are perfectly acceptable. Threaded connections up to 3" may be acceptable for fluids which are not hazardous. At 3" connections are becoming unwieldy and it may be difficult to apply sufficient torque, without damaging components, to tighten joints. Most equipment users will not accept threaded connections larger than 3".

It must be remembered, when using threaded connections, that the pipe or fittings must rotate. This can be a problem when trying to align flanges or bends. The correct orientation of parts/pipes may compromise the joint tightness and sealing. Also, sufficient joints must be included to enable the complete assembly to be assembled and disassembled.

Cutting threads in pipes reduces the effective wall thickness. The effect of the reduction in wall thickness on the working stresses should be evaluated. The thread itself can be considered as a "stress raiser" in situations where fatigue may occur.

Threaded connections are sometimes described as "gas" or "GAZ" connections. This terminology should not be used as it is ambiguous. Some industries will interpret ¼" gas as ¼" BSPT and others as ¼" NPT.

9.5.1 Parallel threads

Parallel threads are used for those applications where sealing is performed independently. The parallel thread forms the mechanical link and provides a location for a seal or the compressive sealing force. Two parallel threads are in popular use:

- a Whitworth form 55° thread
- an American 60° thread

British Standard Pipe {thread} (Parallel), BSP(P), see definition in Chapter 1, is popular for process applications. An inch thread for nominal bore pipework, it is currently specified internationally by BS EN ISO 288 and DIN 3852 Pt 2. All Standards quote dimensions in millimetres.

The American thread is used principally in hydraulic fluid power. Although generally known as an SAE thread it is, in reality, a UNF (Unified National Fine {thread of America}) fastener thread, again see Chapter 1 for a definition. This thread is standardised in American, German and international Standards; SAE J475, DIN 2353 and ISO 725 respectively.

Parallel threads can be sealed by two different methods depending upon the machining of the female adaptor. The most common method is by a flat washer compressed against a perpendicular face. Washers can be fibre, metal or combined. A thick metal washer with an elastomer inner seal is very popular. The alternative is to use an "O" ring in a plain bore just larger than the od of the thread.

9.5.2 Tapered threads

The tapered pipe thread is used predominantly in process applications. The process seal is formed by the locking action of the tapered thread form. These threads can be used without sealants, in the "dry" condition, with some fluids. The user may specify a specific sealing product. When the joint is made permanently it can be seal welded to prevent leakage. Many users do not approve of this construction and specifically exclude seal welding.

Tapered pipe threads are standardised in ANSI B1.20.1, BS 21, DIN 2999 and DIN 3858, and ISO 7-1.

9.5.3 Thread sealants

As mentioned in Section 9.5.2, tapered threads can be sealed dry but the usual practice is to apply a sealant. One of the most popular options is to use PTFE tape. The tape, about 12 to 20 mm wide, is supplied in rolls. The tape is applied to the male thread which is then screwed in and tightened. The PTFE deforms within the thread space to make the seal. PTFE can be used for temperatures between -157° and 260°C.

Liquid and paste sealants can be divided into two main groups:

- setting
- non-setting

Setting compounds cure and become solid and act as an adhesive as well as a sealant. This type of sealant can be very useful in situations where vibration levels are high and screwed connections tend to loosen. Hard setting sealants resist disassembly.

Non-setting compounds tend to thicken, that is to say the viscosity increases dramatically but the compound does not solidify. Resistance to vibration is increased but disassembly is not noticeably impaired. Table 9.40 indicates typical properties of liquid and paste sealants.

Type	Cure time	Operating temperature °C	Acceptable process fluids
Hard setting		0 to 150	Petroleum products, water, steam, dilute mineral acids
Non setting		-40 to 150	Hydrocarbons, natural gas
Silicone based non setting	15 mins	-54 to 204	Mineral oil (HAP)*, phosphate esters, silicate esters, non-aromatic petroleum, dilute alkalis, water
Urethane based hard setting	6 hours	-29 to 93	Ammonia, chlorine gas, hydrogen sulphide, aliphatic hydrocarbons, ethers, some dilute acids

*(HAP) High Analine Point

Table 9.40 Properties of thread sealants

9.5.4 Threaded fittings

Complex pipe arrangements can be constructed using threaded pipe fittings. Most fittings have taper female threads to accept pipes. Fittings are available in a wide range of configurations including straight, tee, cross and elbow. Screwed elbows are very short radius bends. High fluid velocities should be avoided to reduce flow turbulence effects.

Threaded fittings are manufactured in standard pressure ratings. Low pressure fittings, 55 barg, may be made of cast iron; high pressure fittings, up to 750 barg, can be forged steel. Malleable iron is a popular material. Some system builders try to limit their inventory size by standardising on a pressure rating which will fulfil all their process requirements. Low pressure systems are built with higher pressure rated fittings when not actually required. This practice can be dangerous if low pressure piping is used with high pressure fittings.

Most pipework does not carry pressure rating information. Most fittings are stamped with a pressure rating and sometimes also a hydrotest pressure. Low pressure piping can be dangerously over stressed if hydrotested to the high pressure fitting value. From a purely safety aspect, piping systems should be built with consistent pressure ratings. High pressure fittings require much more space than low pressure fittings. Modern equipment is required to be compact and light. Using over-rated fittings works against the basic concept of saving space and material.

Some grades of non-metallic pipe, CPVC and UPVC, are suitable for screwed connections. Threaded fittings are produced as an alternative to cemented connections and this may be an attractive option in some situations.

9.5.5 Threaded unions

Threaded unions are used in threaded pipe systems to enable easy disassembly without the problems associated with cleaning and remaking a threaded connection. The pipework is screwed into both halves of the union which are themselves held together by a union nut. Low pressure fittings, up to 55 barg, are usually of malleable iron. These fittings generally have a cone seating for alignment and sealing. Higher pressure fittings in steel may have a parallel spigot for alignment and either a fibre washer or "O" ring for sealing.

Steel threaded unions, for pipes up to 1½" nb, using an "O" ring face seal can operate at pressures up to 980 barg. These fittings require no assembly clearance. Once the union nut is removed the fitting can be separated radially without any axial movement. Viton™ is used for the standard "O" ring material.

Threaded unions, $\frac{1}{16}$" to 1", are manufactured specifically for high vacuum applications. In 316 stainless steel these unions use a parallel spigot for alignment and to locate the flat ring gasket. Gaskets can be of various materials to suit the particular application:

- silver plated stainless steel
- silver plated nickel
- copper
- aluminium
- TFE

Typical leakage rates are quoted as 4E-9 standard cc/s, 0.000 000 004 cc/s.

9.5.6 Hammer lug unions

Hammer lug unions are a very heavy duty version of the screwed union. The union nut is not equipped with hexagon flats but lugs to allow "flogging". Hammer lug unions are produced in many varieties for various applications. Most are associated with oilfield operations but not all are high pressure. Pressure ratings can be as low as 35 barg and up to 1 034 barg.

The nipples can be attached to the pipe by various methods:

- threaded
- socket weld
- butt weld

Low pressure unions are manufactured in sizes from 1" to 10". The highest pressure unions are made specifically to match API thick wall pipe in sizes from 1" to 4".

Fittings are manufactured to complement the hammer lug unions. Closed die forgings or forged solid blanks are machined to produce elbows, tees and crosses. Swivel joints are also available in several combinations, to allow pipework to be aligned and oriented with fixed connections.

9.6 Welded connections

When pipework is assembled which will not require disassembly, welding should be considered as a suitable connection method. Welded fittings are mass-produced in popular materials to allow complete systems to be built. The low carbon grades of austenitic stainless steels, 304L and 316L, should be considered in situations when post weld heat treatment is not viable.

Socket weld fittings, see Figure 9.9 in Section 9.9.3, are very popular for pipe systems up to DN80. Socket weld fittings should not be used with extremely hazardous fluids or fluids which have a tendency for crevice corrosion. The external weld is difficult to radiograph and the integrity may be questionable.

Butt weld construction, see also Figure 9.9, offers the highest, verifiable integrity. Butt weld connections are easy to radiograph and repair if defects are found. Butt welding is time consuming and setting up can be awkward. However, it should be used when the possibility of fluid leakage must be eliminated. Butt welding can be performed on pipework of any diameter and any wall thickness. When carried out by skilled craftsmen in accordance with properly considered procedures the butt weld connection has the same pressure rating as the parent pipework.

Both socket and butt welding can be accomplished by automatic welding machines which greatly improves the basic quality level.

9.7 Soldered connections

Also known as capillary connections, soldered connections are used extensively in domestic and commercial water systems. Soldered connections are also suitable for low pressure steam, gas, air and some refrigeration systems. When a permanent joint is required a soldered connection may be ideal. Tables 9.41 and 9.42 indicate the pressure and temperature capabilities of standard soft and silver soldered connectors when used with copper pipe.

Size od mm	Maximum working pressure barg		
	30°C	65°C	110°C
6 to 54	16	10	6
67	10	6	4
76 to 108	10	5	1

Table 9.41 Maximum operating conditions for soft soldered joints

Size od mm	Maximum working pressure barg		
	30°C	65°C	110°C
6	242	151	61
8	202	126	51
12	158	99	40
15 to 54	69	43	17

Table 9.42 Maximum operating conditions for silver soldered joints

The figures quoted in the two Tables apply to general purpose applications, not designed to a specific code.

Soldered connections are made in two styles:

- capillary — which is manufactured with a solder insert
- plain or end feeding — which requires the user to supply the solder when the joint is hot

The capillary fitting is very popular because it is very easy to produce a good joint provided everything is clean. The end feed soldered connection offers higher operating conditions but does require a better level of skill to make a good joint. Both these systems can be used very effectively in systems compatible with copper and the type of solder used. Many styles of soldered connectors are available to enable the construction of complex systems. Connectors can allow transition to other joint types such as threaded and flanged.

9.8 Cemented connections

Cemented connections, sometimes called cold solvent welding, are used with non-metallic pipe and tube and are similar in concept to capillary soldered connections. The pipe/tube is inserted in a socket; the cement bonds the pipe/tube od to the id of the socket. This type of assembly is used with:

- ABS
- CPVC
- PVDF
- UPVC

Curing time is usually a few minutes. This type of connection is not as robust as welding on metal pipes. Non-metallic pipes and their connections can be prone to fatigue problems, just like some metallic connections, but the failures occur at much lower stress levels. Junctions between non-metallic and metallic systems can be particularly troublesome if the metallic system lacks flexibility and imposes undue forces and moments.

9.9 Flanges

9.9.1 Manufacture

Flanges are widely used in many industries for diverse applications. The number of flange standards is equally numerous. All standards are not equal or equivalent. The flange specification used must be appropriate for the application. Pipework handling hazardous fluids must conform to a prescribed quality which must be supported by material and testing certification.

Flanges are the most popular connection method for pipe sizes over DN80, but are not the only one possible. Other methods should be considered during the design process. Flanges can be difficult to split after years of hot operation. If joints are not required to be broken routinely then permanent connections should be considered, such as welding.

Flanges can be manufactured by many methods. The following list starts with the highest quality:

- forged to finished shape
- forged
- cast then "HIPed" (HIP, Hot Isostatically Pressed)
- powder metallurgy
- machined from solid
- cast
- machined from plate

Many specifications do not allow flanges to be made from plate. Low grade carbon steel plate can be riddled with laminations and inclusions which considerably weaken the material. This problem is completely cured by using grain refined steel. Higher grade steels and stainless steels do not suffer this problem.

Many specifications do not allow flanges to be cast. This is particularly so as operating pressure increases. 200 barg is a typical benchmark for specifying wrought flanges. Cast flanges may be prohibited when handling hazardous fluids. The problem of cast flanges must be considered in conjunction with size. Cast iron, malleable iron and SG iron are perfectly adequate for high pressure in the smaller sizes. The problem is how to define "small" and "large".

Most flanges are standardised by a recognized authority such as ISO, BSI or ANSI. However special proprietary flange designs are available. These tend to be smaller and lighter and allow a flanged connection to be used where space is very restricted. The complete specification for such flanges, includ-

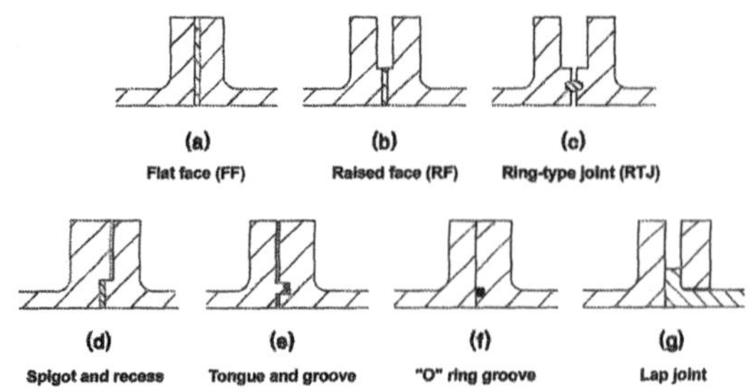

Figure 9.8 Popular flange styles

ing the relationship between design pressure and operation at emergency conditions with relief valve over-pressure and the hydro-test pressure, should be reviewed before committing designs to be manufactured. Flanges may be called upon to transmit large axial forces, torque or bending moments. Some small, light proprietary flanges are only rated for loads associated with the internal pressure.

9.9.2 Flange facings

Flange designs have evolved as industrial processes place higher demands on fluid handling. Higher pressures and temperatures make flange sealing more difficult. Hazardous fluids require a guaranteed high integrity seal. Figure 9.8 shows flange styles that are used in many industries to cope with specific problems.

(a) flat face (FF)
(b) raised face (RF)
(c) ring-type joint (RTJ)
(d) spigot and recess
(e) tongue and groove
(f) "O" ring groove
(g) lap joint

Notice that FF, RF and RTJ use two identical flanges. The other flange styles utilise a matching pair.

9.9.2.1 Flat face (FF)

Flat face flanges are used for low pressure applications and with cast iron or non-metallic valves and fittings. The gasket seals across the full flange face width surrounding the bolting. No part of the flange is subjected to direct bending as a result of the bolting loads. The wide flange faces allow the use of soft gasket materials such as rubber and cork but care must still be taken to avoid over-torqueing.

Corrugated brass or copper gaskets with plenty of graphite compound are used on low pressure steam applications. It is possible to use a steel RF flange with a cast iron FF flange but skill and judgement are necessary. Alignment of the flanges is accomplished solely by the bolting. Maximum misalignment is equal to the bolt clearance. Flat face flanges are most often used with mirror, rough, smooth or stock finish.

9.9.2.2 Raised face (RF)

The raised face flange is the preferred flange style for general applications when the operating pressure is too high for flat faces and cast iron or non-metallic materials. The reduced gasket area promotes a higher gasket seal pressure. Cast iron is not quite ductile enough to be generally successful in the raised face style. Rough treatment and over-torqueing bolting results in broken cast iron flanges. Steel and other metals possess sufficient ductility to withstand the imposed bending stresses. Alignment of the flanges is controlled solely by the bolting. Raised face flanges can be supplied in all the surface finishes; concentric serrated, mirror, rough, smooth, spiral serrated and stock finish.

9.9.2.3 Ring-type joint (RTJ)

The ring-type joint flange is a version of the raised face flange used for high pressure or hazardous fluid applications. A "soft" metal ring gasket is located in the two grooves and slightly deformed to create the seal, see Section 9.10.3. The metal ring aligns both flanges and the bolting only supplies the axial load. The flanges do not close face-to-face and a gap ensures the ring is always compressed. Standard flange designs for pressures up to 145 barg cover diameters up to DN900. The highest pressure ratings, 775 barg, are limited to flanges up to DN300.

The metal gasket rings can only be used once. The cost of rings makes these flanges expensive to use and limits them to systems requiring the highest level of integrity.

Special versions of ring-type joint flanges are used for vacuum applications. These flanges use socket weld connections for ¾" and 1½" od tube. The rings are produced in copper and, nickel, both suitable for temperatures up to 537°C, and Viton™ which can be used up to 232°C.

Note: Do not confuse RTJ with RJT. RJT is an approved hygienic threaded union.

9.9.2.4 Spigot and recess

Spigot and recess flanges are a modification of the raised face flange. One flange has a raised face which engages the recess in the mating flange. The gasket is constrained within the recess. When the gasket is compressed, the flange faces do not close. In some applications it may be desirable to allow the faces to close to control gasket compression. This would be particularly effective with soft gaskets such as rubber or cork.

The spigot and recess combination is a development of the raised face flange as it provides positive alignment and prevents any possibility of gasket blow-out. The range of surface finishes normally used is restricted to mirror, rough and smooth. Spigot and recess flanges can be used on low pressure applications with soft gaskets or high pressure applications.

9.9.2.5 Tongue and groove

The tongue and groove flange configuration is a development of the spigot and recess flange style. The gasket is completely constrained within the groove. Gasket blow-out is prevented and also gasket loss into the process is eliminated. The flange pair can be designed to limit gasket compression or can be designed so the faces remain open. Flange alignment is controlled by the tolerancing of the tongue and the groove. Facing surface finish is usually limited to "mirror" and "smooth". This is not a popular flange style and its use is very limited.

9.9.2.6 "O" ring groove

The "O" ring groove flange principle is significantly different to other flanges. Flanges using flat gaskets or metal rings must apply considerable loads to the gasket/ring to provide a sealing interface pressure. Depending upon the design method used the interface pressure may be considerably higher than the fluid operating pressure. The gasket interface pressure load may be the major load component supplied by the bolting. "O" rings are invariable very soft and can be deformed easily between the fingers. The flange load necessary to compress most elastomer "O" rings sufficiently is insignificant. "O" ring flange designs can be much lighter than other flange styles. Facing surface finishes are always as good as possible, typically to "mirror" quality.

Standard flange designs are available for 414 barg. Elastomer "O" rings have been used for static seals for pressures over 2 000 barg. PTFE is a very popular "O" ring material; it is not an elastomer but a thermoplastic. Special grooves may be required when using PTFE. Metal "O" rings can also be used. "O" ring groove flanges are capable of sealing the highest pressures. Flange alignment is controlled by the bolting clearance.

Spigot and recess and tongue and groove flange styles can be adapted to use "O" rings for sealing.

Note: **Flange orientation** is important. Spigot and recess, tongue and groove, and "O" ring groove flanges can be used in complex piping installations to prevent incorrect assembly. Because the flange assembly consists of a mating pair it is difficult or impossible to fit components, such as non-return valves, backwards.

9.9.2.7 Lap joint

The lap joint is a type of flanged joint as well as a type of facing. If lap joint flanges are considered for an application then the type and nature of the seal must also be considered.

The lap joint uses a separate flange to carry the bolting and apply the axial load. Lap joint connections can be considered when the piping material is exotic, expensive or non-metallic. Exotic materials are very costly, so if the overall size of a component can be reduced, a considerable cost saving can be achieved. Figure 9.8 (g) shows a lap joint flange connected to a flat face flange. This type of connection would occur at vessels or other components where the lap joint principle could not be applied.

In theory, any of the common sealing methods could be applied to the lap joint flange face. If a metal-to-metal joint or plain gasket is proposed, then alignment may be a problem. In these cases alignment would be controlled by the bolting clearance and the clearance between the loose flange and the lap insert. The use of a ring-joint would ensure good alignment. Alternatively, an extension of the "spigot and recess" principle, using the od of the lap insert, could be used to maintain good alignment.

9.9.3 Flange-pipe connections

Flanges must be attached to the pipework. This can be achieved by various means, each with advantages and disadvantages. The following Sections discuss the detailed implications of mounting flanges. The flange-pipe connections shown in Figure 9.9 are used to cope with specific problems:

- (a) threaded
- (b) parallel thread
- (c) circlip
- (d) weld-on collar
- (e) slip-on (SO)
- (f) socket weld (SW)
- (g) butt weld (WN)
- (h) lap joint

9.9.3.1 Threaded

This is the most common method of fitting flanges using a threaded pipe. The thread is a tapered pipe thread and the

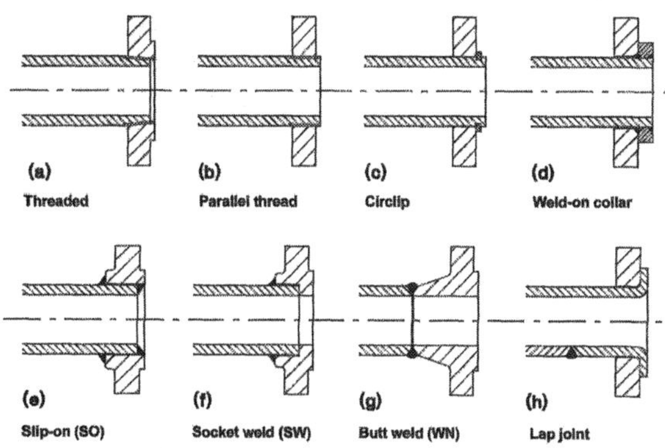

Figure 9.9 Flange-pipe connections

flange is rotated until tight. The pipe does not extend to the front of the flange. Some flanges may have the bore diameter at the raised face reduced to increase the face width. This aspect of design is not covered by some standards. The thread can be sealed by PTFE tape or compound. In some cases the thread can be seal welded at the back. Many piping specifications prohibit seal welding of threaded components. Threaded flanges can be difficult to seal and orientate exactly and should not be used on systems when leaks cannot be tolerated.

Figure 9.9 (b) shows an alternative method of using a threaded flange. A parallel thread is used and the sealing is performed on the pipe end and is a much better approach than 9.9 (a). This type of connection can be improved by using a thread diameter slightly larger than the normal raised face diameter. This allows standard gaskets to be used and is a very useful technique for fitting flanges to materials which are difficult or costly to weld. The flange does not have to be compatible with the fluid because it is not in contact. The flange can be screwed and shrunk in position or pinned.

9.9.3.2 Circlip

A circlip can be used as a shear-ring to attach a flange. A groove in the pipe wall locates the circlip and prevents the flange from moving forward. The flange can be held in the correct orientation by a light shrink fit or pinned. This type of attachment is very effective when the pipe od is increased to the raised face diameter. The flange can be of any suitably strong material.

9.9.3.3 Weld-on collar

The weld-on collar is similar in principle to the circlip but the flange is loose and the pipe material must be suitable for welding. The collar is attached by two welds. The front face must be machined to provide the full "raised face" seal. This is not a common flange attachment method but is perfectly adequate.

9.9.3.4 Slip-on (SO)

Slip-on flanges are very popular for a wide range of applications involving "safe" fluids. The flange is held in position by two fillet welds, one in the bore and one on the back of the flange. This type of flange is not acceptable for hazardous fluids because it is difficult to assess accurately the weld quality. Radiographs are very difficult to interpret. It is available in a wide range of materials, sizes and pressure ratings and is the first choice for a welded flange.

Slip-on flanges which are attached in the normal way are not acceptable for hygienic applications. The gap between the two internal fillet welds can accumulate product, is difficult to clean and cannot be sterilised effectively. This problem can be cured by adding extra weld to the internal fillet so that it can be machined back to produce a square internal corner. Gaskets must have a bore exactly the same as the pipe so that there is no gap to accumulate product and sterilising is made easy.

9.9.3.5 Socket weld (SW)

Socket weld flanges are very popular in the smaller sizes. They are very easy to fit and weld but not used for hazardous fluids because the weld is difficult to radiograph. They are not used with fluids which cause general corrosion or crevice corrosion problems. They are available in many materials and pressure ratings in sizes up to about DN150.

9.9.3.6 Butt weld (WN)

The butt weld flange is the ideal welded flange for normal materials. It is more generally called a "welding neck" flange. The single, full penetration, butt weld is easy to radiograph ensuring high integrity. The flange requires careful alignment with the pipe during initial "tacking". Available in many materials, pressure ratings and sizes, the flange connection is ideally suited for hazardous fluids and hygienic applications.

9.9.3.7 Lap joint

The lap joint flange can be achieved by two methods. When the pipe material is exotic the "lap" is formed by shaping the pipe material after the loose flange has been fitted. If the piping material is readily weldable a "lap" end can be butt welded to the pipe. Both are shown in Figure 9.9 (h). This is a high integrity joint reserved for special applications. The "lap" ends are produced in various lengths by a few fittings manufacturers.

Lap joints are very useful for non-metallic piping and components. Most materials are much better in compression than in tension. The lap joint pipe is held in compression by the loose flange which is in tension due to the bolt load. The lap joint is very effective for joints between non-metallic piping components. The loose flange can be of any appropriate metal. Non-metallic flat face flanges may have a metallic backing flange.

This type of construction can be to accomplish the transition from flanged pipework to compression fittings. Usually the transition involves exposing one screwed thread to the fluid. The lap joint removes this weakness by having the compression coupling machined directly on the lap end. A DN15 flanged connection can be converted to 0.25", 0.375", 0.5" or 10 mm od tube.

Special Note: Friction welding

Friction welding is popular for mass-produced products. Butt weld components can be attached to pipework or valve bodies by friction welding. Any welding used must be able to be inspected for integrity.

Special Note: Flange descriptions

The facing, the pipe connection, the size and the rating, are often connected in the form of an abbreviation. This is most often seen in bills-of-materials, take-off lists, schedules and piping isometrics. The most common combinations are:

- 3"-150LB SOFF — slip-on flat face
- 3"-150LB SORF — slip-on raised face
- 1"-300LB SWFF — socket weld flat face
- 2"-600LB SWRF — socket weld raised face
- 3"-1500LB WNRF — weld neck raised face
- 2"-2500LB WNRTJ — weld neck ring-type joint

9.9.4 Pressure-temperature ratings

Flanges are designed for a specific pressure at a designated temperature. In each flange Standard a range of pressure ratings is defined. Flanges become larger and heavier as the pressure rating increases. Figure 9.10 shows the difference in sizes for ISO 7005 steel flanges for nominal pressures from 10 barg

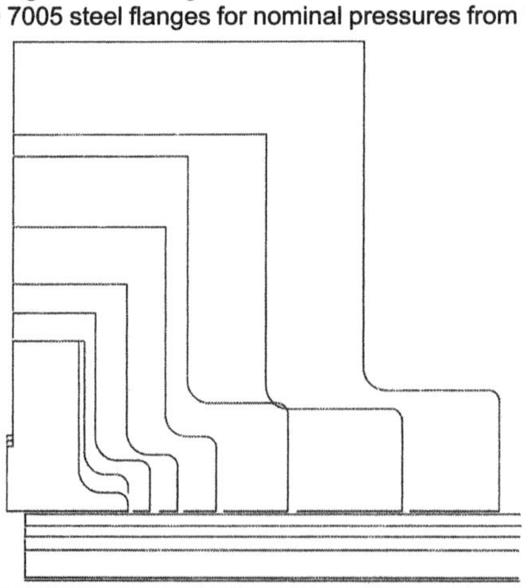

Figure 9.10 ISO 7005 DN200 steel flanges, scale 1:2.5

to 420 barg, PN10, 16, 25, 50, 110, 150, 260 and 420 and pipe sch 10S, 40, XX and 160. Standard steel pipe is shown to illustrate the change in pipe thickness as operating pressures increase.

NOTE: Sch 160 pipe should only be used up to approx 315 barg.

Flanges are affected by temperature. At higher temperatures, the material becomes more elastic and deforms more for the same stress. Different materials are affected by temperature to a varying extent. Flange pressures are tabulated with temperature depending upon the material. Some pressure-temperature relationships are shown graphically in Chapter 13, Section 13.4. Non-metallic materials are affected by temperature. Pressure-temperature ratings for non-metallic pipe are shown in Sections 9.2.6 and 9.2.7.

The standards committees who design flanges do not use the same design philosophies. The pressure-temperature relationship for one flange cannot be applied to another flange. The stress-temperature relationships in pressure vessel standards, such as PD 5500 or ASME VIII, may be helpful.

9.9.5 Special flange connections

In some situations it may be desirable to use a flange connection similar to a ring-type joint but without the gap between the flange faces. MSS has a standard design SP-65 for pipework from DN20 to DN150, see Figure 9.11. The gasket is a lense ring which completely fills the gap between the pipe ends; the seal can be directly on the pipe as shown. MSS approves the use of flanges attached by parallel threads. These flanges are rated at 689 barg at up to 38°C reducing to 276 barg at 454°C. Standard materials are in accordance with ANSI B16.5 groups 1 and 2 except that 304L and 316L are specifically excluded. Standard bolting is B7/2H. See Section 9.10 for suitable gasket materials.

9.9.6 Bolting

The highest possible flange specification can be rendered worthless if the bolting is unsuitable. Flange bolting must be capable of supplying the compressive load to compress the gasket and withstand external forces and moments applied through the pipework. The type of bolting required is dependent upon the process operating conditions and the surrounding environmental conditions. Materials which are easy to check are preferable to materials which require laboratory analysis. Alloy steels which are heat treated to improve the strength are easily checked by hardness. Stainless bolting should not be rusting. The grade of material should be marked on studs and nuts; it is not unknown for the identification to be incorrect. If it's easy to check, check it! There are many suitable materials available, see Tables 13.28 and 13.29 in Chapter 13.

Most bolting is tightened by "torqueing", a technique in which a known torque is applied by a torque or impact wrench. This is the easiest method and the most inaccurate when using plain spanners. The correct torque for a bolt is dependent very much upon friction at both thread and nut face. Lubricated threads are normally specified but the variation in friction with different lubricants is very significant. Tests by a leading turbine manufacturer have shown a range of coefficients of friction from 0.06 to 0.57 depending upon the lubricant used. Important flanges should have the bolting tightened by hydraulic tensioning. A multiple head unit should always be used, two being the absolute minimum, so that opposites can be tightened simultaneously.

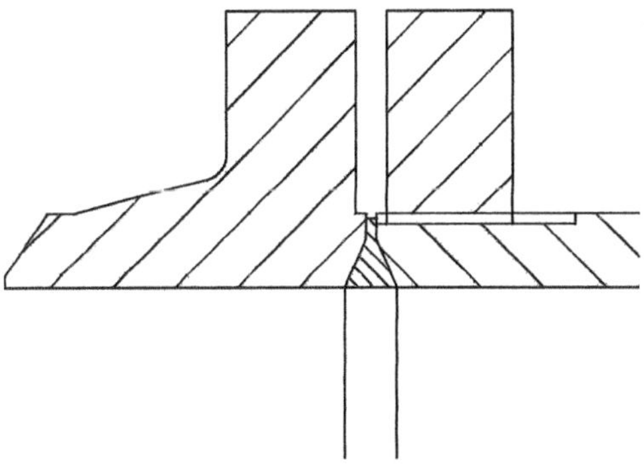

Figure 9.11 High pressure flanges with lense ring gasket

9.10 Gaskets

Effective flange sealing can play an important role in minimising site leakage. It is obviously essential to maintain a good seal on controlled fluids and volatile organic compounds (VOCs). Even a water leak can be costly if the water is treated or potable. Any leak can prove costly simply by considering the possible clean-up costs. Gasket effectiveness depends upon correct bolt tightening; not only the torque but also the order of tightening.

Flanges can seal directly metal-to-metal. In some extreme cases this may be the only option. However the quality requirements of surface finish and face parallelism are such that this option should only be used when unavoidable. Gaskets allow for manufacturing and assembly inaccuracies but only to a limited extent.

Explosive decompression is a problem discussed in Chapter 11 in connection with spindle/stem sealing. Non-metallic gaskets can suffer from similar problems. The potential risk and probable cost of subsequent damage should be evaluated at the design stage.

9.10.1 Non-metallic gaskets

The most common non-metallic gaskets is the flat gasket. 3 mm thick gaskets are used for low pressure applications. 1.5 mm gaskets are used for medium pressure and 0.8 mm and thinner for high pressure. Thick gaskets are generally made from one material, not a mixture. Thinner gaskets are made of compressed fibre held together with a binder. They have very little elasticity and are not able to accommodate flange alignment or surface irregularity problems.

As operating pressures increase, the parallelism and rigidity of the flange faces must increase accordingly. Gasket materials must be selected which are compatible with the process fluid and also the operating conditions. Compressed fibre gaskets may have a porous structure which can allow small molecules to migrate. This type of leakage may be very small but it may also be a hazardous gas and vapour. Expanded PTFE, expanded to provide the resilience solid PTFE lacks, may be available to reduce greatly gas and vapour leakage.

Material	Thickness mm	Max pressure barg	Max temperature °C	Comments
Cork + CR	0.8 to 10.0	low	120/135	General engineering applications with light flanges
IR	0.8 to 5.0	low	80/105	Water and hygienic applications
PTFE	1.5, 3.0	80	-200 to +260	Steam, oil, chlorine, caustics. alcohols, refrigerants

Table 9.43 Flat thick gasket materials

Tables 9.43 and 9.44 indicate the range of materials in common use and the appropriate operating limitations.

Fibre	Binder	Thickness mm	Max pressure barg	Max temp °C	Comments
Cellulose	Glycerine compound	1.0, 1.5, 3.2	8	120	Alcohols, hydrocarbons, vegetable oils, water
Glass	NR	0.4, 0.8, 1.5, 3.0	83	350	Gases, steam, water
Glass	NBR	0.4, 0.8, 1.5 3.0	140	475	General purpose, potable water chlorine
Synthetic	NBR	0.4, 0.8, 1.0	100	420	Steam, oils, fuels,refrigerants
ARAMID™	NBR	0.4, 0.8, 1.0	100	370	Wet steam, solvents, gases, fuel
ARAMID™	SBR	0.4, 0.8, 1.0	100	370	Water, wet steam, gases, ammonia
Asbestos	SBR	0.2, 0.4, 0.8, 1.0, 1.5	130	550	Air, gases, steam, water
Asbestos	NBR	0.4, 0.8, 1.0, 1.5	250	590	Water, paper pulp, hygienic
Asbestos	NBR	0.2, 0.4, 0.8, 1.0, 1.5	130	500	Oil, petrol, refrigerants
Asbestos	SBR	0.4, 0.8, 1.0, 1.5	125	400	Hygienic applications
Asbestos	SBR	0.8, 1.5, 3.0	30	350	Oils, water, general use
Asbestos	CR	0.2, 0.4, 0.8, 1.0, 1.5	125	550	Fuel oils, petrol
Asbestos	PVC blend	0.4, 0.8, 1.5, 3.0		180	Acid services
	Filled PTFE	1.5, 2.0, 3.0	60	260	Wide chemical compatibility

Table 9.44 Flat compressed fibre gasket materials

Figure 9.8 (f) shows a flanged connection sealed by an "O" ring. Standard flanges use elastomer "O" rings for pressures up to 414 barg. "O" rings can be used for higher pressures if care is exercised. A good surface finish is essential, typically 1.6 μm or better. Non-metallic "O" rings are available in many compounds, both elastomers and thermoplastics. Refer to Table 11.1 in Chapter 11 and Table 13.20 in Chapter 13 for more information.

PTFE has very good chemical resistance but poor elastic properties. This deficiency can be overcome by coating a more resilient material with PTFE. Silicone rubber and Viton™ are the most common core elastomers used.

In some applications the "O" ring manufacturer may recommend the use of an "O" ring lubricant. It is advisable to take the manufacturers' advice but check for chemical compatibility with the process fluid. Some manufacturers only have silicone based lubricants.

9.10.2 Combined gaskets

The flat gaskets described in Section 9.10.1 can be strengthened by incorporating fibres which have more rigidity than asbestos or other flexible fibres. Wire mesh can be included to stiffen the gasket and help to resist blow-outs. Common combinations of wire mesh reinforced flat gaskets are listed in Table 9.45.

Fibre	Binder	Reinforcing	Thickness mm	Max pressure barg	Max temp °C	Comments
Asbestos	SBR	Carbon steel wire mesh	0.5, 0.8, 1.0, 1.5	200	550	Suitable for cyclic processes
Synthetic	NBR	Carbon steel wire mesh	1.0, 2.0, 3.0	100	350	Gases, solvents, steam
Asbestos	NBR	Stainless steel wire mesh	0.5, 1.0, 2.0, 3.0	210	600	Suitable for cyclic processes or vibration
Graphite		Stainless steel wire mesh	0.8, 1.5	140	-200 to 500	to 1000°C in non-oxidising environments

Table 9.45 Reinforced flat compressed fibre gasket materials

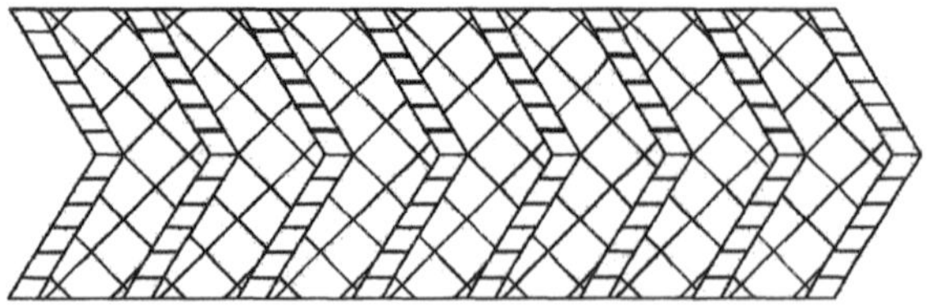

Figure 9.12 A cross-section through a spiral wound gasket

9.10.2.1 Spiral wound gasket

Another style of gasket, the spiral wound gasket, successfully combines the benefits of metallic and non-metallic materials to form a versatile seal, see Figure 9.12.

A narrow strip of metal is bent to form a chevron. A strip of non-metallic material is placed in the vee formed by the chevron. The two strips are wound together over a former and then successively over themselves to form a ring. The spiral wound gasket is very popular in many industries which use raised face, spigot and recess, and tongue and groove flanges. Standard designs handle pressures up to 350 barg and temperatures to 1000 °C.

The basic design can be improved for raised face flanges; a light metal ring is attached to the od to centralise the gasket inside the bolt circle. The spiral wound gasket can be made in many material combinations but the 316/asbestos pairing is mainly standard. For other materials available see Table 9.46.

Metal strip materials	Filler
AISI 310 super stainless steel	PTFE
AISI 321 austenitic stainless steel	Lead
Monel 400™	Aluminium
Inconel 600™	Graphite
Incoloy 800™	
Titanium	
Nickel	

Table 9.46 Alternative materials for spiral wound gaskets

9.10.2.2 Metal cased gasket

Metallic and non-metallic materials can be combined in another form. The soft non-metallic material can be completely surrounded by a thin metal casing. The finished gasket can take two forms, plain or corrugated. The non-metallic filling is usually asbestos. No risk to the process fluid or operating personnel is present because the asbestos is totally enclosed. Several materials are used for the metal casings:

- aluminium
- copper
- brass
- carbon steel
- 5Cr alloy steel
- austenitic stainless steel
- Monel™

9.10.3 Metallic gaskets

Metallic gaskets are available in several forms to suit the application and the flange type used. Thin flat gaskets are the simplest and can be supplied in various materials:

- aluminium

- copper
- soft iron
- carbon steel
- 5Cr alloy steel
- austenitic stainless steel
- Monel™

Flange surface finish and parallelism must be very good. High pressures are possible up to the temperature limit of the flanges. Gaskets can be used dry, or coated with a suitable sealant or lubricant during assembly.

The thin flat gasket can be shaped to allow sealant to be trapped during compression. A thin corrugated metal gasket, with appropriate sealants, can be used for low pressure applications when the parallelism of the mating flanges is difficult to guarantee. These gaskets, usually aluminium, copper or brass, are popular for low pressure steam applications.

9.10.3.1 Serrated gasket

To overcome the high quality requirements of the mating flanges, a modified flat gasket style, called the serrated gasket, was developed for high fluid pressures. The basic gasket is thick, about 4 mm. Each face is machined with a set of concentric grooves which cover the entire face. Sealing is accomplished by crushing the peaks of the grooves. This style of gasket has been used for pressures up to 689 barg and up to the temperature limit of the flanges. Serrated gaskets are usually made of austenitic stainless steel but more exotic materials can be supplied. Finished gaskets can be supplied "as-machined" or coated to assist sealing and corrosion protection. Typical coatings include:

- cadmium plate
- PTFE
- flexible graphite

The serrated gasket can be supplied with an outer support ring to centralise the gasket within the bolt circle.

Ring-type joint rings are available in two forms; octagonal and oval. The octagonal ring has tapered portions at the ends which are the same angle as the groove. The oval ring is semi-circular at the ends and parallel in the middle. The high pressure flanges shown in Figure 9.11 use a metal lense ring as a gasket. Table 9.47 lists the materials commonly used for ring joints and lense rings.

Gasket material	Max Hardness BHN	Max Hardness R_B
Soft iron (1)	90	56
Low carbon steel	120	68
5Cr 0.5Mo steel	130	72
AISI 410	170	86
AISI 304	160	83
AISI 316	160	83
AISI 347	160	83
AISI 310		
AISI 321		
High tensile steel		
2.25Cr alloy steel		
Monel™		
Incoloy™		
Inconel™		
Hastelloy™		
Duplex stainless steels		
Titanium		
Nickel		

(1) Sometimes called ARMCO iron, after the American steel manufacturer

Table 9.47 Materials for ring joints and lense rings

Materials which may be prone to rusting can be electroplated with zinc.

Figure 9.8 (f) shows a flange design for "O" rings. "O" rings can be metallic as well as the more common elastomer types. Solid metal "O" rings require special groove machining and are used for certain applications such as engine cylinder heads. Hollow metal "O" rings can be used in a similar manner to elastomer "O" rings.

Metallic hollow "O" rings are made in three styles:

- plain
- self-energised
- pressurised

Plain "O" rings are sealed with a vacuum or atmospheric pressure nitrogen. Plain rings are used for low positive pressure or vacuum applications.

Self-energised rings are not sealed and a series of very small holes in the bore allow process fluid to pressurise the ring. Pressurised rings are sealed with a positive internal pressure using an inert gas.

Metal "O" rings cover the widest range of operating pressures from full vacuum to over 6 000 barg. Continuous operating temperatures range from -269° to +980°C, up to 1 650°C intermittently.

Typical "O" ring materials include:

- aluminium
- copper
- AISI 304 stainless steel
- AISI 316 stainless steel
- AISI 321 stainless steel
- AISI 347 stainless steel
- Inconel 600™
- Inconel X-750™
- Alloy 718

The surface finish of the mating faces must be good, 0.4 µm. The surface finish can be relaxed slightly if a plated or coated "O" ring is used. Popular coatings include:

- PTFE
- copper
- lead
- silver
- nickel
- gold

Metallic "O" rings can be used in modified spigot and recess and tongue and groove flanges. The dimensions must be adjusted so that when the flange faces touch, the correct compression is applied to the "O" ring.

The "C" ring is a modified version of the metallic "O" ring. The "C" ring is an "O" ring with one side completely open. For flange seals the open side would be in bore to allow self-energising. All the previous comments regarding materials and coatings for metal "O" rings apply. The choice between an "O" ring and a "C" ring should be left to the manufacturer.

Metallic gaskets do provide an excellent seal when fitted and used properly. Most seal rings work harden during the tightening process and cannot be used again. New rings must be made available when joints are broken.

9.10.4 Liquid gaskets

Flanged connections can be sealed metal-to-metal but a liquid gasket can be a great assistance. They can accommodate up to 0.5 mm gaps at low pressure and withstand temperatures up to 260°C. Most liquid gaskets are based on silicone compounds

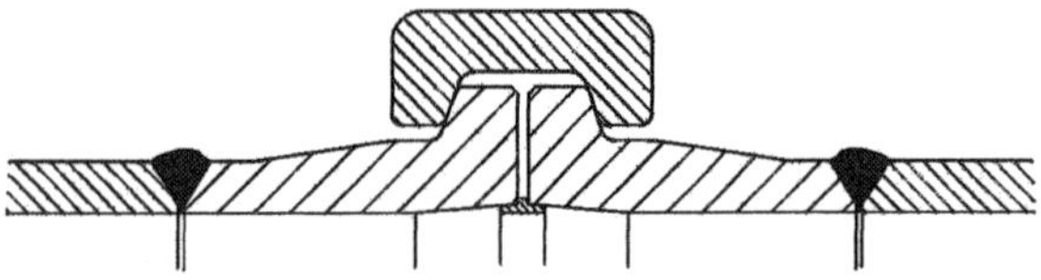

Figure 9.13 A typical process clamped connection

or PTFE dispersions. Modern liquid gaskets do not set like glue, effectively joining components. Disassembly can be carried out without damage. This gasket type should be considered when flanged connections must be broken regularly and can be cleaned effectively before reassembly.

9.11 Clamped connections

Clamped connections are a high integrity connection used for extreme operating conditions, hygienic applications and/or rapid assembly/disassembly. The clamped connection is more rigid than a normal flanged connection and able to resist fatigue better due to vibration. The basic concept relies on the wedging action of opposed tapers to hold the hubs together, see Figure 9.13.

The connection shown uses a shaped metal seal ring in the bore for two separate functions; sealing and alignment. The crush on the seal ring is controlled by the outer parallel ring which maintains a preset gap between the hubs. The clamp can be in two halves and bolted together, similar to a roller chain with a single tension screw. The hubs are always butt weld as shown.

Hygienic versions, in 316L, use an elastomer or thermoplastic seal ring of approved material; silicone rubber, EPDM, PTFE or Viton™. Encapsulated PTFE may be available for some applications. The crush fit in the bore removes the possibility of crevices where product could decay or debris accumulate. Hygienic fittings are usually rated for 10 barg.

Hubs and seal rings can be made in a wide range of materials. Material/fluid compatibility is not a problem. Standard process designs are suitable for pressures over 1 400 barg and temperatures up to 530°C. Exotic materials provide higher ratings. Clamped connections are much lighter and smaller than conventional flange designs. Process temperature changes of 140°C per hour do not induce thermal shock stresses.

The clamped connector profile can be machined directly on to components to avoid welding. Many heavy-duty valves offer these connections as an alternative to flanges. The clamped connection can be used for bonnets and glands when the gland is non-adjustable.

Special versions of connectors are available for applications such as orifice flanges where the orifice can be incorporated in the seal ring.

The clamped connection is a versatile alternative to flanges. However, it is essential to check the bore size of the seal ring. Some high pressure rings have quite small bores compared with the hub bore. Low pressure versions of the clamped connection, using elastomer seals trapped in shaped grooves, are available. The elastomer seal ring is quite effective at maintaining the alignment but does not provide the rigidity of a metal seal ring in the bore. The clamp does help with alignment. These connectors are available in stainless steel for butt welding to inch and metric tube. Sizes range from about 25 mm to 250 mm.

9.12 Grooved connections

The grooved pipe connection system is a simple method of connecting pipes when the working pressure is less than 69 barg for small pipes and 31 barg for the largest. Standard components can be used for pipes up to DN600.

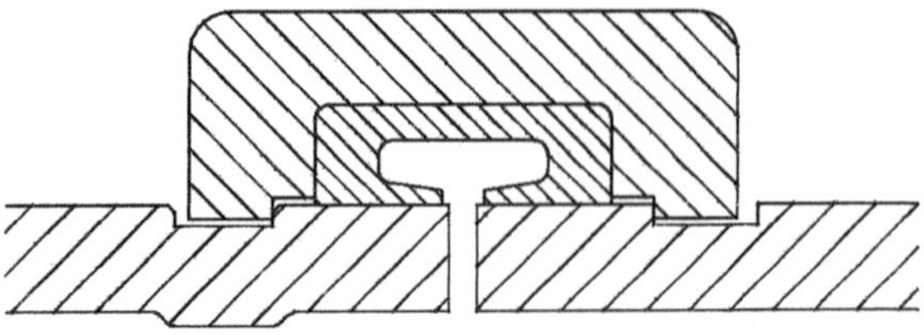

Figure 9.14 A typical grooved connection

The grooved system utilises a groove, cut or rolled, near the end of the pipe to locate a split clamp which prevents axial movement of the pipes relative to each other. An elastomer seal is compressed around the pipe od to make a seal. Figure 9.14 shows a typical joint with a cut and rolled groove.

The process fluid is only in contact with the pipe and the elastomer seal. The grooved system can be used with steel, galvanised, stainless and copper pipe. A selection of elastomer seals is available to provide a wide range of chemical compatibility. Popular materials include EPDM, nitrile rubber, neoprene, Viton™ and silicone rubber.

Rolled grooves are preferred but cut grooves are acceptable. The sizing of the grooves and the clamp allows the two pipes a degree of freedom regarding alignment. Axial, radial and angular misalignment can be accommodated. This system is ideal for piping installations where pipe positions and routings cannot be strictly controlled. The split clamp is usually secured by two bolts but toggle versions are an alternative.

Many pipe fittings, valves and other accessories such as filters, are available with the groove system fitted. A complete system can be built without welding.

The groove piping system is popular for building services applications including:

- domestic water
- hot water, (110°C max)
- chilled water
- air conditioning
- fire water systems
- compressed air
- fuel oil

Accessories are available to allow inter-connection to existing flanged and threaded connection systems. The grooved piping concept provides a simple, effective and approved system for carbon steel piping.

9.13 Compression fittings

Compression fittings are a high integrity connection for small pipes and tubes. High pressure fittings with metallic ferrules, 750 barg maximum, can accept tubes with ods up to 2" or 38 mm. Low pressure fittings with metallic ferrules, up to about 12 barg, can accept tube up to 108 mm. Low pressure fittings, in some cases, can use elastomer ferrules.

There are two basic design concepts:

- The original — using a ferrule with a sharp edge in the bore which cut into the tube od when the ferrule was compressed. The ferrule became part of the tube.
- The modified design — uses the ferrule to swage the tube od rather than cut it.

Both styles co-exist and have found favour in different industries. The swage style can only work effectively up to a specified maximum wall thickness. The cutting ferrule design can work with all thickness. Compression fittings can be applied directly to square cut tube ends and connections made in minutes without special tools or welding.

Some designs permit positive inspection to ensure correct compression of the ferrule and tightening of the union nut. The gap between the body and the union nut can be checked with a "GO–NO-GO" gauge. This simple procedure can detect loose connections and prevent over-torqueing.

Leakage problems can occur at the transition from compression fittings to other systems or components. A high integrity system should not rely on tapered pipe threads for sealing. The initial system design must clarify the type of connections used throughout the system.

The process fluid is in direct contact with the fitting body, the ferrule and the pipework. All components in contact with the process fluid must resist corrosion if leaks are to be avoided. The body, the nuts and the ferrule are exposed to the environment. The environment must also be considered when making material selections.

Compression fittings are not necessarily designed to a Standard or code. If strict compliance with a specific code is mandatory then fitting pressure ratings may need to be reduced. For "coded" systems it is important to check the fittings stress levels early in the design rather than later.

Compression fittings are manufactured for all types of piping connection situations. The list in Table 9.48 indicates some of the wide range of standard configurations available. Tube is used to indicate the compression connection; some, but not all, can be replaced by pipe.

Straight tube to tube	Straight tube to socket weld
Bulkhead tube to tube	Elbow tube to socket weld
Elbow tube to tube	Straight tube to butt weld
Tee tube to tube	Elbow tube to butt weld
Cross tube to tube	Straight tube to flare fitting
Straight tube to male pipe thread	Bulkhead tube to flare fitting
Elbow tube to male pipe thread	Straight tube to pipe flange
45° set tube to male pipe thread	Straight tube to pipe lap joint
Straight tube to female pipe thread	Straight tube to hygienic pipe flange
Elbow tube to female pipe thread	Straight tube to non-metallic tube

Table 9.48 Standard configurations of compression fittings

Various threaded connections can be used; NPT taper, BSPP and BSPT, and SAE with "O" ring seals.

9.13.1 Single ferrule

One of the most popular low pressure compression fitting designs, uses brass bodies and nuts and a plain copper ferrule. This style of compression fitting, see Figure 9.15, is used extensively with light gauge copper tube, to BS EN 1057, for domestic and commercial water systems. Light gauge stainless steel tube can also be used if there is no risk of galvanic corrosion. The illustration shows a compression fitting with a tapered pipe thread. The plain copper ferrule is distorted slightly due to compression by the union nut. This fitting is very good for water systems and compressed air. The connection may be broken and remade many times; the use of a good lubricant with M_2S is recommended. The pipe can be reclaimed by splitting the ferrule, carefully, with a sawcut.

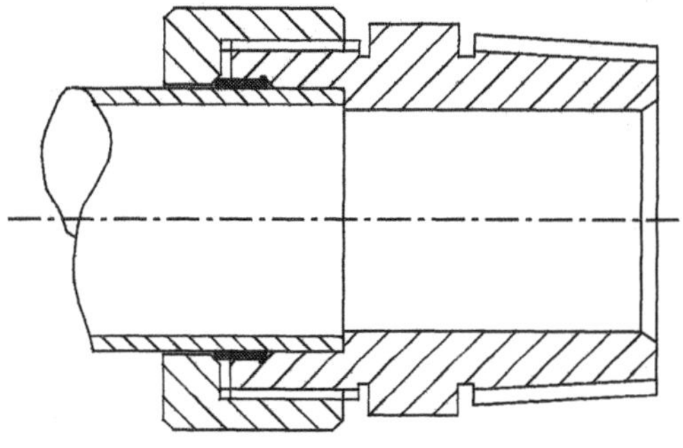

Figure 9.15 A low pressure compression fitting

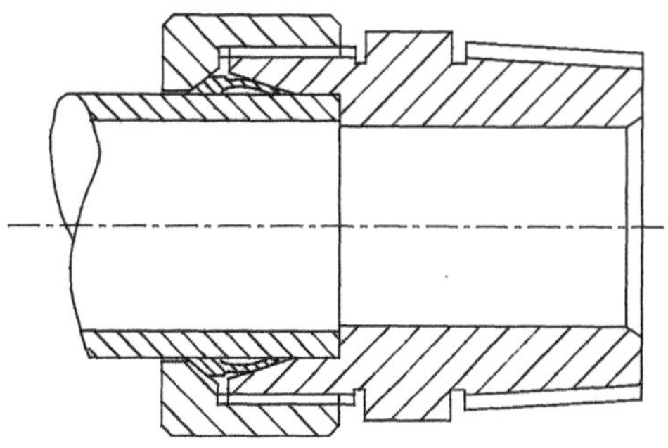

Figure 9.16 A high pressure compression fitting

This type of fitting is available for tube ods from 6 mm to 60.6 mm. Larger sizes, up to 108 mm, use a bolted union rather than threaded. A slightly modified version utilises a shaped brass ferrule; this too distorts without cutting into the pipe wall. Both these styles of ferrule are symmetrical and cannot be fitted incorrectly.

Higher pressure compression fittings create a very good seal by forcing the ferrule to cut into the tube od, see Figure 9.16. The specially shaped ferrule has a barb in the bore. As the ferrule is forced into the body taper the ferrule bore shrinks and the barb bites into the tube. A very effective seal is formed and the ferrule is permanently attached to the pipe. Other designs use two barbs on the ferrule. These joints can be remade many times but the complete pipe is not recoverable. The pipe end must be sawn off just behind the ferrule.

This style of compression fitting is intended for use with steel and stainless steel tube and pipe. Couplings are made to correspond to the ods of schedule pipe and to BS EN 10255. Sizes are generally up to 1.5" od or 1.9" od for 1.5" nb pipe. Pressure ratings are generally to 345 or 414 barg. Carbon steel fittings

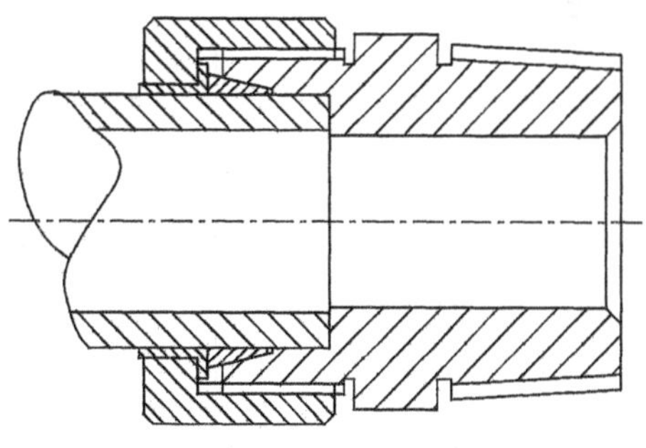

Figure 9.17 A typical double ferrule compression fitting

Figure 9.18 A double ferrule stud coupling
Courtesy of Swagelok Company

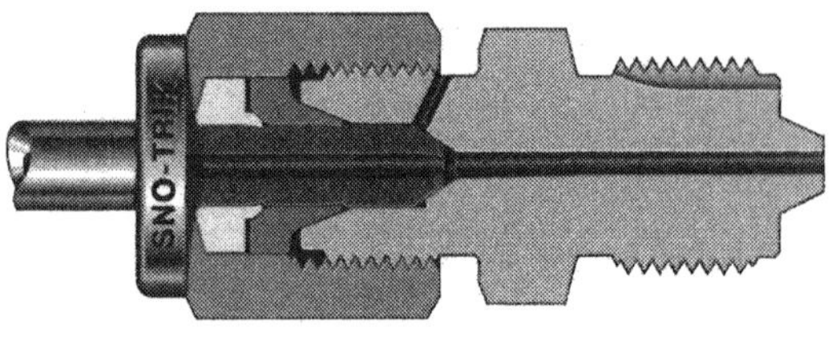

Figure 9.19 A high pressure double ferrule stud coupling
Courtesy of Swagelok Company

are zinc, cadmium or dull chrome plated for protection. The most popular material for stainless fittings is 316.

9.13.2 Double ferrule

Double ferrule fittings were developed to prevent possible problems of the ferrule and tube rotating, driven by the union nut, when fitting initially. The second ferrule is used to provide a thrust bearing for the sealing ferrule, see Figure 9.17. Figure 9.18 shows an exploded view of a particular fitting.

Double ferrule fittings are intended for process applications and are produced in a variety of materials:

- aluminium
- brass
- steel
- stainless steel
- Alloy 20
- titanium

Non-metallic materials such as nylon and PTFE are also available. Many manufacturers can supply sizes up to 1" od or 28 mm. A few can supply couplings up to 2". Many users specify double ferrule fittings whenever compression fittings are used. The double ferrule construction is considered the best choice for hazardous fluids.

Special double ferrule couplings are available for very high pressures, up to 4136 barg. Couplings are produced in a limited range suitable for heavy gauge thick wall tube with ods of $\frac{1}{4}$", $\frac{3}{8}$" and $\frac{9}{16}$". Unlike other compression fittings the tube end must be profiled, not cut square. Figure 9.19 shows a cross section through an assembled coupling. Notice the tell-tale bleed hole for leakage.

Couplings are available with one metallic ferrule and one non-metallic ferrule. The metallic ferrule is swaged to the tube to provide the mechanical grip and prevent the tube moving relative to the coupling. The non-metallic ferrule is usually an "O" ring and only provides the sealing. Couplings for 2" or 38 mm tube use union nuts and larger couplings up to 125 mm use a flanged ring tightened with caphead screws. All sizes are rated at 350 barg.

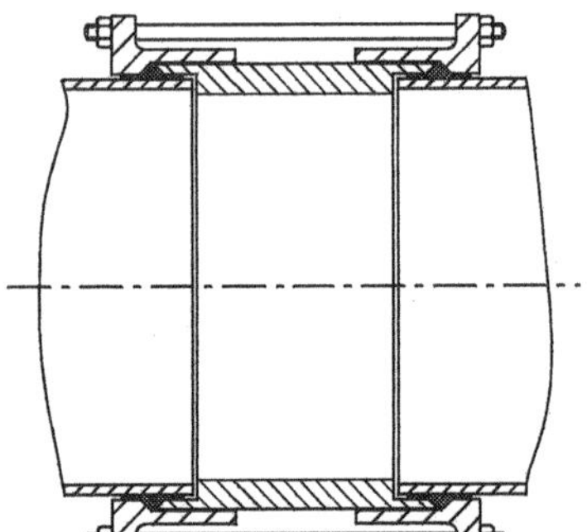

Figure 9.20 Large diameter, low pressure compression fitting

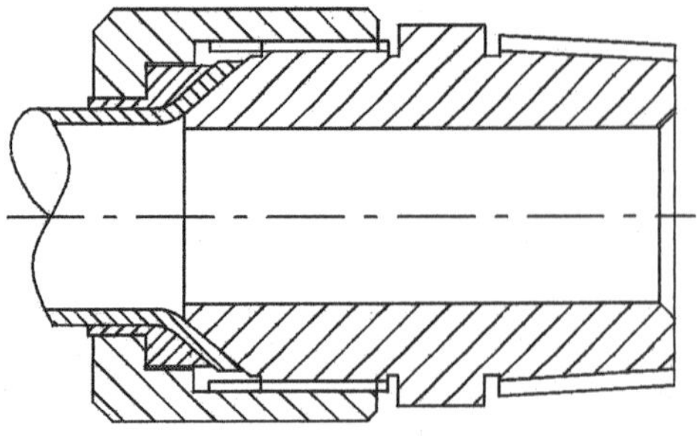

Figure 9.21 A typical flared connection

9.13.3 Non-metallic ferrule

The single and double ferrule fittings described in Sections 9.13.2 and 9.13.1 can be supplied with non-metallic ferrules for special connections. Nylon, PTFE and polyimide are typical materials.

Compression fittings for large diameter, low pressure piping are specifically manufactured with elastomer ferrules, see Figure 9.20. A sleeve is produced with a tapered recess to accept the shaped ferrule which is compressed by a similarly shaped flange. When the tie bolts are tightened, the ferrules are compressed and grip the pipe end. The ferrule completely surrounds the pipe and imparts a degree of alignment flexibility to the joint. Standard metallic materials include carbon steel and malleable iron. Ferrules are available in EPDM, -40° to +90°C, and nitrile rubber, -20° to +100°C.

This type of compression fitting is adaptable to large diameters, DN2 500 having been supplied. Working pressures up to 25 barg are possible. The style of the fitting also allows incorporation into special assemblies. Couplings can be adapted for casting into concrete walls. Couplings can be fitted to other piping components. Special versions can be supplied to connect metallic piping systems to non-metallic systems.

9.14 Flared fittings

Flare fittings rely on the pipe material itself to provide the seal. The pipe end is shaped with a special tool so that the pipe can be clamped directly to a fitting, see Figure 9.21.

The thrust bearing insert between the pipe and the union nut is a design variation which may or may not be present. Flare fittings are popular up 1.5" and 38 mm od tube and are used extensively in refrigeration. Tube wall thickness is important and is usually between 1.6 mm and 4 mm. Small fittings can operate up to about 540 barg at ambient temperature and the largest to 204 barg.

Flared fittings are very similar in concept to compression fittings and the range of fitting configurations is broadly comparable.

9.15 Quick release couplings

This Section deals with automatic isolating couplings. They were probably first designed and used in pneumatics, where the process fluid was safe and working pressures were limited to about 7 barg. Spillage of compressed air during disconnection was not obvious. The design of quick release couplings has progressed significantly but all couplings are not equal. Some couplings have isolating valves in both halves; most only have one. Also, working pressures must be checked carefully. Some couplings will engage and disengage at different pressures. Some high pressure couplings will only engage and disengage after the pressure has been completely removed. This is a very good safety precaution but possibly not always practical. Some coupling designs are not "dry". A small quantity of product is lost each time the couplings are split. This style of design is not acceptable for hazardous fluids.

Quick release couplings allow speedy connection and disconnection of circuits. Many styles of couplings exist which do not automatically isolate the circuits when split. Typical of these are the spring-loaded pin style, the 2 ½" "instantaneous" couplings, as used by water companies and fire brigades.

Quick release couplings use a spring-loaded toggle mechanism to hold the two halves together and to operate the isolating valve(s). Usually pushing the two halves together triggers the toggling mechanism and opens the valve(s).

Single isolating valve couplings are restricted to water and inert gases. Typical sizes range from DN8 to DN15 with working pressures between 35 and 70 barg for all sizes.

Double isolating valve couplings, vent before couple/uncouple, are available in austenitic stainless steel as standard. These couplings will operate with pressures up to 220 barg. Sizes range from DN15 to DN100. These couplings are of the "dry" design. No product is lost when the connection is broken.

Double isolating valve couplings, seal before couple/uncouple, are available in sizes from DN8 to DN40 with pressure ratings up to 600 barg. These couplings are not completely "dry"; a small dead volume of fluid is trapped and lost during uncoupling. The isolating valves are designed for the full rated pressure.

Small couplings, DN8 to DN15, can operate at 414 barg. Coupling or uncoupling is only allowed at 7 barg. Once broken the isolating valves can withstand 70 barg. These couplings are in 316 stainless steel with PTFE seals and seats.

Brass couplings with nitrile seals are available from DN4 to DN80. The smaller couplings are suitable for liquids at 330 barg and gases at 205 barg. The larger sizes are suitable for 130 and 80 barg. These are "dry", connect/disconnect at full pressure couplings. Small couplings, DN8, can be capable of working at over 500 barg on both liquid and gas.

9.16 Approved connections

Approval of connections is divided into two main groups; hygienic and safety. Hygienic couplings are tested for cleanliness and the ability to be cleaned or sterilised. Design guidelines are available but approval is only given after testing. Couplings approved from a safety aspect are usually designed to a code and then tested for acceptability. One pressure vessel code requires a production vessel to be tested to destruction before a design can be accepted. This procedure is obviously very costly when one special vessel is required. Safety approvals may be a prime concern in all forms of transport equipment and military applications.

Hygienic couplings are manufactured in many forms but there are three common features, no crevices, no dead volumes and only approved materials are used. The most common metallic material is 316L stainless steel. Working pressures up to 10 barg are common for pipe ods from ½" to 4". Some designs are available up to 10". Gaskets can be made from a variety of approved compounds:

- silicone rubber
- white nitrile rubber
- black nitrile rubber
- black butyl rubber
- white butyl rubber
- Buna "N"
- EPDM
- PTFE

Proprietary designs for hygienic connections, individual company and trade association, are in abundance:

- WauKesha Cherry-Burrell "S", "Q" and "I"
- ILC
- ISS
- IDF
- RJT
- Tri-Clover Tri-Clamp

Major design standards include:

- BS 4825 Pt 3
- BS 4825 Pt 4
- BS 4825 Pt 5
- DIN 11850
- DIN 11851
- ISO 2852
- ISO 2853
- SMS 1146.

Marine and military connections are subject to rigorous tests before being accepted. Lloyd's Register has a special fire test procedure for couplings and joints. Water pipes are subjected to 800°C and oil pipes to 700°C for 30 minutes and any leakage is grounds for failure.

9.17 Useful references

Mechanics of materials, P G Laurson, WJ Cox, Chapman & Hall; 2nd edition (1947).

Kiwa Water Research, PO Box 1072, 3430 BB Nieuwegein, The Netherlands. www.kiwa.nl.

BS EN 10217-7:2005 Welded steel tubes for pressure purposes. Technical delivery conditions. Stainless steel tubes.

BS EN 12449:1999 Copper and copper alloys. Seamless, round tubes for general purposes.

BS 4127 Specification for light gauge stainless steel tubes, primarily for water applications.

BS EN 10312:2002 Welded stainless steel tubes for the conveyance of aqueous liquids inclu ding water for human consumption. Technical delivery conditions.

BS 5391-1:1976 Specification for acrylonitrile-butadiene-styrene (ABS) pressure pipe. Pipe for industrial uses.

BS 5556 : 1976 (1986) Specification for general requirements for dimension and pressure ratings for pipe of thermoplastics materials.

ASTM D1785-06 Standard Specification for Poly(Vinyl Chloride) (PVC) Plastic Pipe, Schedules 40, 80, and 120.

ASTM F441/F441M-02 Standard Specification for Chlorinated Poly(Vinyl Chloride) (CPVC) Plastic Pipe, Schedules 40 and 80.

ASTM D2241 Standard Specification for Poly(Vinyl Chloride) (PVC) Pressure-Rated Pipe (SDR Series).

DIN 8077 -Draft Document - Polypropylene (PP) pipes - PP-H, PP-B, PP-R, PP-RCT – Dimensions.

DIN 8078 -Draft Document - Polypropylene (PP) pipes - PP-H, PP-B, PP-R, PP-RCT- General quality requirements and testing.

DIN 8061 Unplasticized polyvinyl chloride pipes - General quality requirements and testing.

BS 5118 Noise control on construction and open sites, Part 4 -Code of practice for noise and vibration control applicable to piling operations.

BS AU 110a:1995 Specification for rubber hoses and hose assemblies for automotive air pressure braking systems.

BS 3169:1986 Specification for first aid reel hoses for fire-fighting purposes. (Replaced by BS EN 694:2001 , BS EN 1947:2002)

BS EN 694:2001 Fire-fighting hoses. Semi-rigid hoses for fixed systems.

BS EN 1947:2002 Fire-fighting hoses. Semi-rigid delivery hoses and hose assemblies for pumps and vehicles.

SAE 100R1AT Hydraulic applications requiring 1-braid hose at higher EN/DIN pressures.

DIN 20023 Rubber hoses and hose assemblies (replaced by the European EN 856 standard) - was issued in 1978 specifying the 4SP type hose; the 4SH design being added in 1987.

Noise in valves

10

10.1 The noise problem

Noise can be measured very accurately. The processing power of modern PCs allows virtually instantaneous analysis of noise sources and resolves the noise signal into discrete octave bands or discrete frequencies.

The human ear is one of the most sensitive sensory organs; it is very complex. However it is not equally sensitive at all frequencies. Certain octave bands have more effect than others. At low frequencies there is a transition from hearing to "feeling" when noise becomes vibration. At high frequencies the ear becomes less sensitive and after a "cut-off" frequency the ear does not respond at all. The cut-off frequency varies with age; young people can hear frequencies up to 20 kHz, older people may not hear anything above 12 kHz. Continuous loud background noise can damage the ear permanently and reduce the sensitivity or cause deafness.

The basic premise behind noise legislation is to eliminate hearing damage and to reduce overall noise pollution in the environment. Legislation requires equipment to emit an overall sound pressure level of less than 85 dB(A) at a distance of one metre from the equipment. Some equipment purchasers specify 80 dB(A) as the limiting value.

Valves in pipework can be very difficult to isolate acoustically for accurate measurement. Metal pipework is a very good conductor and radiator of sound waves. Noise from other equipment, such as pumps and compressors, can be radiated from the pipe around a valve and give the false impression that the valve is noisy. Flow conditions created by other equipment can cause valve noise. Valves can be noisy but the overall acoustic environment must be seen in context.

10.2 Noise analysis

As trends in process plant design continue with larger manufacturing units, there is an inherent increase in process pressures and mass flow rates, which must be safely handled by valves at acceptable levels of noise and vibration.

Measurements taken on site during test programmes have indicated that the prediction formula in use is generally accurate to within 5 dB. However, the interpretation of test results is by no means straightforward since numerous factors not covered by the formula can affect the overall noise level generated. Such factors include:

- the effects of background noise
- pipework configuration
- transmission loss
- process conditions

The original research on the generation of hydrodynamic noise was carried out by F M Partridge in California and reported at the Royal Society by M J Lighthill.

Process machinery, such as compressors and pumps, can inject noise into the piping system; this may appear at valves as vibration, or excite valve components to resonate. Centrifugal rotodynamic machines inject a strong signal based on the vane passing frequency, see Equation 10.1.

$$\text{vpf} = \frac{\text{rpm} \times N_i \times N_c}{60} \qquad \text{Equ 10.1}$$

where:

vpf	=	vane passing frequency (Hz)
rpm	=	revolutions per minute (minute^{-1})
N_i	=	number of impeller vanes (non-dimensional)
N_c	=	number of casing vanes (non-dimensional)

The number of casing vanes can be as few as one. The fundamental equation is based on the ideal case where the fluid passing through the impeller acts as a homogeneous mass. Certainly in high power pumps this is not the case and the vane passing frequency should be multiplied by two or three.

Axial flow machines which have many casing or stator vanes inject a signal which is dominated by the blade-rate frequency, see Equation 10.2.

$$\text{brf} = \frac{\text{rpm} \times N_r \times N_s}{60 \times k} \qquad \text{Equ 10.2}$$

where:

brf	=	blade-rate frequency (Hz)
N_r	=	number of rotating blades (non-dimensional)
N_s	=	number of stationary blades (non-dimensional)
k	=	the largest common factor of N_r and N_s (non-dimensional)

Rotary positive displacement machines are built in many configurations. The primary noise frequency will be a function of the number of "elements" passing the process connections.

Reciprocating machines produce noise over a wide spectrum. The dominant frequency is dependent upon the number of cylinders and whether the machine is single or double acting. All integer harmonics will be present in the signal. Harmonics which are multiples of the number of cylinders will be strongest. The fundamental harmonic is based on "rpm" alone. Harmonics less than the product of the number of cylinders and the "rpm" should be weak; except in the case of cavitating pumps when these harmonics can be significant.

Changes in pipe velocity can lead to increased noise levels. For turbulent flow, some manufacturers use the approximation shown in Equation 10.3. If cavitation is present, Equation 10.4 should be used.

Turbulent Noise increase =

$$= 60 \log_{10}\left\{\frac{\text{higher velocity}}{\text{lower velocity}}\right\} \qquad \text{Equ 10.3}$$

Cavitation noise increase =

$$= 120 \log_{10}\left\{\frac{\text{higher velocity}}{\text{lower velocity}}\right\} \qquad \text{Equ 10.4}$$

10.3 Sources of valve noise

Valve noise can be attributed to two main sources:

- A product of mechanical vibration
- A result of fluidic action

10.3.1 Mechanical vibration

Noise caused by the vibration of valve components is normally unpredictable and is usually of secondary concern to the process engineer. The noise may even be considered beneficial since it provides a warning of abnormal conditions within the valve. Table 10.1 shows the various types of mechanical vibration noises which are associated with valves and their possible causes.

Problems associated with this type of vibration/noise have been overcome in many cases by:

- improved design
- recognition of the service limitations of particular valve types
- close attention to working clearances in guides
- the balancing of forces on the valve element

- selecting a suitable actuator

Frequency range	Description	Typical causes
20 to 100 Hz	Oscillatory movement of valve stem/spindle/plug/disc	Inadequate actuator power Actuator instability High fluid velocity Wear
500 to 1 500 Hz	Lateral movement of stem/spindle/plug in its guides	Fluid impingement on plug Excessive guide clearances
3 kHz to 7 kHz	Resonant vibration of valve component (discrete frequency)	Flimsy fabrications Flexible components

Table 10.1 Typical valve mechanical vibration frequencies

10.3.2 Fluid noise

Noise caused by fluidic action may be considered under two sections:

- aerodynamic noise — due to compressible fluids
- hydrodynamic noise— due to incompressible fluids

NOTE: It is worth restating that all fluids are compressible. Liquids are much less compressible than gases but if the pressure is high and the differential pressure great, the compressibility effects should not be ignored until evaluated. Liquid working pressures of 350 barg are commonplace. A popular application pressurises methanol to 520 barg. Liquified natural gases, which are much more compressible than water, are pumped routinely.

10.3.2.1 Aerodynamic noise

This type of noise represents the most serious source of valve noise resulting from the high velocity flow of the fluid. The high velocity results in aerodynamic turbulence, which is generated in the fluid downstream of the trim, as a result of the constricted jet mixing with the slower moving stream. The turbulence generated is carried downstream until pipeline conditions cause it to appear as noise. Further generation of aerodynamic noise can occur when "choked flow" conditions occur at the valve outlet in which case the outlet noise tends to dominate the valve noise.

Methods of attenuating aerodynamically generated noise levels include:

- using low noise trims
- fitting baffle plate silencers
- fitting in-line silencers
- fitting acoustic lagging or an enclosure
- using thick-walled pipework

Generally, noise can be reduced by avoiding the use of high recovery valves.

To prevent the valve outlet becoming the source of noise, a minimum recommended outlet diameter is calculated. This is based on the diameter necessary to produce half sonic velocity. (See Water hammer in Chapter 1.) Entrained gas in liquids or solids can drastically reduce the sonic velocity and produce erratic results.

10.3.2.2 Hydrodynamic noise

The most common source of hydrodynamic noise is cavitation, when part of the liquid undergoes two changes in its state. As the liquid flows through the valve trim, there is an increase in its velocity accompanied by a decrease in liquid pressure, see Figure 10.1. If the liquid pressure falls below the vapour pressure at the inlet temperature, part of the liquid will change its state to become vapour. Downstream of the trim there will be a certain amount of pressure recovery, dependent upon the valve type. If the pressure returns to a value above the vapour pressure the vapour bubbles implode as the vapour condenses to liquid, resulting in cavitation.

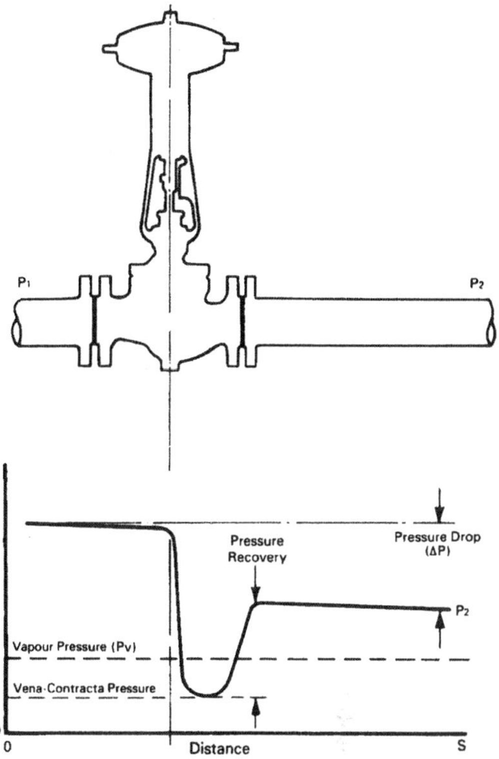

Figure 10.1 Valve pressure profile during cavitation

Cavitation noise can generate sound pressure levels in excess of 100 dB(A). In addition to the noise, the implosion of vapour bubbles results in extremely high pin-point pressures which may destroy the molecular structure of the metal and result in considerable damage. Severe cavitation and vaporisation can also result in "choked flow". In this condition, further increases in pressure drop do not produce a corresponding increase in flow rate. To anticipate these problems, the sizing formulae (see Chapter 8), take into account flashing and cavitation by limiting the pressure drop used in the sizing calculation.

10.3.3 Pipework noise

To prevent the valve outlet becoming a source of noise it is necessary to limit the outlet fluid velocity. The following formulae provide the basis for determining the minimum outlet diameter for low noise operation, based upon the diameter to produce 0.3 sonic velocity.

Metric units:

Gas

$$d = 1.63\sqrt{\frac{Q\sqrt{G}}{P_2}}\,\text{mm} \qquad \text{Equ 10.5}$$

Steam

$$d = 2.19\sqrt{\frac{W}{P_2}}\,\text{mm} \qquad \text{Equ 10.6}$$

For inch pipework:

Gas

$$d = 0.041\sqrt{\frac{Q\sqrt{G}}{P_2}}\,\text{inches} \qquad \text{Equ 10.7}$$

Sound pressure level prediction graph for compressible fluids

Use of the graph

A Determination of sound pressure level given valve type and process data

From the intersection of the valve trim design line and $\Delta P/P_1$ line, draw a vertical line to intersect the pressure drop line. From this point draw a horizontal line to intersect the C_v line. From this point draw a line parallel to the sound pressure level lines and read off the overall S.P.L. in dBA.

B Determination of valve trim design given process data and the maximum permissible sound pressure level

From the sound pressure level valve, draw a line parallel to the S.P.L. lines to intersect the C_v line. From this point draw a horizontal line to intersect the pressure drop line. Contuinue the line vertically to intersect the $\Delta P/P_1$ line. From this point draw a horizontal line and read off the valve trim design necessary to meet the maximum permissible S.P.L.

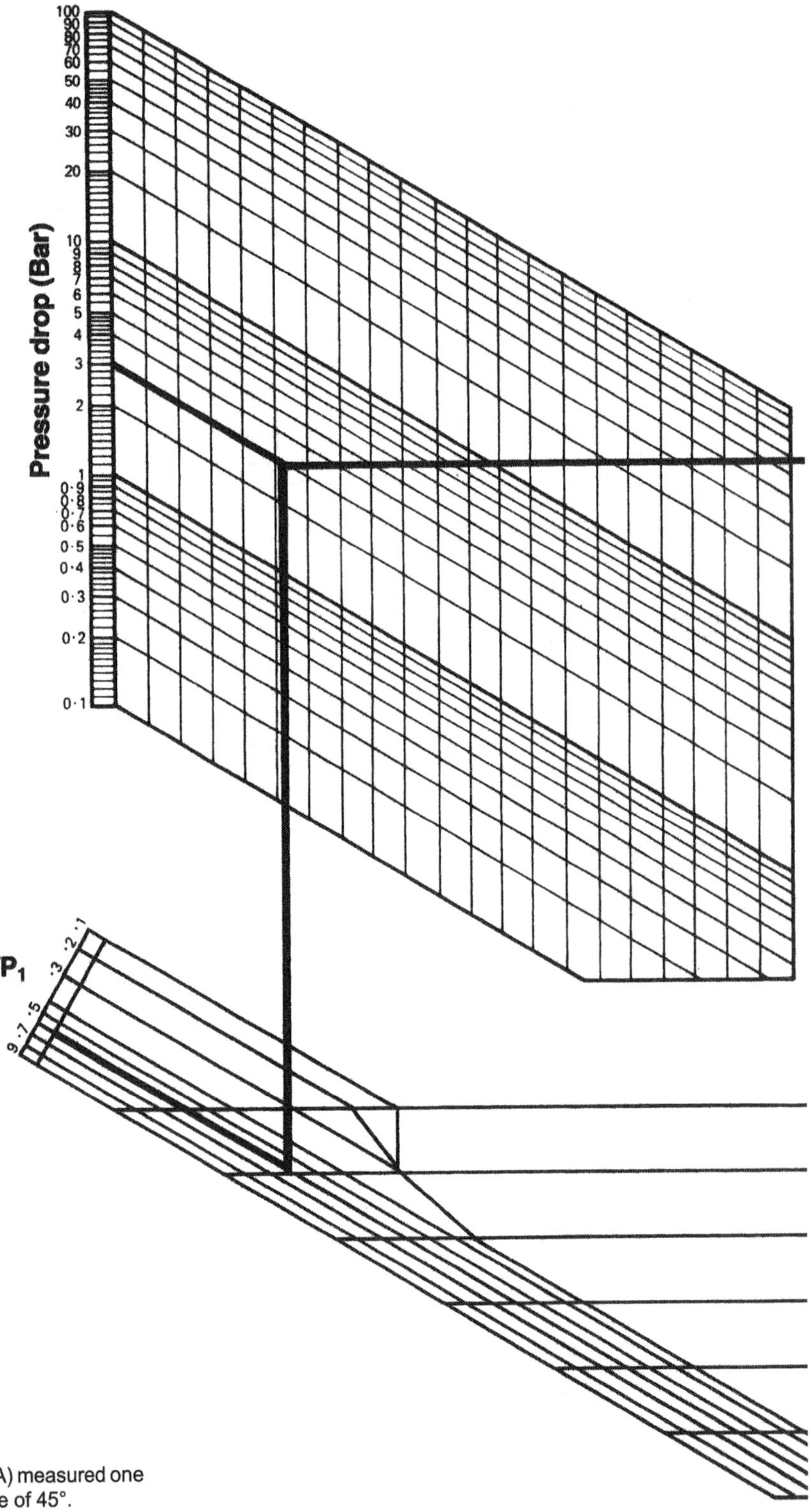

Providing overall sound pressure level in dB(A) measured one metre downstream from the valve at an angle of 45°.

Figure 10.2 Sound pressure level prediction graph for compressible fluids

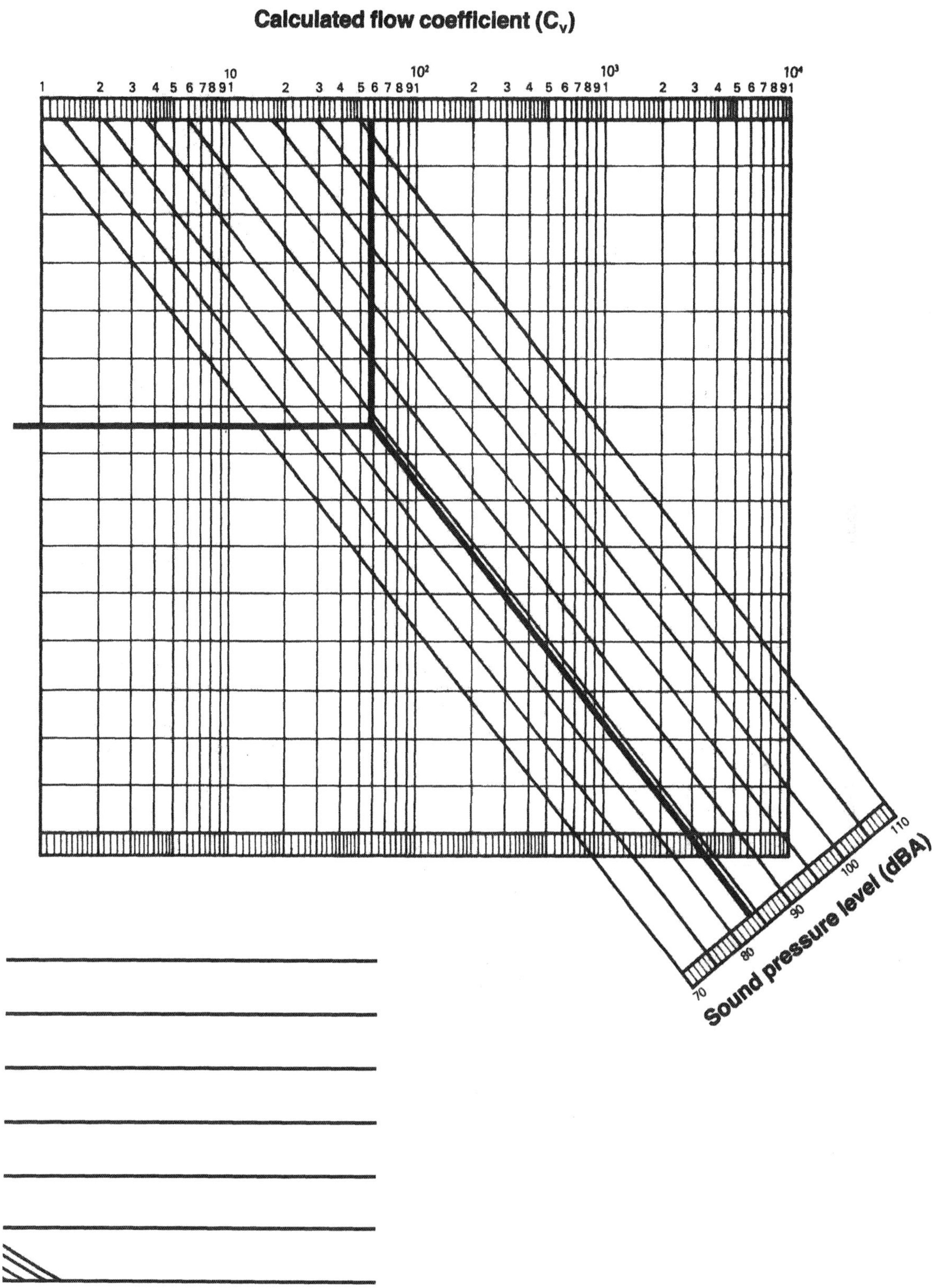
Calculated flow coefficient (C_v)
1 2 3 4 5 6 7 8 9 1
10
10^2
10^3
10^4
70
80
90
100
110
Sound pressure level (dBA)

Steam

$$d = 0.222\sqrt{\frac{W}{P_2}} \text{ inches} \qquad \text{Equ 10.8}$$

where:

Q	=	volume flow rate (m^3/h)
G	=	relative density (non-dimensional)
P_2	=	outlet pressure (bara)
W	=	mass flow rate (kg/h)

10.4 Noise prediction

The graph in Figure 10.2 enables a prediction of the sound pressure level generated by a control valve on compressible fluid service. It expresses the sound pressure level in dB(A) measured one metre from the downstream side of the valve at an angle of 45°. For a complete understanding of the sound pressure level generated by the valve, a prediction of the octave band analysis may be necessary.

The table in Figure 10.3 indicates the average departure from the overall predicted noise level in each of the 11 standard octave bands. An octave band analysis is obtained by subtracting the figures given in the table from the overall sound pressure level.

The noise reductions by low recovery trims are based on results from work carried out by the former Hopkinsons Ltd - now part of Weir Valves & Controls.

10.5 Noise attenuation

Having determined the sound pressure level of a valve, either by measurement or prediction it may prove necessary to attenuate the generated noise. This may be done:

- at the source
- by path treatment

10.5.1 Source treatment

In the prevention of high noise level generation, treatment at the source represents the logical approach. If a valve is installed which generates acoustic power at a level lower than the legislation requirement, then there will be no noise problem. For a number of years, multi-path trim for control valves, see Figure 10.4 and Figure 6.14 in Chapter 6, has been specified for low noise generation on high pressure drop applications where noise would have been of concern. The body of fluid here is broken down into many small jets.

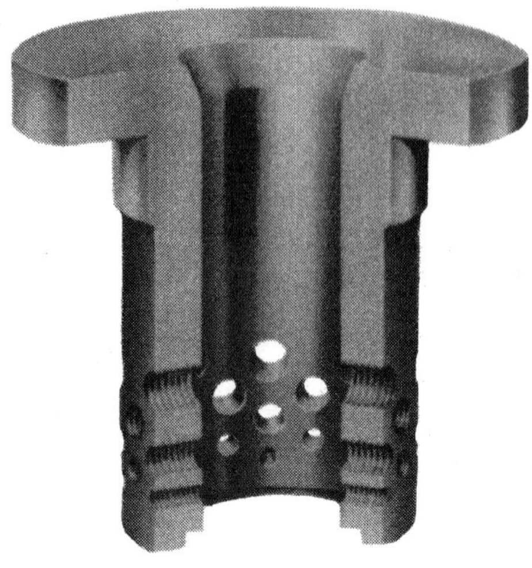

Figure 10.4 Multi-path control valve trim

However, increasingly higher service conditions, coupled with more stringent noise limitations, have created the need for more advanced low noise equipment. Manufacturers have developed additional styles of trims to achieve sound pressure levels some 25 dB lower than a conventional valve on compressible fluid service.

Noise caused by turbulent flow is particularly sensitive to flow path geometry, a fact which is the basis of noise reduction. As fluid flows from the inside of the cage through the cascade trim (see Figure 10.5), to the outside, adiabatic expansion of the fluid takes place with friction reheat. Pressure energy is therefore reduced with the minimum generation of acoustic power and the minimum of pressure recovery. The valve body is of adequate size to ensure low velocities at both inlet and outlet. By

Octave band centre frequency (Hz)	31.5	63	125	250	500	1K	2K	4K	8K	16K	31.5K
Difference to be subtracted from overall S.P.L. (dBA)	15	15	17	18	15	8	6	6	8	18	33

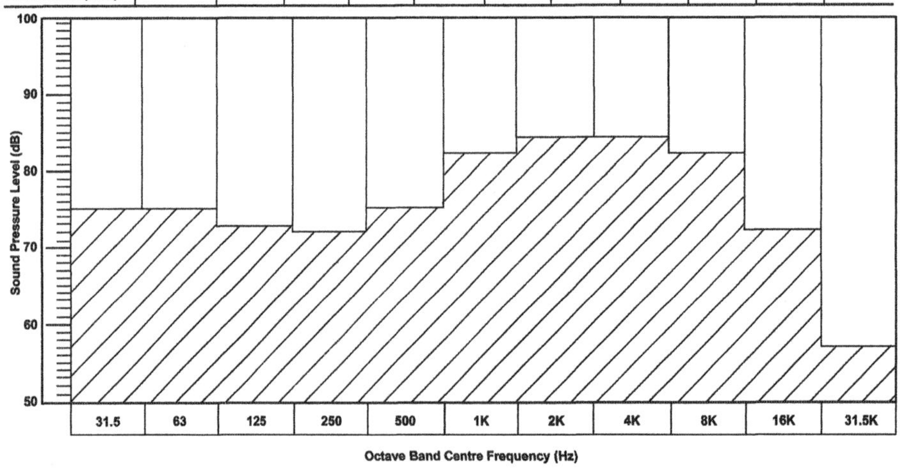

Figure 10.3 Octave band correction values

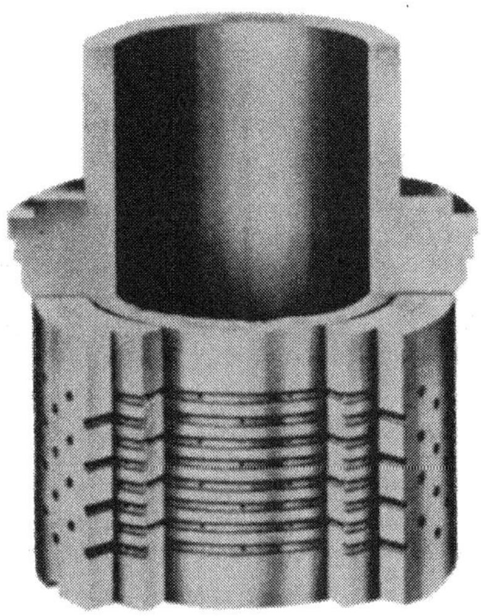

Figure 10.5 A typical cascade trim

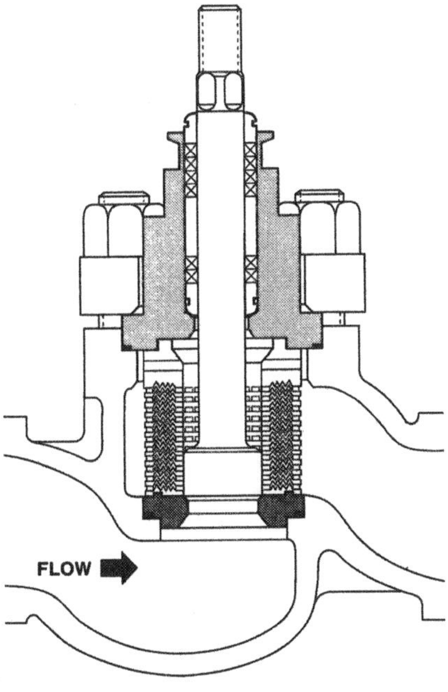

Figure 10.6 An unbalanced Tiger-Tooth™ valve
Courtesy of Flowserve Flow Control

providing a complete guide for the plug throughout the valve travel, the caged design prevents lateral vibration.

An alternative construction style, which also greatly reduces the likelihood of cavitation damage and noise, is shown in Figures 10.6 and 10.7. The expanding liquid, suitable for gas also but, of course, no cavitation damage, is allowed to flow through a multitude of expanding passages which have sudden expansions and sharp corners to destroy pressure and velocity energy. The plug guidance and the high level of overall "stiffness" can be easily seen. The level of noise reduction possible is shown in Figure 10.8. Similar types of trim are available from the so-called DRAG™ valve.

As an additional method of source reduction, a basket type baffle plate silencer, Figure 10.9, may be fitted to the downstream side of the valve. The valve/silencer unit which is designed to ensure low inlet and outlet velocities prevents the generation of high acoustic power by reducing the pressure energy in stages. Sound pressure levels may be reduced by up to 15 dB compared with a conventional valve. The units are designed to pro-

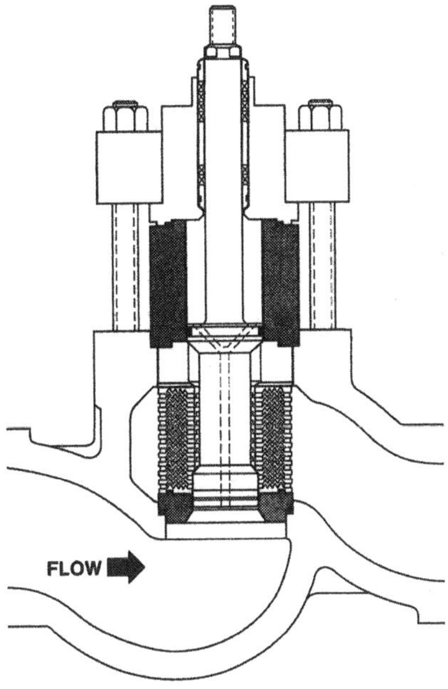

Figure 10.7 A balanced Tiger-Tooth™ valve
Courtesy of Flowserve Flow Control

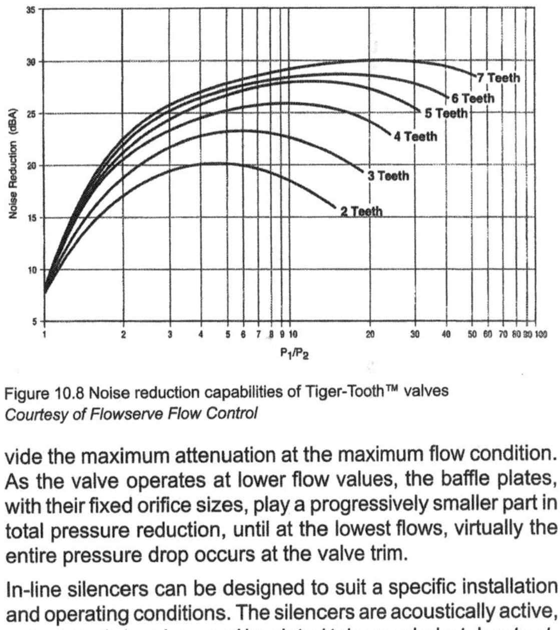

Figure 10.8 Noise reduction capabilities of Tiger-Tooth™ valves
Courtesy of Flowserve Flow Control

vide the maximum attenuation at the maximum flow condition. As the valve operates at lower flow values, the baffle plates, with their fixed orifice sizes, play a progressively smaller part in total pressure reduction, until at the lowest flows, virtually the entire pressure drop occurs at the valve trim.

In-line silencers can be designed to suit a specific installation and operating conditions. The silencers are acoustically active, using tuned chambers and insulated tubes or choke tubes to attenuate a range of frequencies. Figure 10.10 shows the construction of a typical proprietary silencer used with gases and vapours.

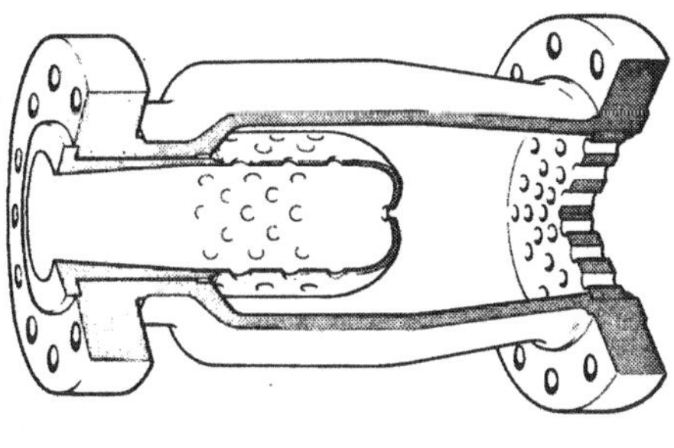

Figure 10.9 Basket type baffle plate silencer

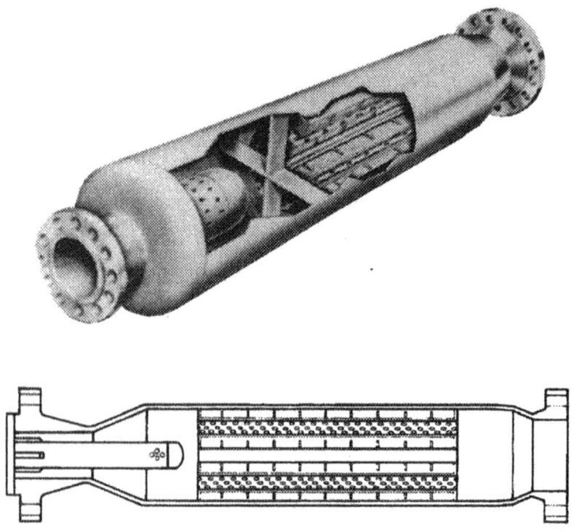

Figure 10.10 In-line acoustic silencer
Courtesy of Flowserve Flow Control

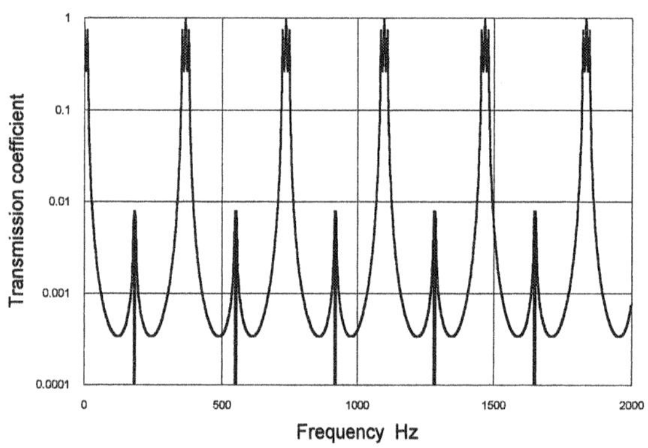

Figure 10.11 Attenuation performance of a simple silencer

Figure 10.11 shows the theoretical transmission coefficient for a simple silencer consisting of two volume bottles connected by a choke tube. The theoretical performance must be modified by an efficiency factor which considers the transition from simple plane wave theory to 3D acoustic waves. Simple silencers of this style, which can be used for liquids, offer attenuation over a wide frequency range. Silencers can be designed so the pass bands (transmission coefficient = 1) avoid significant frequencies in the noise spectrum. In-line silencers can be used with all valve types including the atmospheric outlet of safety relief valves.

10.5.2 Path treatment

In piping systems which are not vented to atmosphere, noise generated in the fluid can only become airborne through vibration of the pipe wall, which, in effect, forms the solid boundary of the fluid. The object of path treatment of high noise levels is to increase the impedance or resistance within the transmission path in order to reduce the acoustic energy transmitted.

Noise generated within a closed system may be carried considerable distances upstream and downstream with minimal loss of intensity. In practice this means that attenuation treatment must be carefully considered for the entire pipeline or the noise may be emitted elsewhere. However, changes in the transmission characteristics of the pipeline will affect the output of noise.

Supports, which are rigidly clamped to the pipework, invariably act as efficient carriers of noise and vibration energy to connected structures, which may themselves convert the vibration to airborne noise. If path treatment is to be fully effective it is essential that rigid supports are replaced with a resilient type along the treated sections. Resilient pipe supports must be selected with great care. The pipe support function cannot be compromised in the interests of noise abatement; both functions must be executed effectively. Poor pipe restraints can lead to pipe failures and the fracture of nozzles on vessels and machinery. Resilient pipe supports must exhibit the required stiffness to fulfil the support role.

Path treatment may be carried out by:

- increasing the pipe wall thickness
- lagging the valve and associated pipework with acoustic insulation
- enclosing the valve and pipework

10.5.3 Pipe wall thickness

The effect of increasing the wall thickness of the downstream pipework is shown in Figure 10.12. Sound pressure level attenuation is related to the pipe schedule and pipe diameter.

10.5.4 Acoustic insulation

Figure 10.13 indicates the sound pressure level (SPL) attenuation which is provided by various thicknesses and densities of acoustic insulation. To obtain the maximum benefit the insulation should be used in conjunction with an outer layer of material such as lead or cement in order to form a double skin.

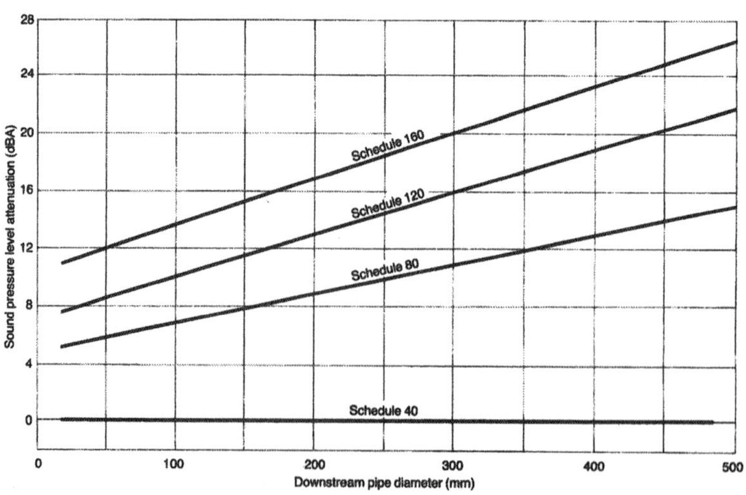

Figure 10.12 Sound pressure level (SPL) attenuation by heavier pipe schedules

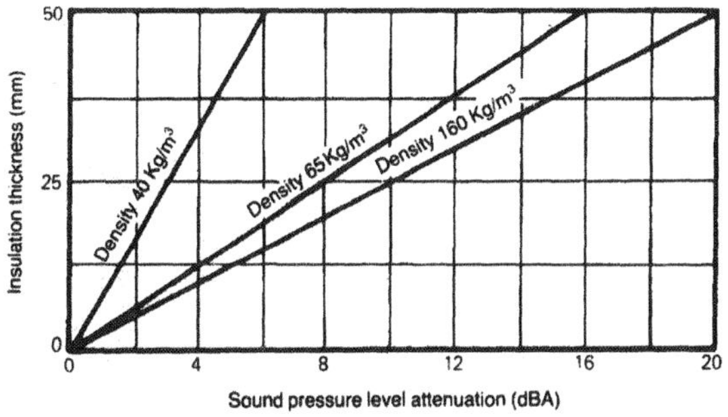

Figure 10.13 Sound pressure level (SPL) attenuation by acoustic insulation

10.5.5 Acoustic enclosures

By enclosing the valve and associated pipework in a suitably designed acoustic enclosure, the transmission of high acoustic energy is reduced. The level of attenuation provided is dependent upon individual design and enclosures are normally tailor-made to meet a specification.

Acoustic enclosures are very effective, very costly and disliked intensely on site. Cursory maintenance, by passing staff, is impossible. Access hatches must be provided for routine maintenance; these are often bulky and must be securely attached to preserve the attenuation. Little indications, such as dripping glands and rusty spindles, go unnoticed until the next routine inspection.

10.5.6 Ear protector zones

The acoustic enclosure problems can be solved by declaring an "Ear Protector Zone". The user defines a specific area of the plant and provides ear protectors at every entrance. Ear protection must be available at the Zone. It is the responsibility of the equipment user to ensure all staff comply with the safety requirements and wear ear protection. This is a very cost-effective way to solve a local problem.

Ear protectors would normally be available anyway. When on-line maintenance or inspection is carried out on insulated or "enclosed" valves, ear protection would be required because of the temporary increase in local noise level.

10.6 Noise calculations

Most valve manufacturers have developed proprietary methods to evaluate valve noise. Typical calculations expect to be within ± 5 dB(A). Most calculations are limited to noise levels greater than 70 dB(A); valves quieter than this are considered to be acceptable. Valve noise is predicted, based on the valve installation being in a free-field, i.e. no reflecting surfaces. If the valve is installed close to hard reflective surfaces, add 3 dB(A) for each close surface. Manufacturers and independent software houses have produced computer programs to predict noise levels.

Isolating and non-return valves can be treated as control valves with very low differential pressures. Valve noise should not be a problem unless velocities are high and fluid forces create mechanical noise in the valve.

Noise from safety relief valves can be predicted using the method in API RP 521. The following formula is used for gases, vapour and steam:

$$L_{100} = L + 10\log_{10}\left\{0.29354W\,k\frac{T}{M}\right\} \qquad \text{Equ 10.9}$$

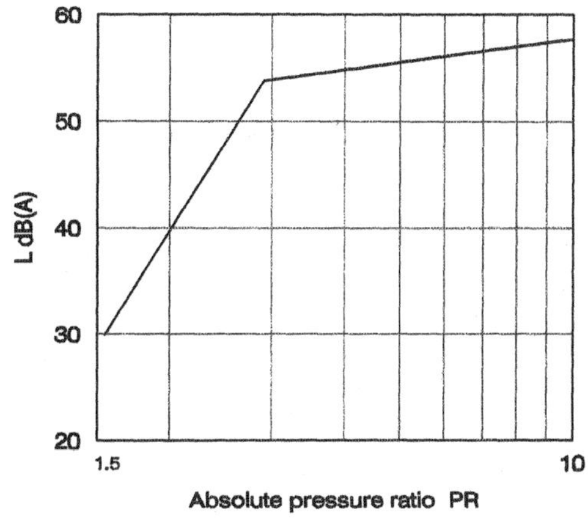

Figure 10.14 Sound pressure level L for Equation 10.9

where:

L_{100} = sound pressure level at 100 ft from point of discharge (dB(A))

L = sound pressure level from Figure 10.14 (dB(A))

PR = absolute relieving pressure divided by absolute back pressure (non-dimensional)

W = maximum relieving capacity (lb/h)

k = ratio of specific heats (non-dimensional)

T = absolute inlet temperature (R)

M = = molecular weight (non-dimensional)

The predicted noise level is based on the relieving capacity of the valve at the specified overpressure, and not the rated capacity of the process. If the noise level is required at a different distance from the valve, use Equation 10.10.

$$L_p = L_{100} - 20\log_{10}\left\{\frac{r}{100}\right\} \qquad \text{Equ 10.10}$$

where:

L_p = sound pressure level at distance r from the point of discharge (dB(A))

r = distance from the point of discharge (feet)

Regulators should be treated as control valves. The chart in Figure 10.1 can be used or manufacturers' proprietary calculations. Standard calculations were given in IEC 60534-8-3 which is now withdrawn.

Examples of sizing and noise calculations are shown in Sections 8.2 and 8.4 in Chapter 8.

10.7 Useful references

Sound Generated Aerodynamically, M J Lighthill, The Bakerian Lecture, 1961, Proceedings of the Royal Society of London. Series A, Mathematical and Physical Sciences, Vol. 267, No. 1329 (May 8, 1962), pp 147-182.

API RP 521 Guide for Pressure-Relieving and Depressuring Systems.

IEC 60534-8-3 Industrial-process control valves - Part 8: Noise considerations - Section 3: Control valve aerodynamic noise prediction method.

Valve stem sealing 11

11.1 Introduction

Valve stem sealing is used to reduce or eliminate product leakage out of the valve. For sub-atmospheric applications the stem seal must reduce or eliminate air leakage into the valve. Many styles of seals have been developed to accommodate the wide range of sizes and operating conditions encountered in modern applications.

Greater public awareness of the environmental impact of leakage from processing plants has led to the development of improved glands, which now also include asbestos free packings. Much of this success was built on the development of "fire-safe" ball valves. Portable electronic sensors permit easy checking of seals in valves with monitoring ports and allow enforcement of environmental regulations.

Most industrialised countries have Standards regulating the emissions of compounds which are hazardous to personnel and/or the environment. This Chapter indicates the range of seal options which may be fitted to valves to achieve the required sealing.

11.1.1 Basic concepts

The majority of stem/spindle sealing is achieved by compressing a soft flexible material between the stem/spindle and the bore of the packing box. The selected material should be softer than the valve parts to reduce wear. The material should also be very flexible so that it can "flow" into the available space in the packing box. At a micro-structure level, the flexible material must be able to flow into the surface finish irregularities of both the stem/spindle and the packing box.

From a valve mechanical design viewpoint the stem/spindle should have a large diameter to be able to:

- resist bending forces
- have low stresses due to actuating forces/torques
- have a very high natural frequency

From a valve sealing viewpoint the stem/spindle should be as small as possible to:

- reduce the area of potential leak paths
- utilise the smallest seal size
- save material around the packing box

The two sets of objectives are in direct conflict and the final design is a compromise to achieve satisfactory performance at a market price level. In general "zero-leakage" is not achieved. Leakage is reduced to very small quantities providing the seal is fitted correctly, maintained properly and operated within specific parameters.

In some cases a physical barrier can be mounted across the complete leakage path. If the barrier is metallic then effectively "zero-leakage" will be achieved, see Section 11.4. If the barrier is elastomeric then a very small leakage will result due to the permeability of the material. As with other types of leakage, the permeability leakage is a function of the differential pressure.

The difficulties of stem/spindle sealing can be overcome by designing a valve which does not require a packing box. See Figures 3.20, 3.22 and 3.23 and Table 3.6 in Chapter 3.

Some rotary valve manufacturers claim better long term performance of rotary seals. It was claimed that reciprocating seals could be degraded by the stem movement. This was illustrated by tests on a number of similar valves and measuring helium leakage through the seals. After 10 thousand operations the linear valve seal was leaking 100 times more than the rotary valve. This is a perfectly acceptable comparison for helium valves. It is not acceptable to extrapolate these test results to other gases or liquids. The molecules of volatile organic compounds (VOCs) are considerably larger than helium and helium tests do not prove increased VOC emissions. For liquids the initial leakage would be very small and after 10 thousand operations the leakage may not be measurably different. Methane can be used as the test gas for quantifying VOC leakage. Methane is the test gas specified by the US Environmental Protection Agency (EPA), for laboratory qualifying tests for valve packing arrangements. Other suitable gases proposed are hexane and isobutylene.

One cause of poor linear valve performance is cited as dust and contamination being carried into the seal during stroking. In poor ambient conditions, both rotary and reciprocating seals can be protected from the environment. In a rotary valve the same section of spindle is in contact with the seal at all times; therefore the wear is concentrated in one place. Also, if moisture is absorbed in the last seal ring local corrosion can be greatly accelerated.

The effect of leaking product through the last ring to the atmosphere can seriously damage both linear and rotary valve seals giving rise to the following:

- does a liquid evaporate to form crystals or deposit solids?
- does a gas/vapour react with moisture in the air?
- does the process fluid harden in the last ring and ruin its sealing capabilities?

There may be many cases where a water, (or compatible liquid/steam), quench can considerably extend the useful seal life. Of course, it can also prevent damaging atmospheric contamination.

Accurate, consistent emission monitoring on site may be a problem with valves not fitted with monitoring ports. Regular routine checks of valve seal leakage are useless if the data is unreliable. It appears necessary to have strict procedures to control all aspects of monitoring and to ensure staff compliance. For outdoor equipment, the position of the detector and wind conditions can produce very variable results.

Hazardous products leaking from the packing box are not the only problem; "safe" products can create problems. Liquids passing through the seal can flash to vapour as the pressure decreases causing "wire drawing", resulting in very short seal life and possible stem/spindle damage. Liquids leaking through the packing can crystallise on contact with the air or vaporise leaving salt or solid deposits. Any solids close to the seal can shorten the effective life. Perhaps the fluid reacts with ambient air in the last packing ring and degrades the seal. This type of attack progressively destroys the complete seal.

The stem/spindle seal is only one potential leak path. Other openings must be sealed effectively and pipe connections must be suitable for the service. For very hazardous fluids the valve body itself must be inspected to ensure integrity. A bellows sealed stem cannot eliminate leakage through a porous valve casting or weld. Similarly, one would not expect to find a valve handling a hazardous fluid fitted with tapered thread process connections or flat flanges using cork gaskets.

It is important to remember that the overall site design may be influenced by product leakage and spillage. Tanks for flammable liquids should be isolated within bunds — low liquid-proof walls or embankments — to prevent the distribution of leakage. Local regulations may require such tanks to be located below ground level. These precautions are also necessary for some gaseous fuels which are heavier than air. The site layout should not include pipe trenches. Individual equipment packages should be mounted on flat structures without pockets which could trap leakage.

A thorough understanding of the products handled is necessary to allow the formulation of a construction philosophy which can be applied consistently. Valves play a small, but vital, part in fluids handling. Fitting "high-spec" valves cannot ensure a safe in-

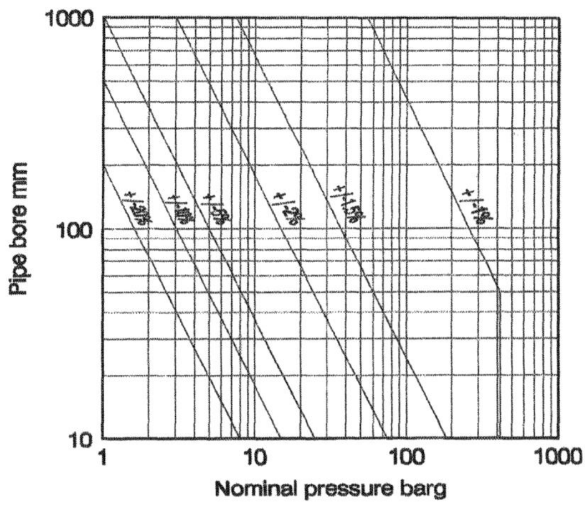

Figure 11.1 Recommended residual pipework pressure pulsations

stallation if other aspects of basic design have been overlooked.

Pressure and temperature ratings quoted in this Chapter are based on steady-state operating conditions. Process conditions are assumed to change slowly enough to allow all components to adjust without seizing or experiencing thermal shock or pressure pulses. Thermal shock and pressure pulses can be very difficult to quantify. Manufacturers of large, high pressure equipment specify temperature changes of less than 50°C per hour for warm-up from cold to operational temperatures. Small, low pressure equipment can have more rapid temperature changes but it is difficult to say how much faster. Piping systems can withstand a certain level of pressure pulsations without suffering undue vibration. The graph shown in Figure 11.1 has been proposed to both CEN and ISO for guidance. If pressure pulsations greater than indicated are to be experienced then the valve manufacturer should be informed.

11.1.2 Explosive decompression

One phenomenon of modern process plants that requires special consideration is explosive decompression. Explosive decompression is the uncontrolled depressurisation of a fluid system due to a failure such as a pipe fracture. High pressure fluid which has been absorbed by elastomers in seals and gaskets expands rapidly and destroys the seal.

Pressurised liquid systems can be damaged by a sudden fall in temperature as liquid flashes to vapour at the atmospheric boiling temperature. Brittle fracture of metallic components is a possibility. If there is a risk of explosive decompression then the full implications must be discussed with the valve manufacturer.

11.2 Single seals

Stem/spindle sealing can be extremely simple. Figure 11.2 shows the simplest possible seal which can be used on both linear and rotary valves.

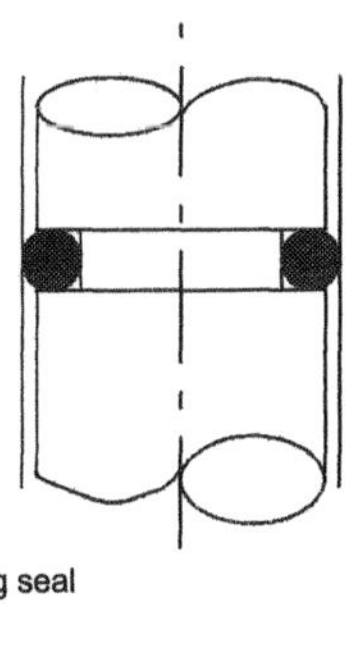

Figure 11.2 Single "O" ring seal

The compression of the "O" ring is controlled by the size of the groove and the packing box bore, a gland, as such, is not required. Wear or corrosion of stem/spindle/packing box results in a loss of compression and hence sealing ability. The surface finish of the valve parts must be good to reduce wear, typically better than 0.4 μm. The plain "O" ring can be replaced by a QUAD™ ring, also called a "lobed" ring, for better sealing qualities. The packing box opening must be machined correctly to provide a good "lead-in" for the "O" ring.

This type of seal is ideal for sites for where the level of maintenance skill is low. No adjustment is possible so the seal life cannot be shortened by over-tightening. Care is required when replacing the seal to apply the correct lubricant specified by the valve manufacturer. The stem/spindle must be removed from the packing box to replace the seal. A modified design allows quicker and easier replacement, see Figure 11.3. The "O" ring is held in position by a non-adjustable gland, replacement involves removing the gland not the stem/spindle. A metal back-up ring in the bottom of the packing box allows clearances to be relaxed but wear to be accommodated. The removal of the groove in the stem/spindle reduces a stress concentration problem.

Figure 11.4 shows a conventional soft packed box with three rings of packing and spherically mounted bolted gland. The top and bottom packing rings are "filled", reinforced with glass or carbon fibre, to increase the ring stiffness and prevent extrusion. Braided packing with reinforced corners can also be used to advantage. This type of packing is adjustable and requires time and skill to bed-in and maintain. Screwed and union glands require less time but are restricted to smaller valves. Bolted glands demand more effort. Soft packing which is not die-formed can compress to about 60% of its original volume. Sufficient gland travel must be available.

Figure 11.5 shows an alternative method of increasing packing stiffness and reducing the likelihood of extrusion. Thinner filled spacers are fitted between the sealing rings. Solid, rather than braided, PTFE rings filled with glass or carbon fibre are popular. Solid PFA or PEEK rings can also be used. Graphite foil or graphite filament spacer rings can be used to assist heat transfer into the packing box wall. Carbon spacers have been used successfully with graphite packing to prevent extrusion.

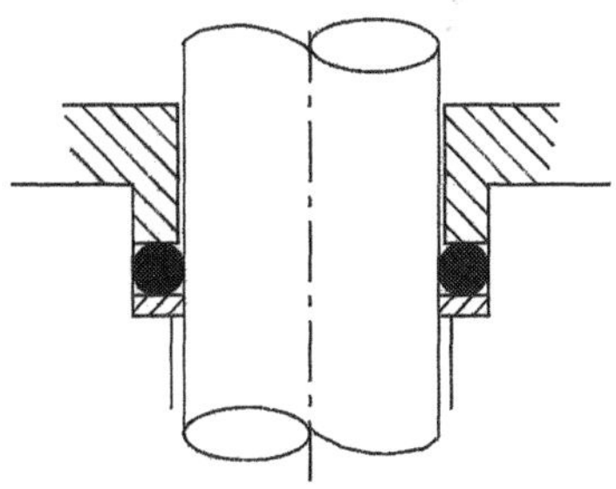

Figure 11.3 Improved single "O" ring seal

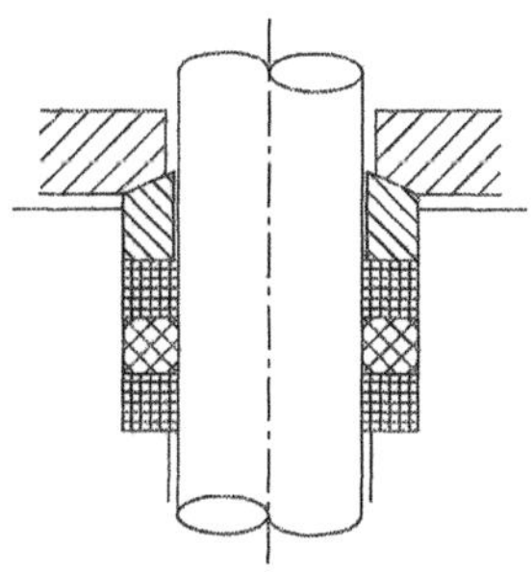

Figure 11.4 Plain soft packed box

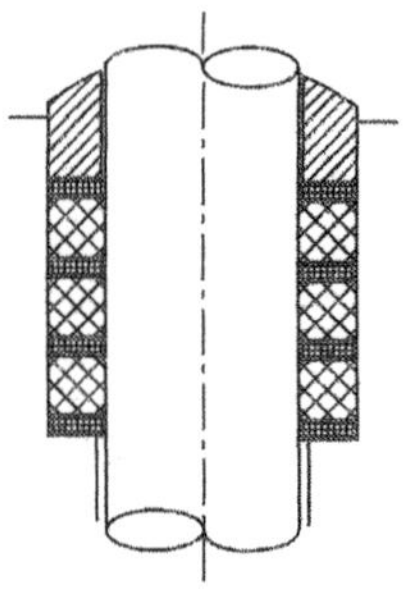

Figure 11.5 Soft packing with stiffening spacers

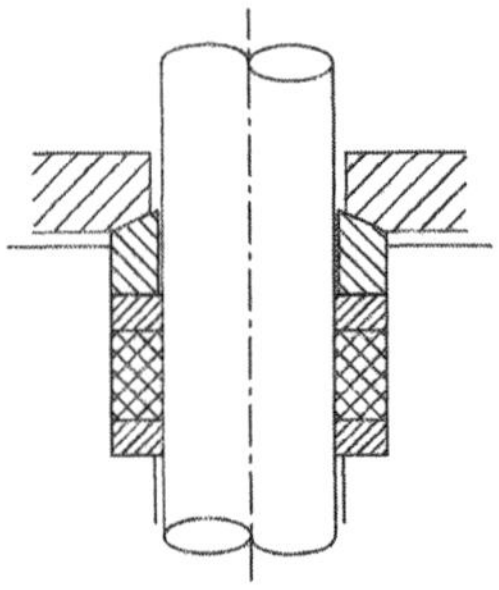

Figure 11.6 Soft packed box with a slug of packing

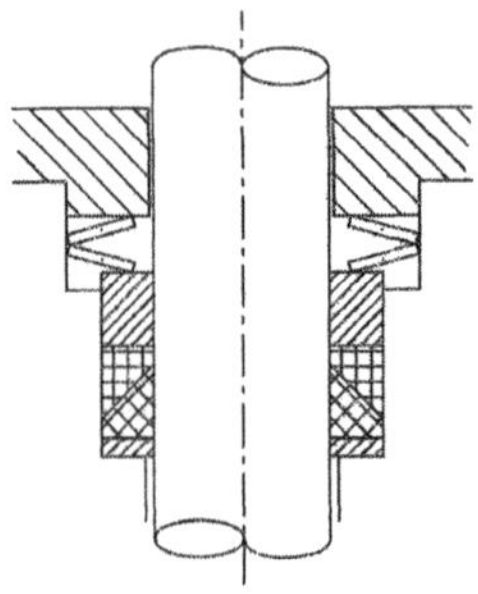

Figure 11.7 Live-loaded packing set

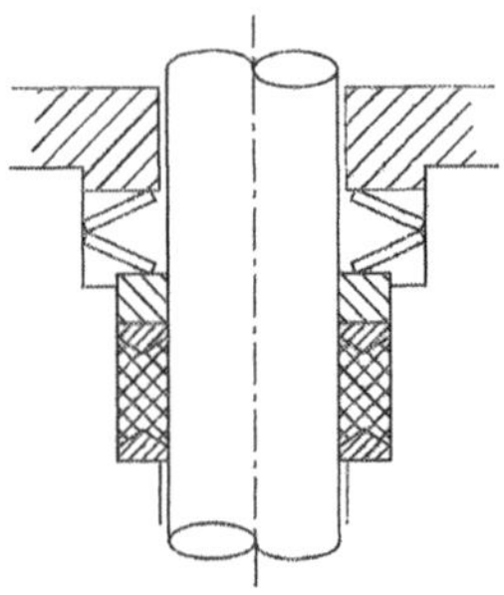

Figure 11.8 Fire-safe packing arrangement

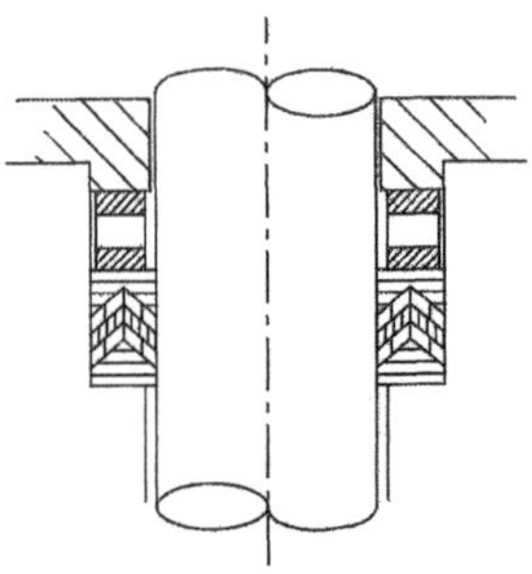

Figure 11.9 Self-energised chevron packing

Figure 11.6 shows another soft packed box but with a "slug" of packing. PTFE and Grafoil™ packing have been used. The top and bottom headers may be quoted as "stainless steel" but this probably means 11/13 Cr. 17-4PH is a better alternative.

Figure 11.7 shows a soft packed box with specially shaped die-formed rings which allows the packing to be "live-loaded". Two disc springs are fitted to provide an axial load on the packing follower. The gland is non-adjustable and locked down solid. Compression of the packing is accommodated by the disc spring deflection. A longer spring stack can be fitted to provide more follower travel. Increased spring load can be applied by doubling up the springs. The specially shaped rings have been produce in PFA, PEEK and UHMWPE to give broad chemical compatibility and temperature ranges.

Figure 11.8 shows another version of the live-loaded packing which has been qualified for fire-safe ball valves. The seal material is reinforced Grafoil™ with metal headers. The maximum process fluid temperature is 121°C but the complete valve assembly has withstood flame temperatures of between 760° and 982°C for 30 minutes. Some graphite packings have created problems when graphite particles have become attached to the stem/spindle. Very good surface finishes reduce the problem but good lubrication and the correct binders are the ultimate solution.

All the packing arrangements described so far have been loaded by pre-compression. Axial forces applied by the gland or a spring have induced the packing to fill the radial space and provide the seal. Figure 11.9 shows a self-energised spring-loaded packing arrangement as fitted to many ball valves. Chevron rings are supported between shaped headers. The fluid pressure under the chevron causes the ring to expand radially and effect a seal on the stem/spindle and the packing box bore. Some initial axial load is required and this is provided by a wave spring.

Several manufacturers supply chevron rings and small detail design differences exist. Some designs have close-fitting chevrons as shown. Other designs include a pocket on the underside of each chevron. The suppliers of the different variations will argue strongly for the benefits of their design. The construction of the headers is probably more important than the chevrons. Plastic or thermoplastic headers are popular, actetal is widely used. Split headers are sometimes used. For critical applications it is worth insisting on solid (unsplit) metal headers. Phosphor bronze, 11-13 Cr or 17-4PH can be used depending upon the application. If the packing box design is good, the headers will not wear and the extra initial cost will easily be recovered. The chevron rings can be endless, that is solid, or split. High pressure applications should always use endless rings even though it takes longer to fit them.

The pressure rating of the "chevron set" is dictated by the number of chevron rings used. Low pressure applications may only require one ring. High pressure applications may require five rings. Some designs add extra sealing capabilities into the top header. It is better to have a plain header and fit an extra standard chevron ring. With long, high pressure packing sets the flexibility of the chevron rings may cause problems by allowing the central rings to partially extrude. This effect can be overcome in a similar manner to adding filled spacers in braided packing. The rigidity of chevron rings can be improved by increasing the fibre content or by using a different ring material. Virgin PTFE chevron rings can be supported by glass or carbon fibre filled PTFE rings. PEEK can be used in support rings.

Figure 11.10 shows a spring-loaded chevron packing arrangement also used in ball valves. The packing is fitted with the spindle from inside the valve body. The wave spring is mounted outside the packing box and exerts its load against a thrust washer which is part of the handle assembly and fitted from the outside. There is a subtle difference between this packing ar-

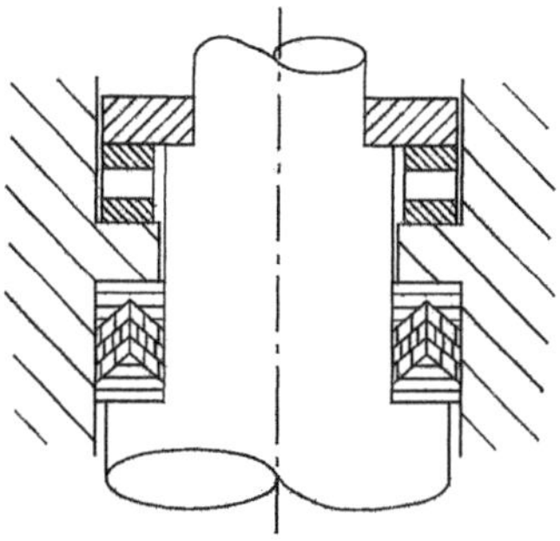

Figure 11.10 Spring-loaded chevron packing

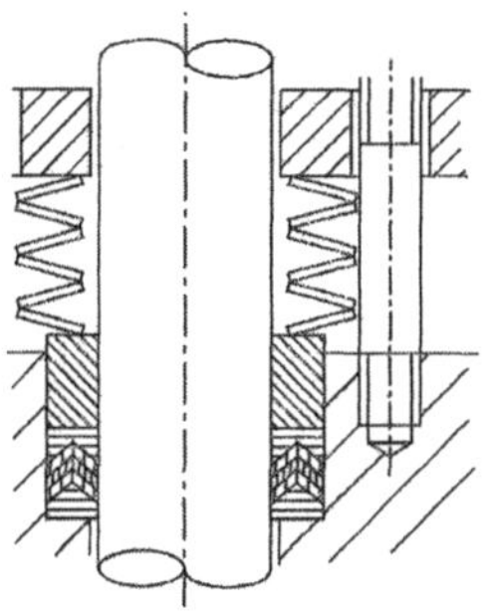

Figure 11.11 Externally loaded chevron packing

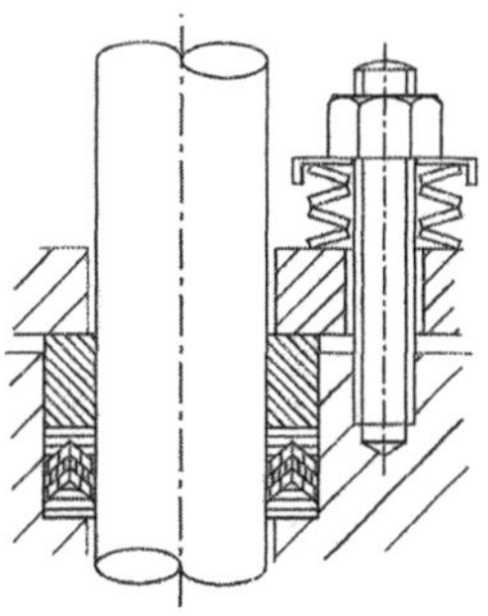

Figure 11.12 Externally loaded gland with chevron packing

rangement and the other spring-loaded styles discussed previously. In Figure 11.10 the spring load assists the internal fluid pressure in applying the sealing force. A fairly light spring can be used because fluid pressure will tend to unload the spring.

All the spring-loaded styles shown previously have used springs working against the internal fluid pressure. Heavy springs were required to prevent the packing moving under the influence of varying fluid pressure. Springs working with the pressure are obviously better from the packing wear point of view because lower loads can be applied at low pressure resulting in less wear.

Legislation regarding undesirable leakage from glands has resulted in extensive publicity for two spring-loaded, "live-loading" options. Figures 11.11 and 11.12 show the two styles promoted by many packing manufacturers and some valve companies. Square braided packing has been advertised as well as chevron. Packing used in valves, where the working pressure cycles over a wide range, may be liable to move in the packing box producing unexpected wear in the packing box bore. As the springs are external they must be heavy and corrosion resistant to ambient conditions. Failure due to stress corrosion would result in high leakage rates. Very poor guidance is provided for the spring stacks. If "all thread" studs are used, as shown in Figure 11.12, the disc springs can become fouled in the threads. No travel stops are fitted to prevent over-tightening or to show the correct loading position. This style of packing can

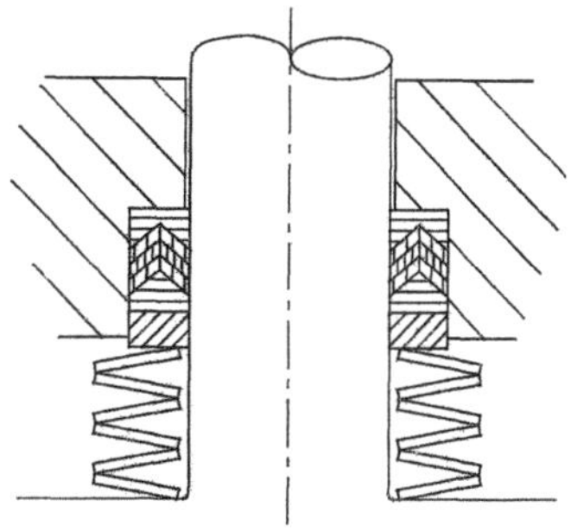

Figure 11.13 Internal live-loading for a tapered plug valve

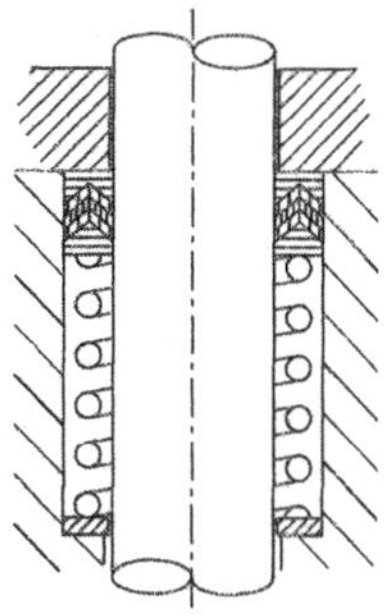

Figure 11.14 Helical spring-loaded chevron packing

be acceptable for site modifications to existing valves, when designed correctly, but not for new valves.

Figure 11.13 indicates one of the correct methods of using spring loading as fitted to a tapered plug valve. The disc spring stack performs two functions simultaneously. The downward reaction of the stack provides the force which seats the plug in the body taper. The spring load compensates for wear. The upward force provides the initial compression for the chevron packing. As the process fluid pressure increases a pressure load is applied to the packing follower which increases the packing compression. Higher internal pressures result in higher sealing forces. Notice that the packing box is plain and surrounded by solid metal. One set of studs retains all the internals within in the valve body. The spring load is controlled by machining tolerances and not adjustable during assembly. The springs are now protected from atmospheric problems but are exposed to the process fluid.

Disc and wave springs are available in a range of materials to cope with many fluid operating conditions:

- high carbon steel
- 11% chrome
- AISI 302
- AISI 316 and 316L
- 17-7PH
- Alloy 20
- Monel™
- Hastelloy C™
- Inconel™

It is very difficult to quantify leakage when using braided packing. Spring-loaded chevron packing removes many of the operating variables. A properly fitted and maintained chevron set should reduce VOC emissions to less than 500 ppm. Figure 11.14 shows the more usual arrangement of spring-loaded packing. Figure 11.15 shows the complete valve bonnet assembly with internally spring-loaded chevron packing, see also Chapter 6, Figures 6.11, 6.12, 6.15, 6.16 and Figure 11.25 in this Chapter. A helical spring can easily exert the light initial

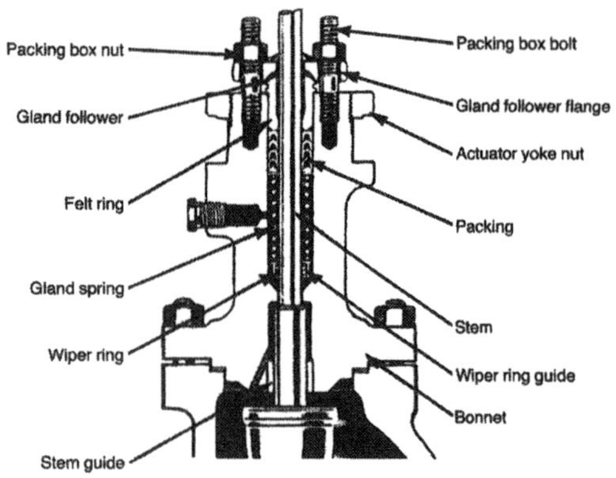

Figure 11.15 Typical bonnet assembly with spring-loaded chevron packing

force necessary. Helical springs do however require considerable space and light disc or wave springs can be fitted as an alternative.

There are many proprietary moulded seal sections which can be used effectively in valves. The ultimate choice depends upon the pressure/temperature rating and the fluid chemical properties.

11.3 Double seals

Double seals are fitted to valves when hazardous fluids are handled or when external lubrication is used. Some users insist on lubrication when braided packing is chosen. Valves operating under vacuum or with high temperature liquid may have a flowing liquid or water quench between the seals. Figure 11.2 shows that an "O" ring can provide a simple but effective seal. Figure 11.16 shows the use of an "O" ring and a lobed ring to provide a double seal. The lobed ring secondary seal could be replaced by an "O" ring or both could be lobed rings.

Figure 11.17 shows a double "O" ring seal as fitted to a butterfly valve. The connection in the gland can be used for lubrication or leakage monitoring. The connection junction with the bore of the packing box must be blended adequately to remove the possibility of "O" ring damage during fitting. When the connection is used for continuous monitoring or venting, the atmospheric "O" ring is loaded by process pressure and therefore has a shorter service life.

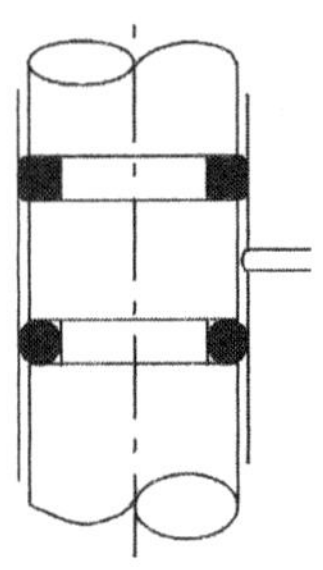

Figure 11.16 Double "O" ring seal

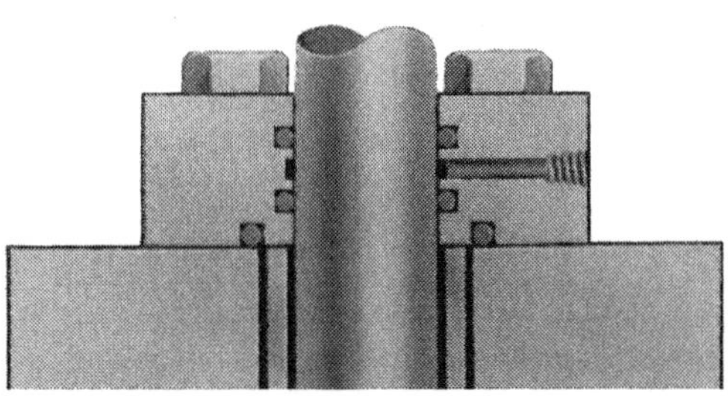

Figure 11.17 Double "O" ring sealed butterfly valve

Some ball valves also use the double "O" ring seal. The spindle is guided and located by a long, close fitting PTFE bush which also acts as a primary seal. One of the "O" rings is fitted as a back-up. In fire-safe valves the second "O" ring is protected by a graphite ring.

The most usual form of double seal is two sets of soft packing, see Figure 11.18. Typically both packing sets are identical to eliminate problems of mis-assembly. The lantern ring can provide extra stem/spindle guidance and should be made of a good bearing material. Figure 11.19 shows the arrangement and performance of a double seal using a graphite ribbon and braided filament for a steam application. When packing utilises a combination of rings it is essential the rings are assembled in the correct order. Figure 11.20 shows a double seal arrangement with spring-loaded chevron packing. Double spring-loaded chevron arrangements should reduce VOC emissions to less than 200 ppm.

It has already been mentioned that spring loading requires considerable space. Figure 11.20 shows the metal bottom header with "pocketed" helical springs; this is the most compact ar-

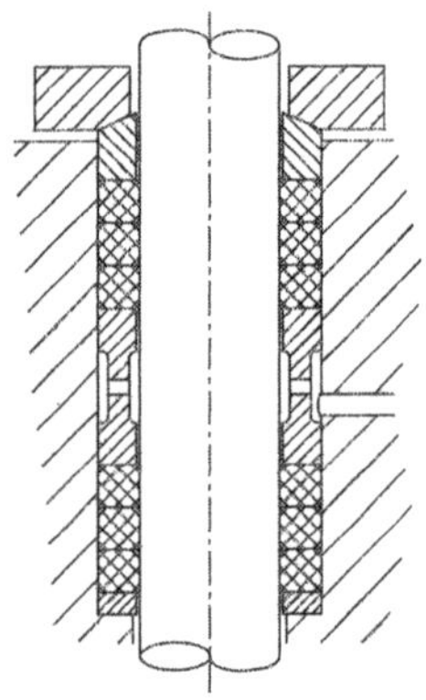

Figure 11.18 Standard double seal arrangement

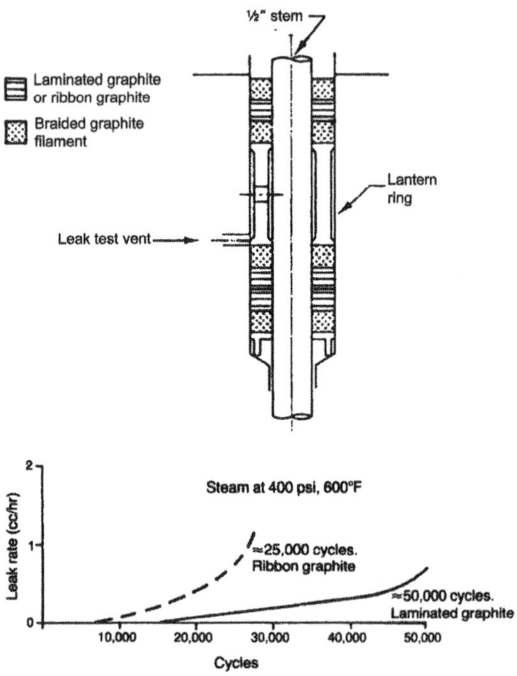

Figure 11.19 Arrangement and performance of double graphite packing

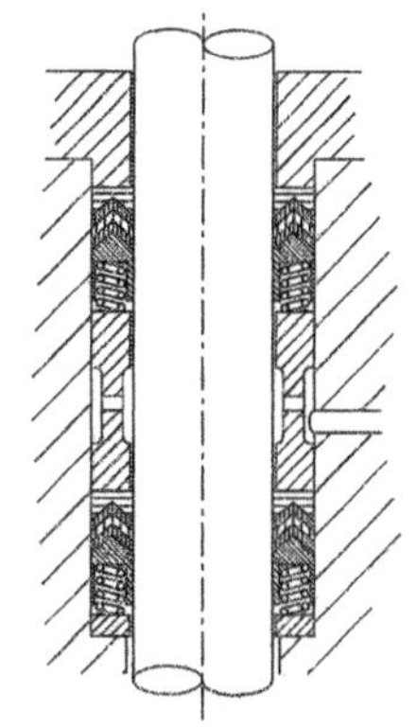

Figure 11.20 Double spring-loaded chevron seals

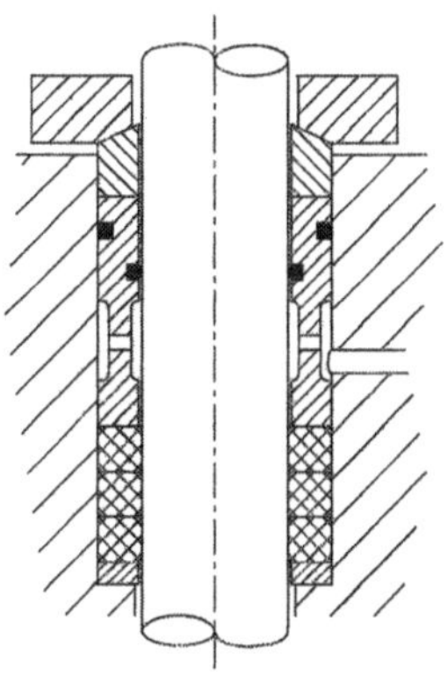

Figure 11.21 Soft packing with "O" ring secondary seals

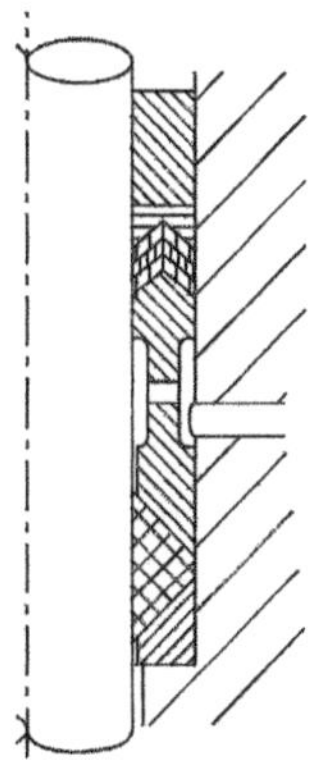

Figure 11.22 A ball valve double seal arrangement

rangement. Metal headers with pocketed springs are superior to plastic headers. It is not essential to use identical packing sets although it is preferable. The lower, inboard set can use stiffer rings or higher temperature compounds. Some arrangements use a single ring as a wiper ring. When the lantern ring connection is used as a vent, or for continuous low pressure monitoring, the second seal can have a much lower pressure rating. Figure 11.21 shows a soft packed adjustable box with "O" rings in the lantern ring. The "O" ring on the outside of the lantern ring is required to prevent leakage. This leakage path is sometimes overlooked.

Figure 11.22 shows an arrangement used in some ball valves. The shaped primary packing is loaded through the secondary chevron packing via a "live-loaded" gland. Other manufacturers reverse the packing arrangement and use graphite for the secondary packing to create a "fire-safe" configuration.

Figure 11.23 shows another seal system for high temperature ball valves. Both seals are made of graphite and act as guide

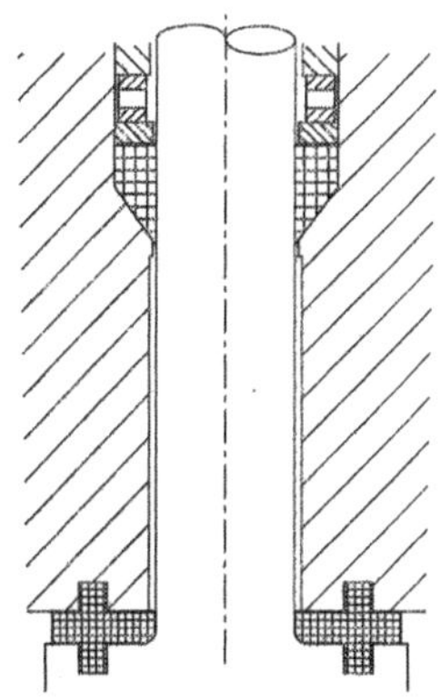

Figure 11.23 High temperature ball valve double seal

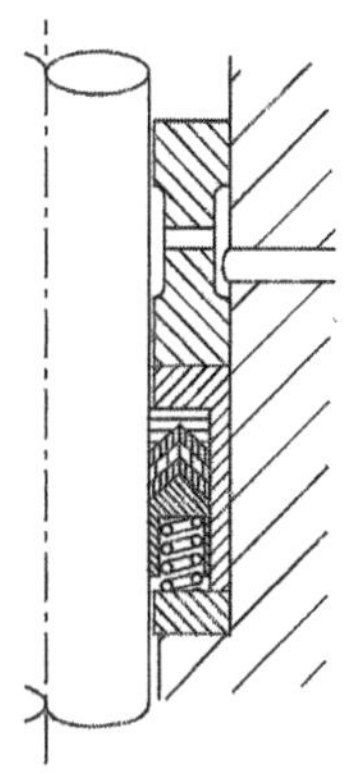

Figure 11.24 Cartridge mounted spring-loaded chevron packing

bearings as well as seals. The wave spring above the secondary seal loads both seals. The primary seal is also loaded by internal fluid pressure which creates an unbalanced force on the spindle.

Apart from Figures 11.21 and 11.23, all the double seal options shown have a serious deficiency — the secondary seal is loaded by process pressure. Even when the primary seal is intact the axial force produced by internal process pressure is transferred to the secondary seal. This means the secondary seal is liable to wear at the same rate as the primary seal. To overcome this problem it is necessary to load the packing sets independently. This is done most easily with spring-loaded chevron packing which can be housed in cartridges, see Figure 11.24. The cartridge length controls the spring load allowing the gland to be bolted down solid. As internal process pressure does not extend beyond the primary seal the secondary seal is only loaded by the springs. The cartridge increases the packing box bore diameter so there are space limitations. Where it can be fitted, double spring-loaded cartridge chevron sets provide the ultimate in safe packing arrangements.

11.4 Special designs

Extreme temperature, high or low, can create sealing problems. Metal valve bodies can operate down to -250°C and to over 1 000°C. Elastomers and plastics, which make the best seals, are limited to continuous maximum temperatures around 260°C. If the packing can be made to operate at a temperature below the process temperature then standard materials can be accommodated. Section 11.3 stated that a lantern ring between two packing sets could be extremely useful. The secondary packing is not in direct contact with flowing process fluid. The top of the packing box is usually surrounded by ambient air which can cool the internals quite effectively. By moving the secondary packing further away from the valve body the cool-

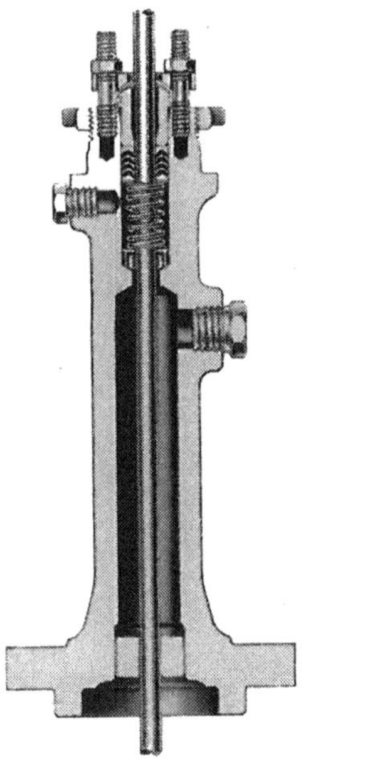

Figure 11.25 An extended bonnet with spring-loaded chevron packing

ing effect can be enhanced. Plain rings can be fitted in the packing box as spacers.

Some manufacturers advocate the use of two lantern rings. Insulating gaskets between lantern rings/spacers or non-metallic lantern rings/spacers can be used to limit the heat transfer along the packing box trim. Ceramic can prove very useful in these situations. Alternatively the bonnet can be extended with an empty cavity, see Figure 11.25. Figure 11.26 indicates predicted and actual test results for the temperature gradient along the bonnet wall and the stem. Standard cast bonnets can be extended by welding in a suitable length of tube, see Figure 11.27.

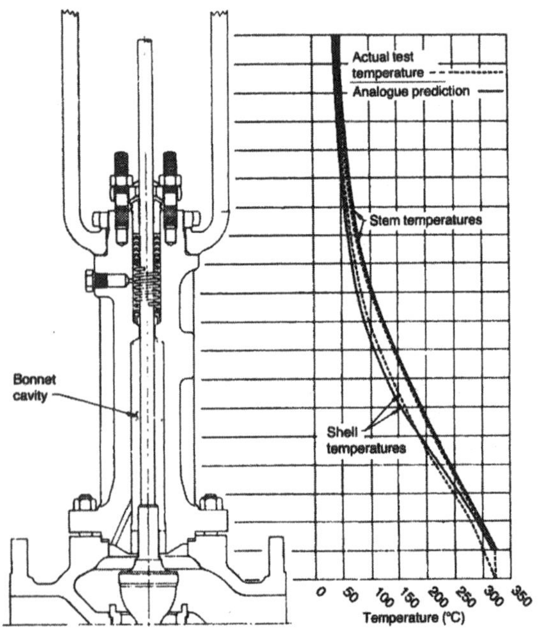

Figure 11.26 Temperature gradient of an extended bonnet

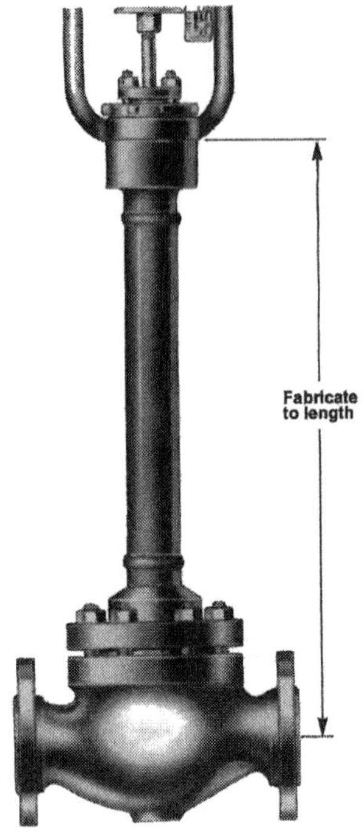

Figure 11.27 An extended cast bonnet

Some valve manufacturers have special bonnet extensions which can be bolted or clamped to existing bonnets. This allows valves to be modified on-site to upgrade the sealing arrangement. The bonnet extension can be fitted with fins to increase the surface area in contact with the ambient air. This technique works for both hot and cold applications. If additional cooling or heating is necessary, a liquid flush can be piped through the lantern ring.

Bellows and diaphragm seals are used for dangerously toxic fluids where leakage would endanger life. These types of seals are also used in cases where the fluid is so expensive that even a slight leakage would result in unacceptable economic losses. The diaphragm and bellows seal completely isolates the fluid from the valve operating mechanism.

Diaphragm seals are limited to smaller plug valve sizes where the plug travel is short. The valve plug is an isolated component within the valve body; it is spring-loaded to lift when the stem is withdrawn. The diaphragm is usually clamped between the valve body and the bonnet flange. The diaphragm can be an elastomer or more usually stainless steel. Elastomers are slightly porous and this must be considered during valve selection. If an elastomer diaphragm is acceptable then a diaphragm valve should also be considered as in chapter 3, Section 3.3.1. Stainless steel diaphragms can be welded to the valve body to eliminate joint problems. Small stainless steel diaphragm sealed valves can operate up to 170 barg.

Bellows seals are not restricted to small valves. The bellows can be made long enough to accommodate the stem travel. Long bellows can be made in sections with stabilising spacers welded between sections. The bellows is formed with both ends open and is attached to the stem and the bonnet, shown in Figure 11.28, (see also Figure 3.16 in Chapter 3). Figure 11.29 shows a globe valve with a bellows sealed bonnet; the fabrication of the bonnet extension is clearly visible.

Whether to attach the top or the bottom of the bellows to the stem is dependent upon the bellows design and the operating conditions. If the top of the bellows is attached to the stem, the

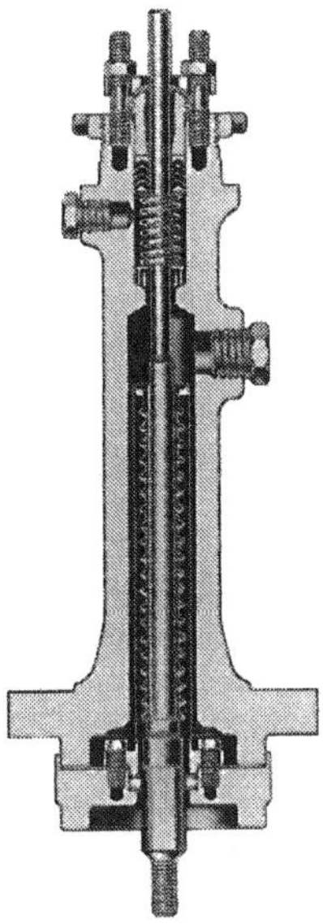

Figure 11.28 A bellows sealed bonnet

Figure 11.29 A globe valve with a bellows sealed bonnet
Courtesy of Dresser-Masoneilan

bellows will be pressurised by the process fluid. If the bottom is attached to the stem the bellows will be surrounded by process fluid. High fluid pressure surrounding the bellows would induce compressive pressure stress which may help with fatigue problems. Conversely, if the bellows is subject to internal process fluid pressure the pressure stresses can be counterbalanced by surrounding the bellows with a pressurised safe fluid.

The best method of bellows attachment is welding which removes leakage doubts. The bellows and associated components should have good "weldability". A bellows can be formed by deep drawing a single piece of metal, or by welding preformed washers together. Bellows can be made with laminations to increase the pressure rating. Bellows are not restricted to linear valves. Manually operated quarter turn valves can be fitted with them to isolate the operating mechanism.

Bellows seals can be used for pressures over 600 barg at ambient temperatures and for temperatures up to 980 °C. As with other metal components, bellows are subject to pressure/temperature ratings, see Chapter 13. Due to the thin metal sections used, high tensile stresses are generated in the bellows which may be prone to failure by fatigue. The manufacturer should be able to predict the number of safe working cycles for the intended operating conditions. In the interests of safety a standard packing box should also be used to back up a bellows seal. It is also important to carry out frequent inspection and maintenance of bellows seals in operation. The bonnet shown in Figure 11.28 includes two tapped bosses; for extremely hazardous fluids a better type of connection should be fitted. One of these connections can be piped to a pressure switch which could raise an alarm if leakage occurs.

NOTE: The actuator on valves fitted with a bellows seal must not be rotated otherwise the bellows will be damaged. Also remember that a bellows acts as a spring and the bellows forces must be taken into account.

Metal bellows do allow a very small quantity of gas to permeate through the metal. The leakage rate with methane has been measured at less than 2.8E-6 mm^3/second for a process pressure of 103 barg. At this leakage rate it would take 11 million years to pass 1 cubic metre. Regulation requirements of measured leakage below 100 ppm can easily be met if weld integrity is assured. The UK Health & Safety Executive (HSE)has identified about 200 fluids and dusts as extremely hazardous. Diaphragm or bellows sealed valves may be the only method of satisfying the regulations and ensuring personnel health.

11.5 Seal materials

The choice of seal material is critical to the ultimate success of any sealing system. The "valve normal" operating conditions are usually well-established and can be used for seal selection. However, operation through process upsets is often forgotten; how much do pressure and temperature change? How quickly and for how long? If a sealing system is intended to survive a short upset excursion and remain effective, the conditions must be evaluated during initial selection. Other non-operational conditions may have a dramatic effect on seal life and performance:

- cleaning with steam
- cleaning with chemicals
- sterilising

The valve seal does not differentiate between operational and non-operational conditions. It is exposed to temperature (do not forget ambient and black bulb temperatures), pressure, fluid and spindle/stem movement, or the lack of movement. The correct material selection can only be made after the valve "life style" has been thoroughly examined. Failure to consider any aspect of "life style" can result in unscheduled stops for maintenance, loss of production and high costs for spare parts.

11.5.1 Braiding materials

Spindle or stem speed is not usually a critical factor in packing material selection. However emergency shut-down (ESD) valves and power-actuated valves for explosion or fire isolation can travel quite fast and it is worth remembering that all packing materials do not have identical dynamic properties. Ceramic filaments are only suitable for the lowest velocities, up to 0.5 m/s. Copper mesh is also very restricted for sliding velocity. All the other materials are suitable for speeds of 3 m/s or faster. Graphite fibre foil has been used at velocities over 30 m/s.

11.5.1.1 Fibres, yarns and filaments

Aluminium

Certain grades of aluminium have very good bearing properties as well as good thermal conductivity. Aluminium mesh has been used as a high temperature packing up to about 500°C for pHs from between about 4 and 7. An aluminium ring in the bottom of the packing box can prevent heat from travelling up the packing by diverting it through the packing box wall.

Aramid

Aramid is a very tough man-made fibre which is used for flak jackets and helmets for armed forces. It is also quite abrasive and must be used with hardened stem/spindles if excessive wear is to be avoided. It has very good chemical resistance and suitable for pH from 2 to 12 or 13 and is very strong and resilient and suitable for pressures between 200 and 350 barg. It is normally braided with a lubricant, graphite, PTFE dispersion or mineral based, to reduce wear problems. Often used as the reinforcing corners with other fibres, it can be suitable for abrasive solids handling and hygienic applications; WRc approved products are available. The temperature range is in the region of -85° to 300°C. When special PTFE dispersions in silicon oil are used the minimum temperature can be as low as -275°C.

Semi-solid complete single ring seals are die-formed from aramid fibres, high temperature binders and flake graphite. These rings are suitable for 70 barg, 400°C and pHs from 4 to 10. Specifically designed for small valves with low usage. The flake graphite can be replaced by PTFE for applications up to 260°C. WRc approved compounds for potable water are available.

In the UK "'WRc Approved" is an independent technical view on whether a product is fit to do the job for which it is claimed or a contractor is fit to provide a specified service. Products and services submitted for approval are assessed by WRc against an existing technical specification (e.g. Water Industry Specification) or an assessment schedule drafted up for the specific product or service.

Asbestos — white

A natural mineral fibre which when treated with lubricating graphite, is still used as a general purpose medium to high pressure packing. Applications up to 480°C include water, steam, oils, alkalis and dilute acids at pressures up to 70 barg. The pH range is generally from 3 or 4 to 14. White asbestos can also use mica as a lubricant which is useful in preventing product discolouration. A PTFE dispersion can be used for lubrication but maximum temperature is reduced to 290°C although better chemical resistance to most solvents is achieved; it is suitable for temperatures down to -70°C. Extra mineral oil can be included for faster applications.

When Inconel™ wire reinforcement is included the maximum temperature is raised to 650°C for pressures up to 210 barg. This packing can use graphite or mica lubrication and is blended with corrosion inhibiters. Although not suitable for oxygen or concentrated acids, it can be used for most other fluids at low to medium pressures.

Semi-solid complete single ring seals are die-formed with asbestos fibres, high temperature binders and flake graphite. Applications range up to 450°C, 70 barg for pH values between 2 and 12. It is suitable for water, steam, oils, distillates and low corrosion fluids.

Asbestos — blue

A natural mineral fibre with good high temperature chemical resistance but much more abrasive than white asbestos. Not often used for modern applications.

Carbon fibre

A very expensive but useful packing material with very good thermal conductivity and low friction. High purity grades are not susceptible to radiation degradation and can be used for nuclear applications. Can be reinforced with Inconel™ wire and graphite lubricated for high pressure applications from -85° to 430°C; over 2 000°C if oxygen contact can be eliminated. Corrosion inhibiters and extra lubricants can be added. Not recommended for use with strong oxidising agents, molten metals or acids. Pressures up to 210 barg with pHs from 0 to 14 can be accommodated.

Ceramic fibre

Alumina silica can be produced as a fine fibre which can be braided into a tough, almost inert packing. Process temperatures up to 1 200°C are possible. Can be combined with Inconel™ wire to produce a thermal shock resistant packing.

Copper

The highest practical thermal conductivity makes copper useful as a thermal barrier. Copper mesh has been used as a high temperature packing up to 800°C.

Cotton

Cotton is the shortest of the natural fibres, less than 50 mm, and produces a soft pliable packing for low to medium pressures, up to 70 barg, non-corrosive applications at temperatures from 65° to 100°C. Cotton can have graphite or grease lubrication incorporated to extend temperature to 120°C; it is a very flexible packing which "beds-in" quickly. The cotton fibre can be coated with rubber compounds to increase the maximum temperature to 130°C. Typically used for water applications, pH 5 to 10, in sewage and mining, cotton impregnated with pure tallow can also be used for hygienic applications up to 50 barg and 120°C.

Flax

A natural fibre from the plant stalk which can be over 1 m in length. Flax has limited chemical resistance. When braided, "raw" flax has a useful temperature range from 0 to 90°C. When lubricated and impregnated with a PTFE dispersion the maximum temperature can increase to 150°C. Suitable for pressures up to 80 barg and a pH range of 5 to 10. Typical applications include sewage and paper making using fresh or seawater.

Flax can also be lubricated by mineral oil or grease with a slight reduction in acceptable pHs, 6 to 9. Flax can be coated with a graphite lubricant prior to braiding. This pre-treatment produces a packing suitable for temperatures up to 120°C at pressures up to 70 barg over a pH range of 5 to 9. Applications are normally restricted to water and seawater. Flax can also be braided with other natural fibres and impregnation with mica dispersion is possible.

Glass fibre

Glass fibres are produced in three different chemical compositions with slightly different properties. Glass fibre is extremely resistant to chemical attack, except concentrated alkalis, and can be used as a universal material for medium and high pressure applications. Inter-strand lubrication, usually a PTFE dispersion, is added to reduce strand chafing. PTFE lubrication limits the maximum temperature to 290°C but the pH range is 4 to 12 for pressures up to 140 or 210 barg. If mineral oil and/or graphite lubrication is used the maximum temperature can be increased to between 480° and 650°C. The use of mineral oil would reduce the range of chemical compatibility.

Graphite fibre

Graphite can be produced as a plain fibre and braided to form a self-lubricating high temperature packing, up to 340°C for general applications. In an inert environment, ambient air excluded, temperatures up to 500°C are possible. Graphite fibres can be braided with PTFE fibres and lubricated for medium pressure universal applications up to 260°C. Pressures up to 400 barg over the full pH range are possible. Typical applications include high pressure steam valves.

Exfoliated graphite

Complete die-form rings can be produced from graphite tape or ribbon which is over 99.5% pure. Free chlorides are guaranteed to be less than 50 ppm with a maximum of 0.1% ash content. This material has broad fluid compatibility; alcohols, alkalis, mineral oils, refrigerants, solvents and water; coupled to low friction at temperatures from -240° to 500°C. High purity graphite is resistant to nuclear radiation. Pressures up to 350 barg with pHs from 0 to 14 are possible. A sacrificial anode can be incorporated to eliminate attack of adjacent metal parts. PTFE inclusions can be added in the bore to assist lubrication of problem fluids but this reduces the maximum allowable temperature to about 300°C.

Combination graphite packing sets

The two graphite forms can be used in combination. The construction of these types of packing is critical in respect of material choice and the order in which the rings are assembled for the various materials. When anti-extrusion rings are necessary they must be fitted correctly. It is, however, possible to double the number of complete valve strokes under test conditions, before the leakage limit is reached if the packing is assembled correctly, as shown in Figure 11.30. This packing arrangement is suitable for operating temperatures of over 600°C for non-oxidising conditions. No lubrication is required, but an extended stuffing box should be used for temperatures above about 425°C.

Graphite honeycomb

A form of packing ring which only compresses about 5 or 6% under full gland load; an ideal seal material for spring loading. The regular structure of the rings promotes very good sealing qualities which allows shorter packing boxes to be used. Tight clearances or anti-extrusion headers are necessary to prevent ring extrusion under pressures up to 300 barg. Special die-formed braided asbestos/graphite packing reinforced with Inconel™ wire has proved successful as end rings. The total packing set shrinkage is only 10 to 15%.

Hemp

A natural material suitable for low pressure, up to 80 barg, non-corrosive applications and pHs from 6 to 9. Typically used for water, sea-water and water/oil emulsions used in low pressure hydraulic systems. Usually mineral oil or grease lubricated and restricted to 120°C.

Jute

A natural vegetable fibre, from 1 to 2 m long, obtained from the plant stalk. Not a popular packing material because of its poor resistance to water.

Polyimide

The thermoset plastic polyimide can be produced as a yarn and braided. When impregnated with a silicone lubricant, a high pressure/temperature abrasion resistant packing is created.

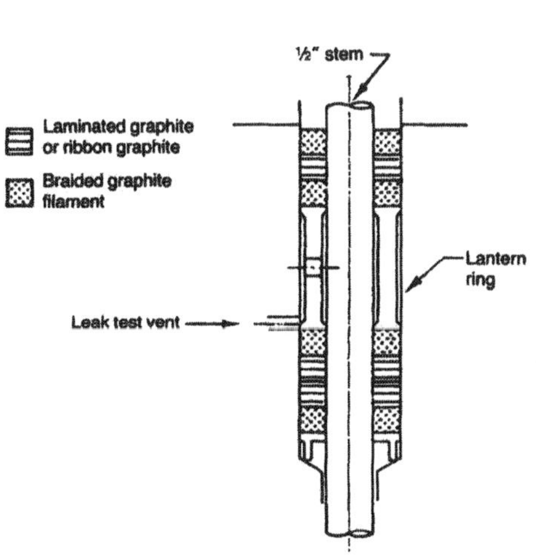

Figure 11.30 Correct assembly of graphite fibre and tape packing sets

Operating limits are 100 barg, -100° to 260°C with pHs from 0 to 12.

PTFE

PTFE can be manufactured as a yarn for braiding. When used with a PTFE lubricant dispersion produces a very flexible packing ideal for hygienic applications from -100° to 280°C. Pressure up to 140 barg with pH values over the full range, 0 to 14, can be handled. Very good for concentrated acids. Close clearances are necessary to avoid packing extrusion. Special braiding techniques coupled with die-forming can increase the maximum pressure to 250 barg. Graphite lubricated PTFE fibre, with or without Aramid reinforced corners, can be used up to 350 barg. Mineral oil can be added during braiding but this reduces the chemical resistance. PTFE packing can also have reinforced corners of polyimide, a thermoset plastic. This style of packing, impregnated with silicone lubricant, can operate with pH values between 0 and 12 at up to 100 barg from -100° to 260°C. Low temperature versions, down to -275°C, can be supplied.

For some high pressure or high temperature applications it may be recommended that PTFE braided rings be separated by glass filled spacer rings to reduce packing ring distortion.

Ramie

A natural fibre which is much stronger than cotton, flax or hemp, wet or dry. The increase in basic strength provides a higher fatigue resistance. The raw fibre is inherently rot resistant and when impregnated with a PTFE dispersion lubricant exhibits good chemical resistance. A soft, conformable packing, with good extrusion resistance, which does not abrade stems or spindles. Suitable for a pH range between 4 and 11 and temperatures between -30° and 120°C. Normal operating pressures are up to 250 barg. By including support rings within the packing the working pressure can be more than doubled.

11.5.2 Moulding and machining materials

11.5.2.1 Elastomers

In the following list, the designation within parentheses is taken from ISO 1629 and that inside square brackets is from ASTM D1418. See Table 11.1 for comparative properties.

Acrylate rubber (ACM) [ACM]

Suitable for high temperature, extreme pressure lubricants containing sulphur, but poor low temperature physical properties and limited resistance to water.

Trade Names: Cyanacryl, Europene AR, Krylene, Krymac 880, Nipol AR, Noxtite PA, Polysar, Thiacril, Vinacryl.

Butyl rubber (IIR) [IIR]

Good resistance to vegetable products is combined with very low gas permeability.

Trade Names: Enjay Butyl, Esso Butyl, Polysar Butyl, Vistanex.

Carboxylated nitrile rubber (XNBR) [XNBR

A modified version of nitrile rubber with better ozone resistance and very good abrasion resistance.

CP rubber (CSM) [CSM]

Very good resistance to a wide range of chemicals and good abrasion resistance. Low temperature properties can be improved by correct choice of plasticizer.

Trade Names: Hypalon, K 51 Penton

EPDM rubber (EPDM) [EPDM]

A development of **EPR rubber** with very wide temperature limits and suitable for water, steam and vegetable products.

Trade Names: Buna AP, EPDM Terpolymer, Epsyn, Intolan epdm, Keltan, KR grade, Nordel, Polysar EPDM, Royalene, Vistalon

EPR rubber (EPM) [EPM]

Developed especially for phosphate-ester hydraulic fluids but has good resistance to vegetable products.

Trade Names: Dutral, Enjay-EPR, Intolan, Olethene, Prescothene

Fluoroelastomers (FPM) [FKM]

Compounds with excellent resistance to petroleum products, acids, alkalies and many solvents with intermittent temperature rating to 300°C.

Trade Names: Dai El, Fluorel, Kel-F, Noxtite, Tecnoflon, Viton

Hydrogenated acrylonitrile rubber (HNBR) [NEM]

A derivative of **nitrile rubber** with higher strength and better abrasion resistance.

Natural rubber (NR) [NR]

Also called "pure gum rubber", used to be the most popular elastomer because of its resilience, low hysteresis, low temperature properties and wide range of hardnesses. Very good for gases, water and castor based hydraulic fluids. Man-made **polyisoprene** has nearly identical physical properties and is now the most popular elastomer.

Chlorobutadiene rubber (CR) [CR]

Developed to overcome the deficiencies of natural rubber when used with some hydrocarbon compounds. Compatible with many CFC refrigerants and can be used at low temperature when ester plasticizers are used. Better ozone and weather resistance.

Trade Names: Baypren, Butaclor, Chloroprene, Denka, Neoprene™

Nitrile rubber (NBR) [NBR]

Also known as acrylonitrile rubber, nitrile content can vary from 18% to 50%. Higher nitrile content rubbers are suitable for petrol and petroleum products.

Trade Names: Breon, Butacril, Butakon A, Chemigum, Europrene N, Herecol, Hycar, Hyclad, Krynac, Luran, Nipol, Nitriflex, Paracril, Perbunan, Polyblend, Polysar Krynac, Revinex

Perfluorinated fluoroelastomer [FFKM]

Expensive but very good chemical resistance, almost as good as **PTFE** but more flexible.

Trade Names: Chemraz, Kalrez, Simriz

Polyisoprene (IR) [IR]

A man-made replacement for natural rubber with almost identical properties.

Trade Names: Cariflex, Natsyn

Polyurethane

Most popular grades are called polyester (AU) or polyether (EU). Very good low temperature properties and outstanding resistance to abrasion.

Trade Names: Adiprene, Caprolan, Cyanaprene, Desmopan, Duthane, Elastollan, Elastothane, Estane, Jectothane, Molythane, Pellethane, Polyvon, Prescollan, Simputhan, Urepan, Vibrathene, Vulkollan

Styrene butadiene rubber (SBR) [SBR]

Similar in most respects to natural rubber

Trade Names: Buna Hüls, Buna SB, Cariflex S, Carom, Europrene, Intol, Krylene, Krynol, Plioflex, Polysar, Solprene, Synpol

Silicone rubber (MVQ) [VMQ]

Very wide temperature range is possible for some hydrocarbon based and some phosphate-ester based lubricants.

Trade Names: Dow-Corning, Prescolastik, Rhodorsil, Silastic, Silicone, Siloprene

11.5.2.2 Thermoplastics

The designation within parentheses is taken from ISO 1043.1 and that inside square brackets is from ASTM D1600

Ethylene tetrafluoroethylene (ETFE) [ETFE]

Almost as good as **PTFE** but can be injection moulded to produce finished seals.

Trade Name: Tefzel

Perfluoralkoxy copolymer (PFA) [PFA]

Almost as good as PTFE but can be compression and injection moulded to produce finished seals.

Trade Name: Teflon-PFA

Polyamide (PA) [PA]

The highest strength thermoplastic with very good abrasion resistance. Good dry running properties.

Trade Names: Durethan, Dymetrol, Nylon, Rilsan, Ultramid, Vestamid

Polypropylene (PP) [PP]

Very good for hot water and low pressure steam.

Trade Names: Hostalen PP, Novolen

Polytetrafluoroethylene (PTFE) [PTFE]

Almost totally chemically inert from -195° to 232°C. Attacked by molten alkaline metals, such as sodium, and some halogenated compounds. Cannot be moulded, with the exception of some of its very similar alloys. Seals are machined from solid. Its unsurpassed lubricating qualities assist with dry operation. PTFE has poor resilience but excellent chemical resistance. The resilience problem can be overcome by applying a PTFE coating to a more elastic core, such as graphite, natural or synthetic rubber, silicone rubber or Viton™.

Trade Names: Algoflon, Fluon, Halon, Hostaflon, Teflon

Polyvinyl chloride (PVC) [PVC]

Has very good elastic properties and can be used to replace elastomers. Suitable for a wide range of chemicals.

Trade Names: Breon, Hostalit, Plaskon

11.5.2.3 Thermoset plastics

Polyimide (PI) [PI]

A high temperature material with good abrasion resistance and low coefficient of friction. Good resistance to hydrocarbon products but not hot water.

Trade Name: Vespel

11.5.2.4 Engineered plastics

Polyetheretherketone (PEEK) [PEEK]

Very good chemical resistance. Can be reinforced to improve high temperature physical properties.

Trade Name: Arlon

11.5.3 Comparative properties of non-metallic materials

Table 11.1 gives an indication of the general corrosion properties of some popular non-metallic materials. Many of these materials have several distinct compounds with slightly varying chemical properties. The most suitable material can only be selected by reviewing the actual operating conditions with a specific fluid. The addition of fibre fillers can increase the temperature and pressure ratings of seals but at the expense of some flexibility.

	PVC	NR IR	SBR	AU EU	IIR	PP	CR	XNBR	CSM	NBR	EPDM	EPM	HNBR	ETFE	ACM	FKM	VMQ	PTFE	FFKM	PFA	PI	PEEK
Low temperature °C	-35	-60	-50	-50	-30	0	-40	-30	-20	-45	-55	-45	-30	-100	-20	-15	-60	-250	-29	-250	-240	-18
High temperature °C	70	80	90	100	100	104	110	110	120	130	140	140	140	150	160	175	200	230	232/316	260	480	260
Abrasion resistance		E	G	E	F	G	G	E	E	G	G	E	E	G	F	G	P	P	P	P	E	
Tear resistance		E	P/F	E	G		G	E	E	G	G	G	E	F	F	F	P		F			
Gas permeability		H	L	M	L		L	L	L	L	L		L	L	L	L	L	L	L		M/H	
Compression set		E	G	G	P		E	G	F	G	G	F	E	G	P	G	G		G			
Resilience		E	F	G	P		G	F	P	F		P	F		F	F	P	P		P	P	
Coefficient of friction		H		H				L		L	L		L			H		L	L	L	L	L/M
Radiation resist		G	F	G	P		F	F	F	F	F		G		F	P	P	P	G	P	E	E
Weather resist		P	F	E	E		P	P	E	P	E	E	G	E	E	E	E	E	E	E	E	
Ozone resist	G	F	F	E	E		G	P	E	P	E	E	G	E	E	E	E	E	E	E		G
Sunlight ageing		P	P	E	E	P	E	P	E	P	E		G			G	G	E		E		
Mineral oil (LAP*)	P	F	P	G	U	P	F	E	G	E	P	P	E	G	E	E	P	E	E	G	G	G
Mineral oil (HAP**)	P	F	P	P	U	P	G	E	G	E	P	P	E	G	E	E	G	E	E	G	G	G
Silicone fluids	F	G	F	F	F	G	F	F	F	F	G	G	F		G	E	P	P/G	E	P/G		G
Diester fluids		P	P	G	G		P	P	P	P		P	P			G	F	G	G	G		
Phosphate esters		P	P	G	E	P	P	P	P	P	E/G	E/G	P	G	P	P	F	G	E	G		
Silicate esters		P	P	F	P	P	F	F	F	F	P	P	F		P	P	F	G	E	G	G	
Aliphatic hydrocarbons	E	P	P	E	P	G	F	G	G	G	P	P	G	G	P	E	P	E	E	E	G	G
Aromatic hydrocarbons	U	P	P	G	P	F	P	F	P	F	P	P	F	G	P	E	P	E	E	E	G	G
Halogenated solvents	U	P	P	P	P	F	P	P	P	P	P	P	P	E			P	E		E	P	G
Ketones	U	G/F	P	P	P	G	G	P	P	P	G	F	P	P	P	P	P	G	G	E	G	G
Aromatic petroleum	G	P	P	G	P	P	P	G	P	G	P	P	G	F	F	E	P	G	E	G	G	G
Non-aromatic petroleum	G	P	P	G	P	P	G	G	F	G	P		G	F	G	E	G	G	E	G		G
Dilute acid	E	G	G	G	G	E	P	P	E	G	G	E	G		P	G	F	E	E	E	P	G
Concentrated acid	E	F	P	P	F	F	P	P	G	P	G	E	G	G	P	G	P	E	G	E	F	G
Dilute alkali		G	G	P	G	E	G	F	G	G	E	E	G		P	G	F	E	E	E	P	G
Concentrated alkali		F	F	P	G	G	P	P	G	P	G	E	G			P	P	E		E	P	G
Cold water	G	E	E	G	G	G	F	P	G	G	E	E	E	E	U	G	F	E	E	E		G
Hot water	U	F		U	F	P	P	P	F	G	E	E	E	E	U	G	F	E	E	E	P	G
Steam	U	U	U	U	U	U	P	P	P/F	P	E	E	E	E	U	U	P	G	E	E		G

E = excellent, G = good, F = fair, U = unsuitable, H = high, M = medium, L = low (*LAP = low aniline point, **HAP = high aniline point)

Table 11.1 Comparative properties of elastomers

11.5.4 Metal seal rings

Aluminium

Certain grades of aluminium have very good bearing properties as well as good thermal conductivity. Solid aluminium rings can be used as headers or anti-extrusion rings within a soft packing set. An aluminium ring in the bottom of the packing box can prevent heat from travelling up the packing by diverting it through the packing box wall. It is suitable for a wide range of fluids, except alkalis and concentrated acids, from -200° up to about 430/500°C.

Copper

Solid copper rings can be used as headers or anti-extrusion rings. The highest practical thermal conductivity makes copper useful as a thermal barrier. Copper is compatible with many fluids and may be used up to 800°C under ideal conditions.

Lead

Lead foil rings are sometimes used as headers or spacers between other rings to avoid extrusion or prevent ring deformation. The high thermal conductivity allows heat to pass easily into the packing box to be dissipated, rather than along the packing. Lead can be used for a wide range of applications, not for alkalis or concentrated acids, up to about 100/260°C.

11.5.5 Bellows and diaphragms

Many materials are available for diaphragms, see Table 3.5 in Chapter 3. All of the materials listed are non-metallic and the operating pressures are low. Stainless steel can be used for diaphragms but this is limited to small valves. The mechanical actuation of diaphragms, physical contact over a small or large area by a solid component, produces high stresses in the diaphragm which restricts the operating pressure. If the diaphragm can be actuated pneumatically or hydraulically the stress regime is reduced dramatically. Experience with reciprocating pumps indicates that large stainless steel diaphragms could be used successfully at pressures up to 350 barg.

There are many popular materials for bellows, see Table 11.2.

Material	Minimum temperature	Maximum temperature
NBR	-10	100
PVC	-10	60
EPDM	-20	120
Viton™	-7	120
PTFE	-50	150
Brass		150
Phosphor bronze		200
Aluminium	-196	220
AISI 304	-196	680
AISI 316	-196	800
AISI 316L	-196	480
AISI 321	-196	815
Monel™	-196	680
Inconel 600™		980
Inconel 625™	-196	700/980
Inconel 718™		
Incoloy 800™		980
Incoloy 825™		850
Hastelloy B and C™		700/980
Titanium		500/600
Nimonic alloys™		700/900

Table 11.2 Temperature limitations of popular bellows materials

Base	Lubricant	Type	Comments
Vegetable	castor oil	oil	partially insoluble in water
Vegetable	rapeseed oil cotton seed oil	oil	partially insoluble in water insoluble in alcohol at low temperature soluble in ether, carbon disulphide, chloroform, carbon tetrachloride, petrol, benzene, mineral oils limited viscosity range readily oxidised
Animal	beeswax	solid	
Animal	tallow	grease	hygienic applications
Animal	whale oil		
Mineral	hydro-carbons	oil grease	wide viscosity range many compounds with blended additives can oxidise and deposit carbon residue
Mineral	mica	solid	wide temperature range <600 °C
Mineral	graphite	solid	wide temperature range <1 500 °C very good heat transfer almost chemically inert
Mineral	tungsten dioxide		wide temperature range almost chemically inert
Man-made	chlorofluoro-carbons	oil grease	outstanding oxidisation resistance wide temperature range <300 °C environmentally unfriendly
Man-made	silicones	oil grease	wide viscosity range wide temperature range <360 °C almost chemically inert
Man-made	PTFE	solid dispersion	wide temperature range almost chemically inert
Man-made	molybdenum disulphide	solid	wide temperature range

Table 11.3 Packing and seal lubricants

Bellows assemblies in control valves can suffer from resonance leading to premature fatigue failure. Valve flutter may be the root cause. The problem can usually be cured by increasing the bellows stiffness which should move the natural frequency. When considering how much to increase the stiffness remember the actuator has to supply the force to overcome the stiffness; check the actuator sizing.

11.6 Lubrication

Some materials have self-lubricating qualities and do not require external lubricants. PTFE is the most notable and graphite similarly for high temperatures. These materials can run dry with little increase in wear providing the surface finishes of the contacting parts are good and the speed not too fast. Both PTFE and graphite can be used as additives to braided and moulded seals to provide inherent lubricating qualities.

Lubricants added to braided packings can confer other useful properties. Fibre wear is reduced and heat transfer properties may be improved. Overall packing resilience is increased. Many types of solid and semi-solid or greases can be applied to seals during manufacture. Table 11.3 lists the most used compounds.

NOTE: Some manufacturers of braided packing advise that new packing should be soaked in oil overnight before fitting. It is certainly worthwhile reading the instructions the day before repacking a valve.

As mentioned earlier it can sometimes be advantageous to lubricate the packing box. The preferred lubricant is silicone grease, which reduces friction for temperatures up to 260°C. Lubrication is carried out by opening an isolating valve and screwing in a feed screw which presses lubricant into the pack-

Stem diameter mm	ANSI Valve Class	Packing Type				Gland bolt tightening torque Nm
		PTFE		PTFE Impreg-nated asbestos	Graphite*	
		Single	Double			
8	All	94	140	—	—	—
	125			284	—	1.5 - 3
	150			284	563	1.5 - 3
	250			428	—	2 - 4
10	300	170	250	428	855	2 - 4
	600			563	1125	3 - 5
	900			720	1440	4 - 6
	1 500			855	1710	4 - 7
	125			405	—	3 - 5
	150			405	810	3 - 5
	250			518	—	4 - 6
12	300	240	350	518	1035	4 - 6
	600			720	1440	5 - 7
	900			923	2246	6 - 10
	1 500			1125	2250	8 - 12
	2 500			1328	2260	9 - 13
	125			490	—	2 - 4
	150			490	960	2 - 4
16	250	284	210	653	—	3 - 5
	300			653	1306	3 - 5
	600			900	1800	4 - 7
	125			788	—	6 - 10
	150			788	1850	6 - 10
	250			990	—	7 - 11
20	300	350	540	990	1980	7 - 11
	600			1485	2970	11 - 16
	900			1980	3960	14 - 22
	1 500			2475	4950	18 - 27
	2 500			2970	5940	22 - 33
	300			1372	2744	16 - 20
	600			1912	3825	18 - 27
25	900	470	700	2385	4770	22 - 33
	1 500			2950	5900	27 - 41
	2 500			3465	6930	33 - 48
	300			1800	3600	18 - 27
	600			2475	4950	24 - 35
32	900	550	790	3150	6300	30 - 44
	1 500			3865	7650	36 - 54
	2 500			4590	4180	44 - 66
	300			2758	5517	22 - 34
	600			3881	7762	33 - 48
50	900	940	1410	5063	10125	44 - 64
	1 500			6188	12575	54 - 78
	2 500			703	14607	61 - 90

Table 11.4 Typical packing friction forces

ing box, see Figure 11.31. The lantern ring forms a chamber in the packing box into which the lubricant is pressed. The isolating valve is normally closed and is only opened to carry out actual lubrication. The valve also, of course, must remain closed when it is necessary to add more silicone grease.

For valves used with oxygen, petroleum-based lubricants must not be used because of the risk of explosion; the valves must be carefully degreased and protected at all times prior to installation.

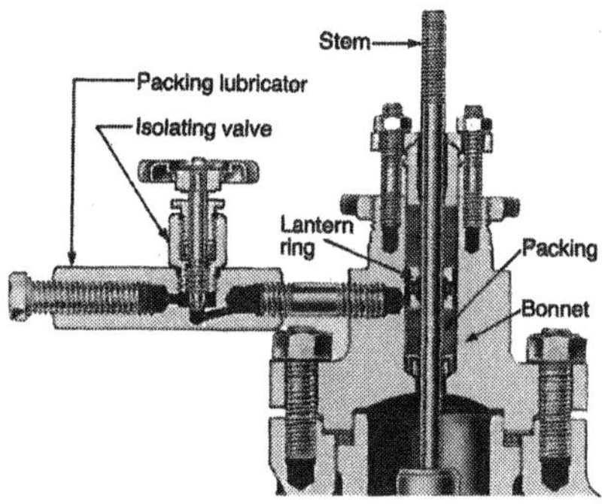

Figure 11.31 A typical packing box lubricator with isolating valve

11.7 Packing friction

The packing inside the packing box is squeezed, either by a gland, spring or fluid pressure, to occupy the space between the stem/spindle and the packing box wall. The packing tends to grip the stem/spindle and a force must be applied to overcome packing friction. Packing friction is a function of the size and the type of packing. Table 11.4 shows typical values of friction forces.

11.8 Packing box assembly

If valves are not going into active service as soon as they are delivered it is beneficial to carry out the following actions:

- Have the packing boxed separately rather than fitted to the valve. Condensation or the residue of hydrotest water can create a chemical reaction between the stem/spindle and the packing. Pitting of the stem/spindle is the usual consequence.
- Valves should be thoroughly dried, left partially open and dessicant placed inside before the blanks are fitted.
- The packing box should be sprayed with moisture displacing preserving wax before the gland is fitted.
- Valves for hygienic applications must be preserved with approved products. The valve should be clearly marked with the date of preservation and an expiry date to initiate re-preservation.

11.9 Useful references

Bellows sealing of valves to meet Clean Air Regulations, Effectiveness and lifetime, J M Pinnington and A J Brown, Valves and actuators for fluid control, BVAMA International Conference, Birmingham, June 1992.

U.S. Environmental Protection Agency (EPA), Ariel Rios Building, 1200 Pennsylvania Avenue, NW, Washington, DC 20460 USA, www.epa.gov.

Health and Safety Executive (HSE), Rose Court, 2 Southwark Bridge, London SE1 9HS UK, Tel: 0845 345 0055, www.hse.gov.uk.

WRc Approval, Frankland Road, Blagrove, Swindon, Wiltshire SL7 2HD UK Tel: 01793 865000, www.wrcapproved.com.

ISO 1629 Rubber and latices – Nomenclature.

ASTM D1418-06 Standard Practice for Rubber and Rubber Latices-Nomenclature.

ISO 1043-1:2001 Plastics - Symbols and abbreviated terms - Part 1: Basic polymers and their special characteristics.

ASTM D1600-99 Standard Terminology for Abbreviated Terms Relating to Plastics.

Actuators 12

12.1 Introduction

The function of an actuator is to adjust the position of the valve to ensure correct control of the process fluid. The valve position may be only "open" or "closed", as in the case of isolating valves, or in any intermediate position for control valves. To operate effectively the actuator must be sufficiently powerful to produce a positive, accurate and quick response to a control signal. In the event of signal failure the actuator may be required to return the valve to a predetermined position or to hold its current position. It is therefore important to specify the correct type and size of actuator in order to meet the demands of the process, reliability and economy.

There are many types of actuators for valves. The pneumatic diaphragm is probably the most widely used and probably the cheapest. But other types of have been developed such as the electro-hydraulic actuator, to improve performance and to use other power sources. Important advances have been achieved in the design of actuators, which are now faster and more accurate. Other types use compressed air more economically. Others interface directly with digital control systems, often called "Smart valves".

When external power is not available the actuator can be driven by the process fluid. This type of valve is called a "regulator", see Chapter 5.

An actuator applies a linear force or a rotary torque to a valve element. Some linear actuators can have the linear motion translated into rotary motion. Figure 12.1 shows the variety of actuator types and their application to linear and rotary valves.

The actuator must be attached securely and rigidly to the valve body. A compact assembly is preferable, to reduce space requirements, the chances of accidental damage and the likelihood of structural vibration problems. Figures 6.3 and 6.4 in Chapter 6 show a globe valve with a diaphragm actuator. Figure 6.20 in Chapter 6 shows the mounting of both diaphragm and piston actuators on rotary valves. All these illustrations show "traditional" actuators, that is actuators powered by compressed air. Air is a readily available, safe commodity. Compressed air is a common utility in process plants. Instrument air, specifically for control purposes, may be derived from normal compressed air supplies or generated independently. Some actuators are fitted with filter-regulators to ensure a stable acceptable supply.

Compressed air, for both general use and instrumentation, should be dried to some extent. Moisture in the air causes rust and may freeze when the temperature is reduced. Water and/or oil droplets can cause flow problems in small diameter orifices and erosion. Moisture in cylinders can cause movement to be erratic. Although water is not a solvent for most industrial lubricants, moisture can remove lubrication from important locations. Moisture or oil accumulation in some assemblies can create a "hydraulic lock" which prevents any movement in one or all directions.

General compressed air systems, as a minimum, will be fitted with a separator, traps and drains to remove condensation. Instrument air quality may require the use of refrigeration or adsorption driers. The dew point of instrument air must be below the minimum temperature experienced anywhere in the air circuit. The Instrumentation, Systems and Automation Society (ISA) recommends a safety margin of 10°C below minimum operational temperature.

General purpose compressed air may have oil entrained to lubricate and extend the useful life of power tools. Oil content should be reduced to 1 ppm for instrument air.

Air entering the compressor will be filtered to prevent solids causing rapid compressor wear. Additional filters may be required in the distribution system to remove any contamination created by the system pipework. Air for actuators and positioners should be filtered to remove particles larger than 5 μm.

Instrument air systems can be very long and pressure losses can vary considerably due to the change in demand. A typical system specification at the point of use would be:

- nominal operating pressure 8.5 barg
- minimum design pressure 4.0 barg
- maximum design pressure 10.5 barg

Air exhausting from actuators must not be forgotten. Personnel protection may be necessary in some instances.

As can be seen from the instrument air pressure range listed above the effectiveness of an air actuated system may vary considerably. If a piston actuator is connected directly to the air supply, the control valve response time will be much shorter with higher pressure air than with lower.

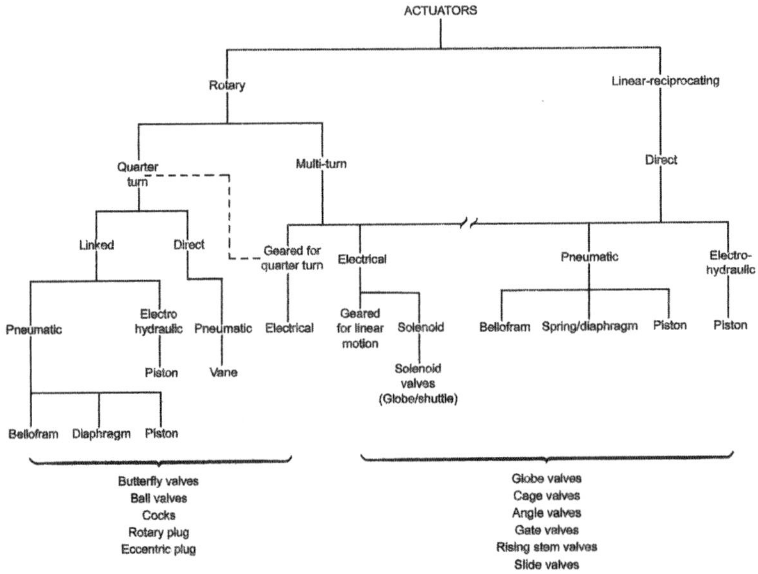

Figure 12.1 Actuator types and applications

Electric powered actuators could solve this problem. Electric supplies are generally very closely regulated. Frequency control of ac supplies is generally better than ±1%. The voltage at the actuator may suffer short term reductions of 20% due to sudden local load changes but these only last for a few milliseconds. If response time is critical for process stability then electric power should be considered. The overall quality of the electricity supply should be quantified before investment is made. Compressed air is usually produced on site and the quality and reliability of supply is within the user's control. Electricity is usually a third-party supply and some extra site equipment may be necessary to improve quality and reliability.

Type of fluctuation	Percentage of complaints
Voltage dips, transients, short interruptions <3 minutes	32%
Long interruptions >3 minutes	24%
Over and under voltage	21%
Surges	13%
Harmonic interference	12%
Flicker	11%
Unbalances	1%

Table 12.1 Typical electric supply problems

A survey held a few years ago, of Italian industrial consumers with 20 kV supplies, illustrates typical supply problems. See Table 12.1.

Motor contactors and electronic control equipment can be sensitive to supply variations. PLCs can shut down if a voltage dip of 10% lasts for longer than 100 ms. Some motor contactors or relays may drop-out with a 30% voltage dip.

Actuators can be suitable for many 50 Hz and 60 Hz supplies:

- 660/575 V 3 ph
- 500/460 V 3 ph
- 415/380 V 3 ph
- 220 V 1 ph
- 110 V 1 ph
- 24 V 1 ph
- 24 V dc

For isolating valves, which only operate occasionally, an intermittent rating, 20% duty cycle with 60 starts per hour, should be adequate. For control applications the actuator should be continuously rated; 1 200 starts per hour are possible.

Actuator stiffness may be important, depending upon the valve type, if process conditions change quickly with wide variations in pressure. A flexible actuator will allow the valve element to move further out of position requiring more actuator counteraction. Mechanical actuation, when a linear valve retains its screwed stem, is the most stiff and the valve element is locked in position.

The local environment around the valve can be important when selecting the actuator type and rating. Air operated actuators are unaffected by hazardous areas created by flammable gases. Electrical equipment can become very costly if particular classification approval is required. Both air and electric may suffer, due to high ambient or black bulb temperatures. Outdoor sites can be subject to black bulb temperatures of 80°C and over the limit of some electronic devices. This is very close to the operating limit of some diaphragm materials.

When high forces are required, an electro-hydraulic actuator may seem an attractive option. The hydraulic fluid specification should be considered when the actuator is located close to very hot equipment, which may be able to ignite drips or spray. Water-oil emulsions can be an alternative to hydraulic oil but seals must be selected carefully. Some oil additives are very costly and small leaks may prove expensive to "top-up". Some applications in marine environments or coastal ambient air, laden with salt, may require material restrictions. Aluminium is particularly unpopular.

The sealing of the enclosure containing the actuator and the positioner can be influenced by the ambient conditions. Electrical equipment is well defined and regulated for physical and electrical protection. Mechanical equipment is sometimes forgotten. Protection from dust and dripping water can be essential. Heaters may be required in mechanical equipment to reduce the operating temperature range and ensure protection from freezing. Lubrication for life can reduce maintenance activities. Actuators and positioners for hygienic applications may need special materials and/or surface finishes and be hoseproof.

12.2 Control signals

Control signals for thermostatic actuators are obviously always temperature related; these are described in detail at the end of Section 6.3.3 in Chapter 6.

12.2.1 Analogue signals

The type and number of control signals available and the extent of monitoring by other systems may significantly influence the choice of actuator. The "traditional" control valve and actuator shown in Figure 6.3, Chapter 6, is disappearing as computers or PLCs take over. The unit shown is rugged and well proven, even if susceptible to instrument air problems, but is an isolated system. If a process problem occurs and the control valve is suspected of a malfunction, an operator or an instrument technician must go to the valve and examine it. Inspection of the local pressure gauges and the travel indicator may immediately indicate the problem enabling speedy remedial action. Otherwise further investigation is required. The problem would be solved by a trained specialist "on-the-spot".

Modern trends are for better instrument communication using expert system software which recognizes the problem and offers a solution, remotely, to the operator in the control room. A maintenance team can then be sent to the valve with a better understanding of the problem and the solution; as well as bringing the spares required for replacement. When the expert system software is competent, the overall plant operating efficiency can be very high.

The valve unit shown in Figure 6.3, Chapter 6, is a completely fluid based system. A pneumatic control signal is piped from the process controller which compares a process fluid signal with the set point. The resulting control signal varies in pressure depending upon the discrepancy between the process variable and the set point. The positioner moves the valve element in proportion to the pneumatic signal. The controller can be mounted close to the valve, remotely or in the control room.

Typical pneumatic control and operating pressures are 0.2 to 1.03 barg (3 to 15 psig). Some diaphragm actuators can work with 0.2 to 1.86 barg (3 to 27 psig) or 0.41 to 2.07 barg (6 to 30 psig). Some piston actuators can work with air pressures up to 10 barg. The control valve can be controlled from a great distance by using a simple electric interface.

A transmitter or transducer measures the process variable and can produce an analogue electrical signal. Transmitters and transducers are available to measure:

absolute pressure	ORP
density	particle counters
dewpoint	pH
differential pressure	relative humidity
dissolved O_2 level	speed (rpm)
flow	temperature
force or weight (load cells)	vacuum
gas analysers	vibration

gauge pressure	viscosity
level	VOC emissions
linear speed	water-in-oil analysers

One of the most popular electric signals is 4-20 mA at 24 V dc. Using an analogue current signal, rather than an analogue voltage signal, removes problems due to wiring resistance and voltage drop. Other analogue electric signals used are:

- 0 to 20 mA dc
- 1 to 5 mA dc
- 10 to 50 mA dc
- 1 to 5 V dc
- 1 to 10 V dc

Temperature detectors, such as thermocouples and resistance temperature detectors, RTDs, produce millivolt dc signals. These signals are standardised in BS EN 60584-1. Part 2 of the Standard specifies applicable tolerances.

Electric transmitters can be "intelligent" and incorporate control methods, such as Proportional Integral Derivative (PID), to allow the use of simple control valves. Indicating displays can be driven by the analogue signal to allow operator information to be shown at convenient locations. Repeaters, small booster amplifiers, can be used with analogue signals to allow signals to travel very great distances. But of course, only one signal is carried by each pair of wire cores. Sending a great amount of different data requires numerous cores. Digital communications can alleviate this problem.

12.2.2 Digital signals

Relatively complex serial digital communications can be transmitted over a simple RS 485 network. The RS 485 protocol is not as well known as RS 232 which is the serial interface fitted to most computers and PLCs. Converters are available to allow the computer RS 232 to communicate directly with RS485 networks. RS 232 links are normally restricted to about 100 m but special modems can extend this to several kilometres using four core cable. Fibre-optic data links can extend the 232 range to many kilometres. Radio 232 links are useful over suitable terrain. The simplest 485 system uses a screened twisted pair for communications. The controller can either transmit or receive data. By using two twisted pairs the 485 can transmit and receive simultaneously. Up to 32 devices can be connected anywhere over a distance of 1.2 km when using the specified cable type. This type of network is also called "multi-drop".

If longer distances are necessary RS 485 can be extended by using line repeaters, although some manufacturers claim 4 km system lengths without repeaters. Some line repeaters also allow the addition of another 16 or 32 devices. The RS 485 protocol can cope with a maximum of 255 devices and transmission speed can be between 100 kbps and 10 Mbps depending upon the cable type and length and the chipset used. Cable lengths shorter than 60 m support the fastest speeds. Cables of the maximum length, 1.2 km, run at the lowest speed. 8 bit binary numbers are transmitted at 10% of the bit speed because of the additional control bits. Multiple 485 networks can be used to control and communicate with many devices. Many devices are fitted with an RS 485 interface as standard or as an option:

- mV and mA transmitters for thermocouples
- transmitters for RTDs
- mV and mA analogue outputs
- multi-channel switch inputs
- relays
- event counters
- frequency measurement
- scanners
- data-loggers
- variable frequency inverters (for motor speed control)
- text displays
- graphic displays
- operator interfaces
- controllers
- PCs
- PLCs

Complex distributed control systems can be constructed using the RS 485 protocol. A number of controllers, with two 485 ports, use one port to communicate with up to 32 local devices; no line repeaters are used. The other 485 port communicates with a supervisory controller or computer. One device can influence the action of any other device, out of a maximum of 961, by communicating via a maximum of three controllers. The supervisory controller can be connected to another 485 network; the maximum number of devices would increase to 29 791. Supervisory controllers can include data-logging and long term data storage. Data history can be invaluable when diagnosing problems and tracing the causes of process upsets. As with all important data, the data is much more valuable than the hardware producing or storing it. Long term storage is a necessity not a luxury.

Devices measuring or supplying analogue signals must convert to or from digital signals. The analogue signal is sampled at a fixed frequency and converted to a digital value. The sampling and conversion takes time. Some devices sample at relatively slow speeds, 1 sample/s to 10 sample/s. This low sampling speed rate may be a problem with some process operations and higher speed devices may be needed. The 485 devices may be fitted with visual displays which show the current variable value or the status of a switch or signal. Measuring devices can also be fitted with high and low alarm outputs independent of the 485 circuits. Settings for the alarm outputs are adjusted remotely by the controller over the 485 network.

Text displays can relay information to operators regarding process conditions or give instructions for manual adjustments. Graphic displays, monochrome or colour, can show one of many stored images. Images can be prepared on a PC and downloaded to the display memory. Operator interfaces can have simple text displays or coloured graphic displays. Keypads for operator input can be from several re-programm -able function keys to full keyboards. Displays with touch sensitive screens can have many keyboard designs and layouts pre-defined and loaded as required by the controller. Operator access can be restricted by password or key-switches. Operator interfaces can include serial, RS 232, or parallel, Centronic, or USB interfaces for printers.

PLCs can be used to provide local intelligence to perform complicated tasks. Local intelligence unloads the main control system and does not "clog up" the 485 with control messages. The PLC program can include instructions for the PLC to report the completion of cycles of operations or to wait, if necessary, for permission to begin specific operations.

The RS 485 protocol is an internationally recognised Standard controlled by Consultative Committee on International Telephony and Telegraphy (CCITT) — now managed by ITU. Many proprietary serial bus protocols, Fieldbus, exist for digital communications, see Table 12.2.

ASI	IoNet
BACnet	ISP
Batibus	LonWorks
Bitbus	Melsecnet

CAB	MODBUS
CANbus	P-net
DeviceNet	Pakscan
Drivecom	Pamring
EIB	PROFIBUS
Esprit	SDS
FIB	SERCOS
Fieldbus Foundation	Seriplex
I/A	Simatic
Instabus	SP50
Interbus-S	World-FIP

Table 12.2 Proprietary communication buses

Three protocols; P-net, PROFIBUS and World-FIP; are defined in a European Standard, EN 50170. The Fieldbus Foundation specification forms the basis of a part of IEC 61158.

Certain protocols have been adopted by industry sectors. The following for example, are popular for building management:

- BACnet
- Konnex, KNX
- CAB
- Esprit
- Instabus
- LonWorks

The popular fieldbuses use a single twisted pair to communicate. The twisted pair forms a radial link to which the devices are attached known as a "multi-drop system". If the fieldbus is broken, the devices on the outer portion are cut off. The integrity of the fieldbus can be improved by using two separate twisted pairs. Some valve manufacturers have developed their own proprietary fieldbus systems which use a ring rather than a radial arm. If the ring is broken in one place, communications are not affected. Devices are only isolated after two breaks in the ring.

Some protocols are slow, 5 kbps, 31.25 kbps is a popular speed and others are very fast, 12 Mbps. Some protocols have very limited sensor numbers and are best suited to small control networks, such as sensors and control valves in a pre-assembled equipment package. Other protocols are capable of connecting many devices.

Digital control can be considered at three basic levels of communications:

- device
- distributed control
- supervisory

The device level covers relatively short distance connections between sensors/actuators and a local controller or PLC. The local controller or PLC can provide local logic for local adjustments or simply act as a "junction box" and pass on instructions from the DCS, distributed control system.

The distributed control level involves the local controllers/PLCs communicating with a remote PLC or computer.

The supervisory level includes the communications between all the remote PLCs or computers to enable the operation of the complete system to be viewed and analysed.

The fieldbuses are generally used for device and distributed control levels although some of the more powerful, PROFIBUS, World-FIP, SP50 and Fieldbus Foundation, can be used at supervisory level. Supervisory level is generally limited to established computer local area networks, LANS, such as:

- DECNet
- Ethernet
- NFS
- Novell
- OSI
- SNA
- TCP/IP
- X.25

Hardware and software systems have been developed to assist with the management of large processes. These "expert" systems interrogate the control buses for strategic data and make recommendations or highlight operational problems. Typical "expert" systems include:

- batch and continuous process optimisation
- decision support
- equipment diagnosis
- plant information
- raw material scheduler
- simulator
- utilities optimisation

Some control system suppliers fit the system with a modem for communications with the supplier's nearest office. The supplier can then interrogate the system at regular intervals to check for minor faults and system efficiency. The system can be set up to phone the supplier under specific conditions so that immediate action can be taken.

Outsourcing of equipment routine maintenance and trouble-shooting has become very popular over the last few years. Equipment suppliers no longer just provide spares to the user. The equipment supplier underwrites the reliability of the equipment by performing all necessary maintenance and adjustment functions, including the supply of replacement parts, for a fixed period for a fixed fee. Telemonitoring allows the supplier to "keep-an-eye" on equipment without frequent site visits. Telemonitoring is of course used for other equipment as well as control valves such as compressor sets, pump sets and turbines.

One form of digital communications, mains borne signalling, has become popular. A high frequency digital signal, about 0.5 V, is superimposed on the 50 or 60 Hz power distribution cabling. The signal is detected by coils around the power cable. This type of signalling can be very cost effective as no control wiring is required. A control system can be added to an existing installation provided power cables are fitted in all areas. Problems can arise due to mains-borne interference if the control signals escape. A blocking filter is required to prevent signal transmissions into uncontrolled areas or off-site. A European Standard, BS EN 50065-1, specifies the frequency bands allocated for various user groups:

- 3-95 kHz electricity suppliers
- 95-148.5 kHz consumers

This type of communication is favoured in building services for lighting, heating and security applications.

12.2.3 Mixed signals

The 4-20 mA analogue current signal is a popular control option. A digital signal can be superimposed on the current signal using a technique called Frequency Shift Keying, (FSK). The digital signal can be used for communications independent of the current signal control functions. Communications can include resetting parameters as well as interrogation. The HART protocol, implements analogue and digital signals using one pair of cores. When device manufacturers have upgraded equipment to support the HART protocol, the user can upgrade an installation without rewiring between measuring sensor or actuators/positioners and controllers. The HART protocol can be used over distances between 1.5 and 3 km depending upon the wiring.

Many sensors and actuator/positioners can be upgraded to the HART protocol. Replacement or modified circuit boards can be fitted to implement the extra facilities. The HART protocol can be used when only the measuring device or actuator/positioner has been upgraded. Device settings can be adjusted in the "field" using a handheld programmer or palmtop computer to send the digital signals.

The HART protocol is very popular; it is much more popular than any of the proprietary digital bus options discussed in 12.2.2.

12.2.4 Electrical safety

Traditional pneumatic control valves, using an analogue process signal, are inherently safe. The process signal is piped to a bellows or Bourdon tube for measurement. Process fluids are contained and pose no threat to personnel or environment. Valves and controllers are unaffected by gas mixtures in the atmosphere providing corrosion is not caused. Gases in the air can have a significant impact on the choice of electrical equipment.

Electric and electronic systems must comply with European standards when used in potentially hazardous areas. The two most popular options are flameproof, FLP, and intrinsically safe, IS. The requirements for FLP equipment are specified in BS EN 60079 and BS EN 50018. IS requirements are specified in BS EN 60079 and EN 50020. Analogue electric signals, 4-20 mA, can comply with both philosophies. Other sensor signals can be fitted with barriers to comply with IS requirements.

The 4-20 mA sensor signal combined with HART digital information can be intrinsically safe. RS 232 serial digital communications can be isolated to provide IS level protection. Some proprietary digital buses; PROFIBUS, World-FIP, LonWorks; can be used as IS systems but data transmissions speeds tend to be drastically reduced, 31.25 kHz may be the maximum. Controllers, PLCs and operator interfaces are available, certified to IS requirements.

Fibre optics permits digital data transmission over longer distances and at higher data rates than electronic communication. Research has shown that broken fibre optics can cause ignition in hazardous areas under some operating conditions. This risk should be evaluated and precautions taken when necessary.

12.2.5 Electromagnetic compatibility

EMC, electromagnetic compatibility, is an international and European standards requirement for all electrical equipment. All electrical equipment has the capability to radiate interference in the form of signals through the mains power supply or through the surrounding air in a manner similar to that of radio transmissions when in use. Electric and magnetic fields both create electromagnetic radiation. The electromagnetic radiation of electric equipment must be below specific requirements and all electric equipment must be capable of operating correctly when subjected to the permissible radiation.

Interference in cables due to radiation can be cured by using screened cable and segregating different services. Equipment may need screening in a similar fashion. The use of screens is only effective when proper earth bonding and complete screen continuity is maintained. The interconnection of screens, cable-to-cable and cable-to-equipment, is extremely important and must be made efficiently. If cable screening cannot be performed effectively then fibre optics should be considered.

Interference transmitted through the mains power supply can be very difficult to eliminate. Electricity supply companies may have more stringent requirements than the normal standards in order to protect other consumers. The sudden voltage drop caused by DOL (direct-on-line) starting of large electric motors is a particular type of interference which has been recognized for a long time. Most supply companies have limits on the motor size which can be started DOL.

Everyday items can produce interference which causes other equipment problems. The following list is not exhaustive:

- fluorescent lights
- dimmers for incandescent lights
- photocell automatic light switches
- electric welders
- heat guns
- cellular (mobile) telephones
- variable frequency inverters

Variable frequency inverters can inject harmonic distortion into the electricity supply. This is not usually a problem with small drives. As drive powers increase, electric power supply companies have set limitations on the permissible distortion. Very large variable frequency drives need specific clearance from the local electric company before installation.

Electrical and magnetic radiation and mains-borne interference can cause serious problems to electric and electronic instrumentation. Operating problems can be expected if adequate precautions are not taken.

International and European standards tend follow national or industrial standards. EN 61000 and the German Standard NAMUR AK 5 should be consulted regarding EMC radiation and immunity.

12.2.6 Choice of signal system

The decision to use one type of system may not be easy. One important factor to be considered is the skill level of site staff and the distance to the nearest manufacturer's service centre. A system which can only be maintained by manufacturers' representatives, can have very high operating costs and may be subject to prolonged shut-downs if representatives are not available. Mechanical equipment may be complex, but can usually be repaired with simple tools by experienced personnel. Electric/electronic equipment may require specialised test equipment if full diagnostic capabilities are not included in the hardware/software. When upgrading or expanding an existing system, retraining of personnel will be necessary to cope with new technology.

Mechanical (pneumatic) systems can be checked with pressure gauges and standard mechanical measuring equipment, such as feeler gauges. Analogue electrical systems can be checked with standard multimeters. Traditional electric analogue controls, 4-20 mA and 1-10 V, can be integrated into modern digital systems. Digital systems require specialised test equipment.

Electronic technology changes rapidly. Any system installed today may become difficult to maintain in two years; spares may not be readily available. The installed equipment may operate satisfactorily, but how long will spares continue to be available? Systems based on proprietary chips may be vulnerable to spares shortage as the manufacturer improves facilities; a guaranteed upgrade path must be assured.

Proprietary software and gateways allow digital and analogue plant and process control systems to be interrogated by management systems, using completely different protocols. Process communication buses can be interconnected with standard computer communication networks. Process information can be collected by accountants and engineers and imported into their favourite database or spreadsheet programs. Management accountants and engineers do not have to learn the process system; a gateway allows them to use familiar sys-

	Diaphragm	Piston	Vane	Electro-mechanical	Electro-pneumatic	Electro-hydraulic
Travel mm	up to 105	unlimited	up to 90°	unlimited	unlimited	unlimited
Speed	slow	medium to very fast	medium	very slow	slow to medium	slow to very fast
Thrust/torque	low to high	low to very high	medium	low to high	low to high	low to very high
Stiffness	low	medium to high	low	high	medium to high	medium to high
Linearity % of travel	1 to 4	1 to 4		1 to 3		0.1 to 1.0
With pneumatic positioner	1 to 2	1 to 2				
Repeatability % of travel	1 to 4	1 to 4		1 to 3		0.1 to 0.5
With electronic "Smart" control	0.25 to 0.5	0.25 to 0.5				
Failure position	all	all	all	stay put (1)	all	all

(1) See fail-safe, Section 12.4.4

Table 12.3 Basic actuator capabilities

tem commands to find the relevant information. Security must be implemented to ensure that access to areas of information is restricted to only those requiring it. Accountants, for example, should not have access to control settings of field instruments or be able to adjust them!

Digital systems can store process variables and information from equipment sensors regarding the health of machinery. Processes can be optimised continuously to achieve the best efficiency for the current conditions. Machinery data can be reviewed automatically for trends and predictions made for loss of efficiency or the latest date when shut-down may be required to prevent serious damage. Electronic systems allow the incorporation of standard and proprietary databases and documentation. CD-ROMs can be used for online access to operating and maintenance information. Maintenance staff "in the field" can use text or graphical displays to read manufacturers' manuals, while standing next to a field device. Documentation updates are simple to implement.

Digital communications enable much more information to be transmitted faster over long distances. Modems and satellites effectively provide worldwide communications. Reliable digital communications can allow processes to operate much more effectively and efficiently. If the support infrastructure is provided to cope with the day-to-day maintenance of a digital system, then digital can be the most cost-effective. If digital maintenance support cannot be provided adequately a more traditional approach must be adopted.

12.3 Types of actuator

There are many types of actuator for use with isolating and process control valves. The least expensive and most widely used is the pneumatic diaphragm actuator with spring return. The primary function of an actuator is to control the position of the valve moving element; gate, plug, ball, butterfly, etc.; between "open" and "closed" to ensure correct control of the process fluid. To do this the actuator must be sufficiently powerful to produce positive, accurate and rapid response to a control signal and be able to return the valve to a suitable predetermined position in the event of signal failure. It is therefore important to specify the correct type and size of actuator in order to meet the demands of accuracy, reliability and economy. Table 12.3 outlines the basic capabilities of the major actuator types.

Diaphragm actuator

The pneumatic diaphragm actuator as shown in Figures 12.2 and 12.3 is the most commonly used in the process industry today for a number of reasons:

- very strong for its size and weight
- simple, inexpensive construction
- no material compatibility problems with dry air
- the petrochemical industry, representing the biggest user, prefers to use compressed air rather than electricity to reduce the fire risk

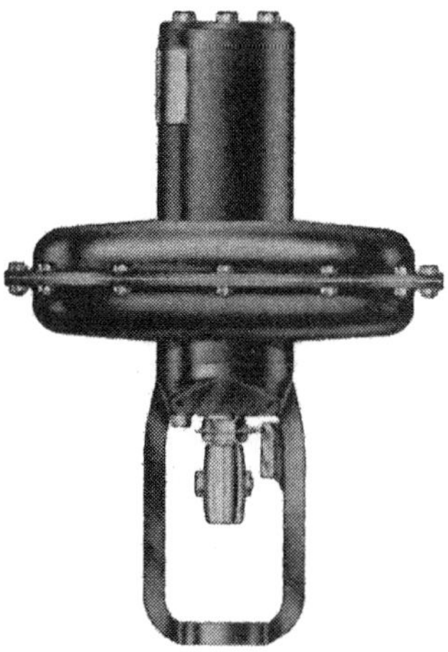

Figure 12.2 A typical diaphragm actuator

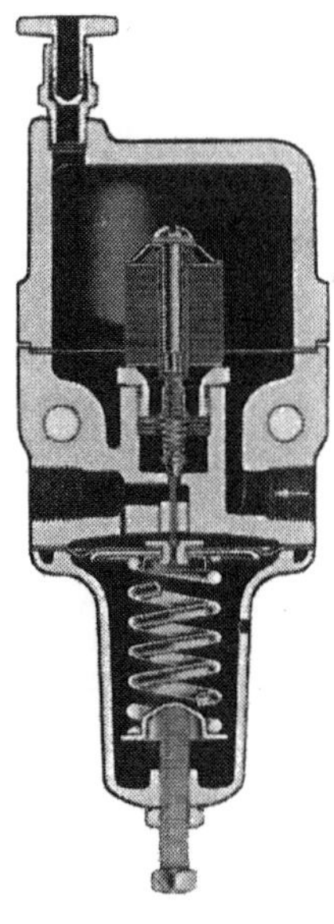

Figure 12.3 A typical reducing pressure filter-regulator

- actuators can be set to return the valve to a pre-determined position, open or closed, in the event of supply air failure
- availability of well-developed associated technology; positioners, regulators, boosters, etc.
- they are the most economical solution for thrusts of up to 10 kN

Diaphragm actuators usually use air pressures between 0.21 and 1.03 barg, 3 and 15 psig, to operate the valve. Figure 12.2 shows a typical actuator with the return spring mounted above the diaphragm. The yoke, for attaching the actuator to the valve body and the stem connector, can be easily seen.

When a pneumatic controller is used to produce the diaphragm actuator pressure it may be necessary to fit a filter-regulator in the air supply to the controller. Figure 12.3 shows such a typical regulator arrangement which consists of:

- a pressure reducing regulator
- a filter
- a catchpot and drain valve for moisture

The regulator is selected for its very large regulating range, 1 : 1000, and very stable pressure, even under varying flow conditions. It is the balance between the pressure under the diaphragm and the spring pressure above the diaphragm which determines the position of the valve plug, and which in turn determines how great the instrument air pressure will be. If the spring is screwed down a long way, then high pressure is required underneath to close the valve, i.e. a high outlet pressure.

Figure 12.4 shows sections through a standard actuator assembled for direct and reverse action. Direct action conventionally is understood to be downward.

In either case the space between the actuator cover and the diaphragm is a completely enclosed volume. In the absence of compressed air the diaphragm is pushed to its uppermost position by the spring acting upon the diaphragm plate. If compressed air is applied to the diaphragm then pressure will increase in the confined space above or below the diaphragm.

Since pressure acts in all directions, then an upward or downward force, the product of pressure and area, will act upon the diaphragm. When this force becomes greater than the spring pre-load, the diaphragm will move up or down, compressing the spring and causing the actuator and valve stems to move up or down as well. Theoretically, when the pressure is 0.21 barg, 3 psig, the actuator is still in its original position, i.e., the air pressure should just balance the spring pre-load. When the pressure is greater than 0.21 barg the stem begins to move. At 0.3 barg for example:

> 0.3 x area = an actuator thrust greater than the spring pre-load.

Increased air pressure produces more thrust until at 1.03 barg, 15 psig, the diaphragm and stem should be at their extreme position.

The actuator spring is designed to balance the applied pressure between 0.21 and 1.03 barg as the stem travels between the initial and extreme positions. This relationship is also known as "the spring range". Spring configurations vary depending upon the size and style of actuator. A single helical compression spring, centred on the stem, is a basic arrangement. Multiple concentric springs may be used to increase the spring load or several springs may be positioned on a pitch circle around the stem.

There are three types of diaphragm actuator:

- direct-acting — air pressure is applied above the diaphragm and pushes the stem downwards
- reverse-acting — air pressure is applied below the diaphragm and pushes the stem upwards
- double-acting — air pressure can be applied to either side of the diaphragm to push the stem in the required direction, no return spring

The type of actuator used depends upon whether the valve is required to open or close in the event of loss of supply air pressure to the actuator, i.e. fail-safe.

Valves controlled by pneumatic diaphragm actuators usually operate relatively slowly because of the time required for air at low-pressure to fill and build up pressure in the space between

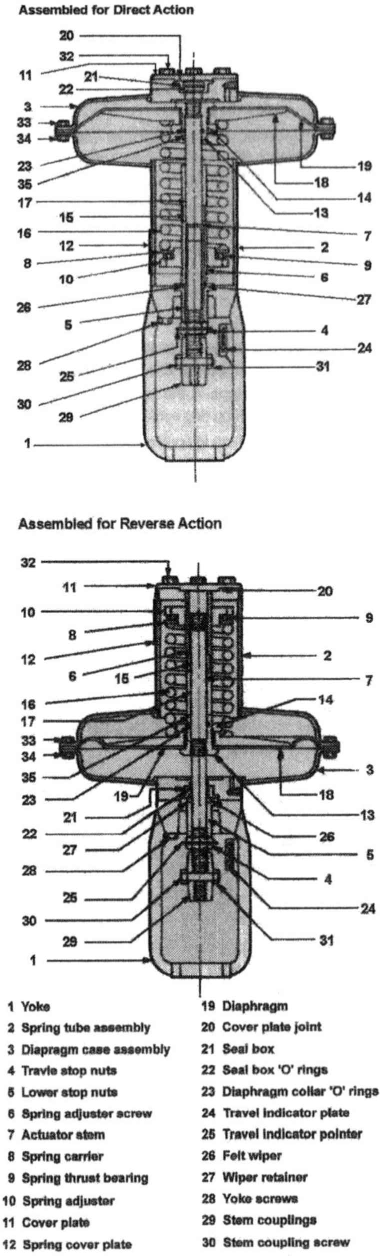

1 Yoke
2 Spring tube assembly
3 Diapragm case assembly
4 Travle stop nuts
5 Lower stop nuts
6 Spring adjuster screw
7 Actuator stem
8 Spring carrier
9 Spring thrust bearing
10 Spring adjuster
11 Cover plate
12 Spring cover plate
13 Diaphragm collar
14 Collar nut
15 Stem guide
16 Spring
17 Travel stop tube
18 Diaphragm button
19 Diaphragm
20 Cover plate joint
21 Seal box
22 Seal box 'O' rings
23 Diaphragm collar 'O' rings
24 Travel indicator plate
25 Travel indicator pointer
26 Felt wiper
27 Wiper retainer
28 Yoke screws
29 Stem couplings
30 Stem coupling screw
31 Stem coupling screw nut
32 Cover plate screws
33 Diaphragm cast screws
34 Diaphragm cast screw nuts
35 Grub screws

Figure 12.4 Direct and reverse-acting diaphragm asemblies

the diaphragm and the cover. The maximum thrust is also limited by the diaphragm area and the maximum air pressure which can be used, usually 1.03 barg, 15 psig, although higher pressure diaphragms, up to 5 barg, 72 psig, are sometimes available depending upon the size.

The typical time to stroke a DN100 valve is 4 to 8 seconds. A DN250 valve may take 20 seconds. These times may be reduced by a factor of four or more by the use of a large air connection on the diaphragm, by booster relays, quick exhaust valves or a combination of these. Some users specify the air pipework between the controller/positioner and the actuator

Figure 12.5 Diaphragm actuators fitted to globe body control valves
Courtesy of Dresser-Masoneilan

must be 10 mm od minimum. This is much larger than the normal air supply pipework to the controller/positioner which is 6 mm od. Some users may specify double ferrule compression fittings, even on air pipes, see Section 9.10.2 in Chapter 9.

When considering fast valve movements the effects of surge and water hammer must not be overlooked. Figure 12.5 shows globe body control valves, with a diaphragm actuator, set up for compressor anti-surge applications. The complexity of the instrumentation and piping is obvious. The solenoid valve vent connection, on top of the diaphragm casing, can be seen clearly.

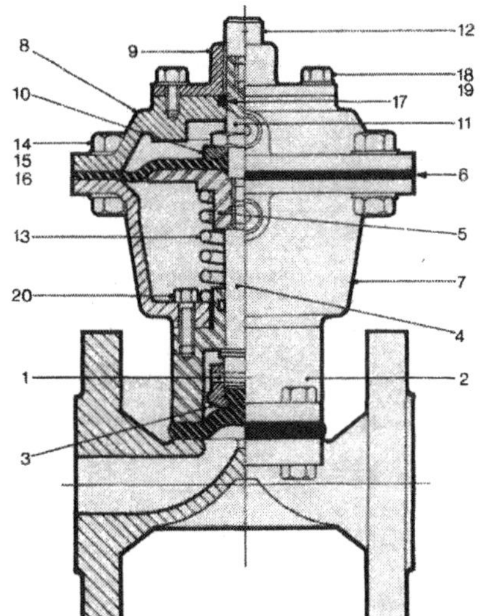

1 Compressor pin
2 Bonnet
3 Compressor
4 Spindle
5 Diaphragm plate
6 Operating diaphragm
7 Cylinder
8 Cover
9 Cover cap
10 Clamp washer
11 Adjusting screw
12 Indicator
13 Spring
14 Cover/cylinder bolt
15 Cover/cylinder nut
16 Cover/cylinder washer
17 'O' ring
18 Cover cap bolt
19 Cover cap washer
20 Cylinder/bonnet bolt

Figure 12.6 A short stroke diaphragm actuator
Courtesy of Crane Process Flow Technologies

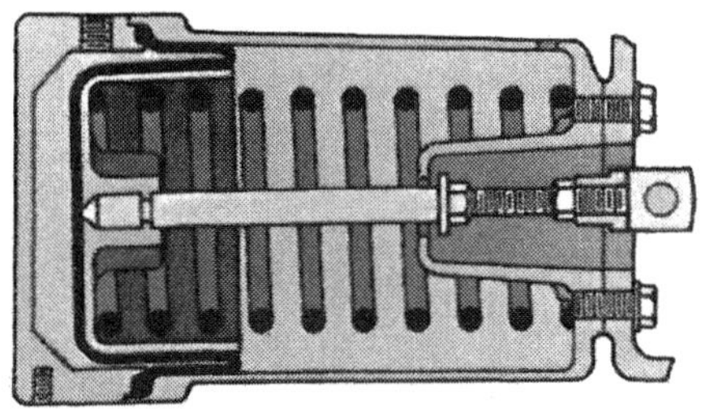

Figure 12.7 A rolling diaphragm actuator

In most spring diaphragm actuators, the moulded, fibre reinforced diaphragm is clamped between the upper and lower diaphragm casings. The diaphragm itself is moulded in the shape of a shallow dish. This limits the stroke that can be obtained and, because a substantial part of the diaphragm is unsupported, it also limits the maximum pressure to about 5 barg, 75 psig, for a 320 square centimetre actuator or to 3 barg, 43 psig, for a 1290 square centimetre actuator. Some diaposition holding is quite flexible due to the volume of air in the diaphragms are relatively well supported by their plates which allow attachment of the stem. Figure 12.6 shows a diaphragm actuator, with fail-safe closing spring return, fitted to a weir type diaphragm valve, only a short stroke is necessary.

Figure 12.7 shows a design that partially overcomes the short stroke deficiency called the "rolling diaphragm". It uses a closely fitting cylinder with very little unsupported area. This design has been applied to conventional actuators for reciprocating valves with 25 to 100 mm stroke, and has a fixed pressure rating of 5 barg, 72 psig, irrespective of size. Figure 12.8 shows a rolling diaphragm actuator fitted to a needle valve. A particular application of the rolling diaphragm is the Camflex™ valve which has a long stroke appropriate for the actuation of a rotary valve.

Diaphragm actuators used for rotary valves are basically similar to those for reciprocating motion valves. Originally, actuators which had been designed for reciprocating valves were adapted to provide rotary motion by fitting a suitable open linkage mechanism. The linkage mechanism had an inherent weakness and could not compensate for angular over-travel of the valve shaft. A properly integrated actuator design is pre-

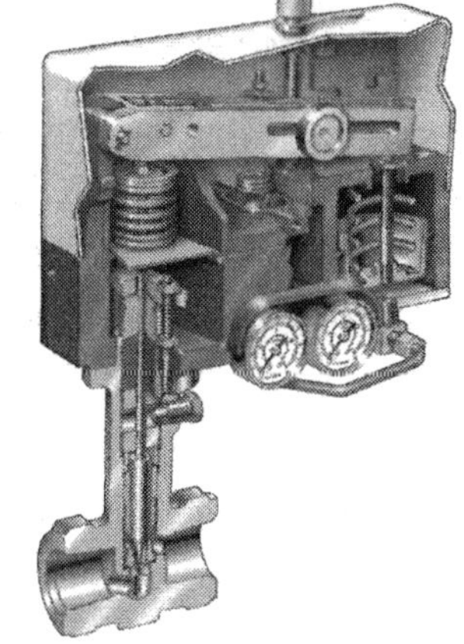

Figure 12.8 A rolling diaphragm actuator fitted to a small valve
Courtesy of Dresser-Masoneilan

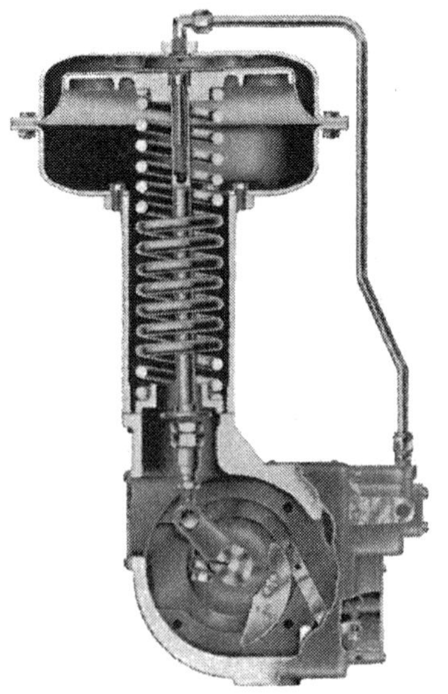

Figure 12.9 A diaphragm actuator fitted to a rotary valve

ferred, as shown in Figure 12.9. This construction has many advantages:

- the mechanism is totally enclosed and therefore protected from damage and the ingress of dirt, dust etc.
- the fitting of a positioner is greatly simplified; see Section 12.5
- the system is much more positive and robust

With this arrangement it is also possible to vary the relationship between signal input and angle of rotation by using different cams in the positioner. Positioner operation and function is dealt with in Section 12.4.2. Rotary actuators may suffer from stiction and require 15 to 25% extra air pressure to break-away.

The diaphragm actuator does have some disadvantages:

- a large proportion of available thrust is used to overcome the spring force
- not quite linear characteristic, within 2% typical
- hysteresis due to diaphragm stiffness, typically 3%
- air permeates through diaphragm and is lost
- diaphragm can fatigue due to flexing
- position holding is quite flexible due to the volume of air in the diaphragm case and the compressibility of air, see **Actuator stiffness** in Chapter 1

The most popular diaphragm material is nylon reinforced nitrile rubber which is suitable for operating temperatures between -40° and 80°C. Viton™ may be available as an option for higher temperatures and pressures and PTFE would comply with hygiene requirements. The diaphragm casing can be in various materials to suit industrial and hygienic requirements:

- pressed steel
- aluminium
- Al Zn alloy epoxy coated
- vinyl ester plastic
- PAS

The stem is normally of steel, but stainless steel is available from a few manufacturers, some of whom allow water to be used. Attaching yokes will be of cast iron, steel or aluminium.

Diaphragm actuator strokes are "industry" standardised to suit the full range of valve sizes:

- 20 mm
- 30 mm
- 54 mm (2 ⅛ ")
- 105 mm (4 ⅛ ")

Travel stops can be adjusted to prevent excessive loads on valve stems. Typical diaphragm areas range from 160 to 2900 cm^2 producing thrusts from under 2 kN to over 125 kN. Diaphragm actuators for rotary drives can produce torques over 1100 Nm.

Piston actuator

Piston actuators are also called "cylinder actuators". Piston actuators are generally used in applications where the stroke of a diaphragm actuator is too short or the thrust too small. They can operate at higher pressures and in some applications a pressure reducing regulator may not be necessary. They are very useful for linear isolating valves such as gate valves. Short stroke piston actuators are available. Figure 12.10 shows an example of a double-acting actuator fitted to a weir type diaphragm valve for isolating applications. Figure 12.11 shows a piston actuator with spring return and a positioner fitted for control valve applications.

Piston actuators can be considered as high pressure versions of the diaphragm actuator. Single acting spring return actuators are available for mostly "on/off" isolating valve applications; the double acting versions can be used on control valves when good position holding is required under varying valve load conditions.

The principle of operation involves creating modulated reciprocating motion by increasing or decreasing air pressure simultaneously on either side of a piston. The air is supplied at high pressure, 6 to 10 barg, which quickly fills the cylinder volume. Piston actuators therefore have an inherent rapid response for

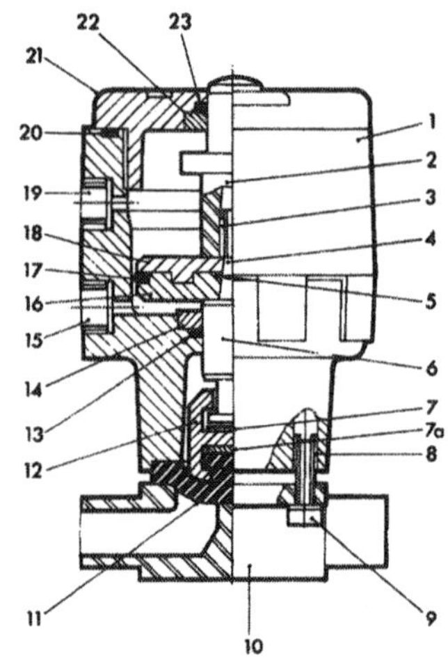

1 Bonnet
2 Indicator
3 Indicator insert
4 Spindle indicator insert
5 Piston inner seal
6 Spindle
7 Thrust disc
7a Thrust disc
8 Bonnet base insert
9 Screw
10 Valve body
11 Diaphragm
12 Compressor
13 Spindle seal
14 Bonnet washer
15 $\frac{1}{8}$" BSP insert (lower port)
16 Lower piston section
17 Piston outer section
18 Upper piston section
19 $\frac{1}{8}$" BSP insert (upper port)
20 Cap seal
21 Cap
22 Cap washer
23 Indicator seal

Figure 12.10 A short stroke piston actuator
Courtesy of Crane Process Flow Technologies

Figure 12.11 A spring return piston actuator with positioner

a given input signal. Actuators with plastic pistons and/or cylinders may be limited to 6 barg air pressure. Figure 12.12 shows a typical double acting piston actuator with an in-line positioner fitted.

The advantages of a piston actuator are:

- it can apply a greater thrust than an equivalent area diaphragm actuator due to higher working pressure
- operates more rapidly than diaphragm actuators because of the smaller working volumes and the higher operating air pressures
- greater inherent "stiffness" (can be 10 times stiffer than a diaphragm), enables higher precision positioning
- greater stiffness promotes better position holding under varying valve thrust conditions
- greater stiffness enhances vibration damping
- it may not need a pressure regulator for on/off isolation applications

Cylinders for longer strokes are generally formed by clamping a plain tube between two end caps. Long studbolts around the periphery of the tube provide the clamping load. The bore of the tube must have a very good surface finish to reduce the wear of the piston seal. Bores are typically ground or honed to produce

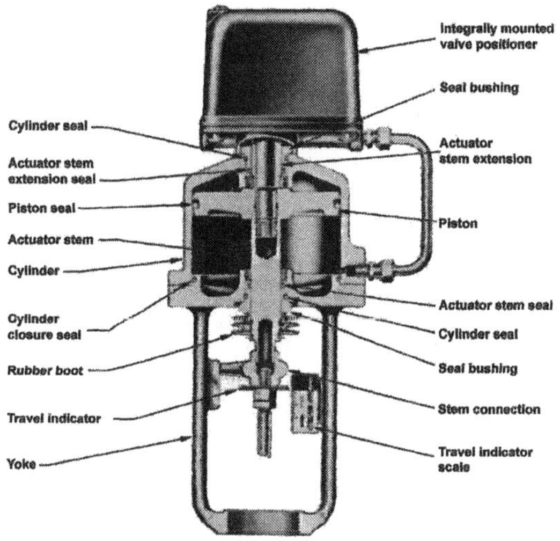

Figure 12.12 A double acting piston actuator with in-line positioner

Figure 12.13 A knife gate valve for isolating applications

a finish between 0.1 and 0.4 μm. One end cap must be fitted with a piston rod seal to prevent the ingress of foreign solids and moisture which could degrade the piston seal. Self-energised "U" rings are used extensively. The "U" ring may be protected by a wiper ring or close fitting plain bush which prevents solids approaching. The other end cap may be fitted with an adjustable travel stop. Piston seals can be dedicated double "U" seals or "O" rings. The basic construction of piston actuators allows virtually unlimited stroke lengths. Figure 12.13 shows a piston actuator fitted to a knife gate valve. Figure 12.14 shows a piston actuator fitted to a mix-proof CIP valve designed for hygienic applications.

Knife gate valves fitted with piston actuators have been used as high speed isolating valves to prevent the spread of fires and explosion damage. Temperature and pressure sensors are used to detect the approach of flame fronts or pressure waves.

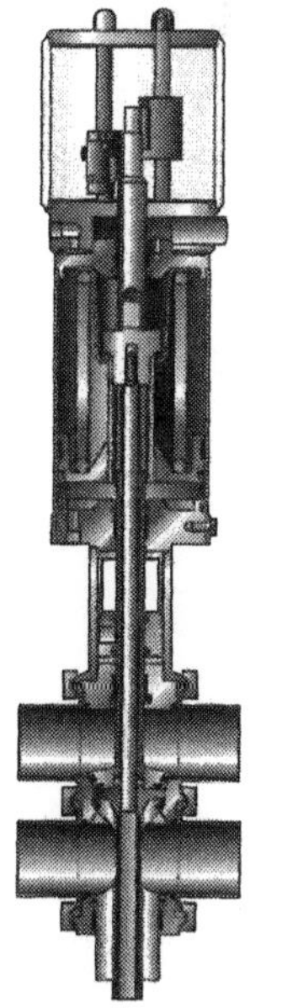

Figure 12.14 A mix-proof diverter valve with CIP facilities
Courtesy of Waukesha Cherry-Burell

An electric signal, which releases high pressure gas to the actuator, is transmitted to a gate valve further along the pipe. To ensure the valve closes as quickly as possible, air or gas is supplied from pressurised cylinders attached to the actuator. To prevent damage to the gate and the valve body, the piston end-of-travel is cushioned in the cylinder to provide rapid deceleration on final closure. Additional deceleration may be provided by physical deformation of travel stops. These gate valves are reset manually after operating conditions have stabilised.

Static and dynamic torques in rotary valves can amount to several thousand Nm. The linear motion of the actuator must be converted to rotary motion efficiently. Various linkage mechanisms have been used in the past. Some mechanisms are now avoided because of the inaccurate control resulting from inherent backlash.

The diaphragm actuator shown in Figure 12.9 solves this problem by allowing for lateral movement of the actuator stem connected directly to an arm attached to the valve shaft. The lateral movement is permitted by diaphragm distortion. Piston distortion is not possible. Lateral movement could be accommodated by attaching the stem to the piston via a spherical bearing. However this would introduce additional friction and possibly lubrication problems. Some manufacturers use a piston which is partially spherical rather than cylindrical. The piston rocks in the cylinder bore, see Figure 12.15.

The "slider crank" arrangement shown in Figures 12.9 and 12.15 are quite effective but the motion conversion is not linear. 1% of piston travel produces varying amounts of spindle rotation depending upon the stroke position. Also, torque at the beginning and end of the stroke is reduced precisely at the places where more torque could be useful to overcome stiction. Quarter turn actuators have used radial arms with slots to accommodate a sliding pin attached to the piston rod. Tandem piston actuators have used the "Scotch yoke" mechanism to convert the motion, but this solution creates bearing problems. Hertz stresses, produced by line contact of a circular pin on a flat surface, can easily cause local plastic deformation which ruins the smooth sliding surface. Cams sliding on the pistons have also been tried. These have the advantage of variable cam profiles to modify the stroke/rotation relationship. One cam can operate between two pistons to provide accurate positioning and return travel without the use of springs. Air pressure is not wasted in the compressing of springs. The cam and piston materials must be "bearing compatible" and sufficient contact area must be provided to reduce wear to acceptable levels.

An alternative method of translating linear piston motion into rotary spindle motion is to use a rack and pinion. With the popularity of quarter turn ball and plug valves, the piston actuator, with rack and pinion, has undergone very rapid development to increase torque and reduce size. Various designs are used, and Figure 12.16 shows from top to bottom, a single piston, a tandem piston and an opposed piston configuration.

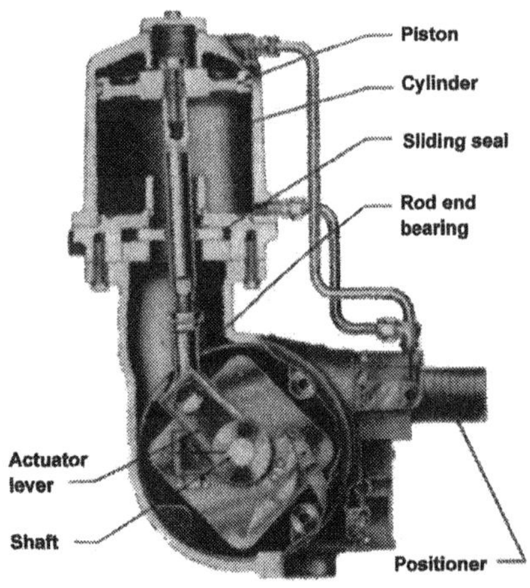

Figure 12.15 A piston actuator for rotary valves

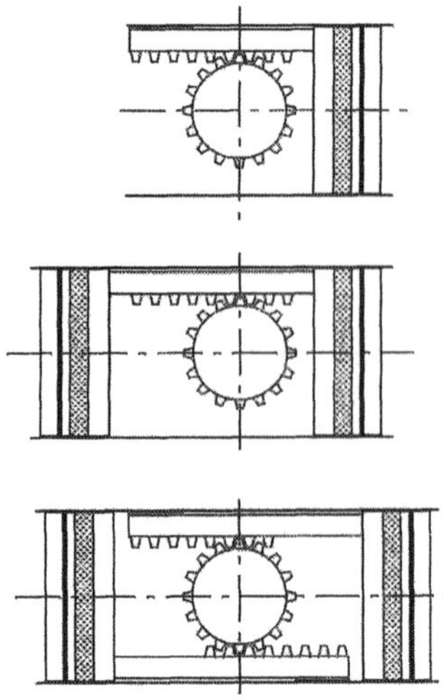

Figure 12.16 Typical rack and pinion piston actuator configurations

The rack and pinion both require support to resist the forces generated by the tooth mesh. Some designs provide a slide along the other side of the spindle; this is good because it supports both the rack and the pinion simultaneously. The pistons are generally fitted with PTFE slide rings, which act as bearings and help to seal. When required, return springs are mounted on the outer faces of the pistons. End caps can have bolts with locknuts for travel stops. A novel idea which takes the multiple piston concept further is shown diagrammatically in Figure 12.17. Some multiple piston actuators are only suitable for compressed air or other clean gases. Figure 12.25 shows an opposed piston actuator powered by hydraulics.

Rack and pinion piston actuators are available for 180° applications. The top and bottom faces of the cylinders are usually provided with standardised interfaces for actuator mounting and mounting accessories on the actuator. Popular Standards include:

- ISO 5211
- DIN 3337
- VDE/VDI 384 NAMUR
- Norbro Standard

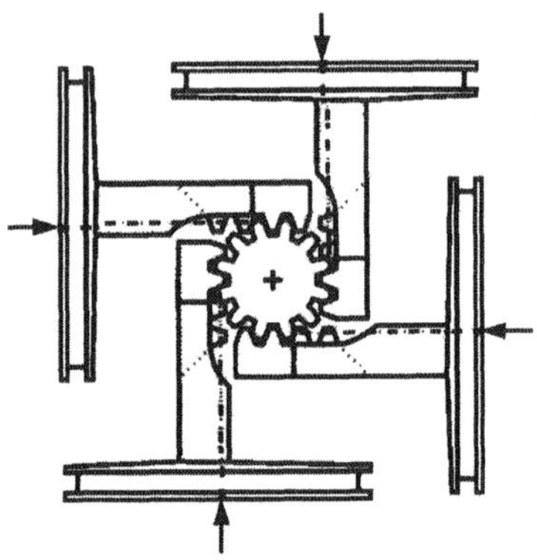

Figure 12.17 A four-piston rack and pinion actuator
Courtesy of Forac Ltd

Figure 12.18 A 90° vane actuator with return spring

Piston actuators are made in a selection of materials to cope with operating conditions, see Table 12.4.

Tube	End caps	Piston	Piston rod Spindle	Piston seal	Rod seal
Glass fibre reinforced epoxy Polymide Aluminium	Aluminium Polymide	Aluminium	Steel Stainless steel	Neoprene	Poly-urethane

Table 12.4 Piston actuator materials

Piston actuators are used with hydraulic oil as well as air. Central hydraulic power packs provide the pressurised oil in a similar manner to that of a central compressor set providing the instrument air. Larger cylinders, up to 600 mm bore, have been used for pressures up to 15 barg. Cylinders up to 200 mm bore have been used on pressures as high as 130 barg. The use of higher pressure fluid allows smaller cylinders to provide high thrusts. Piping sizes can be smaller as the working fluid volume is reduced and pressurised reservoirs can also be smaller. The effective inertia of the piston plus the fluid system is reduced, permitting faster response times and quicker valve operation.

Standardised hydraulic power packs are built for pressures up to 344 barg. One of the best pumps for hydraulic oil which is simple, efficient, reliable and long-lasting is the triple screw pump and this is limited to about 250 barg. This might be a sensible limit when considering higher pressure applications. Table 12.5 indicates the type of output achievable from piston actuators.

Piston actuator type	Fluid	Max output
Single acting spring return	Air 10 barg Oil 12 barg	394 kN 340 kN
Double acting	Air 10 barg Oil 15 barg	134 kN 438 kN
Single acting spring return	Air 10 barg	4 160/2 645 Nm
Double acting		650 000 Nm
Opposed piston spring return	Air 6.0 barg Oil	3 240/1 660 Nm 4 195/2 619 Nm
Opposed piston double acting	Air 6.0 barg Oil	4 955 Nm 3 923 Nm
Four piston double acting	Air 8.2 barg	2 541 Nm

Table 12.5 Piston actuator capabilities

A novel variation on the piston actuator produces rotary motion directly without linkages. Inside the cylinder the piston is connected to a helical gear which meshes with a helical circular rack attached to the cylinder. As the piston slides through the cylinder the gear/rack mesh forces the piston to rotate. The rotary motion is transmitted to the spindle via splines or additional helical gears. The double acting hydraulic piston offers all the usual advantages. Working pressures can be up to 210 barg with output torques, in both directions, up to 76 000 Nm

Vane actuator

The vane actuator is a simple mechanism for quarter turn applications. The vane actuator produces rotary motion directly by applying air pressure to an arm or "vane" attached to the spindle. Because the vane is attached directly to the spindle there is no backlash or lost motion. The arm has seals on all edges. A split casing is assembled around the vane and totally encloses it. When air is applied to one side of the vane a torque is produced which affects the spindle directly, without translation or transmission losses. The torque produced is directly proportional to the differential pressure across the vane. Travel stops can be fitted in the casing edges to prevent over travel and overload in both directions.

Vane actuators can be single and double acting. Clock type torsion springs can be used for spring return applications by adding a spring housing on top of the actuator casing, see Figure 12.18. By adding a spring, vane actuators can fail-open or fail-closed in the event of power failure.

Most actuators require high quality air, see Section 12.1. Vane actuators can accept poor quality air. However, depending upon the type of vane seals used, lubricated air can be acceptable and entrained solids up to 40 μm are possible. Vane actuators can be supplied to operate with hydraulic oil. Gearboxes can be added to extend the operating travel up to 180°.

Vane actuators are generally small, producing torques up to 200 Nm with 6 barg air pressure. They are probably better suited for butterfly valves than for ball or plug valves.

In the context of producing rotary motion, vane motors for compressed air and compressed liquid are available as drivers. These motors are multi-turn drivers in the same sense as electric motors. Vane motors can be used to actuate both linear and rotary valves by fitting a suitable speed reduction and motion translator. Linear valves, for example, can be actuated by using the vane motor to drive the nut in a similar manner to electro-mechanical actuators. Vane motors can be considered when electricity is unavailable or for hazardous areas.

Electro-mechanical actuator

Electro-mechanical actuators are very versatile and can produce all the required motions for valve actuation:

- linear
- part-turn, 90° or 180°
- full turn, 360°
- multi-turn

Electro-mechanical actuators can easily be adapted to suit most types of valve. Electrical safety must be observed when operating in hazardous areas and the ambient temperature must be considered for cooling requirements. Beware of high black bulb temperatures. Each actuator consists of a number of components, see Figure 12.19. Among these are:

- an electric motor, low inertia, high torque; dc or single phase, capacitor-run for small actuators, three phase for larger sizes
- gearing to reduce speed and increase torque
- valve attachment according to ISO 5210
- a change-over mechanism to allow local manual operation; this should be lockable to prevent both methods driving simultaneously

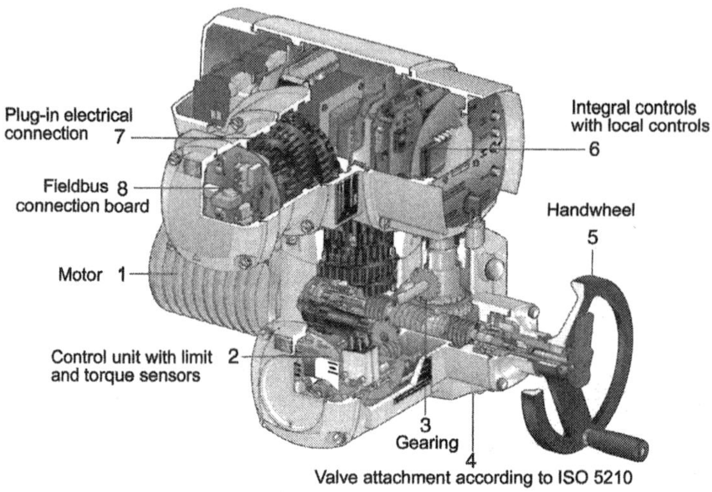

Figure 12.19 Typical electric actuator design
Courtesy of AUMA Riester GmbH & Co KG

- a position transmitter to show the valve position locally
- a torque limiter to prevent motor or valve damage and to prevent over travel, see Figure 12.20

Many accessories can be fitted to enhance valve operation, some important ones include:

- a switch to select local or remote control when local push-buttons are fitted
- an intermediate limit switch to allow valve travel to a predetermined position
- an electronic position transmitter, to allow remote indication or control of valve position, see Figure 12.21

Figure 12.22 shows a typical block diagram for an electro-mechanical actuator indicating the various functions available.

Electro-mechanical actuators for wedge gate valves should be fitted with a "hammer-blow" effect to ensure the valve always opens. Electrical enclosures should have a minimum protection of IP44 and up to IP68. Weatherproofing or tropicalisation should be considered for actuators outdoors. Anti-condensation heaters may be necessary for some climates. Flameproof or explosionproof enclosures are available for hazardous locations. For on/off isolating valve applications the actuator opens the valve to a limit switch. Valve closure should be monitored by a torque setting backed up by a limit switch to prevent damage. Valve opening can be monitored by torque setting as well as the limit switch setting. An indicator should always be fitted if the state of the valve cannot be obviously judged.

Actuators can be fitted with two motors when used for control applications. A large motor can be fitted for rapid isolation duty in the event of control failure; a smaller motor being used for control modulation. Some actuators for isolating applications can have two-speed motors. Normal valve motion is performed at full speed. The initial valve opening and the final travel on closing can be made at 25% speed to avoid surge and water hammer problems.

Electro-mechanical actuators can be the most cost-effective for smaller valves. They are obviously very easy to integrate into digital control system protocols or RS 485, and analogue current signals, 4 to 20 mA. They are the most "stiff" of all the operating mechanisms, offering the highest positional accuracy and holding. Due to the high ratio of gearing used, electro-mechanical actuators are rarely troubled by vibration or "hunting" problems. Immediate motor reversing is not always possible. Some models impose a 2 s delay before restarting the motor.

Motor starting torque may be slightly less than the stall torque when running; 75 to 80% is normal. Specifying the number of starts per hour is critical to long term reliability. For isolating valve applications, the standard 60 starts per hour may be more than acceptable. Control modulating applications should specify much higher starting frequencies; up to 1 200 per hour is not

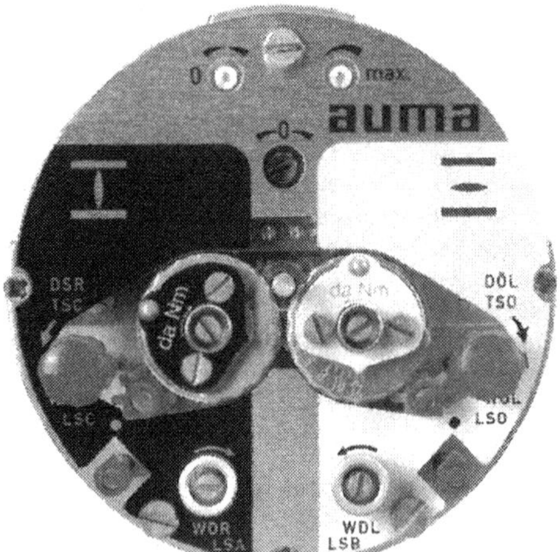

Figure 12.20 Electric torque limiter
Courtesy of AUMA Riester GmbH & Co KG

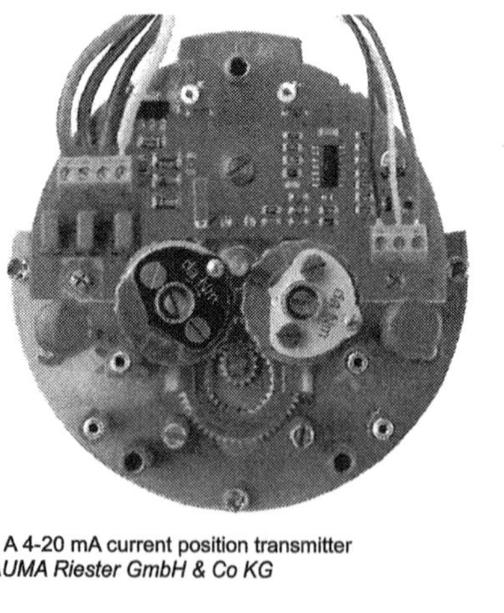

Figure 12.21 A 4-20 mA current position transmitter
Courtesy of AUMA Riester GmbH & Co KG

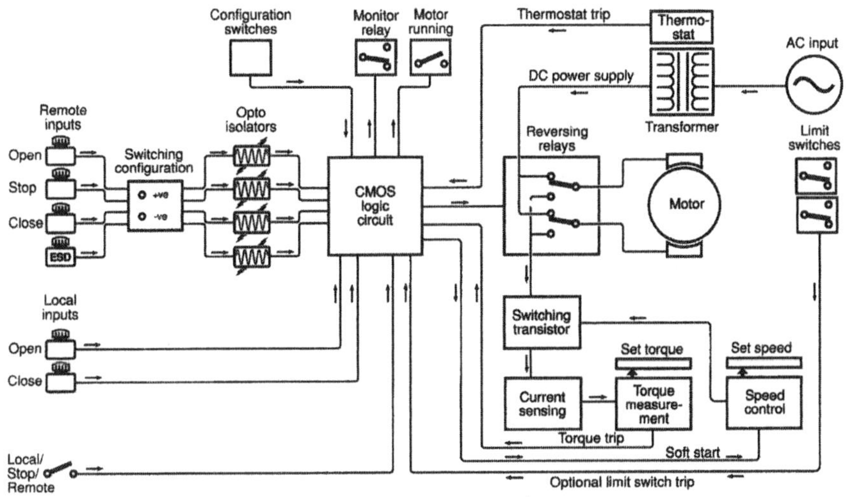

Figure 12.22 A typical electro-mechanical actuator block diagram
Courtesy of Rotork Controls Ltd

uncommon. Actuator cycle times are related to actuator size. For 90° travel a small actuator, 14 Nm, may be adjustable from 5 to 20 seconds. A large actuator, 1080 Nm, may take 1 to 4 minutes. Table 12.6 shows the capabilities of typical electro-mechanical actuators.

Actuator driver	Range degree	Typical output starting/stall torque Nm
12/24 V dc	90 to 360	276/345
115/230 V 1ph ac	90 to 360	276/345
110/240 V 1ph ac	90 to 180	1000
380 V 3ph ac	90 to 180	4000
24 V dc	90	676
110/240 V 1ph ac spring return	90	400
110/240 V 1ph ac 380/660 V 3ph ac	90	1080
110/240 V 1ph ac 380/415 V 3ph ac	multi-turn	10 - 32 000 (4 to 180 rpm)

Table 12.6 Typical electro-mechanical actuator capabilities

Electro-pneumatic actuator

Electro-pneumatic actuation uses compressed air in diaphragm, piston or vane actuators controlled by solenoid valves, see Section 3.3.1 in Chapter 3. This type of actuation is very simple and is ideal for on/off isolating valve applications. Isolating valves can be operated remotely with ease. Both spring return and double acting actuators can be used. Diaphragm actuators may require a pressure reducing regulator to ensure the applied air pressure is not too high. Most piston actuators should operate successfully at plant air pressure. A filter/drier would be a wise precaution. See the start of Section 12.3 for details of actuator capabilities.

Electronics, PLCs, and pressure/flow transducers or transmitters enable this type of actuation to become DIY controllers. The output from a transducer is connected to a PLC. The PLC logic decides whether the value is correct, too small or too big. If necessary the PLC sends signals to solenoid valves to open or close and adjust the valve one way or the other. Speed control can be implemented by using solenoid valves in parallel. Any characterisation of the valve motion required is performed in the PLC logic algorithm. Complicated set pattern operating routines can be performed by the PLC, without operator intervention, triggered by set points in the operating conditions or timed by the PLC. The PLC can also manage alarm and trip set points, as well as variable routines for ESD overrides, based on operating conditions when the ESD signal is received. As PLCs are inexpensive, complex control mechanisms can be implemented inexpensively.

Electro-hydraulic actuator

There are many types of electro-hydraulic actuator. Figures 12.23 and 12.24 show typical examples. The control signal is usually 4-20 mA, and is sensed by the coil of a solenoid, called an "electromagnetic force motor" in Figure 12.24, operating a lever system which opens and closes a small jet that controls the flow of hydraulic oil. When the jet is closed, the pressure in the amplifying relay increases; opening the jet causes the pressure to decrease. These pressures are amplified and then applied to the main piston to impart linear motion to the valve plug in accordance with the input signal. The hydraulic oil pressure used to power the system is provided by a small, constantly running, integral eccentric operated plunger pump, driven by an electric motor. The hydraulic power pack can be integral with the actuator or separate. Figure 12.25 shows an opposed piston quarter turn actuator with a hydraulic power pack. Self-contained electro-hydraulic linear actuators can produce

Figure 12.23 Self-contained electro-hydraulic actuator

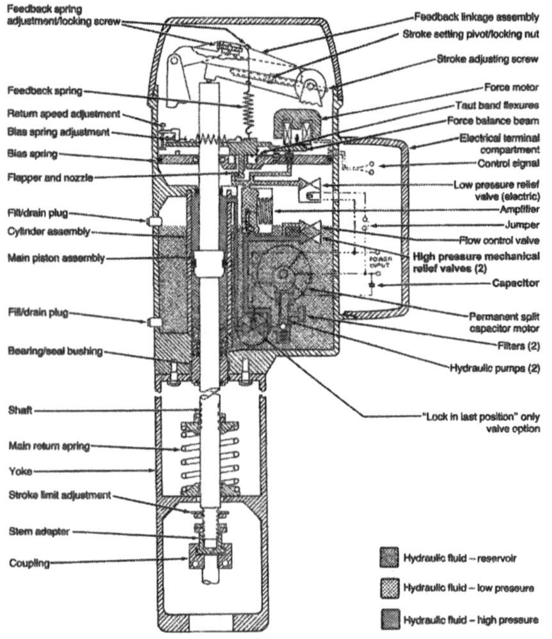

Figure 12.24 Section through self-contained electro-hydraulic actuator

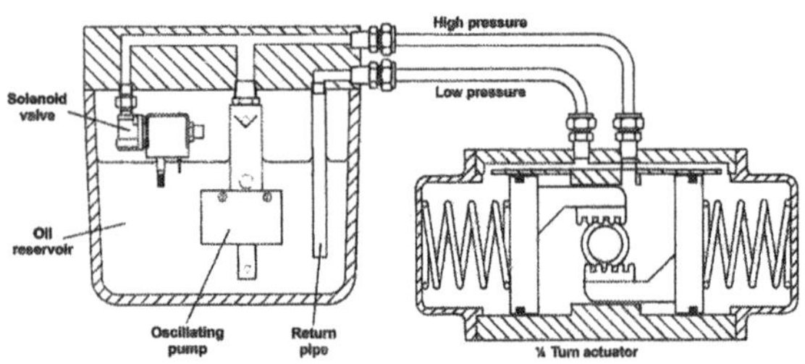

Figure 12.25 A hydraulic opposed piston quarter turn actuator
Courtesy of Rotork Controls Ltd

thrusts of over 330 kN; rotary actuators have torques of over 52 000 Nm.

Electro-hydraulic actuation can be accomplished in a similar manner to electro-pneumatic actuation by using solenoid valves with standard piston actuators. Both types of electro-hydraulic actuation offer very accurate positioning with "stiff" systems. Hydraulic actuation, with pressures up to 130 barg, have been used to obtain faster valve reactions. Speed increases from 4:1 to 15:1 have been claimed. High speed valve operation can create serious problems of surge and water hammer and these effects must be fully investigated before making binding commitments. Thrusts of over 100 kN can be provided by these actuator units. The integral type actuator/power pack is very useful in remote locations where compressed air is not available.

12.4 Accessories

Accessories may be vital for the correct operation of actuators and not necessarily an afterthought which may improve the operation. In some systems, with some actuators, some accessories can be detrimental to overall performance. Accessories must be selected with as much care as the actuator itself. The following guidelines should prove helpful.

12.4.1 Handwheels

A handwheel is often required to operate the valve manually, when starting up a process or in case of emergency; or sometimes a handwheel is used to limit the valve travel. Figure 12.26 shows a handwheel fitted to an eccentric plug valve which is actuated by a rolling diaphragm, controlled by a positioner mounted on the valve spindle.

There are several different ways of mounting handwheels:

- on a separate screwed stem — acting on the end of the actuator stem, which can be used to limit actuator travel or hold the actuator at full travel,
- on a screwed stem — engaged in the spring retainer plate, which can be used to adjust the actuator spring load and increase the load sufficiently to prevent the actuator from moving,
- on a spindle — geared to adjustable travel stops which reset the actuator travel range.

Handwheels should not be used on valves for safety shut-down functions on a process. When specifying handwheels it should be remembered that some mechanisms, for example worm drives, have very large velocity ratios and are capable of producing very large forces if abused. Cheater bars used on handwheels can buckle stems or distort plugs and seats. If a handwheel is too small and cannot be operated without extra le-

Figure 12.26 A handwheel fitted to an eccentric plug valve

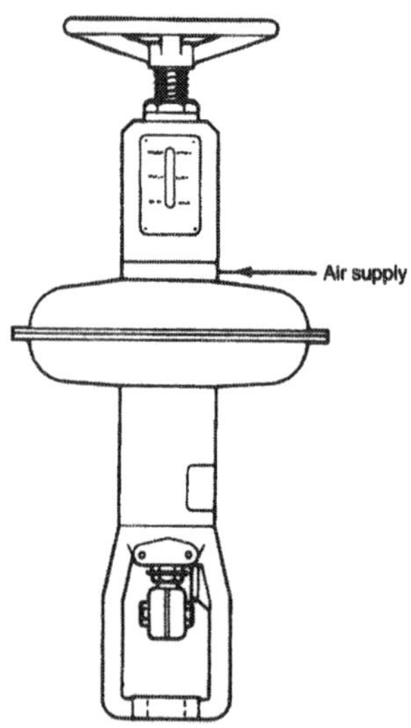

Figure 12.27 A direct-acting diaphragm actuator with handwheel

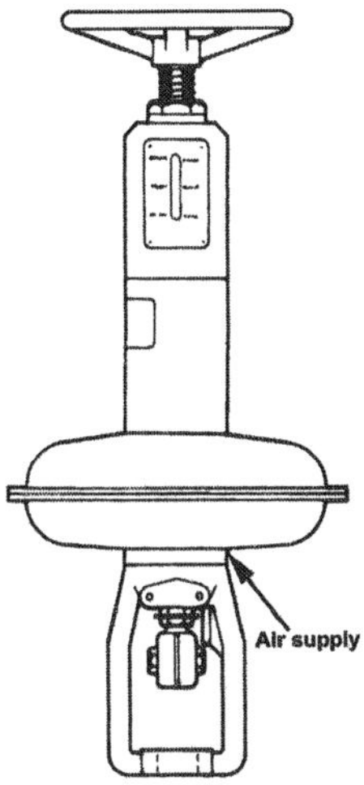

Figure 12.28 Reverse-acting diaphragm actuator with handwheel

verage, something is wrong. Something is worn or binding or the wrong valve/actuator is being used.

Figures 12.27 and 12.28 show handwheels fitted to direct and reverse-acting diaphragm actuators.

Figure 12.29 shows a top-mounted handwheel fitted to a reverse-acting diaphragm actuator on a weir type diaphragm valve. The handwheel can be used to increase the spring load and prevent the pneumatic power opening the valve. The arrangement of multiple spring stacks around the diaphragm plate to provide the correct load can also be seen. Notice that the inner and outer springs are wound opposite hands, to prevent fouling.

The geared type of handwheel can also control the position of the valve stem in both directions. It is much more powerful, and therefore easier to use when operating large valves. Geared handwheels are often preferred in applications requiring frequent manual operation and for valves larger than DN75. When the handwheel is in its intermediate position the valve stem can move freely to the extremes in both directions, see Figure 12.30.

Handwheels for piston actuators are always geared. Usually it is not possible to operate the valve by means of the handwheel until the air pressure has been released. The handwheel mechanism may in fact be damaged if the air pressure is not released. Additional measures taken to avoid such damage are using a torque limiter and bevel gears, rather than the worm drive in the handwheel drive. A bevel gear drive with a shear pin torque limiter, can also be used and is shown in Figure 12.31. Some handwheel mechanism designs use the shear pin as a disengaging drive clutch. Under normal actuated valve operation, the shear pin is not fitted and is held in a safe place on the actuator casing.

It is recommended that block and bleed valves, see Chapter 3, Section 3.3.1, are fitted in all the air lines to facilitate isolation and venting. Fitting block and bleed valves will protect the handwheel mechanism and operators in the event of an unexpected restoration of control air. Once the air lines are isolated and vented the shear pin can be fitted and manual operation can proceed.

Electro-mechanical actuators can be fitted with handwheels. The simplest option is to fit the reduction gearbox with a manual clutch which engages the handwheel and disengages the motor. The clutch is operated by an external lever which can be locked in either position. The drive gearing used ensures the valve will remain in the chosen position. The other option, used on larger actuators, is to use an automatic clutch, operated by

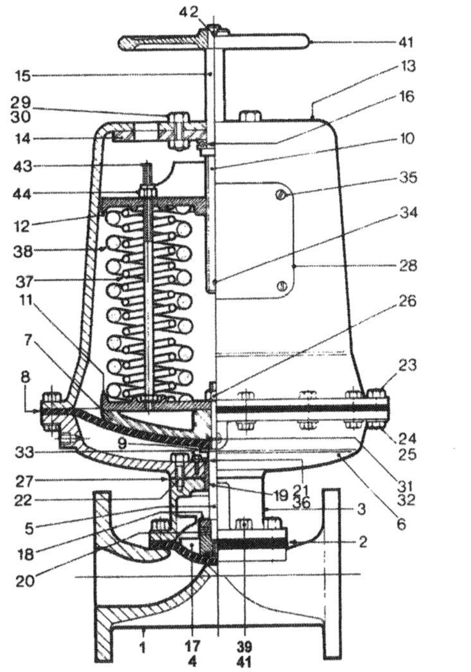

1 Body	16 Thrust race	31 Cylinder plug (1)
2 Line diaphragm	17 Matching piece	32 Cylinder plug (2)
3 Bonnet	18 Spacing collar	33 Spindle bush screw
4 Compressor	19 Spindle bush	34 Spindle pin
5 Spindle	20 Compressor pin	35 Window screen
6 Cylinder	21 'O' ring	36 Bonnet/bush sealing joint
7 Diaphragm plate	22 Bonnet/cylinder bolt	
8 Operating diaphragm	23 Cylinder bolt	37 Spring (1)
9 Clamp washer	24 Cylinder nut	38 Spring (2)
10 Adjusting screw	25 Cylinder washer	39 Body/bonnet stud
11 Lower spring plate	26 Spindle lockout	40 Body/bonnet nut
12 Upper spring plate	27 Bonnet/cylinder joint	41 Handwheel
13 Cover	28 Window	42 Handwheel pin
14 Thrust sleeve	29 Thrust plate bolt	43 Lifting rods
15 Cover plug	30 Thrust plate nut	44 Lifting rod lock nuts

Figure 12.29 A diaphragm operated weir type diaphragm valve with handwheel
Courtesy of Crane Process Fluid Technologies

the handwheel. The initial movement of the handwheel disengages the motor and engages the handwheel. Energising the motor disengages the handwheel and returns the actuator to electric.

Electro-pneumatic actuators can be operated manually if a local air receiver and manual control valves are fitted. Some large actuators, fitted with return springs, can have facilities to manually compress the spring and move the valve. This is a slow process, but useful during commissioning before control systems are on-line. Large valves can be cracked open to allow process priming and venting.

Electro-hydraulic actuators can be operated manually by using a hand operated positive displacement pump to provide the hydraulic pressure. Alternatively, the hydraulic circuit can be fitted with a small accumulator to act as a pressurised reservoir. In the event of power failure, the stored high pressure fluid can be released by manual valves.

12.4.2 Positioners

Positioners are not now considered to be the universal solution to many problems. In fact, in some control circuits the effect of fitting a positioner can be detrimental. Whilst it is an advantage to use positioners in slow systems, it can be a disadvantage in the case of control loops with short reset times. The best solution is to correctly size the actuator and spring and ensure that the valve functions correctly without the need for a positioner.

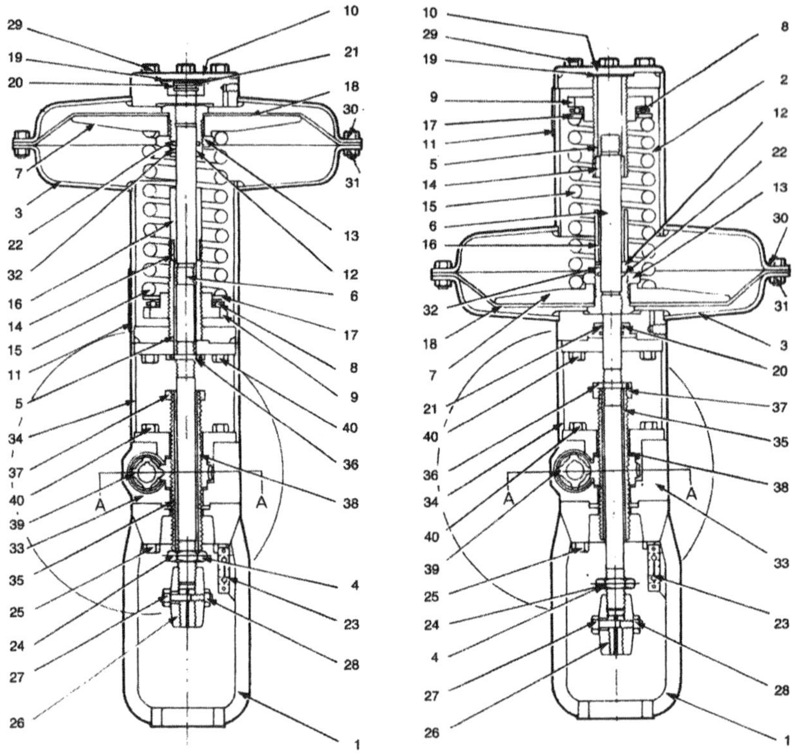

Assembled for direct action

Assembled for reverse action

1	Yoke	21	Seal box 'O' rings
2	Spring tube assembly	22	Diaphragm collar 'O' ring
3	Diaphragm case assembly	23	Travel indicator plate
4	Travel stop nuts	24	Travel indicator pointer
5	Spring adjusting screw	25	Yoke screws
6	Actuator stem	26	Stem coupling
7	Spring carrier	27	Stem coupling screw
8	Spring thrust bearing	28	Stem coupling screw nut
9	Spring adjustor	29	Cover plate screws
10	Cover plate	30	Diaphragm case screws
11	Spring cover plate	31	Diaphragm case screw nuts
12	Diaphragm collar	32	Grub screw
13	Collar nut	33	Gear housing
14	Stem guide	34	Screw housing
15	Spring	35	Screw stem
16	Travel stop tube	36	Actuator stem nut
17	Diaphragm button	37	Screw stem nut assembly
18	Diaphragm	38	Worm gear thrust bearing
19	cover plate joint	39	Grease nipple
20	Seal box	40	Housing screws

Figure 12.30 Geared handwheel fitted to diaphragm actuator

The use of a positioner however should be considered for:

- split-range systems, where a controller controls more than one valve; although it is now preferable to split the controller output electronically, see Section 12.4.5, and control each valve in a range by a full range signal from the control room
- valves where the actuator working pressure is greater than the control signal pressure, higher actuator pressures are used to provide sufficient force to ensure correct valve movement, a frequently used range is 0.4 to 2.0 barg, 6 to 29 psig
- actuators operating at higher pressures to increase actuator "stiffness"
- where it is necessary to achieve the best possible control with a minimum of overshoot and the fastest possible reaction in systems with long pneumatic signal lines between the valve and regulator. In this case it would be better to use electric analogue or digital signal transmission
- where the control loop reacts slowly to changes in valve position, and very accurate positioning is therefore desired

A positioner should also be considered for relatively slow systems, such as mixing/separation, level and temperature control when the volume and mass of the fluids handled are large compared to the process action.

The valve positioner is a feedback mechanism. It allows valves to be positioned precisely in accordance with the output signal of the controller, in the presence of major disturbances caused by unbalanced forces acting on the valve stem, changes in actuator temperature, etcetera. Even a relatively small actuator, that otherwise would have to be bench-set to an unacceptable degree, can be stroked precisely when a valve positioner is used to signal the valve position. Positioners will accept a pneumatic control signal between 0.21 and 1.03 barg. The air supply pressure should be at least 0.4 bar above the maximum outlet pressure and inlet pressures up to 6 barg are fairly standard. The positioner outlet pressure to the actuator will adjustable between 0% and nominally 100% of the supply pressure. Positioner hysteresis should be not greater than 0.5%.

Figure 12.32 indicates, diagrammatically, the arrangement of a standard bellows positioner fitted to a diaphragm actuator.

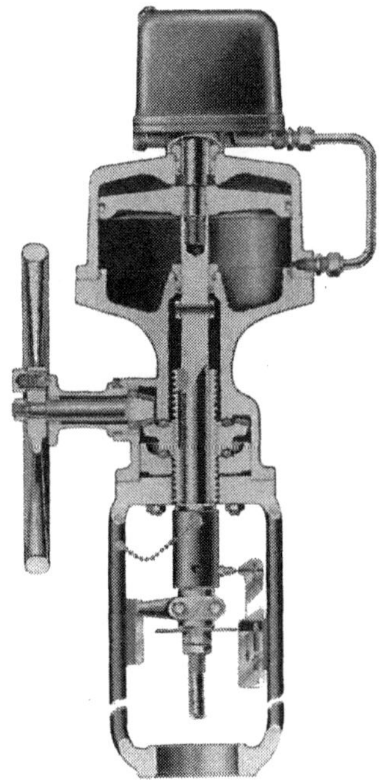

Figure 12.31 Cylinder actuator with bevel geared handwheel and shear pin torque limiter

Figure 12.33 A bellows positioner fitted to a diaphragm actuator

Generally, the positioner housing is attached to the actuator yoke, as shown in Figure 12.33.

The positioner senses the movement of the valve plug by means of a lever, which is connected to the stem between the plug and the actuator, see Figure 12.34. The signal from the controller, 0.21 to 1.03 barg, is transmitted to the positioner bellows via the instrument connection. The bellows expands as the input signal pressure increases, while the internal tension spring tries to restrain movement. Compressed air from the air supply line supplies the relay where it meets both the relay valve trim and a small restrictor jet. The air then passes to a chamber which is totally enclosed by a diaphragm, except for an escape nozzle to atmosphere. This nozzle remains open if the beam and flapper assembly are positioned at the mid-point of its travel.

Whether the beam and flapper assembly are moved to the right or to the left depends upon the bellows signal input setting and the valve position. If for example, the beam and flapper assembly is moved to the left by the position cam or the control signal bellows, the nozzle will then be constricted and more air will flow through the restrictor than through the nozzle, causing pressure to build up in the diaphragm chamber. The relay valve trim presses to the left, closing the exhaust channel and opening the supply pressure line to the top of the actuator diaphragm.

The increased pressure pushes the actuator downwards, thereby moving the cam lever, causing the cam to rotate. This allows the beam and flapper assembly to move away from the nozzle, thus creating a balanced situation where the nozzle releases an equal amount of air, supplied by the restrictor. The relay valve trim then adopts the balanced position, closing both the supply of new air to the top of the actuator diaphragm and also the exhaust channel from the actuator top to atmosphere.

The valve is now balanced and remains in the position dictated by the signal input. If the signal pressure to the bellows is reduced, then the beam and flapper assembly will move away from the nozzle causing the pressure in the diaphragm chamber to fall. The exhaust channel opens and air is released. The pressure in the top of the actuator reduces i.e. the diaphragm pressure is less, whereby the actuator spring causes the valve stem to travel upwards. The cam follows the movement, causing the nozzle opening to correspond with the restriction opening. The relay is in balance and the valve adopts the position

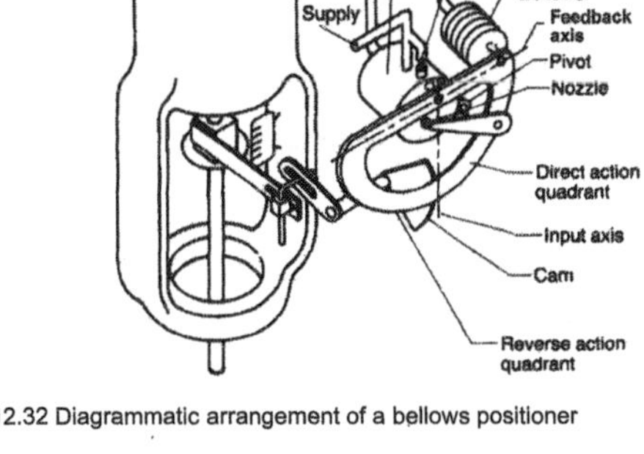

Figure 12.32 Diagrammatic arrangement of a bellows positioner

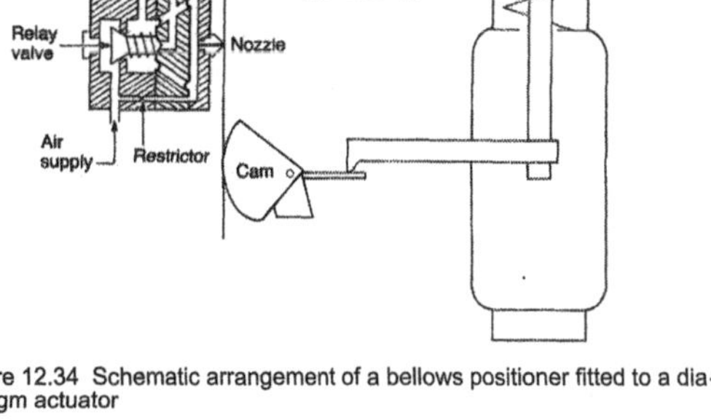

Figure 12.34 Schematic arrangement of a bellows positioner fitted to a diaphragm actuator

Figure 12.35 Positioner internals and pressure gauges

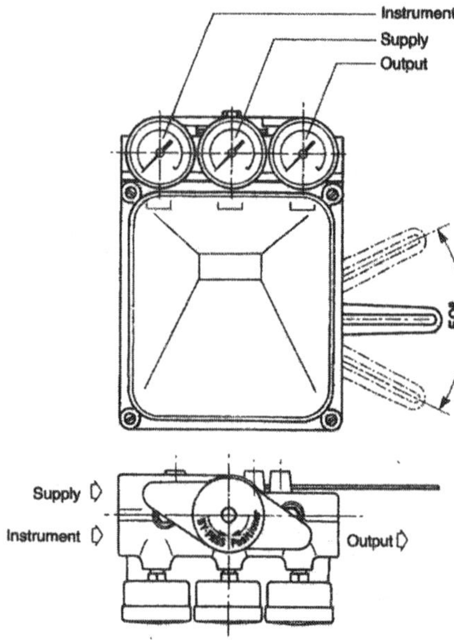

Figure 12.36 Positioner pressure gauges and connections

dictated by the input signal. No air is released from the top of the actuator. The valve is now once again ready to receive a new control signal via the bellows input. The cam can be "characterised", that is the profile modified, to adjust the valve flow characteristics.

The positioner is usually fitted with gauges, see Figures 12.35 and 12.36. The gauges indicate:

- supply air pressure
- signal input pressure from the controller
- signal pressure to the diaphragm

The positioner can be fitted with by-pass valves which allow the output signal of the controller to be transmitted directly to the actuator diaphragm.

NOTE: It is dangerous to use positioners fitted with by-pass valves, in conjunction with control valves operating in split-range arrangements, or where the diaphragm pressure is higher than the controller signal pressure. The use of by-pass valves in such systems is therefore not recommended and they should be removed, if any are fitted, before the valves are commissioned.

Figure 12.37 shows a schematic arrangement of a positioner piped to a piston cylinder. The instrument air pressure signal from the controller, 0.21 to 1.03 barg, 3 to 15 psig, acts on the positioner bellows. The bellows tend to expand as the signal pressure increases and the bell crank pivots about the fulcrum.

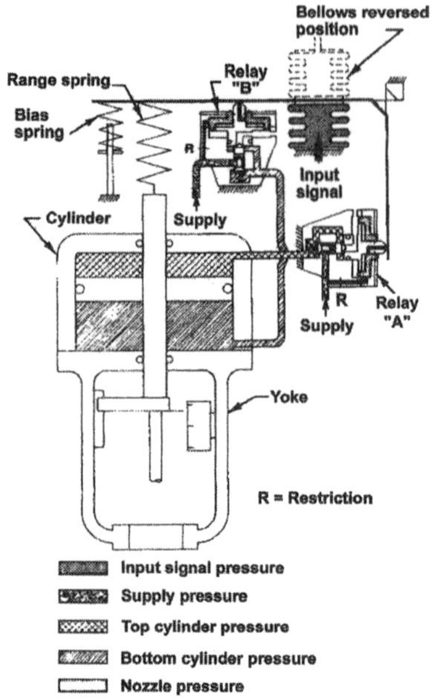

Figure 12.37 Schematic arrangement of a positioner for a piston actuator

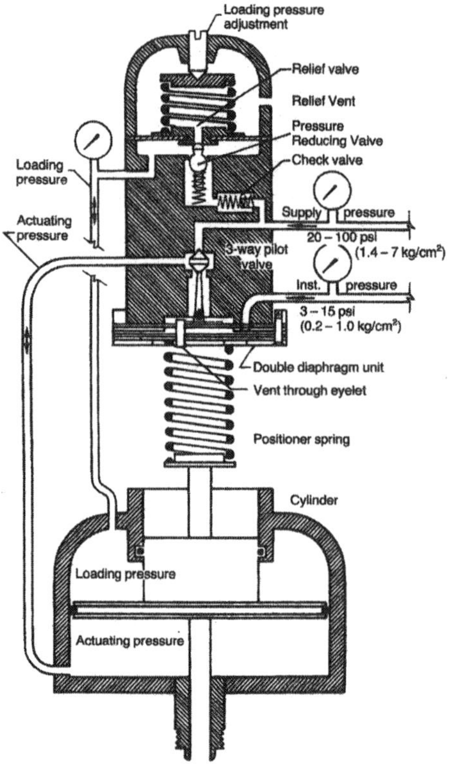

Figure 12.38 An in-line positioner for a piston actuator

This has the effect of closing relay nozzle A and opening relay nozzle B. This causes:

- the pressure to increase in relay A which compels the air pressure above the piston to increase; top cylinder pressure
- the pressure to decrease in relay B which causes a corresponding reduction in air pressure below the piston; bottom cylinder pressure

This "out-of-balance" causes the piston to move downwards in the cylinder until the valve plug is correctly positioned in relation to the input air signal transmitted from the controller. The piston movement is transmitted back to the bell crank via the "range spring", the bottom of which is attached to the upper end of the piston rod. When the axial force produced by the range spring balances the thrust of the control signal bellows, the pressures in both relays A and B are equalised and the piston comes to rest.

Piston actuators can be supplied to either open or close the valve as the control signal pressure increases. This is achieved by reversing the position of the bellows as indicated in Figure 12.37.

Positioners for piston actuators can be built "in-line" on the end of the cylinder as shown diagrammatically in Figure 12.38.

Positioners can be fitted to pneumatically operated rotary valves. The positioner is mounted on the valve spindle itself as shown in Figure 12.39. This arrangement is much simpler than earlier schemes which required a specially designed type of actuator. The basic difference between the two variants is that the cylinder actuator requires a double relay, and air must be supplied both above and below the piston. Figure 12.40 clearly demonstrates the principle of operation.

If the pressure variations in the output of the controller are small, then the spool valve supplying air to the cylinder will remain static. The change in the pressure will be transmitted directly to the rolling diaphragm. If the change in output is large, then the spool valve will either admit, or exhaust, air as required and cause rapid movement of the diaphragm. Feedback of the valve position is provided by the characterised cam attached to the valve spindle, and the feedback spring. The reducing signal input causes the signal module to release air through the exhaust port and causes the opposite sequence of events throughout the system. The diaphragm actuator adopts a new position corresponding to the size of the signal from the controller.

Pneumatic actuators may be arranged to provide "split range" control, which enables full valve travel to be obtained by only using part of the 0.21 to 1.03 barg control signal pressure range. This allows several valves to be controlled by the same controller. Valve positioners may be used to achieve split range operation of control valves. A typical application is the steam range shown in Figure 12.41. When the system is in balance, the amount of steam entering the range at A is sufficient to maintain a constant pressure on the range. If the steam demand at B exceeds the supply, the pressure on the range will tend to reduce. This is sensed by a reverse-acting pressure controller, PIC, in such a way that its output signal will start to increase. As a result more steam will be admitted to the back pressure turbine through valve 1. If there are no other constraints this can continue until the limit of its flow capacity is reached. If this is not sufficient to maintain pressure on the range, extra steam is admitted directly from the boiler through valve 2.

Conversely, if supply exceeds demand, the flow to the back pressure turbine is throttled and, if that is not sufficient to restore the range pressure to its set point, valve 3 is opened to pass the excess steam to a condenser. When the demand decreases the sequence is reversed. Valve 2 begins to close in stages. When this is closed, valve 1 begins to close until the system is once more in balance with a signal to the valves of about 0.5 barg, 8 psig. If the demand is further reduced, the signal output begins to fall below 0.5 barg, valve 3 begins to open in stages and is fully open at 0.2 barg, whereupon excess steam is released from the system. If the demand increases, valve 3 begins to close again until the system is balanced at the pre-set system pressure, where the signal output is at 0.5 barg.

One of the many advantages of such a system is that the complete circuit can be supervised by just one controller which can be adjusted up or down as required to control all valves in the system. In a practical arrangement of this kind there must be adequate relief capacity to cope with instrument failure.

In the classical arrangement, the output signal of the direct-acting pressure controller of 0.21 to 1.03 barg (or 4 to 20 mA if the controller is electronic), is divided into three bands by carefully adjusting the calibration of the positioners on valves 1, 2 and 3:

- valve 3: 0.2 barg (open) to 0.5 barg (closed)
- valve 1: 0.5 barg (closed) to 0.7 barg (open)
- valve 2: 0.7 barg (closed) to 1.0 barg (open)

This arrangement is quite effective, but the calibration is difficult to achieve under operating conditions in the field, and tends to

Figure 12.39 Rolling diaphragm actuator fitted to an eccentric plug valve

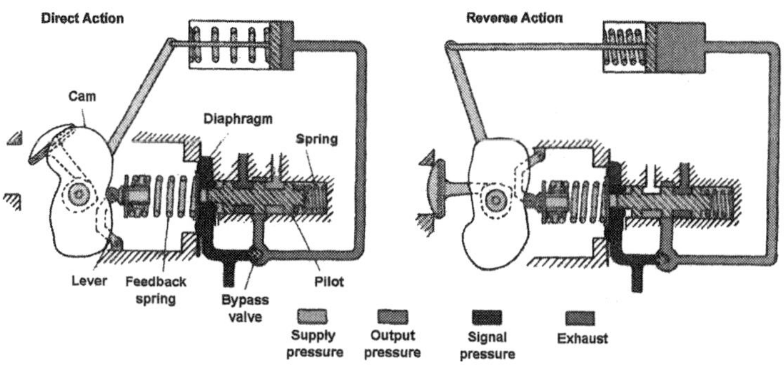

Figure 12.40 Diagrammatic arrangement of positioner for rotary valves

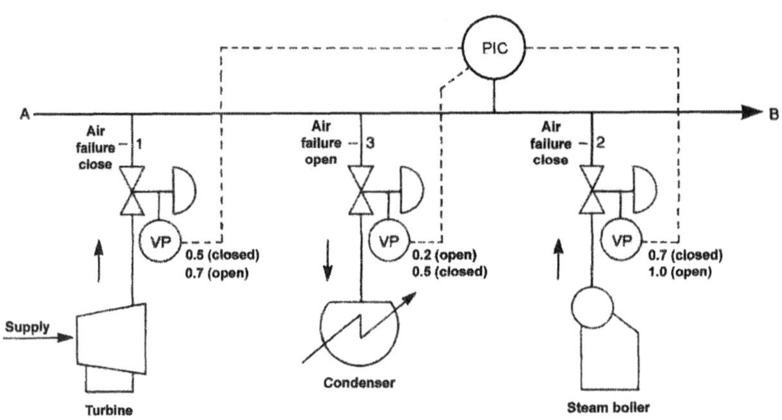

Figure 12.41 Split range control of a steam range

drift under the influence of vibration and extremes of temperature usually experienced on industrial plant out in the open.

The more satisfactory of arranging split range control of, say, a steam range, is to divide the controller output signal electronically by three electronic selector relays in the control room, where long term stability of the settings can be guaranteed to about 0.1%. Each of the three valves is controlled with the full range output, 0.21 to 1.03 barg of the appropriate signal selector relay.

The positioners described so far are all pneumatic/mechanical devices. The pneumatic control signal is compared to the valve stem position and adjustments are made if necessary. The valve/actuator/positioner must be located as close as possible to the controller to reduce piping pressure drop losses. In some complicated process installations and in utility distributions systems, the controller may be a considerable distance from the valve. Distribution systems controlled from a central control room may transmit control signals via radio or fibre optic link. An interface is required between electrical signals and the pneumatic operating system. An electro-pneumatic positioner fulfils this requirement.

The electro-pneumatic positioner is fitted to the actuator yoke and a link is connected to the valve stem in the normal way. The control signal is either 4 to 20 mA or 2 to 10 V. The control signal is compared to the valve position via a cam, which can be "characterised" if necessary. The positioner reduces the air supply pressure to a modulated 0.21 to 1.03 barg control supply which is directed to the diaphragm or piston. These positioners can be single or double acting and set up to be reverse-acting. Split range operation can be accommodated by internal adjustments. Typical electro-pneumatic positioner operating parameters are shown in Table 12.7.

Parameter	Value
Control signal	4 to 20 mA at 8.5 V dc minimum
Supply air pressure	1.4 to 6.9 barg
Output air pressure	0 to 100%
Sensitivity	0.1% span
Linearity	1.5% span
Hysteresis	0.5% span

Table 12.7 Typical electro-pneumatic positioner operating parameters

An alternative electro-pneumatic positioner control for diaphragm actuators can be achieved by using solenoid valves. The driving control signal can be pneumatic or electric; when pneumatic is used it is converted immediately to an electric current signal. The 4 to 20 mA positioner signal is compared electronically with the 4 to 20 mA control signal. If the actuator must compress the spring further, a solenoid valve in the air supply is opened until the signals coincide. If the actuator has travelled too far, a solenoid valve in a vent line is opened and allows the spring to return the actuator until the signals coincide. The positioner Smart electronics are can be configured using the HART protocol over the current loop.

Electrically actuated valves can have positioner facilities added. The stem or spindle is fitted with a potentiometer or a current position transmitter. Every stem/spindle position corresponds to a modulated electric signal between 4 to 20 mA or 2 to 10 V. The electric positioner compares the valve signal to the control signal and transmits an appropriate signal to the electric motor. Some positioners can vary the motor speed in proportion to the error between the two signals. Electric positioners can be used for split range operation in a similar manner to pneumatic positioners.

The most recent designs of positioners are non-contact devices which "watch" the valve stem or spindle. Proximity detectors sense the position of targets on the stem or spindle. Using a digital system with its own on-board microprocessor, they can analyse the valve/actuator performance — control signal, output signal, valve stem position, response time; and create an individual "signature". If the valve/actuator performance deviates from the signature an alarm can be triggered advising pre-emptive maintenance.

12.4.3 Booster relays

A booster relay is a feed forward device and can be used in a similar way to positioners, but for fast rather than slow systems, (see the beginning of Section 12.4.2). Booster relays should be considered for pressure and flow control applications. The typical time to stroke a DN100 valve is 4 seconds. This may be decreased to 1 second or less by the use of a large air connection on the diaphragm, by booster relays, quick exhaust valves or a combination of these.

A booster relay, Figure 12.42, increases the actuator stroke speed by magnifying the volume of air available at a given pressure or by increasing the exhaust capacity. Since the effective areas on the top and bottom of the diaphragm assembly are the same, the relay will reproduce the input as the output in a 1 : 1 ratio. An input pressure on top of the diaphragm assembly forces it down, seating the exhaust face and unseating the supply face of the valve plunger, increasing the output pressure. The output pressure is applied to the bottom of the diaphragm assembly. It will continue to increase until its upward force equals the input pressure's downward force. When these forces are equal, the relay will be in balance and the output pressure will equal the input pressure.

The air ports on the outlet side are larger than the control side to allow greater volumes of air to flow. It is worthwhile increasing the sizes of the actuator connection lines and keeping them as short as possible, to fully utilise the booster relay capacity to a maximum.

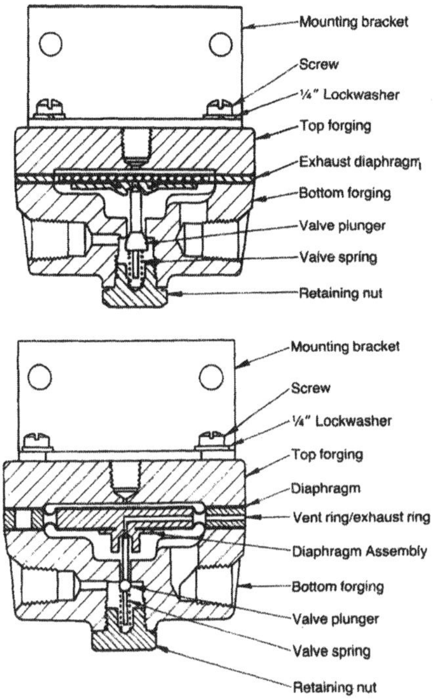

Figure 12.42 Two typical booster relays

12.4.4 Fail-safe, trip valves and lockup

In some situations, fail-safe operation can be accomplished by a return spring either opening or closing the valve once external power has been removed. In the case of double acting and electric/electro actuators, additional equipment is necessary. Holding the valve at its current position or stay-put or lockup is another alternative.

Consider a three-way valve in a temperature control circuit where control is achieved by blending hot and cold water. In the event of air failure, the temperature will rise or fall depending upon how the valve is set to react. Preferably the valve should stay in the same position as it was immediately prior to air failure. Diaphragm actuators can be arranged to lockup. The principle employed is to monitor and position the valve, see Figures 12.43 and 12.44. The trip valve switches when the supply pressure to the reducing regulator falls below a pre-determined value, Figure 12.43, or the trip valve switches when the reduced pressure from the regulator falls below a certain value, Figure 12.44. In either case the actual air pressure controlling the valve is locked whilst the pressure is still sufficient to maintain control over the actuator.

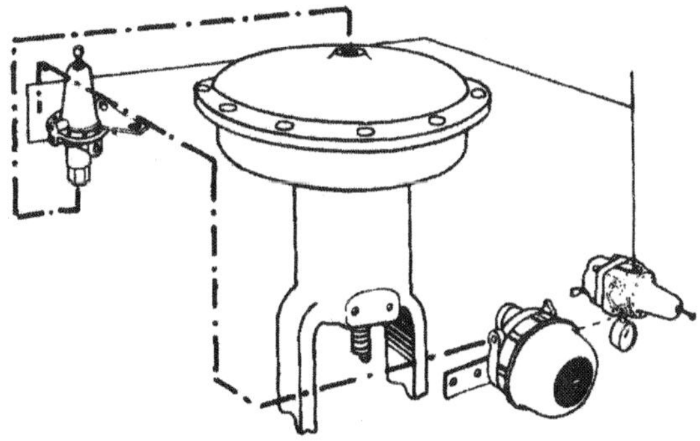

Figure 12.43 A trip valve actuated by supply air pressure

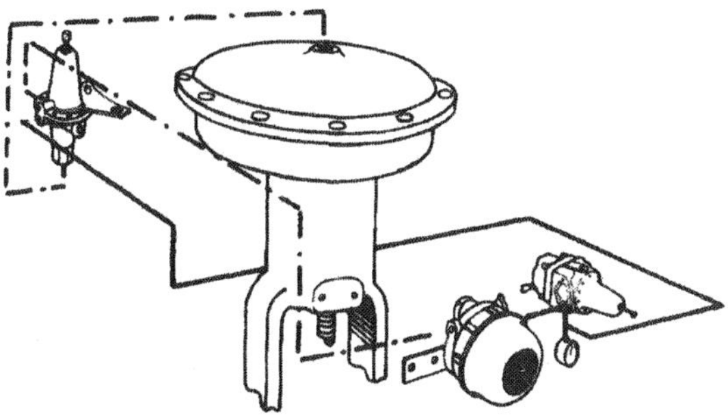

Figure 12.44 A trip valve actuated by pressure reducing regular outlet

The system diagram in Figure 12.45 shows the arrangement for triggering the trip valve from the pressure reducing regulator outlet. The signal travels from the controller to the positioner, from the positioner through the lockup valve, the opening A to B and then to the diaphragm casing. The connection A-B remains open as long as the air supply pressure to D keeps the valve open. The lockup valve spring is set to close the valve A-B when the pressure falls, and this traps the pressure in the diaphragm casing and maintains the valve in the same position. When the pressure rises again, the connection A-B re-opens and the valve can be controlled normally by adjusting the spring D, so that the valve "switches" whilst there is still sufficient pressure in the system to control the valve, even if this is less than would normally be accepted.

Piston actuators can be fitted with a trip valve which can be set to cause the valve to travel to a pre-determined position; open, closed, or stay-put; in the event of air supply failure. The basic principle employed is to add an air reservoir at normal supply pressure, 6 to 7 barg, into the system, connected to the actuator by means of a trip valve or valves. If the air supply pressure falls

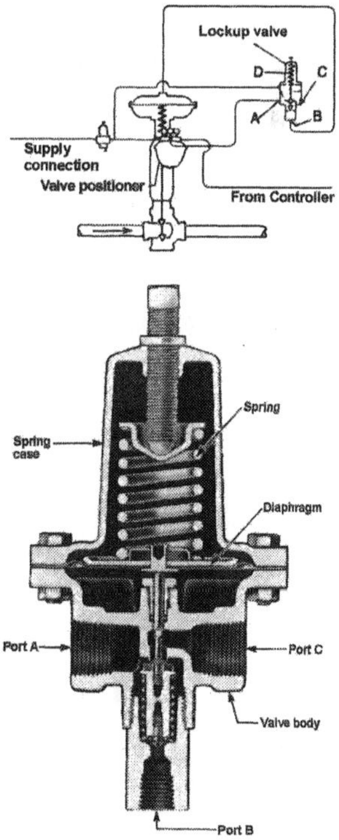

Figure 12.45 Piping schematic and trip valve detail

to less than 75% of normal value, the trip valve is activated and compressed air from the reservoir is supplied to either:

- the top of the cylinder — if the valve stem is required to travel downwards, requiring one trip valve
- the bottom of the cylinder — if the valve stem is required to travel upwards, requiring one trip valve
- the top and bottom of the cylinder — if the valve is to remain stationary, requiring two trip valves

The fail-safe system, Figures 12.46, 12.47 and 12.48, consists of an air reservoir or volume tank, which in the event of air supply pressure failure, can be connected to the top or bottom of the cylinder depending on the function required in the system. The mounting of a trip valve in the connection line to the top or bottom of the actuator determines whether the actuator, and also the valve, will travel up or down. The trip valve operates with the air supply pressure entering the air tank through a check valve. The check valve closes when the air tank is full. The trip valve plug moves to the left and opens for normal operation. If the supply pressure falls to 75% of the normal pressure, the trip valve will react to this as a reduction in the system air supply, whereupon it will move to the right and open the connection between the air tank and the top or bottom cylinder connection depending upon the method of connection.

A three way switching valve or regulator actuates both trip valves to lockup a cylinder when normal air supplies fail. The fact that the valve plug strokes at 75% of the normal working pressure depends upon the ratio of areas in the trip valve. When the system again returns to normal air pressure, both the lockup and fail-safe systems return to their positions for normal operation. If the handwheel must continue to be operated, or if the valve has to be controlled while these systems are engaged, then it is necessary to fit a small needle valve to balance the pressure in each of the lockup valves and to vent the pressure in the fail-safe valve. The needle valve must be closed again before returning to normal operation.

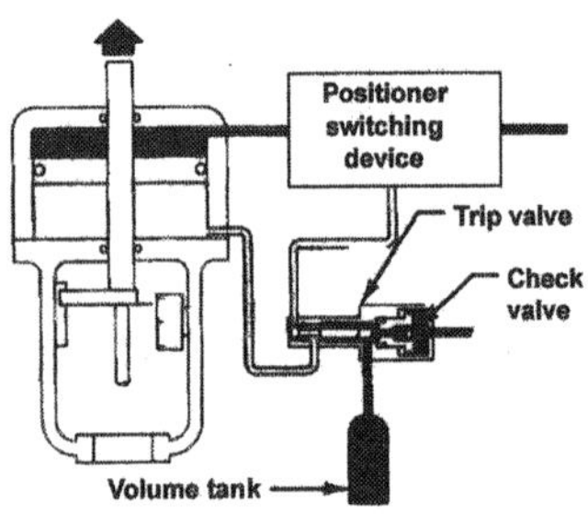

Figure 12.46 A double acting piston configured to fail-safe "up"

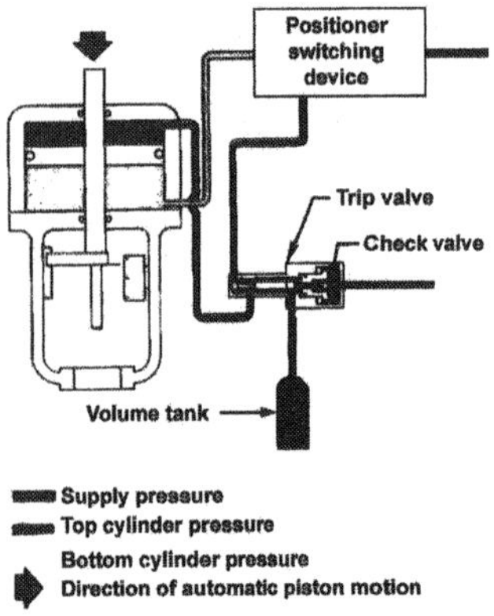

Figure 12.47 A double acting piston configured to fail-safe "down"

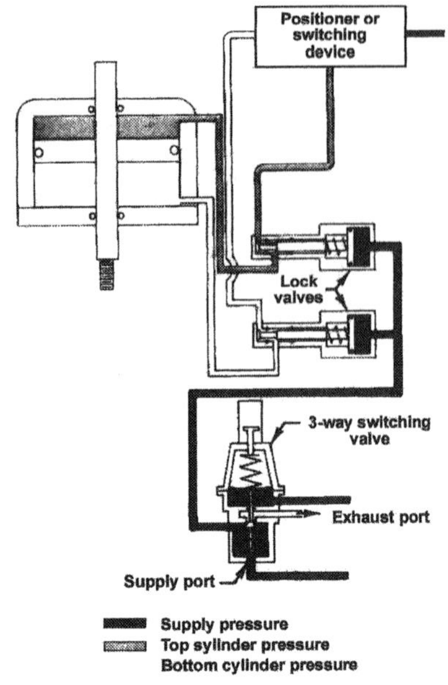

Figure 12.48 A double acting piston configured to lockup

The great advantage of this approach is that the net actuator thrust available is large, since there is no spring force to be overcome, as in other systems.

Electro-mechanical actuators would fail in the stay-put or lockup condition if the power supply failed. The gearing fitted ensures positive position holding. If alternative fail-safe strategies are required then extra equipment must be supplied. If the site is equipped with 24 V dc or 110 V 1 ph ac emergency supplies then the actuator manufacturer may be able to supply a dual voltage system. Some manufacturers, as an option, offer battery back-up systems which can power the valve for a specified number of cycles; main power failure initiates a fail-safe routine. Batteries are greatly affected by ambient temperatures. Low ambient temperature severely reduces the Amp-hour capacity. If actuators and batteries must be located outside, exposed to the elements, accurate data must be available to allow correct selection of accessories. Back-up battery systems may be supplied complete with automatic chargers. Some battery systems disconnect automatically after a preset duration of power failure.

Other electrically dependent actuator systems may adopt similar philosophies to those outlined above.

12.4.5 Pneumatic/hydraulic/electric converters

The converter, Figure 12.49, transmits a pneumatic output signal which is proportional to a dc milliampere input signal. These converters are generally called I/P converters. A combination of input/output ranges are available. Transducers with direct-acting outputs drive down-scale upon input signal failure. The optional "reverse-acting output" drives up-scale upon input signal failure. The converter housing has a drain hole for the coil cavity. This provides a discharge path for water or oil vapours that may be accidentally entrained in the air supply. The drain prevents flooding of the housing and reduces corrosion caused by these undesirable supply conditions. An air purge of the transducer's terminal enclosure can be achieved by plugging the drain hole. This makes the converter suitable for use in corrosive atmospheres.

The input coil and the float are attached to a common centre spindle and make up what is referred to as the "moving coil assembly". This assembly is free to move vertically. The float is to-

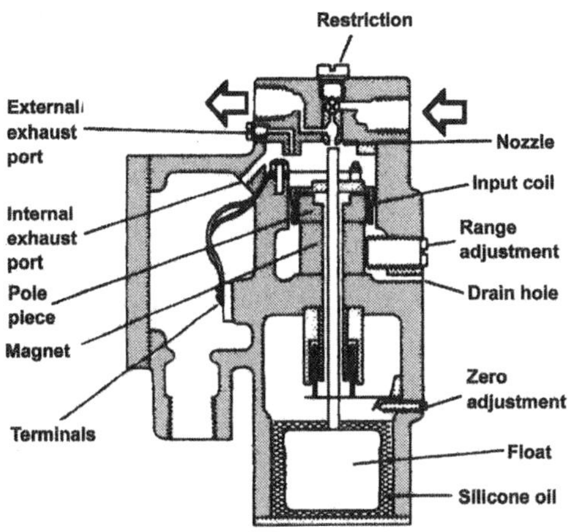

Figure 12.49 A typical current/pressure, I/P converter

tally surrounded by silicone fluid and is sized so that the resultant buoyant force equals the weight of the moving coil assembly. This puts the moving coil assembly into a state of neutral buoyancy or weightlessness, which, together with the viscous damping of the silicone fluid, makes the transducer insensitive to shock and vibration. A permanent magnet provides a magnetic field which passes through the input coil. Current flowing through the coil reacts with the magnetic field to force the moving coil assembly closer to the nozzle.

The free end of the centre spindle serves as a valve plug which restricts the flow of air exhausting from the nozzle. Air is supplied to the nozzle through a restriction. The restriction and the nozzle form a pressure-divided circuit. The pressure in the nozzle, back-pressure, varies according to the restrictive effect imposed by the spindle plug. The nozzle back-pressure is the output of the converter. The nozzle forms a column of air which has a diameter equal to the nozzle diameter. This column of air acts on the spindle plug to oppose and equal the force produced by the coil. The pressure of this column of air, nozzle back-pressure, is determined by the upward force of the moving coil assembly divided by the area of the column of air, i.e. the area of the nozzle.

Thus, the force produced by the coil is continuously balanced by the nozzle back-pressure so that the converter output pressure is, at all times, directly proportional to the coil current. Zero adjustment is accomplished by varying a spring force on the moving coil assembly. Range adjustment is accomplished by changing the gap between the permanent magnet and the end of the range adjustment screw. This screw shunts a portion of the magnetic field. Varying the gap changes the amount of shunting which, in turn, changes the flux density through the coil. I/P converters are quite accurate; linearity should be ± 0.5% and hysteresis 0.3%.

12.4.6 Remote travel indication and limit switches

There are many reasons why it may be necessary to monitor valve position. The type of monitoring can be "end-of-travel indication" which shows the valve has completed a cycle or actual valve position including all intermediate positions. The type of indication required is dependent upon the nature of the control system used. Limit switches can be connected to indicator lamps in the control room and inform operators that a valve is open or closed. This simple system is ideal for isolating valves, and when combined with a flow diagram, can show operators the state of flow paths. Continuous travel indication, perhaps showing the valve position as a percentage of fully open, is useful for control and throttling valves and tells operators what capacity they are operating at:

- does the valve have travel left to allow increased production?
- is it near the limit and production is already at the practical maximum?

Pneumatic positioners described in Section 12.4.2 can be used to indicate the valve position at a location remote from the valve. By reversing the positioner's method of operation it can be used to transmit signals to an indicator, thus providing a reliable basis for remote monitoring. The positioner produces a modulating pneumatic output signal, 0.21 to 1.03 barg, in proportion to the valve travel and can be displayed remotely on a suitably graduated pressure gauge. Positioners using 4 to 20 mA electric travel signals can indicate remotely simultaneously, provided the total applied resistance is within specification.

Valve stem/spindle position can be monitored continuously by proximity detectors or Hall Effect sensors. An electric analogue signal proportional to the valve travel can be transmitted to a suitable indicator. Rotary spindle position can also be continuously monitored via a potentiometer; the user applies a stabilised power supply and measures the output. Standardised resistances of 5 kΩ and 25 kΩ are popular. Current position transmitters, CPTs, which produce a 4 to 20 mA proportional signal from a 15 to 40 V dc supply, can be fitted to some rotary actuators.

End of travel can be indicated remotely by fitting the valve or actuator with limit switches. Limit switches are microswitches which have a "snap" action and can be set accurately to detect a specific physical position. When the valve is closed or wide open a positive indication is given. When neither condition is shown the valve is at an intermediate position. Figure 12.50 shows the installation of cam-operated limit switches on a diaphragm actuator.

Electro-mechanical actuators can sometimes have extra limit switches fitted. These switches can be set at any intermediate position and indicate specific valve positions.

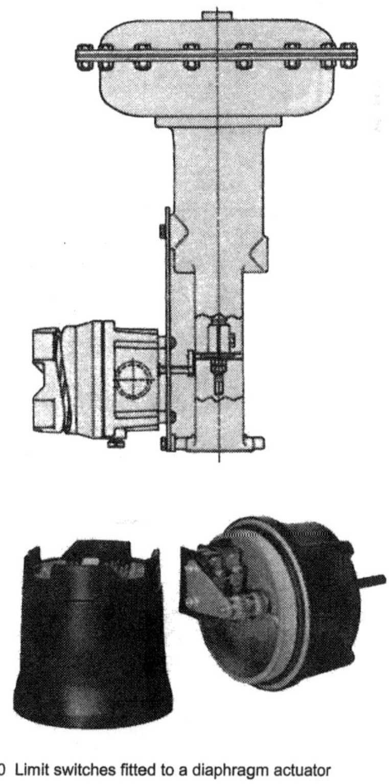

Figure 12.50 Limit switches fitted to a diaphragm actuator

12.4.7 Emergency shut-down overrides

Pneumatically operated diaphragm and piston actuators can have Emergency Shut-Down (ESD) overrides incorporated by using solenoid valves in direct instrument air supplies. Valves can be opened or closed irrespective of any control signals.

Electrical ESD control overrides can be wired directly into many electrically operated actuators. The ESD signal will override any current control signal and open or close the valve as predetermined. ESD signals can by-pass thermal cut-outs in some models.

12.4.8 Dampers

Dampers are sometimes called "snubbers". Large valves working with large pressure drops are subject to large hydrodynamic forces which may produce vibration in the valve stem or spindle. Vibration in a valve can be caused by turbulence and the shedding of eddies and cavitation. There are three ways of counteracting these vibrations:

- select a valve style which is not prone to turbulence/eddies and will not cavitate at the proposed operating conditions
- use an actuator of adequate size and as "stiff" as possible, usually this means choosing at least a double acting piston actuator, hydraulic or electric actuators and avoiding diaphragm actuators
- fit a damper

The damper can be compared to a competition style shock absorber fitted to a rally car; it works in both directions and is adjustable. The damper attenuates the valve stem velocity by using a restriction, or orifice, through which the air/oil is made to flow as the stem moves. The restrictor should be adjustable. Two typical applications of the use of dampers are feed water valves and steam reduction valves for large steam boiler systems.

Dampers cannot be used with booster relays. The desired effects of the two components are diametrically opposed. If stem/spindle vibration is experienced with actuators using booster relays, remove the boosters first. If this does not cure the problem it may be necessary to consider a larger or stiffer actuator.

12.5 Force/torque required in valve operation

The forces and torques to be overcome in a valve are the result of a number of interacting factors which depend upon the actual operating conditions. Unless the valve is balanced, the predominant force is that created by the pressure drop across the valve, which is normally calculated for the valve in the closed position, although the forces can increase dramatically due to the effect of the fluid flowing through the valve.

It is important to know the direction and magnitude of the forces acting on the stem/spindle in order to choose the correct valve/actuator combination which will give the desired travel in response to the actual change in input signal. Some of the forces caused by the pressure drop across the valve are:

- the change in kinetic energy in the fluid flowing through the valve — this is proportional to the pressure drop across the valve
- rubbing friction between the plug and the valve guide, the stem and the packing — the loads varying with temperature and materials used
- vibration/oscillation of the stem/spindle — caused by sudden impulsive forces associated with turbulent flow or cavitation

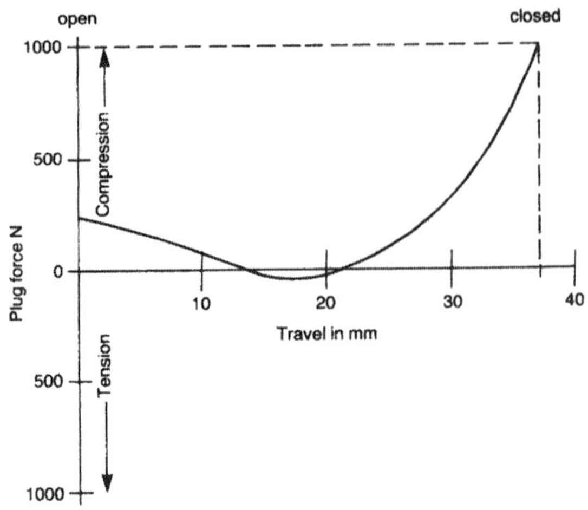

Figure 12.51 Force-stroke diagram for a DN100 double-beat globe valve

Figure 12.51 shows a typical force-stroke diagram for a DN100 top and bottom guided double-beat globe valve with 7 barg differential pressure and 37.5 mm valve travel.

When the valve is closed, the stem is in compression, and a compressive force is exerted on the actuator. This changes to tensile as the valve opens, returning to a considerable compressive force when the valve is in the fully open position. If, for example, this particular actuator were sized to cope only with the pressure drop in the closed position, then the valve would open fully in response to a input signal of less than the full range of 0.21 to 1.03 barg. On the other hand, if the force in the open position were tensile instead of compressive, then the valve would never open completely.

This problem is much more important in the case of unbalanced valves, such as single seat valves. It is often worse when the valve stem is packed with low-friction material such as PTFE. Oscillations can be destructive if the conditions in the valve cause resonance. The proper remedy is to use a stiffer actuator e.g. a positioner or a piston or electro-mechanical type, able to cope with the alternating forces.

12.5.1 Selecting the correct actuator/spring for linear valves

When selecting an actuator and spring for a given set of operating conditions, it is usually possible to refer to pre-calculated tables. This is always carried out by the manufacturers, but an occasional check by the purchaser could be helpful.

Example: To calculate a suitable spring for a direct-acting actuator for a valve with the following specification:

Valve type:	DN50 single seat
Valve action:	Pushdown-to-close
Flow direction:	Flow-to-open
Valve seat:	Steel
Packing:	Single PTFE
Valve travel:	28 mm
Seat circumference:	184.5 mm
Seat area:	2700 mm
Valve stem diameter:	12.5 mm
Valve ΔP:	1 barg max when closed
Fail-safe action:	valve to open

It is assumed that the actuator signal is 0.21-1.03 barg and in this case a positioner is used with an output signal range of 0 to 1.2 barg.

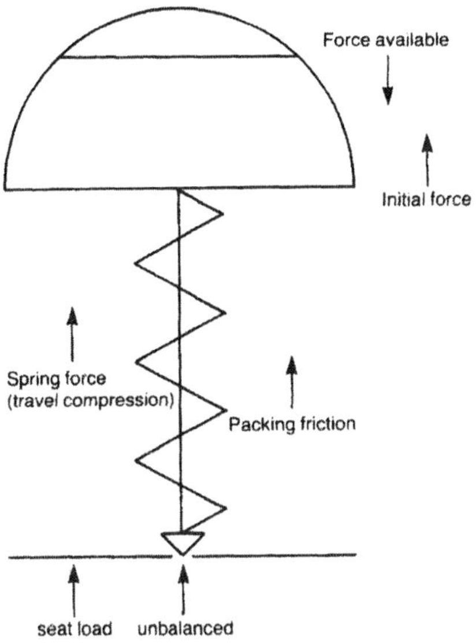

Figure 12.52 Valve/actuator force diagram for open valve

Actuator size:

The actuator should, if possible, be selected from standard sizes which are produced for direct connection to the valve.

In this instance the actuator surface area is 0.0445 m^2.

To calculate the available actuator thrust, see Figure 12.52:

The available actuator thrust is the result of applying air pressure to the top of the diaphragm, in this case:

$$1 \text{ bar} \equiv 100\,000 \text{ N/m}^2$$

$$1 \text{ bar} \times 0.0445 \text{ m}^2 = 4450 \text{ N Force available}$$

To ensure a quick response when the pressure is applied, the spring is given a certain amount of pre-compression, in this case equivalent to 0.2 bar:

$$0.2 \text{ bar} \times 0.0445 \text{ m}^2 = 890 \text{ Initial force}$$

This is an upwards force and should, therefore, be deducted from the maximum available actuator thrust:

$$4450 \text{ N} - 890 \text{ N} = 3560 \text{ N}$$

which is therefore the remaining force available for the downward movement of the actuator stem.

The actuator, however, must have sufficient thrust to:

- overcome the unbalanced force resulting from the pressure drop. The valve seat area is 0.0027 m^2, and the upwards unbalanced force is therefore the product of the area and the differential pressure:

  $$1 \text{ bar} \times 0.0027 \text{ m}^2 = 270 \text{ N Unbalanced force}$$

- impart sufficient force to the valve plug against the seat to prevent leakage, i.e. the pressure required between the plug and the seat (similar to turning a tap off tightly), is 9 N/mm over the circumference of 184.5 mm.

  $$9 \text{ N/mm} \times 184.5 \text{ mm} = 1660 \text{ N Seat load}$$

- overcome packing friction. Packing friction on a 12.5 mm diameter stem with PTFE packing is equivalent to about 225 N, (see also Chapter 11 about packing friction values). The valve stem should be treated to tolerate this friction, otherwise graphite packing is only used where high temperatures necessitate its use.

The actuator must be able to supply the total force required, see Figure 12.53:

Unbalanced forces	270 N
Seat pressure	1660 N
Packing friction	225 N
Total unbalance	2155 N

The resulting force which must be overcome by the spring is therefore:

Net actuator force	3560 N
Minus total unbalance	2155 N

0.2 - 1.0 bar to Diaphragm (0.0445m.)										
Travel	mm (in)	9 (3/8)	11 (7/16)	12 (1/2)	16 (5/8)	20 (3/4)	23 (7/8)	28 (1 1/8)	37 (1 1/2)	Spring rate N/m
Force, N (Force, Lb)		940 (207)								400 000
		1 560 (345)	940 (207)							330 000
		2 200 (552)	1 900 (414)	1 250 (276)						260 000
		2 500 (552)	2 200 (483)	1 560 (345)	940 (207)					240 000
		2 500 (621)	2 200 (483)	1 900 (414)	1 250 (276)					220 000
		2 800 (621)	2 500 (552)	2 200 (483)	1 560 (345)	940 (207)				200 000
		3 100 (690)	2 800 (621)	2 500 (552)	1 900 (414)	1 560 (345)	940 (207)			160 000
			3 100 (690)	3 100 (690)	2 500 (552)	2 200 (483)	1 900 (414)	940 (207)		130 000
					3 100 (690)	2 800 (621)	2 500 (552)	1 900 (414)	940 (207)	100 000
						3 100 (690)	2 800 (621)	2 500 (552)	1 560 (345)	80 000
							3 100 (690)	2 800 (621)	2 200 (483)	70 000
									2 800 (621)	50 000

Table 12.8 Actuator force available

Spring force	1405 N	↑ Spring force travel compression

The spring rate is obtained by dividing the force which the spring is required to balance by the valve stroke length.

$$\frac{\text{Net spring force}}{\text{Stroke}} = \text{Spring rate}$$

$$\frac{1405}{0.028} = 50\ 179\ \text{N/m}$$

A spring should then be selected having a rate as near as possible to, but not greater than, that required. See Table 12.8 which tabulates the actuator force available for various travels with different spring rates fitted to the actuator.

If the ideal spring is not available then it is possible to either:

1) select a larger actuator, or,
2) use a 0.4 to 2.0 barg actuator control signal.

12.5.2 Selecting the correct actuator/spring for a reverse-acting diaphragm

Example: To calculate the spring rate for a reverse-acting actuator for a valve having the following specification:

Valve type:	DN50 single seat
Valve action:	Pushdown-to-close
Flow direction:	Flow-to-open
Valve seat:	Steel
Packing:	Single PTFE
Valve travel:	28 mm
Seat circumference:	184.5 mm
Seat area:	2700 mm
Valve stem diameter:	12.5 mm
Valve ΔP:	3 barg max when closed
Fail-safe action:	valve to close

It is assumed that the actuator signal is 0.21-1.03 barg and in this case a positioner is used with an output signal range of 0 to 1.2 barg.

Actuator size:

The actuator should, if possible, be selected from standard sizes which are produced for direct connection to the valve concerned.

In this case the actuator would have an area of 0.0445 m^2.

To calculate the available actuator thrust, see Figure 12.52:

The available actuator thrust is the result of applying air pressure to the underside of the diaphragm, in this case:

$$1\ \text{bar} \equiv 100\ 000\ \text{N/m}^2$$

$$1\ \text{bar} \times 0.0445\ \text{m}^2 = 4450\ \text{N}\ \text{Force available}\ \uparrow$$

To ensure a quick response when the pressure is applied, the spring is given a certain amount of pre-compression, in this case equivalent to 0.2 barg:

$$0.2\ \text{bar} \times 0.0445\ \text{m}^2 = 890\ \text{N}\ \text{Initial force}\ \downarrow$$

This force acts downwards and should, therefore, be deducted from the maximum available actuator thrust:

$$4450\ \text{N} - 890\ \text{N} = 3560\ \text{N}$$

which is the available force remaining for upward movement of the actuator stem.

The actuator must, however, have sufficient thrust to:

- overcome the unbalanced force resulting from the pressure drop. The valve seat area is 0.0027 m^2 and the upwards unbalanced force is therefore:

  $$3\ \text{bar} \times 0.0027\text{m}^2 = 810\ \text{N}\ \text{Unbalanced}\ \uparrow$$

- impart sufficient thrust on the valve plug against the seat to prevent leakage, i.e. the pressure required between the plug and the seat (similar to turning a tap off tightly), is 9 N/mm over the circumference of 184.5 mm.

  $$9\ \text{N/mm} \times 184.5\ \text{mm} = 1660\ \text{N}\ \text{Seat load}\ \uparrow$$

- In the case of reverse-acting actuators the spring force of 890 N is pressing against the seat. The additional force required to give 9 N/mm seat load is:

  $$1660 - 890 = 770\ \text{N}$$

- If the requirement is less than 890 N to give sufficient sealing force against the seat, the remaining force can be used to overcome the unbalanced forces.
- overcome packing friction. Packing friction on a 12.5 mm diameter stem with PTFE packing is equivalent to about 225 N, (see also Chapter 11 about packing friction values). The valve stem should be treated to tolerate this friction, otherwise graphite packing is only used where high temperatures necessitate its use.

The spring must overcome the sum of the following forces, see Figure 12.53:

Unbalanced forces	810 N
Pressure against seat	770 N
Packing friction	225 N
Total unbalanced	1805 N

The resulting force which must be balanced by the spring is therefore:

Net actuator force	3560 N	
Minus unbalanced force	1805 N	
Spring force	1755 N	Additional spring force required ↓

The spring rate is obtained by dividing the force which the spring is to overcome by the valve stroke length:

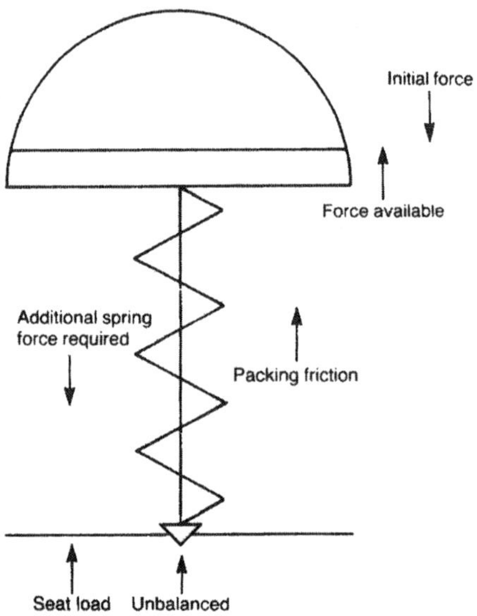

Figure 12.53 Valve/actuator force diagram for closed valve

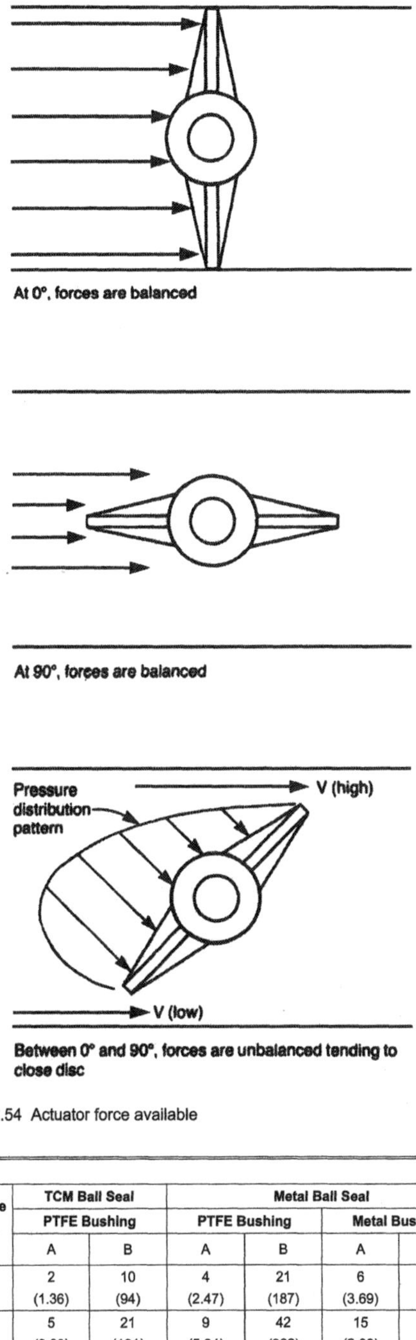

Figure 12.54 Actuator force available

$$\frac{\text{Net spring force}}{\text{Stroke}} = \text{Spring rate}$$

$$\frac{1755}{0.028} = 62\,678\ \text{N/m}$$

A spring should then be selected having a rate as near to, but not greater than, that required. See Table 12.8 which tabulates the actuator force available for various travels with different spring rates fitted to the actuator.

If the ideal spring rate is not available, then it is possible to either:

1) select a larger actuator, or,
2) use a 0.4 to 2 barg actuator control signal and recalculate the spring rate.

In this example the requirement is for a spring rate of about 63 000 N/m. The nearest available spring is rated at 70 000 N/m, the difference between 63 and 70 thousand is not significant. The next available spring, rated at 80 000 N/m would be too stiff and the valve would be unable to open fully.

12.5.3 Actuators for rotary valves

The angles of rotation for this type of valve are either 0° to 90° or 0° to 60°. For control purposes 0° to 60° is usually only found in older valves. 0° to 90° rotation is used for both control and on/off isolating applications. The fluid flow in a butterfly valve illustrated in Figure 12.54 shows:

- at 0° opening, i.e. fully closed, the valve is only subjected to the differential pressure forces
- at 90°, the valve is subjected to the effect of pressure difference acting upon a restriction caused by the shaft and disc profile
- at intermediate angles the liquid flow in a conventional butterfly valve is far greater on one side of the disc than on the other, the actual flow distribution depending upon the valve position

Thus in the zone where the flow is greatest, between the body and the disc, the velocity is relatively high and static pressure is low, whilst in the zone where the flow is small, between the body and the disc, the liquid velocity is relatively low and static pressure is high. The result of the difference in static pressure causes an unbalanced torque in conventional butterfly valves, which is greatest at about 78° open on the standard disc position/torque curve, shown in Figure 12.55.

ANSI Class 150														
Body size mm (in)	TCM Ball Seal		Metal Ball Seal				Flow Ring				C	Travel mm (in)	Yoke size mm (in)	Maximum Tb Nm (in-lb)
	PTFE Bushing		PTFE Bushing		Metal Bushing		PTFE Bushing		Metal Bushing					
	A	B	A	B	A	B	A	B	A	B				
80 (3)	2 (1.36)	10 (94)	4 (2.47)	21 (187)	6 (3.69)	21 (187)	0.39 (.244)	10 (94)	2.35 (1.47)	10.34 (94)	3.68 (2.3)	53.98 (2 1/8)	53.98 (2 1/8)	104.94 (954)
100 (4)	5 (3.08)	21 (191)	9 (5.34)	42 (382)	15 (9.39)	42 (382)	1.30 (.811)	21 (191)	7.79 (4.87)	21.01 (191)	10.4 (6.5)	79.38 (3 1/8)	71.44 (2 13/16)	176 (1 600)
150 (6)	11 (7.19)	48 (432)	20 (12.32)	95 (865)	36 (22.58)	95 (865)	3.28 (2.05)	48 (432)	19.68 (12.3)	48 (432)	35.2 (22)	79.38 (3 1/8)	71.44 (2 13/16)	316.80 (2 880)
200 (8)	20 (12.57)	76 (693)	33 (20.8)	152 (1 386)	68 (42.46)	152 (1 386)	6.93 (4.33)	76 (693)	41.60 (26.0)	76 (693)	73.6 (46)	104.78 (4 1/8)	71.44 (2 13/16)	724.90 (6 590)
250 (10)	34 (21.19)	120 (1 089)	54 (33.73)	236 (2 146)	120 (74.99)	236 (2 146)	13.20 (8.25)	120 (1 089)	79.20 (49.5)	120 (1 089)	144 (90)	104.78 (4 1/8)	71.44 (2 13/16)	1 881.0 (17 100)
300 (12)	53 (33.05)	174 (1 584)	83 (51.86)	349 (3 169)	198 (122.2)	349 (3 169)	22.72 (14.2)	174 (1 584)	135.36 (84.6)	174 (1 584)	249.6 (156)	104.78 (4 1/8)	71.44 (2 13/16)	1 881.0 (17 100)
400 (16)	74 (46.36)	267 (2 426)	120 (75.16)	534 (4 852)	260 (162.8)	234 (4 852)	28 (17.5)	267 (2 426)	168.32 (105.2)	267 (2 426)	472 (295)	104.78 (4 1/8)	71.44 (2 13/16)	1 870 (17 100)
600 (24)	171 (107.1)	518 (4 712)	261 (163)	1 036 (9 418)	671 (419.2)	1 036 (9 418)	81.92 (51.2)	518 (4 712)	491.68 (307.3)	518 (4 712)	1 280 (800)	206.38 (8 1/8)	90.49 127.00 (3 9/16) (5)	3 707 6 248 (33 700) (56 800)

Table 12.9 Characteristic torque factors A, B and C suitable for butterfly, ball and Vee-ball™ valves

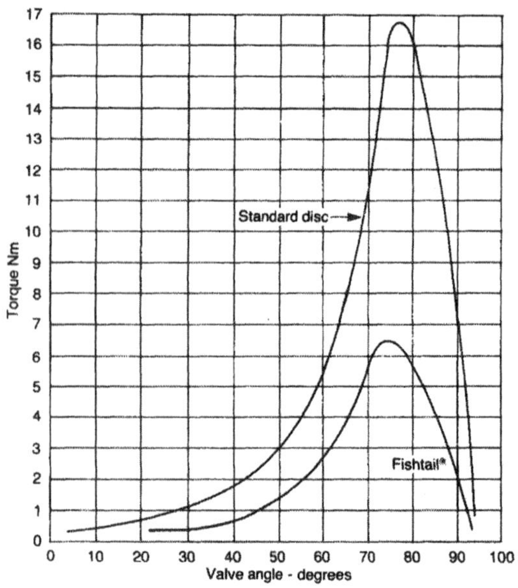

Figure 12.55 Torque exerted by fluid flow over butterfly valves

Figure 12.55 also illustrates the torque requirement for a Fishtail™ valve. In the Fishtail™ valve the disc profile is specially shaped so that the fluid flow distributes itself almost equally on the upper and lower surfaces of the disc. This means that the flow quantity, the flow velocity and static pressure are also more equally distributed about the centre of the disc resulting in a great reduction in the unbalanced forces in the valve.

12.5.4 Sizing the actuator

The torque requirements in a rotary valve are the result of a variety of factors, such as the effects of pressure drop, fluid flow velocity and break-away torque, i.e. the torque required to initially move the valve from its closed position. In practice there are two methods which can be used to determine a suitable size of actuator:

- by calculating the torque resulting from the pressure drop across the valve. Standard tables can be used to assist this calculation, and the following parameters must be known:
  - the size of the valve, its design and construction
  - the proposed actuator, the final choice can be made by repeating the calculation for different size actuators
  - the fail-safe action of the valve in the event of loss of air supply
  - the rated flow conditions
  - pressure conditions in the closed position
  - the air supply pressure to the actuator
- by calculating the torque requirement based on:
  - the break-away torque required

    $Tb = A(\Delta P) + B$
  - the dynamic torque requirement
  - $Td = C(\Delta P)$ at 10°, 60° and 90°

Factors A, B and C for the valve can be obtained from Table 12.9 and multiplied as shown by the pressure drop in the closed position, ΔP_{closed} and the effective pressure drop ΔP_{eff}, for butterfly valves at 10°, 60° and 90° and for ball valves at 70° and 90°.

There are two ways of calculating ΔP_{eff}:

1) if the flow conditions are known, then the actual pressure drop can be calculated,
2) a) for liquids

   $\Delta P = F(P - FP)$

   b) for gases

   use the equation: $\Delta P = 0.14p$

   where "p" is the absolute inlet pressure.

Use the lesser of the two ΔP_{eff} values obtained by applying methods 1) and 2). The actuator size can be chosen from tables quoting actuator output torques.

Table 12.10 shows data for various sizes of piston actuator with 7 barg air supply pressure.

Actuator size	With positioner Nm (in-lb)	Without positioner Nm (in-lb)
30	247.50 (2250)	275 (2500)
40	495 (4500)	550 (5000)
60	990 (9000)	1100 (10,000)
80	2970 (27,000)	3300 (30,000)
100	4400 (40,000)	4950 (45,000)

Table 12.10 Torque capabilities of piston actuators with 7 barg air

12.6 Setting-up

The actuator construction must be compatible with the valve style used. For globe valves, the valve bonnet is machined to accept standard yoke diameters. Ball and rotary valves have standardised bolt patterns to allow the fixing of brackets to attach the actuator. In both cases the actuator is mounted in a standard configuration to allow stems and spindles to be joined and allow full travel.

Linear valves use a standardised split clamp connector to connect the actuator stem to the valve stem. The stem connector, shown in Figure 12.56, fulfils two functions:

- to connect the actuator to the valve
- to provide a means of setting the actuator correctly relative to the valve position

Large forces are transmitted via the stem connector especially when powerful actuators or handwheel mechanisms are fitted. It is therefore essential to ensure that all threads are clean and undamaged and that the connection is properly secured. The connector is tightened after the stem assembly, stem-to-plug and stem-to-actuator, have been adjusted to the correct lengths to permit the valve to travel between its specified limits.

In addition to the stem connector, it is also usual to fit a scale and an indicator to show the position of the valve, see Figure

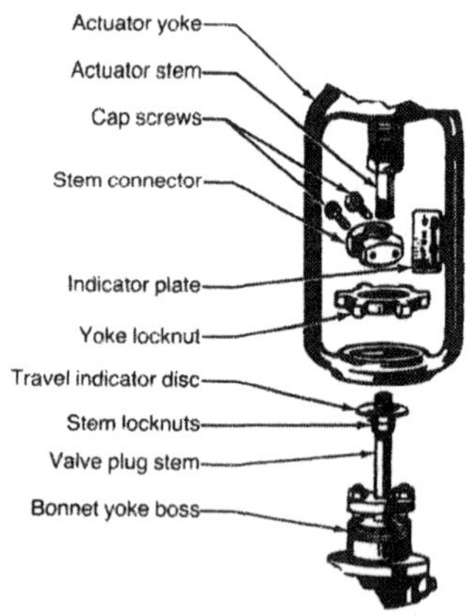

Figure 12.56 A typical stem connector for linear valves

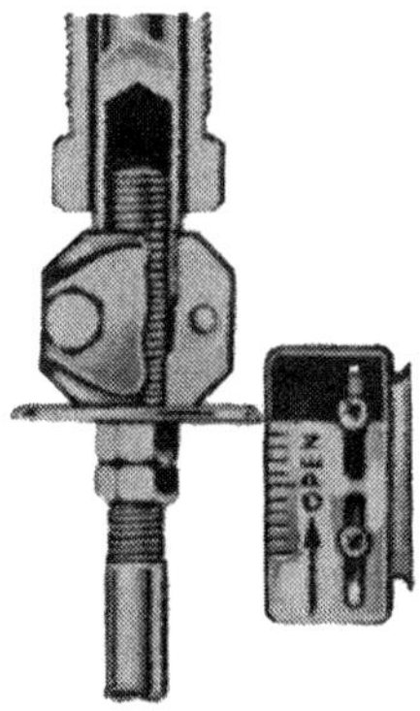

Figure 12.57 Typical stem pointer and travel scale

12.57. The scale is adjusted to the valve stroke and is usually set with the valve in the closed position with the plug resting on the seat. Excessive wear between the plug and the seat will thus be apparent if the indicator does not align with the scale when the valve is closed. Valve stems and their connectors are sometimes protected by a rubber gaiter; alternatively the complete assembly may be encased by a cover mounted on the actuator yoke.

12.6.1 Adjusting the stem connector on linear valves

1) Remove the stem connector in order to push down to close the valve. Ensure that the valve plug is in the closed position, i.e. the seat and plug are in contact with each other.
2) Set the actuator in its corresponding closed position:
a) for direct-acting actuators the valve stroke is downwards from the upper position,
b) for reverse-acting valves the closed position is as it is in the loosely attached position.
3) Move the actuator stem about 2 mm closer to the valve stem and connect both stems by means of the stem connector.
4) Check that the actuator can travel its full stroke. Make sure that the plug touches the seat before the actuator comes to its limit. Minor adjustments, less than 2 mm, can be made by rotating the stem connector slightly, after loosening the cap screws which hold it together.
5) When everything is connected together, screw the indicator plate into a position which corresponds with the valve in closed position.
6) Use a gauge connected to the actuator signal input and a dial indicator on the valve stem, set the valve according to the specification plate.

12.6.2 Adjusting the spindle connector on rotary valves

Rotary valves and actuators can be assembled in many configurations, as shown in Figures 12.58 and 12.59.

The most common alternatives are with the actuator mounted either to the left or to the right with a choice of which direction the valve should travel in the event of air failure. Valves of this type are usually fitted with direct-acting actuators and the arrangement for valve opening or closing in the event of air failure is made by changing the connection between the valve spindle and the lever. In the case of ball valves, the connection point is set at the ideal locations on the valve spindle and the lever, assuming that fine adjustment can be made with the actuator stem.

Fine adjustment on rotary valves can only be made while the valve is not being used in the process, since it must be possible

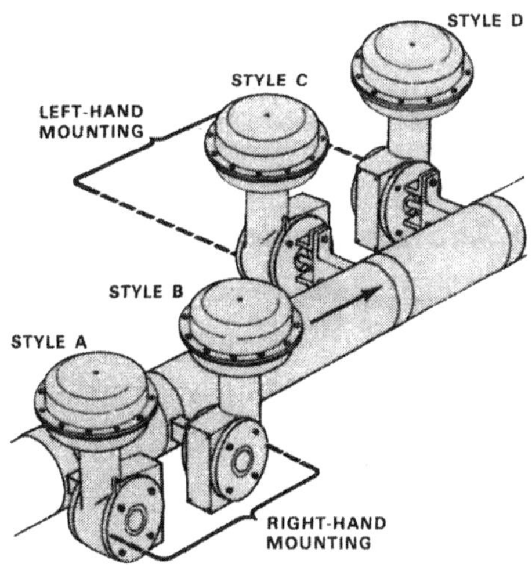

Figure 12.58 Rotary valve and actuator configurations

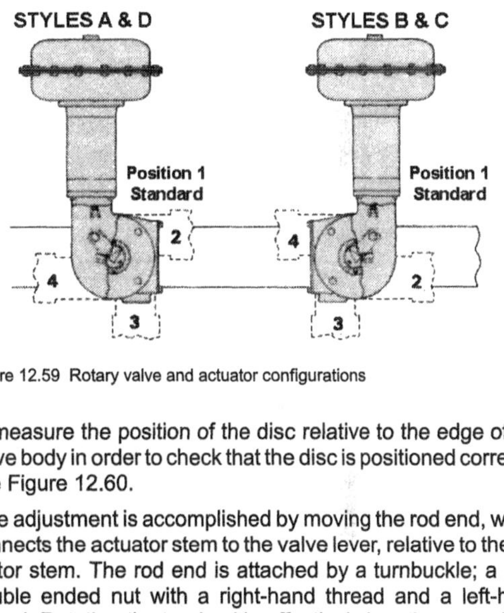

Figure 12.59 Rotary valve and actuator configurations

to measure the position of the disc relative to the edge of the valve body in order to check that the disc is positioned correctly, see Figure 12.60.

Fine adjustment is accomplished by moving the rod end, which connects the actuator stem to the valve lever, relative to the actuator stem. The rod end is attached by a turnbuckle; a long double ended nut with a right-hand thread and a left-hand thread. Rotating the turnbuckle effectively lengthens or shortens the actuator stem.

12.6.3 Bench setting

Bench setting is the method of adjusting the actuator spring compression in a test rig without subjecting the valve to process

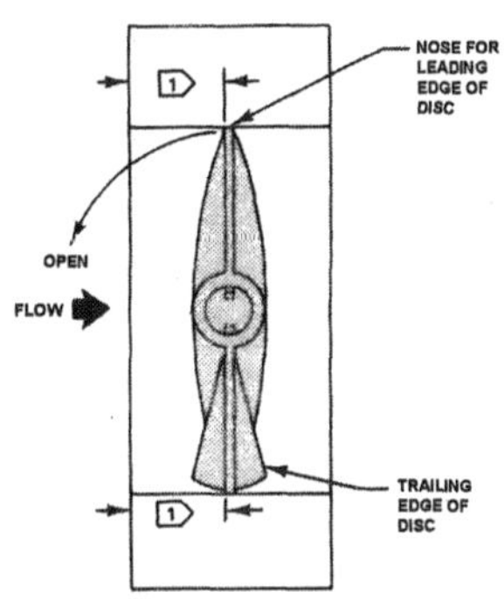

Figure 12.60 Checking disc position for rotary valve actuator adjustment

pressure, in the hope that the valve will function correctly when process conditions are applied.

Most major users do not permit the practice at all or restrict it severely, because the need for bench setting actually signals that the actuator is not large enough for its duty. There have been instances in the past where, in order to gain some competitive advantage, undersized actuators have been supplied.

12.7 Actuator selection

When selecting an actuator, many factors must be considered:

- compatibility with the valve design to be used
- power supply for actuator
    - pneumatic
    - hydraulic
    - electric
    - combined system
- single or double acting
- control signal for actuator
    - pneumatic analogue
    - electric analogue
    - electric digital
    - analogue/digital combination
- force or torque required
- speed of operation
    - slow
    - fast
    - variable
- power required to achieve force/torque and speed
- stiffness of actuator
- fail-safe operation of actuator
    - stay-put
    - close
    - open
- accessories and options required
    - handwheel
    - local powered operation
    - positioners
    - boosters
    - remote indication
    - ESD override facilities
    - damping facilities
- ambient operating conditions
    - electrically hazardous
    - air-borne contamination/corrosion
    - water hazards
    - temperature extremes, including sunlight
    - humidity
    - biological attack
    - vibration, external and valve-generated
    - seismic considerations
- serviceability by local staff and reliability
- initial purchase and installation costs
- operating and maintenance costs

A thorough knowledge and understanding of the process to be controlled and its requirements is essential if the correct choice of actuator is to be made. It is not possible to give general advice on the selection of actuators.

In the past the pneumatic diaphragm actuator has been the most popular for process applications but electric actuators have been used extensively in small "safe" applications. The advent of cheap electronic controls and digital communications means that "electric" actuators are gradually taking over.

NOTE: The size of actuator required for a particular valve can be determined by the supplier. Remember that no control valve can control better than its actuator permits.

12.8 Useful references

The Instrumentation, Systems and Automation Society (ISA), 67 Alexander Drive, Research Triangle Park, NC 27709 USA, Tel: 919 549 8411, www.isa.org.

BS EN 60584-1 and 2 Thermocouples. Reference tables.

BS EN 50065-7:2001 Signalling on low-voltage electrical installations in the frequency range 3 kHz to 148.5 kHz.

BS EN 60079 :2006 Electrical apparatus for explosive gas atmospheres. General requirements.

BS EN 50018:2000 Electrical apparatus for potentially explosive atmospheres. Flameproof enclosure 'd'.

BS EN 50020:2002 Electrical apparatus for potentially explosive atmospheres. Intrinsic safety 'I'.

EN61000-4-3:2002 Electromagnetic compatibility (EMC). Testing and measurement techniques. Radiated, radio-frequency, electromagnetic field immunity test.

EN 50170-A3 AMD 3 General Purpose Field Communication System.

IEC 61158-SER Ed. 1.0 b:2005 Digital data communications for measurement and control - Fieldbus for use in industrial control systems.

VDI/VDE 3845 Mounting for solenoid valves and position signallers to the NAMUR standard

ISO 5211:2001/DIN 3337 Industrial valves — Part-turn actuator attachments.

Valve materials

13

13.1 General

The pressure/temperature ratings quoted in this Chapter are based on steady-state operating conditions. This can be defined as: material failure due to fatigue is unlikely and the material will not be exposed to stresses, additional to the pressure related stresses, caused by thermal shock.

13.1.1 Fatigue

Fatigue is relatively easy to predict and combat. If the internal pressure is likely to change regularly then the stress range and the number of cycles can be calculated. If the number of cycles is large, 7000 is used in some piping and pressure vessel standards as a limitation, then the pressure rating may need to be reduced to compensate. If the pressure range is significant, then fatigue should be investigated as a matter of routine.

13.1.2 Thermal shock

Thermal shock is much more difficult to analyse. The mass of the valve body, the rate of temperature rise of the fluid and the nature of external cooling are all extremely important. If there is any chance the process fluid temperature will change suddenly, greater than 17°C per hour, the valve manufacturer should be consulted. It only takes one severe thermal shock to crack the valve body. Thermal shock can be a very severe problem as illustrated by the recommendations given in the ASME draft code for nuclear pumps and valves, see Figure 13.1. The code was applicable to pumps and valves with ANSI pressure ratings from 300 lb to 2500 lb.

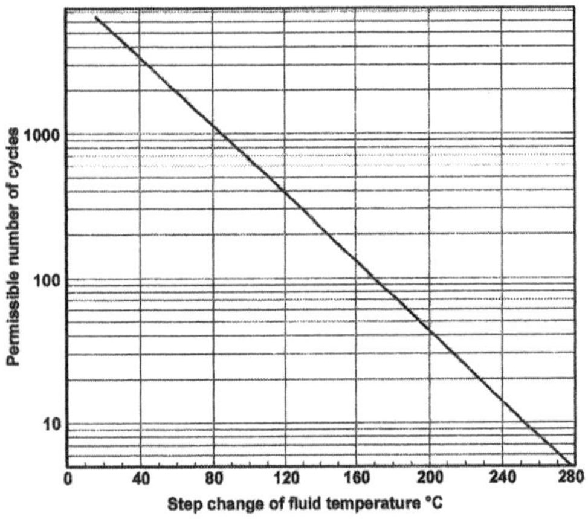

Figure 13.1 Permissible number of rapid temperature excursions

13.1.3 Corrosion

The materials of valve construction do not necessarily have to be totally corrosion resistant to the process fluid, **provided the fluid can be contaminated!** If process fluid purity is essential then all materials in contact must be corrosion resistant. If some contamination is acceptable then the valve materials may be allowed to be attacked. The corrosion rate(s) will determine the useful life of the valve. In order to extend the valve life it may be possible to allocate a corrosion or erosion allowance of sacrificial material. This will be possible when the operating pressure/temperature is below the rated pressure/temperature. Once the corrosion allowance has been used the valve must be replaced or refurbished. Extreme operating environments may also warrant an external corrosion/erosion allowance. Table 13.1 indicates some general guidance on the uses of the basic material groupings.

Material	Applications
Carbon steel	Hot and cold water De-aerated hot water Non-corrosive clean liquids and gases
Grey cast iron	Low pressure applications The graphite film provides good corrosion protection provided it is not damaged by high velocities, aeration of liquids, cavitation or erosion. Clean liquids and gases.
Brass	Low pressure clean water and clean non-corrosive gases.
High tensile steel	High pressure applications with non-corrosive clean liquids and gases.
SG iron Nodular iron Ductile iron	Slightly higher pressures and temperatures than cast iron. Graphite film not as tough. Clean liquids and gases.
11/13Cr steel (1)	De-aerated hot water up to 350°C. Non-oxidising gases up to 650°C. A good replacement for carbon steel for applications over 200°C; better thermal stability, higher pressure capabilities.
Ni-Resist iron	Hot NaOH, seawater, some acids, coke oven gas, coal tar, wet hydrogen sulphide, paper making, hydrocarbons with HCl and H_2S,sewage, low pressure steam.
Gunmetal, bronze	Salt water, seawater, brine and other moderately corrosive aqueous solutions. Aluminium bronze and nickel aluminium bronze better than tin bronze in seawater. Nickel aluminium bronze for high pressures
Austenitic stainless steels	Hot water, hot gases. General corrosive applications at low to medium pressures up to high temperatures. Cryogenic applications.
Higher steel alloys Nickel alloys Titanium	Corrosive or oxidising applications with acids or high temperature/pressure. Chloride compounds or hot flue gases, paper making.
Non-metallic materials	Acids, alkalis, corrosive reagents and solvents at low/moderate temperature and pressure. Can be resistant to erosion.

(1) Many manufacturers consider 11/13Cr steel as stainless steel; it is stain resistant.

Table 13.1 Applications and materials for valves

The selection of materials of construction can be very complicated and may result in compromises to avoid galling and/or galvanic corrosion. The valve manufacturers have very broad experience for solving application problems. Beware of using solutions to unrelated problems. "Wonderful" materials used in other industries may only be average when transferred to valve applications.

13.1.4 Hygienic applications

Valves intended to convey foodstuffs must be designed for effective cleaning and made from approved materials. The dominant materials in common use are:

- 316L stainless steel
- white nitrile rubber
- black butyl rubber
- white butyl rubber
- Buna "N"
- EPDM
- PTFE

European practice for hygienic applications is not identical to American practice. Equipment destined for the USA must comply with the FDA regulations. American equipment which does comply with the FDA requirements is generally acceptable in Europe. EHEDG has committees which consider all aspects of design and use.

13.2 The Pressure Equipment Directive (PED)

The European Pressure Equipment Directive (PED) 97/23/EC became a legal requirement on 29th May 2002. It was implemented in the United Kingdom by the Pressure Equipment Regulations 1999. This Directive applies to the design, manufacture and conformity assessment of pressure equipment and assemblies of pressure equipment with a maximum allowable pressure greater than 0.5 barg.

Among the many exclusions listed in the PED are:

- Pipelines used for the conveyance of fluids or substances
- Water distribution systems
- Equipment coming under the simple pressure vessel directive 87/404/EEC
- Well control equipment
- Casings or machinery where the design is dictated by the static and dynamic operational effects and for which pressure is not a significant design factor.

Equipment is categorised as vessels, steam generators, piping, safety accessories, pressure accessories and assemblies. Equipment size is based on a combination of the internal volume, in litres, or the nominal diameter, in mm, and the maximum allowable pressure, in barg. Fluids are divided into two groups:

Group 1 fluids are designated:

- explosive
- extremely flammable
- highly flammable
- flammable
- very toxic
- toxic
- oxidising

These are designated in Council Directives 67/548/EEC 27 June 1967, 96/54/EEC and 96/56/EEC.

Group 2 fluids comprise all other liquids/gases/vapours including steam.

If a liquid has a vapour pressure greater than 0.5 bar(g) at the maximum allowable temperature then the liquid is treated as a gas. "Assemblies" refers to multiple pieces of equipment assembled by a manufacturer. Assemblies assembled by the user are covered by the "In Use" regulations.

The Essential Safety Requirements prescribe that "adequate safety margins" must be included while considering, among other things, earthquakes, corrosion, erosion and fatigue. The PED gives maximum allowable membrane stresses, for predominantly static loads, based on material composition and strength. The PED does not provide safety margins for fatigue resistance. To provide some consistency the PED should have stated for example, fatigue loading, where the maximum allowable stress should not exceed 80% of the predicted endurance limit for the nominated conditions.

Significant costs may be incurred when implementing material tractability and appraisal of materials by a Notified Body for some applications.

13.3 NACE

The name NACE (originally known as the National Association of Corrosion Engineers) has become synonymous with the requirement to combat sulphide stress cracking, SSC, even though NACE produces much information on all forms of corrosion. The NACE Standard MR-01-75 is applied to equipment which is exposed to hydrogen sulphide. In very broad terms, sulphide stress cracking is a problem for materials which have been hardened and tempered to increase strength. The main thrust of the NACE requirements is to limit material hardness.

In oilfield applications, fluids are divided into two major groups:

- sweet
- sour

Sweet fluids contain no hydrogen sulphide. Sour fluids contain hydrogen sulphide, usually as a trace element. The percentage of hydrogen sulphide and the partial pressure are important factors in the probability of SSC problems.

NACE excludes the use of cast iron for pressure containment and also excludes the use of free-machining varieties of steels.

For carbon steels, the hardness is restricted to 22 HR_C, which is approximately 238 BHN. Welded components must be stress relieved to ensure the heat-affected-zone (HAZ), meets the hardness limit. Alternatively, the carbon content can be restricted to less than 0.18% which effectively removes the hardening possibility. With carbon steels, compliance with NACE does not create problems.

NACE can create major problems with alloy steels. Consider bolting; AISI 4140, ASTM A193 B7 is the most popular bolt/stud material. NACE limits the hardness to 22 HR_C when the normal hardness would be 24 to 32 HRC. The reduction in hardness produces a reduction in strength; the ultimate strength will be reduced from 850/1000 MPa to about 770 MPa maximum. This means a bolted connection will only have approximately 90% of the strength. Equipment working close to the pressure limit may require alternative, better materials.

The austenitic stainless steels cannot be work hardened to increase the strength. Again, this can be a major problem for bolting and similar bar components. SSC is generally a problem experienced by materials in tension. Valve seats, pressed into the body, will be in compression and should not be susceptible to SSC.

When NACE requirements are applied to a valve the design must be checked, not just the materials. Changing from B7 studs to B7M studs can create new problems while eliminating the SSC problem. Table 13.2 lists some popular materials with their respective limiting hardnesses.

Material	Maximum hardness Rockwell "C"
ASTM A216 WCB	22
ASTM A193 B7M	22
ASTM A194 2M	22
ASTM A217 CA-15	22
ASTM A351 CF-8	22
ASTM A351 CF-3	22
ASTM A351 CF-8M	22
ASTM A351 CF-3M	22
PH17-7Mo	30/32
17-4PH	33
ASTM A453 Gr660	35
Monel 400	35
Incoloy 800	35
Incoloy 825	35
Inconel 600	35
ASTM A637 Gr688 (springs)	50
Inconel X-750 (springs)	50

Table 13.2 NACE MR-01-75 hardness limits

Overlays and coatings are acceptable to NACE provided the base material meets the hardness requirements.

Hydrogen sulphide is highly toxic, even at very low concentrations in the ambient air. Tapered thread screwed connections should not be used for sealing pressurised sour fluids. If threads cannot be avoided, they should be protected from the fluid by seals.

13.4 Valve body material

The choice of material is usually based upon requirements for pressure, temperature, resistance to corrosion, abrasion and method of manufacture. A material which is resistant to abrasion may have little resistance to corrosion for a particular fluid.

Standard	Grade	Major constituents	Min tensile strength MPa	Max temp °C	Remarks
ASTM A278	C120		138	230	
ASTM A126	A		145	180	down to -10°C
ASTM A278	C125		172	230	
BS 1452	180		180	191	down to -18°C
ISO 185	200		200	300	down to -10°C
ASTM A278	C130		207	230	
ASTM A126	B		214	230	down to -10°C
BS 1452	220		220	221	down to -18°C
ASTM A278	C135		241	230	
ISO 185	250		250	300	down to -10°C
ASTM A278	C140		276	345	
ASTM A278	C160		414	345	

NOTE: The standards authorities severely limit the low temperature use of grey cast iron. Test results show little change in strength or ductility down to -40°C.

Table 13.3 Grey cast iron for valve bodies

Standard	Grade	Major constituents	Min tensile strength MPa	Max temp °C	Remarks
ISO 1083	350-22		350	350	
ISO 2531	400-5		400	350	
ISO 1083	400-15		400	350	
ASTM A395	414-18		414	≈ 300	
BS 2789	420/12		420	≈ 350	
ASTM A536	65-45-12		448	≈ 350	
ISO 1083	500-7		500	350	
BS 2789	500/7		500	≈ 350	
ASTM A536	80-55-06		550	≈ 350	
ISO 1083	600-3		600	120	
DIN 1693	600-3		600	≈ 120	
BS 2789	600/3		600	≈ 120	
ASTM A536	100-70-03		690	≈ 120	

Table 13.4 Ductile iron for valve bodies

Standard	Grade	Major constituents	Min tensile strength MPa	Max temp °C	Remarks
BS 310	B290/6		290	221	down to -18°C
BS 310	B310/10		310	221	down to -18°C
ISO 5922	B32-12		320	350	down to -10°C
BS 309	W340/3		340	221	down to -18°C
BS 310	B340/12		340	221	down to -18°C
ISO 5922	B35-10		350	350	down to -10°C
BS 309	W410/4		410	221	down to -18°C

Table 13.5 Malleable iron for valve bodies

Standard	Grade	Major constituents	Min tensile strength MPa	Max temp °C	Remarks
BS 3468	L-NiCr 20 2	1.8Cr 20Ni 1Mn 0.3 Cu 1.9Si	170	≈ 650	-50 °C min
ASTM A436	Type 2		172		Good for alkalis and high temperature steam
DIN 1694	GGL-NiCr 20 2				
ISO 2892	L-NiCr 20 2				
BS 3468	S-NiCr 20 2	2.25Cr 20Ni 1Mn 2.25Si	370	≈ 800	7% El 200 BHN
ASTM A439	Type D-2		400		8% El 202 BHN
DIN 1694	GGG-NiCr 20 2				Wear resistant
ISO 2892	S-NiCr 20 2				
BS 3468	S-NiMn 23 4	0.2Cr 22.5Ni 4Mn 2Si	440	Min	Wear resistant
ASTM A571	Type D-2M		417	-196 to -250°C	
DIN 1694	GGG-NiMn 23 4				
ISO	S-NiMn 23 4				

Table 13.6 Ni-resistant cast irons for valve bodies

Standard	Grade	Major constituents	Min tensile strength MPa	Max temp °C	Remarks
ASTM A216	WCB	1Mn 0.6Si 0.3C	485	426	Cast ANSI 1.1
DIN 17245	GS 22Mo4		440	450/500	
BS 1504	161 Gr480	0.25C 0.9Mn 0.4Ni 0.25Cr	480	400	
AFNOR	A48 CM				
ISO 3755	200-400W				Cast ISO 1E0
DIN 1681	GS-38.3		373	300	
ASTM A105		0.35C 0.8Mn 0.35Si	485	426	Forged ANSI 1.1
DIN 17155	30C				
BS 1503	221 Gr490	0.25C 1.2Mn 0.5Ni 0.25Cr	490	480	
ASTM A350	LF2 (1)	0.3C 0.7Mn 0.22Si	485/655	343	Forged ANSI 1.1
BS 1503	221 Gr530	0.27C 1.2Mn 0.5Ni 0.25Cr	530	350	
SEW 680	TTST 41V				
BS 1503	161Gr A	0.25C 0.9Mn 0.4Ni 0.25Cr	490	482	Forged BS 10 Down to -18°C

Table 13.7 Popular carbon steels for valve bodies — (1) Minimum temperature -45°C

Standard	Grade	Major constituents	Min tensile strength MPa	Max temp °C	Remarks
ASTM A182	F12	1Cr 0.55Mo 0.55Mn 0.15C	485	593	Forged ANSI 1.6 Excellent ductility
BS 970 Pt2	709M40 "S"	1Cr 0.3Mo 0.85Mn 0.4C	770	≈ 500	-75 °C min
BS 1503	620	1Cr 0.55Mo 0.55Mn 0.4Ni 0.11C	420	525	
BS 1504 DIN 17245 ASTM A217	621 GS-17 CrMo 55 WC6	1.25Cr 0.55Mo 0.65Mn 0.2Ni 0.1C 1.25Cr 0.5Mo 0.65Mn 0.17C	480 490/635	525 500 590	Cast

Table 13.8 High tensile steel for valve bodies

Standard	Grade	Major constituents	Min tensile strength MPa	Max temp °C	Remarks
ASTM A182 BS 1503	F22 622-490	2.25Cr 1Mo 0.45Mn 0.15C 2.25Cr 1Mo 0.65Mn 0.2Ni 0.11C	515 490	590 580	Forged ANSI 1.10
ISO 2604-1	F34Q				Forged ISO 6E0
ASTM A217 BS 1504 DIN 17006 ISO 4991	WC9 622 GS-35 CrMo V 10 4 C34AH	2.4Cr 1Mo 0.6Si 0.55Mn 2.25Cr 1Mo 0.55Mn 0.2Ni 0.18C 2.4Cr 0.4Mo 0.65Mn 0.35C	485 620 735/1100	590	Cast ANSI 1.10

Table 13.9 2.25% chrome alloy steels for valve bodies

Standard	Grade	Major constituents	Min tensile strength MPa	Max temp °C	Remarks
ASTM A182 ASTM A182 BS 1503 ISO 2604-1	F5 F5a 625 F37	5Cr 0.55Mo 0.45Mn 0.2Ni 0.15C 5Cr 0.55Mo 0.45Mn 0.2Ni 0.25C 5Cr 0.55Mo 0.5Mn 0.2Ni 0.18C	485 620 520	649 649 (1)	Forged ANSI 1.10
ASTM A217 BS 1504 ISO 4991	C5 625 C37H	5.2Cr 0.75Si 0.55Mo 0.55Mn 0.2C 5Cr 0.55Mo 0.55Mn 0.2Ni 0.2C	620 620	649	Cast ANSI 1.13

Table 13.10 5% chrome alloy steels for valve bodies — (1) Final material properties and maximum temperature by agreement

Standard	Grade	Major constituents	Min tensile strength MPa	Max temp °C	Remarks
ASTM A182	F9	9Cr 1Mo 0.75Si 0.45Mn 0.15C	585	649	Forged ANSI 1.14
ASTM A217 BS 1504 ISO 4991	C12 629A C38H	9Cr 1Mo 1Si 0.5Mn 0.2C 9Cr 1.05Mo 0.5Mn 0.5Si 0.2Ni 0.2C	620 620	649 600	Cast ANSI 1.14

Table 13.11 9% chrome alloy steels for valve bodies

Standard	Grade	Major constituents	Min tensile strength MPa	Max temp °C	Remarks
ASTM A182	F6a F6b F6NM	12.5Cr 1Mn 1Si 0.5Mo 0.2Ni 0.15C 12.5Cr 1Mn 1Si 1.5Mo 0.5Ni 0.15C 13Cr 4Ni 0.75Mn 0.5Mo 0.45Si 0.06C	485/760 760/930 760/930		Forged
DIN 17006 Werkstoff	X 10 Cr 13 X 4 CrNi 13 4 1.4001	13Cr 0.8Mn 0.8Si 0.5Ni 0.12C 14Cr 1 Mn 1 Si 0.08 C			
ASTM A217 DIN 17006	CA-15 G-X 5 CrNi 13				Cast
ASTM A487 DIN 17006	CA-6NM G-X 5 CrNi 13 4	 12.75Cr 4.25Ni 1.5Mn 1Si 0.07C			Cast
BS 1503	410S21 410S29 403S17	12.5Cr 1Mn 1Ni 0.8Si 0.12C 12.5Cr 1Mn 1Ni 0.8Si 0.17C 13Cr 1Mn 0.5Ni 0.8Si 0.08C	589 695 464	600 600 400	Forged
BS 1504	420C29A 425C11A	11/13Cr 11/13Cr 4Ni	620 770	600 400	Cast

Table 13.12 11/13% chrome alloy for steels for valve bodies

Standard	Grade	Major constituents	Min tensile strength MPa	Max temp °C	Remarks
ISO 1338	CuPb5Sn5Zn5			220/260	Min -10°C
ISO 1338	CuSn8Pb2			260	Min -10°C
ISO 1338 BS 1400 DIN 1714 ASTM B148 UNS	CuAl10Fe3 AB1 CuAl10Fe A95200 C95200	9.5Al 2.5Fe	450/590	260	Min -10°C
ISO 1338 BS 1400 DIN 1714 ASTM B148 UNS	CuAl10FeNi5 AB2 CuAl10Ni D95500 C95500	9.5Al 5Ni 4.5Fe	640/740	350	Min -10°C

Table 13.13 Cast copper alloys for ISO flanges

Standard	Grade	Major constituents	Min tensile strength MPa	Max temp °C	Remarks
ASTM A705	Type 630	16.7Cr 4Ni 4Cu 1Mn 1Si 0.07C	1310	315	(1)
	S17400		795		(1)
Werkstoff	1.4542	16.5Cr 3.25Ni 4cu 1mn 1Si0.07C			
ASTM A705	Type 631	17Cr 6.1Ni 1Al 1Mn 0.09C	1275	'315	(1)
	S17700		1170		(1)
Werkstoff	1.4568	17Cr 7.1Ni 1Al 1Mn 0.09C			
ASTM A182	F304	19Cr 9.5Ni 1Mn 0.5Si 0.08C	515	815	Forged ANSI 2.1
BS 1503	304S40	18Cr 10Ni 2Mn 1Si 0.07C	510	450	
ASTM A351	CF-8	19.5Cr 9.5Ni 2Si 1.5Mn 0.5Mo 0.08C	485	815	Cast ANSI 2.1
ASTM A351	CF-3	19Cr 10Ni 2Si 1.5Mn 0.5Mo 0.03C	485	427	
BS 1504	801	17Cr 7Ni 2Mn 2Si 0.12C	460	400	
DIN 17440	X5 CrNi 18 9	18.5Cr 9.2Ni 2Mn 1Si 0.07C	500		Min -196°C
DIN 17440	X5 CrNi 19 11	18.5Cr 11.2Ni 2Mn 1Si 0.07C	500		Min -196°C
ISO 4991	C46				
ISO 4991	C47				
ASTM A182	F316	17Cr 12Ni 2.5Mo 1Mn 0.5Si 0.08C	515	815	Forged ANSI 2.2
BS1503	316S40	17.2Cr 12Ni 2.25Mo 2Mn 1Si 0.07C	510	500	
	316S41	17.2Cr 12.7Ni 2.7Mo 2Mn 1Si 0.07C	510	500	
DIN 17440	X5 CrNiMo 18 10	17.5Cr 12Ni 2.25Mo 2Mn 1Si 0.07C	500		Min -60°C
ASTM A351	CF-8M	19.5Cr 10.5Ni 2.5Mo 1.5Si 1.5Mn 0.08C	485	815	Cast ANS 2.2
ASTM A351	CF-3M	19Cr 11Ni 2.5Mo 1.5Si 1.5Mo 0.03C	485	454	
BS 1504	845B	17.5Cr 10Ni 2.5Mo 2Mn 2Si 0.08C	460		
ASTM A182	F304L	19Cr 9.5Ni 1Mn 0.5Si 0.035C	485	454	Forged ANSI 2.3
ASTM A182	F316L	17Cr 12Ni 2.5Mo 1Mn 0.5Si 0.035C	485	454	
BS 1503	304S30	18Cr 11Ni 2Mn 1Si 0.03C	490		
	316S30	17.2Cr 12.5Ni 2.25Mo 2Mn 1Si 0.03C	490		
	316S31	17.2Cr 13Ni 2.25Mo 2Mn 1Si 0.03C	490	450	
DIN 17440	X2 CrNi 18 9	18.5Cr 11.2Ni 2Mn 1Si 0.03C	450		Min -196°C
	X2 CrNiMo 18 10	17.5Cr 12.5Ni 2.25Mo 2Mn 1Si 0.03C	450		Min -60°C

Table 13.14 Stainless steels for valve bodies (1) Age-hardened forgings

Some fluids necessitate the use of special, costly materials or materials which pose manufacturing difficulties in order to prevent severe corrosion damage to the valve. Fortunately the majority of process fluids do not impose excessive requirements with regard to pressure, temperature and corrosion, which is why cast iron and cast carbon steel are the most common materials used.

The following review gives a general indication of cost of materials used for valve bodies. The BS and ASTM designations are quoted for each material. When specifying materials it is always advisable to quote a well known standard designation. Each of the following material specifications is referred to the relevant pressure/temperature rating group in Section 13.7. The specifications quoted are only by way of examples of the most popular materials.

Cast iron, BS 1452, ASTM A126

A relatively cheap material available in various grades used for valve bodies for water, gas and other non-corrosive fluids. Cast iron is easy to cast and to produce intricate castings including relatively thin sections. Very good to machine.

Cast steel, BS 1504-161 Grade 480, ASTM A216 WCB

This is the most popular valve body material and is used for air, saturated and superheated steam, non-corrosive liquids and gases. The cost is about four times greater than cast iron. The addition of alloying elements can, however, increase its pressure and temperature rating far beyond that of cast iron, Maximum recommended temperature is 426°C. It is not the easiest of materials to cast and can be prone to "tearing" when machining.

Chrome-molybdenum steel, BS 1504 625, ASTM A217 Grade C5

A popular alloy for temperatures up to 593°C, used for high pressure steam, oil, gas, petroleum vapour, seawater and other corrosive media. This material is also resistant to erosion. The cost is somewhat greater than for WCB but it is stronger and better at elevated temperatures, up to 649°C.

Stainless steel, type AISI 316, BS 1504-316C16, ASTM A351 CF8M

This material is particularly suited to oxidising and very corrosive fluids. It is suitable for temperatures up to about 800°C. The cost of this material is about two and a half times greater than chrome-molybdenum steel. It can be very difficult to cast and machine. Correct heat treatment after casting is essential. The low carbon version, 316L, although more costly again, can be more economical if weld repairs are required due to the removal of the post-weld heat treatment.

ANSI and ISO group materials which have similar temperature/strength characteristics

The figures describing the pressure capabilities of materials are based on the ASME/ANSI B16.34-2004, ISO 7005-1992 groupings and BS 10-1962. The data shown in the curves in Section 13.7 is accurate but is intended as a quick guide to suitable materials. The shape of the curves highlight the temperatures when materials weaken. The tabulations in the relevant standard should be consulted for specific data. Some pressure ratings vary with size and flange design and some materials are not recommended for the temperatures indicated. The attachment of the flange can also be important.

NOTE: The various flange standard bodies have adopted different design philosophies. ANSI standards allow intermittent operation at +10% design pressure but always pressure test at 150% of design pressure. European and ISO standards do not allow any operation above the design pressure but may only pressure test at 130% of design pressure. The full implications of the choice of standard must be considered during the design/application phase.

Non-metallic seats or linings may not be capable of sustaining the same pressure/temperature rating as the valve body. Man-

Standard	Grade	Major constituents	Min tensile strength MPa	Max temp °C	Remarks
ISO 426-1	CuZn20Al2			200	Min -10°C
ISO 428	CuAl10Fe3	9.5Al 3Fe 1Ni	570	260	Cold worked flanges for -180°C
BS 2872	CA 103				
DIN 17665	CuAl10Fe				
ASTM B124	C62300				
UNS	C62300				
ISO 428	CuAl10Ni5Fe4			260	Min -10°C
ISO 429	CuNi10Fe1Mn	10.5Ni 1.5Fe 0.7Mn	280/310	260	Min -10°C
BS 2872	CN102				
ISO 429	CuNi30Mn1Fe	31Ni 1Mn 0.7Fe	370	350	Min -10°C
BS 2872	CN107				

Table 13.15 Forged copper alloys for ISO 7005 flanges

Standard	Grade	Major constituents	Min tensile strength MPa	Max temp °C	Remarks
BS 1400	LG-4	7Sn 3Pb 2.4Zn	250	≈ 250	
Leaded gunmetal					
BS 1400	G1	10Sn 2.25Zn 1Pb	250/270	≈ 250	
Gunmetal					

Table 13.16 Popular copper alloys for valve bodies

Standard	Grade	Major constituents	Min tensile strength MPa	Max temp °C	Remarks
254 SMO	S31254	20Cr 18Ni 6Mo 0.01C	650	≈ 400	Nitrogen strengthened
					Cast and forged
SAF	2205	22Cr 5.5Ni 3.2Mo 2Mn 0.03C	680		Nitrogen strengthened
ASTM A276	S31803	22Cr 5.5Ni 3Mo 2Mn 1Si 0.03C			Forged
ASTM A182	F51		620		
W 1.4462	X12 CrNiMoN 22 5	22Cr 5.5Ni 3Mo 2Mn 0.03C			
654 SMO	S32654 (Forged)	24Cr 22Ni 7.3Mo 0.01C	750	≈ 400	Nitrogen strengthened
W 1.4465	G-X 2 CrNiMoN 25 25	25Cr 23.5Ni 2.25Mo 2Mn 0.03C			Cast
Zeron	25	25Cr 6.5Ni 2.5Mo 1.5Mn 0.03C	650		Nitrogen strengthened
					Forged
Zeron	100	25Cr 7Ni 3.5Mo 1Mn 0.03C	700		Nitrogen strengthened
ASTM A351	J93380				Cast and forged
ASTM A473	S32760				
W 1.4582	X 4 CrNiMoNb 25 7	25Cr 7Ni 1.6Mo 2Mn 0.06C			
Ferralium	255	25.5Cr 5.5Ni 3Mo 3Cu 2Mn 0.08C	530		Nitrogen strengthened
					Cast
ASTM A182	F XM-19	22Cr 12.5Ni 5Mn 2.2Mo 1Si 0.06C	690	565	Nitrogen strengthened
	S20910				Forged
ASTM A351	CG-6MMN	22Cr 12.5Ni 5Mn 2.2Mo 1Si 0.06C	515	510	Nitrogen strengthened
	J93790				Cast
Carpenter	20Cb-3	20Cr 3.5Cu 2.5Mo 2Mn 1Si 34Ni	600	≈ 875	
ASTM B462					Forged
ASTM A351	CN-7M				Cast
Hastelloy	C	17Mo 16.5Cr 6Fe 4.5W 2.5Co 52Ni	495	≈ 427	Cast
ASTM A494	CW-12M-1				
Monel 400		63Ni 31Cu 1.75Fe	≈ 450	427	Cast
ASTM A494	M30C				
Titanium	Ti0.2Pd	0.2%Pd	385	250/300	Cast
Titanium	Ti-6Al-4V	6Al 4V	900/1300	300	Forged

Table 13.17 Higher corrosion resistance materials

ufacturers' specifications must be reviewed carefully for limitations of operation.

13.4.1 Cast iron and carbon steels

Table 13.3 shows the basic material specifications for grey cast iron used for valves. Table 13.4 shows the grades of ductile iron and Table 13.5 shows the malleable cast irons in popular use. Ductile iron is also called SG iron, which is an abbreviation of spheroidal graphite iron, and nodular cast iron. Ni-Resist cast iron, austenitic spheroidal graphite cast iron, can be useful for high temperature and corrosive applications because of its high nickel content. Wear and abrasion resistance are also good. Type D-2M was developed for cryogenic applications and has excellent anti-galling properties and should be considered for bushings, see Table 13.6.

Figure 13.2 indicates the performance of ISO 7005 cast iron flanges utilising ISO 185 material. Flanges for PN25 and PN40 must use the higher Grade 250 material. Figure 13.3 shows the ratings for BS 10 cast iron flanges. This Standard is widely used for compressed air, gas, water, hot water and steam applications. Figure 13.4 shows the performance of ISO flanges made from ductile cast iron to ISO 2531 Gr 400-5 and ISO 1083 Gr 500-7. Figure 13.5 shows the performance of ISO flanges when cast in malleable cast iron to ISO 5922.

The rating for the most popular steel materials, castings in ASTM A216 WCB and forgings to ASTM A105, are shown in Table 13.6. Also in this category is ASTM A350 LF2, the low temperature steel.

Carbon steels are available in a wide range of chemical compositions which allows selection for specific mechanical properties. Table 13.7 indicates the materials which meet the minimum requirements of the standards. High tensile steel, AISI 4140, is popular for high pressure applications, see Table 13.8. Also in the ANSI 1.9 group is ASTM A271 WC6 which is very similar to BS 1504 621.

Figure 13.7 shows the performance of ISO 7005 steel flanges from cast, forged and plate materials, based on the minimum standard materials Group 1E0. This group includes ISO 3755 200-400 W, ISO 2604-1 F8 and ISO 630 Fe360 B.

Figure 13.8 shows the ratings of BS 10 flanges in carbon steel. Figure 13.9 shows the improved ratings of BS 10 flanges when using alloy steel. Figure 13.10 shows the ratings for high tensile steel forgings, ASTM A182 F12, which is very similar to BS 970 En19 now 709M40. This material is also similar to the forged/cast materials BS 1503-620 and BS 1504-621. The ANSI group includes ASTM A182 F11 and ASTM A217 WC6.

13.4.2 Alloy steels

The low alloy steels include alloys with small additions of chrome and nickel up to the 11/13Cr steels with 4% nickel. The addition of these elements improves the high temperature performance and imparts some corrosion resistance. The addition of chrome and nickel improves the thermal stability of steel and makes these steels popular for applications which suffer wide temperature ranges without getting too hot. Tables 13.9, 13.10 and 13.11 list the popular alloys used.

Alloy steel with 11/13% chrome is popular for many applications such as boiler feed water, see Table 13.12. This group of alloys is not listed in ANSI B16.34, however, it is listed in ASME VIII and BS 5500. Figure 13.11, shows the capabilities of the ANSI group which includes ASTM A217 WC9 which has 2.25% chrome. The ISO equivalent is shown in Figure 13.12, these materials only apply to PN25 and PN40 flanges. Figure 13.13 indicates the performance of the group which includes the 5% chrome steels such as ASTM A217 C5. Figure 13.14 shows the performance of the ANSI group which includes the 9% chrome steels. Figure 13.15 shows the relative performance of 11/13Cr and 12Cr 4Ni cast steels compared to cast 9Cr steels according to BS 5500.

Steels with 5%, 8% and 9% nickel have been developed for low temperature applications down to -196°C. These steels may not be readily available as castings, but forgings to ASTM A522 can be obtained. Such materials may be requested for valves in systems using wrought versions for flanges and pipes (which have standardised their use for pipework and flanges). Nickel steels, BS 1503 503 with 3.5% nickel and BS 1503 509 with 9% nickel, are suitable for operation down to -100°C and -196°C respectively.

13.4.3 Stainless steels

NOTE: There are slight differences in the chemical composition between cast and wrought grades of stainless steel. In some corrosive environments the difference in alloying element proportions may have a significant effect.

Stainless steels are generally accepted as the AISI 300 series. AISI 302 is usually considered as the lowest grade of stainless but is not used significantly for pressure containment. AISI 304 is the most popular of the lower grades, see Table 13.14, Figures 13.16 and 13.17. 304 stainless steel can be used for many applications which do not require the extra capabilities or expense of 316. The most popular grade of stainless steel, overall, is 316, see Table 13.14, Figures 13.18 and 13.19. If welding is required, and post weld heat treatment is difficult or costly then the low carbon grades, 304L and 316L, may be useful if the lower strength and limited temperature range is acceptable, see Figure 13.20.

AISI 304 is an excellent material for cryogenic applications. Standard requirements vary, but 304 is generally acceptable for temperatures down to -196°C to -250°C. The impact strength at -250°C should be at least 20 J. AISI 316 is not so popular for low temperature applications; some codes limit the minimum temperature to -60°C.

13.4.4 Non-ferrous materials

Figure 10.41 indicates the capabilities of ISO 7005 flanges made in copper alloys. The performance shown in the curves in Section 10.6, applies to all the specified alloys, see Tables 13.14 for cast and 13.15 for forged materials, although ISO 426-1, $CuZn_{20}Al_2$ is restricted to 200°C maximum. Table 13.16 shows other copper alloys which are popular valve materials.

Figure 13.22 shows the extension of performance possible when only the $CuAl_{10}Fe_5Ni_5$ and $CuNi_{30}Mn_1Fe$ alloys are used.

13.4.5 Higher corrosion resistance

Higher corrosion resistance can be achieved by increasing the proportions of chrome and nickel. Copper can be an important alloying element. Corrosion resistance can be improved by selecting alternative materials, see Table13.17. Figure 13.23 shows the pressure/temperature rating of the ANSI group which includes Alloy 20 and Carpenter 20™. Excellent resistance to stress corrosion cracking and wide ranging corrosion resistance enables these materials to be used with hot dilute sulphuric acid.

Figure 13.24 indicates the ratings for Monel 400™. This material can be used for seawater, chlorine, hydrofluoric acid, non-oxidising dilute acids and alkalis. It is suitable for the dry gases of chlorine and hydrogen chloride.

Figure 13.25 indicates the performance of Hastelloy C™, Inconel 625™ and Incoloy 825. These materials exhibit excellent high temperature corrosion resistance. Hastelloy C has excellent resistance to pitting, crevice corrosion and stress corrosion cracking. It has very good resistance to hot acids, seawater and brines and is used for wet chlorine, hypochlorites and chlorine dioxide solutions. Inconel 625™ has very good resistance to pitting, crevice corrosion, stress corrosion cracking, corrosion fatigue and erosion corrosion. It also has good performance in high temperature sulphur or chloride compounds. Incoloy 825 has excellent resistance to stress corrosion cracking and good resistance to pitting and crevice corrosion. Applications include sulphuric, phosphoric and nitric acid. Incoloy 825 is a very useful material for high temperature steam applications and wet flue gases.

Titanium has excellent corrosion resistance. Cast titanium valve bodies used to be difficult to source because of the limited foundry capacity within Europe but now castings over 1000 kg are available. Titanium and its alloys are extremely resistant to aqueous chlorine solutions, brine from desalination, chromic acid, hydrochloric acid, nitric acid and seawater.

13.4.6 Non-metallic materials

The range of non-metallic materials for valves is not as comprehensive as the metallic. Non-metallic materials are generally much weaker and the pressure capabilities reduce dramatically as size increases. Operational temperatures are also quite restricted. Table 13.18 lists the most popular materials used for production valves.

Material	Size	Pressure barg	Temperature °C
ABS	DN10 to DN100	15	20
ABS	DN150	12	20
ABS	DN200	9	20
PVC	DN3	3.8	-30 to 50
PVC	DN10 to DN25	15	65
PVC	DN32 to DN100	15	20
PVC	DN150	12	20
PVC	DN200	9	20
CPVC	DN10 to DN25	15	93
uPVC	DN10 to DN150	12	20
PVDF	DN20 to DN65	16	60
PVDF	DN20 to DN65	7.5	80
PVDF	DN20 to DN65	6.0	100

Material	Size	Pressure barg	Temperature °C
PVDF	DN20 to DN65	4.5	120
PVDF	DN20 to DN65	3.5	140
PP	DN15 to DN20 DN32 to DN50	15 12	20 20
PTFE	DN3 to DN8	5	-10 to 90
PTFE	DN8 to DN20	6.9	24 to 150
Kynar	DN15 to DN50	16	20
Kynar	DN15 to DN50	2	138
ETFE	DN3	2	-30 to 70
PEEK	DN3	2.5	0 to 52
Delrin	DN8 to DN40	16	130

Table 13.18 Non-metallic materials for valve bodies

13.5 Trim materials

When using internal materials, always keep within the temperature limitations specified and make sure that the material combinations are suitable. Because the material for the valve trim is often selected with regard to its resistance to abrasion or cavitation, it is essential that the material combination is compatible, i.e. they resist wear and galling in combination.

The process fluid may, or may not, possess lubricating qualities which can assist with both wear and galling. Lube oil or ethylene glycol will assist some combinations whereas steam or paraffin will have little effect. 304 and 316 stainless steel can be surface treated to increase the hardness to about 1100 Hv. The treatment does not impair the corrosion resistance.

The ability to improve the hardness dramatically means both materials can be used as trim and self-mated without the associated risks of galling. Table 13.19 gives a list of material combinations and their suitability.

	304 SST	316 SST	Bronze	Inconel	Monel	Hastelloy B	Hastelloy C	Titanium 75A	Nickel	Alloy 20	Type 416 Hard	Type 440 Hard	17-4PH	Alloy 6 (co-cr)	ENC*	Cr Plate	Al Bronze
304 SST [(1)]	P	P	F	P	P	P	F	P	P	P	F	F	F	F	F	F	F
316 SST [(1)]	P	P	F	P	P	P	F	P	P	P	F	F	F	F	F	F	F
Bronze	F	F	S	S	S	S	S	S	S	S	F	F	F	F	F	F	F
Inconel	P	P	S	P	P	P	F	P	F	F	F	F	F	F	F	F	S
Monel	P	P	S	P	P	P	F	F	F	F	F	F	F	S	F	F	S
Hastelloy B	P	P	S	P	P	P	F	F	S	F	F	F	F	S	F	S	S
Hastelloy C	F	F	S	F	F	F	F	F	F	F	F	F	F	S	F	S	S
Titanium 75A	P	P	S	P	F	F	F	P	F	F	F	F	F	S	F	F	S
Nickel	P	P	S	F	F	S	F	F	P	P	F	F	F	S	F	F	S
Alloy 20	P	P	S	F	F	F	F	F	P	P	F	F	F	S	F	F	S
Type 416 Hard	F	F	F	F	F	F	F	F	F	F	F	F	F	S	S	S	S
Type 440 Hard	F	F	F	F	F	F	F	F	F	F	S	F	S	S	S	S	S
17-4 PH	F	F	F	F	F	F	F	F	F	F	F	S	P	S	S	S	S
Alloy 6 (co-cr)	F	F	F	F	S	S	S	S	S	S	S	S	S	F	S	S	S
ENC [(2)]	F	F	F	F	F	F	F	F	F	F	S	S	S	S	P	S	S
Cr Plate	F	F	F	F	F	S	S	F	F	F	S	S	S	S	S	P	S
Al Bronze	F	F	F	S	S	S	S	S	S	S	S	S	S	S	S	S	P

[(1)] Soft or slightly work-hardened [(2)] Electroless nickel coating
S = Satisfactory F = Fair P = Poor

Table 13.19 Wearing and galling resistance of material combinations

Soft, non-metallic materials are used for valve seats or seat inserts when it is necessary to have a completely leak-free valve. Choice of the correct material requires detailed knowledge of the process fluid and conditions.

Table 13.20 reviews popular materials and operating conditions. When selecting materials it is always advisable to base decisions on previous good results and from the experience of the valve manufacturer.

Material	Pressure barg	Temperature °C
Buna "N"	517	-23 to 121
EP	414	-23 to 150
EP	172	-54 to 163
KALREZ	34	27 to 93
KALREZ	69	32 to 260
Neoprene	414	-23 to 149
Nylon 612		-196 to 180
PCTFE	414	-54 to 204
PEEK	50	-18 to 267
PTFE (virgin)	69	-254 to 204
PTFE (glass filled)	69	-100 to 210
VESPEL	690	-240 to 204
Viton	414	-23 to 204
Viton "E"	517	-54 to 204

Table 13.20 Non-metallic materials for valve seats

Diaphragms can be used to seal the packing box or as an active sealing member of the valve. Diaphragms are usually not-metallic. Table 13.21 lists the usual choices.

Diaphragm material	Operating temperature range °C
Natural rubber	-50/-40 to 90/100
Butyl rubber	-40/-30 to 90/100
EPDM	150
PTFE	175
PTFE/EP rubber	-30 to 150
PTFE/Butyl rubber	-20 to 150
Hypalon (CSM)	-10/0 to 90/100
PTFE/Hypalon	-10 to 120
Fluorocarbon rubber	-5/+5 to 140/150
PTFE/Fluorocarbon rubber	-5 to 175

Table 13.21 Non-metallic materials for diaphragms

A bellows, used for packing box sealing, can be difficult to categorise. It is essentially part of the pressure boundary and could be considered with the valve body. But the construction is usually very light and made of different grades of materials to those of castings and forgings for bodies. A bellows is more easily associated with the trim. Table 13.22 lists the popular material choices.

Springs can play a small, but very important, part in valve operation. Materials range from high carbon steel music wire to high nickel alloys such as Inconel 625™. Table 13.23 lists possible choices.

13.6 Hard facing materials

Metallic hard facing materials are very useful for changing the surface properties either locally or over a complete component. Zones, subject to high velocity fluid causing erosion or subject to cavitation damage, can have exotic alloys placed strategically for protection. One valve manufacturer suggests hard facing trim components for continuous throttling of 10 bar and over. The cost of valve manufacture can be reduced significantly by locating exotic alloys strategically rather than producing complete components.

Metallic alloys can be deposited by various methods:

- welding
- flame spraying using a combustion gas
- electric arc heating with compressed air spray
- plasma spraying
- high velocity oxygen/fuel spraying
- laser beam

Welding and flame spraying are the most used methods. Some alloys can only be deposited by a specific method and the designer has no alternative. Adhesion of the coating to the sub-

13.7 Pressure/temperature ratings for flanges

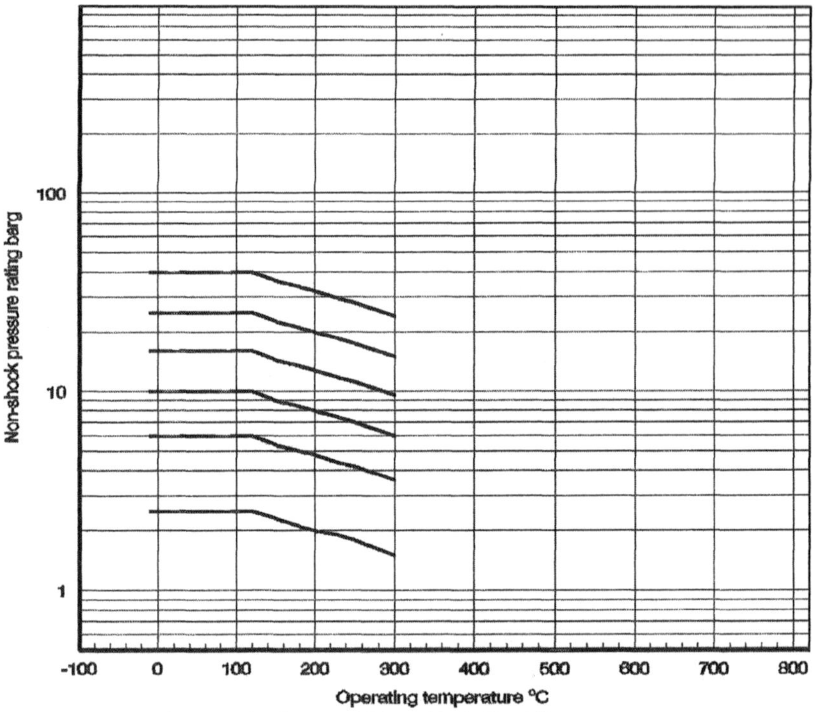

Figure 13.2 Pressure/temperature ratings for ISO 7005 cast iron flanges — PN2.5, 6, 10, 16, 25 and 40

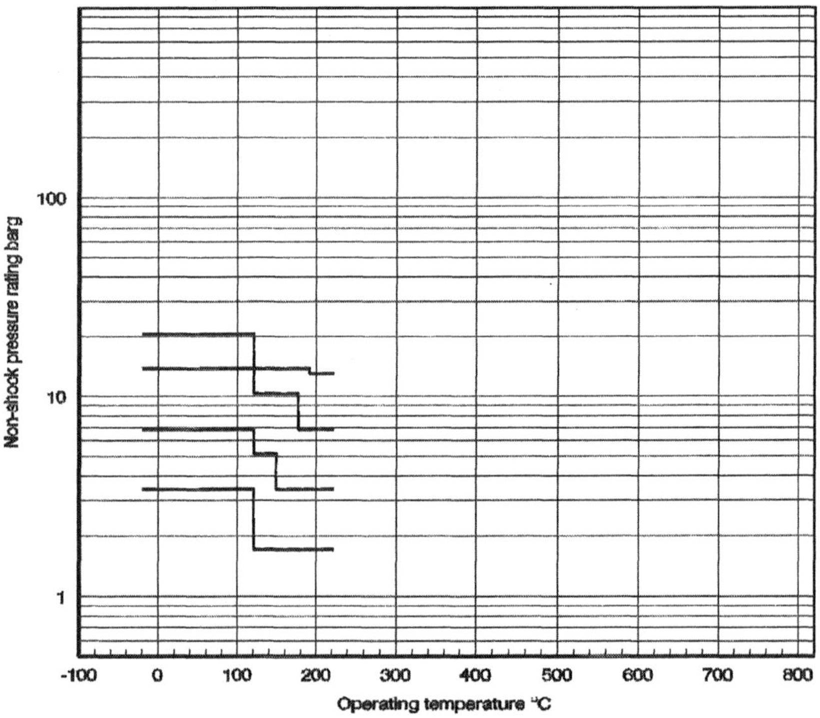

Figure 13.3 Pressure/temperature ratings for BS 10 cast iron flanges — Table A, D, E, F and H

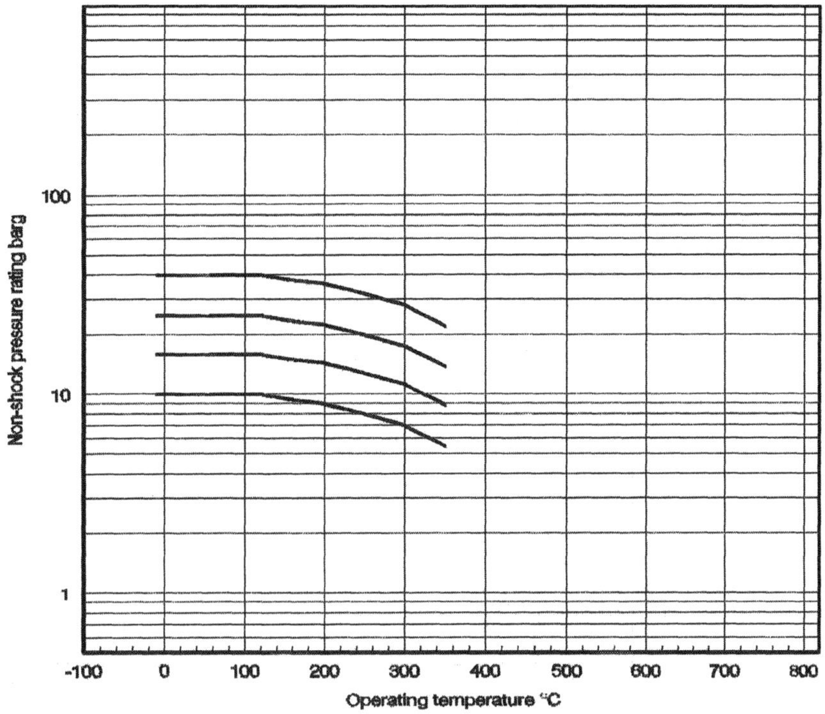

Figure 13.4 Pressure/temperature ratings for ISO 7005 ductile iron flanges — PN10, 16, 25 and 40

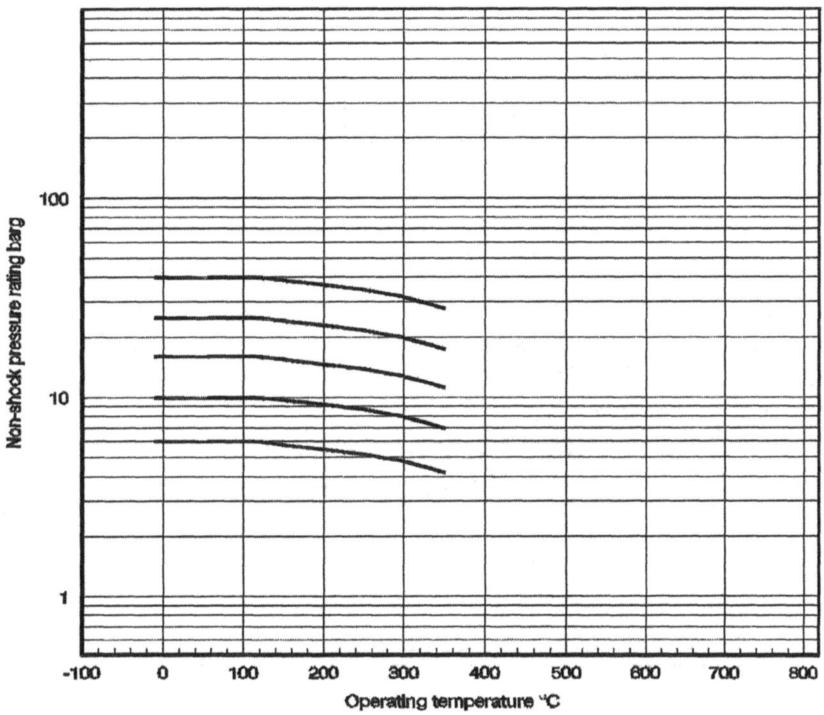

Figure 13.5 Pressure/temperature ratings for ISO 7005 malleable iron flanges — PN6, 10, 16, 25 and 40

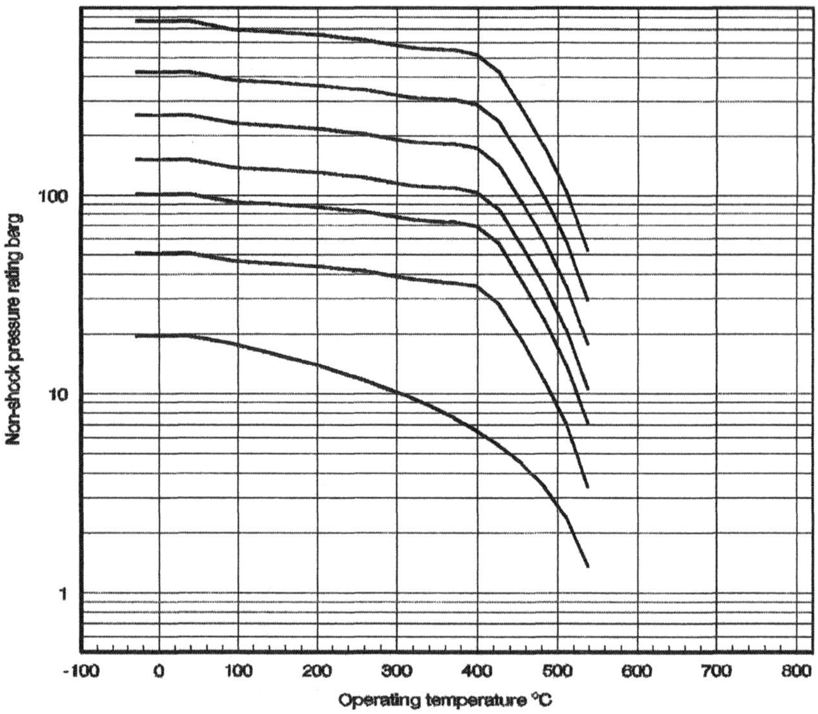

Figure 13.6 Pressure/temperature ratings for ANSI carbon steel flanges — Group 1, 1A, 150lb, 300lb, 600lb, 900lb, 1500lb, 2500lb and 4500lb

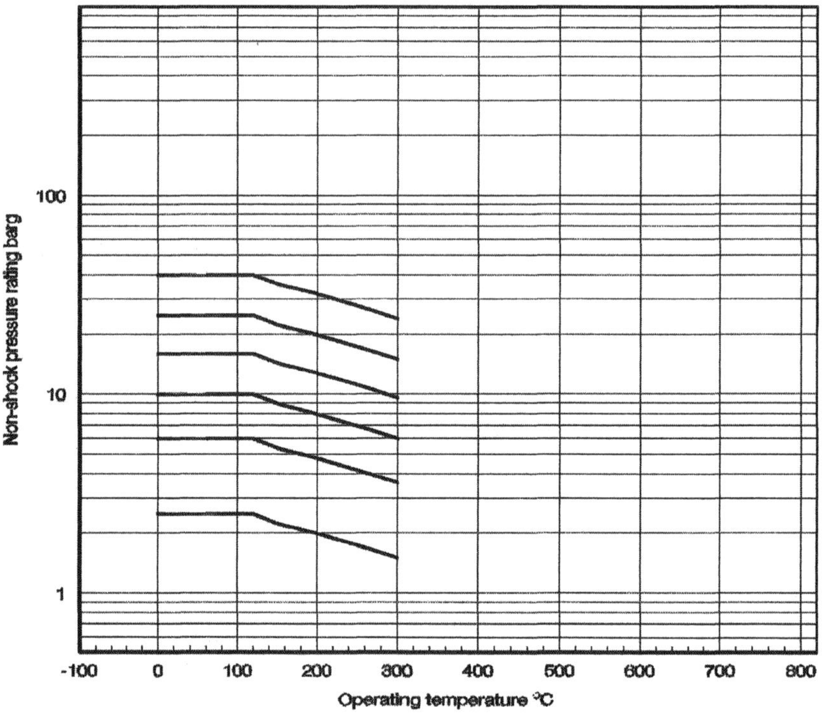

Figure 13.7 Pressure/temperature ratings for ISO 7005 carbon steel flanges — Group 1E0, PN2.5, 6, 10, 16, 25 and 40

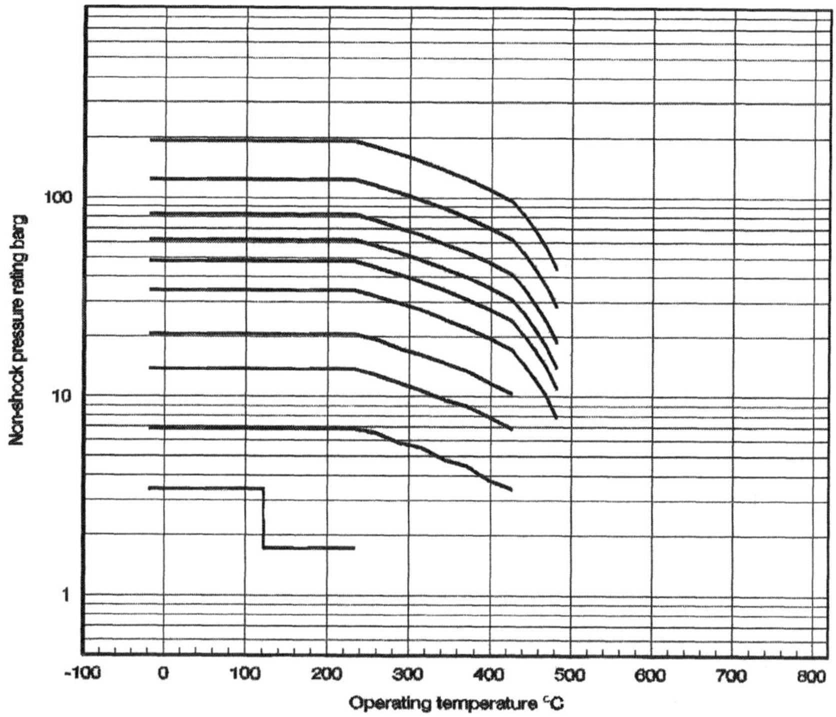

Figure 13.8 Pressure/temperature ratings for BS 10 carbon steel flanges — Table A, D, E, F, H, J, K, R, S and T

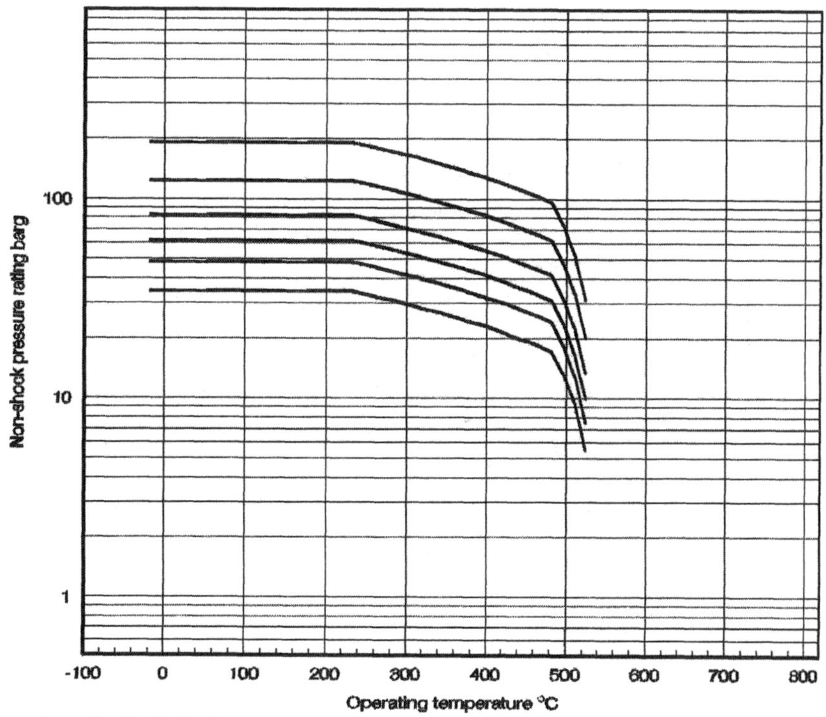

Figure 13.9 Pressure/temperature ratings for BS 10 alloy steel flanges — Table H, J, K, R, S and T

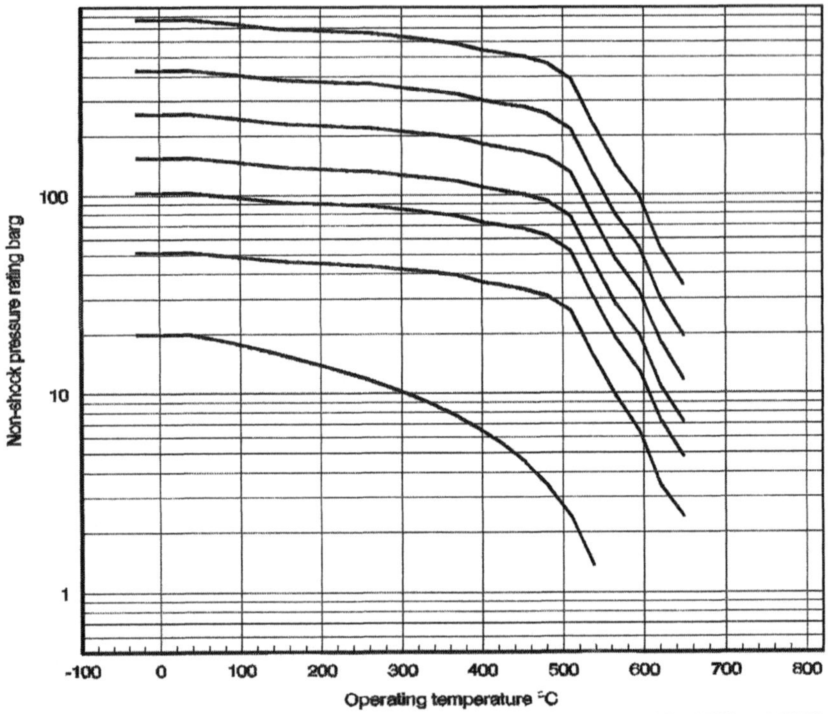

Figure 13.10 Pressure/temperature ratings for ANSI alloy steel flanges — Group 1.9a, 150lb, 300lb, 600lb, 900lb, 1500lb, 2500lb and 4500lb

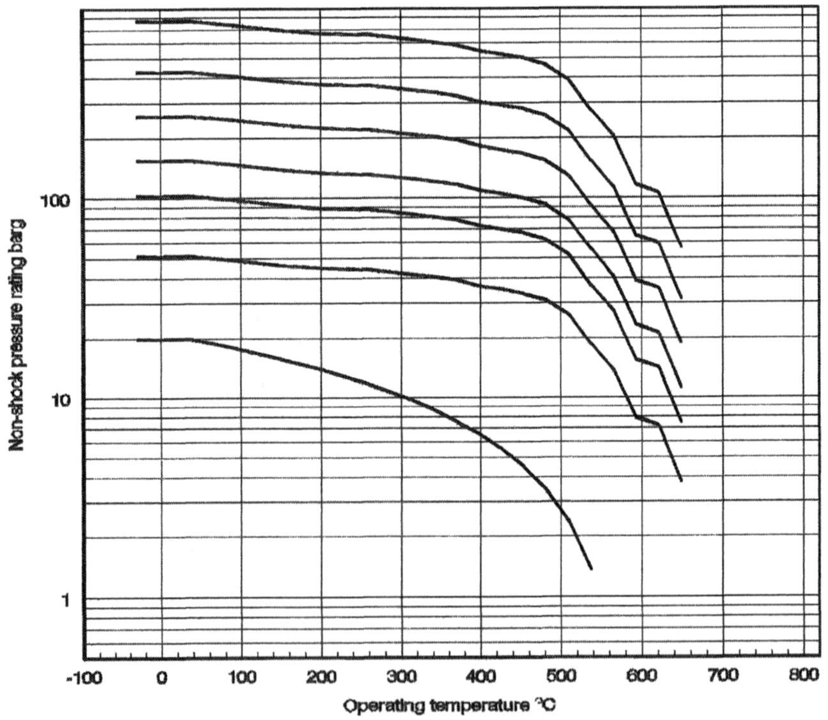

Figure 13.11 Pressure/temperature ratings for ANSI alloy steel flanges — Group 1.10A, 150lb, 300lb, 600lb, 900lb, 1500lb, 2500lb and 4500lb

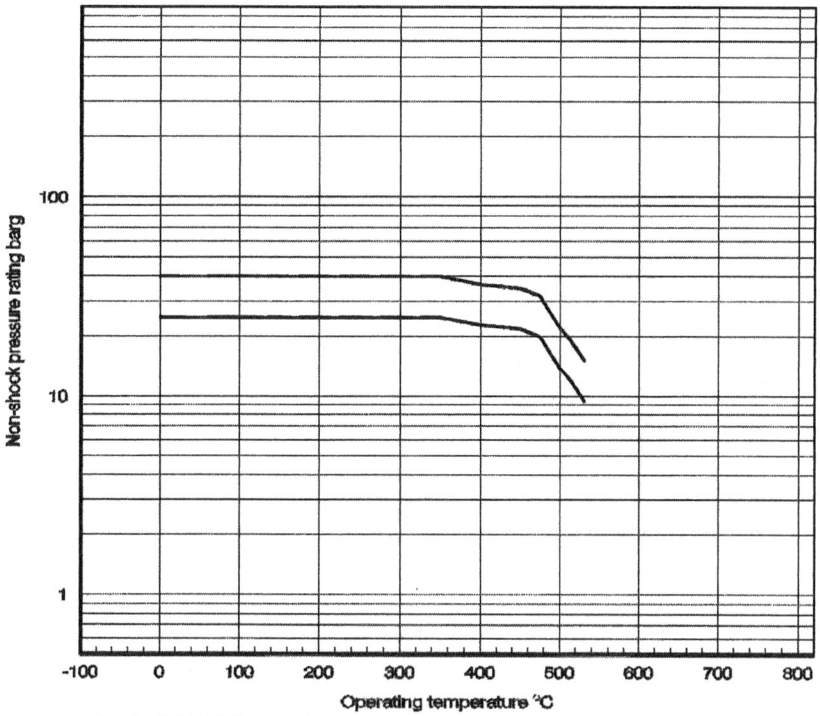

Figure 13.12 Pressure/temperature ratings for ISO 7005 alloy steel flanges — Group 6E0, PN25 and 40

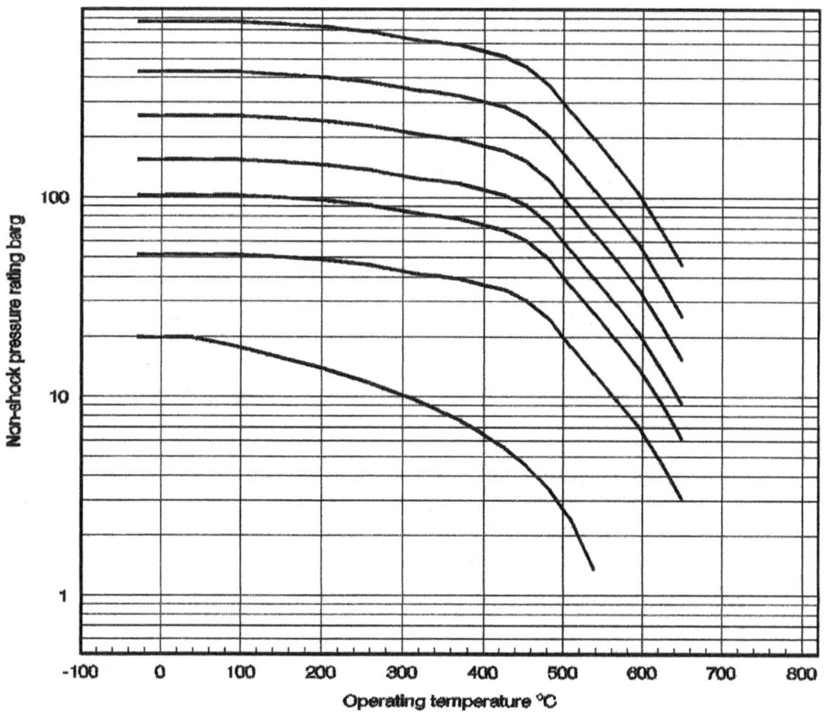

Figure 13.13 Pressure/temperature ratings for ANSI alloy steel flanges — Group 1.13A, 150lb, 300lb, 600lb, 900lb, 1500lb, 2500lb and 4500lb

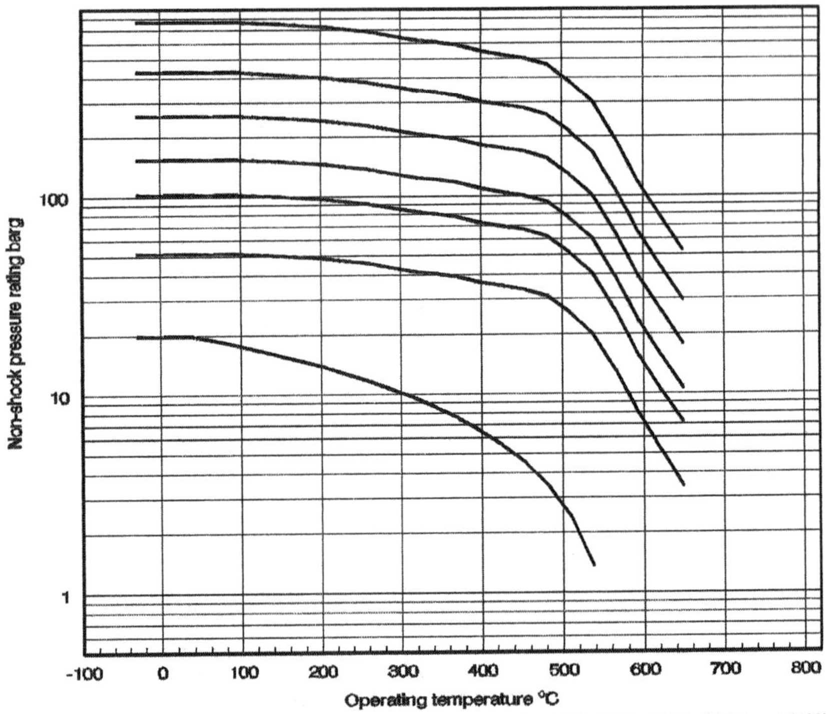

Figure 13.14 Pressure/temperature ratings for ANSI alloy steel flanges — Group 1.14A, 150lb, 300lb, 600lb, 900lb, 1500lb, 2500lb and 4500lb

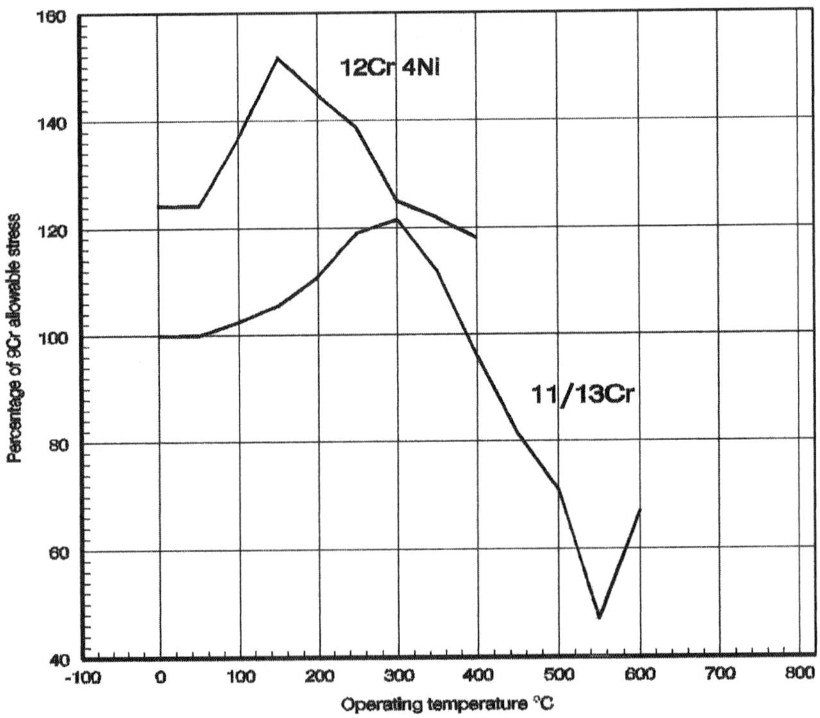

Figure 13.15 Comparative stress/temperature properties of 11/13 Cr and 12Cr 4Ni according to BS 5500

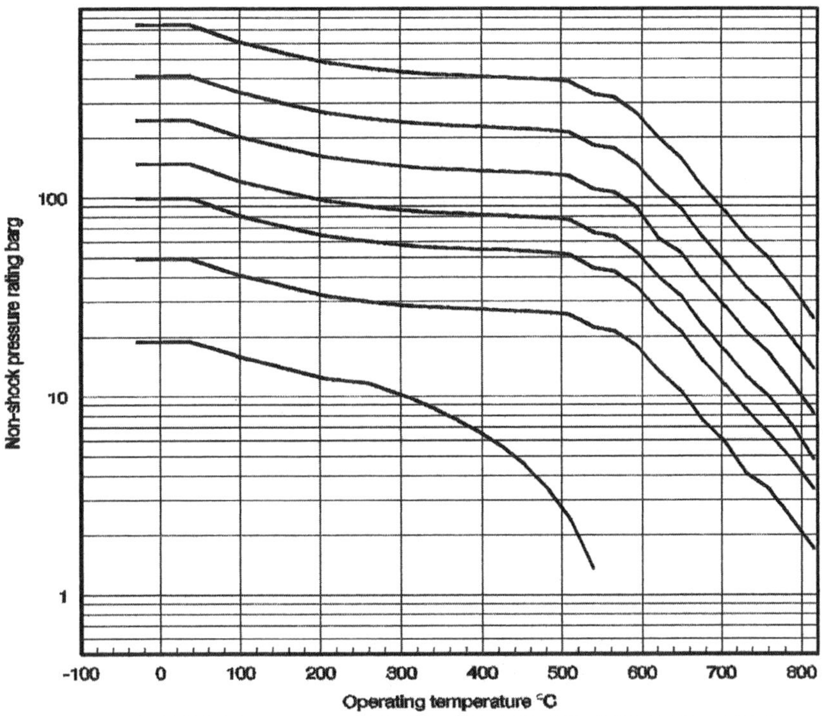

Figure 13.16 Pressure/temperature ratings for ANSI 304 flanges — Group 2.1A, 150lb, 300lb, 600lb, 900lb, 1500lb, 2500lb and 4500lb

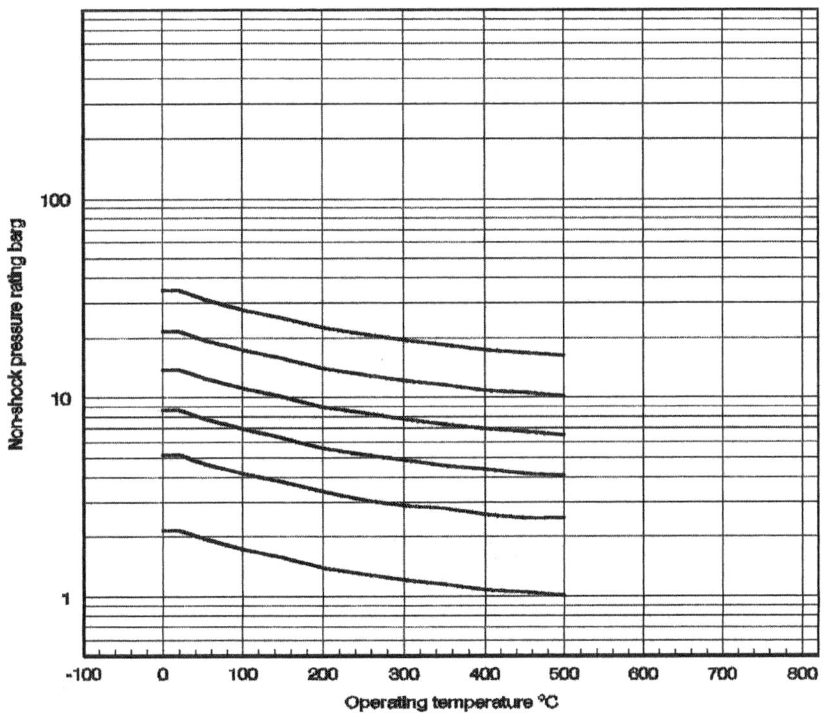

Figure 13.17 Pressure/temperature ratings for ISO 7005 flanges in 304 stainless steel — Group 11E0, PN2.5, 6, 10, 16, 25 AND 40

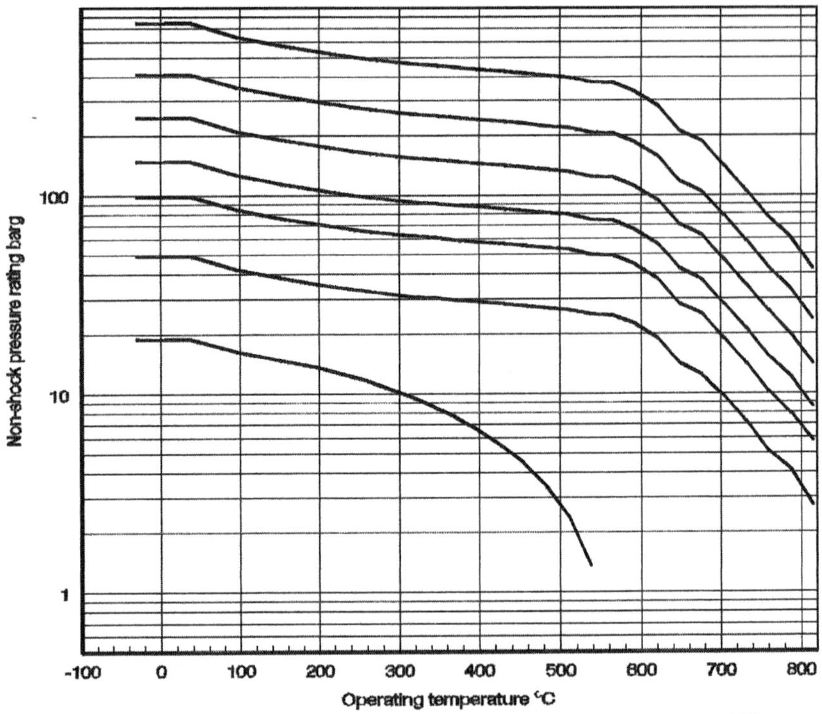

Figure 13.18 Pressure/temperature ratings for ANSI 316 flanges — Group 2.2A, 150lb, 300lb, 600lb, 900, 1500lb, 2500lb and 4500lb

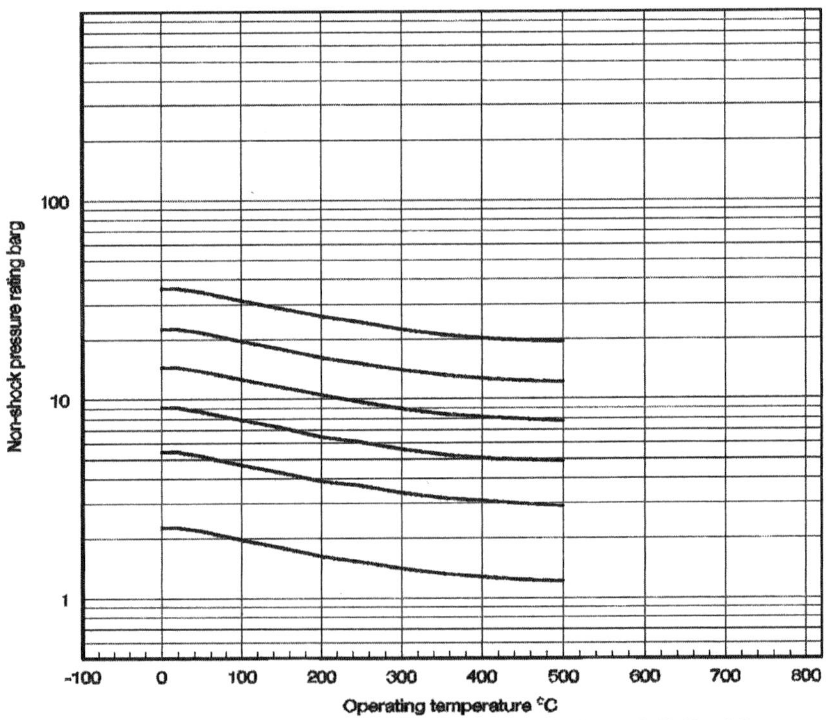

Figure 13.19 Pressure/temperature ratings for ISO 7005 flanges in 316 stainless steel — Group 14E0, PN2.5, 6, 10, 16, 25 and 40

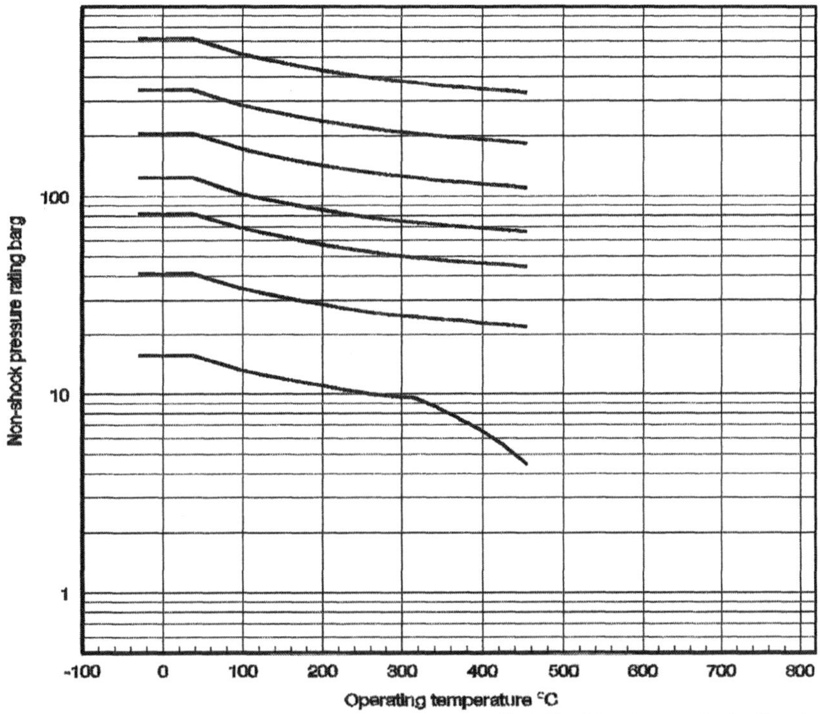

Figure 13.20 Pressure/temperature ratings for 304L and 316L ANSI flanges — Group 2.3A, 150lb, 300lb, 600lb, 900lb, 1500lb, 2500lb and 4500lb

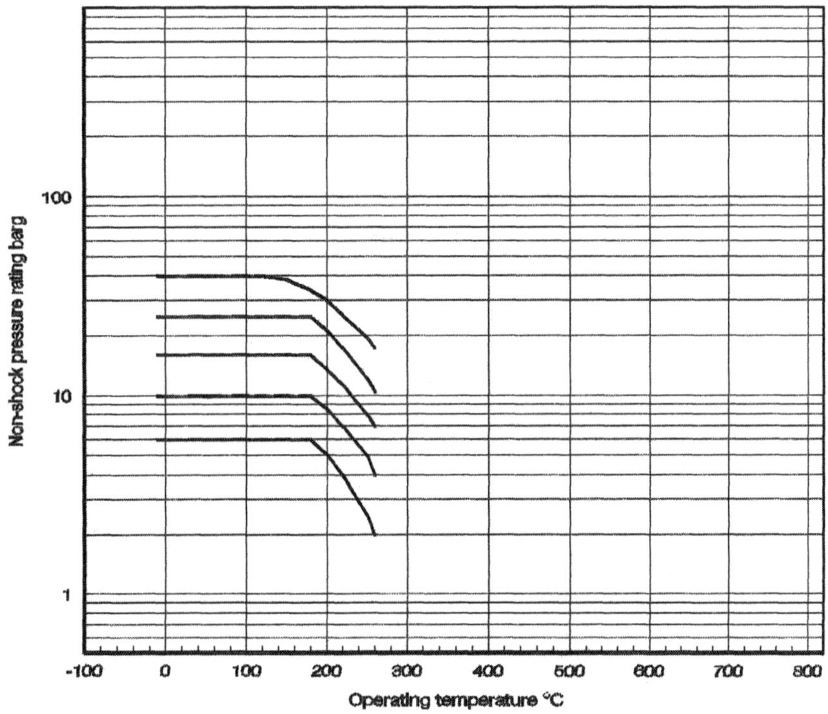

Figure 13.21 Pressure/temperature ratings of ISO 7005 flanges in copper alloys — PN6, 10, 16, 25 and 40

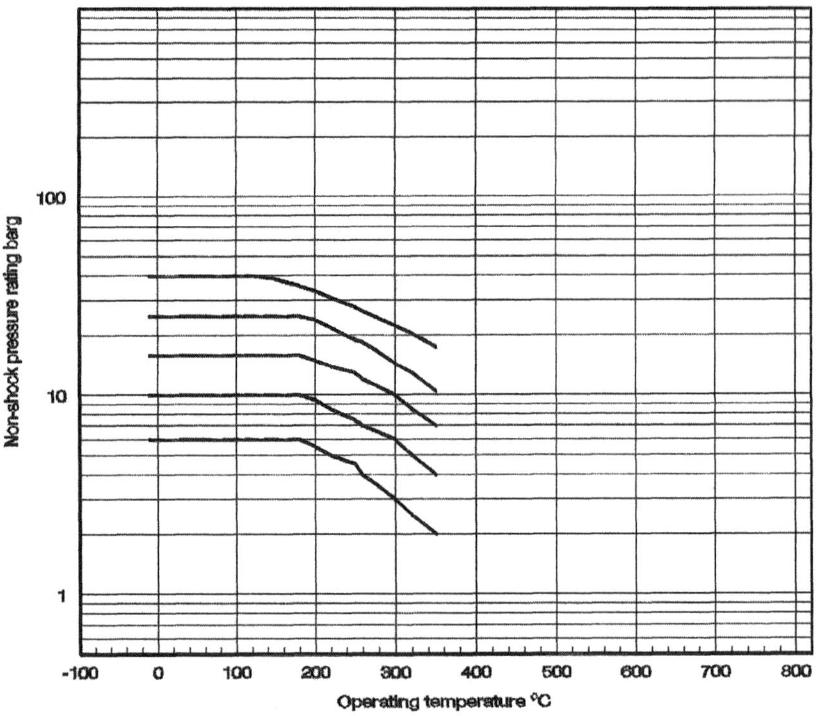

Figure 13.22 Pressure/temperature ratings of ISO 7005 flanges with selected copper alloys — PN6, 10, 16, 25 and 40

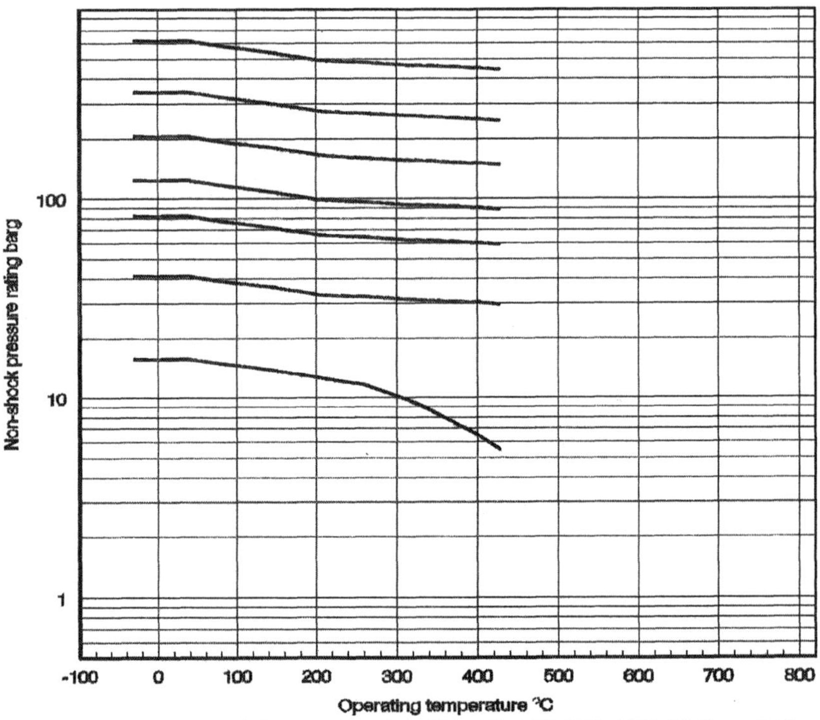

Figure 13.23 Pressure/temperature ratings for Alloy 20 ANSI flanges — Group 3.1A, 150lb, 300b, 600lb, 900lb, 1500lb, 2500lb and 4500lb

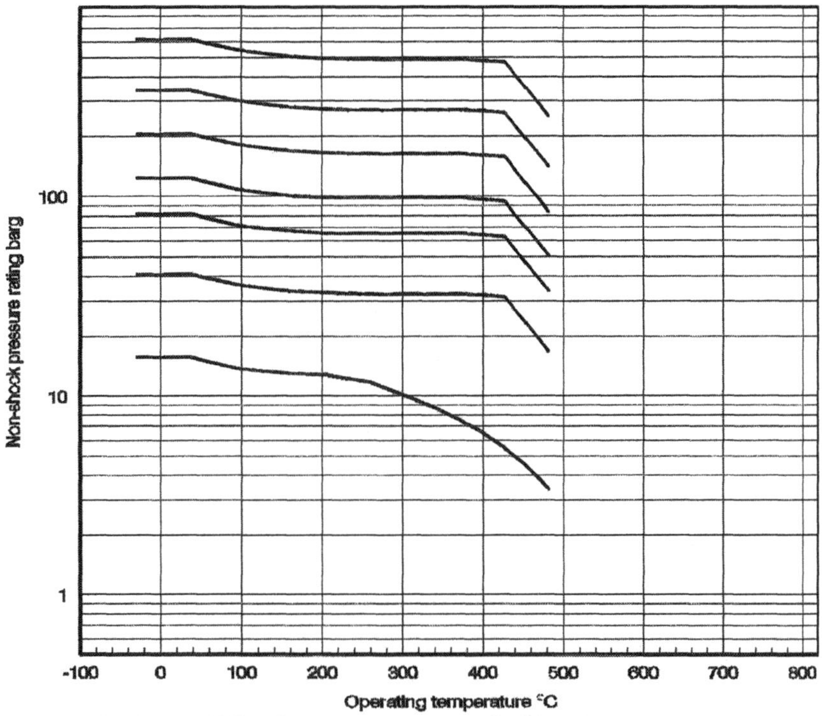

Figure 13.24 Pressure/temperature ratings for Monel 400™ ANSI flanges — Group 3.4A, 150lb, 300lb, 600lb, 900lb, 1500lb, 2500lb and 4500lb

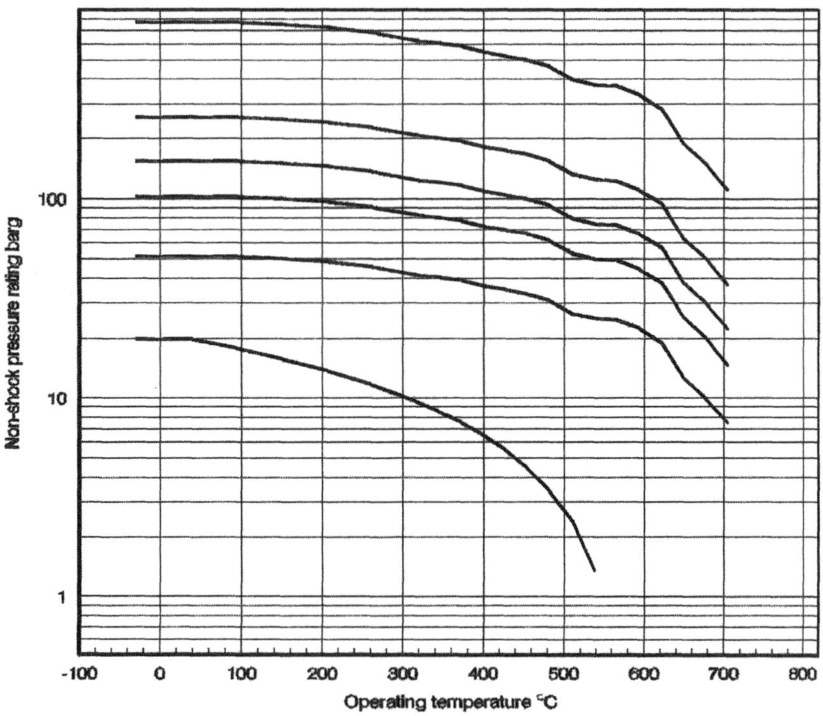

Figure 13.25 Pressure/temperature ratings for Hasteloy C™ ANSI flanges — Group 3.8A, 150lb, 300lb, 600lb, 900lb, 1500lb, 2500lb and 4500lb

Standard	Grade	Major constituents	Min tensile strength MPa	Max temp °C	Remarks
BS 2870 ISO 426/1	CZ103 CuZn20	80Cu 20Zn	265/400	150/200	80/20 brass up to '60 barg
BS 2870 ISO 427 CDA UNS	PB102 CuSn4 510 C51000	95Cu 5Sn 0.2P	310/645	200/250	phosphor bronze up to '70 barg
BS 1449 ASTM A240 DIN 17440 AFNOR A36-209	321S12 321 X10 CrNiTi 18 9 Z6 CNT 18-11	18Cr 10.5Ni 0.08C	510	≈ 500	Standard material, fair corrosion resistance, good mechanical properties
BS 1449 ASTM A240 DIN 17440 AFNOR A36-209	304S15 304 X5 CrNi 18 9 Z6 CN 18-09	18Cr 9.5Ni 0.06C	510	≈ 400	Improved corrosion resistance
BS 1449 ASTM A240 DIN 17440 AFNOR A36-209	304S12 304L X2 CrNi 18 9 Z2 CN 18-10	18Cr 10.5Ni 0.03C	490	≈ 400	Better for welding
BS 1449 ASTM A240 DIN 17440	316S16 316 X5 CrNiMo 18 10	17.5Cr 11.5Ni 2.5Mo 0.07C	540	≈ 400	Better corrosion resistance especially for pitting
BS 1449 ASTM A240 DIN 17440 AFNOR A36-209	316S16 316L X2 CrNiMo 18 10 Z2 CND 17-12	17Cr 12.5Ni 2.5Mo 0.03C	490	450/650	Better for welding
BS 1449 ASTM A240 DIN 17440	347S17 347 X10 CrNiNb 18 9	18Cr 10.5Ni 0.08C	510	≈ 500	Subject to availability of suitable gauge strip
BS 1449 ASTM A167 AFNOR DIN	310S24 310 Z12 CNS25.20 X15 CrNiSi 25 20	24.5Cr 20.5Ni 0.15C	540	≈ 1 000	Subject to availability of suitable gauge strip Good high temperature corrosion resistance
Incoloy 825 UNS	 N08825	42Ni 21.5Cr 3Mo 2Cu	580/720		Excellent resistance to chloride and caustic stress corrosion. Wet flue gas, hp steam, contaminated seawater, acids
Incoloy 800 BS 3072 ASTM B409 DIN	 NA15 X10 NiCrAlTi 3220	32Ni 21Cr 1.5Mn	510/690	≈ 815	High temperature corrosive applications
Inconel 600 BS 3072 ASTM B168 DIN 17750 AFNOR	 NA14 N 06600 NiCr15Fe NC 15 Fe	72Ni 15.5Cr 8Fe	550/690	≈ 450	High temperature strength and oxidation resistance
Monel 400 BS 3072 ASTM B127 DIN 17750 AFNOR	 NA13 N 04400 NiCu30Fe NU 30	63Ni 31Cu 1.75Fe	≈ 450	427	Subject to availability of suitable gauge strip Excellent for chlorine
Nimonic 75 DIN	 NICr20Ti	75Ni 19.5Cr 5Fe	800	≈ 1000	High temperature applications
Nimonic 80A		71Ni 19.5Cr 5fe 2.2Ti 2Co	1190	≈ 1000	High temperature applications
Hastelloy B ASTM B333	 N10001	28Mo 5fe 1.2Co 0.5Cr	≈ 950	≈ 760	Subject to availability of suitable gauge strip Excellent corrosion resistance to acids
Hastelloy C UNS	 N10276	17Mo 16.5Cr 5.7Fe 4.3W	≈ 690	≈ 980	Subject to availability of suitable gauge strip Excellent corrosion resistance to chlorides
Inconel 625 ASTM B443		22Cr 9mo 5fe + No + Ta	827/1034	800/980	Good high temperature physical properties and general corrosion resistance Excellent corrosion fatigue properties
Titanium 115		99.8Ti	400	≈ 350	Very corrosion resistant
Tinanium 260		0.2Pd	386/450	≈ 350	Very corrosion resistant
BS 1470 Aluminium	NS5	3.5Mg 0.5Si 0.5Fe	215	200	Subject to availability of suitable gauge strip Good weldability Cryogenic applications to -190 °C

Table 13.22 Popular materials for bellows

Standard	Grade	Major constituents	Min tensile strength MPa	Max temp °C	Remarks
Phosphor bronze BS2873 ASTM B159	 510A	5Sn 0.2P 0.3Zn 95Cu	670/720	80	Min -40 °C
BS 1409	Gr C	1.15Mn 0.35Si 0.7C	1 080/2 470	150	Min -20 °C
BS 2803	Gr 1	0.75Mn 0.3Si 0.65C	1 240/1 780	170	Min -20 °C
DIN 17221	50 Cr V 4	1.05Cr 0.95Mn 0.25Si 0.5C	1 370	200	
Werkstoff	1.4922	11.5Cr 1Mo 0.55Mn 0.4Ni 0.2C	1 170	550	Min -80 °C
Werkstoff BS 2056 17-7PH	1.4310 301S81	17Cr 7.5Ni 2Mn 1Si 0.15C	1 170	250	Min -200 °C
BS2056	302S25	18.5Cr 8.5Ni 2Mn 0.2Si 0.16C	1 700/2 160	300	Min -196 °C
BS 2056	316S16	18.5Cr 10Ni 3Mo 2Mn 0.2Si 0.12C		300	Min -196 °C
Inconel	X-750	70Ni 15.5Cr 7Fe 2.5Ti 1Mn 1Co	1 030	400/650	Min ≈ -185 °C
Monel BS 3075	K-500 NA 18	63Ni 30Cu 2.8Al 2Fe 1.5Mn 0.6Ti	900/1 240	230	Min -70 °C
Nimonic	90	51Ni 19.5Cr 18Co 5Fe 2.5Ti 1.5Si 1Mn		550/600	

Table 13.23 Spring materials

Standard	Grade	Major constituents	Min tensile strength MPa	Max temp °C	Remarks
Stellite	6	67Co 28Cr 4W 1C			46 HR_c
Stellite	12	62Co 30Cr 8W 1.3C			48 HR_c
Stellite	SF6	54Co 19Cr 13Ni 8W 3Si 1.7B 0.7C			For smaller parts
Colmonoy	6	73Ni 13.5Cr 5Fe 4Si 3B	207	≈ 540	56/61 HR_c Wear resistant under corrosive conditions. Excellent non-galling characteristics
Tungsten carbide	LW-5	25WC 5Ni	48	650/760	69 HR_c Good wear resistance in basic and organic acid solutions. Excellent abrasion resistance in sour fluids. Expensive

Table 13.24 Popular metallic hard face coatings

Standard	Grade	Major constituents	Min tensile strength MPa	Max temp °C	Remarks
Hard chrome plate					> 50 HR_c Widest range of substrate Good adhesion to substrate Non-stick, low coefficient of friction, corrosion resistant
Nickel plate		99Ni	460/540	700	152/190 BHN Corrosion resistant to many acids Anti-fretting Lubrication required for wear resistance
Electroless nickel plate		91Ni 9P			Abrasion/corrosion resistant Excellent adhesion to substrate
Electroless nickel composite		Nickel + Fluorocarbon or PTFE		260/300	Min -80°C 60/70 HR_c Hard self-lubricating surface Low coefficient of friction
Silver plate			1 100/1 800		76/95 BHN Bearing surface for bushes Anti-galling/fretting corrosion resistant

Table 13.25 Electrically or chemically applied coatings

strate varies considerably. Some methods/alloys become an integral part of the component and are almost impossible to remove. Others adhere to the surface by different bonding techniques with varying degrees of success. De-lamination or fracture of very hard coatings can damage other components during valve operation. The bonding of the coating must be assessed very carefully. Another problem is coating porosity. Porosity is not necessarily a problem when the coating is used for erosion or cavitation resistance. When the coating is applied for corrosion resistance any porosity will allow the fluid to attack the substrate. Non-destructive testing, usually dye-penetrant, will highlight the problem.

The edge treatment of the coating is very important. It is usually much better to support the coating fully by the substrate. Coating corners can sometimes be difficult; a high level of operator skill may be essential. Flame sprayed coatings are usually applied to a minimum thickness of about 0.75 mm. Coatings such as tungsten carbide may be applied to a thickness of 0.1 mm.

Coatings can be applied to many substrates. The coefficients of linear expansion should be similar to avoid thermal cracking. When applying hot coatings to austenitic stainless steels it is better to use the low carbon versions to escape problems caused by carbide precipitation within the substrate.

Table 13.24 lists the most popular metallic coatings used. There are many more coatings available.

Hard facing metals can be deposited chemically or electrically. These coatings can be applied for erosion/corrosion resistance or to modify the surface characteristics to prevent galling. Coating thickness can be as thin as 5 μm. Table 13.25 lists the most popular metals.

13.8 Non-metallic coatings

Metallic valve bodies can be lined with various non-metallic compounds to provide corrosion resistance and/or ensure hygienic cleanliness. Coatings may be brushed or sprayed on to the body or moulded in situ.

Coating or lining	Operating temperature range °C
Hard rubber (Ebonite)	-10 to 85
Soft rubber	-10 to 85
Polypropylene (PP)	-10 to 95
Neoprene	-10 to 105
Polyvinylidene fluoride (PVDF)	≈ -40 to 110/130
Butyl rubber	-10 to 120
Halar (ECTFE)	-10 to 120
Nylon 11 (can be FDA or Wrc approved)	≈ -50 to 120/140
Polyvinylidene	-10 to 140
Ethylene tetrafluoroethylene (ETFE)	-10 to 150
Borosilicate glass	-10 to 175
Perfluoroalkoxy (PFA)	-10 to 175
PTFE	-45 to 175
Fluorinated ethylene propylene (FEP)	150
UHMWPE	93

Figure 13.26 Popular non-metallic linings and coatings

Table 13.26 lists the popular options. The chemical inertness of PTFE has been imparted to generic compounds, like FEP and PFA, which are suitable for moulding into complex shapes for liners and coatings. Some liners are moulded to shape and then fitted to the valve body; these are replaceable. Other liners are moulded directly into the valve body and retained by plain or dovetail grooves. These liners are not replaceable, but because the attachment is physical, valves of this style can be used for vacuum applications which tend to separate liners.

Coatings have been specifically developed for valves handling dry solids and powders. Compounds very similar in physical properties to PTFE can be sprayed on metal surfaces and then baked. The coated surface has good "non-stick" characteristics and is protected from corrosion.

Ceramic coatings, applied by dipping, have been developed which combine wear, erosion and corrosion resistance. The surface is chemically treated between coats and may be heat treated finally to achieve the required properties.

When considering the use of a coating or lining the following points should be reviewed to decide acceptability:

- Fracture or porosity — the effect of the process fluid on the body parent material
- Permeability — the molecular leakage of the fluid to the parent material
- Delamination or separation — the effect of the coating/lining on the downstream system

Standard	Grade	Major constituents	Min tensile strength MPa	Max temp °C	Remarks
BS 1506 BS 3692	111 8.8	0.65Mn 0.2Si 0.15C	400	300	Min -20°C
BS 4882 BS 4882 BS 1506 ASTM A193 AISI	B7 L7 621A B7 4140	1.05Cr 0.65Mn 0.27Mo 0.2Si 0.4C	790/860	400	Min -100°C
BS 4882 BS 1506 DIN 17440 ASTM A193 AISI	B6 713 X10 Cr13 B6 410	12.5Cr 1Ni 1Mn 1Si 0.12C	590/700	500	Min 0°C
BS 4882 BS 1506 ASTM A193	B16 661 B16	1.05Cr 0.6Mo 0.55Mn 0.35Si 0.32C	850	525	Min 0°C
BS 4882 BS 4882 BS 1506 DIN 17440 ASTM A193 AISI	B8 B8X 801B X5 CrNi 18 9 B8 304	18.7Cr 8.5Ni 0.2Si 0.08C	540	575	Min -250°C
BS 4882 BS 4882 BS 1506 DIN 17440 ASTM A193 AISI	B8T B8TX 821 Ti X10G NiTi 18 9 B8T 321	18.5Cr 7.5Ni 1.2Mn 0.6Si 0.12C	540	575	Min -250°C
BS 4882 BS 4882 BS 1506 DIN 17440 ASTM A193 AISI	B8C B8CX 821 Nb X10 CrNiNb 18 9 B8C 347	18.5Cr 9Ni 1.2Mn 0.6Si 0.08C	540	575	Min -250°C
BS 4882 BS 4882 BS 1506 DIN 17440 ASTM A193 AISI	B8M B8MX 845 X5 CrNiMo 18 10 B8M 316	17Cr 10Ni 2.75Mo 2Mn 0.6Si 0.08C	540	600	Min -250°C
BS 1473	HB 30-TF	98Al 1Si 0.7Mg 0.6Mn	240	200	Cryogenic applications
BS 1473	HB 15-TF	93Al 4.4Cu 0.8Mn 0.7Si 0.5Mg	430	200	Cryogenic applications

Standard	Grade	Major constituents	Min tensile strength MPa	Max temp °C	Remarks
BS 3692 ISO/R 272 ISO/R 733 ISO 885 ISO888	8			300	Min -20°C
BS 4882 BS 1506 ASTM A194	2H 162 2H	0.8Mn 0.25Si 0.6C	700	450	Min 0°C
BS 4882 BS 4882 BS 1506 ASTM A194	4 L4 240 4	0.65Mn 0.4Mo 0.25Si 0.32C	620	520	Min -100°C
BS 4882 BS 1506 ASTM A194 AISI	7 621A 7 4140	1.05Cr 0.65Mn 0.27M 0.2Si 0.4C	700	565	Min -100°C
BS 4882 BS 1506 DIN 17440 ASTM A194 AIS	8M 845 X5CrNiMo 18 10 8M 316	17Cr 10Ni 2.75Mo 2Mn 0.6Si 0.8C	540	600	Min -250°C
BS 4882 BS 3076 ASTM B637 DIN 17742	80A NA 20 N07080 NiCr20TiAl	75Ni 20Cr 2.3 Ti 1.4Al 1Fe 0.065C	1 000	750	Min -250°C

Table 13.28 Standard nut materials

Standard	Grade	Major constituents	Min tensile strength MPa	Max temp °C	Remarks
ASTM A564	Gr 630 S17400	17Cr 4Ni 4Cu	965	≈ 315	Precipitation hardened
BS 4882	B17B			650	Min -250°C
BS 4882 BS 3076 ASTM B637 DIN 17742	B80A NA20 N07080 NiCr20TiAl	75Ni 20Cr 2.3Ti 1.4Al 1Fe 0.065C	1 000	750	Min -250°C
254 SMO	S31254	20Cr 18Ni 6Mo 0.01C	1 200		Nitrogen strengthened
SAF 2507	S32750	25Cr 7Ni 4Mo 0.03C	910		Nitrogen strengthened
Monel	K-500	66Ni 29Cu 3Al 0.2C	1 000		
Inconel	625	62Ni 21.5Cu 9Mo 2.3Fe 0.03C	1 050		
Incoloy	925	40Ni 27Fe 21Cr 3Mo 2Cu 0.02C	1 200		
Marinel		78Cu 15Ni 4.3Mn 1.5Al 1Fe 0.01C	920		
ASTM A320	L7		930		
Cronifer	2419 MoN	24Cr 17Ni 6Mn 4Mo 0.6Cu 0.03C	1 210	≈ 500	Nitrogen strengthened
ASTM A453	Gr 660 K63198	25.5Ni 14.7Cr 2.1Ti 1.2Mo 0.08C	895	≈ 510	(1)
ASTM A453 UNS	Gr 651 K63198	19.5Cr 9.5Ni 1.3Mo 1.3W 0.31C	655/690	≈ 510	(1)
BS 2874 UNS DIN ISO	CA 106 C61400 CuAl8Fe CuAl8Fe3	90Cu 7Al 2.5Fe	485/550	260	Min ≈ -200°C
BS 2874 UNS DIN ISO	CA 104 C63200 CuAl10Ni CuAl10Fe5Ni5	80Cu 9.7Al 5Fe 5Ni	650/850	370	Min ≈ -10°C
HIDURON 191			725	300	Min -196°C

(1) High strength, high temperature bolting for use with austenitic stainless steel components

Figure 13.29 Special bolting materials

- Galling — the effect of valve components rubbing on the coating/lining
- Abrasion and/or erosion resistance — what will be the consequence of particles flowing with the fluid?

Non-metallic materials are as complex, or even more complex, than metallic materials like stainless steel and duplex stainless steels. New compounds are available to valve designers, almost weekly. The valve designer is reliant on chemists and technologist to advise on chemical compatibility and the suitability of compounds for sealing and bonding to substrates. Permeation through non-metallic materials, plus the absorption of fluids, are the two greatest problems to solve. There is huge variation between materials, and even between different versions of the same material. Compound selection is best guided by testing or field experience.

13.9 Bolting materials

The most popular bolting/stud material for pressurised applications is Grade B7 based on AISI 4140 high tensile steel. The popular grades of stainless steel fasteners, AISI 304 and 316, are equivalent in strength to mild steel. If valve bolting is changed from B7 to stainless then an equivalent strength grade should be used if the pressure/temperature rating is to be maintained.

Care should be taken when selecting materials for corrosive atmospheres, coastline or offshore installations for example, to avoid direct corrosion or galvanic corrosion. Tables 13.27 and 13.28 list standard materials regularly used for bolting and nuts on valves.

Table 13.29 lists special materials which can be used for corrosive applications or high temperatures.

For some corrosive atmosphere applications, marine installations for example, standard fastener materials may be coated with a protective layer. Xylan™ on a phosphate primer has been popular for offshore installations.

13.10 Useful references

The Copper Development Association, Verulam Industrial Estate, 224 London Road, St Albans, AL1 1AQ, UK
Tel: +44 01727 731200, Fax: +44 01727 731216

FDA (Food and Drug Administration), 5600 Fishers Lane, Rockville, MD 20857 USA Tel: 888 463 6332, www.fda.gov.

NACE International, 1440 South Creek Drive, Houston, TX 77084-4906 USA, Tel: 281 228 6200, Fax: 281 228 6300, Email: firstservice@nace.org, www.nace.org.

BSSA, Broomgrove, 59 Clarkehouse Road, Sheffield S10 2LE UK, Tel: 0114 267 1260, Fax: 0114 266 1252, Email: enquiry@bssa.org.uk, www.bssa.org.uk.

BSSA's Stainless Steel Advisory Service (SSAS) UK, Tel: 0114 266 1252 , Fax: +44 0114 267 1265, Email: ssas@bssa.org.uk

MR-01-75/ISO 15156-2:2003 Petroleum and natural gas industries - Materials for use in H2S-containing environments in oil and gas production - Part 2: Cracking-resistant carbon and low alloy steels, and the use of cast irons.

BS 10:1962 Specification for flanges and bolting for pipes, valves and fittings.

ANSI/ASME B16.34-2004 Valves Flanged, Threaded and Welding End.

ISO 7005-1:1992 Metallic flanges - Part 1: Steel flanges.

ISO 5922:2005 Malleable cast iron.

BS PD 5500:2006 Specification for unfired fusion welded pressure vessels.

Instrumentation and ancillaries 14

14.1 Instrumentation

14.1.1 Limit switches

In large complex installations it is often difficult for operating personnel to remember which valves are open and which valves are closed. If an isolating valve is on the suction or discharge side of a compressor or pump, other instrumentation may indicate very quickly that a valve is in the wrong position. If a valve on a tank is left open, the first external indication may be a low level alarm, by which time it may be too late to take action. It would be beneficial for operators in the central control room to see, at a glance, whether valves were open or closed.

Limit switches can be fitted to most process valves. Linear valves used for instrument impulse lines (see Section 3.3.5, Chapter 3), may be too small but quarter turn valves should be adaptable. Limit switches provide an electrical signal to indicate the valve end of travel. Switches can be set to show "closed", "wide open" or two switches can indicate both.

If a valve does not indicate either end of travel then it must be at an intermediate position. Indicator lights on a schematic control panel provide operators with essential information very quickly.

Limit switches can be fitted to manual and actuated valves. Figure 14.1 shows a limit switch assembly fitted to the yoke of a diaphragm actuator.

Traditional limit switches are electro-mechanical devices. Physical contact with the spindle/stem operates a micro-switch which opens or closes at least one pair of contacts. Mechanical switches can be susceptible to vibration which can cause spurious operation or rapid failure due to fatigue. Electro-mechanical limit switches can be replaced by electric inductive switches which can withstand 30 g at frequencies from 10 to 55 Hz. Inductive switches detect a change in the local inductance as a target approaches. Electro-inductive limit switches are not a perfect alternative or a straight exchange. The electro-inductive limit switch requires a 7 to 15 Vdc supply and consumes about 2 mA continuously. Vibration protection is achieved at a cost. Inductive switches can be Ex ia which allows fitting in all environments.

14.1.2 Position sensors

Limit switches provide useful information regarding valve end of travel but cannot indicate that a valve is just cracked open. High differential pressure losses, with possible increased wear rates caused by high velocities, can create operational problems which are difficult to rectify. An indicator which shows valve travel as a percentage, 0 to 100%, would highlight such problems quickly.

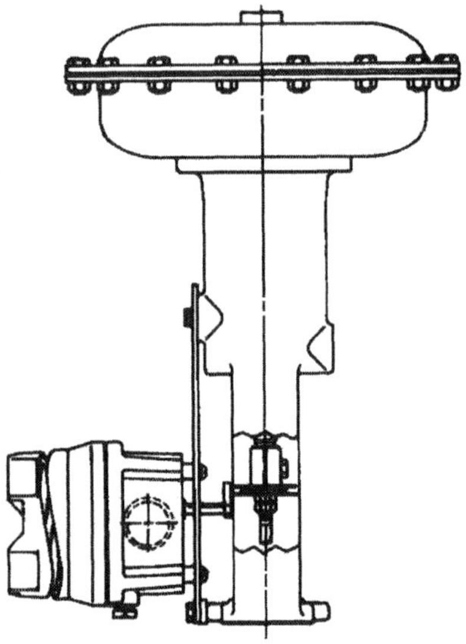

Figure 14.1 A typical limit switch arrangement

Most valves can be fitted with proximity detectors or Hall Effect sensors which continuously monitor the spindle/stem position. An electric analogue signal proportional to the valve travel can be transmitted to a suitable indicator. Rotary spindle position can also be continuously monitored via a potentiometer; the user applies a stabilised power supply and measures the output. Standardised resistances of 5 kΩ and 25 kΩ are popular. Current position transmitters, CPT, which produce a 4 to 20 mA proportional signal from a 15 to 40 V dc supply, can be incorporated into some rotary actuators. Electronic digital systems can also be fitted.

Special Note: Electrical enclosures

Because limit switches and proximity detectors are electrical devices they must be suitable for the working environment. The minimum physical protection should be dust/water-jet proof to protection classification IP65. The electrical characteristics must be suitable to the operating zone and relevant gas groups, see Electrical zones and Physical protection in Chapter 1.

Protection Classification IP 65 to DIN EN 60529 (VDE 0470), offered by an enclosure is shown by the letter IP (Ingress Protection) and two digits. The first digit indicates two factors:

1. Protection for persons
2. Protection for equipment

The second digit indicates the protection against water.

14.1.3 Differential pressure indicators

A differential pressure indicator is a very useful addition to any control valve installation. Valve position can be checked, locally or remotely, but without some knowledge of the flow the valve effectiveness cannot be judged. A differential pressure indicator shows instantly how much throttling the valve is achieving. Also, valve wear or fouling can be assessed if a log is maintained of position and differential pressure.

If a control valve is fitted with a filter it is essential the differential pressure across the filter is measured to assess the filter condition. Silencers fitted after control valves may suffer from an accumulation of solids. Differential pressure measurement across the silencer can be used to assess fouling problems.

14.1.4 Seal leakage measurement

Valve seal leakage may be a concern when hazardous or costly fluids are involved. Seal leakage measurement can be considered from two separate aspects; the rate of leakage or the nature of the leakage. Very small liquid leaks can produce vast quantities of vapour.

Valves with double seals are easy to monitor. The interspace between the seals can be piped to a monitoring system, which can be very simple or complex, to perform multiple measurements. The rate of leakage is an indication of the health of the primary seal. Depending upon the size of the valve and the nature of the leakage the flow can be assessed by various means. Liquid leakage should always be small thus making direct flow rate assessment awkward.

A drain pot, as used with some pump mechanical seals, is ideal. The seal leakage is drained into the pot. The pot is fitted with a level switch to indicate high level. When the drain pot is full it can be drained manually for product recycling. The time taken for it to fill is a direct indication of the leakage rate. As the primary seal wears, the pot will fill quicker. Corrective maintenance can be scheduled when appropriate.

Gas or vapour flow rate can be measured via an orifice and differential pressure; the flow through the orifice being directed to a suitable recovery system. Differential pressure measurement allows the flow rate to be quantified and a leakage/time trend to be calculated. Corrective maintenance can be scheduled before leakage reaches an unacceptable level.

If a valve is handling a pure substance then there is no doubt regarding the exact nature of the leakage. In many process applications the substance handled is a mixture of gases or liquids. The processing of multi-phase mixtures is common and valves are required to handle liquid/gas mixtures successfully. Liquid valves have always been required to handle dissolved gases. Not all the components of a mixture may be expensive or hazardous.

When handling sour water, water leakage would not pose a problem providing the H_2S was left behind. It is more likely the H_2S would leak through the seal before the water. If methane and hydrogen sulphide were processed as a mixture there would be fewer problems with a methane leak. When a valve is handling a mixture it may be very important to identify which component is leaking. The interspace between the seals should be piped to a monitor as well as flow measurement. Monitors can be simple and only identify a single compound, or complicated and identify a range of compounds. The nature of the process, installation and the perceived hazards will determine the type of equipment required.

Seal leakage can be monitored, intermittently, by handheld devices which "sniff" the air local to the valve. This type of measurement is very approximate and, for outdoor valves, very dependent upon wind conditions. Also, comparison of data may not be valid if readings have been taken under different operating conditions. A "hard-wired" installed system can overcome all these problems.

14.1.5 Seat leakage detection

In some situations it may be beneficial to know, with a degree of certainty, whether a closed isolating valve or non-return valve is leaking. Closed non-return valves play a very important part in compressor and pump systems by preventing reverse flow through stand-by units. Continuous leakage through these valves can represent a significant energy cost. Leaking isolating valves can allow unwanted chemical mixtures or admit air to sub-atmospheric vessels.

Fluid flow through very small apertures creates sound at ultrasonic frequencies, typically 50 kHz to 150 kHz. An ultrasound detector can be mounted externally on the valve body to detect the relevant frequency band. Integral electronics convert the sound signal to an electrical signal which can be adjusted to acceptable or unacceptable. When the signal reaches the unacceptable level a solid-state switch closes allowing an alarm signal to be triggered.

14.1.6 Cavitation detection

Cavitation in liquids produces an ultrasonic signature between 100 kHz and 500 kHz. The ultrasound can be detected by special microphones an a similar manner to seat leakage as described in Section 14.1.5. This type of alarm system can be used to extend the valve life providing process conditions can be adjusted to eliminate the cavitation. Alternatively the cavitation alarm can be used to detect unusual operating conditions and indicate the need for investigation.

14.2 Ancillary equipment

14.2.1 Proprietary interlocking systems

Proprietary interlocking systems are fitted to valves and other equipment to prevent unauthorised adjustment and to ensure equipment can only be operated in the correct sequence. The valve manufacturer may fit the interlock or the finished valve may be sent to the lock company to be modified. Reputable locking systems do not infringe the valve manufacturer's warranty. Interlocks can have multiple keys to ensure sequencing.

Interlocking systems are fitted to ensure plant and personnel safety. The interlock system forms an integral part of the plant safety system and is controlled by operators in the central control room or by the safety officer. Keys are located in a locked cabinet and must be logged in and out. The operation and security of an interlocking system is best explained by an example.

Take, for example, a high power centrifugal pump operating with a low specific gravity hydrocarbon liquid. For maintenance purposes the pump is fitted with suction and discharge isolating valves and the casing has vent and drain valves. This pump is normally controlled automatically and starts/stops in response to a control signal. Centrifugal pumps in this situation must be operated correctly to avoid damage to the pump and to prevent the pump initiating a process hazard. This type of pump has some operating restrictions:

a) the pump must not run dry — normal pump operation depends on the process liquid to lubricate the wear rings and stuffing boxes bushes. Pump seizure will result from dry running

b) the pump must not run with the discharge valve closed — when the discharge valve is closed all the pump losses are transferred to the liquid trapped in the pump casing and the liquid temperature rises rapidly. A fire hazard is easily created with flammable process liquids

A proprietary interlocking system can be fitted to the pump system to ensure safe operation. The following sequence is typical when a pump is returned to service after maintenance:

1 The pump system master key, K1, is retrieved from the cabinet in the control room. This key fits the pump drain valve and allows the drain valve to be closed.

2 When the drain valve is completely closed, key K2 is released, which fits the suction valve. The drain valve cannot be opened without K2 being fitted.

3 When K2 is fitted to the suction valve, the valve can be opened to allow the pump to prime and vent. The casing vent valve is already open.

4 When the suction valve is fully opened, K3 is released. When the casing vent valve is closed, K4 is released.

5 Both K3 and K4 must be fitted to the discharge valve before it can be opened. Once the discharge valve is fully open, key K5 is released.

6 Key K5 must be taken to the motor contactor and allows the contactor to be switched from "OFF" to either "HAND" or "AUTO" depending on whether the pump is being put on-line or tested. Key K6 is released from the contactor and must be taken to the cabinet in the control room.

The sequencing of the keys ensures the pump cannot be operated if the suction and discharge valves are closed. Interlocks can be configured to allow partial valve adjustment. An open valve could be closed but not re-opened. This variation can be useful in some operating situations.

The use of interlocking systems provides a very secure operating/safety system. The size and scope of the interlocking system must be considered very early in the project time-scale. Once control logic has been finalised, interlocking should be studied to provide the necessary level of operational safety.

14.2.2 External locking facilities

A very simple and novel method has been developed for locking valves with round handwheels and indicating the degree of hazard involved with the fluid. Circular hinged covers are designed to totally enclose the handwheel. Once fitted the cover is padlocked to prevent removal. Multi-clasps can be fitted to allow different company departments to fit their own padlock. The degree of hazard is indicated by the colour of the cover; green, yellow, blue or red. This type of accessory is extremely useful and quick to implement; no valve modifications are required. Most valves can be modified to accept padlocks. Linear valves can be locked in any position; quarter-turn valves are limited to open or closed. Most standard locking modifications will accept multi-clasps.

Padlocks are very effective for controlling unusual conditions. Interlocking is much better for routine safety control.

14.2.3 Ancillary valves

Valve bodies can have extra small connections for specific duties. Standard cast bodies can have bosses strategically placed to allow extra connections to be used without requiring pattern modifications. The most popular addition is a drain. Depending upon the valve body design, two drains may be necessary. Large valves can be fitted with a by-pass. By-passes can be used for preheating equipment during start-up or to maintain hot standby equipment at a temperature close to operating. A by-pass can also be used to balance a valve by relieving differential pressure loads on guides and seats. By-pass valves can be manual or powered to allow remote operation. Some valves can be fitted with a body cavity relief connection.

Some ball and plug valves can be fitted with facilities to allow the injection of sealing compounds, typically PTFE paste, to seal damaged seats. This is not a common option and may be elusive. Check very early in the inquiry stage if this option is required.

The simplest auxiliary connections for ancillary equipment are screwed. Large valves may have an option for flanged connections. Bosses suitable for screwed connections should be able to accept socket welded connections. The highest integrity connection would be butt welded. Auxiliary connections should have the same level of sealing integrity as the main process connections. If a valve is fitted with RTJ flanges or clamped proprietary connections to eliminate hazardous leaks, it does not make sense to fit a screwed drain plug or screwed auxiliary piping.

In some cases the valve manufacturer may add auxiliary connections by drilling through the main flanges. This approach may contravene the flange design code. The flange thickness may be increased to compensate for the material removed by the hole; this introduces a new problem. Extra long studs or studbolts are required for the connection. This type of valve modification must be considered carefully before proceeding.

Globe or needle valves can be fitted to impulse lines to provide an adjustable attenuation for pressure pulsations. A fixed level of attenuation can be included by adding a snubber. Two basic designs are available. A small circular disc is precision cast with a multitude of fine tortuous flow paths on both sides. The disc is mounted in a holder, which has the pipe connections, and the fluid is forced to flow around both sides of the disc.

The second design is very similar but the disc is formed by sintering stainless steel particles. The fluid is forced to flow through the very fine passages formed between the particles. Snubbers are available in many mounts for $\frac{1}{8}$" to $\frac{1}{2}$" pipes for pressures up to 414 barg.

14.2.4 Heating/cooling jackets

Valve bodies can be fitted with heating or cooling jackets to reduce heat transfer through the valve body. Heating jackets are the most common but cooling can be applied in the same manner. Heating and cooling jackets are fitted to valve bodies on a "case by case" basis. The jacket is usually fabricated and welded to the body. Copper alloy valves may be brazed or silver soldered. These are labour intensive options and tend to be costly.

Heating jackets usually use steam or heat transfer fluids. Cooling can use chilled water or refrigerant or any compatible fluid at the correct temperature. The supply of isolating valves or regulators must be clarified. The requirement for jackets must be established at the beginning of the inquiry stage so the effect on price and delivery can be evaluated.

Heating or cooling of the bonnet or packing box can usually be accomplished without too much difficulty. This may be an option which can be retrofitted at site to cover modified operating conditions.

14.2.5 Inlet filters

Control valves can benefit from the addition of an inlet filter. Most industrial processes are "dirty", that is clean fluids carry very small quantities of, in general, small solids. Large solids should have been removed when the system was flushed prior to commissioning.

Control valves which continuously throttle can produce high velocities which accelerate erosion by very small quantities of solids. One control valve manufacturer advises the use of hard-faced trim for throttling applications of 10 bar and over. Tests with reciprocating pump check valves indicate a doubling of the wear rate due to the presence of 0.25% small solids similar to fine sand. Process upsets can increase the size and quantity of solids entrained by catalyst carry over. Removing solids from the fluid can increase the valve useful life and extend operating time between overhauls.

Inline or "Y" type filters can easily be fitted with the control valve. The filter should be fitted inside the control valve isolating valves, see Figures 17.2 and 17.3 in Chapter 17, which allows the filter to be cleaned whenever the valve is taken off-line. Differential pressure indication must be fitted across the filter. This is the only method of assessing accurately the state of the filter.

Small filters can be built into a globe valve body. These are ideal for low flow rates and intermittent flows such as instrument impulse lines. Filter elements can be made from wire mesh or sintered stainless steel particles. Wire mesh elements can be suitable for 40 μm and larger particles. Sintered element can remove particles as small as 0.5 μm. Brass and stainless steel bodies with $\frac{1}{8}$" to $\frac{1}{2}$" screwed connections or compression fittings can be suitable for pressures up to 414 barg.

14.2.6 Hydraulic dampers

Large actuated valves working with high pressure drops are subject to large fluid forces. Vibration, possibly created by turbulence or eddies, can be transmitted through the valve spindle/stem to the actuator. There are two ways to counteract the vibration:

- fit an actuator with high inherent stiffness; this usually means a piston hydraulic actuator and not a pneumatic diaphragm
- fit a hydraulic damper

Hydraulic dampers, sometimes called "snubbers" incorrectly, are similar in principle to a car shock absorber; a rally or competition version, not a standard saloon version. The valve hydraulic damper works in both directions of travel and is adjustable. The hydraulic damper restricts the speed of movement of the valve spindle/stem. As the spindle/stem moves, oil is forced to pass through a needle valve. The needle valve can be adjusted to increase or decrease the level of damping.

Typical applications include feed water valves and steam valves used in association with power generation. Modern balanced valve trims are designed to minimise out-of-balance forces and flow turbulence.

A hydraulic damper should be fitted solely as a means of speed control.

14.2.7 Silencers

Some valves are prone to produce unacceptable noise. The noise may be continuous or very intermittent. Control valves may produce continuous unacceptable noise. Relief and safety valves should only produce unacceptable noise very occasionally. The noise levels should be predicted at a very early stage, see Chapter 10. Silencers, if required, may take up considerable space and have a major impact on pipework arrangements.

If a control valve is predicted to create unacceptable noise continuously then the first course of action is to consider removing the control valve. It may be beneficial to utilise power recovery rather than destroy the potential energy available.

Relief or safety valves for air and steam, which normally discharge to the atmosphere, are the valves most likely to create noise problems. The larger the valve the bigger the problem.

If noise is likely to be a problem the valve manufacturer should be contacted. It may be possible to implement valve modifications to reduce the noise to acceptable levels. Alternatively, fitting a silencer may be the only option.

14.2.8 Acoustic enclosures

If the pipework adjacent to a valve is to be lagged, for thermal or acoustic insulation, it may be possible to solve a valve noise problem by surrounding the valve with sound absorbent material. A box can be constructed, enclosing the relevant valve parts, to prevent the noise radiating into the surrounding air. Acoustic enclosures only work effectively when noise transmitted through the pipework is not radiated to the local air. Valve manufacturers may supply their own enclosures but are more likely to sub-contract this work to a specialist.

14.3 Useful references

DIN EN 60529 Degrees of protection provided by enclosures (IP code) (IEC 60529:1989 + A1:1999); German version EN 60529:1991 + A1:2000.

NEMA (National Electrical Manufacturers Association),
1300 North 17th Street, Suite 1752, Rosslyn, VA 22209 USA,
Tel: 703 841 3200, Fax: (703) 841-5900, www.nema.org.

Quality, inspection and testing 15

15.1 Introduction

Most valves have the important components manufactured from metal, of which cast iron, carbon steel and stainless steel are the most common. Cast iron forms the basis of all steel products; even though some new steels are made from scrap steel. Iron ore can be converted to iron by three methods:

- blast furnace
- sintering or pelletised/blast furnace
- direct reduction

In a blast furnace, iron ore reacts with hot coke to produce pig iron. The sintering or pelletising process prior to the blast furnace operation is added to enable blending of iron ores, and also to control the size of the blast furnace feed. Sintering, or pelletising, improves the blast furnace operation and reduces energy consumption. Direct reduction produces sponge iron from iron ore pellets by using natural gas. Most iron is produced from sintered iron ore and coke. The steel maker controls the sintering process to produce a consistent iron quality.

Modern blast furnaces are fitted with many instruments and together with computer modelling enable inprocess control. Iron is taken from the blast furnace as finished material for iron foundries. Iron is transferred to the oxygen steel process for conversion to various grades of steel. Iron from direct reduction plants is mixed with scrap steel in an electric arc furnace also to produce various grades of steel.

The valve design engineer is not generally concerned which production method is used. For some alloy steels, used for some arduous applications, the designer may specify vacuum remelted steel for reduced impurities, or may require additives to promote fine grain structure. Principally, the design engineer looks at the physical properties of the material to judge suitability.

Standard tests are applied, solely to assess compliance with the published specifications. Some materials are characterised only by their physical properties or chemical composition, others by both. Grey cast iron is specified by its physical properties. Some low grades of carbon steel are specified by their chemical composition and no physical properties are necessary. Most materials are described by both.

For the physical properties, described in Section 15.1.1, standard test pieces are stretched in a machine which simultaneously measures the increase in length and the applied load. There are several different test piece sizes which give slightly different results. One standard test piece is very small, this fits a machine called a Hounsfield Tensometer. Very small test pieces are useful when samples must be taken from castings or finished parts.

15.1.1 Physical properties

The mechanical properties referred to in Sections 15.1.1.1 to 15.1.1.4 and 15.1.1.6 are a function of the grain direction of the material. Unless specified otherwise all properties relate to the longitudinal direction. Properties in the transverse and through directions may be lower depending upon the physical treatment of the material and its grain structure.

15.1.1.1 Ultimate strength

This is the strength of the material when it fractures. See Chapter 13 for typical values.

15.1.1.2 Limit of proportionality

This is the strength of the material when the relationship between stress and strain ceases to be linear. In low carbon steel this is classified as the "yield point", the onset of plastic deformation, where the material does not return to its original length when the load is removed. Most designs do not stress materials beyond the limit of proportionality.

15.1.1.3 Elongation

How much the material has increased in length when it fractured is known as elongation. Different test pieces have different gauge length, each gauge length gives a slightly different result. Good elongation properties, 15 to 20%, are required for complex components which are highly stressed. Good elongation indicates ductility. Ductility is necessary so that components can deform very slightly to spread the load. A good cast iron may be 4%.

15.1.1.4 Reduction in area

Ductile materials "thin" slightly as they are stretched. When the material fractures, the cross-sectional area of the fracture is less than the original test piece. Reduction in area is reported in most American Standards, but not used very much in Europe.

15.1.1.5 Hardness

The ability of the material to withstand surface indentation. No special test piece is required, and raw material and finished parts can be tested. Several scales of hardness are used — Brinell Hardness Number, Vickers Pyramid Hardness and Rockwell. Approximate conversions are available between scales. In carbon steels, the hardness is directly related to the strength.

Hardness is very important for valves exposed to hydrogen sulphide, H_2S, in sour applications. Welds and HAZ should be checked for compliance to NACE requirements.

15.1.1.6 Impact strength

The ability of the material to withstand shock or impact is important. A special test piece is required to fit the test machine. Most materials lose impact strength as the temperature reduces. Depending upon the material, impact properties should be checked when operating below 0°C. Two different tests are used which give different results, very approximate conversions are available. Charpy is the most popular, Izod is little used. A popular benchmark for offshore equipment is 27 J at the design temperature. It is normal to check three test pieces.

15.1.1.7 Fatigue strength

All the tests defined so far can be performed fairly quickly; "test pieces today, results tomorrow". Fatigue is different. A special test piece is either subjected to repeated tensile loads or repeated bending loads. For repeated tensile loads, the test piece experiences cyclic loads from 0 to +value. A bending test piece is loaded from -value to +value. To find the endurance limit the test piece must not fail. A test piece will appear successful, if it lasts five million cycles, 5 000 000. If the machine runs at 3000 r/min this will take 1667 minutes, 28 hours. Of course, it will not be possible to guess the correct stress so several tests must be run.

Testing for fatigue in air is simple, testing for fatigue in contact with a specific liquid, vapour or gas is more complicated. It is not usual for fatigue strength of materials to be checked on a contract basis. For critical valves, fatigue testing of finished components may be necessary. Most valve designs are not based on fatigue. The valve manufacturer should state if the life of any component is limited by fatigue when operating at the rated conditions.

15.1.1.8 Creep resistance

Creep is the permanent distortion of the material after being subjected to a stress for a long period of time. It is not normally a problem in valves, although non-metallic valves may suffer at temperatures below 250°C. It is usually considered in steam and gas turbines, i.e. hot components. Creep testing is similar to fatigue testing but worse with creep tests running for years. Published research data is used when necessary.

15.1.2 Heat treatment

Many materials require heat treatment to achieve the correct condition or strength. Carbon steels are hardened and tempered to achieve high strength, usually at the expense of ductility. Austenitic stainless steels are stress relieved, softened or solution annealed to modify the physical or chemical properties. The final condition is usually confirmed by taking hardness readings. When components are heat treated to achieve specific physical properties a test piece is heat treated simultaneously. The necessary physical tests are conducted on the test piece.

15.1.3 Chemical composition

When a metallic material is produced as a raw material, its chemical composition is checked. When cast iron is converted to carbon steel in the oxygen process, all the relevant elements are weighed into the converter. Before the steel is poured, the chemical composition is checked. When the steel is poured a sample is cast. The sample is analysed and its chemical properties are the properties of the melt. Certificates will show the name of the steelmaker and the melt, cast or heat number.

The chemical composition may show elements which are not called for in the specification. Low carbon steels may show traces of nickel, chromium and molybdenum. The trace elements are a welcome addition because they tend to enhance the physical properties of the material. Impurities, such as sulphur and phosphorous, will be shown very accurately. The chemical composition of specific components, when necessary, will be traced back to the original melt.

On rare occasions, a sample will be taken from a component and analysed. Modern techniques only require very small samples. It is possible to analyse material without destroying it and some devices are available which can analyse material without removal from the component. Sufficient accuracy is present to differentiate between 304 and 316 stainless steel.

15.1.4 Corrosion resistance

Corrosion resistance of materials is judged from published research. Valve manufacturers carry out long term research on corrosion and develop materials to cope with specific problems. If a valve user wishes to handle a new fluid, of which no valve manufacturer has had previous experience, the user should conduct basic corrosion testing.

15.1.5 Non-destructive testing

Raw material, raw castings and completely finished components can be examined physically to determine the quality of certain aspects of the material. This type of examination falls into two categories:

- surface inspection
- internal inspection

Surface inspection looks for discontinuities in the surface which could be detrimental to the service life of the component. Cracks in the surface create stress raisers which can lead to fatigue failures. Pinholes in the surface may indicate porosity.

Internal inspection can show the integrity of the material and if there are any impurities, inclusions or voids in critical locations. Impurities, inclusions and voids detract from the cross-sectional area available for stressing and create stress raisers. Porosity can lead to leakage problems.

When flaws are detected it has to be decided whether the flaw is serious, if it can be repaired or whether it **should** be repaired. Some national standards, particularly pressure vessel standards, have categories for defects. The manufacturer's requirements may be more or less stringent than published standards. If the flaw is in raw material, a casting or piece of plate, it may be more cost-effective to scrap it rather than expend more time and money on repairs. If the flaw is in a semi-finished piece there may be more incentive to repair. If the flaw is in a finished component there will be strong financial reasons for a repair.

15.1.5.1 Radiographic inspection

Radiography, X-ray, is accepted as the highest grade of internal inspection and the most costly. Radiography is good because a permanent record of the inspection is available at any time for review. It is used mostly for welds, but also critical areas of castings. Components can be taken to fixed machines. Large components or assemblies can be radiographed in situ using radioactive isotopes. Strict safety precautions must be enforced.

15.1.5.2 Ultrasonic inspection

Ultrasonic inspection is popular because it can be conducted with portable equipment. The visual display indicates the position of flaws in respect to the thickness of the material. The size of the flaw must be assessed by an experienced operator. This method is very good for inspecting large flat plates prior to fabrication or forgings.

15.1.5.3 Dye penetrant inspection

Dye penetrant, or "Dye pen" is a surface inspection method which is ideal for locating pinholes and hairline cracks. The surface is first sprayed with dye which is allowed to soak into any surface defects. After a specified time, the dye is washed off and the component cleaned. Chalk is finally sprayed onto the surface. If surface defects exist, the dye trapped in the defect is drawn into the chalk by capillary action. Defects are outlined by dye indications in the chalk. The technique is generally used on finished machined surfaces. A skilled operator can judge the depth of the defect by the size of the "bleed-out".

Dye penetration is particularly useful on welds and surface coatings. Weld integrity is degraded significantly by the presence of surface imperfections. It can indicate if a surface coating has achieved complete coverage without porosity, pinholes or cracks and works on non-metallic coatings as well as metallic.

15.1.5.4 Magnetic particle inspection

Magnetic particle, or "Mag particle" is a surface inspection method which is popular for cast materials. This inspection can only be conducted on materials which can be magnetised by an electric current. The surface is coated with a liquid, bearing small magnetic particles. If the surface contains flaws, the magnetic flux is concentrated around them drawing the magnetic particles towards the flaw.

15.1.6 Repairs

When repairs are undertaken, usually by welding, the repair must be inspected. Initially the faulty material must be removed which can be achieved by any suitable means. The component must be checked to ensure that all faulty material is removed; this is normally by dye penetrant or magnetic particle inspection. The repair is then carried out. The component must then be inspected by the same method which found the original flaw. Valves built specifically for a purchaser may have repairs categorised into "major" and "minor".

15.1.7 Welding

Welding is a skilled occupation. There are many types of welding and the welder carrying out any particular job must be proficient at that particular type. Welders are graded by the types of components they weld, the positions of the welds and by the types of equipment used. Welding pipework is thought to be the most difficult type of welding. Welders for pressure vessels have to be qualified and this is usually assessed by an insurance company. Welding pipework to valve bodies is difficult. Designing weld methods for exotic materials requires very special materials knowledge. All welding for pressure containing parts should be supported by certification.

Some closed die forged valve bodies have flanges attached by friction welding. These welds may require inspection for integrity depending upon the operating conditions and the nature of the process fluid.

15.1.8 Inspection

Inspection of all components is carried out as a matter of course. The degree of inspection is dependent upon the nature and function of the component and the batch size. Mass-produced parts are not as individually scrutinised as a small number of specially-made parts. General dimensional inspection of parts does not normally form part of the purchaser's involvement. In special cases, the purchaser may wish to witness or observe the measurement of specific dimensions.

Valves which must comply with the requirements of The Pressure Equipment Directive (PED) 97/23/EC may be inspected by the Notified Body.

The PED (Pressure Equipment Directive 97/23/EC) applies to the design, manufacture and conformity assessment of pressure equipment and assemblies of pressure equipment with a maximum allowable pressure greater than 0.5 barg.

15.1.9 Assembly

The basic procedure of valve assembly involves the final dimensional check. If components do not have the correct fit or clearance it will be very apparent. Selective assembly of components is frowned upon unless the components are always supplied as a sub-assembly. Spare parts cannot be selectively assembled at site. Assemblies may have fits, clearances and alignment measured and recorded before testing or dispatch. This record is sometimes called a "History Sheet".

15.1.10 Packaging

Packaging is the assembling of the valve and any associated equipment, such as an actuator, into a complete unit. Valves can be very simple manual devices with few parts, or complex assemblies with more than one valve with actuators and controllers. The purchaser may decide to supply actuators and controls independently. When control equipment is supplied, the valve manufacturer is often required to wire all the control elements together.

If a control panel is mounted on the package then switches and transducers can be wired to the panel. If the control panel is mounted separately the valve manufacturer will probably wire the package instrumentation to a terminal box. The purchaser's inspector or instrument engineer may wish to witness the wiring process. Some cable glands are complicated to build up; wiring routes can be better assessed with a full size 3D package and wiring practices inside instruments and terminal boxes may be considered critical.

15.1.11 Pressure testing

Pressure testing of components and assemblies confirms the material integrity and the basic strength of the design. Porosity of castings can be detected easily without lengthy ultrasonic or costly radiographic inspections. Pressure testing does not always reveal thin walls. Whenever possible, components should be pressure tested in a manner which closely represents the pressure effects during normal working. Clamping components in fixtures modifies the elastic behaviour of the component, and does not highlight problems such as distortion of mating flanges.

Mass-produced valves will probably be pressure tested in a clamping test rig. These valves should have undergone rigorous prototype testing which included pressure testing with bolted blank flanges. Finish machined valve bodies can be weighed to verify material content.

Pressure testing does not automatically mean hydrotesting with water. Some tests may be conducted with oil, others with a detergent and water mixture. Some components may be pressurised with compressed air and submerged in a water tank. Pressure testing for packing box seal leakage assessment may be carried out with methane. Critical components for very hazardous fluids may be tested with helium. If the valve must experience fluctuating operating conditions, the manufacturer should be asked how standard steady-state tests can be adapted to more closely reproduce operating conditions.

The pressure to which components may be subjected can be limited by standard flange designs. The purchaser and the manufacturer must ensure that each fully understands the requirements and the limitations of pressure testing. Safety relief valves will generally be subjected to two pressure tests:

- a high pressure for the inlet
- a low pressure for the body and the outlet

15.1.12 Running tests

Running tests are thought by some to be the ultimate quality control test. This may or may not be the case. Many types of test are possible under very diverse operating conditions. Testing at the factory becomes more important as the valve size increases and the more unique the valve. Operational problems are much easier to solve in the factory than at site. The type and scope of tests which can be conducted should be clarified during quotation evaluation.

Running or operational tests should reproduce site conditions as closely as practicable. The choice of test fluid is important. A test with light oil could not be considered representative for a valve destined for boiler feed water. Testing should confirm:

- C_V, K_V or capacity
- compatibility of the trim
- repeatability of seat sealing
- speed of operation
- full travel of spindle/stem
- set points and droop
- stability of the control system

Control and safety relief valves can suffer instability due to the lapping of the plug to the seat. Very good matched surfaces, for low leakage rates, can stick together and resist opening.

A **strip inspection** can be performed after any running tests. Inspecting the trim can confirm the suitability, or otherwise, of the mating materials.

Noise testing can be very difficult. Most valve manufacturers do not have quiet rooms and noise testing is conducted on the normal testbed, sometimes adjacent to the machine shop. Background noise is frequently a problem, as well as pipe wall noise from the inlet and the outlet. Most manufacturers can offer a noise survey — a noise test not complying with any standard, but reporting the background noise at each test point, with the test reading.

15.1.13 Painting

Painting is the last production operation prior to dispatch. Some paint finishes, especially offshore requirements, are quite complicated and have many intermediate stages. The purchaser's inspector may wish to witness some operations.

Paint specifications may require the raw material to be blast cleaned to a bright finish. The surface has obviously to be dry and free from loose fragments. Blast cleaning may have to be performed in a controlled atmosphere. Some inspectors wish to check the environment and see the surface before the primer is applied. Individual coats of some paint finishes have specified thickness, the inspector may wish to measure the paint thickness as coats are applied. The total paint thickness and the finish of the top coat will be the final inspection.

Some valves operate at very high temperatures. The paint finish should be suitable for the operating temperature. The purchaser should be informed of any special procedures necessary to maintain the protective properties of the paintwork.

Some components are machined on all surfaces, eg. forged valve bodies. Valve manufacturers may not wish to blast clean a surface which is machined for design purposes. Paint systems are available for applying to bright machined surfaces, the only subsequent treatment being a thorough degrease.

15.1.14 Preservation and packing

Preservation is the process of protecting the valve in its "as-built" condition until the purchaser installs it. The type of preservation given to any particular valve depends upon several factors:

- valve material combinations
- packing box configuration
- paint finish
- destination
- purchaser

All valves should be dried thoroughly. Traces of hydrotest water can only be damaging. If the internal materials are all corrosion resistant there is no need to use preservative fluids or soft setting waxes. Dessicant must be placed inside the valve prior to sealing the connections. Hygienic valves must be preserved with approved products. It is very annoying and time consuming for the valve user to have to clean thoroughly a valve which has been preserved incorrectly. Connections must be adequately protected from physical damage. Plywood blank flanges are perfectly acceptable if sealed with tape. Screwed connections and plain pipe ends can be protected by plastic plugs and caps; again secured by tape. Any external surfaces, which are not painted or corrosion resistant must be sprayed with a setting preservative.

REMEMBER! 11/13Cr steel is not stainless; it is stain resistant.

The packing should be removed from the packing box and the spindle/stem taped in position. The new packing should be securely attached to the valve. Small valves can be shrink wrapped rather than boxed. Valves built specially for the purchaser may be subject to the purchasers' contract packing requirements.

Poor preservation and packing can ruin the best valves. It is not worth "penny-pinching" on packing because that is the first thing the site personnel see. Instructions on the outside of the packaging should state the packing date and how long the preservation is valid. It is sensible to supply instructions on maintaining the preservation. Instructions should be packed with the valve stating the preservatives used and the correct method(s) of removal.

15.1.15 Purchased equipment

When the valve manufacturer packages the valve to form a complete unit, specifications and purchase orders for the "purchased" equipment will be issued. If the manufacturer is building his standard unit, all the specifications issued will be the valve manufacturer's. If the valve unit is built especially for a user, the user may impose requirements on the "bought out" equipment. It is possible that conflicts may arise between the two sets of specifications. The valve manufacturer is responsible for supplying a complete working package, irrespective of what the purchaser's specifications require.

15.1.16 Functional testing

Valve testing is complicated. It is therefore essential that important decisions regarding the amount of contract equipment to be used and the test arrangement are clarified prior to the test. Another part of testing which must be considered is the functional testing of packaged instrumentation and control equipment.

The following inspection functions are normal:

- pneumatic and electric circuit verification
- pneumatic leak testing
- wiring continuity
- earth and leakage tests
- calibration of instruments
- verification of set points
- simulation of control logic

If the purchaser wishes to exercise control over these activities then proprietary specifications must be issued.

15.1.17 Witnessing

The purchaser, his representative, or the certifying authority may wish to be present when certain activities are performed. The certifying authority is usually an insurance company who wishes to confirm that statutory requirements have been fulfilled.

Viewing activities is approached in two ways:

- **Activities which are interesting or complicated may be observed.** The manufacturer informs the purchaser when the activity will take place, five to ten days notice being normal. The activity proceeds whether the observer is present or not.
- **Critical activities are witnessed**. The manufacturer informs the purchaser/certifying authority when the activity will take place. The activity cannot be performed if the inspector is not present. A hold is placed on that aspect of the contract. Both observing and witnessing cost money and take time. The purchaser has no automatic rights to either. The full extent of external participation in inspection functions must be agreed during the pre-contract negotiations. Changes to external inspection, or the unavailability of inspectors at specific times can lead to extended deliveries and increased costs.

The purchaser may wish to observe/witness activities at the valve manufacturer's sub-suppliers. This can only be done with prior agreement. Also it is important for the purchaser and the valve manufacturer to agree who has to pay for any third party inspection or the certifying authority.

15.1.18 Clarification of specifications

Valve users who have a large population of valves and who purchase many valves every year, will have proprietary specifications covering most aspects of design, operation and maintenance.

The general trend is to take a national or international standard and amend certain clauses. Some clauses are deleted as being irrelevant, others are edited to reduce option choices and some options are modified to become mandatory requirements. When the published Standard is thought to be weak, the proprietary version includes new and rewritten clauses. Compliance with legal regulations is obligatory.

It is possible for the purchaser to impose requirements which are inappropriate for the application. When these problems arise it is essential that the valve manufacturer has access to the purchaser's engineers who understand why specific clauses were written. The manufacturer must have the opportunity for technical discussions with engineers who can change or delete parts of the specification.

With modern business practices there is a great risk that the purchaser may buy the valve as specified, rather than buy the valve he needs. The consequences of this problem is summed up by the following edict:

> "Troubleshooting is making the valve do what is specified. Making the valve do what you want, is up-grading!"

The difference between troubleshooting and up-grading is — who pays for the work!

15.1.19 Certification

It is obvious from the range of testing and inspection operations discussed that the scope for certification is immense. The cost and delivery implications cannot be overlooked. Table 15.1 lists, in the order of popularity, purchaser's favourite documentation.

Component/Assembly	Certification
Valve	Hydrotest Performance Material conformance Material physical Material chemical Material impact Welder qualification Weld procedures Classification approval Noise
Pipework	Hydrotest Material physical Material chemical Welder qualifications Weld procedures Cleanliness
Electrical equipment	Hazardous area Type test Noise (motors) Performance (motors) Vibration (motors)

Table 15.1 Popular certificates

A material conformance certificate is a declaration from the manufacturer that the materials used comply with the specifications. Certificates of conformance which begin "I .... certify that to the best of my knowledge......" are not acceptable.

Valves compliant with the PED regulations must be manufactured from traceable materials. It is not clear whether the purchaser is automatically entitled to have copies of the material certification.

When valves are destined for cold climates, impacted tested material becomes very important. Even structural material is certified. Pipework does not really suffer impact certification problems. It may be easier to use 304 or 316 stainless steel, for which many welders are qualified, than use a specialist nickel alloy.

As fluids become more hazardous and materials more exotic, the requirements for non destructive testing and operator qualifications increase.

Although the quality of the material has already been discussed, what about the quality of the certification? When a material certificate specifies physical properties or chemical composition, how accurate is it, how valid is it and which authority certified it? The first standard to grade certification was DIN 50049. However, this has been superseded by the DIN-adopted European Standard, EN 10204.

The certification is classified as follows:

2.1 **A Certificate of Conformance**. The manufacturer certifies the materials conform to the order specification.

2.2 **A Works Report**. The manufacturer certifies, on the basis of tests performed on the batch, the materials conform to the order specification. No actual certificates are supplied.

2.3 **A Works Certificate**. The manufacturer certifies, on the basis of tests performed on the contract material, the materials conform to the order specification. No actual certificates are supplied.

3.1.A **Certificate issued by the manufacturer's Quality Control or Inspection Departments**, not someone from "manufacturing", stating the test results of the actual material.

3.1.B **Certificate issued by an independent test house**, employed by the manufacturer, stating the test results of the actual material.

3.1.C **Certificate issued by an independent test house**, employed by the purchaser, stating the test results of the actual material.

Certification to 3.1.C is obviously the best from a valve purchaser's viewpoint. It is also the most costly and has the biggest impact on delivery. On all but the most critical applications, certification to 3.1.B has become the standard for large valves built-to-order. The purchaser's inspector can insist on positive material identification on random components to validate certification.

The valve manufacturer retains quality records for a considerable time, 10 years is not uncommon.

Manufacturer's quality could be considered as an extension of certification quality. How much control does the manufacturer have over the manufacturing process? This aspect of Quality Assurance is standardised in BS 5750, EN 29000 and ISO 9000 and the United States ANSI/ASQC Q9000 series which all describe the same set of quality principles.

The basis of manufacturing control is recognised as lying within strict operational instructions. Every employee has a written job description. Routine inspection and manufacturing procedures must be written down and strictly controlled.

All of these documents should form part of the manufacturer's Quality Assurance Manual, QAM. The manuals are issued to all relevant departments. Personnel are not allowed to keep individual copies of often-used procedures. These can become out-of-date and result in incorrect operations. Lines of communications are defined with critical decisions. Special emphasis is given to error reporting and feedback to the designer. Companies who advertise compliance with the standards may have been audited by a third party assessor. Routine inspections are performed at regular intervals to ensure standard compliance.

Compliance with the three standards mentioned is not essential for good manufacturing control. The basic principles can be incorporated into a manufacturing system without strict compliance with the standards. The critical items which are essential for effective control are a Quality Plan, for components or complete assemblies, and the power invested in the Quality Control Department. Inspection at every stage of manufacturing is essential. Reliance on CNC machining is misplaced.

Total Quality Management is a phrase which is used by management to boost the image of companies. If quality plans are not used, if an independent department does not check components/assemblies at each stage of production, what does Total Quality Management have to offer the purchaser?

Self-inspection during manufacturing requires a cultural change within the workforce. This dramatic change to long established working practices is not easy to institute or manage. This type of manufacturing process can work very well when the workforce fully understands the benefits and the implications. A good manufacturing plant can be permanently disabled by an unsuccessful transition. Final assembly is often too late a stage at which to discover that components are incorrect.

Quality Assurance is not intended as a means to higher or better quality levels. Quality Assurance ensures goods are manufactured to the quality level specified. Quality Assurance is intended to produce consistency. Well-regulated manufacturing systems will tend to refine designs and manufacturing techniques as the fault reporting system highlights component or assembly failures; this is an essential cost reduction feature. Improvement in product quality is usually driven by the purchaser, directly or indirectly.

15.1.20 Documentation

All valves and valve units will be supplied with some documentation. The extent of the documentation will depend upon the nature and duty of the valve and the method of purchase. However, some documentation is controlled by legislation and the purchaser must receive this documentation irrespective of the method of purchase.

Safety is controlled in Europe by the UK Machinery Directive and the Pressure Equipment Directive (PED) which are implemented by compliance with harmonised standards. In the US this is covered by ANSI, OSHA and many others. It is the duty of the valve designer to design safe valves. If complete safety cannot be incorporated in the design, documents must be provided which detail the areas of concern and what precautions must be taken by the user. In general these will deal with the fitting of guards, protection against pressurised fluids and isolation of electric supplies prior to working on equipment. Other documentation the purchaser would reasonably expect to receive is discussed later.

Documentation can also apply to manufacturers' published literature. Typically manufacturers show valve flow capacity, C_V or K_V, and list dimensions and materials of construction either by generic term; carbon steel, stainless steel, bronze; or standard specification. Information that should be in publications to endorse the quality and reliability of the product, is also discussed later.

15.2 Mass-produced valves

Mass-produced valves can be purchased by two methods: over-the-counter and by enquiry. Many standard valves are purchased "over-the-counter" from distributors. The use of distributors allows a manufacturer to have his products widely available at many outlets. Some distributors have technical knowledge and skills and can remove some of the skilled back-up required from the manufacturer. A purchaser can review published data, select a valve for an application and purchase a valve over-the-counter. Alternatively, the purchaser can send the valve manufacturer or distributor an enquiry with the duty conditions. From the subsequent quotation, accompanied by standard literature, the purchaser can decide which valve to buy.

Before purchasing a valve, the following questions should be answered, either from reading the published data or direct questions to the distributor or manufacturer:

- Will the valve(s) be subject to all safety requirements such as the PED and ANSI/OSHA?
- Will the valve(s) be CE marked?
- Are the valves designed and built to a standard?
- Are the valves manufactured to a standard?
- Will my assembled valve be subjected to any operational tests?

Some manufacturers do not test all valves. If not all valves are tested, how many valves in a batch are tested and what is a typical batch size? What fluid is used?

- Will my complete assembled valve unit be subjected to any operational tests?
- Will my assembled valve be pressure tested?

If an assembled valve is pressure tested, it is important to know the pressure and how the packing box was sealed.

- To what standard are running tests conducted?
- What Quality Assurance documentation will be supplied with the valve?

The minimum documentation would be a hydrotest certificate and a final inspection release certificate. For a standard valve constructed of 316 stainless steel or a more exotic alloy, it would be reasonable to expect a Works Report, 2.2, as in Section 15.1.19.

Other standard documentation would include:

- Parts lists and assembly drawings
- Installation instructions and guidelines
- Maintenance and safety instructions
- Trouble-shooting guide
- Telephone, fax, e-mail and web address of service department

15.3 Custom-built valves

When a valve is built to the purchaser's specification, the purchase order will have been placed on the basis of quotations received from various manufacturers. The purchaser will have issued an inquiry detailing the operating conditions and any other requirements. The purchaser will have evaluated the quotations based on:

- Valve performance
- Compliance with specifications
- Manufacturing schedule and delivery
- Manufacturer's experience
- Confidence in manufacturer
- Documentation
- Cost

Cost is a difficult variable. If the quotation is evaluated by a contractor, initial cost plus a cursory look at the performance may suffice. A user will evaluate life-cycle costs. Documentation plays an important role in business. Technical documentation plays a critical role in modern plants. The manufacturer should have listed all the documents which would form part of the technical contract. A Vendor Documentation Requirement form, VDR, should have been completed and returned. This form lists all the technical documents and at what time they will be submitted. A preliminary quality plan, QP, would have been submitted with the quotation to indicate the inspection included.

An important part of any quotation is a good installation list; this together with a good quality plan and a good VDR can instil confidence. A quality plan indicates the important inspection functions the manufacturer will conduct to assure an acceptable quality level. The QP also shows the areas where control is taken from the Quality Control Department, QC, and assumed by the manufacturer's Engineering Department or the purchaser's inspector or engineer. In some contracts, a Certifying Authority may also have inspection rights; the CA would normally be an insurance company. The QC department is always in overall control of any contract.

Some problems may be due to design, rather than manufacturing. If a valve body fails hydrotest, and inspection confirms the machining to be correct, the problem will be referred to the Engineering Department. Some components are critical, and any

REF NO.	PROCESS DESCRIPTION OF Q.C. ACTIVITY	Q.C. PROCESS REF	ACCEPTANCE CRITERIA: IN HOUSE	ACCEPTANCE CRITERIA: CLIENT	ACCEPTANCE CRITERIA: ACTUAL	VERIFYING DOCUMENT	IN HOUSE: ENG	IN HOUSE: Q.C.	CLIENT: INSP	CLIENT: ENG	CERTIFYING AUTHORITY
VALVE BODY INSPECTION											
1	VERIFICATION OF MATERIAL COMPLIANCE	ASTM A216WCB	CONFORMS TO SPEC		EN 10204 3.1.B	CERTIFICATE OF MECH & CHEMICAL PROPS.		H	X		
2	VISUAL INSPECTION OF CASTING	SP-55-1985 (R90)	CONFORMS TO SPEC			CERTIFICATE OF VISUAL INSPECTION		H	X		
3	GOODS INWARD INSPECTION	ES 509-018	CONFORMS TO P:O. & SPEC			INSPECTION CARD		R			
4	INSPECTION AFTER MACHINING	TO DRAWING	CONFORMS TO DRAWING			INSPECTION CARD		R			
5	HYDROTEST	SP-61-1992	153 BARG		153 BARG	CERTIFICATE OF HYDROTEST		H	H		3 copies of Certificate
6	INSPECTION/Q.A. REVIEW & RELEASE FOR ASSY	WIM 9-200-8 1	INSPECTION/Q.C. REQUIREMENTS COMPLETE			INSPECTION CARD		R			

H ≡ hold point or witness point, R ≡ document retained at manufacturer's works, X ≡ document included in purchaser's data pack

Figure 15.1 Example of Quality Plan format for a component

problems are referred directly to Engineering. If the purchaser and the manufacturer have agreed to differentiate between minor and major repairs, major repairs will result in a purchaser's hold point. If the valve has difficulty in meeting a requirement on test, both sets of engineers will be involved.

If the preliminary QP is not acceptable to the purchaser, the modifications must be agreed during the pre-contract negotiations. Extra certification and hold points not only affect the cost but also the delivery.

Quality Control Plans can take many forms. Early QPs were produced as drawings in the drawing office. The inspection operations were listed horizontally across the top of the drawing, the components were listed vertically down the left hand side. Different types of dots in the squares indicated various operations and hold points. Because of the style, these plans were often called "pools plans" because of the similarity with football pools coupons. This type of QP is difficult to arrange initially and even worse to modify.

The Energy Industries Council, one of the leading UK trade association for UK companies supplying capital goods and services to the energy industries worldwide, considered the problem of QP's and issued a proposal, *Guideline for the generation of Quality Plans, QAG 01*. The proposal advocated a style of QP, on A4 paper, which could be word processed. The basic idea was to have each component on a separate sheet, organised in a specific format. Complicated components could have their quality requirements extending over several pages.

Figure 15.1 indicates a slightly modified format which has been used very successfully. The references in the "Q.C. Procedure Reference" column; ES and WIM; refer to the manufacturer's standard inspection and workshop instructions. Inspection to national, European or international standards can be performed when appropriate. Components which are not listed in a QP are subject to the manufacturer's standard quality requirements. If the purchaser is considering placing an order, he may ask for a copy of the manufacturer's Quality Control Manual. This would normally be an "Uncontrolled Document". The prospective purchaser would not receive modified or new procedures as they were released.

When a contract is subject to such structured quality requirements, it is usual for the purchaser and the manufacturer to have a meeting, very early after the purchase order placement, to clarify any ambiguous points and present a contract manufacturing schedule. This meeting, sometimes called a "Kick-off meeting" or a "Vendor co-ordination meeting" takes place between the manufacturer's engineers and contract management and the purchaser's engineers and inspectors. If a Certifying Authority is imposed, they must be represented so that their involvement and hold points are correctly interpreted. This meeting confirms the quality requirements and the documentation requirements shown on the VDR. The major and minor repair categories may be finalised on a component by component basis. The manufacturer may agree to submit progress reports during the contract.

As the contract progresses, the manufacturer's contract engineer and the purchaser's inspector can work through the plan and ensure all inspections are completed and that documentation is available.

One important document, which is only used occasionally, is a design review. Valves for very critical applications, very hazardous fluids or subject to legislative control, usually through a Certifying Authority, have important areas of design reviewed by a third party. This review is considered as an additional safety measure and would evaluate important areas of construction such as:

- valve body pressure stressing
- effect of nozzle loads on valve body stressing and deflections
- material combinations
- seat leakage requirements
- spindle/stem sealing capabilities

Valve testing is one of the most important aspects of the quality assurance programme. The preliminary QP with the quotation should have outlined the tests proposed and how they related to the rated operating conditions. The vendor co-ordination meeting should have clarified any missing data. Manufacturer's test facilities do have limitations and all rated operating conditions cannot be accommodated. Discussions with the manufacturer are essential.

The choice of test fluid can cause formidable problems if seat and seal leakage are to be measured. Standard test fluids are water and air. Many process and chemical valves operate with liquids with viscosities over 100 cSt. A good test, however, is one which closely approximates the operating conditions. The test fluid must be agreed with the manufacturer, but one closely approximating the properties of the process fluid would obviously be better than an arbitrary choice based on dogma.

Larger, more complex valve assemblies may have two volumes of documentation supplied with the equipment, an installation and maintenance manual, and a data book. The maintenance manual would contain all the information listed for standard valves. A typical data book would contain:

- Quality plan
- Material certificates
- NACE compliance certificate
- Weld procedures
- Welder qualifications
- Radiographs
- Non-destructive test (NDT) reports
- NDT operator qualifications
- Surface finish verification (for hygienic valves)
- Hydrotest certificates
- Leak test certification
- Instrument certification
- Wiring check results
- Functional test results
- Paint procedure
- Paint inspection certificate
- Overall dimension and weight certificate
- Inspector's release certificate

If the purchaser's inspector had carried out random positive material identification, the material chemical composition certificate would be marked as such next to the inspector's stamp.

15.4 Guidelines for testing and documentation

It is very difficult to generalise about quality requirements for such diverse products as valves, but Tables 15.2 and 15.3 are given for guidance. If necessary, the requirements of the PED **must** be fulfilled.

Quality grade	DN15	DN50	DN100	DN200	DN300	DN500	DN750	DN 1 000
Grade D	320	152	122	85	66	42	24	17
Grade C	448	220	117	135	90	67	37	26
Grade B	584	362	290	246	182	119	57	40
Grade A	920	472	377	255	249	238	88	60

Design pressures shown in barg

Table 15.2 Guidelines for valve quality requirements

Table 15.3 indicates a design pressure associated with a valve diameter. The Table is based on carbon steel valves; the pressures indicated should be adjusted in proportion to the strength of the **ductile** material used. Valves in cast iron would use pressures approximately 35% of the values shown.

Some national governments severely restrict the use of cast iron; pressures in excess of 200 barg may not be permissible irrespective of the valve size. The quality requirements refer to pressure containing and pressure retaining parts. Other tests may be appropriate depending upon the application. The requirements shown are based on a "safe" fluid; hazardous fluids should use requirements one or two levels higher.

Valves for use with hazardous fluids may require additional tests for material, joint and seal integrity. Pressurised tests with air, methane or helium should be considered. Hygienic valves should have the relevant surface finishes checked; the purchaser should ask the manufacturer about his standard procedure.

Quality Grade	Inspection requirements
Grade D	Hydrotest certificate
Grade C	Hydrotest certificate
Grade C	Certificate of compliance, EN 10204-2.2
Grade B	Hydrotest certificate
Grade B	Copy of sizing calculations
Grade B	Copy of Classification Body Approval certificate
Grade B	Surface inspection of pressure containing parts and welds, magnetic particle or dye penetrant
Grade B	Hardness certificate for HAZ and welds for NACE valves
Grade B	Material certificates for pressure containing and pressure retaining parts, EN 10204-3.1.A
Grade A	Hydrotest certificate
Grade A	Copy of sizing calculations
Grade A	Copy of pressure containment calculations
Grade A	Validated Copy of Classification Body Approval certificate
Grade A	Surface inspection of pressure containing parts and welds, magnetic particle or dye penetrant
Grade A	Confirmation of material integrity for pressure containing parts and welds, radiography or ultrasonic inspection
Grade A	Hardness certificate for pressure containing parts, HAZ and welds for NACE valves
Grade A	Material certificates for pressure containing and pressure retaining parts, EN 10204-3.1.B

Table 15.3 Description of quality grades

15.5 Standards

Please refer to Chapter 16, Section 16.24, for relevant standards.

15.6 Useful references

Marchwood Laboratory, Dept. of Materials Engineering, Applied Science Building, University of Wales, Singleton Park, Swansea SA2 8PP, UK, Tel: 01792 205864, www.swan.ac.uk/mateng/test/marchwood.

School of Engineering, Solid Mechanics Group, University of Surrey, Guildford, Surrey GU2 7XH, UK, Tel: 01483 259264, Fax: 01483 306039, www.surrey.ac.uk/MME/Research.

McMaster Steel Research Centre, McMaster University, Steel Research Laboratory, 1280 Main Street West, Hamilton, Ontario L8S 4L8 Canada, Tel: 905 525 9140, www.mcmasteel.mcmaster.ca.

Energy Industries Council, Newcombe House, 45 Notting Hill Gate, London W11 3LQ, UK, Tel: 020 7221 2043, Fax: 020 7221 8813, Email:info@eic-uk.com, www.the-eic.com.

DIN 50049 Inspection Documents for the Delivery of Metallic Products – (Withdrawn, - Replaced by EN 10204).

DIN EN 10204 Metallic products - Types of inspection documents (includes Amendment A1:1995); German version EN 10204:1991 + A1:1995 (DIN-adopted European Standard).

BS 5750-8:1991 Quality systems. Guide to quality management and quality systems elements for services.

ISO 9000 Compendium 9th Edition 2001, International Standards for Quality Management.

ANSI/ASQC Q9000-1-1994 Quality Management and Quality Assurance Standards - Guidelines for Selection and Use.

Machinery Directive - 98/37/EC, (Previously Directive 89/392/EEC as amended by Directives 91/386/EEC and 93/68/EEC. The original Directive and its amendments have been consolidated in the single Directive 98/37/EC)

Standards and specifications 16

16.1 Introduction

Within any civilised community there is legislation which controls relationships between individuals. Successful technology must also be controlled and guided by standards and regulations to ensure its safe application and allow some degree of interchangeability.

Standards were initially applied to the finished dimensions of machine components to ensure easy replacement of worn or broken parts. The scope and activities of standards have been successively extended to cover the complete machine not only with regard to dimensioning but also to such parameters as performance, reliability and safety. Within the valve industry, standardisation has developed in a similar fashion, and today there are a considerable number of standards and regulations. There are also a large number of new standards which are in the process of preparation and old standards in the process of revision.

Some standards have legal authority and can be enforced by the process of law; these are mostly restricted to safety requirements. Most standards are issued for guidance. Comparison between quotations for valves would be difficult or impossible if important items, such as connections, were not standardised. Selection of the best valve would be dubious without very detailed design and construction information on each. The solution to the problem is to issue a specification with the inquiry detailing the important characteristics required. To avoid the necessity of all purchasers writing their own specifications, groups were formed to write a single common specification for certain areas of application.

Because of the diversity of issuing authorities, the style and concept of standards is not uniform. CEN has decided upon rigourous control of all aspects of standards; presentation, structure and content. CEN has produced a manual for all concerned with standard writing and it is worthwhile reviewing for the structure and content requirements.

CEN standards consist of requirements; these are basically statements controlling design, construction and use. Requirements must be verifiable. A competent person, inspecting the equipment/installation/service, must be able to see that a requirement has been met or not. Generalised statements, such as "durable" and "rugged", are subjective and cannot form part of requirements.

In this Chapter, Sections 16.3 - 16.29 focus on the most important standards and specifications relating to valves. They also indicate the sources of other interesting and possibly mandatory requirements.

Valves Manual International concentrates on the process applications of valves. Many standards are published specifically for fluid power valves. These standards are listed when they could be also applicable to process installations, or of general interest.

16.2 Issuing authorities

In principle, a standard or specification can be issued by anyone. In order that the document shall carry some influence, however, the issuing organisation should be one that is recognised by both the manufacturer and the user of the object to be standardised. It is therefore normal practice for new documents to be developed by these three parties. The organisations which issue standards and specifications consist basically of the following categories:

- International standard organisations
- Geographic country group standard organisations
- National standard organisations

International standard organisations require the cooperation and support of many countries. As the world becomes, effectively smaller, more and more international trade transactions will be regulated by international standards.

Geographic country group organisations are made up of countries, usually close to each other, which form an economic trading block. This type of cooperation is becoming popular and could result in an expansion of issuing authorities.

Another type of document is frequently used to assist the purchasing of valves — a specification. A specification is very similar in most respects to a standard but does not necessarily have the same rigorous editorial controls. Specifications are issued by:

- Trade societies and associations
- Corporations, large companies, government departments
- Classification societies (insurance companies)
- Certification authorities (insurance companies)

Trade societies and associations are groups of companies which share a common interest. For example, utility suppliers, water or electricity, could form a trade association with pump and valve manufacturers and firms of consultants which regularly design systems. Together they agree a common approach to the design and operational problems of pumps/valves and write specifications to improve reliability. All members use the same specification for the same items.

Corporations, large companies and government departments issue specifications to control important features, design and/or function. Chemical companies, who develop new manufacturing processes, often issue specifications to regulate size, speed and rating of important equipment.

Classification societies are generally one, or more, insurance companies who issue rules for design and construction. This type of specification is used for marine installations, ships and offshore platforms, and the most notable authorities are: Lloyds Register, Det Norske Veritas and the American Bureau of Shipping. If a design standard requires a component to be 20mm thick but the specification requires 25mm it is the specification which takes precedence. If a ship owner tries to insure a ship, the first question asked is "Can I see your Classification Certificate?"

Certification authorities are very similar to classification societies but work in conjunction with the purchaser on large projects which have some degree of risk or hazard to the purchaser's capital investment or personnel safety. Certification authorities are regularly used for large petrochemical installations and offshore platforms. Specific specifications are not always issued. Experienced inspectors scrutinise and witness all relevant aspects of the equipment; design, raw material, construction, testing, safety; and issue a certificate, if completely satisfied, when the equipment is ready to be dispatched.

The number of issuing authorities can therefore be very large and varied which often presents difficulties when developing new standards. Since there are many conflicting interests, especially when applied to international standards, the result is often a compromise with which none of the involved parties is completely satisfied. The advantages to be gained by the specification and general acceptance of standards and regulations however, usually outweigh the disadvantages.

CEN issues standards to control trade within the EEC and EFTA, (European Free Trade Association). EFTA is an intergovernmental organization. Its members are Iceland, Liechtenstein, Norway and Switzerland. From 1995, it has not been possible to use national standards for trade between member countries; ENs or ISOs are used. ISO is issuing new standards to control world trade. ISO discussed with all standards organisations and trade associations the rationalisation of worldwide standardisation. ISO and CEN have an understanding and try not to duplicate standards.

Standards and specifications usually contain a immunity clause. Strict compliance with any standard does not assure success or guarantee success. Standards are written with the best of intentions with the best information available. It is the responsibility of the user of the standard to ensure the standard's requirements are compatible with the equipment and operation thereof in each individual case. One of the great problems with standards and specifications is that the user purchases the equipment specified; not the equipment needed!

16.2.1 National standards bodies

Some of the most important bodies for issuing standards impacting on the manufacture and use of valves are shown in Table 16.1, in alphabetical order:

	Body	Standards
ANSI	American National Standards Institute	ANSI
ASME	The American Society of Mechanical Engineers	ASME
ASTM	The American Society for Testing Materials	ASTM
BSI	British Standards Institution	BS
CEN	Comité Européen de Normalisation	EN
DIN	Deutsches Institut for Normung	DIN
ISO	International Organisation for Standardisation	ISO

Table 16.1 Standard issuing organisations for valves

There are other standards which do not refer to valves directly but contain essential information for both users and manufacturers. The ATEX regulations, concerning the use of electrical equipment in potentially hazardous environments, were implemented by modifying the existing CEN and IEC standards.

ADR, the European Agreement concerning the International Carriage of Dangerous Goods by Road, is a good example.

NOTE: ADR is a very special Standard. It is not published by a multinational standards authority but it is subject to agreement by many national authorities who publish their own copy.

The standards authorities which are multinational represent the following countries:

ADR Austria, Belgium, Czechoslovakia, Denmark, Finland, France, Germany, Greece, Hungary, Italy, Luxembourg, The Netherlands, New Zealand, Norway, Poland, Portugal, Spain, Sweden, Switzerland, UK, Yugoslavia.

CEN Comité Européen de Normalisation (the European Standards authority) composed by the following National members:

Austria, Belgium, Bulgaria, Cyprus, Czech Republic, Denmark, Estonia, Finland, France, Germany, Greece, Hungary, Iceland, Ireland, Italy, Latvia, Lithuania, Luxembourg, Malta, The Netherlands, Norway, Poland, Portugal, Romania, Slovakia, Slovenia, Spain, Sweden, Switzerland, United Kingdom. CEN is the European regional Standards Authority which issues Standards for the European Union.

CENELEC European Committee for Electrotechnical Standardisation, (a branch of CEN).

ISO Argentina, Australia, Austria, Belgium, Brazil, Bulgaria, Canada, China, Czechoslovakia, Denmark, Egypt, Finland, France, Germany, Greece, Hungary, India, Indonesia, Ireland, Israel, Italy, Japan, Korea (DPR), Korea (REP), Malaysia, The Netherlands, New Zealand, Norway, Pakistan, Poland, Portugal, Romania, Russia, Singapore, South Africa, Spain, Sweden, Switzerland, Turkey, UK, USA, Yugoslavia.

16.2.2 Authorised bodies

Many bodies publish specifications which impact on the manufacturer and use of valves. The most important are listed alphabetically in Table 16.2.

	Body	Standard
API	American Petroleum Institute	API
ARI	Air-Conditioning and Refrigeration Institute (US)	ARI
ASHRAE	American Society of Heating, Refrigeration and Air-conditioning Engineers (of America)	ASHRAE
AWS	American Welding Society	AWS
ASME	American Society of Mechanical Engineers	ASME
AWWA	American Water Works Association	AWWA
CGA	Compressed Gas Association (of America)	CGA
FCI	The Flow Control Institute	FCI
IEEE	The Institution of Electrical and Electronics Engineers (of America)	IEEE
ISA	Instrumentation, Systems and Automation Society	ISA
MSS	Manufacturers Standardization Society of the valve and fittings industry	SP
NACE	National Association of Corrosion Engineers (USA)	MR
NFPA	National Fire Protection Agency (USA)	NFPA
PPI	Plastics Pipe Institute Inc (USA)	
SAE	Society of Automotive Engineers	SAE

Table 16.2 Authorised bodies

16.2.3 Classification societies

These are learned bodies, usually insurance companies, who have rules governing the use of valves and/or associated equipment in certain industries, environments or applications. Typical areas of influence include drinking water, hygienic applications, marine and offshore, oil & gas installations. Table 16.3 indicates some of the most common used societies.

Body	Field of influence
American Society of Mechanical Engineers (ASME)	Stressing of pressure vessels, valve sizing (USA)
American Petroleum Institute (API)	Valve installation and sizing (USA)
Associated Offices Technical Committee (AOTC)	Commercial and industrial compressed air and heating systems (UK)
Bau Institut (BI)	Hygienic applications (Germany)
Beaureau Veritas	Marine, offshore, general process (France)
Det Norske Veritas (DNV)	Marine, offshore, general process (Norway)
Factory Mutual Systems (FM)	Commercial and industrial fire protection systems (UK)
Germanischer Lloyd	Marine, offshore, general process (Germany)
Institute of Electrical Engineers (IEE)	Electrical safety and wiring (UK)
Keurings Instituut voor Waterleiding-Artikelen (KIWA)	Drinking water (The Netherlands)
Lloyd' s Register	Marine, offshore, general process (UK)
Loss Prevention Council (LPC)	Commercial and industrial fire protection systems (UK)
TUV ad-Merkblatt	Pressure vessels (Germany)
Underwriters Laboratory (UL)	Electrical equipment, fire safety (USA)
Underwriters Laboratory (UL)	Electrical equipment, fire safety, fire protection systems (Canada)
Water Research Council (WRc)	Drinking water (UK)

Table 16.3 Classification societies

16.2.4 National governments

National governments appoint organisations to regulate important aspects of valve use. The most important organisations are listed alphabetically in Table 16.4.

Abbreviation	Organisation
DWI	The Drinking Water Inspectorate (UK)
EPA	The Environmental Protection Agency (USA)
FDA	Food and Drink Administration (USA)
HSE	The Health and Safety Executive (UK)
OSHA	Occupational Safety and Health Administration (USA)

Abbreviation	Organisation
TA Luft	Technische Anleitung zur Reinhaltung der Luft (Germany)
USDA	US Department of Agriculture (USA)

Table 16.4 Government appointed organisations

16.2.5 Approved connections

Approval of connections is divided into two main groups:

- hygienic
- safety

Hygienic couplings are tested for cleanliness and the ability to be cleaned or sterilised. Design guidelines are available but approval is only given after testing. Couplings approved from a safety aspect are usually designed to a code and then tested for acceptability. One pressure vessel code requires a production vessel to be tested to destruction before a design can be accepted. This procedure is obviously very costly when one special vessel is required.

Safety approvals may be a prime concern in all forms of transport equipment and military applications.

Hygienic couplings are manufactured in many forms but there are three common features:

- no crevices
- no dead volumes
- approved materials

The most common metallic material is 316 stainless steel. Gaskets can be made from a variety of approved compounds such as:

> silicone rubber, white nitrile rubber, black butyl rubber, white butyl rubber, Buna "N", EPDM, PTFE.

Proprietary designs for hygienic connections are in abundance:

> WauKesha/Cherry Burrell "s", "Q", "I", ILC, ISS, IDF, RJT, SMS 1146, Tri-Clover.

16.3 Components

16.3.1 Gaskets

Gaskets are an important part of bonnet design as well as process connections.

ANSI B16.20 - RTJ rings and grooves

ANSI/ASME B16.21 - Non-metallic flat gaskets for pipe flanges

ANSI/AWWAC111/A21.11 - Rubber gasket joints for ductile-iron and gray-iron pressure pipe and fittings

BS 1806 - "O" rings and housings

BS 3381 - Metallic spiral wound gaskets

BS 4518 - "O" rings and housings

BS 6956 - Joint materials and compounds

FSA - Non-metallic gasketing

FSA - Metallic gaskets

16.3.2 Seals

Proprietary seal specifications greatly outnumber the "standard" seals. There are many standards, too many to list here, which define elastomers for specific fluids and/or specific applications.

ANSI B93.17 - Multiple lip packing sets, methods for measuring stack heights

ANSI B93.62 - Hydraulic fluid power - Reciprocating dynamic sealing devices in linear actuators - Method of testing, measuring and reporting leakage

ANSI/SAE AMS 7261/1D - Phosphonitrilic (FZ) fluoroelastomer sealing rings, high temperature, fluid resistant, 65-75

BS 4371 - Fibrous gland packing

FSA - Chemical compatibility of elastomers for seals

FSA - Compression packings

FSA - Moulded packings

MSS SP-120-2006 - Flexible Graphite Packing System for Rising Stem Steel Valves (Design Requirements)

MSS SP-130-2003 - Bellows Seals for Instrument Valves

MSS SP-132-2004 - Compression Packing Systems for Instrument Valves

16.4 Control and communications

Electronic control and communications will play an increasing role in valve technology in future. Communication "buses" will continue as proprietary standards for the moment.

ANSI/FCI 87-1 - Power signal standard for spring-diaphragm actuated control valves

ANSI/IEEE 961 - FASTBUS, Modular high speed data acquisition and control system

ANSI/ISA S7.3 - Quality standard for instrument air

ANSI/ISA S7.4 - Air pressures for pneumatic controllers, transmitters and transmission systems

EN 50065-1 - Mains bourne signalling systems

EN 61131-1 - Programmable controllers: Part 1: General information - For PLCs and their peripherals. Definitions and principal characteristics ° IEC 61131-1

EN 61131-2 - Programmable controllers: Part 2: General information - For PLCs and their peripherals. Equipment requirements and tests ° IEC 61131-2

EN 61131-3 - Programmable controllers: Part 1: General information - For PLCs and their peripherals. Programming languages ° IEC 61131-3

ISO 1219-1 - Fluid power systems and components - Graphic symbols and circuit diagrams - Part 1: Graphic symbols

ISO 3511-1 - Process measurement control functions and instrumentation - Symbolic representation - Part 1: Basic requirements

ISO 3511-2 - Process measurement control functions and instrumentation - Symbolic representation - Part 2: Extension of basic requirements

ISO 3511-3 - Process measurement control functions and instrumentation - Symbolic representation - Part 3: Detailed symbols for instrument interconnection diagrams

16.5 Design

Many standards provide guidance on the design of systems and valves. Pressure vessel standards are a good source of useful information as well as safety relief valve requirements.

AD A4 - Valve bodies (AD-Merkblätter)

ANSI B1.20.1 - Standards for pipe threads

ANSI B31.10 - Code for pressure piping

ANSI B31.3 - Petroleum refinery piping

ANSI B31.4 - Oil transportation piping

ANSI B31.5 - Refigeration piping systems

ANSI B31.7 - Nuclear power piping

ANSI B31.8 - Gas transmission and distribution piping systems

ANSI/API 14.3.1 - Manual of Petroleum Measurement Standards; Chapter 14.3 Part 1 - General Equations and Uncertainty Guidelines - Concentric, Square-edged Orifice Meters

ANSI/API 14.3.2 - Manual of Petroleum Measurement Standards; Chapter 14.3 Part 2 Specification and Installation Requirements - Concentric - Square-edged Orifice Meters

ANSI/API 14.3.3 - Manual of Petroleum Measurement Standards; Chapter 14.3 Part 3 Natural Gas Applications

ANSI/API 2610, Construction, Operation, Maintenance & Inspection of Terminal and Tank Facilities

ANSI/API 599 - Metal Plug Valves - Flanged, Threaded and Welding End

ANSI/API 600 = ISO 10434 - Bolted Bonnet Steel Gate Valves

ANSI/API 603 - Corrosion-Resistant, Bolted Bonnet Gate Valves - Flanged and Butt-Welding Ends

ANSI/API RP 14E and Installation of Offshore Production Platform Piping Systems

ANSI/ASME B1.1 - Unified inch screw threads, UN and UNR thread form

ANSI/ASME B31.1 - Power piping

ANSI/ASME B31.10 - Pressure piping

ANSI/ASME B31.11 - Slurry transportation piping systems

ANSI/ASME B31.3 - Chemical plant and petroleum refinery piping

ANSI/ASME B31.4 - Liquid transportation systems for hydrocarbons, liquid petroleum gas, anhydrous ammonia and alcohol

ANSI/ASME B31.5 - Refrigeration piping

ANSI/ASME B31.6 - Nuclear power piping

ANSI/ASME B31.8 - Gas transmission and distribution piping systems

ANSI/ASME B31.9 - Building services piping code

ANSI/ASTM A998 - Practice for Structural Design of Reinforcements for Fittings in Factory-Made Corrugated Steel Pipe for Sewers and Other Applications

ANSI/AWWA C150/A21.50 thickness of ductile-iron pipe

ANSI/AWWA C151/A21.51 thickness of centrifugally cast ductile-iron pipe

ANSI/NFPA 1963 - Screw threads and gaskets for fire hose connections

ANSI/SAE AIR 4066 - Hydraulic servo-actuator materials

ANSI/SAE ARP 674A - Groove design, metal "O" ring gaskets

ANSI/SAE HIR 1415A guide for hydraulic systems for submersible vehicles

ANSI/SAE HIR 1496B - Hydraulic system design criteria for advanced marine surface vehicles

API 11V1 - Gas Lift Valves, Orifices, Reverse Flow Valves and Dummy Valves

API 14A = ISO 10432 - Specification for Subsurface Safety Valve Equipment

API 2510 and Construction of LPG Installations

API 521 = ISO 23251 - Guide for Pressure-relieving and De-pressurising Systems

API 594 - Check Valves: Flanged, Lug, Wafer and Butt-welding

API 602 = ISO 15761 - Steel Gate, Globe and Check Valves for Sizes DN 100 and Smaller

API 608 - Metal Ball Valves - Flanged, Threaded and Butt-Welding Ends

API 609 - Butterfly Valves: Double Flanged, Lug- and Water-Type

API 6D = ISO 14313 - Specification for Pipeline Valves

API Bulletin 5C2 - Performance Properties of Casing, Tubing, and Drill Pipe

API Bulletin 5C3 - Formulas and Calculations for Casing, Tubing, Drill Pipe, and Line Pipe Properties

API MPMS Chpt 14.2 - Manual of Petroleum Measurement Standards; Chapter 14.2 Compressibility Factors of Natural Gas and Other Related Hydrocarbon Gases

API MPMS Chpt 14.4 - Manual of Petroleum Measurement Standards; Chapter 14.3 Part 4 Converting Mass of Natural Gas Liquids and Vapors to Equivalent Liquid Volumes

API RP 1111 - Design, Construction, Operation, and Maintenance of Off-shore Hydrocarbon Pipeline and Risers

API RP 11V6 of Continuous Flow Gas Lift Installations Using Injection Pressure Operated Valves

API RP 14B = ISO 10417 - Design, Installation, Repair and Operation of Subsurface Safety Valve Systems

API RP 2N - Planning, Designing, and Constructing Structures and Pipelines for Arctic Conditions

API RP 520 - 1 - Sizing, Selection and Installation of Pressure-relieving Devices in Refineries, Part I Sizing and Selection

API RP 553 - Refinery Control Valves

API RP 86 - Recommended Practice for Measurement of Multiphase Flow

API TR 5TRSR22 - Technical Report in SR22 Supplementary Requirements for Enhanced Leak Resistance

API TR 6F1 - Performance of API and ANSI End Connections in a Fire Test According to API Specification 6FA

API TR 6F2 - Fire Resistance Improvements for API Flanges

API TR 938-C - Use of Duplex Stainless Steels in the Oil Refining Industry

API/IP Spec 1584 - Four-inch Aviation Hydrant System Components and Arrangements

BS 1104 - Acme screw threads

BS 1134 - Assessment of surface texture, guidance and general information

BS 1580 = ISO 263 = ISO/R725 - Unified screw threads

BS 1963 - Pressure operated relay valves for gas burning appliances

BS 3351 - Piping systems for petroleum refineries and petrochemical plants

BS 3643 = ISO 68 - ISO metric screw threads

BS 6920 - Suitability of non-metallic products for use in contact with water intended for human consumption with regards to their effect on the quality of the water

BS MA 18 - Salt water piping systems in ships

DIN 19569 Pt 4 - Sewage treatment plants; principles for the design of structures and technical equipment; specific principles for shut-off devices without body

DIN 2413 - Calculation of wall thickness of steel pipes subjected to internal pressure

DIN 3339 - Valves; body materials

DIN 3790 and sizes of oblique globe valves; 380psig & 150psig

DIN 3840 - Valve bodies; strength calculations for internal pressure

DIN 4762-1 - Surface roughness; terminology

DIN 4768-1 -

EN 10226 - 1 - Pipe threads where pressure tight joints are made on the threads - Part 1: Taper threads - Designation, dimensions and tolerances

EN 10243 - 1 - Steel drop and press forgings - Tolerances on dimensions

EN 10243 - 2 - Steel upset forgings made on horizontal forging machines -Tolerances on dimensions

EN 13445 - all parts - Unfired pressure vessels

EN 20898-2 - Mechanical properties of fasteners - Part 2: Nuts with specified proof load values - Coarse threads

EN ISO 228-1 - Pipe threads where pressure-tight joints are not made on the threads - Part 1: Dimensions, tolerances and designation

EN ISO 898-1 - Mechanical properties of fasteners made of carbon steel and alloy steel - Part 1: Bots, screws and studs

ISA S75.17 - Control valve aerodynamic noise prediction

ISA S75.18 - Control valve stability

ISO 15649 - Petroleum and natural gas industries - Piping

ISO 261 - ISO general-purpose metric screw threads - General plan

ISO 262 - ISO general-purpose metric screw threads - Selected sizes for screws, bolts and nuts

ISO 286 - all parts - ISO system of limits and fits (tolerancing practices)

ISO 3448 - Industrial liquid lubricants - ISO viscosity classification

ISO 7-1 - Pipe threads where pressure-tight joints are made on the threads - Part 1: Dimensions, tolerances and designation

ISO 724 - ISO general-purpose metric screw threads - Basic dimensions

ISO 965 - all parts - ISO general-purpose metric screw threads - Tolerances

ISO R/468 -

MSS SP-60-2004 - Connecting flange joint between tapping sleeves and tapping valves

PD 5500 - Pressure vessels (The withdrawn Standard BS 5500, still used outside the EU)

SSPC SP 6 - Surface preparation specification

16.5 Dimensions and pressure ratings

Dimensions and ratings of valves are the subject of many standards. The standardisation of dimensions is crucial for interchangeability between manufacturers.

ANSI/API 526 - Flanged steel safety relief valves for use in petroleum refineries

ANSI/ARI 720 - Refrigerant access valves and hose connectors

ANSI/ARI 750 - Thermostatic refrigerant expansion valves

ANSI/ARI 760 - Solenoid valves for use with volatile refrigerants

ANSI/ASME A112.21.3M - Hydrants for utility and maintenance use

ANSI/ASME A112.26.1M - Water hammer arresters

ANSI/ASME B16.10 - Face-to-face and end-to-end dimensions of valves

ANSI/ASME B16.33 - Manually operated metallic gas valves for use in gas piping systems up to 125 psig (8.6 barg)

ANSI/ASME B16.34 - Valves - Flanged, threaded and welding end

ANSI/ASME B16.38 - Large metallic valves for gas distribution, manually operated, NPS 2.5 to 12 inch, 125 psig (8.6 barg)

ANSI/ASME B16.40 - Gas shut-offs and valves in gas distribution systems, manually operated, thermoplastic

ANSI/ASME N278.1 - Self-operated and power operated safety related valves, functional specification standard

ANSI/AWWA C500 - Gate valves for water and sewage systems

ANSI/AWWA C501 - Cast iron sluice gates

ANSI/AWWA C502 - Fire hydrants, dry barrel

ANSI/AWWA C503 - Fire hydrants, wet barrel

ANSI/AWWA C504 - Rubber seated butterfly valves

ANSI/AWWA C507 - Ball valves, 6 to 48"

ANSI/AWWA C508 - Swing-check valves for waterworks service, 2" to 24" NPS

ANSI/AWWA C509 - Resilient seated gate valves for water and sewerage systems

ANSI/AWWA C510 - Double check valve backflow prevention assembly

ANSI/AWWA C511 - Reduced pressure principle backflow prevention assembly

ANSI/AWWA C512 - Air release, air/vacuum and combination air valves for waterworks service

ANSI/AWWA C540 - Power actuating devices for valves and sluice gates

ANSI/AWWA C800 - Underground service line valves and fittings

ANSI/CGA E-7 - Medical gas regulators and flowmeters

ANSI/CGA V-1 - Compressed gas cylinder valve outlet and inlet connections

ANSI/CGA V-9 - Compressed gas cylinder valves

ANSI/FCI 74-1 - Spring loaded lift disc check valve standard

ANSI/ISA S75.03 - Face-to-face dimensions for flanged globe-style control valve bodies

ANSI/ISA S75.04 - Face-to-face dimensions of flangeless control valves

ANSI/ISA S75.08 - Installed face-to-face dimensions for clamp or pinch valves

ANSI/ISA S75.12 - Socket-weld end and screwed end globe-style control valves (ANSI Classes 150, 300, 600, 900, 1500, 2500), face-to-face dimensions

ANSI/ISA S75.14 - Face-to-face dimensions for butt-weld end globe-style control valves

ANSI/ISA S75.15 - Butt-weld end globe-style control valves (ANSI Classes 150, 300, 600, 900, 1500 and 2500), face-to-face dimensions

ANSI/ISA S75.16 - Flanged globe-style control valve bodies (ANSI Classes 900, 1500 and 2500), face-to-face dimensions

ANSI/ISA S75.22 - Face-to-centering dimensions for flang-ed globe-style angle control valve bodies

ANSI/NFPA 1964 - Spray nozzles (shut-off and tip) (for fire hose)

ANSI/SAE AS 568A - Aerospace size standard for "O" rings

ANSI/UL 1106 - Marine manually operated shut-off valves for flammable liquids

ANSI/UL 144 - Pressure regulating valves for LP-gas

ANSI/UL 1468 - Direct-acting pressure-reducing and pressure-control valves for fire protection service

ANSI/UL 193 - Alarm valves for fire protection service

ANSI/UL 207 - Refrigerant containing components and accessories

ANSI/UL 312 - Check valves for fire protection service

ANSI/UL 668 - Hose valves for protection service

ANSI/UL 698 - Industrial control equipment for use in hazardous (classified) locations

ANSI/UL 842 - Valves for flammable fluids

BS 1212 - Float operated valves

BS 1414 - Steel wedge gate valves (flanged and butt welding ends) for the petroleum, petrochemical and allied industries (DN25 to DN600, 150LB to 2 500LB)

BS 1415 - Mixing valves; thermostatic and non-thermostatic

BS 1868 - Steel check valves (flanged and butt welding ends) for the petroleum, petrochemical and allied industries (DN15 to DN600, 150LB to 2 500LB)

BS 1873 - Steel globe and globe stop and check valves (flanged and butt welding ends) for the petroleum, petrochemical and allied industries (DN15 to DN400, 150LB to 2 500LB)

BS 2080 - Face-to-face, centre-to-face, end-to-end and centre-to-end dimensions of flanged and butt welding end steel valves for the petroleum, petrochemical and allied industries = ISO 5751

BS 5150 - Cast iron wedge and double disc gate valves for general purposes (DN10 to DN1 000, PN1.6 to PN25)

BS 5151 - Cast iron gate (parallel slide) valves for general purposes (DN40 to DN1 000, PN10 to PN25) [1] ISO 5996

BS 5152 - Cast iron globe and globe stop and check valves for general purposes (DN10 to DN450, PN10 to PN25)

BS 5153 - Cast iron check valves for general purposes (DN10 to DN1 000, PN6 to PN25)

BS 5154 - Copper alloy globe, globe stop and check, check and gate valves for general purposes (DN10 to DN80, PN16 and PN50)

BS 5155 - Specification for butterfly valves (DN40 to DN2 000, up to PN40, 300LB)

BS 5156 - Screw down diaphragm valves for general purposes (DN10 to DN300, PN6, PN10 and PN16)

BS 5157 - Steel gate (parallel slide) valves for general purposes (DN40 to DN600, PN16 to PN100)

BS 5158 - Cast iron and carbon steel plug valves for general purposes (DN10 to DN600, PN10 to PN100)

BS 5159 - Cast iron and carbon steel ball valves for general purposes (DN10 to DN600, PN10 to PN100) [1] ISO 7121

BS 5160 - Specification for flanged steel globe valves, globe stop and check valves for general purposes (DN10 to DN450, PN16 to PN40)

BS 5163 - Double flanged cast iron wedge gate valves for waterworks purposes (DN50 to DN600, PN10 and PN16) [1] ISO 7259

BS 5351 - Steel ball valves for the petroleum, petrochemical and allied industries (DN8 to DN400, 150LB to 900LB)

BS 5352 - Specification for steel wedge gate, globe and check valves, 50 mm and smaller for the petroleum, petrochemical and allied industries (800 pound and 1 500 pound)

BS 5353 - Specification for plug valves (DN8 to DN800, 150LB to 2 500LB)

BS 5793 Pt 3 - Face-to-face dimensions of control valves ° IEC 60534-3-2

BS 5793 Pt 6 - Mounting details for attachment of positioners and actuators ° IEC 60534-6

BS 5840 - Valve mounting details for actuators = ISO 5210/1, /2, ISO 5211/1, /2, /3

BS 6331 - Mounting dimensions for double-acting hydraulic cylinders = ISO 6020/1 [1] ISO 6099, ISO 8133, ISO 8134

BS 6364 - Specification for valves for cryogenic service

BS 7438 - Steel and copper alloy wafer check valves

BS 750 - Specification for underground fire hydrants and surface box frames and covers (16 barg)

DIN 11862 - Drink and dairy fittings; straight-way cocks of stainless steel

DIN 2532/3 - Flanges

DIN 2632/3 - Flanges

DIN 3160 - Valves for refrigerant circuits, PN25

DIN 3161 - Angle valves for refrigerant circuits, PN25

DIN 3163 - Control valves for refrigerant circuits, PN25

DIN 3164 - Position indicators for refrigerant valves

DIN 3202 Pt 2 - Face-to-face and centre-to-face dimensions for butt-weld valves

DIN 3202 Pt 3 - Face-to-face dimensions for wafer type valves

DIN 3202 Pt 4 - Face-to-face and centre-to-face dimensions for screwed valves

DIN 3230 Pt 1 - Specification for valves; enquiry, order and delivery

DIN 3230 Pt 2 - Specification for valves; general requirements

DIN 3230 Pt 3 - Specification for valves; certification of testing

DIN 3230 Pt 5 - Specification for valves; gas distribution and installations; requirements and testing

DIN 3230 Pt 6 - Specification for valves; flammable liquids; requirements and test methods

DIN 3239 Pt 1 - Valve welding ends

DIN 3320 Pt 3 - Safety valves; centre-to-face dimensions for flanged valves, PN40 to PN250 inlet

DIN 3337 - Actuator mounting interface

DIN 3341 - Plate valves for displacement compressors; suction valves, discharge valves; main dimensions, materials, installation

DIN 3352 Pt 1 - Gate valves; general information

DIN 3352 Pt 10 - Gate valves of stainless steel

DIN 3352 Pt 11 - Flanged copper alloy gate valves

DIN 3352 Pt 12 - Socket end copper alloy gate valves

DIN 3352 Pt 13 - Double socket cast iron gate valves with elastomeric seat and inside screw stem

DIN 3352 Pt 2 - Cast iron gate valves, with metallic seat and inside screw stem

DIN 3352 Pt 3 - Cast iron gate valves, with metallic seat and outside screw stem

DIN 3352 Pt 4 - Cast iron gate valves, with elastomeric seat and inside screw stem

DIN 3352 Pt 5 - Steel gate valves, isomorphe series

DIN 3352 Pt 6 - Gate valves of unalloyed and low alloyed steel, with inside screw stem

DIN 3352 Pt 7 - Gate valves of unalloyed and low alloyed steel, with outside screw stem

DIN 3352 Pt 8 - Low temperature steel gate valves, with outside screw stem

DIN 3352 Pt 9 - Gate valves of heat resistant steel

DIN 3354 Pt 1 - Butterfly valves; general data

DIN 3354 Pt 2 - Butterfly valves; soft seat cast iron tight shut-off valves with flanges

DIN 3354 Pt 3 - Butterfly valves; soft seat, steel or cast steel tight shut-off valves with flanges or butt weld ends

DIN 3354 Pt 4 - Butterfly valves; metal seat seal, steel or cast steel tight shut-off valves with flanges or butt weld ends

DIN 3354 Pt 5 - Butterfly valves; concentric tight shut-off, soft seat, wafer or flanged style

DIN 3356 Pt 1 - Globe valves; general information

DIN 3356 Pt 2 - Globe valves; cast iron stop valves

DIN 3356 Pt 3 - Globe valves; carbon steel stop valves

DIN 3356 Pt 4 - Globe valves; high temperature steel stop valves

DIN 3356 Pt 5 - Globe valves; stainless steel stop valves

DIN 3357 Pt 2 - Full bore metal ball valves

DIN 3357 Pt 3 - Reduced bore metal ball valves

DIN 3357 Pt 4 - Full bore non-ferrous metal ball valves

DIN 3357 Pt 5 - Reduced bore non-ferrous metal ball valves

DIN 3357 Pt 6 - Full bore cast iron ball valves

DIN 3357 Pt 7 - Reduced bore cast iron ball valves

DIN 3358 - Linear valve actuator attachment; fixing dimensions

DIN 3359 - Metallic diaphragm valves

DIN 3380 - Gas pressure regulators for supply pres-sures up to 100 bar

DIN 3381 - Safety devices for gas supply installations up to 100 bar; pressure relief governors and safety shut-off devices

DIN 3391 - Valves and multi-functional controls for gaseous fuels; general requirements

DIN 3393 Pt 1 - Combinations of controls for gas appliances; combinations

DIN 3430 - Gas valves, angle pattern ball shut-off valves

DIN 3431 - Gas valves, angle pattern ball shut-off valves with compression fittings

DIN 3432 - Gas valves, straight pattern ball shut-off valves with compressions fittings

DIN 3434 - Gas valves, straight pattern ball shut-off valves with compressions fittings

DIN 3435 - Gas valves, angle pattern ball shut-off valves with compressions fittings

DIN 3436 - Valves for gas installations; nozzles and gaskets

DIN 3437 - Gas stop valves for pressures over 16 bar; requirements and acceptance tests

DIN 3441 Pt 4 - Oblique valves of rigid PVC (unplasticised or rigid); dimensions

DIN 3441 Pt 5 - Unplasticised polyvinyl chloride (PVC-U) valves; PN6 and PN10, wafer type butterfly valves, dimensions

DIN 3441 Pt 6 - Unplasticised polyvinyl chloride (PVC-U) valves; gate valves with inside screw stem, dimensions

DIN 3441 Pt 7 - Valves for water supply systems; requirements and acceptance tests

DIN 3442 Pt 1 - Polypropylene (PP) valves; requirements and testing

DIN 3442 Pt 3 - Polypropylene (PP) diaphragm valves; dimensions

DIN 3543 Pt 3 - PVC tapping valves for plastic pipes; dimensions

DIN 3543 Pt 4 - High density polyethylene (HDPE) tapping valves for HDPE pipes; dimensions

DIN 3544 Pt 1 - High density polyethylene (HDPE) valves; tapping valves; requirements and testing

DIN 477 Pt 1 - Gas cylinder valves rated for test pressures up to 300 bar; types, sizes and outlets

DIN 477 Pt 4 - Compressed gas cylinder valves; swing check valves for camping cylinders

DIN 477 Pt 5 - Gas cylinder valves rated for test pressures up to 450 bar; types, sizes and outlets

DIN 477 Pt 8 - Gas cylinder valves; separate connections for medical gas cylinders up to 3 litres capacity (PIN-index-system)

DIN 4811 Pt 1 - Pressure regulators for LPG; regulators with uncontrolled inlet pressure for outdoor appliances

DIN 4811 Pt 3 - Pressure regulators for LPG; regulators and safety devices with uncontrolled inlet pressure and fixed or adjustable outlet pressure up to 4 bar

DIN 4811 Pt 4 - Pressure regulators for LPG; regulators and safety devices with uncontrolled inlet pressure for appliances with LPG cylinders

DIN 4811 Pt 5 - Pressure regulators for LPG; regulators and safety devices for LPG tanks

DIN 4811 Pt 6 - Pressure regulators for LPG; regulators and safety devices for appliances with controlled inlet pressure

DIN 4811 Pt 7 - Pressure regulators for LPG; regulators for LPG appliances in caravans and sporting boats

DIN 4817 Pt 1 - Shut off valves for liquified gases; definitions, requirements, testing and marking

EN 1074-1 - Valves for water supply; specification for use and appropriate verification tests; general requirements

EN 1074-2 - Valves for water supply; specification for use and appropriate verification tests; isolating valves

EN 1074-3 - Valves for water supply; specification for use and appropriate verification tests; check valves

EN 1213 - Building valves; copper alloy stop valves for potable water supply in buildings; general specification

EN 161 - Automatic shut-off valves for gas burners and gas appliances

EN 19 = ISO 5209 - Marking of general purpose industrial valves

EN 215 - Thermostatic valves for radiators

EN 246 - Sanitary taps; general specifications for flow rate regulators

EN 248 - Sanitary taps; general technical specification for electro-deposited surface nickel chrome coatings

EN 28659 = ISO 8659 = ISO 9393-1 - Unplasticised polyvinyl chloride (PVC-U) valves; requirements and testing

EN 331 - Manually operated ball valves and closed bottom taper plug valves for gas installations for buildings

EN 558-1 - Metal valves for flanged pipe systems; face-to-face and centre-to-face dimensions; general

EN 558-2 - Metal valves for flanged pipe systems; face-to-face and centre-to-face dimensions; PN designated valves

EN 558-3 - Metal valves for flanged pipe systems; face-to-face and centre-to-face dimensions; class designated valves

EN 593 - Industrial valves; metallic butterfly valves for general purposes

EN 88 - Pressure governors for gas appliances for inlet pressures up to 200 mbar

IEC 60534-3-2 - Industrial process control valves; Part 3; dimensions, Section 3; face-to-face dimensions for flangeless control valves except wafer butterfly valves

IEC 60534-5 - Industrial process control valves; marking

IEC 60534-6 - Industrial process control valves; mounting details for attachment of positioners to control valve actuators

IEC 60534-7 - Industrial process control valves; control valve data sheet

ISA S4.01.1 - Face-to-face dimensions of flanged globe valves

ISA S4.01.2 - Face-to-face dimensions of flangeless partial ball valves

ISA S75.03 - Face-to-face dimensions for globe-style control valve bodies (Classes 125, 150, 250, 300 and 600)

ISA S75.04 - Dimensions for flangeless control valves (Classes 150, 300 and 600)

ISA S75.08 - Installed face-to-face dimensions for flanged clamp or pinch valves

ISA S75.12 - Face-to-face dimensions for socket weld-end and screwed-end globe-style control valves (Classes 150, 300, 600, 900, 1500 and 2500)

ISA S75.14 - Face-to-face dimensions for butt weld-end globe-style control valves (Class 4500)

ISA S75.15 - Face-to-face dimensions for butt weld-end globe-style control valves (Classes 150, 300, 600, 900, 1500 and 2500)

ISA S75.16 - Face-to-face dimensions for flanged glode-style control valve bodies (Classes 900, 1500 and 2500)

ISA S75.20 - Face-to-face dimensions for separable flanged, globe-style control valves (ANSI Classes 150, 300 and 600)

ISA S75.22 - Face-to-centreline dimensions for flanged globe-style angle control valve bodies (Classes 150, 300 and 600)

ISO 11191 - Gas cylinder valves; taper threads, gauges

ISO 2503 - Pressure regulators for gas cylinders used in welding, cutting and related processes; terminology, requirements and testing

ISO 4126-1 - Safety relief valves - General requirements

ISO 5145 - Gas cylinder valves; for highest grade gases, sizes, connections, threads

ISO 5210-1 - Multi-turn valve actuator attachment; flange dimensions

ISO 5210-2 - Multi-turn valve actuator attachment; flange and coupling performance characteristic

ISO 5210-3 - Multi-turn valve actuator attachment; dimensions of driving components

ISO 5211-1 - Part-turn valve actuator attachment; flange dimensions

ISO 5211-2 - Part-turn valve actuator attachment; flange and coupling performance characteristics

ISO 7508 - Unplasticised polyvinyl chloride (PVC-U) ball valves and diaphragm valves, dimensions

ISO 8242 - Polypropylene (PP) valves, ball valves, dimensions

ISO 8659 = EN 28659 - Unplasticised polyvinyl chloride (PVC-U) valves; requirements and testing

ISO 9393-1 = EN 28659 - Unplasticised polyvinyl chloride (PVC-U) valves; requirements and testing

MSS SP-101-2001 - Part-turn valve actuator attachment - Flange and driving component dimensions and performance characteristics

MSS SP-102-2001 - Multi-turn valve actuator attachment - Flange and driving component dimensions and performance characteristics

MSS SP-105-2005 - Instrument valves for code applications

MSS SP-108-2002 - Resilient-seated eccentric cast iron plug valves

MSS SP-110-1996 - Ball valves, threaded, socket-welding, solder joint, grooved and flared ends

MSS SP-25-1998 - Standard marking system for valves, fittings, flanges and unions

MSS SP-42-2004 - Class 150 corrosion resistant gate, globe, angle and check valves with flanged and butt-weld ends

MSS SP-67-2002 - Butterfly valves

MSS SP-68-2004 - High pressure offset-seat butterfly valves

MSS SP-70-2006 - Cast iron gate valves, flanged and threaded ends

MSS SP-71-2005 - Cast iron swing check valves, flanged and threaded ends

MSS SP-72-2000 - Ball valves with flanged or butt-welding ends for general service

MSS SP-78-2005 - Cast iron plug valves, flanged and threaded ends

MSS SP-80-2003 - Bronze gate, globe, angle and check valves

MSS SP-81-2006 - Stainless steel, bonnetless, flanged knife gate valves

MSS SP-85-2002 - Cast iron globe and angle valves, flanged and threaded ends

MSS SP-86-2002 - Guidelines for metric data in standards for valves, flanges, fittings and actuators

MSS SP-88-2001 - Diaphragm type valves

MSS SP-99-2005 - Instrument valves

VDE 0730 Pt 2ZB - Electric motor operated appliances for domestic and similar purposes; particular requirements for magnetic valves

VDE/VDI 3845 - Industrial process control valves; interfaces between valves, actuators and auxiliary equipment (NAMUR and ISO 5211)

16.7 Electrical

Electricity is used for both power and instrumentation in connection with valves. There are many different standards, from different areas of the World, covering a wide range of topics including many safety subjects.

ANIA/NFPA 70 - National Electrical Code

IEC 60034-1 - Rotating electrical machines - Part 1: Ratings and performance

IEC 60079 - all parts apparatus for explosive gas atmospheres

IEC 60526 - Degrees of protection provided by enclosures (IP Code)

IEEE 841 - Standard for petroleum and chemical industry - Severe duty totally enclosed fan-cooled squirrel cage induction motors

NEMA 250 - Enclosures for low voltage electrical equipment

16.8 Fasteners

Fasteners are used widely for valve assembly and also for building the valve into the piping system. Fasteners were one of the first engineering components to be standardised but there are a bewildering array of standards currently available. Even though approximately 75% of the World is nominally 'Metric', there is a huge number of 'inch' fasteners still in use. The thread form is very important but the material of the fastener is of equal importance because of the range of strengths available.

EN 10269 - Steels and nickel alloys for fasteners with specified elevated and/or low temperature properties

16.9 Fittings

Fittings for pipes and tubes are very diverse and are covered by a multitude of standards.

ANSI B16.11 - Forged steel pipe fittings - socket weld and threaded (0.125" to 4" nb, ASTM A182 materials), 2000lb to 9000lb ratings

ANSI B16.28 - Wrought steel butt weld short radius elbows and returns (0.5" to 24" nb, ASTM A403 materials)

ANSI B16.9 - Dimensions and tolerances for butt weld fittings

ANSI/ASME B16.11 - Forged steel pipe fittings - socket weld and threaded (0.125" to 4" nb, ASTM A182 materials), 2000lb to 9000lb ratings

ANSI/ASME B16.12 - Cast-iron threaded drainage fittings

ANSI/ASME B16.15 - Cast bronze threaded fittings, Classes 125 and 250

ANSI/ASME B16.22 - Wrought copper and copper alloy solder-joint pressure fittings

ANSI/ASME B16.23 - Cast copper alloy solder-joint drainage fittings - DWV

ANSI/ASME B16.24 - Cast copper alloy pipe flanges and flanged fittings

ANSI/ASME B16.26 - Cast copper alloy fittings for flared copper tubes

ANSI/ASME B16.28 - Wrought steel butt weld short radius elbows and returns (0.5" to 24" nb, ASTM A403 materials)

ANSI/ASME B16.29 - Wrought copper and wrought copper alloy solder-joint drainage fittings - DWV

ANSI/ASME B16.3 - Malleable-iron threaded fittings, Classes 150 and 300

ANSI/ASME B16.32 - Cast copper alloy Solvent® joint fittings for Solvent® drainage systems

ANSI/ASME B16.4 - Cast-iron threaded fittings, Classes 125 and 250

ANSI/ASME B16.42 - Ductile iron pipe flanges and flanged fittings

ANSI/ASME B16.45 - Cast-iron fittings for Solvent® drainage systems

ANSI/ASME B16.9 - Factory made wrought steel butt weld fittings (0.5" to 48" nb, ASTM A403 materials)

ANSI/ASTM A181 - Specification for Carbon Steel Forgings, for General-purpose Piping

ANSI/ASTM A234 - Specification for Piping Fittings of Wrought Carbon Steel and Alloy Steel for Moderate and High Temperature Service

ANSI/ASTM A350 - Specification for Carbon and Low-Alloy Steel Forgings, Requiring Notch Toughness Testing for Piping Components"

ANSI/ASTM A403 - Specification for Wrought Austenitic Stainless Steel Piping Fittings

ANSI/ASTM A420 - Specification for Piping Fittings of Wrought Carbon Steel and Alloy Steel for Low-Temperature Service

ANSI/ASTM A696 - "Standard specification for steel bars, carbon, hot-wrought or cold-finished, special quality, for pressure piping components"

ANSI/ASTM A74 - Specification for Cast Iron Soil Pipe and Fittings

ANSI/ASTM A758 - Specification for Wrought-Carbon Steel Butt-Welding Piping Fittings with Improved Notch Toughness

ANSI/ASTM A774 - Specification for As-Welded Wrought Austenitic Stainless Steel Fittings for General Corrosive Service at Low and Moderate Temperatures

ANSI/ASTM A815 - Wrought Ferritic, Ferritic/Austenitic (Duplex) and Martensitic Stainless Steel Pipe Fitting

ANSI/AWWA C110/A21.10 - Ductile-iron and gray-iron fittings, 3" to 48", for water and other liquids

ANSI/AWWA C153/A21.53 - Ductile-iron compact fittings, 3" to 16", for water and other liquids

ANSI/AWWA C219 - Bolted, sleeve-type couplings for plain-end pipe

ANSI/AWWA C907 - Polyvinyl chloride (PVC) pressure fittings for water - 4" to 8"

ANSI/SAE J513 - Refrigeration tube fittings

API 14L = ISO 16070 - Lock Mandrels and Landing Nipples

API 6H - End Closures, Connectors, and Swivels

BS 1740 Pt 1 - Low pressure screwed pipe fittings (150 lb)

BS 3799 - Steel pipe fittings for the petroleum and allied industries (207 bar(g))

BS 4346 ISO 2035 ISO2043-5 ISO 2048 - Joints and fittings for use with unplasticised PVC pressure pipes

BS 4368 - Carbon and stainless steel compression couplings

BS 864 - Capillary and compression tube fittings

BS AU 153 = ISO 2974 - Fuel injection high pressure pipe fittings

BS AU 68 - Fuel injection low pressure pipe fittings

BS MA 67 - Pipe couplings for use in marine pipework systems; summary and application

DIN 32676 - Food industry fittings; clamp connections for stainless steel tubes

EN 10241 - Threaded steel fittings

EN 10242 - Threaded pipe fitting in malleable cast iron

EN 10253 - 1 - Butt welding pipe fittings wrought carbon steel without specific inspection requirements

EN 10284 - Malleable cast iron fittings with compression ends for plastics piping system

EN 20049 - Malleable cast iron threaded pipe fittings; Part 1: fittings with parallel internal and taper external threads in accordance with ISO 7-1

EN ISO 10806 - Pipework - Non-alloyed and stainless steel fittings for corrugated flexible metallic hoses

EN ISO 8434 - 1 - Metallic tube connections for fluid power and general use - Part 1: 24 Compression fittings

ISO 2016 - Capillary solder fittings for copper tubes - Assembly dimensions and tests

ISO 2037 - Stainless steel pipe and fittings (hygienic)

MSS SP-103-1995 (R00) - Wrought copper and copper alloy insert fittings for polybutylene systems

MSS SP-104-2003 - Wrought copper LW solder joint pressure fittings

MSS SP-107-1991 (R00) - Transition union fittings for joining metal and plastic products

MSS SP-109-2006 - Welded Fabricated Copper Solder Joint Pressure Fittings

MSS SP-111-2005 - Gray-Iron and Ductile-Iron Tapping Sleeves

MSS SP-114-2001 - Corrosion Resistant Pipe Fittings Threaded and Socket Welding, Class 150 and 1000

MSS SP-119-2003 - Belled End Socket Welding Fittings, Stainless Steel and Copper Nickel

MSS SP-123-2006 - Non-Ferrous Threaded and Solder-Joint Unions for Use With Copper Water Tube

MSS SP-124-2001 - Fabricated Tapping Sleeves

MSS SP-129-2003 - Copper-Nickel Socket-Welding Fittings and Unions

MSS SP-43-2001 - Wrought stainless steel butt-welding fittings

MSS SP-51-2003 - Class 150LW Corrosion Resistant Cast Flanged Fittings

MSS SP-75-2004 - Specification for High Test Wrought Butt Welding Fittings

MSS SP-79-2004 - Socket-Welding Reducer Inserts

MSS SP-83-2006 - Class 3000 Steel Pipe Unions, Socket-Welding and Threaded

MSS SP-95-2006 - Swaged Nipples and Bull Plugs

MSS SP-97-2006 - Integrally Reinforced Forged Branch Outlet Fittings

16.10 Flanges

The shape and dimensions of flanges is dealt with by many standards which are sub-divided according to material, type and pressure rating. The following Standards, Table 16.5, are the most popular and cover working pressures up to 420 barg depending upon the material and the temperature.

Standard	Dimensions
ISO 7005	both
BS 4504	metric
ANSI/ASME B16.5 / BS 1560	inch
BS 10	inch
SAE	inch

Table 16.5 Popular flange Standards

ANSI B16.36 - Steel Orifice Flanges

ANSI B16.5 - Steel pipe flanges, flanged valves and fittings (0.5" to 24" nb, 150lb, 300lb, 400lb, 600lb, 900lb, 1500lb, 2500lb)

ANSI/ASME B16.1 - Cast iron flanges

ANSI/ASME B16.36 - Orifice flanges

ANSI/ASME B16.42 - Ductile iron pipe flanges and flanged fittings

ANSI/ASME B16.47 - Large diameter steel flanges

ANSI/ASME B16.5 - Steel flanges

ANSI/ASTM A105 - Specification for Carbon Steel Forgings for Piping Applications (UNS K03504)

ANSI/ASTM A707 - Specification for Forged Carbon and Alloy Steel Flanges for Low-Temperature Service

ANSI/AWWA C207 - Steel pipe flanges for waterworks service, 4" to 144"

API 605 - Large diameter steel flanges (over 24" nb, DN600)

API TR 6AF - Capabilities of API Flanges Under Combinations of Load

API TR 6AF1 - Temperature Derating of API Flanges Under Combination of Loading

API TR 6AF2 - Capabilities of API Integral Flanges Under Combination of Loading

BS 10 - Flanges and bolting for pipes, valves and fittings in carbon, alloy steel and copper (0.5" to 24" nb, cast, plate, bar and forged material)

BS 1560 - Steel pipe flanges and flanged fittings for the petroleum industry

BS 3293 - Carbon steel pipe flanges, over 24", for the petroleum industry

BS 4504 Pt 1 - Flanges and bolting for pipes, valves and fittings, Metric series; Ferrous

BS 4504 Pt 2 - Flanges and bolting for pipes, valves and fittings, Metric series; Copper alloy and composite flanges

EN 1092-1 - Steel flanges - Part 1: PN designated

EN 1092-2 - Cast iron flanges - Part 2: PN designated

ISO 7005-1 - Metallic flanges: steel

ISO 7005-2 - Metallic flanges: cast iron

ISO 7005-3 - Metallic flanges: copper alloy and composite

MSS SP-106-2003 - Cast copper alloy flanges and flange fittings, Class 125, 150 and 300

MSS SP-44-2006 - Steel pipe line flanges

MSS SP-6-2001 - Finishes for Contact Faces of Pipe Flanges and Connecting-End Flanges of Valves and Fittings

MSS SP-65-2004 - High pressure chemical industry flanges and threaded stubs for use with lense gaskets

MSS SP-9-2005 - Spot facing for bronze, iron and steel flanges

16.11 Flow capacity

Regulators and control valves require formalised sizing methods; safety relief valves require a strict methodology for sizing to ensure safe protection of pressurised, or vacuum, systems. Simple isolating valves also need a rational sizing approach to allow logical selection. Correct valve sizing is not necessarily based on the process pipe size.

ANSI/ISA S75.01 - Flow equations for sizing control valves

ANSI/ISA S75.11 - Control valves, inherent flow characteristics and rangeability

BS 6759 Pt 2 - Specification for safety valves for compressed air and inert = ISO 4126 gases

BS 6759 Pt 3[1] ISO 4126 - Specification for safety valves for process fluids

DIN 3320 Pt 1 - Safety valves; definitions, sizing, marking

EN 60534-2-1 to EN 60534-2-3 of control valve

FCI 68-1 - Procedure in rating flow and pressure characteristics of solenoid valves for gas service

FCI 68-2 - Procedure in rating flow and pressure characteristics of solenoid valves for liquid service

FCI 84-1 - Metric definition of the valve flow coefficient Cv

IEC 60534-2 - Industrial process control valves; Part 2; flow capacity; Section 1; sizing equations for incompressible fluid flow under installed conditions

IEC 60534-2-2 - Industrial process control valves; Part 2; flow capacity; Section 2; sizing equations for compressible fluid flow under installed conditions

IEC 60534-2-4 - Industrial process control valves; Part 2; flow capacity, Section 4; inherent flow characteristics and rangeability

ISA S75.01 - Flow equations for sizing control valves

ISA S75.11 - Inherent flow characteristics and rangeability of control valves

16.12 Forces and moments on connections

Very few standards mention the loads which may be applied to connections of valves.

MSS SP-92-1999 - Valve user guide

Rotodynamic pumps in liquid systems have nozzle loads defined for specific applications. Any load applied to a pump may be transmitted through an isolating valve first. Table 16.6 lists allowable forces and moments for steel and alloy steel pumps used in petroleum and allied industries. These pumps would be baseplate mounted and grouted to foundations. Table 16.7 lists allowable loads for heavy duty vertical cast iron pumps which are normally supported by a foundation plinth but are not restrained.

DN	F_X N	F_Y N	F_Z N	F_R N	M_X Nm	M_Y Nm	M_Z Nm	M_R Nm
40	890	710	580	1280	490	230	350	620
50	1330	1070	890	1930	950	470	720	1280
80	1780	1420	1160	2560	1330	680	1000	1800
100	3110	2490	2050	4480	2300	1 180	1760	3130
150	4890	3780	3110	6920	3530	1 760	2580	4710
200	6670	5340	4450	9630	5020	2 440	3800	6750
300	8000	6670	5340	11 700	6100	2 980	4610	8210
350	8900	7120	5780	12 780	6370	3 120	4750	8540
400	10 230	8450	6670	14 580	7320	3 660	5420	9820

Table 16.6 Allowable forces and moments for steel/alloy steel refinery pumps

The "Y" axis is the pipe centre-line, the "Z" axis is vertical and the "X" azis is horizontally transverse across the flange face. Clauses in some trade association specifications effectively allow individual forces and/or moments to be doubled.

DN	F_X N	F_Y N	F_Z N	F_R N	M_X Nm	M_Y Nm	M_Z Nm	M_R Nm
40	550	625	500	975	650	450	525	950
50	750	825	675	1300	700	500	575	1025
80	1125	1250	1025	1975	800	575	650	1175
100	1500	1675	1350	2625	875	625	725	1300
150	2250	2500	2025	3925	1250	875	1025	1825
200	3000	3350	2700	5225	1625	1150	1325	2400
250	3725	4175	3375	6525	2225	1575	1825	3275
300	4475	5000	4025	7825	3025	2150	2475	4450
350	5020	5875	4700	9125	3875	2750	3175	5700
400	5975	6650	5375	10 425	4850	3450	3975	7150

Table 16.7 Allowable forces and moments for cast iron pumps

16.13 Hoses

Hoses are available in many sizes and for many specific applications.

ANSI/NFPA 1961 - Fire hose

ANSI/NFPA 1962 - Care, use and service testing of fire hose, including couplings and nozzles

ANSI/RMA/CGA IP-7 - Specification for rubber welding hose

ANSI/SAE AMS 3386C - Aircraft fuelling hose, acrylonitrile butadiene (NBR) rubber, textile reinforced, collapsing

ANSI/SAE AMS 3387D - Aircraft fuelling hose, acrylonitrile butadiene (NBR) rubber, textile re-inforced, chloroprene covered, non-collapsing

ANSI/SAE AMS 3388C - Aircraft fuelling hose, acrylonoitrile butadiene (NBR) rubber, single wire braid reinforced, non-collapsing

ANSI/SAE AMS 3389C - Aircraft fuelling hose, acrylonitrile butadiene (NBR) rubber, double wire braid reinforced, non-collapsing

API 17J - Unbonded Flexible Pipe

API 17K = ISO 13628-10 and operation of subsea production systems - Part 10: Specification for bonded flexible pipe

API RP 17B = ISO 10420 - Flexible Pipe

BS 3165 - Rubber suction hose for fire fighting purposes

BS 3169 - Specification for first aid reel hoses for fire fighting purposes

BS 3212 - Flexible rubber tubing and hose (including connections where fitted and safety recommendations) for use in LPG vapour phase and LPG/air installations

BS 5122 - Rubber hose for saturated steam

BS 5342[1] ISO 6134 - Specification for rubber hoses for high pressure saturated steam

BS 5842 - Specification for thermoplastic hose assemblies for dock, road and rail tanker use

BS 6391 - Specification for non-percolating layflat delivery hoses and hose assemblies for fire fighting purposes

BS AU 108 - Plain and reinforced hoses of rubber

BS AU 110 - Air pressure rubber brake hose

DIN 2823 - Requirements for chemical hose

DIN 53516 - Abrasion resistance requirements for hose

EN 1360 - Specification for electrically bonded hose and hose assemblies for fuel dispensers

EN 1361 - Specification for aircraft ground fuelling hose and hose assemblies

EN 559 - Rubber hoses for gas welding and allied processes

EN ISO 10380 - Corrugated flexible metallic hose and hose assemblies

EN ISO 10807 - Corrugated flexible metallic hose assemblies for the protection of electrical cables in explosive atmospheres (Withdrawn)

ISO 1403 - Specification for general purpose rubber water hoses

ISO 2398 - Specification for rubber hoses for compressed air

ISO 3861 - Rubber sandblast hose

16.14 Inspection and testing

Manufacturers should be assessed on their in-house process control systems. Great store is set by ISO 9000, EN 29000, BS 5750 and ANSI/ASQC Q 9000. These Standards specify rules about how departments should work, how departments should communicate with each other and most importantly, what to do when things go wrong. Valves for critical applications such as nuclear, pharmaceutical, offshore, are normally purchased from manufacturers who operate one of the standard systems and have been approved by an accredited third party assessor.

All valve manufacturers should operate some type of formalised control system. A traditional system can be as good as

ISO, EN or BS when operated correctly and the Quality Control staff are empowered to scrap substandard components. Inquiries for important valves should always request a Vendor Document Requirement form, VDR, and a Quality Control Plan, QC Plan, to be submitted as part of the quotation. The VDR will show the purchaser what documentation e.g. drawings, manuals, certificates, the manufacturer considers important and will supply with the valve. The QC Plan will show all the inspection operations, hold points and any conditions about which Engineering must be contacted to exert overall control for important decisions. After carefully studying the VDR and the QC Plan, the true character of the manufacturer will be revealed. See Chapter 15 for more detailed information.

ANSI B93.112 - Hydraulic fluid power - Valves - Method for determining the internal leakage

ANSI B93.99 - Hydraulic fluid power - Servo-valves - Test methods

ANSI/API 510 - Pressure Vessel Inspection Code: In-Service Inspection, Rating, Repair, and Alteration

ANSI/API RP 574 - Inspection Practices for Piping System Components

ANSI/API RP 576 - Inspection of Pressure Relieving Devices

ANSI/API RP 578 - Material Verification Program for New and Existing Alloy Piping Systems

ANSI/ASHRAE 102 - Methods of testing non-electric, non-pneumatic thermostatic radiator valves

ANSI/ASHRAE 17 - Method of testing capacity rating of thermostatic refrigerant expansion valves

ANSI/ASME PTC25.3 - Performance test code - Safety and relief valves

ANSI/ASTM A247 - Test Method for Evaluating the Microstructure of Graphite in Iron Castings

ANSI/ASTM A262 - Practices for Detecting Susceptibility to Inter granular Attack in Austenitic Stainless Steels

ANSI/ASTM A275 - Test Method for Magnetic Particle Examination of Steel Forgings

ANSI/ASTM A327 - Test Methods for Impact Testing of Cast Irons

ANSI/ASTM A367 - Test Methods of Chill Testing of Cast Iron

ANSI/ASTM A370 - Test Methods and Definitions for Mechanical Testing of Steel Products

ANSI/ASTM A388 - Practice for Ultrasonic Examination of Heavy Steel Forgings

ANSI/ASTM A438 - Test Method for Transverse Testing of Grey Cast Iron

ANSI/ASTM A609 - Practice for Castings, Carbon, Low-Alloy, and Martensitic Stainless Steel, Ultrasonic Examination Thereof

ANSI/ASTM A745 - Practice for Ultrasonic Examination of Austenitic Steel Forgings

ANSI/ASTM A751 - Test Methods, Practices, and Terminology for Chemical Analysis of Steel Products

ANSI/ASTM A763 - Practices for Detecting Susceptibility to Inter granular Attack in Ferritic Stainless Steels

ANSI/ASTM A799 - Practice for Steel Castings, Stainless, Instrument Calibration, for Estimating Ferrite Content

ANSI/ASTM A800 - Practice for Steel Casting, Austenitic Alloy, Estimating Ferrite Content Thereof

ANSI/ASTM A802 - Practice for Steel Castings, Surface Acceptance Standards, Visual Examination

ANSI/ASTM A926 - Test Method for Comparing the Abrasion Resistance of Coating Materials for Corrugated Metal Pipe

ANSI/ASTM A966 - Test Method for Magnetic Particle Examination of Steel Forgings Using Alternating Current

ANSI/ASTM A993 - Test Method for Dynamic Tear Testing of Cast Irons to Establish Transition Temperature

ANSI/AWS B1.10 - Inspection of welds, - Guide to non-destructive examination

ANSI/AWS B1.11 - Guide for the visual inspection of welds

ANSI/AWS B4.0 - Standard Methods for Mechanical Testing of Welds

ANSI/AWS C3.2 - Standard Method for Evaluating the Strength of Brazed Joints

ANSI/AWS C3.8 - Recommended practices for ultrasonic inspection

ANSI/FCI 75-1 - Test conditions and procedures for measuring electrical characteristics of solenoid valves

ANSI/FCI 79-1 - Standard for proof of pressure ratings for pressure reducing regulators

ANSI/FCI 81-1 - Standard for proof of pressure rating for temperature regulators

ANSI/FCI 82-1 - Recommended methods for testing and classifying the water hammer characteristics of electrically operated valves

ANSI/ISA S75.02 - Control valve capacity test procedure

ANSI/ISA S75.07 - Laboratory measurement of aerodynamic noise generated by control valves

ANSI/ISA S75.13 - Method of evaluating the performance of positioners with analog input signals and pneumatic output

ANSI/ISA S75.19 - Hydrostatic testing of control valves

ANSI/NFPA 26 - Supervision of valves controlling water supplies for fire protection

ANSI/NFPA 291 - Fire flow testing and marking of hydrants

ANSI/NFPA T3.5.16 - Hydraulic fluid power - Flow control valves - Method for measuring and reporting regulating characteristics

ANSI/SAE J1117 - Method of measuring and reporting the pressure differential-flow characteristic of a hydraulic power valve

ANSI/SAE J1227 - Assessing cleanliness of hydraulic fluid power components and systems

ANSI/SAE J1235 - Measuring and reporting the internal leakage of a hydraulic fluid power valve

ANSI/SAE J747 - Hydraulic control valve test procedure

ANSI/SAEAMS 2634 - Ultrasonic inspection of thin-walled metal tubing

ANSI/SAEARP 868B - Pressure drop test method for fuel system components

ANSI/SAEAS 1284A - Standard test procedure and limiting value for shut-off surge pressure of pressure fuel dispensing systems

ANSI/SAEAS 708 - Top visual quality "O" ring packings and gaskets - Surface inspection guide and acceptance standard

ANSI/SAEMA2004A - Thermal shock testing of fluid system piping and fittings

API 598 - Valve Inspection and Testing

API 5B - Threading, Gauging, and Thread Inspection of Casing, Tubing, and Line Pipe Threads

API 607 = ISO 10497 - Testing of Valves - Fire Type-testing Requirements

API 6AV1 - Verification Test of Wellhead Surface Safety Valves and Underwater Safety Valves for Offshore Service

API Publication 1157 - Hydrostatic Test Water Treatment and Disposal Options for Liquid Pipeline Systems

API RP 1110 - Pressure Testing of Liquid Petroleum Pipelines

API RP 11V2 - Gas Lift Valve Performance Testing

API RP 577 - Welding Inspection and Metallurgy

API RP 5A5 = ISO 15463 - Field Inspection of New Casing, Tubing, and Plain-end Drill Pipe

API RP 5B1 - Threading, Gauging, and Thread Inspection of Casing, Tubing, and Line Pipe Threads

API RP 5C5 = ISO 13679 - Recommended Practice on Procedures for Testing Casing and Tubing Connections

API RP 5L3 - Conducting Drop-weight Tear Tests on Line Pipe

API RP 5L8 - Field Inspection of New Line Pipe

API RP 5SI - Purchaser Representative Surveillance and/or Inspection at the Supplier

API RP 5UE - Recommended Practice for Ultrasonic Evaluation of Pipe Imperfections

API RP 6HT - Heat Treatment and Testing of Large Cross Section and Critical Section Components

API TR 17TR1 - Evaluation Standard for Internal Pressure Sheath Polymers for High Temperature Flexible Pipes

API TR 6AM - Material Toughness

API TR 939-A - Research Report on Characterization and Monitoring of Cracking in Wet H2S Service

ASTM D 1598 - Test for time-to-failure of plastic pipe under constant internal pressure

ASTM E446 - Radiography for valve bodies

ASTM E94 - Radiography for valve bodies

BS 4778 Pt 2 - Quality vocabulary - concepts and definitions

BS 4778 Pt 3 - Quality vocabulary - guide to concepts and definitions IEC 60050 (191)

BS 5793 Pt 4 - Inspection and routine testing of control valves IEC 60534-4

BS 5882 - Total quality assurance programme for nuclear installations

BS 5998 - Quality requirements for steel valve castings

BS 6001= ISO 2858 IEC 410 - Sampling procedures and tables for inspection by attributes

BS 6002 = ISO 3951 - Specification for sampling procedures and charts for inspection by variables for percent defective

BS 6755 Pt 1 - Testing of valves - Production pressure testing

DIN 32509 - Hand-operated shut-off valves for welding, cutting and allied processes; design, safety requirements and testing

DIN 32725 - Safety shut-off devices for combustion plants with liquified gas in its liquid state; safety requirements and test

DIN 3357 Pt 1 - Metal ball valves; General requirements and methods of test

DIN 3399 - Gas low-pressure cut-off valves; safety requirements, testing

DIN 3437 - Gas stop valves for pressures over 16 bar; requirements and acceptance tests

DIN 3441 Pt 7 - Valves for water supply systems; requirements and acceptance tests

DIN 3442 Pt 1 - Polypropylene (PP) valves; requirements and testing

DIN 3475 - Valves and fittings of spheroidal cast iron for untreated and potable water; corrosion protection by internal enamelling; quality requirements and testing

DIN 3544 Pt 1 - High density polyethylene (HDPE) valves; tapping valves; requirements and testing

DIN 4811 Pt 2 - Pressure regulators for LPG; pressure regulators with special connections; safety requirements, testing, marking

EN 10002 - 1 - Metallic materials; tensile testing; Part 1: method of test

EN 10002 - 2 - Metallic materials; tensile testing; Part 2: verification of the force measuring system of the tensile testing machines

EN 10002 - 3 - Metallic materials; Tensile testing; Part 3: Calibration of force proving instruments used for the verification of uni-axial testing machines

EN 10002 - 4 - Metallic materials; tensile test; part 4: verification of extensiometers used in uni-axial testing

EN 10002 - 5 - Metallic materials; tensile testing; part 5: method of testing at elevated temperature

EN 10003 - 1 - Metallic materials; Brinell hardness test - Part 1: Test method

EN 10003 - 2 - Metallic materials; Brinell hardness test; -Part 2: Verification of Brinell hardness testing machines

EN 10003 - 3 - Metallic materials; Brinell hardness test; Part 3: Calibration of standardised blocks to be used for Brinell hardness testing machines

EN 10004 - 1 - Metallic materials; hardness test; part 1: Rockwell test (scales A, B, C, D, E, F, G, H, K)

EN 10004 - 2 - Metallic materials; hardness test; Part 2: verification of Rockwell hardness testing machines (scales A, B, C, D, E, F, G, H, K)

EN 10004 - 3 - Metallic materials; hardness test; Part 3: calibration of standardised blocks to be used for Rockwell hardness testing machines (scales A, B, C, D, E, F, G, H, K)

EN 10021 - General technical delivery requirements for steel and iron products

EN 10036 - Chemical analysis of ferrous materials; determination of total carbon in steels and irons; gravimetric method after combustion in a stream of oxygen

EN 10045 - 1 - Charpy impact test on metallic materials, V and U notches; Part 1: Test method

EN 10045 - 2 - Metallic materials; Charpy impact test; part 2: verification of the testing machine (pendulum impact)

EN 10071 - Chemical analysis of ferrous materials; determination of manganese in steels and irons; electrometric titration method

EN 10083 - 1 - Quenched and tempered steels; part 1: technical delivery conditions for special steels

EN 10083 - 2 - Quenched and tempered steels - Part 2: Technical delivery conditions for unalloyed quality steels

EN 10083 - 3 - Quenched and tempered steels - Part 3: Technical delivery conditions for boron steels

EN 10087 - Case hardening steels - Technical delivery conditions

EN 10088 - 1 - Stainless steels - Part 1: List of stainless steels

EN 10088 - 2 - Stainless steels - Part 2: Technical delivery conditions for sheet/plate and strip for general purposes

EN 10088 - 3 - Stainless steels - Part 3: Technical delivery conditions for semi finished products, bars, rods and sections for general purposes

EN 10109 - 1 - Metallic materials - Hardness test - Part 1: Rockwell test (scales A, B, C, D, E, F, G, H, K) and Rockwell superficial test (scales 15N, 30N, 45N, 15T, 30T and 45T)

EN 10109 - 2 - Metallic materials - Hardness test - Part 2: Verification of Rockwell hardness testing machines (scales A, B, C, D, E, F, G, H, K, N, T)

EN 10109 - 3 - Metallic materials - Hardness test - Part 3: Calibration of standardised blocks to be used for Rockwell hardness testing machines (scales A, B, C, D, E, F, G, H, K, N, T)

EN 10136 - Chemical analysis of ferrous materials; determination of nickel in steels and irons; flame atomic absorption spectrometric method

EN 10152 - Electrolytically zinc coated cold rolled steel flat products; technical delivery conditions

EN 10154 - Continuously hot-dip aluminium-silicon (AS) coated steel strip and sheet - Technical delivery conditions

EN 10155 - Structural steels with improved atmospheric corrosion resistance; technical delivery conditions

EN 10163 - 1 - Delivery requirements for surface condition of hot rolled steel plates, wide flats and sections Part 1: general requirements

EN 10163 - 2 - Delivery requirements for surface condition of hot rolled steel plates, wide flats and sections Part 2: plate and wide flats

EN 10163 - 3 - Delivery requirements for surface condition of hot rolled steel plates, wide flats and sections Part 3: sections

EN 10177 - Chemical analysis of ferrous materials; determination of calcium in steels; flame atomic absorption spectrometric method

EN 10178 - Chemical analysis of ferrous materials; determination of niobium in steels; spectrophotometric method

EN 10179 - Chemical analysis of ferrous materials; determination of nitrogen (trace amounts) in steels; spectrophotometric method

EN 10181 - Chemical analysis of ferrous materials; determination of lead in steels; flame atomic absorption spectrometric method

EN 10184 - Chemical analysis of ferrous materials; determination of phosphorus in steels and irons; spectrophotometric method

EN 10188 - Chemical analysis of ferrous materials; determination of chromium in steels and irons; flame atomic absorption spectrometric method

EN 10200 - Chemical analysis of ferrous materials; determination of boron in steel; spectrophotometric method

EN 10201 - Chemical analysis of ferrous materials; determination of silicon in steels and iron; flame atomic absorption spectrometric method

EN 10204 = DIN 50049 - Metallic products. Types of inspection documents

EN 10207 - Steels for simple pressure vessels; technical delivery requirements for plates, strips and bars

EN 10211 - Chemical analysis of ferrous materials - Determination of titanium in steel and iron - Flame atomic absorption spectrometric method

EN 10212 - Chemical analysis of ferrous materials - Determination of arsenic in steel and iron - Spectrometric method

EN 10213 - 1 - Technical delivery conditions for steel castings for pressure purposes - Part 1: General

EN 10213 - 2 - Technical delivery conditions for steel castings for pressure purposes - Part 2: Steel grades for use at room temperature and elevated temperatures

EN 10213 - 3 - Technical delivery conditions for steel castings for pressure purposes - Part 3: Steel grades for use at low temperatures

EN 10213 - 4 - Technical delivery conditions for steel castings for pressure purposes - Part 4: Austenitic and austenitic - ferritic steel grades

EN 10228 - 1 - Non-destructive testing of steel forgings - Part 1: Magnetic particle inspection

EN 10228 - 2 - Non-destructive testing of steel forgings - Part 2: Penetrant testing

EN 10228 - 3 - Non-destructive testing of steel forgings - Part 3: Ultrasonic testing of ferritic or martensitic steel forgings

EN 10228 - 4 - Non-destructive testing of steel forgings - Part 4: Ultrasonic testing of austenitic and austenitic-ferittic stainless forgings

EN 10229 - Evaluation of resistance of steel products to hydrogen induced cracking (HIC)

EN 10232 - Metallic materials; tube (in full section); bend test

EN 10233 - Metallic materials; tube flattening test

EN 10234 - Metallic materials; tube drift expanding test

EN 10235 - Metallic materials; tube flanging test

EN 10236 - Metallic materials; tube; ring expanding test

EN 10237 - Metallic materials; tube; ring tensile test

EN 10246 - 1 - Non-destructive testing of steel tubes - Part 1: Automatic electromagnetic testing of seamless and welded (except submerged arc welded) ferromagnetic steel tubes for verification of hydraulic leak-tightness

EN 10246 - 10 - Non-destructive testing of steel tubes - Part 10: Radiographic testing of the weld seam of submerged arc-weld steel tubes for the detection of imperfections

EN 10246 - 14 - Non-destructive testing of steel tubes - Part 14: Automatic ultrasonic testing for the detection of laminar imperfections for seamless and welded (except submerged arc-welded) steel tubes

EN 10246 - 15 - Non-destructive testing of steel tubes - Part 15: Automatic ultrasonic testing of strip/plate used in the manufacture of welded steel tubes for the detection of laminar imperfections

EN 10246 - 2 - Non-destructive testing of steel tubes - Part 2:

EN 10246 - 3 - Non-destructive testing of steel tubes - Part 3: Automatic eddy current testing of seamless and welded (except submerged arc welded) steel tubes for the detection of imperfections

EN 10246 - 5 - Non-destructive testing of steel tubes - Part 5: Automatic full peripheral magnetic transducer/flux leakage testing of seamless and welded (except submerged arc welded) ferromagnetic steel tubes for the detection of longitudinal imperfections

EN 10246 - 6 - Non-destructive testing of steel tubes - Part 6: Automatic full peripheral ultrasonic testing of seamless steel tubes for the detection of transverse imperfections

EN 10246 - 7 - Non-destructive testing of steel tubes - Part 7: Automatic full peripheral ultrasonic testing of seamless and welded (except submerged arc welded) steel tubes for the detection of longitudinal imperfections

EN 10246 - 8 - Non-destructive testing of steel tubes - Part 8: Automatic ultrasonic testing of the weld seam of electric resistance and induction welded steel tubes for the detection of longitudinal imperfections

EN 10246 - 9 - Non-destructive testing of steel tubes - Part 9: Automatic ultrasonic testing of the weld seam of submerged arc welded steel tubes for the detection of longitudinal and/or transverse imperfections

EN 10254 - Steel closed die forgings - General technical delivery conditions

EN 10276 - 1 - Chemical analysis of ferrous materials - Determination of oxygen in steel and iron - Part 1: Sampling and preparation of steel samples for oxygen determination

EN 1074-1 - Valves for water supply; specification for use and appropriate verification tests; general requirements

EN 1074-2 - Valves for water supply; specification for use and appropriate verification tests; isolating valves

EN 1074-3 - Valves for water supply; specification for use and appropriate verification tests; check valves

EN 12454 - Founding - Visual examination of surface discontinuities - Steel and castings

EN 1369 - Founding; magnetic particle inspection

EN 1370 - Founding; surface roughness inspection by visual tactile comparators

EN 1371-1 - Founding; liquid penetrant inspection; part 1: sand, gravity die and low pressure die castings

EN 1559-1 - Founding - Technical conditions of delivery - Part 1: General

EN 1559-2 - Founding - Technical conditions of delivery - Part 2: Additional requirements for steel castings

EN 1559-3 - Founding - Technical conditions of delivery - Part 3: Additional requirements for iron castings

EN 1559-5 - Founding - Technical conditions of delivery - Part 5: Additional requirements for magnesium alloy castings

EN 1560 - Founding - Designation system for cast iron - Material symbols and material numbers

EN 1561 - Founding - Grey cast irons

EN 1562 - Founding - Malleable cast irons

EN 1563 - Founding - Spheroidal graphite cast irons

EN 1564 - Founding - Austempered ductile cast irons

EN 24159 = ISO 4159 - Ferromanganese and ferrosilicomanganese; determination of manganese content; potentiometric method

EN 24829 - 1 = ISO 4829 -1 - Steel and cast iron; determination of total silicon content; reduced molybdosilicate pectrophotometric method part 1: silicon content between 0,05 and 1%

EN 24829 - 2 = ISO 4829 -2 - Steel and cast iron; determination of total silicon content; reduced molybdosilicate pectrophotometric method part 2: silicon content between 0,01 and 0,05%

EN 24934 = ISO 4934 - Steel and cast iron; determination of sulphur content; gravimetric method

EN 24935 = ISO 4935 - Steel and cast iron; determination of sulphur content; infrared absorption method after combustion in an induction furnace

EN 24937 = ISO 4937 - Steel and iron; determination of chromium content; potentiometric or visual method

EN 24938 = ISO 4938 - Steel and iron; determination of nickel content; gravimetric or titrimetric method

EN 24943 = ISO 4943 - Chemical analysis of ferrous metal; determination of copper content; flame atomic absorption spectrometric method

EN 24946 = ISO 4946 - Steel and cast iron; determination of copper content; 2,2` diquinolyl spectrophotometric method

EN 24947 = ISO 4947 - Steel and cast iron; determination of vanadium content; potentiometric titration method

EN 28233 = ISO 8233 - Thermoplastic valves; torques; test method

EN 28659 = ISO 8659 = ISO 9393-1 - Unplasticised polyvinyl chloride (PVC-U) valves; requirements and testing

EN 29302 - Seamless and welded (except submerged arc-welded) steel tubes for pressure purposes; electromagnetic testing for verification of hydraulic leak-tightness

EN 29303 - Seamless and welded (except submerged arc-welded) steel tubes for pressure purposes; full peripheral ultrasonic testing for the detection of longitudinal imperfections

EN 29658 = ISO 9658 - Steel; determination of aluminium content; flame atomic absorption spectrometric method

EN 917 - Plastic piping systems; thermoplastic valves; test methods for resistance to internal pressure ISO/DIS 9393-1

EN ISO 10280 - Steel and iron - Determination of titanium content - Diantipyrylmethane spectrophotomatric method

EN ISO 10700 - Steel and iron - Determination of manganese content - Flame atomic absorption spectrometric method

EN ISO 3785 - Steel - Designation of test piece axes

EN ISO 945 - Cast iron - Designation of microstructure of graphite

ENV 22605 - 1 = ISO 2605-1 - Steel products for pressure purposes; derivation and verification of elevated temperature properties; Part 1: Yield or proof stress of carbon and low alloy steel products

ENV 22605 - 2 = ISO 2605-2 - Steel products for pressure purposes; derivation and verification of elevated temperature properties; Part 2: Proof stress of austenitic steel products

ENV 22605 - 3 = ISO 2605-3 - Steel products for pressure purposes; derivation and verification of elevated temperature properties; Part 3: Alternative procedure for deriving the elevated temperature yield or proof stress properties when data are limit

IEC 60534-2-3 - Industrial process control valves; Part 2; flow capacity; Section 3; flow capacity - test procedures

IEC 60534-4 - Inspection and routine testing

IEC HD 289 S1 - Routine production tests for electrical equipment

ISA RP75.23 - Considerations for evaluating control valve cavitation

ISA S75.02 - Control valve capacity test procedures

ISA S75.07 - Laboratory measurement of aerodynamic noise generated by control valves

ISA S75.13 - Method of evaluating the performance of positioners with analogue input signals

ISA S75.19 - Hydrostatic testing of control valves

ISO 11203 - Acoustics - Noise emitted by machinery and equipment - Determination of emission sound pressure levels at a work station and at other specified locations from the sound power level

ISO 2503 - Pressure regulators for gas cylinders used in welding, cutting and related processes; terminology, requirements and testing

ISO 2604 - 2 - Steel products for pressure purposes - Quality requirements - Part 2: Wrought seamless tubes

ISO 2604 - 3 - Steel products for pressure purposes - Quality requirements - Part 3: Electric resistance and induction welded tubes

ISO 5208 - Specification for pressure testing requirements for general purpose valves

ISO 7291 - Manifold regulators for welding, cutting and related processes; concepts, requirements and testing

ISO 8233 - Thermoplastic valves; torques; test method EN 28233

ISO 83 - Methods for notched bar tests: The Charpy U-notch impact test on metals

ISO 8501 - all parts - Preparation of steel substrates before application of paints and related products - Visual assessment of surface cleanIness

ISO 8659 = EN 28659 - Unplasticised polyvinyl chloride (PVC-U) valves; requirements and testing

ISO 9393-1 = EN 28659 - Unplasticised polyvinyl chloride (PVC-U) valves; requirements and testing

ISO 9614-2 - Acoustics - Determination of sound power levels of noise sources using sound intensity - Part 2: Measurement by scanning

ISO/R 148 - Methods for notched bar tests: The Charpy V-notch impact test on metals

ISO/R 84 - Methods for notched bar tests: The Izod V-notch impact test on metals

MSS SP-112-1999 - Quality standard for evaluation of cast surface finishes - Visual and tactile method

MSS SP-53-2002 - Quality standard for steel castings and forgings for valves, flanges and fittings and other piping components - Magnetic particle examination method

MSS SP-54-2001 - Quality standard for steel castings for valves, flanges and fittings and other piping components - Radiographic examination method

MSS SP-55-2001 - Quality standard for steel castings for valves, flanges and fittings and other piping components - Visual method

MSS SP-61-2003 - Pressure testing of steel valves

MSS SP-82-1992 - Valve pressure testing methods

MSS SP-93-2004 - Quality standard for steel castings and forgings for valves, flanges and fittings and other piping components - Liquid penetrant examination method

MSS SP-94-2004 - Quality standard for ferritic and martensitic steel castings for valves, flanges and fittings and other piping components - Ultrasonic examination method

VDI 2173 - Fluidic characteristic quantities of control valves and their determination

VDI 2174 - Mechanical characteristic values of control devices for flowing materials and their determination.

VDI 2176 - Fluidic characteristic quantities of throttle valves and their determination

VDI 2177 - Description and inspection of position controllers with pneumatic actuators

16.15 Installation

Some safety relief valves are installed with standard isolating valves in the inlet pipework. This practice is prohibited by many pressure vessel and system "codes" and classification societies. This style of installation is understandable, but not approved, when the process must continue to operate with defective valves.

If isolating valves are to be fitted in this way then valve locking must be strictly observed. Only very senior **operating** personnel must have the authority to take the safety relief valve off-line.

ANSI/ASTM A798 - Practice for Installing Factory-Made Corrugated Steel Pipe for Sewers and Other Applications

ANSI/ASTM A807 - Practice for Installing Corrugated Steel Structural Plate Pipe for Sewers and Other Applications

ANSI/ASTM D3839 - Practice for underground installation of flexible reinforced thermosetting resin pipe and reinforced plastic mortar pipe

ANSI/MSS MSS SP-58 - Pipe hangers and supports - Materials, design and manufacture

ANSI/SAE J1176 - External leakage classification for hydraulic systems

API 686 - Machinery installation and installation design

API RP 520 - 2 - Sizing, Selection, and Installation of Pressure-relieving Devices in Refineries, Part II Installation

BS 3974 - Pipe supports

BS 6759 Pt 1 ISO 4126 - Specification for safety valves for steam and hot water (44kW)

ISA S4.01.1 - Recommended practice for standard control valve manifold design

ISA S4.01.2 - Recommended practice for standard control valve manifold design

ISA S7.3 - Instrument air specification

MSS SP-113-2001 - Connecting Joint between Tapping Machines and Tapping Valves

MSS SP-116-2003 - Service Line Valves and Fittings for Drinking Water Systems

MSS SP-127-2001 - Bracing for Piping Systems Seismic-Wind-Dynamic Design, Selection, Application

MSS SP-45-2003 - By-pass and drain connections

MSS SP-58-2002 - Pipe Hangers and Supports — Materials, Design and Manufacture

MSS SP-69-2003 - Pipe hangers and supports - Selection and application

MSS SP-77-2000 - Guidelines for Pipe Support Contractual Relationships

MSS SP-89-2003 - Pipe hangers and supports - Fabrication and installation practices

VDE 0730 - Liquid hammer damping

16.16 Instrumentation

Instrumentation installed with a valve can be crucial for the success and safety of the installation.

DIN 3398 Pt 1 - Pressure switches for gas; safety requirements, testing

DIN 3398 Pt 2 - Pressure switches for air and flue gases for furnace systems; safety requirements, testing

DIN 3398 Pt 3 - Pressure cut-off switches for gaseous substances; requirements and testing

DIN 3398 Pt 4 - Pressure cut-off switches for liquid fuels and heat transfer oils

16.17 Maintenance and repair

Very few standards or specifications address the problem of repairs.

ANSI/SAE AMS 2694A - Repair welding of aerospace castings

ANSI/API 570 - Piping Inspection Code: Inspection, Repair, Alteration, and Rerating of In-service Piping Systems

API Publication 1132 - Effects of Oxygenated Fuels and Reformulated Diesel Fuels on Elastomers and Polymers in Pipeline/Terminal Component

API RP 11V7 - Repair, Testing and Setting Gas Lift Valves

API RP 14H, Maintenance and Repair of Surface Safety Valves and Underwater Safety Valves Offshore

API RP 2200 - Repairing Crude Oil, Liquefied Petroleum Gas and Product Pipelines

API RP 621 - Reconditioning of Metallic Gate, Globe, and Check Valves

API RP 6DR - Repair and Re-manufacture of Pipeline Valves

API TR 17TR2 - The Ageing of PA-11 in Flexible Pipes

API TR 939-B - Repair and Remediation Strategies for Equipment Operating in Wet H2S Service

16.18 Materials

All valves utilise materials and most trading blocks have their own materials standards. Material manufacturers and valve manufacturers have their own material specifications. The valve user needs an appreciation of the materials standards produced by important organisations.

AMS 4901 - MIL-T-9046, Ti GRADE 4

AMS 4911 - MIL-T-9046, Ti-6AL-4V

AMS 4928 - MIL-T-9046, 6AL-4V

ANSI/ASTM A126 - Specification for Grey Iron Castings for Valves, Flanges, and Pipe Fittings

ANSI/ASTM A128 - Specification for Steel Castings, Austenitic Manganese

ANSI/ASTM A148 - Specification for Steel Castings, High Strength, for Structural Purposes

ANSI/ASTM A153 - Standard specification for zinc coating (hot-dip) on iron and steel hardware

ANSI/ASTM A159 - Specification for Automotive Grey Iron Castings

ANSI/ASTM A182 - Specification for Forged or Rolled Alloy-Steel Pipe Flanges, Forged Fittings, and Valves and Parts for High-Temperature Service

ANSI/ASTM A193 - Alloy Steel and Stainless Steel Bolting materials for high temperature service

ANSI/ASTM A194 - Standard specification for alloy-steel and stainless steel bolting materials for high temperature service

ANSI/ASTM A216 - Specification for Steel Castings, Carbon, Suitable for Fusion Welding, for High- Temperature Service

ANSI/ASTM A217 - Specification for Steel Castings, Martensitic Stainless and Alloy, for Pressure- Containing Parts, Suitable for High-Temperature Service

ANSI/ASTM A227 - Specification for Steel Wire, Cold-Drawn for Mechanical Springs

ANSI/ASTM A228 - Specification for Steel Wire, Music Spring Quality

ANSI/ASTM A229 - Specification for Steel Wire, Oil-Tempered for Mechanical Springs

ANSI/ASTM A230 - Specification for Steel Wire, Oil-Tempered Carbon Valve Spring Quality

ANSI/ASTM A231 - Specification for Chromium-Vanadium Alloy Steel Spring Wire

ANSI/ASTM A232 - Specification for Chromium-Vanadium Alloy Steel Valve Spring Quality Wire

ANSI/ASTM A27 - Specification for Steel Castings, Carbon, for General Application

ANSI/ASTM A276 - Standard specification for stainless steel bars and shapes

ANSI/ASTM A278 - Specification for Grey Iron Castings for Pressure-Containing Parts for Temperatures Up to 350°C

ANSI/ASTM A297 - Specification for Steel Castings, Iron-Chromium and Iron-Chromium-Nickel, Heat Resistant, for General Application

ANSI/ASTM A313 - Specification for Stainless Steel Spring Wire

ANSI/ASTM A336 - Specification for Alloy Steel Forgings for Pressure and High-Temperature Parts

ANSI/ASTM A351 - Specification for Castings, Austenitic, Austenitic-Ferritic (Duplex), for Pressure-Containing Parts

ANSI/ASTM A352 - Specification for Steel Castings, Ferritic and Martensitic, for Pressure-Containing Parts, Suitable for Low-Temperature Service

ANSI/ASTM A389 - Specification for Steel Castings, Alloy, Specially Heat-Treated, for Pressure-Containing Parts, Suitable for High-Temperature Service

ANSI/ASTM A395 - Specification for Ferritic Ductile Iron Pressure-Retaining Castings for Use at Elevated Temperatures

ANSI/ASTM A434 - Standard specification for steel bars, alloy, hot-wrought or cold-finished, quenched and tempered

ANSI/ASTM A436 - Specification for Austenitic Grey Iron Castings

ANSI/ASTM A439 - Specification for Austenitic Ductile Iron Castings

ANSI/ASTM A447 - Specification for Steel Castings, Chromium-Nickel-Iron Alloy (25-12 Class), for High-Temperature Service

ANSI/ASTM A47 - Specification for Ferritic Malleable Iron Castings

ANSI/ASTM A473 - Specification for Stainless Steel Forgings

ANSI/ASTM A479 - Stainless and Heat Resisting Bars and Shapes for use in boilers and other pressure vessels

ANSI/ASTM A48 - Specification for Grey Iron Castings

ANSI/ASTM A484 - Specification for General Requirements for Stainless Steel Bars, Billets, and Forgings

ANSI/ASTM A487 - Specification for Steel Castings Suitable for Pressure Service

ANSI/ASTM A494 - Specification for Castings, Nickel and Nickel Alloy

ANSI/ASTM A518 - Specification for Corrosion-Resistant High-Silicon Iron Castings

ANSI/ASTM A522 - Specification for Forged or Rolled 8 and 9% Nickel Alloy Steel Flanges, Fittings, Valves, and Parts for Low-Temperature Service

ANSI/ASTM A532 - Specification for Abrasion-Resistant Cast Irons

ANSI/ASTM A536 - Specification for Ductile Iron Castings

ANSI/ASTM A541 - Specification for Quenched and Tempered Carbon and Alloy Steel Forgings for Pressure Vessel Components

ANSI/ASTM A560 - Specification for Castings, Chromium-Nickel Alloy

ANSI/ASTM A565 - Specification for Martensitic Stainless Steel Bars, Forgings, and Forging Stock for High-Temperature Service

ANSI/ASTM A571 - Specification for Austenitic Ductile Iron Castings for Pressure-Containing Parts Suitable for Low-Temperature Service

ANSI/ASTM A576 - Standard specification for steel bars, carbon, hot-wrought, special quality

ANSI/ASTM A579 - Specification for Super strength Alloy Steel Forgings

ANSI/ASTM A582 - Standard specification for free-machining stainless steel bars

ANSI/ASTM A592 - Specification for High-Strength Quenched and Tempered Low-Alloy Steel Forged Fittings and Parts for Pressure Vessels

ANSI/ASTM A597 - Specification for Cast Tool Steel

ANSI/ASTM A638 - Specification for Precipitation Hardening Iron Base Superalloy Bars, Forgings, and Forging Stock for High-Temperature Service

ANSI/ASTM A667 - Specification for Centrifugally Cast Dual Metal (Grey and White Cast Iron) Cylinders

ANSI/ASTM A668 - Specification for Steel Forgings, Carbon and Alloy, for General Industrial Use

ANSI/ASTM A703 - Specification for Steel Castings, General Requirements, for Pressure-Containing Parts

ANSI/ASTM A705 - Specification for Age-Hardening Stainless Steel Forgings

ANSI/ASTM A711 - Specification for Steel Forging Stock

ANSI/ASTM A723 - Specification for Alloy Steel Forgings for High-Strength Pressure Component Application

ANSI/ASTM A727 - Specification for Carbon Steel Forgings for Piping Components with Inherent Notch Toughness

ANSI/ASTM A732 - Specification for Castings, Investment, Carbon and Low Alloy Steel for General Application, and Cobalt Alloy for High Strength at Elevated Temperatures

ANSI/ASTM A743 - Specification for Castings, Iron-Chromium, Iron-Chromium-Nickel, Corrosion Resistant, for General Application

ANSI/ASTM A744 - Specification for Castings, Iron-Chromium-Nickel, Corrosion Resistant, for Severe Service

ANSI/ASTM A747 - Specification for Steel Castings, Stainless, Precipitation Hardening

ANSI/ASTM A757 - Specification for Steel Castings, Ferritic and Martensitic, for Pressure-Containing and Other Applications, for Low-Temperature Service

ANSI/ASTM A781 - Specification for Castings, Steel and Alloy, Common Requirements, for General Industrial Use

ANSI/ASTM A823 - Specification for Statically Cast Permanent Mould Grey Iron Castings

ANSI/ASTM A834 - Specification for Common Requirements for Iron Castings for General Industrial Use

ANSI/ASTM A836 - Specification for Titanium-Stabilised Carbon Steel Forgings for Glass-Lined Piping and Pressure Vessels

ANSI/ASTM A837 - Specification for Steel Forgings, Alloy, for Carburising Applications

ANSI/ASTM A842 - Specification for Compacted Graphite Iron Castings

ANSI/ASTM A874 - Specification for Ferritic Ductile Iron Castings Suitable for Low-Temperature Service

ANSI/ASTM A890 - Specification for Castings, Iron-Chromium-Nickel-Molybdenum Corrosion-Resistant, Duplex (Austenitic/Ferritic) for General Application

ANSI/ASTM A897 - Specification for Austempered Ductile Iron Castings

ANSI/ASTM A903 - Specification for Steel Castings, Surface Acceptance Standards, Magnetic Particle and Liquid Penetrant Inspection

ANSI/ASTM A909 - Specification for Steel Forgings, Microalloy, for General Industrial Use

ANSI/ASTM A915 - Specification for Steel Castings, Carbon, and Alloy, Chemical Requirements Similar to Standard Wrought Grades

ANSI/ASTM A921 - Specification for Steel Bars, Microalloy, Hot-Wrought, Special Quality, for Subsequent Hot Forging

ANSI/ASTM A957 - Specification for Investment Castings, Steel and Alloy, Common Requirements, for General Industrial Use

ANSI/ASTM A958 - Specification for Steel Castings, Carbon, and Alloy, with Tensile Requirements, Chemical Requirements Similar to Standard Wrought Grades

ANSI/ASTM A961 - Specification for Common Requirements for Steel Flanges, Forged Fittings, Valves, and Parts for Piping Applications

ANSI/ASTM A965 - Specification for Steel Forgings, Austenitic, for Pressure and High Temperature Parts

ANSI/ASTM A985 - Specification for Steel Investment Casting General Requirements, for Pressure-Containing Parts

API 6A718 - Specification of Nickel Base Alloy 718 (UNS N07718) for Oil and Gas Drilling and Production Equipment

EN 01250 - 4 - Open die steel forgings for general engineering purposes - Part 4: Stainless steels

EN 10028 - 1 - Flat products made of steels for pressure purposes; Part 1: general requirements

EN 10028 - 2 - Flat products made of steels for pressure purposes; Part 2: non - alloy and alloy steels with specified elevated temperature properties

EN 10028 - 3 - Flat products made of steels for pressure purposes; part 3: weldable fine grain steels, normalised

EN 10028 - 4 - Flat products made of steels for pressure purposes; Part 4: Nickel alloy steels with specified low temperature properties

EN 10028 - 5 - Flat products made of steels for pressure purposes; Part 5: weldable fine grain steels, thermomechanically rolled

EN 10028 - 6 - Flat products made of steels for pressure purposes; Part 6: weldable fine grain steels, quenched and tempered

EN 10028 - 7 - Flat products made of steels for pressure purposes; Part 7: Stainless steels

EN 10095 - Heat - resisting steels and alloys

EN 10222 - 1 - Steel forgings for pressure purposes - Part 1: General requirements for open die forgings

EN 10222 - 2 - Steel forgings for pressure purposes - Part 2 : Ferritic and martensitic steels with specified elevated temperature properties

EN 10222 - 3 - Steel forgings for pressure purposes - Part 3: Ferritic and martensitic steels with elevated temperature properties

EN 10222 - 4 - Steel forgings for pressure purposes - Part 4: Nickel steels with specified low temperature properties

EN 10222 - 5 - Steel forgings for pressure purposes - Part 5: Fine grain steels with high proof stress

EN 10222 - 6 - Steel forgings for pressure purposes - Part 6: Austenitic, martensitic and austenitic-ferritic stainless steels

EN 10225 - Weldable structural steels for fixed offshore structures

EN 10267 - Ferritic-pearlitic engineering steels for precipitation hardening from hot-working temperatures

EN 10272 - Stainless steel bars for pressure purposes

EN 10273 - Hot rolled weldable steel bars for pressure purposes with specified elevated temperature properties

EN 10283 - Corrosion resistant steel casting

EN 10285 - Steel tubes and fittings for on and offshore pipelines - External three layer extruded polyethylene based coating

EN 10286 - Steel tubes and fittings for on and offshore pipelines - External three layer extruded polypropylene based coatings

EN 10287 - Steel tubes and fittings for on and offshore pipelines - External fused polyethylene based coatings

EN 10288 - Steel tubes and fittings for on and offshore pipelines - External two layer extruded polyethylene based coatings

ISO 3506 - Corrosion resistant stainles steel fasteners - Specifications

ISO 427 - Wrought copper-tin alloys - Chemical composition and forms of wrought products

ISO 4991 - Steel castings for pressure purposes

ISO 683-13 - Heat-treatable steels, alloy steel and free-cutting steels - Part 13: Wrought stainless steels

ISO 683-18 - Heat-treatable steels, alloy steel and free-cutting steels - Part 18: Bright products of unalloyed and low alloy steels

NACE MR0175 - Sulphide stress cracking resistant metallic materials for oilfield equipment

16.19 Noise

Noise emitted by equipment constitutes a considerable problem when personnel have to be present in the vicinity for long periods. Consideration must be given to the complete system not just the valve. The process pipework may contribute to the overall noise level. Specific areas can be designated "ear protection zones" where ear defenders must be worn. Typical noise level restrictions in current specifications call for 84 dBA.

IEC 994 and EN 60994 give guidance on vibration and pulsation measurement at site for rotodynamic pumps and turbines. Other standards which may be useful include ISO 5343; ANSI S2.17; ANSI S2.40; and API 678. Noise, vibration and pulsations, generated by machinery, can be transmitted through the fluid and the pipe wall and result in significant measurement errors at the valve location.

ANSI/ISA S75.17 - Control valve aerodynamic noise prediction method

EN 60534-8-2 - Method for laboratory measurement of noise ° IEC 60534-8-2

EN 60534-8-3 ° IEC 60534-8-3 - Industrial process control valves; noise considerations; prediction of noise generated by aerodynamic flow

EN 60534-8-4 ° IEC 60534-8-4 - Industrial process control valves; noise considerations; prediction of noise generated by hydrodynamic flow

VDI 3738 - Characteristic noise emission values of technical sound sources; industrial control elements for gas and liquid flows

16.20 Operation

Very few standards or specifications deal with the day-to-day operation of valves. The valve manufacturers' associations do provide some guidance.

API MPMS Chpt 6.7 - Manual of Petroleum Measurement Standards; Chapter 6.7 - Metering Viscous Hydrocarbons

API RP 15TL4 - Care and Use of Fiberglass Tubulars

API RP 5C1 - Care and Use of Casing and Tubing

MSS SP-91-1996 - Guidelines for manual operation of valves

MSS SP-92-1999 - MSS valve user guide

Spec 16C - Choke and Kill Systems

16.21 Painting and corrosion protection

Painting and surface protection is usually the manufacturer's standard or subject to the purchaser's contract requirements.

SP-98-1996 - Protective epoxy coatings for the interior of valves and hydrants

BS 381C - Paint colours

MSS SP-98-2005 - Protective epoxy coatings for the interior of valves and hydrants

API TR 939-D - Stress Corrosion Cracking of Carbon Steel in Fuel Grade Ethanol: Review and Survey & corosion protection

ANSI/ASTM A380 - Practice for Cleaning, Descaling, and Passivation of Stainless Steel Parts, Equipment, and Systems & corrosion protection

16.22 Pipes and tubes

Pipes and tubes are available to many specifications.

ANSI A21.51 - Ductile iron pipe

ANSI B36.10 - Wrought steel pipe dimensions

ANSI B36.19 - Stainless steel pipe dimensions

ANSI/API RP 5L7 - Unprimed Internal Fusion Bonded Epoxy Coating of Line Pipe

ANSI/ASAE S394 - Specifications for irrigation hose and couplings used with self-propelled hose drag agricultural irrigation systems

ANSI/ASME B16.1 - Cast iron pipe flanges and flange fittings, Class 25, 125, 250 and 800

ANSI/ASME B16.25 - Butt-welding ends

ANSI/ASME B16.5 - Steel pipe flanges, flanged valves and fittings (0.5 to 24 nb, 150lb, 300lb, 400lb, 600lb, 900lb, 1500lb, 2500lb)

ANSI/ASTM A1005 - Specification for Steel Line Pipe, Black, Plain End, Longitudinal and Helical Seam, Double Submerged-Arc Welded

ANSI/ASTM A1006 - Specification for Steel Line Pipe, Black, Plain End, Laser Beam Welded

ANSI/ASTM A106 - Specification for Seamless Carbon Steel Pipe for High-Temperature Service

ANSI/ASTM A134 - Specification for Pipe, Steel, Electric-Fusion (Arc)-Welded (Sizes NPS 16 and Over)

ANSI/ASTM A135 - Specification for Electric-Resistance-Welded Steel Pipe

ANSI/ASTM A139 - Specification for Electric-Fusion (Arc)-Welded Steel Pipe (NPS 4 and Over)

ANSI/ASTM A179 - Specification for Seamless Cold-Drawn Low-Carbon Steel Heat-Exchanger and Condenser Tubes

ANSI/ASTM A192 - Specification for Seamless Carbon Steel Boiler Tubes for High-pressure Service

ANSI/ASTM A209 - Specification for Seamless Carbon-Molybdenum Alloy-Steel Boiler and Superheater Tubes

ANSI/ASTM A210 - Specification for Seamless Medium-Carbon Steel Boiler and Superheater Tubes

ANSI/ASTM A213 - Specification for Seamless Ferritic and Austenitic Alloy-Steel Boiler, Superheater, and Heat-Exchanger Tubes

ANSI/ASTM A214 - Specification for Electric-Resistance-Welded Carbon Steel Heat-Exchanger and Condenser Tubes

ANSI/ASTM A249 - Specification for Welded Austenitic Steel Boiler, Superheater, Heat-Exchanger, and Condenser Tubes

ANSI/ASTM A250 - Specification for Electric-Resistance-Welded Ferritic Alloy-Steel Boiler and Superheater Tubes

ANSI/ASTM A252 - Specification for Welded and Seamless Steel Pipe Piles

ANSI/ASTM A254 - Specification for Copper-Brazed Steel Tubing

ANSI/ASTM A268 - Seamless or Welded Ferritic and Martensitic Stainless Steel Tube for general service

ANSI/ASTM A269 - Specification for Seamless and Welded Ferritic and Martensitic Stainless Steel Tubing for General Service

ANSI/ASTM A270 - Specification for Seamless and Welded Austenitic Stainless Steel Sanitary Tubing

ANSI/ASTM A271 - Seamless Austenitic Chromium Nickel Steel Still Tube for use in carrying fluids at elevated temperatures and pressures in various types of heaters

ANSI/ASTM A312 - Specification for Seamless and Welded Austenitic Stainless Steel Pipes

ANSI/ASTM A333 - Specification for Seamless and Welded Steel Pipe for Low-Temperature Service

ANSI/ASTM A334 - Specification for Seamless and Welded Carbon and Alloy-Steel Tubes for Low-Temperature Service

ANSI/ASTM A335 - Specification for Seamless Ferritic Alloy-Steel Pipe for High-Temperature Service

ANSI/ASTM A338 - Specification for Malleable Iron Flanges, Pipe Fittings, and Valve Parts for Railroad, Marine, and Other Heavy-Duty Service at Temperatures Up to 650°F (345°C)

ANSI/ASTM A358 - Specification for Electric-Fusion-Welded Austenitic Chromium-Nickel Alloy Steel Pipe for High-Temperature

ANSI/ASTM A369 - Specification for Carbon and Ferritic Alloy Steel Forged and Bored Pipe for High-Temperature Service

ANSI/ASTM A376 - Specification for Seamless Austenitic Steel Pipe for High-Temperature Central-Station Service

ANSI/ASTM A377 - Index of Specifications for Ductile-Iron Pressure Pipe

ANSI/ASTM A409 - Specification for Welded Large Diameter Austenitic Steel Pipe for Corrosive or High-Temperature Service

ANSI/ASTM A423 - Specification for Seamless and Electric-Welded Low-Alloy Steel Tubes

ANSI/ASTM A426 - Specification for Centrifugally Cast Ferritic Alloy Steel Pipe for High-Temperature Service

ANSI/ASTM A450 - Specification for General Requirements for Carbon, Ferritic Alloy, and Austenitic Alloy Steel Tubes

ANSI/ASTM A451 - Specification for Centrifugally Cast Austenitic Steel Pipe for High-Temperature Service

ANSI/ASTM A498 - Specification for Seamless and Welded Carbon, Ferritic, and Austenitic Alloy Steel Heat-Exchanger Tubes with Integral Fins

ANSI/ASTM A500 - Specification for Cold-Formed Welded and Seamless Carbon Steel Structural Tubing in Rounds and Shapes

ANSI/ASTM A501 - Specification for Hot-Formed Welded and Seamless Carbon Steel Structural Tubing

ANSI/ASTM A511 - Specification for Seamless Stainless Steel Mechanical Tubing

ANSI/ASTM A512 - Specification for Cold-Drawn Buttweld Carbon Steel Mechanical Tubing

ANSI/ASTM A513 - Specification for Electric-Resistance-Welded Carbon and Alloy Steel Mechanical Tubing

ANSI/ASTM A519 - Specification for Seamless Carbon and Alloy Steel Mechanical Tubing

ANSI/ASTM A523 - Specification for Plain End Seamless and Electric-Resistance-Welded Steel Pipe for High-pressure Pipe-Type Cable Circuits

ANSI/ASTM A524 - Specification for Seamless Carbon Steel Pipe for Atmospheric and Lower Temperatures

ANSI/ASTM A53 - Specification for Pipe, Steel, Black and Hot-Dipped, Zinc-Coated, Welded and Seamless

ANSI/ASTM A530 - Specification for General Requirements for Specialised Carbon and Alloy Steel Pipe

ANSI/ASTM A554 - Specification for Welded Stainless Steel Mechanical Tubing

ANSI/ASTM A608 - Specification for Centrifugally Cast Iron-Chromium-Nickel High-Alloy Tubing for Pressure Application at High Temperatures

ANSI/ASTM A618 - Specification for Hot-Formed Welded and Seamless High-Strength Low-Alloy Structural Tubing

ANSI/ASTM A632 - Specification for Seamless and Welded Austenitic Stainless Steel Tubing (Small-Diameter) for General Service

ANSI/ASTM A648 - Specification for Steel Wire, Hard Drawn for Prestressing Concrete Pipe

ANSI/ASTM A674 - Practice for Polyethylene Encasement for Ductile Iron Pipe for Water or Other Liquids

ANSI/ASTM A688 - Specification for Welded Austenitic Stainless Steel Feedwater Heater Tubes

ANSI/ASTM A691 - Specification for Carbon and Alloy Steel Pipe, Electric-Fusion-Welded for High-pressure Service at High Temperatures

ANSI/ASTM A694 - Specification for Carbon and Alloy Steel Forgings for Pipe Flanges, Fittings, Valves, and Parts for High-Pressure Transmission Service

ANSI/ASTM A716 - Specification for Ductile Iron Culvert Pipe

ANSI/ASTM A733 - Specification for Welded and Seamless Carbon Steel and Austenitic Stainless Steel Pipe Nipples

ANSI/ASTM A742 - Specification for Steel Sheet, Metallic Coated and Polymer Precoated for Corrugated Steel Pipe

ANSI/ASTM A746 - Specification for Ductile Iron Gravity Sewer Pipe

ANSI/ASTM A760 - Specification for Corrugated Steel Pipe, Metallic-Coated for Sewers and Drains

ANSI/ASTM A761 - Specification for Corrugated Steel Structural Plate, Zinc-Coated, for Field-Bolted Pipe, Pipe-Arches, and Arches

ANSI/ASTM A762 - Specification for Corrugated Steel Pipe, Polymer Precoated for Sewers and Drains

ANSI/ASTM A771 - Specification for Seamless Austenitic and Martensitic Stainless Steel Tubing for Liquid Metal-Cooled Reactor Core Components

ANSI/ASTM A778 - Specification for Welded, Unanneled Austenitic Stainless Steel Tubular Products

ANSI/ASTM A789 - Specification for Seamless and Welded Ferritic/Austenitic Stainless Steel Tubing for General Service

ANSI/ASTM A790 - Specification for Seamless and Welded Ferritic/Austenitic Stainless Steel Pipe

ANSI/ASTM A796 - Practice for Structural Design of Corrugated Steel Pipe, Pipe-Arches, and Arches for Storm and Sanitary Sewers and Other Buried Applications

ANSI/ASTM A803 - Specification for Welded Ferritic Stainless Steel Feedwater Heater Tubes

ANSI/ASTM A810 - Specification for Zinc-Coated (Galvanised) Steel Pipe Winding Mesh

ANSI/ASTM A813 - Specification for Single- or Double-Welded Austenitic Stainless Steel Pipe

ANSI/ASTM A814 - Specification for Cold-Worked Welded Austenitic Stainless Steel Pipe

ANSI/ASTM A822 - Specification for Seamless Cold-Drawn Carbon Steel Tubing for Hydraulic System Service

ANSI/ASTM A826 - Specification for Seamless Austenitic and Martensitic Stainless Steel Duct Tubes for Liquid Metal-Cooled Reactor Core Components

ANSI/ASTM A847 - Specification for Cold-Formed Welded and Seamless High Strength, Low Alloy Structural Tubing with Improved Atmospheric Corrosion Resistance

ANSI/ASTM A849 - Specification for Post-Applied Coatings, Pavings, and Linings for Corrugated Steel Sewer and Drainage

ANSI/ASTM A851 - Specification for High-Frequency Induction Welded, Unannealed, Austenitic Steel Condenser Tubes

ANSI/ASTM A861 - Specification for High-Silicon Iron Pipe and Fittings

ANSI/ASTM A862 - Practice for Application of Asphalt Coatings to Corrugated Steel Sewer and Drainage Pipe

ANSI/ASTM A865 - Specification for Threaded Couplings, Steel, Black or Zinc-Coated (Galvanised) Welded or Seamless, for Use in Steel Pipe Joints

ANSI/ASTM A872 - Specification for Centrifugally Cast Ferritic/Austenitic Stainless Steel Pipe for Corrosive Environments

ANSI/ASTM A885 - Specification for Steel Sheet, Zinc and Aramid Fiber Composite Coated for Corrugated Steel Sewer, Culvert, and Under drain Pipe

ANSI/ASTM A888 - Specification for Hubless Cast Iron Soil Pipe and Fittings for Sanitary and Storm Drain, Waste, and Vent Piping Applications

ANSI/ASTM A908 - Specification for Stainless Steel Needle Tubing

ANSI/ASTM A928 - Specification for Ferritic/Austenitic (Duplex) Stainless Steel Pipe Electric Fusion Welded with Addition of Filler Metal

ANSI/ASTM A929 - Specification for Steel Sheet, Metallic-Coated by the Hot-Dip Process for Corrugated Steel Pipe

ANSI/ASTM A930 - Practice for Life-Cycle Cost Analysis of Corrugated Metal Pipe Used for Culverts, Storm Sewers, and Other Buried Conduits

ANSI/ASTM A943 - Specification for Spray-Formed Seamless Austenitic Stainless Steel Pipes

ANSI/ASTM A949 - Specification for Spray-Formed Seamless Ferritic/Austenitic Stainless Steel Pipe

ANSI/ASTM A953 - Specification for Austenitic Chromium-Nickel-Silicon Alloy Steel Seamless and Welded Tubing

ANSI/ASTM A954 - Specification for Austenitic Chromium-Nickel-Silicon Alloy Steel Seamless and Welded Pipe

ANSI/ASTM A978 - Specification for Composite Ribbed Steel Pipe, Precoated and Polyethylene Lined for Gravity Flow Sanitary Sewers, Storm Sewers, and Other Special Applications

ANSI/ASTM A984 - Specification for Steel Line Pipe, Black, Plain-End, Electric-Resistance-Welded

ANSI/ASTM A999 - Specification for General Requirements for Alloy and Stainless Steel Pipe

ANSI/ASTM B637 - Seamless Pipe and Tube

ANSI/ASTM B677 - Welded pipe

ANSI/ASTM D3262 - Specification for reinforced plastic mortar sewer pipe

ANSI/ASTM D3295 - Specification for PTFE tubing

ANSI/ASTM D3296 - Specification for FEP-fluorocarbon tubing

ANSI/AWWA C200 - Steel water pipe, 6 and larger

ANSI/AWWA C900 - Polyvinyl chloride (PVC) pressure pipe, 4 to 12, for water distribution

ANSI/AWWA C901 - Polyethylene (PE) pressure pipe and tubing, to 3, for water service

ANSI/AWWA C902 - Polybutylene (PB) pressure pipe and tubing, to 3, for water

ANSI/AWWA C905 - Polyvinyl chloride (PVC) water transmission pipe, nominal diameters 14 to 36

ANSI/AWWA C906 - Polyethylene (PE) pressure pipe and fittings, 4 to 63, for water distribution

ANSI/AWWA C950 - Fibreglass pressure pipe

ANSI/AWWAC115/A21.15 - Flanged ductile-iron pipe with threaded flanges

ANSI/SAEAMS 2223F - Copper and copper alloy seamless tubing

ANSI/SAEAMS 2243F - Tolerances, corrosion and heat resistant steel tubing

ANSI/SAEAMS 2244A - Titanium and titanium alloy tubing

ANSI/SAEAMS 4922 - Seamless hydraulic tubing, titanium alloy, cold worked and precipitation heat treated

ANSI/SAEAMS 4943C - Seamless hydraulic tubing, titanium alloy, annealed

ANSI/SAEAMS 4944C - Seamless hydraulic tubing, titanium alloy, cold worked, stress relieved

ANSI/SAEAMS 5556E - Seamless or welded hydraulic tubing, steel, corrosion and heat resistant, solution heat treated, SAE 30347

ANSI/SAEAMS 5561A - Welded and drawn hydraulic tubing, steel, high pressure

ANSI/SAEAMS 5564 - Hydraulic tubing, steel, high pressure, ultrasonically tested welded or seamless, SAE 30304

ANSI/SAEAMS 5566J - Hydraulic tubing, steel, high pressure, seamless or welded, SAE 30304

ANSI/SAEAMS 5567 - Hydraulic tubing, steel, seamless or welded, corrosion resistant, solution treated, SAE 30304

API 15HR - High Pressure Fiberglass Line Pipe

API 15LE - Polyethylene (PE) Line Pipe

API 15LR - Low Pressure Fiberglass Line Pipe

API 5CT = ISO 11960 - Specification for Casing and Tubing

API 5L - Specification for Line Pipe

API 5L9 - Recommended Practice for External Fusion Bonded Epoxy Coating of Line Pipe

API 5LC - CRA Line Pipe

API 5LCP - Specification for Coiled Line Pipe

API 5LD - CRA Clad or Lined Steel Pipe

API 6A = ISO 10423 - Specification for Wellhead and Christmas Tree Equipment

API RP 5A3 = ISO 13678 - Recommended Practice on Thread Compounds for Casing, Tubing, and Line Pipe

API RP 5C7 - Coiled Tubing Operations in Oil and Gas Well Services

API RP 5L2 - Internal Coating of Line Pipe for Non-Corrosive Gas Transmission Service

BS 1387 = ISO 65 - Steel tubes and tubulars

BS 1600 - Dimensions of steel pipe - for petroleum and allied industries

BS 1973 - Polythene pipe (Type 32) for general purposes including chemical and food industry uses

BS 2871 Pt 2 - Copper tubes for general purposes g ISO/R 196

BS 3059 Pt 2 - Seamless and welded carbon and alloy and austenitic stainless steel boiler and superheater tubes with specific elevated temperature properties

BS 3505= ISO 2505= ISO 3114= ISO 3472/3 - Unplasticised PVC pipe for cold water services

BS 3506 - Unplasticised PVC pipe for industrial purposes

BS 3600 ISO 64= ISO 336 - Dimensions and masses of welded and seamless steel pipes

BS 3601 - Steel pipes and tubes for pressure purposes: carbon steel with specified room temperature properties

BS 3602 ISO 2604 ISO 2605 - Specification for steel pipes and tubes for pressure purposes: carbon and carbon manganese steel with specified elevated temperature properties

BS 3603 - Specification for steel pipes and tubes for pressure purposes: carbon and alloy steel with specified low temperature properties

BS 3604 ISO 2604 ISO 2605 - Specification for steel pipes and tubes for pressure purposes: ferric alloy steel with specified elevated temperature properties

BS 3605 - Seamless and welded austenitic stainless steel pipes and tubes for pressure purposes

BS 4514 - Specification for unplasticised PVC soil and ventilating pipes, fittings and accessories

BS 4772 - Grey iron pipes

BS 4825 - Stainless steel pipe and fittings (hygienic)

BS 4825 Pt 5 - Stainless steel milk pipes and fittings (recessed ring joint type)

BS 4991 ISO 3212-3 - Propylene copolymer pressure pipe

BS 5391 - Specification for ABS pressure pipe

BS 5480 - Specification for GRP pipes and fittings for use with water supply or sewerage

BS 5481 - Specification for unplasticised PVC pipe and fittings for gravity sewers

BS 5556 ISO 161/1 - Specification for general requirements for dimensions and pressure ratings for pipes of thermoplastic materials (metric series)

BS 6323 Pt 3 - Hot finished seamless tube for automobile uses

BS 6323 Pt 4 - Cold drawn seamless tube for automobile uses

BS 6323 Pt 5 - Electric resistance welded tube for automobile uses

BS 6323 Pt 6 - Cold finished erw tube for automobile uses

BS 6572 ISO/R 1164 ISO/R 1166 ISO 2506 - Polythene pipe (Type 32) for above ground use for cold water services

BS 6730 - Black polythene pipes

BS MA 49 - Summary and application of carbon steel tubes for marine pipework systems

BS MA 60 - Copper and copper alloy tubes for marine pipework systems

DIN 1626 - Welded steel pipes in unalloyed and low alloy steels for supply purposes, process plants and tanks

DIN 1629 - Hot finished seamless tubes in unalloyed steels for supply purposes, process plant and tanks, ST52

DIN 17175 - Seamless tubes of heat resistant steels

DIN 17177 - Electric pressure welded steel tubes for elevated temperatures

DIN 2391 - ST35.4 NBK Cold drawn hard, normalised seamless hydraulic tube

DIN 2391 - ST35 BK Cold drawn seamless hard tube

DIN 2391 - ST35 GBK Cold drawn seamless annealed, soft tube

EN 10208 - 1 - Steel pipes for pipe lines for combustible fluids; technical delivery conditions; Part 1: pipes of requirement class A

EN 10208 - 2 - Steel pipes for pipe lines for combustible fluids; technical delivery conditions; Part 2: pipes of requirement class B

EN 10216 - 1 - Seamless steel tubes for pressure purposes; technical delivery conditions - Part 1: non-alloy steel with specified room temperature properties

EN 10217 - 1 - Welded steel tubes for pressure purposes; technical delivery conditions - Part 1: non -alloy steel with specified room temperature properties

EN 10220 - Seamless and welded steel tubes; dimensions and masses per unit length

EN 10224 - Steel tubes and fittings for the conveyance of aqueous liquids including water for human consumption

EN 10240 - Internal and/or external protective coatings for steel tubes - Specification for hot dip galvanised coatings

EN 10255 - Non-alloy steel tubes suitable for welding or threading

EN 1057 - Copper tubes for water, gas and sanitation

EN ISO 1127 - Stainless steel tubes - Dimensions, tolerances and conventional masses per unit length

ISO 4200 - Plain end steel tubes, welded and seamless - General tables of dimensions and masses per unit length

ISO 6708 - Pipework components - Definition and selection of DN (nominal size)

16.23 Pressure vessels

Valves are not pressure vessels but pressure vessel methodology and materials are very important in the valve industry. Pressure vessel standards include a great deal of interesting and useful information.

ANSI/ASME I - Boiler and pressure vessel code - Power Boilers

ANSI/ASME II - Boiler and pressure vessel code - Material Specifications

ANSI/ASME III - Boiler and pressure vessel code - Nuclear Power Plant Components

ANSI/ASME IV - Boiler and pressure vessel code - Heating Boilers

ANSI/ASME IX - Boiler and pressure vessel code - Welding and brazing qualifications

ANSI/ASME V - Boiler and pressure vessel code - Non-destructive examination

ANSI/ASME VI - Boiler and pressure vessel code - Recommended Rules for Care and Operation of Heating Boilers

ANSI/ASME VII - Boiler and pressure vessel code - Recommended Rules for Care of Power Boilers

ANSI/ASME VIII - Boiler and pressure vessel code - Rules for the construction of pressure vessels

ANSI/ASME X - Boiler and pressure vessel code - Fibreglass reinforced Plastic Pressure Vessels

ANSI/ASME XI - Boiler and pressure vessel code - Rules for Inservice Inspection of Nuclear Power Plant Components

ANSI/ASTM A240 - Specification for Heat-Resisting Chromium and Chromium-Nickel Stainless Steel Plate, Sheet, and Strip for Pressure Vessels

ANSI/ASTM A266 - Specification for Carbon Steel Forgings for Pressure Vessel Components

ANSI/ASTM A372 - Specification for Carbon and Alloy Steel Forgings for Thin-Walled Pressure Vessels

ANSI/ASTM A414 - Specification for Steel, Sheet, Carbon, for Pressure Vessels

ANSI/ASTM A508 - Specification for Quenched and Tempered Vacuum-Treated Carbon and Alloy Steel Forgings for Pressure Vessels

ANSI/ASTM A515 - Standard specification for pressure vessel plates, carbon steel, for intermediate and higher temperature service

ANSI/ASTM A516 - Standard specification for pressure vessel plates, carbon steel, for moderate and lower temperature service

ANSI/ASTM A748 - Specification for Statically Cast Chilled White Iron-Gray Iron Dual Metal Rolls for Pressure Vessel Use

ANSI/ASTM A765 - Specification for Carbon Steel and Low-Alloy Steel Pressure-Vessel-Component Forgings with Mandatory Toughness Requirements

ANSI/ASTM A859 - Specification for Age-Hardening Alloy Steel Forgings for Pressure Vessel Components

API Bulletin 2U - Stability Design of Cylindrical Shells

16.24 Qualification

Quality assurance and welding are the two areas where formal certification is usually applied on a regular basis. Both these areas require practical testing and test methods must be formalised to produce a uniform production quality.

ANSI/ASTM A488 - Practice for Steel Castings, Welding, Qualifications of Procedures and Personnel

ANSI/AWS B2.1 - Specification for Welding Procedure and Performance Qualification

ANSI/AWS B2.1-BMG - Base Metal Grouping for Welding Procedure and Performance Qualification

ANSI/AWS B2.2 - Standard for Brazing Procedure and Performance Qualification

ANSI/AWS B2.4 - Specification for Welding Procedure and Performance Qualification for Thermoplastics

ANSI/AWS B5.1 - Specification for the Qualification of Welding Inspectors

ANSI/AWS B5.15 - Specification for the Qualification of Radiographic Interpreters

ANSI/AWS B5.16 - Specification for the Qualification of Welding Engineers

ANSI/AWS B5.17 - Specification for the Qualification of Welding Fabricators

ANSI/AWS B5.2 - Specification for the Qualification of Welding Inspector Specialists and Welding Inspector Assistants

ANSI/AWS B5.4 - Specification for the Qualification of Welder Test Facilities

ANSI/AWS B5.5 - Specification for the Qualification of Welding Educators

ANSI/AWS B5.9 - Specification for the Qualification of Welding Supervisors

ANSI/AWS C1.5 - Specification for the Qualification of Resistance Welding Technician

ANSI/AWS G1.6 - Specification for the Qualification of Plastics Welding Inspectors for Hot Gas, Hot Gas Extrusion, and Heated Tool Butt Thermoplastic Welds

ANSI/AWS QC1 and certification of welding inspectors

API 622 - Type Testing of Process Valve Packing for Fugitive Emissions

API 6FA - Fire Test for Valves

API 6FB - Fire Test for End Connections

API 6FC - Fire Test for Valve With Automatic Backseats

API 6FD - Fire Test for Check Valves

API RP 15S of Spoolable Reinforced Plastic Line Pipe

API RP 591 - Chemical, Oil, and Gas Industries

AWS QC13 - Specification for the Certification of Welding Supervisors

AWS QC15 - Specification for the Certification of Radiographic Interpreters

BS 6755 Pt 2 - Testing of valves - fire type testing

EN 10256 - Non-destructive testing of steel tubes - Qualification and competence of levels 1 and 2 non destructive testing personnel

MSS SP-100-2002 requirements for elastomer diaphragms for nuclear service diaphragm type valves

MSS SP-121-2006 Testing Methods for Stem Packing for Rising Stem Steel Valves

16.25 Safety

The standards and regulations applying to safety are primarily concerned with the way in which valves and systems should be designed and constructed to avoid accidental damage, personnel injury or environmental pollution.

The general principles to be used in the design of complete installations are given in many standards and guidelines. The factors which should be considered include:

- access
- dust/liquid/gas hazards
- fire fighting
- noise
- proximity of adjacent property
- redundancy of equipment
- separation of safety equipment from process equipment.

BS 5908 code of practice for fire precautions in the chemical and allied industries.

BS 8005 Part 2 gives guidance on the design and construction of pumping stations and pumping mains.

The hazards posed by substances are given in:

ADR - European agreement concerning the International carriage of dangerous goods by road.

EH 40/93 - Occupational exposure limits (UK Health and Safety Executive)

ICAO - International civil aviation organization.

IMDG - International maritime dangerous goods code.

NFPA 325M - Fire hazard properties of flammable liquids, gases and volatile solids.

NFPA 704 - Identification of the specific hazards of materials and their severity.

RID - Regulations concerning the International carriage of dangerous goods by rail 2007. RID is part of the Convention concerning International Carriage by Rail (COTIF).

To ensure that a system is capable of withstanding the required nominal pressure it is necessary to pressure test all the equipment which is subjected to internal pressure. The standards dealing with pressure vessels contain regulations concerning pressure testing. For normal land use the valve must normally be tested to a pressure which is at least 30% greater than that of the specified nominal pressure. For marine valves, regulations are laid down by the classification societies which often demand a greater margin of safety.

The risk of explosion is an important consideration from the point of view of safety. There are no special regulations for valves although stringent requirements are specified for electric motors and electrical equipment. There are many standards covering the safety of electrical equipment and these standards contain a wealth of useful information for the mechanical and process engineers designing, selecting and arranging equipment. The following should be reviewed; IEC 34-1; IEC 34-5; BS/EN 60034 Part 5; BS 5345 and some sections of BS 4999.

For marine installations the instructions laid down by the classification societies such as Lloyd's Register must be followed. The safety requirements are of particular importance, since with a ship at sea it cannot be assumed that assistance will be available in the event of an accident. It therefore follows that the safety regulations on board ship are often more stringent than those of an equivalent installation for land use.

In principle all equipment must be type-approved and also have passed special safety inspection controls before installation on board ship is permitted. Similar restrictions are imposed for equipment to operate on offshore platforms.

AD A1 - Safety devices for excess pressure; bursting discs (AD-Merkblätter)

ANSI Z49.1 - Safety in Welding, Cutting, and Allied Processes

API 14C - Recommended practice for Analysis, design, installation and testing of basic surface safety systems for offshore production platforms

API RP 75 - Development of a Safety and Environmental Management Program for Outer Continental Shelf Operations and Facilities

AWS F4.1 - Recommended Safe Practices for the Preparation for Welding and Cutting of Containers and Piping

BS 6759 - Safety valves

DIN 32509 - Hand-operated shut-off valves for welding, cutting and allied processes; design, safety requirements and testing

DIN 32725 - Safety shut-off devices for combustion plants with liquified gas in its liquid state; safety requirements and test

DIN 3447 - Leak test devices for automatic shut-off valves for gaseous fuels; safety requirements

DIN 4811 Pt 2 - Pressure regulators for LPG; pressure regulators with special connections; safety requirements, testing, marking

DIN 8521 Pt 2 - Safety devices to prevent backflow for the application of gases of public gas supply, of air and oxygen up to an overpressure of 0.1 bar, not protecting against flashbacks; safety requirements and testing

EN 60950 - The specification for safety of information technology equipment including electrical business equipment g IEC 60950

EN 61010 - Safety requirements for electrical equipment for measurement, control and laboratory use

VDE 0700 Pt 246 - Safety of household and similar electrical appliances; magnetic valves

The safety of equipment exposed to fire is very important in some applications and installations. Table 16.8 indicates the test requirements for the most used specifications.

Test requirements	API 607	API 6FA	FM 6033	BS 6755 Pt 2
Pretest requirements	none	none	20 000 pressure cycles without leakage	none
Fire conditions	exposure to fire of specified flame temperature for specified duration	exposure to fire of specified flame temperature for specified duration	exposure to fire of specified fuel for specified duration	exposure to fire of specified flame temperature for specified duration
Valve position	closed	closed	closed	closed
Process fluid	water	water	water	water
Process pressure		75% of pressure rating @ 20°C	8.6 barg	75% of pressure rating @ 20°C
Fuel for fire	not specified	not specified	heptane	gas
Flame temperature °C	760 to 871	760 to 980	not specified	760 to 980
Fire duration min	30	30	15	30
Operation after test	open and close	open	none	open
Seat leakage during test (1)	40	400	95	400
External leakage during test (1)	20	100	individual drops	100
Seat leakage after cool-down (1)	40	40 (2)	95	40 (2)
External leakage after cool-down (1)	20	20 (2)	individual drops	20 (2)
External leakage after operation (1)	40	200 (3)	not specified	200 (3)

(1) ml/inch of nominal diameter/min, (2) at low pressure, (3) at test pressure

NOTE: ISO 10497 fire testing ≡ BS 6755 Pt 2 ≡ API 607 3rd edition

Table 16.8 Fire-safe test requirements

Marine and military connections are subject to rigorous tests before being accepted. Lloyd's Register has a special fire test procedure for couplings and joints. Water pipes are subjected to 800°C and oil pipes to 700°C for 30 minutes and any leakage is grounds for failure. Other specifications which may be applicable; OCMA FSV-1 and DIN 3357

16.26 Seat leakage

Seat leakage can be a major problem. Knowing how good a valve was when new enables operational wear to be quantified.

ANSI B147.1 = API 527 - Commercial seat tightness of safety relief valves with metal-to-metal seats

ANSI/API 527 - Commercial seat tightness of safety relief valves with metal-to-metal seats

ANSI/FCI 70-2 - Control valve seat leakage

16.27 Terminology

It is important to use the correct terminology to avoid confusion. However there are a number of standards and specifications to choose from:

ANSI/ASTM A644 - Terminology Relating to Iron Castings

ANSI/AWS A3.0 - Standard Welding Terms and Definitions

ANSI/FCI 86-2 - Regulator terminology

ANSI/ISA S75.05 - Control valve terminology

API 5T1 - Imperfection Terminology

API RP 80 - Guidelines for the Definition of Onshore Gas Gathering Lines

DIN 3211 Pt 1 - Valves; terms and definitions

DIN 3320 Pt 1 - Safety valves; definitions, sizing, marking

EN 10001 - Definition and classification of pig - irons

EN 10020 - Determination and classification of grades of steel

EN 10027 - 1 - Designation systems for steels; part 1: steel names, principal symbols

EN 10027 - 2ation systems for steels; part 2: numerical system

EN 10052 - Vocabulary of heat treatment terms for ferrous products

EN 10079 - Definition of steel products

EN 60534-1 - Specification and terminology for control valves

EN 736-1 - Valves; terminology and definitions of valve types

FCI 70-1 - Standard terminology and definitions for filled thermal systems for remote sensing temperature regulators

FCI 89-2 - Terminology standard for pressure switches and differential pressure switches

FSA - Glossary of terms (for fluid sealing)

IEC 60534-1 - Control valve terminology and general considerations

ISA S51.1 - Process instrumentation terminology

ISA S75.05 - Terminology

ISO 185 - Grey cast iron - Classification

ISO 2503 - Pressure regulators for gas cylinders used in welding, cutting and related processes; terminology, requirements and testing

MSS SP-90-2000 - Guidelines on terminology for pipe hangers and supports

MSS SP-96-2001 - Guidelines on terminology for valves and fittings

16.28 Valves

There are standards and specifications applied to specific valve designs and applications. Valve users can access these documents before deciding on their specific requirements and purchase order preparation.

ANSI B16.10 - Valve dimensions, face and end

EN ISO 4126-1 - Safety devices for protection against excessive pressure - Part 1: Safety valves

MSS SP-115-2006 - Excess Flow Valves for Natural Gas Service

MSS SP-117-2006 - Bellows Steals for Globe and Gate Valves

MSS SP-118-2002 - Compact Steel Globe & Check Valves - Flanged, Flangeless, Threaded and Welding Ends

MSS SP-122-2005 - Plastic Industrial Ball Valves

MSS SP-125-2000 - Gray Iron and Ductile Iron In-Line, Spring-Loaded, Center-Guided Check Valves

MSS SP-126-2000 - Steel In-Line Spring-Assisted Center Guided Check Valves

MSS SP-128-2006 - Ductile Iron Gate Valves

MSS SP-131-2004 - Metallic Manually Operated Gas Distribution Valves

MSS SP-133-2005 - Excess Flow Valves for Low Pressure Fuel Gas Appliances

MSS SP-134-2006 - Valves for Cryogenic Service Including Requirements for Body/Bonnet Extensions

MSS SP-135-2006 - High Pressure Steel Knife Gate Valves

16.29 Welding

Welding performs a critical function in many piping systems. Standards are available covering the important issues.

ANSI B16.25 - Butt weld bevel dimensions

ANSI/AWS C1.4 - Specification for Resistance Welding of Carbon and Low-Alloy Steels

ANSI/AWS C6.2 - Specification for Friction Welding of Metals

ANSI/AWS D10.10 - Recommended practices for local heating of welds in piping and tubing

ANSI/AWS D10.10 - Recommended Practices for Local Heating of Welds in Piping and Tubing

ANSI/AWS D10.11 - Recommended Practices for Root Pass Welding of Pipe Without Backing

ANSI/AWS D10.12 - Guide for Welding Mild Steel Pipe (up to 13 mm wall)

ANSI/AWS D10.4 - Recommended practices for welding austenitic chromium-nickel stainless steel piping and tubing

ANSI/AWS D10.6 - Recommended practice for gas tungsten arc welding of titanium piping and tubing

ANSI/AWS D10.7 - Recommended practices for gas shielded arc welding of aluminium and aluminium alloy pipe

ANSI/AWS D10.8 - Welding of chromium-molybdenum steel piping and tubing

ANSI/AWS D14.4 - Specification for Welded Joints in Machinery and Equipment

ANSI/AWS D18.1 - Specification for Welding of Austenitic Stainless Steel Tube and Pipe Systems in Sanitary (Hygienic) Applications

ANSI/AWS D18.3 - Specification for Welding of Tanks, Vessels, and Other Equipment in Sanitary (Hygienic) Applications

ANSI/AWS G1.1 - Guide to Ultrasonic Assembly of Thermoplastics

ANSI/AWS G2.1 - Guide for the Joining of Wrought Nickel-Based Alloys

ANSI/AWS G2.4 - Guide for the Fusion Welding of Titanium and Titanium Alloys

ANSI/AWWA C206 - Field welding of steel water pipe

API 1104 - Welding of Pipelines and Related Facilities

API 2B - Fabrication of Structural Steel Pipe

API Publication 2201 - Safe Hot Tapping Practices in the Petroleum & Petrochemical Industries

API RP 582 - Recommended Practice and Supplementary Welding Guidelines for the

API RP 5C6 - Welding Connections to Pipe

AWS D11.2 - Guide to welding iron castings

BS 4515 - Specification for process of welding of steel pipelines on land and offshore

EN 287-1 - Approval testing of welders for fusion welding: Steels

EN 287-2 - Approval testing of welders for fusion welding: Aluminium and aluminium alloys

EN 288-1 - Specification and approval of welding procedures for metallic materials: General rules for fusion welding

EN 288-2 - Specification and approval of welding procedures for metallic materials: Specifications and procedures for arc welding

EN 288-3 - Specification and approval of welding procedures for metallic materials: Welding procedure test for arc welding of steels

EN 288-4 - Specification and approval of welding procedures for metallic materials: Welding procedure test for arc welding of aluminium and aluminium alloys

ISO 228-1 - Pipe threads where pressure-tight joints are not made on the threads - Part 1: Dimensions, tolerances and designation

Installation and maintenance 17

17.1 Introduction

Some years ago EDF Group of France, one of Europe's leading energy companies, reported that valves represented 2% of the capital cost of its pressurised water reactor power stations. But valves accounted for 20% of shutdowns and 60% of total maintenance costs. Packing box leakage was a serious problem on hot medium pressure valves working with boric acid in a radioactive environment. A ten year development programme by a packing manufacturer plus user staff training cured the problem, reduced costs to acceptable levels, dramatically increased reliability and reduced maintenance time to 17%. Some packing now lasts for over five years.

An annual report of NBIC, (the American National Board of Boiler and Pressure Vessel Inspectors) some years ago quoted the following figures relating to pipework and valves based on a survey of 6486 accidents:

- piping — 531 accidents
- safety valves — 210 accidents
- pressure control failure — 104 accidents
- safety valve leakage — 38 accidents

It was also reported some years ago that of over 30 control valves tested at one site, serious operating problems were uncovered, see Table 17.1.

Control valve problem	Percentage of valves affected
Positioner	71.0
Bench set	38.7
Travel	35.5
Actuator force	35.5
Friction	35.5
I/P	35.5
Other	6.7

Table 17.1 Typical operating problems

These examples help illustrate that valves can be a source of constant trouble on site. Misrepresentation of the operating conditions may be a significant contributory factor. Purchasing valves to a **price** rather than to a **specification** may be another problem. The consequential costs of valve failures must be considered during the process design and valve selection phases and arrangements made to cope with costly problems. Whole-life cost is likely to be many times greater than the initial purchase cost.

17.2 General

There is one fundamental rule which must be observed before doing **anything** with equipment.

"Read the instruction manual first!"

Some companies have good sources of operational and reliability information from similar installations. This information can be useful for all aspects of valve utilisation. However, it must be remembered that other installations are **similar,** not **identical**. In mechanical engineering it is very difficult, if not impossible, to manufacture two identical components. It is certainly impossible to build two identical process systems and operate them with identical conditions. Information from similar installations is very useful but must be used with care.

It is essential that the site employs, or has very easy access to, personnel experienced with the equipment used. Many valve manufacturers hold training courses at their works explaining all aspects of operation and maintenance. This is a very good investment in the long term. Delays, incurred by waiting for factory staff to travel long distances, can be extremely costly especially when spares can only be ordered after the arrival of a specialist. This type of reliability analysis may restrict the style of purchased equipment.

It is also essential that the site is equipped with workshop facilities and tools appropriate for the equipment in service. Special tools, necessary for routine maintenance, should have been supplied with the valve originally. These tools must be readily available for the appropriate staff.

Some tools themselves require periodic maintenance. Lapping blocks for seats should be checked regularly and reconditioned if necessary. They must be in very good condition if seat leakage is to be minimised. Protective clothing and evacuation/decontamination equipment may be necessary for some hazardous fluid systems. Personnel must be fully aware of the nature of fluid hazards **before** starting work on a system.

Routine inspections, maintenance and repair require consumables which must be available if extra delays are to be avoided. The following items are typical:

- gasket material
- thread tape
- pipe thread sealants
- fastener thread locking compounds
- lubricants, (oil and grease)
- "O" ring lubricant
- anti-seize sprays
- release oil
- standard studs, bolts, studbolts and nuts
- continuously threaded rod
- leak detection sprays or detectors
- water dispersal sprays
- degreasing fluids
- emery tape
- grinding and lapping paste
- paint

Continuously threaded rod is necessary to make special length studs and studbolts. Paint is often overlooked as a standard consumable. Touching-up the paint, after maintenance, can considerably extend the life of external components.

It will also prove very useful to maintain a stock of reliable pressure gauges. These will prove invaluable when trying to discover what a system is actually doing. A portable, electronic flowmeter of the clamp-on type will also be extremely useful.

Operating and instruction manuals should be stored efficiently at site to allow rapid retrieval in case of breakdowns. Manufacturers spend considerable sums of money preparing and printing manuals to promote successful operation of their valves. Copies of the manuals held at company headquarters are of little use to maintenance staff on site.

During normal working a log of operating conditions should be routinely updated. Background operational history will prove invaluable when investigating problems and failures.

17.3 Pre-installation inspection

Before erecting pipe sections, site staff must ensure all relevant equipment is available. Terminations at the beginning and end of the pipe run must be ready to accept the pipe. All intermediate valves and fittings should be present and moved to their final locations. Valves which are "adjustable" such as safety relief valves, regulators and control valves should be tested, if possible, before installation. Some non-return valves are adjustable but pre-installation testing may not be possible due to the wide range of operating possibilities. Testing during commissioning will probably be the only suitable option.

Valve testing should, ideally, be performed as soon as the valve is received at site. There are several good reasons for this recommendation:

- The test will establish whether the valve conforms to its specification in all respects. It is therefore necessary to have the manufacturer's specification available and the process data sheet for the valve, and to confirm every design feature and every function of the valve.
- The tests should also check features not expressly stated in the manufacturer's data sheet, such as seat or gland leakage, friction or hysteresis of the actuator.
- A pressure test at the stated test pressure(s) may be necessary if the factory test certificate has expired. Legal requirements for some systems dictate retesting routinely at specified time intervals.
- The site test record is the basic datum for all subsequent maintenance operations. If the valve is used in corrosive service, a record of body thickness or trim clearances will establish the rate of corrosion of affected parts and their replacement.
- A check of regular and special spare parts will ensure that these can be ordered in time.

It should be evident that testing should be done as early as possible, so that if a defective valve has to be returned to the manufacturer, it will not cause serious delay to a construction programme or a maintenance schedule. After workshop testing, valves should have the inlet and outlet blanked off with plywood and adhesive tape to avoid contamination and entry of dirt or stones, before being returned to storage.

Pipework can be installed initially with safety relief valves, regulators and control valves replaced by spool pieces. This allows the pipe system to be cleaned without damaging valves.

17.3.1 Test equipment

It is possible to obtain proprietary test stands to perform the first three tests listed above; but the equipment is quite simple to construct and many larger sites using "adjustable" valves may prefer to build their own.

A simple test stand should accommodate valve bodies from DN25 to about DN100. Some larger sites may need to test valves up to DN150 but the majority of valves will be DN100 and smaller. Mobile test facilities and staff may be available for large valves. Before building a test stand it is advisable to check the local support companies for suitable test facilities.

The valve should be securely attached to the test stand by its inlet connection, which also acts as a feed for the water supply for pressure and leak testing. The outlet connection of the valve should be blanked off and fitted with a small valve for venting the air from the body. Test stands in which the valve under test is sandwiched between two flanges tightened by a screw jack are not recommended. A mobile hoist of the type used in garages is very convenient for removing actuators and bonnets, etc. The test stand also supports the valve during seat removal, grinding-in and some of the other maintenance operations on the open body described later.

The test stand should be fitted with two air supplies that are adjustable between zero and the highest actuator pressure encountered on site. One of these supplies is used as the main air supply to valve positioners, piston actuators etc., and should be fitted with a simple regulator and gauge to allow it to be set to the correct value. The other supply should be controlled by a precision regulator and the air pressure indicated on an accurate gauge. This gauge, because it is a calibration substandard, should be included in the regular calibration programme on site of all test equipment.

To perform pressure and leak testing at the specified pressures for each valve, a supply of high pressure water is required. Such a supply is most conveniently generated by a direct-acting air-operated reciprocating pump. This type of pump is driven by a regulated supply of compressed air and is capable of furnishing exact test pressures from a few bar to thousands of bar. This pump, with suitable connectors and hoses, will be able to pressure test all valves on site.

Safety relief valves can have the set pressure checked with the high pressure water supply. The flowing characteristics of a valve, overpressure and blowdown, can only be validated with an adequate supply of appropriate fluid.

In the same way in which these tests are carried out before the "adjustable" valve is installed for the first time, they should be performed on each occasion the valve has been removed from the plant for service.

17.4 Installation criteria

17.4.1 Pipework

Valves must be installed correctly if a long, economic service life is to be achieved. Pipe "fall" must ensure adequate venting and draining. Hazardous liquids should be capable of complete drainage to suitable collection points. The pipework must have adequate flexibility to prevent excessive forces and moments being applied to the valve body; distortion can induce seat leakage and promote galling/wear/seizure of trim.

Piping and valves in areas of high fire risk should have the effects of high ambient temperatures fully investigated. Under these circumstances it may be advisable to upgrade flange ratings to provide higher gasket loads or use RTJ metal gaskets or clamped connections with metal seal rings. Safety relief valves, specifically for pressures generated by fire hazards, may be necessary.

Safety relief valves that protect against thermal expansion caused by normal ambient temperature changes may be required in pipe sections locked between two isolating valves. Many high integrity systems utilise socket or butt weld valves. The manufacturer's instructions should be reviewed for disassembly requirements prior to welding and adequate protection must be applied for weld splatter. Butt weld valves can be removed from pipework and replaced after the old welds have been dressed. Socket weld valves and fittings are much more difficult to remove.

Threaded pipework can be damaged by Stilsons™, Footprints™, pipe pliers and chain wrenches during installation. This type of damage can be serious on galvanised pipework and pipework susceptible to fatigue failure. Care must be taken during assembly to prevent damage which can shorten the working life of new equipment. Great care must be taken with the design of threaded pipework systems. It is easy to design a system which cannot be assembled or a system which requires remote connections to be broken to allow the removal of a specific item. These problems may not become apparent until the system assembly stage. Thread paste should be applied to the pipe and not the valve body threads, otherwise paste may enter the valve trim where it will collect dirt and possibly prevent the valve from closing.

For valves with flanged connections, check that the flanges are properly aligned to ensure even gasket contact. Fit the bolts and tighten the nuts in the order shown in Figure 17.1. This will prevent uneven loading on the gasket and risk of flange breakage.

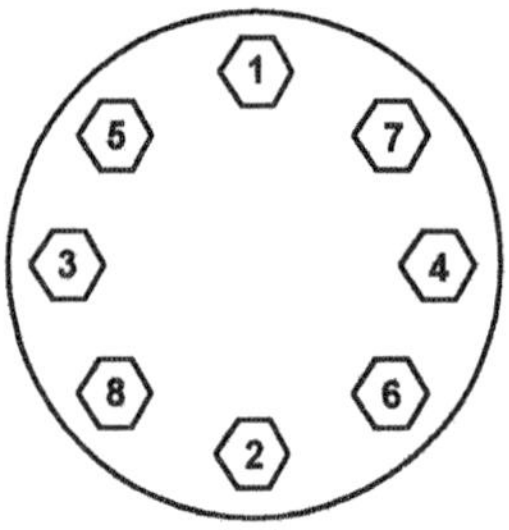

Figure 17.1 Flange bolting tightening sequence

17.4.2 Valve mounting

The physical location of valves within the pipework is important. Valves should not be positioned close to fittings which create severe flow disturbances. Uneven flow patterns can cause unexpectedly high erosion and rapid wear. Regulators and control valves can suffer continuous or intermittent instability due to irregular flow patterns. If a valve must be located close to a bend then a "pulled" large radius pipe bend is preferred to a bend fitting. Valves should not be located at the centre of spans between pipe supports. The increased pipe deflection may lead to vibration problems. The physical location of valves to enable ease of personnel access may be very important, even for automatic valves, since all valves will require inspection at some time. Some, however, can only be maintained/repaired in situ.

Valve internals can be inspected with a boroscope if a small connection is available. Fixed valves with integral seats may require machining or lapping in situ. Access must be considered as an integral part of plant layout, and valves should be located where the complete external condition can be easily seen. Additional instrumentation should be considered when routine inspection visits are not possible.

Some standard valves are not suitable for mounting upside down or in vertical pipe runs. The actual mode of mounting may not be apparent until assembly at site. The manufacturer's installation instructions should be consulted for acceptable mounting positions. Adjustments may be possible to accommodate various positions. Regulators and control valves are best installed so that the actuator is vertically above the valve. The actuator should be adequately supported to avoid vibration problems.

Preservatives on valves should be removed just prior to installation. The manufacturer's instructions will state the appropriate solvents required. Valves for biotechnical and hygienic applications must be cleaned thoroughly with suitably approved solvents.

Piping systems which operate at low or high temperature may be surrounded by substantial layers of insulation and metal protective cladding. The glands of valves should be left exposed for easy, visual inspection and maintenance. The early signals of packing wear, liquid drips and vapour clouds, can be concealed by insulation. Maintenance staff can easily adjust a gland when passing on another job. Stripping insulation for an inspection will only be done as a specific scheduled task. Carbon steel pipework expands approximately 1.2mm per metre per 100°C temperature rise, and stainless steel slightly more, at 1.7mm. The change in pipe length, and consequential forces and moments on connections, must be considered.

Manual valves can generally be mounted with the spindle/stem vertically above the valve or extending horizontally. If the valve is to be mounted in any other orientation check the operating instructions. In general, manual valves with packing boxes should **not** be mounted upside down.

Most solenoid valves must be mounted with the armature vertical, directly above the valve. Some valves will allow a small deviation from the vertical, ±15°, so it is important to check the operating instructions.

Safety relief valves, srvs, **must** be installed correctly if the safety function is to be achieved successfully. The inlet pipe for a srv should be as short as possible. When protecting a vessel or boiler, direct connection to the vessel flange is preferred. If inlet piping is unavoidable then the pipe should be at least one size larger than the valve inlet.

NOTE: Never fit a valve to a connection which is smaller than the valve inlet.

If the pipe length is significant or the viscosity is high check the pressure drop at the maximum flow rate. This should not exceed 3% of the valve set pressure.

Inlet pipes for liquid valves must ensure a good supply of liquid and eliminate the possibility of gas or vapour reducing the valve capacity. Inlet pipes for gas and vapour should "fall" back to the source to avoid the accumulation of condensation and liquid slugs. Hot inlet piping should be insulated up to the srv. Outlet pipework should be free flowing, about 10% friction drop maximum. During pressure drop calculations remember that gas and vapour will expand considerably in the srv outlet pipe. When multiple valves are piped to a common outlet or the outlet pipe is used for other functions, the design must be considered carefully. Do not cost cut on pipe sizes.

Outlet pipes must be self-draining. Liquid or condensation must not be allowed to accumulate on top of the plug and cause corrosion. A separate drain pipe on the main outlet pipe may be necessary to ensure drainage. Some srv designs include pockets within the valve body which cannot be drained. Check valve designs carefully if corrosion is likely to be a major problem. For example — Can a drain pipe be fitted to the valve body?

Bends in the outlet pipe can transmit forces and moments back to the valve body so adequate pipe supports must be fitted. The back pressure at the valve outlet should remain relatively constant over all operating conditions. Variable or significant back pressures require balanced bellows or pilot-operated valves, see Chapter 7. The reaction force caused by the valve operation must be absorbed by the pipework. Additional local supports may be necessary.

Some srvs are installed with standard isolating valves in the inlet pipework. This practice is prohibited by many pressure vessel and system "Codes" and classification bodies. This type of installation is understandable, but not approved, when the process must continue to operate with defective valves. If isolating valves are to be fitted in this way then valve locking must be strictly observed. Only very senior **operating** personnel must have the authority to take the srv off-line.

Srvs are normally mounted with the spring vertically above the plug. Low pressure valves may be mounted upside down when specifically instructed by the manufacturer. Mounting in other positions is at the manufacturer's discretion.

Srvs can be fitted with manual lifting levers to allow lift checking and also to enable solids to be blown off the seat. Good personnel access is necessary to benefit from this facility. Extension chains or rods may be fitted to allow operation from below. Hydraulic or pneumatic cylinders can be fitted to allow remote operation. Some srv designs result in the inlet flange being much thicker than normal and so extra long studs or studbolts may be required.

Gas and steam srvs can produce a lot of noise when operating. Large valves may require large silencers to become socially acceptable. The problem of valve noise must be considered at the design and selection stage. Silencers are likely to be too large to be fitted as an afterthought. See Section 10.6 in Chapter 10 for noise calculations.

Inches							Millimetres					
A*	W**	H**	X**	Actual manifold pipe od	Nominal size Manifold pipe	Nominal size Control valve	Nominal size control valve	Actual manifold pipe od	A*	W**	H**	X**
7¾	27	39	23	1.315	1	1	25	33.4	197	690	990	580
9½	27	39	23	1.900	1½	1½	40	48.3	235	690	990	580
7¾	27	39	23	1.900	1½	1	25	48.3	197	690	990	580
10½	27	39	23	2.375	2	2	50	60.3	267	690	990	580
9¼	27	39	23	2.375	2	1½	40	60.3	235	690	990	580
7¾	27	39	23	2.375	2	1	25	60.3	197	690	990	580
12½	30	42	27	3.500	3	3	80	88.9	317	760	1070	690
10½	30	42	27	3.500	3	2	50	88.9	267	760	1070	690
9¼	30	42	27	3.500	3	1½	40	88.9	235	760	1070	690
14½	35	43	30	4.500	4	4	100	114.3	368	890	1090	760
12½	35	43	30	4.500	4	3	80	114.3	317	890	1090	760
10½	35	43	30	4.500	4	2	50	114.3	267	890	1090	760
14½	45	54	39	6.625	6	4	100	168.3	368	1140	1370	990
12½	45	54	39	6.625	6	3	80	168.3	317	1140	1370	990
18⅝	55	57	46	8.625	8	6	150	219.1	473	1400	1450	1170
14½	55	57	46	8.625	8	4	100	219.1	368	1400	1450	1170

* Actual dimensions from 15A. ** Suggested dimensions.

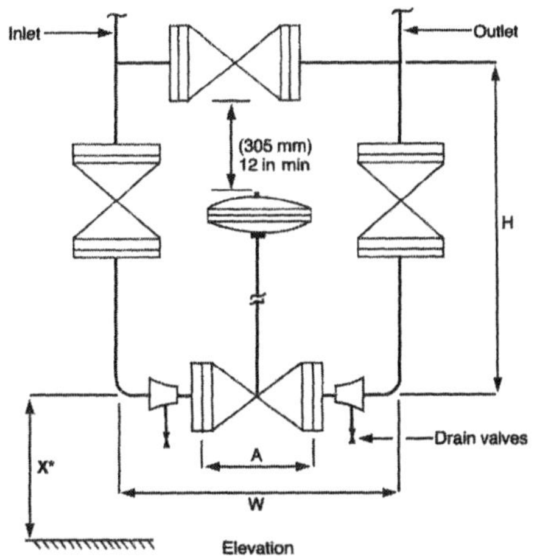

Dimensions shown are suggested piping dimensions and may vary depending upon actual dimensions of components
*** For bottom removal of valve plug**

Figure 17.2 Type I Control valve manifold dimensions for ANSI Class 300LB

Two identical srvs can be fitted with a special double-beat globe valve in the inlet pipework, one working valve and one 100% standby. The standby valve becomes available when the globe valve is changed over. The valve outlets are piped independently. This type of installation allows continuous protection with the facility for standby maintenance and inspection. When the outlet pipe is long a similar type of valve can be fitted to the valve outlets allowing the working valve to utilise the single outlet pipe. The inlet and outlet globe valves can be "ganged" together using a chain drive from one stem to the other. Bursting discs can be used as standby protection while srvs are being maintained. It is advisable to ensure that a stock of replacement discs is always held.

The ISA has recommendations concerning the space which should be allowed for a regulator or control valve in a pipe system. The importance of following these recommendations becomes evident as soon as it is necessary to service the valve or replace components with the valve in situ. The tables in Figures 17.2 and 17.3 refer to cast steel valves although they can also apply to cast iron valves.

Regulators and control valves are frequently used in systems with compressors and pumps which create flow variations and pressure pulsations. Reciprocating and peristaltic machines may be fitted with pulsation dampers to attenuate the effects on pipework. Most compressors and pumps produce pressure pulsations to a greater or lesser extent and dampers or pulsation filters may be fitted to prevent control instability. During regulator/control valve installation, especially with reciprocating machines, check the piping to see if dampers/filters specified have been fitted or are still in storage. Waiting until commissioning may result in long delays.

During system assembly it is sensible to check the pressure ratings of the pipes, fittings, flanges and valves used. The system can only be operated and hydrotested at pressures compatible with the lowest rating.

With large, complex process systems the pipework and valves may be assembled some considerable time before the process is commissioned. It is essential therefore that valves are protected, internally and externally, against corrosion and physical damage. The internal conditions, prior to start-up, can be much more corrosive than the normal operating conditions.

17.5 Commissioning

17.5.1 Pipework cleaning

Foreign particles in the pipeline can destroy valve seats and prevent movement of both plug and rotary valves, stopping them from sealing properly. If they are large enough, foreign

Inches								Millimetres						
A*	W**	H**	X**	Y**	Actual manifold pipe od	Nominal size: Manifold pipe	Nominal size: Control valve	Nominal size control valve	Actual manifold pipe od	A*	W**	H**	X**	Y**
7¾	58	21	23	37	1.315	1	1	25	33.4	197	1470	530	580	940
9½	58	21	23	37	1.900	1½	1½	40	48.3	235	1470	530	580	940
7¾	58	21	23	37	1.900	1½	1	25	48.3	197	1470	530	580	940
10½	58	21	23	37	2.375	2	2	50	60.3	267	1470	530	580	940
9¼	58	21	23	37	2.375	2	1½	40	60.3	235	1470	530	580	940
7¾	58	21	23	37	2.375	2	1	25	60.3	197	1470	530	580	940
12½	67	26	27	39	3.500	3	3	80	88.9	317	1700	660	690	990
10½	67	26	27	39	3.500	3	2	50	88.9	267	1700	660	690	990
9¼	67	26	27	39	3.500	3	1½	40	88.9	235	1700	660	690	990
14½	74	29	30	40	4.500	4	4	100	114.3	368	1880	740	760	1020
12½	74	29	30	40	4.500	4	3	80	114.3	317	1880	740	760	1020
10½	74	29	30	40	4.500	4	2	50	114.3	267	1880	740	760	1020
14½	97	36	39	50	6.625	6	4	100	168.3	368	2460	910	990	1270
12½	97	36	39	50	6.625	6	3	80	168.3	317	2460	910	990	1270
18⅝	109	41	46	52	8.625	8	6	150	219.1	473	2770	1040	1170	1320
14½	109	41	46	52	8.625	8	4	100	219.1	368	2770	1040	1170	1320

* Actual dimensions from 15A. ** Suggested dimensions.

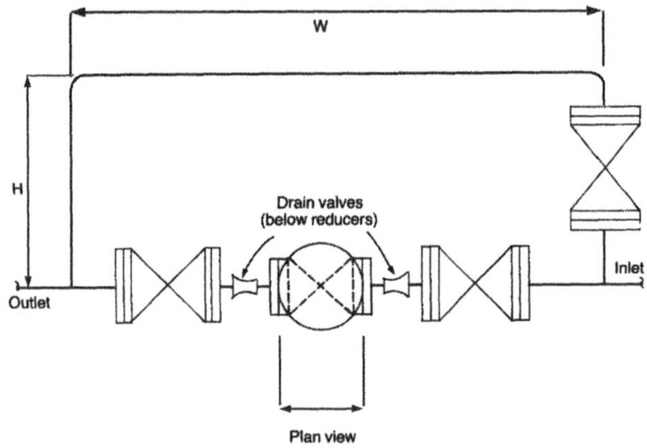

Dimensions shown are suggested piping dimensions and may vary depending upon actual dimensions of components

Figure 17.3 Type VI Control valve manifold dimensions for ANSI Class 300LB

bodies can completely block a control valve. Very small, hard particles can cause rapid wear of the trim clearance. To prevent contamination problems it is necessary to clean all pipework so that it is completely free of all loose weld splatter, slag and other particles. Pipework and valves in solid handling systems should be cleaned to remove "out-of-spec" solids prior to commissioning.

Gas and steam lines in particular, should be carefully cleaned by a series of blowouts, to remove scale and rust which could otherwise damage the valve. This is best done immediately after pressure testing. The progress of the cleaning operation can be checked by blowing against an aluminium target plate.

Systems handling flammable, hazardous or hygienic fluids may require purging of air before the piping can be vented and primed with process fluid. Nitrogen gas, distilled water or proprietary sterilising liquid may be required. Good piping design can ensure that pockets of air or contamination are not overlooked. Vent and drain valves at appropriate locations can indicate the progress of purge fluid. Leak testing of these types of system may be much more important than pressure testing.

17.5.2 Isolating valves

The seat tightness of isolating valves can be checked during purging, leak testing and pressure testing. The pressure decay of isolated pipe sections can be logged while connections and packing boxes are checked. Remember, accurate packing box leakage can only be measured when the pressurised fluid is very similar to the process fluid. Nitrogen leakage may not be very representative of the process gas leakage.

Manual isolating valves should be operated using the handwheel fitted to the valve. If leakage is more than expected/specified **do not use cheater extensions.** If, after inspection, the valve is found to be in good working order and still will not seal properly the actuator mechanism or the valve style must be changed. Cheater bars strain and over-stress components and wear out valves very quickly.

Power-operated isolating valves can have their operating mechanism checked without process fluid in the pipework. The control sequence and the travel limit stops can be verified. Ensure that close control signals actually close the valve. Also verify the effect of power failure and the Emergency Shut Down, ESD, signal when appropriate. The speed of operation of actuated valves can be very important for the stability of the process system. Valves which open or close too quickly can radiate pressure pulsations through the system. The effect of valve speed can only checked properly when the process fluid is pressurised and flowing at rated velocity. Power operated isolating valves should be set to operate as slowly as practicable.

17.5.3 Solenoid valves

Direct-operated solenoid valves can be commissioned without process fluid. Servo-operated solenoid valves require the process fluid to be present and pressurised for correct operation. Open and close signals should be checked for correct settings.

The speed of valve operation may be important. Significant pressure pulsations may be created with larger valves. It may be possible to fit different springs to direct-operating valves so that the speed can be adjusted. Servo-operated valves have an orifice which can be changed to control the speed. In general terms, the slower the operating speed, the better.

Some linear valves operate at very high speed for emergency isolation. These valves will cause severe pressure pulsations and vibration in piping systems which are long and/or of high velocity. Simulated operation for commissioning is best done with pressurised, stationary process fluid; this will avoid the worst effects of the pressure pulsations.

17.5.4 Non-return valves

Non-return valves require flowing fluid for full commissioning. Adjustable valves can be set to open wide at maximum flow rate. When fitted, dampers or dashpots should be adjusted to provide an acceptable closing speed with minimum surge effects. It will be necessary to start and stop the fluid flow. This can take some considerable time with large systems. Listen to valves very carefully, with a stethoscope or screwdriver, while operating. It should be possible to hear doors swinging or chattering due to unstable stable flow conditions or vortices. Doors which move continuously during normal operation will tend to wear quickly. Stronger springs or a different spring rate may solve the problem.

17.5.5 Safety relief valves

Safety relief valves can be adjusted at site to optimise performance. If more than one valve is fitted, isolate all but one and commission in turn. The set pressure will have been fixed and checked by the manufacturer and the spring locked and sealed. Unless process requirements have changed there should be no need to adjust the set point. Many compressor and pump manufacturers set safety relief valves at +10% whereas valve manufacturers prefer +25% because it allows the use of smaller valves. The set pressure for compressor/pump protection **must not** be reset higher without informing the compressor/pump manufacturer. The set pressure can normally be adjusted up to 10% with one spring. Consult the valve manufacturer for adjustments over a wider range and for alternative springs. The set pressure should not be adjusted on-line. Serious accidents have occurred when safety relief valves have become unstable resulting in very high operating pressures. Valves must be adjusted when unpressurised. Only adjust slightly between retries.

The overpressure and blowdown can be adjusted during commissioning provided the operating and fluid conditions are suitable. Standard spring-loaded valves can only be optimised using a fluid which has very similar physical properties to the process fluid. Valves will have one or two adjustments to allow overpressure and blowdown to be set.

Overpressure and blowdown can both be set with the safety relief valve on-line. The operating pressure should be less than 90% of the set pressure when adjustments are made. Set the overpressure first. Move the adjustment by one increment only between retries. Once the overpressure is acceptable, adjust the blowdown. Always replace or tighten locking devices during retries. Resetting the valve to extremes of operating conditions can result in unstable operation.

When the line is clean, spool pieces can be removed and regulators/control valves can be installed. The valve(s) and gaskets should be an easy sliding fit into the gap in the pipe, and upstream and downstream pipes should be carefully aligned and concentric, and the flanges must be parallel. If these precautions are not observed and the pipes are pulled into alignment by bolting up to the valve body haphazardly, the resulting stresses will certainly cause the valve to leak and may cause it to stick or to fail prematurely in service.

17.5.6 Control valves and regulators

Regulators may be totally enclosed with the stem/spindle connection inaccessible without dismantling the valve. These regulators must be verified on the test stand before installation. Regulators of an open construction can be checked in a similar manner to control valves.

Control valves should have been fully checked before installation and it is important to check site records. The actuator and positioner, if fitted, should be disconnected and checked for full stroke. The air supply should be checked before connecting to the positioner. The positioner response to 0%, 25%, 50%, 75% and 100% signals should be measured. The positioner and actuator should be checked for air pressure failure response. If all is in order, reconnect the valve stem and recheck the stroking range.

Regulators and control valves are frequently used in systems with compressors and pumps which create flow variations and pressure pulsations. Check the piping to see if dampers/filters specified have been fitted correctly and, when appropriate, are charged with nitrogen to the correct pressure. The nitrogen charge pressure, with the system at atmospheric pressure, should normally be 70% of the **actual** system operating pressure. Systems which operate over a significant range of pressures may require a different precharge setting.

Initial commissioning may be performed with water, oil, air or nitrogen. Final commissioning and "fine tuning" can only be carried out when rated operating conditions with the proper process fluid are available. Final checks should include listening for flashing and cavitation. Both flashing and cavitation can create serious operating problems if overlooked during valve type selection and sizing.

17.6 Routine maintenance

Regular maintenance operations will include external visual examinations, lubrication and packing adjustment. Staff must be aware of isolation and depressurisation procedures before valves are dismantled. When necessary, decontamination facilities should be clearly available, not concealed at the back of the valve.

It is very important, particularly when using motor actuators with gate valves not primarily designed to be automatically actuated, that the valve gland and the drive-nut are effectively lubricated at all times. If necessary extra lubricators should be fitted.

Valves handling volatile organic compounds, VOCs, or hazardous fluids should have the packing box leakage checked regularly. Some compounds will be regulated by health or environmental legislation. Valves for hazardous or controlled fluids should have sampling ports on the packing box for portable or continuous monitoring. A leakage log for these valves is essential to predict repacking dates.

Some valves are installed but are required to operate or move infrequently. In these situations valves can present seat and spindle/stem sealing problems. The working environment should have been considered during valve type selection, but the ideal valve may not be able to cope perfectly with the operational requirements. The routine maintenance for such valves should include:

- spindle/stem movement
- packing lubrication and adjustment
- seat/plug inspection and relapping

Many sub-contract organisations offer valve maintenance. Routine maintenance, at scheduled times, is very easy to implement as is a major overhaul at annual plant shutdowns. Unscheduled maintenance, such as a service crew passing a valve on the way to another job or called out for emergencies, is more difficult. If a sub-contract maintenance agreement is considered then the use of a local organisation is preferable. Clearly, agreement on rapid response and standard fees must be secured before contracts are awarded.

The staff, facilities and quality assurance programmes of sub-contract organisations must be fully investigated. Factory-trained personnel are essential for complicated valves. High quality machine tools, welding and inspection equipment are necessary for good repair work. The limitations of test equipment must be known before valves are committed for repair. A strict quality assurance programme must be running effectively to ensure all aspects of maintenance and testing achieve the desired objectives. Staff, at all levels, must have very broad engineering experience.

The valve user must be able to quantify the faults and provide acceptance levels for the repairs. Numerical, measurable values must be assigned to important variables. Written descriptions, rather than numerical values, are subject to interpretation. For example:

- "The valve stem in contact with the packing shall be smooth" — is a poor requirement
- "The valve stem in contact with the packing shall have a surface finish between 0.8 and 0.4μm" — is precise and verifiable

When requirements are poorly defined the end result can be less than satisfactory.

A review of the sub-contractor's quality manual will indicate the range of quality levels available for specific tasks. If a sub-contractor cannot meet the quality level for a task then a quotation is unnecessary. Quotations for repair of major valves must be accompanied by an itemised list of operations with associated inspection/quality controls. Extra special attention must be given to the reclamation of worn components by plating, welding or metal spraying. Treatments prohibited on the new valve components should not be allowed during repair. Impregnation of porous castings is frequently prohibited for new valves; there is no sound reason why it should be acceptable as a repair.

Some organisations win contracts by quoting low prices. The contract value is subsequently increased by extras for unexpected work. Maintenance quotations for major valves must include a list of extra costs which covers probable additional work. Extras from the list are then the only additions to the contract considered. Competitive quotations can then be reviewed on a "total" cost basis and so enable repairs rather than replacement to be justified.

The routine maintenance system for safety equipment must operate independently from other maintenance functions. Safety checks, to comply with pressure codes or operating licences, must be performed at prescribed times and cannot be delayed by other jobs. Inspection and maintenance of safety systems must be allocated a much higher priority than other tasks. Environmental health or pollution inspectors can shut down processes if legal requirements have not been met. Computerised management systems can prompt personnel to carry out routine inspections and testing. It is essential to keep very good records of inspection, tests, overhauls, parts failures and replacements.

Specialist companies can offer on-line testing of safety relief valves. The valve is lifted manually, pneumatically or hydraulically to confirm full travel and the capacity. This type of testing is questioned by some valve manufacturers who think it to be inaccurate. It is best to seek the manufacturer's advice before committing contracts. On-line testing of this type may not be acceptable for code or licensing compliance. In any event, on-line testing does not replace routine, regular strip-downs and physical inspection.

Some process systems may require periodic pressure testing to comply with construction codes or legal requirements. The process will have to be shut down to perform the pressure test and the opportunity should be exploited to carry out inspections of critical equipment.

17.6.1 Reconditioning nozzles and discs

Safety relief valves fall into the category of very infrequent operation. They are fitted as a safety system and should not form part of normal control operations. Equipment or machinery operation should not require them to be lifted routinely. Safety relief valves should be used as dedicated safety devices.

Safety relief valves that have lift levers which have not operated, should be lifted manually on a regular basis, therefore check the equipment manufacturer's instructions. Valves should be lifted at least three times a year; some compressor manufacturers recommend once a week. Another recommendation is based on set pressure:

- below 70barg: lift once a week
- above 70 barg: once a year

This last range, however, seems to be illogical.

Nozzles and discs will wear, become eroded or corroded and the safety relief valve will leak. The seat seal must be maintained to ensure proper valve operation. Unless the seat surfaces have been damaged, regular lapping should ensure proper valve operation and good sealing. The seating surfaces are perfectly flat and must be true without irregularities. ANSI/API 527 specifies a surface flatness of within 0.5μm. Standard production methods achieve flatnesses within 0.025 to 0.3μm. The nozzle and disc should not be lapped against each other but individually against a lapping block. If a disc insert is fitted the insert must be removed for lapping.

Safety relief valves that have balance bellows should be dismantled with great care. Damaging the bellows during routine maintenance may prove a costly mistake. Measure and note the position of the spring adjuster prior to dismantling.

Very clean working conditions are necessary for successful lapping. A cast iron or plate glass lapping block can be used for all circular seatings. One side should be used for rough lapping and the other for polishing. Select the "truest" side for polishing. After rough lapping, the nozzle and disc should be polished with the finest compound available. When finished, the seating surfaces should have an even matt finish.

If only the nozzle requires lapping, this can be performed in situ once the top works have been removed. Some applications may require the seating surfaces to be "super-finished" with diamond compound. A dedicated lapping block must be used for this purpose. Some diamond compounds require special thinners; always ensure a supply is available before starting.

When nozzles or discs are badly scratched they must be machined prior to lapping. Nozzles can be machined in situ with a portable machine. When set up properly this is the most accurate method of machining. Standard tools can be used for globe valves as well as safety relief valves. Nozzles can also be machined in a centre lathe. The nozzle can be machined assembled in the body or independently. In either case the set up must be perfectly "true" before starting. Nozzles should be removed by using the tool provided by the manufacturer. If they are tight or seized in with sediment/corrosion, release oil should be tried. If unsuccessful, **gentle** heating of the body may assist.

Discs can be machined in a centre lathe but may be very difficult to grip. Some designs may have centres for machining between centres; this is the easiest option. Again, accurate setting up is essential. Some safety relief valve manufacturers provide remachining drawings showing the new profile and limits of wear. Tools should be ground to the valve manufacturer's instructions and lapped with an oil stone if necessary. Very rigid tooling is essential to produce a smooth finish on the tough materials used. Lap and polish the surfaces before reassembly.

Some safety relief valves include an elastomer seal in the disc or piston to provide good seat tightness. The seal must be inspected regularly for the low leakage to be maintained. Pilot-operated valves have moving parts and seals, and possibly diaphragms, in the pilot assembly. The pilot assembly must be inspected and maintained regularly if the benefits of the pilot system are to be achieved.

All components must be thoroughly cleaned prior to reassembly. It is essential that all traces of lapping/polishing compound are removed. If the spindle has been polished with emery tape, then all traces of emery dust must be removed. Failure to clean parts thoroughly, which have been lapped or polished, will result in rapid wear and loss of seat seal. Sliding bearing surfaces should be lightly lubricated with a lubricant suitable for the service conditions. If the nozzle or disc has been machined, the assembled valve should be tested and reset before returning to production.

When servicing regulators and control valves it is obviously important that the maintenance engineer understands how the valve is constructed and operates. Without such knowledge there is a risk that equipment may be damaged, and worse, that serious damage could be caused to a complete process, with all the consequences of plant shutdown.

All regulator and control valve actuators must be able to react to even small variations in input signal. Any lack of performance is usually obvious from a recorder chart reading of the controlled variable. Before servicing, however, it is important to ensure that the process or the valve is isolated and by-passed, if the repair can be undertaken in situ,e.g., the "stroking" of a valve positioner. If the valve has to be removed from the line, it is usual to do this under a "permit-to-work" system that ensures that the valve is drained or vented and decontaminated of toxic or corrosive substances, before it is removed from site to the workshop area.

Large companies usually carry out their own preventative maintenance programme. They have an established routine for inspection of all larger components as well as the replacement of packings, "O"rings, seals, diaphragms and other non-metallic materials. The advice given in Sections 17.6.2 to 17.6.6 applies to normal maintenance repairs and includes some general instructions on how the work should be carried out. Remember, there are usually specific manufacturers' instructions which must be made available and followed.

If there are a large number of regulator and control valves to be serviced, it may be worthwhile engaging a specialist contractor. Most of the larger manufacturers listed in the Buyers' Guide in Chapter 21 have service departments, service centres or agents that specialise in the overhaul of valves. It is also possible to arrange some expert training of staff, who should then be in a position to undertake regular maintenance or cope with breakdowns with greater confidence. The training should include instruction on how to remove, refit and overhaul an actuator, how to remove a gland packing and repack the box, and how to adjust the stroke of a valve, with or without a valve positioner.

17.6.2 Replacing actuator diaphragms

Before attempting to renew an actuator diaphragm, the compression of the spring should be relieved as much as possible by releasing the spring adjuster. Then two, and only two, of the bolts on opposite sides, holding together the diaphragm chamber, should be replaced by long studs which are fitted with nuts. These should be done up tightly to hold the housing together while the remaining bolts are being removed. When this has

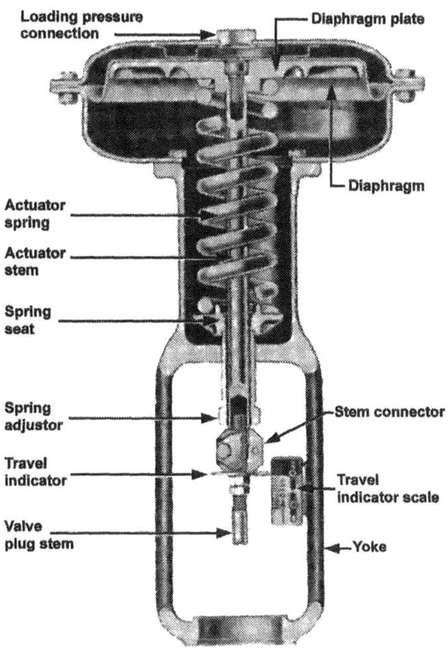

Figure 17.4 A direct-acting diaphragm actuator

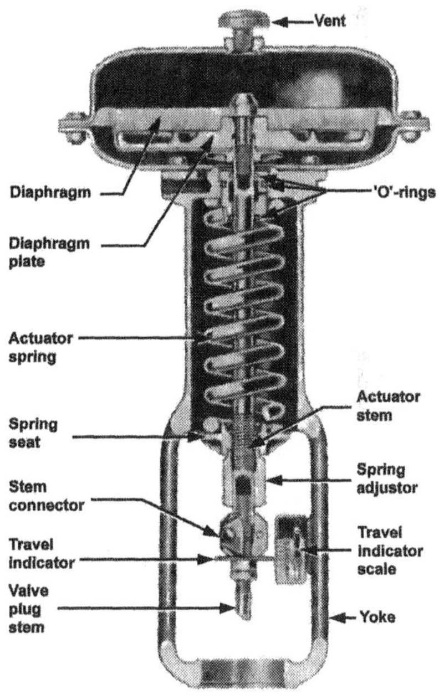

Figure 17.5 A reverse-acting diaphragm actuator

been completed, the stud nuts can be slowly released and the diaphragm housing should come apart.

Then remove the upper casing from the actuator housing. In the case of direct-acting diaphragms, see Figure 17.4, the diaphragm can be lifted directly and replaced with a new one. On reverse-acting actuators, see Figure 17.5, the actuator housing must be disassembled to gain access to the diaphragm.

Most actuators have an injection moulded diaphragm, the profile of which ensures that the surface area remains as constant as possible throughout the valve travel. Suitable replacement diaphragms for the range of standard actuators should be held in the users' stores. When the actuator is reassembled, the

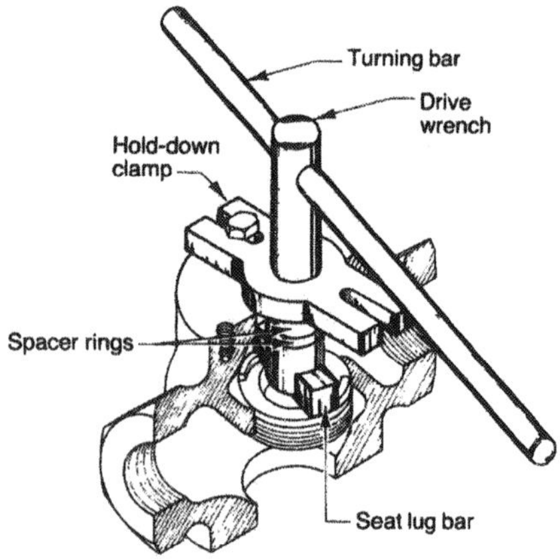

Figure 17.6 Correct use of the seat ring extractor

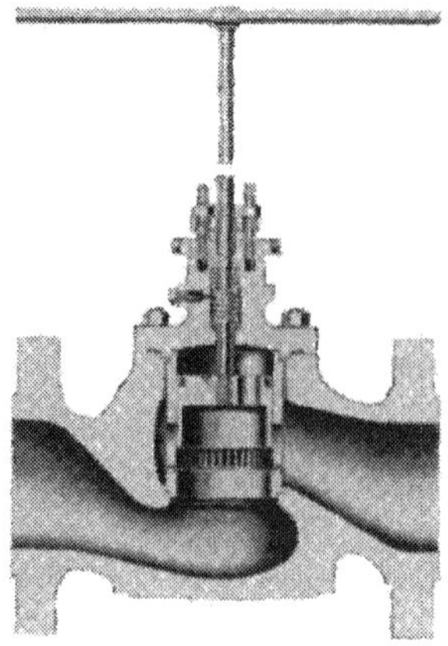

Figure 17.7 Typical set up for valve plug/seat grinding

bolts should be tightened evenly. Check for leakage by using soapy water or a proprietary aerosol.

17.6.3 Replacing seat rings

There are two main types of seat rings in regulators and control valves:

- Cage-guided valves where the seat is retained by the cage
- Screwed-in seat rings

The first type presents no problem.

For screwed-in seat rings, the procedure is as follows:

1 Select an extractor of the correct size.

2 Fit the correct number of spacer rings to keep the seat lug bar down.

3 If the seat cannot be loosened by using the standard turning bar, see Figure 17.6, then it is permissible to add a length of pipe to obtain more torque whilst tapping the drive wrench, lightly, with a hammer. Stubborn seats may need to be soaked with release oil for a couple of hours.

4 Before screwing in the new seat ring, the threads must be cleaned thoroughly and degreased before thread paste can be applied.

5 Tighten the new seat ring using the extractor that was used for removing the old seat.

NOTE: Some seat rings are "tack-welded" to the valve body. This weld must be removed and release oil applied before any attempt is made to unscrew the seat using a seat ring extractor. These extractors can be purchased from the valve manufacturer.

17.6.4 Grinding of valve seats

Some leakage at the valve seat with a metal-to-metal contact with the plug must be expected. There are varying leakage requirements specified according to class, which differ in the quantity of leakage under test conditions.

If the leakage is unacceptable, then the plug must be ground to the seat using a good quality grinding paste. Using a "lift and twist" motion prevents the metal surfaces from scoring each other in case the grinding paste is removed from the surfaces. During this operation the weight of the plug should be supported as appropriate.

To obtain a good grip and good guiding of the valve plug, a special tool should be made and grinding should be carried out with the packing box and packing fitted to the valve body. On no account should any gripping tools be applied to the valve stem, see Figure 17.7

For double-seated valves, where one seat may be ground-in before the other, polishing paste should be applied to the finished seat whilst grinding the other seat. Never allow one seat to be without polishing paste while grinding the other.

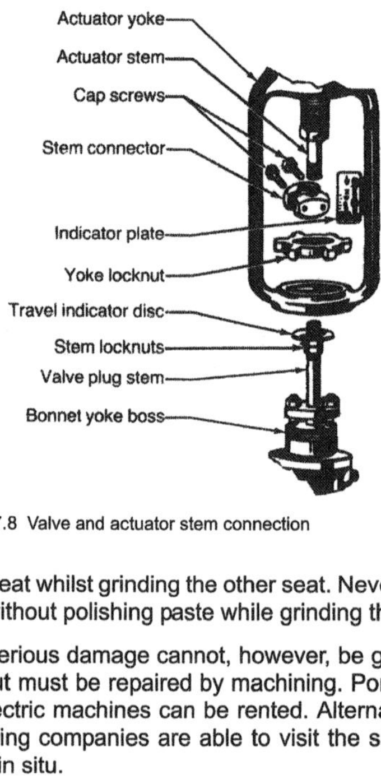

Figure 17.8 Valve and actuator stem connection

More serious damage cannot, however, be ground out in this way, but must be repaired by machining. Portable pneumatic and electric machines can be rented. Alternatively, specialist machining companies are able to visit the site and refurbish valves in situ.

17.6.5 Adjusting the stem connector

17.6.5.1 Linear valves

1 Remove the stem connector and push down the valve stem to close the valve. Ensure that the valve plug is in the closed position - i.e., the seat and plug are in contact with each other, see Figure 17.8.

2 Set the actuator in its corresponding closed position:

For direct-acting actuators, the valve stroke is downwards from the upper position.

For reverse-acting valves, the closed position is as it is in the loosely attached position.

3 Move the actuator stem about 2mm closer to the valve stem and connect both stems by means of the stem connector.

4 Check that the actuator can travel its full stroke. Make sure that the plug touches the seat before the actuator comes to its limit. Minor adjustments, less than 2mm, can be made by screwing the stem connector slightly, after loosening the cap screws which hold it together.

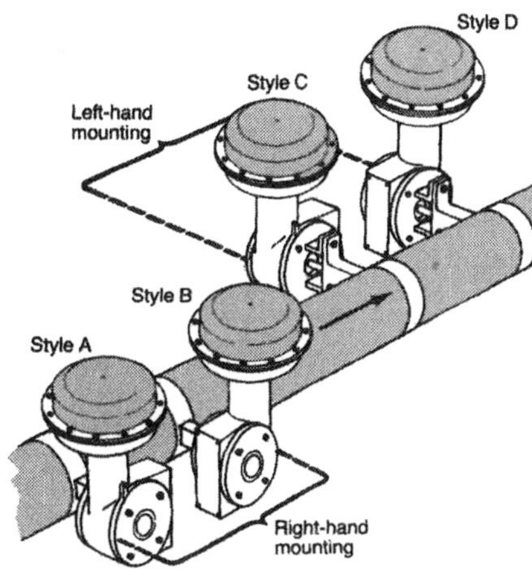

Figure 17.9 Actuator mounting positions

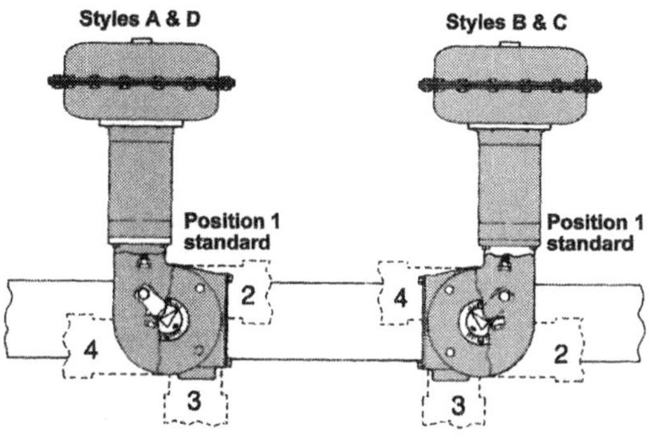

Figure 17.10 Notation for actuator mounting position

5 When everything is connected together, screw the indicator plate into a position which corresponds with the valve in closed position.

6 Using a gauge connected to the actuator signal input and a dial gauge on the valve stem, set the valve according to the specification plate.

17.6.5.2 Rotary valves

Rotary valves and actuators can be assembled together in many ways, as shown in Figures 17.9 and 17.10

The most common alternatives are with the actuator mounted either to the left or to the right with a choice of which direction the valve should travel in the event of air failure.

Valves of this type are usually fitted with direct-acting actuators and the arrangement for valve opening or closing in the event of air failure is made by changing the connection between the valve spindle and lever.

In the case of ball valves, virtually any connection point on the valve spindle is ideal provided fine adjustment can be made with the actuator stem.

Fine adjustment on rotary valves can only be made while the valve is not being used in the process, since it must be possible to measure the position of the disc relative to the bore of the valve body in order to check that the disc is positioned correctly, see Figure 17.11

17.7 Packing boxes

Packing boxes can suffer severe local damage due to wire-drawing. Portable machines can be hired to remachine the packing box in situ. This facility can be extremely useful for large valves in locations which present dismantling problems.

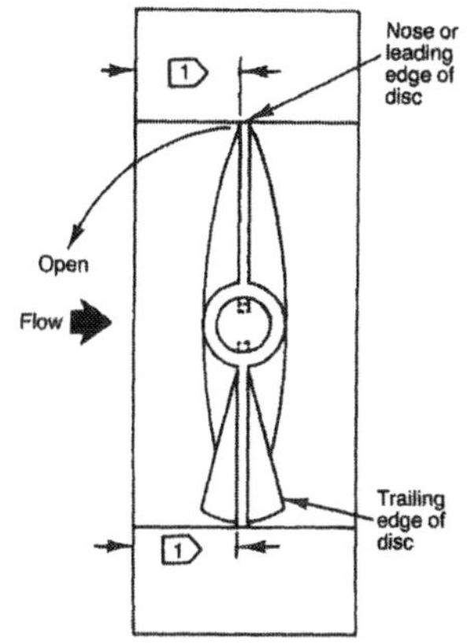

Figure 17.11 Butterfly valve disc position relative to valve bore

17.7.1 Routine packing adjustment

With the exception of chevron packing rings, see Chapter 11, Figures 11.9 to 11.12, all packings are compressed by means of the packing box nuts. This adjustment is extremely critical and can, if not carried out correctly, damage the valve before the process is even started. It is advisable to delegate the job of "running-in" the packing boxes to one person, whereby the packings are compressed in small stages - one third of a turn of the packing box nuts during a period of several hours, until the valves stop leaking, without causing unnecessary stem friction.

The packing box can be likened to a bicycle tyre inner tube, which bursts if it is overloaded. The packing must remain pliable in order to follow the movement of the stem. It must retain its sealing qualities and must not be overloaded by over tightening the packing box nuts.

17.7.2 Replacing spindle/stem packing

The packing around the valve spindle/stem should be replaced if leakage continues, even after adjustment. Packing can also be replaced while carrying out other service work. Chevron rings should be replaced completely when they leak. Check that there is no pressure in the system before removing any packing.

If the packing is of the split type, it is possible to remove it without removing the actuator, but this can easily lead to irreparable damage to the spindle/stem and stuffing box surfaces, unless the proper packing extractor is used.

Do not attempt to remove packing by applying pressure through the lubrication nipple. This can be dangerous and does not usually work, since in most valves one half of the packing is below the lubricating connection.

The following procedure is recommended:

1 Consult the manufacturer's instructions the day before repacking is scheduled. It may be recommended to soak the packing rings in light oil overnight before fitting. Check the availability of approved lubricants and cleaning solvents for biotechnical and hygienic applications.

2 Disconnect the valve stem from the stem connector.

3 Remove the actuator from the valve body.

4 Remove the bonnet and pull out the plug and stem.

5 Insert a copper drift, slightly larger in diameter than the stem, through the bottom of the bonnet and push out the old packing. Use a proprietary packing extractor, shaped somewhat like a flexible corkscrew, to remove the old packing. Discard the old gland packing. If the replacement packing is in the form of chevron rings, check that the original springs and headers are serviceable and in place be-

fore fitting the new packing rings. These should be lightly lubricated to assist assembly.

6 Clean the bonnet and packing box and inspect the spindle/stem. There may be scratches or burrs which can damage the new packing. Spindles and stems that seal against PTFE chevron rings are usually honed to a super finish and must be free from damage of any kind.

NOTE: Never apply tools directly to a valve stem.

7 Check the valve plug, the seat ring and other parts of the valve trim.

8 Reassemble the valve body and fit the bonnet.

9 Assemble the valve body and bonnet in the same way as described in connection with connecting the flanges.

10 Push in the packing rings one by one in the correct order. Use silicone grease, or approved hygienic quality grease, on PTFE packing rings and make sure that they are not damaged by the spindle/stem threads.

For compressed fibre rings fitted to linear control valves, and for all rings fitted to rotary valves, it is essential that each ring should fit the gap between the spindle/stem and the packing box accurately. Because these packings are energised by fluid pressure alone, there is more friction between the ring and the sealing surfaces. So unless the gland is very carefully assembled, the rings will not "bottom" in the box.

It is not good enough to fit the gland follower and to drive the packing to the bottom of the box. This simply does not work, and leads, because the bottom ring is not properly supported, to rapid failure of the packing in service. The only correct way of fitting packing rings of this type, is to measure the depth of the box with a depth gauge and then to fit each ring separately with the aid of a copper drift. When the first ring has bottomed this should be confirmed by checking with the depth gauge. The second and any subsequent rings are fitted in like manner.

The procedure takes rather longer to read about than it does to carry out, but it has been proven time and again in practice, on valves that were once considered difficult, e.g., large butterfly valves on medium and high pressure steam duty.

11 Fit the components which compress the packing.

12 For PTFE packing rings, the gland nuts should be tightened fully; for other packing types, the nuts should only be tightened sufficiently to prevent leakage.

13 Fit the actuator.

17.8 Spare parts

Equipment and valve manufacturers invest considerable sums of money in designing and developing their products. Spare parts are manufactured to the same standard as the new, complete equipment. The sale of spare parts forms an integral part of a manufacturer's income. Small machine shops can easily copy the physical dimensions of a component, but complete "reverse engineering" is impossible. They do not have research and development departments supporting them.

Short-term cost reductions can lead to major cost expenditure when copied parts fail catastrophically and damage other components. Original equipment manufacturers guarantee their spare parts. Guarantees and warranties for the valve itself will normally be invalidated if parts, made by others, are used.

Spares should always be ordered from the original manufacturer of the valve. When ordering spare parts, always state the following:

- valve model
- valve serial number

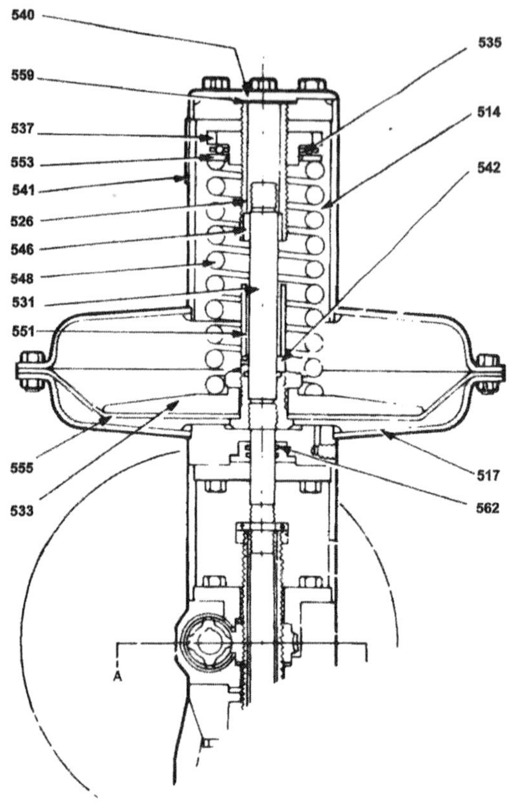

514	Spring tube assembly	541	Spring cover plate
517	Diaphragm case assembly	542	Diaphragm collar
526	Spring adjusted screw	546	Collar nut
531	Actuator stem	548	Stem guide
533	Spring carrier	551	Spring
535	Spring thrust bearing	553	Travel stop tube
537	Spring adjustor	555	Diaphragm button
540	Cover plate	559	Cover plate joint
		562	Seal button

Figure 17.12 A typical parts list

- year of manufacture
- part number and material
- part reference from spare parts list, see Figure 17.12
- quantity
- your valve tag number

When ordering spares for special valves it may be very helpful to include all the data given on the nameplate or a copy of the valve data sheet.

During the manufacturing phase of special valves, a quotation should have been obtained for routine and special spare parts. The quotation would indicate the lead time of any parts which would have to be specifically manufactured. This information should be stored with the other valve data for easy reference.

17.9 Trouble-shooting

When operating problems arise at site, it is not necessarily a valve problem. Calling out a service engineer can be costly when the fault is found to be incorrect installation or "incorrect" operating conditions. The following logical approach can be used initially for all valve types.

Check the installation:

- Was the valve commissioned successfully?
- Is the valve the right way round?
- Is the valve in an acceptable orientation?
- Does the valve move freely?
- Is there any build-up of dirt/wax/corrosion products restricting travel?
- Is all ancillary equipment fitted?
- Is all pipework connected to the correct ports?
- Check for leaks.
- Are the electrics wired correctly?
- Are electric motors running in the correct direction?
- Is automatic control overridden by local settings?
- Have any safety devices tripped?
- Has the valve been serviced recently?
- Have flanged connections been tightened up evenly?
- Are relevant filters and orifices clean?
- Are springs intact?
- If pulsation dampers or accumulators are fitted check gas precharge pressures.

Check the operating conditions:

- Has the valve ever operated successfully?
- Has the valve been operating under data sheet-defined conditions?
- Check operating logs for historical data.
- Is a valve used continuously rated "continuous" or "intermittent"?
- Has there been a recent process upset?
- If instrumentation indicates the valve has always operated within the defined conditions, check when the instruments were last calibrated. Recalibrate if necessary, but note errors and apply to most recent operating conditions.

Actions:

- If installation is correct and operation is within prescribed data sheet limits, read the following paragraphs and the manufacturer's trouble-shooting guide; request a service engineer if problems cannot be resolved.
- If operation is outside prescribed data sheet limits, read the manufacturer's product specification for absolute operational limits. If the valve is unsuitable, check for upgrades or replacement.

Most valves are designed to contain pressurised fluid. Leaks of escaping fluid are relatively easy to detect. Leaks on valves in vacuum services may be much harder to find and so specialist ultrasonic sensors may be necessary.

Process and equipment engineers frequently incorporate safety margins within the operating parameters specified. Adding a percentage to operating pressures ensures that compressors and pumps can cope with all realistic conditions. Equipment manufacturers frown on this practice and would prefer actual conditions to be specified and a safety margin indicated separately. Safety relief valves, regulators and control valves can be undersized by using higher pressures for selection than experienced in normal operation.

Process designers and contractors sometimes overlook trace elements and the possibility of small quantities of solids in process streams. Undocumented trace elements can cause severe corrosion problems which result in:

- packing degradation and fugitive emissions
- gasket failure and leaks
- diaphragm or bellows failure
- seat corrosion and loss of seal
- body corrosion and loss of pressure containment

Unexpected fluid attack and corrosion can cause major operational problems. Very small quantities of abrasive solids can cause rapid valve wear. Research has shown that 0.25% of solids, similar to fine sand, can double the seat/plug wear rate. The wear rate caused by entrained solids is a function of velocity raised to a power of between 2.5 and 5. Small increases in velocity can result in large increases in erosion. Safety relief valves, regulators and control valves, and valves which are liable to operate with high fluid velocities, will be very susceptible to erosion caused by solids.

Erosion can result in:

- spindle/stem wear leading to packing leakage
- spindle/stem wear leading to spindle/stem failure
- seat/plug wear leading to leakage
- seat/plug wear leading to loss of control
- diaphragm wear leading to loss of control
- body wear and loss of pressure containment

Erosion can be alleviated by the use of very hard materials and surface coatings. Hardenable stainless steels, such as 17-4PH and 440C, can be heat treated to resist erosion. Soft metals can be coated with glass, Stellite™, tungsten carbide or ceramic. If solids are specified initially, alloy cast irons can replace soft materials such as carbon steel.

Erosion can produce an increase in corrosion indirectly; erosion corrosion. Mild erosion can remove the protective oxide layer from corrosion resistant materials. Clean metal may be easily attacked by the process fluid. The oxide layer is not allowed to repair itself due to the erosive action of the solids. Rapid material loss ensues.

Mild cavitation can produce similar effects to erosion corrosion. The cavitation implosions remove the protective oxide layer and the corrosive fluid elements remove bulk material. Fully developed cavitation removes the bulk material directly. If the cavitation bubbles implode next to a solid surface, bulk material is plucked from the surface due to the extremely high local pressures created. Cavitation damage has a characteristic "pop-marked" appearance and is usually easy to identify.

Entrained gases in liquid, flashing and cavitation cause a reduction in flow capacity of all liquid valves. The volume of gas/vapour replaces liquid of a much higher density and consequently the mass flow for a given differential pressure is reduced. Flashing and cavitation can be expected in some liquid applications and should have been considered during valve type selection and sizing. Unexpected flashing and cavitation can effectively undersize valves, creating a flow restriction. Process designers and contractors sometimes omit to quantify entrained gas and vapour in liquid streams. The expansion of gas and vapour can create choked flow conditions, restricting mass flow, due to the pressure drop across a valve.

Pressure pulsations, vibration and thermal shock can create very serious operational problems. Equipment and valve ratings are based on steady-state operating conditions. Steady-state is defined as all relevant parts of the valve being at the same temperature, pressure and when appropriate, velocity.

Pressure pulsations and vibration can cause rapid wear in some valves. Valves with loose handwheels and pressure gauges with broken pointers are commonplace. Figure 11.1 in Chapter 11 shows the relationship between pipe size and operating pressure and allowable residual pressure pulsations. This information provides guidance eliminating excessive piping vibration in petrochemical installations. Lightweight stainless steel systems for hygienic applications should use rather lower pulsation values.

Both pressure pulsations and vibration cause moving parts to rattle within their clearances. The continuous movement of parts creates wear, which increases the clearance, which increases the wear rate. Correct valve type selection is the key to the solution but the problem must be recognized at the design stage. Changing the valve type, when possible, may be the only solution.

Most system designers do not understand thermal shock and do not consider it as part of the overall design concept. Ultra safety-conscious designers, such as those involved in nuclear plants, realised the effects could cause major problems and quantified some basic requirements, see Figure 13.1 in Chapter 13. General process equipment is not designed to these requirements, but the implications should be obvious.

Consider a simple example: A hygienic valve conveys yoghurt at -3°C and is then sterilised at 110°C. An almost instantaneous temperature change of 113°C. A heavy nuclear pump or valve would be expected to survive about 500 cycles. A hygienic valve, sterilised twice a day, would not last a year.

Low pressure equipment, such as most hygienic valves, does not suffer as badly as higher pressure equipment. Other equipment in the system, such as compressors or pumps, may still suffer. The mild effects of thermal shock are seizure and jamming. The internal components expand much faster than the guides and pressure containment, and the valve locks up. If valve movement is not forced, the valve will return to normal operation once temperatures have equalised. If the valve is forced, permanent damage may result. The most serious effects of thermal shock are cracking of the pressure containment and the failure of gaskets and seals. If process temperatures change by more than 4°C per minute thermal shock problems are a possibility.

Valves can jam due to slow temperature changes when warmed up properly. Valves which are wide open or closed can suffer from differential expansion problems. All components are warmed, or cooled, slowly to avoid thermal shock but internal clearances of the stem are removed due to the end of travel. Differential expansion/contraction rates result in high tensile/compressive forces being induced in the stem. Linear motion valves are the most common, but butterfly valves can be jammed closed depending upon the seat design. If temperatures are returned to the original value when the valve was opened/closed, normal operation should be possible. When process temperatures vary considerably from shutdown, valves should not be opened wide and sealed on the back seat. The valve can always be opened further while operating.

Solenoid valves are usually designed for clean fluids. Traces of solids may accumulate and prevent full opening or closing. If the solids cannot be filtered out, then the valve must be cleaned regularly.

Spring-loaded safety relief valves, can suffer from chattering, fluttering and instability of operating conditions including low or high popping pressure. Table 17.2 outlines typical causes. Safety relief valves used in steam applications can be very problematic and the manufacturer should be consulted.

Regulators and control valves are frequently used in systems with compressors and pumps which create flow variations and pressure pulsations. Control instability can be caused by undamped pulsations, therefore piping must be checked for correct fitting of dampers/filters and correct gas precharge of bladder/diaphragm devices.

Sympton	Possible cause
Chattering	Corrosion problems Low flow Overpressure too high Oversized valve Poor inlet and outlet piping design Valve inertia (on large valves)
Excessive simmering	Damaged disc or nozzle Incorrect adjustment Scale, wax or corrosion problems
Excessive wear of guides	**Fluttering** Solids in fluid
Excessive wear of guides, disc and nozzle	**Chattering** Solids in fluid
Fluttering	Oversized valve Poor inlet and outlet piping design Valve inertia (on large valves)
High popping pressure	Balance bellows failure High viscosity liquid sticking disc to nozzle Scale, wax or corrosion problems
Leakage from bonnet	Balance bellows failure Damaged disc or nozzle
Long blowdown	Damaged disc or nozzle Incorrrect adjustment Scale, wax or corrosion problems
Low popping pressure	Damaged disc or nozzle Spring creep or relaxation
Seat leakage	Damaged disc or nozzle Incorrect optimisation, short blowdown Lifting lever incorrectly set Operating pressure too high Trapped solids Valve body distortion
Sluggish operation	Damaged disc or nozzle High viscosity product Incorrect adjustment Scale, wax or corrosion problems
Unstable operation	Low flow Poor inlet and outlet piping design

Table 17.2 Typical spring-loaded safety relief valve problems

17.10 Repairs

As valves wear and corrode, or if a serious failure occurs such as cracking due to thermal shock, repairs will be considered. The nature of the repair will be dependent upon the material of the component and the service conditions and whether the repair is structural or cosmetic. Large components, made of exotic materials, will be considered for repair due to the high cost of a complete replacement.

Some repairs can be carried out by machining. Very badly worn packing boxes can be bored oversize and "sleeved" or have the next size larger packing fitted. Any machining which affects pressure containment thicknesses must have the stressing checked by an experienced engineer.

NOTE: Remember, inch packing may be just slightly larger than metric packing.

The most popular types of repair techniques are:

- welding
- surface coatings

Welding can be used for structural and cosmetic repairs. Cracks and loss of surface material can be repaired in many materials provided the damaged area can be cleaned properly and inspected. Welding of cast iron structures can be perfectly acceptable. Welding of cast iron components exposed to process fluid may be more difficult. Cast iron is not welded with cast iron filler but with bronze or nickel alloys. The corrosion resistance of the repaired component may be completely different to the original. The effect of the repair on corrosion resistance must be fully examined prior to the repair being undertaken.

Some alloy steels, used for high pressure valves, are very difficult to weld, requiring accurate pre-heating and post-weld heat treatment. The popular AISI 300 series austenitic stainless steels may need post-weld heat treatment to restore corrosion resistance. This factor is the basis of the widespread use of the low carbon versions, 304L and 316L.

Soldering and brazing should be considered for repairs when welding is not appropriate. Copper and its alloys are mainly used.

Surface coatings are a very popular method of reclaiming worn or corroded components. Very thin coatings can be applied, such as cadmium, nickel and chrome plate, to restore clearances and "fits". Thicker coatings can be applied by flame spraying or, manually, by brush and trowel. Large corroded or eroded valves can be repaired by coating areas with glass beads encapsulated in epoxy adhesive. Typical compounds which can be applied by hand, such as two-pack epoxy putty, can withstand temperatures up to 148°C and withstand a wide range of chemicals.

Surface corrosion can uncover porous areas of material. Porosity can be treated with suitable chemicals depending upon the parent material and the nature of the fault. Whether impregnation is suitable for sealing is dependent upon the size of the pores, the material thickness and the operating pressure. Little data is published, as part of product specifications, on the pressure retaining capabilities of impregnation compounds. If "trial and error" is not feasible it may be wise to remove the porous portion completely and rebuild with sound material.

Impregnation compounds can be two-pack epoxy, anaerobic and heat curing. Two-pack epoxy compounds start to cure when mixed. Anaerobic compounds start to cure when deprived of air. Heat curing compounds only commence curing when a minimum critical temperature has been reached. Some anaerobic and heat curing compounds can be accelerated by chemical catalysts.

Specialist firms have equipment to permit vacuum impregnation. The vacuum ensures air is removed from the porosity, allowing the compound to fill the pores completely. Impregnation compounds have wide chemical resistance and can be suitable for continuous operation at 120°C. Intermittent operation up to 200°C is possible with some.

Non-metallic valve materials can be repaired. As with metallic materials, the exact specification of the material must be known before any repair can be contemplated. Typical repair techniques include:

- solvent welding
- thermal fusion welding
- reinforced adhesive

Patches can be constructed with adhesive and reinforced with metal mesh or glass-fibre ribbon. Whenever possible, trials with solvents and adhesives should be conducted to test the bond strength and the effect of the process fluid.

NOTE: When repairs are made in a material which is different from that of the component, the difference in thermal expansion should be considered. Slight changes in temperature may overstress the bond and result in delamination.

17.11 Useful references

ANSI/API 527 (R2002) Seat Tightness of Pressure Relief Valves.

MSS SP-92-1999, MSS Valve User Guide, Manufacturers Standardization Society.

How to calculate purge gas volumes, D F Schneider, Hydrocarbon Processing 1993, Vol 72; No 11, Gulf Publishing, USA, ISSN 0018-8190.

NBIC (The National Board of Boiler and Pressure Vessel Inspectors), 1055 Crupper Avenue, Columbus, OH 43229 USA, Tel: 614 888 8320, Email: getinfo@nationalboard.org, www.nationalboard.org.

ISA, 67 Alexander Drive, Research Triangle Park, NC 27709 USA, Tel: 919 549 8411, Fax: 919 549 8288, Email: info@isa.org, www.isa.org.

Manufacturers Standardization Society (MSS) of the Valve and Fittings Industry, 127 Park Street N E, Vienna, VA 22180-4602, USA, E-mail: info@mss-hq.com, www.mss-hq.com.

Some applications and solutions 18

18.1 Quarter-turn pneumatic rotary actuators — a comparison

Introduction

In the process industry many liquids, gases or media are regulated by quarter-turn valves i.e. ball valves, butterfly valves or plug valves.

The main purpose of these valves is to open or shut down a flow. However, in some special cases these types of valves are also used to control flow. Typically, these types of valves turn 90° in order to open or shut. The required torque is generated by a either a manual handle, a manual hand wheel, a pneumatic actuator, an electrical actuator or a hydraulic actuator. Reductions in plant manpower has increased the use of actuators.

Pneumatic actuators are a low cost and safe option. Where no compressed air supply air is available, electric or hydraulic actuators can be used .

The types of pneumatic actuators for quarter-turn applications are:

- diaphragm and cylinder actuators
- rack and pinion actuators
- scotch yoke actuators
- vane type actuators

Diaphragm and cylinder actuators

The linear movement of a cylinder is converted to a rotary movement via a crankshaft mechanism. The stroke which is made will be between plus 45°and minus 45°, see Figure 18.1.

A diaphragm actuator consists of two discs with a membrane in between. By applying pressurised air onto the diaphragm, the drive shaft will be moved against the spring force.

The forces depend on the air pressure (max. 6 barg), the diaphragm surface area and the spring constant.

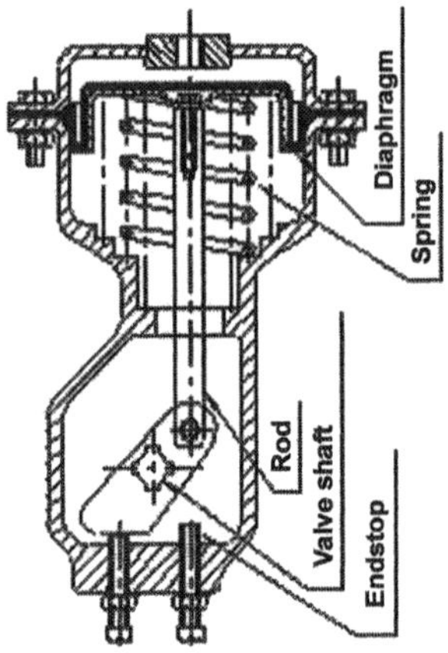

Figure 18.1 Diagrammatic representation of diaphragm actuator
Courtesy of Delan BV

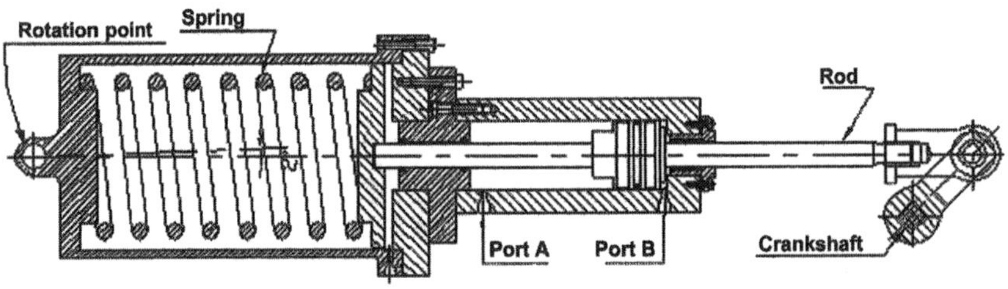

Figure 18.2 Piston air cylinder actuator
Courtesy of Delan BV

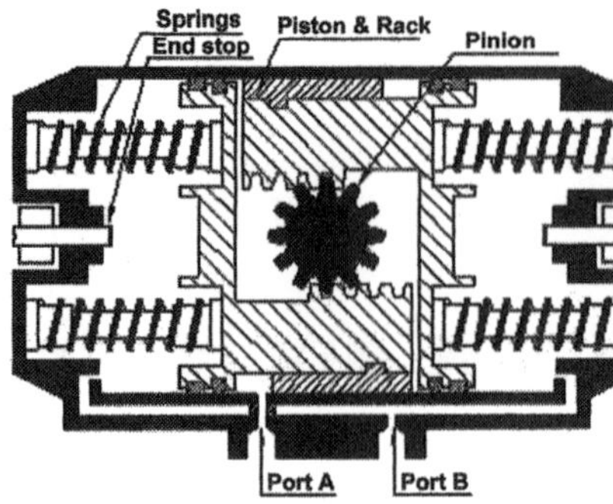

Figure 18.3 Rack and pinion actuator
Courtesy of Delan BV

The benefits of the diaphragm actuator are:

- Few moving parts
- High output torque at low air pressure (because of high diaphragm area)
- Suitability for control applications (little internal friction)

In the cylinder actuator a piston also moves against a spring force. See Figure 18.2. Since the allowable air pressure is higher, the output forces are also higher than those of diaphragm actuators.

Because of the axial force component the actuator should enable a small rotation. These actuator types are mostly used in control applications and will therefore be excluded from a further comparison.

Rack and pinion actuator

The rack and pinion type actuator is the most commonly used. The actuator consists of two cylinders placed opposite each other and connected via a rack and pinion, see Figure 18.3.

Air pressure on Port A will result in a rotation of the pinion. When the pressure is taken away, the pinion will be turned back by the springs. A version without springs will need air pressure on port B to move the actuator to its original position. The maximum air pressure is approximately 10 barg.

Scotch yoke actuator

This actuator also consists of two opposite placed cylinders. However, both pistons are connected to each other by a shaft. This shaft makes a linear movement that is transformed via the yoke to a rotation angle from plus 45° to minus 45°. The yoke will move or turn at pressurisation of both Ports A. When air pressure drops the yoke will be returned by the springs. When no springs are present, air pressure on both Ports B is necessary to turn the yoke. See Figure 18.4.

As with linear cylinders, high output forces can be generated. Both cylinders move at the same time in the same direction resulting in a jolting movement. The torque depends on the starting angle since the distance from the centreline of the valve varies.

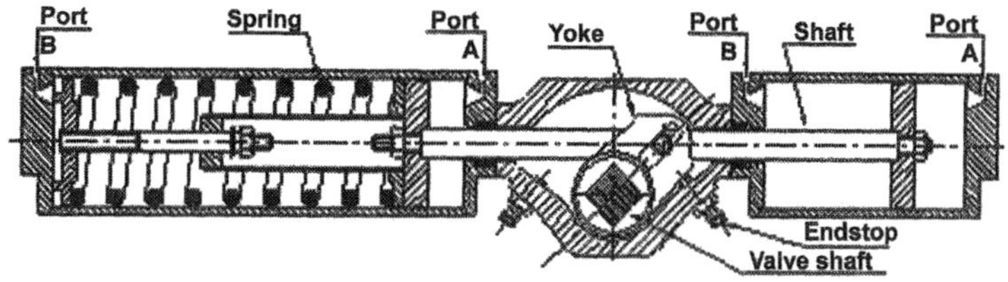

Figure 18.4 Scotch yoke actuator
Courtesy of Delan BV

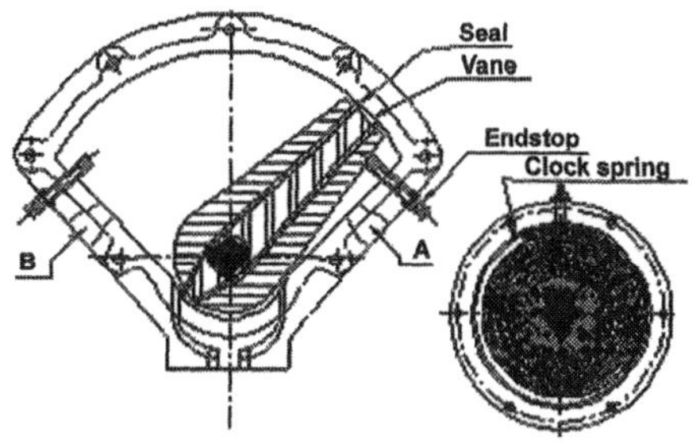

Figure 18.5 Vane actuator
Courtesy of Delan BV

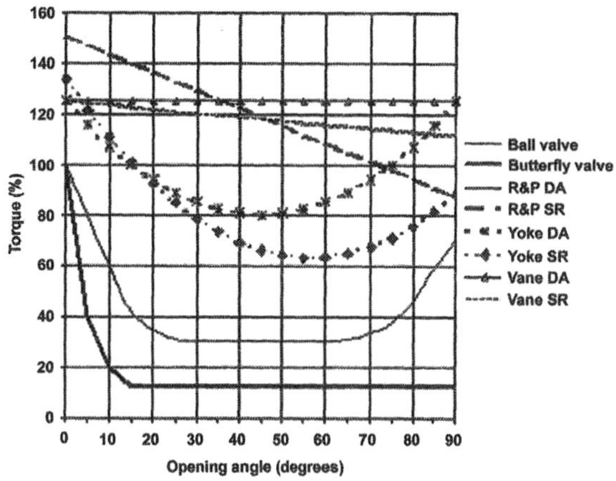

Figure 18.6 Torque and opening angle for various actuators
Courtesy of Delan BV

Vane actuator

In this style of actuator, a vane divides the air chamber in two. The vane consists of a firm body and (on both sides) a seal. The shaft is an integral part of the vane. By applying air pressure on Port A the vane moves a quarter turn. Air pressure on Port B makes the vane move back to its original position. See Figure 18.5.

A separate failsafe spring return unit with a clock spring can be mounted when a safe position is essential for example, in case of air pressure failure.

Double acting or spring return versions

All the actuators can be supplied either as double acting or spring return types. The double acting versions will use air onto Ports A and B to rotate clock and counter clockwise. If air pressure suddenly fails, then the actuator will stop at that last position. The spring return models only use air pressure for one movement. The springs will return the vane to the starting point. If air pressure drops, then the actuator will go to a pre-selected safe and secure position. However, the spring return versions have a lower torque output then the double acting versions. A spring return actuator is larger and has higher air consumption.

A thorough comparison between the "pros and cons" of a double acting model or a spring return model is essential.

Torque

The required torque depends on the specification of the valve, the process and the service duty. The relationship between the torque and the opening angle for a butterfly or a ball valve and the various actuators can be found in Figure 18.6. Opening angle 0° means that the valve is closed

A butterfly valve requires the highest torque upon opening. This torque will decrease on opening until a certain minimal level, because the disk is out of its lining.

With a ball valve the required torque output will increase just before reaching the open position because of an increase in contact area.

From Figure 18.6 it can be seen that a minimum safety margin of 25% should be applied when calculating the required torque output of an actuator. The torque output of a double acting rack and pinion actuator (R&P DA)and a vane actuator (Vane DA) does not depend on the opening.

The curve of a Scotch yoke actuator (Yoke DA) is parabolic. The torque outputs of the spring return actuators is related to the opening angle of the valve. The spring constants are different for a rack and pinion spring return (R&P SR), containing helical springs and the vane failsafe (Vane SR) containing a clock spring. This results in a higher loss of torque for the rack and pinion actuator at a higher opening angle. Taking the safety of 25% into account, the end of stroke becomes critical to the selection of the proper actuator. The minimum opening torque will thus increase to 150%.

Accessories

Limit switch — limit switches give information about the position (2 positions and running) of the actuator and the valve. The switches can be either proximity, micro or pneumatic. Standardised type systems are used to reduce mounting and replacement costs.

Solenoid — air pressure via an electrical signal is ported to the actuator, thus operating the actuator. Double acting actuators need 5/2-solenoids i.e. with 2 positions (open/close) and 5 air ports. Spring return actuators need 3/2-solenoids i.e. with 2 positions and 3 air ports.

Positioner — it is sometimes necessary to move an actuator to an "in-between" position rather than to switch between the open and close position. The positioner consists of a mechanism that compares the measured position with the desired position and, depending on the settings, sends a corrective signal. This corrective signal will be converted into a pneumatic signal and adjusts the position of the actuator. The necessary instrument air pressure is usually 6 barg. The desired position is usually represented as an analogue signal between 4 and 20 mA.

Nowadays the desired position will often be represented as a digital value. Profibus and Fieldbus are examples of this type of communication. It is also possible to exchange other data with the positioner. Hart is an intermediate form which allows other data to be superimposed on the existing signal of between 4-20 mA.

Comparison between actuators

A comparison between the rack and pinion, the Scotch yoke and the vane actuator is difficult. Some choices are historically determined.

- **Materials and corrosion resistance —**

Actuators have to withstand aggressive open air and bad quality pressurised air quality. Although modern compressor installations become better and better there still is the danger of corrosion by water and/or oil in the compressed air. Table 18.1 gives a general comparison.

	Rack and pinion	Scotch yoke	Vane
Casing	Aluminium or glass fibre reinforced polyamide	Steel coated	Zinc alloy coated
Piston	Aluminium	Aluminium	Steel or SS
Bind cap	Aluminium		
Drive shafts	Steel, coated		Steel coated or SS
Seals	"O"-rings	"O"-rings	Shaped Polyurethane
Spring	Coated helical spring	Coated helical spring	Spring steel

Table 18.1 Materials comparisons

Many chemical plants have to deal with a very aggressive environment, resulting in the damaging of the aluminium housing. A special house case or a coating could be a solution. The corrosion resistance of the materials should be investigated per application. A zinc alloy case offers more protection against corrosion.

- **Torques —**

The stated values in Table 18.2 are purely indicative.

	From Nm	To Nm
Spring return Scotch yoke	100	15 000
Spring return rack and pinion	1	4000
Spring return vane	1	7000
Double acting scotch yoke	10	30 000
Double acting rack and pinion	1	15 000
Double acting vane	0.7	19 000

Table 18.2 Torque comparisons at 5.5 barg

- **Cost —**

Generally speaking the price of double acting rack-and pinion actuators is comparable with vane actuators. The spring return rack-and pinion actuators are favourable in price because a simple spring can be incorporated.

- **Stroke speed —**

The stroke speed is set by:

The volume of the actuator and the related amount/quantity/volume of air which needs to be displaced.

The moments of inertia of the moving inner parts.

The restrictions in the air inlet and outlet lines. The solenoids will then become important. See Table 18.3.

	Rack and pinion	Scotch yoke	Vane
Volume	- Lot of dead volume	+/- Dead volume around spring	+ Almost no dead volume
Moment of inertia	+/- Three heavy moving parts	- Lots of moving parts going in one direction	+ One small moving part
Restrictions	- Small, long entrance bores	+/-	+ Almost direct in the chamber

Table 18.3 Stroke speed comparisons

- **Clearance —**

The life cycle of an actuator is affected by backlash. Hysterisis is caused by clearance in control applications which results in an inaccurate control loop. Backlash occurs inside the valve, on the coupling between the shaft of the actuator, the shaft of the valve and in the actuator itself. In control applications there is also backlash in the NAMUR coupling between the actuator and the positioner. Table 18.4 offers some comparisons.

Rack and pinion	Scotch yoke	Vane
- Three moving parts (gear) and internal sideloads	+/- Clearance around lever mechanism	+ Only one moving part

Table 18.4 Clearance comparisons

- **Life cycle —**

The dynamic life cycle is determined by the number of operations. The static life cycle is determined by the resistance against corrosion and time fatiguing factors such as loss of spring force and ageing of grease. With many suppliers life cycle calculations are different, see Table 18.5

	Rack and pinion	Scotch yoke	Vane
Dynamic life	Approximately 2 million movements	Approximately 900 000 movements	Approximately 4 million movements
Static life	Depending on air quality. Springs are firmly pre-tensioned	Depending on air quality. Springs are firmly pre-tensioned	No internal wear. Springs IP65 isolated, not firmly pre-tensioned

Table 18.5 Life cycle comparisons

- **Air consumption —**

Compressed air or instrument air is an important cost aspect in installations. Also, for proper operation the units at the end of the grid should be supplied with enough air at an adequate pressure. As far as stroke speed is concerned, the air volume and consumption of the vane actuator is best.

- **Safety standards —**

An actuator has to comply with the safety standards as stated in the European machinery directives. For example it is prohibited for the actuator to have any moving parts in which fingers or any other part of the body can be trapped/jammed. Most actuators comply with these machinery directives.

- **Explosion-proof —**

Electrical equipment has to comply with ATEX special requirements when installed in an explosive atmosphere. There are also Directives for non-electrical equipment related to static electricity Sparking might occur when standard tooling knocks an apparatus. Aluminium or metal is more vulnerable.

- **User-friendly maintenance —**

	Rack and pinion	Scotch yoke	Vane
Double acting	- Rack should be positioned to the pinion properly	+/- Alignment of the shaft	+/- Special attention to the sealing of the cases
Spring return	- Dismounting might be dangerous when springs com out	+ Although heavy, spring can be mounted safely	+/- The spring should be fixed with a keeper plate

Table 18.6 Maintenance comparisons

Article — Courtesy of Delan BV

18.2 Double isolation in demanding services

Introduction

Demanding applications in oil and gas are rarely kind to valves. High velocities — particularly when opening valves on gas/multi-phase flow against high differential pressure, are a major cause of problems; as are fluids entrained with foreign particles and pipelines affected by scale, corrosion particles and other debris.

For many years the mainstay of the industry was the soft-seated ball valve — relatively inexpensive and with good isolation as a single valve. But in practice the seats proved vulnerable in a number of applications to the point where maintenance costs in terms of seat replacement (and indeed entire valve replacement) became very expensive. Trunnion-mounted metal-seated ball valves were subsequently popular but although these valves offered excellent isolation in the factory, they were often less than satisfactory in service.

The safety, environmental and financial impact of valve leakages meant this situation could not continue and the influence of safety engineers led to a requirement for double isolation valving. The options were two valves in series (which would be potentially costly) or a double block and bleed assembly.

An initial approach to double block and bleed was to use the cavity relief on a trunnion-mounted ball valve to provide verification. However, in this design seats remained vulnerable in many applications. For example, if the first seat started passing, the spring force on the second seat was not always enough to contain the pressure and this second seat could also be prone to failure. Again, with only one obturator if the ball or stem failed, there was no backup and safety engineers now insist on two obturators.

A more successful approach was the split wedge gate valve, which has two obturators and creates the seal mechanically. However care must be taken at valve opening to avoid seat damage by ensuring pressure at the centre of the valve is lower than at either side. In addition both obturators are controlled by the one stem, so a problem here means both seats would fail to seal. The relatively large size and weight of these valves can be a concern and in exotic materials they are very expensive. Additionally, they can often require extra gantries for operational access.

Some modular ball valves have fared well on instrument lines but the seat design generally remains insufficiently robust for the majority of typical service conditions, and they often have a non-standard face-to-face dimension which restricts maintenance interchangeability. Larger twin ball designs are high in weight and cost, offering little advantage over two separate valves.

These criteria have led to the production of the double isolation plug valve (dipv) which combines single isolation technology with low-maintenance attributes, see Figures 18.7 and 18.8.

Selection issues

The bleed To enable the process medium to be safely vented the vent port must be large enough to remove any leakage to ensure there is no pressure on the downstream seat. For example if the first obturator starts passing and a small pressure is created in the cavity between the obturators — not enough to energise the seats — this could creep behind the seats and down the line. This can be a key issue for ball valves but on plug or gate valves the pressure forces the obturator onto the seat — the higher the pressure the greater the seal.

Valve cavities With many media, fine particles can collect in dead areas and settle causing valves to malfunction, particularly in the stem area of gate valves. Such detritus may also settle in the space below the ball in ball valves, causing the ball to graze or score.

Figure 18.7 Double isolation plug valve
Courtesy of Flowserve Flow Control

Figure 18.8 Cutaway view of a double isolation plug valve
Courtesy of Flowserve Flow Control

Seats In abrasive or erosive applications seating materials become progressively harder – the degree of seating contact reduces to allow the seat to take the profile of the ball. This can make sealing problematic or easily prone to damage in service.

With a small seat contact area high loading forces on the seat are generated resulting in a high torque requirement, and if left in one position for extended periods the seat can leave an impression on the ball. In extreme circumstances this can cause deformation of the ball, making it difficult or impossible to turn.

Also, if mechanical means are used to produce the sealing, then force must be applied evenly round the seat and the obturator must not be moved relative to the seat or seat damage can occur.

If the seat becomes scratched or worn there is generally no way of applying extra force to make the seat seal. For this reason some ball valve manufacturers fit valves with emergency sealing injection points to enable emergency sealing to be achieved. However, on the majority of large valves there are a number of injection points which inevitably makes the process time-consuming. Furthermore, if the seat is badly damaged sealant injection is not always effective.

Stem seals Sealing to atmosphere is often the most critical issue for operator safety. Valve stems should be a separate piece to the obturator and allow 1-2° of freedom between stem and obturator so that any side loading is not transferred to the stem or packing. Non-rising stem valves are preferable since external conditions cannot act on the stem and cause packings to deteriorate.

Some packing systems incorporate pressure-energised seals. Though effective when new, the pressure on such seals and the subsequent loading of the seals on the stem tends to increase the torque requirement for valve operation. Over time seals can become worn and there is increased likelihood of stem leakage.

It is recommended that a secondary re-injectable packing system is used to provide a back-up facility. Note this should be comprised of a packing material rather than a sealant. If sealant is used in this area it is often injected at high-pressure and can blow already weak packing.

Maintenance On critical services in oil and gas production, a truly maintenance-free bubble tight shut off valve remains — (no pun intended!) – a "pipe dream". Sadly, maintenance is often ignored until a valve malfunctions, leading to high remedial costs. In many cases hydrocarbon engineers allow valves to pass as repair costs are simply prohibitive and maintenance time is often not available. This may be acceptable on single isolation valves in non-hazardous applications but not with a number of valves. On most valves sealant injection systems are important as they provide temporary respite to facilitate repairs.

Material Critical abrasive, erosive or corrosive applications are increasingly leading to more exotic materials being specified in valve construction. Many engineers are selecting compact flanges and hub ends to reduce costs while neglecting the main cost factor – the weight of the valve itself. The weight and cost saving of using a double valve unit rather than two separate valves with a spool piece is also considerable.

For sour services where standard materials are insufficient, Incolloy 625 is typically specified on all wetted parts — including the stem area, the valve passages, obturators and seat faces. Generally, cost considerations dictate that valve sizes up to six inches are of solid construction, while larger sizes feature weld overlays.

On ball and gate valves where NACE materials are specified seat pockets can be problematic. If fluids are trapped here they can become concentrated and corrosive over time. If this area corrodes, the valve could start passing behind the seats. A typical solution has been to put a weld overlay to protect the body but this incurs considerable extra cost.

Double isolation plug valve design

As its name implies, the double isolation plug valve is two complete valves in a single compact, lightweight face-to-face assembly, with the second valve completely independent of the first, Figure 18.9.

The opposed plug design is preferred in order to maximise the C_v through the valve and minimise the cavity between them. See Figures 18.10 and 18.11. The certainty of isolation and the tiny volume between the plugs means that the size of the vent need be only ½" or ¾" — even on 24" NB valves. The protected seats enable the valve to be opened at full differential pressure with minimal problems and metal-to-metal seats trap the seal-

Figure 18.9 Twin plug assembly
Courtesy of Flowserve Flow Control

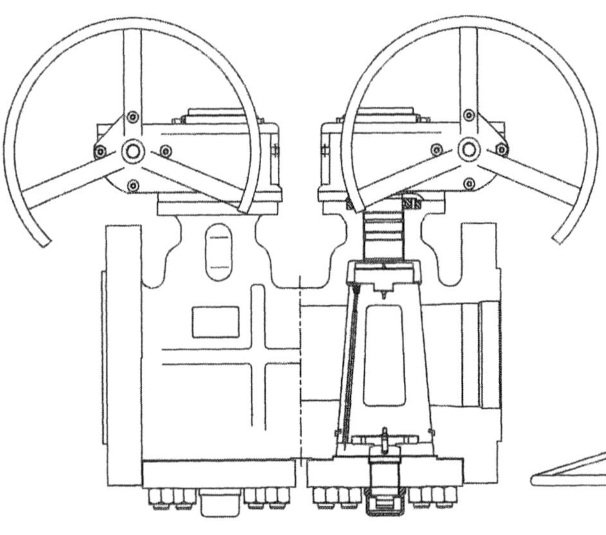

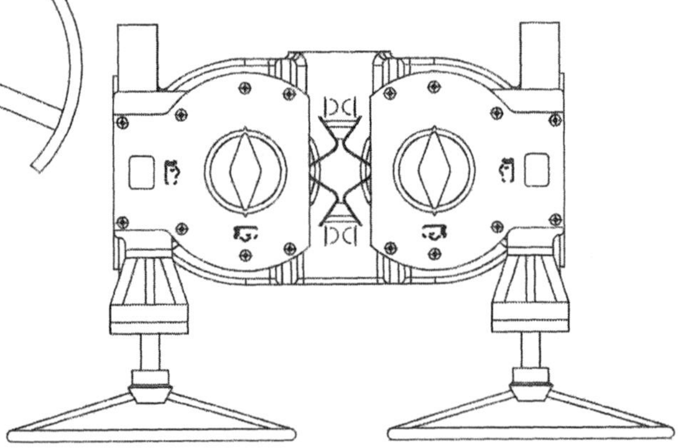

Figure 18.10 Twin plug valves in same orientation
Courtesy of Flowserve Flow Control

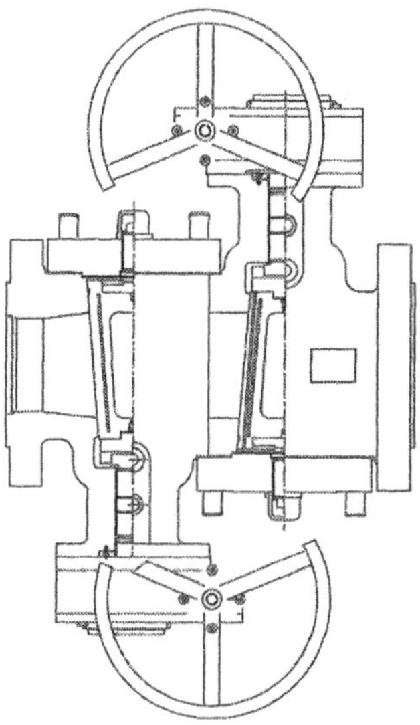

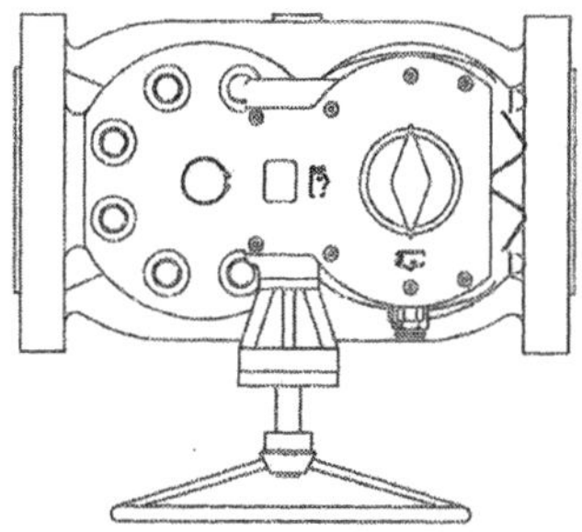

Figure 18.11 Double isolation plug valve with plug orientation inverted
Courtesy of Flowserve Flow Control

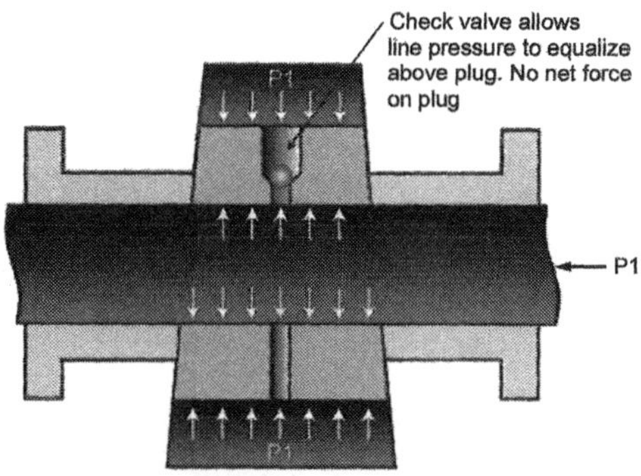

Figure 18.12 Pressure balance plug
Courtesy of Flowserve Flow Control

ant film to give 100% bubble tight performance so that the valve never needs to be taken out of line.

The plug is in contact with the body and a protected pressure balance facility allows pressure in the passage to move to the top of the plug to ensure it can never be locked into its taper. A check valve between the top and bottom of the plug will only open if the pressure is higher at the bottom than at the top to allow pressure equalisation, see Figure 18.12.

This configuration means there is no cavity.

The dipv's large port area ensures minimal pressure drops particularly at higher pressures (60 barg and above) as well as extreme, cost-effectiveness when built in exotic materials – with protected seats and no seat cavities additional extra protection of seat weld overlays is generally not required.

In addition, as the seats of a plug valve are unusually large (typically 12° of cross-sectional area) point loadings do not occur. The dipv seals by trapping a very thin film of sealant between finely lapped surfaces. Since its seats are on the plug they turn out of the line of flow — and so are not prone to damage from foreign particles in the line media, thus achieving long life without the necessity to overhaul the valve.

Maintenance is performed through the dipv's in-line sealant injection system which can be carried out under service conditions so that downtime is minimised. The sealant remains in the grooves and is not lost on the first operation of the valve, but lasts for a number of operations — typically between 10-20 cycles on critical services. The lapped metal-to-metal seats give a good seal and the sealant ensures the bubble tight seal. Service experience in the Middle East has shown that if a plug valve is removed from the line as part of a major overhaul, it only has to be cleaned and re-assembled; no remachining is necessary and expensive valve repairs are avoided. Reports from the field illustrate that even after some years of service, operators are confident they will achieve a bubble tight seal on a double isolation plug valve.

Applications

The double isolation plug valve is now being selected for dirty service and critical isolation applications. These include:

- manifold headers
- test and production separator isolation
- drain valve applications

The double isolation plug valve is also recommended for compressor isolation, produced water and gas injection lines. It is also increasingly being specified for:

- onshore gas applications such as filter isolation
- kicker valves, drain valves for pig launcher and receiver stations

Article — Courtesy of Flowserve Flow Control

18.3 Low noise control for atmospheric venting

Introduction

Modulating vent valves are an interesting solution for applications where any gas or steam needs to be released to the atmosphere. The modulating vent valve prevents noise generation instead of muffling noise at the outlet as a standard vent silencer would do. They can be used in applications involving high pressure drops, temperatures up to 565°C, a low noise requirement, or where space is limited. They are used in various applications such as protecting systems from over pressurisation or discharging gas from systems.

The modulating vent valve is designed simply and ruggedly. It consists of an angle style valve body and uses multi-labyrinth trim technology. True velocity control in severe service applications, is achieved by limiting the fluid velocities at the exit of the

Figure 18.13 Modulating vent valve under final construction
Courtesy of Copes Vulcan, An SPX Process Equipment Operation

trim. Variable flow and pressure regulation is provided by an internal plug, modulated via a valve-mounted actuator on receipt of a system command signal. They provide a compact and economical solution to today's noise problems in atmospheric venting. A valve under final construction is seen in Figure 18.13.

The modulating vent valve principle

The basic principle of a modulating vent valve is to control the velocity of the steam in such a way as to reduce the noise generated by the moving steam. Aerodynamic noise is produced when there is a rapid expansion of the steam from the inlet to the outlet as the pressure is reduced. If the velocity between the inlet and outlet is controlled, such that the steam is allowed to expand gradually, then the noise generated by the expanding steam would be greatly reduced. In a modulating vent valve this controlled expansion is accomplished with a series of expanding right angle turns inside the disk stack. See Figures 18.14 and 18.15.

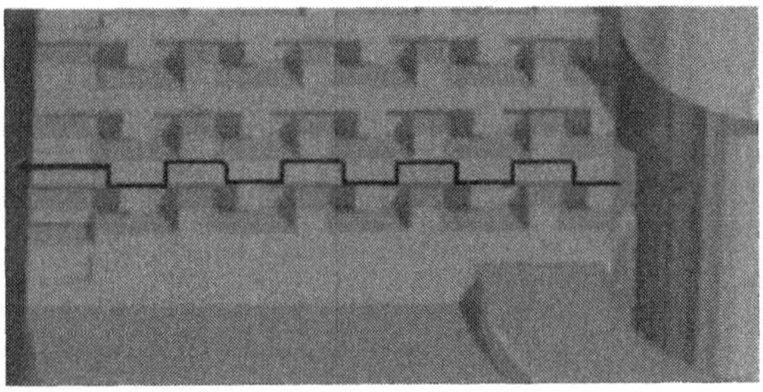

Figure 18.14 Internal view of disk stack
Courtesy of Copes -Vulcan, An SPX Process Equipment Operation

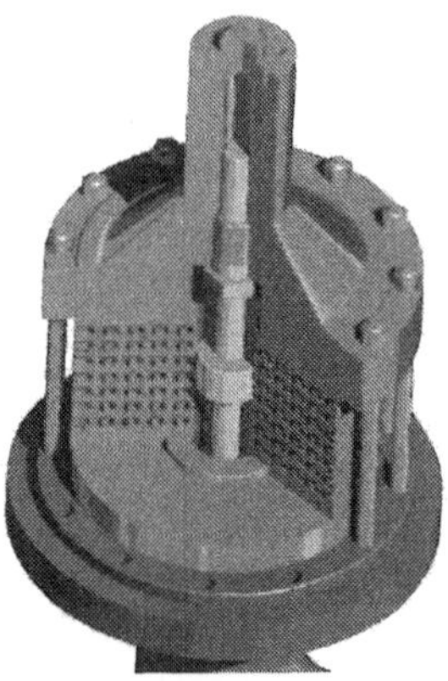

Figure 18.15 Cutaway of disk stack
Courtesy of Copes-Vulcan, An SPX Process Equipment Operation

With each turn, the steam expands gradually as the pressure decreases from the inlet to the outlet. This eliminates the noise source due to rapid expansion. The expansion constant is determined precisely so that choked flow does not occur at any point. The velocity of the steam is controlled so that it is well below Mach 1 at all times.

The noise generated inside the disk stack is high frequency noise. When frequencies are converted to the A-weighted scale, higher frequencies are given a lower dBA value. The higher frequencies are subject to higher atmospheric absorption. As the noise is carried by the steam, part of it is absorbed by the walls of the disk stack at each turn.The disk stack prevents noise generation rather than attenuating noise at the outlet. It is an effective means of reducing noise through velocity control.

Noise level

85 dBA at a distance of 1 metre is a typical noise specification for a control valve. As the valve and vent stack are combined within one valve it is usual to allow for noise degradation over a specified boundary distance. The vent valves should be on an intermittent duty and should be positioned so as to vent the steam to atmosphere without obstructions above the valve. For this reason it is usual to find that the valves are installed high on roof buildings or piping towers.

The vent valve is a hemispherical radiation source with ground reflection. The effects of ground reflection are conservatively estimated at 3 dBA. The sound level heard by an observer in the far field is attenuated by directivity index, distance, and atmospheric conditions, such as molecular absorption, wind, and humidity. Attenuation is due to the inverse square law and atmospheric absorption.

For the purpose of calculation a distance of 12 metres at an angle of 180 degrees from the outlet vent was assumed.

Figure 18.16 Valve without cowling fitted
Courtesy of Copes-Vulcan, An SPX Process Equipment Operation

Figure 18.17 Cutaway view of cowling bolted to main body
Courtesy of Copes-Vulcan, An SPX Process Equipment Operation

Cowling design

The cowling is used to prevent the escaping steam from contacting surrounding machinery and personnel. The purpose of the cowling is to redirect the flow and noise upwards. It does not serve as an attenuator or pressure vessel and therefore it is lightweight and strong, see Figure 18.16. Incorporated into the design of the cowling are the following features:

- **Integral drain connection**

  Integral drain connection is provided to ensure that any rainwater that collects within the cowl can be easily removed. The connection point provided under the cowl can either be left to drain naturally, or can be piped away to drain to a more convenient place.

- **Mesh screen**

  As the valves could be closed for long periods of time, a mesh screen is provided to prevent wildlife and large debris from collecting within the cowling.

The cowling is bolted to the main valve body as a "top-hat" assembly and is provided with lifting eyes to assist with removal. The cowling is then specially painted with a high temperature painting system to provide both corrosion and temperature resistance and is suitable for temperatures up to 550° C. See Figure 18.17.

Body design

The main valve body is of the angle design and incorporates a streamlined flow path into the disk stack assembly, see Figure

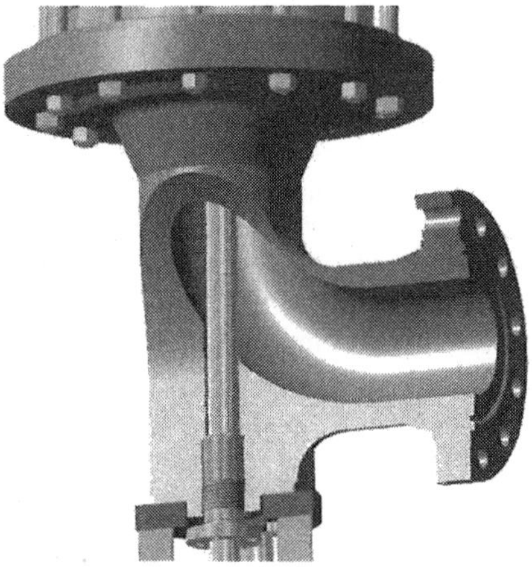

Figure 18.18 Body design
Courtesy of Copes-Vulcan, An SPX Process Equipment Operation

Figure 18.19 Typical Installation
Courtesy of Copes-Vulcan, An SPX Process Equipment Operation

18.18. This streamlining also ensures that when the valve is closed any condensate that forms internally is naturally drained away from the valve internals. An integral body drain is provided for this purpose, or a drain arrangement can be positioned in the upstream piping to the unit providing the piping is naturally sloped away from the valve inlet.

Conclusion

The velocity control concept utilised in the trim design is independent of frequency. Each modulating vent valve is custom-designed to meet the specific requirements of each application. In addition to capacity, it is also sized to provide a low trim exit velocity via multiple outlets for quiet operation. It can be easily positioned with minimal supporting structure.

Most conventional vent silencers being of large size to accommodate the acoustic absorption material are generally 2 to 7 times longer in length and 2 to 4 times larger in diameter to achieve a comparable noise reduction. Thus the supporting structures for conventional silencers need to be designed to support this large weight.

Conventional silencers are generally sections of duct or pipe which have been shaped or treated with the intention of reducing the transmission of sound, while at the same time allowing the free flow of a gas. They are acoustical filters and performance is dependent upon frequency. They are designed to reduce noise levels at a particular frequency; but as the flow rate varies, the frequency at which the highest noise level occurs also varies. When this occurs, silencers become less desirable.

The initial cost of silencers may be low; but frequently, due to the enormous size of the devices, supporting structures are required which can offset the potential savings. Also, space becomes a problem. For high-pressure applications these devices become costly. The modulating vent valve offers an interesting option for applications where any gas or steam needs to be released to the atmosphere, see Figure 18.19.

Article — Courtesy of Copes-Vulcan, An SPX Process Equipment Operation

18.4 Actuation technology designed to assist the operator

Introduction

The widespread adoption of microprocessor technology has resulted in the growth in functionality of field devices. As a result of technological enhancements, and a trend towards decentralisation, aspects of process control systems have been transferred to field devices - particularly actuators. Operators who manage digital control systems now face an increasing number of programmable parameters, control signals and a diverse range of status and diagnostic indicators.

Just under two decades ago, it was typical for field devices to have just a few programming switches to set the desired device configuration and the operator was able to visualise the status of the device based on a few indicator lights.

Since the 1990s, with the availability of enhanced microprocessor technology and, for example, the introduction of fieldbus systems, increasing process control system functionality has been transferred to field devices, particularly actuators. In addition, display functionality has also been significantly enhanced. Today, depending upon the complexity of the device, several dozen parameters can be set via software and the operator is faced with a multitude of status and diagnostic signals.

Device manufacturers must not only consider the enhanced functionality of their devices, but also how the operator can manage this increased complexity and the exponentially increasing available data.

Before expanding on the developments in actuators that have been made to support operators, it is helpful to provide the industrial context within which these enhancements have been made.

Electric actuators play an essential role in flow control applications. As the front end "intelligence" for most valves, actuators convert control signals into mechanical motion and, therefore, fluid control across a wide range of sectors including the water, power and oil/gas industries. Taking the water sector as an example, actuators provide control across a range of application areas including the automation of potable water equipment, the treatment of wastewater and a variety of water and effluent treatment plant control systems. See Figure 18.20

Actuator programming via onboard switches, buttons and displays

When programming actuators at the device itself it is preferable that parameters can be set intuitively and without recourse to

Figure 18.20 Electric actuators in a flow control application
Courtesy of AUMA Riester GmbH & Co KG

Figure 18.21 On site staff benefit from plain text display in national language
Courtesy of AUMA Riester GmbH & Co KG

additional support. This has been primarily achieved by equipping devices with a display that provides clear indications to the operator supported by straightforward controls e.g. push buttons and simple switches. Additionally, these parameters can be set without the need to open the actuator housing.

Further developments in actuator functionality, however, have meant an increasing number of programmable parameters. This places a greater load on the operator with the consequence of potentially longer set-up times and an increased risk of programming errors.

A number of initiatives have been taken to help the user in the programming and management of actuation systems. Designed to avoid the operator becoming over loaded when dealing with a large number of parameters, these developments include:

- An intuitive structure in the programming menu
- Parameters predominantly displayed in plain text in the national language, see Figure 18.21
- Parameters relevant for safe operation clearly displayed in the menu structure
- Operating instructions and display menu structures having a similar logical layout and consistent terminology

It is often prudent to split the operation instructions into two elements:

- Commissioning
- Operating instructions

Commissioning instructions contain comprehensive installation and parameter setting information.

Operating instructions are usefully presented as a manual containing a description of all device parameters written with those responsible for the integration of the field devices in mind, such as PLC software engineers.

Actuator programming via a laptop computer

The development of field device programming via a laptop computer offers the operator some significant benefits compared to using on-device controls and many field device manufacturers offer programming software downloads on their web sites. For example, software can be made available via the actuator company's web site – this can be downloaded as required by operators and installed onto laptop computers running the Windows operating system, see Figure 18.22.

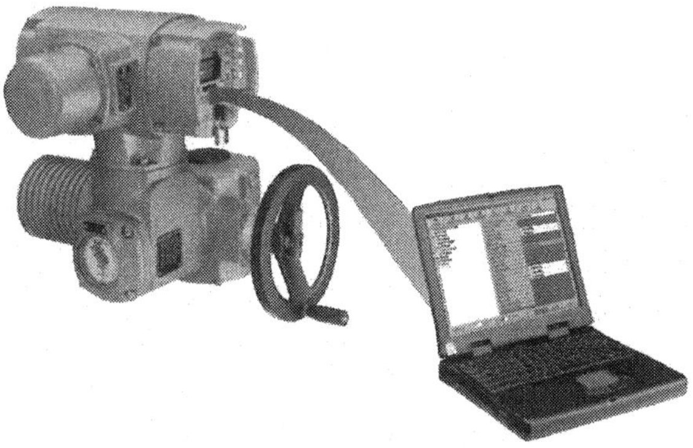

Figure 18.22 Actuation industry developments have facilitated programming via laptop computers
Courtesy of AUMA Riester GmbH & Co KG

Clearer parameter presentation and on-line help

These programs allow a clearer presentation of the device parameters than the display on the device can provide. Additionally, integrated online help functions provide extensive explanation of the parameters and therefore frequently save the operator, who may be out on-site, the time of obtaining and consulting a manual.

Easier on-site device set-up

Databases supplied with programming software ensure that the set up of device functionality does not have to be performed directly at the field device. The parameter settings can be recorded in advance and the laptop physically taken to the field device for downloading.

Simplified documentation

Once downloaded to the field device, the parameter settings can be easily stored in a data record. This is particularly advantageous during maintenance when, for example, a device may need replacing and a new device is required to replicate the operational settings of the existing unit.

Cable-based or wireless connection

Traditionally the link between the programming device, generally a laptop, and the field device has been established with a data cable. However, recent advancements in actuator programming now include the use of wireless technology e.g. Bluetooth device connection.

To many readers wireless connection may appear to be the obvious and ideal solution for data access when using mobile devices. However, in the area of actuation technology with inherent operational and safety implications, there are two main areas to consider:

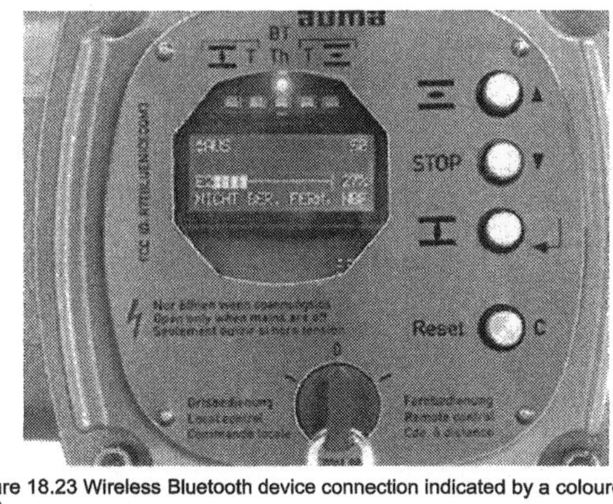

Figure 18.23 Wireless Bluetooth device connection indicated by a coloured LED
Courtesy of AUMA Riester GmbH & Co KG

- Target device identification
- Non-authorised access

In a plant there are often numerous field devices of the same type within range of the programming device using, for example, a Bluetooth interface. It is therefore essential to ensure that the connection is only made to the chosen device. To avoid the risk of incorrect device parameters being programmed, the identification of the target device must be known and when communication has been established, a signal from the specific device, for example an LED, should be provided. To simplify identification, the device should be given a reference such as "Feedwater 1" which is then used in the programming software identification and log-on process. See Figure 18.23.

Depending on the range, it is possible for wireless systems to establish communication from outside the plant to field devices inside the site. To address issues of non-authorised access to the field data, password authentication should be included in the software.

Remote actuator programming via fieldbus

There are well-established protocols for the remote programming of field devices via fieldbus systems from the control room. The EDD/PDM system, which is adopted by Siemens, is just one such protocol. A second approach is the use of manufacturer-independent FDT/DTM technology. FDT is an interface standard for the connection of fieldbus with programming software made available by the device manufacturer. In principle, the FDT/DTM concept is not connected to any particular fieldbus system.

To implement this technology, the following are currently required:

- A Profibus DP-V1 fieldbus protocol
- Provision of field devices that support DP-V1 services
- Device Type Managers (DTMs) for the field devices

In principle, a DTM is a device driver comparable to a printer driver that can be installed on a computer in the control room. From the computer the parameters of all devices that support the DTM are accessible for reading as well as writing. The DTM is a program file (.exe) that installs the programming software. By adopting a DTM solution, manufacturers of field devices are able to add further functionality that benefits the operator – these functions include online help, electronic operation instructions and service information.

Diagnostics

Whether at the device or operating remotely, in the event that a fault occurs it should be possible to identify the cause with the help of diagnostic functions. The operator, who does not work with the device on a daily basis, should be able to interpret independently any diagnostic information without recourse to additional technical support. A plain text display on the device in the national language of the operator is a prerequisite for this. Even if the operator is not in a position to eliminate the failure, important information can be provided to the device manufacturer ensuring that the service technician delivers the correct spare parts.

The same principle applied to parameter programming is used with diagnostics. The programming software allows the display of diagnostic information on a laptop, which is much easier than using the small display on the device. In addition, online help can also be sought to further investigate diagnostic information. In the case of fieldbus systems this can be carried out from the convenience of the control room.

Conclusion

The trend towards decentralisation has considerably increased the complexities of process control system functionality of field

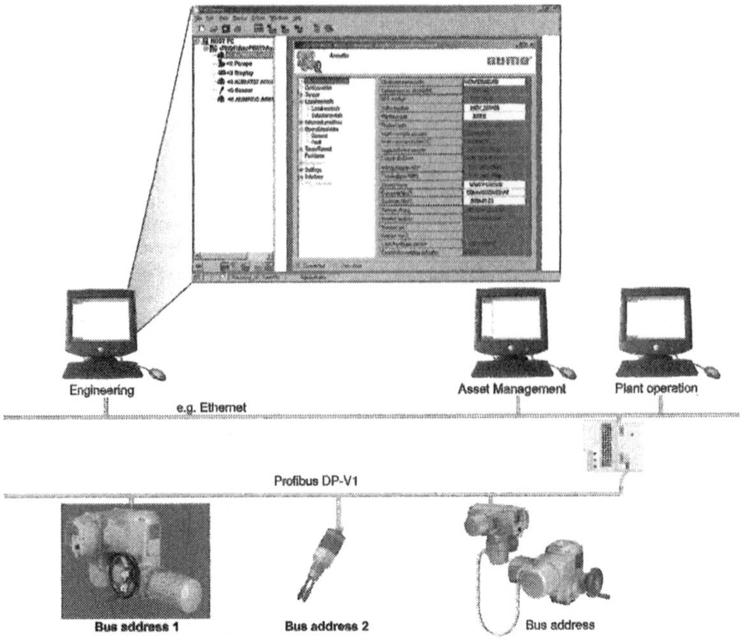

Figure 18.24 Illustration showing levels of accessibility of device data
Courtesy of AUMA Riester GmbH & Co KG

devices. While enhancing the functionality of actuation solutions, forward looking actuator manufacturers have driven developments to support those working in the field who are operating the devices. The initiatives taken support field device programming and have been to the advantage of both those using actuators and the wider actuation industry.

Article — Courtesy of Auma Riester GmbH & Co KG

18.5 Pressure relief with high flow capacity and low cost of ownership

Introduction

In the quest for efficient relief system sizing, it is the goal of the designer to obtain the maximum amount of flow for the smallest financial investment while meeting the often challenging operating requirements of each application. With their full opening characteristics, rupture disc devices provide the best flow option while pressure relief valves (prvs) offer significantly reduced flow but, with automatic reset, avoid immediate downtime. An alternative option uses "buckling pin" technology to control the set pressure of a high flow, in-line relief valve that equals the performance of the best rupture disc devices.

Buckling pin pressure relief valve technology

The buckling pin relief valve (bprv) owes its high in-line flow capacity to a unique hollow rotating disc (see Figure 18.25). Based upon an offset shaft butterfly valve concept, a pressure differential produces a turning moment on the valve shaft in the direction of flow. This moment is converted to a force in the valve external mechanism (see Figure 18.26) where it is restrained by a buckling pin under normal service conditions.

At a predetermined pressure relief point, the buckling pin is designed to activate at the corresponding load, allowing the disc to rotate through 90 degrees, full open. This exposes the hollow disc profile to the flow path, providing the maximum flow area for pressure relief.

Buckling point

Set pressure accuracy of a buckling pin-activated relief valve is assured by the excellent repeatability of buckling pins made from the same controlled heat of raw material. Euler's Law determines that a pin that is held in compression will buckle at a

Figure 18.25 View of the hollow rotating disc
Courtesy of BS & B Safety Systems (UK) Ltd

Figure 18.26 Buckling pin relief valve external mechanism
Courtesy of BS & B Safety Systems (UK) Ltd

Figure 18.27 Buckling pin cartridge assembly
Courtesy of BS & B Safety Systems (UK) Ltd

particular load. This buckling point is controlled by three variables:

- Diameter of the pin
- Length of the pin
- Modulus of elasticity (material) of the pin

Installation

Precise control of buckling pin dimensions within each lot results in buckling pin relief valve accuracy that, it is claimed, equals or exceeds that of rupture disc devices. The buckling pin relief valve takes a centuries old, highly reliable technology, and applies it to a modern engineering application.

Buckling pins can be installed in a cartridge form (see Figure 18.27) allowing a replacement part to be handled conveniently in a similar manner to that of an electrical cartridge fuse. This offers the optimum ease and integrity of installation.

Resetting

Resetting is straightforward, carried out typically in the field by the user. The design basis for the bprv provides a unique combination of high flow capacity, in-line valve technology designed for minimum reset downtime. Reset is by hand and typically takes less than 5 minutes to accomplish even on very large nominal size valves and without the need for specialised tools. The buckling pin which activates the buckling pin relief valve is housed in an enclosure external to the process. The enclosure protects both the pin and mechanism from environmental conditions.

Operating conditions

The application of buckling pin relief valve technology is particularly suited to the following conditions:

- Continuous processes — Minimum downtime in the event of activation
- Hazardous service —The operator does not have to remove the valve from the piping system for routine replacement pin installation, minimizing the risk of exposure to the process
- Large line size — There is no requirement for special lifting equipment, extensive manpower and the resultant extended downtime to install a spare buckling pin
- Inaccessible locations — The buckling pin replacement part is small and convenient for installation – the option of a buckling pin cartridge further simplifies installation in hard to reach locations
- Remote installations — The small buckling pin replacement part can be easily accommodated in limited storage space
- Frequent operation — The cost of a replacement buckling pin is very modest compared to the cost of larger rupture disks and post discharge relief valve testing

Summary

Buckling pin relief valve technology allows a plant or facility to operate at a high productivity level while increasing safety for operators. An attractive overall cost of ownership combined with certification to international codes and standards such as the European Pressure Equipment Directive, (PED) and ASME Section VIII, provides the pressure relief system designer with a new and efficient means of compliance with challenging project objectives.

Article — Courtesy of BS & B Safety Systems (UK) Ltd

18.6 The use of repair intelligence

Valve performance improvement

The improvement of control valve life and performance is a goal well worth striving for, especially when one considers the implications of a valve malfunction or complete breakdown in terms of production and maintenance costs as well as the safety implications of certain malfunctions.

Control valve designers involved in this challenge, have to grapple with combinations of the latest materials and material conditions, innovate in terms of trim design and body design and work within more and more stringently applied commercial, legislative and production restraints.

A large part of the research and development process of valve life and performance improvement is testing and modelling, in an attempt to predict actual operational performance in the field. These activities form a process which is at best indicative and cannot practically be broad enough to cover the whole gamete of operating conditions the valve design will, in reality, experience in its installed environment.

An alternative and perhaps more pragmatic approach to life and performance improvement of specific valve applications is now described. This approach uses actual field results, in the form of intelligence gathered through repair and overhaul activities.

Never intended to encroach on the product improvement activities of valve manufacturers, this approach is essentially concerned with the numerically few but commercially problematic applications. These are the applications where the latest product advances have failed to give an acceptable performance due to their severity or uniqueness, and where no general range of products can possibly hope to provide an optimised solution.

The Repair Intelligence Circle

There exists a wealth of repair intelligence concerning the actual performance of valve designs in the field and on specific applications. The gathering of, and proactive utilisation of this intelligence requires a unique mix of valve availability, tailored data acquisition systems and the right skill competencies. When brought together effectively with feedback and operational data, these form the Repair Intelligence Circle.

Intelligence gathering

A prerequisite to effective collation of performance data is the availability for inspection of valves of particular designs, operating on specific applications. This is important in order to distinguish a "one-off" failure from the more concerning recurring failure modes. Also the scope and depth of gathered intelligence grows in relation to the magnitude of historical data available as well as fresh information being recorded on a continuous basis.

Intelligence pertaining to valve performance in the field is available from a number of sources, these typically include:

- Plant or contractor maintenance documentation
- Operators logs of plant trips and valve malfunctions
- Operators experiences of valve related issues
- Diagnostic reports
- Repair Intelligence

With the exception of repair-related intelligence, it is rare that each individual, or for that matter any combination of information sources, will yield the right kind of information necessary to make a definitive analysis and step-change in valve performance.

On the other hand whilst the information gathered through repair activities can often be utilised in isolation, the process is greatly enhanced through consideration of information from the other sources.

Collation of information from the valve at the repair stage, requires that the valve be handled and processed in an entirely different manner to that of the traditional repair shop.

The valve needs to be assessed before stripping, if pertinent to the performance issue at hand. Some "as received" testing may be required, this may take the form of shell and leak testing as well as operational tests for stroke time, hysteresis etc. Diagnostic footprinting is an ideal way of gathering this type of data.

Figure 18.28 Severe erosion of drilled cage
Courtesy of Severn Unival Ltd, Technical Products and Services Group

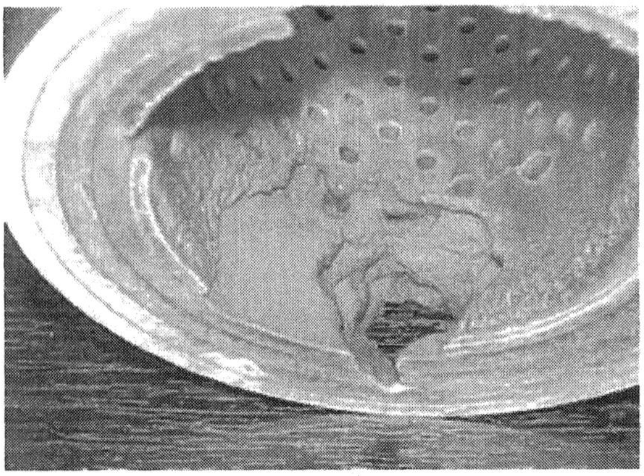

Figure 18.29 Cavitation of ceramic muti-cage trim
Courtesy of Severn Unival Ltd, Technical Products and Services Group

During stripping, the relative position of components will need to be noted and recorded as will information relevant to the investigation, such as:

- Obvious severe damage to components from cavitation, erosion, galling, corrosion and fatigue. (See Figures 18.28 and 18.29.)
- More subtle cause and effect of malfunction, fretting of surfaces, seal displacement, uneven wear, the nature and magnitude of debris ingress

Digital photography and collection of sample materials assist greatly in the recording of this valuable information, however as with many detailed investigations there is no substitute for an experienced technician employed in an ongoing and active system of repair intelligence gathering.

Analysis

Successful analysis of the repair intelligence and the gathered information from the other various sources, is the cornerstone of performance improvement. It is unlikely that an effective solution will be forthcoming if the failure mode is not as fully understood as can practically be achieved.

The skill competencies required for the analysis phase are varied and include those of materials and design, application knowledge and experience of valve failure mechanisms.

Experience suggests that a team approach is best employed for this activity, ensuring the support of all relevant technical specialists is fully utilised.

Numerous theories can and will arise during this phase. Each one should be weighed against the evidence at hand, often further information and data will need to be sought in order to fully consider a line of inquiry.

Key to this stage of the process is the ability to call upon information and data from past failure analysis activities. This ability comes only through the availability of valve repair intelligence gathered over a significant number of years of pro-active intelligence gathering and recording.

The various theories and mechanisms should be challenged and interrogated before being accepted or discounted. More that one possible scenario can often exist and the experience of the specialists involved in the analysis, will be needed to disseminate the information at hand, as well as the previously gathered intelligence, with a view to identifying the "best fit" scenarios.

Determining a solution, reporting and follow-up

Any solution proposed by the analysis team, who are now collectively wearing the different "hat" of a valve solution provider, must address all of the identified elements of the failure scenario or scenarios.

This stage of the improvement process requires the skill competencies of all of the various parties involved but will rely heavily on the design and materials experts.

In providing an effective solution it is imperative that data relating to the operation of previous solutions of a similar nature and having similar features is available.

Any innovative development will include features not fully tried and tested in the field. The presence of data pertaining to the performance of similar designs or elements of the design greatly reduces the unknown factors of a proposed solution. This then allows focussed consideration of the innovative features, in the knowledge that the other elements of the solution are field-proven and will perform.

The presentation of the proposed solution to the client should include a detailed report of all stages of the process. In this way the logic behind the proposal can be demonstrated and understood.

Ideally tests can be arranged to assess the improved performance, or in the case of longer term issues, inspections may be arranged at opportune times in order to assess the ongoing effectiveness of the solution.

The gathering of information relating to the performance of the installed solution and its further development is a fundamental part of the Repair Intelligence Circle; it is this data that feeds back into the analysis of failures and the determination of further effective solutions.

Further opportunities

Other benefits of the database that naturally builds from these activities, lie in the area of criticality studies and spares inventory. Data pertaining to the actual field performance of valve designs on specific applications can be fed into criticality studies as an element of the overall criticality review and ranking of the valve. This then will increase the potential for effective management of spares inventory when spares holdings are linked to valve criticality.

The utilisation of data applicable to a particular plant or facility with regard to performance of valves on specific applications, is a valuable tool in the process of specification and procurement of equipment for plant modification and expansion. Where repair intelligence has been used to improve valve performance on the existing plant, it follows that its extended use in the specification process of new equipment will prevent old lessons being re-learned.

Conclusion

The use of repair intelligence in the manner described here is a feasible and pragmatic approach to performance improvement of valves on specific applications. It is also a valuable source of data on the actual field performance of design features which may be later incorporated in the improvement of product ranges.

It requires all of the elements described to be available and utilised. However, when these are available and managed effectively, significant improvements can and have been seen to be achievable, with obvious desirable benefits to the operator.

Article — Courtesy of Severn Unival Ltd, Technical Products and Services Group

18.7 Designed for continuous use — valves with a permeation-resistant lining PFA-P

Background

Linings made of the well-known thermoplastic fluoroplastic PFA (perfluoroalkoxy) have been used as an alternative to high-alloy, expensive metals for valves, control valves, pumps containers etc. PFA has superseded PTFE, which is processed in a pressure sintering process, in these applications to a large extent. PFA's success is due to several important advantages over PTFE. PFA is processed in a transfer moulding process and as a result the lining wall thicknesses can be accurately defined and reproduced. PFA is almost transparent and therefore permits much more reliable quality control. Moreover, thanks to its dense molecular structure, PFA has generally much lower permeation rates than PTFE with the same wall thicknesses but it has the same chemical and thermal resistance properties.

Barrier effect of PFA

With media which especially tend to permeate, such as chlorine, bromine or fluorine compounds even the good barrier effect of pure PFA with the usual wall thickness of ± 3 mm is often not sufficient to achieve satisfactory service lives of the equipment. If the chemical penetrates the lining, this may cause corrosion on the pressure-bearing metal body and perhaps even failure of the unit. Experience shows that an increase in the PFA lining wall thickness to 5–6 mm already produces substantially longer service lives. This is implemented, for example, in ball and globe control valve bodies. There is a quadratic relationship between the breakthrough time, i.e. the time the medium takes to permeate through the plastic and emerge again, and the wall thickness. In most applications the breakthrough time is higher than the useful life of the units because the temperature and pressure gradients driving the permeation process decrease significantly towards the outside. This is also seen in pump housings where the PFA linings of 5-6 mm normally are particularly thick-walled anyway.

However, in the case of internal wetted components, such as shut-off elements, stems and pump shafts, inner magnetic assembly linings and cans, the wall thicknesses cannot be increased, or only to a very limited extent, for functional reasons. This is exactly where permeation can cause a too early failure of the unit.

Higher permeation resistance

The aim was therefore to provide a lining material even more resistant to permeation. Material engineers concentrated their efforts on creating a PFA variation which could be subjected to thermoplastic processing but with a much higher permeation resistance and with the same chemical resistance and a temperature resistance from -60 to +200°C. With the compound PFA-P (the "-P" stands for permeation resistance), a material is now available which satisfies these general conditions. The carrier polymer PFA is enriched with an extremely corrosion-resistant filler. This filler extends the diffusion paths and acts as a diffusion barrier. The surface resistance of PFA-P as regards electric conductivity corresponds to that of PFA. As a lining material, it covers the same pressure and temperature ranges as pure PFA for valves, globe control valves, pumps etc., i.e. from -60 to + 200°C and from vacuum up to 25 barg. Both the PFA base material and agents used in the production process are FDA-compliant.

In order to confirm the permeation resistance, the permeation of chlorine gas was examined under operating conditions as an example. In this test, samples made of PFA-P and pure PFA with different thicknesses were compared with each other at temperatures of up to 150°C and pressure differences of up to 7 barg.

The result is shown in Figure 18.30.The permeation through PFA-P is roughly half of that compared with pure PFA. This effect is particularly noticeable with the critical elevated operating temperatures. Tests with the extremely mobile helium as a test medium also showed a similar reduction in permeation. The permeation rates of solvents tend to be considerably lower but experience shows that they follow the same trend and so PFA-P should produce substantial improvements here, too.

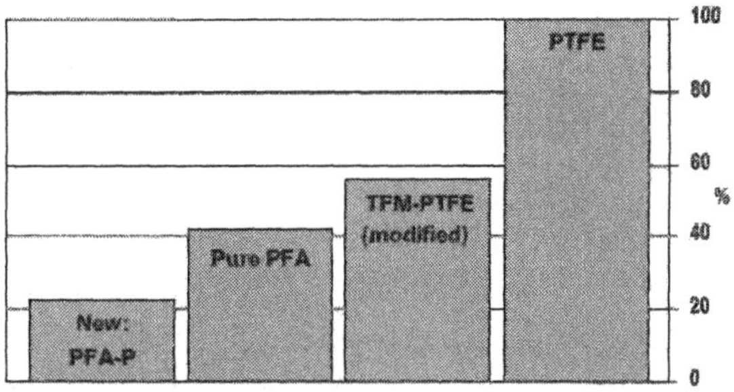

Figure 18.30 Qualitative comparison of the permeability of fully fluorinated plastics
Courtesy of ITT Richter Chemie-Technik GmbH

Figure 18.31 Ball valve dismantled with a PFA-P lining
Courtesy of ITT Richter Chemie-Technik GmbH

Practical example

A ball valve lined with PFA-P, was tested in an application in which monochloroacetic acid was present at approximately 150°C, as an alternative to special valves made of special materials. Even after 1.5 years of continuous use this ball valve showed no signs of failure. And it is considerably cheaper than the special valve previously used. The ball valve lined with PFA-P covers a broad operating spectrum with a temperature range of -60 to +200°C and a pressure range from 1 mbarg up to 16 barg. See Figure 18.31.

Depending on the problem in question, a decision has to be made as to whether a PFA-P lining is necessary for the entire unit or only for critical components. In many cases it is sufficient to line individual components with PFA-P instead of with PFA or PTFE, while the main part of the valve still remains lined with standard PFA.

The valve was type-tested and complied with the German Clean Air Act and was also approved for the transport of hazardous goods in tanks, GGVSE/ADR/RID/TRT24. Depending on the application, the valve can be equipped with a PFA or Al_2O_3 shut-off ball or a cavitation-free ball. With the V-control ball and the special hysteresis-free coupling between the ball valve stem and the actuator, it becomes a complete control valve for k_v 0.8 to 400 (C_v 0.7 to 345).

Summary

The suitability for use with highly permeating media makes PFA-P-lined pumps and valves interesting for a large group of users. Moreover, the chemical and physical properties relevant to applications permit the use of PFA-P where pure PFA has proved successful. In addition, as well as the ball valves described, control valves, sight glasses, sampling valves, vessel drain valves and safety valves can also be produced with a PFA-P lining. For many plant operators PFA-P leads to a significant prolongation of the service life.

In many cases it will already be sufficient to line individual components with PFA-P instead of with PFA or PTFE while the main part of the valve still remains lined with standard pure PFA. A general "all or nothing" policy is not advisable. The problem in question governs whether a PFA-P lining is necessary for the entire unit or only for critical components.

Article — Courtesy of ITT Richter Chemie-Technik GmbH

18.8 Valve sealing solutions — according to TA-Luft

TA-Luft is the national German regulation for leakage and pollution (Literal English translation is "Clean Air"). All industrial plants in Germany have to be upgraded to fulfil the European emission legislation (IPPC-Directive 96/61/EC) and the national requirements TA-Luft 2002 [1, 2] by 2007. For valve applications in particular, the spindle and the housing seals are affected. This article describes modern packing and gasket systems which are capable of performing according to the increased demands of these standards for the reduction of fugitive emissions. The performance of these sealing solutions was tested on internal packing test stands and certified by independent external test institutes.

For the end user not only is the quality of the sealing component important, but also the proper condition of the refurbished valve. Additionally, competent installation and a performance check of the repaired valve are necessary to ensure optimal, trouble-free service. Based on practical application experience gained during several shut-downs in refineries and chemical plants over the last years, guidelines will be proposed to ensure that old and used valves can be refurbished to perform according to TA-Luft.

Introduction

Due to the forthcoming emission legislation in Europe most chemical and process industry plants have to adhere to strict legislation when replacing sealing elements during shutdowns. Besides the need for proper sealing solutions there are requirements with regard to the correct installation and condition of the equipment, i.e. valve or flange connections. All aspects have to be covered to ensure a tight and emissions minimising performance.

In this environment the manufacturer strives to not only supply a packing set or gasket to the customer but to give additional service for a problem-free upgrade of existing valves to Integrated Pollution Prevention Control (IPPC) and TA-Luft requirements. This means as well as the provision of the right products, the company will work with the customer and help in the transition work. Besides delivering the sealing solution the company trains and educates most customers about the requirements and consequences of the IPPC and TA-Luft legislation. The company is also actively involved with the European Sealing Association to provide BAT (Best Available Technology) guidance notes for sealing technology [3] to the European IPPC bureau.

Sealing Solutions for TA-Luft

The company has developed a specific range of packing sets and gasket solutions to specifically cover the refurbishment and upgrade of used valves to meet the new regulations. A wide range of packings and gaskets were tested using in-house test stands as well as field tests with selected customers to arrive at the best and most versatile solutions.

For the retrofit of valves there are two main solutions based on the temperature range of up to 250°C and for temperatures exceeding this limit which will be described in more detail below.

Sealing set for temperatures up to 250 °C

For the specific requirements of fugitive emission sealing a new packing set based on PTFE impregnated non-woven material was developed, shown in Figure 18.32. By the use of these patented materials, a very dense and gas tight packing set could be produced [4]. The set is made of three intermediate rings of PTFE impregnated meta-aramid non-woven material. The end rings consist of carbon non-woven material with a PTFE/graphite impregnation. The set has very good thermal stability and reduced cold flow compared to pure PTFE packings, due to the fibre reinforcement. The strong but flexible end ring material counteracts loss of gland compression force and protects against packing extrusion into clearances.

The set is suitable for standard on/off valves as well as for control valves due to its low coefficient of friction. The graph in Fig-

Figure 18.32 Packing set for fugitive emission sealing
Courtesy of Burgmann Industries GmbH & Co KG

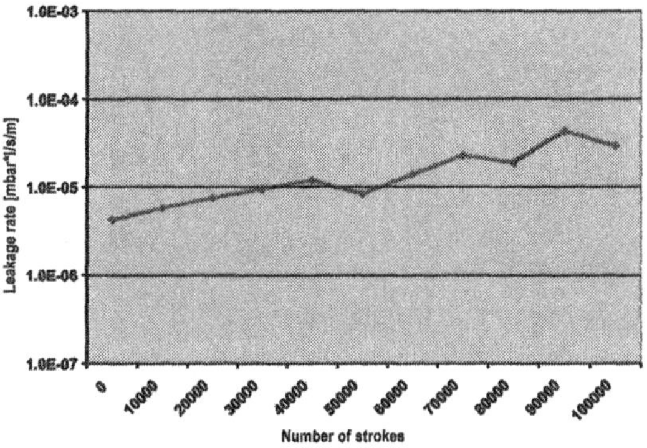

Figure 18.33 Leakage rate over spindle cycles
Courtesy of Burgmann Industries GmbH & Co KG

ure 18.33 shows the results of a test using an in-house packing test stand. The packing was tested at 40 barg with helium according to VDI 2440 [5] at room temperature. Over 100,000 cycles with a stroke length of 10 mm the leakage rate was lower than the TA-Luft limit throughout the test. The measured frictional force showed a slight reduction due to some gland pressure loss.

A field-test with the set in a control valve in a gas refinery (26.4 barg, ambient temperature, mixed hydrocarbons, 4 spindle cycles per day) showed no detectable leakage after 30 months of operation. The leakage was measured on-site by sniffing with a FID (Flame lonisation Detector). The measurement of previously packed graphite sealed valves showed leakage values of up to 5,000 ppm.

A modified PTFE gasket is reinforced by a special filler to reduce cold flow and creep is recommended as a gasket material for valve refurbishment, in combination with the packing set. It combines good chemical resistance with very low gas permeability.

Sealing set for temperatures above 250°C

The packing set shown in Figure 18.34 has been specially developed for the sealing of older valves to meet fugitive emission standards.

This set consists of four different components, each with their own specific function, to guarantee good performance in less than ideal, re-worked, conditions. The end rings are made from braided expanded graphite with carbon filament yarn corners. These give protection against gap extrusion even when valve clearances are large. High density expanded graphite discs follow, which act as a permeation barrier. The central ring of the set is made of coated high density expanded graphite adapter rings and a low density expanded graphite sealing ring with a special friction reducing impregnation. The central ring is "wedge-shaped" externally and can adapt to the spindle surface easily even in reworked valves were the actual dimensions might differ significantly from the nominal stuffing box dimensions.

Figure 18.34 Packing set for older valves for fugitive emissions
Courtesy of Burgmann Industries GmbH & Co KG

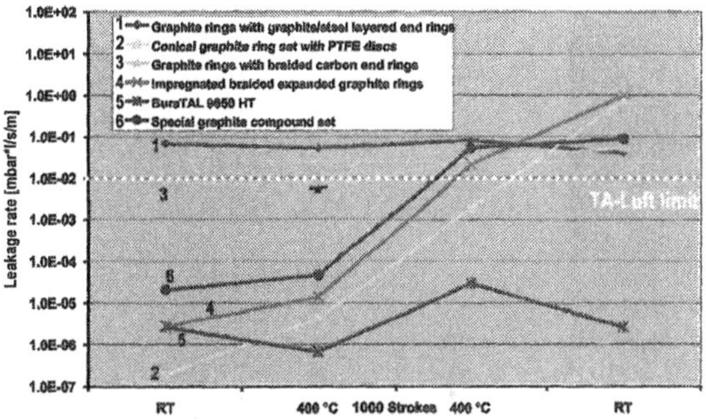

Figure 18.35 Leakage performance of packing set
Courtesy of Burgmann Industries GmbH & Co KG

Figure 18.35 shows the leakage performance of the set during tests using an in-house test stand against competing packing sets from other manufacturers, testing with helium at 40 barg. The packing set was the only one which performed consistently below the TA-Luft limits, without additional spring loading over 1000 spindle cycles. All other packing sets showed higher gland pressure losses and failed to meet or sustain the TA-Luft limit.

In combinations this gasket material is used for standard flange sealing applications. It is made from expanded graphite with an internal impregnation and reinforced with an expanded stainless steel inlay. The gasket can be used as a standard for all flange connections. For specific requirements, solutions based on spiral wound gaskets or grooved gaskets are also available.

All of the above described sealing solutions have also been tested and certified by independent and internationally recognised test institutes, which are approved to issue relevant certification. All the TA-Luft sets described have been certified independently by the MPA (Material Prufanstalt) in Stuttgart.

Life cycle cost of high performance sealing sets

The products described so far are "high grade" products, and require a higher initial outlay in comparison to standard packings, usually supplied in length form. Grade is a ranking concept for the relative standing of competing products [6]. It is based on planned and recognisable differences and features, which may or may not exist. Price is often used by consumers to judge the differences between product grades. A consumer may judge a product to be over-priced if its grade is too high for their needs. A customer must be able to see the value of all the features the product has to offer, in order to justify the purchase of a "high grade" sealing set. It is not enough just to prove that the features exist or are effective.

High grade products cannot compete with low grade products such as cheap packing supplied in length form simply on acquisition costs, so the focus must be on the sustaining or "advantage" costs. These costs combined, with the disposal costs, represent the life cycle cost of the product.

The sustaining costs associated with a packing set are related to issues such as:

- Product leakage
- Maintenance and retightening regime
- Environmental pollution and resulting costs/fines
- System downtime due to seal failure

If a customer considers the full extent of each product's life cycle cost, the "high grade" sealing product is no longer seen to be over-priced compared with a low grade solution.

Experience based on valve refurbishment

Over the last few years much experience has been gained by working together with large chemical and petrochemical plants, both in Germany and throughout Europe, during shut-downs and upgrade projects to meet the IPPC and TA-Luft standards. The co-operation between relevant departments and companies involved proved important in the successful completion of the upgrade.

How organised the end customer is, and how many parties are involved in the plant refurbishment is critical. In general, the maintenance department of the end user manages the project. Often sub-contractors carry out the valve and pipeline refurbishment work. The third party involved is the seal supplier/manufacturer. In most cases the seal supplier works closely with the valve repair contractor during the upgrade and in the planning phase with the customer's maintenance department.

Valve measurements

The first checks made are to see if the valve documentation data is correct. Normally the maintenance department has detailed documentation such as inventory lists, drawings and spare part information. All changes of valve components should also be documented. But in some instances the data is incomplete. If this is the case the following steps have to be taken.

Valve dimensional measurements

A physical survey to establish the stuffing box dimensions has to be undertaken either by the customer or by the seal supplier's personnel. In some cases the valve is not accessible during production so an assumption has to be made based on valve OEM information or similar valves in the plant. The unique identification and marking of the surveyed valves will be imperative for the installation of the replacement packing sets. If the valve is accessible but still in operation the spindle dimension is measured (or the thread). For the housing dimension the gland dimension is taken and a slight addition is calculated. The third unknown factor is the stuffing box depth. If it is not known how many rings are used the maintenance company should be able to turn adapter rings on-site as and when required.

Reworking of valves

If a valve has to be reworked the principle component will be the spindle in most cases. Due to damage, the spindle may have to be replaced or machined slightly smaller. In the case of a smaller spindle the packing set should be able to adapt to the changed dimensions. If this is not possible new sets made from appropriate tooling have to be manufactured. It should also be checked that the permissible clearances are not exceeded after the rework. If this is the case appropriate metal disks should be used which can bridge the bigger clearances. Similar measurements apply for the enlargement of the stuffing box bore due to corrosion problems.

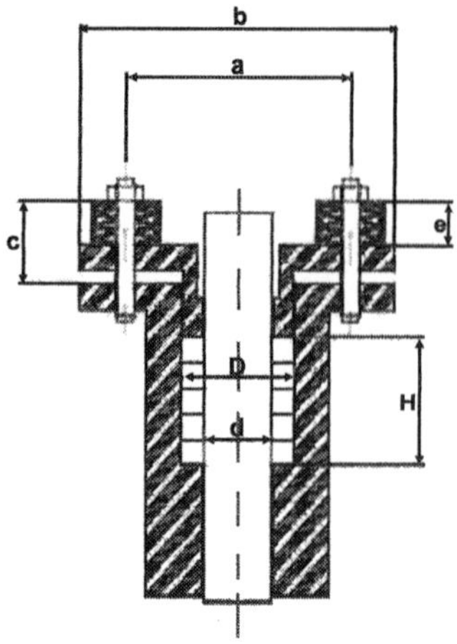

Figure 18.36 Spring-loaded gland arrangement
Courtesy of Burgmann Industries GmbH & Co KG

Determination of spring loading of valve bolts

If the valve is to be fitted with disc springs at the gland bolts, all the dimensions for the spring assembly have to be measured. In some cases longer gland bolts are needed to accommodate the disc spring solution. The protection of the springs by an additional bushing is also recommended to guide the spring stack and to protect the springs against contamination. Figure 18.36 shows a spring-loaded gland arrangement. The gland bolt should be tightened until only a one-millimetre gap between the gland and the bottom of the bushing is achieved. This set-up is very "user friendly" because the proper gland forces are applied and the possibility of over- or under-tightening the set is reduced. Additionally there is a visual indication when the volume loss of the packing set has increased and the springs have to be re-tightened.

There should always be an economic evaluation to see if spring loading is necessary. It is recommended if the valve is hard to access for re-tightening or if long maintenance free periods are required. Also if there are a high number of spindle movements or large temperature fluctuations during operation a spring-energised system can offer more consistent axial force retention.

Installation of sealing sets

Maintenance plan

Detailed installation information and maintenance plans should always be supplied with all sealing sets. All installation forces will be specified to achieve the best sealing performance. If torque measurement is not possible, a specific diagram for the compression-related tightening will be supplied. This information will often be incorporated into customer standards. Experience shows that the maintenance information needs to be easy to understand. Diagrams are more informative than formulae or tables. If the instructions are too complicated they will not be followed and errors can occur, resulting in poor seal performance.

Delivery of sets

The correct packing sets and gaskets can be supplied at an agreed delivery time for the upgrade as a result of the preliminary check and establishment of all valve dimensions. But it is not always possible to check every valve and there will even be dimensional variations when some of the valves are opened. To react quickly in this event, a specialist with the refurbishment company will liase with a project co-ordinator at the seal supplier who can ensure that missing seals or new dimensions can be manufactured immediately to ensure the correct function of the valve in time for the plant re-start.

Differences and problems

In situations where there are problems or dimensional deviations, the experience and skill of the maintenance personnel is crucial. In addition to the specific maintenance and installation instructions, the seal supplier should also provide training sessions for end users or refurbishment companies. A thorough theoretical and practical training will ensure that not only standard situations but also potential problems are covered.

It is also important that there are project leaders at all partners involved in the upgrade. Only these people can organise a quick response and consequent seal replacement in emergencies. A large inventory of tools to cover a wide range of dimensions is also essential. In addition, certain packing set sizes should be available from stock. In many cases tooling or specific packing sets can be adapted to cover slightly different valve dimensions.

Re-work or new valve?

In many instances it is questionable if an old and worn valve should be reworked or replaced by a new one. Experience has shown that most valves can be upgraded to fugitive emission requirements if the work is carefully done. But for smaller valve sizes there is always an economic decision which has to be made. Experience at bigger plants has led to a general rule being adopted: stainless steel valves smaller than DN 50 and carbon steel valves smaller than DN 80 for fugitive emission applications should be replaced by new valves. In general replacement proved to be more economic than refurbishment.

Strategy for valve upgrades

As a general rule based on the above experience there is an easy strategy for valve upgrades. It has been proven over the past few years by upgrades performed with large and small end users in the process industry. The following steps outline the procedure:

1. A competent seal supplier is contacted with the request for reference applications and pricing
2. Technical advice should be sought from the seal supplier. There can be substantial cost savings by using standardised solutions across the plant where possible
3. An offer is made by the seal supplier and the planning phase by end user begins
4. Sealing sets and solutions are ordered
5. Technical details are confirmed and on-site measurements taken as necessary
6. Maintenance or contractor personnel are trained in the correct installation techniques, if necessary
7. Sealing sets are manufactured and delivered
8. Installation by the customer or a service company with the support of the seal supplier begins
9. The seal supplier should be ready to respond rapidly to unforeseen problems
10. Plant can be restarted on schedule!

In all installations there will be additional costs incurred to meet the upcoming environmental legislation. But in most of the cases valves can be refurbished without creating excessive costs using the Best Available Technology (BAT). On the other hand, future costs may be saved due to fewer product losses, longer seal life and additional operational safety.

Conclusion

The experience to date has shown that with a carefully planned approach the upgrading of existing plants to meet the requirements of IPPC or TA-Luft can be fulfilled without any problems in most cases. The co-operation between all partners involved in the upgrade is important. Besides the plant user, these are the seal supplier as well as the maintenance department or specialist maintenance company. Additional measures like planned LDAR (Leak Detection & Repair) programmes will provide further benefits. Emission reductions after valve upgrades can also be directly monitored.

The use of suitable packing rings or packing sets is important to guarantee problem-free operation according to TA-Luft or IPPC guidelines during the lifetime of the valve. The seals should be approved by a recognised test institute to guarantee their technical capability. Also, the proper preparation of the valve as well as the right installation is equally important. Additional measures, like spring loading of the gland, ensure additional performance for trouble free sealing. Despite the increased maintenance costs to meet these new legal requirements there will be cost reductions due to longer service live and increased maintenance intervals. Additionally, reduction of product losses will give benefits in regard to the increased efficiency of the plant.

References used

1. Council Directive 96/61/EC of 24 September 1996 concerning integrated pollution prevention and control, Official Journal L 257,10/10/1996, pp 26 - 40.
2. Erste Allgemeine Verwaltungsvorschrift zum Bundes-Immissionsschutzgesetz (Technische Anleitung zur Reinhaltung der Luft - TA Luft) Entwurf 23.11.2001
3. Sealing Technology - BAT Guidance Notes, ESA Publication No. 014/05, June 2005
4. Th Klenk, R Vogel, *Special non-woven materials for sealing technology - Alternative to braided packings*, BHRG Sealing Conference, York, 2003
5. VDI 2440, Emissionsminderung Mineralölraffinerien, Kommission Reinhaltung der Luft (KRdL) im VDI und DIN - Normenausschuss, 2000-11
6. H. Paul Barringer, *Why you need practical reliability details to define life cycle costs for your products and competitors products*, 16th International Titanium Annual Conference, New Orleans, October 2000

Article — Courtesy of Burgmann Industries GmbH & Co KG

18.9 Final elements in a safety instrumented system (ESD or HIPPS)

Introduction

This article describes the functionality of — and requirements for — isolation valves used in critical safety applications.

ESD history

A well-known term for a safety shutdown valve is: ESD valve, an abbreviation of Emergency Shut Down valve. This valve is typically operated by the ESD system. The designation clearly indicates the function of the valve and therefore it is widely used. ESD is even is repeated in the tag number of the valve.

The function of the valve is to close upon a demand from the shut-down system and thus the valve has a safety function. The term ESD system dates from the time that the safety system consisted of a first layer of defence (ESD system) and a secondary layer, which was the safety relief system (safety relief valves). In case of failure of the process control system, the ESD system would try to shutdown the process. If this failed the relief valves would trigger and reduce the pressure (or temperature) to an acceptable level.

Figure 18.37 Control and safety system consisting of an axial control valve and axial isolation HIPPS valve according to IEC 61508
Courtesy of Mokveld Valves BV

Changing environment

With the present mindset regarding the environment this is of course not a satisfactory solution. Apart from this the industry became aware that the existing *prescriptive* standards describing the above safety system (such as DIN 3381/EN14382), were sometimes over-engineered. Therefore the industry started to design safety systems in a different way. The idea was to make sure that the source of the problem was isolated (rather than relieving the problem to atmosphere). This requires highly reliable equipment while a failure will result in damage to equipment or even endanger the safety of the personnel. In order to isolate the source of the problem a HIPPS system was installed. See Figure 18.37. The High Integrity Pressure Protection System would close the HIPPS valves.

For the design of the safety systems the industry was using guidelines from the prescriptive standards together with company standards. This led to new international *performance-based* standards describing the methods to judge the required safety level for a specific process, and set the requirements of the safety system to achieve that level. These standards are IEC 61508 and IEC 61511.

In these new standards the term ESD system no longer exists. The IEC standards use the term Safety Instrumented System (SIS) for the complete safety loop wherein the valves performing the closure of the source of the problem are called "Final Elements".

And this explains the lengthy title of this article. In a period of approximately 10 years the name as well as the function of the safety shutdown valves has evolved. 10 years ago a relatively simple valve was sufficient for the ESD function while a failure would not directly result in an unsafe situation. However, failure of a final element or - even worse - a complete SIS could well result in damage to equipment or in the loss of lives.

This, of course, changes the requirements for isolation valves used in safety applications. It is of the *utmost importance* that there is a clear understanding what type of safety shut-down valve we are looking at. Or - when it is part of a Safety Instrumented System - that it is clear what safety level is required.

Without going into detail on the IEC standards (and thus with the risk of reducing a complete standard to some one-liners!) the following gives a rough indication of the requirements.

IEC Requirements

IEC 61508 and IEC 61511 are *performance-based* standards. The keywords of these standards are:

- Verification
- Validation
- Documentation

During the HAZOP (Hazard and Operability Study) parts of a project are identified where (for instance) over-pressure could occur. When a control valve reduces the pressure for a lower rated section piping, a means of protection is required. As part of the HAZOP the consequences in case of the over-pressure shall be defined. This shall then lead to what is called a required Safety Integrity Level (SIL) ranging from 0 to 4. The standards advise not to use a SIL4 but they advise to try to improve the basic design and thus avoid the SIL4 requirement. A SIL3 would typically mean that the over-pressure situation might result in an accident with several casualties. The same study may also be performed for the environmental damage in the case of failure to control the process.

Once the required SIL is defined a Safety Instrumented System can be designed. The design of the SIS shall then be such that the requirements of this SIL are met. The standards set limits for the reliability and architecture of the SIS for each SIL. Mainly the last part changes the approach towards the safety shut-down valves in general and more specific to what is referred to generally as HIPPS valves, the final elements in a SIS requiring a SIL3 or sometimes even SIL2.

Failure of these final elements will have serious consequences to personnel or environment and thus the reliability shall be *verified* thoroughly. Several variables of the application may have influence on this reliability and are therefore to be defined prior to the verification of the reliability. Some of these variables are:

- Power supply for the actuator
- Clean or unclean fluids (e.g. gas containing particles)
- The stroking time in relation to the size

ESD valves typically are ball valves, which normally have a closing time of 1-2 seconds per inch. Figure 18.38 shows two axial on-off valves and a ball valve next to each other. This illustrates the different valve solutions for HIPPS/ESD applications. In case the process requires a closing time of 2 seconds, for a 10" valve the ball valve would have to close 5 to 10 times faster than its normal closing time. Increasing the closing speed with a

Figure 18.38 One ball valve for ESD and two axial on-off valves for HIPPS
Courtesy of Mokveld Valves BV

factor 5 to 10 has considerable impact on the mechanical design and thus the reliability of the final element:

- The torque transmitting parts shall be designed to allow faster acceleration and deceleration
- Parts subject to friction shall be adapted while the risk of galling rapidly increases at higher speed
- The actuator shall be able to release its pressure fast enough to allow the rapid movement

Certification in accordance with prescriptive standards, like for example the DIN 3381/EN14382, can be of assistance to assure that the mechanical design of the final element is fit for purpose. It shall then of course be verified that the correct type of equipment is specified and supplied while these standards define different types of safety shut-down valves.

The IEC 61508 and 61511 do not indicate requirements regarding the mechanical design of the final elements, however, the requirement is for "*dependable*" failure rates. *Dependable* means reliability data based on the mechanical design as under evaluation and based on references for this specific application.

A certification in accordance with IEC 61508 and IEC 61511 (usually for SIL3 or SIL2) exists, however this should not be accepted lightly. It is the responsibility of the engineer to *verify* if the data in the certificate can be used for his specific application. SIL certificates shall therefore not be used without reading and *verifying* the report behind the certificate. E.g. failure data based on test cycles in general cannot be used because final elements are not cycling constantly. The final element in a SIS is changing its position only on demand (or test) and we may hope that the safety system does not have regular or even constant demands. For mechanical systems (like valves, actuators and solenoids) there is a huge difference in behaviour between equipment that is constantly switching or staying in the same position for long periods. Therefore the failure rates are not identical. Figure 18.39 shows an integrated SIS for SIL3.

These *verifications* shall take place during the engineering phase and prior to purchasing. For the final element it shall be *verified* that the failure rate as mentioned is applicable for the specific combination of valve and actuator. Most certificates and failure data apply either to the valve only or to the actuator only, which does not give any information of the two parts combined.

In the certificates often a SIL2 or SIL3 is mentioned. However the SIL level shall be considered for the complete loop only. A required Safety level (for the complete loop) is defined and it shall be *validated* that this level is obtained with the different

Figure 18.39 Integrated SIS with redundant pressure switches suitable for SIL3 application
Courtesy of Mokveld Valves BV

components of this loop. The *validation* of the required SIL is not obtained by piling up certificates but by calculating the Probability of Failure on Demand of the complete loop. It is for these reasons that IEC 61508 and IEC 61511 do not recognise certificates and that depending on the SIL level, a third party validation may be required. The different considerations, *verifications* and *validations* during the complete lifecycle of the equipment shall be well *documented* for future reference.

Summary

It can be concluded that over the last years, as a result of IEC 61508 and IEC 61511, the approach towards ESD, or HIPPS valves, has changed completely. The safety requirements have changed and as such what we now call final elements in a Safety Integrated System (rather than ESD valves) require specific attention in the engineering and purchasing phase as well as during the operational phase.

Article — Courtesy of Mokveld Valves BV

18.10 Eliminating plugging of boiler feedpump recirculation valves

Boiler feedpump recirculation

Boiler feedpump recirculation is an extremely arduous duty for a control valve, possibly one of the most severe duties in any power plant. Water passing through the feedpump picks up heat which can lead to cavitation - cavitation is the formation and subsequent collapse of vapour bubbles in the liquid flow stream. If the vapour collapse occurs near a solid component such as a pump impeller or the pipe wall the collapse mechanism can cause damage, damage so severe that components are literally worn away. Protecting the pump against cavitation is possible by maintaining a minimum flow through the pump to minimise any temperature rise. This is accomplished by installing a recirculation system around the feedpump.

In a typical feedpump recirculation system (Figure 18.40), the feedpump recirculation valve takes feedwater from the boiler feedpump discharge and recirculates it to the deaerator in conventional drum style boilers or to the low pressure drum in combined cycle units. Depending on the size and type of plant, the valve may be reducing an inlet pressure of 379 barg to about 10 barg. This large pressure drops cause high-energy cavitation that will very quickly destroy a standard control valve trim. In some plants, the recirculation line runs to the condenser, which creates an even higher pressure drop for the valve to handle, as a result the valve has to handle both cavitating and flashing conditions.

A main cause of failure of the feedpump recirculation valve is due to particulate in the flow stream. If pipes are not cleaned properly following construction then weld slag and other debris can become lodged in the narrow passages of today's drilled hole or stacked disk cages (Figure 18.41). In older plants, there

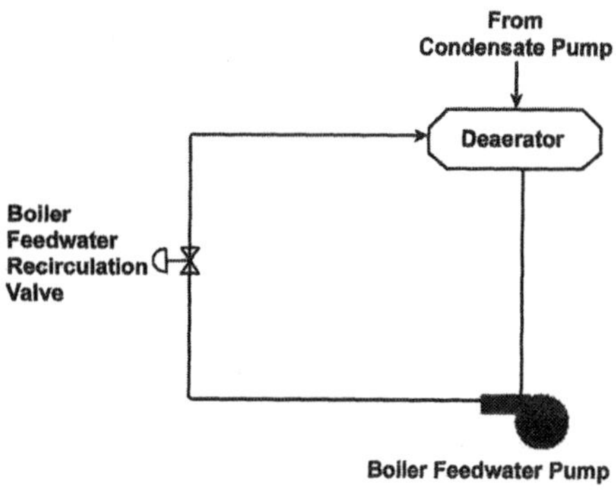

Figure 18.40 A typical feedpump recirculation system
Courtesy of Emerson Process Management

Figure 18.41 Weld slag and other debris lodged in drilled hole or stacked disk cages
Courtesy of Emerson Process Management

can also be issues regarding iron oxide build up or even particulate from degrading upstream equipment. Even if the particulate is not large enough to become lodged in the cages it can cause severe erosion damage to the throttling and shutoff surfaces of the valve trim.

Damage to the shutoff surfaces leads to decreased unit performance as flow that should be producing steam is routed back through the main feedpumps, leading to excess pumping requirements as well as causing severe damage to the valve body. Recent experiences have shown that leaking recirculation valves can cost a plant up to $275,000 each year in lost capacity and maintenance costs.

Dirty Service Trim

A unique solution to this issue is achieved with the use of the patented multi-stage, anti-cavitation control valve trim, or Dirty Service Trim (DST) (Figure 18.42). The DST eliminates cavitation in applications with entrained particulate that could plug the inlet passages or cause severe erosion damage to conventional anti-cavitation trim. The DST design uses a combined axial and radial flow path that features large openings enabling flowing particulate up to 20mm in diameter to pass through.

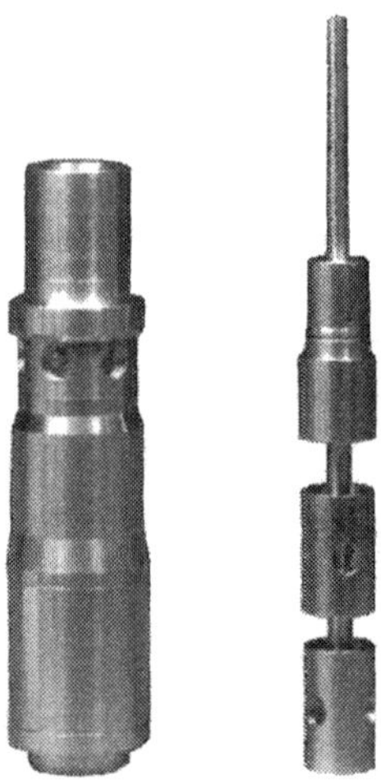

Figure 18.42 Dirty Service Trim DST
Courtesy of Emerson Process Management

This design incorporates a protected seating surface separating the shutoff and throttling areas. All significant pressure drop is taken downstream of the seating surface so as a result, the seating surfaces are not worn away by throttling control action. This results in improved shutoff performance over time. In addition the throttling areas are not required to have the superior surface conditions otherwise needed for tight shutoff.

As can seen in Figure 18.43, the DST staged pressure reduction design takes the majority of the pressure drop in the initial stages of the trim, dramatically reducing the available energy of the fluid leaving the final stage. This eliminates the possibility of high-energy fluid impingement on the valve body, and in cases where there is entrained particulate, erosion damage to the valve body and downstream equipment is eliminated.

In standard anti-cavitation trims (Figure 18.44) there can be a flow between the plug and the cage, past the seals, this flow takes the full pressure drop and as such can cause high velocity impingement on the seating surfaces resulting in poor control and loss of shutoff. In the DST trim clearance flows are subjected to a staged pressure drop eliminating this problem.

Applications

A power plant in Michigan, USA had to bring their 450 MW unit down approximately every other month because of plugging to the small (1.27mm dia) holes in the recirculation valves anti-cavitation trim. A DST solution was installed in the existing valve and it has operated successfully for in excess of seven years without any need for cleaning or maintenance.

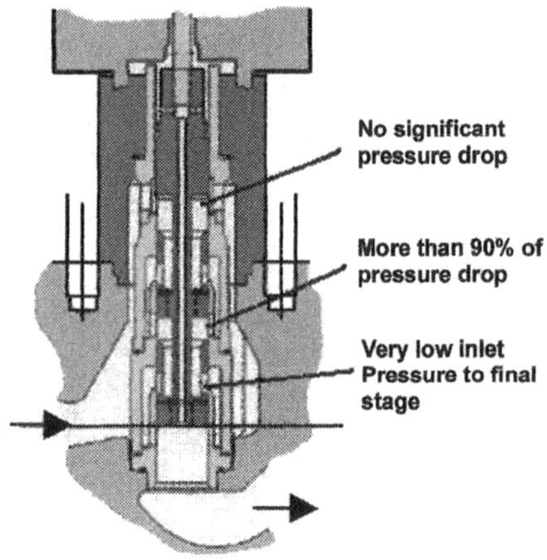

Figure 18.43 Staged pressure reduction design
Courtesy of Emerson Process Management

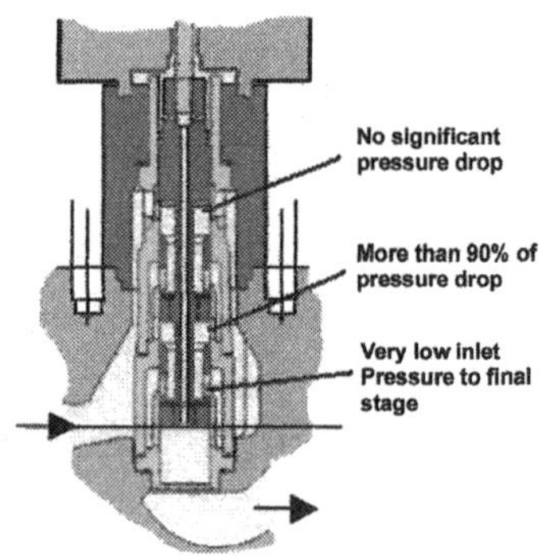

Figure 18.44 Standard anti-cavitation trim
Courtesy of Emerson Process Management

A DST solution was also retrofitted to three existing valves in a nuclear power plant. The plant was attempting to increase the flow rate through the condensate recirculation valves, but the existing valves could not provide the necessary capacity. Because of its large open flow paths, the DST solution provided the required capacity and cavitation protection, while other offerings required the installation of larger valves.

Additional problem applications

There are many other applications that require cavitation protection while being able to pass large particulates. One such application is water injection recycle valves in gas processing plants. These valves often have as much as a 310 barg pressure drop but can be required to pass a sludge-like material.

Figure 18.45 shows before and after pictures of a trim set removed from a gas plant in Syria. DST was installed in the valves in 1997 after the previous trim was plugging every month. The pictures seen were taken four years later when the valves were opened for the first time since installation of the DST solution. The picture on the top shows a large build-up of sludge, but the valve was still able to pass the required amount of flow. The picture on the bottom shows the same trim set after it was cleaned for reinstallation. Notice the damage-free trim components.

Hydrotreatment service is yet another critical application requiring a control valve to address issues with severe erosion and entrained particulate. The two main applications are the hot high pressure separator letdown and cold high pressure separator letdown.

Figure 18.45 Before and after pictures of a trim set removed from a gas plant
Courtesy of Emerson Process Management

In these applications, cavitation, flashing and outgassing can occur during the pressure reduction phase that separates hydrogen and recycles gases from the main product streams. When flashing or outgassing occurs in a control valve, erosion problems are introduced due to entrained liquid droplets in the high velocity gas. In many cases, erosion damage can be caused by catalyst fines entrained in the process stream. Corrosion damage is also a concern because of the presence of H_2S and NH_3.

The DST counteracts the possible erosion damage by incorporating multiple expanding stages that compensate for the volumetric expansion of the entrained gases. Each stage reduces the velocity of the fluid. Proper material selection will also eliminate the occurrence of any chemical attacks.

Conclusion

As demonstrated by these additional examples, the use of large expanding flow passages stages the pressure drop required to eliminate damaging cavitation while also reducing the amount of available energy exiting the trim. By using larger passages, issues with plugging and valve component erosion are eliminated while generally providing more capacity than conventional drilled hole or stacked disk technologies. Along with these issues, tight shutoff can be maintained due to the protected seating surfaces.

There are many issues that can arise when applying control valves to severe service duties, particularly when there is the potential for cavitation. These issues are not always understood or appreciated by end users or specifiers. In order to ensure these issues are correctly addressed it is necessary to approach a supplier that offers more than one anti-cavitation technology, this will help to ensure that the solution you get will give trouble free operation.

Article — Courtesy of Emerson Process Management

18.11 Steam trap isolation ensuring uninterrupted steam flow

High pressure and high capacity steam traps are common where there is a requirement for robust, long life service in a high pressure, high capacity environment such as refineries, power plants and other supercritical plant applications.

In these circumstances continuous operation is paramount during routine maintenance. A special packaged unit has therefore been developed which enables the steam trap to be isolated from the system thus allowing routine maintenance to be carried out without interruption of the steam flow.

Figure 18.46 shows the packaged unit, which is available in pressures from about 10 barg (150 psig) up to 170 barg (2500 psig). It combines a steam trap, two bonnet-less globe valves and a bypass pipe connection specifically designed for applications where operation of the plant is normally 24 hours a day, seven days a week. This fabricated one-piece unit allows

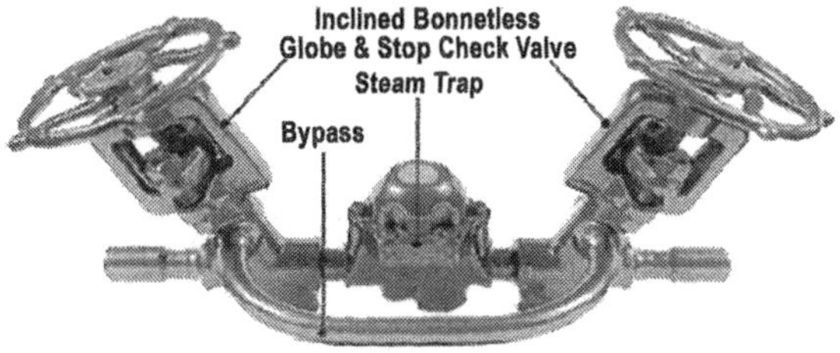

Figure 18.46 The packaged unit
Courtesy of Velan Valves Ltd

maintenance to be carried out to the trap in complete safety by bypassing the trap and using the valves. It also offers a double protection against back pressure which can result in significant losses of steam and also can lead to back up of condensate in the system which in turn can lead to high levels of inefficiency. Problems such as water hammer can easily damage expensive plant equipment such as turbine blades and in some cases can cause death and injury to site personnel due to pipe ruptures.

The trap is forged and also fitted with a fitted integral strainer, automatic air vents and check valves. The internals are all stainless steel and the seats are 316 Stellited for longer life cycling. The trap can also be fitted in any position - horizontal, vertical and angle. The unit is also designed to cope with all degrees of superheat above saturation temperature with no loss of thermal pull on closure. In fact the more superheat the tighter the shut off. Another advantage is that it will start to open 4.4°C below saturation and will be fully open 22°C below saturation to drain condensate.

Conventional installations

In a normal pipe condensate drainage system, configuration around a high pressure, medium or low pressure steam mains would probably resemble something similar to the illustration in Figure 18.47. As can be seen the plant requires equipment and material such as a Y-strainer, 3 valves, a trap, check valves and piping, elbows, tees, unions etc.

Fabrication and heat treatment of the welds on site adds increased time and labour to the job.

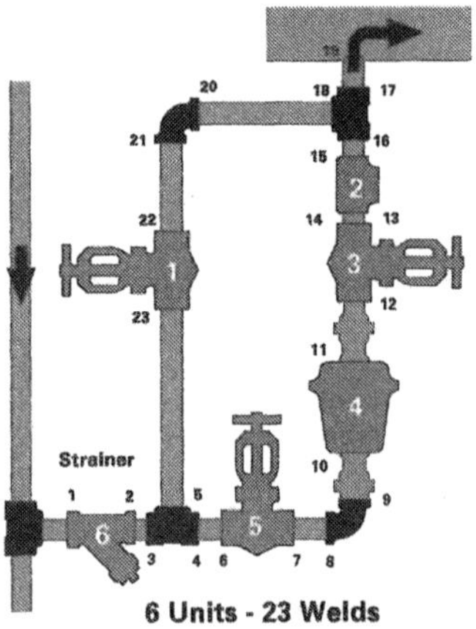

Figure 18.47 Conventional bypass set up (3 valves) and bucket trap
Courtesy of Velan Valves Ltd

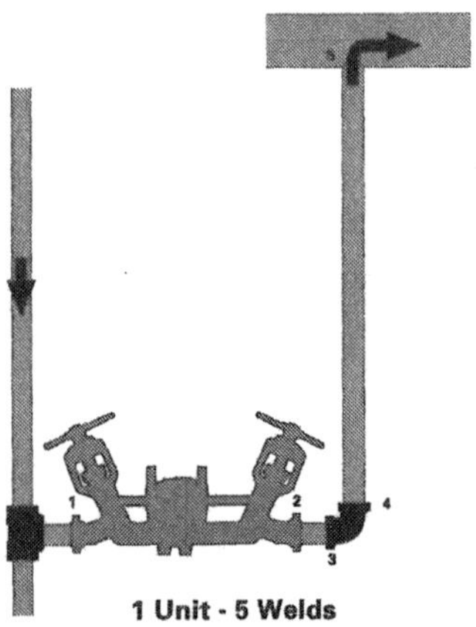

Figure 18.48 Packaged set up with bypass (2 valves only)
Courtesy of Velan Valves Ltd

The equipment used is probably manufactured by several different suppliers therefore leading to increased inventory, paperwork, spares and records etc.

The packaged system

The packaged system developed is shown in Figure 18.48 It is available in various materials A105 N, F22, F91. It is ISO 9001 accredited and is designed to ASME VIII boiler and pressure vessel code. Fabrication into a similar installation can be completed in a short time compared to the many many hours required in conventional systems using all of the loose equipment and material listed earlier. There are only five welds to heat treat and, since only one unit is supplied by one supplier, no extra inventory or records for other equipment is required.

How it works

Automatic steam trap operation (Figure 18.49A)

In normal mode the inlet and outlet valves are in the closed position and this allows the trap to operate normally. The trap will shut off steam and drain condensate automatically when required. The inlet and outlet valves are both in the top closed position offering double protection against bypass leakage. As long as the trap has been sized correctly to handle the conden-

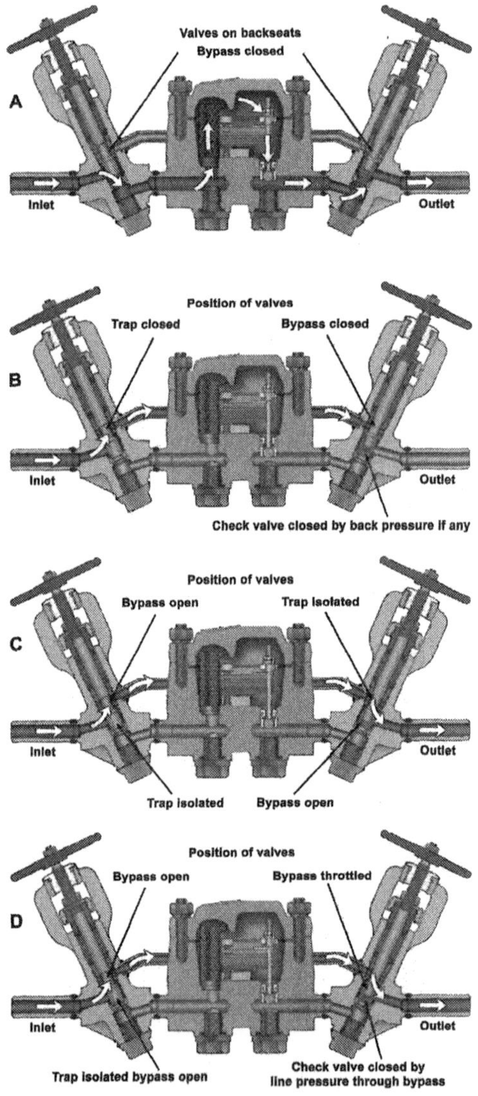

Figure 18.49 Method of operation of packaged unit
Courtesy of Velan Valves Ltd

sate flow, the unit can be left in this position during start up and normal plant operation.

With the bypass valves normally closed, the Stellite and lapped discs and seats assure 100% tightness, thus reducing packing maintenance costs. On start up the trap can also be isolated as the bypass can handle three times the condensate capacity of the trap and then be switched to trap operation during normal operation. It can also be actuated to ease operation.

Complete shut off – No flow, trap closed for service (Figure 18.49B)

The inlet valve is in the bottom position. The trap is now sealed off by the inlet valve and the bypass is closed by the outlet valve. The valve is protected by the check valve portion of the outlet valve.

Bypass open – Full flow, Trap isolated for service (Figure 18. 49C)

This is the preferred service position. With the inlet and outlet valves in bottom position, the trap is protected against back pressure by the check valve disc in the outlet valve and the check valve in the steam trap. The trap is therefore safely isolated for service without interruption to flow or plant operation.

Throttled bypassing – Trap isolated for service (Figure 18.49D)

The inlet valve is in the bottom position and the outlet valve is in the intermediate position. The trap is sealed off by the inlet valve and flow through the bypass is restricted by the position of the outlet valve. The floating check valve of the outlet valve protects the trap from back pressure.

Summary

This patented packaged unit is capable of providing substantial savings in component parts and installation costs and is being used extensively in installations such as: powers stations, refineries, and marine and similar supercritical applications where continuous operation is essential during routine maintenance.

Article — Courtesy of Courtesy of Velan Valves Ltd

18.12 Increased valve actuator functionality in the digital age

Introduction

The valve actuator plays a key role in the reliable, efficient and safe operation of fluids handling processes, particularly now that plants are predominantly automated and computerised (Figure 18.50). The duty of the valve actuator can be divided into three areas:

- Valve operation
- Communication with the controller
- Diagnostics

Using an intelligent electric actuator as an example, we can see how the advent of the digital age has increased the functionality and usefulness of the actuator in all three areas.

Valve operation

The basic function of the actuator is to operate the valve mechanically. Simplicity is the answer to achieving long term reliability here, in this case utilising a single worm and wheel arrangement in an oil bath.

Most valve actuators are inactive for over 90% of the time and during these periods their function is to communicate the position of the valve and their availability for operation. However, when commanded to change the valve position the actuator must respond with accuracy and reliability, especially since some valves only open or close for safety reasons and others are required to be precisely positioned to correct the flow in the process stream. It is also essential that the actuator communicates a signal to confirm that the requested action is being performed, or generates an alarm if it is not or cannot be completed. All these functions are generated by the control and instrumentation elements inside the actuator, which must be protected from the outside environment to ensure long term reliability.

The majority of valve locations are exposed to environmental challenges to a greater or lesser degree, often involving the effects of water — spraying, flooding and even submersion. It is essential therefore that the electrical components in particular are protected from the corrosive effects of the outside atmosphere. The actuator utilises "O" ring seals and its terminal compartment is separately sealed (double-sealed), so that even with the terminal cover removed water cannot reach the internals of the actuator.

Reliability is further enhanced by the "non-intrusive" actuator design. With this technology actuators now use an intrinsically safe, hand held setting tool – rather like a TV programmer – with a secure, dedicated wireless infra-red link to perform all the switch setting and commissioning functions that were previously only achieved by removing electrical covers. See Figure 18.51.

There are significant advantages inherent in this technology. Using the menu on the actuator display screen it is quick and easy to commission the valve, in any weather – day and night – without jeopardising the actuator's integrity and with complete safety even in hazardous areas. The same tool may also be used to download this data and upload it to other actuators with similar commissioning requirements to simplify and speed up the operation, or to download actuator operating data for analysis, as explained later.

Figure 18.50 Valve actuator installation on a petroleum tank farm in China
Courtesy of Rotork Controls Ltd

Figure 18.51 Intelligent valve actuator and "non-intrusive" hand held setting tool
Courtesy of Rotork Controls Ltd

Figure 18.52 Master station, capable of controlling up to 240 actuators on a single two-wire loop
Courtesy of Rotork Controls Ltd

"Non-intrusive" commissioning and communication would not be possible using conventional electro-mechanical switch mechanisms, limit switch counters and torque sensing devices, so it is the replacement of these devices with solid state electronic developments that has facilitated this advancement in reliability and user-friendliness.

Communication with the controller

Digital technology and the advancement of solid-state electronics have given the valve actuator the ability to store and communicate a proliferation of control and indication data – see Figure 18.52. Traditionally, communicating this data to the control room was performed by "hard-wiring", whereby a separate wire or two was required for each control or monitoring function. Typically, up to 26 separate terminal connections were necessary on an electric actuator, but the development of digital bus control technology has radically changed the situation.

Digital bus systems replace the multitude of conventional hard wired control cables with a single loop or strand of two or four wire cable linking multiple numbers of actuators together. Basically, each actuator is given a unique address code in order to recognise or ignore the data instructions that are transmitted along the bus from the controlling element.

The above is an over-simplified description to be universally applicable to the many and diverse systems and protocols that have been introduced, but with all of them installation costs are dramatically reduced. Furthermore, digital communication fully exploits the information gathering potential inherent in intelligent actuators whilst enabling rapid plant upgrades and simplifying the work involved in altering plant functionality. Engineering work is greatly simplified and commissioning times are reduced.

The last two decades have witnessed the development and introduction of many public or "Open" digital systems including Modbus, Profibus, DeviceNet and Foundation Fieldbus. These various standards have been passed into the public domain, documented and controlled by the IEC and other authorities in the USA, Europe and other parts of the world. The advantage promoted by open systems is the ability to mix and match equipment from several different manufacturers on the same bus highway. For example, a Profibus network will often support actuators, flow sensors and motor drives on the same data highway.

There are, in addition, a number of proprietary systems that are designed for a particular item of plant equipment. The advantage of a proprietary system is that the network can be optimised to exactly match the needs of the elements attached to it. For example, they can cope with extremely long data highways – up to 20km – without the need for any repeating devices, facilitating installation in the spacious environments of refineries, tank farms and large sewage treatment plants where actuators are typically used. They can also cope with the removal of power from any of the actuators on the system at any time, and protect against cable failures by providing dual communication paths (both directions) at all times. Open systems can do this, but only if they are extensively modified from their existing design.

Figure 18.53 Uploading commissioning data on an actuator installed in a UK water filtration plant
Courtesy of Rotork Controls Ltd

There are therefore many factors that need to be taken into account when deciding between the use of a proprietary or open system of digital control. These are often predetermined by industry preferences. For example the water and waste treatment industry has adopted the Profibus protocol in many areas whilst Foundation Fieldbus is predominant in oil and gas production and processing industries. In fact a large number of tank farm installations have adopted two-wire digital control technology. In all cases, using the technology has improved functionality through the ability to both control and receive/exchange/update operating data with the valve actuators in the plant.

Diagnostics

Unlike valve operation and communication, diagnostics is an area of valve actuation functionality that has been introduced, rather than enhanced, by digital technology. The inclusion of a data logger within the actuator enables an event-by-event history of valve activity to be generated, including the torque profiles produced during each opening and closing of the valve. These can be compared with the valve torque signature profile logged during the commissioning process to identify the trend of valve operating wear.

Using the hand held setting tool, the user can extract and store this information, which can be downloaded and viewed on a PC running appropriate software. Alternatively, a laptop running the software can be directly linked to the actuator on site. See Figure 18.53. The data collected is analysed to identify potential valve problems, tight spots and changes in the torque profiles which enable preventative maintenance to be planned and unexpected interruptions to plant operation avoided. These abilities also enable plant utilisation to be optimised by the implementation of "just in time" valve maintenance schedules.

Case studies

Two case studies illustrate how users can quickly and conveniently investigate the operation of actuated valves for purposes such as preventative maintenance, fault diagnosis and asset management.

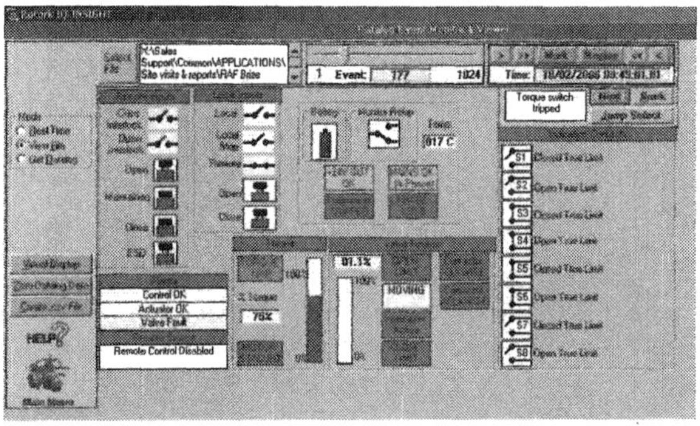

Figure 18.54
Courtesy of Rotork Controls Ltd

In the first, a plug valve operated by an actuator on an aviation refuelling line is displaying an intermittent failure to fully close.

Analysing the data logger event log confirms that the actuator sometimes torque trips before reaching the fully close position. The coloured screen shows a torque trip event (Event 177) on 18 February 2006 at 00.49 hours. (See Figure 18.54.)

Analysing the data logger torque log shows the valve torque profile. The relatively high closing torque profile indicates a valve problem, confirmed by the valve maker. (See Figure 18.55.)

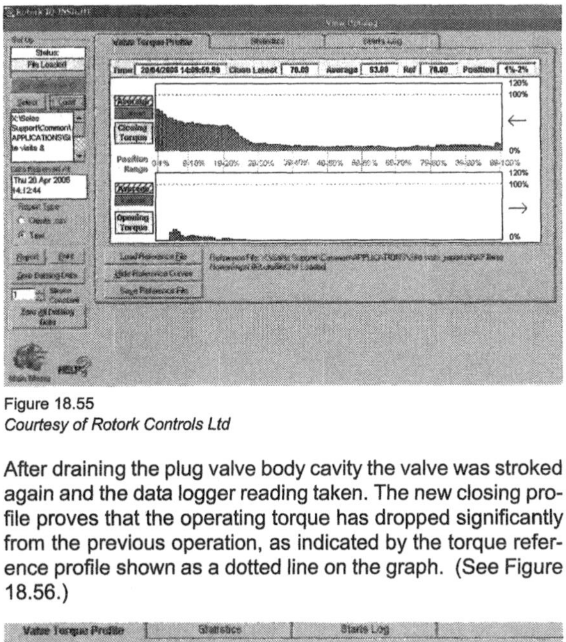

Figure 18.55
Courtesy of Rotork Controls Ltd

After draining the plug valve body cavity the valve was stroked again and the data logger reading taken. The new closing profile proves that the operating torque has dropped significantly from the previous operation, as indicated by the torque reference profile shown as a dotted line on the graph. (See Figure 18.56.)

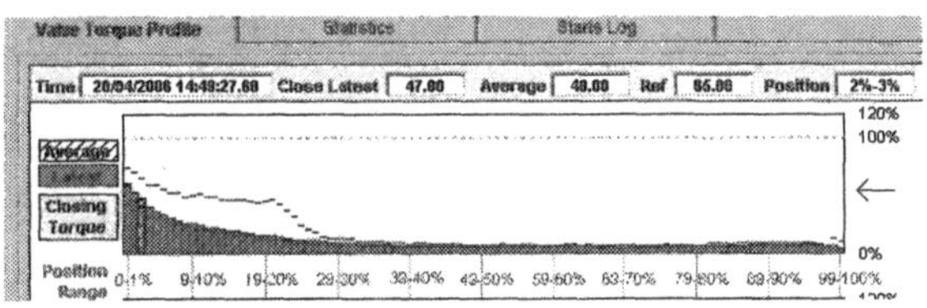

Figure 18.56
Courtesy of Rotork Controls Ltd

The second case study involves a modulating actuator on a steam control valve in a power station. Valve and actuator operational parameters are investigated for process control analysis and maintenance requirements.

Analysing the data logger starts log shows the opening and closing operation (starts) against the valve position. This indicates where the process control requires the valve to be positioned by the modulating actuator over the operational cycle of the boiler unit. (See Figure 18.57.)

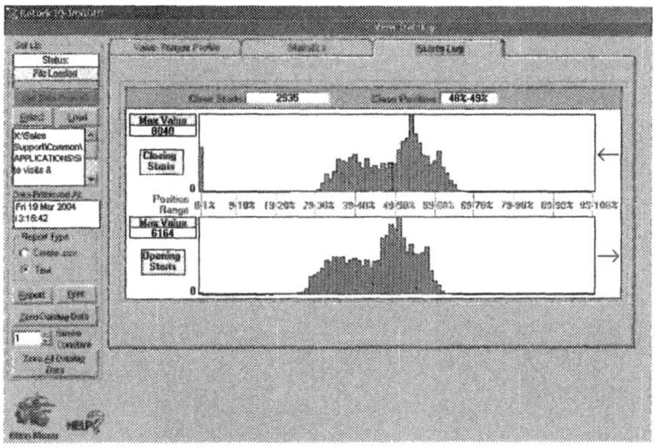

Figure 18.57
Courtesy of Rotork Controls Ltd

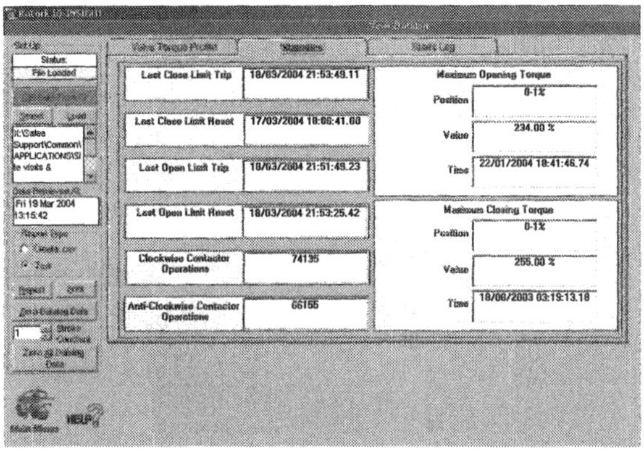

Figure 18.58
Courtesy of Rotork Controls Ltd

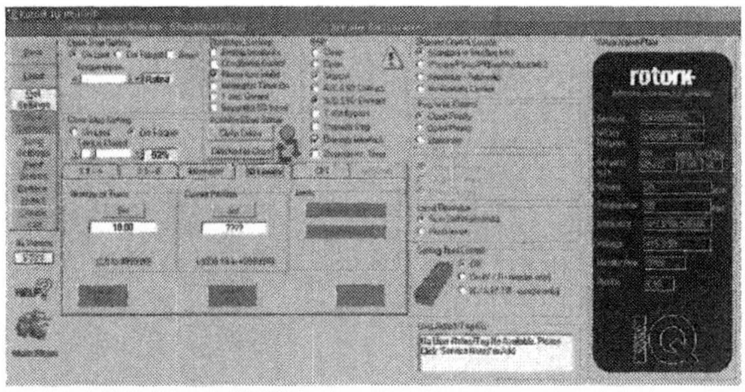

Figure 18.59
Courtesy of Rotork Controls Ltd

The statistics tab of the data logger shows that the actuator has completed over 74,000 close positioning starts and over 66,000 open positioning starts in the nine months since installation. This is well within the maintenance cycle life of the actuator, indicating that it can continue without requiring attention. (See Figure 18.58.)

The actuator configuration is downloaded and saved to allow its type and setup to be recorded as part of the station's asset management system. The file can be viewed for maintenance purposes and uploaded to replacement actuators to replicate the setup configuration. (See Figure 18.59.)

Article — Courtesy of Rotork Controls Ltd

Valve and actuator selection 19

19.1 General operating conditions

To consider any type of valve for any application the following information must be known:

- flow
- temperature
- pressure
- differential pressure (when throttling)
- constituents of the fluid
- properties of the fluid
- ambient conditions

19.1.1 Pressure

When the operating pressure is sub-atmospheric — a vacuum, the packing box configuration and the connection seals should be considered carefully. Leakage of air into a process system is much harder to detect than process fluid leaking out.

When the process system is pneumatic vacuum, air leakage is an inconvenience and represents increased energy consumption.

When the process system is a liquid supply to a pump suction the introduction of air can result in loss of prime, cavitation type symptoms and loss of capacity.

When the process fluid is hazardous the introduction of air can be dangerous.

When the process system is hygienic the induction of atmospheric air can result in contamination with consequential loss of production, and revenue.

Constituents of the fluid refers to solids, trace elements and other phases. Valves selected on the basis of clean fluid may wear very rapidly if abrasive solids are present. There is no acceptable level of solids in clean fluid. The naked eye can only detect particles over 40mm. Particles between 5 and 40mm can cause a major wear problem.

Wear and erosion due to solids is exacerbated by fluid velocity; the material removal rate will increase in proportion to the fluid velocity raised to a power between 2.5 and 5. Small increases in velocity result in a large increase in wear. Wear problems due to undeclared solids will be very apparent in valves used for throttling. Large solids will present problems when trying to close the valve.

Trace elements can pose corrosion problems. Minute quantities of corrosive elements can attack valve materials. Very important valve functions and seat and spindle/stem sealing, can be quickly ruined by unexpected corrosion. Erosion of the protective oxide layer combined with trace element corrosion can prove costly to rectify.

19.1.2 Constituents of the fluid

Liquids are usually considered as single phase. Most users ignore the fact that liquids can include dissolved and entrained gases. Gases and vapours may also include liquid droplets. Dissolved or entrained gases may be hazardous to personnel or corrosive, or both. Hydrogen sulphide, H_2S, is a common example of a very toxic gas which is also corrosive. Air dissolved in liquids can be a source of oxygen for corrosion processes. Dissolved and entrained gases can become very important in throttling and expansion processes when the capacity of the valve can be seriously impaired. Liquid droplets in high velocity gas and vapour streams can act in a very similar manner to solids and cause severe erosion.

19.1.3 Properties of the fluid

The properties of the fluid can be of crucial importance when selecting the correct type of valve. The vapour pressure, when operating close to the boiling point, can be important for valve sizing and trim selection. If the pressure differential across the valve is high enough to make flashing or cavitation a possibility then the valve selection must be made accordingly. Flashing can reduce the valve liquid capacity dramatically. Cavitation can result in rapid valve destruction. Unexpected cavitation in liquid systems results in the removal of metal from important locations within the valve. Initial problems include loss of seat sealing and control accuracy. Material can also be removed from the inside of the valve body leading to pressure containment problems. Special valve trims are produced to cope with cavitation. If cavitation is not suspected, and not specified, a completely new valve may be required to allow successful, economic operation.

The properties of fluids can be changed by the inclusion of solids. A high proportion of solids can change a Newtonian fluid into a non-Newtonian emulsion, slurry or pulp. The viscous performance of the "fluid" in the piping system will have a direct bearing on the operating pressures of the valve.

19.1.4 System operating scenarios

Any variations in the operating conditions must be quantified. For any specific variation, the other fluid parameters must be stated. The duration of normal operation must be specified. Changes in operating conditions are always assumed, by manufacturers, to be slow and slow enough not to cause thermal shock, instability or surge. Rapid changes in temperature can create distortion. This may result in valves jamming wide open or closed. Heavy process valves can suffer body failure due to cracking. Quick acting valves can create flow surges which result in water hammer; very high internal pressures which can exceed the hydrostatic test pressure. **Any** rapid changes in operating conditions must be evaluated; this includes warming up and cooling down before and after process cycles.

There is no standardisation on valve operating regimes. But standards for pumps, however, provide definitions for duty cycles, and the following descriptions may prove useful:

continuous — running for a minimum of 8 hours per day most days, without stopping, at any of the specified operating conditions,

light — running for 3 hours or more, but less than 8 hours, per day most days, without stopping, at any of the specified operating conditions,

intermittent — running for less than 3 hours per day most days, without stopping, at any of the specified operating conditions,

cyclic — if the flow, head/pressure, temperature, viscosity, density or mixture proportions, change regularly in response to a pre-programmed sequence, the purchaser should describe the proposed sequence and the rates of change of the parameters,

irregular — does not operate on most days. The purchaser should describe the predicted operational sequence.

Valves fitted to systems which operate irregularly would seem to have a better chance of lasting longer. This may prove to be a very costly oversimplification. Components in systems with stagnant liquids can corrode very quickly. Stagnant water can promote the growth of micro-organisms which can seriously impair valve functions. Systems which partially drain between cycles can endure very high corrosive conditions.

System operating scenarios must include the possibilities of extreme control functions, i.e. all control valves closing to minimum flow or opening to maximum flow. This type of control sequence can be a problem with regulators and thermostatic valves which operate automatically without external control inputs or instrumentation outputs. Thermal "run-away" or overheating can be experienced when all thermostatic valves close.

19.1.5 Economics

When selecting valves for a range of applications on one site, select the valve which is best for the application, taking into account whole-life costs. Try to standardise on certain sizes, even if this is not always possible, to reduce the spares inventory. Even when different types and sizes of valve are used it is possible to standardise on packing.

The skill level of local personnel must be considered when evaluating actuated and control valves. Simple electric and pneumatic systems should be within the capabilities of all maintenance crews. Complicated hydraulic and electronic systems may require specialist training and extra equipment or even the presence of a service engineer from the manufacturer. Maintenance skill requirements should be considered within the total whole-life costing. Payment for factory service engineers plus the possible loss of production due to travelling time may rule-out complicated systems.

The benefits of complicated systems, including modern communication buses, must be accurately costed. The increased productivity possible, together with better process control, resulting in a more consistent product is extremely attractive. The benefits may be sufficiently large, if target reliability is achieved, to pay for the specialist training and equipment or a factory service engineer. All the fringe benefits, including self-diagnostics and reporting, must be taken into account.

There may be certain cases when the properties of the fluid are uncertain. When this problem arises it is best to discuss the effects of variations, on valve performance and reliability, with the valve manufacturer or a consultant. Values must be placed on uncertain parameters. If this is left to the discretion of valve manufacturers, they will chose values which are best for their particular valve. The user must analyse the problem and chose reasonable values to describe a safe operating "envelope".

19.1.6 Ambient conditions

The conditions external to the valve are often overlooked. Valves can operate in warm, air-conditioned pump rooms or on cellar decks of offshore oil platforms, in the spray zone, with temperatures rarely above 5°C. Valves operating in the desert can be subject to direct sunlight, black bulb temperatures of 85°C daily, with occasional sandstorms to remove paint, lubricant and corrosion products. External corrosion/erosion of the spindle/stem can seriously impair sealing. Material selections and style of construction must be suitable for the environment as well as the process. The valve "overall life-style" must be considered.

19.1.7 System designer

The system designer shoulders the ultimate responsibility for valve selection and the selection of all the other equipment in an interconnected system. It is up to the system designer to define all the pertinent operating conditions for the valve "environment" to ensure all the interconnected equipment is compatible and to arrange all the equipment so that the piping designer can fit appropriate pipes between adjacent equipment. The ultimate success of any system is dependent upon the skill and expertise of the system designer.

19.2 Valve selection according to fluid

19.2.1 Direct selection

For many applications certain types of valve have become the standard and these valves, with few exceptions, are the most suitable. Applications can be designated by fluid or industry. Table 19.1 is a direct selection table for specific fluid properties and refer to the general valve types described throughout *Valves Manual International*. Check with Chapters 3 to 7 for descriptions of the specific valve types. Suppliers and manufacturers are listed in the Buyers' Guide, Chapter 21.

The applications refer to fluids which are generally not corrosive. Corrosive applications require detailed fluid knowledge to allow suitable valve materials to be selected.

Table 19.1 can result in too many alternative categories being offered. In this situation the selection should be reinforced by more information. In particular flow, pressure and temperature can be important.

With approved valves for low pressure fluids with a low concentration of small solids, special attention must be paid to the use of the correct materials.

For approved valves for high pressure fluids with a high concentration of small solids, special attention must be paid to the selection of appropriate materials bearing in mind the low strength of popular austenitic stainless steels.

Valves for fluids with large solids, irrespective of concentration and pressure, belong to the same group listed at the end of the Table. Special attention must be paid to the possibilities of impact damage. Special valves and air-locks are manufactured specifically for system which transport light solids using air. Read Section 3.3.6 in Chapter 3.

19.3 Check lists for valve purchase specification

Many problems and unnecessary costs can be avoided if the following check lists are used. Irrespective of the method of purchase, "off-the-shelf" or by enquiry, the following data should be collated and reviewed by all interested parties. Although more than 100 check items are included under the main groups, they cannot be considered as entirely exhaustive.

In some specialist applications attention must be paid to additional factors. Above all, the check lists should be used when planning and purchasing valves for unusual, critical or difficult fluids. Particular attention should be paid to valves operating at an extreme condition. For general valves, in most industries, a check-list and specification can be drawn up quickly. Terms described as "rated" relate to the valve's normal operating conditions.

The data collated must be accurate. When accuracy is not possible, the data presented must be suitably qualified. There must be no ambiguity. The data collected applies to the valve, not necessarily the system. In variable flow systems the operating conditions at the valve may vary; **do not** specify only the maximums. Every parameter which has a maximum value also has a minimum value. The minimum may be much more important than the maximum; consider -40° to +30°C. The lower value is very significant for many material options; the maximum value is acceptable for virtually all materials. When the transition between values is not slow the rate of change must be quantified. Some valves may be prone to jam open or closed, due to differential thermal expansion, if operating temperatures vary over a wide range. Do not assume everyone knows exactly what **you** are going to do with a valve.

NOTE: Valves are not guaranteed for undefined operating conditions. Data presented to manufacturers or distributors must not require interpretation. A manufacturer or distributor must **not** be required to "read-between-the-lines" to extract important information.

When users and contractors work on large projects, some decisions taken early in the life of the project are forgotten. All relevant data should be reviewed when preparing the final valve specification.

Fluid designation	Valve type	Prime considerations
Clean fluid Low pressure clean Not approved	**Isolating** — all except **Fire hydrant**	
	Non-return — all	
	Safety relief — all except **Shear pin, Buckling pin**	
	Regulator — all	
	Control — all	
Clean fluid Low pressure clean Approved	**Isolating** — all except **Fire hydrant**	Material compatibility with fluid and any cleaning requirements imposed by regulations
	Non-return — all	
	Safety relief — all except **Shear pin, Buckling pin**	
	Regulator — all	
	Control — all	
Clean fluid High pressure clean Not approved	**Isolating — Wedge gate, Conduit gate, Straight globe, Angle globe, Three-port globe, Straight needle, Angle needle, Solenoid, One-piece body ball, Two-piece body ball, Three-piece body ball, Trunnion-mounted ball, Multi-port ball, Parallel plug, Tapered plug, Lubricated plug, Eccentric plug, Multi-port plug, Line blind, Sample, Flush tank, Float, Automatic air release, Combined block and bleed, Combined stop & non-return, Valve manifold**	
	Non-return — Twin disc, Piston, Ball, Controlled closure	
	Safety relief — all except **Shear pin, Buckling pin, Vacuum break valve**	
	Regulator — all	
	Control — all except **Pinch**	
Clean fluid High pressure clean Not approved	**Isolating — Wedge gate, Conduit gate, Straight globe, Straight needle, Angle needle, Solenoid, Parallel plug, Tapered plug, Lubricated plug, Eccentric plug, Multi-port plug, Sample, Flush tank, Float, Automatic air release, Combined block and bleed, Combined stop & non-return**	Material compatibility with fluid and any cleaning requirements imposed by regulations
	Non-return — Twin disc, Piston, Ball, Controlled closure	
	Safety relief — all except **Vacuum break valve**	
	Regulator — all	
	Control — all except **Pinch, Ball, Characterised Ball, Butterfly, Eccentric rotating plug**	
Dirty fluid Low pressure clean Low concentration of small solids Not approved	**Isolating — Parallel gate, Wedge gate, Conduit gate, Weir type diaphragm, Full-bore diaphragm, Pinch, Straight needle, One-piece body ball, Two-piece body ball, Three piece body ball, Trunnion-mounted ball, Multi-port ball, Butterfly, Eccentric disc butterfly, Parallel plug, Tapered plug, Lubricated plug, Eccentric plug, Multi-port plug, Line blind, Sample, Flush tank, Float**	
	Non-return — all except **Piston, (Ball best)**	
	Safety relief — all are suitable when soft seats are fitted	
	Regulator — Review application with manufacturer	
	Control — Review application with manufacturer	
Dirty fluid High pressure Low concentration of small solids Not approved	**Isolating — Parallel gate, Wedge gate, Conduit gate, One-piece body ball, Two-piece body ball, Three-piece body ball, Trunnion-mounted ball, Multi-port ball, Parallel plug, Tapered plug, Lubricated plug, Eccentric plug, Multi-port plug, Line blind**	Renewable soft seat inserts
	Non-return — Twin disc, Ball, Controlled closure	
	Safety relief — all with soft seats except **Vacuum break valve**	
	Regulator — Review application with manufacturer	
	Control — Review application with manufacturer	
Dirty fluid Low pressure High concentration of small solids Not approved	**Isolating — Conduit gate, Weir type diaphragm, Full-bore diaphragm, Pinch, Three-piece body ball, Trunnion-mounted ball, Multi-port ball, Butterfly, Tapered plug, Lubricated plug, Eccentric plug, Multi-port plug, Line blind, Sample, Flush tank**	Material selection
	Non-return — Piston	
	Safety relief — Shear pin, Buckling pin + bursting disc	
	Regulator — Review application with manufacturer	
	Control — Review application with manufacturer	
Dirty fluid High pressure High concentration of small solids Not approved	**Isolating** — all except **Butterfly, Fire hydrant**	Material selection Strength and abrasion resistance
	Non-return — all	
	Safety relief — all except **Shear pin, Buckling pin**	
	Regulator — all	
	Control — all	

Table 19.1 Direct selection of valves for clean and dirty fluids

19.3.1 Fluid properties

Full description

trade name, chemical formula, concentration

hazardous/flammable/toxic

controlled by legislation

abrasive/corrosive

all constituents including trace elements

viscosity

viscosity characteristic with shear (rheology)

vapour pressure or condensing temperature

solidifying temperature

crystallisation temperature

sg/Cp/pH

allowable leakage rate to atmosphere

allowable leakage rate across seat

Fluid contaminants

dissolved/entrained gases in liquids

- release pressure

entrained liquid droplets in gas/vapour

solids

- size/concentration/distribution
- hardness

- rigid/deformable/friable
- abrasiveness (Miller Number or similar)

Suitable lubricants

user recommendations

prohibited lubricants, contaminants or diluents

Other fluids to be considered

chemical cleaning agents

LP steam cleaning

sterilising solutions

19.3.2 Operating conditions

Pressure/temperature

min/rated/max

maximum rate of change

- pressure pulsations/water hammer and associated pipework vibration

safety margins/design conditions

Flow

min/rated/max

maximum rate of change

- modulating valves preferred to pop-action

valve C_v/K_v/characteristic

Duty cycle

speed of operation

continuous/light/intermittent/cyclic

time between routine maintenance

time between re-lubrication

standby cycles

- full of product or flushed and empty

Cleaning cycle

fluid or mechanical

Process upsets

consider the effects on fluid and contaminants. Specify duration if longer than 10/20 minutes.

19.3.3 Environment

Site conditions

indoor/outdoor/onshore/offshore/sheltered/sunshade

min/max ambient temperature

max black bulb temperature

min/max relative humidity

altitude

hazardous area classification

atmospheric pollution

permissible gas concentration in air (ppm)

biological attack

conditions during transportation

- special boxing requirements

19.3.4 Materials

User material recommendations

materials in contact with process fluid state annual corrosion rate, mm/year

NACE requirements

Approved materials

prohibited materials

Physical properties

strength

ductility

impact resistance

creep resistance

erosion resistance

cavitation resistance

thermal shock resistance

Chemical properties

corrosion resistance

corrosion allowance

predicted life

19.3.5 Valve mechanical requirements

Pressure rating

process connections
(inlet and outlet ratings may be different)

pressure rating across closed valve

hydrotest pressure(s)

Connections

type, threaded, flanged, welding, compression, clamp

special surface finish on flanges

additional body connections; vent, drain, by-pass, instrumentation

packing box connections; lube, quench, leakage, instrumentation

Options

special internal surface finish

heat tracing or cooling

extended bonnet for heating/cooling

extension for local manual operation

local manual operation of automatic/actuated valves

insulation

in situ inspection/maintenance
replacing packing on-line

fire-safe

anti-static

enclosed bonnet/yoke

noise level

19.3.6 Valve pneumatic requirements

Power

air pressure and consumption

cleanliness required

lubricated or non-lubricated

Control

distance from controller

booster or positioner required

19.3.7 Valve electrical requirements

Power electrics

ac; voltage, phases, frequency, current/power required

dc; voltage, current/power required

emergency power supplies; availability, type

number of starts per hour (for motors)

Instrumentation

power supplies required, current/power required

Classification

electrical area classification

physical protection

certification

19.3.8 Valve instrumentation requirements

Indicators

pressure gauges for process

pressure gauges for pneumatics/hydraulics

Transmitters

facilities for remote indication

alarms and trips

Switches

limit switches

travel indicators

pressure switch for bellows failure

alarms and trips

Special sensors

packing box leakage measurement

packing box leakage analysers

ultrasound detectors

- seat leakage/flashing/cavitation

Communications

standard/proprietary interface to PLC

standard/proprietary interface to DCS

ESD override facility

HART/field bus/RS 485/RS 232

19.3.9 Erection and installation

Location

space available

access route

lifting facilities

supports

space for ancillaries or auxiliaries

space for expansion/improvements

Position

special orientation

availability of facilities

- electric supply
- pneumatic supply
- communications
- drainage for safe products
- drainage for hazardous products
- vent path for steam

19.3.10 Maintenance

Documentation

parts lists and cross-sectional drawings

lubrication schedule

routine maintenance instructions

repair instructions

trouble-shooting guide for all equipment

list of rapidly wearing parts with predicted life

quotation for commissioning spares

quotation for standard stocked spares

location of manufacturer's nearest service centre

location of manufacturer's nearest spares stockist

Site facilities

instrumentation to indicate valve operating conditions

special tools

staff training

19.3.11 Quality assurance

General

valves designed and built to a standard

valves designed and built to hygienic regulations

valves manufactured to a standard

fully assembled valve pressure tested

fully assembled valve/actuator tested

Fully assembled valve third party type tested

typical batch size of mass-produced valves

number of valves in batch tested

tolerance band on published data

Documentation

pressure test certificate(s)

material certificates

weld procedures

welder qualifications

inspector qualifications

non destructive testing reports

19.3.12 Quotation assessment

Commercial

quotation validity

length of guarantee

point of delivery

time of delivery

price

currency

Technical

standard product or prototype

technical support

size/capacity/features

check actual velocity through ports

check effect of valve pressure drop on compressors/pumps

requirements for auxiliary or ancillary equipment

energy consumption

drawings

quality level

availability for site support

19.4 Actuator selection

The first point which must be considered, and about which a definite decision must be made, is the choice of power supply.

Actuators usually respond to a remote signal without local operator intervention. A suitable power supply must be available. The different types of power supply are not equal in popularity.

Based on valves already installed, power supplies are ranked as follows:

- pneumatic
- electric
- hydraulic

Clean, dry instrument air is the most popular power supply and can be used with many actuator types. Electric actuators are very versatile and can be adapted to most situations. Hydraulic actuators can be supplied from a central system or have a dedicated power pack attached directly to the actuator.

In some remote valve installations, power may not be available. Contact with the "outside world" may be limited to a 4-20mA control signal or an "on/off" signal. The process fluid must be considered as the first choice for actuator power supply. Control valves can be replaced by regulators.

Safe gases, at suitable pressures, can be filtered to provide an equivalent "pneumatic" system provided seals and elastomers are compatible. Safe liquids can similarly provide hydraulic capabilities. If the pneumatic demand is low then a supply of bottled gas may be adequate. The situation may be different with hazardous process fluids. When sufficient differential pressure always exists across the valve, the exhaust fluid can be returned to the process and so eliminate the local hazard. If a very low, or virtually no differential exists during normal operation, other solutions must be investigated.

A turbine or hydraulic motor in the process line can produce hydraulic power or compressed air directly. Alternatively, electricity can be generated. When the process fluid is flammable it can be used as fuel for an engine or gas turbine which can produce compressed air, hydraulic power and/or electricity. The local generation of electricity, in very remote installations, is very helpful. Extremely long control circuits are unnecessary. Control communications can be achieved by radio; VHF or UHF; or microwave link. Also any type of power supply for the actuator can be derived from the electric supply.

The power supply, available and/or preferred, and the ambient environment, place severe constraints on the actuator selection. Pneumatic may be best choice for electrical hazardous areas. Hydraulic may not be a wise choice for very hot environments, such as metal manufacturing plants. Safe hydraulic fluids and water-oil emulsions are available but are not compatible with all equipment. Pneumatic or hydraulic actuators may not work well in environments subject to vibration.

The actuator must be compatible with the valve type used. The most popular actuator, the pneumatic diaphragm, will not be compatible with many long stroke linear valves and a piston actuator will be more appropriate. A diaphragm actuator may not be suitable for high thrusts. The thrust in both directions must be considered for double-acting actuators. Rotary actuators must be capable of the full rotation necessary for the valve. Travel stops may be essential to ensure wide open and closed alignment.

The speed of valve movement may be critical to process operation and safety. Safety relief valves and bursting discs operate very quickly to prevent overpressure. Some linear valves close very quickly to prevent the expansion of flame fronts. High speed valve operation may result in pressure pulsations and water hammer with consequential damage. Slight damage to pipe supports may be acceptable when ensuring overall plant safety. Any damage, due to routine valve operations, is not tolerable. Valve speed should be selected, or be capable of adjustment on site, to provide acceptable control functions without initiating damage. Normal, steady-state process design procedures do not have safety margins to include fatigue failures. Actuator speed variation for different functions is possible.

In remote installations, where power may be restricted or very costly, actuator efficiency may be very important. Actuators which operate against return springs are not energy efficient. This aspect of actuator selection should be considered in conjunction with the fail-safe option. If spring-loading is not required for close or open options then a double-acting stay-put actuator should be used.

A "rigid" actuator may be essential for very accurate positioning and for valves subject to wide pressure variations. Electric and hydraulic should be reviewed first.

Standard accessories, designed to increase the range of functions, may sway the choice of actuator in favour of a specific type. Increased safety, by incorporating an ESD override, may be an important requirement for valves handling hazardous fluids.

One aspect of actuation and control that can not be overlooked is "complexity". A great deal of modern equipment uses solid-state electronics. This requires specialist knowledge and equipment for servicing. Normal handtools and the average multimeter are inadequate. The purchaser must consider the local staff and skill levels and the facilities available when contemplating using the most up-to-date equipment, and what costs and time delay would be involved when relying on factory service engineers?

19.5 Check lists for actuator purchase specification

The beginning of Section 19.3 outlines the benefits of good check lists, so it is important to read this first. A comprehensive specification saves time and money.

19.5.1 Power supply

pneumatic
- other clean gas

hydraulic
- site supply or integral power pack
- preferred liquid

electric

combined system

(qualify pressure, temperature, cleanliness, max flow rate, voltage, frequency, max current)

see also Section 19.3.2 Duty cycle

19.5.2 Actuator type

linear or rotary
- travel
- thrust/torque

speed of operation
- fast/slow/adjustable
- different preset speeds for opening/closing
- water hammer protection

single or double acting

19.5.3 Environment

See Section 19.3.3

19.5.4 Control signal

process fluid

pneumatic

electric analogue

electric digital

analogue/digital combination

19.5.5 Additional features

rigidity of positioning

response to power failure
- fail open/fail closed/stay-put

handwheel

positioners

boosters

remote indication

ESD override facilities

feedback to control system/DCS/PLC

- HART/field bus/RS 485/RS 232

damping facilities

19.6 Purchasing

Chapter 16 covers valve standardisation, including all the important aspects of valve design and application. Chapter 15 covers quality assurance and testing. These chapters should be reviewed for applicability to specific valves and applications. If a valve is to be inspected by the purchaser's representative, acceptability must be determined by "pass" or "fail" criteria. An inspector cannot quantify the state of a valve after one or more years of operation.

The purchase documents should contain references to legislation and regulations, and to the requirements affecting design, to the properties of the fluid, as well as to the operating conditions of the valve. There is no point in making references to obscure local regulations. Manufacturers do not have easy access to all standards and regulations. If, however, particular, local regulations must be observed, a copy of these should be supplied with the inquiry.

It is important that the inquiry and purchase order list all operating conditions for the valve. Safety margins should be shown separately as such.

Users may consider the purchase of refurbished, second-hand valves. This approach is reasonable provided the refurbished valve meets all the requirements. "Out-of-spec" valves, used as an interim measure while waiting for new replacements, can prove very costly in the long term.

NOTE: Remember the site cost-saving edict:

"Trouble-shooting is making the valve do what you specify.

Making the valve do what you want is up-grading!"

Fluid properties & conversions 20

20.1 Introduction

This Chapter brings together useful data on fluid properties and conversions to assist the user in selecting a valve or actuator and in making initial, preliminary design decisions.

The information is provided in three parts comprising:-

Liquid properties: —

A comprehensive Table of liquid properties is provided in Section 20.2.2. this gives important and useful data for valve applications. The data outlined in the Liquid Table is described in detail in Section 20.2.1.1.

Supplementary diagrams covering viscosity, density and solubility are included in Section 20.2.3.

Thermodynamic properties of gases:—

General gas and vapour data:- Useful general information such as molecular weight, specific gas constant etcetera, is given in Section 20.3.1.

Table of physical constants:- Frequently used physical constants and properties of gases and other selective components are listed in tabular form, (see Section 20.3.2). Immediately following this is a detailed list of notes and references and explanations for the values given in the tables, (Section 20.3.3).

Pressure/enthalpy charts:- Calculations can conveniently be performed with sufficient accuracy for most purposes, by using pressure/enthalpy charts. These, sometimes referred to as Mollier charts, are charts on which the following lines are plotted for the gas in question:

- the saturation line - the line dividing the liquid and gaseous states of the gas
- lines of constant temperature
- lines of constant entropy
- lines of constant specific volume

All these lines are plotted on a basis of pressure against enthalpy (total heat).

In Section 20.4 pressure/enthalpy charts are provided for some of the more common gases used in industry. (This data is reproduced with kind permission of the Gas Processors Suppliers Association, USA.)

Units and conversions:—

SI, The International System of Units:- The modern SI-system has been adopted in legislation by practically every country in the world. There are, however, extensive changeover problems and many different kinds of data are still only available in units which do not conform to the SI-system. Section 20.6 attempts to bridge the difficulties experienced when dealing with quantities expressed in older units.

Almost without exception it is recommended that the quantities be converted directly to SI units; calculation can then be carried out using the coherent SI-system. However, it is important that all results of calculations should be reviewed by someone fully conversant with the system of units and the practical values encountered.

20.2 Liquid properties

20.2.1 Introduction

The Liquid Table in Section 20.2.2 gives useful data for valve applications. The characteristics of liquids are not given to that degree of accuracy which, in special cases, may be required in order to specify other process equipment.

Note: The most common liquids, water, oils and certain suspensions are dealt with in more detail in Sections 2.2, 2.3 and 2.4 of Chapter 2. Compressibility and wave

Substance	Entered in table as
Acetic ether	Ethylacetate
Alcohol	Ethyl alcohol
Alum	Aluminium sulphate
Ammonium hydroxide	Ammoniac
Benzol	Benzene
Black liquor	Sulphate liquor
Bromite	Silver bromide
Butanol	Butyl alcohol
Carbitol	Ethyldiglycol
Caustic soda	Sodium hydroxide
Cellosolve	Sodium chloride
Cellosolve acetate	Ethylglycolacetate
Chlorhydric acid	Hydrochloric acid
Chloroacetic acid	Monochloroacetic acid
Chlorobenzene	Monochlorobenzene
Chrome alum	Potassium chrome sulphate
Chrome oxide	Chromic acid
Cooking salt	Sodium chloride
Copper vitriol	Copper sulphate
Cyancalium	Potassium cyanide
Dichloromethane	Methylene chloride
Diethylether	Ether
Dphonylether	Diphenyl oxide
Ethanol	Ethyl alcohol
Ferrichloride	Iron (III) chloride
Ferrosulphate	Iron (II) sulphate
Fixerbad	Sodium tiosulphate
Formaline	Formaldehyde
Fumaric acid	Malein acid
Glauber salt	Sodium sulphate
Hartshorn salt	Ammonium carbonate
Hydrofluoric acid	Hydrofluoric acid
Methanol	Methyl alcohol
Methylaldehyde	Formaldehyde
Naphthalene chloride	Chloronaphthalone
Natron saltpetre	Sodium nitrate
Oleic acid	Fatty acid
Palmitic acid	Fatty acid
Perhydrol	Hydrogen peroxide
Phenol	Carbolic acid
Potash	Potassium carbonate
Potash nitrate	Potassium nitrate
Potash salts	Potassium hydroxide
Pulping liquor	Calcium bisulphite liquor
Radical vinegar	Concentrated acetic acid
Sal ammoniac	Ammonium chloride
Saltlake	Sodium chloride
Slaked lime	Calcium hydroxide
Soda	Sodium carbonate
Soda lye	Sodium hydroxide
Sodium borate	Borax
Styrol	Styrene
Sublimate	Chloride of mercury
Sulphite liquor	Calcium bisulphite
Tannin	Tannic acid
Trichloroethanyl	Chloral
Urea	Carbamide
Water glass	Sodium silicate

Table 20.1 Trade names and synonyms used in the Liquid Table in Section 20.2.2

speed are not included in the Liquid Table. These characteristics are stated for a number of liquids in Section 2.1.5.

The Liquid Table contains the following data:

- **Substance:** the liquid properties are arranged in alphabetical order according to the name of the substance. The trade names with synonyms are given in Table 20.1. The letter D indicates a dangerous (flammable or otherwise hazardous) liquid within the terms of the UK Petroleum (Inflammable liquids and Other Dangerous Substances) Order 1947 - Amended 1968 SI 570. The letter T indicates a high level of toxicity.
- **Chemical formula:** for identification of substance. In practice a substance unfortunately can not be assigned to a single formula because of contamination. Liquids containing contaminants which are insoluble or solid have the worst pumping characteristics, often causing troublesome wear on pumps and shaft seals.
- **Solubility in H_2O:** attention is drawn in the Liquid Table to various diagrams referring to the solubility of the substance in water. The solubility is expressed as a concentration in percentage by weight at varying temperatures. When pumping, precipitation often occurs when the concentration exceeds that stated by the solubility curve for the given temperature. Conversely, there will be precipitation if the temperature drops below the stated solubility temperature.
- **Viscosity:** is stated in mm^2/s (cSt) at +20°C. Viscosity below $5mm^2/s$ indicates that the liquid is easy flowing like water. Viscosity below $1mm^2/s$ indicates that the liquid has poor lubrication qualities and poor frictional dampening properties for acoustic and pressure waves. The Table also refers to diagrams showing dependence of viscosity on temperature. The viscosity will also be used for the determination of frictional losses in the pipework. This depends primarily on whether the flow is laminar or turbulent, see Chapter 3. In the laminar region, the pipe frictional losses are proportional to the viscosity, whereas its effect can be neglected in the turbulent region for the internal roughness of pipes which occur in practice.

  Certain liquids are non-Newtonian and require special care when determining the equivalent (apparent) viscosity.
- **Vapour pressure (absolute):** is stated in kiloPascals (kPa) at +20°C. See curves in Figure 20.3 which show the temperature dependence of vapour pressure. Note that boiling point at atmospheric pressure (101.325 kPa 760 mm Hg) gives another point on the vapour pressure curve. The curves in Figure 20.3 are set out schematically and may deviate from the exact values. This is especially applicable to aqueous solutions - curve 13. All references to this curve apply to chemicals in solution in water. Liquid pressure near to the vapour pressure at the actual liquid temperature can cause cavitation in a pump or pipe.
- **Concentration in H_2O:** the concentration of the substance in water expressed as percentage by weight.
- **Density:** is stated in kg/m^3 at +25°C. Reference is made to various figures for other temperatures and concentrations. The density is used among other things for conversion of pressure to head and for the calculation of the power required for the pump.
- **pH region:** is an expression of acidity or alkalinity and is grouped within pH regions 0 - 4, 4 - 6, 6 - 9 and 9 - 14. This group classification assists in the choice of pump material, the pH regions coinciding with commonly used pump material resistivity to corrosion.
- **Melting point:** note that many contaminants gradually begin to degrade at temperatures immediately above the melting point which can have a harmful effect on the function of the pump.
- **Fire hazard class** stated in accordance with Chapter 2, Section 2.1.7.

NOTE: Where there is no information in the Liquid Table, this means that the actual liquid property is not known and it does not mean that the information omitted is of no general interest in valve technology.

20.2.2 Liquid Table

Substance		Chemical formula	Viscosity mm^2/s at 20 °C	Vapour pressure kPa		Conc. % in H_2O	Density			pH/ region	Boiling point °C at 101.3 kPa (760 mm Hg)	Melting point °C	Fire hazard class- (Chapter 2 Section 2.1.7)	Solubility in H_2O at various temps. See Fig no.	Comments
				at 20 °C	at various temps. See Fig 20.3 curve no.		at 25 °C	at various temps. See Fig no.	at various conc. See Fig no.						
Abietinic acid		$C_{20}H_{30}O_2$										172			
Acetaldehyde	D	CH_3CHO	< 1	100	10		780/20°C	20.4			21	-125	1		
Acetic acid	D	CH_3COOH	1.2	3.5	19		1050	20.4	20.6	0 - 4	119		2b		
					19	5-100				0 - 4					
					19	50	1060			0 - 4					
Acetic acid anhydride	D	$(CH_3CO)_2O$			20		1080			0 - 4	136	- 73			
Acetone	D	CH_3COCH_3	< 1	30	14		790	20.4			57	- 95	1		E-mod and speed of sound - Table 2.1
Acetyl chloride		CH_3COCl					1100				51	- 112	1		
Alkyd solution in paint naphtha			Fig 20.1				900				150		2b		Non-Newtonian - Figure 20.1
Alkyd solution in xylene	D		Fig 20.1	0.9			1000				139		2a		Non-Newtonian - Figure 20.1
Alkyd tixotrope sol. in paint naphtha			Fig 20.1				900				150		2b		Non-Newtonian - Figure 20.1
Allyl alcohol	D	CH_2CHCH_2OH		2.3			850				97		1		
Ally chloride	D	CH_2CHCH_2CL	<1				940				45		1		
Aluminium		Al					2700					660			
Aluminium chloride		$AlCl_3$	< 5	2.2	13	5	1030		20.7	0 - 4				20.10	
				2.2		10	1090			0 - 4					
Aluminium nitrate		$Al(NO_3)_3$		2.2		10	1050		20.7	0 - 4				20.10	
Aluminium sulphate		$Al_2(SO_4)_3$		2.2	13	10	1110		20.5	0 - 4				20.10	
Ammonium alum		$NH_4Al(SO_4)_2$				10	1050			0 - 4					
Ammonium bromide		NH_4Br			13	5	1030		20.7	4 - 9					
Ammonium carbonate		$(NH_4)_2CO_3$			13										
				2.2	13	10	1030			4 - 6					
Ammonium chloride		NH_4Cl			13	10 - 50				4 - 6					
				1.8		26	1070		20.7	4 - 6					
Ammonium fluoride		NH_4F		2.3		6	1030								
				2.2		14	1060			9 - 14					
Ammonium hydroxide		NH_4OH	< 5		1			20.4		9 - 14					
			< 5	9.3		10	960			9 - 14					
			< 5	34.5		25	910								
Ammonium nitrate		NH_4NO_3			13	10 - 20			20.7	6 - 9					
				2.2	13	10	1040			6 - 9					
				0.5	13	60	1230			6 - 9					
Ammonium oxalate		$(COONH_4)_2H_2O$			13	30	1040			6 - 9					
Ammonium perchlorate		NH_4ClO_4				10	1040			6 - 9				20.10	
Ammonium persulphate		$(NH_4)_2S_2O_8$				10	1060			4 - 6					
Ammonium sulphate		$(NH_4)_2SO_4$				0 - conc			20.7	4 - 6				20.10	
				2.2		10	1060			4 - 6					
				0.7		50	1280			4 - 6					
Amyl acetate		$CH_3COOC_5H_{11}$	< 5	0.7	20		880				149	- 78	2b		
Amyl alcohol		$C_5H_{11}OH$	6				810				133	- 78	2b		
Amyl chloride		$C_5H_{11}Cl$					890				109		1		
Amyl polycaptan		$C_5H_{11}SH$					850				100				
Aniline	D	$C_6H_5NH_2$	6	0.1	24		1020	20.4			184	- 6	3		
Aniline hydrochloride		$C_6H_5NH_2HCl$				5				0 - 4					
Antimony		Sb				20	1090			0 - 4		630			
Anthracene oil		$C_{14}H_{10}$					1250				226				
Arsenic acid		H_3AsO_4					2500		20.6						
						10	1070			0 - 4					
Asphalt solution in naphtha			1500 - 7500				900				150		2b		Non-Newtonian
Aviation fuel			< 1	25			720				40		1		Tables - 2.1, 2.17, 2.18
Barium chloride		$BaCl_2$		2.2	13	10	1090		20.8					20.10	
				1.8		26	1280			6 - 9					
Beer					13		1010			4 - 6	100				
Benzene	T	C_6H_6	< 1	13	16		880	20.4			80	5.5	1		
Benzene sulphonic acid		$C_6H_5SO_3H$										525			
Benzoic acid		C_6H_5COOH					1270				250	122			

20.2.2 Liquid Table

Substance		Chemical formula	Viscosity mm²/s at 20 °C	Vapour pressure kPa at 20 °C	Vapour pressure kPa at various temps. See Fig 20.3 curve no.	Conc. % in H_2O	Density at 25 °C	Density at various temps. See Fig no.	Density at various conc. See Fig no.	pH/ region	Boiling point °C at 101.3 kPa (760 mm Hg)	Melting point °C	Fire hazard class- (Chapter 2 Section 2.1.7)	Solubility in H_2O at various temps. See Fig no.	Comments
Benzyl alcohol		C_7H_8O	5	< 0.1			1050	20.4			205		3		
Biphenyl		$C_6H_5C_6H_5$					1990				256	70			
Borax		$Na_2B_4O_7$			13		2370					741			
				2.2	13	3.5	1030								
Borax acid		H_3BO_3			14	10				4 - 6					
						50				4 - 6					
Bromine	D	Br_2	< 1/0 °C		12		3120/20 °C	20.4			59	- 7.2			See Section 2.3 and Table 2.1
Burner oils	D		Fig 2.11				840/20 °C						3		See Section 2.3 and Table 2.1
	E		Fig 2.11				940/20 °C						3		See Section 2.3 and Table 2.1
	F		Fig 2.11				950/20 °C						3		
Butane		C_4H_{10}			8	50	600/0 °C	20.4			- 0.5	138			
Butyl acetate	D	$CH_3COOC_4H_9$	< 1	2	19		880				125	- 77	2a		
Butyl alcohol	D	C_4H_9OH	< 5	0.9	18		810	20.4			118	- 90	2a		
Butyl glycol		$C_6H_{14}O_2$	< 5	< 0.1			900				171		3		
Butyric acid		C_3H_2COOH			21		960	20.4			164	6.5			
Calcium bisulphate		$Ca(HSO_3)_2$			13	25	1040			4 - 6					
Calcium chloride		$CaCl_2$			13			20.7							
Calcium hydroxide		$Ca(OH)_2$			13	5	1060		20.5	9 - 14					
Calcium nitrate		$Ca(NO_3)_2$							20.7						
Camphor		$C_{10}H_{16}O$					990				209	176			
Carbamide		$(H_2N)CO(NH_2)$					1320					132			
						50	1115			4 - 6					
Carbolic acid	D	C_6H_5OH	11				1070	20.4			43				
Carbon disulphide		CS_2	< 1	48	12		1262	20.4			46	- 112	1		
Carbon tetrachloride	T	CCl_4	< 1	20	15		1600				77	- 23			
Castor oil			Table 2.17	(0)			960								
Chloral		CCl_3CHO					1520				98	- 57			
Chloramine		$CH_3C_6H_4SO_2NCl$										183			
Chloride of lime	D	$Ca(ClO)_2\ 4H_2O$					2350								
Chloride of mercury	T	$Hg\ Cl_2$			13	0.1	1000			4 - 6					
							7150								
Chlorine	D	Cl_2			3			20.4			- 35	- 103			
Chlornaphthalene		$C_{10}H_7Cl$	2.5				1200				263	- 25			
Chloroform	D	$CHCl_3$	< 1	31	14		1490	20.4			61	2			E-mod and speed of sound - Table 2.1
Chlorosulphonic acid		$HClSO_3$	< 5		21		1280				158	- 80			
Chromic acid	D	H_2CrO_4	< 5				2700			0 - 4					
			< 5			10 - 50				0 - 4					
Citric acid		$C_6H_8O_7H_2O$			13		1550			0 - 4		153			
Copper (II) chloride		$CuCl_22H_2O$				10	3390					620		20.10	
							1090			0 - 4					
Copper cyanide		$CuCN$			13		2920								
Copper nitrate		$Cu(NO_3)_26H_2O$			13		2070							20.10	
		$Cu(NO_3)_23H_2O$					2320					114			
				2.2	13	10	1090								
Copper sulphate	D	$CuSO_4$					3610							20.10	
					13	10	1160			0 - 4					
				2.2	13	18	1210			0 - 4					
Corn oil			Table 2.17												
Cresol		$C_6H_4(CH_3)OH$			25		1040			0 - 4	191	30			
Cyclohexanol	D	C_6HOH	71	0.3			950	20.4			161		1		
Cyclohexanone	D	$C_6H_{10}O$	< 5	0.7			950				156		2b		
Diacetone alcohol	D		150	0.1	23		940				170	- 43	2b		
Dichloroacetic acid		$CHCl_2$-$COOH$	< 5				1550	20.4		0 - 4	192	- 11			
Dichloroethane	D	CH_2Cl-CH_2Cl		31.2	14		1180	20.4			59		1		
Dichloroethylene	D	$C_2H_2Cl_2$		32			1250	20.4			60	- 81	1		

20.2.2 Liquid Table

Substance		Chemical formula	Viscosity mm²/s at 20 °C	Vapour pressure kPa at 20 °C	Vapour pressure kPa at various temps. See Fig 20.3 curve no.	Conc. % in H_2O	Density at 25 °C	Density at various temps. See Fig no.	Density at various conc. See Fig no.	pH/ region	Boiling point °C at 101.3 kPa (760 mm Hg)	Melting point °C	Fire hazard class- (Chapter 2 Section 2.1.7)	Solubility in H_2O at various temps. See Fig no.	Comments
Dichlorohydrin		$(CH_2Cl_2)_2CHOH$					1360				176				
Dichlorophenoxyacetic acid		$Cl_2C_6H_3O$								0 - 4					
		CH_2COOH													
Diesel fuel			Fig 2.11				830				210		3		
Diethyl carbonate		$(C_2H_5O)_2CO$			19		980				128	- 48	2b		
Dioctylphthalate			82				980								
Dioxan	D	$O_2(CH_2)_4$	2	3.9			1040	20.4			101	12	1		
Dipentene							850				180	- 70	3		
Diphenyloxide		$C_6H_4OC_6H_4$	4	0.03			1070				253	27	3		
Engine oil SAE 5W-50			Fig 2.12				900/20 °C						3		See Section 2.3 and Table 2.1
Ether	D	$C_4H_{10}O$	< 1	74			710	20.4			35		1		
Etherdiethylene		C_2H_4O		150			870				11	- 113			
Ethyl acetate	D	$CH_3COOC_2H_5$	< 1	13.8	16		900	20.4			77		1		
Ethyl alcohol		C_2H_5OH	< 5	8			790	20.4			78		1		
Ethyldiglycol		$HO(CH_2)_4$	< 5	0.02			990				202		3		
		$O_2C_2H_5$													
Ethyl chloride		C_2H_5Cl	< 1	133	9		890	20.4			12	- 139	1		
Ethyl glycol		$C_4H_{10}O_2$	2.3	0.5			930				135		2b		E-mod and speed of sound - Table 2.1
Ethyl glycol acetate		$C_6H_2O_3$	< 5	0.2			970				150		2b		
Ethylene diamine	D	H_2NCH_2	1.5	1.4			900			9 - 14	117	11			
		CH_2NH_2													
Ethylene glycol	D	$C_2H_6O_2$	18	< 0.13	24		1110	20.4			198		3		E-mod and speed of sound - Table 2.1
Fluorine	T	F									- 188	- 223			
Formaldehyde	D	HCHO			5		820				- 20				
Formic acid	D	HCOOH	1.5	5.3	18	20 - 50	1220	20.4			101				
Freon		CF_2Cl_2					1470				- 29	- 155			
Fruit juices					13										
Furfural	D	C_5H4O_2	1.4	0.3	22		1160	20.4			162		3		
Furfurole	D	C_5H6O_2		0.13	23		1160	20.4			171		3		
Gallic acid		$C_6H_2(OH)_3COOH$					1700			0 - 4					
Glycerine		$CH_2OHCHOH$	1200				1260	20.4			290				
		CH_2OH													
Ground nut oil			Table 2.17				910								See Tables 2.17, 2.18
Hexane		C_6H_{14}	< 1	30	14		660		20.4		69	- 90	1		
Hydrobromic acid		$Hbr \cdot H_2O$	< 1				1780		20.6	0 - 4		- 89			
				0.6		50	1520			0 - 4					
Hydrochloric acid	D	HCl		1.3	13	20	1100		20.6	0 - 4	110				
				1.3	13	36	1180			0 - 4					
Hydrofluoric acid	T	HF					990	20.4	20.6		20	- 83			
						1	1005			0 - 4					
				2.2		10	1030			0 - 4					
				1		40	1130			0 - 4					
						75	1240			0 - 4					
Hydrogen		H_2									- 253	-259			
Hydrogen peroxide		H_2O_2			21		1460				151	- 89			
				2.2	21	10	1040								
Hydrogen sulphide	D	H_2S					950/-50 °C	20.4			- 60	- 86			
Hydriodic acid		$HI.H_2O$					1700		20.6		127				
Industrial oils as per ISO -															
Cutting oil			Table 2.17	(0)			900/20 °C						3		See Section 2.3 and Table 2.1
Gear oil			Table 2.16	(0)			905/20 °C						3		

20.2.2 Liquid Table

Substance		Chemical formula	Viscosity mm²/s at 20 °C	Vapour pressure kPa at 20 °C	Vapour pressure kPa at various temps. See Fig 20.3 curve no.	Conc. % in H_2O	Density at 25 °C	Density at various temps. See Fig no.	Density at various conc. See Fig no.	pH/ region	Boiling point °C at 101.3 kPa (760 mm Hg)	Melting point °C	Fire hazard class- (Chapter 2 Section 2.1.7)	Solubility in H_2O at various temps. See Fig no.	Comments
Hydraulic oils			Table 2.17	(0)			880/20 °C						3		
Turbine oil			Table 2.17	(0)			885/20 °C						3		
Iodine	D	I_2					4940	20.4			184	114			
Iodoform		CHI_3					4010					120		20.10	
Iron (II) chloride		$FeCl_2$				10 - 50				0 - 4					
Iron (III) chloride		$FeCl_3$	1	2.2	13	10	1085			0 - 4					
			15	0.7	13	50	1550			0 - 4					
Iron nitrate		$Fe(NO_3)_3$		2.2		10	1080			0 - 4					
				1.8		25	1230			0 - 4					
Iron (II) sulphate		$FeSO_4\ 7H_2O$	< 5		13		1050								
Iron (III) sulphate		$Fe_2(SO_4)_3$			13	30 - 50				0 - 4					
Isobutylalcohol		$(CH_3)_2CHCH_2OH$	< 5	1.2			800				108		2a		
Isopropylalcohol		$CH_3CHOHCH_3$			16		780				82	- 89	1		
Isopropylnitrate					17		1190				100				
Kerosene			< 5	0.1			780				170		2b		See Tables 2.17, 2.18
Lacquer			200 - 500				900				150		2b		Non-Newtonian, see Section 2.4
Lactic acid		$H_6C_3O_3$			10	1020				0 - 4					
Latex			1000 - 4000			50	1050			6 - 9	100				Non-Newtonian, see Section 2.4
Lead		Pb					11350					327			
Lead acetate		$(CH_3COO)_2Pb\ 3H_2O$				10				6 - 9				20.10	
Lead nitrate		$Pb(NO_3)_2$				30				0 - 4				20.10	
Linseed oil			Table 2.17			930									
Liquid resin			20 - 300		(0)		970								
Lubricating oil SAE 75W-140			Table 2.16	(0)			910/20 °C								See Section 2.3 and Table 2.1
Magnesium carbonate		$MgCO_3$					2960								
Magnesium chloride		$MgCl_2\ GH_2O$			13		2320					708		20.10	
				2.2	13	10 - 50				4 - 6					
				2.2	13	10	1080			4 - 6					
				2.2	13	15	1130			4 - 6					
				2.2	13	25	1150			4 - 6					
				2.2	13	30	1280			4 - 6					
Magnesuim sulphate		$MgSO_4.7H_2O$			13		1680							20.10	
				2.2	13	10	1100			4 - 6					
				2.2	13	20	1300			4 - 6					
Maleic		$(HCCOOH)_2$			13		1590					130			
						50	1300			0 - 4					
Manganese chloride		$MnCl_2$			13		2980					650			
						10	1060			4 - 6					
Manganese chloride hydrate		$MnCl_2\ 4H_2O$					2010					58			
Manganese sulphate		$MnSO_4\ 7H_2O$					2090								
						30	1220			4 - 6					
Mercury	T	Hg					13600				357	- 39			E-mod and speed of sound - Table 2.1
Methyl acetate	D	$CH_3CO_2CH_3$	< 1		12		930				58		1		
Methyl alcohol	D	CH_3OH	< 1	13	14		790	20.4			65		1		E-mod and speed of sound - Table 2.1
Methylene chloride		CH_2Cl_2	< 1	72.4	11		1320	20.4			40				
Methylethylketone		C_4H_8O		15	16		810	20.4			80		1		
Methyl glycol		$C_3H_8O_2$	1.6	1.7			970				125		2b		
Methylchloride		CH_3Cl					1790				- 24		1		
Milk, fresh					13		1020								
Milk, sour					13		1020			4 - 6					
Molasses			100 - 50000												Non-Newtonian, see Section 2.4
Monochlorobenzene		C_6H_5Cl	1	1.6	20		1110	20.4			132		2a		
Monochloracetic acid		CH_2ClCO_2H	2.2 (50 °C)		24		1410			0 - 4	189	63			
Mustard			approx 5000		13										Non-Newtonian, see Section 2.4

20.2.2 Liquid Table

Substance		Chemical formula	Viscosity mm²/s at 20 °C	Vapour pressure kPa at 20 °C	Vapour pressure kPa at various temps. See Fig 20.3 curve no.	Conc. % in H_2O	Density at 25 °C	Density at various temps. See Fig no.	Density at various conc. See Fig no.	pH/ region	Boiling point °C at 101.3 kPa (760 mm Hg)	Melting point °C	Fire hazard class- (Chapter 2 Section 2.1.7)	Solubility in H_2O at various temps. See Fig no.	Comments
Naphthalene		$C_{10}H_8$			27		1150				210	80			
Naphthalene sulphonic acid		$C_{10}H_7SO_3H$					1450					102			
						10	1030			0 - 4					
Nickel nitrate		$Ni(NO_3)_2$ $6H_2O$			13		2050				137	57			
					13	10	1050			4 - 6					
Nickel sulphate		$NiSO_4$ $7H_2O$			13		1950								
					13	10	1060			4 - 6					
Nitric acid	D	HNO_3			16		1500		20.6	0 - 4	83	- 42			
					16	1	1004			0 - 4					
					16	5	1030			0 - 4					
					16	10	1050			0 - 4					
				2	16	20	1120			0 - 4					
				1.5	16	40	1250			0 - 4					
					16	65	1480			0 - 4					
					16	80	1452			0 - 4					
Nitro benzine	D	$C_6H_5NO_2$	2	0.2			1200	20.4			211		3		
Nonane		C_9H_20	1	0.7			720	20.4			151	- 54	2b		
Octane		C_8H_{18}	< 1	2			700	20.4			126		1		
Octanol		C_6H_{13} CHOH CH_3	11				830				194	- 15			
Oleum		H_2SO_4 + 13%SO_3					1910			0 - 4					
Olive oil			Table 2.17	(0)			910								
Oxalic acid		$C_2H_2O_4$ $2H_2O$			13		1650					101			
					13	10 - 50				0 - 4					
					13	0.5	1000			0 - 4					
					13	10	1020			0 - 4					
					13	15	1030			0 - 4					
Ozone		O_3									- 112	- 192			
Paraffin							770				240	45 - 55			
Pentane		C_5H_{12}	< 1	67	11		620	20.4			36	- 130	1		
Peracetic acid		CH_3COOOH					1230				105				
Petrol			< 1				730				40		1		
Phenoldicarbonate							1130				312	80			
Phenolphthalein		$C_{20}H_{14}O_4$					1280					261			
Phosphoric acid	D	H_3PO_4			13		1840			0 - 4					
					13	10	1050			0 - 4					
				3	13	20	1120			0 - 4					
				1.3	13	40	1370			0 - 4					
					13	70	1530			0 - 4	181				
					13	85	1690			0 - 4					
Phthalic acid		$C_6H_4(CO_2H)_2$					1600					206			
Phthalic anhydride		$C_6H_4(CO)_2O$					1530					130			
Potassium bi-carbonate		$KHCO_3$			13		2170							20.9	
					13	10				6 - 9					
Potassium bi-chromate		$K_2Cr_2O_7$					2680								
						25	1050			0 - 4					
Potassium bi-sulphate		$KHSO_4$					2320					214			
						5	1035			0 - 4					
Potassium bromide		KBr			13		2750		20.8		1435	730			
Potassium carbonate		K_2CO_3					2420					891·		20.9	
				2.2	13	20	1190			9 - 14					
Potassium chlorate	D	$KClO_3$			13		2320					356		20.9	
Potassium chloride		KCl			13		1980		20.4			776		20.9	
						20	1130			6 - 9					
						10 - 30				6 - 9					
Potassium chromate	D	K_2CrO_4					2730					968		20.9	

20.2.2 Liquid Table

Substance		Chemical formula	Viscosity mm²/s at 20 °C	Vapour pressure kPa at 20 °C	Vapour pressure kPa at various temps. See Fig 20.3 curve no.	Conc. % in H_2O	Density at 25 °C	Density at various temps. See Fig no.	Density at various conc. See Fig no.	pH/ region	Boiling point °C at 101.3 kPa (760 mm Hg)	Melting point °C	Fire hazard class- (Chapter 2 Section 2.1.7)	Solubility in H_2O at various temps. See Fig no.	Comments
Potassium chrome sulphate		$KCr(SO_4)\ 12H_2O$			13		1830								
Potassium cyanate	D	KOCN					2060								
Potassium cyanide	T	KCN			13		1520				635				
					13	5 - 10				9 - 14					
Potassium fluoride	T	$KF\ 2H_2O$					2450				156	41			
Potassium hydroxide	D	KOH					2040		20.5		360			20.9	
				2.2	13	10 - 90				9 - 14					
				2.2	13	30	1290			9 - 14					
				2.2	13	50	1510			9 - 14					
Potassium iodide		KI					3130		20.8						
Potassium nitrate		KNO_3					2110							20.9	
				2.2		10	1080	20.7							
Potassium oxalate		$K_2C_2O_4\ H_2O$					2130								
						15	1170			0 - 4					
Potassium perchlorate		$KClO_4$					2520							20.9	
Potassium permanganate		$KMnO_4$					2700								
				2.2	13	20	1040			0 - 4					
Potassium persulphate		$K_2S_2O_8$					2480								
						1 - conc				0 - 4					
Potassium silicate		K_2SiO_3										976		20.9	
Potassium sulphate		K_2SO_4			13		2660								
Propane		C_3H_8			2		580	20.4			- 42	- 190	1		
Propanol		C_3H_7OH	2.8	2.8			800	20.4			97	- 126	1		
Propionic acid		C_2H_5COOH			21		990				141	- 21			
						25	1030			0 - 4					
Pyridine	D	C_5H_5N		3.6	19		980				116	- 42			
Pryogallol	D	$C_6H_3(OH)_3$					1460				309				
						10	1030			0 - 4					
Quinine		$C_{20}H_{24}N_2O_2$										57			
Salicylic acid		$C_6H_4OHCO_2H$					1440				211	159			
						conc	1000			0 - 4					
Seawater - 4% NaCl			1	2.3	13		1020			6 - 9	100				See Section 2.2.4 and Table 2.1
Sebacic acid		$C_8H_{16}(COOH)_2$					915				240	55			
Silver bromide		AgBr					6470				432				
Silver chloride		AgCl					5560				455				
Silver nitrate	D	$AgNO_3$			13		4350		20.8		212				
Size			500 - 3000							6 - 9	100				Non-Newtonian, see Section 2.4
Sodium acetate		$NaC_2H_3O_2\ 3H_2O$			13	7	1450				123	58		20.9	
					13		1050			6 - 9					
Sodium arsenate	T	$Na_2HAsO_4\ 7H_2O$					1880								
Sodium bi-carbonate		$NaHCO_3$			13		2160								
					13	5	1040			9 - 14					
Sodium bi-sulphate		$NaHSO_4\ H_2O$			13		2100					58			
					13	10	1080			0 - 4					
Sodium bi-sulphite		$NaHSO_3$			13		1480								
				2.2	13	10	1100								
Sodium bromide		$NaBr\ 2H_2O$					2180								
						5 - 10				4 - 6				20.9	
Sodium carbonate		Na_2CO_3			13		2530		20.8			850		20.9	
				2.2	13	10 - 50				9 - 14					
				2.2	13	10	1150			9 - 14					
Sodium chlorate		$NaClO_3$					2480					250		20.9	
						28	1410			4 - 6					
Sodium chloride		NaCl			13		2170		20.8		1413	801		20.9	
				2.2	13	1 - conc				6 - 9					

20.2.2 Liquid Table

Substance	Chemical formula	Viscosity mm^2/s at 20 °C	Vapour pressure kPa		Conc. % in H_2O	Density			pH/ region	Boiling point °C at 101.3 kPa (760 mm Hg)	Melting point °C	Fire hazard class- (Chapter 2 Section 2.1.7)	Solubility in H_2O at various temps. See Fig no.	Comments
			at 20 °C	at various temps. See Fig 20.3 curve no.		at 25 °C	at various temps. See Fig no.	at various conc. See Fig no.						
Sodium di-chromate	$Na_2Cr_2O_7 2H_2O$			13		2520								
Sodium fluoride T	NaF					2550					988		20.9	
					5	1050			4 - 6				20.9	
Sodium fluorsilicate	Na_2SiF_6					2680								
Sodium hydrogen difluoride	$NaHF_2$					2080								
Sodium hydrosulphate	$NaHSO_4$					2440								
	$NaHSO_4 H_2O$					2100					58			
Sodium hydroxide D	$NaOH$			13		2130		20.5		1390	318		20.9	
			2.2	13	10 - 70				9 - 14					
			2.2	13	30	1330			9 - 14					
			2.2	13	50	1530			9 - 14					
Sodium hypochlorite D	$NaOCl$		2.3	13	5	1020			6 - 9					
Sodium nitrate	$NaNO_3$			13		2260		20.7			307		20.9	
				13	4	1030			6 - 9					
Sodium perchlorate	$NaClO_4H_2O$					2020					130		20.9	
					10	1070			4 - 6					
Sodium peroxide	Na_2O_2					2810								
					10	1110			6 - 9					
Sodium phosphate primary	$NaH_2PO_4 12H_2O$					1910					60			
			2.2		10	1070								
Sodium phosphate secondary	$Na_2HPO_4 12H_2O$					1520								
			2.3		50	1250			0 - 4					
Sodium phosphate tertiary	$Na_3PO_4 12H_2O$					1620								
			2.2		conc	1110			9 - 14					
Sodium silicate	Na_2SiO_3			13		2400		20.7			1088			
Sodium sulphate	$Na_2SO_4 10H_2O$			13		1460					32		20.9	
			2.2	13	1	1000			4 - 6					
			2.2	13	5	1020			4 - 6					
Sodium sulphide	$Na_2S 9H_2O$			13		1420								
			2.2	13	20	1070			6 - 9					
Sodium sulphite	$Na_2SO_3 7H_2O$			13		1530							20.9	
Sodium thiosulphate	$Na_2S_2O_3 5H_2O$			13		1730					40			
			2.2		1-conc				4 - 6					
Soya oil		Table 2.17	(0)			920								
Stearic acid	$CH_3(CH_2)_{16}COOH$					840				380	- 70			
Styrene	C_8H_8	< 1	0.95	21		910				145		2b		
Sulphate liquor 20% DS*		Fig 2.20		13		1090								*DS - Dry substance, content by weight - see equations 2.13 to 2.20
40% DS*		Fig 2.20				1195								
50% DS*		Fig 2.20				1250								
60% DS*		Fig 2.20				1305								
Sulphur	S					2060	20.4			445 *	113			
Sulphur dioxide	SO_2					1380	20.4			- 10	- 73			
	$SO_2 H_2O$			5				20.6	0 - 4					
Sulphuric acid D	H_2SO_4			13		1030		20.6	0 - 4	338				
Sulphurous acid D	H_2SO_3			13		1840		20.6	0 - 4					
				13	10	1070			0 - 4					
				13	15	1100			0 - 4					
			2.1	13	20	1140			0 - 4					
				13	30	1220			0 - 4					
				20	40	1300			0 - 4					
				20	50	1400			0 - 4					
				20	60	1500			0 - 4					
			0.1	20	70	1610			0 - 4					
				20	80	1730			0 - 4					
				20	90	1820			0 - 4					
Sweetened juice				13										

20.2.2 Liquid Table

Substance		Chemical formula	Viscosity mm^2/s at 20 °C	Vapour pressure kPa		Conc. % in H_2O	Density			pH/ region	Boiling point °C at 101.3 kPa (760 mm Hg)	Melting point °C	Fire hazard class- (Chapter 2 Section 2.1.7)	Solubility in H_2O at various temps. See Fig no.	Comments
				at 20 °C	at various temps. See Fig 20.3 curve no.		at 25 °C	at various temps. See Fig no.	at various conc. See Fig no.						
Tannic acid		$C_{76}H_{52}O_{46}$				5 - 50				0 - 4					
						10	1035			0 - 4					
Tartaric acid		$C_4H_6O_6$			13	10 - 50				0 - 4					
Tetrachloroethane		$CHCl_2CHCl_2$	1.1	0.8	20		1590	20.4			146				
Tetrachloroethylene	D	C_2Cl_4	< 1	2.4	19		1620	20.4			121				
Tetralin		$C_{10}H_{12}$		0.1			970				207		3		
Tin		Sn					5750				2260	232			
Tin (II) chloride		$SnCl_2\ 2H_2O$					2710					38			
Tin (IV) chloride		$SnCl_4$					3950				652	246			
Tincture of iodine			< 5	8			905				78		1		
Thioglycollic acid		CH_2SHCO_2H					1330				120	- 17			
Tuolene	D	$C_6H_5CO_3$	< 1	3.5	18		860	20.4			110		1		
Trichloroacetic acid	D	CCl_3CO_2H					1620			0 - 4	198	56			
Trichloroethylene	D	C_2HCl_3	< 1	7.5	16		1460	20.4			87				
Turpentine							860				156		2b		E-mod and speed of sound - Table 2.1
Unslaked lime		CaO					3250				2850	2580			
						10				9 - 14					
Valtran			Table 2.17	(0)			920								
Vinylacetate		$C_4H_6O_2$		15.5			930				73		1		
Vinegar							990			0 - 4					
Water			1	2.3	13		997			6 - 9	100				See Section 2.2 and Table 2.1
Xylene	D	$C_6H_4(CH_3)_2$	< 1	1			860	20.4			139		2a		
Zinc		Zn					7140					419			
Zinc chloride		$ZnCl_2$			13		2910					283			
				2.2	13	5	1030			4 - 6					
				2.2	13	20	1150			4 - 6					
				2.2	13	30	1220			4 - 6					
				2.2	13	40	1420			4 - 6					
				2.2	13	60	1750			4 - 6					
Zinc cyanide		$Zn(CN)_2$					1850								
Zinc sulphate		$ZnSO_4\ 7H_2O$			13		1960					100		20.10	

20.2.3 Supplementary diagrams

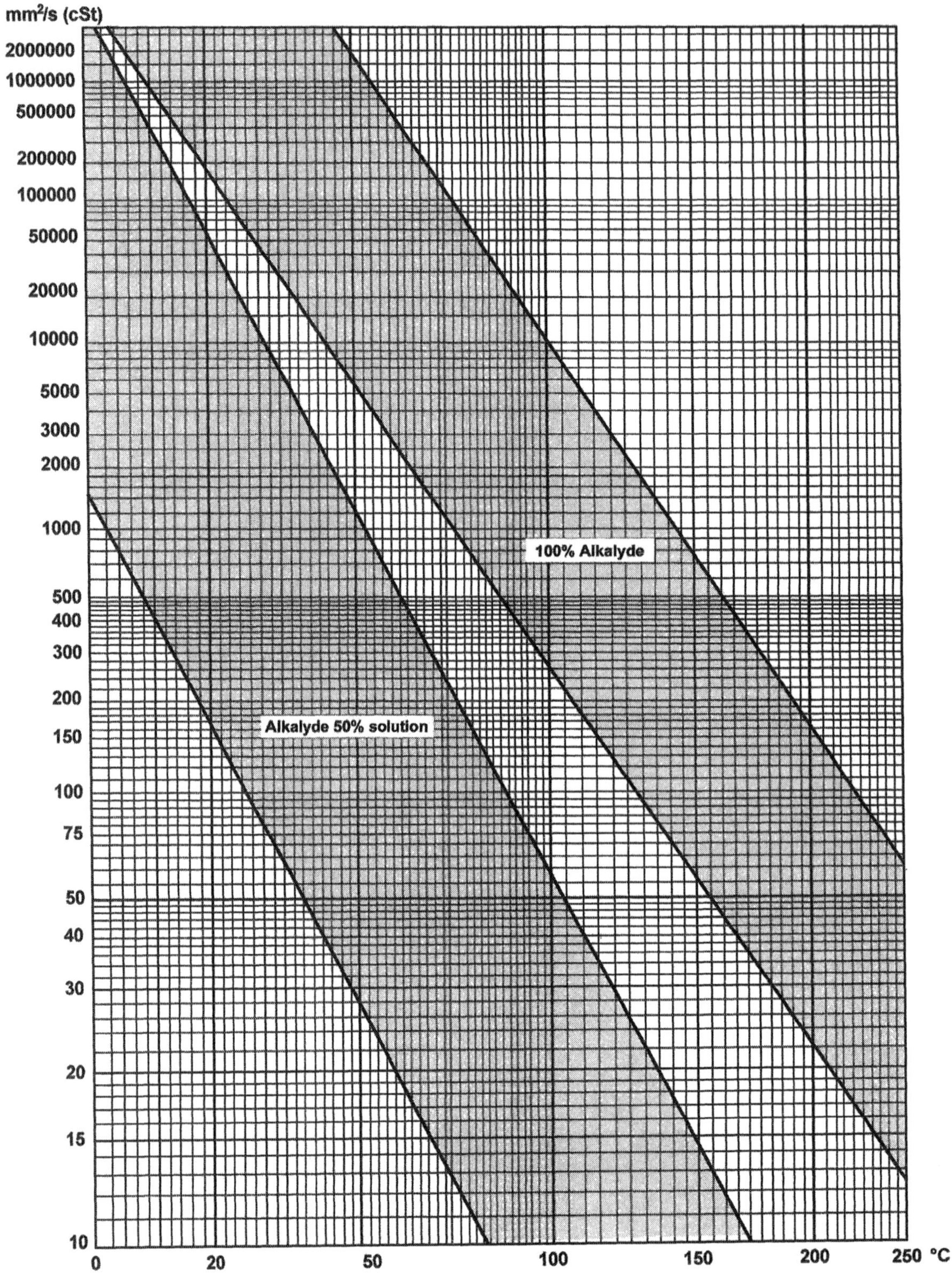

Figure 20.1 Viscosity of alkydes at various temperatures

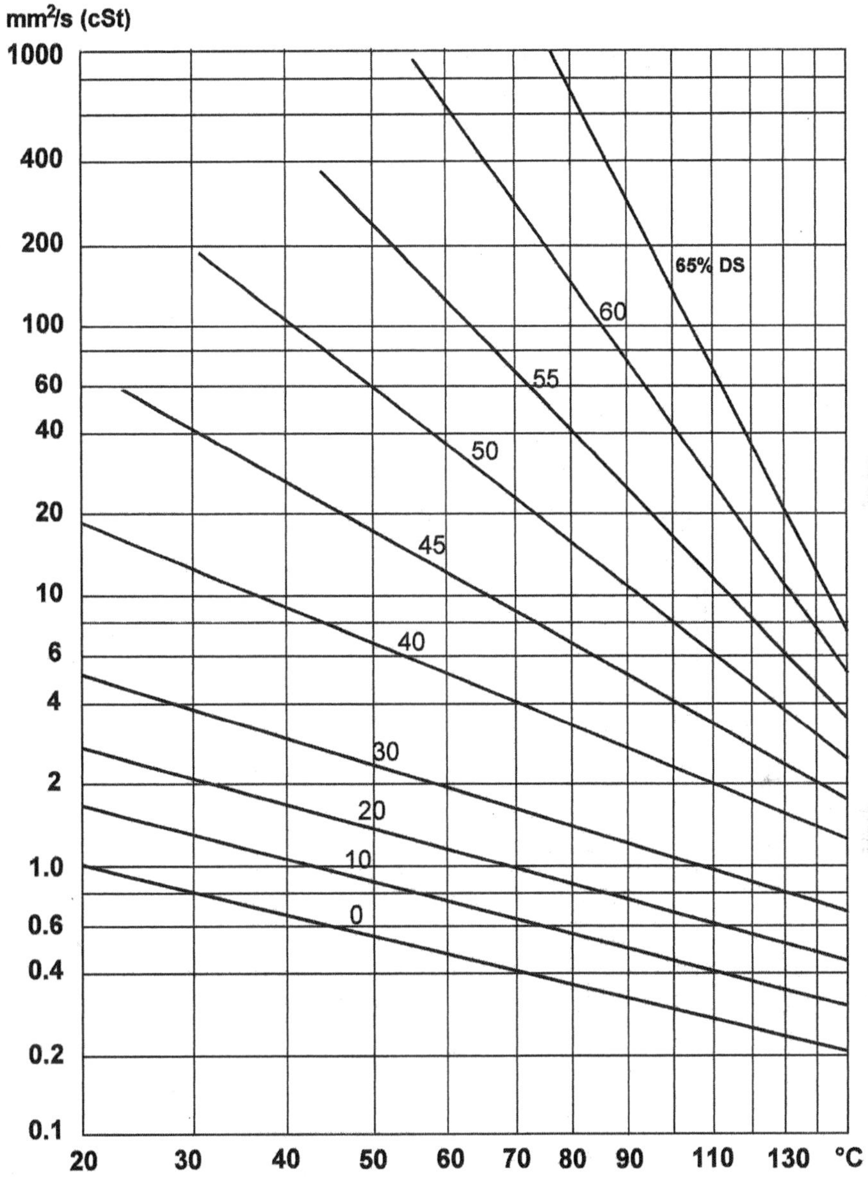

Figure 20.2 Viscosity temperature diagram for black liquor

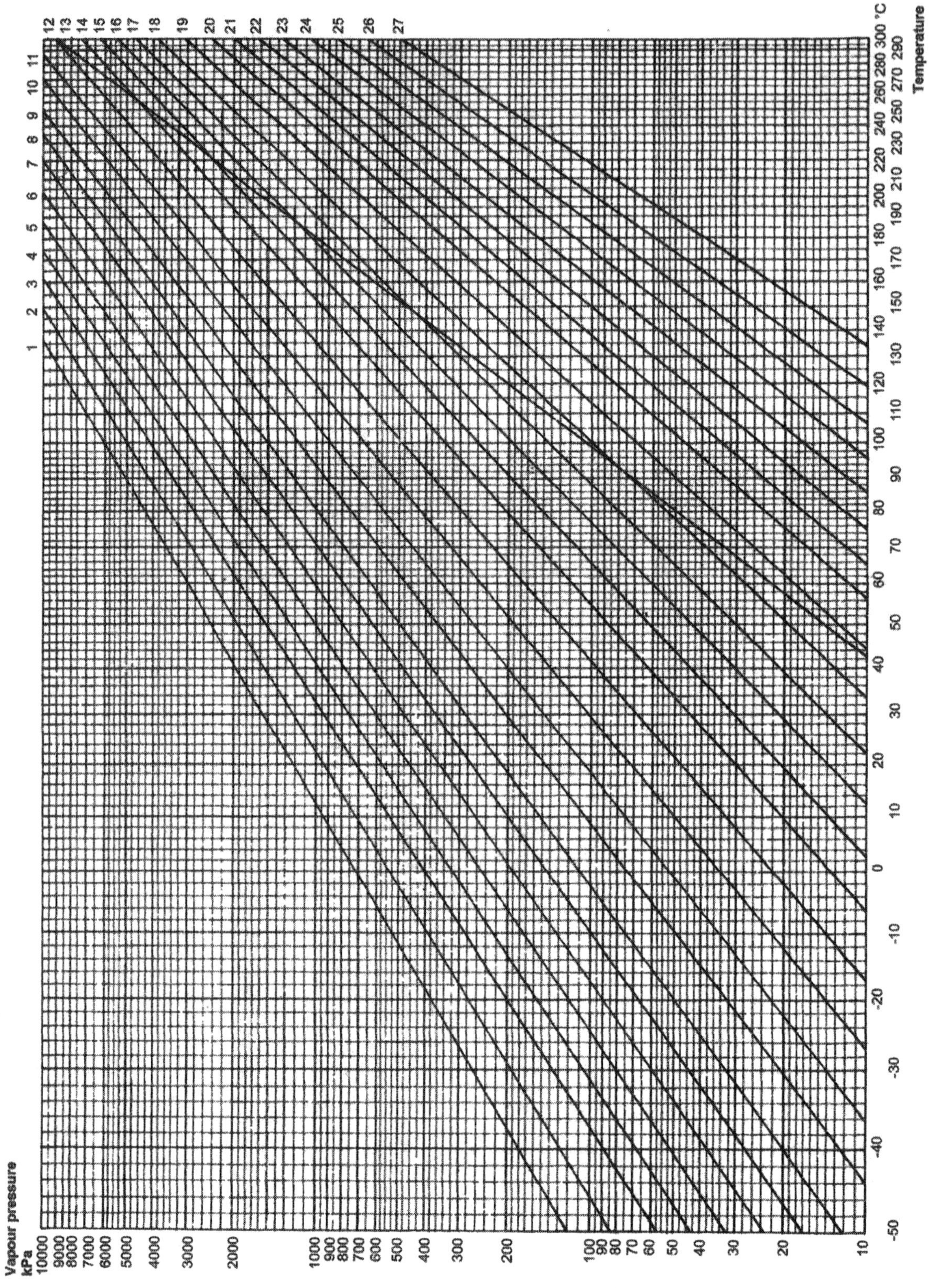

Figure 20.3 Vapour pressure diagram (see Liquid Table, Section 20.2.2 for cross references)

Substance	Temperature °C									
	-100	-75	-50	-25	0	20	50	100	150	200
Acetaldehyde						783				
Acetic acid						1049	1018	960	896	827
Acetone	920	893	868	840	812	791	756			
Ammonia			695		636	609	561	458		
Aniline					1039	1022	996	951		
Benzene					900	879	847	793	731	661
Benzylalcohol					1061	1045	1022			
Bromide					3188	3120				
Butane	698	676	652	627	601	579	542	468	296	
Butanol					825	810				
Butyric acid					977	958	927			774
Carbon bi-sulphide	1432		1362		1292	1262				
Carbon dioxide			1154	1070	925	772				
Chlorine	1717		1595		1469	1411	1314	1111	810	
Chlorobenzene					1128	1106	1074	1019	960	896
Chloroform			1618		1526	1490	1433			
Cyclohexane						779	750	700	645	578
Cyclohexanone					963	952	925			
Cyclohexene				842	830	811	780	732		
Dichloroacetic acid						1552				
Dichloroethane					1207	1176				
Dichloroethylene						1250				
Dioxane						1030	1010			
Ethanol					806	789	763	716	649	557
Ether (Diethylether)	842	816	790	764	736	714	676	611	518	
Ethyl acetate					924	901	864	797	721	621
Ethylene glycol					1128	1112				
Ethyl chloride					919	892	846			
Formic acid						1220	1184			
Furfurol					1181	1160	1128			
Glycerine					1273	1261	1242	1209		
Hexane		742	721	700	678	659	631	580	520	438
Hydrochloric acid	1235		1076		920					
Hydrofluoric acid	1660		1123		1002	987				
Hydrogen sulphide	1170		980		870					
Iodine	5060		5010		4960	4940			3780	
Methanol					810	792	765	714	650	553
Methyl acetate					959	934	894	822	734	610
Methylene chloride					1362	1326				
Methylethylketone					826	803				
Nitrobenzene					1223	1203	1174			
Nonane			769	751	733	718	694	653	609	
Octane		860	757	738	719	703	678	635	588	532
Octanol					842	829				
Pentane	737	715	693	670	646	626	596	533	460	
Phenol					1092	1071	1050			
Propane	646	619	590	560	528	501	450			
Propanol						804	779	733	674	592
Styrol						910				
Sulphur	2100		2080		2070	2060			1780	1740
Sulphur dioxide			1557		1435	1383	1296	1118	768	
Tetrachloroethane					1626	1593				
Tetrachloroethylene					1656	1621				
Toluole					885	868	839	739	737	672
Trichlorethylene						1463				
Xylene					881	864	838	795		678

Figure 20.4 Density for liquids at various temperatures

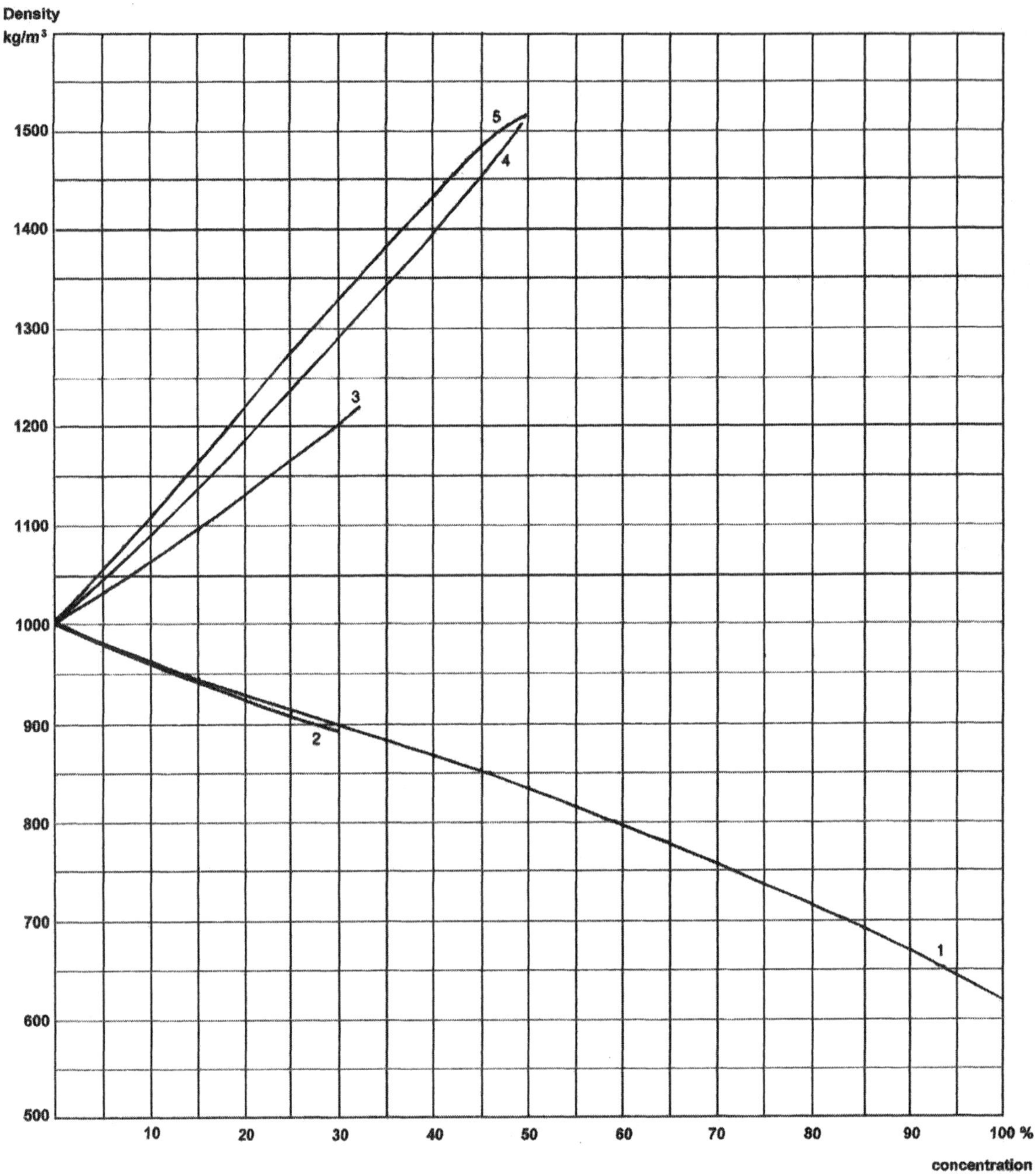

1. Ammonia solution at 15 °C
2. Ammonium hydroxide solution 20 °C
3. Calcium hydroxide solution 20 °C
4. Potassium hydroxide solution 15 °C
5. Sodium hydroxide solution 15 °C

Figure 20.5 Density against concentration for various bases

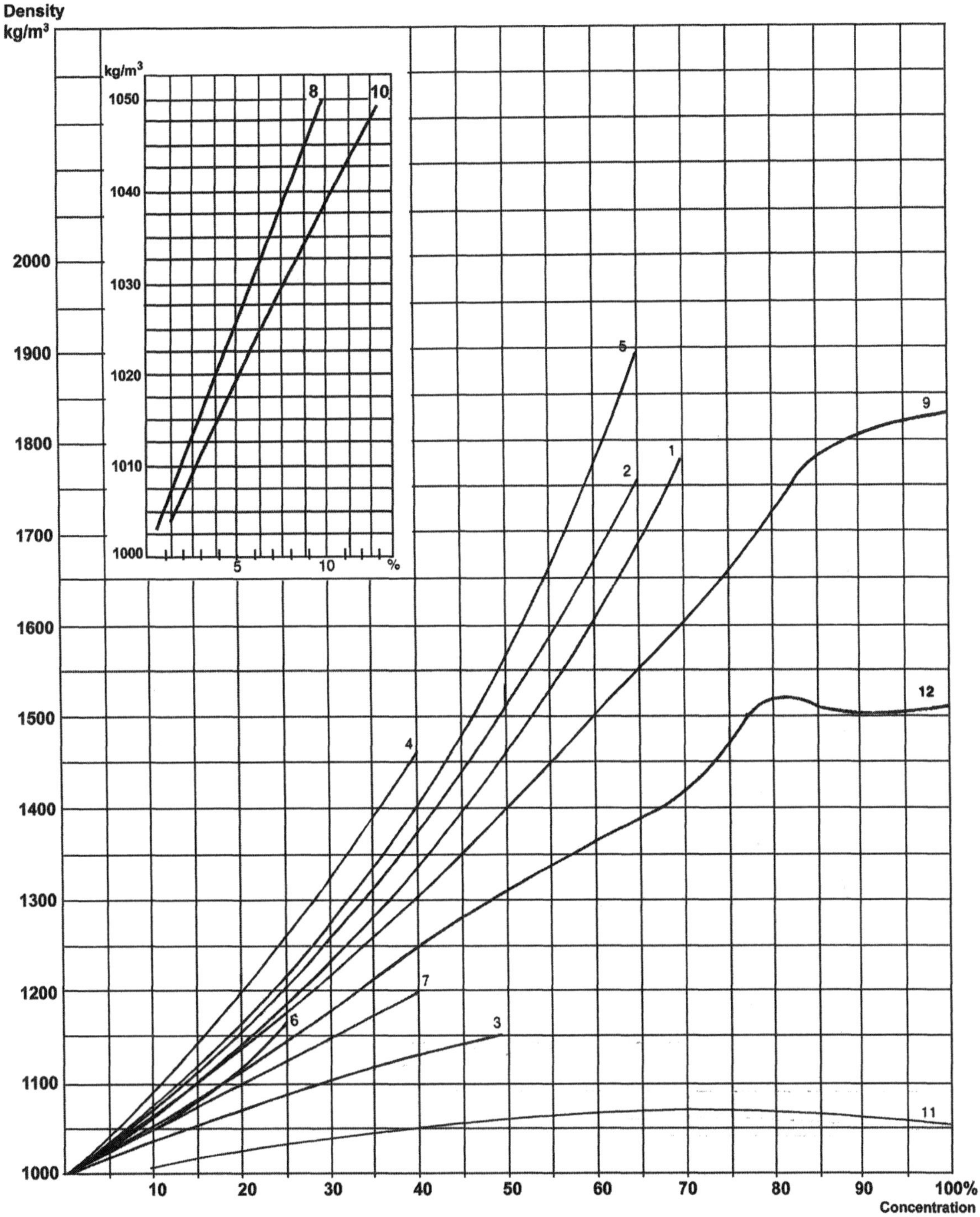

1. Arsenic acid	at	15°C	7. Hydrochloric acid	at 20°C
2. Hydrobromic acid		20°C	8. Sulphur dioxide solution	15°C
3. Hydrofluoric acid		20°C	9. Sulphuric acid	20°C
4. Iodic acid		18°C	10. Sulphurous acid	15,5°C
5. Hydroidinic acid		20°C	11. Acetic acid	20°C
6. Chloric acid		18°C	12. Nitric acid	18°C

Figure 20.6 Density against concentration of various acids

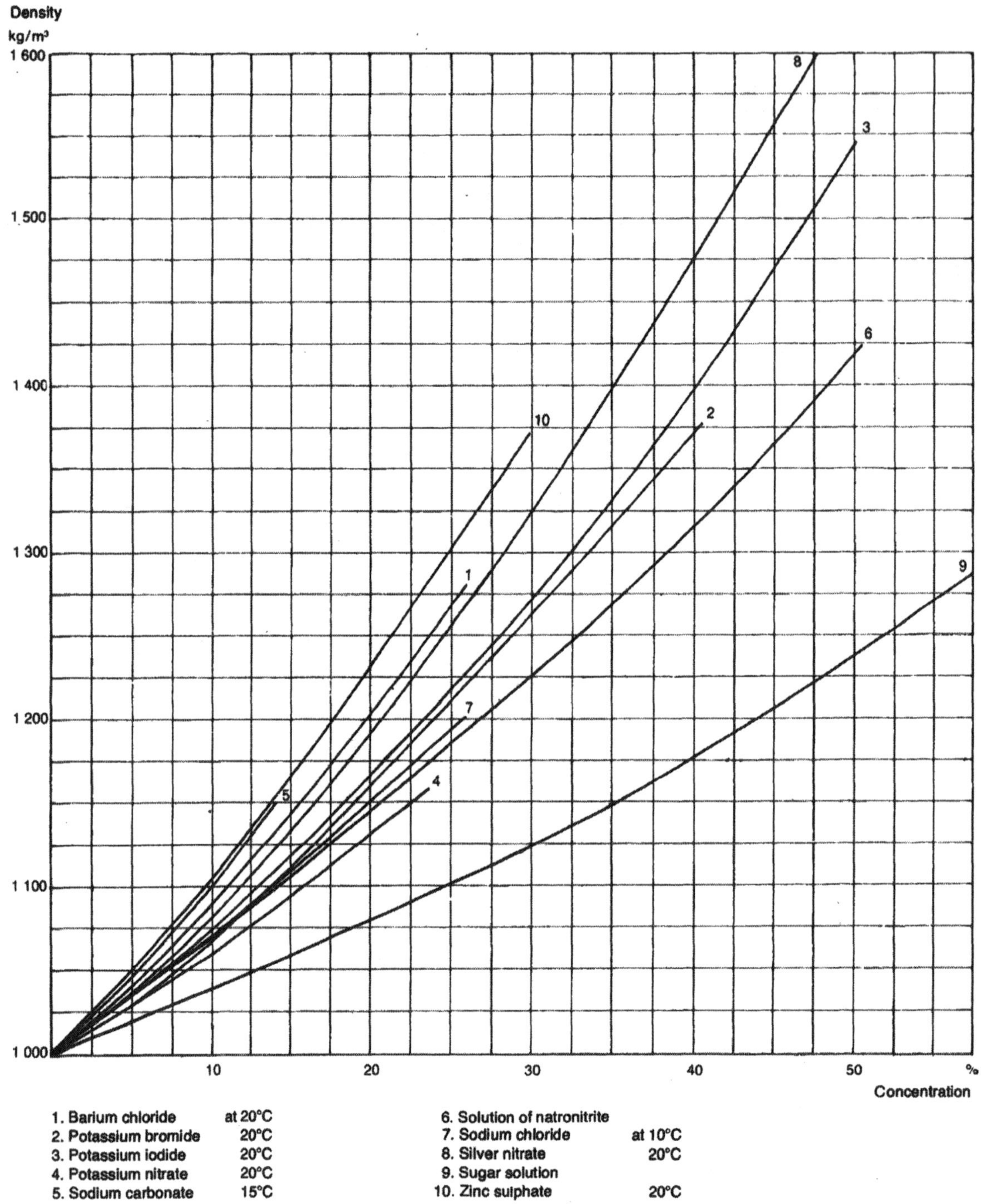

Figure 20.7 Density against concentration for various salts

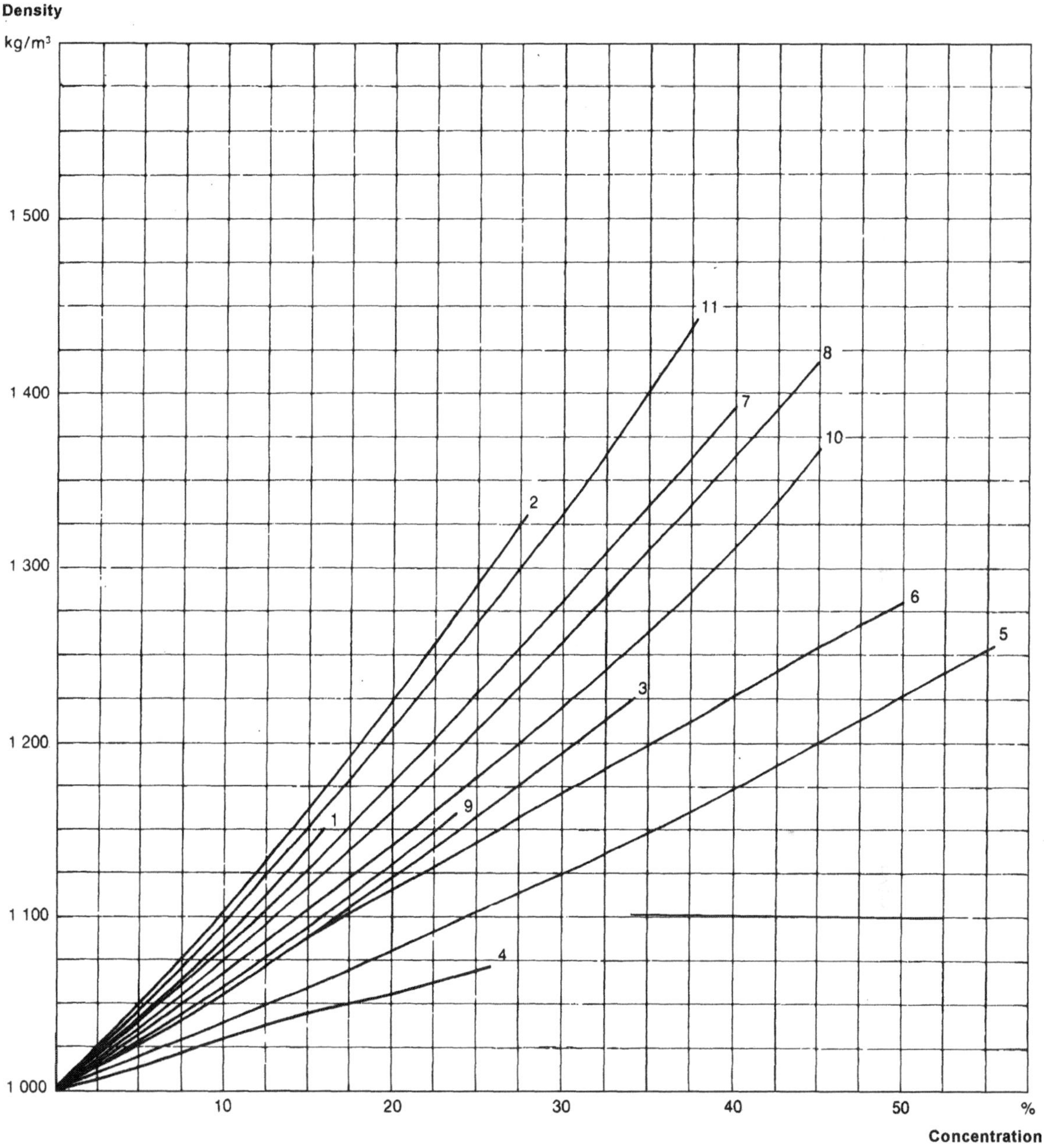

1. Aluminium chloride	at 18°C	7. Calcium chloride	at 20°C
2. Aluminium sulphate	19°C	8. Calcium nitrate	18°C
3. Ammonium bromide	18°C	9. Potassium chloride	20°C
4. Ammonium chloride	20°C	10. Sodium nitrate	20°C
5. Ammonium nitrate	20°C	11. Sodium silicate	20°C
6. Ammonium sulphate	20°C		

Figure 20.8 Density against concentration for various salts

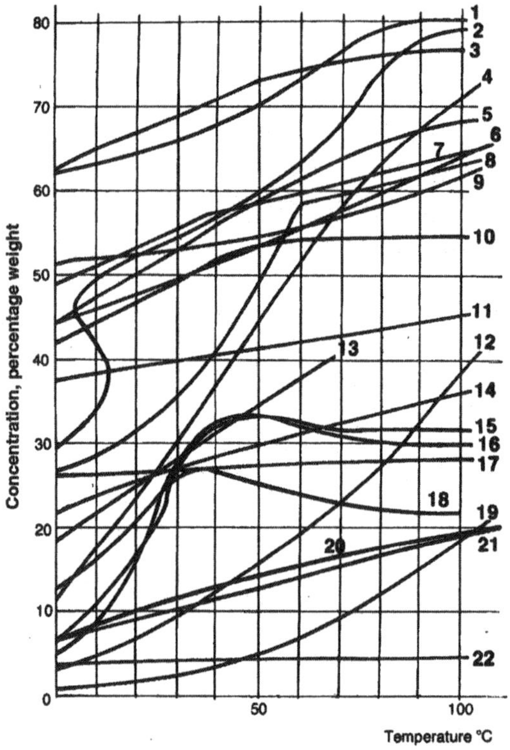

	Curve no.
Potassium bicarbonate	13
Potassium hydroxide	7
Potassium carbonate	9
Potassium chlorate	12
Potassium chloride	14
Potassium chromate	11
Potassium nitrate	4
Potassium perchlorate	19
Potassium sulphate	21
Sodium acetate	8
Sodium bicarbonate	20
Sodium bromide	10
Sodium dichromate	1
Sodium fluoride	22
Sodium hydroxide	2
Sodium carbonate	15
Sodium chlorate	5
Sodium chloride	17
Sodium nitrate	6
Sodium perchlorate	3
Sodium sulphate	16
Sodium sulphite	18

Figure 20.9 Solubility of salts in water at various temperatures

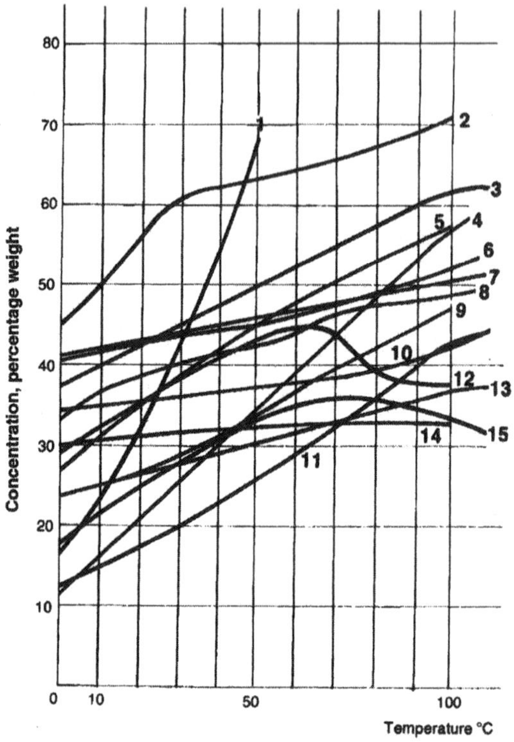

	Curve No.
Aluminium chloride	14
Aluminium nitrate	3
Aluminium sulphate	9
Ammonium perchlorate	4
Ammonium sulphate	7
Barium chloride	13
Lead acetate	1
Lead nitrate	5
Iron (II) chloride	8
Copper (II) chloride	6
Copper (II) nitrate	2
Copper (II) sulphate	11
Magnesium chloride	10
Magnesium sulphate	15
Zinc sulphate	12

Figure 20.10 Solubility of salts in water at various temperatures

20.3 Thermodynamic properties of gases

20.3.1 General gas and vapour data

Gas or vapour	Chemical formula	Molecular weight	Specific gas constant (1)	Specific heat & constant pressure c_p (2)				Ratio of specific heat γ (3)				Hazar-dous (4)
				-40 °C	15 °C	65 °C	150 °C	-40 °C	15 °C	65 °C	150 °C	
Acetaldehyde	CH_3CHO	44.05	0.189	-	-	-	-	@30 °C>	1.14	-	-	PFR
Acetic acid	CH_3COOH	60.05	0.138	-	-	-	-	-	1.15	@136 °C>	1.15	PFR
Acetylene	C_2H_2	26.04	0.319	1.48	1.66	1.79	1.96	1.31	1.26	1.24	1.21	PFR
Air (dry)	(5)	28.96	0.287	1	1	1	1.02	1.4	1.4	1.24	1.39	
Anhydrous ammonia	NH_3	17.03	0.488	-	2.12	2.2	2.33	-	1.31	1.3	1.27	PF
Argon	Ar	39.95	0.208	0.523	0.523	0.523	0.523	1.67	1.67	1.67	1.67	
Benzene	C_6H_6	78.11	0.106	-	1.01	1.21	1.51	-	1.12	-	1.08	PF
Boron trifluoride	BF_3	67.82	0.123	-	-	-	-	-	1.2	-	-	PF
Butadiene	C_4H_6	54.09	0.154	-	1.44	1.63	1.93	-	1.12	1.11	1.09	PFR
Butane	C_4H_{10}	58.12	0.143	-	1.64	1.86	2.23	-	1.094	1.08	1.07	PF
Butylene	C_4H_8	56.11	0.148	-	1.49	1.7	2.03	-	1.11	1.1	1.08	PF
Carbon dioxide	CO_2	44.01	0.189	0.791	0.841	0.892	1.06	1.34	1.304	1.28	1.25	P
Carbon disulphide	CS_2	76.13	0.109	-	-	-	-	-	1.21	-	-	PF
Carbon monoxide	CO	28.01	0.297	1.04	1.04	1.04	1.05	1.4	1.4	1.4	1.4	PF
Carbon tetrachloride	CCl_4	153.82	0.054	-	-	-	-	-	1.11	-	-	P
Chlorine	Cl_2	70.91	0.117	-	0.481	0.49	0.51	-	1.355	-	-	PR
Cyclohexane	C_6H_{12}	84.16	0.099	-	1.21	1.46	1.9	-	1.089	@80 °C>	1.08	PF
Cyclopentane	C_5H_{10}	70.13	0.118	-	1.13	1.37	1.78	-	1.117	-	-	PF
Cyclopropane	C_3H_6	42.08	0.198	-	-	-	-	-	1.11	-	-	PF
Demethylamine	C_2H_7N	45.08	0.184						1.15			PF
Diethylene glycol	$C^4H_{10}O_3$	106.17	0.078						1.07			PF
Diemthyl ether	C_2H_6O	46.07	0.180						1.11			PFR
Dowtherm A™		165	0.050						1.05			
Dowtherm E™		147	0.056						1			
Ethane	C_2H_6	30.07	0.276	1.53	1.72	1.92	2.27	1.22	1.193	1.17	1.14	PF
Ethyl alcohol	C_2H_6O	46.07	0.180	-	1.28	-	-	1.29	1.13	@90 °C>	1.13	F
Ethyl chloride	C_2H_5Cl	64.52	0.129	1.316	1.49	1.66	1.98	-	1.19	-	-	PF
Ethylene	$C_2H_6O_2$	28.05	0.296	1.31	1.51	1.7	2.0	1.25	1.243	1.21	1.17	PFR
Ethylene glycol	$2(CH_3O)$	62.07	0.134	-	-	-	-	-	1.09	-	-	PF
Ethylene oxide	C_2H_4O	44.05	0.189	-	-	-	-	-	1.2	1.19	1.15	PFR
Glycerine	$C_3H_8O_3$	92.1	0.090	-	-	-	-	-	1.06	-	-	PF
Helium	He	4	2.078	5.193	5.193	5.193	5.193	1.66	1.66	-	-	
Heptane	C_7H_{16}	100.21	0.083	-	@98 °C >	1.98	2.2	-	@98 °C >	1.04	1.04	PF
Hexane	C_6H_{14}	86.18	0.096	-	1.66	1.87	2.23	@69 °C>	1.06	1.06	1.05	PF
Hydrogen	H_2	2.02	4.116	13.92	14.27	14.41	14.49	1.42	1.41	1.40	1.40	F
Hydrogen chloride	HCl	36.46	0.228	-	0.812	0.813	0.814	-	1.41	-	-	PFR
Hydrogen sulphide	H_2S	34.08	0.244	0.975	0.996	1.02	1.05	1.34	1.33	1.32	1.3	PF
Isobutane	C_4H_{10}	58.12	0.143	-	1.62	1.86	2.23	-	1.094	1.08	1.07	PF
Isobutylene	C_4H_8	56.1	0.148	-	1.55	1.75	2.06	-	1.11	1.09	1.08	PF
Isopentane	C_5H_{12}	72.15	0.115	@28 °C>	1.68	1.84	2.21	@28 °C>	1.07	1.07	1.06	PF
Isoprene	C_5H_8	68.12	0.122	-	1.49	-	-	-	1.09	-	-	PFR
Isopropyl alcohol	C_3H_8O	60.1	0.138	-	-	-	-	-	1.09	-	-	PF
KLEA 61™ (R407B) (6)		102.9	0.081	-	0.795	0.875	0.995					
KLEA 66™ (R407C)		86.2	0.096	-	0.975	1.06	1.18					
KLEA 134a™		102.03	0.081	0.711	0.814	0.901	1.04					
KLEA 407A™		90.1	0.092	-	0.804	0.885	1.01					
Krypton	Kr	83.8	0.099						1.68			
Methane	CH_4	16.04	0.518	2.12	2.21	2.34	2.61	1.33	1.308	1.29	1.25	PF
Methyl alcohol	CH_4O	32.04	0.259	-	1.13	-	-	-	1.203	-	-	PF
Methyl chloride	CH_3Cl	50.49	0.165	@25 °C>	0.833	-	-	@25 °C>	1.29	-	-	PF
Methyl mercapton	CH_4S	48.11	0.173						1.2			PF
Monomethylamine	CH_5N	31.06	0.268						1.02			PF
Naphthalene	$C_{10}H_8$	128.17	0.065						1.07			PF
Natural gas	(7)	17.4/19.0	0.478						1.27			PF
n-Butane	C_4H_{10}	58.12	0.143	-	1.64	1.86	2.23	-	1.094	1.08	1.07	PF
n-Decane	$C_{10}H_{22}$	142.29	0.058	-	1.62	1.85	2.21	-	1.03	-	-	F
Neon	Ne	20.18	0.412	1.03	1.03	1.03	-	1.66	1.66	1.66		
n-Heptane	C_7H_{16}	100.21	0.083	-	@98 °C>	1.98	2.2	-	@98 °C>	1.04	1.04	PF
n-Hexane	C_6H_{14}	86.17	0.096	-	@69 °C>	1.85	2.2	-	@69 °C>	1.06	1.05	PF
Nitrogen	N_2	28.01	0.297	1.04	1.04	1.04	1.05	1.4	1.4	1.4	1.4	
Nitric acid	NO	30.01	0.277	1	0.996	0.996	1	1.38	1.39	1.39	1.38	P
Nitrous oxide	N_2O	44.01	0.189	-	0.879	-	-	-	1.303	@100 °C>	1.28	
n-Octane	C_8H_{18}	114.22	0.073	-	@125 °C>	1.88	2.19	-	@125 °C>	1.04	1.03	F
n-Nonane	C_9H_{20}	128.25	0.065	-	1.67	-	-	-	1.04	-	-	F
n-Pentane	C_5H_{12}	72.15	0.115	@36 °C>	1.72	1.85	2.21	@36 °C>	1.074	1.07	1.06	PF

Gas or vapour	Chemical formula	Molecular weight	Specific gas constant (1)	Specific heat & constant pressure c_p (2)				Ratio of specific heat γ (3)				Hazar-dous (4)
				-40 °C	15 °C	65 °C	150 °C	-40 °C	15 °C	65 °C	150 °C	
Octane	C_8H_{18}	114.23	0.073	-	1.57	1.9	2.19	-	@125 °C>	1.04	1.03	F
Oxygen	O_2	32	0.26	0.913	0.917	0.925	0.946	1.4	1.4	1.39	1.38	PFR
Pentane	C_5H_{12}	72.15	0.115	@36 °C>	1.72	0.85	2.21	@36 °C>	1.074	1.07	1.06	PF
Phosgene	$COCl_2$	98.92	0.084	0.515	0.569	0.611	0.661	1.19	1.17	1.16	1.14	P
Propadiene	C_3H_4	40.07	0.207	-	-	-	-	-	1.69	-	-	
Propane	C_3H_8	44.1	0.188	1.39	1.63	1.85	2.23	1.16	1.133	1.11	1.09	PF
Propylene	C_3H_6	42.08	0.198	1.27	1.48	1.67	1.98	1.18	1.154	1.14	1.11	PFR
Propylene oxide	C_3H_6O	58.08	0.143	-	-	-	-	-	1.13	-	-	PFR
Refrigerant 11 (8) (9)	CCl_3F	137.38	0.061	@24 °C>	0.561	0.59	0.653	@24 °C>	1.14	1.13	1.1	
Refrigerant 12 (10)	CCl_2F_2	120.92	0.069	@30 °C>	0.607	-	-	@30 °C>	1.14	-	-	
Refrigerant 13 (11)	$CClF_3$	104.47	0.08	0.057	0.628	0.687	0.766	1.17	1.15	1.13	1.12	
Refrigerant 21 (12)	$CHCl_2F$	102.93	0.081	-	0.569	0.62	0.707	-	1.18	1.16	1.13	
Refrigerant 22 (13)	$CHClF_2$	86.48	0.096	-	0.624	0.674	0.762	-	1.2	1.17	1.14	
Refrigerant 113 (14)	$C_2Cl_3F_3$	187.39	0.044	@48 °C>	0.666	0.678	0.749	@48 °C>	1.08	1.08	1.07	
Refrigerant 114 (15)	$C_2Cl_2F_4$	170.93	0.049	-	0.657	0.703	0.787	-	1.09	1.08	1.07	
Refrigerant 115 (16)	C_2ClF_5	154.48	0.054						1.08			
Refrigerant 500		99.29	0.084					@30 °C>	1.13			
Refrigerant 502		111.6	0.075					@30 °C>	1.133			
Styrene	C_8H_8	104.15	0.08						1.076			PFR
Sulphur dioxide	SO_2	64.06	0.13		0.615				1.24			P
Sulphur hexafluoride	SF_6	146.05	0.057						1.09			P
Steam (dry saturated)	H_2O	18.02	0.461		@100 °C>	1.88			1.135	1.135	1.135	
Steam (superheated)	H_2O	18.02	0.461				1.91		1.3	1.3	1.3	
Toluene	C_7H_8	92.14	0.09	-	@110 °C>	1.45	1.59		1.09		1.06	PF
Trimethylamine	C_3H_9N	59.11	0.141						1.18			PF
Triethylene glycol	$C_6H_{14}O_4$	150.18	0.055						1.04			PF
Vinyl chloride	C_2H_3Cl	62.5	0.133						1.19			PFR
Xenon	Xe	131.3	0.063						1.66			
o-Xylene	C_8H_{10}	106.16	0.078	-	1.22	-	-	-	1.069			PF
m-Xylene	C_8H_{10}	106.16	0.078	-	1.16	-	-	-	1.072			PF
p-Xylene	C_8H_{10}	106.16	0.078	-	1.16	-	-	-	1.072			PF

(1) kJ/(kg K)
(2) kJ/(kg K) at 1.01325 bar absolute
(3) At 1.01325 bar absolute
(4) General hazards; F-flammable, P-personnel, R-chemically reactive
(5) Analysis by weight; 75.53% N_2, 23. 14% O_2, 1.28% Ar, 0.05% CO_2. Real air is about 36% relative humidity with γ = 1.395 at 1.01325 bara and 20 °C
(6) New refrigerants to replace CFCs; KLEA 61 for R502, KLEA 66 for R22, KLEA 134a for R12
(7) Approximately 955 Methane, relative density 0.6
(8) This group of refrigerants are called Fluorocarbons or CFCs
(9) Trichloromonofluoromethane
(10) Dichlorodifluoromethane
(11) Chlorotrifluoromethane
(12) Dichloromonofluoromethane
(13) Chlorodifluoromethane
(14) Trichlorotrifluoroethane
(15) Dichlorotetrafluoroethane
(16) Chloropentafluoroethane

20.3.2 Table of physical constants

	See Note No: →		A.	B.		C.	D.				
								Critical constant			
Number	Compound	Formula	Molar mass (molecular weight)	Boiling point, °C 101.3250 kPa (abs)	Vapour pressure, kPa (abs), 40 °C	Freezing point, °C 101.3250 kPa(abs)	Refractive index n_D 15 °C	Pressure kPa(abs)	Temperature, K	Volume, m³/kg	Number
1	Methane	CH_4	16.043	-161.51	(35000)*	-182.45*	1.00040*	4599.	190.56	0.00617	1
2	Ethane	C_2H_6	30.070	-88.59	(6000)*	-182.79*	1.21098*	4880.	305.41	0.00489	2
3	Propane	C_3H_8	44.097	-42.07	1369.8	-187.62*	1.29526*	4240.	369.77	0.00454	3
4	Isobutane	C_4H_{10}	58.123	-11.79	530.89	-159.59	1.3249*	3640.	407.82	0.00446	4
5	n-Butane	C_4H_{10}	58.123	-0.51	379.61	-138.35	1.33626*	3784.	425.10	0.00439	5
6	Isopentane	C_5H_{12}	72.150	27.83	151.31	-159.89	1.34803	3381.	460.35	0.00427	6
7	n-Pentane	C_5H_{12}	72.150	36.05	114.70	-129.71	1.35196	3365.	469.66	0.00434	7
8	Neopentane	C_5H_{12}	72.150	9.50	270.0	-16.58	1.342*	3199.	433.71	0.00420	8
9	n-Hexane	C_6H_{14}	86.177	68.72	37.297	-95.31	1.37736	3030.	506.4	0.00429	9
10	2-Methylpentane	C_6H_{14}	86.177	60.24	50.68	-153.67	1.36601	3010.	497.46	0.00426	10
11	3-Methylpentane	C_6H_{14}	86.177	63.26	45.73	-162.89	1.37120	3120.	504.4	0.00426	11
12	Neohexane	C_6H_{14}	86.177	49.72	73.41	-99.825	1.36314	3080.	488.66	0.00417	12
13	2,3-Dimethylbutane	C_6H_{14}	86.177	57.96	55.34	-128.53	1.36967	3130.	499.86	0.00415	13
14	n-Heptane	C_7H_{16}	100.204	98.37	12.342	-90.55	1.38262	2740.	539.2	0.00426	14
15	2-Methylhexane	C_7H_{16}	100.204	90.03	17.226	-118.26	1.37969	2730.	530.06	0.00420	15
16	3-Methylhexane	C_7H_{16}	100.204	91.85	16.155	—	1.38354	2810.	535.16	0.00403	16
17	3-Ethylpentane	C_7H_{16}	100.204	93.47	15.265	-118.58	1.38828	2890.	540.46	0.00415	17
18	2,2-Dimethylpentane	C_7H_{16}	100.204	79.17	26.32	-123.78	1.37696	2770.	520.36	0.00415	18
19	2,4-Dimethylpentane	C_7H_{16}	100.204	80.47	24.85	-119.21	1.37620	2740.	519.66	0.00417	19
20	3,3-Dimethylpentane	C_7H_{16}	100.204	86.04	20.94	-134.44	1.38592	2950.	536.26	0.00413	20
21	Triptane	C_7H_{16}	100.204	80.86	25.41	-24.56	1.38439	2950.	531.06	0.00397	21
22	n-Octane	C_8H_{18}	114.231	125.65	4.146	-56.76	1.39274	2490.	568.4	0.00420	22
23	Diisobutyl	C_8H_{18}	114.231	109.08	8.417	-91.160	1.38762	2490.	549.96	0.00422	23
24	Isooctane	C_8H_{18}	114.231	99.21	12.966	-107.35	1.38651	2570.	543.86	0.00410	24
25	n-Nonane	C_9H_{20}	128.258	150.78	1.358	-53.48	1.40080	2280.	594.7	0.00433	25
26	n-Decane	$C_{10}H_{22}$	142.285	174.11	0.4814	-29.63	1.40745	2100.	617.7	0.00439	26
27	Cyclopentane	C_5H_{10}	70.134	49.22	73.99	-93.827	1.40081	4508.	511.6	0.00371	27
28	Methylcyclopentane	C_6H_{12}	84.161	71.83	33.75	-142.43	1.40430	3784.	532.75	0.00379	28
29	Cyclohexane	C_6H_{12}	84.161	80.78	24.64	6.550	1.42083	4073.	553.5	0.00366	29
30	Methylcyclohexane	C_7H_{14}	98.188	100.94	12.211	-126.59	1.41806	3471.	572.15	0.00375	30
31	Ethane (Ethylene)	C_2H_4	28.054	-103.73	(9700)*	-169.15*	(1.229)*	5040.	282.34	0.00466	31
32	Propene (Propylene)	C_3H_6	42.081	-47.68	1691.	-185.25*	1.3010*	4665.	365.55	0.00448	32
33	1-Butene (Butylene)	C_4H_8	56.108	-6.23	459.0	-185.35*	1.3390*	4043.	419.92	0.00426	33
34	cis-2-Butene	C_4H_8	56.108	3.72	338.3	-138.90	1.3560*	4243.	435.54	0.00417	34
35	trans-2-Butene	C_4H_8	56.108	0.88	366.5	-105.54	1.3490*	3964.	428.59	0.00424	35
36	Isobutene	C_4H_8	56.108	-6.91	477.4	-140.34	1.3477*	4000.	417.86	0.00425	36
37	1-Pentene	C_5H_{10}	70.134	29.95	141.65	-165.21	1.36522	3513.	464.74	0.00421	37
38	1,2-Butadiene	C_4H_6	54.092	10.84	269.	-136.19	—	(4502)*	(444)*	(0.0043)*	38
39	1,3-Butadiene	C_4H_6	54.092	-4.41	436.1	-108.89	1.3979*	4277*	425.	0.00407	39
40	Isoprene	C_5H_8	68.119	34.05	123.8	-145.95	1.41510	(93850)*	(484)*	(0.0041)*	40
41	Acetylene	C_2H_2	26.038	-84.01*	—	-80.8*	—	6139.	308.31	0.00432	41
42	Benzene	C_6H_6	78.114	80.07	24.38	5.532	1.49472	4898.	562.12	0.00332	42
43	Toluene	C_7H_8	92.141	110.60	7.895	-94.98	1.49133	4106.	591.76	0.00343	43
44	Ethylbenzene	C_8H_{10}	106.167	136.17	2.873	-94.963	1.49052	3606.	617.16	0.00352	44
45	o-Xylene	C_8H_{10}	106.167	144.39	2.051	-25.18	1.50045	3734.	630.29	0.00348	45
46	m-Xylene	C_8H_{10}	106.167	139.09	2.528	-47.86	1.49206	3536.	617.01	0.00354	46
47	p-Xylene	C_8H_{10}	106.167	138.32	2.648	13.26	1.49068	3511.	616.19	0.00357	47
48	Styrene	C_8H_8	104.152	145.23	2.00	-30.63	1.54969	4050.	(646)*	0.00333	48
49	Isopropylbenzene	C_9H_{12}	120.194	152.38	1.47	-96.021	1.48635	3209.	631.1	0.00355	49
50	Methyl alcohol	CH_4O	32.042	64.67	35.44	-97.65	1.32464	8097.	512.60	0.00368	50
51	Ethyl alcohol	C_2H_6O	46.069	78.26	17.903	-114.1	1.35739	6148.	513.88	0.00362	51
52	Carbon monoxide	CO	28.010	-191.45	—	-204.99*	1.00028*	3494.	132.86	0.00329	52
53	Carbon dioxide	CO_2	44.010	-78.464*	—	-56.56*	1.00038*	7374.	304.11	0.00214	53
54	Hydrogen sulphide	H_2S	34.082	-60.266	2859.7	-85.48*	1.00057*	8963.	373.37	0.00288	54
55	Sulphur dioxide	SO_2	64.065	-9.94	630.2	-75.47*	1.00059*	7884.	430.8	0.00190	55
56	Ammonia	NH_3	17.0305	-33.32	1555.	-77.69*	1.00033*	11350.	405.5	0.00425	56
57	Air	N_2+O_2	28.9625	-194.34	—	—	1.00028*	3771.	132.43	0.00323	57
58	Hydrogen	H_2	2.0159	-252.850*	—	-259.347*	1.00013*	1293.	33.0	0.03185*	58
59	Oxygen	O_2	31.9988	-182.954*	—	-218.792*	1.00027*	5043.	154.59	0.00229	59
60	Nitrogen	N_2	28.0134	-195.798	—	-209.997*	1.00028*	3398.	126.21	0.00318	60
61	Chlorine	Cl_2	70.9054	-33.95	1146.	-100.95	1.3740*	7977.	416.86	0.00175	61
62	Water	H_2O	18.0153	99.974*	7.3849	0.000	1.33339	22064.	647.10	0.003106	62
63	Helium	He	4.0026	-268.95	—	—	1.00003*	227.5	5.20	0.01436	63
64	Hydrogen chloride	HCl	36.461	85.14	6547.	-114.17*	1.00040*	8310.	324.68	0.00222	64

NOTE: Numbers in this table do not have accuracies greater than 1 part in 1000; in some cases extra digits have been added to calculated values to achieve internal consistency or to permit recalculation of experimental values. * See the Notes and References in Section 20.3.3

20.3.2 Table of physical constants *(continued)*

	E.			F.	G.	H.	I.			J.		
	Density of liquid 101.3250 kPa(abs), 15 °C						Ideal gas 101.3250 kPa(abs), 15 °C			Specific Heat 15 °C 101.3250 kPa(abs) kJ/(Kg K)		
Number	Relative density (specific gravity) 15 °C/15 °C	Kg/m³ (wt in vacuum)	m³/kmol	Temperature coefficient of density, 1/T	Accentric factor, ω	Compressibility factor of real gas, Z, 15 °C 101.3250 kPa(abs)	Relative density (specific gravity) Air = 1	m³ gas/kg	Volume ratio gas to liquid	C_p^0, Ideal gas	C_p^0, Liquid	Number
1	(0.3)*	(300.)*	(0.05)*	—	0.0108	0.9980	0.55392	1.4738	(442.15)*	2.2040	—	1
2	0.35808*	357.76*	0.084051*	—	0.0972	0.9918	1.0382	0.78635	281.32*	1.7053	4.0283	2
3	0.50776*	507.30*	0.086925*	-0.00303*	0.1515	0.9824	1.5226	0.53619	272.01*	1.6242	2.5918	3
4	0.56349*	562.98*	0.10324*	-0.00220*	0.1852	0.9709	2.0068	0.40680	229.02*	1.6164	2.3879	4
5	0.58459*	584.06*	0.099515*	-0.00197*	0.1981	0.9664	2.0068	0.40680	237.60*	1.6514	2.3949	5
6	0.62491	624.35	0.11556	-0.00162	0.2286	—	2.4912	0.32772	204.61	1.6071	2.2301	6
7	0.63157	631.00	0.11434	0.00161	0.2510	—	2.4912	0.32772	206.79	1.6233	2.2737	7
8	0.59724*	596.70*	0.12092*	-0.00191*	0.1965	0.9579	2.4912	0.32772	195.55*	1.6320	2.2994	8
9	0.66449	663.89	0.12981	-0.00136	0.2990	0.9881	2.9755	0.27438	182.15	1.6151	2.2306	9
10	0.65834	657.75	0.13102	-0.00141	0.2780	0.9848	2.9755	0.27438	180.47	1.6107	2.2053	10
11	0.66953	668.93	0.12883	-0.00137	0.2736	0.9863	2.9755	0.27438	183.54	1.5845	2.1697	11
12	0.65455	653.96	0.13178	-0.00146	0.2334	0.9798	2.9755	0.27438	179.43	1.5982	2.1483	12
13	0.66675	666.15	0.12937	-0.00137	0.2481	0.9840	2.9755	0.27438	182.78	1.5788	2.1458	13
14	0.68846	687.84	0.14568	-0.00122	0.3483	0.9948	3.4598	0.23597	162.30	1.6074	2.2087	14
15	0.68359	682.97	0.14672	-0.00127	0.3312	0.9931	3.4598	0.23597	161.16	1.5955	2.1832	15
16	0.69206	691.44	0.14492	-0.00124	0.3231	0.9935	3.4598	0.23597	163.16	1.5833	2.1338	16
17	0.70327	702.64	0.14261	-0.00125	0.3111	0.9939	3.4598	0.23597	165.80	1.6157	2.1502	17
18	0.67884	678.23	0.14774	-0.00129	0.2870	0.9903	3.4598	0.23597	160.04	1.6155	2.1611	18
19	0.67763	677.02	0.14801	-0.00130	0.3035	0.9907	3.4598	0.23597	159.75	1.6564	2.1930	19
20	0.69721	696.58	0.14385	-0.00095	0.2687	0.9920	3.4598	0.23597	164.37	1.6009	2.0992	20
21	0.69616	695.53	0.14407	-0.00157	0.2501	0.9907	3.4598	0.23597	164.12	1.5765	2.0876	21
22	0.70718	706.54	0.16168	-0.00115	0.3978	0.9977	3.9441	0.20699	146.25	1.6026	2.1923	22
23	0.69845	697.82	0.16370	-0.00121	0.3571	0.9959	3.9441	0.20699	144.44	1.5701	2.1383	23
24	0.69669	696.06	0.16411	-0.00118	0.3043	0.9941	3.9441	0.20699	144.08	1.5974	2.0495	24
25	0.72231	721.66	0.17773	-0.00110	0.4425	0.9990	4.4284	0.18436	133.04	1.5990	2.1855	25
26	0.73452	733.86	0.19389	-0.00102	0.4881	0.9996	4.9127	0.16618	121.95	1.5962	2.1797	26
27	0.75125	750.57	0.093441	-0.00129	0.1938	0.9844	2.4215	0.33714	253.05	1.1329	1.7636	27
28	0.75512	754.44	0.11155	-0.00125	0.2266	0.9905	2.9059	0.28095	211.96	1.2546	1.8451	28
29	0.78385	783.14	0.10747	-0.00119	0.2094	0.9928	2.9059	0.28095	220.02	1.2120	1.8223	29
30	0.77437	773.67	0.12691	-0.00112	0.2352	0.9953	3.3902	0.24081	186.31	1.3329	1.8405	30
31	—	—	—	—	0.0860	0.9935	0.96863	0.84282	—	1.4966	—	31
32	0.52183*	521.36*	0.080714*	-0.00308*	0.1404	0.9843	1.4529	0.56189	292.95*	1.4919	2.3927	32
33	0.60095*	600.41*	0.093449*	-0.00196*	0.1914	0.9697	1.9373	0.42143	253.03*	1.4856	2.2001	33
34	0.62921*	628.64*	0.089253*	-0.00193*	0.2054	0.9662	1.9373	0.42143	264.92*	1.3911	2.2154	34
35	0.61179*	611.24*	0.091794*	-0.00199*	0.2034	0.9665	1.9373	0.42143	257.59*	1.4875	2.2663	35
36	0.60218*	601.64*	0.093258*	-0.00208*	0.1953	0.9698	1.9373	0.42143	253.55*	1.5317	2.2960	36
37	0.64588	645.30	0.10868	-0.00155	0.2313	0.9483	2.4215	0.33714	217.56	1.5027	2.1661	37
38	0.65859*	658.00*	0.082207*	-0.00182*	0.2557	(0.969)	1.8677	0.43172	287.63*	1.4362	2.2621	38
39	0.62787*	627.30*	0.086230*	-0.00198*	0.1956	0.9721	1.8677	0.43172	274.21*	1.4303	2.2347	39
40	0.68665*	686.03*	0.099294*	-0.00148*	0.1745	—	2.3520	0.34711	238.13*	1.4661	2.1718	40
41	—	—	—	—	0.1949	0.9930	0.89902	0.90810	—	1.6627	—	41
42	0.88511	884.31	0.088333	-0.00121	0.2090	0.9938	2.6971	0.30270	267.68	1.0149	1.7147	42
43	0.87236	871.57	0.10572	-0.00106	0.2632	0.9971	3.1814	0.25661	223.66	1.0866	1.6771	43
44	0.87210	871.32	0.12185	-0.00101	0.3027	0.9986	3.6657	0.22272	194.06	1.1604	1.7207	44
45	0.88507	884.27	0.12006	-0.00094	0.3113	0.9989	3.6657	0.22272	196.94	1.2107	1.7409	45
46	0.86934	868.56	0.12223	-0.00099	0.3257	0.9987	3.6657	0.22272	193.44	1.1462	1.6957	46
47	0.86611	865.33	0.12269	-0.00100	0.3214	0.9986	3.6657	0.22272	192.72	1.1480	1.6839	47
48	0.91112	910.30	0.11442	-0.00101	(0.2412)	—	3.5961	0.22703	206.66	1.1149	1.7258	48
49	0.86675	865.97	0.13880	-0.00098	0.3260	0.9991	4.1500	0.19672	170.36	1.2830	1.7589	49
50	0.79666	795.94	0.040257	-0.00117	0.5649	—	1.1063	0.73795	587.37	1.3567	2.4754	50
51	0.79436	793.65	0.058047	-0.00108	0.6446	—	1.5906	0.51324	407.33	1.3829	2.3564	51
52	0.78931*	788.60*	0.035519*	—	0.0477	0.9996	0.9671	0.84417	665.71*	1.0403	—	52
53	0.82268*	821.94*	0.053544*	-0.01049*	0.2667	0.9964	1.5196	0.53726	441.59*	0.83294	—	53
54	0.80262*	801.90*	0.042499*	0.00283*	0.0948	0.9845	1.1767	0.69382	556.37*	0.99794	2.1098	54
55	1.3973*	1396.0*	0.045892*	—	0.2548	0.9801	2.2120	0.36907	515.22*	0.61942	1.3589	55
56	0.61826*	617.70*	0.027571*	—	0.2558	0.9876	0.58802	1.3884	857.61*	2.0791	4.6931	56
57	0.87469*	873.90*	0.033142*	—	—	0.9996	1.0000	0.81639	713.45*	1.0040	—	57
58	0.071064*	71.000*	0.028393*	—	-0.2216*	1.0006	0.06960	1.729	832.78*	14.261	—	58
59	1.1420*	1141.0*	0.028045*	—	0.0216	0.9992	1.1048	0.73893	843.12*	0.91673	—	59
60	0.80933*	808.60*	0.034644*	—	0.0370	0.9997	0.96723	0.84402	682.48*	1.0395	—	60
61	1.4258*	1424.5*	0.049776*	—	0.0878	(0.9875)	2.4482	0.33347	475.02*	0.47612	—	61
62	1.0000	999.10	0.018032	-0.00023	0.3445	—	0.62202	1.3125	1311.3	1.8617	4.1863	62
63	0.12509*	124.98*	0.032026*	—	0.	1.0005	0.13820	5.9074	738.30*	5.1931	—	63
64	0.85377*	853.00*	0.042744*.	-0.00539*	0.1259	0.9923	1.2589	0.64851	553.18*	0.79910	—	64

NOTE: Numbers in this table do not have accuracies greater than 1 part in 1000; in some cases extra digits have been added to calculated values to achieve internal consistency or to permit recalculation of experimental values. * See the Notes and References in Section 20.3.3

20.3.2 Table of physical constants *(continued)*

	See Note No: →	K.					L.	M.					
		Heating value, 15 °C							Flammability limits, vol% in air mixture		ASTM octane number		
		Net		Gross									
Number	Compound	MJ/m³ Ideal gas, 101.3250 kPa(abs)	MJ/kg Liquid (Wt in vacuum)	MJ/m³ Ideal gas, 101.3250 kPa(abs)	MJ/kg Liquid (Wt in vacuum)	MJ/m³ Liquid	Heat of vaporisation 101.3250kPa (abs) at boiling point, kJ/kg	Air required for combustion ideal gas m³(air)/m³(gas)	Lower	Higher	Motor method D-357	Research method D-908	Number
1	Methane	33.949	—	37.708	—	—	510.44	9.548	5.0	15.0	—	—	1
2	Ethane	60.429	47.162*	66.065	51.595*	18459.*	488.86	16.710	2.9	13.0	+0.05	+1.6*	2
3	Propane	86.418	45.955*	93.936	49.986*	25358.*	431.78	23.871	2.0	9.5	97.1	+1.8*	3
4	Isobutane	112.007	45.210*	121.406	49.032*	27604.*	366.46	31.032	1.8	8.5	97.6	+0.1*	4
5	n-Butane	112.396	45.342*	121.794	49.164*	28715.*	386.08	31.032	1.5	9.0	89.6*	93.8*	5
6	Isopentane	138.087	44.899	149.363	48.594	30340.	342.20	38.193	1.3	8.0	90.3	92.3	6
7	n-Pentane	138.380	44.973	149.656	48.668	30710.	357.45	38.193	1.4	8.3	62.6*	61.7*	7
8	Neopentane	137.485	44.739*	148.762	48.434*	28901.*	315.32	38.193	1.3	7.5	80.2	85.5	8
9	n-Hexane	164.399	44.735	177.550	48.342	32094.	334.78	45.355	1.1	7.7	26.0	24.8	9
10	2-Methylhexane	164.076	44.666	177.230	48.275	31753.	322.48	45.355	1.18	7.0	73.5	73.4	10
11	3-Methylpentane	164.186	44.690	177.339	48.299	32309.	325.61	45.355	1.2	7.7	74.3	74.5	11
12	Neohexane	163.658	44.571	176.810	48.180	31508.	305.30	45.355	1.2	7.0	93.4	91.8	12
13	2,3-Dimethylbutane	163.982	44.647	177.135	48.256	32146.	316.56	45.355	1.2	7.0	94.3	+0.3	13
14	n-Heptane	190.389	44.555	205.424	48.102	33087.	317.05	52.516	1.0	7.0	—	—	14
15	2-Methylhexane	190.095	44.503	205.131	48.051	32817.	305.58	52.516	1.0	7.0	46.4	42.4	15
16	3-Methylhexane	190.238	44.534	205.275	48.082	33246.	308.27	52.516	(1.01)	(6.6)	55.8	52.0	16
17	3-Ethylpentane	190.322	44.553	205.356	48.100	33797.	310.57	52.516	(1.00)	(6.5)	69.3	65.0	17
18	2,2-Dimethylpentane	189.625	44.416	204.661	47.963	32530.	291.70	52.516	(1.09)	(6.8)	95.6	92.8	18
19	2,4-Dimethylpentane	189.798	44.452	204.835	48.000	32497.	294.90	52.516	(1.08)	(6.8)	83.8	83.1	19
20	3,3-Dimethylpentane	189.880	44.471	204.915	48.018	33448.	295.60	52.516	(1.04)	(7.0)	86.6	80.8	20
21	Triptane	189.685	44.434	204.720	47.982	33373.	288.41	52.516	(1.08)	(6.8)	+0.1	+1.8	21
22	n-Octane	216.373	44.418	233.284	47.919	33857.	301.23	59.677	0.8	6.5	—	—	22
23	Diisobutyl	215.791	44.330	232.704	47.831	33377.	284.86	59.677	(0.92)	(6.3)	55.7	55.2	23
24	Isooctane	215.726	44.342	232.642	47.843	33302.	269.54	59.677	0.95	6.0	100.0	100.0	24
25	n-Nonane	242.399	44.319	261.191	47.784	34484.	289.26	66.839	0.7	5.6	—	—	25
26	n-Decane	268.393	44.234	289.067	47.669	34983.	278.31	74.000	0.7	5.4	—	—	26
27	Cyclopentane	131.152	43.786	140.546	46.954	35242.	389.25	35.806	(1.48)	(8.3)	84.9*	+0.1	27
28	Methylcyclopentane	156.751	43.656	168.029	46.824	35326.	345.53	42.968	1.0	8.35	80.0	91.3	28
29	Cyclohexane	156.031	43.438	167.306	46.606	36499.	356.46	42.968	1.2	8.35	77.2	83.0	29
30	Methylcyclohexane	181.563	43.358	194.715	46.525	35995.	316.74	50.129	1.1	6.7	71.1	74.8	30
31	Ethane (Ethylene)	55.963	—	59.724	—	—	482.64	14.323	2.7	36.0	75.6	+0.03	31
32	Propene (Propylene)	81.460	(45.38)	87.099	(48.55)	(25312)	437.73	21.484	2.0	11.7	84.9	+0.2	32
33	1-Butene (Butylene)	107.463	44.921*	114.979	48.089*	28873.*	390.68	28.645	1.6	10.	80.8*	97.4	33
34	cis-2-Butene	107.176	44.764*	114.689	47.931*	30131.*	416.16	28.645	1.6	10.	83.5	100.0	34
35	trans-2-Butene	107.021	44.713*	114.538	47.881*	29267.*	405.65	28.645	1.6	10.	—	—	35
36	Isobutene	106.762	44.617*	114.276	47.785*	28749.	394.24	28.645	1.6	10.	—	—	36
37	1-Pentene	133.456	44.624	142.850	47.791	30840.	359.31	35.806	1.3	10.	77.1	90.9	37
38	1,2-Butadiene	104.116	45.074*	109.755	47.538*	31280.*	446.31	26.258	(1.62)	(10.3)	—	—	38
39	1,3-Butadiene	101.874	44.116*	107.513	46.580*	29220.*	430.99	26.258	2.0	12.5	—	—	39
40	Isoprene	127.328	43.803	134.843	46.412	31840.	380.26	33.419	(1.12)	(8.5)	81.0	99.1	40
41	Acetylene	53.159	—	55.038	—	—	819.51	11.935	1.5	100.	—	—	41
42	Benzene	134.051	40.137	139.689	41.843	37002.	393.78	35.806	1.2	8.0	+2.8	—	42
43	Toluene	159.538	40.522	167.056	42.451	36999.	360.10	42.968	1.2	7.1	+0.3	+5.8	43
44	Ethylbenzene	185.552	40.922	194.947	43.014	37479.	335.32	50.129	1.0	8.0	97.9	+0.8	44
45	o-Xylene	185.091	40.808	194.484	42.901	37936.	346.62	50.129	1.0	7.6	100.0	+0.	45
46	m-Xylene	185.020	40.799	194.413	42.892	37254.	342.86	50.129	1.0	7.0	+2.8	+4.0	46
47	p-Xylene	185.049	40.809	194.444	42.901	37124.	339.09	50.129	1.0	7.0	+1.2	+3.4	47
48	Styrene	180.302	40.505	187.816	42.212	38426.	355.52	47.742	1.1	8.0	+0.2	>+3.*	48
49	Isopropylbenzene	211.323	41.192	222.598	43.410	37592.	312.00	57.290	0.8	6.5	99.3	+2.1	49
50	Methyl alcohol	28.605	19.929	32.362	22.703	18070	1076.71	7.161	5.5	44.0	—	—	50
51	Ethyl alcohol	54.038	26.807	59.678	29.701	23572.	840.48	14.323	3.28	19.0	—	—	51
52	Carbon monoxide	11.965	—	11.964	—	—	215.78	2.387	12.50	74.20	—	—	52
53	Carbon dioxide	0.0	—	0.0	—	—	573.28*	—	—	—	—	—	53
54	Hydrogen sulphide	21.905	14.772*	23.784	16.043*	12865.*	548.08	7.161	4.30	45.50	—	—	54
55	Sulphur dioxide	0.0	—	0.0	—	—	389.29	—	—	—	—	—	55
56	Ammonia	13.401	—	16.220	—	—	1371.07	3.581	15.50	27.00	—	—	56
57	Air	0.0	—	0.0	—	—	205.16	—	—	—	—	—	57
58	Hydrogen	10.223	—	12.102	—	—	441.99	2.387	4.00	74.20	—	—	58
59	Oxygen	0.0	—	0.0	—	—	213.03	—	—	—	—	—	59
60	Nitrogen	0.0	—	0.0	—	—	199.08	—	—	—	—	—	60
61	Chlorine	—	—	—	—	—	287.84	—	—	—	—	—	61
62	Water	—	*	1.8792*	0.0	0.	2256.64	—	—	—	—	—	62
63	Helium	—	—	0.0	—	—	—	—	—	—	—	—	63
64	Hydrogen chloride	—	—	—	—	—	442.95	—	—	—	—	—	64

NOTE: *Numbers in this table do not have accuracies greater than 1 part in 1000; in some cases extra digits have been added to calculated values to achieve internal consistency or to permit recalculation of experimental values.* * See the Notes and References in Section 20.3.3

20.3.3 Notes and references to Table of physical constants

Number	Compound	Formula	Molar mass (molecular weight)	Boiling point, °C 101.3250 KP (abs)	Vapour pressure KP (abs), 40 °C	Freezing point, °C 101.3250 kPa (abs)	Refractive index n_D 15 °C	Critical constants: Pressure kPa (abs)	Critical constants: Temperature, °C	Critical constants: Volume, m^3/kg	Number
	See Note No: →		A.	B.		C.	D.				
1	Methane	CH_4	15	4	a,b	c,44	i,44	44	44	4	1
2	Ethane	C_2H_6	15	44	a,b	c,44	h,44	44	44	44	2
3	Propane	C_3H_8	15	44	28	c,44	h,44	44	44	44	3
4	Isobutane	C_4H_{10}	15	44	26	44	h,44	44	44	44	4
5	n-Butane	C_4H_{10}	15	44	27	44	h,44	27	44	44	5
6	Isopentane	C_5H_{12}	15	44	44	44	44	44	44	44	6
7	n-Pentane	C_5H_{12}	15	44	44	44	44	44	44	44	7
8	Neopentane	C_5H_{12}	15	44	44	44	h,44	44	44	44	8
9	n-Hexane	C_6H_{14}	15	44	44	44	44	44	44	44	9
10	2-Methylpentane	C_6H_{14}	15	44	44	44	44	44	44	44	10
11	3-Methylpentane	C_6H_{14}	15	44	44	f,17,23	44	44	44	44	11
12	Neohexane	C_6H_{14}	15	44	44	44	44	44	44	44	12
13	2,3-Dimethylbutane	C_6H_{14}	15	44	44	44	44	44	44	44	13
14	n-Heptane	C_7H_{16}	15	44	44	44	44	44	44	44	14
15	2-Methylhexane	C_7H_{16}	15	44	44	44	44	44	44	44	15
16	3-Methylhexane	C_7H_{16}	15	44	44	f,32	44	44	44	44	16
17	3-Ethylpentane	C_7H_{16}	15	44	44	44	44	44	44	44	17
18	2,2-Dimethylpentane	C_7H_{16}	15	44	44	44	44	44	44	44	18
19	2,4-Dimethylpentane	C_7H_{16}	15	44	44	44	44	44	44	44	19
20	3,3-Dimethylpentane	C_7H_{16}	15	44	44	44	44	44	44	44	20
21	Triptane	C_7H_{16}	15	44	44	44	44	44	44	44	21
22	n-Octane	C_8H_{18}	15	44	44	44	44	44	44	44	22
23	Diisobutyl	C_8H_{18}	15	44	44	44	44	44	44	44	23
24	Isooctane	C_8H_{18}	15	44	44	44	44	44	44	44	24
25	n-Nonane	C_9H_{20}	15	44	44	44	44	44	44	44	25
26	n-Decane	$C_{10}H_{22}$	15	44	44	44	44	44	44	44	26
27	Cyclopentane	C_5H_{10}	15	44	44	44	44	44	44	44	27
28	Methylcyclopentane	C_6H_{12}	15	44	44	44	44	44	44	44	28
29	Cyclohexane	C_6H_{12}	15	44	44	44	44	44	44	44	29
30	Methylcyclohexane	C_7H_{14}	15	44	44	44	44	44	44	44	30
31	Ethane (Ethylene)	C_2H_4	15	44	b	c,44	h,21	44	44	44	31
32	Propene (Propylene)	C_3H_6	15	44	44	44	h,44	44	44	44	32
33	1-Butene (Butylene)	C_4H_8	15	44	44	44	h,44	44	44	44	33
34	cis-2-Butene	C_4H_8	15	44	44	44	h,44	44	44	44	34
35	trans-2-Butene	C_4H_8	15	44	44	44	h,44	44	44	44	35
36	Isobutene	C_4H_8	15	44	44	44	h,44	44	44	44	36
37	1-Pentene	C_5H_{10}	15	44	44	44	44	44	44	44	37
38	1,2-Butadiene	C_4H_6	15	44	44	44		a	a	a	38
39	1,3-Butadiene	C_4H_6	15	44	20	44	h,44	44	44	44	39
40	Isoprene	C_5H_8	15	44	44	44		a	a	a	40
41	Acetylene	C_2H_2	15	d,44	b	c,44		44	44	44	41
42	Benzene	C_6H_6	15	44	44	44	44	44	44	44	42
43	Toluene	C_7H_8	15	44	44	44	44	44	44	44	43
44	Ethylbenzene	C_8H_{10}	15	44	44	44	44	44	44	44	44
45	o-Xylene	C_8H_{10}	15	44	44	44	44	44	44	44	45
46	m-Xylene	C_8H_{10}	15	44	44	44	44	44	44	44	46
47	p-Xylene	C_8H_{10}	15	44	44	44	44	44	44	44	47
48	Styrene	C_8H_8	15	12	12	44	44	43	a,11	43	48
49	Isopropylbenzene	C_9H_{12}	15	44	44	44	44	44	44	44	49
50	Methyl alcohol	CH_4O	15	p,45	45	45	45	45	45	45	50
51	Ethyl alcohol	C_2H_6O	15	45	45	45	45	45	45	45	51
52	Carbon monoxide	CO	15	45	b	c,45	i,45	45	45	45	52
53	Carbon dioxide	CO_2	15	d,3	b	c,45	i,45	45	45	45	53
54	Hydrogen sulphide	H_2S	15	25	25	c,25	i,45	45	45	45	54
55	Sulphur dioxide	SO_2	15	45	18	c,45	i,45	45	45	45	55
56	Ammonia	NH_3	15	1	18	c,45	i,45	45	45	45	56
57	Air	N_2+O_2	15,34	36	b		i,47	36	36	36	57
58	Hydrogen	H_2	15	r,s,10	b	c,r,s	i,45	n,45	n,45	n,45	58
59	Oxygen	O_2	15	r,10	b	c,r,s,10	i,45	45	45	45	59
60	Nitrogen	N_2	15	6	b	c,45	i,45	6	6	6	60
61	Chlorine	Cl_2	15	1	2	c,45	h,45	2	2	45	61
62	Water	H_2O	15	r,10	45	45	45	31	31	31	62
63	Helium	He	15	12	b		i,45	7	7	7	63
64	Hydrogen chloride	HCl	15	1	18	c,45	i,45	45	45	45	64

20.3.3 Notes and references to Table of physical constants *(continued)*

	E. Density of liquid 101.3250 kPa(abs), 15 °C			F.	G.	H.	I. Ideal gas 101.3250 kPa (abs), 15 °C			J. Specific Heat 15 °C 101.3250 kPa (abs) kJ/(Kg K)		
Number	Relative density (specific gravity) 15 °C/15 °C	Kg/m^3 (wt in vacuum)	m^3/kmol	Temperature coefficient of density, 1/T	Acentric factor, ω	Compressibility factor of real gas, Z, 15 °C 101.3250 kPa(abs)	Relative density (specific gravity) Air = 1	m^3 gas/kg	Volume ratio gas to liquid	C_p, Ideal gas	C_p, Liquid	Number
1	b,a	b,a,t	b,a	b,a	4	44			b,a	8	b	1
2	h,29	h,t,29	h,29	h,29	44	44			h	8	41	2
3	h,28	h,28	h,28	h,28	28	44			h	8	28	3
4	h,m,26	h,m,26	h,m,26	h,m,26	26	44			h	8	26	4
5	h,27	h,27	h,27	h,27	27	44			h	8	27	5
6	44	44	44	44	44					8	30	6
7	44	44	44	44	44					8	38	7
8	h,44	h,44	h,44	h,44	44	44			h	8	19	8
9	44	44	44	44	44					8	38	9
10	44	44	44	44	44					8	44	10
11	44	44	44	44	44					8	44	11
12	44	44	44	44	44					8	44	12
13	44	44	44	44	44					8	44	13
14	44	44	44	44	44					44	38	14
15	44	44	44	44	44					44	44	15
16	44	44	44	44	44					44	32	16
17	44	44	44	44	44					44	44	17
18	44	44	44	44	44					44	44	18
19	44	44	44	44	44					44	44	19
20	44	44	44	44	44					44	44	20
21	44	44	44	44	44					44	44	21
22	44	44	44	44	44					44	38	22
23	44	44	44	44	44					44	44	23
24	44	44	44	44	44					44	44	24
25	44	44	44	44	44					44	38	25
26	44	44	44	44	44					44	38	26
27	44	44	44	44	44					8	17,46	27
28	44	44	44	44	p,44					44	17,46	28
29	44	44	44	44	44					8	9	29
30	44	44	44	44	44					44	17,46	30
31	b	b	b	b	44	44			b	8	b	31
32	h,44	h,44	h,44	h,44	44	44			h	8	44	32
33	h	h	h	h	44	44			h	44	44	33
34	h,p	h,p	h,p	h,p	44	44			h	44	44	34
35	h,p	h,p	h,p	h,p	44	44			h	44	p,13	35
36	h,p	h,p	h,p	h,p	44	44			h	44		36
37	44	44	44	44	44					44	44	37
38	h,p	h,p	h,p	h,p	p,44	a			h	44	44	38
39	h,p	h,p	h,p	h,p	20	a			h	44	42,46	39
40	p	p	p	p	44					44	39	40
41	b	b	b	b	p,44	44			b	8		41
42	44	44	44	44	44					44	44	42
43	44	44	44	44	44					44	44	43
44	44	44	44	44	44					44	44	44
45	44	44	44	44	44					44	44	45
46	44	44	44	44	44					44	44	46
47	44	44	44	44	44					44	44	47
48	44	44	44	44	a,40					44		48
49	44	44	44	44	44					44	42	49
50	45	45	45	45	45					45	45	50
51	45	45	45	45	45					45	45	51
52	j,33	j,33	j,33		33				q	8	b	52
53	h,3	h,3	h,3	h,3					h	8		53
54	h,25	h,t,25	h,25	h,25	25	45			h	8	25	54
55	h	h	h		18	45			h	8		55
56	h	h	h		18	45			h	45		56
57	j,36	j,36	j,36						q	35		57
58	j,n,16	j,n,16	j,n,16		n,18	45			q	8	b	58
59	j,37	j,37	j,37		45	45			q	8	b	59
60	j,6	j,6	j,6		6	45			q	8	b	60
61	h	h	h		2	a			h	45		61
62	24	24	24	24	45					44	22	62
63	j	j	j		45	45			q	8	b	63
64	h	h	h	h	45	45			h	45		64

20.3.3 Notes and references to Table of physical constants *(continued)*

Number	See Note No: →	K.					L.	M.	Flammability limits, vol% in air mixture		ASTM octane number		
		Heating value, 15 °C											
		Net		Gross									
	Compound	MJ/m³ ideal gas, 101.3250 kPa(abs)	MJ/kg Liquid (Wt in vacuum)	MJ/m³ ideal gas, 101.3250 kPa(abs)	MJ/kg Liquid (Wt in vacuum)	MJ/m³ Liquid	Heat of vaporisation 101.3250kPa (abs) at boiling point, kJ/kg	Air required for combustion ideal gas m³(air)/m³(gas)	Lower	Higher	Motor method D-357	Research method D-908	Volume
1	Methane	44	b	44	b	b	4		v	v			1
2	Ethane	44	h,44	44	h,44	h,44	29		v	v	e	e,g	2
3	Propane	44	h,44	44	h,44	h,44	28		v	v		e,g	3
4	Isobutane	44	h,44	44	h,44	h,44	26		v	v		e,g	4
5	n-Butane	44	h,44	44	h,44	h,44	27		v	v	g	g	5
6	Isopentane	44	44	44	44	44	44		v	v			6
7	n-Pentane	44	44	44	44	44	44		v	v	g	g	7
8	Neopentane	44	h,44	44	h,44	h,44	44		v	v			8
9	n-Hexane	44	44	44	44	44	44		v	v			9
10	2-Methylpentane	44	44	44	44	44	44		v	v			10
11	3-Methylpentane	44	44	44	44	44	44		v	v			11
12	Neohexane	44	44	44	44	44	44		v	v			12
13	2,3-Dimethylbutane	44	44	44	44	44	44		v	v		e	13
14	n-Heptane	44	44	44	44	44	44		v	v			14
15	2-Methylhexane	44	44	44	44	44	44		v	v			15
16	3-Methylhexane	44	44	44	44	44	44		a,v	a,v			16
17	3-Ethypentane	44	44	44	44	44	44		a,v	a,v			17
18	2,2-Dimethylpentane	44	44	44	44	44	44		a,v	a,v			18
19	2,4-Dimethylpentane	44	44	44	44	44	44		a,v	a,v			19
20	3,3-Dimethylpentane	44	44	44	44	44	44		a,v	a,v			20
21	Triptane	44	44	44	44	44	44		a,v	a,v	e	e	21
22	n-Octane	44	44	44	44	44	44		v	v			22
23	Diisobutyl	44	44	44	44	44	44		a,v	a,v			23
24	Isooctane	44	44	44	44	44	44		v	v			24
25	n-Nonane	44	44	44	44	44	44		v	v			25
26	n-Decane	44	44	44	44	44	44		a,v	a,v			26
27	Cyclopentane	44	44	44	44	44	44		a,v	a,v	g	e	27
28	Methylcyclopentane	44	44	44	44	44	44		v	v			28
29	Cyclohexane	44	44	44	44	44	44		v	v			29
30	Methylcyclohexane	44	44	44	44	44	44		v	v			30
31	Ethane (Ethylene)	44	b	44	b	b	44		v	v		e	31
32	Propene (Propylene)	44		44		44	44		v	v		e	32
33	1-Butene (Butylene)	44	h,44	44	h,44	h,44	44		v	v	g		33
34	cis-2-Butene	44	h,44	44	h,44	h,44	44		a,v	a,v			34
35	trans-2-Butene	44	h,44	44	h,44	h,44	44		a,v	a,v			35
36	Isobutene	44	h,44	44	h,44	h,44	44		a,v	a,v			36
37	1-Pentene	44	44	44	44	44	44		v	v			37
38	1,2-Butadiene	44	h,44	44	h,44	h,44	43		a,v	a,v			38
39	1,3-Butadiene	44	h,44	44	h,44	h,44	43		v	v			39
40	Isoprene	44	44	44	44	44	43		a,v	a,v			40
41	Acetylene	44		44			43		v	v			41
42	Benzene	44	44	44	44	44	44		v	v	e		42
43	Toluene	44	44	44	44	44	44		v	v	e	e	43
44	Ethylbenzene	44	44	44	44	44	44		v	v		e	44
45	o-Xylene	44	44	44	44	44	44		v	v			45
46	m-Xylene	44	44	44	44	44	44		v	v	e	e	46
47	p-Xylene	44	44	44	44	44	44		v	v	e	e	47
48	Styrene	44	44	44	44	44	43		v	v	e	e	48
49	Isopropylbenzene	44	44	44	44	44	44		v	v		e	49
50	Methyl alcohol	45	45	45	45	45	45		v	v			50
51	Ethyl alcohol	45	45	45	45	45	45		v	v			51
52	Carbon monoxide	8	b	8	b	b	45		v	v			52
53	Carbon dioxide						k,45						53
54	Hydrogen sulphide	14	h,45	14	h,45	h,45	25		v	v			54
55	Sulphur dioxide						45						55
56	Ammonia	45		45			45		v	v			56
57	Air						36						57
58	Hydrogen	45		45			n,45		v	v			58
59	Oxygen						37						59
60	Nitrogen						45						60
61	Chlorine						43						61
62	Water	8	u,8	u,8			24						62
63	Helium												63
64	Hydrogen chloride						43						64

Explanation of Notes used in Section 20.3.3

a. Values in parentheses are estimated values.

b. The temperature is above the critical point.

c. At saturation pressure (triple point).

d. Sublimation point.

e. The + sign and number following specify the number of cm^3 of TEL added per gallon to achieve the ASTM octane number of 100, which corresponds to that of Isooctane (2,2,4 - Trimethylpentane).

f These compounds form a glass.

g. Average value from octane numbers of more than one sample.

h. Saturation pressure and 15°C.

I. Index of refraction of the gas.

j. Densities of the liquid at the normal boiling point.

k. Heat of sublimation.

m. Equation 2 of the reference was refitted to give: a = 0.7872957; b = 0.1294083; c = 0.03439519.

n. Normal hydrogen (25% para, 75% ortho).

p. An extrapolated value.

q. Gas at 15°C and the liquid at the normal boiling point.

r. Fixed points on the 1968 International Practical Temperature Scale (IPTS-68).

s. Fixed points on the 1990 International Temperature Scale (ITS-90).

t. Densities at the normal boiling point are: Ethane, 554.0 [291; Propane, 581.0 [28]; Propene, 609.1 [5]; Hydrogen Chloride, 1192 [43]; Hydrogen Sulphide, 949.0 [25];Ammonia, 681.6 [43]; Sulphur Dioxide, 1462 [43].

u. Technically, water has a heating value in two cases: net ((2.466 MJ/kg) when water is liquid in the reactants, and gross (+1.879 MJ/m^3) when water is gas in the reactants. The value is the ideal heat of vaporisation (enthalpy of the ideal gas less the enthalpy of the saturated liquid at the vapour pressure). This is a matter of definition; water does not burn.

v. Extreme values of those reported in *The Air Density Equation and the Transfer of the Mass Unit*, FE Jones, J. Res. Natl. Bur. Stand. (U.S.) 83 (5), 419 (Sep - Oct 1978).

A. Molar mass (molecular mass) is based upon the following atomic weights: C = 12.011; H = 1.00794; O = 15.9994; N = 14.0067; S = 32.066; Cl = 35.4527. The values were rounded off after calculating the molar mass using all significant figures in the atomic weights.

B. Boiling point: the temperature at equilibrium between the liquid and vapour phases at 101.3250 kPa.

C. Freezing point: the temperature at equilibrium between the crystalline phase and the air saturated liquid at 101.3250 kPa.

D. The refractive index reported refers to the liquid or gas and is measured for light of wavelength corresponding to the sodium D-line (589.26 nm).

E. The relative density: ρ(liquid, 15°C)/ρ(water, 15°C). The density of water at 15°C is 999.10 kg/m^3.

F. The temperature coefficient of density is related to the expansion coefficient by: $(\delta\rho/\delta T)_P/\rho = -(\delta\rho V/\delta T)_P/V$, in units of 1/T.

G. Pitzer acentric factor: $\omega = -\log_{10}(P/P_c) - 1$, P at $T = 0.7\ T_c$

H. Compressibility factor of the real gas, Z = PV/RT, is calculated using the second virial coefficient.

I. The density of an ideal gas relative to air is calculated by dividing the molar mass of the of the gas by 28.9625, the calculated average molar mass of air. (See *The Air Density Equation and the Transfer of the Mass Unit*, FE Jones, J. Res. Natl. Bur. Stand. (U.S.) 83 (5), 419 (Sep - Oct 1978), for the average composition of dry air. The specific volume of an ideal gas is calculated from the ideal gas equation. The volume ratio is: V(ideal gas)/V(liquid in vacuum).

J. The liquid value is not rigorously C_P, but rather it is the heat capacity along the saturation line C_S defined by: $C_S = C_P - T\ (\delta V/\delta T)_P(\delta P/\delta T)_S$. For liquids far from the critical point, $C_S = C_P$.

K. The heating value is the negative of the enthalpy of combustion at 15°C and 101.3250 kPa (abs.) in an ideal reaction (one where all gases are ideal gases). For an arbitrary organic compound, the combustion reaction is:

$C_nH_mO_hS_jN_k$ (s,l,or,g) + (n + m/4 - h/2 + j) O_2(g) $\rightarrow$

n CO_2(g) + m/2 H_2O (g or l) + k/2 N_2(g) + j SO_2(g),

where s, l and g denote respectively solid, liquid and ideal gas. For gross heating values, the water formed is liquid; for net heating values, the water formed is ideal gas. Values reported are on a dry basis.

To account for water in the heating value, see GPA 2172. The MJ/kg liquid column assumes a reaction with the fuel in the liquid state, while the MJ/m^3 ideal gas column assumes the gas in the ideal gas state. Therefore, the values are not consistent if used in the same calculation, e.g. a gas plant balance.

L. The heat of vaporisation is the enthalpy of the saturated vapor at the boiling point at 101.3250 kPa minus the enthalpy of the saturated liquid at the same conditions.

M. Air required for the combustion of ideal gas for compounds of formula $C_nH_mO_h,S_jN_k$ is: V(air)/V(gas) = (n + m/4 (h/2 + j)/0.20946.

Comments

Units:

reported values are in SI units based on the following:

mass: kilogram, kg

length: metre, m

temperature: International Temperature of 1990 (ITS-90), where 0°C = 273.15 K.

Other derived units are:

volume: cubic metre, m^3

pressure: Pascal, Pa (1 Pa = N/m^2)

energy: Joule, J

Gas constant, R:

8.314510 J/(kmol)

0.008314510 m^3(kPa/(kmol)

1.987216 cal_{th}/kmol)

1.985887 Btu(I.T.)/(R(lbmol)

Conversion factors:

1 m^3 = 35.31467 ft^3 = 264.1721 gal.

1 kg = 2.204623 lbm

1 kg/m^3 = 0.06242795 lbm/ft^3 =0.001 g/cm^3

1 kPa = 0.01 bar = 0.009869233 atm = 0.1450377 psia

1 atm = 101.3250 kPa = 14.69595 psia = 760 Torr

1 kJ = 0.2390057 $kcal_{th}$ = 0.2388459 kcal (I.T.)
= 0.9478172 Btu (I.T.)

References used in Section 20.3.3

D Ambrose, National Physical Laboratory, Teddington, Middlesex, England: Feb. 1980, NPL Report Chem 107.

Ambrose, D Hall, DJ Lee, DA Lewis, G. B.; Mash, C. J. J. Chem. Thermo., 11, 1089 (1979).

Carbon Dioxide International Thermodynamic Tables of the Fluid State-3, S Angus, B Armstrong, KM de Reuck, Eds. Pergamon Press, Oxford, 1976.

Methane International Thermodynamic Tables of the Fluid State-5, Pergamon Press, Oxford, 1978.

Propylene (Propene). International Thermodynamic Tables of the Fluid State-7, S Angus, B Armstrong, KM de Reuck, Eds., Pergamon Press, Oxford, 1980.

Nitrogen International Thermodynamic Tables of the Fluid State-6, S Angus, B Armstrong, KM de Reuck, Eds., Pergamon Press, Oxford, 1979.

Helium. International Thermodynamic Tables of the Fluid State-4, S Angus,KM de Reuck, RD McCarthy, Eds., Pergamon Press, Oxford, 1977.

Heating Values of Natural Gas and its Components, GT Armstrong, TL Jobe, NBSIR 82-2401, May 1982.

JG Aston, GJ Szasz, HL Finke, J. Am. Chem. Soc., 65, 1135 (1943).

CR Barber, Metrologia 5, 35 (1969).

Styrene, Its Polymers, Copolymers and Derivatives, RH Boundy, RF Boyer, Eds., A.C.S. monograph No. 115, Reinholt, N.Y., 1952.

P Chaiyavech, M Van Winkle, J. Chem. Eng. Data, 4, 53 (1959).

J Chao, KR Hall, J Yao, Thermochimica Acta, 64, 285 (1983).

CODATA Task Group on Key Values for Thermodynamics, CODATA Special Report No. 7, 1978.

Commission on Atomic Weights and Isotopic Abundances, Pure and Appl. Chem. 63, 975 (1991).

A Tabulation of the Properties of Normal Hydrogen from Low Temperature to 300 K and from 1 to 100 Atmospheres, JW Dean, NBS Tech. Note 120, November 1961.

DR Douslin, HM Huffman, J. Am. Chem. Soc., 68, 1704 (1946).

The Vapor Pressure of 30 Inorganic Liquids Between One Atmosphere and the Critical Point, D.G. Edwards, Univ. of Calif., Lawrence Radiation Laboratory, UCRL-7167. June 13, 1963.

EDSU, Engineering Sciences Data, Engineering Sciences Data Unit, EDSU International Ltd, London.

JL Flebbe, DA Barclay, DB Manley, J. Chem. Eng. Data, 27, 405 (1982).

AW Francis, J. Chem. Eng. Data, 5, 534 (1960).

DC Ginnings, GT Furukawa, J. Am. Chem. Soc. 75, 522 (1953).

Recommended Reference Materials of the Realization of Physicochemical Properties, G Girard, Chapter 2, K.N. Marsh, Ed.; Blackwell Sci. Pub., London, 1987.

AR Glasgow, ET Murphy, CB Willingham, FD Rossini, J. Res. NBS, 37, 141 (1946).

Hydrogen Sulfide Provisional Thermochemical Properties from 188 to 700 K at Pressures to 75 MPa, RD Goodwin, NBSIR 831694, October 1983.

Thermophysical Properties of Isobutane from 114 to 700 K at Pressures to 70 MPa, RD Goodwin, WM Haynes, NBS Tech. Note 1051, January 1982.

Thermophysical Properties of Normal Butane from 135 to 700 K at Pressures to 70 MPa RD Goodwin, WM Haynes, NBS Monograph 169, April 1982.

Thermophysical Properties of Propane from 85 to 700 K at Pressures to 70 MPa, RD Goodwin, WM Haynes, NBS Monograph 170, April 1982.

Thermo- physical Properties of Ethane, 90 to 600 K at Pressures to 700 bar, RD Goodwin, HM Roder, GC Straty, NBS Tech. Note 684, August 1976.

GB Guthrie, HM Huffman, J. Am. Chem. Soc., 65, 1139 (1943).

NBS/NRC Steam Tables, L Haar, JS Gallagher, GS Kell, Hemisphere Publishing Corporation, Washington, 1984.

HM Huffman, GS Park, SB Thomas, J. Am. Chem. Soc., 52, 3241 (1930).

Thermodynamic Property Values for Gaseous and Liquid Carbon Monoxide from 70 to 300 Atmospheres, JG Hust, RB Stewart, NBS Technical Note 202, Nov. 1963.

The Air Density Equation and the Transfer of the Mass Unit, FE Jones, J. Res. Natl. Bur. Stand. (U.S.) 83 (5), 419 (Sep - Oct 1978).

Gas Tables: (SI Units), JH Keenan, J Chao, J Kaye, John Wiley and Sons, Inc., New York, 1983.

The Matheson Unabridged Gas Data Book, Matheson Gas Products; New York, 1974.

Thermophysical Properties of Oxygen from the Freezing Liquid Line to 600 R for Pressures to 5000 Psia, RD McCarty, LA Weber, NBS Technical Note 384, July 1971.

JF Messerly, GB Guthrie, SS Todd, HL Finke, J. Chem. Eng. Data, 12, 338 (1967).

JF Messerly, SS Todd, GB Guthrie, J. Chem. Eng. Data, 15, 227 (1970).

Computer Aided Data Book of Vapor Pressures. Ohe, Data Book Publishing Co., Tokyo, Japan, 1976.

Measurements of the Specific Heats, Cs, and Cv, of Dense Gaseous and Liquid Ethane, HM Roder, J. Res. Nat. Bur. Stand. (U.S.) 80A, 739 (1976).

RB Scott, CH Meyers, RD Rands, FG Brickwedde, N Bekkedahl, J. Res. NBS, 35, 39 (1945).

The Chemical Thermodynamics of Organic Compounds, DR Stull, EF Westrum, GC Sinke, John Wiley & Sons, Inc., New York, 1969.

TRC Thermodynamic Tables (Hydrocarbons), Thermodynamics Research Center, Texas A&M University System, College Station, Texas.

TRC Thermodynamic Tables (Non-Hydrocarbons), Thermodynamics Research Center, Texas A&M University System, College Station, Texas.

Tables on the Thermophysical Properties of Liquids and Gases, NB Vargaftik, Hemisphere Publishing Corporation, Washington D.C., 1975.

Handbook of Chemistry and Physics, RC Weast, The Chemical Rubber Co., Cleveland, Ohio, 1969.

20.4 Pressure/enthalpy charts

20.4.1 Nitrogen

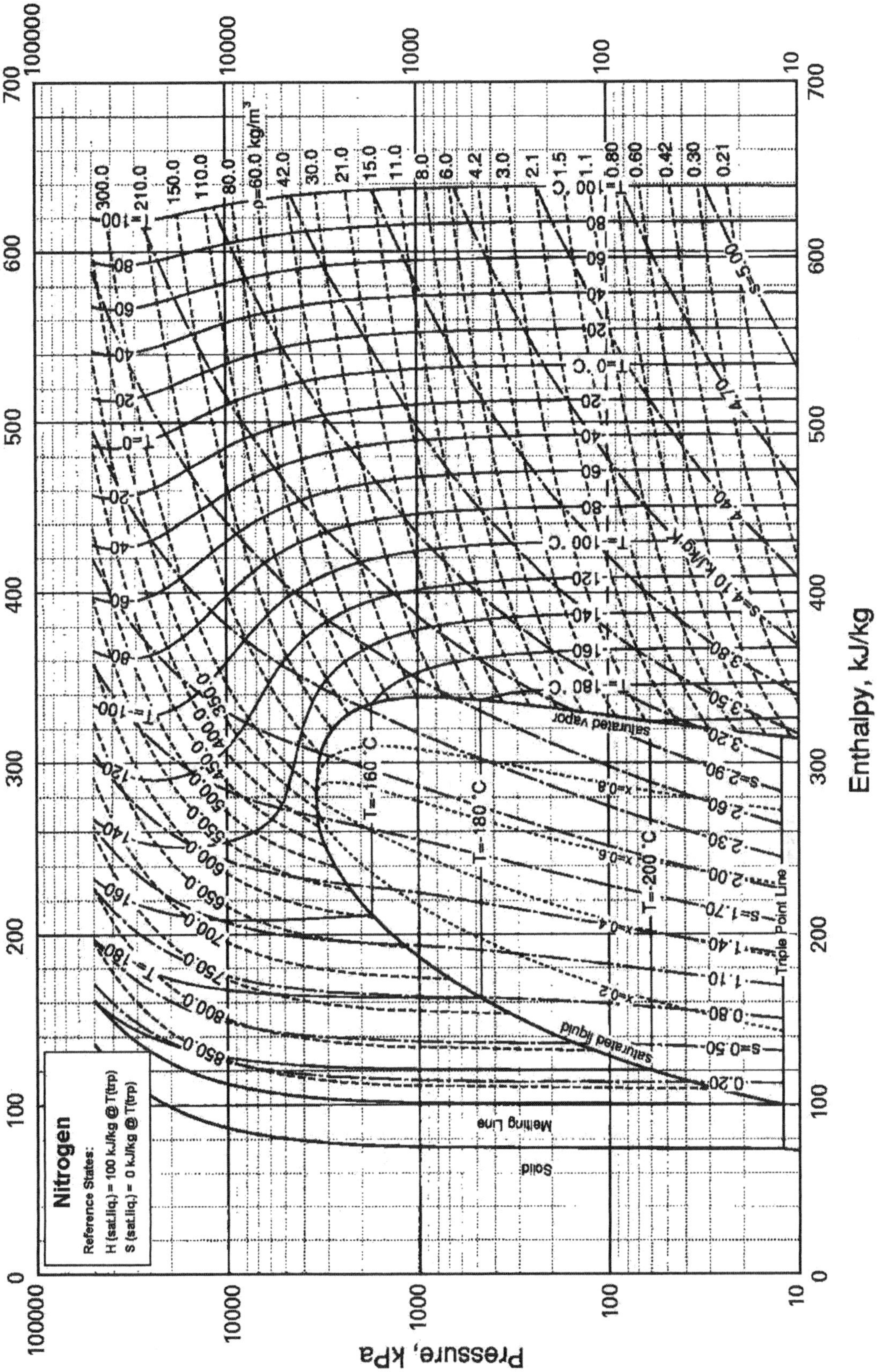

J F Ely, CEPR Department, Colorado School of Mines

20.4.2 Carbon dioxide

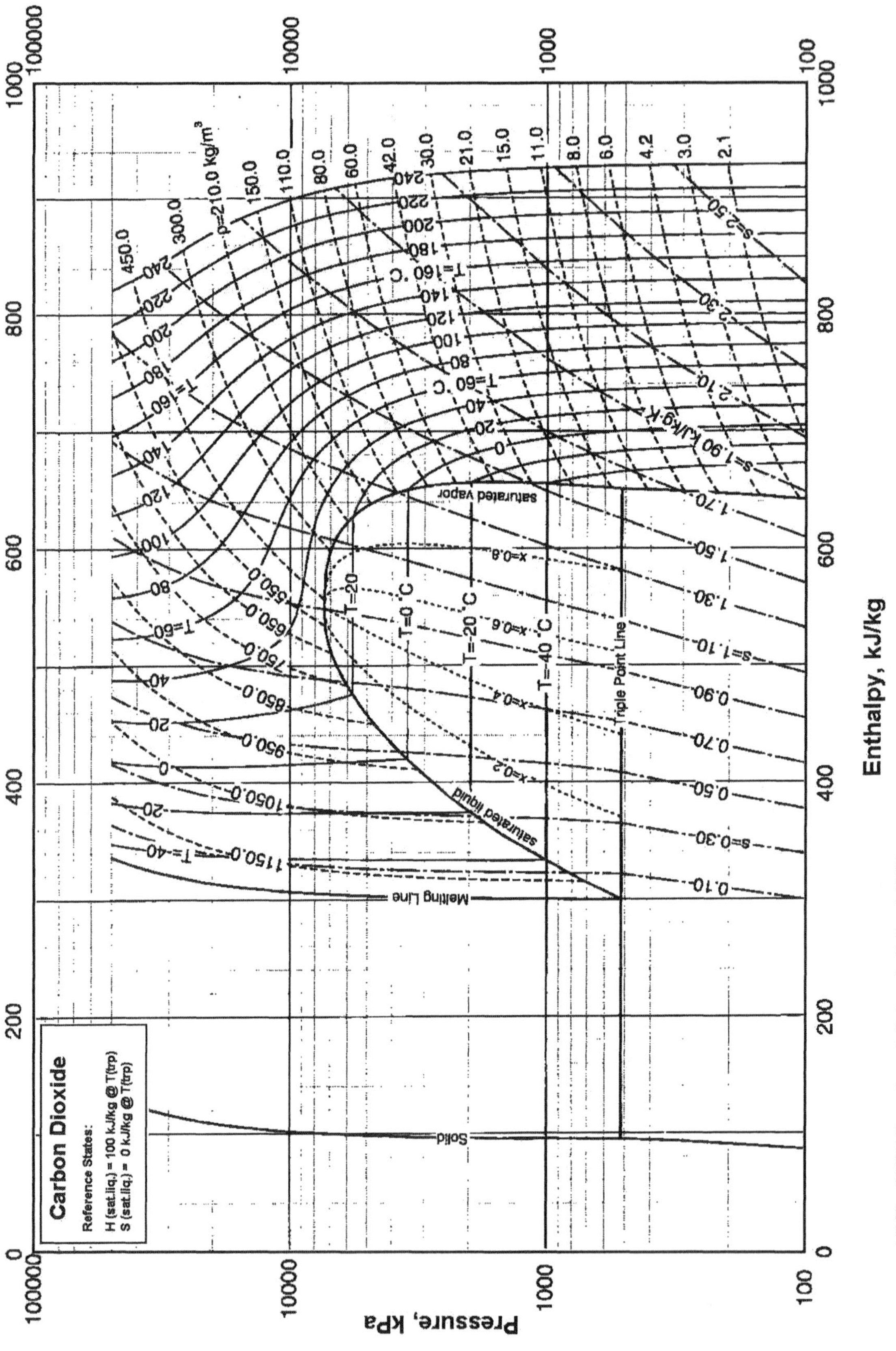

20.4.3 Methane

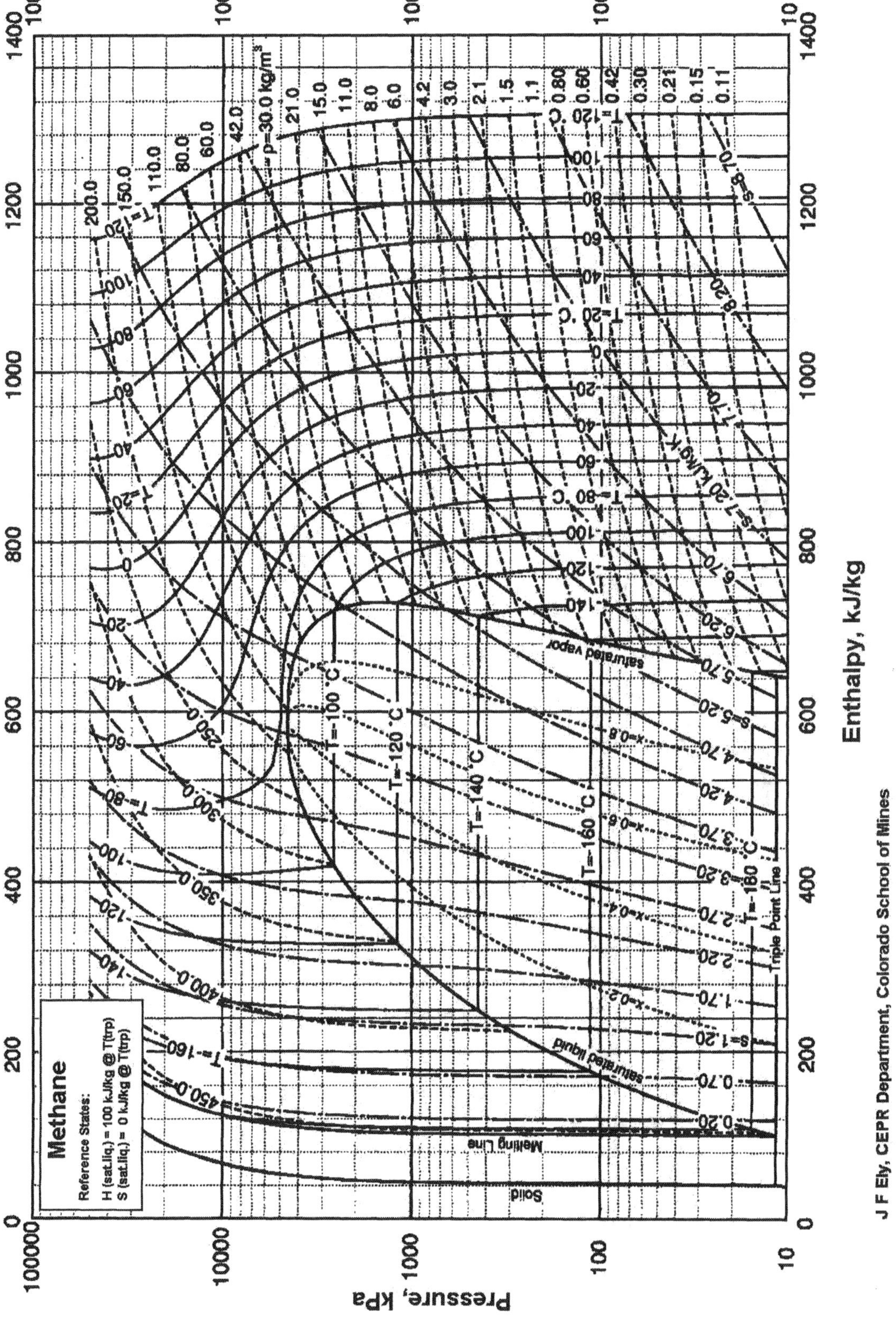

20.4.4 Ethane

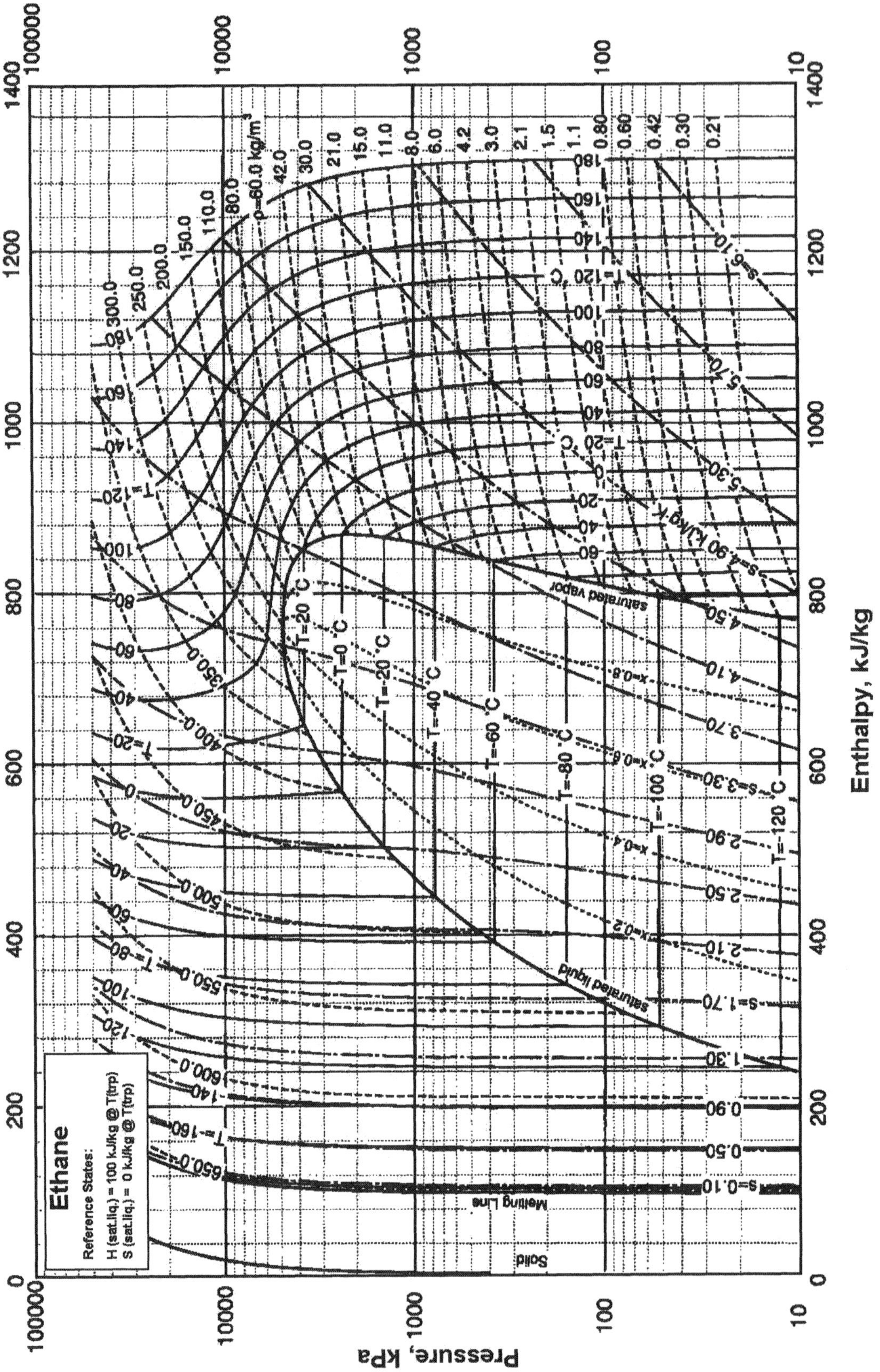

J F Ely, CEPR Department, Colorado School of Mines

20.4.5 Ethylene

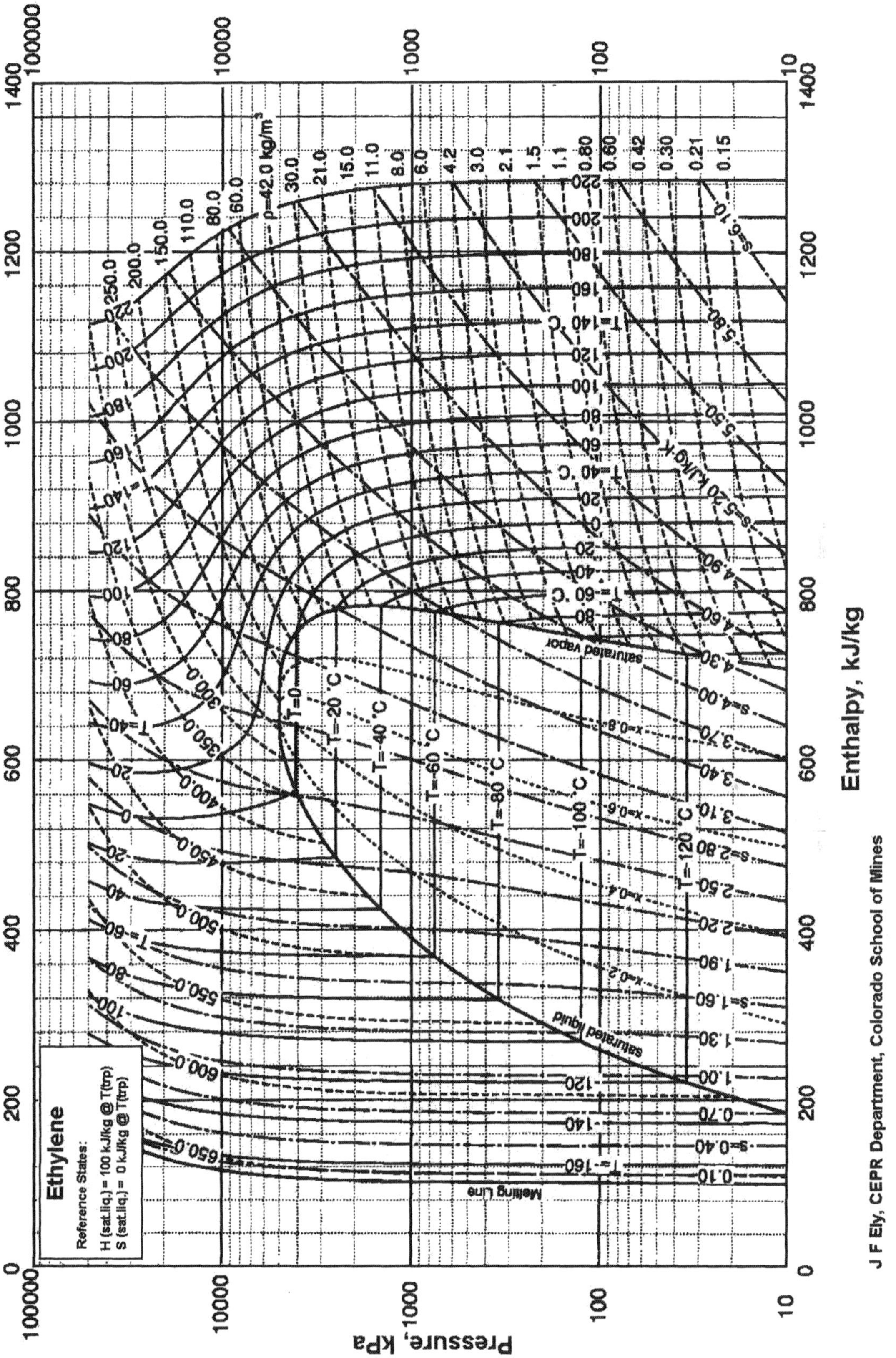

20.4.6 Propane

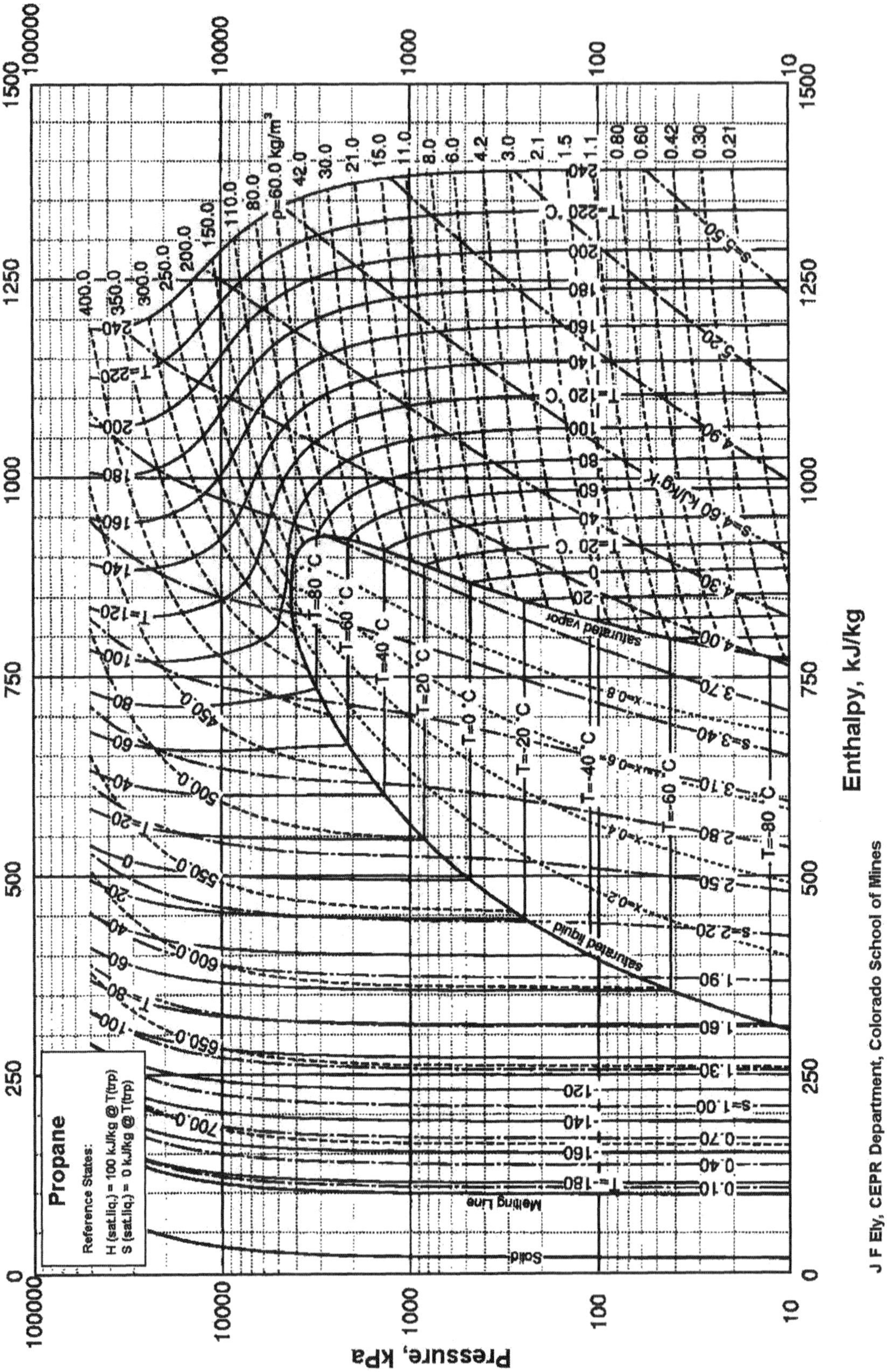

20.4.7 Propylene

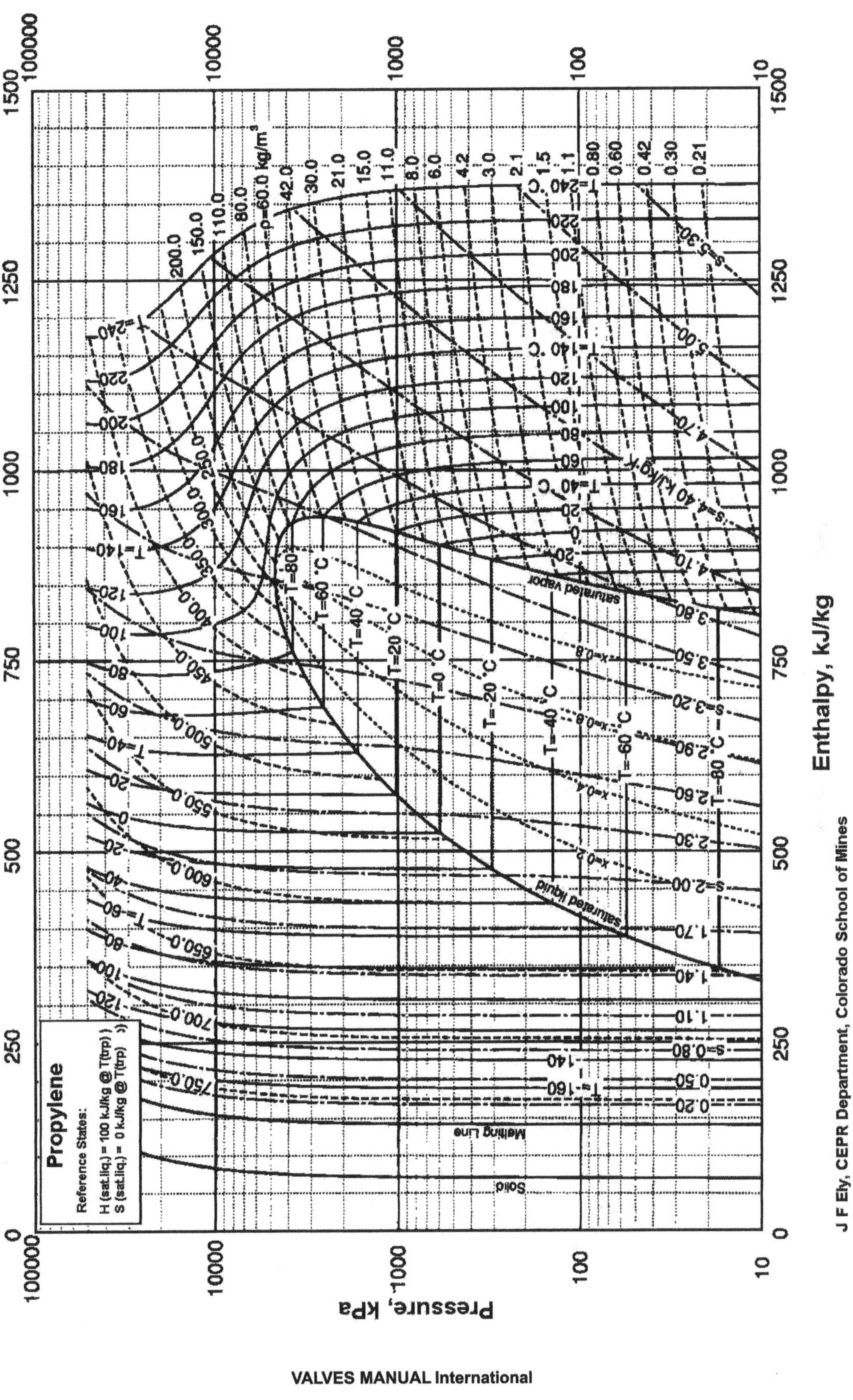

20.4.8 Iso-butane

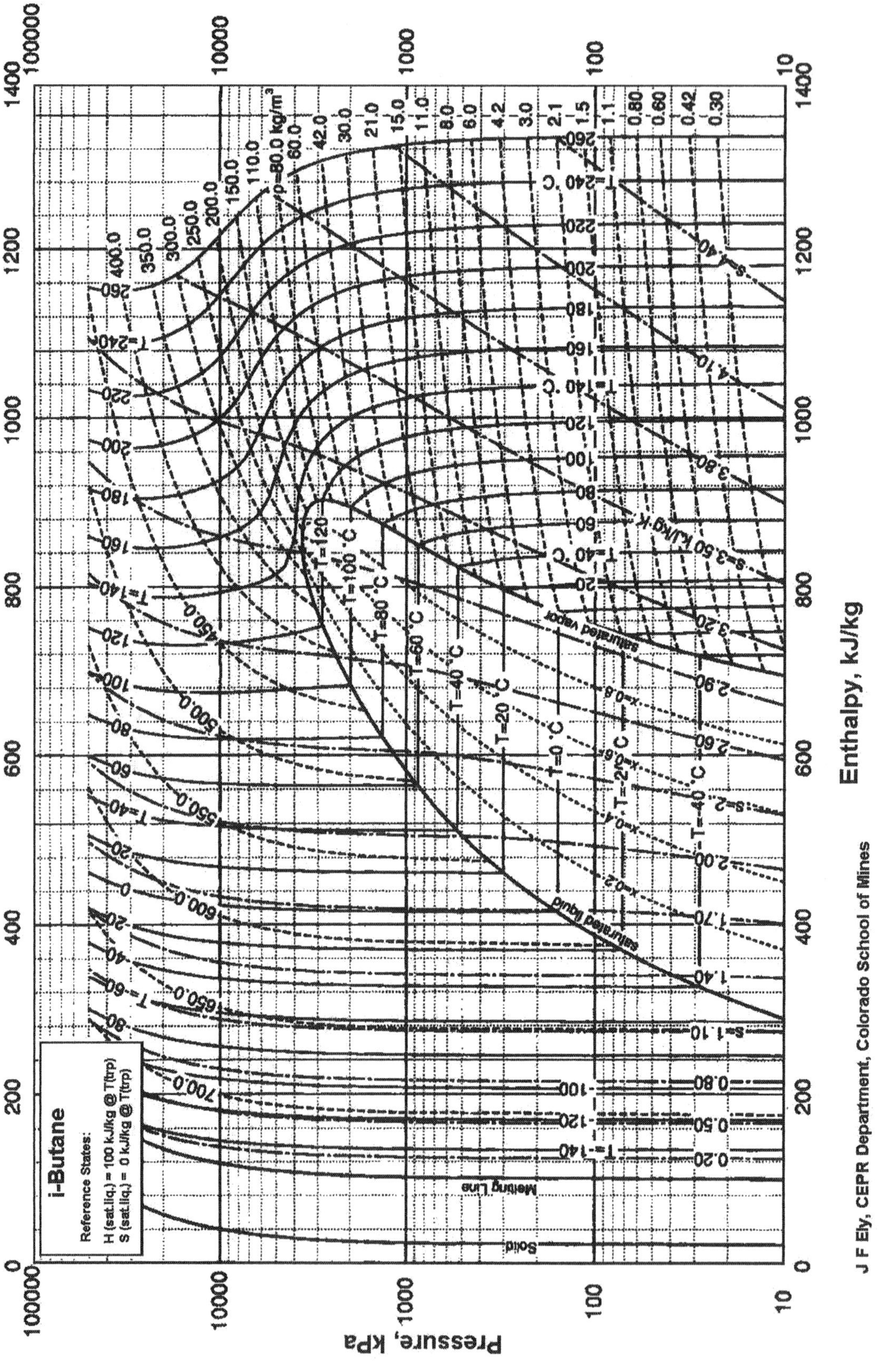

20.4.9 Normal-butane

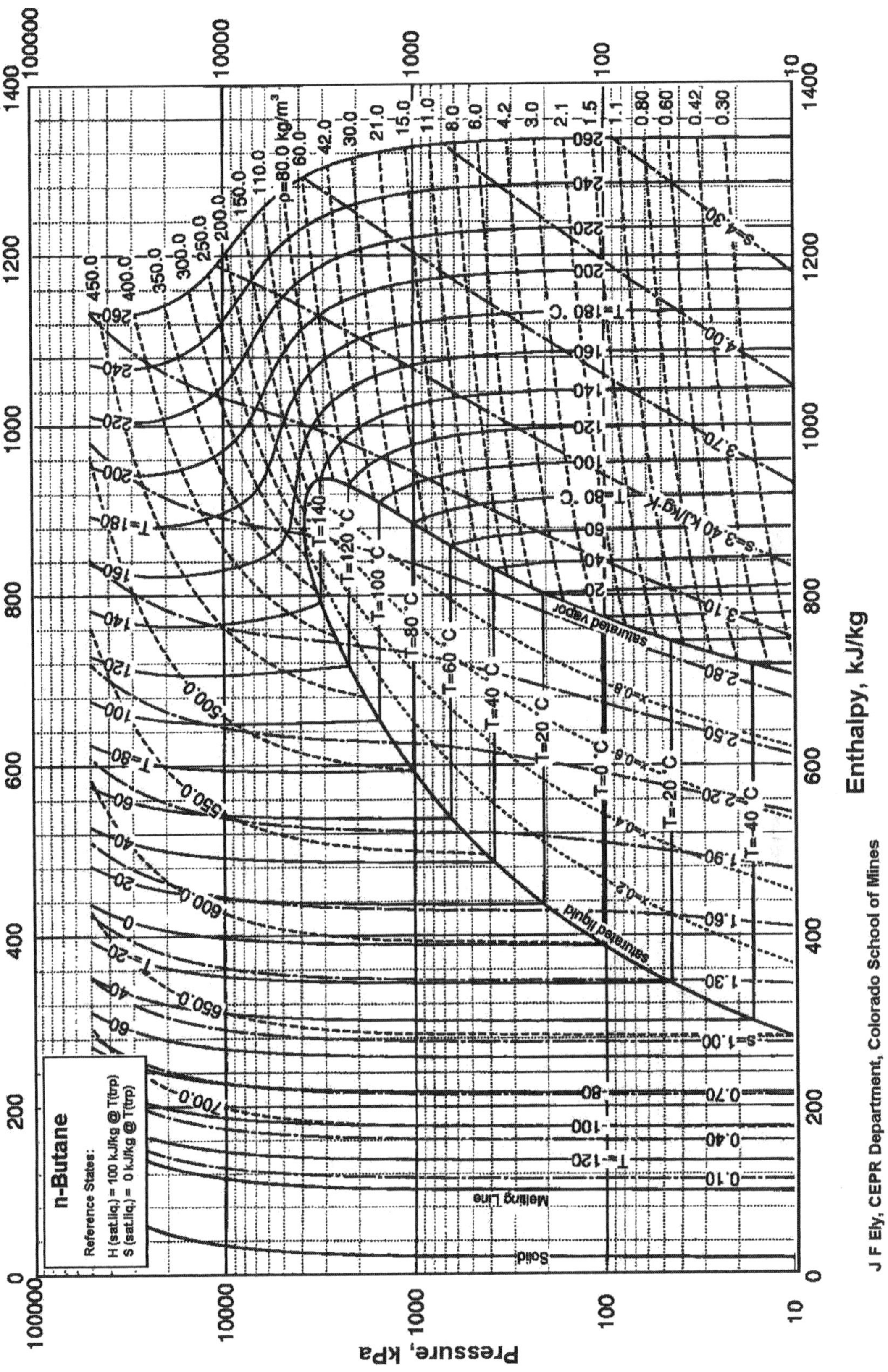

20.4.10 Iso-pentane

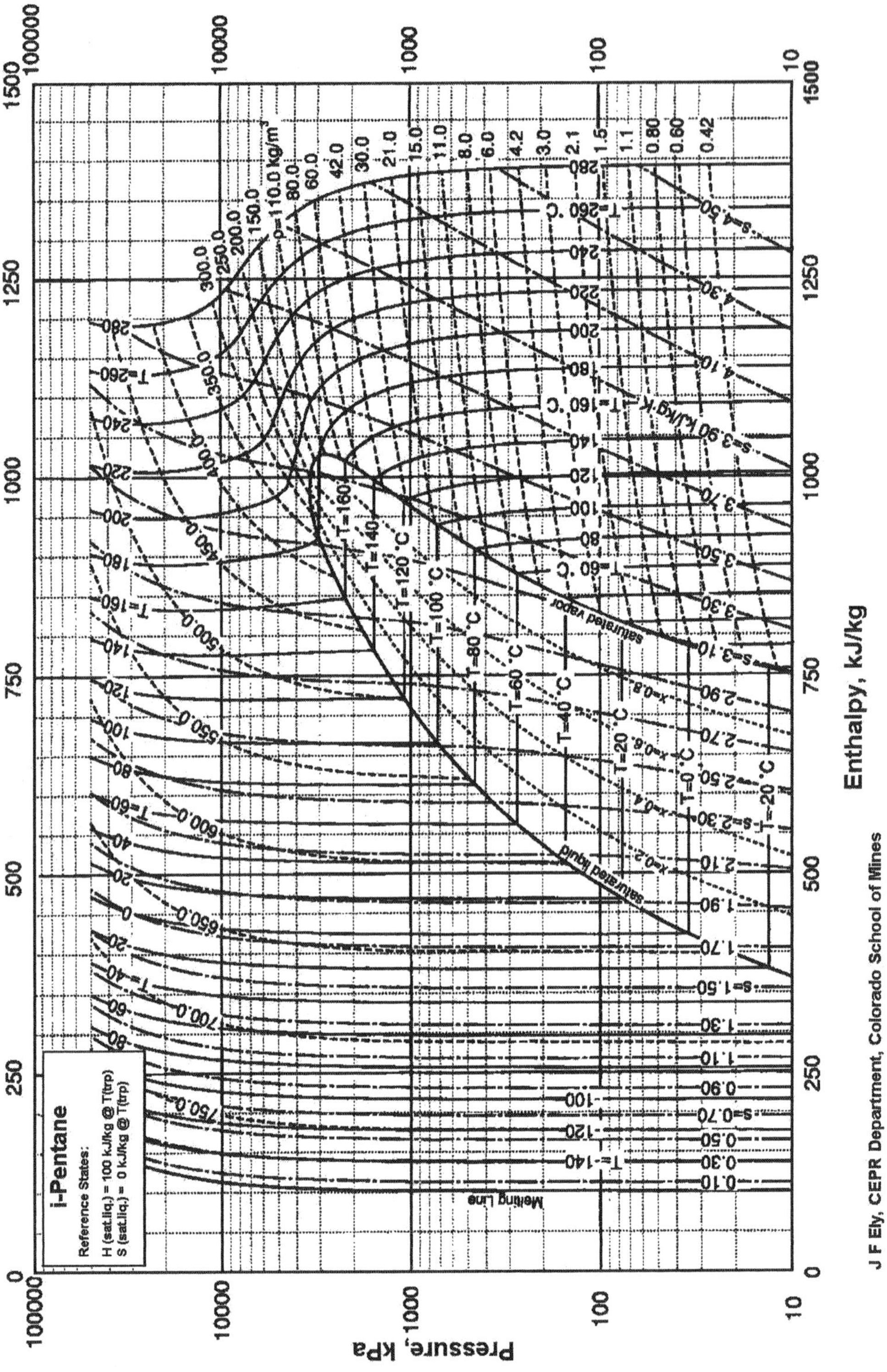

20.4.11 Normal pentane

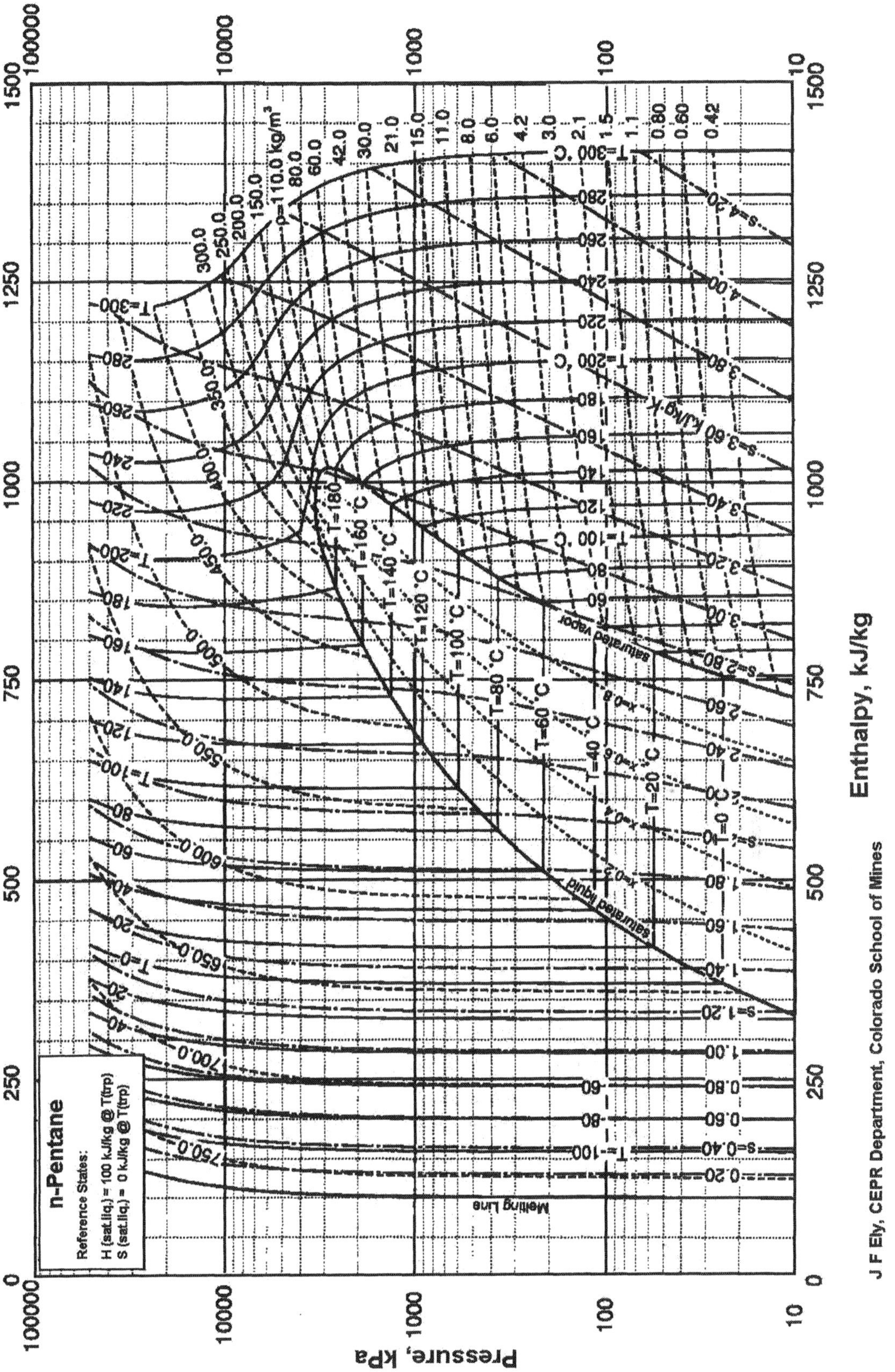

20.4.12 Oxygen

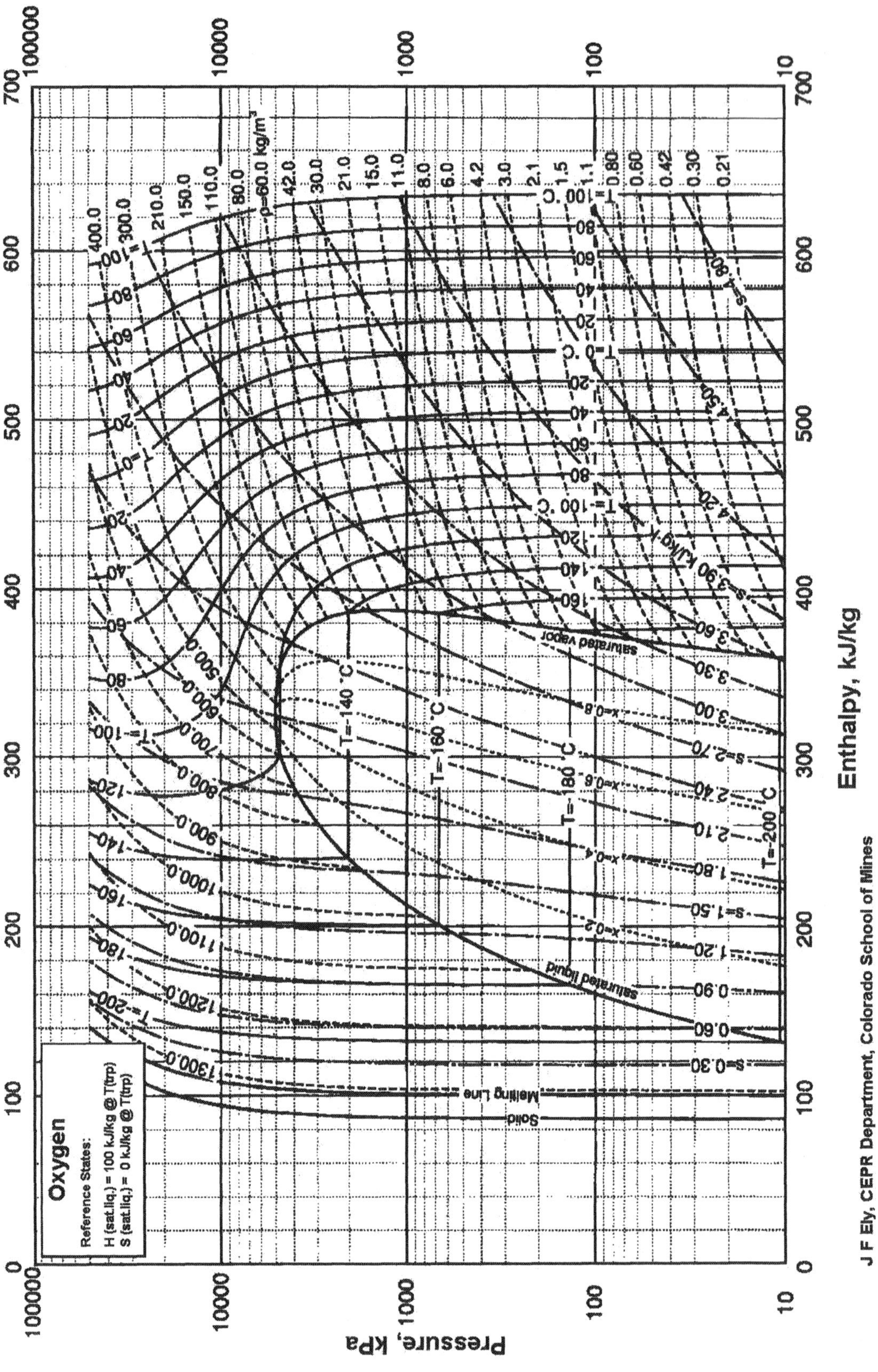

J F Ely, CEPR Department, Colorado School of Mines

20.5 SI, The International System of Units

Système International d'Unités,(international abbreviation, SI) the international measurement unit system is based on an earlier metric system and is being used more extensively worldwide. The SI system is systematically constructed to cover in practice all scientific, technical and daily requirements and is subject to international agreement. This means that it is now possible to apply the SI system uniformly throughout the world. A measurement system which is suitable for all technical and scientific purposes has to fulfil many requirements. Some of the basic requirements which SI satisfies are consistency, consequential applicability, coherence, the convenient expression of multiples and sub-multiples over a wide range of numerical values and accuracy.

- Consistency means that each unit shall represent one, and only one, quantity.
- Consequential applicability means that each quantity shall be measured in one, and only one, unit.
- Coherence means that all units for every quantity shall be compatible so as to eliminate the need for arbitrary conversion factors in calculations involving related quantities.
- Convenient expression of multiples and sub-multiples means the convenient multiplication of units to enable the use of practical numerical values within a particular application.
- Accuracy means that the base units shall be precisely derived and defined. Six of the seven base units are thus determined from distinct precisely defined physical phenomena, the seventh, the kilogram, is determined by one standard body which is held in Paris.

In 1971 the EEC, now the EC (European Community) ratified a Directive, 71/354/EEC, on units which committed all member states to amend legislation to authorise SI units within 18 months of that date and to implement all provisions of the Directive within a further five years. An amending Directive, 76/770/EEC, legislates the obligations. The Units of Measurement Directives place non-SI units into four chapters A to D.

Chapter A prescribes those units which are for permanent use and member states are obliged to authorise them in their laws by 21 April, 1978.

Chapter B contains a list of all units which member states have undertaken to cease to authorise in their laws with effect from 31 December, 1977.

Chapter C contains a list of units which member states have undertaken to cease to authorise in their laws with effect from 31 December, 1979.

Chapter D covers remaining units and some other units and will be reviewed before 31 December, 1979.

The formal content of the SI is determined and authorised by the General Conference of Weights and Measures (CGPM) and, for more detailed descriptions of the System reference should be made to BS 3763 and SI - The International System of Units published by the UK Office of Public Sector Information (formerly called HMSO). However, the basic advice to industry on the use of SI is now contained in ISO 1000, BS 5555.

The SI system includes two classes of units:

- Base units
- Derived units

The system also includes a number of Non-SI units, the "bar" for example, which are retained for use together with the SI units and their multiples because of their practical importance or because of their specialized fields. Multiples and sub-multiples of SI units and recognised Non-SI units are formed by attaching a prefix, micro, milli, kilo, Mega etc., to the unit symbol.

20.5.1 Brief history of unit systems

Although often called the metric system, the SI system essentially has more basic units and overcomes problems encountered during the development of a consistent system of units to serve all science and engineering functions. The metric system was introduced at the beginning of the nineteenth century based on the unit of length being the metre. The unit of mass followed and together with the unit of time formed the basic metric system. In 1873 the British Association for the Advancement of Science agreed on the use of the centimetre, gramme and second as the basic units for scientific work (the CGS system) but engineering within the United Kingdom had been well established using the British units of feet, pounds and seconds (the FPS system). As electrical experiments took place it was the metric system that became the basis for units peculiar to the electrical sciences and many basic electrical units were added to the CGS system.

An international authority on the metric system was established in 1875 with the Bureaux International des Poids et Mesures at Sèvres defining the units of length, mass and time as the metre, kilogramme and second respectively (the MKS system) as these units were more convenient than the CGS system. The basic units of length, mass and time are insufficient to cover electrical units and consequently units employing the permeability and permittivity of free space became necessary. Confusion also arose because of the links with the CGS system with, in particular, the use of p (the ratio of the circumference of a circle to its diameter) appearing in equations which were usually not associated with circles.

To overcome the complications the International Electrotechnical Commission (IEC) rationalized the units in 1950 and adopted as a fourth basic unit the unit of electrical current, the ampere (the MKSA system). This had been suggested about a half century earlier by an Italian professor called Giorgi and this system of units was consequently named the Giorgi System. Although this system of units covered electrical engineering it did not cover all branches of science and consequently the Conference Générale des Poids et Measures (CGPM) in 1954 agreed a rationalized and coherent system of units which became in 1960 the SI system.

After some subsequent additions the system now has seven basic units as follows:

length	metre (m)
mass	kilogramme (kg)
time	second (s)
electric current	Ampere (A)
light	candela (cd)
temperature	Kelvin (K)
substance	mole (mol)

In addition the following supplementary units are in use:

plane angle	radian (rad)
solid angle	steradian (sr)

The supplementary units are both ratios and therefore have no basic units.

The SI system is based around the seven basic units and the two supplementary units together with derived units for the more commonly used quantities and a series of prefixes used for the formation of multiples and submultiples.

Actual temperatures are normally ex-pressed in Celsius units (°C) and temperature differences in Kelvin units (K).

20.5.2 Method of expressing symbols and numbers

The following rules apply to symbols for units

- The symbol should be lower case unless the unit is derived from a proper name in which case the symbol should be upper case (or the first letter upper case if more than one letter), for example, metre - m, Ampere - A, Hertz - Hz
- The symbol should not contain a final full stop, for example, m not m
- The symbol should remain unaltered in the plural, for example, m not ms
- If multiple symbols are required they should be separated by a space if confusion can occur, for example, kg m/s^2 not kgm/s^2. If a pair of units are each represented by a single letter they are not separated if the absence of a space is not likely to cause confusion, for example Nm
- The symbol should be reduced to its simplest expression, for example, W not J/s, Js^{-1}, kg m^2/s^3 or kg m^2 s^{-3}
- Where more than one symbol is required and division is involved use a solidus or superscript, do not use more than one solidus, for example, use m/s^2 or m s^2, do not use m/s/s

Large numbers should be written with the digits grouped in threes and with a '.' or ',' used to denote the decimal place, for example, 12 345 678, 0.000 012, 0,012 3. Use 10 000 not 10,000 to denote ten thousand, for example. One exception to this rule is that numbers with only four digits and without a decimal point do not normally have the space after the first digit, for example 1234 not 1 234.

Alternatively very large or very small numbers can be represented in exponential form, that is a number multiplied by a factor in the form of 10 to a positive or negative power. For example 100 is equivalent to 10^2 and 0.001 is equivalent to 10^{-3}. Therefore 1 234 000 can be expressed in the form 1.234 x 10^6 and 0.0123 can be expressed as 1.23 x 10^{-2}.

There is a further group of specialised units which are primarily for use within astronomy and physics.

20.5.3 Multiples of SI units

The prefixes in Table 20.2 are used to form names and symbols of multiples and subdivisions of the SI units. The symbol of a prefix is considered to be combined with the unit symbol for the base unit, supplementary unit or derived unit to which it is directly attached, forming with it a symbol for a new unit which can be provided with a positive or negative exponent and which can be combined with other unit symbols to form symbols for compound units.

Factor by which the unit is multiplied	Prefix		Example
	name	symbol	
10^{24}	yotta	Y	
10^{21}	zetta	Z	
10^{18}	exa	E	
10^{15}	peta	P	
10^{12}	tera	T	1 terrajoule = 1TJ
10^{9}	giga	G	1 gigawatt = 1GM
10^{6}	mega	M	1 megavolt = 1 MV
10^{3}	kilo	k	1 kilometre = 1 km
10^{2}	hecto	h	1 hectogram = 1 hg
10^{1}	deca	da	1 decalumen = 1 dalm
10^{-1}	deci	d	1 decimetre = 1 dm
10^{-2}	centi	c	1 centimetre = 1 cm
10^{-3}	milli	m	1 milligram = 1 mg
10^{-6}	micro	μ	1 microgram = 1 μm
10^{-9}	nano	n	1 nanohenry = 1 nH
10^{-12}	pico	p	1 picofarad = 1 pF
10^{-15}	femto	f	1 femtometre= 1 fm
10^{-18}	atto	a	1 attosecond = 1 as

Table 20.2 Multiples of SI units

Whenever possible units should be multiples or submultiples of 3 - hecto, deca, deci and centi therefore should not normally be used.

The prefix symbol should appear immediately before the basic symbol, for example mA for milliampere. It should be noted that mm^3 means (0.001 m)^3 not 0.001 m^3 and that mm^{-1} means (10^{-3} m)$^{-1}$ not 10^{-3} m^{-1}. The use of dk should be avoided as this may cause confusion, for example, dkg is decagramme not deci killogramme.

It is normal practice to use millimetre (mm) as the unit of length on engineering drawings.

20.5.4 Derived units

These are expressed in terms of base units and/or supplementary units by multiplication and division according to the laws of physics relating the various quantities, see Table 20.3.

Quantity	Name of derived SI unit	Symbol	Expressed in terms of base or supplementary units
frequency	Hertz	Hz	1 Hz = 1/s
force	Newton	N	1 N = 1 kg m/s^2
pressure, stress	Pascal	Pa	1 Pa = 1 N/m^2
energy, work, heat	Joule	J	1 J = 1 Nm
power	Watt	W	1 W = 1 J/s
electric charge, quantity of electricity	Coulomb	C	1 C = 1 A s
electric potential	Volt	V	1 V = 1 J/c = 1 W/A
electric capacitance	Farad	F	1 F = 1 C/V
electric resistance	Ohm	Ω	1 Ω = V/A
electric conductance	Siemens	S	1 S = 1/Ω
magnetic flux	Weber	Wb	1 Wb = 1 V s
magnetic flux density	Tesla	T	1 T = 1 Wb/m^2
inductance	Henry	H	1 H = 1 Wb/A
luminous flux	Lumen	lm	1 lm = 1 cd sr
illuminance	Lux	lx	1lx = 1 lm/m^2
radioactivity	Becquerel	Bq	1 Bq = 1/s
absorbed dose	Gray	Gy	1 Gy = 1 J/kg

Table 20.3 Some derived SI units having special names

Non-SI units. There are certain units not included in SI which cannot, for a variety of reasons, be eliminated, despite the fact that these can, in principle, be expressed in SI units. Some of the non-SI units which may be used together with the SI units and their multiples and are recognised by the CIPM, Comité International des Poids et Mesures, are shown in Table 20.4.

Quantity	Name of unit	Unit symbol	Definition
Time	minute	min	1 min = 60 sec
Time	hour	h	1 h = 60 min
Time	day	d	1 d = 24 h
Plane angle	degree	°	1° = (p/180) rad
Plane angle	minute	′	1 ′ = (1/60)°

Quantity	Name of unit	Unit symbol	Definition
Plane angle	second	"	1 " = (1/60)
Volume	litre	l	1 l = 1 dm^3
Mass	tonne	t	1 t = 10^3 kg
Pressure of fluid	bar	bar	1 bar = 10^5Pa

Table 20.4 Non-SI units

Conductance is sometimes used as the reciprocal of resistance for which the unit is the Siemen (S), also referred to as mho.

20.6 Conversion factors for SI units

A number of electrical terms evolved from the CGS system may still be encountered as well as units in the FPS system and other special units. In general it is best to convert units to the SI system before inserting quantities into equations as this will avoid problems involving conversion factors in the equations themselves. The engineering units which may be encountered in other systems of units are listed in Tables 20.5 to 20.7 with the conversion factor to give SI units.

Quantity	Unit	Conversion factor	SI unit
Length	cable	219.456	m
	chain (Gunter's)	20.116 8	
	chain (Ramden's)	30.48	
	fathom	1.828 8	
	feet	0.304 8	
	furlong	201.168	
	inch	0.025 4	
	micron	1 x 10^{-6}	
	mile (nautical Brit)	1.853 184 x 10^3	
	mile (nautical Int)	1.852 x 10^3	
	mile (statute)	1.609 34 x 10^3	
	mil	2.54 x 10^{-5}	
	rod	5.029 2	
	yard	0.914 4	
Area	acre	4.046 856 x 10^3	m^2
	are	100	
	centare	1	
	hectare	1 x 10^4	
Volume	barrel (Brit)	0.163 65	m^3
	barrel (US petrol)	0.158 98	
	barrel (US dry)	0.115 63	
	barrel (US liquid)	0.119 24	
	gallon (Brit) (Imp)	4.545 9 x 10^{-3}	
	gallon (US dry)	4.404 8 x 10^{-3}	
	gallon (US liquid)	3.785 3 x 10^{-3}	
	litre	1 x 10^{-3}	
	ounce (Brit fluid)	2.841 225 x 10^{-5}	
	ounce (US fluid)	2.957 373 x 10^{-5}	
	pint (Brit) (Imp)	5.682 4 x 10^{-4}	
	pint (US dry)	5.505 95 x 10^{-4}	
	pint (US liquid)	4.731 63 x 10^{-4}	
	quart (Brit) (Imp)	1.136 49 x 10^{-3}	
	quart (US dry)	1.101 19 x 10^{-3}	
	quart (US liquid)	9.463 2 x 10^{-4}	

Table 20.5 Conversion factors for length, area and volume

To obtain the quantity in SI units, the original units should be multiplied by the conversion factor, for example, to convert a length of 10 inch to metre units multiply by 0.0254 giving 0.254 m. To convert from one non-SI unit to another non-SI unit, first convert to the SI unit by multiplying by the conversion factor and then convert to the other non-SI unit by dividing by the conversion factor for the other unit. If multiple units are involved, convert each separately by multiplying by the appropriate conversion factor for any units with positive powers and dividing by the appropriate conversion factors for any units with negative powers. For example, to convert lb/in^3 (or lb in^{-3}) to SI units multiple by 0.453 6 to convert lb to kg and divide by 0.025 4^3 to convert in^{-3} to m^{-3} giving 27 680 kg m^{-3} (or 27680kg/m^3).

To convert Fahrenheit to Celsius or centigrade subtract 32 and then multiply by 5/9.

To convert Fahrenheit to Rankine add 459.67

To convert Celsius and centigrade to Kelvin add 273.15

To convert Rankine to Kelvin multiply by 5/9

For temperature conversion to Kelvin (K) the following should be noted:

Absolute zero =	0 Kelvin
	0 Rankine
	-273.15 Celsius
	-273.15 centigrade
	-459.67 Fahrenheit
Ice point =	273.15 Kelvin
	491.67 Rankine
	0 Celsius
	0 centigrade
	32 Fahrenheit

Quantity	Unit	Conversion factor	SI unit
Energy and Work	British Thermal Unit (Btu)	1.054 35 x 10^3	J
	Btu (IST)	1.055 04 x 10^3	
	Btu (mean)	1.055 87 x 10^3	
	Btu (39°F)	1.059 66 x 10^3	
	Btu (60°F)	1.054 68 x 10^3	
	calorie, gm (cal, gm)	4.184	
	cal, gm (mean)	4.19	
	cal, gm (15°C)	4.186	
	cal, gm (20°C)	4.182	
	calorie, kg (cal, kg)	4.184 x 10^3	
	cal, kg (mean)	4.19 x 10^3	
	centigrade heat unit (15°C)	1.898 3 x 10^3	
	erg	1 x 10^{-7}	
	foot-pound	1.355 82	
Force	dyne	1 x 10^{-5}	N
	kilogramme force	9.807	
	poundal	0.138 255	
Mass	hundredweight (long)	50.802 3	kg
	hundredweight (short)	45.359 2	
	ounce (troy)	0.031 103	
	ounce (avdp)	0.028 35	
	pounds (troy)	0.373 241	
	pounds (avdp)	0.453 592	
	slugs	14.593 9	
	ton (long)	1.106 046 x 10^3	
	ton (short)	9.071 847 x 10^2	
	ton (metric) (tonne)	1 x 10^3	
Power	cheval-vapeur	735.499	W
	horsepower	746	
	horsepower (boiler)	9.809 5 x 10^3	
	horsepower (metric)	735.499	
Pressure	atmosphere (atm)	1.103 25 x 10^5	Pa
	bar	1 x 10^5	
	inch of mercury (32°F) (in Hg)	3.386 4 x 10^3	
	inch of mercury (60°F) (in Hg)	3.376 1 x 10^3	
	inch of water (4°C) (in H_2O)	0.248 7	
	mm of mercury (0°C)	133.313	
	torr	133.313	
Velocity	knot	0.514 4	m/s
Viscosity (dynamic)	centipoise (cp)	1 x 10^{-3}	Ns/m^2
	poise (p)	0.1	
Viscosity (kinematic)	centistoke (cSt)	1 x 10^{-6}	m^2/s
	Stoke (St)	1 x 10^{-4}	

Table 20.6 Conversion factors for quantities related to mechanics

Quantity	Unit	Conversion factor	SI unit
Electric charge	Faraday (chem) Faraday (phys)	9.649×10^4 9.652×10^4	C
Magnetic flux	line Maxwell	1×10^{-8} 1×10^{-8}	Wb
Magnetic flux density	Gauss	1×10^{-4}	T
Magnetizing force	Oersted	79.577 47	A-turns/m
Magnetomotive force	Gilbert	0.795 77	A-turns

Table 20.7 Conversion factors for quantities

Detailed conversion information is now given for a number of different quantities.

20.6.1 Plane angle

Quantity designation:

SI unit: radian, rad.

Normal multiple units: mrad, μrad.

Example: 2 rad = 2 × 57.2958 = 114.5916°

rad	... g gon, grade	... ° degree	... ′ minute	... ″ second	angular mil
1	63.6620	57.2958	3.43775 × 103	0.206265 × 106	1.00268 × 103
$15.7080 \cdot 10^{-3}$	1	0.9	54	3.24 × 103	15.75
$17.4533 \cdot 10^{-3}$	1.11111	1	60	3.6 × 103	17.5
$0.290888 \cdot 10^{-3}$	$18.5185 \cdot 10^{-3}$	$16.6667 \cdot 10^{-3}$	1	60	0.291667
$4.84814 \cdot 10^{-6}$	$0.308642 \cdot 10^{-3}$	$0.277778 \cdot 10^{-3}$	$16.6667 \cdot 10^{-3}$	1	$4.86111 \cdot 10^{-3}$
$0.997331 \cdot 10^{-3}$	$63.4921 \cdot 10^{-3}$	$57.1429 \cdot 10^{-3}$	3.42857	205.714	1

grade (g) (or gon), 1 g = 1 gon = μ/200 rad.

1° = /180 rad.

For some purposes, the angular mil is taken to be 10^{-3} rad. The figures shown here are based on the concept that an angular mil is equal to 360/6400 degrees.

NOTE: 1 grade (... g) = 1 /100 of a right angle

20.6.2 Length

Quantity designation l.

SI unit: metre (m).

Normal multiple units: km, cm, mm, μ.

Example: 3 in = 3 × 25.4 × 10^{-3} = 76.2 × 10^{-3} m

metre m	inch in	foot ft	yard yd	mile
1	39.370	3.2808	1.0936	$0.62137 \cdot 10^{-3}$
$25.4 \cdot 10^{-3}$	1	$83.333 \cdot 10^{-3}$	$27.778 \cdot 10^{-3}$	$15.783 \cdot 10^{-6}$
0.3048	12	1	0.33333	$0.18939 \cdot 10^{-3}$
0.9144	36	3	1	$0.56818 \cdot 10^{-3}$
$1.6093 \cdot 10^{3}$	$63.36 \cdot 10^{3}$	$5.28 \cdot 10^{3}$	$1.76 \cdot 10^{3}$	1
$1.852 \cdot 10^{3}$	$72.913 \cdot 10^{3}$	$6.0761 \cdot 10^{3}$	$2.0254 \cdot 10^{3}$	1.1508

1 nautical mile = 6080 ft = 1853.184 m

1 Å, 1 Ångström = 10^{-10} m

1 astronomic unit = 0.1496 · 10^{12} m

1 light year = 9.4605 · 10^{15} m

1 parsec = 30.857 · 10^{12} m

20.6.3 Area

Quantity designation: A.

SI unit: square metre (m^2).

Normal multiple units: km^2, dm^2, cm^2, mm^2.

Example: 4 ft^2 = 4 ·92.903 ·10^{-3}= 0.371612 m^2

m^2	in^2	ft^2	yd^2	acre	square mile $mile^2$
1	$1.5500 \cdot 10^{3}$	10.764	1.1960	$0.24710 \cdot 10^{-3}$	$0.38610 \cdot 10^{-6}$
$0.64516 \cdot 10^{-3}$	1	$6.9444 \cdot 10^{-3}$	$0.77161 \cdot 10^{-3}$	$0.15942 \cdot 10^{-6}$	$0.24910 \cdot 10^{-9}$
$92.903 \cdot 10^{-3}$	144	1	0.11111	$22.957 \cdot 10^{-6}$	$35\text{-}870 \cdot 10^{-9}$
0.83613	$1.296 \cdot 10^{3}$	9	1	$0.20661 \cdot 10^{-3}$	$0.32283 \cdot 10^{-6}$
$4.0469 \cdot 10^{-3}$	$6.2726 \cdot 10^{6}$	$43.56 \cdot 10^{3}$	$4.84 \cdot 10^{3}$	1	$1.5625 \cdot 10^{-3}$
$2.5900 \cdot 10^{6}$	$4.0145 \cdot 10^{9}$	$27.878 \cdot 10^{6}$	$3.0976 \cdot 10^{6}$	640	1

1 acre = 100 m^2

1 hectare = 100 acres = 10000 m^2

20.6.4 Volume

Quantity designation: V.

SI unit: cubic metre (m^3).

Normal multiple units: dm^3, cm^3, mm^3.

Non SI unit: litre (l): 1 l = 0.001 m^3 = 1 dm^3.

Normal multiple units: cl, ml.

Example:

5 US gallon = 5 · 3.7854 · 10^{-3} = 18.927 · 10^{-3} m^3 = 18.927 l

m^3	in^3	ft^3	yd^3	UK gallon	US gallon
1	$61.024 \cdot 10^{3}$	35.315	1.3080	219.97	264.17
$16.387 \cdot 10^{-6}$	1	$0.57870 \cdot 10^{-3}$	$21.434 \cdot 10^{-6}$	$3.6046 \cdot 10^{-3}$	$4.3290 \cdot 10^{-3}$
$28.317 \cdot 10^{-3}$	$1.728 \cdot 10^{3}$	1	$37.037 \cdot 10^{-3}$	6.2288	7.4805
0.76456	$46.656 \cdot 10^{3}$	27	1	168.18	201.97
$4.5461 \cdot 10^{-3}$	277.42	0.16054	$5.9461 \cdot 10^{-3}$	1	1.2010
$3.7854 \cdot 10^{-3}$	231	0.13368	$4.9511 \cdot 10^{-3}$	0.83268	1

gross (register) tonnage used in shipping
1 ton = 100 ft^3 = 2.83168 m^3

1 UK fluid ounce, fl oz = 28.4131 cm^3

1 US fluid ounce, fl oz = 29.5735 cm^3

20.6.5 Time

Quantity designation: t.

SI unit: second (s).

Normal multiple units: ms, μs, ns.

Non SI units: day (d), hour (h), minute (min).

Example: 100000 s = 100000/3600 = 27.778 h

s	min	h	d (day)	week
1	16.6667 · 10^{-3}	0.277778 · 10^{-3}	11.5741 ·10^{-6}	1.65344 · 10^{-6}
60	1	16.6667 · 10^{-3}	0.694444 · 10^{-3}	99.2063 · 10^{-6}
3.6 · 10^{3}	60	1	41.6667 · 10^{-3}	5.95238 · 10^{-3}
86.4 · 10^{3}	1.44 · 10^{3}	24	1	0.142857
604.8 · 10^{3}	10.08 · 10^{3}	168	7	1

1 tropical year = 31556925.974 s = 365.24219878 d

1 sidereal year = 31558150 s

1 calendar year = 365 d = 8760 h

20.6.6 Linear velocity

Quantity designation: v.

SI unit: metre per second (m/s).

Normal multiple units: km/h.

m/s	km/h	ft/s	mile/h
1	3.6	3.2808	2.2369
0.27778	1	0.91134	0.62137
0.3048	1.0973	1	0.68182
0.44704	1.6093	1.4667	1
0.51444	1.852	1.6878	1.1508

1 knot = 1 nautical mile per hour = 1.853 km/h

20.6.7 Linear acceleration

Quantity designation: a.

SI unit: metre per second squared (m/s^2).

m/s^2	cm/s^2	ft/s^2	in/s^2	g
1	100	3.2808	39.37	0.10197
10 × 10-3	1	32.808 · 10-3	393.7 · 10^{-3}	1.0197 · 10^{-3}
0.3048	30.48	1	12	31.081 · 10^{-3}
25.4 × 10^{-3}	2.54	83.33 · 10^{-3}	1	2.59 · 10^{-3}
9.80665	980.665	32.174	386.09	1

20.6.8 Angular velocity

Quantity designation: ω.

SI unit: radian per second (rad/s).

The SI and Imperial units are identical.

Angular velocity is normally calculated from N revolutions/s by $2\pi N$

20.6.9 Angular acceleration

Quantity designation: a.

SI unit: radian per second squared (rad/s^2)

The SI and Imperial units are identical.

20.6.10 Mass

Quantity designation: m.

SI unit: kilogram (kg).

Normal multiple units: μg, g, mg.

Non SI unit: tonne (t) = 1000 kg.

kg	lb (pound)	oz (ounce)	hundred-weight cwt (UK)	ton (UK)
1	2.2046	35.274	19.684 · 10^{-3}	0.98421 · 10^{-3}
0.45359	1	16	8.9286 · 10^{-3}	0.44643 · 10^{-3}

kg	lb (pound)	oz (ounce)	hundred-weight cwt (UK)	ton (UK)
14.594	32.174	514.79	0.28727 · 10^{-3}	14.363 · 10^{-3}
28.350 × 10^{-3}	62.5 × 10^{-3}	1	0.55804 · 10^{-3}	27.902 · 10^{-6}
50.802	112	1.792 · 10^{3}	1	50 · 10^{-3}
1.0161 · 10^{3}	2.24 · 10^{3}	35.84 · 10^{-3}	20	1

oz = ounce, also called ounce avoirdupois

1 ounce troy = 31.1035 · 10^{-3} kg

cwt = hundredweight

USA cwt = 100 lb USA ton = 2000 lb

20.6.11 Density

Also called specific weight.

Quantity designation: ρ.

SI unit: kilogram per cubic metre (kg/m^3)

Non SI units: kg/dm^3, g/cm^3.

kg/m^3	g/cm^3	lb/in^3	lb/ft^3
1	10^{-3}	36.127 · 10^{-6}	62.428 · 10^{-3}
10^{3}	1	36.127 · 10^{-3}	62.428
27.680 · 10^{3}	27.680	1	1.728 · 10^{3}
16.019	16.019 · 10^{-3}	0.57870 · 10^{-3}	1

The term specific gravity or relative density is also used and is the ratio of the mass of a given volume of substance to the mass of an equal volume of water at temperature of 4 °C and a pressure of 101.325 kPa absolute. The density of water at 4 °C and 101.325 kPa absolute is 1000.02 kg/m^3.

20.6.12 Force

Quantity designation: F.

SI unit: newton (N).

Normal multiple units: MN, kN.

N	dyn	Kilogram-force, kgf kilopond,	pound-force lbf
1	0.1 · 10^{6}	0.10197	0.22481
10 · 10^{-6}	1	1.0197 · 10^{-6}	2.2841 · 10^{-6}
9.8066	0.98066 · 10^{6}	1	2.2046
4.4482	0.44482 · 10^{6}	0.45359	1

20.6.13 Torque

Quantity designation: T.

SI unit: Newton metre (Nm).

Normal multiple units: MNm, kNm.

Nm	kgf m	lbf × in	lbf × ft
1	0.10197	8.8508	0.73756
9.8066	1	86.796	7.2330
0.11299	11.521 · 10^{-3}	1	83.333 · 10^{-3}
1.3558	0.13826	12	1

Torque, power and speed are related by the formula: $P = 2\pi NT$

20.6.14 Pressure, stress

Quantity designation: p, σ, τ.

SI unit: Pascal (Pa), 1 Pa = 1 N/m^2.

Normal multiple units: GPa, MPa, kPa and for stress: MN/m^2, N/m^2, N/mm^2.

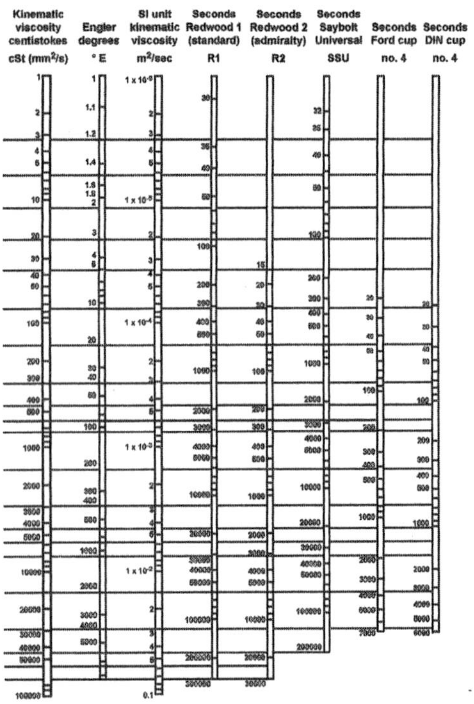

Figure 20.11 Nomogram for conversion of Kinematic viscosity to other units of viscosity, in Section 20.6.16

Pa	bar	kgf/cm² technical atmos	kgf/mm²	Torr (≈ mm Hg)	standard atm	lbf/in² (psi)
1	$10 \cdot 10^{-6}$	$10.197 \cdot 10^{-6}$	$0.10197 \cdot 10^{-6}$	$7.5006 \cdot 10^{-3}$	$9.8692 \cdot 10^{-6}$	$0.14504 \cdot 10^{-3}$
$100 \cdot 10^{3}$	1	1.0197	$10.197 \cdot 10^{-3}$	750.06	0.98692	14.504
$98.066 \cdot 10^{3}$	0.98066	1	$10 \cdot 10^{-3}$	735.56	0.96784	14.223
$9.8066 \cdot 10^{6}$	98.066	100	1	$73.556 \cdot 10^{3}$	96.784	$1.4223 \cdot 10^{3}$
133.32	$1.3332 \cdot 10^{-3}$	$1.3595 \cdot 10^{-3}$	$13.595 \cdot 10^{-6}$	1	$1.3158 \cdot 10^{-3}$	$19.337 \cdot 10^{-3}$
$101.32 \cdot 10^{3}$	1.0132	1.0332	$10.332 \cdot 10^{-3}$	760	1	14.696
$6.8948 \cdot 10^{3}$	$68.948 \cdot 10^{-3}$	$70.307 \cdot 10^{-3}$	$0.70307 \cdot 10^{-3}$	51.715	$68.046 \cdot 10^{-3}$	1

The preferred pressure unit for pump applications is the "bar".

1 mm H_2O ≈ 9.81 Pa

1 in H_2O ≈ 249.09 Pa

1 in Hg ≈ 3386.4 Pa

1 ata = 1 technical atmosphere (absolute)

1 atu = 1 technical atmosphere (gauge)

20.6.15 Dynamic viscosity

Quantity designation: η.

SI unit: Pascal second Pa s.

Normal multiple units mPa s (= cP).

N s/m²	N s/mm²	P (poise)	cP m Pa s
1	10^{-6}	10	10^{3}
10^{6}	1	10×10^{6}	10^{9}
0.1	0.1×10^{-6}	1	100
10^{-3}	10^{-9}	10×10^{-3}	1

20.6.16 Kinematic viscosity

Quantity designation: μ.

SI unit: square metres per second (m^2/s).

Normal multiple units: mm^2/s.

For conversion to other units of viscosity see nomogram in Figure 20.11.

m²/s	mm²/s cSt	St (Stoke)
1	10^{6}	$10 \cdot 10^{3}$
10^{-6}	1	$10 \cdot 10^{-3}$
$0.1 \cdot 10^{-3}$	100	1

	cSt x	°E x	R1 x	SSU x
Kinematic viscosity cSt	1	7.58	0.247	0.216
Engler degrees °E	0.132	1	0.0326	0.0285
Seconds, Redwood 1, R1	4.05	30.7	1	0.887
Seconds, Saybolt Universal SSU	4.62	35.11	1.14	1

The above factors apply for values above 60cSt

20.6.17 Energy

Quantity designations: E, W, Q, L, U depending upon the type of energy.

SI unit: Joule (J).

Normal multiple units: TJ, GJ, MJ, kJ, mJ.

Since 1 J = 1 Nm = 1 Ws then Nm and Ws can also be used for all types of energy. The unit Joule should, however, be used for expressing all types of energy.

1 erg = 0.1×10^{-6} J.

J Joule	kWh kilowatt hour	kgf m kilogram-force metre	kcal kilo-calorie	ch h metric horse-power hour	ft. lbf (foot pound-force)	Btu (British thermal unit)
1	$0.27778 \cdot 10^{-6}$	0.10197	$0.23885 \cdot 10^{-3}$	$0.37767 \cdot 10^{-6}$	0.73756	$0.94782 \cdot 10^{-3}$
3.6 · 106	1	$0.36710 \cdot 10^{6}$	859.85	1.3596	$2.6552 \cdot 10^{6}$	$3.4121 \cdot 10^{3}$
9.8066	$2.7241 \cdot 10^{-6}$	1	$2.3423 \cdot 10^{-3}$	$3.7037 \cdot 10^{-6}$	7.2330	$9.2949 \cdot 10^{-3}$
$4.1868 \cdot 10^{3}$	$1.163 \cdot 10^{-3}$	426.94	1	$1.5812 \cdot 10^{-3}$	$3.0880 \cdot 10^{3}$	3.9683
$2.6478 \cdot 10^{6}$	0.73550	$0.27 \cdot 10^{6}$	632.42	1	$1.9529 \cdot 10^{6}$	$2.5096 \cdot 10^{3}$
1.3558	$0.37662 \cdot 10^{-6}$	0.13826	$0.32383 \cdot 10^{-3}$	$0.51206 \cdot 10^{-6}$	1	$1.2851 \cdot 10^{-3}$
$1.0551 \cdot 10^{3}$	$0.29307 \cdot 10^{-3}$	107.59	0.25200	$0.39847 \cdot 10^{-3}$	778.17	1

20.6.18 Power

Quantity designation: P.

SI unit: Watt (W).

Normal multiple units: GW, MW, kW, mW, μW.

W	kgf m/s	kcal/s	kcal/h	ch metric horse-power	hp horse-power	ft lbf/s	Btu/h
1	0.10197	0.23885 · 10^{-3}	0.85985	1.3596 · 10^{-3}	1.3410 · 10^{-3}	0.73756	3.4121
9.8066	1	2.3423 · 10^{-3}	8.4322	13.333 · 10^{-3}	13.151 · 10^{-3}	7.2330	33.462
4.1868 · 10^{3}	426.94	1	3.6 · 10^{3}	5.6925	5.6146	3.0880 · 10^{3}	14.286 · 10^{3}
1.163	0.11859	0.27778 · 10^{-3}	1	1.5812 · 10^{-3}	1.5596 · 10^{-3}	0.85779	3.9683
735.50	75	0.17567	632.42	1	0.98632	542.48	2.5096 · 10^{3}
745.70	76.040	0.17811	641.19	1.0139	1	550	2.5444 · 10^{3}
1.3558	0.13826	0.32383 · 10^{-3}	1.1658	1.8434 · 10^{-3}	1.8182 · 10^{-3}	1	4.6262
0.29307	29.885 · 10^{-3}	69.999 · 10^{-6}	0.25200	0.39847 · 10^{-3}	0.39302 · 10^{-3}	0.21616	1

20.6.19 Flow

Quantity designation: q_V.

SI unit: cubic metre per second (m^3/s).

Non-SI units: l/s, ml/s, m^3/h.

I gallon/ min Igpm	US gallon/ min USgpm	barrel/ day bpd	m^3/s	m^3/h	l/s	l/min
1	1.2009	41.175	75.768 · 10^{-6}	272.76 · 10^{-3}	75.768 · 10^{-3}	4.5461
0.83268	1	34.286	63.09 · 10^{-6}	227.12 · 10^{-3}	63.09 · 10^{-3}	3.7854
24.286 · 10^{-3}	29.167 · 10^{-3}	1	1.84 · 10^{-6}	6.6244 · 10^{-3}	1.84 · 10^{-3}	110.41 · 10^{-3}
13.198 · 10^{3}	15.85 · 10^{3}	543.44 · 10^{3}	1	3600	1000	60000
3.6662	4.4029	150.95	277.78 · 10^{-6}	1	277.78 · 10^{-3}	16.667
13.198	15.85	543.44	1.0 · 10^{-3}	3.6	1	60
219.97 · 10^{-3}	264.17 · 10^{-3}	9.0573	16.667 · 10^{-6}	60 · 10^{-3}	16.667 · 10^{-3}	1

1 barrel = 42 US gallon

20.6.20 Temperature

Absolute temperature: Quantity designation: T.

SI unit: Kelvin (K).

Temperature: Quantity designation: t.

unit degree Celsius (Centigrade) (°C).

	Kelvin*** scale	Celsius scale	Rankine scale	Fahrenheit** scale	Physical relation-ship
Relative temperature value	0 K	- 273.15 °C	0 °R	- 459.67 °F	Absolute zero
	273.15 K	0 °C	491.67 °R	32 °F	Melting point of ice*
	273.16 K	0.01 °C	491.688 °R	32.018 °F	Triple point of water*
	373.15 K	100 °C	671.67 °R	212 °F	Boiling point of water*
Relative temperature intervals (diffs)	1 K 0.55556 K	1 °C 0.55556 °C	1.8 °R 1 °R	1.8 °F 1 °F	

* For defined conditions

** Value in °C = $\frac{1}{1.8}$ (value in °F -32) = (value in K -273.15)

*** Value in K = $\frac{5}{9}$ × (value in °R)

20.7 Other conversion factors

20.7.1 Hardness

Hardness is not defined within the SI system. The following table can be used for conversion between the popular systems used.

Tensile strength	Vickers hardness (F ≥ 98 N)	Brinell hardness	Rockwell hardness	
N/mm^2	HV	BHN	HR_B	HR_C
200	63	60		
210	65	62		
220	69	66		
225	70	67		
230	72	68		
240	75	71		
250	79	75		
255	80	76		
260	82	78		
270	85	81	41	
280	88	84	45	
285	90	86	48	
290	91	87	49	
300	94	89	51	
305	95	90	52	
310	97	92	54	
320	100	95	56	
330	103	98	58	
335	105	100	59	
340	107	102	60	
350	110	105	62	
360	113	107	63.5	
370	115	109	64.5	
380	119	113	66	
385	120	114	67	
390	122	116	67.5	
400	125	119	69	
410	128	122	70	
415	130	124	71	
420	132	125	72	
430	135	128	73	
440	138	131	74	
450	140	133	75	
460	143	136	76.5	
465	145	138	77	
470	147	140	77.5	
480	150	143	78.5	
490	153	145	79.5	
495	155	147	80	
500	157	149	81	
510	160	152	81.5	
520	163	155	82.5	
530	165	157	83	
540	168	160	84.5	
545	170	162	85	
550	172	163	85.5	
560	175	166	86	
570	178	169	86.5	
575	180	171	87	
580	182	172		
590	184	175	88	
595	185	176		
600	187	178	89	
610	190	181	89.5	
620	193	184	90	
625	195	185		
630	197	187	91	
640	200	190	91.5	
650	203	193	92	
660	205	195	92.5	
670	208	198	93	

Tensile strength	Vickers hardness (F ≥ 98 N)	Brinell hardness	Rockwell hardness	
675	210	199	93.5	
680	212	201		
690	215	204	94	
700	219	208		
705	220	209	95	
710	222	211	95.5	
720	225	214	96	
730	228	216		
740	230	219	96.5	
750	233	221	97	
755	235	223	97.5	
760	237	225		
770	240	228	98	
780	243	231		21
785	245	233		
790	247	235	99	
800	250	238	99.5	22
810	253	240		
820	255	242		23
830	258	245		
835	260	247		24
840	262	249		
850	265	252		
860	268	255		25
865	270	257		
870	272	258		26
880	275	261		
890	278	264		
900	280	266		27
910	283	269		
915	285	271		
920	287	273		28
930	290	276		
940	293	278		29
950	295	280		
960	299	284		
965	300	285		
970	302	287		30
980	305	290		
990	308	293		
995	310	295		31
1000	311	296		
1010	314	299		
1020	317	301		32
1030	320	304		
1040	323	307		
1050	327	311		33
1060	330	314		
1070	333	316		
1080	336	319		34
1090	339	322		
1095	340	323		
1100	342	325		
1110	345	328		35
1120	349	332		
1125	350	333		
1130	352	334		
1140	355	337		36
1150	358	340		
1155	360	342		
1160	361	343		
1170	364	346		37
1180	367	349		
1190	370	352		
1200	373	354		
1210	376	357		38
1220	380	361		
1230	382	363		39
1240	385	366		
1250	388	369		

Tensile strength	Vickers hardness (F ≥ 98 N)	Brinell hardness	Rockwell hardness	
1255	390	371		
1260	392	372		40
1270	394	374		
1280	397	377		
1290	400	380		
1300	403	383		41
1310	407	387		
1320	410	390		
1330	413	393		42
1340	417	396		
1350	420	399		
1360	423	402		43
1370	426	405		
1380	429	408		
1385	430	409		
1390	431	410		
1400	434	413		44
1410	437	415		
1420	440	418		
1430	443	421		
1440	446	424		
1450	449	427		45
1455	450	428		
1460	452	429		
1470	455	432		
1480	458	435		46
1485	460	437		
1490	461	438		
1500	464	441		
1510	467	444		
1520	470	447		
1530	473	449		47
1540	476	452		
1550	479	455		
1555	480	456		
1560	481	457		
1570	484	460		48
1580	486	462		
1590	489	465		
1595	490	466		
1600	491	467		
1610	494	470		
1620	497	472		49
1630	500	475		
1640	503	478		
1650	506	481		
1660	509	483		
1665	510	485		
1670	511	486		
1680	514	488		50

Values based on DIN 50150.

20.7.2 Material toughness

Material toughness is not defined by SI. Most materials are assessed by conducting impact testing. Two differing test methods can be used with various sizes and styles of test specimen. The following table can be used as a **guide** to the relative toughness of the various tests:

Charpy V notch kgm/cm^2	Charpy V notch ft lb	Charpy V notch Joule	Izod ft lb
0.4	2.3	3.1	2.5
0.9	5.2	7	6.4
1.5			10.8
2.2			16
3.1	18	24.4	21.5
4.1	23.8	32.2	27.8

Charpy V notch kgm/cm^2	Charpy V notch ft lb	Charpy V notch Joule	Izod ft lb
5.2	30	40.6	34.1
6.5	37.7	51	40.4
8.0	46.4	92.9	46.7
9.4	54.5	73.9	53
10.9	63	85.4	59.3
12.6			65.6
14.1	82	111	71.9
15.8			78.2
17.7	102	138	84.5
19.4			90.8
21.1	122	165	97.1
23.0	134	182	103.4

20.8 Normal quantities and units used in valve technology

Quantity		Recommended unit	
Name	Symbol	Name	Units
Area	A	square metre square centimetre square millimetre	m^2 cm^2 mm^2
Compressibility	χ	reduction in unit volume per bar differential pressure	bar^{-1}
Density	ρ	kilogram per cubic metre gram per cubic decimetre	kg/m^3 g/dm^3
Energy	E	Joule kiloJoule MegaJoule	J kJ MJ
Flow (1)	q_v	litre per second litre per minute litre per hour cubic metre per hour	l/s l/min l/h m^3/h
Force	F	Newton kiloNewton MegaNewton	N kN MN
Frequency	f	Hertz kiloHertz MegaHertz	Hz kHz MHz
Head (suction, discharge, differential)	H	metre	m
Length	l	metre millimetre micron	m mm µm
Mass (weight)	m	tonne kilogram gram	tonne kg g
MIP (Minimum inlet pressure)	MIP	bar (gauge)	bar
Moment of inertia	J	kilogram metre squared	kgm^2
NPIP (Net positive inlet pressure)	NPIP	bar	bar
NPSH (Net positive suction head)	NPSH	metre	m
Power	P	Watt kiloWatt MegaWatt	W kW MW
Pressure (absolute or gauge)	p	bar	bar
Shaft speed	n	revolutions per second revolutions per minute revolutions per hour	r/s r/min r/h
Stress	σ τ	MegaNewton per square metre	MN/m^2
Temperature	T	degree Celsius	°C
Time	t	second minute hour	s min h
Torque	T	Newton metre kiloNewton metre MegaNewton metre	Nm kNm MNm
Velocity	v	metre per second	m/s
Viscosity, dynamic	η	Poise centiPoise	P cP
Viscosity, kinematic	μ	Stoke centiStoke	St cSt

(1) Pumps handle liquid by volume.

The user must convert mass flow to volume flow.

20.9 Useful references

Hydraulic Institute, 9 Sylvan Way, Parsippany NJ, 07054 USA Tel: 973 267 9700, Fax: 973 2679055, www.pumps.org.

VDMA (German Engineering Federation), Lyoner Straße 18, D-60528 Frankfurt, Germany, Tel: 069 66 03-0, www.vdma.org.

American Bureau of Shipping, ABS, 16855 Northchase Drive,Houston, TX 77060 USA Tel: 281 877 5800, Fax: 281 877 5803, www.eagle.org.

Det Norske Veritas DNV, Veritasveien 1, N-1322 Høvik Norway, Tel: 067 57 99 00, Fax: 067 57 99 11, www.dnv.com.

Lloyd's Register (LR), 71 Fenchurch Street, London EC3M 4BS, UK, Tel: 020 7709 9166, Fax: 020 7488 4796, www.lr.org.

Bureau International des Poids et Mesures (BIPM), Pavillon de Breteuil, F-92312 Sèvre Cedex, France, www.bipm.fr.

Office of Public Sector Information (OPSI), Admiralty Arch, North Side, The Mall, London, SW1A 2WH, UK, Tel: 01603 723011, www.opsi.gov.uk.

The Stationery Office Ltd (TSO) UK, Tel: 0870 242 2345, Email: esupport@tso.co.uk, www.tsoshop.co.uk.

EFTA (European Free Trade Association), 9-11, rue de Varembé, CH-1211 Geneva 20 Switzerland, Tel: 022 332 2626, Fax: 022 332 2699, www.secretariat.efta.int.

BVAA (British Valve & Actuator Association) 9 Manor Park, Banbury, Oxfordshire, OX16 3TB, UK, Tel: 01295 221270, Fax: 01295 268965, www.bvaa.org.uk.

European Hygienic Engineering & Design Group (EHEDG), Avenue Grand Champ, 148 B-1150 Brussels, Belgium, Tel: 32 02 7617408, www.ehedg.org.

NAFEM (North American Association of Foodservice Equipment Manufacturers), 161 North Clark Street, Suite 2020, Chicago, IL 60601, USA, Tel: 312 821 0201, Fax: 312 821 0202, www.nafem.org.

Buyers' guide 21

21.1 Introduction

The Buyers' Guide summarises the broad valve types and actuators, based on the individual Chapters in the book devoted to each type, namely:

- Isolating valves — Chapter 3
- Non-return valves — Chapter 4
- Regulators — Chapter 5
- Control valves — Chapter 6
- Safety relief valves — Chapter 7
- Actuators — Chapter 12

The Guide has been grouped in this way to simplify representation of the major generic valve groups. This is due to the difficulty in trying to impose tight boundaries on various valve type descriptions used in various parts of the world. Although each Chapter describes in detail all the valve types that have been placed within that generic group, it is strongly recommended that direct contact with the relevant companies is made – initially perhaps via their websites - to ensure that their product ranges are clearly identified from the broad classification and that these details can be clarified wherever necessary.

This is followed by ancillary products and services. Trade or brand names are comprehensively listed too. It is preceded by the names and addresses and contact details of all companies appearing in the Guide. They are listed alphabetically, by country.

Company names and addresses, Section 21.2

Company information, listed alphabetically by country.

Isolating valves, Section 21.3

Companies listed alphabetically by country.

Non-return valves, Section 21.4

Companies listed alphabetically by country.

Regulators, Section 21.5

Companies listed alphabetically by country.

Control valves, Section 21.6

Companies listed alphabetically by country.

Safety relief valves, Section 21 7

Companies listed alphabetically by country.

Actuators, Section 21.8

Companies listed alphabetically by country.

Ancillary products and services, Section 21.9

Companies listed alphabetically by product and/or service.

Trade or brand names, Section 21.10

Companies listed alphabetically by trade or brand name.

21.2 Company names and addresses

Detailed company information, listed alphabetically by country

AUSTRALIA (AU)

John Valves
Creswick Road
Ballarat
VIC 3350
Australia
Tel: +61 3 5333 0777
Fax: +61 3 5338 1771
Email: john.ferris@johnvalves.com.au
Web: www.johnvalves.com.au

Tyco Flow Control
114 Albatross Road
Nowra
New South Wales 2541
Australia
Tel: +61 2 4448 0300
Fax: +61 2 4423 3232
Web: www.tycoflowcontrol.com

AUSTRIA (AT)

Brunnbauer-Armaturen Produktions GmbH
Akaziengasse 36
PO Box 63
A-1234 Wien
Austria
Tel: +43 1 699 96000
Fax: +43 1 699 9640
Email: brunnbauer@brunnbauer.at
Web: www.brunnbauer.at

CCI - Valve Technology GmbH
Lemböckgasse 63/1
Kobe city
A 1230 Wien
Austria
Tel: +43 1 869 2740
Fax: +43 1 865 3603
Email: scarter@ccivalve.com
Web: www.ccivalve.com

EPC GmbH
IZ Nö Süd, Strasse 1
Objekt 50
A-2355 Wiener Neudorf
Austria
Tel: +43 2236 614660
Fax: +43 2236 6146630
Email: epc@epc.at
Web: www.epc.at

BELGIUM (BE)

Belgium Ventiel NV
Blokhuisstraat 24
Industriepark Noord 1
B 2800 Mechelen
Belgium
Tel: +32 1529 4070
Fax: +32 1520 1413
Email: sales@belven.com
Web: www.belgiumventiel.com

Dresser Valves Europe
Boulevard du Souverain 207 B2
Vorstlaan
B-1160 Brussels
Belgium
Tel: +32 2 344 0970
Fax: +32 2 344 1123
Web: www.masoneilan.com

Icarus SA
Parc Industriel des Hauts Sarts, Zone 3,
Rue des Alouettes 100
B-4040 Herstal
Belgium
Tel: +32 4 240 0101
Fax: +32 4 240 0640
Email: valves@icarus-be.com
Web: www.icarus-be.com

KSB Belgium SA
Zoning Industriel Sud
B-1301 Wavre
Belgium
Tel: +32 10 435 227
Fax: +32 10 435 255
Email: laurent.jonniaux@ksb.com
Web: www.ksbgroup.com

Truflo Rona - Truflo International
Parc Industriel des Hauts Sarts
B-4040 Herstal
Belgium
Tel: +32 4 240 6886
Fax: +32 4 248 0246
Email: sales@truflorona.com
Web: www.truflointernational.com

BRAZIL (BR)

Dresser Industria e Comercio Ltda
Divisao Masoneilan
Rua Funchal, 129 Conj 5A
04551 060 Sao Paulo SP
Brazil
Tel: +55 11 2146 3600
Fax: +55 11 2146 3610
Web: www.masoneilan.com

Emerson Process Management
Avenue Hollingsworth 325
Sorocaba
San Paulo 18087-105
Brazil
Tel: +55 15 238 3788
Fax: +55 15 228 3300
Web: www.emersonprocess.com/fisher

CANADA (CA)

Cameron's Valves & Measurement Group
For most current contact information go to www.c-a-m.com

Dresser–Masoneilan DI Canada Inc
Harrington Court, 2nd Floor
Burlington
Ontario L7N 3P3
Canada
Tel: +1 905 335 3529
Fax: +1 905 336 7628
Web: www.masoneilan.com

Floval Equipment Ltd
250 Rayette Road, Unit 1
Concord
Ontario L4K 2G6
Canada
Tel: +1 905 669 4500
Fax: +1 905 669 4905
Email: sales@floval.com
Web: www.floval.com

GGosco Engineering Inc
4-1272 Speers Road
Oakville
Ontario L6L 5T9
Canada
Tel: +1 905 825 2627
Fax: +1 905 825 4051
Email: sales@ggosco.com
Web: www.ggosco.com

Newman Hattersley - Truflo International
48-2400 Lucknow Drive
Mississauga
Ontario L5S 1T9
Canada
Tel: +1 905 678 1240
Fax: +1 905 673 9255
Email: inquiries@newmanhattersley.com
Web: www.truflointernational.com

Spirax Sarco Canada Ltd
383 Applewood Crescent
Concord
Ontario L4K 4J3
Canada
Tel: +1 905 660 5510
Fax: +1 905 660 5503
Email: info@spirax.ca
Web: www.spirax-sarco.com

Velan Inc
7007 Côte de Liesse
Montreal
Quebec H4T 1G2
Canada
Tel: +1 514 748 7743
Fax: +1 514 748 8635
Email: sales@velan.com
Web: www.velan.com

CHINA (CN)

Baoyi Group Co Ltd
Oubei Third Bridge Industrial Estate
Wenzhou 325105
China
Tel: +86 577 6737 5888
Fax: +86 577 6731 5333
Email: baoyi@baoyi.com
Web: www.baoyi.com

China Shenjiang Valve Co Ltd
Qiligang Industrial Zone
Yueqing
Wenzhou 3250605
Zhejiang
China
Tel: +86 577 6267 1099
Fax: +86 577 6267 2751

Email: chsv@sjv.cn
Web: www.valvesworld.com

Danfoss (Tianjin) Ltd
1407 Jin Wan Mansion
358 Nanjing Road, Nankai District
Tianjin 301700
China
Tel: +86 22 2750 1402
Fax: +86 22 2517 1401
Web: www.danfoss.com.cn

Dresser Flow Solutions
Suite 1703, Capital Mansion
6 Xinyuannan Road Chaoyang District
Beijing 100004
China
Tel: +86 10 8486 4515
Fax: +86 10 8486 5305
Web: www.masoneilan.com

Jiangsu Yandian Valve Co Ltd
8 Caoyang Road
Shangzhuang Town, Yancheng City
Jiangsu 224023
China
Tel: +86 515 862 1172
Fax: +86 515 862 1208
Email: sales@ydfvalve.com
Web: www.ydfvalve.com

Kenda Industrial Corporation
508, 820#, Xiahe Road
Xiamen 361004
China
Tel: +86 592 229 4111
Fax: +86 592 229 4999
Email: kenda@kendavalve.com
Web: www.kendavalve.com

Metso Automation (Shanghai) Co Ltd
Section A,168 Fu Te Dong Yi Road
Shanghai Waigaoqiao Trade Zone
Shanghai 200131
China
Tel: +86 21 5866 8383
Fax: +86 21 5866 1001
Email: info.automation@metso.com
Web: www.metsoautomation.com

Neway Valve (Suzhou) Co Ltd
No 999 Xiangjiang Road
Suzhou 215129
China
Tel: +86 512 666 51365
Fax: +86 512 666 51360
Email: neway@neway.com.cn
Web: www.newayvalve.com

Shanghai Flow Valve & Fitting Co Ltd
504A Zhongda Square
989 Dongfang Road
Shanghai 200122
China
Tel: +86 21 6875 7387
Fax: +86 21 5831 9620
Email: shhfl@online.sh.cn
Web: www.flow-valve.com

Shanghai Halin Control Valve Co Ltd
208 Lane 800
GaoJing Road, Qingpu District
Shanghai 201702
China
Tel: +86 21 5988 8520
Fax: +86 21 5988 8557
Email: sales@halin.cn
Web: www.halin.cn

Shanghai Karon Valves Machinery Co Ltd
9B, 686 Gubei Road
Shanghai 200336
China
Tel: +86 21 6241 5197
Fax: +86 21 6241 3949
Email: exportsales@karon-valve.com
Web: www.karon-valve.com

Shanghai Neles-Jamesbury Valve Co Ltd
333 Qinqiao Road
Jinqiao Export Processing Zone, Pudong
Shanghai 201206
China
Tel: +86 21 6100 6611
Fax: +86 21 6104 3200
Email: snj@snjvalve.com
Web: www.sn-j.com

SKVC (Group) Corporation
10 Qu Xiang San Road
Suzhou 215001
China
Tel: +86 512 6251 6088
Fax: +86 512 6251 307970
Email: info@skvc.cn
Web: www.skvc.cn

Sufa Technology Industry Co Ltd
501 Zhuijiang Road
Suzhou 215011
China
Tel: +86 512 6753 3655
Fax: +86 512 6753 2707
Email: sales@sufavalve.com
Web: www.sufavalve.com

Suhou Viza Valve Co Ltd
20D King's Tower
12 Shishan Road
Suzhou 215011
China
Tel: +86 512 6701 5219
Fax: +86 512 6808 1920
Email: sales@vizavalve.com
Web: www.vizavalve.com

Suzhou Sufa Da Valve Co Ltd
2114 Renmin Road
Suzhou 215001
China
Tel: +86 512 6753 3655
Fax: +86 512 6753 2587
Email: sales@sufada.com
Web: www.sufada.com

Xiamen Jia Da Quan Valves & Fittings Co Ltd
903 Lianhua Building
188 Jiahe Road
Xiamen 361009
China
Tel: +86 592 552 7803
Fax: +86 592 551 8127
Email: xjiada@chinavalve.net.cn
Web: www.china-valve-fitting.com

Xiamen Landee Industries Co Ltd
Suite 2201, Huiteng Metropolis
No 321 Jiahe Road
Xiamen 361012
China
Tel: +86 592 520 4188
Fax: +86 592 520 4189
Email: landee@landee.cn
Web: www.landee.cn

Zhejiang Beize Valve Manufacturing Co Ltd
3051 Yongqiang Road
Wenzhou 325025
Zhejiang
China
Tel: +86 577 8899 8160
Fax: +86 577 8899 7822
Email: bzvalve@188.com
Web: www.valve-suppliers.com

Zhejiang Chaoda Valve Co Ltd
Madao, Oubei
Yongjia
Zhejiang 325105
China
Tel: +86 577 6731 9955
Fax: +86 577 6731 1151
Email: chaoda@chinavalve.com
Web: www.chinavalve.com

Zhejiang Dafulong Valve Co Ltd
Zhangbao Industrial Zone
Oubei
Wenzhou 325105
China
Tel: +86 577 6731 5866
Fax: +86 577 6731 5844
Email: dafulong@mail.wzptt.zj.cn
Web: www.dafulon.com

CZECH REPUBLIC (CZ)

Arako spol sro
Hviezdoslavova 18
CZ-746 01 Opava
Czech Republic
Tel: +420 553 694 111
Fax: +420 553 694 777
Email: arako@arako.cz
Web: www.arako.cz

Armatura Krnov spol sro
Bruntálská 5
CZ-794 01 Krnov
Czech Republic
Tel: +420 554 616 852
Fax: +420 554 616 861
Email: prodej@armaturkakrnov.cz
Web: www.armaturkakrnov.cz

Armatury Group as
Bolatická 39
CZ-74721 Kravare
Czech Republic
Tel: +420 553 680 111
Fax: +420 553 680 333
Email: kravare@agroup.cz
Web: www.armaturygroup.cz

I B C Praha spol sro
Karlštejnská 9
CZ-252 25 Jinocany
Czech Republic
Tel: +420 251 006 111
Fax: +420 251 006 222

Email: armatury@ibcpraha.cz
Web: www.ibcpraha.cz

Jihomoravská Armaturka spol sro
Lipová alej 3087/1
PO Box 123
CZ-695 01 Hodonin
Czech Republic
Tel: +420 518 318 111
Fax: +420 518 354 003
Email: sales@jmahod.cz
Web: www.jmahod.cz

LDM spol sro
Litomyšlská 1378
CZ-560 02 Ceská Trebová
Czech Republic
Tel: +420 465 502 511
Fax: +420 465 533 101
Email: sale@ldm.cz
Web: www.ldm.cz

Mostro as
Komoranská 1900/63
CZ-14314 Praha 4–Modrany
Czech Republic
Tel: +420 261 312 111
Fax: +420 261 312 333
Email: mostro@mostro.cz
Web: www.mostro.cz

MSA as
Hlucinska 41
CZ-747 22 Dolni Benesov
Czech Republic
Tel: +420 653 881 111
Fax: +420 653 651236
Email: s@msa.cz
Web: www.msa.cz

ZPA Pecky as
Tr 5 Kvetna 166
PL-289 11 Pecky
Czech Republic
Tel: +420 321 785 141
Fax: +420 321 785 165
Email: sales@zpa-pecky.cz
Web: www.zpa-pecky.cz

DENMARK (DK)

Broen A/S
Skovvej 30
DK-5610 Assens
Denmark
Tel: +45 64712095
Fax: +45 64712495
Email: broen@broen.com
Web: www.broen.com

Danfoss A/S
Nordborgvej 81
DK-6430 Nordborg
Denmark
Tel: +45 7488 2222
Fax: +45 7488 0949
Email: danfoss@danfoss.com
Web: www.danfoss.com

FINLAND (FI)

Larox Flowsys Oy
Marssitie 1
PO Box 338
FIN-53101 Lappeenranta
Finland
Tel: +358 201 113311
Fax: +358 201 113300
Email: info@larox.fi
Web: www.larox.fi

Metso Automation Oy
Levytie 6
FIN-00880 Helsinki
Finland
Tel: +358 20 483150
Fax: +358 20 483151
Email: info.automation@metso.com
Web: www.metsoautomation.com

Naval Oy
PO Box 32
FIN-23801 Laitila
Finland
Tel: +358 2 85091
Fax: +358 2 856506
Email: pgustafsson@flowserve.com
Web: www.naval.fi

Sento Oy
Satamakatu 3
PO Box 13
FIN-24101 Salo
Finland
Tel: + 358 2 7277200
Fax: + 358 2 7277201
Email: hogfors.salo@sento.fi
Web: www.sento.fi

FRANCE (FR)

A K Müller France
14 Espaces Multiservices
Boulevard Courcerin
F - 77183 Croissy Beaubourg
France
Tel: +33 1 64 62 95 14
Fax: +33 1 64 62 95 12
Email: info@akmuller.fr
Web: www.akmuller.fr

Banides et Debeaurain
Route d'Eu
BP 43
F-76470 Le Tréport
France
Tel: +33 2 35 86 05 00
Fax: +33 2 35 50 62 56
Email: export@banides-debeaurain.fr
Web: www.banides-debeaurain.fr

Bayard SA
ZI - 4 avenue Lionel Terray
F-69881 Meyzieu
France
Tel: +33 4 37 44 24 24
Fax: +33 4 37 44 23 00
Email: edutel@tyco-valves.com
Web: www.bayard.fr

Buracco SA
10 Rue de Verdun
BP 6
F-71301 Montceau-les-Mines
France
Tel: +33 3 85 67 31 00
Fax: +33 3 85 57 06 82
Email: buracco@buracco.com
Web: www.buracco.com

Cameron's Valves & Measurement Group
For most current contact information go to www.c-a-m.com
France

Danfoss Socla SA
Water Controls Division
365 Rue du Lieutenant Putier
F-71530 Virey-le-Grand
France
Tel: +33 3 85 97 42 42
Fax: +33 3 85 41 97 42
Email: contact@danfoss.com
Web: www.watervalves.com

Defontaine SA
3 Rue Louis Renault
BP 329
F-44803 Saint-Herbain Cedex
France
Tel: +33 2 40 67 89 89
Fax: +33 2 40 67 89 00
Email: info@defontaine.com
Web: www.defontaine.com

Emerson Process Management
Rue Paul Baudray
PO Box 10145
F-68701 Cernay
France
Tel: +33 3 89 37 64 00
Fax: +33 3 89 75 65 18
Email: info.fr@emersonprocess.com
Web: www.emersonproces.com/fisher

Guichon Valves
Rue Jacqueline Auriol
BP 401
F-73104 Aix-les-bains Cedex
France
Tel: +33 4 79 61 40 00
Fax: +33 4 79 35 24 02
Email: info@guichon.com
Web: www.guichon.com

H+ Valves SA
Impasse les "Saillants"
Chessy les mines
F-69380 Lozanne
France
Tel: +33 4 78 43 91 52
Fax: +33 4 78 43 91 01
Email: gen@hvalves.com
Web: www.hvalves.com

Huot Groupe Saint-Mihiel SA
2 Rue de la Marsoupe
BP 36
F-55301 Saint Mihiel Cedex
France
Tel: +33 3 29 91 66 55
Fax: +33 3 29 90 20 17
Email: commercial@huot.fr
Web: www.huot-sa.com

Masoneilan Dresser Produits Industriels
Energy 5 130/190 Boulevard de Verdun
F-92413 Courbevoie
France
Tel: +33 1 49 04 90 00

Fax: +33 1 49 04 90 10
Web: www.masoneilan.com

Mecafrance SA
Rue du Marais
F-80400 Ham
France
Tel: +33 03 2381 4300
Fax: +33 03 2381 4520
Email: mecafrance.info@meca france-sa.com
Web: www.mecafrance-sa.fr

Meca-Inox
1 Rue Jean-Pierre Timbaud
PO Box 77
F-95101 Argenteuil Cedex
France
Tel: +33 1 30 25 79 80
Fax: +33 1 30 25 79 81
Email: infos@meca-inox.tm.fr
Web: www.meca-inox.com

Metso Automation SAS
6-8 Rue du Maine
F 68271 Wittenheim
France
Tel: +33 3 89 50 64 00
Fax: +33 3 89 50 64 40
Email: info.automation@metso.com
Web: www.metsoautomation.com

Perolo SA
69-71 Rue des Macons
F-33390 Blaye
France
Tel: +33 5 57 42 67 00
Fax: +33 5 57 42 67 02
Email: perolo@perolo.com
Web: www.perolo.com

Randex SA
153 Rue du Placyre
PO Box 334
F-38509 Voiron
France
Tel: +33 4 76 05 23 48
Fax: +33 4 76 65 60 38
Email: info@randex.fr
Web: www.randex.fr

Roforge
Parc d'activités Stelytec
PO Box 19
F-42401 Saint Chamond Cédex 1
France
Tel: +33 4 77 22 55 77
Fax: +33 4 77 31 45 48
Email: web@roforge.fr
Web: www.roforge.fr

SACCAP
112-114 Avenue de Vendôme
F-41034 Blois
France
Tel: +33 2 54 52 33 20
Fax: +33 2 54 52 33 22
Email: info@saccap.fr
Web: www.saccap.fr

SAPAG Industrial Valves
2 rue du Marais
F-80400 Ham
France
Tel: +33 3 23 81 43 00
Fax: +33 3 23 81 18 69
Email: info@sapag-valves.com
Web: www.sapag-valves.com

Sart von Rohr SAS
25 Rue de la Chapelle
PO Box 2
F-68620 Bitschwiller Les Thann
France
Tel: +33 3 89 37 79 50
Fax: +33 3 89 37 79 51
Email: sartventes@sart-von-rohr.fr
Web: www.sart-von-rohr.fr

SNRI
Route du Treuil
F-16700 Ruffec
France
Tel: +33 5 45 29 60 00
Fax: +33 5 45 31 12 91
Email: snri@www.snri.fr
Web: www.snri.fr

Spirax Sarco SA
PO Box 61
F-78193 Trappes
France
Tel: +33 1 30 66 43 43
Fax: +33 1 30 66 11 22
Email: courrier@fr.spiraxsarco.com
Web: www.spiraxsarco.com

Strahman Valves France
Savoie Hexapôle
F-73420 Mery
France
Tel: +33 4 79 35 78 00
Fax: +33 4 79 35 78 20
Email: sales@mg-valves.com
Web: www.mg-valves.com

Tyco Flow Control
For EMEA see Tyco Flow Control, The Netherlands
Web: www.tycoflowcontrol.com

Valpes
ZI Centr'Alp
89 Rue des Étangs
F-38430 Moirans
France
Tel: +33 4 76 35 06 06
Fax: +33 4 76 35 14 34
Email: info@valpes.com
Web: www.valpes.com

Vanatome
ZI de la Brassière
PO Box 10
F 26241 Saint Vallier
France
Tel: +33 4 75 23 67 08
Fax: +33 4 75 23 67 12
Email: info@vanatome.com
Web: www.vanatome.com

Vannes Lefebvre
Avenue de Lassus
PO Box 45
F-59481 Haubourdin
France
Tel: +33 3 20 17 51 00
Fax: +33 3 20 17 51 01
Email: vannes@lefebvre.fr
Web: www.lefebvre.fr

Velan SAS
90 Rue Challemel Lacour
F-69367 Lyon Cedex 7
France
Tel: +33 4 78 61 67 00
Fax: +33 4 78 72 12 18
Email: sales@velan.com
Web: www.velan.fr

Weir Valves & Controls France SAS
ZI Du Bois Rigault
Rue Jean Baptiste Grison
F-62880 Vendin Le Vieil
France
Tel: +33 3 21 79 54 50
Fax: +33 3 21 28 62 00
Email: sarasin@weirvalvefr.com
Web: www.weirvalve.com

Weir Valves & Controls SEBIM SAS
ZI La Palunette
Chateauneuf Les Martigues
F-13220 Bouches du Rhône
France
Tel: +33 4 42 07 00 95
Fax: +33 4 42 07 11 77
Email: coseb@weirvalvefr.com
Web: www.weirvalve.com

GERMANY (DE)

A u K Müller GmbH & Co KG
Dresdener Strasse 162
D-40595 Düsseldorf
Germany
Tel: +49 211 73910
Fax: +49 211 7391281
Web: www.akmueller.de

A+R Armaturen GmbH
Altenhagener Strasse 4a
D-32107 Bad Salzuflen
Germany
Tel: +49 5208 91020
Fax: +49 5208 910290
Email: postoffice@ar-armaturen.com
Web: www.ar-armaturen.com

Adams Armaturen GmbH
Baukauer Strasse 55
D 44653 Herne
Germany
Tel: +49 2323 2090
Fax: +49 2323 209285
Email: info@adams-armaturen.de
Web: www.adams-armaturen.de

AKO Armaturen & Separations GmbH
Adam-Opel-Strasse 5
D 65468 Trebur-Astheim
Germany
Tel: +49 6147 91590
Fax: +49 6147 915959
Email: ako@ako-armaturen.de
Web: www.ako-armaturen.de

AKO Regelungstechnik GmbH & Co KG
Porschestrasse 16
D 51381 Leverkusen
Germany
Tel: +49 2171 3940
Fax: +49 2171 394429
Email: info@akomail.de
Web: www.ako-regelungstechnik.de

ARCA-regler GmbH
Kempener Strasse 18
D 47918 Tönisvorst
Germany
Tel: +49 2156 77090
Fax: +49 2156 770955
Email: sale@arca-valve.com
Web: www.arca-valve.com

ARI-Armaturen
Mergelheide 56-60
D 33758 Schloss Holte-Stukenbrock
Germany
Tel: +49 5207 9940
Fax: +49 5207 994158
Email: info@ari-armaturen.com
Web: www.ari-armaturen.com

Armaturenfabrik Franz Schneider GmbH & Co KG
Bahnhofplatz 12
D 74226 Nordheim
Germany
Tel: +49 7133 101129
Fax: +49 7133 101180
Email: u.frank@as-schneider.com
Web: www.as-schneider.com

Armaturen-und Metallwerke Zöblitz GmbH
Bahnhofstrasse 16
D 09517 Zöblitz
Germany
Tel: +49 37363 4800
Fax: +49 37363 48090
Email: info@armaturen-zoeblitz.de
Web: www.armaturen-zoeblitz.de

Armaturenwerk Altenburg GmbH
Am Weissen Berg 30
D 04600 Altenburg/Thüringen
Germany
Tel: +49 3447 8930
Fax: +49 3447 81110
Email: info@awa-armaturenwerk.de
Web: www.awa-armaturenwerk.de

ASP Armaturen
Gambacher Weg 1a
D-84347 Pfarrkirchen
Germany
Tel: +49 8561 985470
Fax: +49 8561 9836589
Web: www.asp-armaturen.de

AUMA Riester GmbH & Co KG
Aumastrasse 1
D-79379 Müllheim
Germany
Tel: +49 7631 8090
Fax: +49 7631 809250
Email: riester@auma.com
Web: www.auma.com

AWP Kälte-Klima-Armaturen GmbH
Gewerbegebiet Ost
Strasse A1
D-17291 Prenzlau
Germany
Tel: +49 3984 85590
Fax: +49 3984 85 5918
Email: info@awpvalves.com
Web: www.awpvalves.com

AZ-Armaturen GmbH
Waldstrasse 7
D-78087 Mönchweiler
Germany
Tel: +49 7721 75040
Fax: +49 7721 750413
Email: info@az-armaturen.de
Web: www.az-armaturen.de

Bar GmbH
Auf der Hohl 1
D-53547 Dattenberg
Germany
Tel: +49 2644 96070
Fax: +49 2644 960735
Web: www.bar-gmbh.com

Berluto Armaturen GmbH
Tempelsweg 16
D 47918 Tönisvorst
Germany
Tel: +49 2151 70380
Fax: +49 2151 700763
Email: info@berluto.de
Web: www.berluto.de

Bopp & Reuther Sicherheits und Regelarmaturen GmbH
Carl-Reuther-Strasse 1
D-68305 Mannheim
Germany
Tel: +49 621 7491662
Fax: +49 621 7491310
Email: sales@sr.boppureuther.com
Web: www.sr.boppureuther.com

Braunschweiger Flammenfilter GmbH
Industriestrasse 11
D 38110 Braunschweig
Germany
Tel: +49 5307 8090
Fax: +49 5307 7824
Email: office@protego.de
Web: www.protego.de

Burgmann Industries GmbH & Co. KG
Ässere Sauerlacher Strasse 6-10
D-82502 Wolfratshausen
Germany
Tel: +49 8171 230
Fax: +49 8171 231095
Email: info@de.eagleburgmann.com
Web: www.eagleburgmann.com

Bürkert GmbH & Co KG
Christian-Bürkert-Strasse 13-17
D-74653 Ingelfingen
Germany
Tel: +49 7940 10111
Fax: +49 7940 10448
Email: info@de.buerkert.com
Web: www.burkert.com

Cera System Vershleissschutz GmbH
Heinrich-Hertz-Strasse 2 4
D-07629 Hermsdorf
Germany
Tel: +49 36601 9190
Fax: +49 36601 91990
Email: helmut.burghardt@cerasystem.de
Web: www.cerasystem.de

CH Zikesch Armaturentechnik GmbH
Laubenhof 13
D-45326 Essen
Germany
Tel: +49 20183 4080
Fax: +49 20134 2771
Email: info@zikesch.de
Web: www.zikesch.de

Christian Bollin Armaturenfabrik GmbH
Westerbachstrasse 290-294
D-65936 Frankfurt
Germany
Tel: +49 6934 1021
Fax: +49 6934 3985
Email: bollin@bollin.de
Web: www.bollin.de

Daume Regelarmaturen GmbH
Jathostrasse 8
D-30916 Isernhagen bei Hannover
Germany
Tel: +49 511 9021 0
Fax: +49 511 9021417
Email: mail@daume-regelarmaturen.de
Web: www.daume-regelarmaturen.de

Drehmo GmbH
Industriestrasse 1
D 57482 Wenden
Germany
Tel: +49 2762 612311
Fax: +49 2762 612359
Email: drehmo@drehmo.com
Web: www.drehmo.com

Dresser Valves Europe GmbH
Heiligenstrasse 75
D-41751 Viersen
Germany
Tel: +49 2162 81700
Fax: +49 2162 8170200
Email: dresservalves@dpi.fr
Web: www.masoneilan.com

E Georg Lüdecke Armaturen GmbH
Heinrich-Hauck-Strasse 2
D-92224 Amberg
Germany
Tel: +49 9621 76820
Fax: +49 9621 768299
Email: info@luedecke.de
Web: www.luedecke.de

Ebro Armaturen Gebr Bröer GmbH
Karlstrasse 8
D-58135 Hagen
Germany
Tel: +49 2331 9040
Fax: +49 2331 904111
Email: post@ebro-armaturen.com
Web: www.ebro-armaturen.com

Ehlers GmbH
Industriestrasse 13
D-52134 Herzogenrath
Germany
Tel: +49 2407 5760
Fax: +49 2407 576101
Email: aachen@eriks.de
Web: www.eriks-deutschland.de

EMG Automation GmbH
Industriestrasse 1
D 57482 Wenden
Germany
Tel: +49 2762 6120

Fax: +49 2762 612237
Email: info@emg-automation.com
Web: www.emg-automation.de

END-Armaturen GmbH & Co KG
Oberbecksener Strasse 78
D-32547 Bad Oeynhausen
Germany
Tel: +49 5731 79000
Fax: +49 5731 7900199
Web: www.end.de

Erhard Armaturen GmbH & Co KG
Meeboldstrasse 22
D 89522 Heidenheim
Germany
Tel: +49 7321 3200
Fax: +49 7321 320525
Email: export@erhard.de
Web: www.erhard.de

Ernst Schmieding GmbH & Co KG
Natorper Strasse 55
D 59439 Holzwickede
Germany
Tel: +49 2301 1890
Fax: +49 2301 18965
Email: schmieding@tyco-valves.com
Web: www.schmieding.de

FIRST GmbH
PO Box 1545
D-42908 Wermelskirchen
Germany
Tel: +49 2196 887780
Fax: +49 2196 8877849
Email: info@firstgmbh.de
Web: www.firstgmbh.com

Fischer Mess- und Regeltechnik GmbH
Bielefelder Strasse 37a
D-32107 Bad Salzuflen
Germany
Tel: +49 5222 9740
Fax: +49 5222 7170
Email: info@fischermesstechnik.de
Web: www.fischer.ag

Flowserve Flow Control GmbH
Manderscheidtstrasse 19
D-45141 Essen
Germany
Tel: +49 201 89195
Fax: +49 201 8919600
Email: ukfcinfo@flowserve.com
Web: www.flowserve.com

Flowserve Flow Control GmbH
Rudolf-Plantk-Strasse 2
D-76275 Ettlingen
Germany
Tel: +49 7243 1030
Fax: +49 7243 103222
Email: ukfcinfo@flowserve.com
Web: www.argus-valves.com

Foxboro Eckardt GmbH
Pragstrasse 82
D-70376 Stuttgart
Germany
Tel: +49 711 5020
Fax: +49 711 502597
Email: salessupport@foxboro-eckardt.de
Web: www.foxboro-eckardt.com

Frank GmbH
Starkenburgstrasse 1
D 64546 Mörfelden-Waldorf
Germany
Tel: +49 6195 9260
Fax: +49 6105 92649
Email: c.habedank@frank-gmbh.de
Web: www.frank-gmbh.de

Franz Dürholdt GmbH & Co KG
Industriearmaturen
Friedrich-Engels-Allee 259
D-42285 Wuppertal
Germany
Tel: +49 202 280860
Fax: +49 202 2808640
Email: info@duerholdt.de
Web: www.duerholdt.de

Friatec AG
Theodor Jansen Strasse
D-66386 St Ingbert
Germany
Tel: +49 6894 5930
Fax: +49 6894 593179
Email: info@jansen-valves.de
Web: www.friatec.de

Fritz Barthel Armaturen GmbH & Co KG
Klopstockplatz 5-7
D 22765 Hamburg
Germany
Tel: +49 4039 82020
Fax: +49 4039 820277
Email: post@barthel-armaturen.de
Web: www.barthel-armaturen.de

Fromme Armaturen GmbH & Co KG
Haupstrasse 12
D-38275 Haverlah
Germany
Tel: +49 5341 338411
Fax: +49 5341 338413
Email: info@fromme-armaturen.com
Web: www.fromme-armaturen.de

G Bee GmbH
Bahnhofstrasse 157
D 74321 Bietigheim-Bissengen
Germany
Tel: +49 7142 5980
Fax: +49 7142 53519
Email: info@g-bee.de
Web: www.g-bee.de

Gebr Tuxhorn GmbH & Co KG
Westfalenstrasse 36
D 33647 Bielefeld
Germany
Tel: +49 521 448080
Fax: +49 521 4480844
Email: a.koenemann@tuxhorn.de
Web: www.tuxhorn.de

Gefa Processtechnik GmbH
Germaniastrasse 28
D 44379 Dortmund
Germany
Tel: +49 231 610090
Fax: +49 231 6100980
Email: gefa@gefa.com
Web: www.gefa.com

Geier Armaturen GmbH & Co KG
Am Rosenhügel 21-27
D-42553 Velbert
Germany
Tel: +49 2053 93040
Fax: +49 2053 7732
Email: info@geier-armaturen.de
Web: www.geier-armaturen.de

GEMÜ Gebrüder Müller Apparatebau GmbH & Co. KG
Fritz-Müller-Strasse
D-74653 Ingelfingen-Creisbach
Germany
Tel: +49 79 401230
Fax: +49 79 40123192
Email: info@gemue.de
Web: www.gemue.de

Georg Schünemann GmbH
Buntentorsdeich 1
D 28201 Bremen
Germany
Tel: +49 421 559090
Fax: +49 421 5590940
Email: info@sab-bremen.de
Web: www.sabfilter.de

Gerhard Götze KG
Robert-Mayer-Strasse 21
D-71636 Ludwigsburg
Germany
Tel: +49 7141 4889460
Fax: +49 7141 4889488
Email: info@goetze-armaturen.de
Web: www.goetze-armaturen.de

Gestra GmbH
Münchener Strasse 77
D-28215 Bremen
Germany
Tel: +49 421 35030
Fax: +49 421 3 03393
Email: gestra.ag@flowserve.com
Web: www.gestra.de

Gossler Fluidtec GmbH
Borsigstrasse 4-6
D 21462 Reinbek
Germany
Tel: +49 407 2709200
Fax: +49 407 27095200
Email: info-gft@gossler.de
Web: www.gossler.de

GS Anderson GmbH
Hesslingsweg 71
D 44309 Dortmund
Germany
Tel: +49 231 92560
Fax: +49 231 9256150
Email: info@anderson-dortmund.com
Web: www.anderson-dortmund.com

Gustav Mankenberg GmbH
Spenglerstrasse 99
D 23556 Lübeck
Germany
Tel: +49 451 879750
Fax: +49 451 8797599
Email: gm@mankenberg.de
Web: www.mankenberg.de

Handtmann Armaturenfabrik GmbH & Co KG
Arthur-Handtmann-Strasse 23

D-88400 Biberach
Germany
Tel: +49 7351 3420
Fax: +49 7351 3422762
Email: info@handtmann.de
Web: www.handtmann.de

Hans Sasserath & Co KG
Mühlenstrasse 62
D 41352 Korschenbroich
Germany
Tel: +49 2161 61050
Fax: +49 2161 610520
Email: sasserath@syr.de
Web: www.syr.de

Hartmann Valves GmbH
PO Box 2028
D-31303 Burgdorf-Ehlershausen
Germany
Tel: +49 5085 98010
Fax: +49 5085 980140
Email: w.michaeli@hartmann-valve.com
Web: www.hartmann-valve.com

Herose GmbH
Elly-Heuss-Knapp-Strasse 12
D-23843 Bad Oldesloe
Germany
Tel: +49 4531 5090
Fax: +49 4531 509120
Email: info@herose.de
Web: www.herose.de

Hofer Hochdrucktechnik GmbH
Ruhrorter Strasse 45
D 45478 Mülheim an der Ruhr
Germany
Tel: +49 208 469960
Fax: +49 208 4699611
Email: info@andreas-hofer.de
Web: www.andreas-hofer.de

Holter Regelarmaturen GmbH & Co KG (HORA)
Helleforthstrasse 58-60
D-33758 Schloss Holte-Stukenbrock
Germany
Tel: +49 5207 89030
Fax: +49 5207 88037
Email: mail@hora.de
Web: www.hora.de

Honeywell GmbH
Hardhofweg
D-74821 Mosbach
Germany
Tel: +49 6261 810
Fax: +49 6261 81309
Email: info.imc.de@honeywell.com
Web: www.honeywell.de

IBK Wiesehahn GmbH
PO Box 200154
D 46244 Bottrop
Germany
Tel: +49 2045 89030
Fax: +49 2045 890320
Email: info@ibk.de
Web: www.ibk.de

IMI Norgren Buschjost GmbH & Co KG
Detmolder Strasse 256
D-32545 Bad Oeynhausen
Germany
Tel: +49 5731 7910
Fax: +49 5731 791179
Email: mail@buschjost.de
Web: www.buschjost.com

Johnson Controls Regelungstechnik GmbH
Bonsiepen 13
D-45143 Essen
Germany
Tel: +49 201 240 0
Fax: +49 201 2400351
Web: www.jci.com

Keulahütte GmbH, Krauschwitz
Geschwister-Scholl-Strasse 15
D 02957 Krauschwitz
Germany
Tel: +49 35771 54 0
Fax: +49 35771 54220
Email: keulahuette@vem-group.com
Web: www.keulahuettekrauschwitz.de

Kieback & Peter GmbH & Co KG
Tempelhofer Weg 50
D 12347 Berlin
Germany
Tel: +49 30 600 950
Fax: +49 30 600 95164
Email: info@kieback-peter.de
Web: www.kieback-peter.de

Klaus Union GmbH & Co KG
Blumenfeldstrsse 18
D-44795 Bochum
Germany
Tel: +49 234 45950
Fax: +49 234 432387
Email: info@klaus-union.de
Web: www.klaus-union.de

Klinger GmbH
Richard-Klinger Strasse 17
D 65510 Idstein
Germany
Tel: +49 6126 40160
Fax: +49 6126 401611
Email: mail@klinger-gmbh.de
Web: www.klinger-gmbh.de

KSB Armaturen GmbH
Johann-Klein-Strasse 9
D-67227 Frankenthal
Germany
Tel: +49 6233 862578
Fax: +49 6233 863401
Web: www.ksb.de

Kühme Armaturen GmbH
Am Vorort 14
D-44894 Bochum
Germany
Tel: +49 234 29800
Fax: +49 234 2980210
Email: info@kuehme.de
Web: www.kuehme.de

Laun GmbH Maschinen- und Apparatebau
Waldstrasse 71
D-65451 Kelsterbach
Germany
Tel: +49 6107 986690
Fax: +49 6107 9866920
Email: info@laun-online.com
Web: www.laun-online.com

Leser GmbH & Co KG
Wendenstrasse 133-135
D-20537 Hamburg
Germany
Tel: +49 4025 165100
Fax: +49 4025 165500
Email: sales@leser.com
Web: www.leser.com

Linde AG
Linde Engineering Division, MAPAG Works
Von-Holzapfel-Strasse 4
D-86497 Horgau
Germany
Tel: +49 8294 86950
Fax: +49 8294 869581
Email: mapag@linde-le.com
Web: www.mapag.de

Lorch GmbH
Heilbronner Strasse 5
D-70771 Leinfelden-Echterdingen
Germany
Tel: + 49 711 22720426
Fax: + 49 711 22720492
Email: vertrieb@lorch.de
Web: www.lorch.de

Ludwig Mohren KG
Weststrasse 30-34
D 52074 Aachen
Germany
Tel: +49 241 88770
Fax: +49 241 874154
Web: www.ludwigmohren.de

M&S Armaturen GmbH
Industriestrasse 24-26
D-26446 Friedeburg
Germany
Tel: +49 4465 8070
Fax: +49 4465 80740
Email: info@ms-armaturen.de
Web: www.ms-armaturen.de

Mecafrance (Deutschland) GmbH
Körnerstrasse 22
D 53175 Bonn Bad-Godesberg
Germany
Tel: +49 228 935500
Fax: +49 228 935506
Email: mfd-bonn@mecafrance.de
Web: www.mecafrance.de

Mittelmann Sicherheits-technik GmbH & Co KG
Schillerstrasse 50
D 42489 Wülfrath
Germany
Tel: +49 2058 90102
Fax: +49 2058 901211
Email: verkauf@mittelmann.com
Web: www.mittelmann.com

Neumo GmbH & Co KG
Henry-Ehrenberg-Platz
D-75438 Knittlingen
Germany
Tel: +49 7043 360
Fax: +49 7043 36130
Email: info@neumo.de
Web: www.neumo.de

Niezgodka GmbH
Bargkoppelweg 73
D 22145 Hamburg
Germany
Tel: +49 4067 94690
Fax: +49 4067 96959
Email: ni@niezgodka.de
Web: www.niezgodka.de

Noval Industriearmaturen GmbH
Porsestrasse 19
D-39104 Magdeburg
Germany
Tel: +49 391 405550
Fax: +49 391 4055549
Email: noval@noval.de
Web: www.noval.de

OHL Gutermuth Industriearmaturen GmbH
Helmeshäuser Strasse 9
D 63674 Altenstadt
Germany
Tel: +49 6047 800620
Fax: +49 6047 800629
Email: og@ohl-gutermuth.de
Web: www.ohl-gutermuth.de

Parker Hannifin GmbH & Co KG
Herl Refrigerating Specialties
Wankelstrasse 40
D 50996 Köln
Germany
Tel: +49 2236 39000
Fax: +49 2236 390039
Email: herl.info@parker.com
Web: www.herl.de

Perrin GmbH
Siemenstrasse 1
D 61130 Nidderau
Germany
Tel: +49 6187 9280
Fax: +49 6187 928251
Email: postoffice@perrin.de
Web: www.perrin.de

Pesch Armaturen GmbH & Co
Otto-Hahn-Strasse 23
D-50997 Köln
Germany
Tel: +49 2236 891601
Fax: +49 2236 891608
Email: wp@peschgruppe.de
Web: www.peschgruppe.de

Pfaudler-Werke GmbH
Pfaudler Strasse
D-68723 Schwetzingen
Germany
Tel: +49 6202 850
Fax: +49 6202 22412
Email: sales@pfaudler.de
Web: www.pfaudler.de

Phönix Armaturen-Werke Bregel GmbH
Am Stadtbruch 6
D 34471 Volkmarsen
Germany
Tel: +49 5693 9880
Fax: +49 5693 988178
Email: info@phoenix-armaturen.de
Web: www.phoenix-armaturen.de

Phönix Messtechnik GmbH
Salzschlirfer Strasse 13
D-60386 Frankfurt
Germany
Tel: +49 69 41674220
Fax: +49 69 41674229
Email: sales@phoenix-mt.comd
Web: www.phoenix-mt.com

Pister-Kugelhähne GmbH
Vogesenstrasse 37
D 76461 Muggensturm
Germany
Tel: +49 7222 50020
Fax: +49 7222 500250
Email: verkauf@pister-gmbh.com
Web: www.pister-gmbh.com

PS Automation GmbH
Philipp-Krämer-Ring 13
D-67098 Bad Dürkheim
Germany
Tel: +49 6322 60030
Fax: +49 6322 600320
Email: info@ps-automation.com
Web: www.ps-automation.com

Reineke Mess- und Regeltechnik GmbH
Von-Ebner-Eschenbach-Strasse 5
D-44807 Bochum
Germany
Tel: +49 234 95950
Fax: +49 234 9595200
Email: info@reineke-online.info
Web: www.reineke-online.com

Remote Control
Gutenbergstrasse 22
D-41564 Kaarst-Büttgen
Germany
Tel: +49 2131 795760
Fax: +49 2131 7957615
Email: info@remotecontrol.de
Web: www.remotecontrol.de

Rheinhütte GmbH & Co
Rheingaustrasse 96-100
D-65203 Wiesbaden
Germany
Tel: +49 611 6040
Fax: +49 611 604328
Email: info@rheinhuette.de
Web: www.rheinhuette.de

Richter Chemie-Technik GmbH
PO Box 10 06 09
D 47883 Kempen
Germany
Tel: +49 2152 1460
Fax: +49 2152 146190
Email: richter-info@richter-ct.com
Web: www.itt-richter.de

RITAG Ritterhuder Armaturen GmbH & Co
Industriestrasse 7-9
D 27711 Osterholz-Scharmbeck
Germany
Tel: +49 4791 92090
Fax: +49 4791 920985
Email: contact@ritag.com
Web: www.ritag.de

Rotech GmbH
Im Katzentach 16-18
D 76275 Ettlingen
Germany
Tel: +49 7243 59310
Fax: +49 7243 593131
Email: info@rotech.de
Web: www.rotech.de

Rötelmann GmbH
In der Lacke 10
D 58791 Werdohl
Germany
Tel: +49 2392 91910
Fax: +49 2392 919114
Email: info@roetelmann.de
Web: www.roetelmann.de

Rudolf von Scheven GmbH
Brinkerstrasse 1-7
D 45549 Sprockhövel
Germany
Tel: +49 2324 9740
Fax: +49 2324 77320
Email: info@von-scheven.de
Web: www.von-scheven.de

SAFI GmbH
Deutschland-Kunstoffarmaturen-Vertriebs
Jahnstrasse 29
D-64319 Pfungstadt
Germany
Tel: +49 6157 81919
Fax: +49 6157 82025
Email: info@safi-d.de
Web: www.safi-d.de

Samson AG Mess- und Regeltechnik
Weismüllerstrasse 3
D 60314 Frankfurt
Germany
Tel: +49 69 40090
Fax: +49 69 40091507
Email: samson@samson.de
Web: www.samson.de

Sauter-Cumulus GmbH
Hans-Bunte-Strasse 15
D-79108 Freiburg
Germany
Tel: +49 761 51050
Fax: +49 761 5105234
Email: sauter-cumulus@de.sauter-bc.com
Web: www.sauter-cumulus.com

Schöneberg Armaturen GmbH
Markplatz 18
D-76356 Weingarten
Germany
Tel: +49 7244 70250
Fax: +49 7244 702540
Email: office@schoeneberg-armaturen.de
Web: www.schoeneberg-armaturen.de

Schroedahl-Arapp Spezialarmaturen GmbH & Co KG
Schoenenbacher Strasse 4
D-51580-Reichshof-Mittelagger
Germany
Tel: +49 2265 99270
Fax: +49 2265 992747
Email: info@schroedahl.de
Web: www.schroedahl.com

Schubert & Salzer Control Systems GmbH
Bunsenstrasse 38

D 85053 Ingolstadt
Germany
Tel: +49 841 96530
Fax: +49 841 9653405
Email: infoschubert-salzer.com
Web: www.schubert-salzer.com

SchuF-Armaturen und Apparatebau GmbH
An der Guldenmühle 8 10
D-65817 Eppstein/Frankfurt
Germany
Tel: +49 6198 571100
Fax: +49 6198 571200
Email: verkauf@schuf.de
Web: www.schuf.de

Schwietzke Armaturen GmbH
Weusterstrasse 13
D 46240 Bottrop
Germany
Tel: +49 2041 74770
Fax: +49 2041 747789
Email: marketing@schwietzke.de
Web: www.schwietzke.de

Siebeck-Bitter GmbH Armaturen und Metallwerk
Am Sandbach 35-41
D 40878 Ratingen
Germany
Tel: +49 2102 45060
Fax: +49 2102 450650
Email: info@siebeck-bitter.de
Web: www.siebeck-bitter.de

Siekmann Econosto GmbH & Co KG
Freigrafenweg 2
D 44357 Dortmund
Germany
Tel: +49 231 93750
Fax: +49 231 9375199
Email: info@siekmann-econosto.de
Web: www.siekmann-econosto.de

Siemens AG
Power Generation
Mellinghofer Strasse 55
D-45473 Mülheim an der Ruhr
Germany
Tel: +49 208 4560
Web: www.powergeneration.siemens.com

SIPOS Aktorik GmbH
Donaustrasse 36
D 90451 Nürnberg
Germany
Tel: +49 911 632840
Fax: +49 911 63284111
Email: info@sipos.de
Web: www.sipos.de

Spirax Sarco GmbH
PO Box 10 20 42
D-78420 Konstanz
Germany
Tel: +49 7531 58060
Fax: +49 7531 580622
Email: vertrieb@de.spiraxsarco.com
Web: www.spirax-sarco.com

Stahl-Armaturen PERSTA GmbH
Mülheimer Strasse 18
D 59581 Warstein
Germany
Tel: +49 2902 76202
Fax: +49 2902 76703
Email: info@persta.com
Web: www.persta.com

Strack GmbH
Im Hasenwinkel 18
D 39179 Barleben
Germany
Tel: +49 39203 898930
Fax: +49 39203 898959
Email: info@strack-valve.de
Web: www.strack-valve.de

Südmo Components GmbH
Industriestrasse 7
D 73469 Riesbürg
Germany
Tel: +49 9081 8030
Fax: +49 9081 803158
Email: info@sudmo.de
Web: www.suedmo.de

TLV Euro Engineering GmbH
Daimler-Benz-Straße 16-18
D-74915 Waibstadt
Germany
Tel: +49 7263 91500
Fax: +49 7263 91550
Email: info@tlv.com
Web: www.tlv.com

Tuchenhagen GmbH
Am Industriepark 2-10
D-21514 Büchen
Germany
Tel: +49 4155 49240
Fax: +49 4155 492428
Email: fc-sales@tuchenhagen.de
Web: www.tuchenhagen.de

Tyco Flow Control
For EMEA See Tyco Flow Control, The Netherlands
Web: www.tycoflowcontrol.com

Uni-Geräte GmbH
Holtumsweg 13
D 47652 Weeze
Germany
Tel: +49 2837 913420
Fax: +49 2837 1444
Email: info@uni-geraete.de
Web: www.uni-geraete.de

VAG-Armaturen GmbH
Carl-Reuther-Strasse 1
D-68305 Mannheim
Germany
Tel: +49 621 7490
Fax: +49 621 7492153
Email: info@vag-armaturen.com
Web: www.vag-armaturen.com

W Bälz & Sohn GmbH & Co
Koepffstrasse 5
D-74076 Heilbronn
Germany
Tel: +49 7131 15000
Fax: +49 7131 150021
Email: mail@baelz.de
Web: www.baelz.de

Warex Valve GmbH
Stauverbrink 2
D-48308 Senden-Bösensell
Germany
Tel: +49 2536 99580
Fax: +49 2536 995829
Email: info@warex-valve.com
Web: www.warex-valve.com

Welland & Tuxhorn AG
Gütersloher Strasse 257
D-33649 Bielefeld
Germany
Tel: +49 521 94180
Fax: +49 521 9418170
Email: info@welland-tuxhorn.de
Web: www.welland-tuxhorn.de

Werner Böhmer GmbH
PO Box 91 12 20
D 45537 Sprockhövel
Germany
Tel: +49 2324 70010
Fax: +49 2324 700179
Email: boehmer@boehmer.de
Web: www.boehmer.de

Wilhelm Schley GmbH & Co
Carl-Zeiss-Strasse 4
D 22946 Trittau
Germany
Tel: +49 4154 80810
Fax: +49 4154 82184
Email: info@wilhelm-schley.com
Web: www.wilhelm-schley.com

WT Armatur GmbH & Co KG
Industriestrasse 5
D 67133 Maxdorf
Germany
Tel: +49 6237 92800
Fax: +49 6237 928050
Email: info@wta.eu
Web: www.wt-armatur.de

Z & J Technologies GmbH
Bahnstrasse 52
D-52355 Düren
Germany
Tel: +49 2421 6910
Fax: +49 2421 691200
Email: postoffice@zjtechnologies.de
Web: www.zjtechnologies.de

Zwick Armaturen GmbH
Egerstrasse 25
D-58256 Ennepetal
Germany
Tel: +49 2333 98565
Fax: +49 2333 98566
Email: info@zwick-gmbh.de
Web: www.zwick-gmbh.de

HUNGARY (HU)

DKG-East
Vár u 9
H-8800 Nagykanizsa
Hungary
Tel: +36 93 313140
Fax: +36 93 310160
Email: bodnarcs@dkgeast.hu
Web: www.dkgeast.hu

INDIA (IN)

Advance Valves Group
142 A & B, Noida Special Economic Zone
Phase 2
Noida 201 301
India
Tel: +91 120 2462374
Fax: +91 120 2462376
Email: global@advancevalves.com
Web: www.advancevalves.com

Amco Industrial Valves
6 Thandava Murthy Street
Royapuram
Chennai 600 013
India
Tel: +91 44 2590 1646
Fax: +91 44 2590 1646
Email: info@amcovalves.com
Web: www.amcovalves.com

Audco India Ltd
Mountpoonamallee Road
Manapakkam
Chennai 600 089
India
Tel: +91 44 234 2323
Fax: +91 44 234 5055
Email: arundev@larsentoubro.com
Web: www.ailvalves.com

AV Valves Ltd
16 Industrial Estate
Nunhai
Agra 282 006
India
Tel: +91 562 34521891
Fax: +91 562 345217
Email: avvalves@sancharnet.in
Web: www.altaindia.com

BDK Process Equipment Inc
47/48 Gokul Road
Karnataka
Hubli 580 030
India
Tel: +91 836 2331499
Fax: +91 836 2330799
Email: bdkhubli@bdkindia.com
Web: www.bdkindia.com

Bharat Heavy Electricals Ltd (BHEL)
BHEL House
Siri Fort
New Delhi 110049
India
Tel: +91 11 26001010
Fax: +91 11 26493021
Email: query@bhel.com
Web: www.bhel.com

Canle Valves PVT Ltd
18/3-4 Menambedu Road
Padi
Chennai 600 050
India
Tel: +91 44 6357320
Fax: +91 44 6357320
Email: canle@vsnl.com

Dresser Valve India Pvt Ltd
305/306, "Midas", Sahar Plaza
Mathurdas Vasanji Road
J B Nagar, Andheri East
Mumbai 400059
India
Tel: +91 22 8354790
Fax: +91 22 8354791
Web: www.masoneilan.com

Excel Hydro-Pneumatics Pvt Ltd
51 First Floor, Raja Industrial Estate
PK Road, Muland (W)
Mumbai 400 080
India
Tel: +91 2258 3291
Fax: + 91 22564 5538
Email: excel@bom2.vsnl.net.in
Web: www.excelhydro.com

Fainger Leser Valves PVT Ltd
136/137, Sanjay Building No 3
Mittal Industrial Estate
Marol, Andheri (E)
Mumbai 400 059
India
Tel: +91 22 850 1692
Fax: +91 22 850 4470
Email: info@fainger.com
Web: www.fainger.com

Fisher Sanmar Ltd
147 Karapakkam Village
Chennai 600 096
India
Tel: +91 44 24504300
Fax: +91 44 24501913
Email: engg@sanmargroup.com
Web: www.sanmargroup.com

Flowel Valves
B-102
Shubhlaxmi Tower
Opp Sanghvi High School, Naranpura
Ahmedabad 380013
India
Tel: +91 79 7414077
Fax: +91 79 27436555
Web: www.flowel_valves.tripod.com

Fouress Engineering (India) Ltd
Mahalaxmi Chambers 22
22 Bhulabhai Desai Road
Mumbai 400 026
India
Tel: +91 22 4964400
Fax: +91 22 4937544
Email: hq@fouressindia.com
Web: www.fouressengg.com

Indian Valve Pvt Ltd
53 Industrial Estate
Satpur
Nashik 422007
India
Tel: +91 253 2350170
Fax: +91 253 2351673
Email: indvalve_nsk@sancharnet.in
Web: www.ivc-india.com

Intervalve (India) Ltd
212/2 Hadapsar
Off Soli Poonawalla Road
Hadapsar
Pune 411 028
India
Tel: +91 20 2699 3900
Fax: +91 20 2699 3921
Email: intervalve@poonawallagroup.com
Web: www.poonawallagroup.com

Juneja Metal Works
Variana Village, PO Basti Guzan
Karpurthala Road
Jalandhar 144 002
India
Tel: +91 181 2650321
Fax: +91 181 2650460
Email: info@junejametalworks.com
Web: www.junejametalworks.com

Kirloskar Brothers Ltd
Udyog Bhavan
Tilak Road
Pune 411 002
India
Tel: +91 20 2444 0770
Fax: +91 20 2444 0824
Email: kblin@kbl.co.in
Web: www.kbl.co.in

Rotex Automation Ltd
114 New India Industrial Estate
Off Mahakali Caves Road, Andheri (East)
Mumbai-400 093
India
Tel: +91 22 5695 2161
Fax: +91 22 5692 3783
Email: sales-bby@rotexindia.com
Web: www.rotexindia.com

Shreeraj Industries
101 Samruddhi, Sattar Taluka Society
Opp High Court Lane
Ahmedabad-380 014
India
Tel: +91 79 754 2813
Fax: +91 79 754 0847
Email: info@Shreerajindia.com
Web: www.shreerajindia.com

Tyco Flow Control
For Asia See Tyco Flow Control, Singapore
Web: www.tycoflowcontrol.com

Virgo Valves & Controls Ltd
277 Hinjewadi Phase II
Maan (Mulshi)
Pune 411 057
India
Tel: +91 20 66744000
Fax: +91 20 66744021
Email: alondhe@virgoengineers.com
Web: www.virgoengineers.com

ISRAEL (IL)

Egmo Ltd
PO Box 1111
Nahariya 22110
Israel
Tel: +972 4 9855130
Fax: +972 4 9855175
Email: marktg@egmo.co.il
Web: www.egmo.co.il

Habonim Industrial Valves & Actuators
Kfar Hanassi
M P Hevel Corazim 12305
Israel
Tel: +972 4 6914911
Fax: +972 4 6914902

Email: sales_international@habonim.com
Web: www.habonim.com

ITALY (IT)

ABV Srl
Via Bichi
I-55100 San Marco (Lucca)
Italy
Tel: +39 0583 464296
Fax: +39 0583 462111
Email: info@abvvalves.com
Web: www.abvvalves.com

AC MO SpA
Via Tommaso da Modena 28
I-31056 Roncade (TV)
Italy
Tel: +39 0422 840220
Fax: +39 0422 840923
Email: info@acmospa.com
Web: www.acmospa.com

Aerre Inox Srl
Lottizzazione Gerola
I-26010 Fiesco (Cremona)
Italy
Tel: +39 0374 370 828
Fax: +39 0374 370 833
Email: info@aerreinox.it
Web: www.aerreinox.it

Air Torque SpA
Via alla Campagna 1
I-24060 Costa di Mezzate (Bergamo)
Italy
Tel: +39 035 682299
Fax: +39 035 687791
Email: info@airtorque.it
Web: www.airtorque.it

Airaga Rubinetterie SpA
Valduggia Frazione Zuccaro
I-13010 Zuccaro (VC)
Italy
Tel: +39 0163 47851
Fax: +39 0163 48189
Web: www.airaga.com

Alfa Valvole Srl
Viale del Lavoro 19
I-20010 Casorezzo (MI)
Italy
Tel: +39 02 90296206
Fax: +39 02 90296292
Email: alfavalvole@alfavalvole.it
Web: www.alfavalvole.it

AVF Astore Valves and Fittings Srl
Via Tangoni 26
I-16030 Casarza Ligure (GE)
Italy
Tel: +39 0185 47811
Fax: +39 0185 466894
Email: astore@astore.it
Web: www.astore.it

Bardiani Valvole SpA
Via G di Vittorio 50/52
I-43045 Fornovo di Taro
Italy
Tel: +39 0525 400044
Fax: +39 0525 3408
Email: bardiani@bardiani.com
Web: www.bardiani.com

Besa Ing Santangelo SpA
Via delle Industrie Nord 1/A
I-20090 Settala Milano
Italy
Tel: +39 0295 37021
Fax: +39 0295 379342
Email: info@besa.it
Web: www.besa.it

Caleffi SpA
S R 229, n 25
I-28010 Fontaneto d'Agogna (NO)
Italy
Tel: +39 0322 8491
Fax: +39 0322 863305
Email: info@caleffi.it
Web: www.caleffi.it

Calobri Srl
Via Giacomo Leopardi 10
I-20019 Settimo Milanese (MI)
Italy
Tel: +39 02 3282951
Fax: +39 02 3281558
Email: info@calobri.com

Carraro Srl
Via Enrico Fermi 22
I-20090 Segrate (MI)
Italy
Tel: +39 02 2699121
Fax: +39 02 26922452
Email: info@carrarovalvole.it
Web: www.carrarovalvole.it

Cesare Bonetti SpA
Via Cesare Bonetti 17
I-20024 Garbagnate Milanese
Italy
Tel: +39 02 990721
Fax: +39 02 9952483
Email: bont.post@bont.it
Web: www.cesare-bonetti.it

Cimberio SpA
Via Torchio 57
I-28017 San Maurizio d'Opaglio (NO)
Italy
Tel: +39 0322 923001
Fax: +39 0322 967216
Email: info@cimberio.it
Web: www.cimberio.com

Colves Srl
Via F Serpero 4/F4
I-20060 Masate (MI)
Italy
Tel: +39 02 957 64357
Fax: +39 02 957 64342
Web: www.colves.com

Dafram SpA
Strada Statale 78, Km 6
I-62010 Urbisaglia (MC)
Italy
Tel: +39 733 51191
Fax: +39 733 50196
Email: sales@dafram.it
Web: www.dafram.it

Effebi SpA
Via Giuseppe Verdi 68
I-25060 Bovezzo (Brescia)
Italy
Tel: +39 030 211010
Fax: +39 030 2110302
Email: effebi@effebi.it
Web: www.effebi.it

Enolgas Bonomi SpA
Via Europa 227
I-25062 Concesio (Brescia)
Italy
Tel: +39 030 2184 311
Fax: +39 030 2184 385
Email: enolgas@enolgas.it
Web: www.enolgas.it

Eurovalve Srl
Via Camicie Rosse 11
I-20090 Opera (MI)
Italy
Tel: +39 02 57601845
Fax: +39 02 57601700
Email: sales@eurovalve.it
Web: www.eurovalve.it

Ferrero Rubinetterie Srl
Via Dogliani 84
I-12060 Farigliano (CN)
Italy
Tel: +39 0173 746001
Fax: +39 0173 76405
Email: info@ferrero-valves.com
Web: www.ferrero-valves.com

FIP - Formatura Iniezione Polimeri SpA
Pian di Parata
I-16015 Casella (GE)
Italy
Tel: +39 010 96211
Fax: +39 010 9621209
Email: info@fipnet.it
Web: www.fipnet.it

FNC SpA
Via Maestri del lavoro
Zona Industriale via per Gorla
I-21040 Cislago (VA)
Italy
Tel: +39 02 9667121
Fax: +39 02 96382186
Email: info@fncitalia.it
Web: www.fncitalia.it

Fratelli Fortis Srl
Via Dr Bellosta 34
I-28017 San Maurizio d'Opaglio (NO)
Italy
Tel: +39 0322 96215
Fax: +39 0322 967276
Email: info@.fortis.it
Web: www.fortis.it

Fratelli Pettinaroli SpA
Via Pianelli 38
I-28017 San Maurizio d'Opaglio (NO)
Italy
Tel: +39 0322 96217
Fax: +39 0322 96545
Web: www.pettinaroli.com

Galli & Cassina SpA
Via Drizza 30/32
I-20020 Solaro (MI)
Italy
Tel: +39 02 967 99632

Fax: +39 02 967 99699
Email: e.patriarchi@gallicassina.it
Web: www.gallicassina.it

GEI Gruppo Energia Italia Srl
Via Pennella 94
I-38057 Pergine Valsugana (TN)
Italy
Tel: +39 0461 532552
Fax: +39 0461 533117
Email: gei@geivalvole.it
Web: www.geivalvole.it

Ghibson Italia Srl
Via Cassola 6/9
I-40050 Monteveglio (BO)
Italy
Tel: +39 051 835711
Fax: +39 051 830344
Email: info@ghibson.it
Web: www.ghibson.it

Greiner SpA
Via Montesuello 212
I-25065 Lumezzane (BS)
Italy
Tel: +39 030 8927511
Fax: +39 030 8927590
Email: emilio.lena@greiner.it
Web: www.greiner.it

GT Attuatori Srl
Viale Europa 17
I-20090 Cusago (MI)
Italy
Tel: +39 02 90390322
Fax: +39 02 90390368
Email: info@gtrevisan.it
Web: www.gtrevisan.it

Indra Srl
Via Novara 10/B-C
I-20013 Magenta (MI)
Italy
Tel: +39 02 97298663
Fax: +39 02 97291855
Email: indra@.indra.it
Web: www.indra.it

Italprotec Sas
Via 1 Maggio 11
I-20040 Cavenago Brianza (MI)
Italy
Tel: +39 02 95 339500
Fax: +39 02 95 335075
Email: enquiry@italprotec.com
Web: www.italprotec.com

Italvalv Snc
Strada del Corriere 27
I-15060 Sant Antonio de Basaluzzo (AL)
Italy
Tel: +39 0143 489491
Fax: +39 0143 489329
Email: italvalv@italvalv.it
Web: www.italvalv.it

Italvalvole Sas
Via Amendola 125
I-13836 Cossato (Biella)
Italy
Tel: +39 015 980641
Fax: +39 015 926297
Email: info@italvalvole.it
Web: www.italvalvole.it

IVR Valvole SpA
Via Brughiera III
1 - Loc Piano Rosa - S S 142, km 37
I-28010 Boca (NO)
Italy
Tel: +39 0322 888811
Fax: +39 0322 888893
Email: e.savioli@ivrvalvole.it
Web: www.ivrvalvole.it

La TecnoValvo Srl
Via Boito 5 (entrance Viale Lombardia 10)
I-20021 Bollate (MI)
Italy
Tel: +39 02 3503508
Fax: +39 02 3503494
Email: info@latecnovalvo.com
Web: www.latecnovalvo.com

LA TIS Service Srl
Via Lago d'Iseo 2/4/6/8
I-24060 Bolgare (BG)
Italy
Tel: +39 035 835 4811
Fax: +39 035 835 4888
Email: osvaldo.pandini@latis-service.com
Web: www.latis-service.com

LCM Italia Srl
Via Vesuvio 25
I-21010 Cardano al Campo (Varese)
Italy
Tel: +39 0331 730658
Fax: +39 0331 263313
Email: infomi@cheropiping.com
Web: www.cheropiping.com

LVF SpA
Via Giuseppe Mazzini 6
I-24060 San Paolo d'Argon (BG)
Italy
Tel: +39 035 4255 211
Fax: +39 035 959 210
Email: info@lvf.it
Web: www.lvf.it

Maran E Peracini Srl
Via Valduggia 25
I-13011 Borgosesia (VC)
Italy
Tel: +39 0163 26757
Fax: +39 0163 22464
Email: info@maranperacini.com
Web: www.maranperacini.com

MEI Valvole Industriali Srl
Via Stipeti
I-55060 Coselli Lucca
Italy
Tel: +39 0583 403265
Fax: +39 0583 403138
Email: info@meivalvole.it
Web: www.meivalvole.it

Nencini SpA
Loc Belvedere
PO Box 211
I-53034 Colle Val d'Elsa (SI)
Italy
Tel: +39 0577 930880
Fax: +39 0577 930883
Email: nencini@nencini.com
Web: www.nencini.com

Nicolini Claudio srl
Via C Treves 68
I-20090 Trezzano sul Naviglio (Milano)
Italy
Tel: +39 02 484 00722
Fax: +39 02 484 00726
Email: info@nicolinivalves.it
Web: www.nicolinivalves.it

Novasfer Srl
Via G Marconi 12/ A-B-C Fraz Carzago
I-25080 Calvagese della Riviera (BS)
Italy
Tel: +39 030 6809011
Fax: +39 030 6800172
Email: info@novasfer.it
Web: www.novasfer.it

Omal Automation
Via San Lorenzo 70
I-25069 Villa Carcina (BS)
Italy
Tel: +39 0308 900145
Fax: +39 0308 900423
Email: omal@omal.it
Web: www.omal.com

OMB Valves SpA
Via Europa 7
I-24069 Cenate Sotto (BG)
Italy
Tel: +39 035 4256711
Fax: +39 035 942638
Email: info@ombvalves.com
Web: www.ombvalves.com

Orsenigo Srl
Via 1 Maggio
I-24060 Bolgare (BG)
Italy
Tel: +39 035 442 3392
Fax: +39 035 442 3294
Email: orsenigo@orsenigosrl.com
Web: www.orsenigosrl.com

Orton - Truflo International
Via Grilli 2/A
San Nicolo A Trebbia
I-29010 Rottofreno (PC)
Italy
Tel: +39 0523 762511
Fax: +39 0523 762512
Email: sales@ortonvalve.com
Web: www.truflointernational.com

Penta Srl
Via Torricelli 27
I-25080 Molinetto di Mazzano (BS)
Italy
Tel: +39 030 2629175
Fax: +39 030 2629176
Email: info@pentavalves.it
Web: www.pentavalves.it

Petrolvalves Srl
Viale G Borri 42
I-21053 Castellanza (VA)
Italy
Tel: +39 033 1334111
Fax: +39 0331 675830
Email: pvcomm@petrolvalves.it
Web: www.petrolvalves.com

Pibiviesse SpA
Via Bergamina 24

I-20014 Nerviano (MI)
Italy
Tel: +39 0331 408711
Fax: +39 0331 408800
Email: info@pibiviesse.it
Web: www.pibiviesse.it

Rastelli Rubinetterie SpA
Regione Monticelli 10/14
I-28045 Invorio (NO)
Italy
Tel: +39 0322 255431
Fax: +39 0322 255117
Email: exportrq@rastelli.it
Web: www.rastelli.it

RIV Rubinetterie Italiane Valvole SpA
Via delle Acacie
8 - Zona Industriale D4
I-28075 Grignasco (NO)
Italy
Tel: +39 0163 4151
Fax: +39 0163 411914
Email: riv-vg@riv-vg.com
Web: www.riv-vg.com

Rubinetterie Bresciane Bonomi SpA
Via Industriale 30
I-25065 Lumezzane SS (BS)
Italy
Tel: +39 030 8250011
Fax: +39 030 8920465
Email: rb@bonomi.it
Web: www.bonomi.it

Saint-Gobain Condotte SpA
Via Romagnoli 6
I-20146 Milano
Italy
Tel: +39 02 42482
Fax: +39 02 4243257
Email: info@sgcondotte.com
Web: www.sgcondotte.com

Servovalve
Via Quasimodo 27
I-20020 Santo Stefano Ticino (MI)
Italy
Tel: +39 0297 48461
Fax: +39 0297 484646
Email: servovalve@servovalve.it
Web: www.servovalve.it

SIRCA International Srl
Via Trieste 8
I-20060 Tressano Rosa (MI)
Italy
Tel: +39 02 92010204
Fax: +39 02 92010216
Email: sirca@tin.it
Web: www.sircainternational.com

Sitindustrie Equipment Srl
VALVOmetal
Via per Orlonghetto 3
I-13018 Valduggia (VC)
Italy
Tel: +39 0163 4361
Fax: +39 0163 48113
Email: valvometal@sitindustrie.com
Web: www.sitindustrie.com

Starline SpA
Via F Baracca 30
I-24060 S Paolo D'Argon (Bergamo)
Italy
Tel: +39 035 958041
Fax: +39 035 958413
Email: starline@starline.it
Web: www.starline.it

SteriValves Srl
Via Enrico Mattei 393
I-55100 Mugnano Lucca
Italy
Tel: +39 0583 440 560
Fax: +39 0583 440 559
Email: info@sterivalves.com
Web: www.sterivalves.com

Technical Srl
Via Toscana 9
I-20060 Vignate (MI)
Italy
Tel: +39 02 9593991
Fax: +39 02 9560273
Email: technical@technical.it
Web: www.technical.it

Thermomess srl
Via E Montale 6
I-20020 Robecchetto con Induno (MI)
Italy
Tel: +39 0331 875697
Fax: +39 0331 875297
Email: info@thermomess.com
Web: www.thermomess.com

Tyco Flow Control
For EMEA See Tyco Flow Control, The Netherlands
Web: www.tycoflowcontrol.com

Valbia Srl
Via A Canossi 31
PO Box 4
I-25060 Brozzo (BS)
Italy
Tel: +39 030 89 69 411
Fax: +39 030 86 10 014
Email: valbia@bonomi.it
Web: www.valbia.it

Valpres Srl
Via A Gitti 11
PO Box 40
I-25060 Marcheno VT (BS)
Italy
Tel: +39 030 89 69 311
Fax: +39 030 89 60 239
Email: valpres@bonomi.it
Web: www.valpres.it

Valvitalia SpA
Via Tortona 69
I-27055 Rivanazzano (PV)
Italy
Tel: +39 0383 945 911
Fax: +39 0383 945 962
Email: info@valvitalia.com
Web: www.valvitalia.com

Valvoindustria Ing Rizzio SpA
Via Circonvallazione 10
I-13018 Valduggia (VC)
Italy
Tel: +39 0163 47891
Fax: +39 0163 47895
Email: info@vironline.com
Web: www.vironline.com

Valvole Hofmann
Via Muzzano 31
I-13897 Occhieppo Inferiore (Biella)
Italy
Tel: +39 015 2593403
Fax: +39 015 2593844
Email: valvolehofmann@valvolehofmann.com
Web: www.valvolehofmann.com

Velan Srl
Via delle Industrie 9/5
I-20050 Mezzago (MI)
Italy
Tel: +39 039 624111
Fax: +39 039 6883357
Email: sales@velan.it
Web: www.velan.com

Watts Cazzaniga SpA
Via Parco
I-20046 Biassono (MI)
Italy
Tel: +39 039 49861
Fax: +39 039 4986222
Email: info@wattscazzaniga.it
Web: www.wattsindustries.com

JAPAN (JP)

ABB KK
6-2-2 Takatsukadai Nishi-ku
Kobe 651-22
Japan
Tel: +81 78 991 5920
Fax: +81 78 991 5900
Web: www.abb.co.jp

Bailey Japan Co Ltd
511 Baraki
Izunokuni-shi
Shizuoka 410 2193
Japan
Tel: +81 559 49 3311
Fax: +81 559 49 1114
Web: www.bailey.co.jp

Bunkaboeki Kogyo Co Ltd
1-11 Saga 1-Chome
Koto-ku
Tokyo 135-0031
Japan
Tel: +81 3 5620 1894
Fax: +81 3 5620 1795
Email: info@bbk.co.jp
Web: www.bbk.co.jp

CCI KK
6-2-2 Takatsukadai, Nishi-ku
Kobe City
Hyogo 651-2271
Japan
Tel: +81 726 41 7641
Fax: +81 726 41 7667
Email: matsuo@ccijapan.co.jp
Web: www.ccivalve.com

Eagle Industry Co Ltd
1-12-15 Shiba-Daimon
Minato-ku
Tokyo 105-8587
Japan

Tel: +81 3 3432 4771
Fax: +81 3 3438 2370
Email: info@ekk.co.jp
Web: www.ekk.co.jp

Fujikin
1-4-8 Shibata, Kita-ku
Osaka 530
Japan
Tel: +81 6 6372 7141
Fax: +81 6 6375 0697
Web: www.fujikin.co.jp

Fukui Seisakusho Co Ltd
6.1-Chome
Shodai Tajika Hirakata-shi
Osaka 573 1132
Japan
Tel: +81 72 857 4521
Fax: +81 72 857 3764
Email: fki@fkis.co.jp
Web: www.fkis.co.jp

Furukawa Kogyo Co Ltd
2-3, Marunouchi 2-Chome
Hikone-shi
Chiyoda-ku
Japan
Tel: +81 3 3212 6570
Fax: +81 3 3212 6578
Email: info@furukawakk.co.jp
Web: www.furukawakk.co.jp

Fushiman Co Ltd
4-6-12 Omori-minami
Ota-ku
Tokyo 143-0013
Japan
Tel: +81 3 3741 8951
Fax: +81 3 3741 7664
Email: gyomu@fushiman.co.jp
Web: www.fushiman.co.jp

Hatanaka Special Valve Industries Co Ltd
1-3-11 Fukagawa Koto-ku
Tokyo 135-0033
Japan
Tel: +81 3 3820 5671
Fax: +81 3 5639 5115
Email: info@hsv.co.jp
Web: www.hsv.co.jp

Heiwa Valve Co Ltd
5-2-11 Nishi Gotanda
Shinagawa ku
Tokyo 141-0031
Japan
Tel: +81 3 3493 5855
Fax: +81 3 3493 5858
Email: mail@heiwa-valve.co.jp
Web: www.heiwa-valve.co.jp

Hirata Valve Industry Co Ltd
3-2-3 Hisamoto
Takatsu ku
Kawasaki 213-8691
Japan
Tel: +81 44 833 2311
Fax: +81 44 822 5101
Web: www.hvi.co.jp

Hirose Valve Industry Co Ltd
2-34 Yasukiyocho
Hikone-shi
Shiga 522-0082
Japan
Tel: +81 749 23 2020
Fax: +81 749 23 2027
Web: www.hirose-valves.co.jp

Hisaka Works Ltd
4-2-14 Fushimimachi
Chuo-ku
Osaka 541-0044
Japan
Tel: +81 6 6391 8940
Fax: +81 6 6391 8941
Email: valve@hisaka.co.jp
Web: www.hisaka.co.jp

Hitachi Valve Ltd
200 Obuke
Asahicho, Mie-Gun
Mie Prefecture 510-8102
Japan
Tel: +81 593 77 4711
Fax: +81 593 77 2714
Email: info@hitachi-valve.co.jp
Web: www.hitachi-valve.co.jp

Ichinose Company Ltd
4-7-6 Itachibori
Nishi ku
Osaka 550-0012
Japan
Tel: +81 6 6539 3112
Fax: +81 6 6536 3165
Web: www.insins.co.jp

Ihara Science Corporation
13-17 Oi 4 Chome
Shinagawa-ku
Tokyo 140-0014
Japan
Tel: +81 3 5742 2701
Fax: +81 3 5742 2233
Web: www.ihara-sc.co.jp

Ishida Valve Manufacturing Co Ltd
Fujishima Building, 4-13-4 Shiba
Minato-ku
Tokyo 108 0014
Japan
Tel: +81 3 3455 5271
Fax: +81 3 3455 8690
Email: t-ishida@ishida-valve.co.jp
Web: www.ishida-valve.com

Ito Koki Co Ltd
10-4 Hakodono-Cho
Higashiosaka City
Osaka 579 8037
Japan
Tel: +81 729 85 2521
Fax: +81 729 82 2210
Email: honsha@itokoki.co.jp
Web: www.itokoki.co.jp

Kane Kogyo Co Ltd
2036 Okusa
Komaki City
Aichi-Pref 485-0802
Japan
Tel: +81 568 79 2476
Fax: +81 568 79 6422
Email: overseasales@kkk-kane.co.jp
Web: www.kkk-kane.co.jp

Kaneko Sangyo Co Ltd
10-6 Shiba 5-Chome
Minato-ku
Tokyo 108-0014
Japan
Tel: +81 3 3455 1411
Fax: +81 3 3456 5820
Email: yasuyuki.sakamoto@kaneko.co.jp
Web: www.kaneko.co.jp

Kitamura Valve Mfg Co Ltd
2500 Daimon
Urawa shi
Saitama 377-0963
Japan
Tel: +81 48 878 1111
Fax: +81 48 878 5289
Email: info@tfcj.co.jp
Web: www.ktm-valve.co.jp

Kitz Corporation
1-10-1 Nakase
Mihama ku
Chiba 261-8577
Japan
Tel: +81 43 299 0111
Fax: +81 43 299 0121
Email: kitzheadquarters@kitz.co.jp
Web: www.kitz.co.jp

Kubota Corporation
1-2-47 Shikitsu-higashi
Naniwa-ku
Osaka 556-8601
Japan
Tel: +81 6 6648 2111
Fax: +81 6 6648 3862
Email: kubota@kubota.co.jp
Web: www.kubota.co.jp

Kurimoto Ltd
1-9 Shimbashi 4-Chome
Minato-Ku
Tokyo 104-0005
Japan
Tel: +81 3 3436 8191
Fax: +81 3 3436 8026
Web: www.kurimoto.co.jp

KVC Co Ltd
208-153 Nagahara-Cho
Sakaimachi Ayanokoji, Shimogoyo-ku
Kyoto 600-8073
Japan
Tel: +81 75 361 3361
Fax: +81 75 371 9341
Email: kvc@kvc-kyoto.co.jp
Web: www.kvc-kyoto.co.jp

Maezawa Industries Inc
7-2 2 Yaesu
Chuo-ku
Tokyo 104 8351
Japan
Tel: +81 3 3281 5521
Fax: +81 3 3274 4157
Email: prir_info@maezawa.co.jp
Web: www.maezawa.co.jp

Mihana Seisakusho Co Ltd
4270-8 Kokubu Higanjyo-cho
Kashiwara City
Osaka 582-0023
Japan

Tel: +81 729 76 0387
Fax: +81 729 76 0386
Email: mihana-v@maple.ocn.ne.jp
Web: www.mihana-v.co.jp

Miyairi Corporation
Hatchobori Kagayaki Building
3-9-7 Hatchobori, Chuo-ku
Tokyo 104-0032
Japan
Tel: +81 3 3553 7375
Fax: +81 3 3552 3674
Email: mtvalve@miyairicorp.co.jp
Web: www.miyairicorp.co.jp

Miyawaki Inc
2-1-30 Tagawakita
Yodagawa-ku
Osaka 532-0021
Japan
Tel: +81 6 6302 5549
Fax: +81 6 6302 5595
Email: export@miyawaki.net
Web: www.miyawaki.net

Motoyama Eng Works Ltd
5-2, Aza Kameoka
Ohhira-mura, Kurokawa-gun
Miyagi 981-3697
Japan
Tel: +81 22 344 4546
Fax: +81 22 344 4533
Email: mew@motoyama-cp.co.jp
Web: www.motoyama-cp.co.jp

Nihon Klingage Co Ltd
1-14-5 Ginza
Chuo-ku
Tokyo 104-0061
Japan
Tel: +81 3 3567 7002
Fax: +81 3 3535 3712
Email: nakajima@klingage.co.jp
Web: www.klingage.co.jp

Nihon Koso Co Ltd
1-16-7 Nihombashi
Chuo-ku
Tokyo 103-0027
Japan
Tel: +81 3 5202 4300
Fax: +81 3 5202 4301
Web: www.koso.co.jp

Niigata Masoneilan Co Ltd (NIMCO)
20th Floor, Marive East Tower
WBG 2 6 Nakase, Mihama ku Chiba shi
Chiba 261 7120
Japan
Tel: +81 43 297 9222
Fax: +81 43 299 1115
Web: www.masoneilan.com

Nippon Ball Valve Co Ltd
5-650 Ohtoriminami Machi
Sakai-shi
Osaka 593-8325
Japan
Tel: +81 722 71 8421
Fax: +81 722 73 3462
Email: info@nbv.co.jp
Web: www.nbv.co.jp

Nippon Daiya Valve Co Ltd
1-3-22 Hiromachi
Shinagawa-Ku
Tokyo 140-0005
Japan
Tel: +81 3 3490 4801
Fax: +81 3 3490 7950
Email: tokyo111@ndv.co.jp
Web: www.ndv.co.jp

Nippon Valqua Industries Ltd
Shinjuka-Mitsui Building
2-1-1 Nishi-Shinjuku, Shinjuku ku
Tokyo 163-0406
Japan
Tel: +81 3 5325 3421
Fax: +81 3 5325 3436
Web: www.valqua.co.jp

Okano Valve Mfg Company
1-14 Nakamachi
Moji ku
Kitakyushu 800-8601
Japan
Tel: +81 93 372 9244
Fax: +81 93 382 1200
Email: okano-kaiei@themis.ocn.ne.jp
Web: www.okano-valve.co.jp

Okumura Engineering Corporation
446-1, Otani, Hino-cho
Gamo-Gun
Shiga 529-1608
Japan
Tel: +81 748 52 2131
Fax: +81 748 52 5025
Web: www.okm-net.co.jp

Sankyo Seisakusho Co Ltd
Yotsubashi Shinkosan Building 702, 5-2
1-chome Kitahorie, Nishi-ku
Osaka 550-0014
Japan
Tel: +81 6 6541 9081
Fax: +81 6 6541 9098
Web: www.sankyo-s-s.co.jp

Shida Engineering Co Ltd
4-1-17 Osaki
Shinagawa ku
Tokyo 141-0032
Japan
Tel: +81 3 3492 2037
Fax: +81 3 3492 2039
Email: mail@shida-valve.com
Web: www.shida-valve.com

Shimizu Alloy Manufacturing Co Ltd
928 Higashi-Nonamicho
Hikone-shi
Shiga 522-0033
Japan
Tel: +81 749 23 3131
Fax: +81 749 22 0687
Email: honshaeigyo@shimizugokin.co.jp
Web: www.shimizugokin.co.jp

Shimizu Iron Works Co Ltd
4-16 Yasukiyocho
Hikone-shi
Shiga 522-0082
Japan
Tel: +81 749 22 2551
Fax: +81 749 22 2542
Web: www.shimizu-valve.co.jp

Shoritsu Seisakusho Co Ltd
2-7-21 Matsushima
Edogawa-ku
Tokyo 132-0031
Japan
Tel: +81 3 3654 9211
Fax: +81 3 3651 5688
Email: shoritsu@sfv.co.jp
Web: www.sfv.co.jp

Showa Valve Co Ltd
155-9 Koizumi
Hikone
Shiga 522-0043
Japan
Tel: +81 749 22 4545
Fax: +81 749 26 1785
Email: export@showavalve.co.jp
Web: www.showavalve.co.jp

TIX Corporation – IKS Division
6-21 Kita-Shinagawa, 5-Chome
Shinagawa-ku
Tokyo 141-0001
Japan
Tel: +81 3 3495 7961
Fax: +81 3 3495 7960
Email: kodama@tix.co.jp
Web: www.tix.co.jp

Toa Valve Company Ltd
12-1, 5-Chome, Nishi-Tachibana-cho
Amagasaki
Hyogo 660-0054
Japan
Tel: +81 6 6416 8865
Fax: +81 6 6419 5641
Email: sales1@toavalve.co.jp
Web: www.toavalve.co.jp

Toko Valex Co Ltd
4-2-17 Matsushima
Edogawa-ku
Tokyo 132-0031
Japan
Tel: +81 3 3655 5161
Fax: +81 3 3655 6769
Email: info@toko-valex.co.jp
Web: www.toko-valex.co.jp

Tomoe Valve Co Ltd
3-11-11 Shinmachi
Nishi-ku
Osaka 550 0013
Japan
Tel: +81 6 6110 2101
Fax: +81 6 6110 2105
Email: bu1@tomoe.co.jp
Web: www.tomoe.co.jp

Toyo Valve Co Ltd
TT-2 Building, 8-1, 3-Chome
Nihonbashi-Ningyocho
Chuo-ku
Tokyo 103 0013
Japan
Tel: +81 3 3249 5314
Fax: +81 3 3249 5303
Email: toyo@toyovalve.co.jp
Web: www.toyovalve.co.jp

TV Valve Co Ltd
4-33-8 Omori-Higashi
Ota-ku

Tokyo 143-0012
Japan
Tel: +81 3 3763 4311
Fax: +81 3 3763 4317
Email: info@tv-valve.com
Web: www.tv-valve.com

Utsue Valve Co Ltd
2-1-13 Kitamura
Taisho ku
Osaka 551-0032
Japan
Tel: +81 6 6552 3161
Fax: +81 6 6551 3245
Web: www.utsue-valve.co.jp

V TEX Corporation
Taniguchi Building 28-11
Minami Ohi 6-Chome
Shinagawa-ku
Tokyo 140-0013
Japan
Tel: +81 3 3765 4161
Fax: +81 3 3765 4168
Web: www.vtex.co.jp

Venn Co Ltd
2-13 Tamagawa 2-Chome
Ohta-ku
Tokyo 146-0095
Japan
Tel: +81 3 3759 0170
Fax: +81 3 3759 1414
Email: info@venn.co.jp
Web: www.venn.co.jp

Yoshitake Inc
7-3 Futano-Cho
Mizuho ku
Nagoya 467-0861
Japan
Tel: +81 52 881 7199
Fax: +81 52 881 7201
Email: ys-admin@yoshitake.co.jp
Web: www.yoshitake.co.jp

KUWAIT (KW)

Dresser Flow Solutions - Middle East Operations
10th Floor, Al Rashed Complex
Fahad Salem Street, P O Box 242
Safat, 13003
Kuwait
Tel: +965 9061157
Fax: +965 3987879
Web: www.masoneilan.com

MALAYSIA (MY)

Dresser Flow Solutions
Business Suite, 19A 9 1, Level 9
UOA Centre, No.19, Jalan Pinang
50450 Kuala Lumpur
Malaysia
Tel: +60 3 2161 0322
Fax: +60 3 2163 3612
Web: www.masoneilan.com

MEXICO (MX)

Dresser Valve de Mexico SA
Henry Ford No 114, Esq Fulton
Fraccionamiento Industrial San Nicolas
54030 Tlalnepantla
Mexico
Tel: +52 5 310 9863
Fax: +52 5 310 5584
Web: www.masoneilan.com

THE NETHERLANDS (NL)

Delan BV
PO Box 218
2181 AE Hillegom
The Netherlands
Tel: +31 252 528870
Fax: +31 252 528470
Email: info@delan.org
Web: www.delan.org

DMN Westinghouse
PO Box 6
2210 AA Noordwijkerhout
The Netherlands
Tel: +31 252 361 800
Fax: +31 252 375 972
Email: dmn@dmn-nwh.nl
Web: www.dmn.info

HP Valves Oldenzaal BV
Zutphenstraat 1
7575 EJ Oldenzaal
The Netherlands
Tel: +31 541 519555
Fax: +31 541 522045
Email: sales@hpvalves.com
Web: www.hpvalves.com

Mokveld Valves BV
Nijverheidsstraat 67
PO Box 227
2800 AE Gouda
The Netherlands
Tel: +31 182 597500
Fax: +31 182 517977
Email: info@mokveld.com
Web: www.mokveld.nl

Noxon Stainless BV
PO Box 6096
5700 ET Helmond
The Netherlands
Tel: +31 492 582111
Fax: +31 492 538970
Email: info@noxon.nl
Web: www.noxon.nl

Red Point Alloys BV
Radonstraat 2
2718 TA Zoetermeer
The Netherlands
Tel: + 31 79 3632070
Fax: +31 79 3611088
Email: info@redpoint.nl
Web: www.redpoint.nl

Tyco Flow Control
Mijkenbroek 22
4824 AB Breda
The Netherlands
Tel: +31 76 5434100
Fax: +31 76 5434399
Web: www.tycoflowcontrol.com

Ventil Test Equipment BV
PO Box 161
2260 AD Leidschendam
The Netherlands
Tel: +31 70 3209327
Fax: +31 70 3203737
Email: info@ventil.nl
Web: www.ventil.nl

Wouter Witzel EuroValve
PO Box 54
NL-7580 CZ Losser
The Netherlands
Tel: +31 535 369536
Fax: +31 535 369500
Email: info@wweurovalve.nl
Web: www.wweurovalve.nl

NORWAY (NO)

Bryne Mekanikk AS
Nordlysvegen 1
PO Box 98
N-4349 Bryne
Norway
Tel: +47 5177 2140
Fax: +47 5177 2141
Email: post@bm.no
Web: www.bm.no

Danfoss A/S
Nordborgvej 81
N-6430 Nordborg
Norway
Tel: +45 7488 2222
Fax: +45 7488 0949
Email: contact@danfoss.com
Web: www.danfoss.com

Kongsberg Esco A/S
PO Box 85
N-3602 Kongsberg
Norway
Tel: +47 926 45700
Fax: +47 3273 2999
Email: post@esco.no
Web: www.esco.no

Norske Ventiler AS
PO Box 115
N-5346 Ågotnes
Norway
Tel: +47 563 15450
Fax: +47 563 15451
Email: info@norskeventiler.no
Web: www.norskeventiler.no

Oceaneering Rotator AS
PO Box 24
Hagen 20
N-4685 Nodeland
Norway
Tel: +47 38185500
Fax: +47 381821 52
Email: rotator@oceaneering.com
Web: www.rotator.no

Scan Armatur AS
PO Box 43
N-5346 Ågotnes
Norway
Tel: +47 5631 2900
Fax: +47 5631 2910
Email: mail@scanarmatur.no
Web: www.scanarmatur.no

Steinsvik Maskinindustri AS
Frakkagjerd
N-5563 Førresfjorden
Norway
Tel: +47 52 754700
Fax: +47 52 754701
Email: nortech@nortechinc.com
Web: www.nortechinc.com

Westad Industri AS
Heggenveien
N-3360 Geithus
Norway
Tel: +47 32 789500
Fax: +47 32 789510
Email: pgroe@westad.com
Web: www.westad.com

POLAND (PL)

Tehaco Sp Zoo
ul. Nowy Swiat 4
PL-80-299 Gdan´sk
Poland
Tel: +48 58 554 5929
Fax: +48 58 552 7228
Email: tehaco@tehaco.com.pl
Web: www.tehaco.com.pl

Varimex-Valves Co Ltd
Radna 6/8
PL-00-341 Warsaw
Poland
Tel: +48 22 828 1531
Fax: +48 22 828 1532
Email: biuro@varimex-valves.pl
Web: www.varimex-valves.pl

Zetkama Fabryka Armatury Przemyslowej SA
ul Slaska 24
PL-57-300 Klodzko
Poland
Tel: +48 74 865 2100
Fax: +48 74 865 2101
Email: head.office@zetkama.com.pl
Web: www.zetkama.com.pl

PORTUGAL (PT)

Velan Válvulas Industriais Lda
Avenue Ary dos Santos
P-1679-018 Famoes
Portugal
Tel: +351 21 934 7800
Fax: +351 21 934 7809
Email: sales@velan.com
Web: www.velan.pt

QATAR (QA)

Delta (Doha) Corporation
PO Box 4926
Doha
Qatar
Tel: +974 4784880
Fax: +974 4784881
Email: crp@qatar.net.qa
Web: www.deltadoha.com

ROMANIA (RO)

Ario SA
10 Industriei Street Bistrita
RO-420063 Bistrita, BN County
Romania
Tel: +40 263 234 160
Fax: +40 263 234 008
Email: info@ario.ro
Web: www.ario.ro

Rominserv Valves IAIFO
B-dul Mihai Viteazul 58
RO-450090 Zalau
Romania
Tel: +40 260 607 300
Fax: +40 260 661 594
Email: dan.folescu@rompetrol.com
Web: www.iaifo.ro

RUSSIA (RU)

Armagus JSC
4 Rudnitskoy Street
Gus-Khrustalniy 601501
Russia
Tel: +7 9241 23241
Fax: +7 9241 28940
Email: sales@armagus.ru
Web: www armagus.ru

Armalit-1 JSC
2 Trefoleva Street
St Petersburg 198097
Russia
Tel: +7 812 2525936
Fax: +7 812 2524200
Email: market@armalit1.ru
Web: www.armalit1.ru

Arzil JSC
18 Chugurin Street
Georgievsk
Stavropolsky Region 357806
Russia
Tel: +7 87951 24435
Fax: +7 87951 24435
Email: nizhnik@mail.ru

Blagoveshchensky Valve Plant JSC
1 Sedova Street
Blagoveshchensk
Bashkortostan 453430
Russia
Tel: +7 3476 621357
Fax: +7 3476 621378
Email: baz@ufanet.ru
Web: eng.pktba.ru

DS Controls
Nekhinskaya Street, 61
173021 Veliky Novgorod
Russia
Tel: +7 8162 15 7898
Fax: +7 8162 15 7921
Web: www.masoneilan.com

Fitting-P
PO Box 244
Naberezhny Chelny
Tatarstan 423810
Russia
Tel: +7 8552 550028
Email: market@arm99.chelny.ru
Web: www.fitting.ru

JSK PKTBA
75 Pobedy Prospekt
Penza 440060
Russia
Tel: +7 8412 457501
Fax: +7 8412 457800
Email: vd@pktba.ru
Web: www.ptpa.ru

PYCT-95
17 Butlerova Street
Moscow 117342
Russia
Tel: +7 4953 301077
Fax: +7 4953 301077
Email: contact@roost.ru
Web: www.roost.ru

Tulaelectroprivod JSC
958 Plechanovo
Leninsky
Tula 301114
Russia
Tel: +7 8767 96709
Fax: +7 876 796717
Email: privod@tula.net
Web: www.tulaprivod.ru

Tyazhpromarmatura JSC
60 Nekrasova Street
Alexin
Tulsky 301368
Russia
Tel: +7 48753 42980
Fax: +7 48753 27120
Email: aztpa@aleksin.tula.net
Web: www.aztpa.ru

Zeim
Yakovlev 1
Cheboksary 428 020
Russia
Tel: +7 8352 305221
Fax: +7 8352 305111
Email: sales@zeim.ru
Web: www zeim.ru

SAUDI ARABIA (SA)

Dresser Al Rushaid Valve & Instrument Co Ltd
(Darvico)
PO Box 10145
Jubail Industrial City 31961
Saudi Arabia
Tel: +966 3 341 0278
Fax: +966 3 341 7624
Web: www.masoneilan.com

SINGAPORE (SG)

Alton International (S) Pte Ltd
80 Tuas Avenue 1
Singapore 639525
Singapore
Tel: +65 8618586
Fax: +65 8619708
Email: enquiry@alton.com.sg
Web: www.alton.com.sg

Dresser Singapore Pte Ltd
16 Tuas Avenue 8
Singapore 639231
Singapore
Tel: +65 668616100
Fax: +65 668617172
Web: www.masoneilan.com

Emerson Process Management
1 Pandan Crescent
Singapore 128461
Singapore
Tel: +65 67778211
Fax: +65 67770947
Email: enquiries@ap.emerson process.com
Web: www.emersonproces.com/fisher

Remote Control Asia Pte Ltd
No 9 Kaki Bukit Road 1
03-03 Eunos Technolink
Singapore 415938
Singapore
Tel: +65 68487150
Fax: +65 67465815
Email: info@remotecontrol.com.sg
Web: www.remotecontrol.com.sg

Tyco Flow Control
45 Tuas Avenu 9
Singapore 639189
Singapore
Tel: +65 8611655
Fax: +65 8616997
Web: www.tycoflowcontrol.com

SOUTH AFRICA (ZA)

AMD Rotolok
PO Box 56314
Pinegowrie 2123
South Africa
Tel: +27 11 674 1166
Fax: +27 11 674 4372
Email: info@amd-rotolok.co.za
Web: www.amd-rotolok.co.za

Atval (Pty) Ltd
22 Makriel Road
Wadeville Ext 2, Germiston
Gauteng
South Africa
Tel: +27 11 827 3554
Fax: +27 11 827 0879
Email: atval@mweb.co.za
Web: www.atval.co.za

AZ Valves South Africa
PO Box 3862
Kempton Park 1620
South Africa
Tel: +27 11 397 3665
Fax: +27 11 397 3803
Email: azvalves @icon.co.za
Web: www.azvalves.com

Dynamic Fluid Control Pty Ltd
32 Lincoln Road Industrial Sites
Benoni South 1501
Johannesburg
South Africa
Tel: +27 11 748 0200
Fax: +27 11 421 2749
Email: dfc@dfc.co.za
Web: www.dfc.co.za

Gunric Valves
PO Box 255
Florida
Johannesburg 1709
South Africa
Tel: +27 11 474 6180
Fax: +27 11 474 6136
Email: sales@gunric.com
Web: www.gunric.com

Insamcor (Pty) Ltd
7-9 Simba Street, Sebenza
Edenvale
Gauteng
South Africa
Tel: +27 11 609 8610
Fax: +27 11 609 7681
Email: info@insamcor.co.za
Web: www.insamcor.co.za

Mine Line (Pty) Ltd
128 Impala Road
Doornkop
Johannesburg
South Africa
Tel: +27 11 765 2635
Fax: +27 11 765 2650
Email: mineline@mweb.co.za
Web: www.mineline.co.za

Mitech
Corner Brine Avenue & Horn Street
Chloorkop Ext 23
Gautang
South Africa
Tel: +27 11 230 1300
Fax: +27 11 230 1333
Email: info@za.spiraxsarco.com
Web: www.mitech.co.za

Premier Valves
PO Box 11735
Randhart 1457
South Africa
Tel: +27 11 908 3760
Fax: +27 11 908 3700
Email: info@premiervalves.co.za
Web: www.premiervalves.co.za

Spirotech International
17 Resnick Street
Factoria
Krugersdorp
South Africa
Tel: +27 11 955 4690
Fax: +27 11 955 4697
Email: sales@spirotech.co.za
Web: www.spirotech.co.za

Thermal Valve Manufacturers (Pty) Ltd
PO Box 75971
Garden View
Johannesburg 2047
South Africa
Tel: +27 11 873 9483
Fax: +27 11 873 8855
Email: sales@t-v-m.com
Web: www.savama.co.za

SOUTH KOREA (KR)

A-Sung Plastic Valve Co Ltd
A-Sung Building, 4th Floor
Kuro-Dong, Kuro-Ku
Seoul 152-060
South Korea
Tel: +82 2 671 19005
Fax: +82 2 671 5687
Email: asung@asungvalve.com
Web: www.asungvalve.com

Goodwin Korea Co Ltd
#382-45 Wonchang-Dong
SeoGu
Incheon
South Korea
Tel: +82 32 579 6313
Fax: +82 32 579 6314
Email: goodwinkorea@goodwin.co.kr
Web: www.goodwin.co.kr

KumKang Valve Mfg Co Ltd
209-10, Jang-Dong
Dalseo-Gu
Daegu 704-190
South Korea
Tel: +82 53 581 4500
Fax: +82 53 581 4505
Email: kuka@kuka.co.kr
Web: www.kuka.co.kr

PK Valve
(641-370) 192-1 Shinchon-Dong
Changwon KN
South Korea
Tel: +82 55 268 3777
Fax: +82 55 268 0281
Email: mst@pkvalve.co.kr
Web: www.pkvalve.co.kr

Samjin Precision Co Ltd
1-31 Daehwa-Dong
Daeduk-ku
Taejeon 306-802
South Korea
Tel: +82 42 672 3600
Fax: +82 42 626 3142
Email: samjin@samjinvalve.com
Web: www.samjinvalve.com

Shin Han Control Valve Co Ltd
407 Kumkang B/D
14 35 Yeoido-Dong, Yeoungdeungpo-ku
Seoul 150 010
South Korea
Tel: +82 2 784 3603
Fax: +82 2 784 3606
Email: shcv@shcontrol.co.kr
Web: www.shcontrol.co.kr

Velan Ltd
1060-4 Shingil-Dong
Ansan City
Kyunggi-Do 425-833
South Korea
Tel: +82 31 491 2811
Fax: +82 31 491 2813
Email: sales@velan.com
Web: www.velan.com

SPAIN (ES)

Babcock Power España SA
Carretera Ugarte-Galindo s/n
E-48510 Val de Trapaga, Vizcaya
Spain
Tel: +34 944 728 000
Fax: +34 944 728 032
Email: bpe@bpe.com.es
Web: www.bbe.es

BAC Valves SA
PO Box 13
Tapis 126
E-17600 Figueres
Spain
Tel: +34 972 677 052
Fax: +34 972 509 040
Email: sales@bacvalves.com
Web: www.bacvalves.com

Beligcast Internacional SL
Barrio Zabalondo 31
31-ZIP Code 93
E 48100 Mungia Vizcaya
Spain
Tel: +34 944 889 120
Fax: +34 944 889 125
Email: belgicast@tyco-valves.com
Web: www.belgicast.es

Centork Valve Control SL
Porto 110 Ipintza
Txatxamendi 24-26
E-20100 Lezo (Guipuzcoa)
Spain
Tel: +34 943 316 137
Fax: +34 943 223 657
Email: info@centork.com
Web: www.centork.com

JC Fábrica de Válvulas SA
C/Cantàbria 2
Poligono Industrial Les Salines
E-08830 Sant Boi de Llobregat (Barcelona)
Spain
Tel: +34 936 548 686
Fax: +34 936 548 687
Email: jc@jc-valves.com
Web: www.jc-ballvalves.com

Kitz Corporation of Europe SA
Ramón Viñas 8
E-08930 Sant Adrià de Besòs
Barcelona
Spain
Tel: +34 934 621 408
Fax: +34 934 620 349
Email: kitzeurope@kitzeurope.com
Web: www.kitzeurope.com

Orbinox SA
Poligono Industrial s/n
E-20270 Anoeta
Spain
Tel: +34 943 698 030
Fax: +34 943 653 066
Email: orbinox@orbinox.com
Web: www.orbinox.com

Pekos Valves SA
Rec del Molinar 9
Polígono Industrial El Circuit
E-08160 Montmeló (Barcelona)
Spain
Tel: +34 935 799 370
Fax: +34 935 799 244
Email: valvulas@pekos.es
Web: www.pekos.es

Válvulas VS SL
Pol Santelices Nave B-2
E 48550 Muskiz
Spain
Tel: +34 944 801 6422
Fax: +34 944 802 385
Email: comercial@valvulasvs.com
Web: www.valvulasvs.com

VYC Industrial SA
Transversal 179
E-08225 Terrassa (Barcelona)
Spain
Tel: +34 937 357 500
Fax: +34 937 358 135
Email: info@vycindustrial.com
Web: www.vycindustrial.com

SWEDEN (SE)

CCI Sweden
PO Box 603
SE-661 29 Säffle
Sweden
Tel: +46 533 689600
Fax: +46 533 689601
Email: lars.wall@ccivalve.se
Web: www.ccivalve.com

NAF AB
Gelbgjutaregatan 2
SE 581 87 Linköping
Sweden
Tel: +46 13 316100
Fax: +46 13 136054
Email: info@naf.se
Web: www.naf.se

Remote Control Sweden
PO 80
Kontrollvägen 15
SE-791 22 Falun
Sweden
Tel: +46 23 58700
Fax: +46 23 58745
Email: info@remotecontrol.se
Web: www.remotecontrol.se

SWITZERLAND (CH)

Biar SA
5 rue des Epineys
CH 1948 Lourtier
Switzerland
Tel: +41 27 779 1111
Fax: +41 27 779 1112
Email: contact@biar.com
Web: www.sampling-valves.com

CCI Switzerland
Im Link II
PO Box 65
CH-8404 Winterthur
Switzerland
Tel: +41 52 264 9500
Fax: +41 52 264 9520
Email: marcel.schmid@ccivalve.ch
Web: www.ccivalve.com

ChemValve AG
Sternenhaldenstrasse 13
CH-8712 Stäfa
Switzerland
Tel: +41 1 928 1788
Fax: +41 1 928 1787
Email: mrb@bluewin.ch
Web: www.chemvalve.ch

CLA-VAL SA
Chemin des Mésanges 1
CH-1032 Romanel/Lausanne
Switzerland
Tel: +41 21 643 1555
Fax: +41 21 643 1550
Email: cla-val@cla-val.ch
Web: www.cla-val.ch

Famat SA
Chemin des Jordils 40
CH 1025 St-Sulpice
Switzerland
Tel: +41 21 695 2626
Fax: +41 21 695 2627
Email: info@famat.com
Web: www.famat.ch

Georg Fischer Rohrleitungssysteme (Schweiz) AG
Ebnatstrasse 101
CH 8201 Schaffhausen
Switzerland
Tel: +41 52 631 3026
Fax: +41 52 631 2897
Email: ch.ps@georgfischer.com
Web: www.piping.georgfischer.ch

InterApp AG
Grundstrasse 24
CH 6343 Rotkreuz
Switzerland
Tel: +41 41 798 2233
Fax: +41 41 798 2234
Email: info@ch.interapp.net
Web: www.interapp.net

SISTAG Absperrtechnik
Alte Kantonsstrasse
CH-6274 Eschenbach
Switzerland
Tel: +41 41 449 9944
Fax: +41 41 448 3431
Email: info@sistag.ch
Web: www.sistag.ch

Tyco Flow Control
For EMEA see Tyco Flow Control, The Netherlands
Web: www.tycoflowcontrol.com

Valv AG
Alte Rheinstrasse
CH 9451 Kriessern
Switzerland
Tel: +41 71 757 5959
Fax: +41 71 757 5960
Email: info@valv.ch
Web: www.valv.ch

von Rohr Armaturen AG
Fichtenhagstrasse 4
CH 4132 Muttenz 2
Switzerland
Tel: +41 61 467 9120
Fax: +41 61 467 9121

Email: info@von-rohr.ch
Web: www.von-rohr.ch

TAIWAN (TW)

Apex Controls Co Ltd
15 Jen-Yi Street
Wu-Jih Hsiang Taichung 414
Taiwan
Tel: +886 4 23388440
Fax: +886 4 23388441
Email: sales@apexcontrols.com.tw
Web: www.apexcontrols.com.tw

Bola-Tek Manufacturing Co Ltd
PO Box 35-78
Taichung 404
Taiwan
Tel: +886 4 22348000
Fax: +886 4 22349000
Email: bolatek@ms18.hinet.net
Web: www.bola-tek.com.tw

Bueno Enterprise Co Ltd
17 Gongye Road
Guantian Industrial District
Tainan
Taiwan
Tel: +886 6 6935311
Fax: +886 6 6935336
Email: bueno@buenoco.jp
Web: www.buenoeco.com

EMICO Eayuan Metal Industrial Co Ltd
3F-8, No 665, Sec 2
Wu-Chuan W Road
Taichung
Taiwan
Tel: +886 4 3832828
Fax: +886 4 3847711
Email: emico@ms4.hinet.net
Web: www.emico.com.tw

Haitima Corporation
8F, 201 Tiding Boulevard, Sec 2
Neihu Chiu
Taipei 114
Taiwan
Tel: +886 2 26585800
Fax: +886 2 26582266
Web: www.haitima.com.tw

HCHTS Industries Corporation (HICO)
8F-15, 409 Lin-Shen North Road
Taipei 104
Taiwan
Tel: +886 2 25818803
Fax: +886 2 25614751
Email: hico@hicovalve.com
Web: www.hicovalve.com

John Valve Mfg Factory Co Ltd
Room C, 17Fl,
270 Chung Ming South Road
Taichung 414
Taiwan
Tel: +886 4 23780078
Fax: +886 4 23754978
Email: sales@johnvalve.com.tw
Web: www.johnvalve.com.tw

Kingdom Precision Casting Co Ltd
14-F-4, 120 Chung Cheng 1st Road
Kaohsiung 80284
Taiwan
Tel: +886 7 7713121
Fax: +886 7 7235165
Email: sales@kicasting.com
Web: www.kicasting.com

Mars Valve Co Ltd
83, Sec 1
Chung-De 8th Road
Taichung
Taiwan
Tel: +886 4 22463808
Fax: +886 4 22470912
Email: mars@marsvalve.com.tw
Web: www.marsvalve.com.tw

Super Inox Corporation
22 Wu-Chuan 2nd Road
Hsin-Chuang
Taipei Hsien
Taiwan
Tel: +886 3 2115015
Fax: +886 3 2115025
Email: inox@ms16.hinet.net
Web: www.inox.com.tw

Value Valves Co Ltd
9 Chung-Shan Road
Tu-Cheng Industrial District
Taipei Hsien
Taiwan
Tel: +886 2 22698000
Fax: +886 2 22686600
Email: david@valuevalves.com.tw
Web: www.valuevalves.com

Valve-Tek Manufacturing Co Ltd
105 Cheng Kuang Street
Taiping
Taichung
Taiwan
Tel: +886 4 22312299
Fax: +886 4 22313399
Email: valvetek@tcts.seed.net.tw
Web: www.valve-tek.com.tw

Velan-Valvac
PO Box 2020
Taichung
Taiwan
Tel: +886 4 2792649
Fax: +886 4 2750855
Email: sales@velan.com
Web: www.velan.com

Wyeco Auto Valves Co Ltd
4F, No 98, Section 3
Chien Kuo North Road
Taipei
Taiwan
Tel: +886 2 25025166
Fax: +886 2 25012863
Email: wysal@ms15.hinet.net
Web: www.wyeco.com.tw

Yi Ming Machinery Ind Co Ltd
276 Jung Shiau Road
Pali
Taipei Hsien
Taiwan
Tel: +886 2 86303236
Fax: +886 2 26100589
Email: valve@mail.mold.net.tw
Web: www.valve.com.tw

Zipson Steel Industrial Co Ltd
9 Chen Kong 2 Street
Minhsiung Industrial Park
Chia-Yi
Taiwan
Tel: +886 2 29573466
Fax: +886 2 29573472
Email: loogoon@ms22.hinet.net
Web: www.taiwanvalve.com.tw

TURKEY (TR)

Asteknik Valve
Mustafa Kemal Pasa
Organize Sanayi Bölgesi
TR-16500 M K Pasa / Bursa
Turkey
Tel: +90 224 632 5260
Fax: +90 224 632 5267
Email: asteknik@asteknikvana.com
Web: www.asteknikvana.com

Ayvaz Endüstriyei Mamuller Sanayi ve Ticaret AS
Atatürk Sanayi Bölgesi
Istasyon Mah Mustafa Inan Cad No 36
TR-34860 Hadimköy-Istanbul
Turkey
Tel: +90 212 771 0145
Fax: +90 212 771 2560
Email: info@ayvaz.com
Web: www.ayvaz.com

Burçelik Valve Company
Ata Center
Ahi Evren Caddesi No 1 K
TR-380870 Maslak-Istanbul
Turkey
Tel: +90 212 286 6640
Fax: +90 212 286 6644
Email: burcelik@burcelik.com.tr
Web: www.burcelik.com.tr

Duyar Vana Makina Sanayi ve Ticaret AS
Ugur Mumcu Mah 2347 Sok No 7
TR-34120 G O Pas a Istanbul
Turkey
Tel: +90 212 668 1808
Fax: +90 212 594 7342
Email: info@duyarvalve.com
Web: www.duyarvalve.com

Gedik Döküm ve Vana Sanayi ve Ticaret AS
Yayalar cad No 78
TR-81520 Pendik-Istanbul
Turkey
Tel: +90 216 307 1262
Fax: +90 216 307 2868
Email: dokum@gedikdokum.com
Web: www.gedikdokum.com

InterValf
Yali Yolu Sokak 56, Kat 1
Bostanci
TR-34740 Istanbul
Turkey
Tel: +90 216 410 6900
Fax: +90 216 410 7736
Email: info@intervalf.com
Web: www.intervalf.com

UNITED ARAB EMIRATES

Dresser Flow Solutions - Middle East Operations
P O Box 61302, Roundabout 8
Units JA01 & JA02, Jebel Ali Free Zone
Dubai
United Arab Emirates
Tel: +971 4 883 8752
Fax: +971 4 8838 038
Web: www.masoneilan.com

Emerson FZE
PO Box 17033
Jebel Ali Free Zone
Dubai
United Arab Emirates
Tel: +971 4 883 5235
Fax: +971 4 883 5312
Web: www.emersonproces.com/fisher

UNITED KINGDOM (UK)

A K Müller (UK) Ltd
Terminus Road
Chichester
West Sussex
PO19 8DW
United Kingdom
Tel: +44 1243 523145
Fax: +44 1243 523149
Email: valves@akmuller.co.uk
Web: www.akmuller.co.uk

Abacus Valves Ltd
Unit 10 Omega Park
Wilsom Road
Alton
Hampshire
GU34 2YU
United Kingdom
Tel: +44 1420 544121
Fax: +44 1420 543032
Email: sales@abacus-valves.co.uk
Web: www.abacusvalves.com

Adanac Valve Specialities Ltd
14 Windmill Avenue
Woolpit Business Park
Woolpit
Suffolk
IP30 9UP
United Kingdom
Tel: +44 1359 240404
Fax: +44 1359 240406
Email: info@adanac.co.uk
Web: www.adanac.co.uk

Advanced Component Technology Ltd
Unit 4
Ryefield Way
Silsden
Yorkshire
BD20 0EF
United Kingdom
Tel: +44 8707 576664
Fax: +44 8707 576665
Email: sales@adcomtec.co.uk
Web: www.adcomtec.co.uk

Advanced Valve Technologies
Thanet Reach, Millennium Way
Broadstairs
Kent
CT10 2QQ
United Kingdom
Tel: +44 1843 600000
Fax: +44 1843 600333
Email: avt@advalve.com
Web: www.advalve.com

Aeon Pipe Systems Ltd
Belasis Business Centre
Coxwold Way
Belasis Hall Technology Park
Billingham
Teeside
TS23 4EA
United Kingdom
Tel: +44 1642 345690
Fax: +44 1642 345691
Email: group.communications@aeon-online.com
Web: www.aeon-online.com

Alco Valves (UK) Ltd
Mission Works
Birds Royd Lane
Brighouse
West Yorkshire
HD6 1LQ
United Kingdom
Tel: +44 1484 710511
Fax: +44 1484 713009
Email: uk@alco-valves.com
Web: www.alco-valves.com

Anson Ltd
Team Valley Trading Estate
Gateshead
Tyne & Wear
NE11 0JW
United Kingdom
Tel: +44 191 482 0022
Fax: +44 191 487 8835
Email: sales@anson.co.uk
Web: www.anson.co.uk

Auld Valves Ltd
Cowlairs Industrial Estate
Finlas Street
Glasgow
G22 5DQ
United Kingdom
Tel: +44 141 557 0515
Fax: +44 141 558 1059
Email: sales@auldvalves.com
Web: www.auldvalves.com

BEL Valves
St Peters
Newcastle-Upon-Tyne
NE6 1BS
United Kingdom
Tel: +44 191 265 9091
Fax: +44 191 276 3244
Email: eric.gardiner@belvalves.co.uk
Web: www.belvalves.co.uk

Bestobell Valves Ltd
President Park
President Way
Sheffield
S4 7UR
United Kingdom
Tel: +44 114 224 0000
Fax: +44 114 278 4974
Email: sales@bestobellvalves.com
Web: www.bestobellvalves.com

Bifold Fluidpower Ltd
Greenside Way
Middleton
Manchester
M24 1SW
United Kingdom
Tel: +44 161 345 4777
Fax: +44 161 345 4780
Email: sales@bifold-fluidpower.co.uk
Web: www.bifold-fluidpower.co.uk

BiS Valves Ltd
Unit 17- 22 Kingfisher Park
West Moors
Wimborne
Dorset
BH21 6US
United Kingdom
Tel: +44 1202 896322
Fax: +44 1202 896718
Email: info@bisvalves.co.uk
Web: www.bisvalves.co.uk

Blackhall Engineering Ltd
Bradford Road
Brighouse
West Yorkshire
HD6 4DJ
United Kingdom
Tel: +44 1484 713717
Fax: +44 1484 400155
Email: sales@blackhall.co.uk
Web: www.blackhall.co.uk

Bray Controls (UK)
16-18 Fountain Crescent
Inchinnan Business Park
Inchinnan
Renfrewshire
PA4 9RE
United Kingdom
Tel: +44 141 812 5199
Fax: +44 141 812 6199
Email: mark.stevenson@bray.com
Web: www.bray.com

Brian Donkin Valves Ltd
Derby Road
Chesterfield
Derbyshire
S40 2EB
United Kingdom
Tel: +44 1246 500029
Fax: +44 1246 297455
Email: enquiries@donkinvalves.com
Web: www.donkinvalves.com

Broady Flow Control Ltd
English Street
Kingston Upon Hull
Humberside
HU3 2DU
United Kingdom
Tel: +44 1482 619600
Fax: +44 1482 619700
Email: paul.robinson@broady.co.uk
Web: www.broady.co.uk

Brooksbank Valves Ltd
Sackville Street
Skipton
Yorkshire
BD23 2PS
United Kingdom

Tel: +44 1756 792346
Fax: +44 1756 792347
Email: sales@brooksbank.co.uk
Web: www.brooksbank.co.uk

BS & B Safety Systems (UK) Ltd
Adamson House, Towers Business Park
Wilmslow Road
Manchester
M20 2YY
United Kingdom
Tel: +44 161 955 4202
Fax: +44 161 955 4282
Email: sales@bsb-systems.co.uk
Web: www.bsb.ie

Bürkert Fluid Control Systems
Brimscombe Port Business Park
Brimscombe
Stroud
GL5 2QF
United Kingdom
Tel: +44 1453 731353
Fax: +44 1453 731343
Email: sales.uk@burkert.com
Web: www.burkert.co.uk

Callow Engineering Ltd
Carlton Works, Lansdown Road
Monton, Eccles
Manchester
M30 9JP
United Kingdom
Tel: +44 161 787 7170
Fax: +44 161 707 5344
Email: enquires@jfletcher.co.uk
Web: www.jfletcher.co.uk

Cameron's Valves & Measurement Group
For most current contact information go to www.c-a-m.com

Colson Industries Ltd
Park Road
Elland
HX5 9HN
United Kingdom
Tel: +44 1422 377999
Fax: +44 1422 377333
Email: sales@colson.co.uk
Web: www.colson.co.uk

Comid Engineering Ltd
Townfield Works
Greenacres Road
Oldham
OL4 2AB
United Kingdom
Tel: +44 161 6249592
Fax: +44 161 6271620
Email: sales@comid.co.uk
Web: www.comid.co.uk

Conflow Ltd
President Park
President Way
Sheffield
S4 7UR
United Kingdom
Tel: +44 114 224 0000
Fax: +44 114 278 4978
Email: info@conflow.com
Web: www.conflow.com

Copes-Vulcan
An SPX Process Equipment Operation
Road Two, Winsford Industrial Estate
Winsford
Cheshire
CW7 3QL
United Kingdom
Tel: +44 1606 552041
Fax: +44 1606 558275
Email: copes.vulcan@processequipment.spx.com
Web: www.spxprocessequipment.com

Crane Fluid Systems
Nacton Road
Ipswich
IP3 9QH
United Kingdom
Tel: +44 1473 277300
Fax: +44 1473 277301
Email: ghatton@crane-ltd.co.uk
Web: www.cranefs.com

Crane Process Flow Technologies
Grange Road
Cwmbran
Gwent
NP44 3XX
United Kingdom
Tel: +44 1633 486666
Fax: +44 1633 486777
Email: peter.quinn@craneflow.com
Web: www.saundersvalves.com

Curtiss-Wright Flow Control UK Ltd
Napoleon House
Gore Cross Business Park
Corbin Way
Bridport
Dorset
DT6 3UX
United Kingdom
Tel: +44 1308 422256
Fax: +44 1308 427760
Email: salessp@curtisswright.com
Web: www.solentpratt.com

Davis Pneumatic Systems Ltd
Huxley Close
Newnham Industrial Estate
Plymouth
Devon
PL7 4BQ
United Kingdom
Tel: +44 1752 336421
Fax: +44 1752 345828
Email: sales@davispneumatic.co.uk
Web: www.davispneumatic.co.uk

Delta Fluid Products Ltd
Delta Road
St Helens
WA9 2ED
United Kingdom
Tel: +44 1744 458615
Fax: +44 1744 453675
Email: enquiry@deltafluidproducts.com
Web: www.deltafluidproducts.co.uk

DI UK Ltd
Unit 4, Suite 1.1, Nobel House
Grand Union Office Park, Packet Boat Lane
Uxbridge
Middlesex
UB8 2GH
United Kingdom
Tel: +44 1895 454 900
Fax: +44 1895 454 919
Web: www.masoneilan.com

DMI Young & Cunningham Ltd
West Chirton Industrial Estate
Gloucester Road
North Shields
Tyne & Wear
NE29 8RQ
United Kingdom
Tel: +44 191 2704690
Fax: +44 191 2704691
Email: ken.hurst@yandc.co.uk
Web: www.dmiuk.co.uk

DPL
Dairy Pipe Lines
Shire Hill
Saffron Walden
Essex
CB11 3AX
United Kingdom
Tel: +44 1799 522885
Fax: +44 1799 525916
Email: dairypipelines@dpluk.co.uk
Web: www.dpluk.co.uk

Dynafluid Ltd/Gresswell Valves
Units D1 &D2
Halesfield 21
Telford
TF7 4NX
United Kingdom
Tel: +44 1952 580946
Fax: +44 1952 582546
Email: p.carlisle@dynafluid.com
Web: www.dynafluid.com

Dynamic Controls Ltd
Union Street
Royton
Oldham
OL2 5JD
United Kingdom
Tel: +44 161 633 3933
Fax: +44 161 633 4113
Email: sales@dynamiccontrols.uk.com
Web: www.dynamiccontrols.co.uk

EIM Controls UK Ltd
6-7 Galaxy House
New Greenham Park
Newbury
Berkshire
RG19 6HW
United Kingdom
Tel: +44 1635 817315
Fax: +44 1635 817460
Email: sales.international@eim-co.com
Web: www.eim-co.com

Elfab Ltd
Alder Road
Northshields
Tyne & Wear
NE29 8SD
United Kingdom
Tel: +44 191 293 1234
Fax: +44 191 293 1200
Email: sales@elfab.com
Web: www.elfab.com

ESME Valves Ltd
Coronation Road
Basingstoke
Hampshire
RG21 2ET
United Kingdom
Tel: +44 1256 464646
Fax: +44 1256 841597
Email: info@esme-valves.co.uk
Web: www.esme-valves.co.uk

Flo-Dyne Ltd
Asheridge Business Centre
Asheridge Road
Chesham
Buckinghamshire
HP5 2PT
United Kingdom
Tel: +44 1494 770088
Fax: +44 1494 770099
Email: enquiries@flo-dyne.net
Web: www.flo-dyne.net

Flowguard Ltd
Watford Bridge
New Mills
Stockport
SK22 4HJ
United Kingdom
Tel: +44 1663 745976
Fax: +44 1663 742788
Email: swheatley@flowguard.com
Web: www.flowguard.demon.co.uk

Flowserve Flow Control (UK) Ltd
Burrell Road
Haywards Heath
Sussex
RH16 1TL
United Kingdom
Tel: +44 1444 314400
Fax: +44 1444 314401
Email: ukfcinfo@flowserve.com
Web: www.flowserve.com

Flowtech Inc Ltd
53 Martin Avenue
Fareham
Hampshire
PO14 2RZ
United Kingdom
Tel: +44 1329 662656
Fax: +44 1329 661146
Email: mail@flowtech-uk.com
Web: www.flowtech-uk.com

Forac Ltd
Unit 9, Riverbank Business Centre
Old Shoreham Road
Shoreham-by-sea
West Sussex
BN43 5FL
United Kingdom
Tel: +44 1273 467100
Fax: +44 1273 467101
Email: sales@forac.co.uk
Web: www.forac.co.uk

FST Northvale Korting Ltd
Uxbridge Road
Leicester
LE4 7ST
United Kingdom
Tel: +44 116 2665911
Fax: +44 116 2610050
Email: sales@northvalekorting.co.uk
Web: www.northvalekorting.co.uk

GA Valves Sales Ltd
Birds Royd Lane
PO Box 5
Brighouse
West Yorkshire
HD6 3UD
United Kingdom
Tel: +44 1848 711985
Fax: +44 1848 719848
Email: sales@gavalves.co.uk
Web: www.gavalves.co.uk

Goodwin International Ltd
Goodwin House
Leek Road
Hanley
Stoke-on-Trent
ST1 3NR
United Kingdom
Tel: +44 1782 220000
Fax: +44 1782 208060
Email: goodwinplc@goodwin.co.uk
Web: www.goodwin.co.uk

Greene Tweed & Co Ltd
Ruddington Fields
Ruddington
Nottinghamshire
NG11 6JS
United Kingdom
Tel: +44 115 9315777
Fax: +44 115 9315888
Email: mktng@gtweed.com
Web: www.gtweed.com

H H Valves Ltd
Unit 1 Eastworks
Lamberhead Industrial Estate
Wigan
WN5 8EG
United Kingdom
Tel: +44 1942 218111
Fax: +44 1942 224800
Email: gd@hhvalves.com
Web: www.hhvalves.com

Hale Hamilton (Valves) Ltd
Cowley Road
Uxbridge
Middlesex
UB8 2AF
United Kingdom
Tel: +44 1895 236525
Fax: +44 1895 231407
Email: sales@halehamilton.com
Web: www.halehamilton.com

Henry Technologies Ltd
Mossland Road
Hillington Park
Glasgow
G52 4XZ
United Kingdom
Tel: +44 141 882 4621
Fax: +44 141 810 9199
Email: pallen@henrytech.co.uk
Web: www.henrytech.co.uk

Hobbs Valve Ltd
Unit L, Trecenydd Business Park
Caerphilly
South Wales
CF83 2RZ
United Kingdom
Tel: +44 29 2086 1099
Fax: '+44 29 2086 1073
Email: info@hobbsvalve.co.uk
Web: www.hobbsvalve.co.uk

Hoerbiger Holding UK Ltd
1649 Pershore Road
Stirchley
Birmingham
B30 3DR
United Kingdom
Tel: +44 121 433 3636
Fax: +44 121 4597794
Email: info@hoerbiger.com
Web: www.hoerbiger.com

IMC UK Ltd
Patrick Gregory Road
Wolverhampton
WV11 3DZ
United Kingdom
Tel: +44 1902 305678
Fax: +44 1902 305678
Email: a.robertson@imc-uk.com
Web: www.alconsolenoids.com

Instrumentation Products Division, Parker Hannifin Ltd
Riverside Road
Barnstaple
EX31 1NP
United Kingdom
Tel: +44 1271 313131
Fax: +44 1271 373636
Email: ipd@parker.com
Web: www.parker.com

Invensys APV
23 Gatwick Road
Crawley
West Sussex
RH10 9JB
United Kingdom
Tel: +44 1293 527777
Fax: +44 1293 552640
Web: www.apv.com

JL Controls Ltd
The Tanneries
Titchfield
Hampshire
PO14 4AR
United Kingdom
Tel: +44 1329 844406
Fax: +44 1329 844408
Email: mail@e-valves.co.uk
Web: www.flowtech-uk.com

K Controls Ltd
Process House, Stone close
Horton Road
West Drayton
Middlesex
UB7 8JU
United Kingdom
Tel: +44 1895 449601
Fax: +44 1895 448586
Email: sales@k-controls.co.uk
Web: www.k-controls.co.uk

Kemutec Powder Technologies Ltd
Springwood Way

Macclesfield
Cheshire
SK10 2ND
United Kingdom
Tel: +44 1625 412000
Fax: +44 1625 412001
Email: sales@kemutec.com
Web: www.kemutec.com

Kinetrol Ltd
Trading Estate
Farnham
Surrey
GU9 9NU
United Kingdom
Tel: +44 1252 733838
Fax: +44 1252 713042
Email: sales@kinetrol.com
Web: www.kinetrol.com

Klinger Ltd
The Klinger Building, Wharfedale Road
Euroway Trading Estate
Bradford
BD4 6SG
United Kingdom
Tel: +44 1274 688222
Fax: +44 1274 688549
Email: enquiries@klingeruk.co.uk
Web: www.klingeruk.co.uk

Koso Kent Introl Ltd
Armytage Road Industrial Estate
Brighouse
West Yorkshire
HD6 1QF
United Kingdom
Tel: +44 1484 710311
Fax: +44 1484 407407
Email: control.valves@kentintrol.com
Web: www.kentintrol.com

Leeds Valve Company Ltd
Town End
Gildersome
Leeds
LS27 7HE
United Kingdom
Tel: +44 113 252 5051
Fax: +44 113 252 5081
Email: sales@leedsvalvenet
Web: www.leedsvalve.com

LG Ball Valves Ltd
Units 5 & 6, Westgate Trading Estate
Aldridge
Walsall
WS9 8EX
United Kingdom
Tel: +44 1922 459999
Fax: +44 1922 458688
Email: sales@lgball-valves.co.uk
Web: www.lgball-valves.co.uk

M V Fluids Handling
Marven House
1 Field Road
Reading
RG1 6AP
United Kingdom
Tel: +44 118 950 3707
Fax: +44 118 939 3222
Email: info@mvfh.co.uk
Web: www.mvfh.co.uk

Midland-ACS
Patrick Gregory Road
Wolverhampton
WV11 3DZ
United Kingdom
Tel: +44 1902 305678
Fax: +44 1902 305676
Email: sales@imc-uk.com
Web: www.midland-acs.com

Oliver Valves Ltd
Parkgate
Knutsford
Cheshire
WA16 8DX
United Kingdom
Tel: +44 1565 632636
Fax: +44 1565 654089
Email: sales@valves.co.uk
Web: www.valves.co.uk

Peter Smith Valve Co Ltd
Occupation Road
Cinderhill Road
Bulwell
Nottingham
NG6 8RX
United Kingdom
Tel: +44 115 927 2831
Fax: +44 115 977 0233
Email: petersmithvalve@hotmail.co.uk
Web: www.petersmithvalve.co.uk

Posi-flate
14 Carters Lane
Kiln Farm
Milton Keynes
MK11 3ER
United Kingdom
Tel: +44 1908 564455
Fax: +44 1908 564615
Email: sales@posiflate.co.uk
Web: www.posiflate.com

Pyropress Engineering Company Ltd
Bell Close, Newnham Industrial Estate
Plympton
Plymouth
PL7 4JH
United Kingdom
Tel: +44 1752 339866
Fax: +44 1752 336681
Email: info@pyropress.com
Web: www.pyropress.com

Remote Control Ltd
Unit 40, Trent Valley Works
Station Road
Rugeley
WS15 3HB
United Kingdom
Tel: +44 1889 576888
Fax: +44 1889 577676
Email: info@remotecontrol.co.uk
Web: www.remotecontrol.co.uk

RGS Electro-Pneumatics Ltd
West End Business Park, Blackburn Road
Oswaldtwistle
Accrington
Lancashire
BB5 4WZ
United Kingdom
Tel: +44 1254 872277
Fax: +44 1254 390133
Email: sales@rgs-e-p.co.uk
Web: www.rgs-e-p.co.uk

Rotolok Ltd UK
1 Millennium Place
Tiverton Business Park
Tiverton
Devon
EX16 6SB
United Kingdom
Tel: +44 1884 232232
Fax: +44 1884 232200
Email: sales@rotolok.co.uk
Web: www.rotolok.co.uk

Rotork Controls Ltd
Brassmill Lane
Bath
BA1 3JQ
United Kingdom
Tel: +44 1225 733200
Fax: +44 1225 333467
Email: sales@rotork.co.uk
Web: www.rotork.com

Sabre Instrument Valves Ltd
Golf Road
Hale
Altrincham
Cheshire
WA15 8AH
United Kingdom
Tel: +44 161 925 4000
Fax: +44 161 925 4001
Email: valves@sabreuk.com
Web: www.sabre-valves.com

Safety Systems UK Ltd
Sharp Street
Worsley
Manchester
M28 3NA
United Kingdom
Tel: +44 161 790 7741
Fax: +44 161 703 8451
Web: www.safetysystemsuk.com

Saint-Gobain Pipelines
Lows Lane
Stanton-by-Dale
Ilkerston
Derbyshire
DE7 4QU
United Kingdom
Tel: +44 115 930 5000
Fax: +44 115 944 9513
Web: www.saint-gobain-pipelines.co.uk

Schubert & Salzer UK Ltd
Unit 2a Larchwood House
Orchard Street
Redditch
Worcestershire
B98 7DP
United Kingdom
Tel: +44 1527 62795
Fax: +44 1527 64343
Email: info.ukschubert-salzer.com
Web: www.schubert-salzer.com

Score (Europe) Ltd
Glenugie Engineering Works
Peterhead
Aberdeenshire

AB42 0YX
United Kingdom
Tel: +44 1779 480000
Fax: +44 1779 481100
Email: customersupport@score-group.com
Web: www.score-europe.org

Seetru Ltd
Albion Dockside Works
Bristol
BS1 6UT
United Kingdom
Tel: +44 117 927 9204
Fax: +44 117 929 8193
Email: sales@seetru.com
Web: www.seetru.co.uk

Severn Glocon Ltd
St Luke Street
Southgate Street
Gloucester
GL1 5RE
United Kingdom
Tel: +44 1452 318900
Fax: +44 1452 318901
Email: sales@severnglocon.co.uk
Web: www.severnglocon.com

Severn Unival Ltd
Milford Buildings
Milford Street
Huddersfield
HD1 3DY
United Kingdom
Tel: +44 1484 518 080
Fax: +44 1484 518 087
Email: helenhood@severnunival.co.uk
Web: www.severnglocon.com

Shaw Valves
Dowker Street
Milnsbridge
Huddersfield
HD3 4JX
United Kingdom
Tel: +44 1484 651177
Fax: +44 1484 645854
Email: sales@taylorshaw.co.uk
Web: www.taylorvalves.co.uk

Shipham Valves Ltd
Hawthorn Avenue
Hull
HU3 5JX
United Kingdom
Tel: +44 1482 323163
Fax: +44 1482 224057
Email: sales@shipham-valves.com
Web: www.shipham-valves.com

Smart Valve & Controls
Uxbridge Road
Leicester
LE4 7ST
United Kingdom
Tel: +44 116 268 8120
Fax: +44 116 261 0050
Email: sales@smartvalves.co.uk
Web: www.smartvalves.co.uk

Spirax-Sarco Ltd
Charlton House
Cheltenham
GL53 8ER
United Kingdom
Tel: +44 1242 521361
Fax: +44 1242 573342
Email: info@spiraxsarco.com
Web: www.spiraxsarco.com

Taylor Shaw
Dowker Street
Milnsbridge
Huddersfield
HD3 4JX
United Kingdom
Tel: +44 1484 651177
Fax: '+44 1484 645854
Email: sales@taylorshaw.co.uk
Web: www.taylorshaw.co.uk

TC Fluid Control Ltd
Valves & Steam Division
Broadgate Broadway Business Park
Chadderton
Oldham
OL9 9XA
United Kingdom
Tel: +44 161 684 7488
Fax: +44 161 684 7487
Email: energy@tc-fluidcontrol.com
Web: www.tc-fluidcontrol.com

Thompson Valves - Truflo International
17 Balena Close
Creekmoor
Poole
Dorset
BH17 7EF
United Kingdom
Tel: +44 1202 697521
Fax: +44 1202 605385
Web: www.truflointernational.com

Tomoe Valve Ltd
Clearwater Road
Queensway Meadows Industrial Estate
Newport
Gwent
NP19 4ST
United Kingdom
Tel: +44 1633 636800
Fax: +44 1633 636801
Email: sales@tomoetritec.co.uk
Web: www.tomoeeurope.com

Transmark Fcx
Heaton House, Riverside Drive
Hunsworth Lane
Bradford
BD19 4DH
United Kingdom
Tel: +44 1274 700000
Fax: +44 1274 700111
Email: enquiries@transmark-fcx.com
Web: www.transmark-fcx.com

TUV NEL Ltd
East Kilbride
Glasgow
G75 0QF
United Kingdom
Tel: +44 1355 220222
Fax: +44 1355 272999
Email: info@tuvnel.com
Web: www.tuvnel.com

Tyco Flow Control
For EMEA see Tyco Flow Control, The Netherlands
Web: www.tycoflowcontrol.com

Vector International
Baglan Industrial Park
Port Talbot
West Glamorgan
SA12 7BY
United Kingdom
Tel: +44 1639 822 555
Fax: +44 1639 822 623
Web: www.vectorint.com

Velan Valves Ltd
Unit 1, Lakeside Business Park
Pinfold Road
Thurmaston
Leicester
LE4 8AS
United Kingdom
Tel: +44 116 269 5172
Fax: +44 113 269 3695
Email: sales@velan.com
Web: www.velan.co.uk

Warren-Morrison Valves Ltd
Peel Street
Northam
Southampton
SO14 5QS
United Kingdom
Tel: +44 2380 223752
Fax: +44 2380 223110
Email: sales@wmvalves.com
Web: www.wmvalves.com

Weir Valves & Controls UK Ltd
Britannia Works
Birkby
Huddersfield
West Yorkshire
HD2 2UR
United Kingdom
Tel: +44 1484 820820
Fax: +44 1484 820484
Email: info@weirvalve.com
Web: www.weirvalve.com

Xomox Ltd
Riverside Road
Pottington Industrial Estate
Barnstaple
Devon
EX31 1XL
United Kingdom
Tel: +44 1271 31111
Fax: +44 1271 322022
Email: uk@xomox.de
Web: www.xomox.de

YPS Valves Ltd
Grangefield Industrial Estate
Pudsey
LS28 6QW
United Kingdom
Tel: +44 113 256 7725
Fax: 44 113 255 1275
Email: info@yps-valves.co.uk
Web: www.yps-valves.co.uk

UNITED STATES (US)

American Valve Inc
PO Box 35229
Greensboro
NC 27425
USA
Tel: +1 336 668 0554
Fax: +1 336 668 4070
Email: sales@americanvalve.com
Web: www.americanvalve.com

AMRI Inc
2045 Silber Road
Houston
TX 77055
USA
Tel: +1 713 682 0000
Fax: +1 713 682 0080
Email: info@amrivalves.com
Web: www.amrivalves.com

AOP Industries Inc
2101 South Broadway
Moore
OK 73160
USA
Tel: +1 405 912 4446
Fax: +1 405 912 4440
Email: sales@aopind.com
Web: www.aopind.com

Asahi/America Inc
35 Green Street
PO Box 653
Malden
MA 02148-0134
USA
Tel: +1 781 321 5409
Fax: +1 781 321 4421
Email: nsmith@asahi-america.com
Web: www.asahi-america.com

ASCO
50-60 Hanover Road
Florham Park
NJ 07932
USA
Tel: +1 973 966 2000
Fax: +1 973 966 2448
Email: info-valve@asco.com
Web: www.ascovalve.com

AUMA Actuators Inc
100 Southpointe Boulevard
Canonsburg
PA 15317
USA
Tel: +1 724 743 2862
Fax: +1 724 743 4711
Web: www.auma-usa.com

Automation Technology Inc
PO Box 310
Industry
TX 78944
USA
Tel: +1 979 357 2570
Fax: +1 979 357 2571
Web: www.atiactuators.com

Balon Corporation
3245 S Hattie
Oklahoma City
OK 73129
USA
Tel: +1 405 677 3321
Fax: +1 405 677 3917
Email: sales@balon.com
Web: www.balon.com

Bernard Controls Inc
15740 Park Row
Suite 100
Houston
TX 77084
USA
Tel: +1 281 578 6666
Fax: +1 281 578 2797
Email: bsales@bernardcontrols.com
Web: www.bernardcontrols.com

Bray Controls USA
13333 Westland East Boulevard
Houston
Texas 77041
USA
Tel: +1 281 894 5454
Fax: +1 281 894 9499
Email: steve.drollinger@bray.com
Web: www.bray.com

Cameron's Valves & Measurement Group
3250 Briarpark Drive
Houston
TX 77042
USA
Tel: +1 281 261 3754
Fax: +1 281 261 3754
Web: www.c-a-m.com

Cashco Inc
607 West 15th Street
Ellsworth
KS 67439
USA
Tel: +1 785 472 4461
Fax: +1 785 472 3539
Email: sales@cashco.com
Web: www.cashco.com

CCI (Control Components Inc)
22591 Avenida Empresa
Rancho Santa Margarita
CA 92688
USA
Tel: +1 949 858 1877
Fax: +1 949 858 1878
Web: www.ccivalve.com

Check-All Valve Mfg Co
1800 Fuller Road
West Des Moines
IA 50265
USA
Tel: +1 515 224 2301
Fax: +1 515 224 2326
Email: sales@checkall.com
Web: www.checkall.com

Circle Seal Controls Inc
2301 Wardlow Circle
Corona
CA 92880-2881
USA
Tel: +1 909 270 6200
Fax: +1 909 270 6201
Email: sales@circlesealcontrols.com
Web: www.circle-seal.com

CLA-VAL Automatic Control Valves
1701 Placentia Avenue
Costa Mesa
CA 92627-4475
USA
Tel: +1 949 722 4800
Fax: +1 949 548 5441
Email: claval@cla-val.com
Web: www.cla-val.com

Conbraco Industries Inc
PO Box 247
Matthews
NC 28106
USA
Tel: +1 704 841 6000
Fax: +1 704 841 6021
Email: conbraco@conbraco.com
Web: www.conbraco.com

Conflow Inc
270 Meadowlands Boulevard
Washington
PA 15301
USA
Tel: +1 724 746 0200
Fax: +1 724 746 0940
Email: sales@conflowusa.com
Web: www.conflow.com

Conval Inc
265 Field Road
Somers
CT 06071
USA
Tel: +1 860 763 0761
Fax: +1 860 763 3557
Email: sales@conval.com
Web: www.conval.com

CPC-Cryolab
12501 Telecom Drive
Tampa
FL 33635
USA
Tel: +1 813 978 1000
Fax: +1 813 977 3329
Email: sales@cpc-cryolab.com
Web: www.cpc-cryolab.com

Crane Nuclear Inc
2825 Cobb International Boulevard
Kenneshaw
GA 30152
USA
Tel: +1 770 429 4720
Fax: +1 770 429 4753
Email: tburky@cranevs.com
Web: www.cranenuclear.com

DeltaValve USA
857 West, South Jordan Parkway
Suite 100
South Jordan
UT 84095
USA
Tel: +1 801 984 1000
Fax: +1 801 984 1001
Email: customercare@deltavalve.com
Web: www.deltavalve.com

DeZURIK Water Controls
250 Riverside Avenue North
Sartell
MN 56377-1743
USA
Tel: +1 320 259 2000
Fax: +1 320 259 2227
Email: info@dezurikwater.com
Web: www.dezurikwater.com

DFT Durabla Fluid Technology Inc
140 Sheree Boulevard
Exton
PA 19341-0566
USA
Tel: +1 610 363 8903
Fax: +1 610 524 9242
Email: dft@dft-valves.com
Web: www.dft-valves.com

Dresser Inc
15455 Dallas Parkway
Addison
TX 75001
USA
Tel: +1 972 361 9800
Fax: +1 972 361 9903
Web: www.dresser.com

Dresser–Masoneilan
85 Bodwell Street
Avon
MA 02322-1190
USA
Tel: +1 508 586 4600
Fax: +1 508 941 5497
Email: marketing@masoneilan.com
Web: www.masoneilan.com

DynaQuip Controls
10 Harris Industrial Park
St Clair
MO 63077
USA
Tel: +1 800 545 3636
Fax: +1 636 629 5528
Web: www.dynaquip.com

DynaTorque Inc
2076 Northwoods Drive
Muskegon
MI 49442
USA
Tel: +1 231 788 7025
Fax: +1 231 788 7030
Email: info@dynatorque.com
Web: www.dynatorque.com

EIM Controls Inc
13840 Pike Road
Missouri City
TX 77489
USA
Tel: +1 281 499 1561
Fax: +1 281 499 8445
Email: sales.usa@eim-co.com
Web: www.eim-co.com

Emerson Process Management
205 South Center Street
Marshalltown
IA 50158
USA
Tel: +1 515 754 3011
Fax: +1 515 754 2830
Web: www.emersonprocess.com/fisher

Enertech
2950 Birch Street
Brea
CA 92621
USA
Tel: +1 714 528 2301
Fax: +1 714 528 0128
Email: enertech@curtisswright.com
Web: www.enertechnuclear.com

Everest Valve Company
PO Box 24295
Houston
TX 77229
USA
Tel: +1 713 923 8696
Fax: +1 713 923 6270
Email: info@everestvalveusa.com
Web: www.everestvalveusa.com

Everlasting Valve Company Inc
108 Somogyi Court
South Plainfield
NJ 07080
USA
Tel: +1 908 769 0700
Fax: +1 908 769 8697
Email: info@everlastingvalveusa.com
Web: www.everlastingvalveco.com

Farris Engineering
Curtiss-Wright Corporation
10195 Brecksville Road
Brecksville
OH 44141
USA
Tel: +1 440 838 7690
Fax: +1 440 838 7699
Email: farris@farris.curtisswright.com
Web: www.cwfc.com

FC Kingston
Storm Manufacturing Group Inc
23201 Normandie Avenue
Torrance
CA 90501-5050
USA
Tel: +1 310 326 8287
Fax: +1 310 326 8310
Email: sales@fckingston.com
Web: www.kingstonvalves.com

FEBCO – Division of Watts Water Technologies Inc
3816 South Willow Avenue
Fresno
CA 93725
USA
Tel: +1 559 441 5300
Fax: +1 559 441 5301
Web: www.febcoonline.com

Flomatic Corporation
15 Pruyn's Island Drive
Glens Falls
NY 12801
USA
Tel: +1 518 761 9797
Fax: +1 518 761 9798
Email: flomatic@flomatic.com
Web: www.flomatic.com

Flow-Quip Inc
7440 E 46th PL
Tulsa
OK 74145-6305
USA
Tel: +1 918 663 3313
Fax: +1 918 663 9034
Email: sales@flowquip.com
Web: www.flowquip.com

Flowsafe Inc
S3865 Taylor Road
Orchard Park
NY 14127
USA
Tel: +1 716 667 2585
Fax: +1 716 667 2580
Email: info@flowsafe.com
Web: www.flowsafe.com

Flowserve Flow Control
1350 North Mountain Springs Parkway
Springville
UT 84663-0903
USA
Tel: +1 801 489 8611
Fax: +1 801 489 2681
Email: ukfcinfo@flowserve.com
Web: www.flowserve.com

Flowserve Flow Control
1978 Foreman Drive
Cookeville
TN 38501
USA
Tel: +1 931 432 4021
Fax: +1 931 432 5518
Email: ukfcinfo@flowserve.com
Web: www.flowserve.com

Flowserve Flow Control
1511 Jefferson Street
Sulphur Springs
TX 75482
USA
Tel: +1 903 885 4691
Fax: +1 903 439 3411
Email: ukfcinfo@flowserve.com
Web: www.flowserve.com

GA Industries Inc
9025 Marshall Road
Cranberry Township
PA 16066
USA
Tel: +1 724 776 1020
Fax: +1 724 776 1254
Email: ga@gaindustries.com
Web: www.gaindustries.com

Gilmore Valve Company
1231 Lumpkin Road
Houston
TX 77043
USA
Tel: +1 713 468 8778
Fax: +1 713 468 8791
Email: sales@gilmorevalve.com
Web: www.gilmorevalve.com

Groth Corporation
13650 N Promenade Boulevard
Stafford

TX 77477
USA
Tel: +1 281 295 6800
Fax: +1 281 295 6999
Email: sales@grothcorp.com
Web: www.grothcorp.com

H O Trerice Company
12950 W Eight Mile Road
Oak Park
MI 48237
USA
Tel: +1 248 399 8000
Fax: +1 248 399 7246
Email: info@trerice.com
Web: www.trerice.com

Harold Beck & Sons Inc
2300 Terry Drive
Newtown
PA 18940
USA
Tel: +1 215 968 4600
Fax: +1 215 968 6483
Email: sales@haroldbeck.com
Web: www.haroldbeck.com

Henry Pratt Company
401 South Highland Avenue
Aurora
IL 60506-5563
USA
Tel: +1 630 844 4000
Fax: +1 630 844 4124
Email: info@henrypratt.com
Web: www.henrypratt.com

Hex, a division of Richards Industries
3170 Wasson Road
Cincinnati
OH 45209
USA
Tel: +1 513 533 5600
Fax: +1 513 871 0105
Email: hex@richardsind.com
Web: www.hexvalve.com

High Pressure Equipment Company
P O Box 8248
1222 Linden
Erie
PA 16505
USA
Tel: +1 814 838 2028
Fax: +1 814 838 6075
Email: sales@highpressure.com
Web: www.highpressure.com

Hoke Inc
405 Centura Court
PO Box 4866
Spartanburg
SC 29303
USA
Tel: +1 864 574 7966
Fax: +1 864 587 5608
Email: rtaylor@circortech.com
Web: www.hoke.com

Homestead Valve
a division of Olson Technologies, Inc
160 Walnut Street West
Allentown
PA 18102
USA
Tel: +1 610 770 1100
Fax: +1 610 770 1108
Email: sales@homesteadvalve.com
Web: www.homesteadvalve.com

Hunt Valve Company Inc
1913 East State Street
Salem
OH 44460
USA
Tel: +1 330 337 9535
Fax: +1 330 337 3754
Web: www.huntvalve.com

Hydroseal Valve Company
1500 S E 89th Street
Oklahoma City
OK 73149
USA
Tel: +1 800 654 4842
Fax: +1 405 631 5034
Email: greg.barber@circorenergy.com
Web: www.hydroseal.com

Indelac Controls Inc
6810 Powerline Drive
Florence
KY 41042
USA
Tel: +1 859 727 7890
Fax: +1 859 727 4070
Email: dvoges@indelac.com
Web: www.indelac.com

ITT Engineered Valves
33 Centerville Road
Lancaster
PA 17603
USA
Tel: +1 717 509 2200
Fax: +1 717 509 2336
Web: www.engvalves.com

ITT Pure-Flo
33 Centerville Road
Lancaster
PA 17603
USA
Tel: +1 717 509 2200
Fax: +1 717 509 2216
Web: www.ittpureflo.com

JMC Valve Inc
9009 Pinehill Lane Suite 216
Houston
TX. 77041
USA
Tel: +1 713 939 9559
Fax: +1 713 939 9560
Email: sales@jmcvalve.com
Web: www.jmcvalve.com

Johnson Controls Inc
5757 N Green Bay Avenue
Milwaukee
WI 53209
USA
Tel: +1 414 524 1200
Email: webmaster@jci.com
Web: www.jci.com

Jordan Controls Inc
5607 W Douglas Avenue
Milwaukee
WI 53218
USA
Tel: +1 414 461 9200
Fax: +1 414 461 1024
Email: info@jordancontrols.com
Web: www.jordancontrols.com

Jordan Valve, a division of Richards Industries
3170 Wasson Road
Cincinnati
OH 45209
USA
Tel: +1 513 533 5600
Fax: +1 513 871 0105
Email: info@richardsind.com
Web: www.jordanvalve.com

Kalsi Engineering Inc
745 Park Two Drive
Sugar Land
T77478-2885
USA
Tel: +1 281 240 6500
Fax: +1 281 240 0255
Email: kalsi@kalsi.com
Web: www.kalsi.com

Kerotest Manufacturing Corporation
5500 Second Avenue
Pittsburgh
PA 15207
USA
Tel: +1 412 521 5942
Fax: +1 412 521 5990
Email: kerotest@kerotest.com
Web: www.kerotest.com

KF Industries
1500 SE 89th Street
Oklahoma City
OK 73149-5249
USA
Tel: +1 405 631 1533
Fax: +1 405 631 5034
Email: kfinfo@kfvalves.com
Web: www.kfvalves.com

Kitz Corporation of America
10750 Corporate Drive
Stafford
TX 77477
USA
Tel: +1 281 491 7333
Fax: +1 281 491 9402
Email: tomh@kitz.com
Web: www.kitz.com

Leslie Controls Inc
12501 Telecom Drive
Tampa
FL 33637
USA
Tel: +1 813 978 1000
Fax: +1 813 978 7543
Email: sales@lesliecontrols.com
Web: www.lesliecontrols.com

M & J Valves
19191 Hempstead Highway
Houston
TX 77065
USA
Tel: +1 281 469 0550

Fax: +1 281 894 1332
Web: www.spxprocessequipment.com

Mallard Control
4970 Washington Boulevard
Beaumont
TX 77707
USA
Tel: +1 409 842 5392
Fax: +1 409 842 8165
Email: greg.barber@circirenergy.com
Web: www.mallardcontrol.com

Marwin Valve,
a division of Richards Industries
3170 Wasson Road
Cincinnati
OH 45209
USA
Tel: +1 513 533 5600
Fax: +1 513 533 7343
Email: marwin@richardsind.com
Web: www.marwinvalve.com

Mastergear
5466 Rockton Road
South Beloit
IL 61080
USA
Tel: +1 815 389 4004
Fax: +1 815 389 8383
Email: sales@mastergearworldwide.com
Web: www.mastergearworldwide.com

Maxon Corporation
201 East 18th Street
Muncie
IN 47302
USA
Tel: +1 765 284 3304
Fax: +1 765 286 8394
Web: www.maxoncorp.com

Metrex Valve Corporation
505 South Vermont Avenue
Glendora
CA 91741
USA
Tel: +1 800 266 4027
Fax: +1 626 335 1514
Email: info@metrexvalve.com
Web: www.metrexvalve.com

Metso Automation USA Inc
44 Bowditch Drive
Shrewsbury
MA 01545
USA
Tel: +1 508 852 0200
Fax: +1 508 852 8172
Email: kathy.martin@metso.com
Web: www.metsoautomation.com

Milwaukee Valve Company
16550 West Stratton Drive
New Berlin
WI 53151
USA
Tel: +1 262 432 2700
Fax: +1 262 432 2701
Web: www.milwaukeevalve.com

MOGAS Industries Inc
14330 East Hardy Street
Houston
TX 77039
USA
Tel: +1 281 449 0291
Fax: +1 281 590 3412
Email: mogas@mogas.com
Web: www.mogas.com

Moog Flo-Tork
1701 North Main Street
PO Box 68
Orrville
OH 44667-0068
USA
Tel: +1 330 682 0010
Fax: +1 330 683 6857
Web: www.flo-tork.com

Mueller Steam Specialty
1491 NC Highway 20 West
St Pauls
NC 28384
USA
Tel: +1 910 865 8241
Fax: +1 910 865 8245
Web: www.muellersteam.com

Newmans Valve
1300 Gazin Street
Houston
TX 77020
USA
Tel: +1 713 675 8631
Fax: +1 713 675 1589
Email: itlsales@newmansvalve.com
Web: www.newcovalves.com

NIBCO Inc
1516 Middlebury Street
Elkhart
IN 46516-4740
USA
Tel: +1 574 295 3000
Fax: +1 574 295 3307
Email: international@nibco.com
Web: www.nibco.com

Nicholson Steam Trap
150 Coldenham Road
Waldem
NY 12586-2000
USA
Tel: +1 845 778 5566
Fax: +1 845 778 7123
Email: sales@nicholsonsteamtrap.com
Web: www.nicholsonsteamtrap.com

Nil-Cor
12241 Rockhill Avenue NE
Alliance
OH 44601
USA
Tel: +1 330 823 0500
Fax: +1 330 821 6722
Email: info@nilcor.com
Web: www.nilcor.com

Nortech
16420 Parks Ten Place, Suite 580
Houston
TX 77084
USA
Tel: +1 281 578-6900
Fax: +1 281 578 6991
Email: nortech@nortechinc.com
Web: www.nortechinc.com

OCV Control Valves
7400 East 42nd Place
Tulsa
OK 74145-4744
USA
Tel: +1 918 627 1942
Fax: +1 918 622 8916
Email: sales@controlvalves.com
Web: www.controlvalves.com

Parker Hannifin Corporation
Instrumentation Products Division USA (IPD)
1005 A Cleaner Way
Huntsville
AL 36265 9681
USA
Tel: +1 256 881 2040
Fax: +1 256 881 5072
Email: ipdsales@parker.com
Web: www.parker.com

Parker, Fluid Control Division
95 Edgewood Avenue
New Britain
CT 06051
USA
Tel: +1 860 827 2300
Fax: +1 860 827 2384
Email: skinnervalve@parker.com
Web: www.parker.com

PBM Inc
1070 Sandy Hill Road
Irwin
PA 15642
USA
Tel: +1 724 863 0550
Fax: +1 724 864 9255
Email: info@pbmvalve.com
Web: www.pbmvalve.com

Polyjet Valves
3816 South Willow Avenue
Fresno
CA 93725
USA
Tel: +1 559 441 5300
Fax: +1 539 441 5301
Email: polyjet@polyjetvalves.com.
Web: www.spxvalves.com

Posi-flate
1125 Willowlake Boulevard
St Paul
MN 55110
USA
Tel: +1 651 484 5800
Fax: +1 651 484 7015
Email: info@posiflate.com
Web: www.posiflate.com

Powell Valves
2503 Spring Grove Avenue
Cincinnati
OH 45214
USA
Tel: +1 513 852 2000
Fax: +1 513 852 2997
Email: bwebster@powellvalves.com
Web: www.powellvalves.com

Process Development & Control
1075 Montour West Industrial Park
Coraopolis

PA 15108
USA
Tel: +1 724 695 3440
Fax: +1 724 695 8635
Email: sales@pdcvalve.com
Web: www.pdcvalve.com

QTRCO
13120 Theis Lane
Tomball
TX 77375
USA
Tel: +1 281 516 0277
Fax: +1 281 516 0288
Email: ed@qtrco.com
Web: www.qtrco.com

Red Valve Company
700 North Bell Avenue
Carnegie
PA 15106
USA
Tel: +1 412 279 0044
Fax: +1 412 279 7878
Email: valves@redvalve.com
Web: www.redvalve.com

Red-White Valve Corporation
20600 Regency Lane
Lake Forest
CA 92630
USA
Tel: +1 949 859 1010
Fax: +1 949 859 7200
Email: info@rwvcusa.com
Web: www.redwhitevalveusa.com

Remote Control Inc
PO Box 355
386 Dry Bridge Road
North Kingstown
RI 02852
USA
Tel: +1 401 294 1400
Fax: +1 401 294 3388
Email: info@remotecontrol.us
Web: www.rciactuators.com

Rexa Koso America Inc
4 Manley Street
West Bridgewater
MA 02379
USA
Tel: +1 508 584 1199
Fax: +1 508 584 2525
Email: ghynes@rexa.com
Web: www.kosoamerica.com

Richards Industries
3170 Wasson Road
Cincinnati
OH 45209
USA
Tel: +1 513 533 5600
Fax: +1 513 533 2590
Email: info@richardsind.com
Web: www.richardsind.com

Rockwood-Swendeman
150 Coldenham Road
PO Box 230
Walden
NY 12586-2035
USA
Tel: +1 845 778 5566
Fax: +1 845 778 7123
Email: sales@rockwoodswendeman.com
Web: www.rockwoodswendeman.com

Ross Valve Manufacturing Co Inc
6 Oakwood Avenue
PO Box 595
Troy
NY 12181
USA
Tel: +1 518 274 0961
Fax: +1 518 274 0210
Email: sales@rossvalve.com
Web: www.rossvalve.com

Rotolok Valves Inc
Industrial Ventures II
2711 Gray Fox Road
Monroe
NC 28110
USA
Tel: +1 704 282 4444
Fax: +1 704 282 4242
Email: sales@rotolok.com
Web: www.rotolok.com

Saint-Gobain Performance Plastics Corporation
7301 Orangewood Avenue
Garden Grove
CA 92841
USA
Tel: +1 714 630 5818
Fax: +1 714 688 2614
Web: www.microelectronics.saint-gobain.com

Samson Controls Inc
4111 Cedar Boulevard
TX 77520-8588
USA
Tel: +1 281 383 3677
Fax: +1 281 383 3690
Web: www.samsoncontrols.com

SMG Valves
7693 Manchester Highway
Morrison
TN 37357
USA
Tel: +1 931 668 5377
Fax: +1 931 668 5378
Email: sales@smg-global.com
Web: www.smg-global.com

Spence Engineering Co Inc
150 Coldenham Road
Walden
NY 12586
USA
Tel: +1 845 778 5566
Fax: +1 845 778 1072
Email: sales@spenceengineering.com
Web: www.spenceengineering.com

Spirax Sarco Inc
1150 Northpoint Boulevard
Blythewood
SC 29016
USA
Tel: +1 803 714 2000
Fax: +1 803 714 2222
Web: www.spiraxsarco.com

SVF Flow Controls Inc
13560 Larwin Circle
Santa Fe Springs
CA 90670
USA
Tel: +1 562 802 2255
Fax: +1 562 802 3114
Email: international@svf.net
Web: www.svf.net

Swagelok Company
31500 Aurora Road
Solon
OH 44139
USA
Tel: +1 440 349 5934
Fax: +1 440 349 5843
Web: www.swagelok.com

Target Rock
a Division of Curtiss-Wright
Flow Control Corporation
1966E Broadhollow Road
East Farmingdale
NY 11735
USA
Tel: +1 516 293 3800
Fax: +1 516 293 6144
Email: tr_sales@curtisswright.com
Web: www.curtisswright.com

Therm-Omega-Tech Inc
353 Ivyland Road
Warminster
PA 18974-2205
USA
Tel: +1 215 674 9992
Fax: +1 215 674 8594
Email: valves@thermomegatech.com
Web: www.thermomegatech.com

Tru-Tech Industries
PO Box 838
Mars
PA 10646
USA
Tel: +1 724 625 2955
Fax: +1 724 625 3553
Email: sales@tru-tech.net
Web: www.tru-tech.net

Tyco Flow Control
9700 West Gulf Bank Road
Houston
TX 77040
USA
Tel: +1 713 466 1176
Web: www.tycoflowcontrol.com

US Para Plate Corporation
2301 Wardlow Circle
Corona
CA 92880
USA
Tel: +1 951 270 6200
Fax: +1 951 270 6201
Email: sales@usparaplate.com
Web: www.usparaplate.com

Valcor Engineering Corporation
2 Lawrence Road
Springfield
NJ 07081
USA
Tel: +1 973 467 8400

Fax: +1 973 467 8382
Email: valcorscientific@valcor.com
Web: www.valcor.com

Val-Matic Valve & Manufacturing Corporation
905 Riverside Drive
Elmhurst
IL 60126
USA
Tel: +1 630 941 7600
Fax: +1 630 941 8042
Email: valves@valmatic.com
Web: www.valmatic.com

Valvtechnologies Inc
5904 Bingle Road
Houston
TX 77092
USA
Tel: +1 713 860 0400
Fax: +1 713 860 0499
Email: sales@valv.com
Web: www.valv.com

Valv-Trol Company
1340 Commerce Drive
PO Box 2259
Stow
OH 44224-1000
USA
Tel: +1 330 686 2800
Fax: +1 330 686 2820
Email: sales@valv-trol.com
Web: www.valv-trol.com

Velan Valve Corporation
94 Avenue C
Williston
VT 05495-9732
USA
Tel: +1 802 863 2562
Fax: +1 802 862 4014
Email: sales@velan.com
Web: www.velan.com

Victaulic Company
PO Box 31
4901 Kesslersville Road
Easton
PA 18044-0031
USA
Tel: +1 610 559 3300
Fax: +1 610 250 8817
Email: victaulic@victaulic.com
Web: www.victaulic.com

Warren Controls Inc
2600 Emrick Boulevard
Bethlehem
PA 18020-8010
USA
Tel: +1 610 317 0800
Fax: +1 610 317 2989
Email: sales@warrencontrols.com
Web: www.warrencontrols.com

WauKesha Cherry-Burell
611 Sugar Creek Road
Delavan
WI 53115
USA
Tel: +1 262 728 1900
Fax: +1 262 728 4904
Email: info@processequipment.spx.com
Web: www.spxprocessequipment.com

Weir Valves & Controls USA Inc
285 Canal Street
Salem
MA 01970
USA
Tel: +1 978 744 5690
Fax: +1 978 741 3626
Email: sales@weirvalveusa.com
Web: www.weirvalve.com

Wey Valve Inc
Highway 6 North
Nettleton
MS 38858
USA
Tel: +1 601 963 2020
Fax: +1 601 963 2025
Email: info@weyvalve.com
Web: www.weyvalve.com

Wood Group Pressure Control
3250 Briarpark Drive
Suite 100
Houston TX 77042
USA
Tel: +1 832 325 4200
Fax: +1 832 325 4350
Web: www.wgpressurecontrol.com

Xomox Corporation
4444 Cooper Road
Cincinnati
OH 45242-5686
USA
Tel: +1 513 745 6000
Fax: +1 513 745 6086
Email: sales@xomox.com
Web: www.xomox.com

Zimmermann & Jansen Inc
620 North Houston Avenue
PO Box 3365
Humble
TX 77347
USA
Tel: +1 281 446 8000
Fax: +1 281 446 8126
Email: zj.central@zjinc.com
Web: www.zjinc.com

Zy-Tech Global Industries Inc
10600 Corporate Drive
Stafford
TX 77477
USA
Tel: +1 281 565 1010
Fax: +1 281 565 3171
Email: sales@zy-tech.com
Web: www.zy-tech.com

21.3 Isolating valves

Companies listed alphabetically by country — See Chapter 3 for product categories

AUSTRALIA
John Valves
Tyco Flow Control

AUSTRIA
Brunnbauer-Armaturen Produktions GmbH
CCI - Valve Technology GmbH

BELGIUM
Belgium Ventiel NV
Icarus SA
KSB Belgium SA
Truflo Rona - Truflo International

BRAZIL
Emerson Process Management

CANADA
Cameron's Valves & Measurement Group
GGosco Engineering Inc
Newman Hattersley - Truflo International
Spirax Sarco Canada Ltd
Velan Inc

CHINA
Baoyi Group Co Ltd
China Shenjiang Valve Co Ltd
Danfoss (Tianjin) Ltd
Jiangsu Yandian Valve Co Ltd
Kenda Industrial Corporation
Metso Automation (Shanghai) Co Ltd
Neway Valve (Suzhou) Co Ltd
Shanghai Flow Valve & Fitting Co Ltd
Shanghai Halin Control Valve Co Ltd
Shanghai Karon Valves Machinery Co Ltd
Shanghai Neles-Jamesbury Valve Co Ltd
SKVC (Group) Corporation
Sufa Technology Industry Co Ltd
Suhou Viza Valve Co Ltd
Suzhou Sufa Da Valve Co Ltd
Xiamen Jia Da Quan Valves & Fittings Co Ltd
Xiamen Landee Industries Co Ltd
Zhejiang Beize Valve Manufacturing Co Ltd
Zhejiang Chaoda Valve Co Ltd
Zhejiang Dafulong Valve Co Ltd

CZECH REPUBLIC
Arako spol sro
Armatura Krnov spol sro
Armatury Group as
I B C Praha spol sro
Jihomoravská Armaturka spol sro
LDM spol sro
Mostro as
MSA as

DENMARK
Broen A/S

FINLAND
Larox Flowsys Oy
Metso Automation Oy
Naval Oy
Sento Oy

FRANCE
Banides et Debeaurain
Bayard SA
Buracco SA
Cameron's Valves & Measurement Group
Danfoss Socla SA
Defontaine SA
Emerson Process Management
Guichon Valves
Mecafrance SA
Meca-Inox
Metso Automation SAS
Perolo SA
Randex SA
Roforge
SACCAP
SAPAG Industrial valves
SNRI
Spirax Sarco SA
Strahman Valves France
Tyco Flow Control
Valpes
Vanatome
Vannes Lefebvre
Velan SAS

GERMANY
A+R Armaturen GmbH
Adams Armaturen GmbH
AKO Armaturen & Separations GmbH
ARCA-regler GmbH
ARI-Armaturen
Armaturenfabrik Franz Schneider GmbH & Co KG
Armaturen-und Metallwerke Zöblitz GmbH
Armaturenwerk Altenburg GmbH
ASP Armaturen
AWP Kälte-Klima-Armaturen GmbH
AZ-Armaturen GmbH
Berluto Armaturen GmbH
Bürkert GmbH & Co KG
Cera System Vershleissschutz GmbH
Christian Bollin Armaturenfabrik GmbH
Daume Regelarmaturen GmbH
E Georg Lüdecke Armaturen GmbH
Ebro Armaturen Gebr Bröer GmbH
Ehlers GmbH
END-Armaturen GmbH & Co KG
Erhard Armaturen GmbH & Co KG
Ernst Schmieding GmbH & Co KG
Flowserve Flow Control GmbH
Flowserve Flow Control GmbH
Foxboro Eckardt GmbH
Frank GmbH
Franz Dürholdt GmbH & Co KG
Friatec AG
Fritz Barthel Armaturen GmbH & Co KG
Fromme Armaturen GmbH & Co KG
G Bee GmbH
Gebr Tuxhorn GmbH & Co KG
Gefa Processtechnik GmbH
Geier Armaturen GmbH & Co KG
GEMÜ Gebrüder Müller Apparatebau GmbH & Co. KG
Georg Schünemann GmbH
Gestra GmbH
Gossler Fluidtec GmbH
GS Anderson GmbH
Gustav Mankenberg GmbH
Handtmann Armaturenfabrik GmbH & Co KG
Hartmann Valves GmbH
Herose GmbH
Hofer Hochdrucktechnik GmbH
Honeywell GmbH
IMI Norgren Buschjost GmbH & Co KG
Keulahütte GmbH, Krauschwitz
Klaus Union GmbH & Co KG
Klinger GmbH
KSB Armaturen GmbH
Kühme Armaturen GmbH
Laun GmbH Maschinen- und Apparatebau
Linde AG
Ludwig Mohren KG
M&S Armaturen GmbH
Mecafrance (Deutschland) GmbH
Mittelmann Sicherheitstechnik GmbH & Co KG
Neumo GmbH & Co KG
Noval Industriearmaturen GmbH
OHL Gutermuth Industriearmaturen GmbH
Parker Hannifin GmbH & Co KG
Perrin GmbH
Pesch Armaturen GmbH & Co
Phönix Armaturen-Werke Bregel GmbH
Pister-Kugelhähne GmbH
Rheinhütte GmbH & Co
Richter Chemie-Technik GmbH
RITAG Ritterhuder Armaturen GmbH & Co
Rötelmann GmbH
Rudolf von Scheven GmbH
SAFI GmbH
Samson AG Mess-und Regeltechnik
Schöneberg Armaturen GmbH
Schubert & Salzer Control Systems GmbH
SchuF-Armaturen und Apparatebau GmbH
Schwietzke Armaturen GmbH
Siebeck-Bitter GmbH Armaturen und Metallwerk
Siekmann Econosto GmbH & Co KG
Siemens AG
Spirax Sarco GmbH
Stahl-Armaturen PERSTA GmbH
Strack GmbH
Südmo Components GmbH
TLV Euro Engineering GmbH
Tuchenhagen GmbH
Tyco Flow Control
Uni-Geräte GmbH
VAG-Armaturen GmbH
Warex Valve GmbH
Welland & Tuxhorn AG
Werner Böhmer GmbH
Wilhelm Schley GmbH & Co
WT Armatur GmbH & Co KG
Z & J Technologies GmbH
Zwick Armaturen GmbH

HUNGARY
DKG-East

INDIA
Advance Valves Group
Amco Industrial Valves
Audco India Ltd
AV Valves Ltd
BDK Process Equipment Inc
Bharat Heavy Electricals Ltd (BHEL)
Canle Valves PVT Ltd
Excel Hydro-Pneumatics Pvt Ltd
Flowel Valves
Fouress Engineering (India) Ltd
Indian Valve Pvt Ltd
Intervalve (India) Ltd
Juneja Metal Works
Kirloskar Brothers Ltd
Rotex Automation Ltd
Shreeraj Industries
Virgo Valves & Controls Ltd

ISRAEL
Egmo Ltd
Habonim Industrial Valves & Actuators

ITALY
ABV Srl
AC MO SpA
Aerre Inox Srl
Airaga Rubinetterie SpA
Alfa Valvole Srl
AVF Astore Valves and Fittings Srl
Bardiani Valvole SpA
Caleffi SpA
Calobri Srl
Carraro Srl
Cesare Bonetti SpA
Cimberio SpA
Colves Srl
Dafram SpA
Effebi SpA
Enolgas Bonomi SpA
Eurovalve Srl
Ferrero Rubinetterie Srl
FIP - Formatura Iniezione Polimeri SpA
FNC SpA
Fratelli Fortis Srl
Fratelli Pettinaroli SpA
Galli & Cassina SpA
GEI Gruppo Energia Italia Srl
Ghibson Italia Srl
Greiner SpA
GT Attuatori Srl
Indra Srl
Italprotec Sas
Italvalv Snc
Italvalvole Sas
IVR Valvole SpA
La TecnoValvo Srl
LA TIS Service Srl
LCM Italia Srl
LVF SpA
Maran E Peracini Srl
MEI Valvole Industriali Srl
Nencini SpA
Nicolini Claudio srl
Novasfer Srl
OMB Valves SpA
Orsenigo Srl
Orton - Truflo International
Penta Srl
Petrolvalves Srl
Pibiviesse SpA
Rastelli Rubinetterie SpA
RIV Rubinetterie Italiane Valvole SpA
Rubinetterie Bresciane Bonomi SpA
Saint-Gobain Condotte SpA
SIRCA International Srl
Sitindustrie Equipment Srl
Starline SpA
SteriValves Srl
Technical Srl
Thermomess srl
Tyco Flow Control
Valpres Srl
Valvitalia SpA
Valvoindustria Ing Rizzio SpA
Velan Srl

JAPAN
Bunkaboeki Kogyo Co Ltd
CCI KK
Fujikin
Fukui Seisakusho Co Ltd
Furukawa Kogyo Co Ltd
Fushiman Co Ltd
Hatanaka Special Valve Industries Co Ltd
Heiwa Valve Co Ltd
Hirata Valve Industry Co Ltd
Hirose Valve Industry Co Ltd
Hisaka Works Ltd
Hitachi Valve Ltd
Ichinose Company Ltd
Ihara Science Corporation
Ishida Valve Manufacturing Co Ltd
Ito Koki Co Ltd
Kane Kogyo Co Ltd
Kaneko Sangyo Co Ltd
Kitamura Valve Mfg Co Ltd
Kitz Corporation
Kubota Corporation
Kurimoto Ltd
KVC Co Ltd
Maezawa Industries Inc
Miyairi Corporation
Nihon Klingage Co Ltd
Nihon Koso Co Ltd
Nippon Ball Valve Co Ltd
Nippon Daiya Valve Co Ltd
Nippon Valqua Industries Ltd
Okano Valve Mfg Company
Okumura Engineering Corporation
Sankyo Seisakusho Co Ltd
Shimizu Alloy Manufacturing Co Ltd
Shimizu Iron Works Co Ltd
Shoritsu Seisakusho Co Ltd
Showa Valve Co Ltd
TIX Corporation – IKS Division
Toa Valve Company Ltd
Tomoe Valve Co Ltd
Toyo Valve Co Ltd
Utsue Valve Co Ltd
Yoshitake Inc

THE NETHERLANDS
DMN Westinghouse
HP Valves Oldenzaal BV
Mokveld Valves BV
Noxon Stainless BV
Red Point Alloys BV
Tyco Flow Control
Wouter Witzel EuroValve

NORWAY
Danfoss A/S
Kongsberg Esco A/S
Norske Ventiler AS
Scan Armatur as
Steinsvik Maskinindustri AS
Westad Industri AS

POLAND
Tehaco Sp Zoo
Varimex-Valves Co Ltd
Zetkama Fabryka Armatury Przemyslowej SA

PORTUGAL
Velan Válvulas Industriais Lda

QATAR
Delta (Doha) Corporation

ROMANIA
Ario SA
Rominserv Valves IAIFO

RUSSIA
Armagus JSC
Armalit-1 JSC
Arzil JSC
Blagoveshchensky Valve Plant JSC
Fitting-P
JSK PKTBA
PYCT-95
Tyazhpromarmatura JSC
Zeim

SINGAPORE
Alton International (S) Pte Ltd
Emerson Process Management
Tyco Flow Control

SOUTH AFRICA
AMD Rotolok
Atval (Pty) Ltd
AZ Valves South Africa
Dynamic Fluid Control Pty Ltd
Gunric Valves
Insamcor (Pty) Ltd
Mine Line (Pty) Ltd
Mitech
Premier Valves
Spirotech International
Thermal Valve Manufacturers (Pty) Ltd

SOUTH KOREA
A-Sung Plastic Valve Co Ltd
KumKang Valve Mfg Co Ltd
PK Valve
Samjin Precision Co Ltd
Velan Ltd
Spain
Babcock Power España SA
BAC Valves SA

Beligcast Internacional SL
JC Fábrica de Válvulas SA
Kitz Corporation of Europe SA
Orbinox SA
Pekos Valves SA
Válvulas VS SL
VYC Industrial SA

SWEDEN

NAF AB

SWITZERLAND

Biar SA
CCI Switzerland
ChemValve AG
Famat SA
Georg Fischer Rohrleitungssysteme (Schweiz) AG
InterApp AG
SISTAG Absperrtechnik
Tyco Flow Control
Valv AG

TAIWAN

Apex Controls Co Ltd
Bola-Tek Manufacturing Co Ltd
Bueno Enterprise Co Ltd
EMICO Eayuan Metal Industrial Co Ltd
Haitima Corporation
HCHTS Industries Corporation (HICO)
John Valve Mfg Factory Co Ltd
Kingdom Precision Casting Co Ltd
Mars Valve Co Ltd
Super Inox Corporation
Value Valves Co Ltd
Valve-Tek Manufacturing Co Ltd
Velan-Valvac
Wyeco Auto Valves Co Ltd
Yi Ming Machinery Ind Co Ltd
Zipson Steel Industrial Co Ltd

TURKEY

Asteknik Valve
Ayvaz Endüstriyei Mamuller Sanayi ve Ticaret AS
Burçelik Valve Company
Duyar Vana Makina Sanayi ve Ticaret A S
Gedik Döküm ve Vana Sanayi ve Ticaret AS
InterValf

UNITED ARAB EMIRATES

Emerson FZE

UNITED KINGDOM

Adanac Valve Specialities Ltd
Advanced Valve Technologies
Aeon Pipe Systems Ltd
Alco Valves (UK) Ltd
Anson Ltd
BEL Valves
Bestobell Valves Ltd
Bifold Fluidpower Ltd
Blackhall Engineering Ltd
Bray Controls (UK)
Brian Donkin Valves Ltd
Brooksbank Valves Ltd
Bürkert Fluid Control Systems
Cameron's Valves & Measurement Group
Colson Industries Ltd
Conflow Ltd
Copes-Vulcan
Crane Fluid Systems
Crane Process Flow Technologies
Curtiss-Wright Flow Control UK Ltd
Delta Fluid Products Ltd
DMI Young & Cunningham Ltd
DPL
Dynamic Controls Ltd
Flowserve Flow Control (UK) Ltd
Flowtech Inc Ltd
GA Valves Sales Ltd
Goodwin International Ltd
H H Valves Ltd
Hale Hamilton (Valves) Ltd
Henry Technologies Ltd
Hobbs Valve Ltd
IMC UK Ltd
Instrumentation Products Division, Parker Hannifin Ltd
Invensys APV
JL Controls Ltd
Kemutec Powder Technologies Ltd
Leeds Valve Company Ltd
LG Ball Valves Ltd
M V Fluids Handling
Oliver Valves Ltd
Peter Smith Valve Co Ltd
Posi-flate
RGS Electro-Pneumatics Ltd
Rotolok Ltd UK
Sabre Instrument Valves Ltd
Safety Systems UK Ltd
Saint-Gobain Pipelines
Schubert & Salzer UK Ltd
Shaw Valves
Shipham Valves Ltd
Spirax-Sarco Ltd
Taylor Shaw
TC Fluid Control Ltd
Thompson Valves- Truflo International
Tomoe Valve Ltd
Transmark Fcx
Tyco Flow Control
Velan Valves Ltd
Warren-Morrison Valves Ltd
Weir Valves & Controls UK Ltd
Xomox Ltd
YPS Valves Ltd

USA

American Valve Inc
AMRI Inc
AOP Industries Inc
Asahi/America Inc
ASCO
Balon Corporation
Bray Controls USA
Cameron's Valves & Measurement Group
Cashco Inc
CCI (Control Components Inc)
Circle Seal Controls Inc
Conbraco Industries Inc
Conflow Inc
Conval Inc
CPC-Cryolab
Crane Nuclear Inc
DeltaValve USA
DeZURIK Water Controls
DynaQuip Controls
Emerson Process Management
Enertech
Everest Valve Company
Everlasting Valve Co
FC Kingston
Flomatic Corporation
Flowserve Flow Control
GA Industries Inc
Gilmore Valve Company
H O Trerice Company
Henry Pratt Company
Hex, a division of Richards Industries
High Pressure Equipment Company
Hoke Inc
Homestead Valve
Hunt Valve Company Inc
ITT Engineered Valves
ITT Pure-Flo
JMC Valve Inc
Jordan Valve, a division of Richards Industries
Kerotest Manufacturing Corporation
KF Industries
Kitz Corporation of America
Leslie Controls Inc
M & J Valves
Marwin Valve, a division of Richards Industries
Metso Automation USA Inc
Milwaukee Valve Company
MOGAS Industries Inc
Mueller Steam Specialty
Newmans Valve
NIBCO Inc
Nicholson Steam Trap
Nil-Cor
Nortech
Parker Hannifin Corporation
Parker, Fluid Control Division
PBM Inc
Posi-flate
Powell Valves
Process Development & Control
Red Valve Company
Red-White Valve Corporation
Rexa Koso America Inc
Richards Industries
Rockwood-Swendeman
Ross Valve Manufacturing Co Inc
Rotolok Valves Inc
Saint-Gobain Performance Plastics Corporation
SMG Valves
Southern Manufacturing Inc
Spence Engineering Co Inc
Spirax Sarco Inc
SVF Flow Controls Inc
Swagelok Company
Target Rock
Therm-Omega-Tech Inc
Tru-Tech Industries
Tyco Flow Control
US Para Plate Corporation
Valcor Engineering Corporation
Val-Matic Valve & Manufacturing Corporation

Valvtechnologies Inc
Velan Valve Corporation
Victaulic Company
WauKesha Cherry-Burell
Weir Valves & Controls USA Inc
Wey Valve Inc
Xomox Corporation
Zimmermann & Jansen Inc
Zy-Tech Global Industries Inc

21.4 Non-return valves

Companies listed alphabetically by country — See Chapter 4 for product categories

AUSTRALIA
John Valves
Tyco Flow Control

AUSTRIA
Brunnbauer-Armaturen Produktions GmbH
CCI - Valve Technology GmbH

BELGIUM
Belgium Ventiel NV
Icarus SA
Truflo Rona - Truflo International

CANADA
Cameron's Valves & Measurement Group
Floval Equipment Ltd
Newman Hattersley - Truflo International
Spirax Sarco Canada Ltd
Velan Inc

CHINA
Baoyi Group Co Ltd
China Shenjiang Valve Co Ltd
Jiangsu Yandian Valve Co Ltd
Kenda Industrial Corporation
Neway Valve (Suzhou) Co Ltd
Shanghai Flow Valve & Fitting Co Ltd
Shanghai Halin Control Valve Co Ltd
Shanghai Karon Valves Machinery Co Ltd
Sufa Technology Industry Co Ltd
Suhou Viza Valve Co Ltd
Suzhou Sufa Da Valve Co Ltd
Xiamen Jia Da Quan Valves & Fittings Co Ltd
Xiamen Landee Industries Co Ltd
Zhejiang Beize Valve Manufacturing Co Ltd
Zhejiang Chaoda Valve Co Ltd
Zhejiang Dafulong Valve Co Ltd

CZECH REPUBLIC
Armatura Krnov spol sro
I B C Praha spol sro
Jihomoravská Armaturka spol sro
Mostro as
MSA as

FRANCE
Bayard SA
Buracco SA
Cameron's Valves & Measurement Group
Danfoss Socla SA
Defontaine SA
Meca-Inox
Randex SA
Roforge
SACCAP
SAPAG Industrial Valves
SNRI
Spirax Sarco SA
Vanatome
Vannes Lefebvre
Velan SAS

GERMANY
Adams Armaturen GmbH
ARI-Armaturen
Armaturenwerk Altenburg GmbH
ASP Armaturen
AWP Kälte-Klima-Armaturen GmbH
Berluto Armaturen GmbH
Cera System Vershleissschutz GmbH
CH Zikesch Armaturentechnik GmbH
E Georg Lüdecke Armaturen GmbH
Ebro Armaturen Gebr Bröer GmbH
Ehlers GmbH
END-Armaturen GmbH & Co KG
Erhard Armaturen GmbH & Co KG
Ernst Schmieding GmbH & Co KG
Flowserve Flow Control GmbH
Frank GmbH
Franz Dürholdt GmbH & Co KG
Fromme Armaturen GmbH & Co KG
Gebr Tuxhorn GmbH & Co KG
Gefa Processtechnik GmbH
GEMÜ Gebrüder Müller Apparatebau GmbH & Co. KG
Georg Schünemann GmbH
Gestra GmbH
GS Anderson GmbH
Herose GmbH
Honeywell GmbH
Keulahütte GmbH, Krauschwitz
Klaus Union GmbH & Co KG
KSB Armaturen GmbH
Linde AG
Mittelmann Sicherheitstechnik GmbH & Co KG
Noval Industriearmaturen GmbH
Parker Hannifin GmbH & Co KG
Phönix Armaturen-Werke Bregel GmbH
Pister-Kugelhähne GmbH
Richter Chemie-Technik GmbH
RITAG Ritterhuder Armaturen GmbH & Co
Rötelmann GmbH
SAFI GmbH
Schubert & Salzer Control Systems GmbH
Schwietzke Armaturen GmbH
Siebeck-Bitter GmbH Armaturen und Metallwerk
Siekmann Econosto GmbH & Co KG
Spirax Sarco GmbH
Stahl-Armaturen PERSTA GmbH
Strack GmbH
Südmo Components GmbH
Tyco Flow Control
VAG-Armaturen GmbH
Wilhelm Schley GmbH & Co
WT Armatur GmbH & Co KG
Z & J Technologies GmbH

INDIA
Advance Valves Group
Amco Industrial Valves
Audco India Ltd
AV Valves Ltd
BDK Process Equipment Inc
Bharat Heavy Electricals Ltd (BHEL)
Canle Valves PVT Ltd
Fouress Engineering (India) Ltd
Indian Valve Pvt Ltd
Intervalve (India) Ltd
Juneja Metal Works
Kirloskar Brothers Ltd
Shreeraj Industries

ITALY
AC MO SpA
Aerre Inox Srl
Caleffi SpA
Calobri Srl
Cesare Bonetti SpA
Cimberio SpA
Colves Srl
Enolgas Bonomi SpA
Ferrero Rubinetterie Srl
Fratelli Fortis Srl
GEI Gruppo Energia Italia Srl
Ghibson Italia Srl
Greiner SpA
Italprotec Sas
La TecnoValvo Srl
LA TIS Service Srl
LVF SpA
Maran E Peracini Srl
MEI Valvole Industriali Srl
Nencini SpA
Novasfer Srl
OMB Valves SpA
Orton - Truflo International
Petrolvalves Srl
Rastelli Rubinetterie SpA
Rubinetterie Bresciane Bonomi SpA
Saint-Gobain Condotte SpA
SIRCA International Srl
Sitindustrie Equipment Srl
Technical Srl
Tyco Flow Control
Valvitalia SpA
Watts Cazzaniga SpA

JAPAN
Bunkaboeki Kogyo Co Ltd
CCI KK
Eagle Industry Co Ltd
Fujikin
Furukawa Kogyo Co Ltd
Hatanaka Special Valve Industries Co Ltd
Hirata Valve Industry Co Ltd
Hirose Valve Industry Co Ltd
Hitachi Valve Ltd
Ichinose Company Ltd
Ihara Science Corporation

Ishida Valve Manufacturing Co Ltd
Ito Koki Co Ltd
Kitz Corporation
Kubota Corporation
Kurimoto Ltd
KVC Co Ltd
Maezawa Industries Inc
Miyairi Corporation
Miyawaki Inc
Nippon Valqua Industries Ltd
Okano Valve Mfg Company
Shida Engineering Co Ltd
Shimizu Alloy Manufacturing Co Ltd
Shimizu Iron Works Co Ltd
Shoritsu Seisakusho Co Ltd
Showa Valve Co Ltd
Toa Valve Company Ltd
Toyo Valve Co Ltd
Utsue Valve Co Ltd
Venn Co Ltd

THE NETHERLANDS

HP Valves Oldenzaal BV
Mokveld Valves BV
Noxon Stainless BV
Red Point Alloys BV

NORWAY

Norske Ventiler AS
Scan Armatur AS

POLAND

Varimex-Valves Co Ltd
Zetkama Fabryka Armatury Przemyslowej SA

PORTUGAL

Velan Válvulas Industriais Lda

ROMANIA

Ario SA
Rominserv Valves IAIFO

RUSSIA

Arzil JSC
Blagoveshchensky Valve Plant JSC
Fitting-P
JSK PKTBA
PYCT-95

SINGAPORE

Alton International (S) Pte Ltd
Tyco Flow Control

SOUTH AFRICA

Atval (Pty) Ltd
Dynamic Fluid Control Pty Ltd
Gunric Valves
Mine Line (Pty) Ltd
Premier Valves
Thermal Valve Manufacturers (Pty) Ltd

SOUTH KOREA

A-Sung Plastic Valve Co Ltd
Goodwin Korea Co Ltd
PK Valve
Velan Ltd

SPAIN

Babcock Power España SA
Beligcast Internacional SL
Orbinox SA
Válvulas VS SL
VYC Industrial SA

SWEDEN

NAF AB

SWITZERLAND

CCI Switzerland
Famat SA
Georg Fischer Rohrleitungssysteme (Schweiz) AG
SISTAG Absperrtechnik
Tyco Flow Control

TAIWAN

Apex Controls Co Ltd
EMICO Eayuan Metal Industrial Co Ltd
HCHTS Industries Corporation (HICO)
John Valve Mfg Factory Co Ltd
Kingdom Precision Casting Co Ltd
Super Inox Corporation
Value Valves Co Ltd
Valve-Tek Manufacturing Co Ltd
Yi Ming Machinery Ind Co Ltd

TURKEY

Ayvaz Endüstriyei Mamuller Sanayi ve Ticaret AS
Duyar Vana Makina Sanayi ve Ticaret AS
Gedik Döküm ve Vana Sanayi ve Ticaret AS

UNITED KINGDOM

Abacus Valves Ltd
Adanac Valve Specialities Ltd
Alco Valves (UK) Ltd
BEL Valves
Bestobell Valves Ltd
BiS Valves Ltd
Blackhall Engineering Ltd
Bray Controls (UK)
Cameron's Valves & Measurement Group
Conflow Ltd
Crane Fluid Systems
Crane Process Flow Technologies
DMI Young & Cunningham Ltd
DPL
Dynafluid Ltd/Gresswell Valves
Dynamic Controls Ltd
Flowserve Flow Control (UK) Ltd
FST Northvale Korting Ltd
GA Valves Sales Ltd
Goodwin International Ltd
Hale Hamilton (Valves) Ltd
Henry Technologies Ltd
Hoerbiger Holding UK Ltd
Invensys APV
K Controls Ltd
Oliver Valves Ltd
Peter Smith Valve Co Ltd
Saint-Gobain Pipelines
Schubert & Salzer UK Ltd
Shaw Valves
Shipham Valves Ltd
Spirax-Sarco Ltd
Taylor Shaw
Transmark Fcx
Tyco Flow Control
Weir Valves & Controls UK Ltd
Xomox Ltd
YPS Valves Ltd

USA

AOP Industries Inc
Asahi/America Inc
Balon Corporation
Cameron's Valves & Measurement Group
CCI (Control Components Inc)
Check-All Valve Mfg Co
Circle Seal Controls Inc
Conbraco Industries Inc
Conflow Inc
Conval Inc
CPC-Cryolab
Crane Nuclear Inc
DFT Durabla Fluid Technology Inc
Enertech
Everest Valve Company
FC Kingston
FEBCO – Division of Watts Water Technologies Inc
Flomatic Corporation
Flowserve Flow Control
GA Industries Inc
Gilmore Valve Company
Henry Pratt Company
High Pressure Equipment Company
Hunt Valve Company Inc
Hydroseal Valve Company
ITT Engineered Valves
KF Industries
Kitz Corporation of America
M & J Valves
Mallard Control
Milwaukee Valve Company
Mueller Steam Specialty
Newmans Valve
NIBCO Inc
Nicholson Steam Trap
Nil-Cor
Parker Hannifin Corporation
Parker, Fluid Control Division
Powell Valves
Red Valve Company
Red-White Valve Corporation
Saint-Gobain Performance Plastics Corporation
Spence Engineering Co Inc
Spirax Sarco Inc
Swagelok Company
Target Rock
Tyco Flow Control
US Para Plate Corporation
Valcor Engineering Corporation
Val-Matic Valve & Manufacturing Corporation
Valvtechnologies Inc
Valv-Trol Company
Velan Valve Corporation
Victaulic Company
WauKesha Cherry-Burell
Weir Valves & Controls USA Inc
Xomox Corporation
Zimmermann & Jansen Inc

21.5 Regulators

Companies listed alphabetically by country — See Chapter 5 for product categories

BELGIUM
Dresser Valves Europe

BRAZIL
Dresser Industria e Comercio Ltda
Emerson Process Management

CANADA
Dresser–Masoneilan DI Canada Inc

CHINA
Dresser Flow Solutions
Shanghai Halin Control Valve Co Ltd
Suzhou Sufa Da Valve Co Ltd

CZECH REPUBLIC
LDM spol sro

FINLAND
Larox Flowsys Oy

FRANCE
A K Müller France
Danfoss Socla SA
Emerson Process Management
H+ Valves SA
Masoneilan Dresser Produits Industriels
Metso Automation SAS
Sart von Rohr SAS
SNRI
Spirax Sarco SA
Tyco Flow Control
Valpes
Weir Valves & Controls France SAS

GERMANY
A u K Müller GmbH & Co KG
ARCA-regler GmbH
ARI-Armaturen
CH Zikesch Armaturentechnik GmbH
Gustav Mankenberg GmbH
IMI Norgren Buschjost GmbH & Co KG
Hans Sasserath & Co KG
Herose GmbH
Holter Regelarmaturen GmbH & Co KG (HORA)
Leser GmbH & Co KG
Schubert & Salzer Control Systems GmbH
Siekmann Econosto GmbH & Co KG
Tyco Flow Control
VAG-Armaturen GmbH

INDIA
Bharat Heavy Electricals Ltd (BHEL)
Dresser Valve India Pvt Ltd
Fisher Sanmar Ltd
Fouress Engineering (India) Ltd
Indian Valve Pvt Ltd
Tyco Flow Control

ITALY
AC MO SpA
Besa Ing Santangelo SpA
Carraro Srl
GEI Gruppo Energia Italia Srl
Indra Srl
Orton - Truflo International
Rastelli Rubinetterie SpA
Rubinetterie Bresciane Bonomi SpA
SIRCA International Srl
Tyco Flow Control
Valvoindustira Ing Rizzio SpA

JAPAN
Fujikin
Hirata Valve Industry Co Ltd
Ichinose Company Ltd
Kane Kogyo Co Ltd
Kubota Corporation
Motoyama Eng Works Ltd
Nihon Koso Co Ltd
Niigata Masoneilan Co Ltd (NIMCO
Okano Valve Mfg Company
Sankyo Seisakusho Co Ltd
Shida Engineering Co Ltd
Toko Valex Co Ltd
V TEX Corporation
Venn Co Ltd
Yoshitake Inc

KUWAIT
Dresser Flow Solutions - Middle East Operations

MALAYSIA
Dresser Flow Solutions

MEXICO
Dresser Valve de Mexico SA

THE NETHERLANDS
Tyco Flow Control

NORWAY
Danfoss A/S

RUSSIA
DS Controls

SAUDI ARABIA
Dresser Al Rushaid Valve & Instrument Co Ltd

SINGAPORE
Dresser Singapore Pte Ltd
Emerson Process Management
South Korea
Shin Han Control Valve Co Ltd
Tyco Flow Control

SOUTH AFRICA
Atval (Pty) Ltd

SWITZERLAND
Tyco Flow Control

UNITED ARAB EMIRATES
Dresser Flow Solutions - Middle East Operations
Emerson FZE

UNITED KINGDOM
Alco Valves (UK) Ltd
Auld Valves Ltd
Bestobell Valves Ltd
Bifold Fluidpower Ltd
BS & B Safety Systems (UK) Ltd
Curtiss-Wright Flow Control UK Ltd
DI UK Ltd
Dynamic Controls Ltd
ESME Valves Ltd
Hale Hamilton (Valves) Ltd
Hoerbiger Holding UK Ltd
Midland-ACS
Safety Systems UK Ltd
Schubert & /salzer UK Ltd
Spirax-Sarco Ltd
TC Fluid Control Ltd
Thompson Valves-Truflo International
Tyco Flow Control
Weir Valves & Controls UK Ltd

USA
Cameron's Valves & Measurement Group
Conflow Inc
Emerson Process Management
Groth Corporation
High Pressure Equipment Company
Jordan Valve, a division of Richards Industries
Mallard Control
Parker Hannifin Corporation
Richards Industries
Saint-Gobain Performance Plastics Corporation
Samson Controls
Spirax Sarco Inc
Swagelok Company
Target Rock
Therm-Omega-Tech Inc
Tyco Flow Control

21.6 Control valves

Companies listed alphabetically by country — See Chapter 6 for product categories

AUSTRALIA
Tyco Flow Control

AUSTRIA
CCI - Valve Technology GmbH
EPC GmbH

BELGIUM
Dresser Valves Europe
Icarus SA

BRAZIL
Dresser Industria e Comercio Ltda
Emerson Process Management

CANADA
Dresser–Masoneilan DI Canada Inc
GGosco Engineering Inc
Spirax Sarco Canada Ltd

CHINA
Danfoss (Tianjin) Ltd
Dresser Flow Solutions
Kenda Industrial Corporation
Metso Automation (Shanghai) Co Ltd
Shanghai Flow Valve & Fitting Co Ltd
Shanghai Karon Valves Machinery Co Ltd
Sufa Technology Industry Co Ltd
Xiamen Jia Da Quan Valves & Fittings Co Ltd
Zhejiang Beize Valve Manufacturing Co Ltd

CZECH REPUBLIC
I B C Praha spol sro
LDM spol sro

FINLAND
Metso Automation Oy

FRANCE
A K Müller France
Bayard SA
Danfoss Socla SA
Emerson Process Management
Masoneilan Dresser Produits Industriels
Metso Automation SAS
SAPAG Industrial valves
Sart von Rohr SAS
Spirax Sarco SA
Vanatome
Velan SAS

GERMANY
A u K Müller GmbH & Co KG
Adams Armaturen GmbH
AKO Regelungstechnik GmbH & Co KG
ARCA-regler GmbH
ARI-Armaturen
AWP Kälte-Klima-Armaturen GmbH
Bar GmbH
Berluto Armaturen GmbH
Cera System Vershleissschutz GmbH
CH Zikesch Armaturentechnik GmbH
Daume Regelarmaturen GmbH
Dresser Valves Europe GmbH
Ebro Armaturen Gebr Bröer GmbH
END-Armaturen GmbH & Co KG
Ernst Schmieding GmbH & Co KG
Fischer Mess-und Regeltechnik GmbH
Flowserve Flow Control GmbH
Foxboro Eckardt GmbH
Franz Dürholdt GmbH & Co KG
Georg Schünemann GmbH
Gestra GmbH
GS Anderson GmbH
Gustav Mankenberg GmbH
Herose GmbH
Holter Regelarmaturen GmbH & Co KG (HORA)
Honeywell GmbH
IMI Norgren Buschjost GmbH & Co KG
Johnson Controls Regelungstechnik GmbH
Keulahütte GmbH, Krauschwitz
Kieback & Peter GmbH & Co KG
Klinger GmbH
Kühme Armaturen GmbH
Linde AG
Lorch GmbH
Neumo GmbH & Co KG
Niezgodka GmbH
OHL Gutermuth Industriearmaturen GmbH
Parker Hannifin GmbH & Co KG
Phönix Armaturen-Werke Bregel GmbH
Phönix Messtechnik GmbH
Reineke Mess- und Regeltechnik GmbH
Richter Chemie-Technik GmbH
Rotech GmbH
Samson AG Mess-und Regeltechnik
Sauter-Cumulus GmbH
Schroedahl-Arapp
Spezialarmaturen GmbH & Co KG
Schubert & Salzer Control Systems GmbH
SchuF-Armaturen und Apparatebau GmbH
Schwietzke Armaturen GmbH
Spirax Sarco GmbH
Südmo Components GmbH
TLV Euro Engineering GmbH
Tuchenhagen GmbH
Tyco Flow Control
Uni-Geräte GmbH
VAG-Armaturen GmbH
W Bälz & Sohn GmbH & Co
Welland & Tuxhorn AG
WT Armatur GmbH & Co KG

INDIA
Dresser Valve India Pvt Ltd
Fisher Sanmar Ltd
Fouress Engineering (India) Ltd
Indian Valve Pvt Ltd

ITALY
Caleffi SpA
Carraro Srl
Eurovalve Srl
FIP - Formatura Iniezione Polimeri SpA
Fratelli Fortis Srl
GEI Gruppo Energia Italia Srl
Italvalv Snc
Italvalvole Sas
LA TIS Service Srl
Nencini SpA
Nicolini Claudio srl
Novasfer Srl
Orton - Truflo International
Saint-Gobain Condotte SpA
Valvitalia SpA
Valvole Hofmann
Watts Cazzaniga SpA

JAPAN
ABB KK
Bailey Japan Co Ltd
CCI KK
Eagle Industry Co Ltd
Fujikin
Fushiman Co Ltd
Hirata Valve Industry Co Ltd
Hisaka Works Ltd
Hitachi Valve Ltd
Ihara Science Corporation
Kaneko Sangyo Co Ltd
Kitz Corporation
Kubota Corporation
Kurimoto Ltd
Maezawa Industries Inc
Miyairi Corporation
Motoyama Eng Works Ltd
Nihon Koso Co Ltd
Niigata Masoneilan Co Ltd (NIMCO
Nippon Ball Valve Co Ltd
Nippon Daiya Valve Co Ltd
Nippon Valqua Industries Ltd
Okano Valve Mfg Company
Okumura Engineering Corporation
Sankyo Seisakusho Co Ltd
Shida Engineering Co Ltd
Shimizu Alloy Manufacturing Co Ltd
Shimizu Iron Works Co Ltd
Toa Valve Company Ltd
Toko Valex Co Ltd
TV Valve Co Ltd
Utsue Valve Co Ltd

KUWAIT
Dresser Flow Solutions - Middle East Operations

MALAYSIA
Dresser Flow Solutions

MEXICO
Dresser Valve de Mexico SA
The Netherlands
Mokveld Valves BV

NORWAY
Bryne Mekanikk AS
Danfoss A/S
Kongsberg Esco A/S
Oceaneering Rotator AS
Scan Armatur as

POLAND
Varimex-Valves Co Ltd
Zetkama Fabryka Armatury Przemyslowej SA
Romania
Rominserv Valves IAIFO

RUSSIA
Armagus JSC
DS Controls
JSK PKTBA
PYCT-95

SAUDI ARABIA
Dresser Al Rushaid Valve & Instrument Co Ltd

SINGAPORE
Dresser Singapore Pte Ltd
Emerson Process Management
Tyco Flow Control

SOUTH AFRICA
Dynamic Fluid Control Pty Ltd
Mitech
Premier Valves
Thermal Valve Manufacturers (Pty) Ltd
Shin Han Control Valve Co Ltd

SPAIN
VYC Industrial SA

SWEDEN
CCI Sweden
NAF AB

SWITZERLAND
CCI Switzerland

CLA-VAL SA

Georg Fischer Rohrleitungssysteme (Schweiz) AG

SISTAG Absperrtechnik

von Rohr Armaturen AG

TAIWAN

Value Valves Co Ltd

Wyeco Auto Valves Co Ltd

UNITED ARAB EMIRATES

Dresser Flow Solutions - Middle East Operations

Emerson FZE

UNITED KINGDOM

A K Müller (UK) Ltd

Auld Valves Ltd

Bestobell Valves Ltd

BiS Valves Ltd

Blackhall Engineering Ltd

Bray Controls (UK)

Bürkert Fluid Control Systems

Callow Engineering Ltd

Conflow Ltd

Copes-Vulcan

Curtiss-Wright Flow Control UK Ltd

DI UK Ltd

DMI Young & Cunningham Ltd

Dynamic Controls Ltd

ESME Valves Ltd

Flowserve Flow Control (UK) Ltd

FST Northvale Korting Ltd

GA Valves Sales Ltd

Invensys APV

Koso Kent Introl Ltd

Leeds Valve Company Ltd

Midland-ACS

Schubert & Salzer UK Ltd

Severn Glocon Ltd

Spirax-Sarco Ltd

TC Fluid Control Ltd

Transmark Fcx

Xomox Ltd

USA

AMRI Inc

Bray Controls USA

Cameron's Valves & Measurement Group

Cashco Inc

CCI (Control Components Inc)

Circle Seal Controls Inc

CLA-VAL Automatic Control Valves

CPC-Cryolab

DeZURIK Water Controls

DFT Durabla Fluid Technology Inc

Dresser–Masoneilan

Emerson Process Management

Enertech

Everest Valve Company

FC Kingston

Flomatic Corporation

Flowserve Flow Control

GA Industries Inc

H O Trerice Company

Homestead Valve

Hunt Valve Company Inc

Johnson Controls Inc

Jordan Valve, a division of Richards Industries

Kitz Corporation of America

Leslie Controls Inc

M & J Valves

Mallard Control

Marwin Valve, a division of Richards Industries

Maxon Corporation

Metso Automation USA Inc

OCV Control Valves

Parker, Fluid Control Division

PBM Inc

Polyjet Valves

Powell Valves

Process Development & Control

Red Valve Company

Rexa Koso America Inc

Richards Industries

Spence Engineering Co Inc

Spirax Sarco Inc

SVF Flow Controls Inc

Therm-Omega-Tech Inc

Tyco Flow Control

Valvtechnologies Inc

Valv-Trol Company

Warren Controls Inc

Weir Valves & Controls USA Inc

Xomox Corporation

21.7 Safety relief valves

Companies listed alphabetically by country — See Chapter 7 for product categories

AUSTRIA

Brunnbauer-Armaturen Produktions GmbH

BELGIUM

Icarus SA

CANADA

Spirax Sarco Canada Ltd

CHINA

Metso Automation (Shanghai) Co Ltd

Xiamen Jia Da Quan Valves & Fittings Co Ltd

FRANCE

Banides et Debeaurain

Bayard SA

H+ Valves SA

Metso Automation SAS

Spirax Sarco SA

Weir Valves & Controls France SAS

Weir Valves & Controls SEBIM SAS

GERMANY

AKO Regelungstechnik GmbH & Co KG

ARI-Armaturen

Armaturenfabrik Franz Schneider GmbH & Co KG

Armaturen-und Metallwerke Zöblitz GmbH

Bopp & Reuther Sicherheits und Regelarmaturen GmbH

Braunschweiger Flammenfilter GmbH

Dresser Valves Europe GmbH

END-Armaturen GmbH & Co KG

Ernst Schmieding GmbH & Co KG

Franz Dürholdt GmbH & Co KG

Gebr Tuxhorn GmbH & Co KG

Georg Schünemann GmbH

Gestra GmbH

Hans Sasserath & Co KG

Herose GmbH

Hofer Hochdrucktechnik GmbH

Klaus Union GmbH & Co KG

Leser GmbH & Co KG

Lorch GmbH

Parker Hannifin GmbH & Co KG

Phönix Messtechnik GmbH

Pister-Kugelhähne GmbH

Richter Chemie-Technik GmbH

Samson AG Mess-und Regeltechnik

Sauter-Cumulus GmbH

SchuF-Armaturen und Apparatebau GmbH

Spirax Sarco GmbH

TLV Euro Engineering GmbH

Tyco Flow Control

Wilhelm Schley GmbH & Co

INDIA

Bharat Heavy Electricals Ltd (BHEL)

Fainger Leser Valves PVT Ltd

Tyco Flow Control

ITALY

Bardiani Valvole SpA

Besa Ing Santangelo SpA

FNC SpA

Italvalvole Sas

Saint-Gobain Condotte SpA

Technical Srl

JAPAN

Fukui Seisakusho Co Ltd

Fushiman Co Ltd

Hatanaka Special Valve Industries Co Ltd

Kaneko Sangyo Co Ltd

Mihana Seisakusho Co Ltd

Motoyama Eng Works Ltd

Sankyo Seisakusho Co Ltd

Toa Valve Company Ltd

V TEX Corporation

Yoshitake Inc

THE NETHERLANDS

Mokveld Valves BV

NORWAY

Scan Armatur as

POLAND

Varimex-Valves Co Ltd

QATAR

Delta (Doha) Corporation

RUSSIA

Armagus JSC

Armalit-1 JSC

Blagoveshchensky Valve Plant JSC

SPAIN
VYC Industrial SA

TURKEY
Duyar Vana Makina Sanayi ve Ticaret AS

UNITED KINGDOM
Auld Valves Ltd
Bestobell Valves Ltd
BiS Valves Ltd
Blackhall Engineering Ltd
Broady Flow Control Ltd
BS & B Safety Systems (UK) Ltd
Delta Fluid Products Ltd
DPL
Dynafluid Ltd/Gresswell Valves
ESME Valves Ltd
GA Valves Sales Ltd
Hale Hamilton (Valves) Ltd
Henry Technologies Ltd
Invensys APV
Oliver Valves Ltd
Pyropress Engineering Company Ltd
Safety Systems UK Ltd
Seetru Ltd
Shaw Valves
Spirax-Sarco Ltd
TC Fluid Control Ltd
Warren-Morrison Valves Ltd
Weir Valves & Controls UK Ltd

USA
Asahi/America Inc
Cameron's Valves & Measurement Group
Circle Seal Controls Inc
Conbraco Industries Inc
CPC-Cryolab
Dresser Inc
Farris Engineering
FC Kingston
Flowsafe Inc
Gilmore Valve Company
Groth Corporation
High Pressure Equipment Company
Hydroseal Valve Company
Mallard Control
Maxon Corporation
Metrex Valve Corporation
Nicholson Steam Trap
Parker Hannifin Corporation
Rockwood-Swendeman
Spence Engineering Co Inc
Spirax Sarco Inc
Valv-Trol Company
WauKesha Cherry-Burell
Weir Valves & Controls USA Inc
Wood Group Pressure Control

21.8 Actuators

Companies listed alphabetically by country — See Chapter 12 for product categories

AUSTRALIA
Tyco Flow Control

AUSTRIA
CCI - Valve Technology GmbH

BELGIUM
Belgium Ventiel NV
KSB Belgium SA

BRAZIL
Emerson Process Management

CANADA
GGosco Engineering Inc
Spirax Sarco Canada Ltd

CHINA
Shanghai Halin Control Valve Co Ltd
Shanghai Neles-Jamesbury Valve Co Ltd

CZECH REPUBLIC
ZPA Pecky as

FINLAND
Metso Automation Oy

FRANCE
Defontaine SA
Emerson Process Management
Meca-Inox
Randex SA
Sart von Rohr SAS
Spirax Sarco SA
Tyco Flow Control
Valpes
Velan SAS

GERMANY
ARCA-regler GmbH
ARI-Armaturen
AUMA Riester GmbH & Co KG
AZ-Armaturen GmbH
Bar GmbH
Daume Regelarmaturen GmbH
Drehmo GmbH
Ebro Armaturen Gebr Bröer GmbH
EMG Automation GmbH
END-Armaturen GmbH & Co KG
Flowserve Flow Control GmbH
Franz Dürholdt GmbH & Co KG
Gefa Processtechnik GmbH
GEMÜ Gebrüder Müller Apparatebau GmbH & Co. KG
GS Anderson GmbH
Holter Regelarmaturen GmbH & Co KG (HORA)
Linde AG
PS Automation GmbH
Remote Control
Richter Chemie-Technik GmbH
Rudolf von Scheven GmbH
Samson AG Mess-und Regeltechnik
Siekmann Econosto GmbH & Co KG
SIPOS Aktorik GmbH
Spirax Sarco GmbH
Südmo Components GmbH
Tuchenhagen GmbH
Tyco Flow Control
Uni-Geräte GmbH

INDIA
BDK Process Equipment Inc
Fouress Engineering (India) Ltd
Rotex Automation Ltd
Virgo Valves & Controls Ltd

ISRAEL
Egmo Ltd
Habonim Industrial Valves & Actuators

ITALY
Air Torque SpA
Caleffi SpA
Colves Srl
Cimberio SpA
FIP - Formatura Iniezione Polimeri SpA
Fratelli Pettinaroli SpA
Galli & Cassina SpA
GEI Gruppo Energia Italia Srl
Ghibson Italia Srl
GT Attuatori Srl
Italvalvole Sas
IVR Valvole SpA
LA TIS Service Srl
Omal Automation
Orton - Truflo International
Saint-Gobain Condotte SpA
Servovalve
SIRCA International Srl
Technical Srl
Valbia Srl
Valvitalia SpA
Valvoindustria Ing Rizzio SpA

JAPAN
CCI KK
Kitz Corporation
Miyairi Corporation
Nihon Koso Co Ltd
Showa Valve Co Ltd
Toko Valex Co Ltd
Tomoe Valve Co Ltd
Toyo Valve Co Ltd

THE NETHERLANDS
Delan BV
Mokveld Valves BV
Noxon Stainless BV

NORWAY
Oceaneering Rotator AS

RUSSIA
JSK PKTBA
PYCT-95
Tulaelectroprivod JSC
Tyazhpromarmatura JSC
Zeim

SINGAPORE
Alton International (S) Pte Ltd
Emerson Process Management
Remote Control Asia Pte Ltd
Tyco Flow Control

SOUTH AFRICA

Atval (Pty) Ltd
Dynamic Fluid Control Pty Ltd
Gunric Valves
Insamcor (Pty) Ltd
Mitech
Thermal Valve Manufacturers (Pty) Ltd

SOUTH KOREA

Velan Ltd

SPAIN

Centork Valve Control SL
Kitz Corporation of Europe SA

SWEDEN

NAF AB
Remote Control Sweden

SWITZERLAND

CCI Switzerland
Georg Fischer Rohrleitungssysteme (Schweiz) AG
Valv AG
Taiwan
Wyeco Auto Valves Co Ltd
Zipson Steel Industrial Co Ltd
United Arab Emirates
Emerson FZE

TAIWAN

Apex Controls Co Ltd
John Valve Mfg Factory Co Ltd
Kingdom Precision Casting Co Ltd
Velan-Valvac
Wyeco Auto Valves Co Ltd
Zipson Steel Industrial Co Ltd

UNITED KINGDOM

Advanced Component Technology Ltd
BEL Valves
Bray Controls (UK)
Bürkert Fluid Control Systems
Copes-Vulcan
Crane Process Flow Technologies
Davis Pneumatic Systems Ltd
DPL
EIM Controls UK Ltd
Flowserve Flow Control (UK) Ltd
Flowtech Inc Ltd
Forac Ltd
Kinetrol Ltd
Midland-ACS
Posi-flate
Remote Control Ltd
RGS Electro-Pneumatics Ltd
Rotork Controls Ltd
Severn Glocon Ltd
Smart Valve & Controls
Spirax-Sarco Ltd
Transmark Fcx
Xomox Ltd

USA

AMRI Inc
AOP Industries Inc
Asahi/America Inc
AUMA Actuators Inc
Automation Technology Inc
Bernard Controls Inc
Bray Controls USA
CCI (Control Components Inc)
DeZURIK Water Controls
DynaQuip Controls
DynaTorque Inc
EIM Controls Inc
Emerson Process Management
Enertech
FC Kingston
Flow-Quip Inc
Flowserve Flow Control
GA Industries Inc
H O Trerice Company
Harold Beck & Sons Inc
Henry Pratt Company
Hunt Valve Company Inc
Indelac Controls Inc
ITT Engineered Valves
Johnson Controls Inc
Jordan Controls Inc
Jordan Valve, a division of Richards Industries
KF Industries
Kitz Corporation of America
Leslie Controls Inc
Marwin Valve, a division of Richards Industries
Mastergear
Metso Automation USA Inc
Milwaukee Valve Company
Moog Flo-Tork
NIBCO Inc
Parker, Fluid Control Division
PBM Inc
Posi-flate
Powell Valves
Process Development & Control
QTRCO
Remote Control Inc
Rexa Koso America Inc
Richards Industries
Samson Controls Inc
Spirax Sarco Inc
SVF Flow Controls Inc
Therm-Omega-Tech Inc
Tru-Tech Industries
Tyco Flow Control
Valvtechnologies Inc
Velan Valve Corporation
Wood Group Pressure Control
Xomox Corporation

21.9 Ancillary products and services

Companies listed alphabetically by product and/or services

BURSTING DISCS

BS&B Safety Systems UK Ltd UK
Elfab Ltd UK

COMPUTER SOFTWARE

FIRST GmbH DE
Warren Controls Inc US

CONNECTIONS

Clamped connectors

M V Fluids Handling UK
NIBCO Inc US
Swagelok Company US
Vector International UK
Victaulic Company US

Expansion Joints

Ayvaz AS TR
Banides et Debeaurain F
Duyar Vana Makina Sanayi ve Ticaret A S TR
Haitima Corporation TW
Huot Groupe Saint-Mihiel SA F
IBK Wiesehahn GmbH DE
Nippon Valqua Industries Ltd JP
Red Valve Company US
Vector International UK
Venn Co Ltd JP
Yoshitake Inc JP

Fittings

AVF Astore Valves and Fittings Srl I
Instrumentation Products Division, Parker Hannifin Ltd UK
Lorch GmbH DE
NIBCO Inc US
Niezgodka GmbH DE
Noxon Stainless BV NL
Parker Hannifin Corporation US
Randex SA F
Rastelli Rubinetterie SpA I
Red Point Alloys BV NL
RIV Rubinetterie Italiane Valvole SpA I
Saint-Gobain Condotte SpA I
Saint-Gobain Pipelines UK
Seetru Ltd UK
Südmo Components GmbH DE
Super Inox Corporation TW
Taylorshaw UK
TC Fluid Control Ltd UK
Vanatome F
Vector International UK
Victaulic Company US
Xiamen Jia Da Quan Valves & Fittings Co Ltd CN
Xiamen Landee Industries Co Ltd CN

Gaskets & seals

BS & B Safety Systems (UK) Ltd UK
Burgmann Industries GmbH & Co. KG DE
Greene Tweed & Co Ltd UK
IBK Wiesehahn GmbH DE
Klinger Ltd UK
Nippon Valqua Industries Ltd JP
Vector International UK

Flanges

Noxon Stainless BV NL
Red Point Alloys BV NL
Vector International UK
Xiamen Jia Da Quan Valves & Fittings Co Ltd CN
Xiamen Landee Industries Co Ltd CN
YPS Valves Ltd UK
Zhejiang Chaoda Valve Co Ltd CN

Flare fittings

Vector International UK

Screwed connections

Hale Hamilton (Valves) Ltd UK
RGS Electro-Pneumatics Ltd UK
Warren-Morrison Valves Ltd UK
YPS Valves Ltd UK

Tubes

Noxon Stainless BV NL

INSTRUMENTATION

Flow

Kemutec Powder Technologies Ltd UK
Larox Flowsys Oy FI
Midland-ACS UK
Phönix Messtechnik GmbH DE
Pyropress Engineering Company Ltd UK
Südmo Components GmbH DE
Zhejiang Chaoda Valve Co Ltd CN

Level

Copes-Vulcan Ltd UK
Phönix Messtechnik GmbH DE
Pyropress Engineering Company Ltd UK
Seetru Ltd UK
VYC Industrial SA ES
Warren Controls Inc US

Pressure

BS & B Safety Systems (UK) Ltd UK
FST Northvale Korting Ltd UK
Hale Hamilton (Valves) Ltd UK
M V Fluids Handling UK
Midland-ACS UK
Pyropress Engineering Company Ltd UK
Thompson Valves- Truflo International UK

Silencers

Flo-Dyne Ltd UK
Midland-ACS UK
RGS Electro-Pneumatics Ltd UK
Rotech GmbH DE

Temperature

FST Northvale Korting Ltd UK
M V Fluids Handling UK
Midland-ACS UK
Pyropress Engineering Company Ltd UK

PIPES & HOSES

Ayvaz AS TR
Larox Flowsys Oy FI
Nippon Valqua Industries Ltd JP
Red Point Alloys BV NL
Saint-Gobain Condotte SpA I
Saint-Gobain Performance Plastics Corp US
Südmo Components GmbH DE
Swagelok Company US
Victaulic Company US
Zhejiang Chaoda Valve Co Ltd CN

SERVICES

Actuator repair

Crane Nuclear Inc US
Score (Europe) Ltd UK
Smart Valve & Controls UK
Thermal Valve Manufacturers (Pty) Ltd SA
Tyco Flow Control NI
Tyco Flow Control US
Tyco Flow Control SG
Tyco Flow Control AU
Ventil Test Equipment BV NL
DMI Young & Cunningham Ltd UK
Rotex Automation Ltd IN
TUV NEL Ltd UK

Consulting engineering

JSK PKTBA RU
Kalsi Engineering Inc US
Score (Europe) Ltd UK
Severn Unival Ltd UK
Smart Valve & Controls UK

Valve reconditioning

Comid Engineering Ltd UK
JSK PKTBA RU
Premier Valves ZA
Rötelmann GmbH DE
Score (Europe) Ltd UK
Seetru Ltd UK
Severn Glocon Ltd UK
Smart Valve & Controls UK
Thermal Valve Manufacturers (Pty) Ltd ZA
Tyco Flow Control NL
Tyco Flow Control US
Tyco Flow Control SG
Tyco Flow Control AU

Valve repair

Comid Engineering Ltd UK
Crane Nuclear Inc US
Dresser Flow Control Solutions UK
JSK PKTBA RU
Premier Valves ZA
Score (Europe) Ltd UK
Seetru Ltd UK
Severn Unival Ltd UK
Smart Valve & Controls UK
Thermal Valve Manufacturers (Pty) Ltd ZA
Tyco Flow Control NL
Tyco Flow Control US
Tyco Flow Control SG
Tyco Flow Control AU
Ventil Test Equipment BV NL
Z & J Technologies GmbH DE

PULSATION DAMPERS

Flowguard Ltd UK
Flo-Dyne Ltd UK

21.10 Trade or brand names

Companies listed alphabetically by Trade or brand name

ACRIS
AMRI Inc US

AIL
Audco India Ltd I

Alco
Alco Valves (UK) Ltd UK

Anderson Greenwood
Tyco Flow Control NL

Apollo
Conbraco Industries Inc US

ATI
Automation Technology Inc US

Attwood & Morrill
Weir Valves & Controls USA Inc US

AVID
Tyco Flow Control NL

Bailey
Safety Systems UK Ltd UK

Barton
Cameron's Valves & Measurement Group US

Batley
Weir Valves & Controls UK Ltd UK

Baumann
Emerson Process Management UK

BBK
Bunkaboeki Kogyo Co Ltd JP

Beck
Harold Beck & Sons Inc US

BEL
BEL Valves UK

Bestobell
Bestobell Valves Ltd UK

Biffi
Tyco Flow Control NL

BioConnect®
Neumo GmbH & Co KG DE

BioControl®
Neumo GmbH & Co KG DE

Birflo
Bestobell Valves Ltd UK

Birkett
Safety Systems UK Ltd UK

BIS
BiS Valves Ltd UK

Blakeborough Controls
Weir Valves & Controls UK Ltd UK

Bray McCannaick
Bray Controls UK Ltd UK

Broady
Broady Flow Control Ltd UK

Brownall
Delta Fluid Products Ltd UK

BZVF
Zhejiang Beize Valve Manufacturing Co Ltd CN

Caldon
Cameron's Valves & Measurement Group US

Cam-Centric®
Val-Matic Valve & Manufacturing Corporation US

Cameron
Cameron's Valves & Measurement Group US

CCIMS®
Instrumentation Products Division, Parker Hannifin Ltd UK

Chem-Plug®
SMG Valves US

Cheroflow
LCM Italia Srl I

Clarkson
Tyco Flow Control NL

Cliff Mock
Cameron's Valves & Measurement Group US

Conflow
Bestobell Valves Ltd UK

Conval
FIRST GmbH DE

Crosby
Tyco Flow Control NL

DANAIS
AMRI Inc US

Darvico
Dresser Al Rushaid Valve & Instrument Co Ltd ZA

Definox
Defontaine SA F

DELTA
Invensys APV UK

Demco
Cameron's Valves & Measurement Group US

Descote
Tyco Flow Control NL

Diamond
Dairy Pipe Lines UK

DPL
Dairy Pipe Lines UK

Dublok
Sabre Instrument Valves Ltd UK

EGMO
Egmo Ltd IL

El-O-Matic
Emerson Process Management UK

EPIC
Severn Glocon Ltd UK

F C Kingston
FC Kingston US

FIP
FIP - Formatura Iniezione Polimeri SpA I

FireLock EZ™
Victaulic Company US

Fishtail™
Emerson Process Management UK

FloCheck
Tyco Flow Control NL

Flow Group Ltd
Bestobell Valves Ltd UK

Foster
Cameron's Valves & Measurement Group US

Gachot
Tyco Flow Control NL

GEMÜ
GEMÜ Gebrüder Müller Apparatebau GmbH & Co. KG DE

General
Cameron's Valves & Measurement Group US

Gloceal
Severn Glocon Ltd UK

Golden-Anderson
GA Industries Inc US

Goodwin
Goodwin International UK

Gourd
Hitachi Valve Ltd JP

Gresswell
Dynafluid Ltd/Gresswell Valves UK

Grove
Cameron's Valves & Measurement Group US

Hancock
Dresser – Masoneilan US

Hancock
Tyco Flow Control NL

Hattersley Heaton
H H Valves Ltd UK

Henry
Henry Technologies Ltd UK

HICO
HCHTS Industries Corporation TW

Hindle
Tyco Flow Control NL

HiP
High Pressure Equipment Company US

Homestead Reg
Homestead Valve US

Hopkinsons
Weir Valves & Controls UK Ltd UK

HORA
Holter Regelarmaturen GmbH & Co KG DE

IAT	Delta Fluid Products Ltd	UK
Inox	Super Inox Corporation	TW
INTELLI+	Bernard Controls Inc	US
INTELLI-POCKET	Bernard Controls Inc	US
Intervalve	Tyco Flow Control	NL
IQ	Rotork Controls Ltd	UK
IQT	Rotork Controls Ltd	UK
ISORIA	AMRI Inc	US
IVC	Indian Valve Pvt Ltd	IN
John	John Valves	AU
Joumatic	ASCO	US
Kajaani comistency	Metso Automation Oy	FI
Karon	Shanghai Karon Valves Machinery Co Ltd	CN
KeyPort®	SMG Valves	US
Keystone	Tyco Flow Control	NL
Kite	Score (Europe) Ltd	UK
KKK	Kane Kogyo Co Ltd	JP
Korting	FST Northvale Korting Ltd	UK
KTM	Tyco Flow Control	NL
KUKA	KumKang Valve Mfg Co Ltd	KR
Kunkle	Tyco Flow Control	NL
L&M Valve	Tyco Flow Control	NL
Larner-Johnson®	Blackhall Engineering Ltd	UK
LESER	Leser GmbH & Co KG	DE
Leser	Seetru Ltd	UK
Limitorque	Flowserve Flow Control	US
Linaflow	Warren-Morrison Valves Ltd	UK
MAC	Weir Valves & Controls UK Ltd	UK
Magna™	Posi-flate	UK
MAMMOUTH	AMRI Inc	US
Marston	Safety Systems UK Ltd	UK
Marvac	Safety Systems UK Ltd	UK
Masoneilan	Dresser – Masoneilan	US
MCF	Tyco Flow Control	NL
Memoryseal®	Velan Valves Ltd	UK
MetsoDNA	Metso Automation Oy	FI
MG Valves	Strahman Valves France	F
Minimatic	FST Northvale Korting Ltd	UK
Morin	Tyco Flow Control	NL
Mucon	Kemutec Powder Technologies Ltd	UK
NABIC	Delta Fluid Products Ltd	UK
Navco	Cameron's Valves & Measurement Group	US
Neles	Metso Automation Oy	FI
NeoTecha	Tyco Flow Control	NL
Newco	Newmans Valve	US
Nil-Cor®	Nil-Cor	US
Norval	FST Northvale Korting Ltd	UK
NuFlo	Cameron's Valves & Measurement Group	US
Nutron	Cameron's Valves & Measurement Group	US
OIC	Newmans Valve	US
Orbit	Cameron's Valves & Measurement Group	US
Paper IQ	Metso Automation Oy	FI
Permaseal®	SMG Valves	US
Phastite®	Instrumentation Products Division	UK
Piping King	Velan Valves Ltd	UK
PVEG	Larox Flowsys Oy	FI
Q	Rotork Controls Ltd	UK
Raimondi	Tyco Flow Control	NL
RAVEN™	Copes-Vulcan	UK
Redflex ®	Red Valve Company	US
Red-White Valves	Toyo Valve Co Ltd	JP
Rexa	Rexa Koso America Inc	US
Rhodes	Delta Fluid Products Ltd	UK
Ring-O	Cameron's Valves & Measurement Group	US
Rovalve	Tyco Flow Control	NL
Samson	Samson AG Mess-und Regeltechnik	DE
SARASIN - RSBD	Weir Valves & Controls France SAS	FI
Sarasin-RSBD	Weir Valves & Controls UK Ltd	UK
SEBIM	Weir Valves & Controls SEBIM SAS	F
Securaseal®	Velan Valves Ltd	UK
Sensodec	Metso Automation Oy	FI
Shaw Valves	Taylor Shaw	UK
Shipham	Bestobell Valves Ltd	UK
Sinclair Collins™	Parker, Fluid Control Division	US
Skilmatic	Rotork Controls Ltd	UK
Skilmatic	Rotork Controls Ltd	UK
Skinner Valve™	Parker, Fluid Control Division	US
SMG	SMG Valves	US
SNJ	Shanghai Neles-Jamesbury Valve Co Ltd	CN
Sperryn	Delta Fluid Products Ltd	UK
Spiraflo	Spirax-Sarco Ltd	UK
Spiratec	Spirax-Sarco Ltd	UK
Stepseat	LG Ball Valves Ltd	UK
STERICOM®	Tuchenhagen GmbH	DE

Supero
Okano Valve Mfg Company JP

Swingclear
LG Ball Valves Ltd UK

SYR
Hans Sasserath & Co KG DE

TBV
Cameron's Valves & Measurement Group US

T-Dynamo
Tomoe Valve Co Ltd JP

Techlok DuoSeal
Vector International UK

Techno
Cameron's Valves & Measurement Group US

TEHACO®
Tehaco Sp Zoo PL

Texstream
Cameron's Valves & Measurement Group US

Thornhill Craver
Cameron's Valves & Measurement Group US

Tork-Mate™
Posi-flate UK

Torqseal
Velan Valves Ltd UK

Triangle
Tyco Flow Control NL

TRICENTRIC®
Weir Valves & Controls USA Inc US

TRI-CHECK
Zwick Armaturen GmbH DE

TRI-CON
Zwick Armaturen GmbH DE

Truflo
CCI (Control Components Inc) US

TruSeal
Cameron's Valves & Measurement Group US

TVT
Hobbs Valve Ltd UK

Valbia
Valbia Srl I

Valmet Consistency
Metso Automation Oy FI

Valvac
Velan Valves Ltd UK

Valvcon
Metso Automation Oy FI

Valve-Retrofits
Rötelmann GmbH DE

ValveWorks®
Warren Controls Inc US

VALVOmetal
Sitindustrie Equipment Srl I

Valvtron
Tyco Flow Control NL

Vanessa
Tyco Flow Control NL

Varec
Tyco Flow Control NL

VARIBELL®
EPC GmbH AT

VARIVENT®
Tuchenhagen GmbH DE

Vee-Ball®
Emerson Process Management UK

Ventil
Ventil Test Equipment BV NL

VESTA
Tuchenhagen GmbH DE

Virgo
Virgo Valves & Controls Ltd IN

VIZA
Suhou Viza Valve Co Ltd CN

Wade
Delta Fluid Products Ltd UK

Weber
Transmark Fcx UK

Wheatley
Cameron's Valves & Measurement Group US

WKM
Cameron's Valves & Measurement Group US

WMV
Warren-Morrison Valves Ltd UK

Wyeco
Wyeco Auto Valves Co Ltd TW

Yarway
Tyco Flow Control NL

Z & J
Z & J Technologies GmbH DE

Zetkama
Zetkama Fabryka Armatury Przemyslowej SA PL

Zikesch
CH Zikesch Armaturentechnik GmbH DE

Zimmermann & Jansen
Z & J Technologies GmbH DE

Zipson
Zipson Steel Industrial Co Ltd TW

Index

D

E

U

V

FSC
www.fsc.org
MIX
Papier | Fördert gute Waldnutzung
FSC® C083411

Kolibri 360 GmbH
Ferdinand-Jühlke-Straße 7
99095 Erfurt, Deutschland
produktsicherheit@kolibri360.de